evolution THIRD EDITION

COMPANION WEBSITE
sites.sinauer.com/evolution3e

The **evolution** Companion Website provides you with a range of valuable study and review tools to help you master the material presented in the textbook. Available free of charge, the site is designed to help you understand the concepts and learn the terminology introduced in each chapter, analyze real-world research, and work with simulations of evolutionary systems.

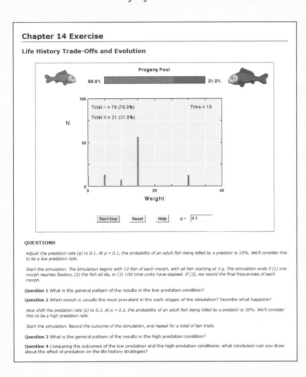

Chapter 14 Exercise

Life History Trade-Offs and Evolution

QUESTIONS

Adjust the predation rate (p) to 0.1. At p = 0.1, the probability of an adult fish being killed by a predator is 10%. We'll consider this to be a low predation rate.

Start the simulation. The simulation begins with 10 fish of each morph, with all fish starting at 5 g. The simulation ends if (1) one morph reaches fixation, (2) the fish all die, or (3) 100 time units have elapsed. If (3), we record the final frequencies of each morph.

Question 1 What is the general pattern of the results in the low predation condition?

Question 2 Which morph is usually the most prevalent in the early stages of the simulation? Describe what happens?

Now shift the predation rate (p) to 0.3. At p = 0.3, the probability of an adult fish being killed by a predator is 30%. We'll consider this to be a high predation rate.

Start the simulation. Record the outcome of the simulation, and repeat for a total of ten trials.

Question 3 What is the general pattern of the results in the high predation condition?

Question 4 Comparing the outcomes of the low predation and the high predation conditions, what conclusion can you draw about the effect of predation on the life history strategies?

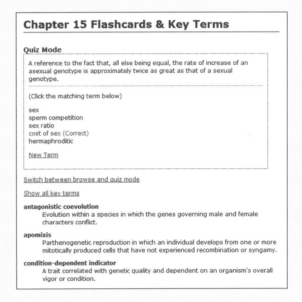

Chapter 15 Flashcards & Key Terms

Quiz Mode

A reference to the fact that, all else being equal, the rate of increase of an asexual genotype is approximately twice as great as that of a sexual genotype.

(Click the matching term below)

sex
sperm competition
sex ratio
cost of sex (Correct)
hermaphroditic

New Term

Switch between browse and quiz mode

Show all key terms

antagonistic coevolution
Evolution within a species in which the genes governing male and female characters conflict.

apomixis
Parthenogenetic reproduction in which an individual develops from one or more mitotically produced cells that have not experienced recombination or syngamy.

condition-dependent indicator
A trait correlated with genetic quality and dependent on an organism's overall vigor or condition.

Features of the Companion Website

Data Analysis Exercises: These inquiry-based exercises challenge you to think as a scientist and to analyze and interpret experimental data. Based on real papers and experiments, these exercises involve answering questions by analyzing the data from the experiments.

Simulation Exercises: These exercises include interactive modules that allow you to explore some of the dynamic processes of evolution. Each exercise poses questions that you answer by running a simulation and observing and analyzing the outcomes.

Online Quizzes: For each chapter of the textbook, the site includes a multiple-choice quiz that covers all the main topics presented in the chapter. Your instructor may assign these quizzes, or they may be made available to you as self-study tools. (Instructor registration is required for student access to the quizzes.)

Flashcards & Key Terms: Flashcard activities help you learn and review the many new terms introduced in the textbook. Each chapter's set of flashcards includes all of the key terms introduced in the chapter.

Chapter Summaries: Concise overviews of the important concepts and topics covered in each chapter.

Chapter Outlines: A convenient outline of each chapter's sections and sub-sections.

Glossary: A complete online version of the glossary, for quick access to definitions of important terms.

evolution
THIRD EDITION

evolution

THIRD EDITION

DOUGLAS J. FUTUYMA

Stony Brook University

Chapter 20, "Evolution of Genes and Genomes"
by Scott V. Edwards, Harvard University

Chapter 21, "Evolution and Development"
by John R. True, Stony Brook University

SINAUER ASSOCIATES, INC. • *Publishers*
Sunderland, Massachusetts U.S.A.

On the Cover

A tropical American leafcutter ant (*Atta cephalotes*) carries a leaf fragment to the colony's nest, as a smaller worker, shown in silhouette, rides along to guard against parasitic flies. The ants use leaves to cultivate a fungus that is the ants' sole food, and which grows only in leafcutter nests. The ant and the fungus have evolved dependence on each other. This mutualistic relationship is threatened by another fungus that attacks the fungal gardens. The ants carry on their body surface, and nourish with special glandular secretions, bacteria that produce an antibiotic that controls the fungal weed. The intricate adaptations of these species to one another illustrate the outcome of coevolution.

evolution, THIRD EDITION

Copyright © 2013 by Sinauer Associates, Inc. All rights reserved. This book may not be reproduced in whole or in part without permission from the publisher.

Sinauer Associates, Inc., 23 Plumtree Road, Sunderland, MA 01375 USA
Phone: 413-549-4300
FAX: 413-549-1118
Email: publish@sinauer.com, orders@sinauer.com
Website: www.sinauer.com

Library of Congress Cataloging-in-Publication Data

Futuyma, Douglas J., 1942-
    Evolution / Douglas J. Futuyma, Stony Brook University. -- Third edition.
       pages cm
    Includes bibliographical references and index.
    ISBN 978-1-60535-115-5 (casebound)
1. Evolution (Biology) I. Title.
QH366.2.F87 2013
576.8--dc23   2012047209

Printed in the U.S.A
7 6 5 4 3 2 1

To Tom and Theresa, Jeanne and Todd, Paul and Lindsay,
and to Annabelle, Matthew, Phoebe, and Audrey

Brief Contents

Contents

Preface

Following Watson and Crick's determination of the structure of DNA in 1953, the growth and self-confidence of molecular biology led not only to unparalleled understanding of fundamental biological processes, but also to divergence between molecular and cellular biology, on one hand, and what was often termed "whole-organism biology" on the other. Evolutionary biology was practiced largely by whole-organism biologists, and although they often proclaimed that evolution was the central, if not sole, unifying theory of biology, such unification, or even broad ratification of the claim, could be hard to discern.

Biology has by no means fully emerged from this schism, especially given the increasing, and necessary, specialization in training and research in the life sciences. But the separation is breaking down; the unification is happening. Like all the other life sciences, evolutionary biology has been transformed by molecular understanding and methods, and now speaks much the same language as the molecular-mechanistic disciplines. Evolutionary biology is becoming increasingly mechanistic as it explores such frontiers as the evolution of genomes, metabolomes, biochemical pathways, complex systems, and the developmental processes that translate DNA sequences into phenotypes. Conversely, scientists who work in molecular, genomic, cell, and developmental biology increasingly recognize that the interpretive framework, the principles, and the analytical methods of evolutionary biology are not only useful, but also critically important for their fields: evolutionary science is being integrated throughout biology. At the same time, the disciplines of organismal biology, including paleobiology, ecology, functional morphology and physiology, animal behavior, and systematics, continue to be central to evolutionary science and are largely framed by evolutionary principles. Together with Scott Edwards and John True, I have attempted to take a look at these many relationships between evolutionary biology and diverse biological disciplines.

An updated edition of *Evolution* is warranted because of the extraordinary pace of change of evolutionary biology, as new analytical and laboratory methods are developed and our understanding of biological mechanisms advances. Growing evidence of function for much of the nontranslated DNA that dominates eukaryotic genomes will have important implications for our understanding of molecular evolution; growing debate and evidence concern the possible evolutionary significance of epigenetic modification of genes; comparative genomics is revealing how new genetic and developmental functions have evolved. Phylogenetics and evolutionary genomics have become more powerful with the development of new analytical methods; a historical, phylogenetic perspective is reshaping community ecology; extraordinary new fossils add to our knowledge of life's history.

One of the most important changes in the field has been a more intense focus on the social and practical implications of evolutionary biology. To be sure, the possible implications of human evolution for interpreting human diversity and behavior have been obvious and controversial for 150 years, and biologists have long understood that evolutionary processes have important applications in health, agronomy, and many other fields. But these applications have now become a significant dimension of evolutionary research, treated in both traditional journals and new journals such as *Evolutionary Applications* and *Evolution, Medicine, and Public Health*. We must, moreover, address the opposition to evolution, most pronounced in the United States, that is but one face of a widespread skepticism about science that has truly dangerous implications—as the debate in the United States about global climate change shows.

I have maintained the structure of the previous edition of this book. It begins with phylogeny as a framework for inferring evolutionary history, and with history as the natural perspective for evolutionary biology. This leads into macroevolutionary patterns, which I think intrigue many beginning students and which provide opportunity for introducing some of the most important evidence for evolution—a critically important aspect of any undergraduate course in the subject. They also constitute many of the *explananda* of evolutionary biology: the observations that a theory of evolution must account for. Most of the subsequent chapters treat evolutionary processes and how they explain a wide variety of phenomena, from genomes to ecological interactions among species. The final chapters return to macroevolution, now viewed through the lens of evolutionary processes, and to the evidence for evolution, the nature of science and the failings of creationism, and some of the social applications of evolutionary biology. Human evolution is treated throughout the text rather than as a separate topic, an approach that reveals the social and personal relevance of the various topics in evolutionary science and points to the commonality of humans with the rest of the biological world. Every chapter has been updated, and the approach and coverage in a few differ substantially from the previous editions.

The literature in almost all areas of evolutionary biology has grown so voluminous, and the methods in many so altered, that it has become nearly impossible to follow—or to understand fully—all of them. So I again thank Scott Edwards (Harvard University) and John True (Stony Brook University), for contributing revised chapters on genome evolution and evolutionary developmental biology, respectively, topics that I doubt I could have competently addressed.

Innumerable colleagues and students have contributed indirectly to this book, through their lectures, publications, conversations, and questions. Among the many who have contributed more directly by providing advice, references, answers to my questions or corrections of my errors are David Begun, Michael Bell, Jerry Coyne, Liliana Dávalos, N. Delaney, Daniel Dykhuizen, Walter Eanes, Bruce Grant, Mark Kirkpatrick, Jeffrey Levinton, Joanna Masel, Hugh McGuinness, Mohamed Noor, Sally Otto, Joshua Rest, Marty Schoenhals, Brad Shaffer, Ellen Simms, Chris Simon, Jason Weir, John Wiens, and Jerry Wilkinson. I owe special thanks to Scott Edwards and John Wiens for helping to improve the treatment of phylogenetic methods. I appreciate all the advice I've received, even if I have not acted on it, and I take full responsibility for the errors or inadequacies that will surely come to light. The many colleagues who contributed photos are acknowledged in the figure captions. Aman Gill helped with some illustrations. Rob DeSalle and the American Museum of Natural History have generously provided work space in that marvelous institution, Michael Purugganan and New York University provided hospitality in the last stages of the book's preparation, and the Department of Ecology and Evolution at Stony Brook continues to provide support and intellectual sustenance. I apologize to the many colleagues who provided input, and whom I have inadvertently omitted. Scott Edwards adds his thanks to Feng-Chi Chen, Allan Drummond, and Harold Zakon for contributions to Chapter 20. John True would like to thank Prof. Jer-Ming Hu of National Taiwan University for productive discussions leading to a more thorough treatment of plant evolutionary developmental biology in Chapter 21.

Since 1977, I have been fortunate to work with the Sinauer team, which is surely unexcelled in competence and devotion to quality. I owe special thanks to Laura Green, Jefferson Johnson, David McIntyre, Norma Roche, and Chris Small, and above all to Andy Sinauer.

DOUGLAS J. FUTUYMA
STONY BROOK, NEW YORK
FEBRUARY 2013

To the Student

The great geneticist François Jacob, who won the Nobel Prize in Biology and Medicine for discovering mechanisms by which gene activity is regulated, wrote in 1973 that "there are many generalizations in biology, but precious few theories. Among these, the theory of evolution is by far the most important." Why? Because, he said, evolution explains a vast range of biological information and unites all of the biological sciences, from molecular biology to ecology. "In short," he wrote, "it provides a causal explanation of the living world and its heterogeneity."

Jacob did not himself do research on evolution, but like most thoughtful biologists, he recognized its pivotal importance in the biological sciences. Evolution provides an indispensable framework for understanding phenomena ranging from the structure and size of genomes to the ecological interactions among different species.

Evolutionary biology is not only a framework for interpreting biological phenomena; it also has important philosophical implications (as the persistent opposition to teaching evolution testifies), and it has diverse practical applications. If you use a DNA-sequencing service to find out about your ancestry, you are benefiting directly from evolutionary science. Just in the last year, evolutionary studies have addressed the management of commercial fisheries, methods to control disease-carrying mosquitoes and other pests, ways to retard the evolution of antibiotic resistance by pathogenic microbes, evolutionary changes in invasive species, and the prospects of extinction versus adaptation to human-induced climate change. The rapidly developing field of evolutionary medicine includes research on medically important genetic variation within and among human populations; consequences of modern environments to which humans have not adapted (such as effects on autoimmune diseases, allergies, cardiovascular health, and cancer); ageing; pathogen evolution of resistance to antibiotics; and studies of how other species have adapted to organisms or conditions that cause pathology in humans. (Why do other primates not develop AIDS, even though they carry viruses related to HIV?)

Evolution, therefore, is as relevant to our lives as genetics and physiology—which is to say that everyone should know something about it. For anyone who envisions a career based in the life sciences—whether as physician or as biological researcher—an understanding of evolution is indispensable.

How can you best learn about evolution? Because all organisms, and all their characteristics, are products of a history of evolutionary change, the scope of evolutionary biology is far greater than any other field of biological science. For this reason, courses in evolution generally do not emphasize the details of the evolution of particular groups of organisms—the amount of information would be simply overwhelming. Instead, evolution courses emphasize the *general principles* of evolution, the *processes of evolutionary change* that apply to most or all organisms, and the *most common patterns of change*, those that have characterized many different groups. Above all, it is important to learn *how evolutionary hypotheses have been tested*, in other words, what the evidence is for (or against) these postulated causes and patterns of evolutionary change.

Evolutionary biology largely concerns events that happened in the past, so it differs from most other biological disciplines, which analyze the properties and functions of organisms' characteristics without reference to their history. However, in both evolutionary biology and other biological disciplines, we often must make inferences about invisible objects and processes (e.g., DNA replication). We make inferences about these things by (1) posing informed hypotheses about what they are and how they work, then (2) generating predictions (making deductions) from these hypotheses about data that we can actually obtain, and finally (3) judging the validity of each hypothesis by the match between our observations and what we expect to see if the hypothesis were true. In this book, we often approach a topic first by describing a hypothesis about a process or pattern in evolution, often based on a verbal or mathematical model, and then presenting an example of evidence, in which observations are compared with the predictions that the hypothesis makes. In some cases, we describe further elaborations or implications of the evolutionary process, and illustrate some of the kinds of biological observations it helps to explain.

You may find that the emphasis in your course differs from what you have experienced in other biology courses. I suggest that you pay special attention to chapters or passages where the fundamental principles and methods are introduced, be sure you understand them before moving on, and reread these passages after you have gone through one or more later chapters. (You may want to revisit Chapters 2, 9, 10, and 12–13 in particular.) Be sure to emphasize understanding, not memorization (except of some particularly important information, such as the geological time scale and major events in the history of life). Be aware that the material in this book builds cumulatively; almost every concept, principle, or major technical term introduced in any chapter is used again in later chapters. Evolutionary biology is a unified whole; just as carbohydrate metabolism and amino acid synthesis cannot be divorced in biochemistry, so it is for topics as seemingly different as the phylogeny of species and the theory of genetic drift.

In every field of science, the unknown greatly exceeds the known. Thousands of research papers on evolutionary topics are published each year, and many of them raise new questions even as they attempt to answer old ones. No one, least of all a scientist, should be afraid to say "I don't know" or "I'm not sure," and that refrain will sound fairly often in this book. To recognize where our knowledge and understanding are uncertain or lacking is to see where research may be warranted, or where exciting new research trails might be blazed. I hope that some readers will find evolution so rich a subject, so intellectually challenging, so fertile in insights, and so deep in its implications that they will adopt evolutionary biology as a career. But all readers, I hope, will find in evolutionary biology the thrill of understanding and the excitement of finding both answers and intriguing new questions about the living world, including ourselves. *Felix qui potuit rerum cognoscere causas*, wrote Virgil: happy is the person who can learn the nature of things.

Media and Supplements

to accompany **evolution** THIRD EDITION

eBOOK

Evolution, Third Edition is available in several ebook formats. Please visit the Sinauer Associates website for more information: www.sinauer.com.

FOR STUDENTS

Companion Website (sites.sinauer.com/evolution3e)

Evolution's Companion Website features review and study tools to help students master the material presented in the textbook. Access to the site is free of charge, and requires no passcode. The site includes:

- *Chapter Outlines and Summaries*: Concise overviews of the important topics covered in each chapter.

- *Data Analysis Exercises*: Expanded for the Third Edition, these inquiry-based problems are designed to sharpen the student's ability to reason as a scientist, drawing on data from real experiments and published papers.

- *Simulation Exercises*: Interactive modules that allow students to explore many of the dynamic processes of evolution and answer questions based on the results they observe.

- *Online Quizzes*: Quizzes that cover the major concepts introduced in each chapter. (Instructor registration is required for student access to the quizzes.)

- *Flashcards & Key Terms*: Easy-to-use activities that help students learn the key terminology introduced in each chapter by browsing terms and quizzing themselves.

- *Glossary*: A complete online version of the glossary, for quick access to definitions of important terms.

FOR INSTRUCTORS

Instructor's Resource Library

The *Evolution* Instructor's Resource Library includes a variety of resources to help instructors in developing their courses and delivering their lectures. The IRL includes:

- *Textbook Figures and Tables*: All of the figures (including photographs) and tables from the textbook are provided as JPEGs (both high- and low-resolution), reformatted and relabeled for optimal readability when projected.

- *PowerPoint Presentations*: For each chapter, all figures and tables are provided in a ready-to-use PowerPoint presentation, making it easy to quickly insert figures into lecture presentations.

- *Answers* to the textbook end-of-chapter Problems and Discussion Topics.

- *Quiz Questions*: All of the questions from the Companion Website's online quizzes are provided in Microsoft Word format.

- *Data Analysis and Simulation Exercises*: All of the exercises from the Companion Website are provided as Word documents, for use in class or as assignments.

Online Quizzing

Available via the Companion Website, instructors have access to a set of online quizzes that can be used either as assigned homework or as a self-study tool. Instructors choose how they want the quizzes to be used by their students, and can quickly see results via the quiz administration site. (Instructor registration is required for student access to the quizzes.)

Evolutionary Biology

Biologists, looking broadly at living things, are stirred to ask thousands of questions. Why do whales have lungs, and why do snakes lack legs? Why do some insects attack only a single species of plant while others eat many different species? Why do salamanders have more than ten times as much DNA as humans? Why are there 19,000 species of orchids and more than 300,000 species of beetles? Why do people differ in facial features, skin color, attitudes, behavior? What accounts for the astonishing variety of organisms? What accounts for their intricate, multitudinous adaptations?

In one of the most breathtaking ideas in the history of science, Charles Darwin proposed that "all the organic beings which have ever lived on this earth have descended from some one primordial form." From this idea, it follows that every characteristic of every species—the trunk of an elephant, the number and sequence of its genes, the catalytic abilities of its enzymes, the structure of its cells and organs, its physiological tolerances and nutritional requirements, its life span and reproductive system, its capacity for behavior—is the outcome of an evolutionary history. The evolutionary perspective illuminates every subject in biology, from molecular biology to ecology. Indeed, evolution is the unifying theory of biology. "Nothing in biology makes sense," said the geneticist Theodosius Dobzhansky, "except in the light of evolution."

A plethora of adaptations
The hands of this big-eared bat (*Plecotus townsendii*) are modified to serve as wings, the tail and hind legs support a membrane used to catch insects, and the ears are highly modified for echolocation, the bat's use of "sonar" to detect insect prey and the structure of its habitat. The only scientific explanation of adaptations is the theory of evolution by natural selection.

What Is Evolution?

The word "evolution" comes from the Latin *evolvere*, "to unfold or unroll"—to reveal or manifest hidden potentialities. Today "evolution" has come to mean, simply, "change." The word is sometimes used to describe changes in individual objects such as stars. **Biological (or organic) evolution**, however, is *change in the properties of groups of organisms over the course of generations*. The development, or ONTOGENY, of an individual organism is not considered evolution: individual organisms do not evolve. Groups of organisms, which we may call **populations**, do evolve: they undergo *descent with modification*. Populations may become subdivided, so that several populations are derived from a *common ancestral population*. If different changes transpire in the several populations, the populations DIVERGE—that is, they become different from each other. As these processes have continued over vast periods of time, they have resulted in the emergence of millions of different forms of life, all having descended, with modification, from the "universal common ancestor" of all known living things.

The changes in populations that are considered evolutionary are those that are passed via the genetic material from one generation to the next. Over the course of many generations, many such changes may accumulate. Thus biological evolution may be slight or substantial: it embraces everything from slight changes in the proportions of different forms of a gene within a population to the alterations that led from the earliest organism to corals, grasshoppers, tomatoes, and humans.

No more dramatic example of evolution by natural selection can be imagined than today's crisis in antibiotic resistance. Before the 1940s, most people in hospital wards did not have cancer or heart disease. They suffered from tuberculosis, pneumonia, meningitis, typhoid fever, syphilis, and many other kinds of bacterial infection—and they had little hope of being cured (**Figure 1.1**). Infectious bacterial diseases condemned millions of people in the developed countries to early death. People in developing countries bore not only these burdens, but also diseases such as malaria and cholera, even more heavily than they do today.

By the 1960s, the medical situation had changed dramatically. The discovery of antibiotic drugs and subsequent advances in their synthesis led to the conquest of most bacterial diseases, at least in developed countries. The sexual revolution in the 1970s was encouraged by the confidence that sexually transmitted diseases such as gonorrhea and syphilis were merely a temporary inconvenience that penicillin could cure. In 1969, the Surgeon General of the United States proclaimed that it was time to "close the book on infectious diseases."

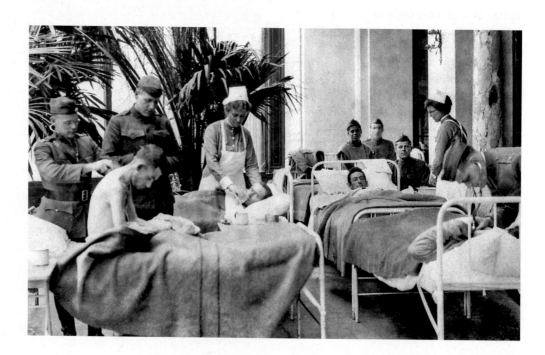

FIGURE 1.1 A tuberculosis ward at a U.S. Army base hospital in France during World War I. Until recently, it was thought that antibiotics, which came into widespread use after World War II, had conquered this devastating bacterial disease. (Photo courtesy of the National Library of Medicine.)

He was wrong. Today we confront not only new infectious diseases such as AIDS, but also a resurgence of old diseases such as tuberculosis. The same bacteria are back, but now they are resistant to penicillin, ampicillin, erythromycin, vancomycin, fluoroquinolones—all the weapons that were supposed to have vanquished them. Almost every hospital in the world treats casualties in this battle against changing opponents, but in the process, they are unintentionally making those opponents stronger. In this and many other ways, humans are instigating an explosion of evolutionary change (Palumbi 2001).

For example, *Staphylococcus aureus*, a bacterium that causes many infections in surgical patients, has evolved resistance to a vast array of antibiotics, starting with penicillin and working its way through many others. Almost the only effective remedy that remains against *S. aureus* is vancomycin, but it, too, is becoming less effective. Likewise, drug-resistant strains of *Neisseria gonorrheae*, the bacterium that causes gonorrhea, have steadily increased in abundance, and many strains of the pneumonia and cholera bacteria are highly resistant to antibiotics. The same thing is happening with other kinds of pathogens. Throughout the tropics, the microorganism that causes malaria is now resistant to chloroquine and is becoming resistant to other drugs as well. HIV, the human immunodeficiency virus that causes AIDS, rapidly evolved resistance to AZT and other drugs that at first were effective against it. People now survive with HIV by taking a combination of three drugs, but there is great concern about how long it will be until virus populations evolve resistance to these drugs as well.

As the use of antibiotics increases, so does the incidence of bacteria that are resistant to those antibiotics; thus any gains made are almost as quickly lost (**Figure 1.2**). Why is this happening? Do the drugs cause drug-resistant mutations in the bacteria's genes? Do the mutations occur even without exposure to drugs—that is, are they present in unexposed bacterial populations? How many mutations result in resistance to a drug? How often do such mutations occur? Do the mutations spread from one bacterium to another? Are they spread only among bacteria of the same species, or can they pass between different species? How is the growth of the organism's population affected by such mutations? Can the evolution of resistance be prevented by using lower doses of drugs? Higher doses? Combinations of different drugs? Can an individual recover from infection by drug-resistant organisms by faithfully following a physician's prescription, or will this work only if everyone else is just as conscientious?

The principles and methods of evolutionary biology have provided some answers to these questions and have shed light on many other problems that affect society. Evolutionary biologists and other scientists trained in evolutionary principles have traced the transfer of HIV to humans from chimpanzees and mangabey monkeys (Gao et al. 1999; Korber et al. 2000). They have studied the evolution of insecticide resistance in disease-carrying and crop-destroying insects. They have helped to devise methods of nonchemical pest control and have laid the foundations for transferring genetic resistance to diseases and insects from wild plants to crop plants. Evolutionary principles and knowledge are being used in biotechnology to design new drugs and other useful products, and in medical genetics to identify and analyze inherited diseases as well as variation in susceptibility to infectious diseases. In the fields of computer science and artificial intelligence, "evolutionary computation" uses principles taken directly from evolutionary theory to solve mathematically intractable practical problems, such as constructing complex timetables or processing radar data.

The importance of evolutionary biology goes far beyond its practical uses, however. An evolutionary framework provides answers to many questions about ourselves. How do we account for human variation—the fact that almost everyone is genetically and phenotypically unique? Are there human races, and if so, how do they differ, and how and when did they develop? What accounts for behavioral differences between

FIGURE 1.2 Evolution of drug resistance. An increase in the consumption of a penicillin-like antibiotic in a community in Finland between 1978 and 1993 was matched by a dramatic increase in the percentage of isolates of the bacterium *Moraxella catarrhalisis* from middle-ear infections in young children that were resistant to the drug. (After Levin and Anderson 1999.)

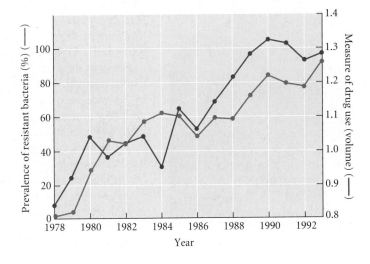

men and women? How did exquisitely complex, useful features such as our hands and our eyes come to exist? What about apparently useless or even potentially harmful characteristics such as our wisdom teeth and appendix? Why do we age, undergo senescence, and eventually die? Why are medical researchers able to use monkeys, mice, and even fruit flies and yeasts as models for processes in the human body? Moreover, as soon as Darwin published *On the Origin of Species*, the evolutionary perspective was perceived to bear on long-standing questions in philosophy. If humans, with all their mental and emotional complexity, originated by natural processes, where do ethics and moral precepts find a foundation and origin? What, if anything, does evolution imply about the meaning and purpose of life? Must one choose between evolution and religious belief?

Before Darwin

Darwin's theory of biological evolution is one of the most revolutionary ideas in Western thought, perhaps rivaled only by Newton's theory of physics. It profoundly challenged the prevailing world view, which had originated largely with Plato and Aristotle. Foremost in Plato's philosophy was his concept of the *eidos*, the "form" or "idea," a transcendent ideal form imperfectly imitated by its earthly representations. For example, the reality—the "essence"—of the true equilateral triangle is only imperfectly captured by the triangles we draw or construct. Likewise, the horse (or any other species) has an immutable essence, but each individual horse has imperfections. In this philosophy of **essentialism**, variation is accidental imperfection.

Plato's philosophy of essentialism became incorporated into Western philosophy largely through Aristotle, who developed Plato's concept of immutable essences into the notion that species have fixed properties. Later, Christians interpreted the biblical account of Genesis literally and concluded that each species had been created individually by God in the same form it has today. (This belief is known as "special creation.") Christian theologians and philosophers elaborated on Platonic and Aristotelian philosophy, arguing that since existence is good and God's benevolence is complete, He must have bestowed existence on every creature of which He could conceive. Because order is superior to disorder, God's creation must follow a plan: specifically, a gradation from inanimate objects and barely animate forms of life through plants and invertebrates and up through ever "higher" forms of life. Humankind, being both physical and spiritual in nature, formed the link between animals and angels. This "Great Chain of Being," or *scala naturae* (the scale, or ladder, of nature), must be permanent and unchanging, since change would imply that there had been imperfection in the original creation (Lovejoy 1936).

As late as the eighteenth century, natural history was justified partly as a way to reveal the plan of creation so that we might appreciate God's wisdom. Carolus Linnaeus (1707–1778), who established the framework of modern taxonomy in his *Systema Naturae* (1735), won worldwide fame for his exhaustive classification of plants and animals, undertaken in the hope of discovering the pattern of the creation. Linnaeus classified "related" species into genera, "related" genera into orders, and so on. To him, "relatedness" meant propinquity in the Creator's design.

Belief in the literal truth of the biblical story of creation started to give way in the eighteenth century, when a philosophical movement called the Enlightenment, largely inspired by Newton's explanations of physical phenomena, adopted reason as the major basis of authority and marked the emergence of science. The foundations for evolutionary thought were laid by astronomers, who developed theories of the origin of stars and planets, and by geologists, who amassed evidence that Earth had undergone profound changes, that it had been populated by many creatures now extinct, and that it was very old. The geologists James Hutton and Charles Lyell expounded the principle of **uniformitarianism**, holding that the same processes operated in the past as in the present and that the observations of geology should therefore be explained by causes that we can now observe. Darwin was greatly influenced by Lyell's teachings, and he adopted uniformitarianism in his thinking about evolution.

Carolus Linnaeus

In the eighteenth century, several French philosophers and naturalists suggested that species had arisen by natural causes. The most significant pre-Darwinian evolutionary hypothesis, representing the culmination of eighteenth-century evolutionary thought, was proposed by the Chevalier de Lamarck in his *Philosophie Zoologique* (1809). Lamarck proposed that each species originated individually by spontaneous generation from nonliving matter, starting at the bottom of the chain of being. A "nervous fluid" acts within each species, he said, causing it to progress up the chain. Species originated at different times, so we now see a hierarchy of species because they differ in age (**Figure 1.3A**).

Lamarck argued that species differ from one another because they have different needs, and so use certain of their organs and appendages more than others. The more strongly exercised organs attract more of the "nervous fluid," which enlarges them, just as muscles become strengthened by work. Lamarck, like most people at the time, believed that such alterations, acquired during an individual's lifetime, are inherited—a principle called **inheritance of acquired characteristics**. The theory of evolution based on this principle is called **Lamarckism**. In the most famous example of Lamarck's theory, giraffes must have stretched their necks to reach foliage above them. Hence their necks were lengthened; longer necks were inherited, and over the course of generations, their necks got longer and longer. This could happen to any and all giraffes, so the entire species could have acquired longer necks because it was composed of individual organisms that changed during their lifetimes (see Figure 1.4A).

Lamarck's ideas had little impact during his lifetime, partly because they were criticized by respected zoologists, and partly because after the French Revolution, ideas issuing from France were considered suspect in most other countries. Lamarck's ideas of how evolution works were wrong, but he deserves credit for being the first to advance a coherent theory of evolution.

Jean-Baptiste Pierre Antoine de Monet, Chevalier de Lamarck

Charles Darwin

Charles Robert Darwin (February 12, 1809–April 19, 1882) was the son of an English physician. He briefly studied medicine in Edinburgh, then turned to studying for a career in the clergy at Cambridge University. He apparently believed in the literal truth of the Bible as a young man. He was passionately interested in natural history and became a companion of the natural scientists on the faculty. In 1831, at the age of 22, his life was forever changed when he was invited to serve as a naturalist and captain's gentleman companion on the H.M.S. *Beagle*, a ship the British navy was sending to chart the waters of South America.

The voyage of the *Beagle* lasted from December 27, 1831, to October 2, 1836. The ship spent several years

(A) Lamarck's theory

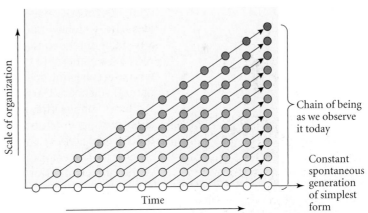

FIGURE 1.3 (A) Lamarck's theory of organic progression. Over time, species originate by spontaneous generation, and each evolves up the scale of organization, establishing a *scala naturae*, or chain of being, that ranges from newly originated simple forms of life to older, more complex forms. In Lamarck's scheme, species do not originate from common ancestors. (B) Darwin's theory of descent with modification, represented by a phylogenetic tree. Lineages (species) descend from a common ancestor, undergoing various modifications in the course of time. Some (such as the leftmost lineages) may undergo less modification from the ancestral condition than others (the rightmost lineages). (A after Bowler 1989.)

(B) Darwin's theory

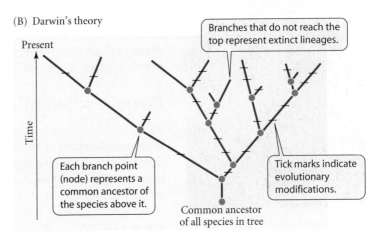

Charles Robert Darwin

traveling along the coast of South America, where Darwin observed the natural history of the Brazilian rain forest and the Argentine pampas, then stopped in the Galápagos Islands, which lie on the equator off the coast of Ecuador. In the course of the voyage, Darwin became an accomplished naturalist, collected specimens, made innumerable geological and biological observations, and conceived a new (and correct) theory about the formation of coral atolls.

Soon after Darwin returned, the ornithologist John Gould pointed out that Darwin's specimens of mockingbirds from the Galápagos Islands were so different from one island to another that they represented different species. Darwin then recalled that the giant tortoises, too, differed from one island to the next. These facts, and the similarities between fossil and living mammals that he had found in South America, triggered his conviction that different species had evolved from common ancestors.

Darwin's comfortable finances enabled him to devote the rest of his life exclusively to his biological work (although he was chronically ill for most of his life after the voyage). He set about amassing evidence of evolution and trying to conceive of its causes. On September 28, 1838, at the age of 29, he read an essay by the economist Thomas Malthus, who argued that the rate of human population growth is greater than the rate of increase in the food supply, so that unchecked growth must lead to famine. This essay was the inspiration for Darwin's great idea, one of the most important ideas in the history of thought: natural selection. Darwin wrote in his autobiography that "being well prepared to appreciate the struggle for existence which everywhere goes on from long-continued observation of the habits of animals and plants, it at once struck me that under these circumstances favourable variations would tend to be preserved and unfavourable ones to be destroyed." In other words, if individuals of a species with superior features survived and reproduced more successfully than individuals with inferior features, and if these differences were inherited, the average character of the species would be altered.

Mindful of how controversial the subject would be, Darwin then spent 20 years developing his theory, amassing evidence, and pursuing other researches before publishing his ideas. In 1844, he wrote a private essay outlining his theory, and in 1856, he finally began a book he intended to call *Natural Selection*. He never completed it, for in June 1858, he received a manuscript from a young naturalist, Alfred Russel Wallace (1823–1913). Wallace, who was collecting specimens in the Malay Archipelago, had independently conceived of natural selection. Darwin had extracts from his 1844 essay presented orally, along with Wallace's manuscript, at a meeting of the major scientific society in London, and he set about writing an "abstract" of the book he had intended. The 490-page "abstract," titled *On the Origin of Species by Means of Natural Selection, or The Preservation of Favoured Races in the Struggle for Life*, was published on November 24, 1859; it instantly made Darwin, by now 50 years old, both a celebrity and a figure of controversy.

For the rest of his life, Darwin continued to read and correspond on an immense range of subjects, to revise *The Origin of Species* (it had six editions), to perform experiments of all sorts (especially on plants), and to publish many more articles and books, of which *The Descent of Man* is the most renowned. Darwin's books reveal an irrepressibly inquisitive man, fascinated with all of biology, creative in devising hypotheses and in bringing evidence to bear upon them, and profoundly aware that every fact of biology, no matter how seemingly trivial, must fit into a coherent, unified understanding of the world.

Darwin's Evolutionary Theory

The Origin of Species contains two major theses. The first is Darwin's theory of **descent with modification**. It holds that all species, living and extinct, have descended, without interruption, from one or a few original forms of life (**Figure 1.3B**). Species that diverge from a common ancestor are at first very similar, but accumulate differences over great spans of time, so that they may come to differ radically from one another. Darwin's conception of the course of evolution is profoundly different from Lamarck's, in which the concept of common ancestry plays almost no role.

Alfred Russel Wallace

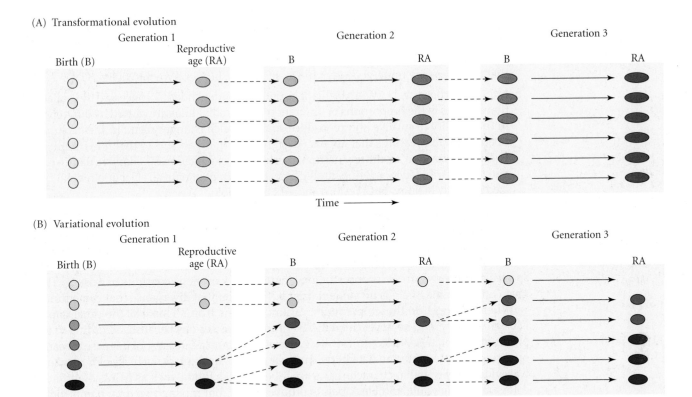

(A) Transformational evolution

Generation 1 — Birth (B) — Reproductive age (RA)

Generation 2 — B — RA

Generation 3 — B — RA

Time ⟶

(B) Variational evolution

Generation 1 — Birth (B) — Reproductive age (RA)

Generation 2 — B — RA

Generation 3 — B — RA

Time ⟶

FIGURE 1.4 A diagrammatic contrast between transformational and variational theories of evolutionary change, shown across three generations. Within each generation, individuals are represented at birth and later in their lives, at reproductive age. The individuals in the left column in each generation are the offspring of those in the right column of the preceding generation. (A) In transformational evolution, individuals are altered during their lifetimes, and their progeny are born with those alterations. (B) In variational evolution, hereditarily different forms at the beginning of the history are not transformed, but instead differ in their survival and reproductive rates, so that their proportions change from one generation to another.

The second theme of *The Origin of Species* is Darwin's theory of the causal agents of evolutionary change. He summarized his theory of **natural selection** in the following way: "If variations useful to any organic being ever occur, assuredly individuals thus characterized will have the best chance of being preserved in the struggle for life; and from the strong principle of inheritance, these will tend to produce offspring similarly characterized. This principle of preservation, or the survival of the fittest, I have called natural selection." This theory is a VARIATIONAL THEORY of change, differing profoundly from Lamarck's TRANSFORMATIONAL THEORY, in which individual organisms change (**Figure 1.4**).

What is often called "Darwin's theory of evolution" actually includes five theories (Mayr 1982a):

1. *Evolution as such* is the simple proposition that the characteristics of lineages of organisms change over time. This idea was not original with Darwin, but it was Darwin who so convincingly marshaled the evidence for evolution that most biologists soon accepted that it has indeed occurred.

2. *Common descent* is a view of evolution that is radically different from the scheme Lamarck had proposed (see Figure 1.3). Darwin was the first to argue that species had diverged from common ancestors and that all of life could be portrayed as one great family tree representing actual ancestry.

3. *Gradualism* is Darwin's proposition that the differences between even radically different organisms have evolved incrementally, by small steps through intermediate forms. The alternative hypothesis is that large differences evolve by leaps, or SALTATIONS, without intermediate forms.

4. *Populational change* is Darwin's hypothesis that evolution occurs by changes in the proportions of individuals within a population that have particular inherited characteristics (see Figure 1.4B). This concept was a completely original idea that contrasts both with the sudden origin of new species by saltation and with Lamarck's account of evolutionary change by transformation of individuals. The "populational thinking" that Darwin

introduced contrasts with Platonic essentialism, which viewed variation from the ideal "type" to be unimportant imperfection. For Darwin, the average was a statistical abstraction; there exist only varied individuals, and there are no fixed limits to the variation that a species may undergo (Mayr 1982a; Hey 2011; but see Winsor 2006).

5. *Natural selection* was Darwin's brilliant hypothesis, independently conceived by Wallace, that changes in the proportions of different types of individuals are caused by differences in their ability to survive and reproduce, and that such changes result in the evolution of **adaptations**, features that appear "designed" to fit organisms to their environment. Because it provided an entirely natural, mechanistic explanation for adaptive design that had been attributed to a divine intelligence, the concept of natural selection revolutionized not only biology, but Western thought as a whole.

Darwin proposed that the various species descended from a common ancestor evolve different features because those features are adaptive under different "conditions of life"—different habitats or habits. Moreover, the pressure of competition favors the use of different foods or habitats by different species. He believed that no matter how extensively a species has diverged from its ancestor, new hereditary variations continue to arise, so that given enough time, there is no evident limit to the amount of divergence that can occur.

Where, though, do these hereditary variations come from? This was the great gap in Darwin's theory, and he never filled it. The problem was serious because, according to the prevailing belief in BLENDING INHERITANCE, variation should decrease, not increase. Because offspring are often intermediate between their parents in features such as color or size, it was widely believed that characteristics are inherited like fluids, such as paints of different colors. Blending white and black paints produces gray, but mixing two gray paints yields more gray, not black or white. Darwin never knew that Gregor Mendel had solved the problem in a paper that was published in 1865, but not widely noticed until 1900. Mendel's theory of PARTICULATE INHERITANCE proposed that inheritance is based not on blending fluids, but on particles that pass unaltered from generation to generation—so that variation can persist. The concept of "mutation" in such particles (later called genes) was developed only after 1900 and was not clarified until considerably later.

The Origin of Species is extraordinarily rich in insights and implications. Darwin drew on an astonishingly broad variety of information, from observed variation in domesticated species to embryology to geographic patterns in the distribution of species, to support his hypotheses. And he showed, or at least glimpsed, how research in every biological subject— taxonomy, paleontology, anatomy, embryology, biogeography, physiology, behavior, ecology—could be advanced and reinterpreted in the light of evolution.

Philosophical Issues

Thousands of pages have been written about the philosophical and social implications of evolution. Darwin argued that every characteristic of a species can vary and can be altered radically, given enough time. Thus he rejected the essentialism that Western philosophy had inherited from Plato and Aristotle and put variation in its place. Darwin also helped to replace a static conception of the world—one virtually identical to the Creator's perfect creation—with a world of ceaseless change. It was Darwin who extended to living things, including the human species, the principle that change, not stasis, is the natural order. In contrast to traditional views that elevated the human species to a special position, distinct from other living things, Darwin began the trend to see humans as part of the natural world, a species of animal (though a very remarkable species, to be sure!) subject to the same processes as others, including natural selection.

Darwin has been credited with making biology a science, for he proposed to replace supernatural explanations in biology with purely natural causes. His theory of random, purposeless variation acted on by blind, purposeless natural selection provided a revolutionary new kind of answer to almost all questions that begin with "Why?" Before Darwin, both philosophers and people in general answered questions such as "Why do plants have flowers?" or "Why are there apple trees?"—or diseases, or sexual reproduction—by imagining the possible purpose that God could have had in creating them. This kind of explanation was made completely superfluous by Darwin's theory of natural selection. The adaptations of organisms—long cited as the most conspicuous evidence of intelligent design in the universe—could be explained by purely mechanistic causes. For evolutionary biologists, the pink color of a magnolia's flower has a *function* (attracting pollinating insects), but not a *purpose*. It was not designed in order to propagate the species, much less to delight us with its beauty, but instead came into existence because magnolias with brightly colored flowers reproduced more prolifically than magnolias with less brightly colored flowers. The unsettling implication of this purely material explanation is that, except in the case of human behavior, we need not invoke, nor can we find any evidence for, any design, goal, or purpose anywhere in the natural world.

It must be emphasized that all of science has come to adopt the way of thought that Darwin applied to biology. Astronomers do not seek the purpose of comets or supernovas, nor chemists the purpose of hydrogen bonds. The concept of purpose plays no part in scientific explanation.

Ethics, Religion, and Evolution

In the world of science, the reality of evolution has not been in doubt for more than a hundred years, but evolution remains an exceedingly controversial subject in the United States and a few other countries. The **creationist movement** opposes the teaching of evolution in public schools, or at least demands "equal time" for creationist beliefs. Such opposition arises from the fear that evolutionary science denies the existence of God, and consequently, that it denies any basis for rules of moral or ethical conduct.

Our knowledge of the history and mechanisms of evolution is certainly incompatible with a literal reading of the creation stories in the Bible's Book of Genesis—as it is incompatible with hundreds of other creation myths people have devised. A literal reading of some passages in the Bible is also incompatible with the principles of physics, geology, and other natural sciences. But does evolutionary biology deny the existence of a supernatural being or a human soul? No, because science, including evolutionary biology, is silent on such questions. By its very nature, science can entertain and investigate only hypotheses about material causes that operate with at least probabilistic regularity. It cannot test hypotheses about supernatural beings or their intervention in natural events.

Evolutionary biology has provided natural, material causes for the diversification and adaptation of species, just as the physical sciences did when they explained earthquakes and eclipses. The steady expansion of the sciences, to be sure, has left less and less to be explained by the existence of a supernatural Creator, but science can neither deny nor affirm such a being. Indeed, some evolutionary biologists are devoutly religious, and many nonscientists, including many priests, ministers, and rabbis, hold both religious beliefs and belief in evolution (see Chapter 23). A commitment to METHODOLOGICAL NATURALISM in scientific investigation (i.e., the assumption that only natural causes operate) does not require a commitment to PHILOSOPHICAL NATURALISM (a belief that only natural processes and causes exist). Most people use methodological naturalism in everyday affairs, as when we assume that there is a material cause when our car or computer malfunctions.

Wherever ethical and moral principles are to be found, it is probably not in science, and surely not in evolutionary biology. Opponents of evolution have charged that evolution by natural selection justifies the principle that "might makes right," and certainly more than one dictator or imperialist has invoked the "law" of natural selection to justify atrocities. But evolutionary theory cannot provide any such precept for behavior. Like any other science, it describes how the world is, not how it should be. The supposition that what is "natural" is "good" is called by philosophers the NATURALISTIC FALLACY.

Various animals have evolved behaviors that we give names such as cooperation, monogamy, competition, infanticide, and the like. Whether or not these behaviors ought to be, and whether or not they are moral, is not a scientific question. The natural world is amoral—the concepts of "moral" and "immoral" simply do not apply outside the realm of human behavior. Despite this, the concepts of natural selection and evolutionary progress were taken as a "law of nature" by which Marx justified class struggle, by which the Social Darwinists of the late eighteenth and early nineteenth centuries justified economic competition and imperialism, and by which the biologist Julian Huxley justified humanitarianism (Hofstadter 1955; Paradis and Williams 1989). All these ideas are philosophically indefensible instances of the naturalistic fallacy. Infanticide by lions and langur monkeys does not justify infanticide in humans, monogamy in penguins is irrelevant to human norms, and evolution provides no basis for human ethics.

Evolutionary Theories after Darwin

Although *The Origin of Species* raised enormous controversy, by the 1870s most scientists accepted the historical reality of evolution by descent, with modification, from common ancestors. This theory provided a new framework for exploring and interpreting the history and diversification of life, a project that was especially promoted by the German zoologist Ernst Haeckel. Thus the late nineteenth and early twentieth centuries were a "golden age" of paleontology, comparative morphology, and comparative embryology, during which a great deal of information on evolution in the fossil record and on relationships among organisms was amassed (Bowler 1996). But the consensus did not extend to Darwin's theory of the cause of evolution, natural selection. For about 60 years after the publication of *The Origin of Species*, all but a few faithful Darwinians rejected natural selection, and numerous theories were proposed in its stead. These theories included neo-Lamarckian, orthogenetic, and mutationist theories (Bowler 1989).

Ronald A. Fisher

NEO-LAMARCKISM includes several theories based on the old idea of inheritance of modifications acquired during an organism's lifetime. Such modifications might have been due, for example, to the direct effect of the environment on development (as in plants that develop thicker leaves if grown in a hot, dry environment). In a famous experiment, the German biologist August Weismann (1834–1914) cut off the tails of mice for many generations and showed that this mutilation had no effect on the tail length of their descendants. Extensive subsequent research has provided no evidence that specific hereditary changes can be induced by environmental conditions under which they would be advantageous.

Theories of ORTHOGENESIS, or "straight-line evolution," held that the variation that arises is directed toward fixed goals, so that a species evolves in a predetermined direction without the aid of natural selection. Some paleontologists held that such trends need not be adaptive and could even drive species toward extinction. None of the proponents of orthogenesis ever proposed a mechanism for it.

MUTATIONIST theories were advanced by some geneticists who observed that discretely different new phenotypes can arise by a process of mutation. They supposed that such mutant forms constituted new species and thus believed that natural selection was not necessary to account for the origin of species. Mutationist ideas were advanced by Hugo de Vries, one of the biologists who "discovered" Mendel's neglected paper in 1900, and by Thomas Hunt Morgan, the founder of *Drosophila* genetics. The last influential mutationist was Richard Goldschmidt (1940), an accomplished geneticist who nevertheless erroneously argued that evolutionary change within species is entirely different in kind from the origin of new species and higher taxa. These, he said, originate by sudden, drastic changes that reorganize the whole genome. Although most such reorganizations would be deleterious, a few "hopeful monsters" would be the progenitors of new groups.

J. B. S. Haldane

The Evolutionary Synthesis

These anti-Darwinian ideas were refuted in the 1930s and 1940s by the geneticists, systematists, and paleontologists who reconciled Darwin's theory with the facts of genetics (Mayr and Provine 1980; Smocovitis 1996). The consensus they forged is known as the **evolutionary synthesis**, or **modern synthesis** and its chief principle, that adaptive evolution is caused by natural selection acting on particulate (Mendelian) genetic variation, is often referred to as **neo-Darwinism**. Ronald A. Fisher and John B. S. Haldane in England and Sewall Wright in the United States developed a mathematical theory of population genetics, which showed that mutation and natural selection together cause adaptive evolution: mutation is not an alternative to natural selection, but is rather its raw material. The study of genetic variation and change in natural populations was pioneered in Russia by Sergei Chetverikov and continued by Theodosius Dobzhansky, who moved from Russia to the United States. In his influential book *Genetics and the Origin of Species* (1937), Dobzhansky conveyed the ideas of the population geneticists to other biologists, thus influencing their appreciation of the genetic basis of evolution.

Other major contributors to the synthesis included the zoologists Ernst Mayr, in *Systematics and the Origin of Species* (1942), and Bernhard Rensch, in *Evolution Above the Species Level* (1959); the botanist G. Ledyard Stebbins, in *Variation and Evolution in Plants* (1950); and the paleontologist George Gaylord Simpson, in *Tempo and Mode in Evolution* (1944) and its successor, *The Major Features of Evolution* (1953). These authors argued persuasively that mutation, recombination, natural selection, and other processes operating within species (which Dobzhansky termed **microevolution**) account for the origin of new species and for the major, long-term features of evolution (termed **macroevolution**).

Sewall Wright

Fundamental principles of evolution

The principal claims of the evolutionary synthesis are the foundations of modern evolutionary biology. Although some of these principles have been extended, clarified, or modified since the 1940s, most evolutionary biologists today accept them as substantially valid. These, then, are the fundamental principles of evolution, to be discussed at length throughout this book. (Qualifications to some of these statements are discussed in the chapters cited.)

1. *The phenotype* (the observed characteristics of an individual) *is different from the genotype* (the set of genes in an individual's DNA). Phenotypic differences among individual organisms may be due partly to genetic differences and partly to direct effects of the environment.

2. *Environmental effects on an individual's phenotype do not alter the genes passed on to its offspring.* In other words, acquired characteristics are not inherited. (See Chapter 9.)

3. *Hereditary variations are based on particles—genes—that retain their identity as they pass through the generations; they do not blend* with other genes. This is true of both discretely varying traits (e.g., brown vs. blue eyes) and continuously varying traits (e.g., body size, intensity of pigmentation). Variation in continuously varying traits is based on several or many discrete, particulate genes, each of which affects the trait slightly (POLYGENIC IN-HERITANCE).

Ernst Mayr

4. *Genetic variation arises by random mutation and recombination.* Genes mutate, usually at a fairly low rate, to equally stable alternative forms, known as ALLELES. The phenotypic effects of such mutations can range from undetectable to very great. The variation that arises by mutation is amplified by recombination among alleles at different loci.

5. *Evolutionary change is a populational process.* It entails, in its most basic form, a change in the relative abundances (proportions, or **frequencies**) of individual organisms with different genotypes (hence, often, with different phenotypes) within a popu-

G. Ledyard Stebbins, George Gaylord Simpson, and Theodosius Dobzhansky

lation. (A population is a group of individuals of the same species that occupy a speci-fied geographic space and may interact with each other.) One genotype may gradually replace other genotypes over the course of generations. Replacement may occur in only certain populations or in all the populations that make up a species.

6. *Change in genotype frequencies may be random or nonrandom.* The rate of mutation is too low for mutation by itself to shift a population from one genotype to another. Instead, the change in genotype frequencies within a population can occur by either of two principal processes: random fluctuations (GENETIC DRIFT) or nonrandom changes due to the superior survival or reproduction of some genotypes compared with others (i.e., natural selection). Natural selection and random genetic drift can operate simultaneously.

7. *Natural selection can account for both slight and great differences among species.* Even a low intensity of natural selection can (under certain circumstances) bring about substantial evolutionary change over time. Adaptations are traits that have been shaped by natural selection.

8. *Natural selection can alter populations beyond the original range of variation* by increasing the frequency of alleles that, by forming new combinations with other genes, give rise to new phenotypes.

9. *Populations can accumulate considerable genetic variation.* Mutations can accumulate in natural populations. Hence many populations contain enough genetic variation to evolve rapidly when environmental conditions change, rather than having to wait for new favorable mutations.

10. *Populations of a species in different geographic regions differ in characteristics that have a ge-netic basis.* These differences are often adaptive and must therefore be the consequence of natural selection.

11. *The differences between different species, and between different populations of the same spe-cies, are often based on differences at several or many genes,* many of which have a small phenotypic effect. This pattern supports the hypothesis that the differences between species evolve by rather small steps.

12. *Species are groups of interbreeding or potentially interbreeding individuals that do not exchange genes with other such groups.* Species are not defined simply by phenotypic differences. Rather, different species of sexually reproducing organisms represent distinct, separately evolving "gene pools."(See Chapter 17.)

13. *Speciation (the origin of two or more species from a single ancestor species) usually occurs by the genetic differentiation of geographically segregated populations.* Because of the geo-graphic segregation, interbreeding does not prevent incipient genetic differences from developing. Fully distinct species have genetic differences that prevent free interbreeding even if they are no longer geographically separated. (See Chapter 18.)

14. *Higher taxa arise by the prolonged, sequential accumulation of small differences, rather than by the sudden mutational origin of drastically new "types."* This principle is supported by the observation that there are often gradations in phenotypic characteristics among species assigned to the same genus, to different genera, and to different families or other higher taxa. (See Chapters 4 and 22.)

15. *All organisms form a great "tree of life" (or PHYLOGENY) that has developed by the branching of common ancestors into diverse lineages,* chiefly through speciation. (See Chapter 2.) All forms of life appear to have descended from a single common ancestor in the remote past.

Evolutionary Biology since the Synthesis

Since the evolutionary synthesis, a great deal of research has tested and elaborated on its basic principles, and these principles have withstood the tests. But progress in both evolutionary studies and other fields of biology has required some modifications and many extensions of the basic principles of the evolutionary synthesis, and it has spurred additional theory to account for biological phenomena that were unknown in the 1940s.

Motoo Kimura

Since James Watson and Francis Crick established the structure of DNA in 1953, advances in genetics, molecular biology, and molecular and information technology have revolutionized the study of evolution. Molecular biology has provided tools for studying a vast number of evolutionary topics, such as mutation, genetic variation, species differences, development, and the phylogenetic history of life, in greater depth and detail than ever before. New mathematical theory has been developed to provide evolutionary interpretations of newly discovered molecular and genetic phenomena.

As molecular and computational technology has become more sophisticated and available, new fields of evolutionary study have developed. Among these fields is MOLECULAR EVOLUTION (analysis of the processes and history of change in genes). The NEUTRAL THEORY OF MOLECULAR EVOLUTION has been particularly important in this field. This hypothesis, especially as developed by Motoo Kimura (1924–1994), holds that most of the evolution of DNA sequences occurs by genetic drift rather than by natural selection, but it provides a foundation for detecting effects of natural selection on DNA sequences. Because the entire GENOME——the full DNA complement of an organism—can now be sequenced, molecular evolutionary studies have expanded into EVOLUTIONARY GENOMICS, which is concerned with variation and evolution in multiple genes or even entire genomes. EVOLUTIONARY DEVELOPMENTAL BIOLOGY, based partly on molecular genetics, is devoted to understanding how the evolution of developmental processes underlies the evolution of morphological features at all levels, from cells to whole organisms.

The advances in these new fields are complemented by vigorous research, new discoveries, and new ideas about long-standing topics in evolutionary biology, such as the evolution of adaptations and of new species. Since the mid-1960s, evolutionary theory has expanded into areas such as ecology, animal behavior, and reproductive biology. Detailed theories that explain the evolution of particular kinds of characteristics such as life span, ecological distribution, and social behavior were pioneered by the

Ernst Mayr, George C. Williams, John Maynard Smith

evolutionary theoreticians William Hamilton and John Maynard Smith in England and George Williams in the United States. The study of macroevolution has been renewed by provocative interpretations of the fossil record and by new methods for studying phylogenetic relationships. Research in evolutionary biology is progressing more rapidly than ever before.

Since Darwin's time, research on evolution, and in biology more broadly, has transformed evolutionary biology. Were Darwin to reappear today, he would not understand the great majority of scientific papers about evolution. *Modern evolutionary biology does not equal Darwinism*, and any antievolutionary critiques of Darwin's ideas are completely irrelevant to our understanding of evolution today.

How Evolution Is Studied

Evolutionary biology is a more historical science than most other biological disciplines, for it seeks to determine what the history of life has been and what has caused those historical events. It complements studies of the PROXIMATE CAUSES (immediate, mechanical causes) of biological phenomena—the subject of cell biology, neurobiology, and many other biological disciplines—with analysis of the ULTIMATE CAUSES (the historical causes, especially the action of natural selection) of those phenomena. If we ask what causes a male bird to sing,

William D. Hamilton

the proximate causes include the action of testosterone or other hormones, the structure and action of the singing apparatus (syrinx), and the operation of certain centers in the brain. The ultimate causes lie in the history of events that led to the evolution of singing in the bird's remote ancestors. For example, past individuals whose genes inclined them to sing may have been more successful in attracting females or in driving away competing males, and thus may have transmitted their genes to more descendants than did their less tuneful competitors. These two kinds of answers to the question are both valid, and together they form a more complete explanation of bird song than either does alone.

Occasionally we can document an evolutionary change as it occurs or piece together records to reconstruct a recent change, just as we do when studying human history. Usually, however, we must *infer* evolutionary history and its causes by interpreting less direct evidence. Some historical events are inferred from fossils, the province of paleontology. Other evolutionary events are inferred from comparisons among living organisms or by studying their phylogenetic relationships, which provide a framework that enables us to draw conclusions about the historical evolution of their phenotypic characteristics and even their genes.

The causes of evolution, such as genetic drift and natural selection, are often studied by comparing data, such as patterns of variation in genes, with THEORETICAL MODELS. They are also studied by the methods of EXPERIMENTAL EVOLUTION, in which laboratory populations of rapidly reproducing organisms adapt to an environment (e.g., a stressful temperature) designed by an investigator. The adaptive reasons for certain characteristics (e.g., bird song) may be inferred from experimental and other functional studies, from their "fit" to a theoretical design (e.g., the heart fits a "pump" design), or by comparing many populations or species to see if the characteristic is correlated with a specific environmental factor or way of life.

Especially when we make inferences about history, or about past causes of change such as natural selection, we do not see the changes occurring, nor observe the causes in action. But *throughout science, causes* (and even the existence of objects) *are not seen; rather, they are inferred*. All of chemistry, for example, concerns invisible atoms and orbitals that govern the association of atoms into molecules. These theoretically postulated entities and their behavior have been confirmed because the theory that employs them makes predictions (hypotheses) that have been matched by observed data. We know that DNA replicates semiconservatively not because anyone has ever seen DNA do that, but because the outcome of a famous experiment matched the prediction made by the semiconservative replication hypothesis.

This hypothetical-deductive method, which was employed by Darwin, has been a powerful tool throughout the sciences since his time and is the basis of much evolutionary research. For example, would you predict that the DNA in mitochondria carries more mutations that are harmful to males than to females? There is no obvious biochemical reason to expect this, but evolutionary theory makes such a prediction. The mitochondria of both males and females are inherited from the mother; the mitochondria in males are not transmitted in sperm and are thus at a "dead end." If a mutation reduces the survival or reproduction of females, it is unlikely to be transmitted to subsequent generations, but the transmission of a mutation will not be affected if it is similarly harmful to males only because males do not transmit the gene anyway. So, male-deleterious mitochondrial mutations are expected to accumulate. This prediction, from the theory of natural selection at the level of the gene, has been verified: mitochondrial variants affect male, but not female, fertility in humans and other animals, and they cause variation in reproductive gene expression in male fruit flies (Innocenti et al. 2011). This example illustrates how evolutionary hypotheses can be tested, and it also shows how they can predict and reveal aspects of biology we would not otherwise have expected.

Evolution as Fact and Theory

Is evolution a fact, a theory, or a hypothesis? Biologists often speak of the "theory of evolution," but they usually mean by that something quite different from what most nonscientists understand by that phrase.

In science, a **hypothesis** is an informed conjecture or statement of what might be true. Most philosophers (and scientists) hold that we do not know anything with absolute certainty. What we call facts are hypotheses that have acquired so much supporting evidence that we act as if they were true. A hypothesis may be poorly supported at first, but it can gain support to the point that it is effectively a fact. For Copernicus, the revolution of Earth around the Sun was a hypothesis with modest support; for us, it is a hypothesis with such strong support that we consider it a fact. Occasionally, an accepted "fact" may need to be revised in the face of new evidence; for example, humans have 46 chromosomes, not 48 as once thought.

In everyday use, "theory" refers to an unsupported speculation. Like many words, however, this term has a different meaning in science. A **scientific theory** is a mature, coherent body of interconnected statements, based on reasoning and evidence, that explain some aspect of nature—usually many aspects. Or, to quote the *Oxford English Dictionary*, a theory is "a scheme or system of ideas and statements held as an explanation or account of a group of facts or phenomena; a hypothesis that has been confirmed or established by observation or experiment, and is propounded or accepted as accounting for the known facts; a statement of what are known to be the general laws, principles, or causes of something known or observed." Thus atomic theory, quantum theory, and the theory of plate tectonics are elaborate schemes of interconnected ideas, strongly supported by evidence, that account for a great variety of phenomena. "Theory" is a term of honor in science; the greatest accomplishment a scientist can aspire to is to develop a valid, successful new theory.

Given these definitions, evolution is a fact. But the fact of evolution is explained by evolutionary theory.

In *The Origin of Species*, Darwin propounded two major hypotheses: that organisms have descended, with modification, from common ancestors; and that the chief cause of modification is natural selection acting on hereditary variation. Darwin provided abundant evidence for descent with modification, and hundreds of thousands of observations from paleontology, geographic distributions of species, comparative anatomy, embryology, genetics, biochemistry, and molecular biology have confirmed this hypothesis since Darwin's time. Thus the hypothesis of descent with modification from common ancestors has long had the status of a scientific fact.

The explanation of how modification occurs and how ancestors give rise to diverse descendants constitutes the theory of evolution. We now know that Darwin's hypothesis of natural selection on hereditary variation was correct, but we also know that there are more causes of evolution than Darwin realized and that natural selection and hereditary variation themselves are more complex than he imagined. A body of ideas about the causes of evolution, including mutation, recombination, gene flow, isolation, random genetic drift, the many forms of natural selection, and other factors, constitutes our current theory of evolution, or "evolutionary theory." Like all theories in science, it is a work in progress, for we do not yet know the causes of all of evolution, or of all the biological phenomena that evolutionary biology will have to explain. In evolutionary biology, as in every other scientific discipline, there are "core" principles that have withstood skeptical challenges and are highly unlikely to require revision, and there are "frontier" areas in which research actively continues. Some widely held ideas about frontier subjects may prove to be wrong, as some examples in this book will illustrate. But the uncertainty at the frontier does not undermine the core. The main tenets of evolutionary theory—descent with modification from a common ancestor, in part caused by natural selection—are so well supported that almost all biologists confidently accept evolutionary theory as the foundation of the science of life.

Go to the
EVOLUTION
Companion Website at
sites.sinauer.com/evolution3e
for quizzes, data analysis and simulation exercises, and other study aids.

Summary

1. Evolution is the unifying theory of the biological sciences. Evolutionary biology aims to discover the history of life and the causes of the diversity and characteristics of organisms.

2. Charles Darwin's major work, *On the Origin of Species*, published in 1859, contains two major hypotheses: first, that all organisms have descended, with modification, from common ancestral forms of life, and second, that the chief agent of modification is natural selection.

3. Darwin's hypothesis that all species have descended with modification from common ancestors is supported by so much evidence that it has become as well established a fact as any in biology. His theory of natural selection as the chief cause of evolution was not broadly supported until the evolutionary synthesis that occurred in the 1930s and 1940s.

4. Evolutionary biology makes important contributions to other biological disciplines and to social concerns in areas such as medicine, agriculture, computer science, and our understanding of ourselves.

5. The implications of Darwin's theory, which revolutionized Western thought, include the ideas that change, rather than stasis, is the natural order; that biological phenomena, including those seemingly designed, can be explained by purely material causes rather than by divine creation; and that no evidence for purpose or goals can be found in the living world, other than in human actions.

6. Like other sciences, evolutionary biology cannot be used to justify beliefs about ethics or morality. Nor can it prove or disprove theological hypotheses such as the existence of a deity. Many people hold that although evolution is incompatible with a literal interpretation of some passages in the Bible, it is compatible with religious belief.

7. The evolutionary theory developed during and since the evolutionary synthesis consists of a body of principles that explain evolutionary change. Among these principles are the following: that genetic variation arises by random mutation and recombination; that the proportions of alleles and genotypes within a population change over time; that such changes in the proportions of genotypes may occur either by random fluctuations (genetic drift) or by nonrandom, consistent differences among genotypes in survival or reproduction rates (natural selection); and that as a result of different histories of genetic drift and natural selection, populations of a species may diverge and become reproductively isolated species.

Terms and Concepts

adaptation

creationist movement

descent with modification

essentialism

evolution (biological evolution; organic evolution)

evolutionary synthesis (modern synthesis)

frequency

hypothesis

inheritance of acquired characteristics

Lamarckism

macroevolution

microevolution

natural selection

neo-Darwinism

population

scientific theory

uniformitarianism

Suggestions for Further Reading

The readings at the end of each chapter include major works that provide a comprehensive treatment and an entry into the professional literature. The references cited within each chapter also serve this important function.

No one should fail to read at least part of Darwin's *On the Origin of Species by Means of Natural Selection, or The Preservation of Favoured Races in the Struggle for Life*, in either the first edition (1859) or the sixth edition (1872), in which Darwin deleted "On" from the title. After some adjustment to the Victorian prose, you will be enthralled by the craft, detail, completeness, and insight in Darwin's arguments. It is an astonishing book.

Among biographies of Darwin, *The Reluctant Mr. Darwin: An Intimate Portrait of Charles Darwin and the Making of His Theory of Evolution,* by David Quammen (W. W. Norton, New York, 2006), is enthralling and is a "must." The best scholarly biographies of Darwin include Janet Browne's superb two-volume work, *Charles Darwin: Voyaging* and *Charles Darwin: The Power of Place* (Knopf, New York, 1995 and 2002, respectively); and *Darwin,* by A. Desmond and J. Moore (Warner Books, New York, 1991), which emphasizes the role played by the religious, philosophical, and intellectual climate of nineteenth-century England on the development of his scientific theories.

Important works on the history of evolutionary biology include P. J. Bowler, *Evolution: The History of an Idea* (University of California Press, Berkeley, 1989); E. Mayr, *The Growth of Biological Thought: Diversity, Evolution, and Inheritance* (Harvard University Press, Cambridge, MA, 1982), a detailed, comprehensive history of systematics, evolutionary biology, and genetics that bears the personal stamp of one of the major figures in the evolutionary synthesis; and E. Mayr and W. B. Provine (eds.), *The Evolutionary Synthesis: Perspectives on the Unification of Biology* (Harvard University Press, Cambridge, MA, 1980), which contains essays by historians and biologists, including some of the major contributors to the synthesis.

Books that expose the fallacies of creationism and explain the nature of science and of evolutionary biology include R. T. Pennock, *Tower of Babel: The Evidence against the New Creationism* (MIT Press, Cambridge, MA, 1999); B. J. Alters and S. M. Alters, *Defending Evolution: A Guide to the Creation/*

Evolution Controversy (Jones and Bartlett, Sudbury, MA, 2001); M. Pigliucci, *Denying Evolution: Creationism, Scientism, and the Nature of Science* (Sinauer Associates, Sunderland, MA, 2002); and E. C. Scott, *Evolution versus Creationism: An Introduction*, second edition (University of California Press, Berkeley, 2009). The evidence for evolution is presented in two outstanding books, *Why Evolution Is True*, by J. A. Coyne (Viking, New York, 2009), and *The Greatest Show on Earth: The Evidence for Evolution*, by Richard Dawkins (Free Press, New York, 2009).

Darwin's birth, in 1809, was marked in 2009 by many bicentennial celebrations and books. Among these is *Evolution Since Darwin: The First 150 Years*, edited by M. A. Bell et al. (Sinauer Associates, Sunderland, MA, 2010), a collection of essays by historians and evolutionary biologists who summarize the state of knowledge and current research directions in all the subfields of evolutionary biology.

Websites

Several excellent websites provide good introductions to evolution; most of them also include material on teaching evolution and on creationism.

"Understanding Evolution" (http://evolution.berkeley.edu) is an excellent site developed by the Museum of Paleontology at the University of California, Berkeley.

"Evolution" (www.pbs.org/wgbh/evolution) is based on a superb series of WGBH/NOVA programs on the Public Broadcasting System.

The National Academy of Sciences of the U.S.A., the members of which are leaders in science, has a website devoted to evolution (www.nationalacademies.org/evolution) and has published an excellent 70-page booklet, *Science, Evolution, and Creationism* (2008), that can be accessed for free through the website or purchased at low cost (order at www.nap.edu).

Problems and Discussion Topics

1. How does evolution unify the biological sciences? What other principles might do so?

2. Analyze and evaluate Ralph Waldo Emerson's couplet,

   Striving to be man, the worm
   Mounts through all the spires of form.

   What pre-Darwinian concepts does it express? What fault in it will a Darwinian find?

3. Some scientists vigorously rejected Darwin's ideas when *On the Origin of Species* was published. Richard Owen, perhaps the most respected biologist in England, raised a number of objections to Darwin's ideas, including the following: "Are all the recognised organic forms of the present date, so differentiated, so complex, so superior to conceivable primordial simplicity of form and structure, as to testify to the effects of Natural Selection continuously operating through untold time? Unquestionably not. The most numerous living beings … are precisely those which offer such simplicity of form and structure, as best agrees … with that ideal prototype from which … vegetable and animal life might have diverged." How might Darwin, or you, argue against Owen's logic?

4. In the same spirit as the preceding question, find and analyze other arguments made by Owen or other critics of Darwin such as François Pictet, Adam Sedgwick, St. George Jackson Mivart, or Louis Agassiz. Some of their essays are reprinted in *Darwin and His Critics* by D. L. Hull (1973). Many of their criticisms were the same as those being raised by modern-day creationists.

5. Discuss how a creationist versus an evolutionary biologist might explain some human characteristics and the implications of their differences. Sample characteristics: eyes; wisdom teeth; individually unique friction ridges (fingerprints); five digits rather than some other number; susceptibility to infections; fever when infected; variation in sexual orientation; limited life span.

6. Since 2001, the complete DNA sequences of the genomes of many people have been determined. If humans, along with all other forms of life, have evolved from a common ancestor, what evidence of this would be expected in the human genome? In what ways might the history and processes of evolution help us to interpret and make sense of human genome sequence data?

7. How might the evolution of antibiotic resistance in pathogenic bacteria be slowed down or prevented? What might you need to know in order to achieve this aim?

8. Drawing on sources available in a good library, discuss how the "Darwinian revolution" affected one of the following fields: philosophy, literature, psychology, or anthropology.

9. Should both evolution and creationism be taught as alternative theories in science classes?

The Tree of Life: Classification and Phylogeny

About 2000 million (2 billion) years ago, a bacterium, not unlike the *Escherichia coli* bacteria in our intestines, took up residence within the single cell of another bacteria-like organism. This partnership flourished, since each partner evidently provided biochemical services to the other. The "host" in this partnership developed a modern nucleus, chromosomes, and mitotic spindle, while the "guest" bacterium within it evolved into a mitochondrion. This ancestral eukaryote gave rise to diverse unicellular descendants. Some descendant lineages subsequently became multicellular when the cells they produced by mitosis remained together. These cells evolved mechanisms of regulating gene expression that enabled groups of them to form different tissues and organs. One such lineage became the progenitor of green plants, another of the fungi and animals (**Figure 2.1**).

Between about 1000 million (1 billion) and 600 million years ago, a single animal species gave rise to two species that became the progenitors of two quite astonishingly different groups of animals: one evolved into the starfishes, sea urchins, and other echinoderms, and the other into the chordates, including the vertebrates. Most of the species derived from the earliest vertebrate are fishes, but one of these species would become the ancestor of the tetrapod (four-legged) vertebrates. About 150 million years after the first land-dwelling tetrapod evolved, some of its descendants stood on the brink of mammalhood. After another 125 million years or so, the mammals had diversified into many groups, including the first primates, adapted to life in trees. Some primates became small, some evolved prehensile tails, and one became the ancestor of large, tailless apes. About 14 million

How do we classify organisms? Despite its outward appearance, this rock hyrax (*Procavia capensis*) is not a rodent, but is instead most closely related to elephants and manatees. These herbivores live in social groups, in which some members act as sentries and warn of approaching predators. As in elephants and manatees, the male's testes are retained within the body. (Photo by the author.)

This chapter and others require familiarity with the fundamental aspects of molecular genetics, such as DNA and its transcription and translation. If you need to refresh your knowledge of codons, exons, pyrimidines, and so forth, please refer to a textbook of introductory biology.

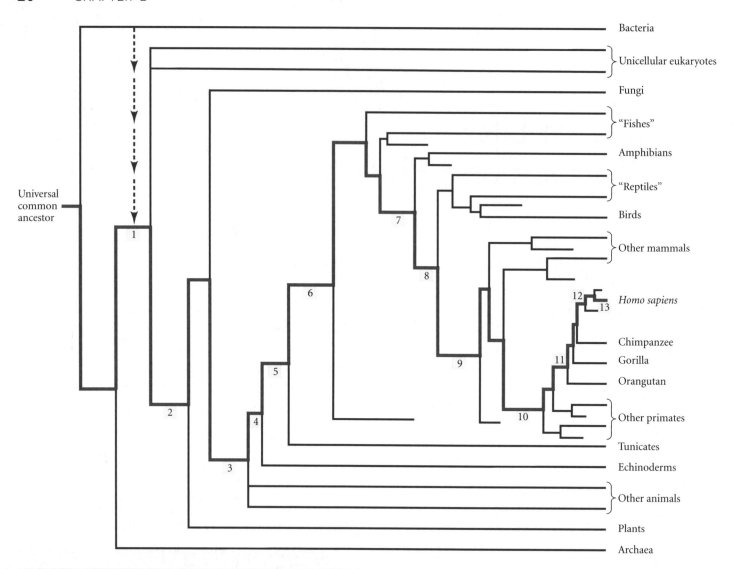

Universal common ancestor

Bacteria
Unicellular eukaryotes
Fungi
"Fishes"
Amphibians
"Reptiles"
Birds
Other mammals
Homo sapiens
Chimpanzee
Gorilla
Orangutan
Other primates
Tunicates
Echinoderms
Other animals
Plants
Archaea

1. Origin of eukaryotes: a symbiotic bacterium becomes the mitochondrion
2. Multicellularity: cell and tissue differentiation
3. Animals: internal digestive cavity; muscles
4. Deuterostomes: embryonic blastopore develops into anus
5. Chordates: notochord; dorsal nerve cord
6. Vertebrates: bony skeleton
7. Tetrapods: limbs
8. Amniotes: amniotic egg; other water-conserving features
9. Mammals: unique jaw joint; bones of middle ear; milk
10. Primates: binocular vision; arboreality
11. Anthropoid apes: loss of tail
12. Hominins: bipedalism
13. *Homo sapiens* spreads from Africa

FIGURE 2.1 Tracing the path (in red) of evolution to *Homo sapiens* from the universal ancestor of all life, in the context of the tree of life. Some of the major events are shown here as character changes; their position on the tree has been inferred from the distribution of character states across the phylogeny. Such evolutionary histories are often fascinating.

years ago, one such ape gave rise to the Asian orangutan on the one hand and an African descendant on the other. The African descendant split into the gorilla and another species. About 5 to 6 million years ago, that species, in turn, divided into one lineage that became today's chimpanzees and another that underwent rapid evolution of its posture, feet, hands, and brain: our own quite recent ancestor.

This is an astonishing story, but it is our best current understanding of some of the high points in our history, in which, metaphorically speaking, the human species developed as one twig in a gigantic tree, the great tree of life. With the passage of time, a species is likely to "branch"— to give rise to two species that evolve different modifications of some of their features. Those species branch in turn, and their descendants may be altered further still. By this process of branching and modification, repeated innumerable times over the course of many millions of years, many millions of kinds of organisms have evolved from a single ancestral organism (**Figure 2.2**).

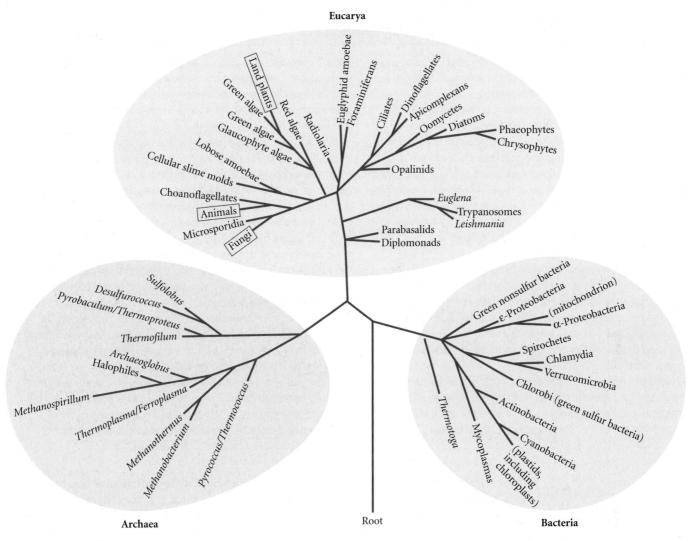

Eucarya

Land plants
Green algae
Green algae
Glaucophyte algae
Red algae
Radiolaria
Euglyphid amoebae
Foraminiferans
Ciliates
Dinoflagellates
Apicomplexans
Oomycetes
Diatoms
Phaeophytes
Chrysophytes
Lobose amoebae
Cellular slime molds
Opalinids
Choanoflagellates
Animals
Euglena
Trypanosomes
Leishmania
Microsporidia
Fungi
Parabasalids
Diplomonads

Sulfolobus
Desulfurococcus
Pyrobaculum/*Thermoproteus*
Thermofilum
Archaeoglobus
Halophiles
Methanospirillum
Thermoplasma/*Ferroplasma*
Methanothermus
Methanobacterium
Pyrococcus/*Thermococcus*

Green nonsulfur bacteria
ε-Proteobacteria
(mitochondrion)
α-Proteobacteria
Spirochetes
Chlamydia
Verrucomicrobia
Chlorobi (green sulfur bacteria)
Actinobacteria
Cyanobacteria (plastids, including chloroplasts)
Thermotoga
Mycoplasmas

Archaea   Root   **Bacteria**

FIGURE 2.2 The tree of life. This estimate of the relationships among some of the major branches is based mostly on DNA sequences, especially those of genes encoding ribosomal RNA. Of the three empires of life, Archaea and Eucarya appear to have the most recent common ancestor. The majority of taxa in the tree are unicellular; the land plants, animals, and fungi comprise the major multicellular taxa. (After Baldauf et al. 2004.)

Evolutionary biologists have developed methods of "reconstructing" or "assembling" the tree of life—of estimating the **phylogeny**, or genealogical relationships, of organisms (i.e., which species share a recent **common ancestor**, which share distant ancestors, and which share even more remote ancestors). The resulting portrayal of relationships is not only fascinating in itself (have you ever thought of yourself as related to a starfish, a butterfly, or a mushroom?), but is also an important foundation for understanding many aspects of evolutionary history, such as the pathways by which various characteristics have evolved.

A few points in our brief sketch of human origins, such as the timing of certain events, have been learned from the fossil record. However, most of this history has been determined from studying not fossils, but living organisms. In this chapter, we will become acquainted with some of the methods by which we can infer phylogenetic relationships, and we will see how our understanding of those relationships is reflected in the classification of organisms. In the following chapter, we will examine some common evolutionary patterns that these approaches have helped to elucidate.

Classification

PHYLOGENETIC ANALYSIS—the study of relationships among species—has historically been closely associated with the classification and naming of organisms (known as TAXONOMY). Both are among the tasks of the field of SYSTEMATICS.

Box 2A Taxonomic Practice and Nomenclature

Standardized names for organisms are essential for communication among scientists. To ensure that names are standardized, taxonomy has developed rules of procedure.

Most species are named by taxonomists who are experts on that particular group of organisms. A new species may be one that has never been seen before (e.g., an organism dredged from the deep sea), but many unnamed species are sitting in museum collections, awaiting description. Moreover, a single species often proves, on closer study, to be two or more very similar species. A taxonomist who undertakes a REVISION—a comprehensive analysis—of a group frequently names new species. A species name has legal standing if it is published in a journal or even in a privately produced publication that is publicly available.

The name of a species consists of its genus name and its specific epithet; both are Latin or latinized words. These words are always *italicized* (or underlined), and the genus name is always capitalized. In entomology and certain other fields, it is customary to include the name of the AUTHOR (the person who conferred the specific epithet); for example, the corn rootworm beetle *Diabrotica virgifera* LeConte.

Numerous rules govern the construction of species names (e.g., genus name and specific epithet must agree in gender: *Rattus norvegicus*, not *Rattus norvegica*, for the brown rat). It is recommended that the name have meaning [e.g., *Vermivora* ("worm eater")] *chrysoptera* ("golden-winged") for the golden-winged warbler; *Rana warschewitschii*, "Warschewitsch's frog"], but it need not. Often a taxonomist will honor another person by naming a species after him or her. (Naming a species after oneself is gauche and just isn't done.)

The first rule of nomenclature is that no two species of animals, or of plants, can bear the same name. (It is permissible, however, for the same name to be applied to both a plant and an animal genus; for example, *Alsophila* is the name of both a fern genus and a moth genus.) The second is the rule of PRIORITY: the valid name of a taxon is the oldest available name that has been applied to it. Thus it sometimes happens that two authors independently describe the same species under different names; in this case, the valid name is the older one, and the younger name is a junior SYNONYM. Conversely, it may turn out that two or more species have masqueraded under one name; in this case, the name is applied to the species that the author used in his or her description. To prevent the obvious ambiguity that could arise in this way, it has become standard practice for the author to designate a single specimen (the TYPE SPECIMEN, or HOLOTYPE) as the "name-bearer" so that later workers can determine which of several similar species rightfully bears the name. The holotype, usually accompanied by other specimens (PARATYPES) that exemplify the range of variation, is deposited and carefully preserved in a museum or herbarium.

In the early 1700s, European naturalists believed that God must have created species according to some ordered scheme, as we saw in Chapter 1. It was therefore a work of devotion to discover "the plan of creation" by cataloguing the works of the Creator and discovering a "natural," true, classification. The scheme of classification that was adopted then, and is still used today, was developed by the Swedish botanist Carolus Linnaeus (1707–1778). Linnaeus introduced BINOMIAL NOMENCLATURE, a system of two-part names consisting of a genus name and a specific epithet (such as *Homo sapiens*). He proposed a system of grouping species in a HIERARCHICAL CLASSIFICATION of groups nested within larger groups (such as genera nested within families) (**Box 2A**). The levels of classification—such as kingdom, phylum, class, order, family, genus, and species—are referred to as TAXONOMIC CATEGORIES, whereas a particular group of organisms assigned to any of these levels is a **taxon** (plural: **taxa**). Thus the rhesus monkey (*Macaca mulatta*) is placed in the genus *Macaca*, in the family Cercopithecidae, in the order Primates; *Macaca*, Cercopithecidae, and Primates are taxa that exemplify the taxonomic categories genus, family, and order, respectively. Several "intermediate" taxonomic categories, such as superfamily and subspecies, are sometimes used in addition to the more familiar and universal ones.

In assigning species to **higher taxa** (those above the species level), Linnaeus used features that he imagined represented propinquity in God's creative scheme. For example, he defined the order Primates by the features "four parallel upper front [incisor] teeth; two pectoral nipples," and on this basis included bats among the Primates. But without an evolutionary framework, naturalists had no objective basis for classifying mammals by their teeth rather than by, say, their color or size. Indeed, because we now use an evolutionary framework, bats are not included in the same order as primates today.

Classification took on an entirely different significance after *On the Origin of Species* was published in 1859. In the Galápagos Islands, as we saw in Chapter 1, Darwin had observed that different islands harbored similar, but nevertheless distinguishable, mockingbirds.

Box 2A *(continued)*

In revising a genus, a taxonomist usually introduces changes in the taxonomy. Here are some examples of such changes:

- Species that were placed by previous authors in different genera may be brought together in the same genus because they are shown to be closely related. These species retain their specific epithets, but if they are shifted to a different genus, the author's name is written in parentheses.

- A species may be removed from a genus because it is determined not to be closely related to the other members, and placed in a different genus.

- Forms originally described as different species may prove to be the same species and be synonymized.

- New species may be described.

The rules for naming higher taxa are not all as strict as those for species and genera. In zoology (and increasingly in botany), names of subfamilies, families, and sometimes orders are formed from the name of the type genus (the first genus described). Most family names of plants end in -aceae. In zoology, subfamily names end in -inae and family names in -idae. Thus *Columba* (Latin for "pigeon"), the genus of the familiar pigeon, is the type genus of the family Columbidae and the subfamily Columbinae; *Rosa* (rose) is the type genus of the family Rosaceae. Endings for categories above family are standardized in some, but not all, groups (e.g.,

-formes for orders of birds, such as Passeriformes, the perching birds that include *Passer*, the house sparrows). Names of taxa above the genus level are not italicized, but are always capitalized. Adjectives or colloquial nouns formed from these names are not capitalized: thus we may refer to murids or to murid rodents (in the family Muridae) without capitalization.

Systematists today rely on phylogeny when classifying organisms. A **monophyletic** taxon is one that includes all the named descendants of a particular common ancestor (for example, the traditional class Aves, which includes all birds, is monophyletic). A **paraphyletic** taxon includes some, but not all, of the descendants from a particular ancestor. (The traditional class Reptilia is paraphyletic because it did not include the birds, which share a common ancestor with dinosaurs and crocodiles.) A **polyphyletic** taxon includes species from two different ancestors, each of which has descendants that are placed in different taxa. [The falcons, hawks, and eagles have been included in the order Falconiformes, but DNA evidence suggests that falcons are more closely related to parrots and songbirds, and that hawks and eagles may be more closely related to a clade that includes owls (Hackett et al. 2008). Thus Falconiformes may prove to be polyphyletic.] Most systematists today hold the opinion that classifications should consist of monophyletic taxa only and thus reflect phylogenetic relationships.

He came to suspect that the different forms of mockingbirds had descended from a single ancestor, acquiring slight differences over time. But this thought, logically extended, suggested that that ancestor itself had been modified from an ancestor further back in time, which could have given rise to yet other descendants—various South American mockingbirds, for instance. By this logic, some remote ancestor might have been the progenitor of all species of birds, and a still more remote ancestor the progenitor of all vertebrates.

Darwin therefore gave meaning to the notion of "closely related" species: they are descended from a relatively recent common ancestor, whereas more distantly related species are descended from a more remote (i.e., further back in time) common ancestor. The features that species share, such as the vertebrae that all vertebrates possess, were not independently bestowed on each of those species by a Creator, but were inherited from the ancestral species in which the features first evolved. With breathtaking daring, Darwin ventured that all species of organisms, in all their amazing diversity, had descended by endless repetition of such events, through long ages, from perhaps only one common ancestor—the universal ancestor of all life.

In Darwin's words, all species, extant and extinct, form a great "Tree of Life," or phylogenetic tree, in which closely adjacent twigs represent living species derived only recently from their common ancestors, whereas twigs on different branches represent species derived from more ancient common ancestors (**Figure 2.3**). He expressed this metaphor in some of his most poetic language:

> The affinities of all the beings of the same class have sometimes been represented by a great tree. I believe this simile largely speaks the truth. The green and budding twigs may represent existing species; and those produced during former years may represent the long succession of extinct species. At each period of growth all the growing twigs have tried to branch out on all sides, and to overtop

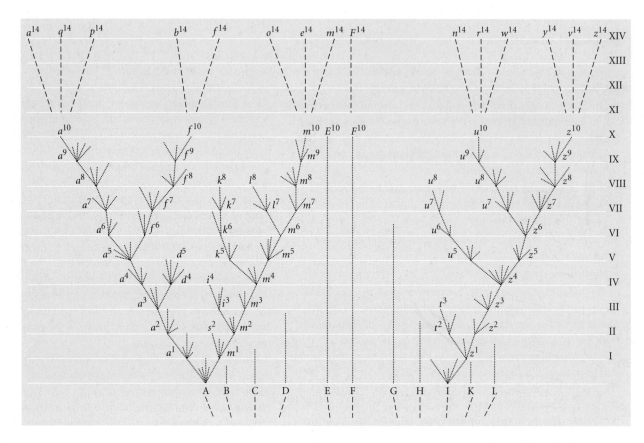

FIGURE 2.3 Darwin's representation of hypothetical phylogenetic relationships, showing how lineages diverge from common ancestors and give rise to both extinct and extant species. Time intervals (between Roman numerals) represent thousands of generations. Darwin omitted the details of branching for intervals X through XIV. Extant species (at time XIV) can be traced to ancestors A, F, and I; all other original lineages have become extinct. Distance along the horizontal axis represents degree of divergence (as, for example, in body form). Darwin recognized that rates of evolution vary greatly, showing this by different angles in the diagram; for instance, the lineage from ancestor F has survived essentially unchanged. (From Darwin 1859.)

and kill the surrounding twigs and branches, in the same manner as species and groups of species have at all times overmastered other species in the great battle for life. The limbs divided into great branches, and these into lesser and lesser branches, were themselves once, when the tree was young, budding twigs; and this connection of the former and present buds by ramifying branches may well represent the classification of all extinct and living species in groups subordinate to groups. Of the many twigs which flourished when the tree was a mere bush, only two or three, now grown into great branches, yet survive and bear the other branches; so with the species which lived during long-past geological periods, very few have left living and modified descendants. From the very first growth of the tree, many a limb and branch has decayed and dropped off; and these fallen branches of various sizes may represent those whole orders, families, and genera which have now no living representatives, and which are known to us only in a fossil state. As we here and there see a thin, straggling branch springing from a fork low down in a tree, and which by some chance has been favoured and is still alive on its summit, so we occasionally see an animal like the Ornithorhynchus or Lepidosiren,* which in some small degree connects by its affinities two large branches of life, and which has apparently been saved from fatal competition by having inhabited a protected station. As buds give rise by growth to fresh buds, and these, if vigorous, branch out and overtop on all sides many a feebler branch, so by generation I believe it has been with the great Tree of Life, which fills with its dead and broken branches the crust of the earth, and covers the surface with its ever-branching and beautiful ramifications.

Ornithorhynchus, the duck-billed platypus, is a primitive, egg-laying mammal. *Lepidosiren* is a genus of living lungfishes, a group that is closely related to the ancestor of the tetrapod (four-legged) vertebrates, and which is known from ancient fossils.

Under Darwin's hypothesis of descent with modification, a hierarchical classification reflects a real historical process that has produced organisms with true genealogical relationships, close or distant in varying degree. Different genera in the same family share fewer characteristics than do species within the same genus because each has departed further from their more remote common ancestor; different families within an order stem from a still more remote common ancestor and retain still fewer characteristics in common. Classification, then, can portray, to some degree, *the real history of evolution*.

Inferring Phylogenetic History

Phylogenetic trees

Two of the major processes in the evolution of a group of organisms are **anagenesis**, which is evolutionary change of features within a single lineage, and **cladogenesis**, or branching of a lineage into two or more descendant lineages (from the Greek *clados*, "branch"). Following cladogenesis, anagenesis in each lineage results in their becoming more different from each other (**divergence**, or **divergent evolution**). A **phylogenetic tree** (or phylogeny, or CLADOGRAM) is a representation of a postulated history of cladogenesis, of the branching of ancestral lineages into diverse descendant lineages (**Figure 2.4**). Two other kinds of evolutionary events are sometimes also represented in such diagrams: extinction (e.g., taxon F in Figure 2.4) and reticulation, which occurs when two lineages merge or form a "hybrid" descendant, so that the tree has a netlike structure (from the Latin *rete*, "net"). We will focus on branching trees here and consider reticulation later.

The phylogenetic tree in Figure 2.4 has three "terminal taxa," human (A), chimpanzee (B) and bonobo (C), all living species. [The same tree could also represent three higher taxa, such as lizards (A), crocodiles (B), and birds (C).] Each segment in the tree is a **lineage**, or BRANCH, which may split at an internal BRANCH POINT (such as D), representing the formation of two descendant lineages (B and C) from their common ancestor. All the descendants of any one ancestor form a **clade**; thus B and C form a clade (also called a MONOPHYLETIC GROUP) that is "nested" within the larger clade A + B + C. Two clades that originate from a common ancestor are called **sister groups**. (If B and C are species, they are sister species.)

The tree in Figure 2.4 represents the *genealogical relationships* among the taxa, meaning the temporal order of branching by which they have originated from the common ancestor (E in this case), which lies at the root of the tree. Thus a tree has an implicit *time scale* from past (at the root) to more recent time (e.g., the present). This time scale may be relative, implying only the order of branching, but in some cases an absolute time scale is used, and

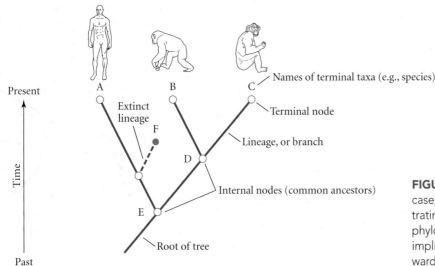

FIGURE 2.4 A phylogenetic tree of three taxa, in this case, human (A), chimpanzee (B), and bonobo (C), illustrating major phylogenetic terms. The time scale in most phylogenetic trees is a relative one, but the tree always implies the passage of time from the root of the tree toward the tips of the branches.

(A)

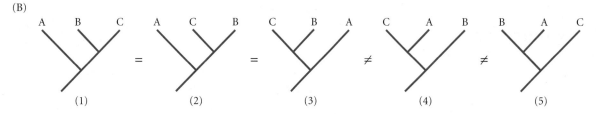

FIGURE 2.5 Different representations of phylogenies. (A) These four trees carry exactly the same information. (B) Trees 1, 2, and 3 carry exactly the same information, namely, that species B and C are more closely related than either is to A. Trees 4 and 5 represent different possible relationships among the three species.

(B)

(1)   (2)   (3)   (4)   (5)

branch points are drawn to match the dates at which the branching events are thought to have occurred. All phylogenetic trees, however, are intended to convey one crucial aspect of evolution: the order of branching, which defines which species are more closely, and which more distantly, related. *Two species are more closely related to each other than to a third species if they are derived from a more recent common ancestor.* By analogy, two siblings are more closely related to each other than they are to a cousin because they share more recent common ancestors (their parents) with each other than with their cousin (a grandparent).

Note that *closeness of relationship is not the same as similarity*: a person might more closely resemble her cousin than her sister with respect to eye color or many other features, but she is still more closely related to her sister. Likewise, two closely related species may be less similar to each other than one is to a more distantly related species. A phylogeny portrays relationship (common ancestry), not similarity.

A phylogenetic tree may be drawn in any of several different ways as a matter of convenience or convention. **Figure 2.5A** illustrates four equivalent trees that differ in the orientation of the implied time axis and in their use of angular or rectangular junctions. Furthermore, the clades arising from a branch point may be rotated without any change in the diagram's meaning, as shown by trees 1, 2, and 3 in **Figure 2.5B**. However, trees 4 and 5 in this figure represent relationships different from those variously portrayed in trees 1–3.

The length of the branches in a phylogenetic tree may or may not have any meaning, depending on what information a researcher means to convey. If the tree conveys only branching order, the relative lengths of branches have no significance. If the tree is accompanied by an absolute time scale, however, the positions of branch points and the lengths of branches are estimates of real time. In some phylogenetic trees, the length of a branch indicates an estimate of the number of evolutionary changes (e.g., DNA nucleotide substitutions) that have transpired in that evolving lineage.

The ROOT of a phylogenetic tree is the ancestral lineage that first branches and gives rise to two or more descendants. So far, we have considered rooted trees, but it will be useful to consider diagrams called "UNROOTED trees" as well: branching diagrams in which the position of the ancestor (the root) relative to the various species is not indicated a priori. This concept is helpful when we seek to determine which of several possible phylogenetic relationships among species is most likely to be the true one. Tree 1 in **Figure 2.6A**, in which the root lies between species D and the others, can be portrayed as the unrooted tree 2, in which the dot indicates the position of the lineage that is ancestral to all the species in tree 1. An unrooted tree (tree 2) can be converted to any of several hypothetical phylogenies by "rooting" it at different branches (e.g., trees 3 and 5 result from rooting tree 2 at different branches). Rooting a tree confers a temporal direction on the relationships among the taxa.

Often, biologists are interested in analyzing the relationships among certain taxa (perhaps A, B, and C), which they refer to collectively as an **ingroup**. They will include **outgroups** in the analysis: one or more other taxa (such as D) that they are quite sure

(A)

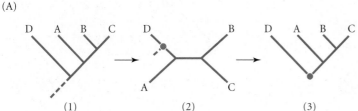

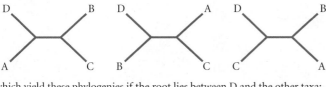

(B)   For 4 taxa, there are 3 possible unrooted trees:

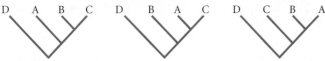

which yield these phylogenies if the root lies between D and the other taxa:

For 4 taxa, there are 3 possible unrooted trees;
For 5, there are 15;
For 10, there are 2,027,025!

FIGURE 2.6 Phylogenetic analyses often use unrooted trees, which are then converted to rooted trees. (A) A rooted tree (1) can be shown as an unrooted tree (2). An unrooted tree can be converted into a rooted tree by placing the root along a branch. Placing the root as shown in tree 2 produces the rooted tree 3; placing it as shown in tree 4 produces a different rooted tree (5). (B) For four taxa, there are three possible unrooted trees. If we know where to root them (e.g., between D and the other taxa), there are three corresponding rooted trees.

(based on prior evidence) are more distantly related to the ingroup than the members of the ingroup are to each other. All the possible phylogenetic trees considered plausible will be trees rooted between the ingroup and the outgroups (as in **Figure 2.6B**).

Data for inferring phylogenies

Although relationship among species does not equal similarity, certain similarities and differences in the characteristics of species constitute the evidence used for inferring their phylogenetic relationships. Each trait of an organism (e.g., the number of toes on a hind limb) is called a **character**; a character may have various **character states** (e.g., five toes in humans, three in rhinoceroses, one in horses). The variation in a character may be DISCRETE (as in toe number) or CONTINUOUS (e.g., length of a toe). As we will see, determining phylogenetic relationships from such evidence is beset by various difficulties. Research on how best to infer phylogenies despite various pitfalls has resulted in analytical methods that are still being improved today.

Phenotypic characters that have proved useful for phylogenetic analyses of various organisms have included not only external and internal morphological features, but also differences in behavior, cell structure, biochemistry, and chromosome structure. Phylogenetic study has been revolutionized, and placed on a far firmer basis, by DNA sequencing, which can reveal variation at hundreds or even thousands of base pair positions. At each such position ("site"), the identity of the nucleotide base (A, T, C, or G) may be considered a character. (We note only one base, such as A, rather than the pair A–T, because specifying the complementary base adds no information.) DNA sequences often provide more data than morphological or other phenotypic features, and they have enabled systematists to find the evolutionary relationships of many organisms that could not formerly be classified because they have so few variable phenotypic features (**Figure 2.7**). Moreover, the four possible states of each character in a DNA sequence are much less ambiguous than many phenotypic features (for example, there can be gradations of size or coloration).

A character is **homologous** in two species if they inherited it from their common ancestor. The character might be similar or identical in the two species (e.g., number of eyes in

FIGURE 2.7 Phylogenetic analysis, based on DNA sequences, has revealed the relationships of some formerly puzzling organisms. (A) *Rafflesia arnoldii* (left), a parasite with the largest flowers of any plant, is now known to be related to a group of families that includes spurges such as *Euphorbia sikkimensis* (right). (B) The fungus cultivated underground by leafcutter ants (left) has been identified as a member of the Lepiotaceae, a family that includes mushrooms such as *Macrolepiota procera* (right). (B, photo of ants courtesy of U. Mueller, Texas/Austin).

(A)

(B)

Fungus

various mammals), or it might have different character states. Similarly, a character state is homologous in two species if that state was inherited from their common ancestor (e.g., one toe in horses and zebras). *Homologous character states that are shared among species provide good evidence of common ancestry if they evolved only once and have been retained by all the descendants of the ancestor in which they first evolved.*

Figure 2.8 shows a hypothetical phylogenetic tree for a group of five taxa, each illustrating evolutionary change in a different character (in this case, a different DNA site). In Figure 2.8A and B, nucleotide substitutions at two DNA sites conform to the criteria for homology. For character 1, A (adenine) is an ANCESTRAL character state, and C (cytosine) is a DERIVED character state; for character 2, T is ancestral and G is derived. If taxa 1, 2, and 3 constitute an ingroup, and if we know that taxa 4 and 5 are outgroups, then these shared derived states (nucleotide substitutions) provide evidence that the ingroup taxa (3, 2, and 1) are indeed a clade (as shown by character 1) and that taxa 1 and 2 are more closely related to each other than they are to taxon 3 (as shown by character 2). As a real example, the presence of vertebrae has long been considered evidence that the various vertebrate animals share a common ancestor that is not shared by other animals. There is good reason to think that vertebrae evolved only once, and that all the descendants of the vertebrate common ancestor possess vertebrae.

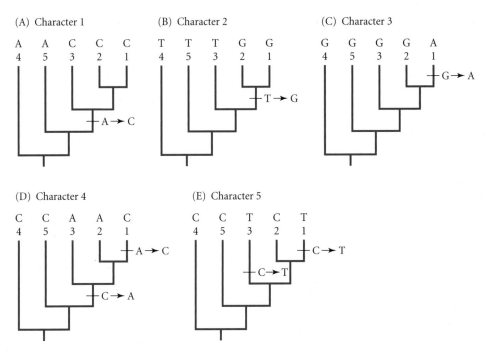

FIGURE 2.8 Phylogenetically informative and uninformative similarities among species, illustrated by nucleotide substitutions (labeled at top) at five sites in a corresponding DNA sequence from five species (1–5). The trees show the actual history of branching that led to these species. (A) Species 1–3 share a C (cytosine) at this DNA site, whereas species 4 and 5 have A (adenine). The simplest historical explanation of this pattern is that C was substituted for A in the common ancestor of 1–3. C is a *derived character state* that is *homologous* in these species. A is an *ancestral character state*. Character state C provides evidence that species 1–3 form a clade. (B) At this site, species 1 and 2 share a G (guanine). The simplest explanation is substitution of G for T (thymine) in the common ancestor of 1 and 2. This pattern provides evidence that species 1 and 2 form a clade, nested within the clade 1–3. (C) Species 1 differs from all other species at this site, because the G → A substitution occurred after species 1 diverged from its nearest relative. This character does not tell us that species 1 is more closely related to 2 than to any other species: it is *uninformative*. Likewise, the similarity of 2 and 3 to 4 and 5 does not mean they form a clade; they simply have not undergone evolution at this DNA site. A character is phylogenetically informative only if it has at least two states, each shared by at least two taxa. (D) The ancestral C was replaced by A in the ancestor of 1–3, but then the A was replaced by C in species 1: a case of *evolutionary reversal* (resulting in *homoplasy*). This character is actually *misleading*: it can be interpreted as evidence that species 1 is more closely related to 4 and 5 than to 2 and 3, and that species 2 and 3 are each other's closest relatives, sharing the derived character state A. (E) Another example of a misleading homoplasy: *convergent evolution*. C has been replaced by T twice, independently in species 3 and 1. This could lead us to think that these species are closest relatives, when in fact they are not.

Not all similarities provide useful phylogenetic information, however, as shown in Figure 2.8 C–E. In Figure 2.8C, character 3 has not evolved at all, except in taxon 1. State G is homologous among all the other taxa, but two members of the ingroup (species 2 and 3) are more similar to each other than to species 1 only because they have not evolved: they share the ancestral character state G. The difference between them and species 1 is consistent with any of the three possible relationships among these three species. We can conclude that differences in the rate of evolution among different lineages can lead to patterns of similarity versus difference that are uninformative or misleading. In particular, *ancestral character states that are shared between species are less likely to provide reliable phylogenetic information than derived states*. For example, lizards and crocodiles, which share scales and quadrupedal (four-legged) locomotion, are more similar to each other than they are to feathered, bipedal birds. But scales and quadrupedal locomotion are ancestral traits; birds have evolved faster in these respects. Actually, birds and crocodiles are closer relatives than are crocodiles and lizards. (On the other hand, the derived states—feathers, bipedalism—are evidence that ostriches, swallows, and other birds form a clade of descendants from a common ancestor.)

A character state is **homoplasious** if it has independently evolved two or more times, and so does not have a unique origin. Such similarities are phylogenetically misleading because the taxa that share such a character state have not all inherited it from their common ancestor. The causes of homoplasy include evolutionary reversal, convergent evolution (convergence), and parallel evolution (parallelism).

In Figure 2.8D, character 4 shows an EVOLUTIONARY REVERSAL in nucleotide substitutions, from C to A and back to C. If our only evidence of relationships among species 1, 2, and 3 were this character, we could be misled into thinking that species 2 and 3 were the closest relatives. Similarly, the earliest chordates lacked paired limbs, whereas most later vertebrates have paired fins or legs, inherited from a Silurian ancestor. However, moray eels, caecilians, snakes, and several lineages of lizards have reverted to the limbless state.

In CONVERGENCE or PARALLELISM, a similar or identical character state evolves independently in two or more lineages (Figure 2.8E). The loss of limbs in the several distantly related vertebrate groups named above illustrates convergent evolution.

Similarly, bipedal locomotion evolved independently in humans and in theropod dinosaurs (including *Tyrannosaurus* and the birds). (All of these causes of homoplasy are discussed further in Chapter 3.)

Inferring phylogenies: The method of maximum parsimony

Even before Darwin, taxonomists held that some traits provided a firmer foundation for classification than others. Animals with vertebrae formed a more "natural" group than, say, all black animals (crows, bears, beetles, etc.). After Darwin, taxonomists recognized that evolution explains why this should be so, and they named taxa both on the basis of postulated common ancestry and on the basis of evolutionary divergence. For example, vertebrates were placed in the class Reptilia if their embryos have an amnion, which was taken as evidence of common ancestry; but birds were placed in a separate class (Aves), even though they have an amnion, because their morphology is so different, having evolved to fit a very different way of life (flight). Taxonomists had not formulated objective criteria for inferring evolutionary relationships; nevertheless, many of the classifications they produced, based on morphological features, have proved to reflect relationships that are supported by modern research.

Many systematists trace the modern practice of inferring phylogenetic relationships to the German entomologist Willi Hennig. Hennig (1966) pointed out that, as we have seen, taxa may be similar because they share (1) uniquely derived character states, (2) ancestral character states, or (3) homoplasious character states. Hennig asserted that only similarities that are due to uniquely derived character states (which he called **synapomorphies**) provide evidence for the clades—the nested monophyletic groups—that make up a phylogenetic tree. Thus, for example, we are confident that the tetrapod limb, the amnion, and the feather each evolved only once. We conclude that all tetrapods (vertebrates with four limbs rather than fins) form a single clade, that the amniotes* form a monophyletic group within the tetrapods, and that all feather-bearing animals (birds) form a single monophyletic group within the amniotes (**Figure 2.9**). However, this does not mean that all vertebrates without feathers form a single branch, and indeed they do not. The lack of feathers is an ancestral character state that does not provide evidence that featherless animals are all more closely related to one another than to birds. (Fishes, lizards, and frogs all lack feathers, but so, after all, do all invertebrates.) Hennig thus articulated a *scientific basis* for determining phylogenetic relationships—that is, the real history of evolution. He and his followers also promulgated an *opinion* on how organisms should be classified: every named taxon should be a monophyletic group (i.e., a clade; advocates of this position are called CLADISTS).

Hennig's principle, that uniquely derived character states define monophyletic groups, poses two difficulties: First, how can we tell which state of a character is derived? Second, how can we tell whether it is uniquely derived or homoplasious? It might be supposed that the fossil record would tell us what the ancestor's characteristics were, and sometimes it is indeed very helpful. But as we will see, the relationships among fossils and living species have to be interpreted, just like those among living species alone. Besides, the great majority of organisms have a very incomplete fossil record (as described in Chapter 4).

All the methods devised to infer phylogenies require evidence based on many characters, not just one. Ideally, each character will provide independent evidence, so that each monophyletic group in the phylogeny will be "supported" by a lot of independent evidence. An early method devised to achieve this goal depends on the concept of parsimony. **Parsimony** is the principle, dating at least from the fourteenth century, that the simplest explanation,

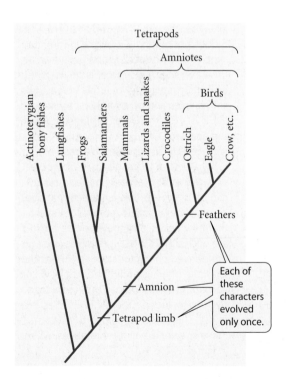

FIGURE 2.9 A phylogeny of some groups of vertebrates, showing monophyletic groups (tetrapods, amniotes, birds) whose members share derived character states that evolved only once (synapomorphies).

---

*AMNIOTES are those vertebrates—reptiles, birds, and mammals—characterized by a major adaptation for life on land: the amniotic egg, with its tough shell, protective embryonic membranes (chorion and amnion), and a membranous sac (allantois) for storing embryonic waste products.

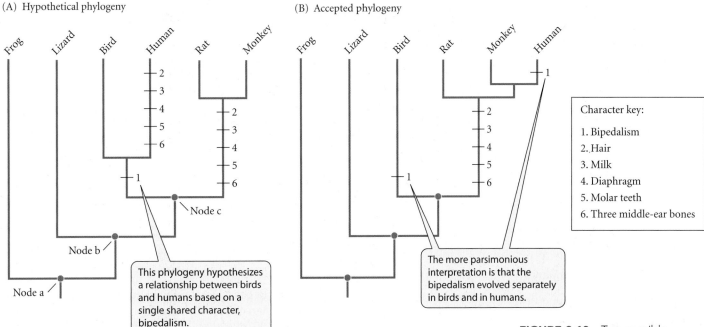

(A) Hypothetical phylogeny

(B) Accepted phylogeny

This phylogeny hypothesizes a relationship between birds and humans based on a single shared character, bipedalism.

The more parsimonious interpretation is that the bipedalism evolved separately in birds and in humans.

Character key:

1. Bipedalism
2. Hair
3. Milk
4. Diaphragm
5. Molar teeth
6. Three middle-ear bones

FIGURE 2.10 Two possible hypotheses for the phylogenetic relationships of humans. Frog and lizard are outgroups, which lack the six characters listed in the box. That implies that possession of these characters is a derived condition. (A) A hypothetical phylogeny postulating a close relationship between humans and birds, based on a shared derived character: bipedal (two-legged) posture and locomotion. Tick marks show the changes in several characters that are implied by each phylogeny. This phylogeny implies that characters 2–6 all evolved twice. (Alternatively, they could have evolved in the common ancestor C, but then have been lost in the bird lineage. In either case, ten evolutionary changes must be postulated for these five characters.) (B) The accepted phylogeny, in which humans are most closely related to other mammals. Characters 2–6 are considered uniquely derived synapomorphies of mammals. The accepted phylogeny requires fewer (six) evolutionary changes than the phylogeny in (A), and is therefore a more parsimonious hypothesis.

requiring the fewest undocumented assumptions, should be preferred over more complicated hypotheses that require more assumptions for which evidence is lacking. The method of **maximum parsimony** is among the simplest methods of phylogenetic analysis.

Any tree that is drawn to represent the phylogeny of a set of species is a hypothesis, one of many conceivable hypotheses (i.e., possible trees). (As noted in Figure 2.6B, the number of possible trees for even a modest number of species is astronomical.) The method of maximum parsimony is based on the principle that among the various phylogenetic trees that can be imagined for a group of taxa, *the best estimate of the true phylogeny is the one that requires us to postulate the fewest evolutionary changes.* Notice in Figure 2.8 that whereas the trees with strictly homologous character states (A and B) show only a single state change, the trees with homoplasious changes (D and E) show multiple changes in a single character. *Thus the tree with the fewest evolutionary changes, according to the parsimony principle, has the fewest homoplasious changes in character states: it requires us to postulate the fewest "extra" evolutionary changes for which we have no other, independent, evidence.* If we are quite sure that the common ancestor of birds and mammals did not have milk and other typical mammalian features (as we deduce from their absence in other vertebrates, i.e., outgroups), then the fact that birds and humans are both bipedal will not mislead us into supposing that humans are more closely related to birds than to other mammals, for this hypothesis requires us to postulate homoplasious evolution (i.e., multiple origins) of milk and other mammalian characters (**Figure 2.10**).

The steps in a maximum parsimony analysis, then, are these (**Figure 2.11**):

1. Select a set of ingroup species (e.g., species 1, 2, and 3 in Figure 2.11) whose phylogenetic relationships we wish to determine, and one or more (preferably more) outgroup species (e.g., species 4 in Figure 2.11) that we are sure, from prior information, are less closely related to the ingroup species than the ingroup species are to each other. (For example, the ingroup species could be several primates, with species from other orders of mammals serving as outgroups.)

2. Determine the state of each of many characters in each species, and enter those data in a data matrix (as in Figure 2.11A).

3. Draw an unrooted tree joining all the species and plot the state changes of the first character on this tree, entering the fewest changes that can account for the differences in character

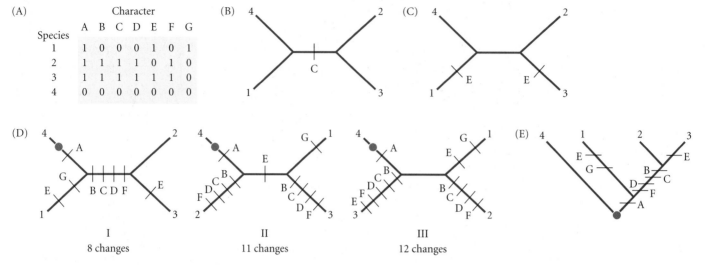

FIGURE 2.11 Steps in a phylogenetic analysis using the maximum parsimony method. (A) A data matrix of seven characters (A–G) scored in each of four species (1–4). Each character can have two different states, denoted as 0 and 1. Species 4 is an outgroup; we are interested in finding the relationships among the ingroup species 1, 2, and 3. (B) On one of the possible unrooted trees that can be drawn for these four species, the change in character C is marked in the position that requires the smallest number of changes, namely, between species 1 + 4 (with state 0) and 2 + 3 (with state 1). (C) When the same is done for character E, it proves necessary to postulate two changes in state (as marked) on this tree. (D) Here are shown all three possible unrooted trees for these four species, and the position of state changes in all characters is marked on each tree. Tree I has the fewest evolutionary changes across all characters, so this is the most parsimonious unrooted tree. (E) Because D is an outgroup, meaning that we are confident that it is more distantly related to the other species than they are to one another, we root tree I between D and the other species, yielding a most parsimonious phylogeny in which species 2 and 3 are the closest relatives.

state among the species (Figure 2.11B). Repeat this step for all the characters (e.g., Figure 2.11C), and count the total number of character state changes (see tree I in Figure 2.11D).

4. Repeat step 3 for all possible unrooted trees that can be drawn for the species (trees I–III in Figure 2.11D). For more than a few species, there will be a huge number of possible trees, so use one of several available computer programs to accomplish this gargantuan task.

5. Choose the "shortest" tree, i.e., the one that requires the fewest character state changes to explain the data in the character state matrix. Now root that tree between the outgroups and the ingroup to form a time-directed conventional phylogenetic tree (Figure 2.11E).

This tree is the "best estimate" of the true phylogeny, according to the assumptions of the maximum parsimony method. Of course, there may be many trees, with somewhat different arrangements, that are equally short, or nearly so, in which case the investigator may present a CONSENSUS TREE. For example, suppose that among several nearly equally short phylogenetic arrangements of twenty species, three (or more) species always form a monophyletic group, but with different arrangements of those three species with respect to one another. There is no basis on which to choose among these arrangements, so the three species may be represented by three branches from a single ancestral NODE. Instead of forming a DICHOTOMY (simple fork), these branches form an "unresolved POLYTOMY" on the consensus tree.

Parsimony analysis of DNA sequences: An example

The parsimony method was developed for use with either phenotypic (e.g., morphological) or molecular sequence data, but it originated at a time when molecular data were sparse and expensive to obtain. One example of a most parsimonious phylogeny based on DNA sequences is a pioneering analysis of human relationships to other primates (**Figure 2.12**) by Morris Goodman and his associates, who sequenced more than 10,000 base pairs of a sequence that includes a pseudogene of hemoglobin (**Figure 2.13**). (A **pseudogene** is a formerly functional gene that has lost its function and may no longer be transcribed, but which accumulates mutations that have no effect on the organism.) Figure 2.13A illustrates some short portions of the entire sequence in humans, three other apes, and two outgroups (species of monkeys), showing some derived character states that unite humans with successively more distantly related species. Notice that there are several gaps (marked by asterisks) in the sequences of *Homo* and *Pan* as compared with those of the other species. The sequences of all the species have been ALIGNED to match the existing nucleotides to the greatest possible extent. (This matching may be done by means of an algorithm that maximizes some criterion that describes more or less likely alignments, according to specific rules.) During evolution, both insertions and deletions of base pairs may occur, giving rise

(A) Gibbon (B) Orangutan (C) Gorilla (D) Chimpanzee

FIGURE 2.12 Members of the primate superfamily Hominoidea. (A) White-handed gibbon (*Hylobates lar*), family Hylobatidae. (B) Orangutan (*Pongo pygmaeus*). (C) Lowland gorilla (*Gorilla gorilla*). (D) Common chimpanzee (*Pan troglodytes*).

FIGURE 2.13 Evidence for phylogenetic relationships among primates, based on the ψη-globin pseudogene. (A) Portions of the sequence in six primates. *Macaca*, an Old World monkey, and *Ateles*, a New World monkey, are successively more distantly related outgroups with reference to the Hominoidea. Sequences are identical except as indicated. Using *Ateles* and *Macaca* as outgroups, positions 3913, 6375, and 8468 exemplify synapomorphies of the other four genera, and position 8230 provides a synapomorphy of *Gorilla*, *Pan*, and *Homo*. Synapomorphies of *Pan* and *Homo* include base pair substitutions at positions 5365, 6367,

and 8224, and deletions at 3903–3906 and 8469–8474 (red asterisks).* Unshared derived states include 3911 and 3913 (*Macaca*), 8230 (*Pongo*), 6374 (*Gorilla*), 5361 (*Pan*), and 6374 (*Homo*). (B) The most parsimonious phylogeny based on the ψη-globin sequence, using *Ateles* as an outgroup. The minimal number of changes is indicated along each branch. A tree that split up the *Homo–Pan–Gorilla* group would have 65 more changes, and one that split *Homo* and *Pan* would have 8 more changes. The figure includes one of several proposed classifications for humans and apes. (A after Goodman et al. 1989; B after Shoshani et al. 1996.)

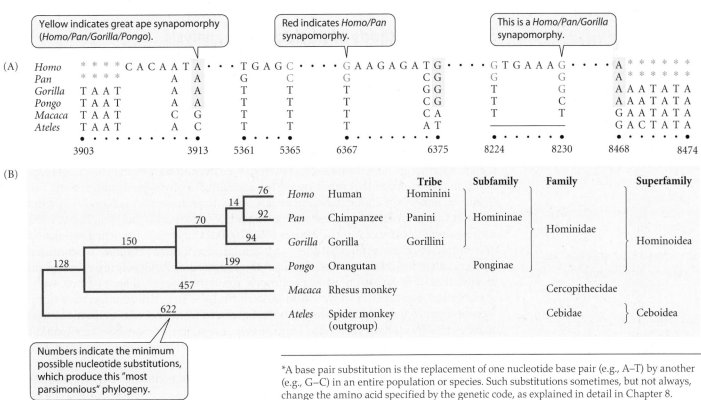

*A base pair substitution is the replacement of one nucleotide base pair (e.g., A–T) by another (e.g., G–C) in an entire population or species. Such substitutions sometimes, but not always, change the amino acid specified by the genetic code, as explained in detail in Chapter 8.

TABLE 2.1 Percentage of divergence between DNA sequences of the ψη-globin pseudogene among orangutan (*Pongo*), gorilla (*Gorilla*), chimpanzee (*Pan*), and human (*Homo*)[a]

	Gorilla	Pan	Homo
Pongo	3.39	3.42	3.30
Gorilla		1.82	1.69
Pan			1.56
Homo			0.38

Source: Data from Bailey et al. 1991.

[a]The percentage of divergence between two human sequences is given in the lower right cell. Divergences between *Homo* and other species are calculated using the average of these two sequences. Values are not corrected for multiple substitutions.

to mismatches (INDELS) that necessitate alignment in order to establish which base pairs are homologous among the different sequences. The two indels in Figure 2.13A (indicated by asterisks) are actually additional evidence of the common ancestry of human and chimpanzee.

In traditional classifications, the primate superfamily Hominoidea consists of three families: the gibbons (Hylobatidae), the human family (Hominidae), and the great apes (Pongidae). The great apes are the two orangutan species (*Pongo*) in southeastern Asia; the two gorilla species (*Gorilla*) in Africa; and the chimpanzee and bonobo (*Pan*), also in Africa. From anatomical evidence, it has long been accepted that the Hominoidea is a monophyletic group and that the Hylobatidae are more distantly related to the other species in that group than those species are to one another.

Physically, *Pongo*, *Pan*, and *Gorilla* appear more like one another than like *Homo* (see Figure 2.12). Hence the traditional view was that Pongidae is monophyletic and that *Homo* branched off first. However, molecular data have definitively shown that *Homo* and *Pan* are each other's closest relatives (Ruvolo 1997). The ψη-globin pseudogene that Goodman et al. studied shows that human, chimpanzee, and gorilla are much more similar to one another than to orangutan (**Table 2.1**). Fourteen synapomorphies unite chimpanzee and human as sister groups, including deletions of short sequences such as sites 8469–8474. In contrast, only three sites (none of which is shown in Figure 2.13) support the hypothesis that chimpanzee and gorilla are the closest relatives. The phylogenetic tree in Figure 2.13B, which shows chimpanzee and human as closest relatives and gorillas as sister group to this pair, is eight steps (evolutionary changes) shorter than a tree that separates chimpanzee from human. Many other molecular data sets support this conclusion. Such phylogenetic studies have led taxonomists to alter the classification of apes and humans by including all the great apes with humans in the family Hominidae (see Figure 2.13B).

Statistical methods of phylogenetic analysis

As DNA sequence data became more readily available, other methods of inferring phylogenies could be applied that have certain advantages over maximum parsimony. (DNA sequence data may not be available, however, if a study includes taxa known only from ancient fossils, or when DNA cannot be obtained from museum specimens of rare or otherwise unobtainable species.) Whereas there are no general "rules" of morphological evolution that might, for example, indicate where in a phylogeny certain characters are especially likely to be homoplasious and therefore misleading, comparisons of DNA sequences among and within species have enabled evolutionary biologists to develop *models of sequence evolution* that can increase the reliability of phylogenetic inferences.

For example, over a long period of evolutionary time, multiple nucleotide substitutions (MULTIPLE HITS) are likely to occur at any single DNA site within an evolutionary lineage, such as a replacement of A by G and later of G by T. If the lineage branched repeatedly into descendant species during that time, the G might have been retained by some descendants, and we would have evidence, from the presence of A, G, and T in various related species, that there had been two substitutions. But if the lineage did not branch, we would see only the derived T, and we could not tell that the first substitution ever occurred. Moreover, since there are only three possible changes from one nucleotide to another, the changes could well have been A to G and back to A, and in that case too, there would be no evidence that there had been any evolution at all. A model of sequence evolution can take this problem into account.

If a branch in a phylogeny shows very few changes in sequence, we can assume that changes have been so rare that multiple hits at any one site are improbable. But if a branch

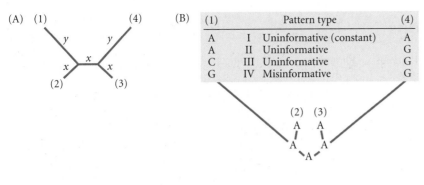

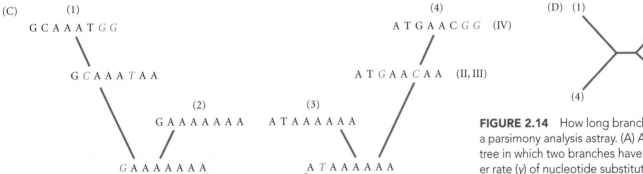

FIGURE 2.14 How long branches can lead a parsimony analysis astray. (A) An unrooted tree in which two branches have had a higher rate (*y*) of nucleotide substitution, and are therefore longer, than the other branches, in which the rate (*x*) has been lower. (B) The same tree, showing four patterns of nucleotide similarity or difference for a single DNA site. Class II, in which one lineage has evolved a unique difference, is uninformative for parsimony analysis. So is class III, in which both lineages 1 and 4 have evolved unique nucleotide substitutions. Class IV, showing convergent evolution (A to G), is misleading. All these classes of change are likely to affect more nucleotide positions in lineages with a higher rate of sequence evolution. (C) This tree illustrates the patterns in part B with an eight-nucleotide sequence. Nucleotide substitutions during evolution along the branches are highlighted in red. (D) This unrooted tree is most parsimonious, but it is incorrect. Maximum likelihood analysis takes account of the differences in branch length caused by class II and III substitutions and is more likely to find the correct phylogeny. (A, B, C after Swofford et al. 1996.)

shows many changes in DNA sequence, multiple hits are more likely, and the number of detectable changes may be less than the number that actually occurred. If there have been many undetected changes in each of two branches on a tree, then much of the history of characters that were shared with other branches will have been lost, and the characters (nucleotides) that once were shared will have been replaced by nucleotides that are uninformative about relationships, or may even be misleading because they have arisen independently. This is especially likely for LONG BRANCHES, those on which there appear to have been many nucleotide substitutions (**Figure 2.14**). A branch may be long either because the rate of DNA sequence evolution in the particular gene used has been exceptionally high, or because the lineage is a long-lived branch that appears not to have split into descendants over a long time. (It may really not have split, or its descendants may have become extinct, or they may exist but simply not be included in the study.) Thus two long branches may have accumulated not only uninformative nucleotide substitutions, but also convergent substitutions that make the species appear more closely related than they actually are. In other words, "long branches attract." Under these conditions, maximum parsimony is likely to yield the wrong phylogenetic tree.

Joseph Felsenstein, who first described this problem, adapted a well-known statistical method, **maximum likelihood (ML)**, to the problem of estimating phylogenies with high statistical confidence (Felsenstein 1981). This approach and the Bayesian method (**Box 2B**) are now the most widely used methods for estimating phylogenies from molecular sequence data.

In using maximum likelihood, one postulates, or estimates from the data itself, a *model* of the evolution of DNA sequences and asks which tree, of all possible trees, maximizes the chance (likelihood) of the observed character states having evolved. One of the important ways in which the ML approach differs from maximum parsimony is that the branch lengths estimated for ML trees are usually much more accurate than those found in a parsimony analysis, due to better accounting for

Box 2B More Phylogenetic Methods

Many methods have been proposed for inferring phylogenies (Felsenstein 2004). Some are computerized algorithmic methods that calculate a single tree from the data using a specific and repeatable protocol. Among these, the NEIGHBOR-JOINING METHOD, which does not assume equal rates of DNA sequence evolution among lineages, is most frequently used. It proceeds by joining the species that are most similar overall (e.g., in DNA sequence), then joining these clusters to the next most similar species to form larger clusters, and repeating until all species have been attached to a cluster. Most practitioners, however, prefer TREE-SEARCHING METHODS, which examine a large sample of the huge number of trees that are possible for even a few taxa. These methods use various search routines to maximize the chance of finding the shortest trees that are compatible with the data.

Some tree-searching programs, such as those based on maximum parsimony, save many of the trees that are examined so that the shortest tree found can be compared with those that are nearly as short. We are then interested in knowing if the shortest tree (or particular groupings within that tree) is reliable or if it differs from other short trees only by chance. This statistical problem is often addressed by a procedure called BOOTSTRAPPING, in which many random subsamples of the characters (e.g., nucleotide sites) are used for repeated phylogenetic analyses. Our confidence in the reliability of a particular grouping is greater if this grouping is consistently found using these different data sets (bootstrap samples).

Two powerful and widely used tree-searching methods are MAXIMUM LIKELIHOOD (ML), which is described in the main text, and the Bayesian method. The increasingly popular Bayesian method differs from ML by maximizing the probability of observing a particular tree, given the model and the data (Huelsenbeck et al. 2001). The procedure, briefly summarized, is this. Begin with a possible tree (perhaps the one found by neighbor-joining) and a slightly different tree, and compare their likelihoods (L) by means of a likelihood ratio test (see main text). Then compare the better of the two with a third tree, slightly altered from the second. Repeat this procedure several million times. This procedure builds up a sequential collection of trees that have increasing L values. Eventually, the improvements in L become so small that L levels off, meaning that further minor variations of the tree topology have about the same likelihood. All the preceding trees, with lower L (together referred to as "burn-in"), are discarded, leaving a set of very similar, highly likely trees. A consensus tree formed from these trees is then the best estimate of the real phylogeny.

multiple hits. Unlike maximum parsimony, which cannot place substitutions that occur in a single species, ML can estimate where on the tree such substitutions occur.

In ML, one calculates, using an evolutionary model, the likelihood of the collected data given a proposed phylogenetic tree; for example, the probability that the tree would have produced the observed distribution of character states (nucleotides) among the species at a single nucleotide position in the DNA sequence. This procedure is repeated, using that postulated tree, for each nucleotide position. The probability of any combination of two or more independent events equals the product of their separate probability values. Therefore, the likelihood that a particular tree produced the observed DNA sequences in each of the species equals the product of all the individual per-site probabilities (and is a very small number). Then the entire procedure is repeated, using another possible tree, and the likelihood of the data, given that tree, is calculated. This process is repeated for all possible trees. (Needless to say, this procedure requires a huge amount of computation and is possible only with the use of high-speed computers.) The tree with the greatest likelihood (L) is taken to be the best estimate of the evolutionary history: both the history of branching (cladogenesis) and the history of changes at each nucleotide site along each branch of the tree (anagenesis). The likelihoods of any two trees can be compared, to see if one is statistically significantly greater, by a simple test of their ratio (L_1/L_2). (In practice, the ratio of the logarithmic values is used.)

What is the model of DNA sequence evolution on which the ML method is based? The many possible models vary in realism, and therefore in complexity. All the models that are commonly used are "time-reversible": they assume that the probability of a change from nucleotide i to nucleotide j in a time interval is the same as that of a change back from j to i in the same length of time. The simplest model (the Jukes–Cantor model) assumes that all twelve possible substitutions between the four nucleotide states are equally probable. This assumption can be evaluated from the data: one compares pairs of similar DNA sequences, perhaps from closely related species, and estimates the proportion of sites that display each possible substitution. In fact, the rate of change between two states (α) can be estimated from the data.

The Jukes–Cantor assumption usually proves to be unrealistic, so more complex models are generally used. The Kimura two-parameter model, for example, allows for one rate (α) for each possible change (transition) between purines or between pyrimidines and another rate (β) for each possible change (transversion) between a purine and a pyrimidine (**Figure 2.15**). The most realistic and complex model used by most researchers at this time is the GTR + Γ + 1 (General Time-Reversible + Gamma + 1) model, which allows a different probability for each of the six possible time-reversible nucleotide substitutions per unit time; the fraction of sites that remain unchanged; and the possibility that the rates of change differ at different sites in the sequence. (Certain sites might change very slowly and be almost invariant among all the species in a study, whereas others might be highly variable, displaying a higher rate of change.) The Γ represents a frequency distribution of these different rates among different sites. All these rates can be estimated from comparisons among DNA sequences in a maximum likelihood analysis.

Figure 2.16 illustrates the maximum likelihood phylogeny of some primates, including humans, that Page and Goodman (2001) obtained from an analysis of two nuclear noncoding DNA sequences, using a model of sequence evolution similar to the Kimura two-parameter model. The relationships found in this analysis are the same as those found in many other studies, including the maximum parsimony analysis described earlier (see Figure 2.13).

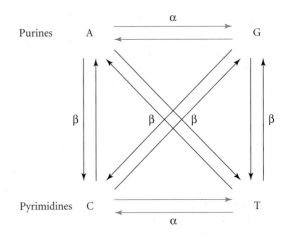

FIGURE 2.15 A two-parameter model in which the rate of transition (α) differs from the rate of transversion (β). Transitions (red arrows) are substitutions of one purine (A, G) or of one pyrimidine (C, T) for the other. Transversions (black arrows) are substitutions of a purine for a pyrimidine, or vice versa. Some models allow all of the six time-reversible substitutions shown here to have different rates.

Evaluating phylogenetic hypotheses

How can we judge the validity of our phylogenetic inferences? The phylogenetic tree that we obtain from a particular set of data is a phylogenetic *hypothesis* that is *provisionally* accepted (as is any scientific statement). Additional data may lead us to modify or abandon that hypothesis, or they may lend further support to it.

One way to test the validity of phylogenetic methods is to apply them to phylogenies that are absolutely known. Many investigators have simulated evolution on a computer, allowing computer-generated lineages to branch and their characters to change according to various models of the evolutionary process. (For example, characters might change at random at different average rates, or one of two species derived from a common ancestor might undergo changes faster than the other.) The investigators then see whether or not the various phylogenetic methods can use the final characteristics of the simulated lineages to yield an accurate history of their branching.

Perhaps more convincing to the skeptic are a few studies in which phylogenetic methods have been applied to experimental populations of real organisms that have been split into separate lineages by investigators (creating artificial branching events) and allowed to evolve. In one such study, David Hillis and coworkers (1992; Cunningham et al. 1998) successively subdivided lineages of T7 bacteriophage and exposed them to a mutation-causing chemical, which caused them to accumulate DNA sequence differences rapidly over the course of about 300 generations. The investigators then scored the eight resulting lineages (**Figure 2.17**) for sequence differences and performed a phylogenetic analysis of the data. For this many populations, there are 135,135 possible dichotomous trees (in which each lineage branches

FIGURE 2.16 Relationships among hominoid primates, based on a maximum likelihood analysis of sequences of two genes. The numbers above the branches are branch lengths inferred from the ML analysis. The pointers to each node show the estimated time (Mya) of each divergence, based on the estimated branch lengths, calibrated by fossil evidence giving a minimal age of the divergence between hominoid primates and Old World monkeys. This analysis, like many others, estimates the divergence between human and chimpanzee lineages at between 5 and 6 million years ago. (After Page and Goodman 2001.)

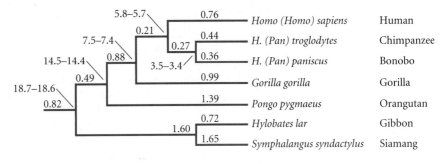

FIGURE 2.17 The true phylogeny of the experimental populations of T7 bacteriophage studied by Hillis et al. An estimate of the phylogeny based on sequence differences among the populations at the end of the experiment was exactly the same as the true phylogeny. (After Hillis et al. 1992.)

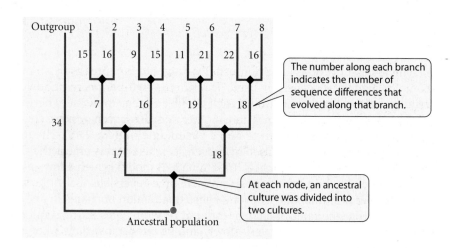

The number along each branch indicates the number of sequence differences that evolved along that branch.

At each node, an ancestral culture was divided into two cultures.

FIGURE 2.18 Relationships among major groups of living vertebrates, as estimated from morphological characters (left) and from DNA sequences (right). On the whole, these two sources of information provide similar estimates of the phylogeny. The relationships of turtles are unclear from morphological data. Among these taxa, the two sources of data conflict only with respect to the relationships of the turtles. (After Meyer and Zardoya 2003; van Rheede et al. 2006; Crawford et al. 2012.)

into two others), but the phylogenetic analysis the investigators used correctly found the one true tree.

There are two major ways to judge the reliability of a phylogeny inferred by any method. First, as is the case throughout science, a statistical test may be used to judge whether one hypothesis (tree) fits the data better than another; both maximum likelihood and Bayesian methods (see Box 2B) enable this approach. Second, as is also true throughout science, *the chief way of confirming a phylogenetic hypothesis is to see if it agrees with multiple, independent sources of data*. These sources might be different, unlinked gene sequences. Alternatively,

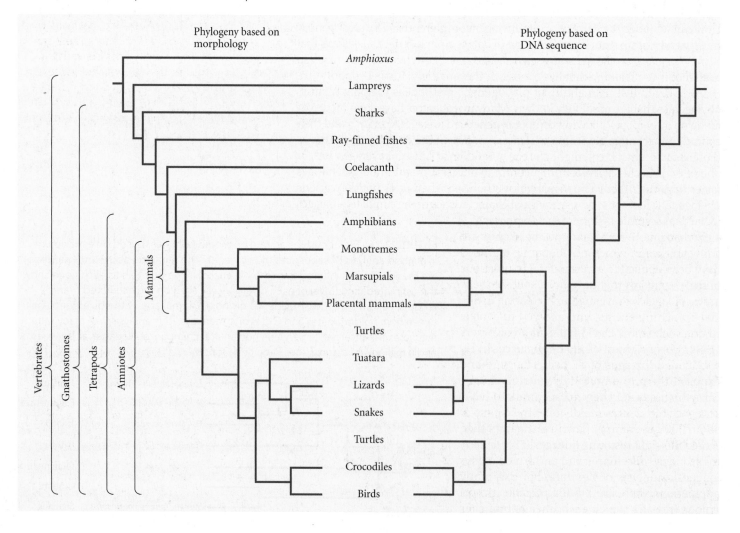

morphological features and DNA sequences, which evolve largely independently of each other and thus provide independent phylogenetic information, might be compared (see below and Chapter 20). With some exceptions, these two kinds of data usually yield similar estimates of phylogeny. For instance, the phylogenetic relationships among higher taxa of vertebrates inferred from DNA sequences are the same, with few exceptions, as those inferred from morphological features (**Figure 2.18**). In the same vein, the genome sequence of sea urchins, determined in 2006 (Sodergren et al. 2006), confirmed that among invertebrate phyla, echinoderms are the closest relatives of the chordates (including humans)—a conclusion reached in the 1890s by comparisons of embryonic development.

Estimating Time of Divergence

Early in the study of molecular evolution, evidence arose that the rate of sequence evolution might be fairly constant over long spans of time, so that sequence differences could serve as a **molecular clock** to estimate the time at which lineages diverged. For example, Charles Langley and Walter Fitch (1974) found that the number of nucleotide substitutions between gene sequences in various pairs of mammals (inferred indirectly from the amino acid sequences of the encoded proteins, since DNA sequencing had not yet been invented) was linearly related to the time since those lineages had diverged from their common ancestors, as estimated from the age of the oldest known fossils of these mammal groups (**Figure 2.19**).

It may also be possible to determine whether sequence evolution is fairly constant even without information on divergence time. One method of doing so is the RELATIVE RATE TEST (Wilson et al. 1977). We know that the time that has elapsed from any common ancestor (i.e., any branch point on a phylogenetic tree) to each of the living species derived from that ancestor is *exactly the same*. Therefore, if lineages have diverged at a constant rate, the number of changes along all paths of the phylogenetic tree from one descendant species to another through their common ancestor should be about the same (**Figure 2.20**). In the example of the hominoid ψη-globin pseudogene (see Table 2.1), the percentage of sites that differ between the orangutan and the several African apes shows a narrow range, from 3.30 (to human) to 3.42 (to chimpanzee). These numbers are so close that they indicate a fairly constant rate of divergence, although the human lineage appears to have slowed down somewhat.

The relative rate test, when applied to DNA sequence data from various organisms, has shown that rates of sequence evolution are often quite similar among taxa that are fairly closely related, but may differ considerably among distantly related taxa. For example, sequence evolution among hominoid primates (apes, including humans) has been slower than among other primates. Moreover, there exists considerable intrinsic variation in the rates of mutation and other processes that affect the rate of sequence evolution, so there always exists an appreciable "estimation error" around even the best estimates of divergence time from DNA sequences. Many methods have been developed to identify and correct for variations in rates of sequence evolution and other sources of error (Welch and Bromham 2005).

In the simplest case, a molecular evolutionary clock (to the extent that one exists) can enable us to *estimate the absolute time* since

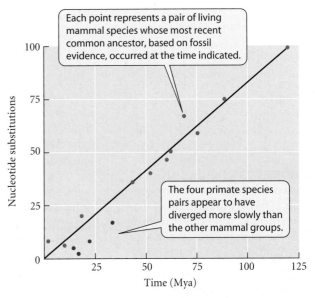

FIGURE 2.19 This plot of base pair differences against time since divergence constitutes some of the earliest evidence that the rate of sequence evolution might be approximately constant. Each point represents a pair of living mammal species whose most recent common ancestor, based on fossil evidence, occurred at the time indicated on the x-axis. (The fossil would indicate the minimal age of the lineage to which a living species belongs.) The y-axis shows the number of base pair differences between the species, inferred from the amino acid sequences of seven proteins. The four green circles represent pairs of primate species, which appear to have diverged more slowly than other mammal groups. (After Langley and Fitch 1974.)

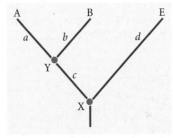

FIGURE 2.20 The relative rate test for constancy of the rate of molecular divergence. Sequences are obtained for living species A and B and for outgroup species E. Y and X represent ancestral species. Lowercase italic letters represent the number of character differences (e.g., nucleotide changes) along each branch. The "genetic distance" between A and E is $D_{AE} = a + c + d$. That between B and E is $D_{BE} = b + c + d$. If the rate of nucleotide substitution is constant, then $a = b$, so $D_{AE} = D_{BE}$. If rate constancy holds throughout the tree, the distance between any pair of species that have species X as a common ancestor will equal that between any other such pair of species.

different taxa diverged—if we can determine how fast the clock is "ticking"; that is, the rate (r) at which nucleotide substitutions occur. Assuming a molecular clock,

$$D = 2rt$$

where D is the proportion of base pairs that differ between the two sequences, r is the rate of divergence per base pair per million years (My), t is the time (in My) since the species' common ancestor, and the factor 2 represents the two diverging lineages. Then the time when two sequences started to diverge is $t = D/2r$ million years ago (Mya). If there exist dated fossils from either of the two clades, the oldest such fossil specifies the *minimal age of divergence*. (The clades might have diverged before then, but older fossils may not yet have been discovered.)

In order to estimate the rate of molecular evolution, we plot (using parsimony or maximum likelihood) where each nucleotide change occurred on our estimated phylogeny. For example, Figure 2.13B shows 76 changes between *Homo* and its common ancestor with *Pan*, 14 changes between that common ancestor and the *Gorilla* branch, and 70 changes between that common ancestor and the *Pongo* branch. Since these hominoids diverged from the lineage leading to Old World monkeys (Cercopithecidae, represented by the rhesus monkey, *Macaca*), 76 + 14 + 70 + 150 = 310 base pair changes have accrued in the lineage leading to *Homo*. The *Macaca* lineage has accrued 457 changes since its common ancestor with *Homo*. The oldest fossils of cercopithecoid monkeys are dated at 25 million years ago, providing a *minimal* estimate of time since divergence between the *Macaca* and the hominoids. Therefore, the rate of substitutions per base pair for the rhesus monkey lineage is apparently 457/10,000 base pairs sequenced/25 My = 1.83×10^{-3} per My, or 1.83×10^{-9} per year. From the common ancestor to *Homo*, the average rate has been 310/10,000/25 My = 1.24×10^{-3} per My. The average rate at which substitutions have occurred in each lineage is therefore r = (457 + 310)/2 = 383.5/10,000/25 My, or 1.53×10^{-3} per My. Suppose, then, that the proportion of base pairs that differ between the ψη-globin pseudogene sequences of two primate species were 0.0256. Then $t = D/2r = 0.0256/([2][0.00153]) = 8.3$ My, our best estimate of when the two species diverged from their common ancestor.

This approach, however, suffers from two major uncertainties. We are already aware of one: the rate has been calibrated by use of a single fossil reference point, which provides only one estimate of a minimal age. It would be better to use multiple reference points—fossils of a number of different clades in the phylogeny—to increase the information used for calibration.

The second uncertainty comes from the assumption that the number of observed nucleotide differences between taxa (e.g., between *Homo* and *Macaca* in the example) is the number that has actually occurred. We know, both from theory and from data, that as nucleotide substitutions occur over time, more and more of them are likely to be multiple hits that erase evidence of previous substitutions. Eventually, substitutions are so likely to be multiple hits that the percentage of divergent nucleotides no longer increases: it reaches a plateau (**Figure 2.21**). Some genes, and some nucleotide positions within genes, undergo substitution faster than others and will be especially likely to cause an underestimate of the number of differences that have actually occurred (i.e., the "real" branch length, in contrast to the observed branch length). Consequently, it is important to use a model of nucleotide substitution, such as those already described, to calculate r from estimated branch lengths (numbers of substitutions) rather than observed branch lengths.

For example, Steiper and Young (2006) estimated the time of divergence of major clades of primates, based on a DNA sequence of 59,764 base pairs in thirteen species of primates and six other (outgroup) mammals. They started from an accepted phylogeny of these species, based on earlier studies by other authors. They used a maximum likelihood method to estimate branch lengths, and they

FIGURE 2.21 The proportions of base pairs at different codon positions in the DNA sequences of the mitochondrial gene *COI* that differ between pairs of vertebrate species, plotted against the time since their most recent common ancestor (estimated from the fossil record). Sequence differences evolve most rapidly at third-base positions and most slowly at second-base positions within codons (as we will see in Chapter 8). Divergence at third-base positions increases rapidly at first, then levels off as a result of multiple substitutions at the same sites. Thus these positions provide no phylogenetic information for taxa that diverged more than about 75 million years ago (Mya); more slowly evolving sites in the sequence (such as second-base positions) are useful for analyzing older relationships. The species used for this analysis include two cypriniform fish species, one frog, one bird, one marsupial, two rodents, two seals, two baleen whales, and *Homo*. (After Mindell and Thacker 1996.)

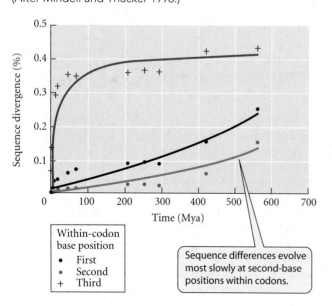

Within-codon base position
- First
- Second
+ Third

Sequence differences evolve most slowly at second-base positions within codons.

estimated rates of sequence evolution by a Bayesian method, using rate calibrations based on paleontological estimates of four divergence times in the phylogeny. These calibrations included the divergence between baboons and macaques, represented by a 5.5-My-old macaque fossil, and the human–chimpanzee divergence, based on a hominin fossil (*Sahelanthropus*) that is 6 to 7 My old. Their analysis incorporated uncertainty both in the calibration ages and the branch lengths, and so provided a range of ages for each node in the phylogeny (**Figure 2.22**).

FIGURE 2.22 Results of a study of divergence times for some lineages of primates, based on ML estimates of branch lengths and calibration of the rate of sequence evolution using several fossils. Asterisks denote calibrated nodes. Calibration using younger fossils tended to yield younger estimates of divergence time than that using older fossils, as shown by the range of ages indicated at each branch point. The authors' best estimates of divergence time are in boldface. The labels at left are geological ages (see Chapter 4). (After Steiper and Young 2006.)

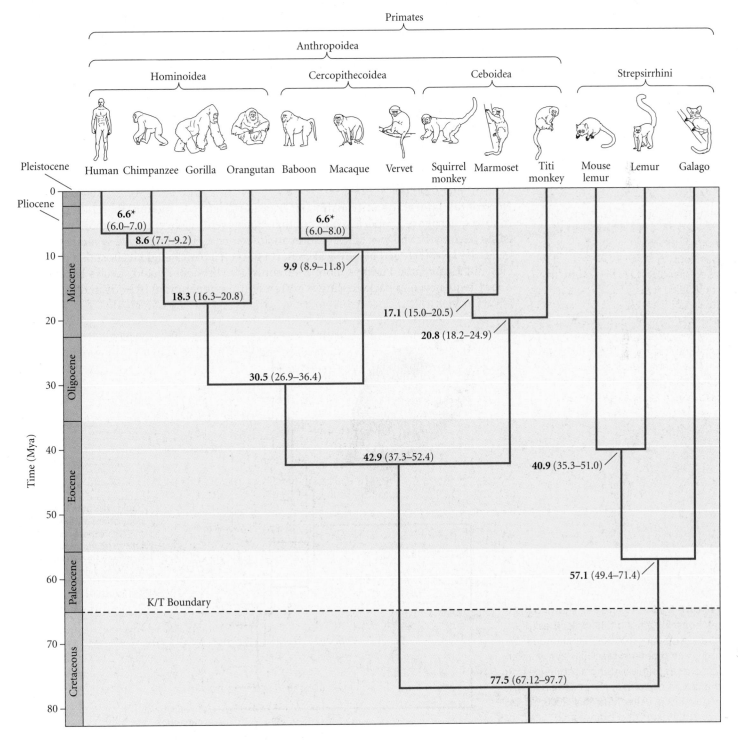

Gene Trees and Species Trees

So far, we have been concerned with inferring phylogenetic trees of species. Because genes replicate, different copies of a gene, within a single species or more than one species, have a history of descent from common ancestral genes, just as species do. Using the same principles that we use for inferring phylogenetic relationships among species, we can infer the historical relationships among variant DNA sequences of a gene (**haplotypes**) and construct a phylogeny of genes, often called a **gene tree** or a **gene genealogy**. This is exactly what was done in the examples of phylogenetic analysis of species relationships based on DNA sequences: the relationships among homologous sequences (genes) were assumed to be the same as the relationships among the species from which the genes were obtained. As we will see in later chapters, gene trees are often constructed for different haplotypes (sequences) of a gene from a single species. For example, the phylogeny of haplotypes of the mitochondrial cytochrome *b* gene in MacGillivray's warbler (**Figure 2.23**) shows that the genetic variants within Mexican populations of this species are more closely related to one another than they are to haplotypes in birds from the United States.

Even when a phylogeny of species (a species tree) is based on gene trees, *an accurately estimated gene tree may imply the wrong species phylogeny*. We take note here of two common causes of this disparity.

Horizontal gene transfer

Horizontal (**lateral**) **gene transfer** (HGT or LGT) refers to the incorporation of a gene from one lineage (e.g., species) into another. Such events violate our assumption that lineages have a strict history of separation and divergence. HGT can have several causes. A gene may be transferred by physical contact between a parasite and its host or a prey species and its predator. Perhaps more commonly, two or more related species hybridize at a low frequency, and backcrossing between F_1 (first-generation) hybrids and the parent

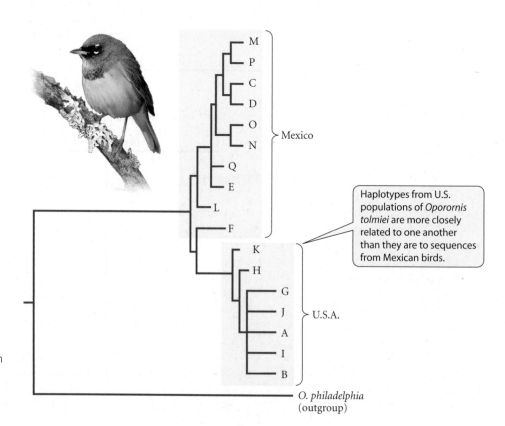

FIGURE 2.23 A gene tree showing the relationships among the haplotypes (different sequences) of the mitochondrial cytochrome *b* gene in MacGillivray's warbler (*Oporornis tolmiei*), using a sequence from the mourning warbler (*O. philadelphia*) as an outgroup. (After Milá et al. 2000; art © Tim Zurowski/Photolibrary.com.)

Haplotypes from U.S. populations of *Oporornis tolmiei* are more closely related to one another than they are to sequences from Mexican birds.

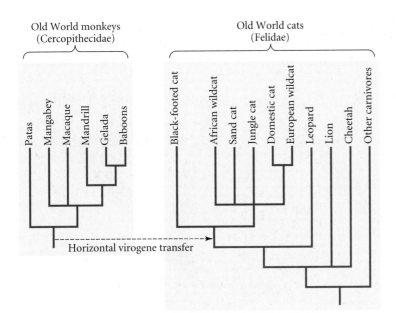

FIGURE 2.24 Phylogenies of some Old World monkeys and cats (Felidae). Both the monkeys and one group of small cats (blue branches) have a "virogene" with such a similar sequence that horizontal transfer from the ancestor of these monkeys to the ancestor of the small cats is the most reasonable interpretation. (After Li and Graur 1991.)

species introduces genes from one species into the population of the other. Because of natural selection, certain of the introduced genes may become prevalent and others not, a phenomenon called INTROGRESSION (see Chapter 17).

HGT can be detected because the gene tree for the transferred gene will differ from the tree inferred from most other genes. For example, a certain gene has been found only in one group of closely related species of cats and in Old World monkeys. This gene would imply that these cats are more closely related to monkeys than they are to other cats, so it is clearly incongruent with the phylogeny indicated by many other genes (**Figure 2.24**). The gene may have been transferred from monkeys to cats by a virus. Similarly, although most genes studied in the parasitic plant *Rafflesia* (see Figure 2.7) show that it is related to spurges, it has a mitochondrial gene closely related to that of its host, a member of the grape family, implying that the parasite acquired this gene from its host (Davis and Wurdack 2004).

Lateral gene transfer is very important in the evolution of bacteria, in which genes are transferred among distantly related species by phage, by uptake of naked DNA released from dead cells, and by conjugation (Ochman et al. 2000). Phylogenetic incongruence and several other lines of evidence have shown that lateral transfer has provided various bacteria with genes for antibiotic resistance, for the ability to invade hosts and cause disease, and for adaptation to extreme environments such as hot springs.

Incomplete lineage sorting

Incomplete lineage sorting is a slightly challenging concept, but it has some important implications (see Chapter 17). Suppose that two successive branching events occur, giving rise first to species A and then to species B and C (**Figure 2.25**). Now suppose that the ancestral species that gave rise to A, B, and C carries two different haplotypes of the gene we are studying (a POLYMORPHISM). Perhaps, just by chance, haplotype 1 becomes FIXED in species A (that is, the other haplotype is lost from that species by genetic drift; we will see how this can happen in Chapter 10), and haplotype 2 is fixed in the common ancestor of B and C, and then diverges by further mutations into haplotypes 2a in species B and 2b in species C. Then the gene tree will reflect the species phylogeny (Figure 2.25A). But if such sorting of haplotypes into the diverging species does not occur, and the common ancestor of B and C is also polymorphic (i.e., if there is incomplete sorting of gene lineages), then the haplotypes may become fixed, by chance, in the three species in such a way that the most closely related species do not inherit the most closely related haplotypes (Figure 2.25B,C).

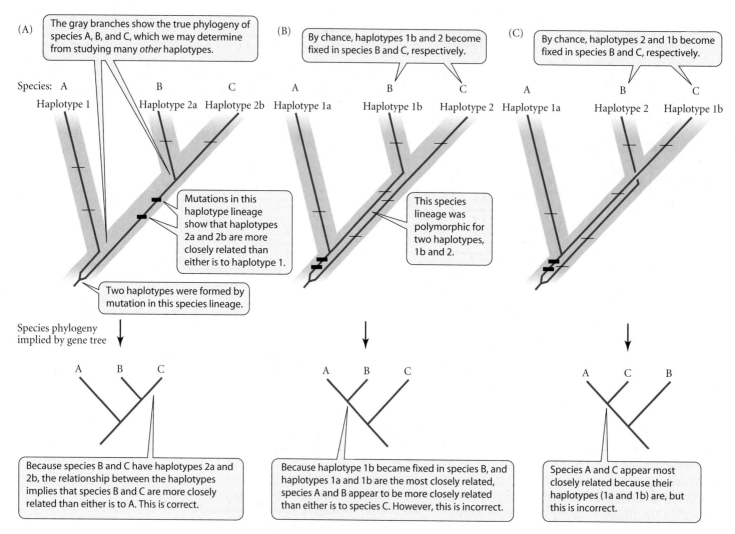

FIGURE 2.25   A gene tree (red lines) may or may not reflect the true phylogeny of the species from which the genes are sampled (gray outer envelopes). Species A, B, and C have arisen by two successive branching events. (A) Haplotypes 2a and 2b form a monophyletic group, as indicated by two shared base pair substitutions (thick tick marks), and the species phylogeny implied by the gene tree matches the actual phylogeny of the species. Thin tick marks show other mutations that distinguish the three haplotypes. (B) Haplotypes 1a and 1b form a monophyletic group and falsely imply that species A and B are sister species. The common ancestor of species B and C was polymorphic for two gene lineages (1b and 2), and the gene lineages that became fixed in the sister species B and C are not sister lineages. (C) The common ancestor of species B and C is again polymorphic for gene lineages 2 and 1b, but the fixation of these genes is the reverse of diagram B. The gene tree falsely implies that species A and C are most closely related. (After Maddison 1995.)

A phylogeny based on these haplotypes will therefore imply incorrect relationships among the species.

Figure 2.26 provides a real example, in which sequences of six independently segregating genes were taken from several individuals of each of four species of grasshoppers (Carstens and Knowles 2007). At each gene locus, the relationships among haplotypes differ, in many different ways, from the best estimate of the species phylogeny based on all the available data. Each species has inherited several gene lineages from its common ancestor with other species. Similarly, about 30 percent of nucleotide positions in a comparison of humans, chimpanzees, and gorillas show incomplete lineage sorting because

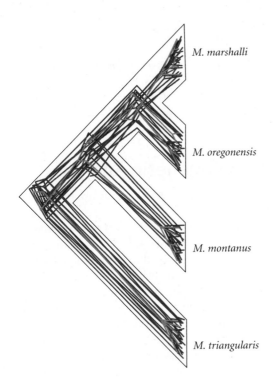

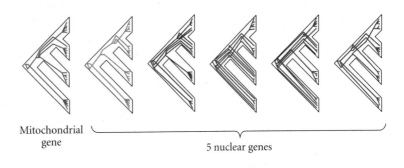

FIGURE 2.26 A phylogeny of four species of grasshoppers (*Melanoplus*) inferred from multiple samples of each of six genes in each species. Trees for the individual genes (at right) show relationships among haplotypes that vary in many different ways from the best estimate of the species phylogeny (black outer lines), indicating that each of these four species inherited several gene lineages from the common ancestor of all four species. The diagram at left shows all six gene trees nested within the species phylogeny. (From Carstens and Knowles 2007, courtesy of L. L. Knowles.)

Mitochondrial gene

5 nuclear genes

of the short time between the branching events that gave rise to these closely related species (Scally et al. 2012).

It is often difficult to distinguish the effects of incomplete lineage sorting from those of horizontal gene transfer due to hybridization, both of which are most likely to occur between recently diverged species or populations. Methods are being developed to reliably infer phylogenies from DNA sequences despite these problems (Degnan and Rosenberg 2009; Knowles and Kubatko 2010). The greater the number of independent genes used, the more reliable the phylogeny is likely to be. Phylogenetic studies of some taxa, such as primates, now employ dozens or hundreds of gene sequences that have been deposited in the database GenBank, and some studies use the sequence of entire genomes. For example, the orders of mammals form four major clades, but the branches connecting these major groups are short, indicating that these lineages diverged during a relatively brief time (**Figure 2.27**). In 2007, two research groups independently analyzed relationships among representative species from each clade for which most of the genome had been sequenced. One research group (Wildman et al. 2007) used 1698 gene sequences, summing to 1,443,825 base pairs; the other (Hallström et al. 2007) used 1961 gene sequences (1,569,141 base pairs). Both groups obtained the same strongly supported phylogeny.

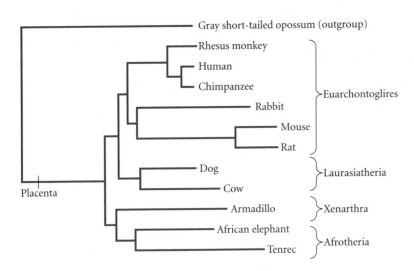

FIGURE 2.27 Relationships among 11 species of placental (eutherian) mammals, which represent four major clades. This phylogeny, based on 1698 gene sequences, shows that Xenarthra is the sister group of Afrotheria, rather than Laurasiatheria + Euarchontoglires, as some investigators had postulated. Hallström et al. obtained the same tree, based on a somewhat different set of 1961 genes. (After Wildman et al. 2007.)

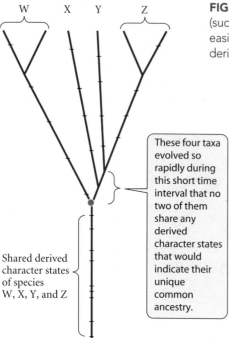

These four taxa evolved so rapidly during this short time interval that no two of them share any derived character states that would indicate their unique common ancestry.

Shared derived character states of species W, X, Y, and Z

FIGURE 2.28 Rapid evolutionary radiation. Each tick mark indicates an evolutionary change (such as a nucleotide substitution). The relationships among clades W, X, Y, and Z cannot be easily determined because they all diverged within such a short interval of time that few or no derived character states evolved in the common ancestors of any subset of these taxa.

Some Other Aspects of Phylogenetic Analysis

Several aspects of evolution that we have considered only in passing are important considerations in phylogenetic analysis.

As we have seen, *the process of evolution often erases the traces of prior evolutionary history.* For instance, teeth provide important features for determining relationships among many mammals, but they cannot be used to assess the relationships of toothless anteaters or baleen whales, whose ancestors lost teeth. In DNA sequences, as already discussed, multiple substitutions at a site over the course of time erase earlier synapomorphies and cause the amount of sequence divergence between taxa eventually to level off at a plateau, which is attained sooner in rapidly evolving than in slowly evolving sequences (see Figure 2.21). For this reason, slowly evolving sequences are required to assess relationships among taxa that diverged in the distant past. Conversely, rapidly evolving DNA sequences are useful for phylogenetic analyses of taxa that have diverged recently, because they will have accumulated more differences than slowly evolving sequences and will therefore provide more phylogenetic information. (Sequences that evolve as fast as the gray curve in Figure 2.21 will be useful for taxa that diverged less than 200 Mya, but older divergences will be more accurately estimated by using sequences that evolve at rates like those shown by the black and red curves.)

These considerations are particularly important in analyzing "bursts" of divergence of many lineages during a short period. Such bursts are called EVOLUTIONARY RADIATIONS, or ADAPTIVE RADIATIONS because the lineages have often acquired different adaptations (see Chapter 3). When successive branching (speciation) events occur over a short time, few nucleotide substitutions (or other changes) occur in the interval between successive nodes, except in rapidly evolving sequences (**Figure 2.28**). Moreover, incomplete lineage sorting is more likely over short than long intervals between branch points. The phylogeny of an adaptive radiation is even more difficult to resolve if the burst occurred long ago, because the rapidly evolving genes required to resolve relationships among lineages that arose during a short interval will have experienced multiple hits from the time of the radiation until the present, and so will have lost phylogenetic information. Although ancient adaptive radiations are difficult to resolve phylogenetically, this same difficulty reveals that an evolutionary radiation took place.

We have already noted that hybridization (interbreeding) can result in horizontal transfer of some genes between species. But hybridization between two ancestral species has also given rise to entirely new species, quite often in plants and occasionally in animals (see Chapter 18). In such cases, part of the phylogeny will be RETICULATED (netlike) rather than strictly branching, and some genes in the hybrid population will be most closely related to genes in each of two other species (**Figure 2.29**).

(A) Phylogeny based on sequence of gene 1

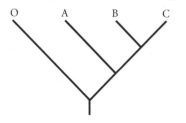

(B) Phylogeny based on sequence of gene 2

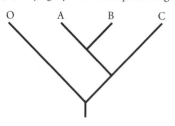

(C) Reticulate phylogeny

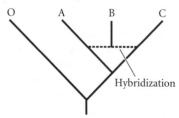

Hybridization

FIGURE 2.29 Hybridization and reticulate evolution. (A, B) Phylogenies inferred for species A, B, C, and an outgroup (O), based on the sequences of two different genes. (C) If many genes show each of these two patterns, the most plausible inference is that species B has arisen by hybridization between species A and C, and that the history of this taxon can thus best be portrayed as reticulate (netlike) evolution. (After Sang and Zhong 2000.)

Applications and Extensions of Phylogenetics

Phylogenetic information has many practical applications. It is often important, for example, to learn where a new infectious disease originated in order to trace the path and mechanism of its spread. Gene trees of the infectious agent can provide critically important information. For example, the deadliest form of human malaria is caused by the parasite *Plasmodium falciparum*, which is restricted to humans. Because related species of parasites are often found in related hosts, it has long been suspected that *P. falciparum* arose from an ancestor that parasitized another primate, but which primate has been a controversial question. A recent comprehensive analysis indicates that human *P. falciparum* evolved from one of the malarial *Plasmodium* strains carried by gorillas, rather than the *Plasmodium* species of chimpanzees, as had been previously thought (**Figure 2.30**; Liu et al. 2010).

Another of the many contexts in which phylogenetics is useful is in biological pest control. For example, plants introduced from one part of the world to another sometimes become economically important weeds. One approach to weed control is to import herbivorous insects or other natural enemies of the plant from its region of origin. Candidates for biological control, however, must be screened to be sure that they will not attack crops or other important nontarget species. For practical reasons, these screening tests are usually limited to those nontarget plant species thought to be at greatest risk. And the species at greatest risk are likely to be those that are most closely related to the weed, because insects with narrow food plant ranges usually attack closely related plants. This research clearly requires knowledge of phylogenetic relationships among plants. For example, American ragweed (*Ambrosia artemisiifolia*), which not only invades cropland but also disperses highly allergenic pollen, has become a damaging weed in Europe. American species of ragweed-specialist insects have been screened on sunflowers (*Helianthus annuus*) because this crop plant is in the same clade as ragweed (although they look very different). At least one of these insects may be unsuitable for controlling ragweed because it can survive on sunflowers (Gerber et al. 2011).

The applications of phylogenetic methods extend beyond biology. No one doubts that French, Spanish, and the other romance languages evolved from Latin, an example of nongenetic CULTURAL EVOLUTION. Students of cultural evolution are increasingly using phylogenetic methods, borrowed from evolutionary biology, to analyze the history of languages and other cultural traits. A recent study used Bayesian phylogenetic techniques, together with a "linguistic clock," to test alternative hypotheses about the origin of the Semitic language; the investigators concluded that it originated in the early Bronze Age (about 5750 years ago) in the Near East (Kitchen et al. 2009). "Language trees" can provide clues to

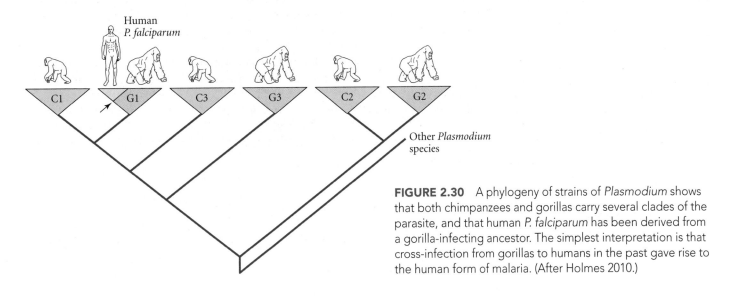

FIGURE 2.30 A phylogeny of strains of *Plasmodium* shows that both chimpanzees and gorillas carry several clades of the parasite, and that human *P. falciparum* has been derived from a gorilla-infecting ancestor. The simplest interpretation is that cross-infection from gorillas to humans in the past gave rise to the human form of malaria. (After Holmes 2010.)

human population history: a phylogenetic analysis of 87 languages provided evidence that the Indo-European language family arose between 7800 and 9800 years ago and spread with agriculture from Anatolia (present-day Asian Turkey) into Europe (Gray and Atkinson 2003). Phylogenetic methods are being used to shed light on many other aspects of cultural history and cultural adaptation (Mace and Holden 2005; Mace and Jordan 2011).

Summary

1. A phylogeny is the history of the events by which species or other taxa have successively originated from common ancestors. It may be depicted by a phylogenetic tree, in which each branch point (node) represents the division of an ancestral lineage into two or more lineages. Closely related species have more recent common ancestors than distantly related species. The group of species descended from a particular common ancestor is a monophyletic group; a phylogenetic tree portrays nested sets of monophyletic groups. Phylogenetic trees, inferred from the character states of the taxa, represent evolutionary relationships and provide a framework for analyzing many aspects of evolution.

2. Biologists are aware of many reasons for possible errors in inferring phylogenetic relationships, and they continue to develop analytical methods to improve the reliability of phylogenetic inferences. For example, overall similarity among organisms is not the best indicator of phylogenetic relationships. Two species may be more similar to each other than to a third because they retain ancestral character states (whereas the third has diverged), because they independently evolved similar character states (homoplasy), or because they share derived character states that evolved in their common ancestor. Only unique shared derived character states are evidence of phylogenetic relationship. Thus a monophyletic group is marked by uniquely derived character states shared by the group's members.

3. Phylogenetic relationships can be obscured by different rates of evolution among lineages and by homoplasy. Several methods are used to estimate phylogenies in the face of these misleading features. The method of maximum parsimony, according to which the best estimate of the phylogeny is the tree that requires one to postulate the smallest number of evolutionary changes to account for the differences among species, is frequently used. DNA sequences are increasingly used to infer phylogenies. They can be analyzed by powerful statistical methods, such as maximum likelihood and Bayesian analysis.

4. A phylogenetic tree is a statement about evolutionary relationships, and like all scientific statements, it is a hypothesis. We gain confidence in the validity of the hypothesis when new data, such as additional characters, support it. Uncertainties remain about the phylogenetic relationships among many taxa, but there are also many well-supported phylogenies.

5. Evolutionary processes may make it difficult to infer phylogenetic relationships. For example, shared derived character states may be erased by subsequent evolution, as when successive base pair substitutions occur at the same site in a DNA sequence. The nucleotide differences between the sequences taken from two species will then represent only a fraction of the nucleotide changes that actually transpired. The magnitude of the missing fraction can be estimated by using models of DNA sequence evolution.

6. The rate of evolution of DNA sequences can be shown in some cases to be fairly constant (providing an approximate "molecular clock"), such that sequences in different lineages diverge at a roughly constant rate. The absolute rate of sequence evolution can sometimes be calibrated if the ages of fossils of some lineages are known and if the rate of sequence evolution is measured not simply by the observed number of differences between sequences from different species, but by the observed number plus the missing fraction estimated by an evolutionary model. The estimated rate of sequence evolution can then be used to estimate the absolute age of some evolutionary events, such as the origin of other taxa.

7. If multiple lineages originate from a common ancestor within a short time, the relationships among them may be unresolved because there was not enough time for derived character states to evolve between successive branching events. Moreover, an accurately estimated gene tree (the phylogeny of homologous DNA sequences from different species) may differ from the phylogeny of the species themselves because of horizontal gene transfer between species or because of a process called incomplete lineage sorting. With enough data, these difficulties can usually be surmounted.

8. In some cases, a phylogeny of species (species tree) is not strictly dichotomous (branching), but may include reticulation (joining of separate lineages into one). This can occur if some species have originated by hybridization between different ancestral species.

Terms and Concepts

anagenesis	cladogenesis	gene tree (gene genealogy)	homology (homologous)
character	common ancestor	haplotype	homoplasy (homoplasious)
character state	divergence (divergent evolution)	higher taxon	
clade			

horizontal (lateral) gene
    transfer
incomplete lineage
    sorting
ingroup
lineage

maximum likelihood
    (ML)
molecular clock
monophyletic
outgroup

paraphyletic
parsimony (maximum
    parsimony)
phylogenetic tree
phylogeny

polyphyletic
pseudogene
sister group
synapomorphy
taxon (plural: taxa)

Suggestions for Further Reading

An introduction to phylogenetic methods using molecular data is given by B. G. Hall in *Phylogenetic Trees Made Easy: A How-to Manual for Molecular Biologists* (third edition, Sinauer Associates, Sunderland, MA, 2007), which includes instructions for several widely used tree-construction software programs.

For deep coverage of phylogenetic analysis, see J. Felsenstein, *Inferring Phylogenies* (Sinauer Associates, Sunderland, MA, 2004).

J. Sullivan and P. Joyce review model selection in phylogenetics in "Model selection in phylogenetics" (*Annu. Rev. Ecol. Evol.*

Syst. 36: 445–4660, 2005) and B. Rannala and Z.-H. Yang review phylogenetic inference using whole genomes in "Phylogenetic inference using whole genomes" (*Annu. Rev. Genomics Hum. Genet.* 9: 217–231, 2008).

Tree Thinking: An Introduction to Phylogenetic Biology, by D. A. Baum and S. D. Smith (Roberts and Company, Greenwood Village, CO, 2012), is a comprehensive introduction to the concepts, methods, and uses of phylogenetics in biology, for non-specialists.

Problems and Discussion Topics

1. Suppose we are sure, because of numerous characters, that species 1, 2, and 3 are more closely related to one another than to species 4 (an outgroup). We sequence a gene and find ten nucleotide sites that differ among the four species. The nucleotide bases at these sites are

   (Species 1) GCTGATGAGT (Species 2) ATCAATGAGT

   (Species 3) GTTGCAACGT (Species 4) GTCAATGACA

   Estimate the phylogeny of these taxa by plotting the changes on each of the three possible unrooted trees and determining which tree requires the fewest evolutionary changes.

2. There is evidence that many of the differences in DNA sequence among species are not adaptive. Other differences among species, both in DNA and in morphology, are adaptive (as we will see in Chapters 12 and 20). Do adaptive and nonadaptive variations differ in their utility for phylogenetic inference? Can you think of ways in which knowledge of a character's adaptive function would influence your judgment of whether or not it provides evidence for relationships among taxa?

3. It is possible for two different genes to imply different phylogenetic relationships among a group of species. What are the possible reasons for this? If there is only one true history of formation of these species, what might we do in order to determine which (if either) gene accurately portrays it? Is it possible for both phylogenetic trees to be accurate even if there has been only one history of species divergence?

4. Explain why rapidly evolving DNA sequences are useful for determining relationships only among taxa that evolved from quite recent common ancestors, and why slowly evolving sequences are useful for resolving relationships only among taxa that diverged much longer ago.

5. Given the principle stated in Problem 4, suppose you determine the DNA sequence of a particular gene from each of several animal species. (Perhaps, for the sake of argument, these species include horse, sheep, giraffe, human, and kangaroo [an outgroup].) How could you tell if this gene has evolved at a rate (not too fast or slow) that would make it useful for estimating the relationships among these animals?

6. Suppose the DNA sequence in the gene you used in Problem 5 has evolved much faster in some lineages (say, horse and human) than in others. Could that affect the estimate of the phylogeny? How? Is there any way you could tell if there was indeed a big difference in the rate of sequence evolution among lineages?

7. What should a biologist do if she or he finds that different methods of analyzing the same data (say, the maximum parsimony method and the maximum likelihood method) provide different estimates of the relationships among certain taxa? What should she do if the different analytical methods give the same estimate, but the estimate differs depending on which of two different genes has been sequenced? (*Hint:* Your answers do not depend on knowing how maximum likelihood works.)

8. Do the quandaries described in Problem 7 ever occur? Choose a group of organisms that interests you, find recent phylogenetic studies of this group, and see whether such problems have been encountered. (You can do this using keywords, such as "phylogeny" and "[taxon name, e.g., deer]," in any of several literature-search engines that your instructor can point you to.)

9. What is the evidence that incomplete lineage sorting has affected DNA variation in humans, chimpanzees, and gorillas? How could the authors of the study described in the text (Scally et al. 2012) tell that it had occurred?

Patterns of Evolution

I n order to classify organisms, systematists compare their characteristics. These comparisons, especially when coupled with the methods of phylogenetic analysis introduced in the previous chapter, are an indispensable basis for inferring the often fascinating history of evolution of various groups of organisms. Beyond that, they enable us to infer the history of changes in the characteristics of organisms, even in the absence of a fossil record (Pagel 1999). In fact, they are the only way of inferring the history of evolution of features that leave little or no fossil trace, such as DNA sequences, biochemical pathways, and behaviors. Phylogenetic and systematic studies enable us to describe past changes in genes, genomes, biochemical and physiological features, development and morphology, and life histories and behavior, as well as associated changes in geographic distribution, habitat associations, and ecological interactions among different species. From such studies of many groups of organisms, it has been possible to find common themes, or *patterns of evolution*.

Because organisms are so diverse, there are few universal "laws" of biology of the kind that are known in physics (Mayr 2004). However, we can make generalizations about what kinds of evolutionary changes have been prevalent—and developing such general statements is one of the chief tasks of science. Furthermore, general patterns of change are the most important phenomena for evolutionary biology to explain. For instance, we know that the size of the genome—the amount of DNA—in various organisms varies greatly. The search for an explanation of genome size might become more interesting (and perhaps easier) if we determined in which clades genome size increased and in which it decreased, rather than knowing only that the genome of one species is larger than that of another.

An evolutionary trend This Australian lizard, *Lerista punctatovittata* (family Scincidae), illustrates the limb reduction that has evolved independently in several lineages of lizards, including the ancestor of the snakes. This species has tiny, vestigial forelegs, as seen in this photograph, but other species of *Lerista* display a complete range of reduction of legs and digits, from fully functional legs to none at all.

Phylogenetic and comparative studies provide insights into almost every aspect of evolution and help us to understand evolutionary processes (Futuyma 2004). This chapter describes a few of the most important patterns that have emerged from a long history of systematic and phylogenetic analyses, concentrating on morphological and genomic characteristics. In later chapters, we will encounter phylogenetic insights into many other topics.

Inferring the History of Character Evolution

One of the most important uses of phylogenetic information is to reconstruct the history of evolutionary change in interesting characteristics by "mapping" character states on the phylogeny and inferring the state in each common ancestor (see Chapter 2). The simplest methods employ maximum parsimony: we assign to ancestors those character states that require us to postulate the fewest homoplasious evolutionary changes for which we lack independent evidence. This method enables us to infer when (i.e., on which branch or segment of a phylogeny) changes in characters occurred, and thus to trace their history.

Humans, for example, have nonopposable first (great) toes, while the orangutan, gorilla, and chimpanzee have opposable first toes (like our thumbs). In **Figure 3.1**, we consider two possible evolutionary histories. The common ancestors are labeled A_1, A_2, and A_3 in order of increasing recency. If we assume that A_1 and A_2 had opposable toes and that A_3, the immediate common ancestor of chimpanzees and humans, had nonopposable toes, we have to postulate two changes, with the chimpanzee reverting to the ancestral state (Figure 3.1A). If, however, we assume that A_3, like A_1 and A_2, had opposable toes, we need to infer only one evolutionary change; namely, the shift to nonopposable toes in the human lineage (Figure 3.1B). This is the more parsimonious hypothesis, so our best estimate is that the common ancestor of humans and chimpanzees had opposable toes.

Figure 3.2 illustrates a more complex and important application of this approach. The host species that a parasite, such as a virus, inhabits can be considered a character of the parasite. Based on their nucleotide sequences, strains of the human immunodeficiency virus (HIV) that cause AIDS are classified into two major groups, HIV-1 and HIV-2, which differ in disease etiology. When AIDS surfaced in the 1980s, the question of where it had come from was a major issue, not least because of fundamentalist religious leaders who shouted that it was God's punishment of homosexual men. Science, of course, provides more useful and testable hypotheses, and so biologists soon showed, by nucleic acid sequencing, that HIV is related to lentiviruses carried by other primates. Phylogenetic analysis of viruses from various primates provided strong evidence that HIV-1 has evolved at least twice from the virus found in chimpanzees, and that HIV-2 has been acquired from the sooty mangabey monkey. Infection of humans probably occurred through the butchering of these animals, which are eaten in parts of Africa.

In this book, we will encounter many examples of evolutionary inferences from phylogenetic trees. These inferences are fundamentally important in the study of molecular evolution. They have even been used to infer the amino acid sequences and functions of

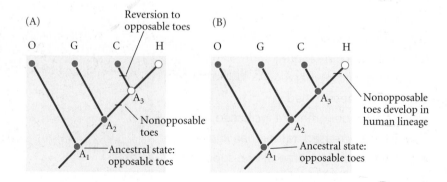

FIGURE 3.1 Two possible histories of change of a character (opposable versus nonopposable toes) in the Hominoidea (O, orangutan; G, gorilla; C, chimpanzee; H, human). (A) If nonopposable toes (open circles) are hypothesized for A_3, the common ancestor of chimpanzee and human, two state changes (tick marks) must be postulated. (B) If opposable toes are hypothesized for A_3, only one change need be postulated. It is therefore most parsimonious to conclude that humans evolved from an ancestor with opposable toes.

FIGURE 3.2 A phylogeny of strains of human immunodeficiency virus (HIV-1 and HIV-2) and simian immunodeficiency viruses (SIV) from chimpanzees and various African monkeys. The internal branches are colored to show the most parsimonious interpretation of the host species of ancestral virus lineages (green, monkeys; red, chimpanzee; gold, human). For example, the different strains of HIV-1 are most closely related to different strains of chimpanzee SIV and are most parsimoniously interpreted as having evolved from the chimpanzee virus. Strains of HIV-1 fall into several groups that apparently originated as separate transfers from chimpanzees to humans. HIV-2 probably originated by at least one transfer from sooty mangabeys. (After Hahn et al. 2000.)

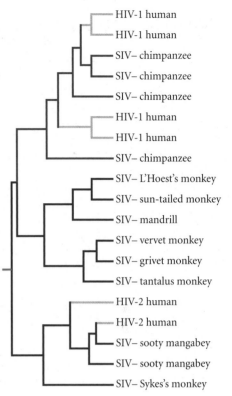

ancestral proteins, which have then been experimentally synthesized (Thornton 2004). For example, by comparing DNA sequences of the visual pigment protein rhodopsin in diverse vertebrates, Belinda Chang and colleagues (2002) inferred the most probable amino acid sequence of rhodopsin in the common ancestor of living archosaurs (crocodilians and birds). This reconstruction provides our best estimate of the protein in dinosaurs, the archosaurs from which birds evolved. The research team synthesized the postulated rhodopsin and expressed it in a mammalian cell line in tissue culture. They found that the protein performed its proper function and that its greatest light absorption was slightly red-shifted compared with the rhodopsins of living archosaurs.

Some Patterns of Evolutionary Change Inferred from Systematics

Many important patterns and principles of evolution have been based on systematic studies over the course of the last century or more and have been clarified by modern phylogenetic studies. Indeed, systematic studies of living organisms have provided a vast amount of evidence for the reality of evolution (**Box 3A**). In this section, we will illustrate some

---

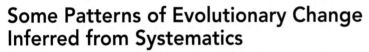

Box 3A Evidence for Evolution

Systematists have always classified organisms by comparing characteristics among them. The early systematists had amassed an immense body of comparative information even before *The Origin of Species* was published. Their data suddenly made sense in light of Darwin's theory of descent from common ancestors; indeed, Darwin drew on much of this information as evidence for his contention that evolution has occurred. Since Darwin's time, the amount of comparative information has increased greatly, and today it includes data not only from the traditional realms of morphology and embryology, but also from cell biology, biochemistry, and molecular biology.

All of this information is consistent with Darwin's hypothesis that living organisms have descended from common ancestors. Indeed, innumerable biological observations are hard to reconcile with the alternative hypothesis, that species have been individually created by a supernatural being, unless that being is credited with arbitrariness, whimsy, or a devious intent to make organisms *look* as if they have evolved. From the comparative data amassed by systematists,

we can identify several patterns that confirm the historical reality of evolution and which make sense only if evolution has occurred.

1. **The hierarchical organization of life**. Before Linnaeus, there had been many attempts to classify species, but those early systems just didn't work. One author, for instance, tried to classify species in complicated five-parted categories—but organisms simply don't come in groups of five. But organisms do fall "naturally" into the hierarchical system of groups-within-groups that Linnaeus described. A historical process of branching and divergence will yield objects that can be hierarchically ordered, but few other processes will do so. For instance, languages can be classified in a hierarchical manner, but elements and minerals cannot.

2. **Homology.** Similarity of structure despite differences in function follows from the hypothesis that the characteristics of organisms have been modified from the characteristics of their ancestors, but it is hard to reconcile with the hypothesis of intelligent

(continued)

Box 3A *(continued)*

design. Design does not require that the same bony elements form the frame of the hands of primates, the digging forelimbs of moles, the wings of bats, birds, and pterodactyls, and the flippers of whales and penguins (see Figure 3.3). Modification of pre-existing structures, not design, explains why the stings of wasps and bees are modified ovipositors and why only females possess them. All proteins are composed of "left-handed" (L) amino acids, even though the "right-handed" (D) optical isomers would work just as well if proteins were composed only of those. But once the ancestors of all living things adopted L amino acids, their descendants were committed to them; introducing D amino acids would be as disadvantageous as driving on the right in Great Britain or on the left in the United States. Likewise, the nearly universal, arbitrary genetic code (see Chapter 8) makes sense only as a consequence of common ancestry.

3. **Embryological similarities.** Homologous characters include some features that appear during development, but would be unnecessary if the development of an organism were not a modification of its ancestors' ontogeny. For example, tooth primordia appear and then are lost in the jaws of fetal anteaters. Some terrestrial frogs and salamanders pass through a larval stage within the egg that has the features of typically aquatic larvae, but hatch ready for life on land. Early in development, human embryos briefly display branchial pouches similar to the gill slits of fish embryos.

4. **Vestigial characters.** The adaptations of organisms have long been, and still are, cited by creationists as evidence of the Creator's wise beneficence, but no such claim can be made for the features, displayed by almost every species, that served a function in the species' ancestors, but do so no longer. Cave-dwelling fishes and other animals display eyes in every stage of degeneration. Flightless beetles retain rudimentary wings, concealed in some species beneath fused wing covers that would not permit the wings to be spread even if there were reason to do so. In *The Descent of Man*, Darwin listed a dozen vestigial features in the human body, some of which occur only as uncommon variations. They included the appendix, the coccyx (four fused tail vertebrae), rudimentary muscles that enable some people to move their ears or scalp, and the posterior molars, or wisdom teeth, that fail to erupt, or do so aberrantly, in many people. At the molecular level, every eukaryote's genome contains numerous nonfunctional DNA sequences, including pseudogenes: silent, untranscribed sequences that retain some similarity to the functional genes from which they have been derived (see Chapter 20).

5. **Convergence.** There are many examples, such as the eyes of vertebrates and cephalopod molluscs, in which functionally similar features actually differ profoundly in structure (see Figure 3.4). Such differences are expected if structures are modified from features that differ in different ancestors, but are inconsistent with the notion that an omnipotent Creator, who should be able to adhere to an optimal design, provided them. Likewise, evolutionary history is a logical explanation (and creation is not) for cases in which different organisms use very different structures for the same function, such as the various modified structures that enable vines to climb (see Figure 3.12).

6. **Suboptimal design.** The "accidents" of evolutionary history explain many features that no intelligent engineer would be expected to design. For example, the paths followed by food and air cross in the pharynx of terrestrial vertebrates, including humans, so that we risk choking on food. The human eye has a "blind spot," which you can find at about 45° to the right or left of your line of sight. It is caused by the functionally nonsensical arrangement of the axons of the retinal cells, which run forward into the eye and then converge into the optic nerve, which interrupts the retina by extending back through it toward the brain (see Figure 3.4).

7. **Geographic distributions.** The study of systematics includes the geographic distributions of species and higher taxa. This subject, known as biogeography, is treated in Chapter 6. Suffice it to say that the distributions of many taxa make no sense unless they have arisen from common ancestors. For example, many taxa, such as marsupials, are distributed across the southern continents, which is easily understood if they arose from common ancestors that were distributed across the single southern land mass that began to fragment in the Mesozoic.

8. **Intermediate forms.** The hypothesis of evolution by successive small changes predicts the innumerable cases in which characteristics vary by degrees among species and higher taxa. Among living species of birds, we see gradations in beaks; among snakes, some retain a vestige of a pelvic girdle and others have lost it altogether. At the molecular level, the difference among DNA sequences for the same protein ranges from almost none among very closely related species through increasing degrees of difference as we compare more remotely related taxa.

For each of these lines of evidence, hundreds or thousands of examples could be cited from studies of living species. Even if there were no fossil record, the evidence from living species would be more than sufficient to demonstrate the historical reality of evolution: all organisms have descended, with modification, from common ancestors. We can be even more confident than Darwin and assert that all organisms that we know of are descended from a single original form of life.

patterns of evolution, drawing both on the long tradition of comparative morphology and on contemporary studies at the molecular level.

Most features of organisms have been modified from pre-existing features

One of the most important principles of evolution is that *the features of organisms almost always evolve from pre-existing features of their ancestors*; they do not arise de novo, from nothing. The wings of birds, bats, and pterodactyls are modified forelimbs (**Figure 3.3**); they do not arise from the shoulders (as in angels), presumably because the ancestors of

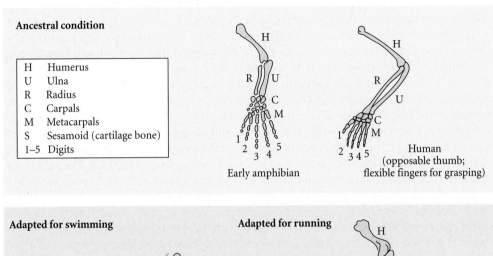

Ancestral condition

H	Humerus
U	Ulna
R	Radius
C	Carpals
M	Metacarpals
S	Sesamoid (cartilage bone)
1–5	Digits

Early amphibian

Human
(opposable thumb;
flexible fingers for grasping)

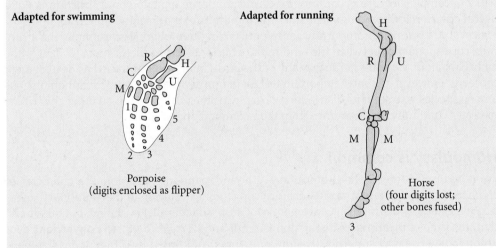

Adapted for swimming

Porpoise
(digits enclosed as flipper)

Adapted for running

Horse
(four digits lost;
other bones fused)

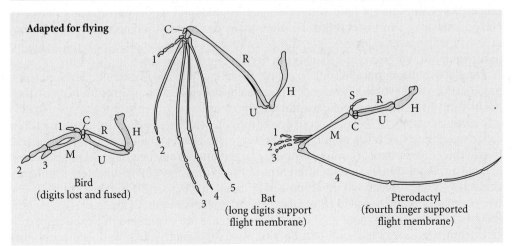

Adapted for flying

Bird
(digits lost and fused)

Bat
(long digits support
flight membrane)

Pterodactyl
(fourth finger supported
flight membrane)

FIGURE 3.3 The forelimb skeletons of some tetrapod vertebrates. Compared with the "ground plan," as seen in the early amphibian, bones have been lost or fused (e.g., horse, bird) or modified in relative size and shape. Modifications for swimming evolved in the porpoise, for running in the horse, and for flight in the bird, bat, and pterodactyl. All the bones shown are homologous among these organisms except for the sesamoid bones (S) in the pterodactyl; these bones have a different developmental origin from the rest of the limb skeleton. (After Futuyma 1995.)

these animals had no shoulder structures that could be modified for flight. The middle-ear bones of mammals evolved from jaw bones of reptiles (see Chapter 4). Likewise, existing proteins have been modified from ancestral proteins and have new functions (see Chapter 20). Genes encoding the sodium channel proteins of animals, which generate action potentials in the nervous system, are homologous to a gene in single-celled choanoflagellates (the sister group to animals), in which the protein acts as an ion selectivity filter (Liebeskind et al. 2011). In other words, related organisms have homologous characters that have been inherited, and sometimes modified to serve a different function, from an equivalent organ in the common ancestor. Homologous characters generally have similar genetic and developmental underpinnings, although these foundations have sometimes undergone substantial divergence among species (see Chapter 21). DNA sequencing has revealed far more homologous genes among very distantly related organisms, in different phyla or even kingdoms, than biologists had expected.

A *character* may be homologous among species (e.g., "toes"), but a given *character state* may not be (e.g., a certain number of toes). The pentadactyl (five-toed) state is homologous in humans and crocodiles (as far as we know, both have an unbroken history of pentadactyly as far back as their common ancestor), but the three-toed state in guinea pigs and rhinoceroses is not homologous, for this condition has evolved independently in these animals by modification from a five-toed ancestral state.

A character (or character state) is *defined* as homologous in two species if it has been derived from their common ancestor, but *diagnosing* homology—that is, determining whether or not characters of two species are homologous—can be difficult. The most common criteria for *hypothesizing* homology of anatomical characters are correspondence of *position* relative to other parts of the body and correspondence of *structure* (the parts of which a complex feature is composed). Correspondence of shape or of function is not a useful criterion for homology (consider the forelimbs of a horse and an eagle). Embryological studies are often important for hypothesizing homology. For example, the structural correspondence between the hindlimbs of birds and crocodiles is more evident in the embryo than in the adult because many of the bird's bones become fused as development proceeds. However, the most important test for judging whether a character is homologous among species is to see if its distribution on a phylogenetic tree (based on other characters) indicates continuity of inheritance from their common ancestor.

Homoplasy is common

We noted in Chapter 2 that **homoplasy**—the independent evolution of a character or character state in different taxa—includes convergent evolution and evolutionary reversal. Some authors distinguish convergent evolution from parallel evolution, but others feel that this is not a meaningful distinction (Arendt and Reznick 2008). In **convergent evolution (convergence)**, superficially similar features are formed by different developmental pathways. The eyes of both vertebrates and cephalopod molluscs (such as squids and octopuses) have a lens and a retina, but their many profound differences indicate that they evolved independently: for example, the axons of the retinal cells arise from the cell bases in cephalopods, but from the cell apices in vertebrates (**Figure 3.4**).

Parallel evolution (parallelism), by contrast, is thought to involve similar developmental modifications that evolved independently (often in closely related organisms, because they are likely to have similar developmental mechanisms to begin with). For example, in several diverse crustacean lineages, from one to as many as three pairs of feeding structures, called maxillipeds, develop not on the head segments that bear the animals' mouthparts, but on their anteriormost thoracic segments, which ancestrally bear legs. Michael Averoff and Nipam Patel (1997) studied differences in the expression of two "master" regulatory genes (Hox genes, which will be discussed in detail in Chapter 21) in these animals. The two genes they studied, called *Ultrabithorax* (*Ubx*) and *abdominal A* (*abdA*), determine how each segment develops in all arthropods. These genes are expressed in leg-bearing thoracic segments, but not in mouthpart-bearing head segments. Averoff and Patel found that

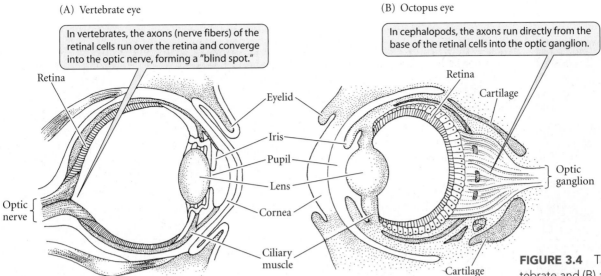

(A) Vertebrate eye

In vertebrates, the axons (nerve fibers) of the retinal cells run over the retina and converge into the optic nerve, forming a "blind spot."

Retina

Optic nerve

(B) Octopus eye

In cephalopods, the axons run directly from the base of the retinal cells into the optic ganglion.

Retina

Cartilage

Eyelid

Iris

Pupil

Lens

Cornea

Ciliary muscle

Optic ganglion

Cartilage

FIGURE 3.4 The eyes of (A) a vertebrate and (B) a squid or octopus (cephalopod mollusc) are an extraordinary example of convergent evolution. Despite their many similarities, note the several differences, including interruption of the retina by the optic nerve in the vertebrate, but not in the cephalopod. (From Brusca and Brusca 1990.)

the evolutionary transformation of crustacean legs into maxillipeds corresponds exactly with a loss of expression of *Ubx* and *abdA* in the thoracic segment(s) involved (**Figure 3.5**), and that this has happened in parallel in the several lineages that have maxillipeds.

There is increasing evidence that a particular gene may contribute to similar evolutionary changes in distantly related organisms. For example, mutations in the melanocortin-1-receptor (*Mc1r*) gene cause the difference between dark and pale populations of both the pocket mouse *Chaetodipus intermedius* and the beach mouse *Peromyscus polionotus*, which belong to

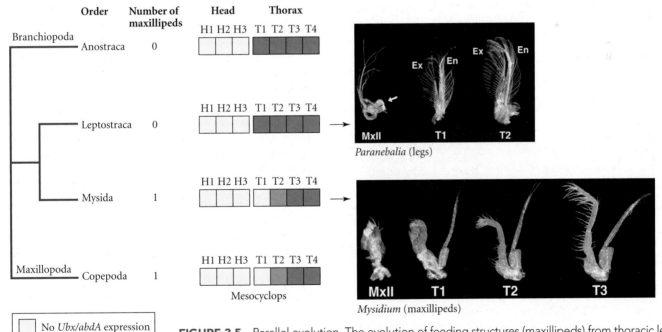

No *Ubx/abdA* expression

Weak expression

Strong expression

FIGURE 3.5 Parallel evolution. The evolution of feeding structures (maxillipeds) from thoracic legs in crustaceans is marked by parallel reduction or loss of expression of the genes *Ubx* and *abdA* in the thoracic segments that bear those structures. The top photo (*Paranebalia*, order Leptostraca) shows the ancestral condition: there is a mouthpart (the maxilla, MxII) on head segment H3, and legs (each with two branches, En and Ex) on thoracic segments T1 and T2. In *Mysidium* (order Mysida), in the bottom photo, thoracic segment T3 has a normal leg (En branch in yellow), but the appendages on T2 and (especially) T1 show maxilla-like modifications of the En branch (green and red, respectively). The copepod *Mesocyclops* (not pictured) has similar modifications of gene expression and morphology. (After Averoff and Patel 1997.)

(A)

(B)

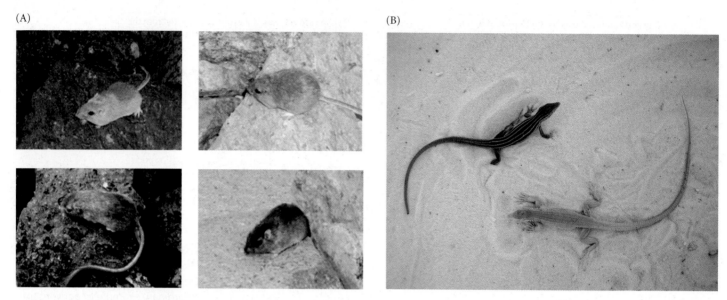

FIGURE 3.6 Convergent (parallel) evolution based on mutations of the same gene, *Mc1r*. (A) Dark (bottom) and pale (top) populations of the pocket mouse *Chaetodipus intermedius* occupy dark rocky outcrops (left) and pale desert soils (right), respectively. The difference is believed to reduce visibility to predators (compare upper right and lower left photos with those at upper left and lower right). (B) A similar difference, also probably due to mutations of *Mc1r*, is seen among populations of the whiptail lizard *Aspidoscelis inornata* that are associated with different desert soils. (A courtesy of Michael Nachman; B courtesy of Erica Bree Rosenblum.)

distantly related families, and mutations of the same gene appear to underlie dark versus pale forms of several lizards and birds (**Figure 3.6**). However, some dark populations of both species of mice owe their coloration not to a mutation of this gene, but to other genes that have not yet been identified. Thus, even within a single species, similar phenotypes may have either similar or different genetic and developmental foundations (Nachman et al. 2003; Hoekstra et al. 2006; Arendt and Reznick 2008).

Evolutionary reversals constitute a return from an "advanced," or DERIVED, character state to a more "primitive," or ANCESTRAL, state (Porter and Crandall 2003). For example, winged insects evolved from wingless ancestors, but many lineages of insects have lost their wings in the course of subsequent evolution. It has long been thought that complex characteristics, once lost, are unlikely to be regained, a principle known as DOLLO'S LAW. However, recent phylogenetic and experimental evidence casts some doubt on this principle (Collin and Miglietta 2008). Birds have not had teeth for at least 70 million years, but tooth primordia develop in embryos of chickens with the *ta*^{2} mutation (which, however, results in death before hatching; Harris et al. 2006). The ancestral life history pattern of salamanders includes an aquatic larval stage with features such as gills and parts of the skeleton that differ from the adult stage. The aquatic larval stage has been lost in the evolution of the wholly terrestrial plethodontine subfamily of salamanders, but phylogenetic analysis showed that it has been regained in one lineage of this subfamily, the dusky salamanders (*Desmognathus*;

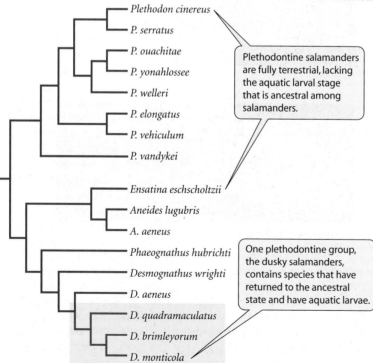

Plethodontine salamanders are fully terrestrial, lacking the aquatic larval stage that is ancestral among salamanders.

One plethodontine group, the dusky salamanders, contains species that have returned to the ancestral state and have aquatic larvae.

FIGURE 3.7 A phylogeny, based on DNA sequences, of part of the salamander family Plethodontidae, showing that species of *Desmognathus* with aquatic larvae (red) are nested within a large group of taxa that lack the larval stage (blue). Most outgroups (i.e., other salamander families) possess the larval stage, which is an ancestral feature of salamanders generally. (After Chippindale et al. 2004.)

Figure 3.7). However, the terrestrial plethodontines develop certain features of the aquatic larval skeleton as they develop within the egg, even though they have adult features when they hatch (Kerney et al. 2012). This observation suggests that the aquatic larval stage of the dusky salamanders results from an elaboration of larval features that had never been entirely lost in their ancestors.

Homoplasious features are often (but not always) adaptations by different lineages to similar environmental conditions. In fact, a correlation between a particular homoplasious character in different groups and a feature of those organisms' environment or niche is often the best initial evidence of the feature's adaptive significance. For example, a long, thin beak has evolved independently in at least six different lineages of nectar-feeding birds. Such a beak enables these birds to reach nectar in the bottoms of the long, tubular flowers in which they often feed (**Figure 3.8**). Likewise, long tubular flowers have evolved independently in many lineages of bird-pollinated plants.

MIMICRY, a condition in which features of one species have specifically evolved to resemble those of another species, provides especially interesting examples of convergent evolution. Two of the most common kinds of mimicry are BATESIAN MIMICRY and MÜLLERIAN MIMICRY, named after the nineteenth-century naturalists who first described the phenomena. Both of these are defenses against predators. A Batesian mimic is a palatable or innocuous animal that has evolved resemblance to an unpalatable or dangerous animal (the MODEL); for instance, many harmless flies have bright yellow and black color patterns like those of wasps, and some palatable butterfly species closely resemble other species that are toxic (see Figure 9.1B). Predators that learn, from unpleasant experience, to avoid the model will

(A)

(B)

(C)

(D)

FIGURE 3.8 Four bird groups in which similar bill shape has evolved independently as an adaptation for feeding on nectar. (A) A South American purple honeycreeper (*Cyanerpes caeruleus*), family Thraupidae. (B) An iiwi (*Vestiaria coccinea*), one of the many Hawaiian honeycreepers, family Fringillidae. The Hawaiian honeycreepers are a well-studied example of adaptive radiation. (C) Hummingbirds, family Trochilidae. This violet sabrewing (*Campylopterus hemileucurus*) is from Costa Rica. (D) Sunbirds, family Nectariniidae. The greater double-collared sunbird (*Nectarinia afra*) is native to South Africa.

also tend to avoid attacking the mimic. In Müllerian mimicry, two or more unpalatable or dangerous species have evolved similar characteristics, and a predator that associates the features of one with an unpleasant experience avoids attacking both species. There are many examples of mimicry "rings" that involve several species of distantly related unpalatable butterflies (Müllerian mimics; see Figure 19.12) as well as palatable butterflies (Batesian mimics) that share the same color pattern. Some species represent intermediate conditions between the Batesian and Müllerian extremes (Mallet and Joron 1999).

Rates of character evolution differ

Different characters evolve at different rates, as is evident from the simple observation that any two species differ in some features but not in others. Some characters, often called **conservative characters**, are retained with little or no change over long periods among the many descendants of an ancestor. For example, humans retain the pentadactyl limb that first evolved in early amphibians (see Figure 3.3). All amphibians and reptiles have two aortic arches, and all mammals have only the left arch. Body size, in contrast, evolves rapidly; within orders of mammals, it may vary at least a hundredfold. The rate of DNA sequence evolution varies among genes, among segments within genes, and among the three positions within amino acid–coding triplets of bases (codons).

Evolution of different characters at different rates within a lineage is called **mosaic evolution (Figure 3.9)**. It is one of the most important principles of evolution, for it says that a species

Heliocidaris tuberculata (indirect development)
Adult

Larva

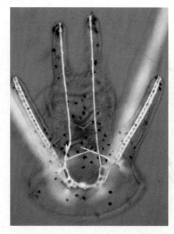

Heliocidaris erythrogramma (direct development)
Adult

Larva

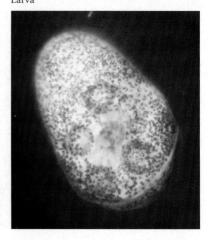

FIGURE 3.9 An example of mosaic evolution. Adults of the closely related sea urchins *Heliocidaris tuberculata* and *H. erythrogramma* are very similar and can even be hybridized in the laboratory; hence most adult features have evolved slowly or not at all. In contrast, larval characteristics have evolved rapidly. *H. tuberculata* has a pluteus larva that is typical of most sea urchins, whereas *H. erythrogramma* has evolved a nonfeeding, direct-developing larval form. (Adult photos © David Harasti; larvae photos courtesy of Rudolf A. Raff.)

evolves not as a whole, but piecemeal: many of its features evolve quasi-independently. (There are important exceptions; for example, features that function together may evolve in concert.) This principle largely justifies the practice of analyzing evolutionary mechanisms not in terms of whole organisms, but in terms of changes in individual features, or even individual genes underlying such features.

Every species is a mosaic of PLESIOMORPHIC (ancestral, or "primitive") and APOMORPHIC (derived, or "advanced") characteristics. For example, the amphibian lineage leading to frogs split from the lineage leading to mammals before the mammalian orders diversified, so in terms of *order of branching*, frogs are an older branch than cows or humans. In that sense, frogs might be assumed to be more primitive. But relative to early Paleozoic amphibians, frogs have both ancestral features (e.g., five toes on the hind foot, multiple bones in the lower jaw) and features (e.g., lack of teeth in the lower jaw) that are more advanced than those of many mammals, in that they have changed further from the ancestral state. Moreover, numerous differences among frog species have evolved in the recent past. For example, one genus has direct development without a tadpole stage, and another gives birth to live young. Humans likewise have both "primitive" characters (e.g., five digits on hands and feet; teeth in the lower jaw) and some that are "advanced" compared with frogs (e.g., a single lower jawbone). *Because of mosaic evolution, it is inaccurate or even wrong to consider one living species more "advanced" than another.*

Evolution is often gradual

Darwin argued that evolution proceeds by small successive changes (GRADUALISM) rather than by large "leaps" (SALTATIONS). Whether or not evolution is always gradual is unknown, and the issue is much debated (see Chapter 22). Many higher taxa that diverged in the distant past (e.g., the animal phyla; many orders of insects and of mammals) are very different and are not bridged by intermediate forms, either among living species or in the fossil record. However, the fossil record does document intermediates in the evolution of some higher taxa, as we will see in Chapter 4.

Gradations among living species are commonplace and provide support for gradualism. For example, the length and shape of the bill differs greatly among species of sandpipers, but the most extreme forms are bridged by species with intermediate bills (**Figure 3.10**). Many lineages of plants in hot, dry environments have independently evolved C_4 photosynthesis, which differs from the more common C_3 condition by an additional biochemical pathway and by modification of the size and chloroplast content of the bundle sheath cells that surround the leaf veins. Some species in the family Molluginaceae have an intermediate condition, and a phylogenetic analysis showed that the fully developed C_4 adaptation has evolved through the sequential addition of several of these features (**Figure 3.11**; Christin et al. 2011).

Change in form is often correlated with change in function

One of the reasons a homologous character may differ so greatly among taxa is that its form may have evolved as its function has changed. The sting of a wasp or bee, for example, is a modification of the ovipositor that other members of the Hymenoptera use to insert eggs into plant or arthropod hosts (that is why only female wasps and bees sting). In the many groups

FIGURE 3.10 Variation in the shape and length of the bill among sandpipers (Scolopacidae). The two vertical series are drawn to scale and are spaced to match the differences in bill length, which range from 18 mm (bottom left) to 166 mm (upper right). Note the gradations in both curvature and length. The phylogenetic relationships among these species are not well resolved, but the variation shows how very different bills could have evolved through small changes. (After Hayman et al. 1986.)

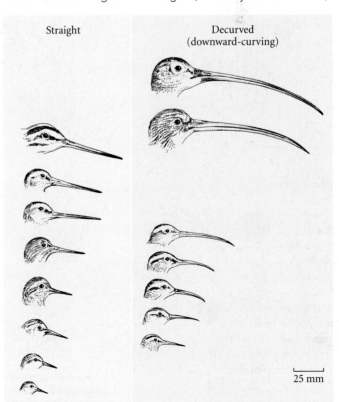

Straight   Decurved (downward-curving)

25 mm

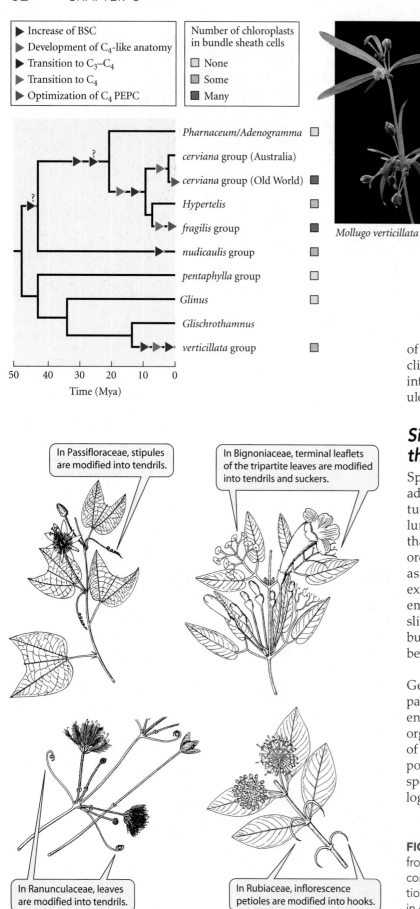

Legend:

- ▶ Increase of BSC
- ▶ Development of C$_4$-like anatomy
- ▶ Transition to C$_3$–C$_4$
- ▶ Transition to C$_4$
- ▶ Optimization of C$_4$ PEPC

Number of chloroplasts in bundle sheath cells
- ☐ None
- ☐ Some
- ■ Many

Pharnaceum/Adenogramma ☐
cerviana group (Australia)
cerviana group (Old World) ■
Hypertelis ☐
fragilis group ■
nudicaulis group ☐
pentaphylla group ☐
Glinus ☐
Glischrothamnus
verticillata group ☐

Time (Mya): 50 40 30 20 10 0

Mollugo verticillata

FIGURE 3.11 Stepwise evolution of the C$_4$ photosynthetic phenotype in the plant family Molluginaceae. The bundle sheath cells (BSC), which surround a leaf vein, increased in size in two lineages (blue arrowheads). The bundle sheath cells took on a characteristic C$_4$ anatomy (green arrowheads), including chloroplasts that became more numerous in more "advanced" C$_4$ species. Metabolic changes occurred as lineages transitioned to C$_3$–C$_4$ and C$_4$ photosynthesis (purple and orange arrowheads). At about the same time, evolutionary changes in PEPC (red arrowheads), a key enzyme of the C$_4$ pathway, occurred. Question marks indicate alternative scenarios for the transition to C$_3$–C$_4$ photosynthesis. (After Christin et al. 2011.)

of plants that have independently evolved a vinelike climbing habit, the structures that have been modified into climbing organs include roots, leaves, leaflets, stipules, and inflorescences (**Figure 3.12**).

Similarity among species changes throughout ontogeny

Species are often more similar as embryos than as adults. Karl Ernst von Baer noted in 1828 that the features common to a higher taxon (such as the subphylum Vertebrata) often appear earlier in development than the specific characters of lower-level taxa (such as orders or families). This generalization is now known as VON BAER'S LAW. Probably the most widely known example is the similarity of many tetrapod vertebrate embryos, all of which display pharyngeal clefts (gill slits), a notochord, segmentation, and paddlelike limb buds before the features typical of their class or order become apparent (**Figure 3.13**).

One of Darwin's most enthusiastic supporters, the German biologist Ernst Haeckel, reinterpreted such patterns to mean that "ontogeny recapitulates phylogeny"; that is, that the development of the individual organism (ONTOGENY) repeats the evolutionary history of the adult forms of its ancestors. Haeckel thus supposed that by studying embryology, one could read a species' phylogenetic history, and therefore infer phylogenetic relationships among organisms directly. By

In Passifloraceae, stipules are modified into tendrils.

In Bignoniaceae, terminal leaflets of the tripartite leaves are modified into tendrils and suckers.

In Ranunculaceae, leaves are modified into tendrils.

In Rubiaceae, inflorescence petioles are modified into hooks.

FIGURE 3.12 Structures modified for climbing in vines from different plant groups show that structures can become modified for new functions, and that different evolutionary paths to the same functional end may be followed in different groups. (After Hutchinson 1969.)

the end of the nineteenth century, however, it was already clear that Haeckel's dictum, referred to as the BIOGENETIC LAW, rather seldom holds (Gould 1977). For example, the pharyngeal clefts and branchial arches of embryonic mammals and reptiles never acquire the form typical of adult fishes. Moreover, various features develop at different rates, relative to one another, in descendants than in their ancestors, and embryos and juvenile stages have stage-specific adaptations of their own (such as the amnion in the embryo). Thus the biogenetic law is certainly not an infallible guide to phylogenetic history. Embryological similarities provided Darwin with some of his most important evidence of evolution, however, and they continue to shed important light on how characteristics have been transformed during evolution.

Development and Morphological Evolution

Until a few decades ago, classification and phylogenetic studies relied chiefly on analyses of morphological characters, including their change during embryonic development. In the course of their work, systematists and comparative morphologists documented many common patterns of evolution that they expressed in terms of underlying developmental changes. Some patterns of developmental change in morphology are individualization, heterochrony, allometry, heterotopy, and changes in complexity (Rensch 1959; Müller 1990; Raff 1996; Wagner 1996). Today one of the most active research areas concerns the genetic and developmental basis of such evolutionary changes (such as the evolution of crustacean appendages portrayed in Figure 3.5). This research is described in more detail in Chapter 21.

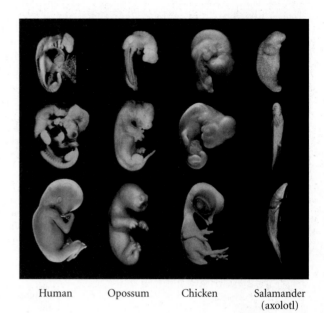

| Human | Opossum | Chicken | Salamander (axolotl) |

FIGURE 3.13 Micrographs show the similarities—and differences—among several vertebrate embryos at different stages of development. Each begins (at top) with a similar basic structure, although they acquire this structure at different stages and sizes. As the embryos develop, they become less and less alike. (Adapted from Richardson et al. 1998; photo courtesy of M. Richardson and R. O'Rahilly.)

Individualization

The bodies of many organisms consist of MODULES—distinct units that have distinct genetic specifications, developmental patterns, locations, and interactions with other modules (Raff 1996). Some such modules are repeated at various sites on the body, and are termed SERIALLY HOMOLOGOUS if they are arrayed along the body axis (as vertebrae are) and HOMONYMOUS if they are not. In some cases, serially homologous or homonymous features lack distinct individual identities (e.g., leaves of many plants, teeth of many fishes) and may be considered aspects of a single character. An important evolutionary phenomenon is the acquisition of distinct identities by such modules, called **individualization** (Müller and Wagner 1996; Wagner 1996), which in turn is an important basis for mosaic evolution. For instance, the teeth of most reptiles are uniform, but they became individualized (differentiated into incisors, canines, premolars, and molars) during the evolution of mammals. Distinct tooth identity was later lost during the evolution of the toothed whales (**Figure 3.14**).

Heterochrony

Heterochrony (Gould 1977; McKinney and McNamara 1991) is broadly defined as *an evolutionary change in the timing or rate of developmental events* (from the Greek *heteros*, "different," and *chronos*, "time"). Many phenotypic changes appear to be based on such changes in timing, although several other developmental mechanisms can produce similar changes (Raff 1996).

Relatively global heterochronic changes, affecting many characters simultaneously, are illustrated by cases in which the time of development of most SOMATIC features (those other than the gonads and related reproductive structures) is altered relative to the time of

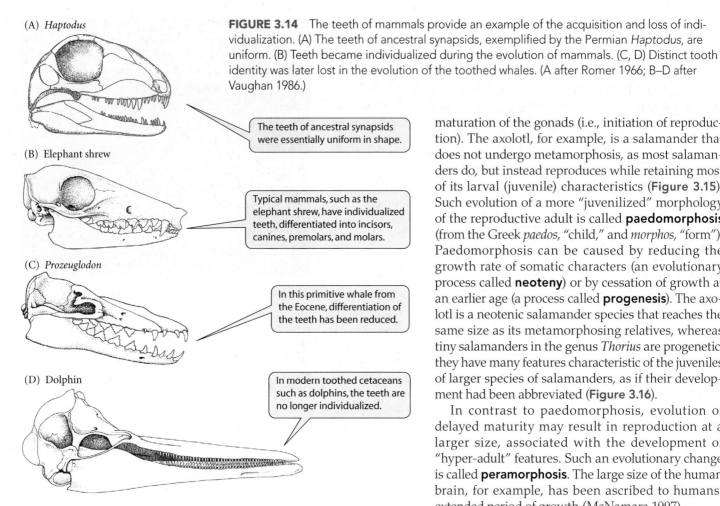

(A) *Haptodus*

The teeth of ancestral synapsids were essentially uniform in shape.

(B) Elephant shrew

Typical mammals, such as the elephant shrew, have individualized teeth, differentiated into incisors, canines, premolars, and molars.

(C) *Prozeuglodon*

In this primitive whale from the Eocene, differentiation of the teeth has been reduced.

(D) Dolphin

In modern toothed cetaceans such as dolphins, the teeth are no longer individualized.

FIGURE 3.14 The teeth of mammals provide an example of the acquisition and loss of individualization. (A) The teeth of ancestral synapsids, exemplified by the Permian *Haptodus*, are uniform. (B) Teeth became individualized during the evolution of mammals. (C, D) Distinct tooth identity was later lost in the evolution of the toothed whales. (A after Romer 1966; B–D after Vaughan 1986.)

maturation of the gonads (i.e., initiation of reproduction). The axolotl, for example, is a salamander that does not undergo metamorphosis, as most salamanders do, but instead reproduces while retaining most of its larval (juvenile) characteristics (**Figure 3.15**). Such evolution of a more "juvenilized" morphology of the reproductive adult is called **paedomorphosis** (from the Greek *paedos*, "child," and *morphos*, "form"). Paedomorphosis can be caused by reducing the growth rate of somatic characters (an evolutionary process called **neoteny**) or by cessation of growth at an earlier age (a process called **progenesis**). The axolotl is a neotenic salamander species that reaches the same size as its metamorphosing relatives, whereas tiny salamanders in the genus *Thorius* are progenetic: they have many features characteristic of the juveniles of larger species of salamanders, as if their development had been abbreviated (**Figure 3.16**).

In contrast to paedomorphosis, evolution of delayed maturity may result in reproduction at a larger size, associated with the development of "hyper-adult" features. Such an evolutionary change is called **peramorphosis**. The large size of the human brain, for example, has been ascribed to humans' extended period of growth (McNamara 1997).

Allometry

Allometric growth, or **allometry**, refers to the *differential rate of growth of different parts or dimensions of an organism* during its ontogeny. For example, during human postnatal growth, the head grows at a slower rate than the body as a whole, and the legs grow at a faster rate. Changes in the growth rates of individual characters—that is, "local" heterochrony—appear to have played an extremely important role in evolution. For example, many evolutionary changes can be described as if local heterochronies had altered the *shape* of one or more characters: an increased rate of elongation of the digits "accounts for" the shape of a bat's wing compared with the forelimbs of other mammals (see Figure 3.3); an elephant's tusks are incisor teeth that have grown much faster than the other teeth.

FIGURE 3.15 Paedomorphosis in salamanders. (A) The tiger salamander (*Ambystoma tigrinum*), like most salamanders, undergoes metamorphosis from an aquatic larva (left; note the presence of gills) to a terrestrial adult (right). (B) The adult axolotl (*Ambystoma mexicanum*), with gills and tail fin, resembles the larva of its terrestrial relative. The axolotl remains aquatic throughout its life.

(A)

(B)

(A) Paedomorphic (*Thorius*)

(B) Nonpaedomorphic

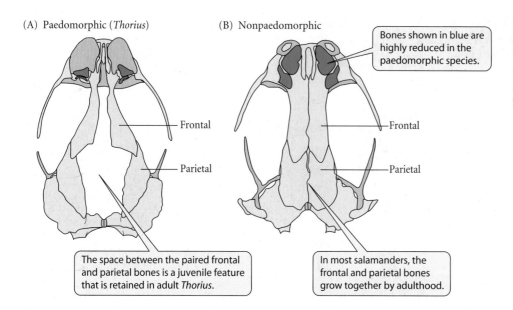

Bones shown in blue are highly reduced in the paedomorphic species.

Frontal

Parietal

Frontal

Parietal

The space between the paired frontal and parietal bones is a juvenile feature that is retained in adult *Thorius*.

In most salamanders, the frontal and parietal bones grow together by adulthood.

FIGURE 3.16 Comparison of the skulls of the progenetic dwarf salamander *Thorius* and a typical nonprogenetic relative, *Pseudoeurycea*. The skull of adult *Thorius* has a number of juvenile features. (After Hanken 1984.)

Allometric growth is often described by the equation

$$y = bx^a$$

where y and x are two measurements, such as the height and width of a tooth or the size of the head and the body. (In many studies, x is a measure of body size, such as weight, because many structures change disproportionately with overall size.) The ALLOMETRIC COEFFICIENT, a, describes their relative growth rates. If $a = 1$, growth is ISOMETRIC, meaning that the two structures or dimensions increase at the same rate, and shape does not change. If y increases faster than x, as for human leg length relative to body size or weight, $a > 1$ (positive allometry); if it increases relatively slowly, as for human head size, $a < 1$ (negative allometry; **Figure 3.17A**). These curvilinear relationships between y and x often appear more linear if transformed to logarithms, yielding the equation for a straight line: $\log y = \log b + a \log x$ (**Figure 3.17B**). For example, antler size in deer increases allometrically with body mass. The largest species of deer, the extinct Irish elk, had antlers much larger, relative to body mass, than those of any other deer (**Figure 3.18**).

(A) Arithmetic plots

(B) Logarithmic plots

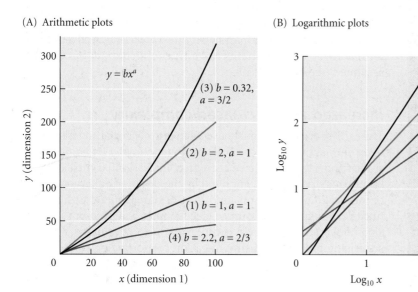

$y = bx^a$

(3) $b = 0.32$, $a = 3/2$

(2) $b = 2$, $a = 1$

(1) $b = 1$, $a = 1$

(4) $b = 2.2$, $a = 2/3$

FIGURE 3.17 Hypothetical curves showing various allometric growth relationships between two body measurements, y and x, according to the equation $y = bx^a$. (A) Arithmetic plots. Curves 1 and 2 show isometric growth ($a = 1$), in which y is a constant multiple (b) of x. Curves 3 and 4 show positive ($a > 1$) and negative ($a < 1$) allometry, respectively. (B) Logarithmic plots of the same curves have a linear form. The slope differences depend on a. Curves 1 and 2 have slopes equal to 1.

(A)

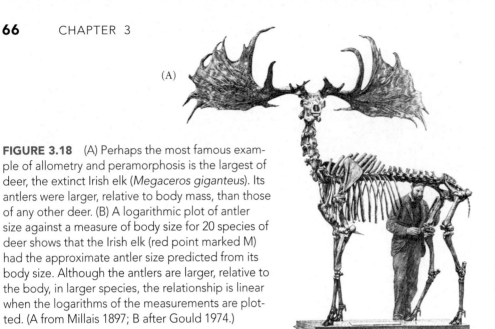

(B)

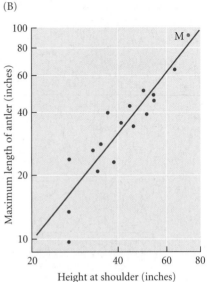

FIGURE 3.18 (A) Perhaps the most famous example of allometry and peramorphosis is the largest of deer, the extinct Irish elk (*Megaceros giganteus*). Its antlers were larger, relative to body mass, than those of any other deer. (B) A logarithmic plot of antler size against a measure of body size for 20 species of deer shows that the Irish elk (red point marked M) had the approximate antler size predicted from its body size. Although the antlers are larger, relative to the body, in larger species, the relationship is linear when the logarithms of the measurements are plotted. (A from Millais 1897; B after Gould 1974.)

Heterotopy

Heterotopy is an evolutionary change in the position within an organism at which a phenotypic character is expressed (from the Greek *topos*, "place"). Studies of the distribution of gene products have revealed many heterotopic differences among species in sites of gene expression. For instance, certain species of deep-sea squid have organs on the body that house light-emitting bacteria. The light is diffracted through small lenses made up of the same two proteins that compose the lenses of the squids' eyes (Raff 1996).

Heterotopic differences among species are very common in plants. For example, the major photosynthetic organs of most plants are the leaves, but photosynthesis is carried out in the stem in cacti and many other plants that occupy dry environments. In many unrelated species of lianas (woody vines), roots grow along the aerial stem (**Figure 3.19**). In some lianas these roots serve as holdfasts, while in others they grow down to the soil from the canopy high above.

The bones of vertebrates provide many examples of heterotopy. For example, many phylogenetically new bones have arisen as SESAMOIDS—bones that develop in tendons or other connective tissues subject to stress (Müller 1990). Many dinosaurs had bony tendons in the tail, and the giant panda (*Ailuropoda melanoleuca*) is famous for having a "thumb" that is not a true jointed digit, but a single sesamoid (see Figure 22.10).

Increased and decreased complexity

The earliest organisms must have had very few genes and must have been very simple in form. Both phylogenetic and paleontological studies show that there have been great increases in complexity during the history of life, exemplified by the origin of eukaryotes, then of multicellular organisms, and later of the elaborate tissue organization of plants and animals (see Chapter 5). Given our impression of overall increases in complexity during evolution, it can be surprising to learn that simplification of morphology—reduction and loss of structures—is one of the most common trends within clades (Adamowicz and Purvis 2006). The primitive "ground plan"

FIGURE 3.19 *Monstera deliciosa* is a creeping vine native to Central America. The species has evolved exposed roots that grow from an aerial stem—an example of heterotopy.

Stem

Roots

(A) *Eusthenopteron* (lobe-finned fish)

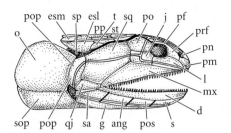

(B) *Milleretta* (early amniote)

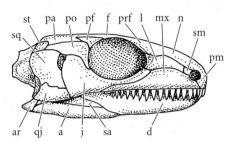

(C) *Canis* (modern mammal)

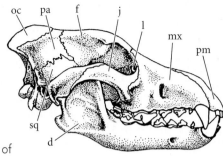

FIGURE 3.20 An example of reduction and loss of structures during evolution. The number of bones in the skull is higher in early lobe-finned fishes (such as the Devonian *Eusthenopteron*), from which amniotes were derived, than in their early amniote descendants (such as the Permian *Milleretta*). Among the later amniotes are placental mammals, such as the domestic dog (*Canis*), in which the skull has fewer elements still. The reduction in the number of bones in the lower jaw is particularly notable. (After Romer 1966.)

of flowers, for example, includes numerous sepals, petals, stamens, and carpels, but the number of one or more of these elements has been reduced in many lineages of flowering plants, and many clades have lost petals or sepals, or both, altogether. The number of digits in tetrapod vertebrates has been reduced many times (consider the single toe of horses shown in Figure 3.3), but has increased only once (in extinct ichthyosaurs). Early lobe-finned fishes had far more skull bones than their amniote descendants (**Figure 3.20**). Many such changes can be ascribed to increased functional efficiency.

Phylogenetic Analysis Documents Evolutionary Trends

The term **evolutionary trend** refers to repeated changes of a character in the same direction, either within a single lineage or, often, in many lineages independently. For example, Justen Whittall and Scott Hodges (2007) documented both kinds of trends in a phylogenetic analysis of columbines (*Aquilegia*; **Figure 3.21**). During the

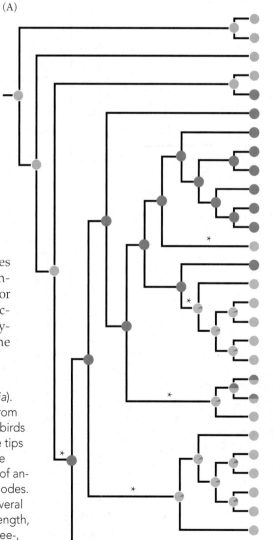

(A)

(B)

A. laramiensis

A. formosa

A. longissima

FIGURE 3.21 Evolutionary trends in columbines (*Aquilegia*). (A) A phylogenetic analysis of several species shows shifts from pollination by bees (blue circles) to pollination by hummingbirds (red) and then by hawkmoths (lime green). The circles at the tips of the branches represent different species; the colors of the circles at internal nodes show the inferred pollination habit of ancestors; and the asterisks mark shifts between pollination modes. (B) These pollinator shifts are associated with changes in several flower characteristics, including nectar spurs of increasing length, as illustrated by representative species of (top to bottom) bee-, hummingbird-, and hawkmoth-pollinated columbines. (After Whittall and Hodges 2007; photos courtesy of Justen Whittall.)

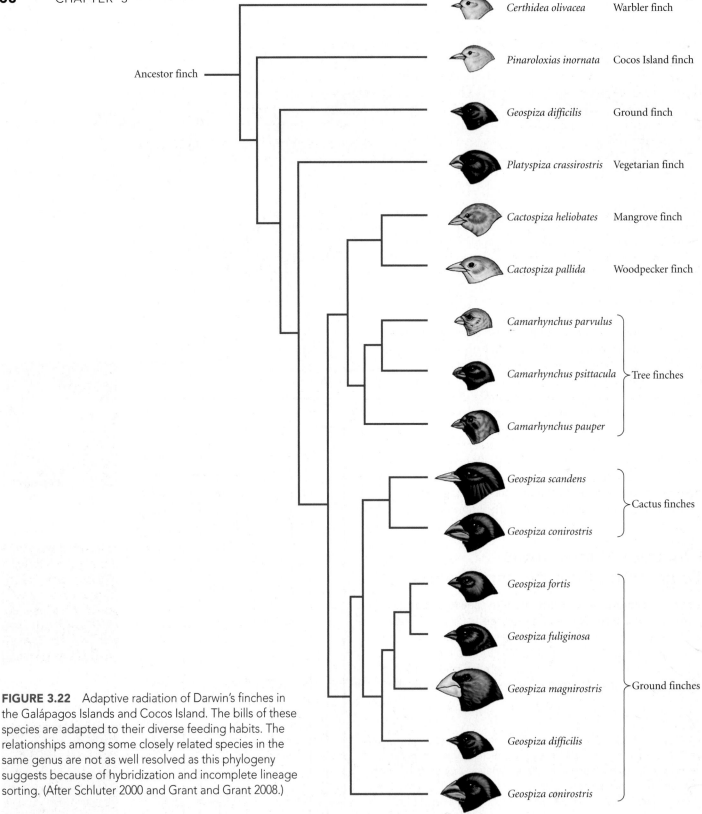

FIGURE 3.22 Adaptive radiation of Darwin's finches in the Galápagos Islands and Cocos Island. The bills of these species are adapted to their diverse feeding habits. The relationships among some closely related species in the same genus are not as well resolved as this phylogeny suggests because of hybridization and incomplete lineage sorting. (After Schluter 2000 and Grant and Grant 2008.)

diversification of columbine species, there have been repeated shifts from pollination by short-tongued bumblebees to longer-tongued hummingbirds, and from hummingbirds to still longer-tongued hawkmoths. In concert with these repeated shifts, the plants have consistently evolved longer nectar spurs, tubular extensions of the petals from which the pollinator extracts nectar. Whittall and Hodges suggest that consistent shifts to longer-tongued, rather than shorter-tongued, pollinators occur because short-tongued animals cannot reach the nectar in long spurs and avoid such flowers. Shorter-tongued insects may visit a greater variety of plant species more indiscriminately, wasting the columbine's pollen by depositing it on other species, so it may be advantageous to exclude these promiscuous visitors. Other common evolutionary trends among flowering plants include shifts from low to high chromosome number and DNA content, many to few flower parts (e.g., stamens or carpels), separate to fused flower parts, radial to bilateral symmetry of the flower, animal to wind pollination, and woody to herbaceous structure. We will analyze the kinds and causes of evolutionary trends in Chapter 22.

Many Clades Display Adaptive Radiation

Evolutionary radiation, as described in the previous chapter, is divergent evolution of numerous related lineages within a relatively short time. In most cases, the lineages become modified for different ways of life, and the evolutionary radiation may be called an **adaptive radiation** (Schluter 2000). The characteristics of the members of an evolutionary radiation usually do not show a trend in any one direction. Evolutionary radiation, rather than sustained, directional evolutionary trends, is probably the most common pattern of long-term evolution.

Several adaptive radiations have been extensively studied and are cited in many evolutionary contexts. The most famous example is the adaptive radiation of Darwin's finches in the Galápagos Islands. These finches, which are all descended from a single ancestor that colonized the archipelago from South America, differ in the morphology of the bill, which provides adaptation to different diets (**Figure 3.22**). Different species of *Geospiza* feed on seeds that vary in size and hardness. Other genera include *Camarhynchus*, which excavates insects from wood, *Certhidea*, which feeds on nectar and insects, and *Cactospiza*, in which one species has the unique habit of using twigs and cactus spines as tools to extricate insects from crevices.

Relatively few kinds of animals and plants have colonized the Hawaiian Islands, but many of those that have done so have given rise to adaptive radiations. For example, more than 800 species of drosophilid flies (fruit flies) occur in the islands. In morphology and sexual behavior, they are more diverse than all the drosophilids in the rest of the world combined (Carson and Kaneshiro 1976). Many of the species have bizarre modifications of the mouthparts, legs, and antennae, which are associated with unusual mating behavior. Among plants, the Hawaiian silverswords and their relatives, three closely related genera of plants in the sunflower family, occupy habitats ranging from exposed lava rock to wet forest, and their growth forms include shrubs, vines, trees, and creeping mats (**Figure 3.23**). Despite these great differences, most silverswords can produce fertile hybrids when crossed (Carlquist et al. 2003).

FIGURE 3.23 Some members of the Hawaiian silversword alliance: closely related species with different growth forms. (A) *Argyroxiphium sandwicense*, a rosette plant that lacks a stem except when flowering (as it is here). (B) *Wilkesia gymnoxiphium*, a stemmed rosette plant. (C) *Dubautia menziesii*, a small shrub.

(A) *Argyroxiphium sandwicense*   (B) *Wilkesia hobdyi*   (C) *Dubautia menziesii*

FIGURE 3.24 A sample of the ecologically diverse Cichlidae of the African Great Lakes. (A) *Pseudotropheus zebra* combs algae from rock surfaces. (B) *Serranochromis robustus* is an active predator on other cichlids. (C) *Cyrtocara moorii* feeds on small invertebrates stirred up by other cichlid species that dig in sand. (D) *Chilotilapia euchilus* uses its enlarged lips to suck invertebrates out of rock crevices.

The cichlid fishes in the Great Lakes of eastern Africa have undergone perhaps the most spectacular adaptive radiation (Seehausen 2006a; Koblmüller et al. 2008). Lake Victoria has more than 200 species, Lake Tanganyika at least 140, and Lake Malawi more than 500—perhaps as many as 1000. The species within each lake vary greatly in coloration, body form, and the form of the teeth and jaws (**Figure 3.24**), and they are correspondingly diverse in feeding habits. There are specialized feeders on insects, detritus, rock-encrusting algae, phytoplankton, zooplankton, molluscs, baby fishes, and large fishes. Some species feed on the scales of other fishes, and one has the gruesome habit of plucking out other fishes' eyes. Adaptive differences between the teeth of certain closely related species exceed the differences between entire families of other fishes. The species in each lake are a monophyletic group and have diversified very rapidly.

Patterns of Evolution in Genes and Genomes

So far, we have seen how phylogenetic analyses can reveal historical patterns of evolution in the phenotypic characteristics of organisms. The same methods are now used to probe the evolution of DNA sequences, such as genes, and of the structure of genomes. This section presents just a few examples of evolutionary patterns at the molecular level; more instances will arise in subsequent chapters (see Chapter 20).

Convergent evolution

Many of the evolutionary patterns that have been found for morphological features are also observed at the molecular level. Among the many examples of convergent evolution of proteins is prestin, which is located in the membrane of hair cells of the cochlea in the inner ear of mammals and is thought to affect frequency selectivity in hearing. When Ying Li et al. (2010) performed maximum likelihood phylogenetic analysis of the prestin gene sequence in various mammals, they found that the sequence of the dolphin (a cetacean) clustered with those of echolocating bats (**Figure 3.25A**), rather than with the pig and cow, known to be the closest relatives of cetaceans. Toothed whales, including dolphins, are the only mammals other than bats that use echolocation (sonar). The convergent similarities in sequence are located primarily in intracellular portions of the protein (**Figure 3.25B**), suggesting that these parts of the protein may be especially important for echolocation.

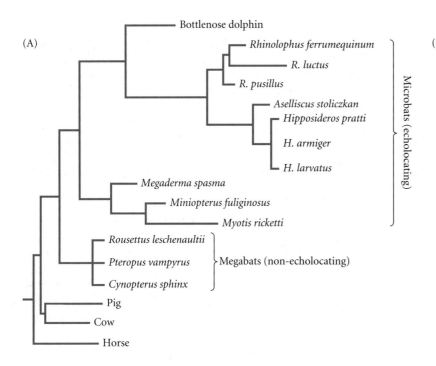

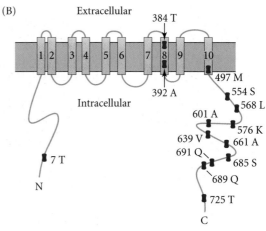

FIGURE 3.25 Evidence of convergence of the prestin gene. (A) A phylogeny of some mammals based on this gene erroneously indicated that the bottlenose dolphin is related to echolocating bats, whereas it is known to be related to artiodactyl hoofed mammals (pig and cow). Large bats (megabats) do not echolocate. (B) A diagram of the prestin protein, which is located in the hair cells of the cochlea. Some portions lie in the cell membrane and others within the cell. The convergent amino acids are marked by their number in the protein sequence and by a letter designating their identity. (After Li et al. 2010).

Genome size

Genome sizes are frequently measured in picograms (pg) of DNA; 1 pg is roughly equivalent to 1 Gb (1 gigabase, or 1 billion base pairs) of actual sequence. As data on genome sizes from hundreds of organisms were compared, a curious pattern emerged. It was expected that physiologically and behaviorally complex organisms, such as mammals, would have more complex, and therefore larger, genomes than simpler organisms. On a very broad scale, this holds true (**Figure 3.26**): viral and bacterial genomes are tiny compared with those of eukaryotes, particularly those of mammals, amphibians, and some plants. But major groups of eukaryotes, such as vertebrates and flowering plants, seem not to differ as had originally been expected. For example, the haploid genome of the puffer fish is about 0.5 Gb, and human and mouse genomes are both about 3 Gb—but salamanders have enormous haploid genome sizes of up to 50 Gb. Moreover, genome size varies more than tenfold among species of salamanders, and about twofold even between species in the same genus. Similarly, flowering plants have a range of genome sizes spanning three orders of magnitude, from 0.1 to 100 Gb, encompassing the range seen across all vertebrates. The amount of DNA per genome was called the C-VALUE, and so lack of correspondence between genome size and phenotypic complexity in eukaryotes was dubbed the C-VALUE PARADOX.

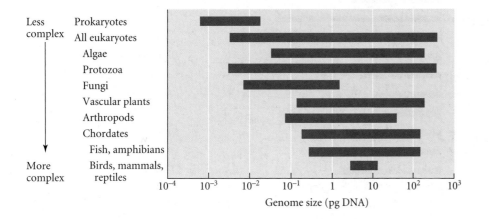

FIGURE 3.26 Genome size variation. The bars indicate the range of genome size (amount of DNA in picograms; 1 pg of DNA is approximately equivalent to 1 billion base pairs) for particular clades. Taxa are arranged from top to bottom in order of supposedly increasing organismal complexity. Within eukaryotes, there is little relationship between genome size and organismal complexity. This "C-value paradox" may be a result of great variation among lineages in the amount of repetitive (noncoding) DNA. (After Gregory 2001.)

This seeming paradox was resolved in the 1960s by Roy Britten, Mary Lou Pardue, and others, who showed that eukaryotic genomes usually contain a great deal of apparently noninformative, highly repetitive DNA that varies greatly in amount among species. Many of the highly repetitive sequences are thought not to be useful to the organism. Today entire genomes are being sequenced and protein-coding genes are being identified, so that it is possible to compare the number of functional genes among species, instead of simply amounts of DNA. The phylogenetic transition from unicellular to multicellular eukaryotes seems to have been marked by a considerable increase in gene number, but there is surprisingly little variation among animals and plants. It may be disconcerting to find that humans have only two-thirds as many genes as a small crustacean (*Daphnia*) and less than half as many as rice (**Figure 3.27**).

Comparative studies are beginning to shed some light on the evolution of genome size. For example, ENDOSYMBIOTIC MICROORGANISMS—parasites or mutualists that live within eukaryotic host organisms—generally have smaller genomes than their free-living relatives. Nancy Moran and colleagues have extensively studied the evolution of reduced genomes in *Buchnera*, a bacterial clade that has been an intracellular symbiont of aphids for about 200 My (Moran 2003; van Ham et al. 2003; see Figure 19.3). Genome comparisons have revealed that the common ancestor of *Buchnera* species lost over 2000 genes, compared with its relative *E. coli*. Because *Buchnera*'s host aphids provide many essential nutrients and metabolic functions for the symbionts (see Figure 16.26C), natural selection for retaining many genes has been relaxed.

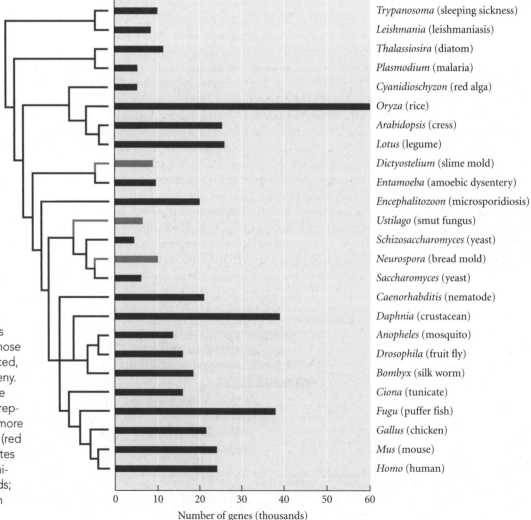

FIGURE 3.27 Numbers of genes estimated for some eukaryotes whose genomes have been fully sequenced, arranged on a provisional phylogeny. Multicellular eukaryotes with tissue organization (plants and animals, represented by blue branches) have more genes than unicellular eukaryotes (red branches) or multicellular eukaryotes that lack pronounced tissue organization (some fungi and slime molds; green branches). (Data from Lynch 2007 and Colbourne et al. 2011.)

Trypanosoma (sleeping sickness)
Leishmania (leishmaniasis)
Thalassiosira (diatom)
Plasmodium (malaria)
Cyanidioschyzon (red alga)
Oryza (rice)
Arabidopsis (cress)
Lotus (legume)
Dictyostelium (slime mold)
Entamoeba (amoebic dysentery)
Encephalitozoon (microsporidiosis)
Ustilago (smut fungus)
Schizosaccharomyces (yeast)
Neurospora (bread mold)
Saccharomyces (yeast)
Caenorhabditis (nematode)
Daphnia (crustacean)
Anopheles (mosquito)
Drosophila (fruit fly)
Bombyx (silk worm)
Ciona (tunicate)
Fugu (puffer fish)
Gallus (chicken)
Mus (mouse)
Homo (human)

Number of genes (thousands)
0   10   20   30   40   50   60

Duplicated genes and genomes

Gene duplication is a mutational event (see Chapter 8) by which a new gene (say, β) arises as a copy of a pre-existing gene (α), so that a single gene locus in an ancestor is represented by two loci in the descendant. These two genes will subsequently undergo different evolutionary changes in sequence and can therefore be distinguished. If two species (1 and 2) both inherit the duplicated pair (α, β) from their common ancestor, the relationships among the genes represent two forms of homology, and so warrant different terms. The genes that originate from an ancestral gene duplication are **paralogous**, whereas the genes that diverge from a common ancestral gene by phylogenetic splitting at the organismal level are **orthologous** (**Figure 3.28**). The phylogenetic relationships among the orthologous and paralogous genes in two or more species can be determined by standard phylogenetic methods.

This process may occur repeatedly over evolutionary time, generating a **gene family**. In the human genome, for example, the twelve members of the globin gene family include genes that encode myoglobin and several α- and β-hemoglobin chains (**Figure 3.29**). The origin of myoglobin and hemoglobin by duplication of an ancestral globin gene occurred during the ancestry of the vertebrates, all of which have both genes. The α- and β-hemoglobins arose by gene duplication in the ancestor of jawed vertebrates, all of which have a functional hemoglobin composed of both α- and β-chains, whereas the jawless vertebrates (e.g., lamprey) have only a single hemoglobin chain. The origin of the other globin genes can be similarly traced based on their sequences and phylogenetic distribution.

Many genes are duplicated as parts of large chromosomal blocks (paralogous regions, or "paralogons"), which often contain hundreds of genes. Indeed, **polyploidy**, whereby the entire nuclear genome is duplicated, occurs commonly in plants and occasionally in animals. The Hox genes, which play a major role in the embryonic development of animals, are represented by one set of genes in

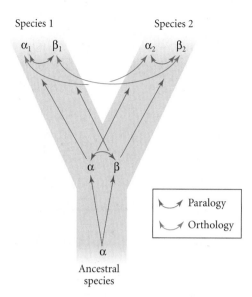

FIGURE 3.28 Orthology and paralogy in gene families. When an ancestral gene (α) undergoes duplication, the two resulting genes (α + β) have a paralogous relationship to one another (blue arrow). A speciation event after duplication results in divergence of the ancestral set of two paralogous genes. Within the genomes of the two diverged species, the α + β still have a paralogous relationship (blue arrows). However, the copies of α in species 1 and species 2 are orthologous (red arrows) because the two genes are related to one another via speciation, not duplication. Likewise, the copies of β in species 1 and 2 are orthologous.

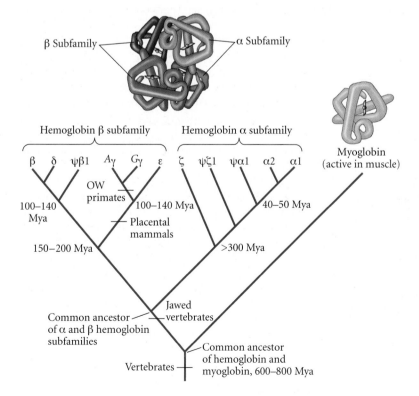

FIGURE 3.29 The phylogeny of genes in the globin family in the human genome. Myoglobin consists of a single protein unit, whereas mammalian hemoglobins consist of four subunits, two each from the α and β subfamilies. Each branch point on the tree denotes a gene duplication event; some of these events are marked with estimates of when the duplication occurred. The origin of hemoglobin and myoglobin from a common ancestral gene occurred in the ancestor of all vertebrates, but the α and β hemoglobin subfamilies originated by duplication in an ancestor of the jawed vertebrates. The duplication of the β hemoglobin into two genes occurred in the ancestor of placental mammals, since the $A_\gamma/G_\gamma/\epsilon$ genes are lacking in monotremes and marsupials. In some instances, one of the pair of genes formed by duplication became a nonfunctional pseudogene, symbolized by ψ. (After Li 1997 and Hartwell et al. 2000.)

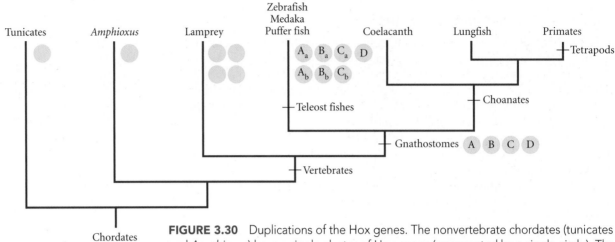

FIGURE 3.30 Duplications of the Hox genes. The nonvertebrate chordates (tunicates and *Amphioxus*) have a single cluster of Hox genes (represented by a single circle). The vertebrates have four Hox clusters, implying that the single cluster has undergone two duplications. Three of the four clusters of the ancestral gnathostome (jawed vertebrate) were duplicated in the ancestor of the teleost fishes. (After Panopoulou and Poustka 2005.)

Go to the

EVOLUTION

Companion Website at

sites.sinauer.com/evolution3e

for quizzes, data analysis and simulation exercises, and other study aids.

nonvertebrate chordates, four sets in most vertebrates, and seven sets in teleost (bony) fishes (**Figure 3.30**). Some other groups of genes show similar patterns. It appears that the entire genome underwent two successive duplications in the ancestor of vertebrates and further duplication in the ancestor of the teleosts (Kuraku et al. 2009).

Summary

1. Phylogenetic analyses have many uses in addition to describing the branching history of life. An important one is inferring the history of evolution of interesting characteristics by "mapping" changes in a character onto a phylogeny that has been derived from other data. Such systematic studies have yielded information on common patterns and principles of character evolution.

2. New features almost always evolve from pre-existing characters.

3. Homoplasy, which is common in evolution, is often a result of similar adaptations in different lineages. It includes convergent evolution, parallel evolution, and reversal.

4. For the most part, different characters evolve piecemeal, at different rates. This phenomenon is called mosaic evolution. Conservative characters are those that are retained with little or no change over long periods; other characters may evolve rapidly and vary widely across a single lineage.

5. Differences among related species illustrate that large differences often evolve gradually, by small steps. Although this pattern is common, it may not be universal.

6. Changes in structure are often associated with change in a character's function. Different structures may be modified to serve similar functions in different lineages.

7. The evolution of morphological features involves changes in their development. Such evolutionary changes include individualization of repeated structures, alterations in the timing (heterochrony) or site (heterotopy) of developmental events, and increases and decreases in structural complexity. Heterochrony can result in changes in the shape of features, often because of allometric growth.

8. Trends in the evolution of a character may be documented by phylogenetic analysis. Evolutionary trends may occur within a single lineage or repeatedly among different lineages.

9. In an adaptive radiation, numerous related lineages arise in a relatively short time and evolve in many different directions as they adapt to different habitats or ways of life. Radiation, rather than directional trends, is perhaps the most common pattern of long-term evolution.

10. Both the size of the genome (amount of DNA) and the number of protein-coding genes have increased and decreased in different lineages. These aspects of the genome do not appear to be strongly correlated with the level of complexity that is usually ascribed to different organisms.

11. One important evolutionary process by which the number of genes in a genome may increase is gene duplication followed by divergence of the duplicated genes. Repeated duplications have given rise to gene families with variable numbers of genes. Polyploidy, the duplication of the entire genome, has occurred frequently in evolution.

Terms and Concepts

adaptive radiation

allometry (allometric growth)

conservative characters

convergent evolution (convergence)

evolutionary reversal (reversal)

evolutionary trend

gene duplication

gene family

heterochrony

heterotopy

homoplasy

individualization

mosaic evolution

neoteny

orthology

paedomorphosis

parallel evolution (parallelism)

paralogy

peramorphosis

polyploidy

progenesis

Suggestions for Further Reading

Most books and other sources cover only some of the many subjects discussed in this chapter.

A broad perspective on the uses of phylogenies in studying evolution and ecology is provided by D. R. Brooks and D. A. McLennan in *The Nature of Diversity: An Evolutionary Voyage of Discovery* (University of Chicago Press, Chicago, 2002) and, briefly, by D. J. Futuyma, "The fruit of the tree of life: Insights into evolution and ecology" [(pp. 25–39 in *Assembling the Tree of Life*, J. Cracraft and M. J. Donoghue (eds.), Oxford University Press, New York, 2004)].

The early history of interpretation of morphological evolution in a developmental context was reviewed by Stephen Jay Gould in *Ontogeny and Phylogeny* (Harvard University Press, Cambridge, MA, 1977). More contemporary perspectives are provided by R. A. Raff in *The Shape of Life: Genes, Development, and the Evolution of Animal Form* (University of Chicago Press, Chicago, 1996).

Many aspects of the evolution of genes and genomes are treated from a phylogenetic perspective by W.-H. Li in *Molecular Evolution* (Sinauer Associates, Sunderland, MA, 1997). S. B. Carroll treats many of this chapter's topics in *The Making of the Fittest: DNA and the Ultimate Forensic Record of Evolution* (W. W. Norton, New York, 2006).

Problems and Discussion Topics

1. In Problem 1 in Chapter 2, you were asked to estimate the phylogeny of species 1, 2, and 3, based on the following DNA sequences:

    (Species 1) GCTGATGAGT (Species 2) ATCAATGAGT

    (Species 3) GTTGCAACGT (Species 4) GTCAATGACA

    Species 4 is an outgroup. Suppose the species are primates that differ in mating system: species 1 and 2 are monogamous, and species 3 and 4 are polygamous. We also happen to know that another polygamous species, 5, is more distantly related to species 1–4 than those species are to one another. Given your best estimate of the phylogenetic history, what has been the probable history of evolution of the mating system?

2. In the absence of a fossil record, how might phylogenetic analysis tell us whether the rate of diversification (increase in number of species) has differed among evolutionary lineages?

3. The antlers of the extinct Irish elk (see Figure 3.18) were so large that some paleontologists proposed that the species became extinct because the antlers simply became too unwieldy and prevented the elk from escaping predators. Other biologists wondered why such big antlers evolved; some suggested that males with larger antlers could win fights with other males and gain greater access to mates. Stephen Jay Gould (1974) calculated that, given the allometric relationship between antler size and body size among species of deer, the Irish elk's antlers were exactly as large as we would expect in such a large deer. If antler size is a result of a growth correlation with body size, is it necessary to postulate an adaptive advantage for the Irish elk's large antlers in order to account for their great size? And would such a correlation imply that natural selection could not prevent the antlers from becoming so large that they caused the species' extinction?

4. Problem 3 illustrates questions about evolutionary processes that arise when patterns of evolution are described. Specify some questions about evolutionary processes that might be raised by each of the following patterns: (a) modification of different structures to achieve similar adaptive functions (e.g., Figure 3.12); (b) evolutionary trends (e.g., Figure 3.21); (c) evolutionary reversal (e.g., salamander life history, Figure 3.7); and (d) reduction or loss of structures or genes during evolution (e.g., Figure 3.20).

5. The first analyses of the human genome offered many insights based on an evolutionary, phylogenetic perspective. After reading these analyses (International Human Genome Sequencing Consortium 2001; Venter et al. 2001) or associated commentaries, describe five questions about or insights into the human genome that could be addressed with the help of phylogenetic analyses.

6. Early in this chapter, the claim was made that phylogenetic information is the basis for describing patterns of evolution, yet some examples of such patterns were presented without phylogenetic trees. Consider the following examples and discuss what phylogenetic evidence or inference was left unstated: (a) The fusion of leg bones during embryonic development of birds is a derived trait, not an ancestral trait, relative to the unfused condition in crocodiles. (b) Pentadactyly (five digits) is homologous in humans and crocodiles. (c) The sting of a wasp is derived from an ovipositor, but modified in both structure and function. (d) Teeth became individualized

in the evolution of mammals, but this pattern was reversed in the whales. (e) Paedomorphosis is a derived (not ancestral) condition in salamanders.

7. The C-value paradox was a paradox only because many biologists assumed that amphibians are more complex than fishes, reptiles more complex than amphibians, and mammals more complex than reptiles. Might the "paradox" disappear if this assumption is false? How might you objectively determine whether or not the assumption is valid?

8. Why would it be surprising if Dollo's law were violated, and complex characteristics that have been lost during evolution (such as birds' teeth) were regained much later?

Evolution in the Fossil Record

Although some of the history of evolution can be inferred from living
organisms, the most direct evidence of that history is found in the fossil
record. Fossils tell us of the existence of innumerable creatures that have
left no living descendants, of great episodes of extinction and diversification, and
of the movements of continents and organisms that explain their present distributions. Only from this record can we obtain an absolute time scale for evolutionary events or evidence of the environmental conditions in which they transpired.
The fossil record also provides many details of which we would otherwise have
no knowledge; for instance, although we can infer some of the evolutionary
changes in the human lineage from comparisons with other primates, research
on fossils verifies some such inferences and reveals other information—such as
the sequence of particular
anatomical changes—that
comparisons with other living
species cannot provide.

The fossil record provides
evidence on two particularly
important themes: phenotypic
transformations in particular
lineages and changes in biological diversity over time. The
first of these themes is the
chief subject of this chapter.

**Members of the hominin
family tree** *Homo sapiens*
is the only surviving species of a
once diverse family that included
(from left) *Australopithecus africanus, Homo habilis,* and *H. erectus.*
The two large skulls to the right
are 'modern' humans, one from a
92,000-year-old Israeli site (upper
right) and one from a 22,000-year-
old French site (lower right).

Some Geological Fundamentals

Rocks at Earth's surface originate as molten material (magma) that is extruded from deep within Earth. Some of this extrusion occurs via volcanoes, but much rock originates as new crust forms at mid-oceanic ridges (**Figure 4.1**). Rock formed in this way is called IGNEOUS (Latin, "from fire") rock. SEDIMENTARY (Latin, "settle," "sink") rock is formed by the deposition and solidification of sediments, which are usually formed either by the breakdown of older rocks or by precipitation of minerals from water. Under high temperatures and pressures, both igneous and sedimentary rocks are altered, forming METAMORPHIC (Greek, "changed form") rocks.

Most fossils are found in sedimentary rocks; they are never found in igneous rocks, and they are usually altered beyond recognition in metamorphic rocks. A few fossils are found in other situations; for example, insects are found in amber (fossilized plant resin), and some mammoths and other species have been found frozen in permafrost.

Plate tectonics

Alfred Wegener first broached the idea of continental drift in 1915, but it was not until the 1960s that both definitive evidence and a theoretical mechanism for continental drift convinced most geologists of its reality. The theory of PLATE TECTONICS has revolutionized geology.

The lithosphere, the solid outer layer of Earth bearing both the continents and the crust below the oceans, consists of eight major and a number of minor PLATES that move over the denser, more plastic asthenosphere below. The heat of Earth's core sets up convection currents within the asthenosphere. At certain locations, such as the mid-oceanic ridge that runs longitudinally down the floor of the Atlantic Ocean, magma from the asthenosphere rises to the surface, cools, and spreads out to form new crust, pushing the existing plates to either side. The plates move at velocities of 5 to 10 centimeters per year. Where two plates come together, the leading edge of one may be forced to plunge under the other (a phenomenon known as SUBDUCTION), rejoining the asthenosphere. The pressure of these collisions is a major cause of mountain building. When a plate moves over a "hot spot" where magma is rising from the asthenosphere, volcanoes may be born, or a continent may be rifted apart. The Great Lakes of eastern Africa lie in such a rift valley;

FIGURE 4.1 Plate tectonic processes. At a mid-oceanic ridge, rising magma creates new lithosphere and pushes the existing plates to either side. When moving lithospheric plates meet, one plunges under the other, frequently causing earthquakes and mountain building. Heat generated by this process of subduction melts the lithosphere, causing volcanic activity.

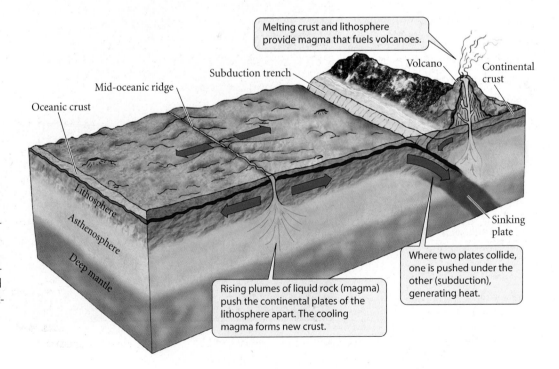

the Hawaiian Islands are a chain of volcanoes that have been formed by the movement of the Pacific plate over a hot spot (see Chapter 6).

Geological time

Astronomers have amassed evidence that the universe originated in a "big bang" about 14 billion years ago (1 billion = 1000 million) and has been expanding from a central point since then. Earth and the rest of the solar system are about 4.6 billion years old, but the oldest known rocks on Earth are about 3.8 billion years old. Living things existed by about 3.5 billion years ago; the first fossil evidence of animal life is about 800 million (0.8 billion) years old (see Chapter 5).

Such time spans are hard for us to comprehend. By analogy, if Earth's age is represented by one year, the first life appears in late March, the first marine animals make their debut in late October, the dinosaurs become extinct and the mammals begin to diversify on December 26, the human and chimpanzee lineages diverge at about 13 hours before midnight on December 31, and the Common Era, dating from the birth of Christ, begins about 13 seconds before midnight.

Absolute ages of geological events can often be determined by **radiometric dating**, which measures the decay of certain radioactive elements in minerals that form in igneous rock. The probability that a radioactive parent atom (e.g., uranium-235) will decay into a stable daughter atom (lead-207) is constant over time. As a result, each element has a specific HALF-LIFE. The half-life of U-235, for example, is about 0.7 billion years, meaning that in each 0.7 billion-year period, half the U-235 atoms present at the beginning of the period will decay into Pb-207. The ratio of parent to daughter atoms in a rock sample thus provides an estimate of the rock's age (**Figure 4.2**). Only igneous rocks can be dated radiometrically, so the age of a fossil-bearing sedimentary rock must be estimated by dating igneous formations above or below it.

Long before radioactivity was discovered—indeed, before Darwin's time—geologists had established the relative ages (i.e., earlier vs. later) of sedimentary rock formations by applying the principle that younger sediments are deposited on top of older ones. Layers of sediment deposited at different times are called **strata**. Different strata have different characteristics, and they often contain distinctive fossils of species that persisted for a short time and are thus the signatures of the age in which they lived. Using such evidence, geologists can match contemporaneous strata in different localities. In many locations, sediment deposition has not been continuous, and sedimentary rocks have eroded; thus any one area usually has a very intermittent geological record, and some time intervals are well represented at only a few localities on Earth. In general, the older the geological age, the less well it is represented in the fossil record because erosion and metamorphism have had more opportunity to take their toll.

The geological time scale

Most of the eras and periods of the **geological time scale** (Table 4.1) were named and ordered before Darwin's time, by geologists who did not entertain the idea of evolution. These geological eras and periods were distinguished, and are still most readily recognized in practice, by distinctive fossil taxa. Great changes in faunal composition—the

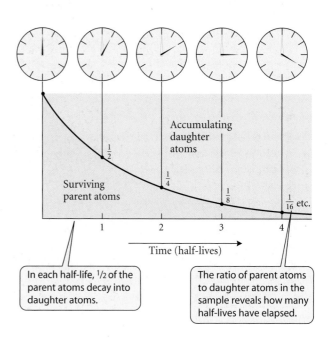

FIGURE 4.2 Radiometric dating. The loss of parent atoms and the accumulation of daughter atoms occurs at a constant rate. In one half-life—a unit of time that is specific to each element—half the remaining parent atoms decay into daughter atoms. The relative amounts of the two elements indicate how many half-lives have elapsed. (After Eicher 1976.)

TABLE 4.1 The geological time scale

Era	Period (abbreviation)	Epoch	Millions of years from start to present	Major events
CENOZOIC	Quaternary (Q)	Recent (Holocene)	0.01	Continents in modern positions; repeated glaciations and lowering of sea level; shifts of geographic distributions; extinctions of large mammals and birds; evolution of *Homo erectus* to *Homo sapiens*; rise of agriculture and civilizations
CENOZOIC	Quaternary (Q)	Pleistocene	2.56	
CENOZOIC	Tertiary (T)	Pliocene	5.3	Continents nearing modern positions; increasingly cool, dry climate; radiation of mammals, birds, snakes, angiosperms, pollinating insects, bony fishes
CENOZOIC	Tertiary (T)	Miocene	23.0	
CENOZOIC	Tertiary (T)	Oligocene	33.9	
CENOZOIC	Tertiary (T)	Eocene	55.8	
CENOZOIC	Tertiary (T)	Paleocene	65.5	
MESOZOIC	Cretaceous (K)		145	Most continents separated; continued radiation of dinosaurs; increasing diversity of angiosperms, mammals, birds; mass extinction at end of period, including last ammonoids and nonavian dinosaurs
MESOZOIC	Jurassic (J)		200	Continents separating; diverse dinosaurs and other reptiles; first birds; archaic mammals; gymnosperms dominant; evolution of angiosperms; ammonoid radiation; "Mesozoic marine revolution"
MESOZOIC	Triassic (Tr)		251	Continents begin to separate; marine diversity increases; gymnosperms become dominant; diversification of reptiles, including first dinosaurs; first mammals
PALEOZOIC	Permian (P)		299	Continents aggregated into Pangaea; glaciations; low sea level; increasing "advanced" fishes; diverse orders of insects; amphibians decline; reptiles, including mammal-like forms, diversify; major mass extinctions, especially of marine life, at end of period
PALEOZOIC	Carboniferous (C)		359	Gondwana and small northern continents form; extensive forests of early vascular plants, especially lycopsids, sphenopsids, ferns; early orders of winged insects; diverse amphibians; first reptiles
PALEOZOIC	Devonian (D)		416	Diversification of bony fishes; trilobites diverse; origin of ammonoids, tetrapods, insects, ferns, seed plants; mass extinction late in period
PALEOZOIC	Silurian (S)		444	Diversification of agnathans; origin of jawed fishes (acanthodians, placoderms, Osteichthyes); earliest terrestrial vascular plants, arthropods
PALEOZOIC	Ordovician (O)		488	Diversification of echinoderms, other invertebrate phyla, agnathan vertebrates; mass extinction at end of period
PALEOZOIC	Cambrian (Є)		542	Marine animals diversify: first appearance of most animal phyla and many classes within relatively short interval; earliest agnathan vertebrates; diverse algae
PROTEROZOIC			2500	Earliest eukaryotes (ca. 1900–1700 Mya); origin of eukaryotic kingdoms; trace fossils of animals (ca. 1000 Mya); multicellular animals from ca. 640 Mya, including possible Cnidaria, Annelida, Arthropoda
ARCHEAN			Lower limit not defined	Origin of life in remote past (first fossil evidence at ca. 3500 Mya); diversification of prokaryotes (bacteria and archaea); photosynthesis generates oxygen, replacing oxygen-poor atmosphere; evolution of aerobic respiration

Source: Gradstein, et al. 2004.

results of mass extinction events—mark many of the boundaries between them. The absolute times of these boundaries are only approximate and are subject to slight revision as more information accumulates.

Phanerozoic time (whose beginning is marked by the first appearance of diverse animals) is divided into three ERAS, each of which is divided into PERIODS. We will frequently refer to these divisions, and to the EPOCHS into which the Cenozoic periods are divided. *Every student of evolution should memorize the sequence of the eras and periods, as well as a few key dates*, such as the beginning of the Paleozoic era (and the Cambrian period, 542 million years ago, or 542 Mya), the Mesozoic era (and Triassic period, 251 Mya), the Cenozoic era (and Tertiary period, 65.5 Mya), and the Pleistocene epoch (2.56 Mya).

The Fossil Record

Some short parts of the fossil record in certain localities provide detailed evolutionary histories, and some groups of organisms, such as abundant planktonic protists with hard shells, have left an exceptionally good record. In some respects, such as the temporal distribution of many higher taxa (e.g., phyla and classes), the fossil record is adequate to provide a reasonably good portrait (Benton et al. 2000). In other respects, the fossil record is very incomplete (Jablonski et al. 1986). Consequently, the origins of many taxa have not been well documented. We know that the fossil record is incomplete because continuing exploration constantly yields new discoveries; for instance, a huge new dinosaur, perhaps the most massive terrestrial animal in Europe and one of the largest animals known, was discovered in Spain in 2006 (Royo-Torres et al. 2006). Most of the discoveries that have documented the origin of birds from dinosaurs have been made in Chinese deposits since the early 1990s.

The incompleteness of the fossil record has several causes. First, many kinds of organisms rarely become fossilized because they are delicate, or lack hard parts, or occupy environments—such as humid forests—where decay is rapid. Second, because sediments generally form in any given locality very episodically, they typically contain only a small fraction of the species that inhabited the region over time. Third, if fossils are to be found, the fossil-bearing sediments must become solidified into rock; the rock must persist for millions of years without being eroded, metamorphosed, or subducted; and the rock must then be exposed and accessible to paleontologists. Finally, the evolutionary changes of interest may not have occurred at the few localities that have strata from the right time; a species that evolved new characteristics elsewhere may appear in a local record fully transformed, after having migrated into the area. Paleontologists agree that the approximately 250,000 described fossil species represent much less than 1 percent of the species that lived in the past.

Two of the most basic questions we may ask of the fossil record are, first, whether it provides evidence for evolutionary change—for descent with modification—and second, whether or not evolution is gradual, as Darwin proposed. The answer to the first question is an unequivocal "yes," as the following examples show. Regarding the second question, the fossil record provides many examples of gradual change, but we must admit the possibility that some characteristics have evolved by large, discontinuous changes.

Evolutionary changes within species

A detailed, continuous sedimentary record can be expected only over short geological time spans. Some such records have preserved histories of change within single evolving lineages. In many such cases of anagenesis, character states tend to change gradually (Levinton 2001). Such changes have been particularly well documented in planktonic foraminiferans, abundant unicellular protists with shells of calcium carbonate that drop to the ocean floor in huge numbers. For example, Michal Kucera and Björn Malmgren (1998) found that in the foraminiferan genus *Contusotruncana*, the shell became increasingly cone-shaped over the course of about 3 million years (**Figure 4.3**).

In another detailed study, Michael Bell and his colleagues (1985) studied Miocene fossils of a stickleback fish, *Gasterosteus doryssus*, in strata that were laid down annually for

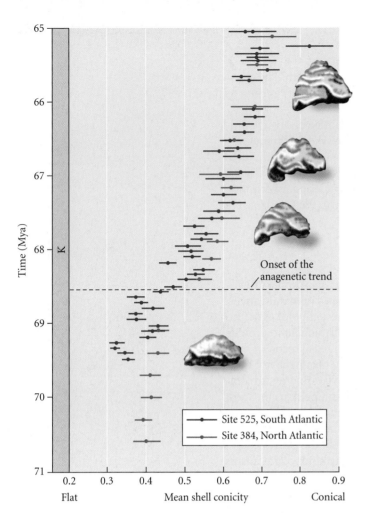

FIGURE 4.3 Gradual evolution of shell shape in the foraminiferan *Contusotruncana*. The diagram shows changes in mean conicity from older (at bottom) to younger (at top) samples over the course of 3.5 My. The circles represent the mean of each sample from a specific time interval, and the horizontal bars indicate variation among individuals within samples. Note that the change (anagenetic trend) began after more than 2.5 My of constant shape (stasis). The ancestral, flatter form is named *C. fornicata*, and the later, more conical phenotype is named *C. contusa*. (After Kucera and Malmgren 1998.)

110,000 years. They took samples from layers that were on average about 5000 years apart. Three of the five characters they studied changed more or less independently and gradually (**Figure 4.4A**). Two other characters changed abruptly at one point in time. Those shifts were caused by the extinction of the local population and immigration of another population—apparently a different species—in which those features were different (**Figure 4.4B**). An interval of 5000 years between samples is a long time by normal human standards, but is unusually fine-scaled by the standards of geological time. The gradual evolution that Bell and colleagues documented would not have been evident if the samples had been spaced hundreds of thousands of years apart, as is more typical of paleontological data.

Origins of higher taxa

In this section, we present several examples of macroevolutionary change—the origin of higher taxa over long periods of geological time. These examples illustrate several important principles, such as homology, change of structure with function, and mosaic evolution, that were introduced in Chapter 2. Moreover, they give the lie to claims by antievolutionists that the fossil record fails to document macroevolution. Anyone educated in biology should be able to counter such charges with examples such as these.

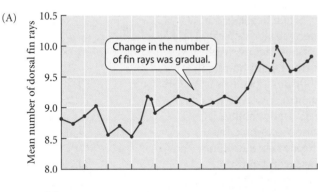

(A)

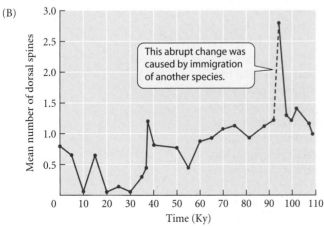

(B)

FIGURE 4.4 Changes in the mean values of characters in fossil sticklebacks, *Gasterosteus doryssus*. (A) In a study of five characters, three characters, including the number of dorsal fin rays, showed independent, gradual change. (B) Dorsal spine number was one of two characters displaying abrupt changes in value. Values are given for intervals of 5000 years over a span of 110,000 years. (After Bell et al. 1985; photo courtesy of M. Bell.)

Gasterosteus doryssus

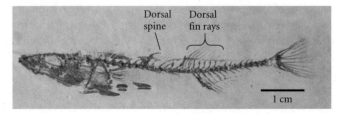

(A) Hypothesized ancestor

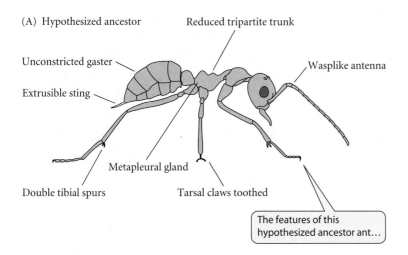

Reduced tripartite trunk

Unconstricted gaster

Wasplike antenna

Extrusible sting

Metapleural gland

Double tibial spurs

Tarsal claws toothed

The features of this hypothesized ancestor ant…

(B) *Sphecomyrma freyi*

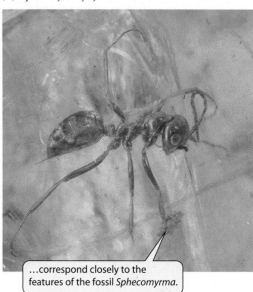

…correspond closely to the features of the fossil *Sphecomyrma*.

FIGURE 4.5 A fossil can help confirm an evolutionary hypothesis. (A) E. O. Wilson and colleagues hypothesized the features that the ancestor of ants should have had, based on comparisons between living ants and the wasps from which they are thought to have arisen. (B) A mid-Cretaceous ant, *Sphecomyrma freyi*, fossilized in amber. Its morphological features closely match those that had been hypothesized by the investigators. (A after Wilson et al. 1967; B courtesy of E. O. Wilson.)

Recall from Chapter 3 that it is often possible to infer the ancestral characteristics of a taxon from phylogenetic comparisons among living species; we are quite sure, for example, that humans descended from primates that had a tail and opposable toes (see Figure 3.3). Consequently, we can often recognize a fossil as a transitional stage in the origin of a taxon because even before it is found, some of its critical features are predictable. For instance, E. O. Wilson, F. M. Carpenter, and W. L. Brown hypothesized what features the ancestor of ants should have had, based on comparisons between primitive species of living ants and related families of wasps. Several years after Wilson and colleagues published their hypothesis, Cretaceous ants (*Sphecomyrma*) preserved in amber—older than any previously known fossil ants—were found, and they corresponded to the predicted morphology in almost all their features (**Figure 4.5**). It is on the basis of such predictions that intermediate stages in the evolution of higher taxa are recognized.

The early stages of a higher taxon are often members of a **stem group**, a clade from which a modern **crown group** later evolves, but which does not have all the defining features of the crown group. The relationships among extinct forms, and their relationships to living forms, are inferred by phylogenetic analysis. Such analyses are usually based on morphological features, for DNA sequences can seldom be extracted except from quite young fossils. It is usually not possible to tell whether a fossil candidate (such as *Sphecomyrma*) is *the* ancestor of the crown group or is a member of the stem group closely related to the ancestor. But this distinction is less important than the ability of such fossils to elucidate the stages in the evolution of characteristics.

THE ORIGIN OF TETRAPODS Sarcopterygii, or lobe-finned fishes, appeared in the early Devonian, about 408 Mya. They include lungfishes and coelacanths, a few of which are still alive, as well as extinct groups such as osteolepiforms, which had distinctive tooth structure and skull bones (**Figure 4.6A**). Osteolepiforms, the stem group that gave rise to the crown group Tetrapoda (four-legged vertebrates), had a tail fin and fleshy paired fins, which had a central axis of several large bones to which lateral bones and slender, jointed rays (radials) articulated. The head could not be flexed relative to the body, and the braincase had a joint between the anterior and posterior sections, as it does in living sarcopterygians. The first definitive tetrapods, such as *Ichthyostega* from the very late Devonian of Greenland, had the same tail fin and distinctive teeth and skull, but the gill cover bones at the rear of the skull

(A)

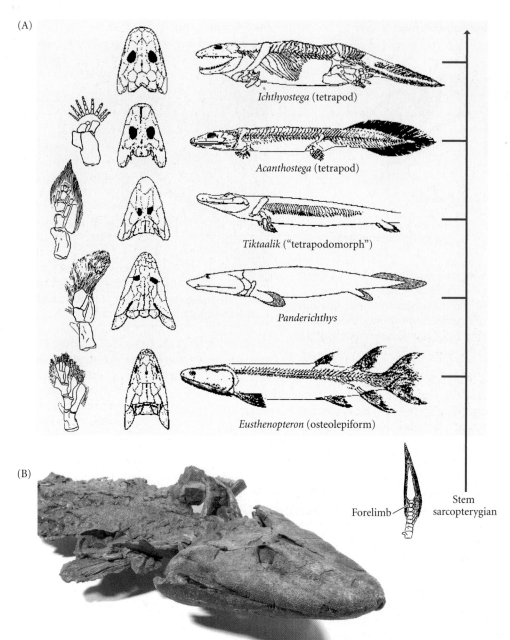

(B)

(C)

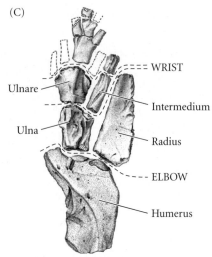

FIGURE 4.6 (A) The lineage leading from stem sarcopterygian fishes to early tetrapods (*Acanthostega*, *Ichthyostega*), including the recently discovered intermediate form *Tiktaalik*. Although *Tiktaalik* and the tetrapods have a flatter skull than the fish *Eusthenopteron*, the structure (as seen from above, at left) is very similar, except that the gill cover bones at the rear have been lost. Among the drawings of forelimbs at far left, note the intermediate structure of the forelimb of *Tiktaalik*. (B) An articulated skeleton of *Tiktaalik*. (C) Drawing of the pectoral fin, or forelimb, of *Tiktaalik*, showing positions of joints and the homologues of the limb bones of tetrapods. Tetrapod digits probably evolved from the many small bones (radials) at the end of the fin. (A after Ahlberg and Clack 2006 and Shubin et al. 2006; C after Shubin et al. 2006.)

had been lost, and the head could now be moved on a more flexible neck. Most importantly, they had larger pectoral and pelvic girdles and fully developed tetrapod limbs that bore more than five digits (unlike almost all later tetrapod vertebrates; Clack 2002).

Clearly, ichthyostegids show a mosaic of sarcopterygian and tetrapod features. Until recently, only a few fossils, such as *Panderichthys*, provided evidence of intermediate steps in the transition from fin to limb (see Figure 4.6A). Hoping to fill in more of this evolutionary sequence, Neil Shubin and colleagues explored likely Devonian deposits in northern Canada and found just what they sought: a rich fossil deposit of a new "tetrapodomorph" sarcopterygian that they named *Tiktaalik roseae* (Daeschler et al. 2006; Shubin et al. 2006). Like ichthyostegids, *Tiktaalik* had a flat, mobile head and elongate snout and lacked gill cover bones; it also had overlapping ribs, which would have provided the support that the body of a partly terrestrial animal requires (**Figure 4.6B**). Most importantly, the pectoral (shoulder) girdle and fins of *Tiktaalik* are intermediate between the sarcopterygian and tetrapod conditions. The humerus, ulna, radius, and wrist bones are clearly homologous to

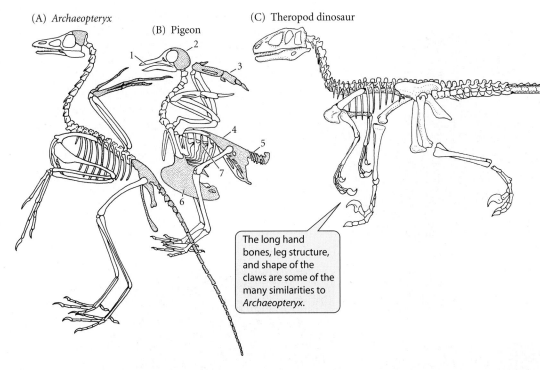

(A) *Archaeopteryx*

(B) Pigeon

(C) Theropod dinosaur

The long hand bones, leg structure, and shape of the claws are some of the many similarities to *Archaeopteryx*.

FIGURE 4.7 Skeletal features of (A) *Archaeopteryx*, (B) a modern bird (pigeon), and (C) a dromaeosaurid theropod dinosaur, *Deinonychus*. Compared with *Archaeopteryx*, the modern bird has (1) no teeth; (2) an expanded braincase with fused bones; (3) fusion and reduction of the three digits of the hand; (4) fusion of the pelvic bones and several vertebrae into a single structure; (5) fewer tail vertebrae, several of which are fused; (6) a greatly enlarged, keeled sternum (breastbone); and (7) horizontal processes that strengthen the rib cage. Theropod dinosaurs share many features with *Archaeopteryx*, the structure of the hand and leg being perhaps most evident in this figure. (A, B after Colbert 1980; C after Ostrom 1976.)

those of early tetrapods and reveal a critical feature: the limb could be flexed at the elbow and wrist (**Figure 4.6C**). All the anatomical details of girdle, limbs, and ribs show that *Tiktaalik* could hold its body off the ground—it could do push-ups. Its skull also has several intermediate features, including a less mobile braincase joint (Downs et al. 2008). Its mode of breathing was intermediate between that of lungfishes and that of terrestrial tetrapods. These features are more pronounced in recently discovered fossils of *Ventastega*, a very early tetrapod that is intermediate between *Tiktaalik* and *Ichthyostega*. Its limbs and girdles resemble those of *Ichthyostega*, and its skull is like that of *Tiktaalik*, except that some bones are closer in pattern to *Ichthyostega* and it has lost the braincase joint (Ahlberg et al. 2008). *Tiktaalik*, *Ventastega*, and their relatives are transitional forms that make the distinction between fishes and the earliest tetrapods difficult to draw.

THE ORIGIN OF BIRDS In recent years, it has become clear that birds are dinosaurs. Not long ago, birds (placed until recently in the class Aves) were defined by their feathers. But because of the many extraordinary fossils that have been discovered in China, the distinction between birds and dinosaurs has now become arbitrary (Chiappe and Dyke 2002; Xu et al. 2003; Norell and Xu 2005; Chiappe 2007; Xu et al. 2011). Birds are clearly theropod dinosaurs, the group that includes *Tyrannosaurus* and members of the Dromaeosauridae (e.g., *Velociraptor* and *Deinonychus*).

Archaeopteryx (see Figure 4.8D) was the first feathered dinosaur (or early bird) to be discovered (in 1860, in Upper Jurassic strata in Germany). It has only a few of the many modifications of the skeleton of modern birds (**Figure 4.7A,B**). Except for the feathers that enabled it to fly, its features—such as teeth, fully developed hands with long, clawed fingers, elongate tail, and leg structure—closely resemble those of other theropod dinosaurs, many of which were similarly small in size (**Figure 4.7C**). It was to be expected, then, that feathers, which had been considered the defining feature of birds, would be found in some other theropods.

Since 1996, Xing Xu, Mark Norell, and many other researchers have described an astonishing variety of feathered theropod dinosaurs found in China (**Figure 4.8**). In some, such as the compsognathid *Sinosauropteryx*, the feathers are filaments that coat the body. Other dinosaurs, such as the oviraptorosaur *Caudipteryx*, had long, broad feathers on the hands

(A)

(B)

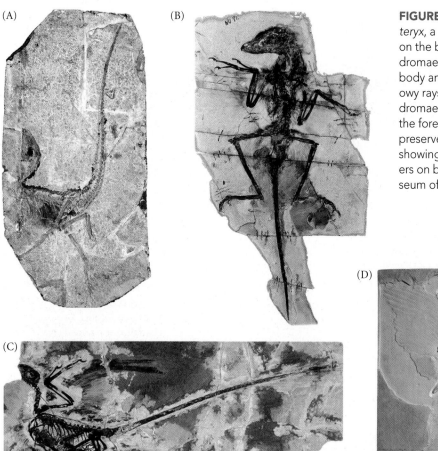

(C)

(D)

FIGURE 4.8   Feathered dinosaurs. (A) *Sinosaurop-teryx*, a compsognathid with filamentous feathers on the body (about 125 Mya). (B) *Sinornithosaurus*, a dromaeosaurid with short, filamentous feathers on the body and longer filaments on the arms, visible as shadowy rays (125 Mya). (C) *Microraptor gui*, a four-winged dromaeosaurid with long feathers at the back of both the forelimbs and hindlimbs (120–110 Mya). (D) A well-preserved specimen of *Archaeopteryx lithographica*, showing the wing feathers and the long tail with feathers on both sides. (A, B © Mick Ellison, American Museum of Natural History; C from Xu et al. 2003.)

and tail. At least two extraordinary four-winged dinosaurs, *Microraptor gui* and *Anchiornis huxleyi* (**Figure 4.9**), had long feathers on all four limbs. Feathers are now thought to have characterized the entire theropod clade (perhaps even *Tyrannosaurus*), and they have been described in a very distantly related dinosaur (an ornithischian; see Figure 5.23) as well. Some of the features of modern birds, such as hollow limb bones, evolved in theropods long before *Archaeopteryx*, and other characters, such as fusion of the tail vertebrae, evolved later. Later still, one lineage evolved such features as the keeled breastbone, loss of teeth, and loss of claws on the hands that typify the 10,000 species of living dinosaurs.

We are not sure of the adaptive function of hollow bones in theropods or the body feathers of *Sinosauropteryx*—though perhaps the feathers provided insulation, which may well have been their original function. Certainly, however, these features, in modified form, later became useful in flight. (Hollow bones lighten the load, and some feathers became modified as airfoils.) These features provide an example of evolutionary change of a structure's function and of PREADAPTATION, possession of a feature that fortuitously plays a different, useful role at a later time (see Chapter 11).

THE ORIGIN OF MAMMALS   The origin of mammals from the earliest amniotes (see Figure 2.9) is one of the most fully documented examples of the evolution of a major taxon (Sidor and Hopson 1998; Kemp 2005). Although some features of living mammals, such as hair and mammary glands, do not become fossilized, mammals do have diagnostic

(A)

(B)

5 cm

FIGURE 4.9 Reconstruction of the plumage of two Jurassic four-winged dinosaurs, with long feathers fringing both forelimbs and hindlimbs. (A) *Anchiornis huxleyi*, from the late Jurassic, ca. 155 Mya. (B) The late Jurassic *Microraptor gui* (120–110 Mya) had iridescent plumage, with rainbow-like reflections. The form and coloration of feathers in both species probably had social and sexual functions, as in living birds. The colors were reconstructed from pigment-containing organelles in the fossilized feathers. (A by M. A. DiGiorgio from Q. Li et al. 2010; B from Q. Li et al. 2012, courtesy of Mick Ellison.)

skeletal features, and it is important to learn these in order to understand how mammals evolved. Here we will describe only changes in the skull and jaw, but the evolution of the postcranial skeleton has also been well described.

- The lower jaw consists of several bones in reptiles, but only a single bone (the dentary) in mammals (see Figure 3.20).

- The primary (and in all except the earliest mammals, the exclusive) jaw articulation is between the dentary and the squamosal bones, rather than between the articular and the quadrate bones, as in other tetrapods.

- Early amniotes have a single bone (the stapes, or stirrup) that transmits sound, whereas mammals have three such bones (hammer, anvil, stirrup) in the middle ear.

- Mammals' teeth are differentiated into incisors, canines, and multicusped (multipointed) cheek teeth (premolars and molars), whereas most other tetrapods have uniform, single-cusped teeth.

Other features that distinguish most mammals from other amniotes include an enlarged braincase, a large space (temporal fenestra) behind the eye socket, and a secondary palate that separates the breathing passage from the mouth cavity.

Soon after the first amniotes originated, during the Carboniferous, they gave rise to the Synapsida, a group characterized by a temporal fenestra that probably provided space for enlarged jaw muscles to expand into when contracted (**Figure 4.10A**). The temporal fenestra became progressively enlarged in later synapsids (**Figure 4.10B–D**).

Permian synapsids in the order Therapsida (Figure 4.10B) had large canine teeth, and the center of the palate was recessed, suggesting that the breathing passage was partially separated from the mouth cavity. The hind legs were held rather vertically, more like a mammal than a reptile.

Cynodont therapsids, which lived from the late Permian to the late Triassic, represent several steps in the approach toward mammals. The rear of the skull was compressed, giving it a doglike appearance (Figure 4.10C,D); the dentary was enlarged relative to the other

(A) Synapsid (*Haptodus*)

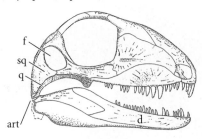

Synapsids had large jaw muscles, multiple bones in the lower jaw, and single-cusped (single-pointed) teeth.

(B) Therapsid (*Biarmosuchus*)

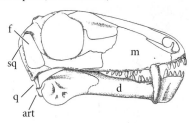

Synapsids of the order Therapsida had large canine teeth, large maxilla bones (m), and long faces.

(C) Early cynodont (*Procynosuchus*)

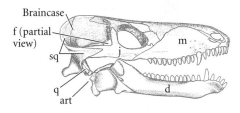

In cynodont therapsids, the side of the braincase was vertical and the large temporal fenestra was lateral to it.

(D) Cynodont (*Thrinaxodon*)

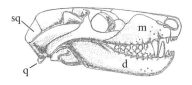

The cynodont dentary (the major jaw bone) became enlarged, and the cheek teeth had multiple cusps.

(E) Advanced cynodont (*Probainognathus*)

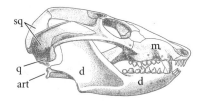

In advanced cynodonts, complex cusp patterns enhanced chewing and the dentary formed an articulation with the squamosal.

(F) *Morganucodon*

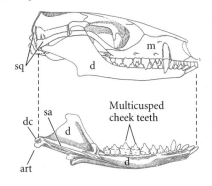

Multicusped cheek teeth

Morganucodon was almost a mammal, with typical mammalian teeth and a lower jaw composed almost entirely of the dentary. The jaw had a double articulation with the skull.

FIGURE 4.10 Skulls of some stages in evolution from early synapsids to early mammals. (A) A pelycosaur, *Haptodus*. Note temporal fenestra (f), multiple bones in lower jaw, single-cusped teeth, and articular/quadrate (art/q) jaw joint. (B) An early therapsid, *Biarmosuchus*. Note enlarged temporal fenestra. (C) An early cynodont, *Procynosuchus*. The side of the braincase is now vertically oriented, separated by a large temporal fenestra from a lateral arch formed by the jugal (j) and squamosal (sq). Note the enlarged dentary (d). (D) A later cynodont, *Thrinaxodon*. Note multiple cusps on rear teeth, large upper and lower canine teeth, and greatly enlarged dentary with a vertical extension to which powerful jaw muscles were attached. (E) An advanced cynodont, *Probainognathus*. The cheek teeth had multiple cusps, and both the articular and dentary bones of the lower jaw articulated with the skull. (F) *Morganucodon* was almost a mammal. Note multicusped cheek teeth (including inner cusps) and double articulation of lower jaw, including articulation of a dentary condyle (dc) with the squamosal (sq). (After Futuyma 1995; based on Carroll 1988 and various sources.)

bones of the lower jaw; the cheek teeth had a row of several cusps, and a bony shelf formed a secondary palate that was incomplete in some cynodonts and complete in others. The quadrate was smaller and looser than in previous forms and occupied a socket in the squamosal.

In the advanced cynodonts of the middle and late Triassic (**Figure 4.10E**), the cheek teeth had not only a linear row of cusps, but also a cusp on the inner side of the tooth. This innovation begins a history of complex cusp patterns in the cheek teeth of mammals, which are modified in different lineages for chewing different kinds of food. The lower jaw of these cynodonts had not only the old articular/quadrate articulation with the skull, but also an articulation between the dentary and the squamosal, marking a critical transition between the ancestral condition and the mammalian state.

In *Morganucodon* of the late Triassic and very early Jurassic (**Figure 4.10F**), the teeth are typical of mammals. *Morganucodon* has both a weak articular/quadrate hinge and a fully developed mammalian articulation between the dentary and the squamosal. The articular and quadrate bones are sunk into the ear region, and together with the stapes, they closely approach the condition in modern mammals, in which these bones transmit sound to the inner ear.

Hadrocodium, from the early Jurassic, carries the trend from *Morganucodon* to the very brink of mammalhood (Luo et al. 2001). This tiny animal is very similar to *Morganucodon*, but the articular and quadrate bones are fully separated from the jaw joint and fully lodged in the middle ear, and the lower jaw consists entirely of the dentary. *Morganucodon*, *Hadrocodium*, and later mammals also show sequential steps in enlargement of the brain, chiefly in the olfactory bulb and the neocortex (Rowe et al. 2011).

This description touches on only the highlights of a complex history. For example, several of the changes in the lower jaw and middle ear occurred independently in different cynodont lineages (Luo 2011). Nonetheless, the fossil record shows that most mammalian characters (e.g., posture, tooth differentiation, skull changes associated with jaw musculature, secondary palate, brain size, reduction of the elements that became the small bones of the middle ear) evolved gradually. Evolution was mosaic, with different characters "advancing" at different rates. No new bones evolved; in fact, many bones have been lost in the transition to modern mammals (Sidor 2001), and all the bones that persist are modified from those of the stem amniotes (and in turn, from those of early tetrapods and even lobe-finned fishes). Some major changes in the form of structures are associated with changes in their function. The most striking example is the articular and quadrate bones, which serve for jaw articulation in all other tetrapods, but became the sound-transmitting middle-ear bones of mammals (Luo 2011). Because the evolution of mammals from synapsids, over the course of more than 130 My, has been gradual, there is no cutoff point for recognizing mammals: the definition of "Mammalia" is arbitrary.

THE ORIGIN OF CETACEANS   Whales and dolphins, traditionally distinguished as the order Cetacea, evolved from terrestrial ancestors. Among living mammals, their closest relatives, according to molecular phylogenetic analyses, are hippopotamuses (Gatesy et al. 1999). Thus cetaceans fall within the order Artiodactyla—even-toed hoofed mammals— along with camels, pigs, and ruminants such as cattle and antelopes. The group as a whole is now often called Cetartiodactyla.

Compared with basal (primitive) mammals, living cetaceans are greatly modified, owing to their adaptations for aquatic life. All share a uniquely shaped tympanic bone that encloses the ear, one of several modifications for hearing in water; a nasal opening far back on top of the skull; stiff elbow, wrist, and finger joints, all enclosed in a paddlelike flipper; a rudimentary pelvis (sometimes associated with a hindlimb rudiment) that is disconnected from the vertebral column; and a lack of the fused, differentiated sacral (lower back) vertebrae that land mammals have. Toothed whales have a large cavity (foramen) in the lower jaw (mandible) that contains a sound-transmitting pad of fat.

Philip Gingerich, J. G. M. Thewissen, and colleagues have discovered many Eocene fossils, mostly in Pakistan, that document many details of the evolution of cetaceans from about 50 to 35 Mya, including features such as the transition to aquatic locomotion (**Figure 4.11**) and to underwater hearing (Gingerich et al. 2001; Thewissen and Williams 2002; Nummela et al. 2004). The closest known relatives of cetaceans appear to be the Raoellidae, small Eocene artiodactyls that share the distinctive cetacean tympanic bone and certain features of the teeth, and which evidently were semiaquatic, like hippos (Figure 4.11A). The oldest known cetacean, *Pakicetus* (53–48 Mya), was a terrestrial or semiaquatic animal with the distinctive cetacean tympanic bone. The slightly younger *Ambulocetus* (48–47 Mya; Figure 4.11B) was adapted for life in shallow coastal waters. It had short hind legs but large feet, with separate digits that bore small hooves. The mandibular foramen was larger than in pakicetids, starting a steady increase in size. *Ambulocetus* was a predator that had long jaws and teeth with somewhat reduced cusps. In the protocetids (e.g., *Rodhocetus*, 49–39 Mya; Figure 4.11C), the fusion between sacral vertebrae was reduced, tooth form was simpler, and the nasal opening was farther back from the tip of the snout. Protocetids had an artiodactyl ankle bone and small hooves at the tips of the toes, but the pelvis and hindlimbs of some of the aquatic protocetids were too small and weak to bear their weight. The enlarged mandibular foramen shows that sound was transmitted by a fat pad from the lower jaw to the ear. Complete adaptation to life in water is shown by the dorudontine basilosaurids, of about 35 Mya (Figure 4.11D), in which the teeth were even simpler, the nostrils were farther back, and the forelimbs were flipperlike, with an almost inflexible wrist and elbow. The pelvis and hindlimbs were completely nonfunctional. The small pelvis was disconnected from the vertebral column, and the hind feet and legs barely protruded from the surface of the body. Dorudontines probably had a horizontal tail fin (fluke) and were, indeed, a small step away from modern cetaceans.

FIGURE 4.11 Reconstruction of stages in the evolution of cetaceans from terrestrial artiodactyl ancestors. (A) Eocene raoellids, perhaps the sister group of Cetacea, were terrestrial but show some evidence of semiaquatic life. (B) The amphibious *Ambulocetus*. (C) The middle-Eocene protocetid *Rodhocetus* had the distinctive ankle bones of artiodactyls, but had numerous cetacean characters. (D) *Dorudon*, of the middle to late Eocene, had most of the features of modern cetaceans, although its nonfunctional pelvis and hindlimb were larger. (E) A modern toothed whale, the harbor porpoise, *Phocoena phocoena*. The nostrils, forming a blowhole, are far back on the top of the head, accounting for the peculiar shape of the skull. (A after Thewissen et al. 2007; B–D after Gingerich 2003 and de Muizon 2001; E, art by Nancy Haver.)

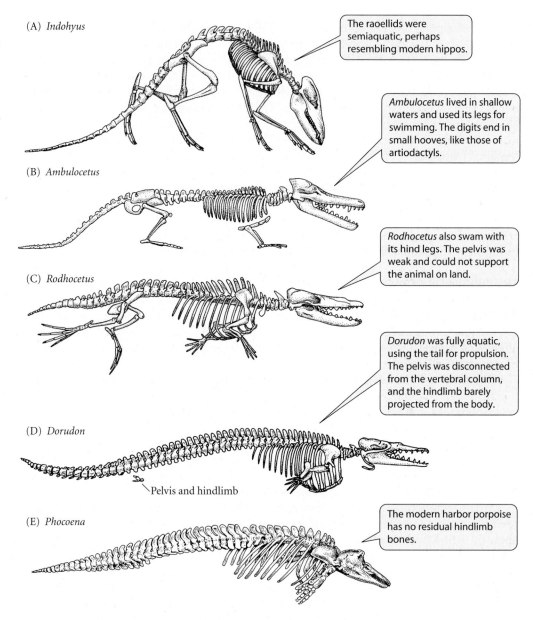

(A) *Indohyus*

The raoellids were semiaquatic, perhaps resembling modern hippos.

(B) *Ambulocetus*

Ambulocetus lived in shallow waters and used its legs for swimming. The digits end in small hooves, like those of artiodactyls.

(C) *Rodhocetus*

Rodhocetus also swam with its hind legs. The pelvis was weak and could not support the animal on land.

Dorudon was fully aquatic, using the tail for propulsion. The pelvis was disconnected from the vertebral column, and the hindlimb barely projected from the body.

(D) *Dorudon*

Pelvis and hindlimb

(E) *Phocoena*

The modern harbor porpoise has no residual hindlimb bones.

The Hominin Fossil Record

The Hominidae, as generally conceived today, includes the common ancestor of humans and living great apes (see Figure 2.13), the extinct members of the stem group to which their common ancestor belonged, and all the descendants of that ancestor. These descendants include the African apes (chimpanzee, bonobo, and gorilla) and their sister group, referred to as "hominins." DNA sequence differences imply that the chimpanzee and human lineages diverged 5 to 6 Mya (see Chapter 2). No chimpanzee fossils have been found, but diverse Miocene apes, undoubtedly members of the hominid stem group, have been described, as have unequivocal fossil hominins. Among these, there is strong evidence of general, more or less unidirectional trends in many characters, such as cranial capacity, a measure of brain size by volume (**Figure 4.12**). Modern humans clearly evolved through many intermediate steps from ancestors that in most anatomical respects were apelike. Thus much of the broad sweep of human evolution has been superbly documented.

There is disagreement, however, about how many distinct hominin species and genera should be recognized, in part because fossil specimens are too few, and too widely separated in time and space, to characterize variation within species compared with that between species. Some taxa are known from one or a few fragmentary bones that provide insufficient information for phylogenetic analysis. Moreover, the differences among various hominins are quantitative (differences in degree) and often rather slight, so it is difficult to determine whether specific earlier populations were ancestral to later ones or were collateral relatives of the actual ancestral lineage. Hence, even if the overall pattern of evolution is clear, the specific phylogenetic relationships among hominin taxa may not be.

All early hominins have been found in Africa, as have several forms that are considered hominins by some, but not all, paleoanthropologists (**Figure 4.13**). One such form, *Ardipithecus ramidus* (**Figure 4.14A**), from 4.4-My-old deposits in Ethiopia, was described in detail in 2009. It may be a key link between hominins and their common ancestor with apes such as chimpanzees. Tim White and collaborators (2009) reported that *Ardipithecus* has many apelike features, such as a small brain (the same size as a chimpanzee's) and adaptations for climbing, but that it also has hominin features, such as small canine teeth (enlarged for fighting in male apes) and features of the foot and pelvis that indicate that it could walk upright. Their proposal that *Ardipithecus* is a member of the hominin lineage is still being debated.

Among the undisputed hominins, the species represented by the most extensive and informative early fossil material, about 3.5 My old, has been named *Australopithecus afarensis* (**Figure 4.14B**). *Australopithecus anamensis* (4.2–3.9 Mya), known from more fragmentary material, is probably an earlier member of the *afarensis* lineage. That these forms are not far removed from the common ancestor of humans and chimpanzees is indicated by their many "primitive," or ancestral, features (**Figure 4.15A,B**). They had a lower face projecting far beyond the eyes, relatively large canine teeth, long arms relative to the legs, a small brain volume (about 400 cc), and curved bones in the fingers and toes

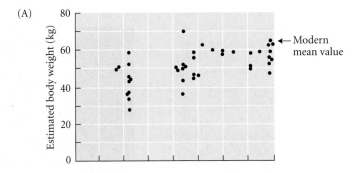

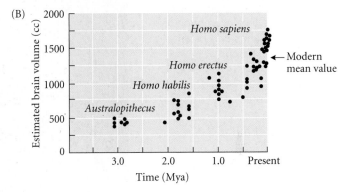

FIGURE 4.12 Estimated body weights (A) and brain volumes (B) of fossil hominins. There has been a steady, fairly gradual increase in brain volume, even though body size has not increased very much in the last 2 million years. The arrows indicate modern averages. (After Jones et al. 1992.)

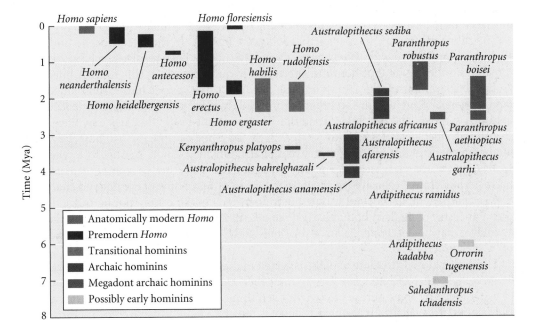

FIGURE 4.13 The approximate time spans of named hominin taxa in the fossil record. Each time span indicates either the range of dated fossils or the range of estimated dates for taxa known from few specimens. Researchers do not all agree on whether or not some of these forms are distinct species. (After Wood and Baker 2011.)

FIGURE 4.14 (A) *Ardipithecus ramidus* as it may have appeared in life. The small braincase, long fingers, and opposable big toe are ancestral features, shared with other African apes, but the bipedal posture is a hominin feature. (B) Skeletal remains of the Pliocene hominin *Australopithecus afarensis.* This famous specimen, nicknamed "Lucy," is unusually complete. This key fossil hominid shows that bipedal stance and locomotion preceded the evolution of substantially increased brain size.

(A)

(B)

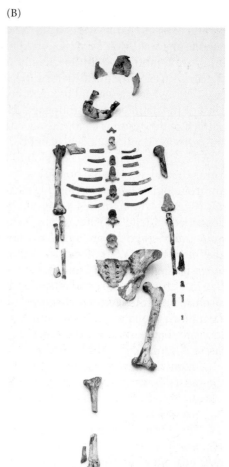

(which imply that they climbed trees). However, the structure of the pelvis and hindlimb clearly shows that *anamensis* and *afarensis* were bipedal. In fact, "fossilized" footprint traces have been found in rock formed from volcanic ash near an *afarensis* site in Tanzania. Bipedalism seems to have been the first distinctively human trait to have evolved.

The number of hominin species that followed *afarensis* in the late Pliocene and early Pleistocene and the relationships among them have not yet been resolved (see Figure 4.13). Most authorities agree that hominin species were quite diverse at this time. A lineage of "robust" australopithecines (*Paranthropus*), of which three species have been named, had large molars and premolars and other features adapted for powerful chewing; they probably fed on tubers and hard plant material. The robust australopithecines may have made stone tools, the oldest of which are 2.6 to 2.3 My old. They became extinct, however, without having contributed to the ancestry of modern humans. A more slender form was *Australopithecus africanus*, which is generally thought to have descended from *A. afarensis*, but had a greater cranial capacity (about 450 cc) (**Figure 4.15C**). Some of the derived characters of *africanus* resemble those of robust australopithecines, so it may not be in the line of direct ancestry of modern humans. *Australopithecus sediba*, described in 2010 and dated at 1.98 Mya, may be descended from *A. africanus* and is claimed to have humanlike features of the pelvis and hand.

The earliest fossils that are usually assigned to the genus *Homo* range from about 1.9 to 1.5 Mya (i.e., early Pleistocene). Originally called *Homo habilis*, they are variable enough to be assigned by some authors to two or even three species, *H. habilis*, *H. ergaster*, and *H. rudolfensis*. *Homo habilis* (in the broad sense) is the epitome of a discovered "missing link"

FIGURE 4.15 Frontal and lateral reconstructions of the skulls of a chimpanzee and some fossil hominins. (A) *Pan troglodytes*, the chimpanzee. Note large canine teeth, low forehead, prominent face, and brow ridge. (B) *Australopithecus afarensis*. Some of the same features as in the chimpanzee are evident. (C) *Australopithecus africanus* has smaller canines and a higher forehead. (D) *Homo habilis*. The face projects less, and the skull is more rounded, than in earlier forms. (E) *Homo erectus*. Note the still more vertical face and rounded forehead. (F) *Homo neanderthalensis*. The rear of the skull is more rounded than in *H. erectus*, and cranial capacity is greater. (A, B after Jones et al. 1992; C–F after Howell 1978.)

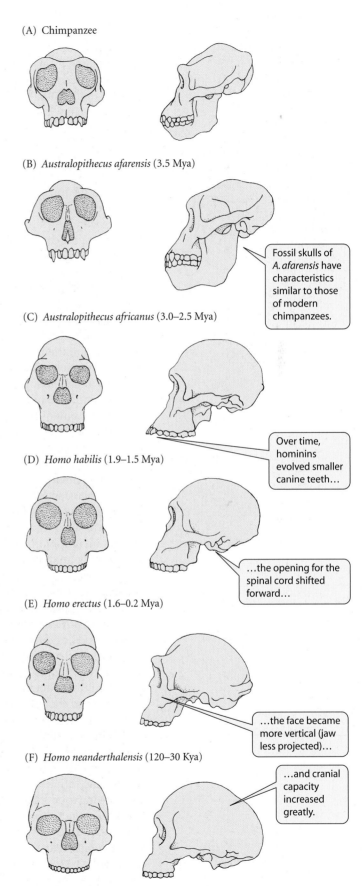

(A) Chimpanzee

(B) *Australopithecus afarensis* (3.5 Mya)

Fossil skulls of *A. afarensis* have characteristics similar to those of modern chimpanzees.

(C) *Australopithecus africanus* (3.0–2.5 Mya)

Over time, hominins evolved smaller canine teeth…

(D) *Homo habilis* (1.9–1.5 Mya)

…the opening for the spinal cord shifted forward…

(E) *Homo erectus* (1.6–0.2 Mya)

…the face became more vertical (jaw less projected)…

(F) *Homo neanderthalensis* (120–30 Kya)

…and cranial capacity increased greatly.

(Figure 4.15D), a "transitional hominin" that could as well be assigned to *Australopithecus* (Wood and Baker 2011). The oldest specimens are very similar to *Australopithecus africanus*, and the younger ones grade into the later form *Homo erectus*. Compared with *Australopithecus*, *Homo habilis* more nearly resembles modern humans in its greater cranial capacity (610 to nearly 800 cc), flatter face, and shorter tooth row. Although the limbs retain apelike proportions that suggest an ability to climb, the structure of the leg and foot indicates that its bipedal locomotion was more nearly human than that of the australopithecines. *Homo habilis* is associated with stone tools (referred to as Olduwan technology) and with animal bones that bear cut marks and other signs of hominin activity.

Later hominin fossils, from about 1.6 Mya to about 200 Kya (thousand years ago), are referred to as *Homo erectus*. Most authorities think that *habilis*, *erectus*, and *sapiens* are a single evolutionary lineage. In most respects, *erectus* from the middle Pleistocene onward has fairly modern human features in its skull, in its postcranial anatomy, and in indications of its behavior. The skull is rounded, the face projects less than in earlier forms, the teeth are smaller, and the cranial capacity is larger, averaging about 1000 cc and evidently increasing over time (Figure 4.15E). At least 1 Mya (perhaps as far back as 1.7 Mya), *erectus* spread from Africa into Asia, extending eastward to China and Java. Throughout its range, *erectus* is associated with stone tools, termed Acheulian technology, that are more diverse and sophisticated than the Olduwan tools of *H. habilis*. The use of fire was widespread by half a million years ago.

Starting about 400 or 300 Kya, *Homo erectus* grades into forms with more *sapiens*-like features, with mean cranial capacity increasing from about 1175 cc at 200 Kya to its modern mean of 1400 cc. The best known of these populations are the Neanderthals of Europe and southwestern Asia, first described from the Neander Valley (Thal)* of western Germany. Neanderthals had dense bones, thick skulls, and projecting brows (Figure 4.15F), but contrary to the popular image of a stooping brute, Neanderthals walked fully erect, had brains as large as or even larger than ours (up to 1500 cc), and had a fairly elaborate culture that included a variety of stone tools (Mousterian culture)

---

*The German word *Thal* ("valley") is pronounced, and today is spelled, *Tal*, and the name of the hominin is sometimes spelled "Neandertal" in English.

and may have included ritualized burial of the dead. Their remains extend from about 230 to 30 Kya.

A research group led by Svante Pääbo has extracted DNA from Neanderthal bones and has sequenced the genome, which proved to differ considerably from that of living humans (Green et al. 2010). For this reason, Neanderthals, once considered a subspecies of *Homo sapiens*, are now regarded as a distinct species, *Homo neanderthalensis*. Then Pääbo's group sequenced DNA from a single finger bone from a deposit in a Siberian cave and were astonished to find that it differs so greatly from *neanderthalensis* (also found in the cave) that it represents a different population, and probably a different species (Reich et al. 2010). The bone is thought to be about 50,000 years old, and its DNA is matched by DNA taken from a single molar tooth in the same deposit. Some features of the tooth differ from both modern and Neanderthal molars. This population, named for the cave, is referred to as "Denisovan." Additional evidence of the unexpectedly high diversity of *Homo* species in the recent past came to light when fossil material of a meter-tall "dwarf" hominin, named *Homo floresiensis*, was found on the Indonesian island Flores, dated at 18 Kya.

Modern *Homo sapiens*, anatomically virtually indistinguishable from today's humans, appeared earlier in Africa (ca. 170 Kya) than elsewhere. Modern humans overlapped with Neanderthals in the Middle East for much of the Neanderthals' history, but abruptly replaced them in Europe about 40 Kya. By 12 Kya, and possibly earlier, modern humans had spread from northeastern Asia across the Bering Land Bridge to northwestern North America, and thence rapidly throughout the Americas. (The history of the spread of modern *Homo sapiens,* and the genetic relationships between modern humans and the Neanderthals and Denisovans, are being revealed by variation in DNA sequences, a story that will be continued in Chapter 6.)

"Upper Paleolithic" culture emerged about 40 Kya. The earliest of several successive cultural "styles" in Europe, the Aurignacian, is marked by stone tools more varied and sophisticated than those of the Mousterian culture. Moreover, culture became more than simply utilitarian: art, self-adornment, and possible mythical or religious beliefs are increasingly evident from about 35 Kya onward. Agriculture, which resulted in an enormously increased human population density and began the human transformation of Earth, is about 11,000 years old. There is, at least at present, no way of knowing which (if any) of these cultural advances were associated with genetic changes in the capacity for reason, imagination, and awareness, but they are not paralleled by any increase in brain size or other anatomical changes.

Throughout hominin evolution, different hominin features evolved at different rates (mosaic evolution). On average, brain size (cranial capacity) increased throughout hominin history, although not at a constant rate, and there were progressive changes, from *afarensis* to *africanus* to *erectus* to *sapiens*, in many other features, such as the teeth, face, pelvis, hands, and feet. The very fuzziness of the taxonomic distinctions among the named forms attests to the mosaic and gradual evolution of hominin features. Although many issues remain unresolved, the most important point is fully documented: modern humans evolved from an apelike ancestor.

Why these changes occurred—what advantages they may have provided—is the subject of much speculation, but little evidence (Lovejoy 1981; Fedigan 1986). What evidence exists is indirect, consisting mostly of inferences from studies of other primates, contemporary cultures, and anatomy and artifacts.

The erect posture and bipedal locomotion are the first major documented changes toward the human condition. A plausible hypothesis is that bipedalism freed the arms for carrying food back to the social unit, especially to an individual's mate and offspring. Food sharing occurs in chimpanzees, which have a complex social structure that includes matrilocal family groups and "friendships." Chimpanzees also make and use a variety of simple tools, such as stone and wooden hammers they use to crack nuts, and twigs fashioned to "fish" termites out of their nests. The advantages gained by using a greater variety of tools may have selected for greater intelligence and brain size. However, many authors, beginning with Darwin, have emphasized that social interactions, such as learning how

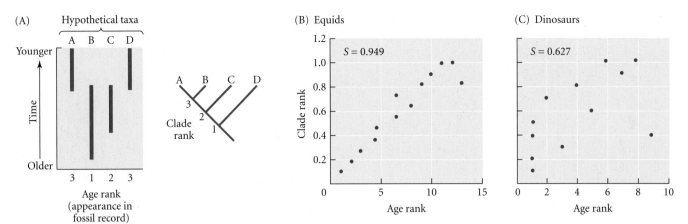

FIGURE 4.16 Correlations between clade rank—the relative order of branching from the base of a phylogenetic tree—and age rank—the relative order of first appearance of the clades in the fossil record. (A) Illustration of how age rank (left) and clade rank (right) are determined for four hypothetical clades, A–D. In this example, clade rank and age rank are not correlated. For example, the phylogeny shows the D lineage to be as old as its sister group (A, B, C), but it is known only from relatively recent fossils. (B) Correlation between clade rank and age rank for 13 clades in the horse family, Equidae. (C) Same, for 12 clades of dinosaurs. *S* is the Spearman rank correlation coefficient, which ranges from 0 (no correlation) to 1.0 (for perfectly correlated variables). (After Norell and Novacek 1992.)

to provide parental care, forming cooperative liaisons with other group members, detecting cheaters in social exchanges, and competing for resources within and among groups, would place a selective premium on intelligence, learning, and communication—thus selecting for greater intelligence and a larger brain.

Phylogeny and the Fossil Record

In inferring phylogenetic relationships among living taxa, we conclude that certain taxa share more recent common ancestors than others. If such statements are correct, then there should be some correspondence between the relative times of origin of taxa, as inferred from phylogenetic analysis, and their relative times of appearance in the fossil record. We can expect this correspondence to be imperfect because of the great imperfection of the fossil record; for example, a group that originated in the distant past might be recovered only from recent deposits. Moreover, although a lineage may have branched off early, it may not have acquired its diagnostic characters until much later. The synapsid clade, for example, did not acquire the diagnostic characters of mammals until long after it had diverged from other reptiles. In many taxa, nevertheless, there is a strong overall correspondence between phylogenetic branching order and order of appearance in the fossil record (Norell and Novacek 1992; Benton and Hitchin 1997). Just by phylogenetic analysis of living species, we infer that the common ancestors of the different orders of mammals, of mammals and reptiles, of these groups and amphibians, and of all tetrapods and sarcopterygian fishes are sequentially older. The sequence in which these groups appear in the fossil record matches the phylogeny. **Figure 4.16** shows how the correlation between phylogenetic branching order (clade rank) and the sequence of appearance of the branches in the fossil record (age rank) can be determined and graphs this correlation for two sample lineages.

Evolutionary Trends

The fossil record presents many instances of evolutionary trends. One of the best known is a steady increase in average body size in the horse family (Equidae) over the course of almost 50 My (**Figure 4.17A**). Tooth height also increased, on average, in this lineage. In the mid-Miocene, one lineage of equids evolved high-crowned (hypsodont) molariform teeth, an adaptation for feeding on abrasive grasses. The tooth crown became steadily higher in this lineage (**Figure 4.17B**). Commonly, some lineages buck the overall trend and undergo reversal; for example, several equid lineages became smaller. Certain evolutionary changes appear never to have been reversed, however. Ever since *Morganucodon*, for example, mammals have had a single lower jawbone and have never reverted to the multiple bones of their ancestors.

Many taxa in the fossil record display PARALLEL TRENDS. For example, the horse family is only one of many animal clades in which average body size has increased—a pattern called **Cope's rule**. Similarly, multiple lineages often evolve through similar phenotypic

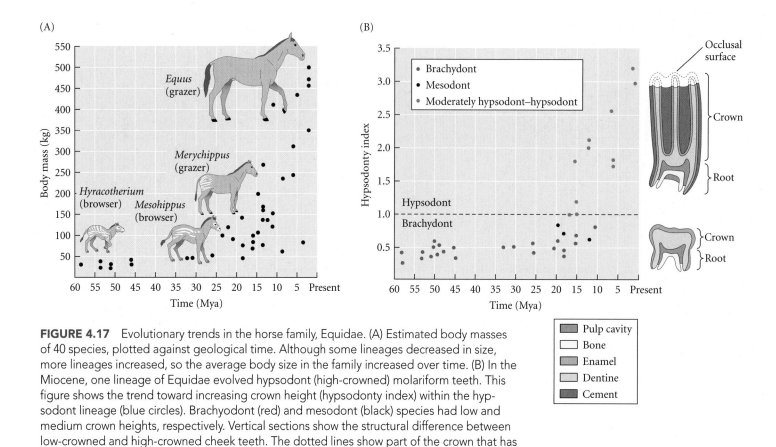

FIGURE 4.17 Evolutionary trends in the horse family, Equidae. (A) Estimated body masses of 40 species, plotted against geological time. Although some lineages decreased in size, more lineages increased, so the average body size in the family increased over time. (B) In the Miocene, one lineage of Equidae evolved hypsodont (high-crowned) molariform teeth. This figure shows the trend toward increasing crown height (hypsodonty index) within the hypsodont lineage (blue circles). Brachydont (red) and mesodont (black) species had low and medium crown heights, respectively. Vertical sections show the structural difference between low-crowned and high-crowned cheek teeth. The dotted lines show part of the crown that has worn away. (A after MacFadden 1986; B after Strömberg 2006.)

stages, called **grades**. For example, balanomorph barnacles, which first appeared in the Cretaceous, are enclosed by a cone-shaped skeleton made up of a number of slightly overlapping plates. Ancestrally, there were eight plates, but as the balanomorphs diversified in the Cenozoic, the proportion of genera with eight plates steadily declined as several lineages independently evolved through six-plate, four-plate, and even one-plate grades (**Figure 4.18**). Shells with fewer plates, and hence fewer lines of vulnerability between them, may provide greater protection against predatory snails (Palmer 1982).

Punctuated Equilibria

Although we have described paleontological examples of gradual transitions through intermediate states, these kinds of transitions are by no means universally found in the fossil record. Intermediate stages in the evolution of many higher taxa are not known, and many closely related species are separated by smaller but nonetheless distinct gaps in the fossil record (**Figure 4.19A**). Most paleontologists have followed Darwin in ascribing these gaps to the incompleteness of the fossil record. In 1972, however, Niles Eldredge and Stephen Jay Gould proposed a more complicated, and much more controversial, explanation, which they called **punctuated equilibria**. Their hypothesis applies to the abrupt appearance of closely related species, not to higher taxa.

"Punctuated equilibria" refers to both a *pattern* of change in the fossil record and a *hypothesis* about evolutionary processes. A common pattern, Eldredge and Gould said, is one of long periods in which species exhibit little or no detectable phenotypic change, interrupted by rapid shifts from one such "equilibrium" state to another; that is, **stasis** that is "punctuated" by rapid change (**Figure 4.19C**). They contrasted this pattern with what they called **phyletic gradualism**, the traditional notion of slow, incremental change (**Figure 4.19B**).

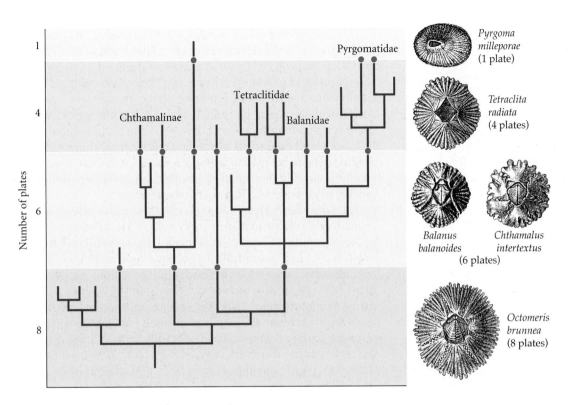

FIGURE 4.18 A parallel trend. The phylogeny of balanomorph barnacles shows the reduction in the number of shell plates that occurred during the Cenozoic in several independent lineages. The vertical axis is not time, but grade of organization (plate number). The drawings are from an extensive monograph on barnacles by Charles Darwin. (After Palmer 1982; drawings by G. Sowerby, from Darwin 1854.)

Pyrgoma milleporae (1 plate)

Tetraclita radiata (4 plates)

Balanus balanoides (6 plates)

Chthamalus intertextus

Octomeris brunnea (8 plates)

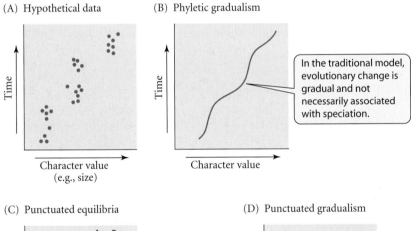

(A) Hypothetical data

(B) Phyletic gradualism

In the traditional model, evolutionary change is gradual and not necessarily associated with speciation.

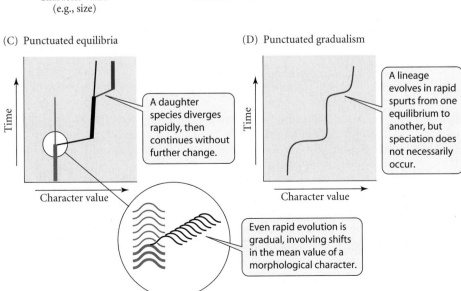

(C) Punctuated equilibria

A daughter species diverges rapidly, then continues without further change.

Even rapid evolution is gradual, involving shifts in the mean value of a morphological character.

(D) Punctuated gradualism

A lineage evolves in rapid spurts from one equilibrium to another, but speciation does not necessarily occur.

FIGURE 4.19 Three models of evolution, as applied to a hypothetical set of fossils. (A) Hypothetical values for a character in fossils from different time periods. These data might correspond to any of the models shown in panels B–D. (B) The traditional "phyletic gradualism" model. (C) The "punctuated equilibria" model of Eldredge and Gould, in which morphological change occurs in new species. Morphological evolution, although rapid, is still gradual, as shown in the inset. (D) The "punctuated gradualism," or "punctuated anagenesis," model of Malmgren et al. (1983).

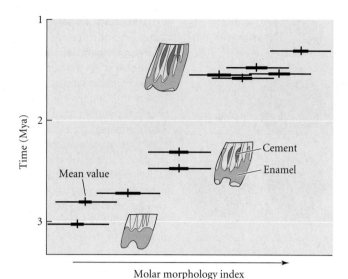

FIGURE 4.20 Phyletic gradualism: change in a molar of the grass-feeding vole *Mimomys*. Grass wears down the molar surface, so it is advantageous to have a high (hypsodont) tooth with enamel (pink) and cement (brown) forming grinding ridges at the tooth's surface. An index of change in these several tooth features shows a gradual increase over more than 1.5 My. Horizontal bars show variation around the mean, which is indicated by vertical marks. Enamel, cement, and tooth height all increased. These hypsodont teeth are similar in structure and function to those of grass-feeding horses (see Figure 4.17). (After Chaline and Laurin 1986.)

The fossil record offers examples of both gradual and punctuated patterns. Several features of the molar teeth of a lineage of grass-feeding voles (*Mimomys occitanus*) changed gradually and directionally in the late Pliocene and Pleistocene (**Figure 4.20**; Chaline and Laurin 1986). However, this pattern is rarely seen in such fine detail. Gene Hunt (2007a) analyzed data for more than 250 phenotypic traits in fossil lineages and concluded that most fit a model of either stasis or random fluctuations that result in a gradual net change over time (**Figure 4.21**). Consistent directional change was seldom recorded, perhaps because it often occurs too quickly to be preserved in a coarse fossil record.

The hypothesis that Eldredge and Gould introduced is that characters evolve primarily in concert with true speciation—that is, the branching of an ancestral species into two species (see Figure 4.19C). They based their hypothesis on a model, known as "founder effect speciation" or "peripatric speciation," proposed by Ernst Mayr in 1954, which we will consider in Chapter 18. The thrust of that model is that new species appear suddenly in the fossil record because they evolved in small populations separated from the ancestral species and then, fully formed, migrated into the region where the fossil samples were taken. The evolutionary change they underwent may have been gradual, but it was rapid and took place "off stage."

In perhaps the best example conforming to Eldredge and Gould's hypothesis, species of *Metrarabdotos*, a Miocene genus of bryozoans, or "moss animals," persisted with little change for several million years, while new species appeared abruptly, without evident intermediates (**Figure 4.22**; Cheetham 1987). However, many lineages in the fossil record quickly evolve between

(A) Directional evolution

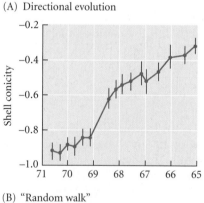

(B) "Random walk"

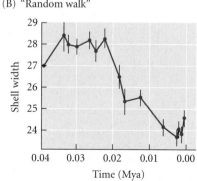

(C) Punctuated change

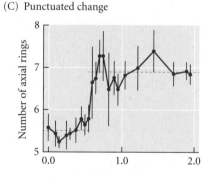

(D) Stasis

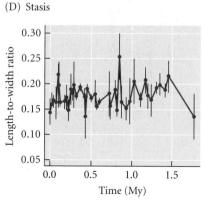

FIGURE 4.21 Examples in the fossil record that fit three models of evolution. (A) Directional evolution: steady change in shell conicity in the foraminiferan *Contusotruncana* (cf. Figure 4.3). (B) A random walk: random fluctuations in shell width of the land snail *Mandarina*. (C) A punctuated change between two intervals of stasis (dashed lines): the number of axial rings in the posterior segment of an Ordovician trilobite, *Flexicalymene*. (D) Stasis in the length-to-width ratio of the lower first molar of an Eocene mammal, *Cantius*. Note that the value is not actually constant, but instead fluctuates. (A, B after Hunt 2007a; C, D after Hunt 2010.)

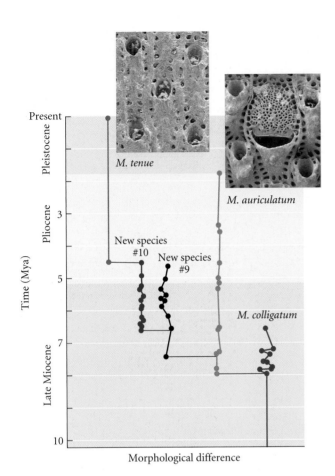

FIGURE 4.22 Punctuated equilibria: the phylogeny and temporal distribution of a lineage of bryozoans (*Metrarabdotos*). The horizontal distance between points represents the amount of morphological difference between samples. The general pattern is one of abrupt shifts to new, rather stable morphologies. (After Cheetham 1987; photos courtesy of A. Cheetham.)

long-stable states, but do not undergo speciation (**Figure 4.19D**). This pattern has been called **punctuated gradualism** or **punctuated anagenesis**.

Eldredge and Gould's hypothesis that bursts of evolution are correlated with speciation would not have been so controversial if they had not further proposed that, except in populations that are undergoing speciation, morphological characters generally *cannot* evolve, due to internal genetic "constraints." This proposition is contradicted by considerable evidence from populations of living species (see Chapters 9 and 13), and Eldredge and Gould's hypothesis that evolutionary change requires speciation is not widely accepted (Hunt 2010).

Rates of Evolution

The rate of evolutionary change varies greatly among characters, among evolving lineages, and within the same lineage over time. Although one may describe the change in, say, the height of a horse's tooth in millimeters (mm) per million years, an increase of 1 mm represents far greater change if the original tooth was 5 mm high than if it was 50 mm high. Therefore, evolutionary rates are usually described in terms of proportional, rather than absolute, change by using the logarithm of the measurement rather than the original scale of measurement. J. B. S. Haldane, one of the pioneers of the evolutionary synthesis, proposed a unit of measurement of evolutionary rate that he called the DARWIN, which he defined as a change by a factor of 2.718 (the base of natural logarithms) per million years.

Most morphological characters display very low rates of evolution in fossil lineages, reflecting the pattern of stasis to which Eldredge and Gould called attention. Usually, the feature is not absolutely constant, but instead fluctuates (see Figures 4.4 and 4.21D). Even the highest rates of change in fossil lineages are usually far lower than those observed over the course of a few centuries (or less) in species that have been transported to new regions or otherwise affected by human-induced environmental change (see Chapter 13).

FIGURE 4.23 Measures of the rate of character evolution depend on the time interval. Each point shows an estimate of the evolutionary rate of a character, plotted against the time interval over which evolution occurred. The data fall into four classes, as described on the figure. Lineages that changed so greatly that they would show "exaggerated change" are not plotted because their relationships would often not be recognized, and so they would not be compared. The rate scale is logarithmic, so rates in class IV are extremely low compared with those in classes I and II. (After Gingerich 1983.)

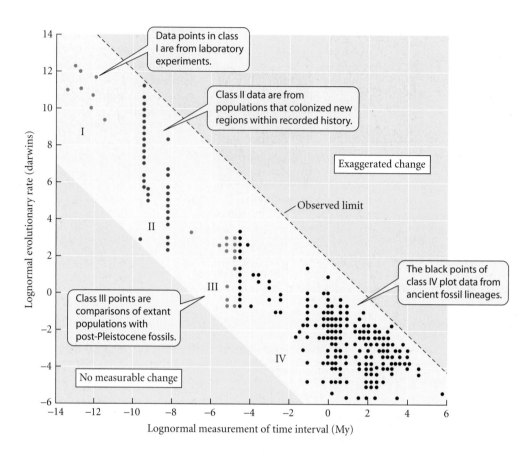

Go to the
EVOLUTION
Companion Website at
sites.sinauer.com/evolution3e
for quizzes, data analysis and simulation exercises, and other study aids.

In general, the longer the time interval over which rates of evolution are measured, the lower they are (**Figure 4.23**), because as the time interval increases, it encompasses longer periods of stasis relative to the brief episodes of rapid adaptive evolution. If characteristics evolved in a single direction for thousands or millions of years, even at extremely low rates, organisms would be vastly more different from one another than they are, and mice would long since have become bigger than the largest dinosaurs. Evolutionary changes can be very rapid, but they are not sustained at high rates for very long.

Summary

1. Although many evolutionary histories are well known, the fossil record of most lineages of organisms is very incomplete.

2. The origins of many higher taxa, such as tetrapods, birds, mammals, cetaceans, and the genus *Homo*, have been documented in the fossil record. These examples show mosaic evolution and gradual change in individual features. The decision of whether to classify intermediate fossils in one taxon or another is often arbitrary.

3. Changes in the form of characters are sometimes associated with major changes in their function.

4. The relative times of origin of taxa, as inferred from phylogenetic analysis, often correspond to their relative times of appearance in the fossil record.

5. Evolutionary trends are evident in the fossil record, but such trends may be reversed in related lineages.

6. Detailed records of change within individual species show that although characters commonly remain relatively unchanged for long periods, when changes do occur they are rapid and may pass gradually through intermediate steps. The term "punctuated equilibria" refers both to the pattern of stasis with rapid shifts to new phenotypes, and to the hypothesis, not widely accepted, that most changes in morphology occur in association with the evolution of new species (i.e., splitting of lineages).

Terms and Concepts

Cope's rule
crown group
geological time scale

grade
phyletic gradualism
punctuated equilibria

punctuated gradualism
(punctuated anagenesis)
radiometric dating

stasis
stem group
strata

Suggestions for Further Reading

The second edition of S. M. Stanley's *Earth System History* (W. H. Freeman, New York, 2005) is a thorough introduction to geological processes, Earth history, and major events in the history of life from a paleontologist's perspective. Other useful works on paleontology include R. L. Carroll's comprehensive and abundantly illustrated *Vertebrate Paleontology and Evolution* (W. H. Freeman, New York, 1988); and *Paleobiology: A Synthesis*, edited by D. E. G. Briggs and P. R. Crowther (Blackwell Publishing, Oxford, 1990), which contains a collection of brief, authoritative essays on numerous topics.

Punctuated Equilibrium by Stephen Jay Gould (Belknap Press of Harvard University Press, Cambridge, MA, 2007) is excerpted from his much longer book, *The Structure of Evolutionary Theory* (Belknap Press, Cambridge, MA, 2002), published shortly before his death. For criticism of the hypothesis of punctuated equilibrium, see J. S. Levinton's *Genetics, Paleontology, and Macroevolution* (Cambridge University Press, Cambridge, 2001).

Excellent treatments of some of this chapter's topics, written for a general audience, include *Your Inner Fish: A Journey into the 3.5-Billion-Year History of the Human Body* by Neil Shubin (Allen Lane/Pantheon, New York, 2008), on anatomical homologies between humans and other vertebrates; *Evolution: What the Fossils Say and Why It Matters* by D. R. Prothero (Columbia University Press, New York, 2007), a useful sourcebook for counteracting creationism; *Gaining Ground: The Origin and Evolution of Tetrapods* by J. A. Clack (Indiana University Press, Bloomington, 2002); and *Glorified Dinosaurs: The Origin and Early Evolution of Birds* by L. M. Chiappe (John Wiley, New York, 2007).

Problems and Discussion Topics

1. "Time averaging" refers to the condensation of fossil samples from different time intervals into a single sample. Refer to Bell's study of sticklebacks (Figure 4.4). What would the data look like if he had collapsed samples spanning 20,000 years into single samples instead of analyzing separate samples at 5000-year intervals? What conclusions might Bell have drawn about the evolution of dorsal spine number if his only samples had been from 70 Kya and 30 Kya?

2. The rate of evolution of DNA sequences (and other features) is often calibrated by the age of fossil members of the taxa to which the living species belong (see Chapter 8). How do imperfections in the fossil record affect the estimates of evolutionary rates obtained in this way? Is there any way of setting limits to the range of possible rates?

3. An ideal fossil record would enable researchers to distinguish the patterns of phyletic gradualism, punctuated equilibria, and punctuated gradualism (see Figure 4.19). How would you do so? How do imperfections of the fossil record make it difficult to distinguish these patterns?

4. Creationists deny that the fossil record provides intermediate forms that demonstrate the origins of higher taxa. Of *Archaeopteryx*, some of them say that because it had feathers and flew, it was a bird, not an intermediate form. Evaluate this argument.

5. Consider the hypothesis that Eldredge and Gould advanced to explain the pattern they called punctuated equilibria. What would be the implications for evolution if the hypothesis were true versus false?

6. Changes in a phenotypic character in a population are considered evolutionary changes only if they have a genetic basis. Alterations of organisms' features directly by the environment they experience are not evolution. Because we cannot breed extinct organisms to determine whether differences have a genetic basis, how might we decide which phenotypic changes represent evolution and which do not? Consider (a) the difference in dorsal spine number of sticklebacks at 65 versus 60 Kya (Figure 4.4B), (b) the difference in the same character at 70 versus 25 Kya, and (c) the difference in the shape of the rear teeth in *Morganucodon* (Figure 4.10F) compared with *Procynosuchus* (Figure 4.10C). Can we be more confident that the difference is an evolved one in some cases than in others?

7. What are the possible causes of trends such as those illustrated by barnacles and horses in this chapter? How would you assess the validity of each cause you can think of?

8. What might be the possible causes of a history of stasis followed by a rapid change of a character?

A History of Life on Earth

If we could look at Earth of 3,500,000,000 years ago, at about the time of life's beginning, we would see only bacteria-like cells. Among these cells would be our most remote ancestors, unrecognizably different from ourselves. And if we then time-traveled through life's history toward the present, we would see played out before us a drama grander and more splendid than we can imagine: a planetary stage of many scenes, on which emerge and play—and then, most likely, die—millions and millions of species with features and roles more astonishing than any writer could conceive.

This chapter describes some high points in the grand history of life, especially the origin, diversification, and extinction of major groups of organisms. Most of this material treats geological and paleontological evidence, but phylogenetic studies of living organisms have also revealed critically important information that has helped us trace life's history.

A Jurassic scene A carnivorous theropod dinosaur, *Ceratosaurus*, attacks the herbivorous sauropod *Brachiosaurus*, one of the largest known terrestrial animals. The cycads in this scene are seed plants that were far more prominent in the Mesozoic than they are today. Flowering plants were only beginning to evolve at this time.

Patterns in the History of Life

There is much more information in this chapter than you may wish to memorize. You may regard it largely as a source of information, or you may simply enjoy reading a sketch of one of the greatest stories of all time. A number of major events and important points that a well-trained biologist should know are highlighted in italics. As you read this chapter, moreover, notice examples of the following general patterns:

1. Climates and the distribution of oceans and land masses have changed over time, affecting the geographic distributions of organisms.
2. The taxonomic composition of the biota has changed continually as new forms have originated and others have become extinct.
3. At several times, extinction rates have been particularly high (so-called **mass extinctions**).
4. Especially after mass extinctions, the diversification of higher taxa has sometimes been relatively rapid.
5. The diversification of higher taxa has included increases both in the number of species and in the variety of their forms and ecological habits.
6. Extinct taxa have sometimes been replaced by unrelated but ecologically similar taxa.
7. Of the variety of forms in a higher taxon that were present in the remote past, usually only a few have persisted in the long term.
8. The geographic distributions of many taxa have changed greatly.
9. Over time, the composition of the biota increasingly resembles that of the present.

Before Life Began

Most physicists agree that the current universe came into existence about 14 billion (14,000,000,000) years ago through an explosion (the "big bang") from an infinitely dense point. Elementary particles formed hydrogen shortly after the big bang, and hydrogen ultimately gave rise to the other chemical elements. The collapse of a cloud of "dust" and gas formed our galaxy less than 10 billion years ago. Throughout the history of the universe, material has been expelled into interstellar space, especially during stellar explosions (supernovas), and has condensed into second- and third-generation stars, of which the Sun is one. Our solar system was formed about 4.6 billion years ago, according to radiometric dating of meteorites and moon rocks. Earth is the same age as those bodies, but because of geological processes such as subduction (see Figure 4.1), the oldest known rocks on Earth are younger, dating from about 4.06 billion years ago. Earth was probably formed by the collision and aggregation of many smaller bodies, the impact of which contributed enormous heat.

Early Earth formed a solid crust as it cooled, releasing gases that *included water vapor but very little oxygen*. As Earth cooled, oceans of liquid water formed, probably by 4.5 billion years ago, and quickly achieved the salinity of modern oceans. By 4 billion years ago, there were probably many small protocontinents, which gradually aggregated to form large land masses over the next billion years.

The Emergence of Life

The simplest things that might be described as "living" must have developed as complex aggregations of molecules. These aggregations, of course, would have left no fossil record, so it is only through chemical and mathematical theory, laboratory experimentation, and extrapolation from the simplest known living forms that we can hope to develop models of the emergence of life.

"Life" is a difficult concept to define. However, it is generally agreed that an assemblage of molecules is "alive" if it can capture energy from the environment, use that energy to replicate itself, and thus be capable of evolving. In the living things we know, these functions are performed by nucleic acids, which carry information, and by proteins, which replicate nucleic acids, transduce energy, and generate (and in part constitute) the phenotype. These components are held together in compartments—cells—formed by lipid membranes.

Although living or semi-living things might have originated more than once, *we can be quite sure that all organisms we know of stem from a single common ancestor* because they all share certain features that are arbitrary as far as we can tell (Crick 1968; Theobald 2010). For example, organisms synthesize and use only L optical isomers of amino acids as building blocks of proteins*; L and D isomers are equally likely to be formed in abiotic synthesis, but a functional protein can be made only of one type or the other. D isomers could have worked just as well. The genetic code, the machinery of replication and protein synthesis, and basic metabolic reactions are among the other features that are universal among organisms and thus imply that they all stem from a common ancestral form, or "last common ancestor" (LCA).

The most difficult problem in accounting for the origin of life is that in known living systems, only nucleic acids replicate, but their replication requires the action of proteins that are encoded by the nucleic acids. Despite this and other obstacles, progress has been made in understanding some of the likely steps in the origin of life (**Figure 5.1**; Zimmer 2009; Lazcano 2010; Lilley and Sutherland 2011).

First, *simple organic molecules*, the building blocks of complex organic molecules, *can be produced by abiotic chemical reactions*. Such molecules have been found in space and in carbonaceous meteorites. In a famous experiment, Stanley Miller found that electrical discharges in an atmosphere of methane (CH_4), ammonia (NH_3), hydrogen gas (H_2), and water (H_2O) yield amino acids and compounds such as hydrogen cyanide (HCN) and formaldehyde (H_2CO), which undergo further reactions to yield sugars, amino acids, purines, and pyrimidines.

Next, some such simple molecules must have formed polymers that could replicate. *Once replication originated, prebiotic evolution by natural selection could occur*, because variants that replicated more prolifically and more faithfully would increase relative to others. Polymerization may have been facilitated by adsorption to clay particles or by concentration caused by evaporation. The most likely early replicators were short RNA (or RNA-like) molecules. *RNA has catalytic properties, including self-replication*. Some RNA sequences (RIBOZYMES) can cut, splice, and elongate oligonucleotides, and short RNA template sequences can self-catalyze the formation of complementary sequences from free nucleotides. Recent experiments have shown that clay particles with RNA adsorbed onto their surfaces can catalyze the formation of a lipid envelope, which in turn catalyzes the polymerization of amino acids into short proteins. Thus aggregates that have some of the critical components of a protocell, including self-replicating RNA, can be formed by chemical processes alone.

It is now thought that the first steps in the origin of life took place in an "RNA world," in which catalytic, replicating RNAs existed before proteins or DNA. Within this RNA world, evolution occurred, since *natural selection and evolution can occur in nonliving systems of replicating molecules*. When Sol Spiegelman (1970) placed RNAs, RNA polymerase

---

*Some bacteria are known to synthesize a few D amino acids from their L counterparts using a specialized enzyme.

FIGURE 5.1 The approximate timing of some events in the early history of Earth and life. (After Becerra et al. 2007.)

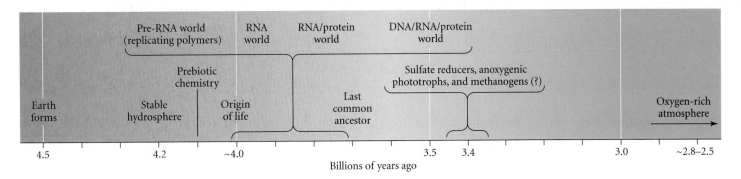

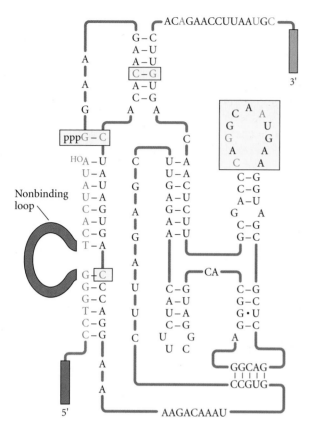

FIGURE 5.2 The sequence and structure of the catalytic RNA, a ligase, that evolved in a simple laboratory system. The oligonucleotide substrate, shown in gray at left, includes residues that bind to the RNA ligase as shown, as well as a nonbinding loop. Mutations that occurred during the experiment are indicated in red. The mutations that were critical for enhanced function are enclosed in boxes. (After Paegel and Joyce 2008).

(a catalytic RNA isolated from a virus, phage Qβ), and nucleotide bases in a cell-free medium, different RNA sequences were replicated by the polymerase at different rates, so that their proportions changed. Tracey Lincoln and Gerald Joyce (2009) extended this approach by mixing two RNAs that catalyzed synthesis of each other's sequences from a pool of nucleotides. In another experiment, a catalytic RNA (RNA ligase) evolved greater efficiency in ligating an oligonucleotide to itself when it was "grown" in an automated system with RNA polymerase enzymes and reagents (**Figure 5.2**; Paegel and Joyce 2008).

Long RNA sequences would not replicate effectively because the mutation rate would be too high for them to maintain any identity. A larger genome might evolve, however, if two or more coupled macromolecules each catalyzed the replication of the other. Replication probably was slow and inexact originally, and only much later acquired the fidelity that modern organisms display. Moreover, many different oligonucleotides undoubtedly could replicate themselves. Before proteins evolved, there were no phenotypes—only genotypes. So *the first "genes" need not have had any particular base pair sequence.* Thus there is no force to the argument, frequently made by skeptics, that the assembly of a particular nucleic acid sequence is extremely improbable (say, 1 chance in 4^{50} for a 50-base-pair sequence).

How protein enzymes evolved is perhaps the greatest unsolved problem. Eörs Szathmáry (1993) has suggested that this process began when cofactors, consisting of an amino acid joined to a short oligonucleotide sequence, aided RNA ribozymes in self-replication. Many contemporary coenzymes have nucleotide components. RNA ribozymes can also catalyze the formation of peptide bonds, so the next step may have been the stringing together of several such amino acid–nucleotide cofactors. Ultimately, the ribozyme evolved into the ribosome, the oligonucleotide component of the cofactor into transfer RNA, and the strings of amino acids into catalytic proteins. Such ensembles of macromolecules, packaged within lipid membranes, may have been precursors of the first cells—although many other features evolved between that stage and the only cells we know. The origin of cells is often considered the first of the major evolutionary transitions in the history of life (**Table 5.1**).

TABLE 5.1 Six major transitions in the history of evolution leading to higher-level formations, or groups

Major transition	Group formed	Group transformation
Separate replicators (genes) → genome within cell	Compartmentalized genomes	Evolution of large, complex genomes
Separate unicells → symbiotic unicell	Eukaryotic cells	Evolution of symbiotic organelle and nuclear genomes; transfer of genes between them; formation of "hybrid genomes"
Asexual unicells → sexual unicells	Zygote (sexually reproducing organism)	Evolution of meiosis and (often obligate) sexual reproduction
Unicells → multicellular organism	Multicellular organisms	Evolution of cell and tissue differentiation and of somatic vs. germ cells
Multicellular organisms → eusocial societies	Origin of societies (in only a few lineages)	Evolution of reproductive and nonreproductive castes (e.g., social insects)
Separate species → interspecific mutualistic associations	Origin of interspecific mutualisms	Evolution of physically conjoined partners (e.g., endosymbioses)

Source: After Bourke 2011, modified from Maynard Smith and Szathmáry 1995.

Precambrian Life

The Archean, prior to 2.5 billion years ago, and the Proterozoic, from 2.5 billion to 542 Mya, are together referred to as Precambrian time. (Table 4.1 lists the divisions of the geological time scale and traces the major events that occurred during each one.) The oldest known rocks (3.8 billion years old) contain carbon deposits that may indicate the existence of life. There is strong evidence of life by 3.0 billion years ago, and debated evidence as far back as 3.5 billion years ago, in the form of bacteria-like microfossils and layered mounds (stromatolites; **Figure 5.3**) with the same structure as those formed today along the edges of warm seas by cyanobacteria (blue-green bacteria).

The early atmosphere had little oxygen, so the earliest organisms were anaerobic. *When photosynthesis evolved* in cyanobacteria and other bacteria, *it introduced oxygen into the atmosphere.* The atmospheric concentration of oxygen increased greatly about 2.4 billion years ago, probably as a result of geological processes that buried large quantities of organic matter and prevented it from being oxidized (Knoll 2003). As oxygen built up in the atmosphere, many organisms evolved the capacity for aerobic respiration, as well as mechanisms to protect the cell against oxidation.

Living things today are classified into three "empires," or "domains": the Eucarya (all eukaryotic organisms) and two groups of prokaryotes, the Archaea and Bacteria (see Figure 2.1). *For about 2 billion years—more than half the history of life—the two prokaryotic empires were the only life on Earth.* Today many Archaea are anaerobic and inhabit extreme environments such as hot springs. [One such species is the source of the DNA polymerase enzyme (Taq polymerase) used for the polymerase chain reaction (PCR) that is the basis of much of modern molecular biology and biotechnology.] The Bacteria are extremely diverse in their metabolic capacities, and many are photosynthetic.

Molecular phylogenetic studies of prokaryotes and of the nuclear genes of eukaryotes show that Eucarya are more closely related to Archaea (possibly lying within the Archaea) than

(A)

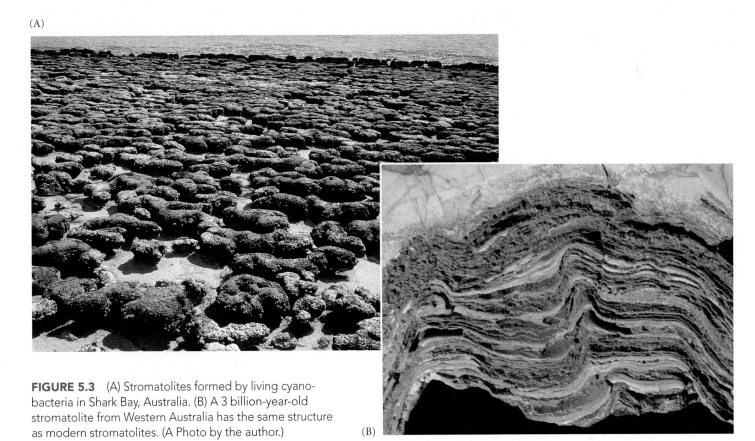

FIGURE 5.3 (A) Stromatolites formed by living cyanobacteria in Shark Bay, Australia. (B) A 3 billion-year-old stromatolite from Western Australia has the same structure as modern stromatolites. (A Photo by the author.)

(B)

to Bacteria (**Figure 5.4**). However, some DNA sequences provide conflicting evidence about relationships both among and within the empires, implying that there was extensive lateral transfer of genes among lineages during the early history of life, when well-defined, integrated genomes did not yet exist (Woese 2000; Fournier et al. 2009). Thus the early history of prokaryotes may have been more like a network than a phylogenetic tree, at least for many genes. The LCA of all modern organisms undoubtedly was the evolutionary outcome of a long history in which RNA gave way to DNA as the genetic material and genes with new functions arose (by gene duplication and mutation; see Chapters 8 and 20) and were mixed into new combinations (Becerra et al. 2007). The prokaryotes that descended from the LCA diversified greatly in their metabolic capacities (Cavalier-Smith 2006): photosynthetic, chemoautotrophic, sulfate-reducing, methanogenic, and other forms soon evolved, and these forms continue today to be the prime movers of the biogeochemical cycles on which ecosystems depend.

A major event in the history of life was the origin of eukaryotes, which are distinguished by such features as a cytoskeleton, a nucleus with multiple chromosomes and a mitotic spindle, and a cell membrane rather than a rigid cell wall. Most eukaryotes undergo meiosis, the basis of highly organized recombination and sexual reproduction. Almost all eukaryotes have mitochondria, and many have chloroplasts.

Mitochondria and chloroplasts are descended from bacteria that were ingested, and later became **endosymbionts**, in protoeukaryotes (Margulis 1993; Maynard Smith and Szathmáry 1995; see Figure 5.4 and Table 5.1). Endosymbiosis has evolved many times in the history of life, and it has been an important source of functional complexity; its role in the origin of eukaryotes is surely the most important instance (Margulis and Fester 1991; Moran 2007). Mitochondria and chloroplasts are similar to bacteria in both their ultrastructure and their DNA sequences (Dyall et al. 2004). Mitochondria are derived from the purple bacteria (a group that includes *E. coli*), and chloroplasts from cyanobacteria that became included within the ancestral cell that gave rise to the green algae and red algae. Unicellular eukaryotic green algae and red algae, in turn, became endosymbionts of at least six other lineages, providing photosynthetic and other functions. These lineages, representing tertiary or quaternary endosymbiosis, include dinoflagellates (unicellular organisms that can be toxic to fish), brown algae, apicomplexans (parasites that include *Plasmodium*, the cause of malaria), and several other groups (Keeling 2004; Lane 2010). Almost nothing is known about the ancestral "host" of the bacteria that became mitochondria. It is unlikely that it had the features (e.g., cell wall) of a bacterium or archaean; instead, it probably had protoeukaryotic features, such as a cell membrane that could engulf smaller organisms and an endoskeleton that prevented the cell from collapsing (Maynard Smith and Szathmáry 1995; Poole and Penny 2006).

Although chemical evidence suggests that eukaryotes may have evolved by about 2.7 billion years ago, the earliest eukaryote fossils are about 1.8 billion years old, which is consistent with

FIGURE 5.4 Some major branches of the tree of life, showing the current understanding of relationships among the three "empires" (Bacteria, Archaea, Eucarya) and among some lineages within each. Red arrows indicate the origin of mitochondria and chloroplasts from bacterial lineages and their endosymbiotic association with early eukaryotic lineages. Green arrows show some of the cases in which green algae became endosymbionts, functioning as chloroplasts within other eukaryote lineages. (After Palmer et al. 2004, Keeling 2007, and Parfrey et al. 2010.)

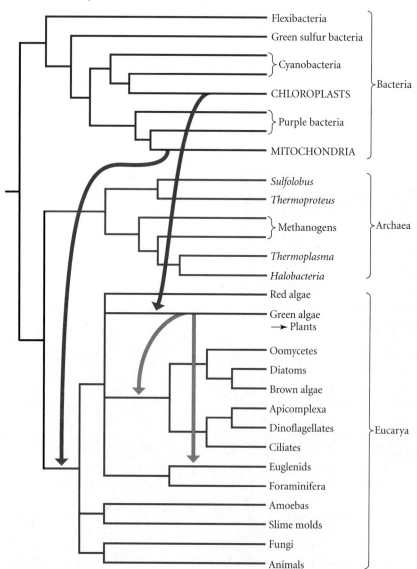

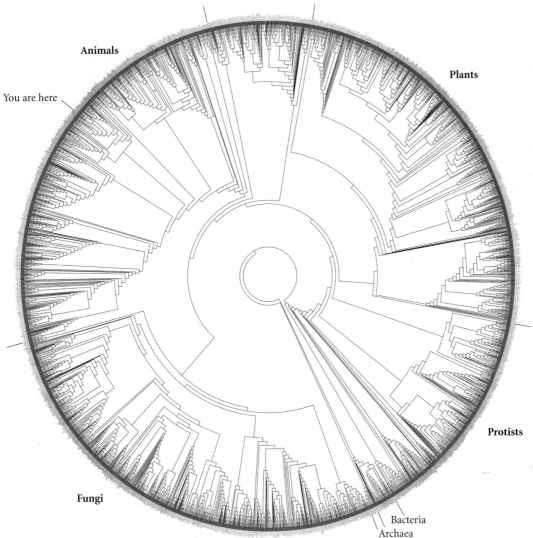

Animals

You are here

Plants

Protists

Bacteria

Archaea

Fungi

FIGURE 5.5 A comprehensive tree of life. This recently compiled phylogeny of thousands of species is based on DNA sequences. The root is in the center, and branches are reflected into a circular figure for the sake of compact display. Note the position of the human species ("You are here"). To zoom in on branches of interest, visit www.zo.utexas. edu/faculty/antisense/Download filesToL.html. (From Hillis 2010, courtesy of D. M. Hillis.)

estimates of the date of the common ancestor of eukaryotes derived from DNA sequence comparisons (Parfrey et al. 2011). Eukaryotes underwent rapid diversification, perhaps triggered by the increasing availability of nitrogen. They include far more lineages than the five kingdoms cited in older textbooks: many lineages of "algae" and "protozoans" are more distantly related to one another than fungi and animals are to each other (see Figure 5.4). These lineages are included in **Figure 5.5**, which portrays a more comprehensive tree of life, showing the phylogenetic relationships among thousands of species (best viewed in the original on the World Wide Web).

For nearly a billion years after their origin, almost all eukaryotes seem to have been unicellular, and most lineages remain so (**Figure 5.6**). *Multicellularity evolved at least six times*: in animals, two groups of fungi, and three groups of "algae," including the green algae from which plants evolved (**Figure 5.7**). The advantage of multicellularity was almost surely the "division of labor" between different cell types with different functions (Grosberg and Strathmann 2007; Michod 2007), which is a prerequisite for large size and for the development of elaborate organ systems. In the evolution of animals and plants, and perhaps in that of the other multicellular lineages as well, the first step seems to have been the evolution of cell adhesion. That step was followed by intercellular bridges that facilitate the movement of nutrients and signaling molecules and by the evolution of new signaling molecules and transcription factors (Knoll 2011). The evolution of tissues and organs depended on these molecules, which control the expression of different genes in different

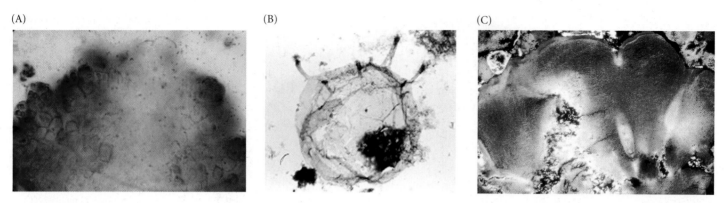

FIGURE 5.6 Some Proterozoic fossils. (A) A 1.5 billion-year-old colony of the cyanobacterium *Eoentophysalis*. (B) A unicellular eukaryote, the acritarch alga *Tappania*, from 1.5 billion-year-old strata in northern Australia. (C) A late Proterozoic (from about 590 Mya) multicellular red alga. (Photos courtesy of A. H. Knoll.)

cells. Of course, levels of transcription and translation are regulated in unicellular organisms, so the basic mechanisms already existed. In multicellular eukaryotes, however, genes may have multiple binding sites for different transcription factors and may thus contribute to diverse, complex developmental pathways (see Chapter 21). Primitive multicellularity, based on adhesion of cells formed by cell division, has evolved in laboratory cultures of yeast (Ratcliff et al. 2012; see Figure 22.14).

The oldest fossils of multicellular animals are about 575 My old. Among the first animal fossils are the creatures known as the EDIACARAN FAUNA of the late Proterozoic (about 635–542 Mya) and early Cambrian. Most of these animals were soft-bodied, lacked skeletons, and appear to have been flat creatures that crept or stood on the sea floor (**Figure 5.8**). They are hard to classify with reference to later animals, but some may have been stem groups of the Cnidaria (corals and relatives) and the Bilateria (bilaterally symmetrical phyla with three embryonic germ layers). Ediacaran animals seem to have lacked features, such as mouthparts or locomotory appendages, that might be used in interacting with other animals, and there is little evidence that they were subject to predation (Xiao and Laflamme 2009).

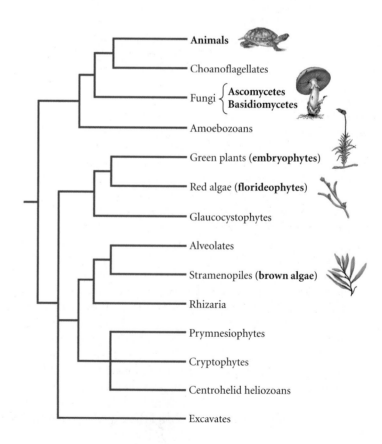

FIGURE 5.7 In this phylogeny of major eukaryotic lineages, the taxa in bold are the six lineages in which multicellularity independently evolved. (After Knoll 2011.)

(A)

(B)

FIGURE 5.8 Members of the Ediacaran fauna. (A) *Tribrachidium heraldicum*. The triradial form of this animal differs from any Phanerozoic animals. (B) The actual relationship of the wormlike *Dickinsonia costata* to later animals is unknown. (B courtesy of Martin Smith.)

Paleozoic Life: The Cambrian Explosion

The Paleozoic era began with the Cambrian period, starting about 542 Mya. For the first 10 My or so, animal diversity was low; then, during a period of about 20 My, almost all the modern phyla and classes of skeletonized marine animals, as well as many extinct groups, appeared in the fossil record. This interval marks the first appearance of brachiopods, trilobites (**Figure 5.9A**) and other classes of arthropods, molluscs, echinoderms (**Figure 5.9B**), and (especially in the Burgess Shale of British Columbia) animals that are hard to classify into later phyla. The Cambrian diversification included the earliest jawless (agnathan) vertebrates: the early Cambrian *Haikouichthys* had eyes, gill pouches, a notochord, segmented musculature, and other features resembling those of larval lampreys (**Figure 5.10A**; Shu et al. 2003), and the late Cambrian conodonts had teeth made of cellular bone (**Figure 5.10B**).

This diversification, perhaps the most dramatic adaptive radiation in the history of life (Valentine 2004), is generally called the **Cambrian explosion** because it transpired over a short time relative to preceding and succeeding intervals. Some paleontologists, however, have pointed out that 20 My is really quite a long time, and so prefer other terms, such as "Cambrian fuse" (Prothero and Buell 2007). By whatever name, the Cambrian advent of animal phyla has long been a major puzzle and source of debate. First, how quickly did evolution actually occur? Several investigators have applied molecular clocks to the DNA sequence divergence among the living animal phyla, but they have obtained widely

(A)
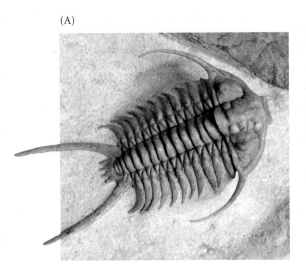

(B)
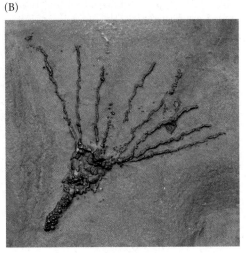

FIGURE 5.9 Two animal groups that first appeared during the Cambrian explosion. Both of these fossils were uncovered in the sandy shales of southern Utah, an area that once was covered by shallow seas. (A) A Cambrian trilobite (*Paraceraurus*). Trilobites, an arthropod group, were very diverse throughout the Paleozoic but became extinct at the end of the Permian. (B) An echinoderm (*Gogia spiralis*) from the early Cambrian. Many groups of echinoderms—which, along with chordates, constitute the deuterostome animals—flourish in the modern fauna.

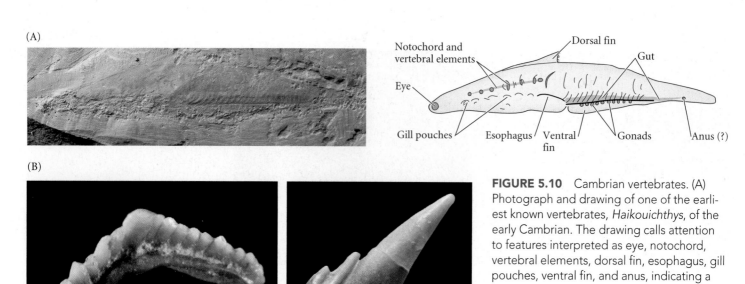

(A)

(B)

FIGURE 5.10 Cambrian vertebrates. (A) Photograph and drawing of one of the earliest known vertebrates, *Haikouichthys*, of the early Cambrian. The drawing calls attention to features interpreted as eye, notochord, vertebral elements, dorsal fin, esophagus, gill pouches, ventral fin, and anus, indicating a postanal tail region characteristic of vertebrates. (B) Bony, toothlike structures of Cambrian conodonts. Conodonts were slender, finless chordates believed to be related to agnathans (jawless vertebrates such as lampreys). (A courtesy of D.-G. Shu, from Shu et al. 2003)

variable estimates of how long ago these lineages originated, ranging from Ediacaran time (ca. 580 Mya) to as far back as 1000 Mya (Smith and Peterson 2002). Since most paleontologists think it unlikely that diverse molluscs, arthropods, and others have eluded detection in Precambrian deposits, they conclude either that the more extreme age estimates are unreliable, or that the Cambrian explosion represents the rapid evolution of shells and skeletons within clades that did not evolve these fossilizable characteristics until long after they diverged from their common ancestors. This second hypothesis is very plausible. For example, the earliest known shell-bearing cephalopod molluscs (squids and their relatives) are late Cambrian nautiloids, but a shell-less cephalopod 30 million years older has recently been found in the Burgess Shale (Smith and Caron 2010).

Molecular phylogenetic studies show that animals are most closely related to the unicellular choanoflagellates (Choanozoa), which have cell adhesion proteins and cell signaling factors like those of animals and form colonies by cell division (Abedin and King 2008; King et al. 2008; **Figure 5.11**). Sponges (phylum Porifera), which have many choanoflagellate-like cells, are thought by many researchers to be the sister group of the other animals. Choanoflagellates, sponges, and metazoans share many genes that underlie cell and developmental processes, suggesting that the "genetic toolkit" of animals evolved during Ediacaran time (Erwin 2009; Knoll 2011). In fact, recent analyses of DNA sequences favor an Ediacaran origin of the bilaterally symmetrical (bilaterian) phyla, which is consonant with the possibly bilaterian nature of some Ediacaran animals (Peterson et al. 2008).

The radially symmetrical Cnidaria (jellyfishes, corals) and the Ctenophora (comb jellies) are basal branches relative to the Bilateria—bilaterally symmetrical animals with a head, often equipped with mouth appendages, sensory organs, and a brain (**Figure 5.12**). According to DNA-based phylogenetic studies, the Bilateria include three major branches: the deuterostomes, in which the blastopore formed during gastrulation becomes the anus, and two groups of protostomes, in which the blastopore becomes the mouth. The largest deuterostome phyla are the Echinodermata (sea urchins and relatives, in which a radially symmetrical adult form has evolved from a bilaterally symmetrical ancestor) and the Chordata (including the vertebrates, tunicates, and amphioxus). The protostomes include two major clades: the Ecdysozoa (arthropods, nematodes, and some smaller phyla) and the Spiralia (molluscs, annelid worms, brachiopods, and a variety of other groups) (see Figure 5.12).

(A)

(B)

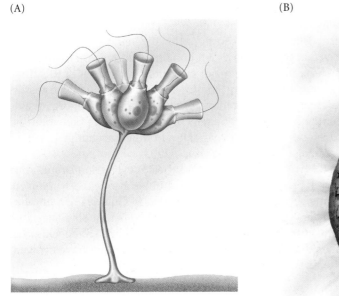

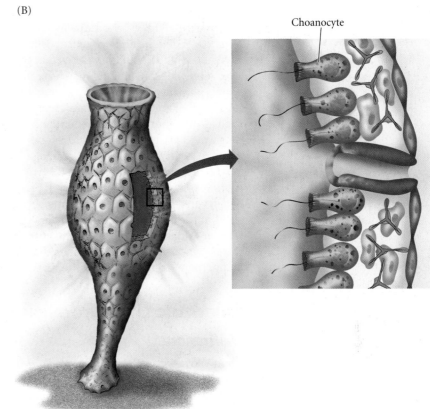

Choanocyte

FIGURE 5.11   (A) A choanoflagellate. (B) A sponge, with a close-up showing choanocytes (collar cells).

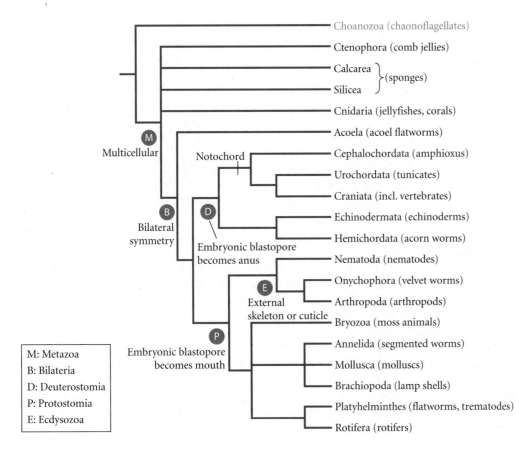

Choanozoa (chaonoflagellates)

Ctenophora (comb jellies)

Calcarea ⎫
⎬ (sponges)
Silicea ⎭

Cnidaria (jellyfishes, corals)

Acoela (acoel flatworms)

Cephalochordata (amphioxus)

Urochordata (tunicates)

Craniata (incl. vertebrates)

Echinodermata (echinoderms)

Hemichordata (acorn worms)

Nematoda (nematodes)

Onychophora (velvet worms)

Arthropoda (arthropods)

Bryozoa (moss animals)

Annelida (segmented worms)

Mollusca (molluscs)

Brachiopoda (lamp shells)

Platyhelminthes (flatworms, trematodes)

Rotifera (rotifers)

M Multicellular

Notochord

B Bilateral symmetry

D Embryonic blastopore becomes anus

E External skeleton or cuticle

P Embryonic blastopore becomes mouth

M: Metazoa
B: Bilateria
D: Deuterostomia
P: Protostomia
E: Ecdysozoa

FIGURE 5.12   An estimate of relationships among some animal phyla, based on the sequences of multiple genes. The bilaterally symmetrical animals (Bilateria) include deuterostome and protostome lineages. The relationships among sponges, Cnidaria, Ctenophora, and Bilateria are uncertain and are shown as a polytomy. The Choanozoa (choanoflagellates) are unicellular forms with structural similarities to animal cells. (After Edgecombe et al. 2011.)

If, as seems almost indubitable, this extraordinary morphological diversity evolved within about 20 My, how and why did so many great changes evolve at that time? A combination of genetic and ecological causes may account for this diversification (Knoll 2003; Marshall 2006). Regulatory genes that govern the differentiation of body parts (such as the Hox genes; see Chapter 21) may have undergone major evolutionary changes at this time, which may have allowed many new genetic combinations to arise. Some of the resulting morphological changes may have led to novel interactions among different organisms, such as predation, that further enhanced diversity by selecting for protective skeletons and new ways of overcoming such defenses. It is possible that environmental changes, such as an increase in atmospheric oxygen, also played a role (Knoll 2003).

The end of the Cambrian (488 Mya) was marked by a series of extinction events. The trilobites, of which there had been more than 90 Cambrian families, were greatly reduced, and several classes of echinoderms became extinct. As Stephen Jay Gould (1989) emphasized, if the early vertebrates had also succumbed, we would not be here today. The same may be said about every point in subsequent time: had our ancestral lineage been among the enormous number of lineages that became extinct, humans would not have evolved, and perhaps no other form of life like us would have, either (see Chapter 22).

Paleozoic Life: Ordovician to Devonian

Marine life

Many of the *animal phyla diversified greatly in the Ordovician* (488–444 Mya), *giving rise to many new classes and orders*. Many of the predominant Cambrian groups did not recover their earlier diversity, so the Ordovician fauna had a very different character from that of the Cambrian. Most Ordovician animals were EPIFAUNAL (i.e., living on the surface of the sea floor), although some bivalves (clams and relatives) evolved an INFAUNAL (burrowing)

(A)

(B)

(C)

(D)

FIGURE 5.13 Ammonoids and nautiloids. Shells housed the squid-like body of these cephalopod molluscs. (A–C) Three of the diverse forms of ammonoid shells. (A) A Jurassic ammonoid, *Craspedites*, showing the intricate sutures that evolved in many later ammonoids. (B) *Australoceras* (Cretaceous), with a very different form of shell. (C) *Kosmoceras* (Jurassic); this specimen was found in a region southeast of Moscow. (D) An orthoconic nautiloid, believed to belong to the same group as the modern *Nautilus*. These animals had non-coiled shells.

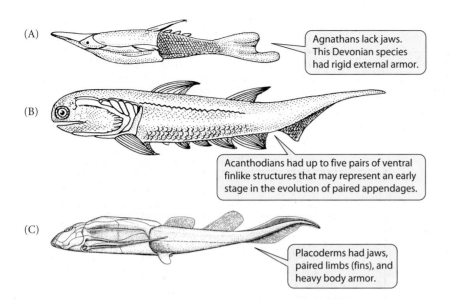

(A)

Agnathans lack jaws. This Devonian species had rigid external armor.

(B)

Acanthodians had up to five pairs of ventral finlike structures that may represent an early stage in the evolution of paired appendages.

(C)

Placoderms had jaws, paired limbs (fins), and heavy body armor.

FIGURE 5.14 Extinct Paleozoic classes of vertebrates. (A) An agnathan (jawless vertebrate), class Heterostraci (*Pteraspis*, Devonian). (B) A gnathostome (jawed vertebrate), class Acanthodii (*Climatius*, Devonian). (C) A placoderm (*Bothriolepis*, Devonian). (A after Romer 1966; B after Romer and Parsons 1986; C after Carroll 1988.)

habit. The major large predators were starfishes and nautiloids (shelled cephalopods; that is, molluscs related to squids). Reefs were built by two groups of corals, with contributions from sponges, bryozoans, and cyanobacteria. The Ordovician ended with a mass extinction that in proportional terms may have been the second largest of all time. It may have been caused by a drop in temperature and a drop in sea level, for there were glaciers at this time in the polar regions of the continents.

Among the groups that survived the mass extinction were the nautiloids, which gave rise to the ammonoids, shell-bearing cephalopods that are among the most diverse groups of extinct animals (**Figure 5.13**). During the Silurian (439–416 Mya), most vertebrates were armored agnathans, jawless fishlike vertebrates that, except for one group, lacked paired fins (**Figure 5.14A**); the only living agnathans are hagfishes and lampreys. *The first known gnathostomes*, marine vertebrates with jaws and two pairs of fins (**Figure 5.14B,C**), *also appeared during the Silurian*. It is during this time that the bony fishes (Osteichthyes) arose. During the Devonian (416–354 Mya), the "age of fishes," two subclasses of bony fishes flourished: the lobe-finned fishes (Sarcopterygii), which included diverse lungfishes and osteolepiforms (see Chapter 4), and the ray-finned fishes (Actinopterygii), which would later diversify into the largest group of modern fishes, the teleosts.

Terrestrial life

Terrestrial plants, including mosses, liverworts, and vascular plants, are a monophyletic group that evolved from green algae (Chlorophyta) (**Figure 5.15**; Judd et al. 2007). Living on land required the evolution of an external surface and spores that are resistant to loss of water, structural support, vascular tissue to transport water within the plant body, and internalized sexual organs, protected from desiccation. The *first known terrestrial organisms are* mid-Ordovician spores and spore-bearing structures (sporangia) of *very small plants*, which were apparently related to today's liverworts (Wellman et al. 2003). By the mid-Silurian, there were small vascular plants, less than 10 centimeters tall, that lacked true roots and had sporangia at the ends of short, leafless, dichotomously branching stalks (**Figure 5.16A**). By the end of the Devonian, terrestrial plants had greatly diversified: there were ferns, club mosses (**Figure 5.16B**), and horsetails (**Figure 5.16C**), some of which were large trees. In the earliest vascular plants, the haploid phase of the life cycle (gametophyte), which produces eggs and sperm, was as complex as the diploid (sporophyte) phase, which produces haploid spores by meiosis. In later plants, the gametophyte became reduced to a small, inconspicuous part of the life cycle. Early plants depended on water for the fertilization of ovules by swimming sperm, as is true of living ferns and some other groups. The spore-bearing plants were joined at the end of the Devonian by the first seed plants (**Figure 5.16D**).

FIGURE 5.15 The phylogeny and Paleozoic fossil record of major groups of terrestrial plants and their closest relatives among the green algae (Chlorophyta). The broad bars show the known temporal distribution of each group in the fossil record. The Coleochaetales and Charales (two groups of green algae), liverworts, mosses, club mosses, Selaginellales, Isoetales, horsetails, ferns, and seed plants have living representatives. (After Kenrick and Crane 1997a.)

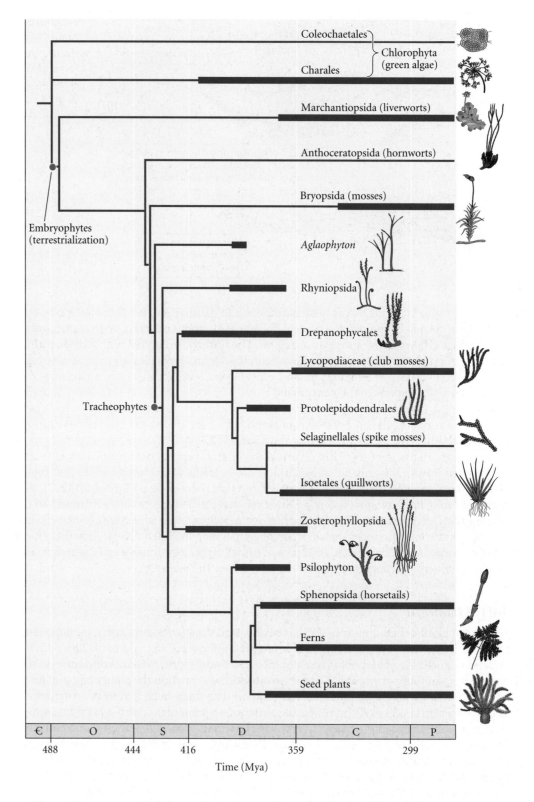

The earliest terrestrial arthropods are known from the Silurian. They fall into two major groups, both of which have marine antecedents. The chelicerates included spiders, mites, scorpions, and several other groups. The earliest mandibulates included detritus-feeding millipedes from the early Devonian, predatory centipedes, and primitive wingless insects, which evolved from crustaceans. *The first terrestrial vertebrates*, the ichthyostegid tetrapods, *evolved from lobe-finned fishes late in the Devonian* (see Chapter 4).

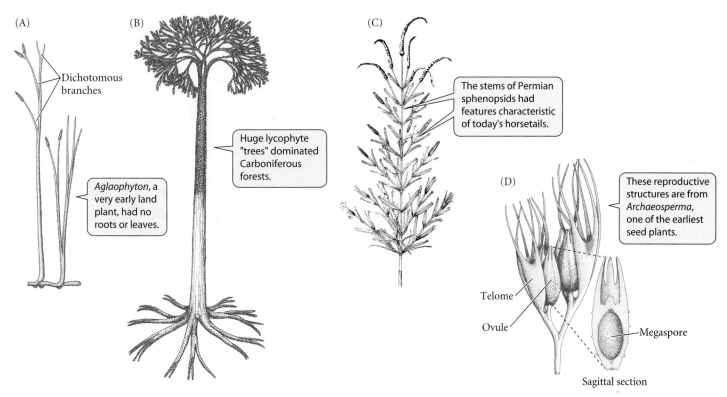

(A) Dichotomous branches

Aglaophyton, a very early land plant, had no roots or leaves.

(B) Huge lycophyte "trees" dominated Carboniferous forests.

(C) The stems of Permian sphenopsids had features characteristic of today's horsetails.

(D) These reproductive structures are from *Archaeosperma*, one of the earliest seed plants.

Telome

Ovule

Megaspore

Sagittal section

FIGURE 5.16 Paleozoic vascular plants, portrayed at different scales. (A) *Aglaophyton*, from the Devonian, was less than 10 centimeters tall. (B) *Lepidodendron*, a Carboniferous lycophyte tree as much as 30 meters tall. (C) Part of the stem of a large sphenopsid tree from the Permian. (D) Reproductive structure of a Devonian seed plant, *Archaeosperma*. (A from Kidston and Lang 1921; B, D from Stewart 1983; C from Boureau 1964.)

Paleozoic Life: Carboniferous and Permian

Terrestrial life

During the Carboniferous (359–299 Mya), land masses were aggregated into the super-continent Gondwana in the Southern Hemisphere and several smaller continents in the Northern Hemisphere. Widespread tropical climates favored the development of extensive swamp forests dominated by horsetails, club mosses, and ferns, which were preserved as the coal beds that we mine today. *The seed plants began to diversify in the late Paleozoic.* Some of them had wind-dispersed pollen, which freed them from depending on water for fertilization. The evolution of the seed provided the embryo with protection against desiccation as well as a store of nutrients that enabled the young plant to grow rapidly and overcome adverse conditions. Bear in mind that none of these plants had flowers.

The first winged insects evolved during the Carboniferous, and they rapidly diversified into many orders, including primitive dragonflies, orthopteroids (roaches, grasshoppers, and relatives), and hemipteroids (leafhoppers and their relatives). In the Permian, the first insect groups with complete metamorphosis (distinct larval and pupal stages) evolved, including beetles, primitive flies (Diptera), and the ancestors of the closely related Trichoptera (caddisflies) and Lepidoptera (moths and butterflies). Some orders of insects became extinct at the end of the Permian. This fossil sequence corresponds to the DNA-based phylogeny of living insects, in which groups without complete metamorphosis are the basal branches and the orders with complete metamorphosis form a terminal clade.

Tetrapod lineages ("amphibians") *diversified in the Carboniferous,* but most became extinct by the end of the Permian. Several late Carboniferous and early Permian lineages (anthracosaurs) are intermediate between amphibians and reptiles, and have been classified as both by different researchers. *Anthracosaurs gave rise to the first known amniotes,* the captorhinomorphs. These primitive amniotes soon gave rise to the *synapsids,* which *included the ancestors of mammals and increasingly acquired mammal-like features* (see Figure 4.10). The first amniotes also gave rise to the diapsids, a major reptilian stock whose descendants dominated the Mesozoic landscape (see Figure 5.21).

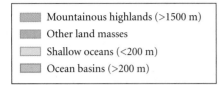

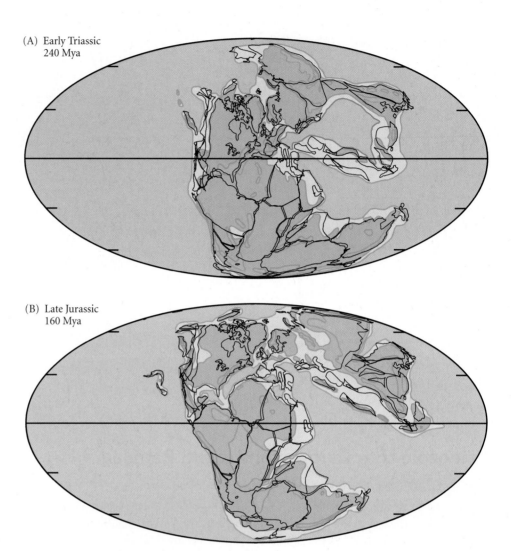

FIGURE 5.17 The distribution of land masses at several points in geological time. (A) In the earliest Triassic, most land was aggregated into a single mass (Pangaea). (B) Eurasia and North America were fairly separate by the late Jurassic. (C) Gondwana had become fragmented into most of the major southern land masses by the late Cretaceous. (D) By the late Oligocene, the land masses were close to their present configurations. The outlines of the modern-day continents are shown in all of the maps; other black lines delineate important plate boundaries. (Maps © 2004 by C. R. Scotese/ PALEOMAP Project.)

The End-Permian mass extinction

During the Permian, the continents approached one another and *formed a single world continent*, **Pangaea** (**Figure 5.17A**), which started to break apart, but then re-formed during the Triassic. Sea level dropped to its lowest point in history, and climates were greatly altered by the arrangement of land and sea. The Permian ended 252 Mya with the *end-Permian mass extinction*, one of the most significant events in the history of life. It is estimated that in this, *the most massive extinction event in the history of Earth so far*, at least 52 percent of the families—and perhaps as many as 96 percent of all species—of skeleton-bearing marine invertebrates became extinct within less than 200,000 years (Shen et al. 2011). Groups such as ammonoids, stalked echinoderms, brachiopods, and bryozoans declined greatly, and major taxa such as trilobites and several major groups of corals became extinct. Some orders of insects and many families of amphibians and mammal-like reptiles became extinct, and the composition of plant communities changed considerably (McElwain and Punyasena 2007). Paleontologists have not yet fully agreed on the cause of this event, but are tending to favor the hypothesis that it was triggered by vast volcanic eruptions in Siberia that covered 7 million square kilometers (2.7 million square miles) with layers of basalt as much as 6500 meters (4 miles) deep. These eruptions are thought to have released poisonous gases such as hydrogen sulfide and vast quantities of carbon dioxide, which in turn caused global warming, aridity, and increased acidity of ocean water (which interferes with the formation of calcium carbonate shells and skeletons). The temperature change, in turn,

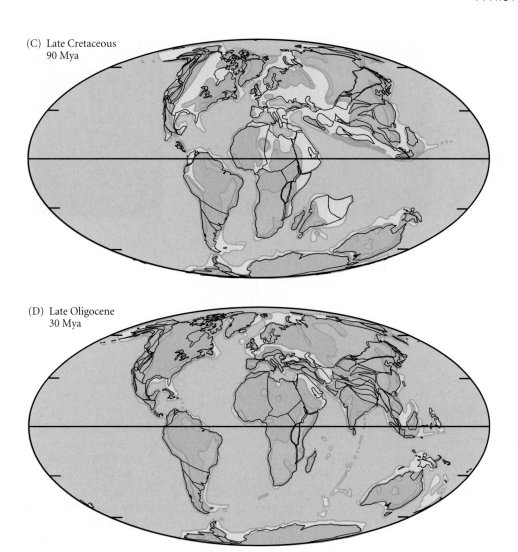

(C) Late Cretaceous
90 Mya

(D) Late Oligocene
30 Mya

may have caused turnover in the water column and the reduced oxygen level that is evident in the geological record from this time. On land, there were massive wildfires, deforestation, and soil erosion (Erwin 2006; Knoll et al. 2007; Shen et al. 2011).

Mesozoic Life

The Mesozoic era, divided into the Triassic (251–200 Mya), Jurassic (200–145 Mya), and Cretaceous (145–65.5 Mya) periods, is often called the "age of reptiles." By its end, Earth's flora and fauna were acquiring a rather modern cast, but this biota was preceded by the evolution of some of the most extraordinary creatures of all time. *During the Mesozoic, Pangaea began to break up*, beginning with the Jurassic formation of the Tethyan Seaway between Asia and Africa, and then the full separation of a northern land mass, called **Laurasia**, from a southern continent known as **Gondwana**. Laurasia began to separate into several fragments during the Jurassic (**Figure 5.17B**), but northeastern North America, Greenland, and western Europe remained connected until well into the Cretaceous. The southern continent, Gondwana, consisted of Africa, South America, India, Australia, New Zealand, and Antarctica. These land masses slowly separated in the late Jurassic and the Cretaceous, but even then the South Atlantic formed only a narrow seaway between Africa and South America (**Figure 5.17C**; see also Figure 6.12A). Throughout the Mesozoic, sea level rose, and many continental regions were covered by shallow epicontinental seas. Although the polar regions

(A)

(B)

(C)

FIGURE 5.18 Features of marine predators and prey that escalated during and after the "Mesozoic marine revolution." (A) The huge claws of modern lobsters represent a trait found in several crustacean groups. Such claws allow some lobsters and crabs to crush and rip mollusc shells. (B) Spines on both bivalves and gastropods (such as this *Murex*) prevent some fishes from swallowing these prey and may reduce the effectiveness of crushing predators. (C) Thick shells and narrow apertures, as in the gastropod *Cypraea mauritiana*, deter predators.

were cool, most of Earth enjoyed warm climates, with global temperatures reaching an all-time high in the mid-Cretaceous, after which substantial cooling occurred.

Marine life

Extinctions continued during the earliest Triassic, but diversity slowly recovered. Many of the marine groups that had been decimated during the end-Permian extinction again diversified; ammonoids (see Figure 5.13), for example, increased from two genera to more than a hundred by the middle Triassic. Planktonic foraminiferans (shelled protists) and modern corals evolved, and bony fishes continued to radiate. Another *mass extinction occurred at the end of the Triassic*, associated with a massive release of carbon into the atmosphere and global warming (Ruhl et al. 2011). Marine biodiversity decreased by about half; groups such as ammonoids and bivalves were devastated, but then recovered and experienced yet another adaptive radiation. The teleosts, today's dominant group of bony fishes, evolved and began to diversify. *During the Mesozoic* and continuing into the early Cenozoic, *predation seems to have escalated* (Vermeij 1987; Huntley and Kowalewski 2007): during this so-called Mesozoic marine revolution, crabs and bony fishes evolved mechanisms for crushing mollusc shells, and molluscs evolved protective mechanisms such as thick shells and spines (**Figure 5.18**).

During the Jurassic and Cretaceous, modern groups of gastropods (snails and relatives), bivalves, and bryozoans rose to dominance; gigantic sessile bivalves (rudists) formed reefs; and the seas harbored several groups of marine reptiles. The *end of the Cretaceous is marked by* what is surely the best-known *mass extinction* (often called the K/T EXTINCTION, using the abbreviations for Cretaceous and Tertiary), in which not only the last of the nonavian dinosaurs, but about 15 percent of marine animal families and up to 47 percent of genera became extinct (Jablonski 1995). Ammonoids, rudists, most marine reptiles, and many families of invertebrates and planktonic protists became entirely extinct. Many paleontologists have concluded that this extinction was caused by the impact of an asteroid or some other extraterrestrial body, although some dispute this hypothesis (see Chapter 7).

Terrestrial plants and arthropods

For most of the Mesozoic, the flora was dominated by gymnosperms (i.e., seed plants that lack flowers). The major groups were the cycads (**Figure 5.19A**) and the conifers and their relatives. The conifers include *Ginkgo*, a Triassic genus that has left a single surviving species as a "living fossil" (**Figure 5.19B,C**). The *angiosperms, or flowering plants* (**Figure 5.19D**), *probably originated in the late Jurassic*, for shrubby and herbaceous angiosperms, some of which were related to certain living families, were fairly diverse in the early Cretaceous (Friis et al. 2010). Many of the anatomical features of angiosperms, including flowerlike structures, evolved individually in various Jurassic groups of gymnosperms, some of which were almost certainly pollinated by insects. The great diversification of flowering plants

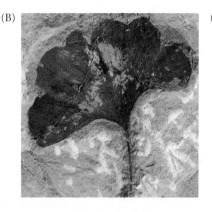

FIGURE 5.19 Seed plants. (A) A living cycad (*Encephalartos* sp.). Abundant and highly diverse during the Mesozoic, approximately 130 species of these gymnosperms survive today. (B) A fossilized *Ginkgo* leaf from the Paleocene. (C) A leaf of the sole surviving ginkgo species (*Ginkgo biloba*). (D) *Protomimosoidea*, a Paleocene/Eocene fossil member of the legume family, an angiosperm group that includes mimosas and acacias. (D courtesy of W. L. Crepet.)

began 20 to 30 My later and continued throughout the Cenozoic, when they achieved ecological dominance over the gymnosperms.

The anatomically most "advanced" groups of insects made their appearance in the Mesozoic (**Figure 5.20**). By the late Cretaceous, most families of living insects, including ants and social bees, had evolved. Throughout the Cretaceous and thereafter, insects and angiosperms affected each other's evolution and may have augmented each other's diversity. As different groups of pollinating insects evolved, adaptive modification of flowers to suit different pollinators gave rise to the great floral diversity of modern plants. *It is largely because of the spectacular increase of angiosperms and insects that terrestrial diversity is greater today than ever before.*

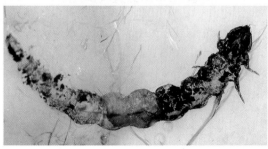

FIGURE 5.20 Some fossil insects. (A) A Jurassic relative of roaches (*Rhipidoblattina*) was a predator, unlike modern roaches. (B) An early Cretaceous beetle from one of the morphologically most primitive beetle families (Cupedidae), which still has a few "living fossil" species. (C) One of the earliest known fossil bees (*Protobombus*, Eocene) was a member of the social bee family, Apidae. (D) Among living moths, the family Micropterigidae has the most ancestral features, such as biting mandibles. This Cretaceous micropterigid larva (100 Mya) is among the earliest lepidopteran fossils. (From Grimaldi and Engel 2005, courtesy of D. Grimaldi.)

Vertebrates

The major groups of amniote vertebrates are distinguished by the number of openings in the temporal region of the skull (at least in the stem members of each lineage; **Figure 5.21**). One such group included marine reptiles that flourished from the late Triassic to the end of the Cretaceous, among them the dolphinlike ichthyosaurs, which gave birth to live young (**Figure 5.22A**).

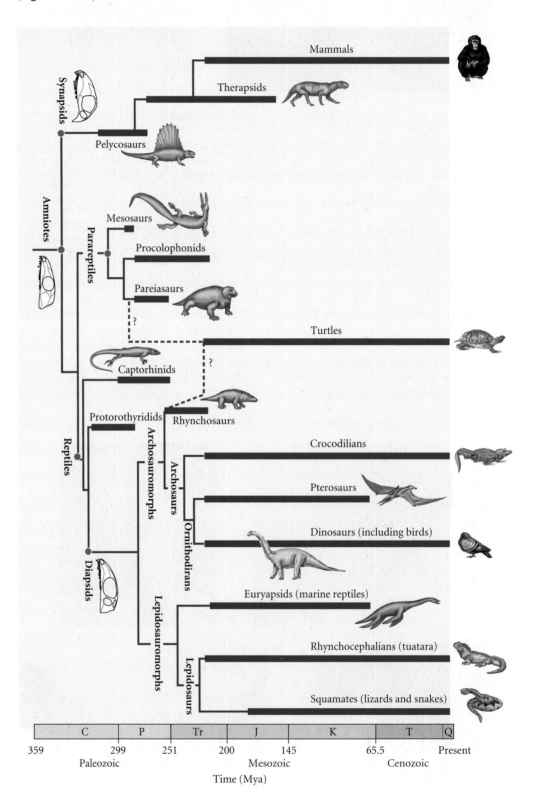

FIGURE 5.21 Phylogenetic relationships and temporal duration (thick bars) of major groups of amniote vertebrates. Some authors define "reptiles" as one of the two major lineages of amniotes, the other being the synapsids, which includes mammals. Recent DNA evidence suggests that turtles are the sister group of the archosaurs and thus are highly modified diapsids (see Figure 2.18). (After Lee et al. 2004.)

FIGURE 5.22 Some Mesozoic reptiles. (A) A marine ichthyosaur (Greek, "fish lizard"), convergent in form with sharks and porpoises. (B) *Lagosuchus*, a Triassic thecodont archosaur ("ruling lizard"), showing the body form of the stem group from which dinosaurs evolved. (C) A pterosaur (Greek, "wing lizard"), one of three flying lineages of vertebrates.

(A) An ichthyosaur

This fossil preserved the outline of the ichthyosaur's skin.

(B) *Lagosuchus talampayensis*

The *diapsids*, with two temporal openings, *became one of the most diverse groups of amniotes.* One major diapsid lineage, the lepidosauromorphs, includes the lizards, which became differentiated into modern suborders in the late Jurassic and into modern families in the late Cretaceous. One group of lizards evolved into the snakes. They probably originated in the Jurassic, but their sparse fossil record begins only in the late Cretaceous.

The archosauromorph diapsids were the most spectacular and diverse of the Mesozoic amniotes. Most of the late Permian and Triassic archosaurs were fairly generalized predators a meter or so in length (**Figure 5.22B**). From this generalized body plan, numerous specialized forms evolved. Among the most highly modified archosaurs are the pterosaurs, one of the three vertebrate groups that evolved powered flight. The wing consisted of a membrane extending to the body from the rear edge of a greatly elongated fourth finger (**Figure 5.22C**). The pterosaurs diversified greatly: one was the largest flying vertebrate known, while others were as small as sparrows.

Dinosaurs evolved from archosaurs related to the one pictured in Figure 5.22B. Dinosaurs are not simply any old large, extinct reptiles, but members of the orders Saurischia and Ornithischia, which are distinguished from each other by the form of the pelvis. Both orders included bipedal forms and quadrupeds that were derived from bipedal ancestors. Both orders arose in the Triassic, but neither order became diverse until the Jurassic.

Dinosaurs eventually became very diverse: more than 39 families are recognized (**Figure 5.23**). The Saurischia included carnivorous, bipedal theropods and herbivorous, quadrupedal sauropods. Among the noteworthy theropods are *Deinonychus*, with a huge, sharp claw that it probably used to disembowel prey; the renowned *Tyrannosaurus rex* (late Cretaceous), which stood 15 meters high and weighed about 7000 kilograms; and the small theropods from which birds evolved. The sauropods, herbivores with small heads and long necks, include the largest animals that have ever lived on land, such as *Apatosaurus* (= *Brontosaurus*); *Brachiosaurus*, which weighed more than 80,000 kilograms; and *Diplodocus*, which reached about 30 meters in length. The Ornithischia—herbivores with specialized, sometimes very numerous, teeth—included the well-known stegosaurs, with dorsal plates that probably

(C) *Pterodactylus*

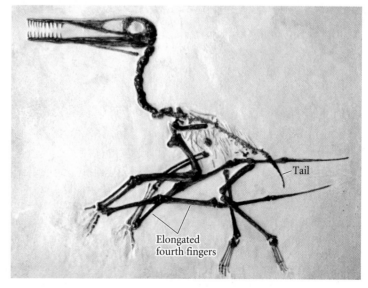

Tail

Elongated fourth fingers

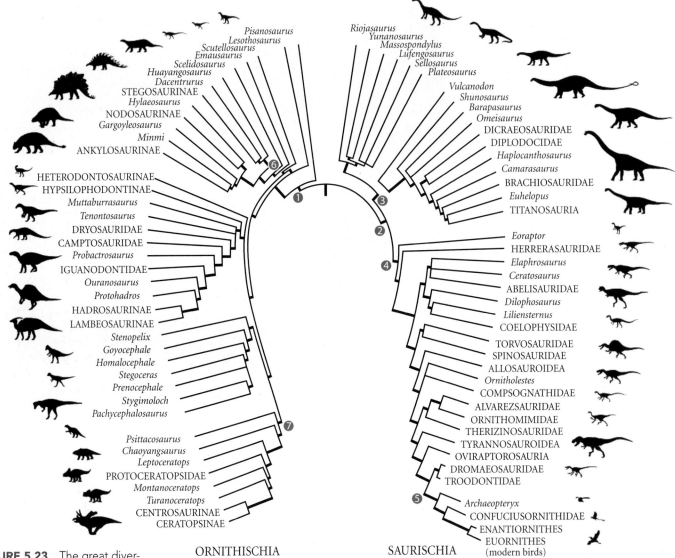

ORNITHISCHIA   SAURISCHIA

FIGURE 5.23 The great diversity of dinosaurs. The root of this proposed phylogeny is central, near the top. The two great clades of dinosaurs, Ornithischia (1; left) and Saurischia (2; right), are curved downward to fit the page. The Saurischia included the Sauropoda (3) and Theropoda (4), of which birds (5) are the only survivors. Ornithischia included stegosaurs (6) and ceratopsians (7). All ornithischian lineages are now extinct. (From Sereno 1999, *Science* 284: 2139. Copyright © AAAS.)

served for thermoregulation, and the ceratopsians (horned dinosaurs), of which *Triceratops* is the most widely known. The extinction of the ceratopsians at the end of the Cretaceous left only one surviving lineage of dinosaurs, which radiated extensively in the late Cretaceous or early Tertiary and today includes about 10,000 species. Aside from these theropod dinosaurs, more familiarly known as birds, the only living archosaurs are the 22 species of crocodilians.

The late Paleozoic *synapsids* (see Figure 5.21), with a single temporal opening, *gave rise to the therapsids*, sometimes called "mammal-like reptiles," which flourished until the middle Jurassic. *The first therapsid descendants that can be considered borderline mammals were the morganucodonts and Hadrocodium of the late Triassic and early Jurassic* (see Figure 4.10F). Although most mammals in later Mesozoic deposits are known only from teeth and jaw fragments, it is clear that several lineages of small mammals proliferated during that time. Among the most diverse were the multituberculates, which originated in the mid-Jurassic, became highly diverse in tooth morphology and probably diet (**Figure 5.24**), and reached their greatest species diversity in the latest Cretaceous and the Paleocene before they became extinct in the Eocene (Wilson et al. 2012). They seem not to have been affected by the K/T extinction or the demise of the dinosaurs.

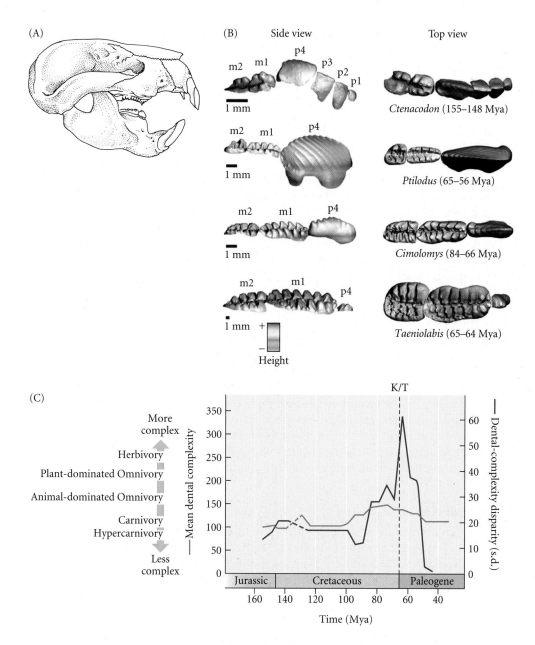

FIGURE 5.24 Multituberculate mammals resembled rodents, but had peculiar, blade-shaped premolar teeth. They became increasingly ecologically diverse during the Cretaceous. (A) Skull of an Eocene species, *Taeniolabis*. (B) Side (buccal) and top-down (occlusal) views of the premolar (p) and molar (m) teeth of four multituberculates from different time periods. The most recent, *Taeniolabis*, had reduced premolars and large, complex molars adapted for a plant diet. (C) The average complexity of multituberculate teeth (brown line) and the variation among genera (blue line) increased to a maximum near the K/T boundary. Changes in number of genera closely followed the plot of average tooth complexity. Dental complexity increases from hypercarnivorous to carnivorous to omnivorous to herbivorous diets (vertical arrow at left). (A from Romer 1966; B and C adapted from Wilson et al. 2012.)

Another Mesozoic mammalian lineage gave rise to the marsupials and the placental (eutherian) mammals in the early Cretaceous. Many of the mammals of this time had the primitive body plan of a generalized mammal, resembling living species such as opossums in overall form. Most of the modern orders of placental mammals are not recognizable in the fossil record until the early Cenozoic.

The Cenozoic Era

The Cenozoic era embraces six epochs, Paleocene through Pleistocene. The last 12,000 years are often distinguished as a seventh epoch (the Holocene, or Recent). Traditionally, the first five epochs (from 65.5 to 2.6 Mya) are referred to as the Tertiary period, and the Pleistocene and Recent (2.6 Mya–present) as the Quaternary period. Some paleontologists divide the era into the Paleogene (65.5–23 Mya) and Neogene (23 Mya–present) periods.

By the beginning of the Cenozoic, North America had moved westward, becoming separated from Europe in the east, but forming the broad Bering Land Bridge between Alaska and Siberia. *The Bering Land Bridge remained above sea level throughout most of the Cenozoic*

(**Figure 5.17D**). *Gondwana broke up* into the separate island continents of South America, Africa, India, and, far to the south, Antarctica plus Australia (which separated in the Eocene). About 18 to 14 Mya, during the Miocene, Africa made contact with southwestern Asia, India collided with Asia (forming the Himalayas), and Australia moved northward, approaching southeastern Asia. *During the Pliocene*, about 3.5 Mya, *the Isthmus of Panama arose, connecting North and South America for the first time.*

This reconfiguration of continents and oceans contributed to major climate changes. *In the late Eocene and Oligocene, there was global cooling and drying*; extensive savannahs (sparsely forested grasslands) formed for the first time, and Antarctica acquired glaciers. Sea level fluctuated, dropping drastically in the late Oligocene (about 25 Mya). In the Miocene, cacti and other plant groups that were adapted to arid conditions, including lineages with C_4 photosynthesis, diversified in several parts of the world (Arakaki et al. 2011). During the Pliocene, temperatures increased to some extent, but toward its end, temperatures dropped, and the Pleistocene epoch, starting about 2.56 million years ago, was marked by a series of about eleven GLACIAL/INTERGLACIAL CYCLES. The most recent such "Ice Age" ended only about 8000 years ago.

Aquatic life

Most of the marine groups that survived the K/T mass extinction proliferated early in the Cenozoic, and some new higher taxa evolved, such as the burrowing sea urchins known as sand dollars. The taxonomic composition of early Cenozoic marine communities was quite similar to that of modern ones. Teleost fishes continued to diversify throughout the Cenozoic, becoming by far the most diverse aquatic vertebrates.

During the Pleistocene glaciations, so much water was sequestered in glaciers that sea level dropped as much as 100 meters below its present level. As many as 70 percent of the mollusc species along the Atlantic coast of North America became extinct in the early Pleistocene; these extinctions were especially pronounced in the tropics.

Terrestrial life

Most of the modern families of angiosperms and insects had become differentiated by the Eocene or earlier, and many fossil insects of the late Eocene and Oligocene belong to genera that still survive. The dominant plants of the savannahs that developed in the Oligocene were grasses (Poaceae), which underwent a major adaptive radiation at this time, and herbaceous plants, many groups of which evolved from woody ancestors. Among the most important of these groups is the family Asteraceae, which includes sunflowers, daisies, ragweeds, and many others. It is one of the two largest plant families today.

Many living orders and families of birds are recorded from the Eocene (55.8–33.9 Mya) and Oligocene (33.9–23 Mya). The largest order of birds, the perching birds (Passeriformes), first displayed its great diversity in the Miocene (23–5.3 Mya). Another great adaptive radiation was that of the snakes, which began an exponential increase in diversity in the Oligocene. Snakes today feed on a great variety of animal prey, from worms and termites to bird eggs and wild pigs, and they include marine, burrowing, and arboreal forms, some of which glide between trees.

The adaptive radiation of mammals

The Cenozoic is sometimes called the "age of mammals" (Prothero 2006). However, the time course of the diversification of placental mammals is controversial. Almost all mammalian fossils that can be assigned to modern orders occur after the K/T boundary (65.5 Mya), and most paleontologists maintain that most of the orders diverged shortly before or after this time (Wible et al. 2007). It has often been suggested, moreover, that the extinction of the large nonavian dinosaurs at the end of the Cretaceous relieved the subjugated mammals from competition and predation and allowed them to undergo adaptive radiation. (The

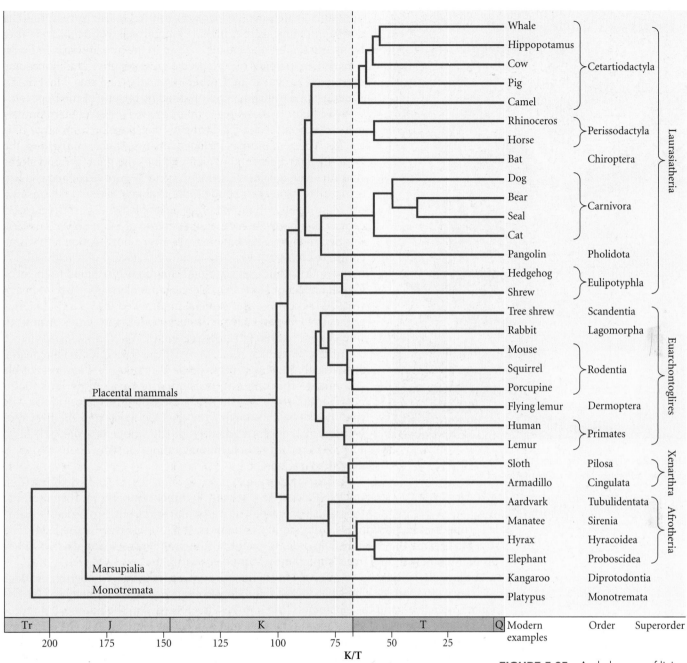

Placental mammals

Marsupialia

Monotremata

	Whale	
	Hippopotamus	
	Cow	Cetartiodactyla
	Pig	
	Camel	
	Rhinoceros	Perissodactyla
	Horse	
	Bat	Chiroptera
	Dog	
	Bear	Carnivora
	Seal	
	Cat	
	Pangolin	Pholidota
	Hedgehog	Eulipotyphla
	Shrew	
	Tree shrew	Scandentia
	Rabbit	Lagomorpha
	Mouse	
	Squirrel	Rodentia
	Porcupine	
	Flying lemur	Dermoptera
	Human	Primates
	Lemur	
	Sloth	Pilosa
	Armadillo	Cingulata
	Aardvark	Tubulidentata
	Manatee	Sirenia
	Hyrax	Hyracoidea
	Elephant	Proboscidea
	Kangaroo	Diprotodontia
	Platypus	Monotremata

Laurasiatheria

Euarchontoglires

Xenarthra

Afrotheria

| Tr | J | K | T | Q |

200 175 150 125 100 75 50 25

K/T

Time (Mya)

Modern examples   Order   Superorder

FIGURE 5.25 A phylogeny of living groups of mammals, based on DNA sequence data. The timing of branch points is based on sequence divergence, calibrated by many different fossils. The data indicate that most orders diverged from one another during the Cretaceous. Some very short branches imply that some groups of orders diverged within a relatively short time, as is characteristic of many adaptive radiations. Some relationships shown here, such as the sister-group relationship of Xenarthra and Afrotheria, are still tentative. (After Meredith et al. 2011.)

multituberculates, however, show little sign of having been subjugated, as noted earlier.) On the other hand, DNA sequence differences, analyzed by methods that use multiple fossils to calibrate the rate of sequence evolution, indicate that divergence among the stem lineages leading to many of the living orders of mammals occurred in a burst of diversification about 80 Mya, during the Cretaceous (**Figure 5.25**; Meredith et al. 2011). The defining features of the living orders, however, may not have evolved until the early Cenozoic, when many of the major lineages within those orders evolved (Meredith et al. 2011). For example, the two major lineages of Carnivora, one including cats and hyenas and the other dogs and weasels, diverged about 60–55 Mya, according to DNA analysis.

Marsupials are known as fossils from all the continents, including Antarctica. Today they are restricted to Australia and South America (except for the North American

FIGURE 5.26   The giant ground sloth *Megatherium* ("great beast") was a Pleistocene representative of the Xenarthra.

opossum, which evolved from South American ancestors). The marsupial families that include kangaroos, wombats, and other living Australian marsupials evolved in the mid-Tertiary. In South America, marsupials experienced a great adaptive radiation; some resembled kangaroo rats, others saber-toothed cats. Most South American marsupials became extinct by the end of the Pliocene.

In addition to marsupials, many groups of placental mammals evolved in South America during its long isolation from other continents. These mammals included an ancient placental group, the Xenarthra (or Edentata), which includes the giant ground sloths that survived until the late Pleistocene (**Figure 5.26**) and the few armadillos, anteaters, and sloths that still persist. At least six orders of hoofed mammals that resembled sheep, rhinoceroses, camels, elephants, horses, and rodents evolved in South America, but declined and became extinct after South America became connected to North America in the late Pliocene. The extinction of many South American mammals has been attributed to the ecological impact of North American mammal groups, such as bears, weasels, peccaries, and camels, that moved into South America at this time as part of what has been called the "Great American Interchange" of terrestrial organisms.

Among placental orders, one of the earliest and, in many ways, structurally most primitive is the Primates. The earliest fossils assigned to this order are so similar to basal eutherians that it is rather arbitrary whether they are called primates or not. The first monkeys are known from the Oligocene, and the first apes (superfamily Hominoidea) from the Miocene, about 22 Mya. The fossil record of hominin evolution, starting about 6 Mya, is described in Chapter 4.

Rodents (Rodentia), related to Primates, are recorded first from the late Paleocene. Perhaps by direct competition, they replaced the multituberculates, a nonplacental but ecologically similar group that had originated from stem mammals in the Jurassic (see Figure 5.24). The rodents became the most diverse order of mammals, in part because of an extraordinary proliferation of rats and mice (superfamily Muroidea) within the last 10 My.

A monophyletic group of orders (Afrotheria), recognized from DNA evidence, includes such very different-looking creatures as the rodentlike hyraxes (see Chapter 2, p. 19), the aquatic manatees, and the elephants. This group is first recorded in the Eocene, but represents one of the earliest splits among the placental orders (see Figure 5.25). The elephants (Proboscidea) underwent the greatest diversification, differentiating into at least 40 genera. Woolly mammoths survived through the most recent glaciation, until about 10,000 years ago, and two genera (the African and Indian elephants) persist today (**Figure 5.27**).

In the Eocene, a group of archaic hoofed mammals gave rise to a great radiation of carnivores and of hoofed mammals (ungulates) in the orders Perissodactyla and Artiodactyla, the early members of which differed only slightly. The Perissodactyla, or odd-toed ungulates, were very diverse from the Eocene to the Miocene, then dwindled to the few extant species of rhinoceroses, horses, and tapirs. The Artiodactyla (now called Cetartiodactyla by some authors) are first known in the Eocene as rabbit-sized animals that had the order's diagnostic ankle bones, but otherwise bore little similarity to the pigs, camels, and ruminants that appeared soon thereafter. In the Miocene, the ruminants began a sustained radiation, mostly in the Old World, that is correlated with the increasing prevalence of grasslands. Among the families that proliferated are the deer, the giraffes and relatives, and the Bovidae, the diverse family of antelopes, sheep, goats, and cattle. During the Eocene, an artiodactyl lineage related to today's hippopotamuses became aquatic and evolved into the cetaceans: the dolphins and whales (see Figure 4.11).

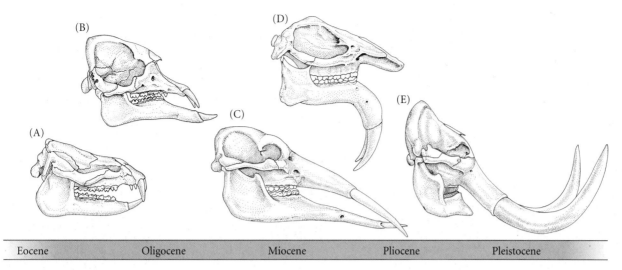

| Eocene | Oligocene | Miocene | Pliocene | Pleistocene |

FIGURE 5.27 Proboscidea, the order of elephants, has only two living genera, but was once very diverse. A few of the extinct forms are (A) an early, generalized proboscidean, *Moeritherium* (late Eocene–early Oligocene); (B) *Phiomia* (early Oligocene); (C) *Gomphotherium* (Miocene); (D) *Deinotherium* (Miocene); and (E) *Mammuthus*, the woolly mammoth (Pleistocene). (After Romer 1966.)

Pleistocene events

Because of its recency and drama, the last Cenozoic epoch, *the Pleistocene, is critically important for understanding today's organisms.*

By the beginning of the Pleistocene, the continents were situated as they are now. North America was connected in the northwest to eastern Asia by the Bering Land Bridge, in the region where Alaska and Siberia almost meet today (**Figure 5.28A**). North and South America were connected by the Isthmus of Panama. Except for those species that have become extinct, Pleistocene species were very similar to, or indistinguishable from, the living species that descended from them.

Global temperatures began to drop during the Pliocene, about 3 Mya, and then, in the Pleistocene itself, underwent violent fluctuations at intervals of about 100,000 years. When temperatures cooled, continental glaciers as thick as 2 kilometers formed at high latitudes, receding during the warmer intervals. *At least four major glacial advances,* and many minor ones, *occurred. The most recent glacial episode,* termed the Wisconsin in North America and the Würm in Europe, *reached its maximum about 20,000 years ago* (see Figure 5.28A), *and the ice melted back between 15,000 and 8000 years ago.* During glacial episodes, *sea level dropped* as much as 100 meters below its present level. This drop exposed parts of the continental shelves, extending many continental margins beyond their present boundaries and *connecting many islands to nearby land masses.* (Japan, for example, was a peninsula of Asia, New Guinea was connected to Australia, and parts of Indonesia were an extension of Southeast Asia; **Figure 5.28B**.) Temperatures in equatorial regions were apparently about as high as they are today, so the latitudinal temperature gradient was much steeper than at present. *The global climate during glacial episodes was generally drier.* Thus mesic and wet forests became restricted to relatively small favorable areas. Grasslands expanded, contributing to the diversification of grazing mammals in Africa and perhaps to the origin of bipedal hominins. During interglacial episodes, the climate became warmer and generally wetter.

These events profoundly affected the distributions of organisms (see Chapter 6). When sea level was lower, many terrestrial species moved freely between land masses that are now isolated; for example, the ice-free Bering Land Bridge was a conduit from Asia to North America for species such as woolly mammoths, bison, and humans. The distributions of many species shifted toward lower latitudes during glacial episodes and toward higher latitudes during interglacial episodes, when tropical species extended far beyond their present limits. Fossils of elephants, hippopotamuses, and lions have been taken from interglacial deposits in England, whereas Arctic species such as spruce and musk ox inhabited the southern United States during glacial episodes. These repeated shifts in geographic distributions resulted from normal processes of dispersal and establishment in new favorable areas, coupled with extinction of populations in areas that became

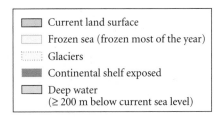

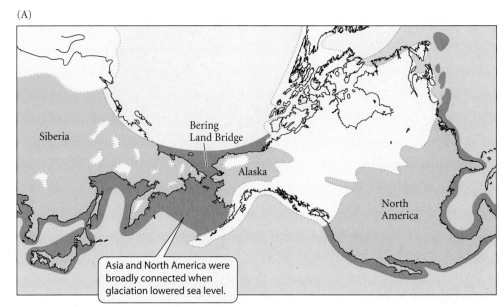

FIGURE 5.28 Pleistocene glaciers lowered sea level by at least 100 meters, so that many terrestrial regions that are now separated by oceanic barriers were connected. (A) Eastern Asia and North America were joined by the Bering Land Bridge. Note the extent of the glacier in North America. (B) Indonesia and other islands were connected to either southeastern Asia or Australia. (After Brown and Lomolino 1998.)

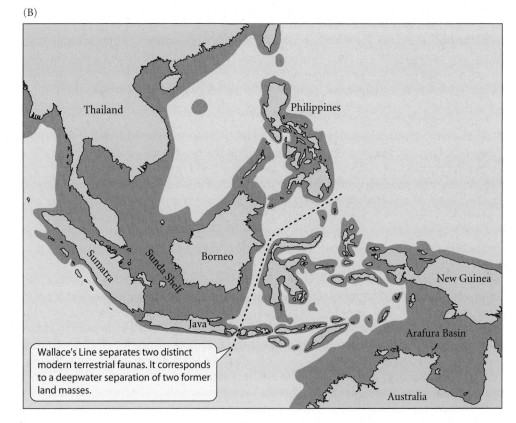

climatically unfavorable to a species. Many species were extirpated over broad areas; for instance, beetle species that occurred in England during the Pleistocene are now restricted to such far-flung areas as northern Africa and eastern Siberia (Coope 1979). Many species that had been broadly, rather uniformly distributed became isolated in separated areas (refuges, or **refugia**) where favorable conditions persisted during glacial episodes. *Some such isolated populations diverged* genetically and phenotypically, in some instances becoming different species. However, the frequent shifts in the distributions of populations may have prevented many of them from becoming different species by mixing them together (see Chapter 18). In some cases, populations have remained in their glacial refugia to this

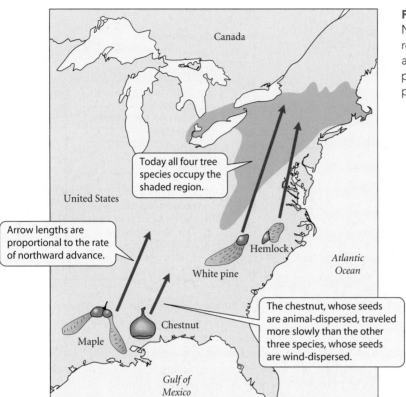

FIGURE 5.29 Different rates of northward spread of four North American tree species from refugia after the most recent glacial episode. After the glacial episode, maple and chestnut moved north from the Gulf region, and white pine and hemlock spread from the mid-Atlantic coastal plain. (After Pielou 1991.)

Today all four tree species occupy the shaded region.

Arrow lengths are proportional to the rate of northward advance.

The chestnut, whose seeds are animal-dispersed, traveled more slowly than the other three species, whose seeds are wind-dispersed.

day, isolated from the major range of their species (see Figure 6.8). On the other hand, many species have rapidly spread over broad areas from one or a few local refugia and have achieved their present distributions only in the last 8000 years or less. From studies of the distribution of fossil pollen of different ages, for example, it has become clear that some of the plant species that are typically associated in communities today moved over the land rather independently, at different rates (Jackson and Williams 2004). Thus *the species composition of ecological communities changed kaleidoscopically* (**Figure 5.29**).

Aside from changes in species' geographic distributions, the most conspicuous effects of the changes in climate were extinctions. At the end of the Pliocene, *many shallow-water marine invertebrate species became extinct, especially tropical species*, which may have been poorly equipped to withstand even modest cooling (Stanley and Campbell 1981; Jackson 1995). No major taxa became entirely extinct, however. On land, the story was different: except in Africa, *a very high proportion of large-bodied mammals and birds became extinct* in the late Pleistocene and Holocene. These animals included mammoths, saber-toothed cats, giant bison, giant beavers, giant wolves, ground sloths, and all the endemic South American ungulates. Archaeological evidence, mathematical population models, and the timing of extinctions relative to human population movements and climate change indicate that both human hunting and climate change were major causes of this "megafaunal extinction" (Martin and Klein 1984; Alroy 2001; Prescott et al. 2012).

The most recent glaciers had hardly retreated when major new disruptions began. The advent of human agriculture about 11,000 years ago began yet another reshaping of the terrestrial environment. For the last several thousand years, deserts have expanded under the impact of overgrazing, forests have succumbed to fire and cutting, and climates have changed as vegetation has been modified or destroyed. At present, under the impact of an exponentially growing human population and its modern technology, species-rich tropical forests face almost complete annihilation, temperate zone forests and prairies have been eliminated in much of the world, marine communities suffer pollution and appalling overexploitation, and global warming caused by combustion of fossil fuels threatens to change

FIGURE 5.30 A comparison of the current possible mass extinction with the five major mass extinctions of the past. The estimated extinction rate [number of extinct species/(number of species × millions of years)] is plotted against the percentage of species estimated to have become extinct in past mass extinction events, or likely to become extinct in the near future. At right, the vertical lines indicate the ranges of extinction rates that could have produced the magnitude of each of the "Big Five" mass extinctions. The curves at left are the estimated extinction rates for all of the species of birds (B), mammals (M), amphibians (A), and reptiles (R) that have become extinct in the last 500 years (blue dots), for the hypothetical extinction of all species currently designated as "critically endangered" (red), and the same for species currently designated as "threatened" (green). If these rates of extinction were to continue as shown by the dashed arrow, 75% of all species would eventually become extinct within 250 to 2270 years, depending on whether extinction rate is calculated from the current number of critically endangered or of critically endangered and threatened species. This extinction magnitude would equal that in the K/T extinction in which the last nonavian dinosaurs perished. (After Barnosky et al. 2011.)

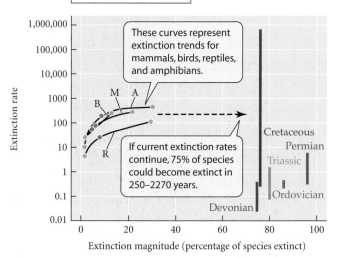

climates and habitats so rapidly that many species are unlikely to adapt (Wilson 1992; Kareiva et al. 1993). An analysis of sample regions covering 20 percent of Earth's surface projected the extinction of 18 to 35 percent of species in those regions, solely as a result of climate change, within just the next 50 years (Thomas et al. 2004). A detailed recent analysis of the numbers of threatened species of terrestrial vertebrates suggests that biological diversity could be reduced by 75 percent (a proportion comparable to the K/T mass extinction) within 900, and perhaps as few as 240, years (**Figure 5.30**; Barnosky et al. 2011; see also Box 7A). Even if these estimates are twice as pessimistic as they should be, it is clear that *one of the greatest ecological disasters, and one of the greatest mass extinctions of all time, is under way, and can be mitigated only if humankind acts decisively and quickly.*

Go to the

EVOLUTION

Companion Website at
sites.sinauer.com/evolution3e

for quizzes, data analysis and simulation exercises, and other study aids.

Summary

1. Evidence from living organisms indicates that all living things are descended from a single common ancestor. Some progress has been made in understanding the origin of life, but a great deal remains unknown.

2. The first fossil evidence of life dates from about 3.5 billion years ago, about a billion years after the formation of Earth. The earliest life forms of which we have evidence were prokaryotes.

3. Eukaryotes evolved about 1.8 billion years ago. Their mitochondria and chloroplasts evolved from endosymbiotic bacteria.

4. Although stem lineages of some modern phyla evolved before the Cambrian period, the fossil record displays an explosive diversification of the animal phyla near the beginning of the Cambrian, about 542 Mya. The causes of this rapid diversification are debated, but may include a combination of genetic and ecological events.

5. Terrestrial plants and arthropods evolved in the Silurian, and insects in the Devonian; tetrapods evolved in the late Devonian from lobe-finned fishes.

6. The most devastating mass extinction of all time occurred at the end of the Permian (about 252 Mya). It profoundly altered the taxonomic composition of Earth's biota.

7. Seed plants and amniotes became diverse and ecologically dominant during the Mesozoic era (251–65.5 Mya). Flowering plants and plant-associated insects diversified greatly from the middle of the Cretaceous onward. A mass extinction (the "K/T extinction") at the end of the Mesozoic included the extinction of the last nonavian dinosaurs.

8. The climate became drier during the Cenozoic era, favoring the development of grasslands and the evolution of herbaceous plants and grassland-adapted animals.

9. Most orders of placental mammals probably originated in the late Cretaceous, but underwent adaptive radiation in the early Tertiary. It is possible that the extinction of nonavian dinosaurs permitted them to diversify.

10. A series of glacial and interglacial episodes occurred during the Pleistocene (the last 2.6 Mya), during which some extinctions occurred and the distributions of species were greatly altered.

11. Humans have caused species extinctions since the spread of agriculture or earlier. Human population growth and technology have had an accelerating impact on biological diversity, and have probably initiated another mass extinction.

Terms and Concepts

Cambrian explosion	Gondwana	mass extinction	refugia
endosymbionts	Laurasia	Pangaea	

Suggestions for Further Reading

S. M. Stanley, *Earth and Life through Time*, second edition (W. H. Freeman, New York, 1993), is a comprehensive introduction to historical geology and the fossil record. In the fourth edition of *Life of the Past* (Prentice-Hall, Upper Saddle River, NJ, 1999), W. I. Ausich and N. G. Lane provide a well-illustrated introduction to the theme of this chapter. A good overview of the origin of life is by A. Lazcano in *Evolution Since Darwin: The First 150 Years*, M. A. Bell et al. (eds.) (Sinauer Associates, Sunderland, MA, 2010), pp. 353–375. J. Maynard Smith and E. Szathmáry, *The Major Transitions in Evolution* (W. H. Freeman, San Francisco, 1995), is an interpretation by leading evolutionary theoreticians of major events, ranging from the origin of life to the origins of societies and languages.

A. K. Behrensmeyer et al. (eds.), *Terrestrial Ecosystems through Time: Evolutionary Paleoecology of Terrestrial Plants and Animals* (University of Chicago Press, Chicago, 1992), presents detailed summaries of changes in terrestrial environments and communities in the past. *After the Ice Age: The Return of Life to North America* by E. C. Pielou (University of Chicago Press, Chicago, 1991) describes the effects of Pleistocene climate change on today's ecology and distribution of species.

Useful books on the evolution of major taxonomic groups include J. W. Valentine, *On the Origin of Phyla* (University of Chicago Press, Chicago, 2004); P. Kenrick and P. R. Crane, *The Origin and Early Diversification of Land Plants* (Smithsonian Institution Press, Washington, D.C., 1997); R. L. Carroll, *Vertebrate Paleontology and Evolution* (W. H. Freeman, New York, 1988); M. J. Benton (ed.), *The Phylogeny and Classification of the Tetrapods* (Clarendon, Oxford, 1988); D. B. Weishampel, P. Dodson, and H. Osmolska, *The Dinosauria* (University of California Press, Berkeley, 1990); and D. R. Prothero, *After the Dinosaurs: The Age of Mammals* (Indiana University Press, Bloomington, 2006).

Problems and Discussion Topics

1. Why, in the evolution of ancestral eukaryotes, might it have been advantageous for separate organisms to become united into a single organism? Can you describe analogous, more recently evolved, examples of intimate symbioses that function as single integrated organisms?

2. Early in the origin of life, as it is presently conceived, there was no distinction between genotype and phenotype. What characterizes this distinction, and at what stage of organization may it be said to have come into being?

3. If we employ the biological species concept (see Chapter 17), when did species first exist? What were organisms before then, if not species? What might the consequences of the emergence of species be for processes of adaptation and diversification?

4. How would you determine whether the morphological diversity of animals has increased, decreased, or remained the same since the Cambrian?

5. Compare terrestrial communities in the Devonian and in the Cretaceous, and discuss what may account for the difference between them in the diversity of plants and animals.

6. Read papers on the different estimates of the timing of either the origin of bilaterian animal phyla or the orders of mammals, and discuss how the difference between paleontological and molecular phylogenetic estimates might be resolved.

7. Animals that are readily classified into extant phyla, such as Mollusca and Arthropoda, appeared in the Cambrian without transitional forms that show how their distinctive body plans evolved. What kinds of evidence would be useful for reconstructing the likely steps in the evolution of such body plans? Consult Valentine 2004 on a phylum of your choice.

8. Are body cavities homologous among those phyla that have them? See Valentine 2004.

9. What is the evidence that the megafaunal extinction in the Pleistocene was partly caused by humans?

10. Discuss the implications of the spread and change of species distributions after the retreat of the Wisconsin glaciers for evolutionary changes in species and for the species composition of ecological communities (see also Chapters 7, 17, and 19).

The Geography of Evolution

Where did humans originate, and by what paths did they spread throughout the world? Why are kangaroos found only in Australia, whereas rats are found worldwide? Why are there so many more species of trees, insects, and birds in tropical than in temperate zone forests?

These questions illustrate the problems that **biogeography**, the study of the geographic distributions of organisms, attempts to solve. The answers often lie in the evolutionary history of the organisms and the history of Earth. For example, geological study of the history of the distributions of land masses and climates often sheds light on the causes of organisms' distributions. Conversely, organisms' distributions have sometimes provided evidence for geological events. In fact, the geographic distributions of organisms were used by some scientists as evidence for continental drift long before geologists agreed that it really happens. In some instances, the geographic distribution of a taxon may best be explained by historical circumstances; in other instances, ecological factors operating at the present time may provide the best explanation. Hence the field of biogeography may be roughly divided into **historical biogeography** and **ecological biogeography**. Historical and ecological explanations of geographic distributions are complementary, and both are important (Myers and Giller 1988; Lomolino et al. 2010).

New World and Old World monkeys All New World monkeys (South and Central America), including the red howler *Alouatta seniculus* (left), belong to the taxon Platyrrhini. Douc langurs (*Pygathrix nemaeus*, below) are Old World monkeys (Africa and Asia) and belong to the entirely distinct taxon Catarrhini.

Biogeographic Evidence for Evolution

Charles Darwin and Alfred Russel Wallace initiated the field of biogeography. Wallace devoted much of his later career to the subject and described major patterns of animal distribution that are still valid today. The distributions of organisms provided both Darwin and Wallace with inspiration and with evidence that evolution had occurred. To us, today, the reasons for certain facts of biogeography seem so obvious that they hardly bear mentioning. If someone asks us why there are no elephants in the Hawaiian Islands, we will naturally answer that elephants couldn't get there. This answer assumes that elephants originated somewhere else: perhaps on a continent. But in a pre-evolutionary world view, the view of special divine creation that Darwin and Wallace were combating, such an answer would not hold: the Creator could have placed each species anywhere, or in many places at the same time. In fact, it would have been reasonable to expect the Creator to place a species wherever its habitat occurred.

Darwin devoted two chapters of *The Origin of Species* to showing that many biogeographic facts that make little sense under the hypothesis of special creation make a great deal of sense if a species (1) has a definite site or region of origin, (2) achieves a broader distribution by dispersal, and (3) becomes modified and gives rise to descendant species in the various regions to which it disperses. (In Darwin's day, there was little inkling that continents might have moved. Today the movement of land masses also explains certain patterns of distribution.) Darwin emphasized the following points:

First, he said, *"neither the similarity nor the dissimilarity of the inhabitants of various regions can be wholly accounted for by climatal and other physical conditions."* Similar climates and habi-

(A)

tats, such as deserts and rain forests, occur in both the Old and the New World, yet the organisms inhabiting them are unrelated. For example, the cacti (family Cactaceae) are restricted to the New World, and the cactuslike plants in Old World deserts are members of other families (**Figure 6.1**). All the monkeys in the New World belong to one anatomically distinguishable group (Platyrrhini), and all Old World monkeys to another (Catarrhini), even though they have similar habitats and diets (see chapter-opening figure).

Darwin's second point is that *"barriers of any kind, or obstacles to free migration, are related in a close and important manner to the differences between the productions [organisms] of various regions."* Darwin noted, for instance, that marine species on the eastern and western coasts of South America are very different.

Darwin's "third great fact" is that *inhabitants of the same continent or the same sea are related, although the species themselves differ from place to place.* He cited as an example the aquatic rodents of South America (the coypu

(B)   (C)

FIGURE 6.1 Convergent growth form in desert plants. These plants, all leafless succulents with photosynthetic stems, belong to three distantly related families. (A) A cactus, *Stenocereus* (Cactaceae), in Oaxaca, Mexico. (B) A carrion flower of the genus *Stapelia* (Apocynaceae). These fly-pollinated succulents can be found from southern Africa to eastern India. (C) A member of the Euphorbiaceae (*Euphorbia candelabrum*) in Ethiopia, Africa. (A, C by the author.)

and capybara), which are structurally similar to, and related to, South American rodents of the mountains and grasslands, not to the aquatic rodents (beaver, muskrat) of the Northern Hemisphere.

"We see in these facts," said Darwin, "some deep organic bond, throughout space and time, over the same areas of land and water, independently of physical conditions. … The bond is simply inheritance [i.e., common ancestry], that cause which alone, as far as we positively know, produces organisms quite like each other."

For Darwin, it was important to show that a species had not been created in different places, but had a *single region of origin*, and had spread from there. He drew particularly compelling evidence from the inhabitants of islands. First, *remote oceanic islands generally have precisely those kinds of organisms that have a capacity for long-distance dispersal* and lack those that do not. For example, the only native mammals on many islands are bats. Second, *many continental species of plants and animals have flourished on oceanic islands to which humans have transported them*. Thus, said Darwin, "he who admits the doctrine of the creation of each separate species, will have to admit that a sufficient number of the best adapted plants and animals were not created for oceanic islands." Third, most of the species on islands are clearly *related to species on the nearest mainland*, implying that that was their source. This is the case, as Darwin said, for almost all the birds and plants of the Galápagos Islands. Fourth, the *proportion of endemic species on an island is particularly high when the opportunity for dispersal to the island is low*. Fifth, *island species often bear marks of their continental ancestry*. For example, as Darwin noted, hooks on seeds are an adaptation for dispersal by mammals, yet on oceanic islands that lack mammals, many endemic plants nevertheless have hooked seeds.

It is a testimony to Darwin's knowledge and insight that all these points hold true today, after a century and a half of research. Our greater knowledge of the fossil record and of geological events such as continental drift and sea level changes has added to our understanding, but has not negated any of Darwin's major points.

Major Patterns of Distribution

The geographic distribution of almost every species is limited to some extent, and many higher taxa are likewise restricted (**endemic**) to a particular geographic region. For example, the salamander genus *Plethodon* is limited to North America, and *Plethodon caddoensis* occupies only the Caddo Mountains of western Arkansas. Some higher taxa, such as the pigeon family (Columbidae), are almost cosmopolitan (found worldwide), whereas others are narrowly endemic (e.g., the kiwi family, Apterygidae, which is restricted to New Zealand).

Wallace and other early biogeographers recognized that many higher taxa have roughly similar distributions, and that the taxonomic composition of the biota is more uniform within certain regions than between them. For example, Wallace discovered a sharp break in the taxonomic composition of animal species among the islands that lie between southeastern Asia and Australia: as far east as Borneo, most vertebrates belong to Asian families and genera, whereas the fauna to the east has Australian affinities. This faunal break has been called WALLACE'S LINE ever since. Based on these observations, Wallace designated several **biogeographic realms** for terrestrial and freshwater organisms that are still widely recognized today (**Figure 6.2**). These realms are the PALEARCTIC (temperate and tropical Eurasia and northern Africa), the NEARCTIC (North America), the NEOTROPICAL (South and Central America), the ETHIOPIAN (sub-Saharan Africa), the ORIENTAL (India and Southeast Asia), and the AUSTRALIAN (Australia, New Guinea, New Zealand, and nearby islands). These realms are more the result of Earth's history than of its current climate or land mass distribution. For example, Wallace's Line separates islands that, despite their close proximity and similar climate, differ greatly in their fauna. These islands are on two lithospheric plates that approached each other only recently, and they are assigned to two different biogeographic realms: the Oriental and the Australian.

Each biogeographic realm is inhabited by many higher taxa that are much more diverse in that realm than elsewhere, or are even restricted to that realm. For example, the endemic

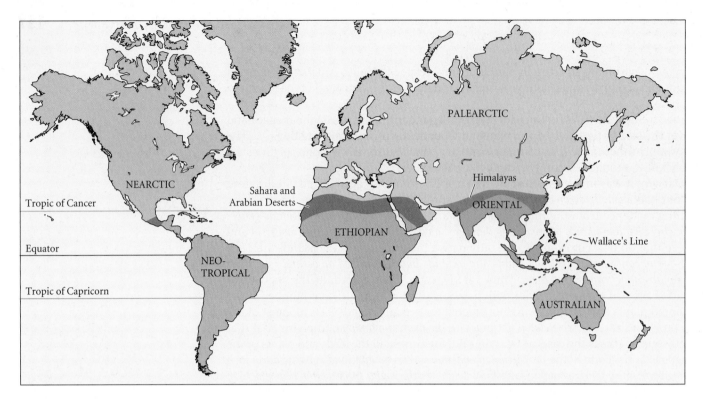

FIGURE 6.2 Biogeographic realms. The biogeographic realms recognized by Wallace are the Palearctic, Ethiopian, Oriental, Australian, Nearctic, and Neotropical.

(or nearly endemic) taxa of the Neotropical realm include the Xenarthra (anteaters and allies), platyrrhine primates (such as spider monkeys and marmosets), many hummingbird groups, a large assemblage of suboscine birds such as flycatchers and antbirds, many families of catfishes, and plant families such as the bromeliads (**Figure 6.3**). Within each realm, most individual species have more restricted distributions; regions that differ markedly in habitat, or that are separated by mountain ranges or other barriers, will have rather different sets of species. Thus a biogeographic realm can often be divided into faunal and floral PROVINCES, or regions of endemism (**Figure 6.4**).

The borders between biogeographic realms (or provinces) cannot be sharply drawn because some taxa infiltrate neighboring realms to varying degrees. In the Nearctic realm

FIGURE 6.3 Examples of taxa endemic to the Neotropical biogeographic realm. (A) An armadillo (order Xenarthra). (B) An anteater (order Xenarthra) hanging from a tree. (C) The chestnut-crowned antpitta (Formicariidae) represents a huge evolutionary radiation of suboscine birds in the Neotropics. (D) This armored catfish belongs to the Callichthyidae, one of many families of freshwater catfishes restricted to South America.

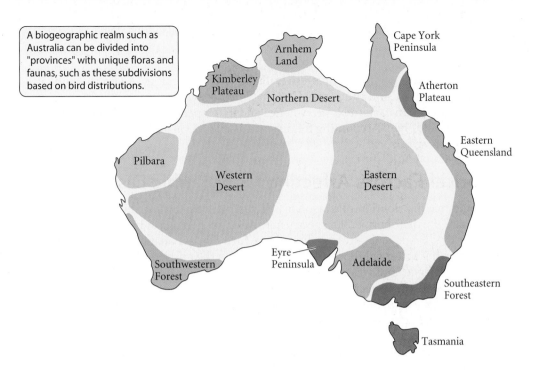

A biogeographic realm such as Australia can be divided into "provinces" with unique floras and faunas, such as these subdivisions based on bird distributions.

Cape York Peninsula

Arnhem Land

Kimberley Plateau

Northern Desert

Atherton Plateau

Pilbara

Western Desert

Eastern Desert

Eastern Queensland

Eyre Peninsula

Adelaide

Southwestern Forest

Southeastern Forest

Tasmania

FIGURE 6.4 Provinces, or regions of endemism, in Australia, based on the pattern of distribution of birds. Distributions of other vertebrates form similar patterns. (After Cracraft 1991.)

(North America), for instance, some species, such as bison, trout, and birches, are related to Palearctic (Eurasian) taxa. But other Nearctic species are related to, and have been derived from, Neotropical stocks: examples include an armadillo, an opossum, and the Spanish moss (*Tillandsia usneoides*), a bromeliad that festoons southern trees.

Some taxa have **disjunct distributions**; that is, their distributions have gaps (**Figure 6.5**). Disjunctly distributed higher taxa typically have different representatives in each area they occupy. For example, many taxa are represented on two or more southern continents, including lungfishes, marsupials, cichlid fishes, and *Araucaria* pines (Goldblatt

(A)

(B)

(C)

(D)

FIGURE 6.5 Examples of disjunct distributions. (A) Among many genera of plants found in both eastern North America and eastern Asia is *Cypripedium*, the lady's-slippers. This is the pink lady's-slipper *Cypripedium acaule*, an orchid in eastern North America. (B) A related lady's-slipper, *Cypripedium franchetii*, growing in Sichuan, China. (C) Boas (family Boidae) are distributed in tropical America and in southern Pacific islands. This is the South American emerald tree boa, *Corallus caninus*. (D) A Pacific boa, *Candoia aspera*, from New Guinea. (A, B by the author.)

1993). Another common disjunct pattern is illustrated by alligators (*Alligator*), skunk cabbages (*Symplocarpus*), and tulip trees (*Liriodendron*), which are among the many genera that are found both in eastern North America and in temperate eastern Asia, but nowhere in between (Wen 1999). Accounting for disjunct distributions has long been a preoccupation of biogeographers.

Historical Factors Affecting Geographic Distributions

The geographic distribution of a taxon is affected by both current and historical factors. The limits to the distribution of a species may be set by geological barriers that it has not crossed or by current ecological conditions to which it is not adapted. In this section, we will focus on the historical processes that have led to the current distribution of a taxon: extinction, dispersal, and vicariance.

The distribution of a species may have been reduced by the extinction of some populations, and that of a higher taxon by the extinction of some constituent species. For example, the horse family (Equidae) originated and became diverse in North America, but it later became extinct there; only the African zebras and the Asian wild asses and horses have survived. (Horses were reintroduced into North America by European colonists.) Likewise, extinction is the cause of the disjunction between related taxa in eastern Asia and eastern North America. During the early Tertiary, many plants and animals spread throughout the northern regions of North America and Eurasia. Their spread was facilitated by a warm, moist climate and by land connections from North America to both Europe and Siberia. Many of these taxa became extinct in western North America in the late Tertiary as a result of mountain uplift and a cooler, drier climate, and were later extinguished in Europe by Pleistocene glaciations (Wen 1999; Sanmartín et al. 2001).

Species expand their ranges by **dispersal** (i.e., movement of individuals). Some species of plants and animals can expand their ranges very rapidly. Within the last 200 years, many species of plants accidentally brought from Europe by humans have expanded across most of North America from New York and New England, and some birds, such as the European starling (*Sturnus vulgaris*) and the house sparrow (*Passer domesticus*), have done the

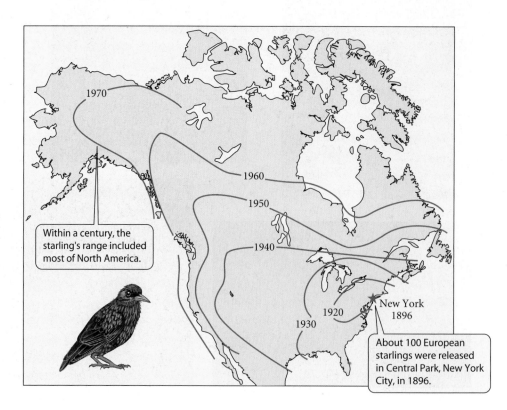

FIGURE 6.6 The history of range expansion of the European starling (*Sturnus vulgaris*) following its introduction into New York City in 1896. (After Brown and Gibson 1983.)

Within a century, the starling's range included most of North America.

New York 1896

About 100 European starlings were released in Central Park, New York City, in 1896.

FIGURE 6.7 A cattle egret (*Bubulcus ibis*) accompanying a water buffalo. This heron feeds on insects stirred up by grazing ungulates both in the Old World and in the New World, to which it recently dispersed.

same within a century (**Figure 6.6**). Other species have crossed major barriers on their own. The cattle egret (*Bubulcus ibis*) was found only in tropical and subtropical parts of the Old World until about 80 years ago, when it arrived in South America, apparently unassisted by humans (**Figure 6.7**). It has since spread throughout the warmer parts of the New World.

If a major barrier to dispersal breaks down, many species may expand their ranges more or less together, resulting in correlated patterns of dispersal (Lieberman 2003). For example, many plants and animals moved between South and North America when the Isthmus of Panama was formed in the Pliocene (see Chapter 5) and between Europe and North America over a trans-Atlantic land bridge in the early Tertiary (Sanmartín et al. 2001).

Vicariance refers to the separation of populations of a widespread species by barriers arising from changes in geology, climate, or habitat. The separated populations diverge, and they often become different subspecies, species, or higher taxa. For example, in many fish, shrimp, and other marine animal groups, the closest relative of a species on the Pacific side of the Isthmus of Panama is a species on the Caribbean side of the isthmus. This pattern is attributed to the divergence of populations of a broadly distributed ancestral species that was sundered by the rise of the isthmus in the Pliocene (Lessios 1998). Vicariance sometimes accounts for the presence of related taxa in disjunct areas.

Dispersal and vicariance are both important processes, and neither can be assumed, a priori, to be the sole explanation of a taxon's distribution. In many cases, dispersal, vicariance, and extinction all have played a role. We have seen, for example, that during the Pleistocene glaciations, species shifted their ranges by dispersal into new regions (see Figure 5.29). Some northern, cold-adapted species became distributed far to the south. When the climate became warmer, these species recolonized northern regions, and southern populations became extinct, except for populations of some such species that survived on cold mountaintops (**Figure 6.8**). (Populations

FIGURE 6.8 The disjunct distribution of a saxifrage (*Saxifraga cernua*) in northern and mountainous regions of the Northern Hemisphere. Relict populations have persisted at high altitudes following the species' retreat from the southern regions that it occupied during glacial periods. (After Brown and Gibson 1983; photo courtesy of Egil Michaelsen and the Norwegian Botanical Association.)

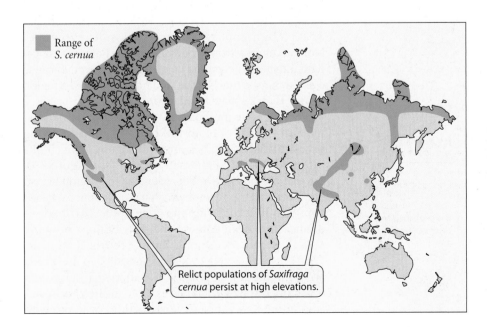

Range of *S. cernua*

Relict populations of *Saxifraga cernua* persist at high elevations.

or species that have been "left behind" in this way may be called RELICTS.) In this case, the vicariant disjunction of populations, caused by the formation of inhospitable intervening habitat, followed dispersal and went hand in hand with extinction.

Testing Hypotheses in Historical Biogeography

Biogeographers have used a variety of guidelines for inferring the histories of distributions. Some of these guidelines are well founded. For example, the distribution of a taxon cannot be explained by an event that occurred before the taxon originated: a genus that originated in the Miocene, for example, cannot have achieved its distribution by continental movements that occurred in the Cretaceous. Some other guidelines are more debatable. Some authors in the past assumed that a taxon originated in the region where it is presently most diverse. But this need not be so, as the horse family shows: although now native only to Africa and Asia, horses are descended from North American ancestors.

The major hypotheses that can be proposed to account for a taxon's distribution are dispersal and vicariance. For example, one might ask whether the marsupials, found today in Australia and South America, dispersed across an ocean from one continent to another, or whether they descended from ancestors on a single land mass that split into these two continents. Phylogenetic analysis plays a leading role in evaluating these hypotheses, but other sources of evidence can be useful as well. For example, an area is often suspected of having been colonized by dispersal if it has a highly "unbalanced" biota—that is, if it lacks a great many taxa that it would be expected to have if it had been joined to other areas. This assumption has been applied especially to oceanic islands that lack forms such as amphibians and nonflying mammals. The fossil record can also provide important evidence (Lieberman 2003)—for instance, it may show that a taxon proliferated in one area before appearing in another—and geological data may describe the appearance or disappearance of barriers. For example, fossil armadillos are limited to South America throughout the Tertiary and are found in North American deposits only from the Pliocene and Pleistocene, after the Isthmus of Panama was formed. This pattern implies that armadillos dispersed into North America from South America. Paleontological data must be interpreted cautiously, however, because a taxon may be much older, and have inhabited a region longer, than a sparse fossil record shows.

Phylogenetic methods are the foundation of most modern studies of historical biogeography. Inferring ancestral distributions from a phylogeny is much like inferring ancestral character states (see Figure 3.1). Several methods have been developed (Ronquist and Sanmartín 2011), most of which use a parsimony-based approach to reconstruct the geographic distributions of ancestors from data on the distributions of living taxa (**Figure 6.9**). Among these, Fredrik Ronquist's (1997) DISPERSAL-VICARIANCE ANALYSIS (DIVA) most fully accounts for the importance of dispersal, although it assumes that vicariance is the "null hypothesis" accounting for changes in distribution, in accord with

FIGURE 6.9 Inferring the biogeographic history of a clade by parsimony-based tree fitting. (A) Species 1–7 are distributed in areas A–F as shown by the dotted lines. The areas are assumed to have arisen by successive fragmentation of an original area by some kind of barrier in the sequence indicated by the area cladogram on the right. For example, areas E and F became separated most recently. (B) When the organism phylogeny (in black) is superimposed on the area cladogram (in brown), we can see that the distribution of the species is explained by vicariance except insofar as extinction or dispersal is required. This analysis assumes that speciation events occurred at the same time as fragmentation of the areas. (After Sanmartín and Ronquist 2004.)

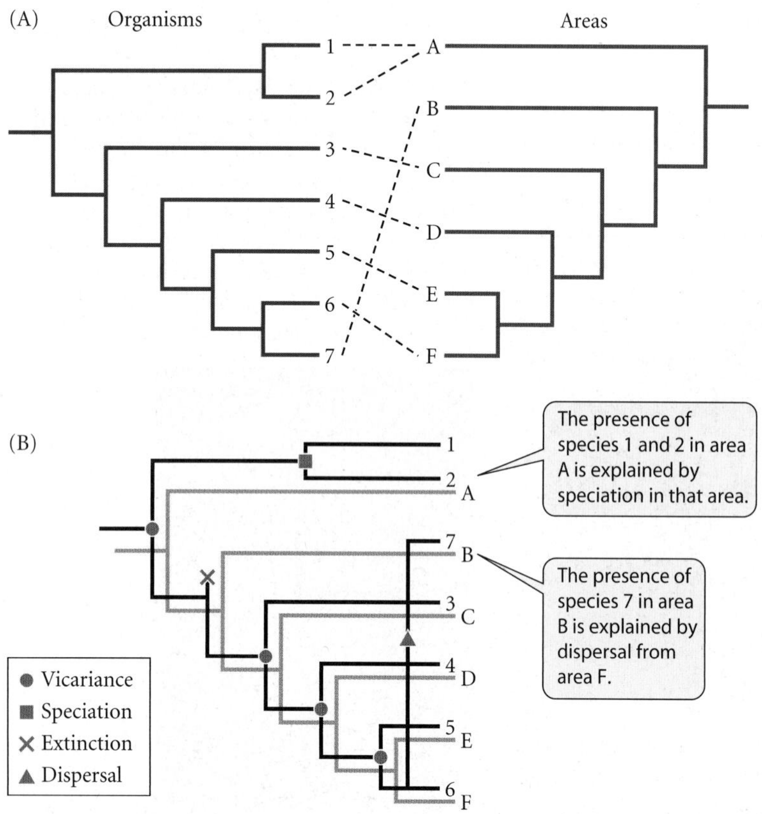

(A) Organisms / Areas

(B)

The presence of species 1 and 2 in area A is explained by speciation in that area.

The presence of species 7 in area B is explained by dispersal from area F.

- Vicariance
- Speciation
- Extinction
- Dispersal

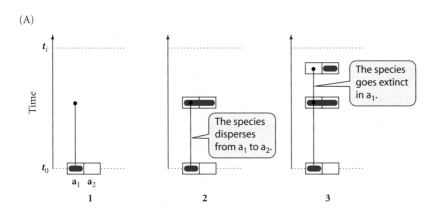

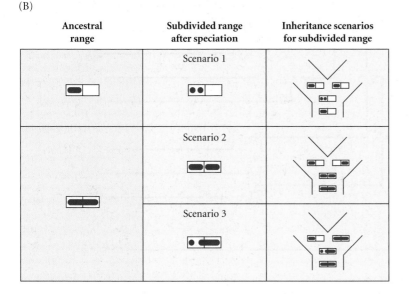

FIGURE 6.10 Simulation of changes in geographic range used in the DEC analysis of geographic distributions. (A) An example of a shift in the range of a single species by dispersal from one area (a_1) to another (a_2), followed by extinction in a_1. Over a series of time intervals, the time to the next such event is randomly generated using predetermined probabilities of dispersal and extinction. Boxes represent geographic areas and shaded ovals represent species ranges. (B) The model supposes several possible ways in which geographic ranges are inherited after speciation. In scenario 1, the ancestral range is a single area that both daughter species inherit. In scenario 2, the ancestor is widespread, and each daughter species occupies only part of the ancestral area. This scenario represents speciation by vicariance. In scenario 3, speciation occurs within an area, and one daughter species inherits only the area where it originated, while the remainder of the lineage inherits the entire ancestral range. These scenarios correspond to models of speciation described in Chapter 18. (After Ree et al. 2005.)

the well-supported principle that new species are generally formed during geographic isolation. Whenever either dispersal or extinction must be invoked in order to explain a distribution, a "cost" is exacted. The historical hypothesis that accounts for the species' distributions with the lowest "cost" is considered the most parsimonious, or optimal, hypothesis. Later, Richard Ree and collaborators (2005) introduced DEC, a statistical method that uses maximum likelihood (see Box 2B) to infer the history of dispersal to a new area, extinction in a formerly occupied area, and vicariance or cladogenesis that best explains the distribution of taxa among areas (**Figure 6.10**).

Examples of historical biogeographic analyses

ORGANISMS IN THE HAWAIIAN ISLANDS The Hawaiian Islands, in the middle of the Pacific Ocean, have been formed as a lithospheric plate has moved northwestward, like a conveyor belt, over a "hot spot," which has caused the sequential formation of volcanic cones. This process has been going on for tens of millions of years, and a string of submerged volcanoes that once projected above the ocean surface lies to the northwest of the present islands. Of the current islands, Kauai, at the northwestern end of the archipelago, is about 5.1 My old; the southeasternmost island, the "Big Island" of Hawaii, is the youngest, and is less than 500,000 years old (**Figure 6.11A**).

Given the geological history of the archipelago, the simplest phylogeny expected of a group of Hawaiian species would be a "comb," in which the most basal lineages occupy Kauai and the youngest lineages occupy Hawaii. This pattern would occur if species

FIGURE 6.11 Dispersal accounts for the distribution of species in the Hawaiian Islands. (A) The present Hawaiian archipelago, showing the approximate dates of each island's formation. (B) A phylogeny of Hawaiian species of *Psychotria* trees shows successive shifts from older islands to younger islands, as expected if younger islands were colonized after they formed. (C) Molecular phylogeny of a Hawaiian genus of crickets (*Laupala*) in the Hawaiian Islands; each branch represents a distinct species. As in *Psychotria*, successively younger groups are found on younger islands. (After Ree et al. 2008 and Mendelson and Shaw 2005; photo courtesy of Kerry Shaw.)

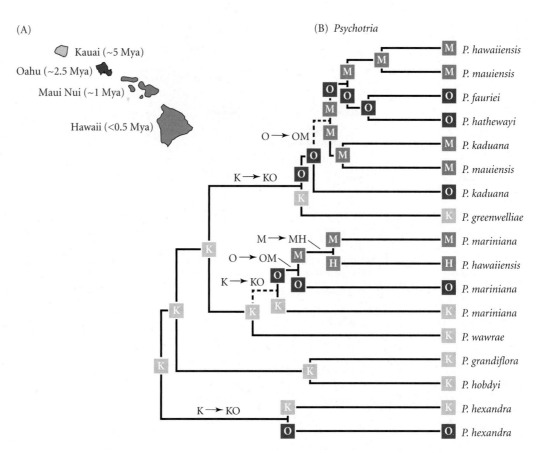

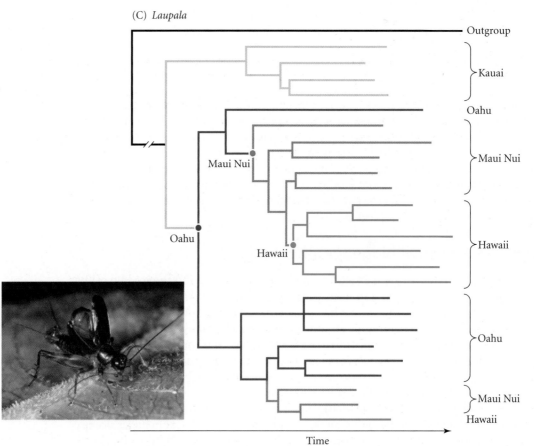

(A)

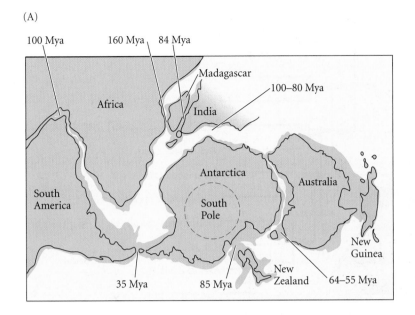

(B)

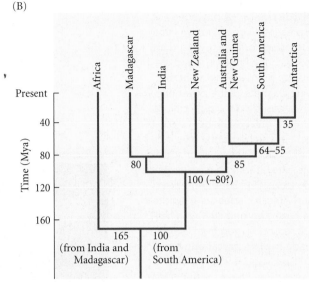

FIGURE 6.12 The breakup of Gondwana. (A) A view of Gondwana in the early Cretaceous (120 Mya), centered on the present South Pole, indicating the approximate times at which connections among the southern land masses were severed. The current configurations of the continents are shown by the black lines; green areas beyond these lines were also exposed land during the early Cretaceous. (B) A branching diagram, sometimes called an area cladogram, that attempts to depict the history of the breakup of Gondwana. (A after Cracraft 2001.)

successively dispersed to new islands as they were formed, did not disperse from younger to older islands, and did not suffer extinction. Ree and Smith (2008) found just this pattern when they used the DEC method to analyze the distribution of trees in the genus *Psychotria* (**Figure 6.11B**). A similar pattern has been found for other Hawaiian taxa, such as the cricket genus *Laupala* (Mendelson and Shaw 2005), analyzed by more traditional methods (**Figure 6.11C**). In all of these cases, dispersal has proceeded from older to younger islands.

GONDWANAN DISTRIBUTIONS   Many intriguing biogeographic problems are posed by taxa that have members on different land masses in the Southern Hemisphere. The simplest hypothesis is, of course, pure vicariance: the breakup of Gondwana isolated descendants of a common ancestor. If the distribution of a taxon among the Southern Hemisphere land masses were caused entirely by vicariance, its phylogeny would be expected to match the AREA CLADOGRAM, a depiction of the sequence by which the land masses became isolated by plate tectonics (**Figure 6.12**).

One of the best examples of a distribution that seems best explained by the fragmentation of Gondwana is the Cichlidae, a large family of freshwater fishes (Sparks and Smith 2004; Genner et al. 2007; Azuma et al. 2008). The species in tropical America and Africa form monophyletic sister groups, which together are more distantly related to species in Madagascar and India (**Figure 6.13**). The times of origin of these clades, estimated by calibrating DNA sequence divergence by many different fish fossils, matches the sequence and times of separation of these land masses in the Cretaceous.

In many cases, however, distributions that suggest vicariance are not so simply explained, and dispersal has been a more important process (de Queiroz 2005; Crisp et al. 2011). For instance, the southern beeches (*Nothofagus*), distributed in southern South America, Australia, New Zealand, and on the island of New Caledonia (near the Philippines), are a classic example of disjunct distribution that has been attributed to the rifting of Gondwana. Estimates of the times of divergence among different lineages, based on fossil-calibrated sequence divergence of several genes, however, were only partly consistent with this hypothesis (Cook and Crisp 2005). The molecular dating of the disjunction between Australian and South American species in two different subgenera coincides well with the time of rifting between these continents, but two other subgenera gave rise to species in both New Zealand and Australia, in both instances more than 30 My after

FIGURE 6.13 A phylogeny of representatives of the fish family Cichlidae from the four land areas in which the family is found. Each branch point is situated at the best estimate of the divergence time, based on multiple calibrations of mtDNA sequence divergence. The horizontal bars indicated the statistical error around the estimated time of divergence of clades. The South American and African clades are estimated to have diverged 89 Mya. Their ancestor diverged from the clade in Madagascar and India/Sri Lanka about 96 Mya. The dashed line marks the separation of Africa from South America 100 Mya. Given the margin of error in divergence estimates, the phylogeny is consistent with vicariant separation of clades caused by the fragmentation of Gondwana. (After Azuma et al. 2008.)

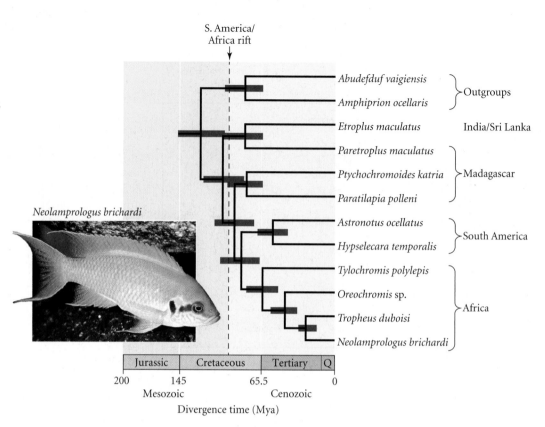

these land masses had separated (**Figure 6.14**). It seems likely that early separation by the breakup of Gondwana was followed by dispersal long afterward.

Another example is provided by the endemic (and highly endangered) biota of Madagascar, the large island east of Africa. Many groups, such as lemurs, occur nowhere else. Together with India, Madagascar broke away from Gondwana more than 120 Mya (see Figure 6.12). India became separated from Madagascar about 88 Mya and collided with southern Asia about 50 Mya. For many years, biogeographers postulated that many of the endemic Madagascan taxa originated by vicariant separation from their relatives in Africa and other southern land masses. However, recent molecular phylogenetic studies indicate

FIGURE 6.14 A simplified phylogeny of the four major lineages (subgenera) of southern beeches (*Nothofagus*), showing branching dates estimated by DNA sequence difference. In subgenera *Fuscospora* and *Lophozonia*, closely related species are found in New Zealand and Australia, even though these land masses separated long before the rift between Australia and South America. The brown bars show estimated times at which Gondwanan land masses separated, antedating the divergence of *Nothofagus* lineages. Consequently, vicariance by continental drift does not explain the disjunct distribution of these plants. (After Cook and Crisp 2005; photo by the author.)

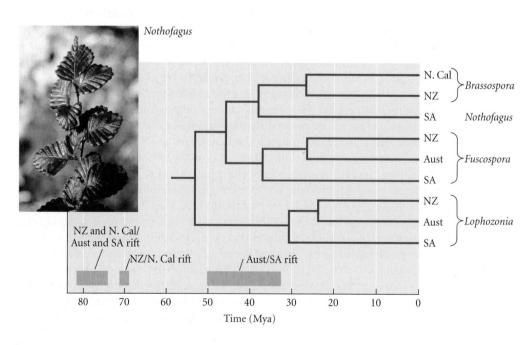

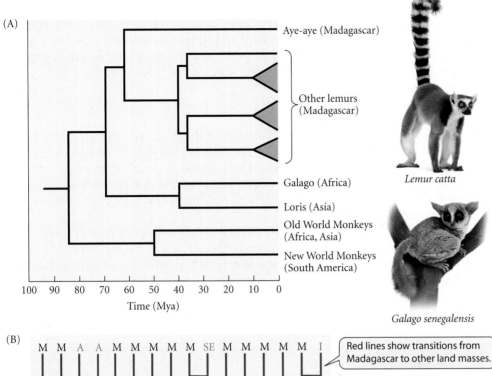

Lemur catta

Galago senegalensis

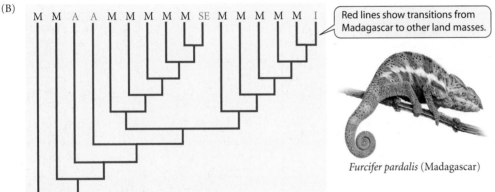

Red lines show transitions from Madagascar to other land masses.

Furcifer pardalis (Madagascar)

FIGURE 6.15 Phylogenetic evidence of the history of two groups of animals in Madagascar. (A) The prosimians of Madagascar (lemurs, including the aye-aye) diverged from the African and Asian prosimians (galagos and lorises) well after Madagascar separated from the other Gondwanan land masses, more than 80 Mya. (B) Chameleons are distributed in Africa (A), India (I), Madagascar (M), and the Seychelles (SE) in the Indian Ocean. Because the phylogenetic distribution over these areas differs from the sequence by which the areas became separated (see Figure 6.12B), the distribution of chameleons is best explained by dispersal from Madagascar, rather than vicariance caused by the breakup of Gondwana. (A after Yoder and Yang 2004; B after Raxworthy et al. 2002.)

that most lineages of Madagascan plants and animals are too young to have been isolated by the division of Gondwana, and that dispersal has played the major role in their distributions (Yoder and Nowak 2006). Madagascar's ants, frogs, snakes, several groups of birds, and all four groups of terrestrial mammals—including the lemurs, a primate group unique to Madagascar—appear to have evolved from ancestors that dispersed over the ocean from Africa (**Figure 6.15A**). Chameleons (slow-moving lizards that catch insects with their extraordinary projectile tongues) have a different, curious history (Raxworthy et al. 2002): they seem to have originated in Madagascar after the breakup of Gondwana and dispersed over water to Africa, India, and the Seychelles islands in the Indian Ocean (**Figure 6.15B**).

The composition of regional biotas

As we have just seen, the geographic history of a clade is often complex and may include both vicariance and dispersal events. Thus the taxonomic composition of the biota of any region is a consequence of diverse events, some ancient and some more recent. For example, the biota of South America has (1) taxa that are remnants of the Gondwanan biota (e.g., anteaters and other xenarthans), some of which are shared with other southern continents (e.g., cichlid fishes; *Araucaria* pines); (2) groups that originated by transoceanic dispersal from Africa during the Tertiary, after South America became isolated by continental drift (e.g., New World monkeys; guinea pigs and related rodents); (3) descendants of species that came from North and Central America after the formation of the Isthmus of Panama in the Pliocene (e.g., llamas; deer); and (4) species that have colonized South America within historical time, whether assisted by humans (e.g., many weeds) or not (e.g., the cattle egret, which apparently arrived from Africa in the 1930s; see Figure 6.7).

Phylogeography

Phylogeography is the description and analysis of the processes that govern the geographic distribution of lineages of genes, especially within species and among closely related species (Avise 2000). These processes include the dispersal of the organisms that carry the genes, so phylogeography provides insight into the past movements of species and the history by which they have attained their present distributions. Such analyses have been carried out on hundreds of species.

Early studies of phylogeography used phylogenetic analyses of polymorphic genes within species (gene trees) to infer population history. However, the genealogy of any one genetic locus may not accurately reflect the history of populations, owing to factors such as incomplete lineage sorting, changes in population size, and natural selection. For this

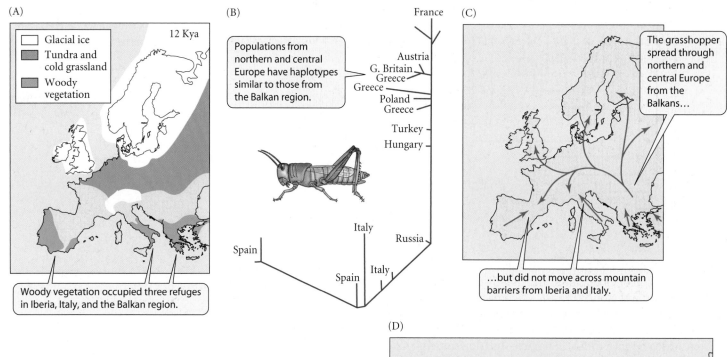

FIGURE 6.16 The recolonization of Europe from glacial refugia as inferred from patterns of genetic variation. (A) Europe during the last glacial maximum (about 12 Kya). (B) Genetic relationships among current populations of the grasshopper *Chorthippus parallelus*. The length of a line segment reflects both the number of mutational differences between haplotypes and the difference in proportions of those haplotypes among populations from the areas indicated. (C) Arrows show the inferred spread of *Chorthippus parallelus* after the glacier melted back. (D) A similar inference of postglacial spread for the hedgehog *Erinaceus europaeus*. Unlike the grasshopper, the hedgehog was able to cross the Alps and the Pyrenees. (A after Taberlet et al. 1998; B after Cooper et al. 1995; C and D after Hewitt 2004.)

reason, modern phylogeography relies on integrating information from multiple loci and comparing it with the results of models based in population genetics (Knowles 2009; Hickerson et al. 2010; see Chapters 9–12 for descriptions of population genetic models).

Pleistocene population shifts

Many authors have used the phylogeographic approach to infer the history of populations during the Pleistocene, when (as we know from paleontological evidence) many northern species occurred far to the south of their present distributions during glacial episodes and moved northward after the glaciers receded (see Chapter 5). Moreover, we know that different species occupied different glacial refugia and had different paths of movement. Many species have left no fossil traces of their paths, but phylogeographic analysis can help to reconstruct them (Taberlet et al. 1998; Hewitt 2000). For example, fossil pollen shows that refugia for woody vegetation in Europe during the most recent glacial period were located in Iberia (Spain and Portugal), Italy, and the Balkans (**Figure 6.16A**), and that the vegetation expanded most rapidly from the Balkans as the glacier retreated. The grasshopper *Chorthippus parallelus*, sampled from throughout Europe, has unique mitochondrial DNA haplotypes in Iberia and in Italy, whereas the haplotypes found in central and northern Europe are most closely related to those in the Balkans (**Figure 6.16B**). Thus we can conclude that this herbivorous insect expanded its range chiefly from the Balkans and did not cross the Pyrenees from Iberia, nor the Alps from Italy (**Figure 6.16C**). A similar analysis of hedgehogs (*Erinaceus europaeus* and *E. concolor*) indicated, in contrast, that these insectivorous mammals colonized northern Europe from all three refugia (**Figure 6.16D**). Often, many species with overlapping distributions prove to have similar phylogeographic patterns, suggesting that geographic patterns of biodiversity have been shaped partly by Earth history (**Figure 6.17**).

Modern human origins

Phylogeography has also been applied to our own distribution. We saw in Chapter 4 that *Homo erectus* was broadly distributed throughout Africa and Asia by about 1 Mya and had evolved into more *sapiens*-like forms by about 300,000 years ago (300 Kya). Until recently, the term "archaic *sapiens*" was applied to these later forms, both those from Africa and those from Europe and Asia. This term includes the populations known as Neanderthals (now recognized as a distinct species, *Homo neanderthalensis*), which persisted until about 29 Kya. The DNA from recently discovered fragmentary remains of the Denisovan population in Siberia, from about 50 Kya, suggests that it was a separate species from *neanderthalensis*. Late Pleistocene fossils of a small-bodied hominin named *Homo floresiensis*, recently found on the island of Flores in Indonesia, are probably also descended from *H. erectus*. Fossils of "anatomically modern humans," or "modern *sapiens*," date from 195 to 170 Kya in Africa, but are not found outside of Africa for another 80,000 years.

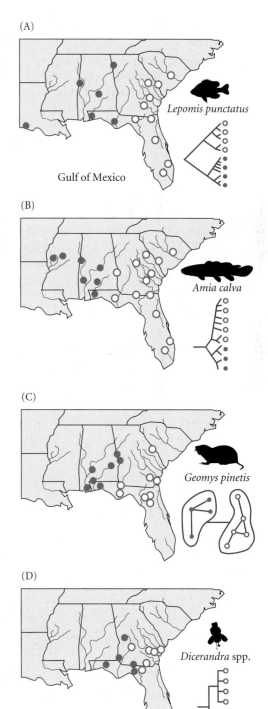

FIGURE 6.17   Patterns of genetic divergence among populations may show that many species have experienced a similar history of subdivision. Gene trees show sharp divisions between eastern and western populations of both freshwater and terrestrial species in the coastal plain of the southeastern United States. These populations may have been isolated and diverged in two refugia during the Pleistocene. (A) Spotted sunfish (*Lepomis punctatus*). (B) Bowfin (*Amia calva*). (C) Pocket gopher (*Geomys pinetis*). (D) A mint, the coastal plain balm *Dicerandra*. (After Soltis et al. 2006).

FIGURE 6.18 Two hypotheses on the origin of modern humans. (A) The multiregional hypothesis posits a single wave of expansion by *Homo erectus* from Africa to parts of Asia and Europe. These various populations (including the populations called Neanderthals) all evolved into modern *H. sapiens*. Dashed lines indicate gene flow among regional populations. (B) The replacement hypothesis proposes that populations of *H. erectus*, derived from African ancestors, gave rise to archaic *sapiens* (including Neanderthals), but that Asian and European populations of archaic *sapiens* became extinct after modern *sapiens* expanded out of Africa in a second wave of colonization.

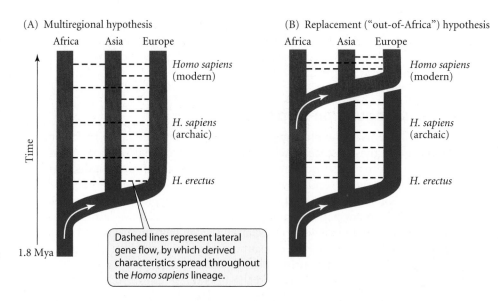

Two major hypotheses for the relationship between modern *Homo sapiens* and "archaic *sapiens*" have been proposed (Satta and Takahata 2002; Relethford 2008). Advocates of the MULTIREGIONAL HYPOTHESIS held that "archaic *sapiens*" populations in Africa and Eurasia all evolved into modern *sapiens*, with gene flow spreading modern traits among the various populations (**Figure 6.18A**). According to this hypothesis, some genetic differences found among today's human populations in Europe and Asia should trace back to genetic differences that developed among populations of *erectus* and "archaic *sapiens*" nearly a million years ago.

In contrast, the REPLACEMENT HYPOTHESIS, or OUT-OF-AFRICA HYPOTHESIS, holds that after "archaic *sapiens*" spread from Africa to Asia and Europe, modern *sapiens* evolved from archaic *sapiens* in Africa, spread throughout the world in a second expansion, and replaced the populations of archaic *sapiens* without interbreeding with them to any substantial extent (**Figure 6.18B**). That is, the modern *sapiens* that evolved from archaic *sapiens* in Africa was reproductively isolated from the Neanderthals and other Eurasian populations of archaic *sapiens*—it was a distinct biological species. According to this hypothesis, the world's populations of archaic *sapiens* became almost entirely extinct as a result of competition or climate change (Finlayson 2005), so most genes in today's populations should be descended from those carried by the modern *sapiens* population that spread from Africa. Because closely related species often do interbreed to some slight extent (see Chapter 17), however, it is possible that modern humans acquired some Neanderthal genes, even if the two forms did not interbreed extensively.

Early genetic studies (Cann et al. 1987; Vigilant et al. 1991), which used mitochondrial DNA (mtDNA), supported (and indeed inspired) the replacement hypothesis. Later, more extensive studies of mtDNA and nuclear genes have generally supported this hypothesis. The gene tree of mtDNA, for example, shows several basal clades of African haplotypes and a derived clade that includes not only several African haplotypes but also all the non-African populations from throughout the world. Since these early studies, a huge amount of data on DNA sequence variation among humans has been collected, and sequences of Neanderthal and Denisovan DNA have further revealed the history of *Homo sapiens*, which proves to be more complex than anyone had envisioned (Stoneking and Krause 2011).

The basic pattern revealed by mtDNA (**Figure 6.19**) has been strongly supported: *humans in Europe, Asia, and the Americas are descended from populations that emerged from Africa*, probably via southwestern Asia (the Middle East), *50,000 to 100,000 years ago*. The amount of genetic variation within populations of hunter-gatherers south of the Saharan Desert, especially in southern Africa, is much greater than in non-African populations, as we expect if colonizing populations carried only a sample of genes from the ancestral

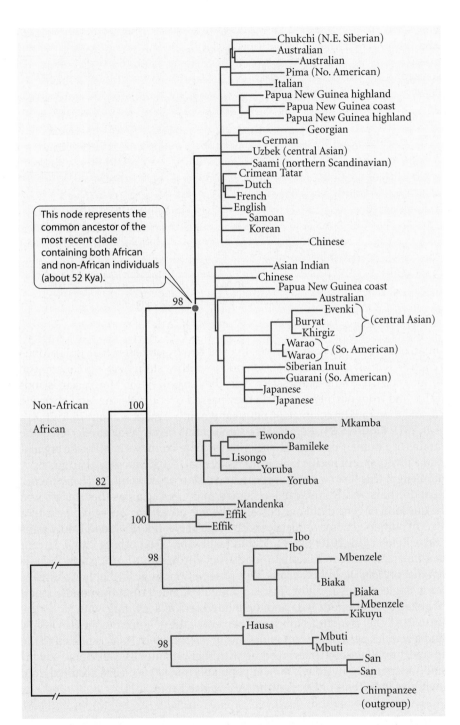

FIGURE 6.19 A gene tree based on complete sequences of mitochondrial genomes from human populations throughout the world. Haplotypes from individuals in Africa (green background) are phylogenetically basal, as expected if the human species originated in Africa, and show high sequence diversity (represented by the lengths of the branches). Haplotypes taken from individuals in the rest of the world (yellow background) form a single clade of very similar haplotypes (denoted by short branches), as expected if these populations had been recently derived from a small ancestral population. Some populations (e.g., Australian) are represented by more than one individual. Numerals represent bootstrap values (see Box 2B). (After Ingman et al. 2000.)

population. The genetic divergence among some African populations, especially of hunter-gatherers, suggests that they have been diverging for up to 115,000 years. In contrast, the genetic differences among non-African populations are very slight, on the whole.

Genetic similarities and differences among human populations have also been used to trace the dispersal of humans from Africa throughout the rest of the world. The amount of DNA sequence variation within human populations is lower the farther they are from Africa—just as we expect from a history of successive colonizations (**Figure 6.20**). The earliest movement from southwestern Asia to the east occurred perhaps 62 to 75 Kya, when humans colonized Australia: an inference from genetics that is consistent with archaeological

FIGURE 6.20 Genetic diversity among individuals within each of many human populations decreases the farther the populations are from eastern Africa (represented by the capital of Ethiopia). More and more distant regions were colonized successively, each by a relatively small number of people, who carried only a sample of the genes in the population from which they had come. The measure of genetic diversity used here is the average heterozygosity of single nucleotide positions throughout the genome (SNPs; see Chapter 9). (After Li et al. 2008.)

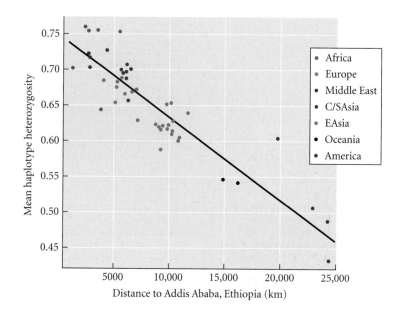

evidence of human occupation of Australia more than 50 Kya (Rasmussen et al. 2011). Much later, perhaps 25 to 38 Kya, in a second expansion out of southwestern Asia, humans spread throughout Europe and Asia. From southeastern Asia, they colonized the Melanesian islands north of Australia, and from northeastern Asia they passed over the Bering land bridge to North America, and then rapidly colonized Central and South America, about 16 to 20 Kya (Rasmussen et al. 2011; **Figure 6.21**). Remarkably, some of this history is also reflected in the phylogeny of strains of a human-pathogenic bacterium sampled from various Pacific populations, and also in a phylogeny of Pacific languages (Gray et al. 2009; Moodley et al. 2009).

Throughout much of this history, the European and Asian descendants of *Homo erectus*, such as the Neanderthals and Denisovans, occupied many of the areas that *sapiens* was colonizing. The question of gene exchange between them is now being answered by extensive comparisons of human genome sequences with those of Neanderthals and Denisovans (Green et al. 2010; Reich et al. 2010; Skoglund and Jakobsson 2011). About 2 to 4 percent of the genome of all non-African humans today, from throughout Europe and Asia, is descended from that of *Homo neanderthalensis*. This observation suggests that non-Africans originated from a single dispersal out of Africa and are descended from the population of *sapiens* that inhabited southwestern Asia and there interbred, at a low rate, with Neanderthals (see Figure 6.21). What is more, as much as 5 percent of the genome in southern Pacific populations, and a smaller percentage in populations in southeastern Asia, is related to the Denisovan genome, implying some interbreeding with that hominin as well. Some advantageous immune system genes in today's human populations may have been acquired from interbreeding with Neanderthals or Denisovans (Abi-Rached et al. 2011).

Geographic Range Limits: Ecology and Evolution

The geographic distribution of a species results not only from the history of its ancestors, but also from current factors, a major subject of ecological biogeography. The border of a species' geographic range is sometimes set by utterly unfavorable conditions, as when the distribution of a terrestrial organism stops at the ocean's edge. But most species have a restricted range within a continent or ocean, and the reasons for these limits are less evident.

Ecological niches

A species can persist only where environmental conditions permit a population growth rate greater than or equal to zero, and this can be the case only if the organism can tolerate each

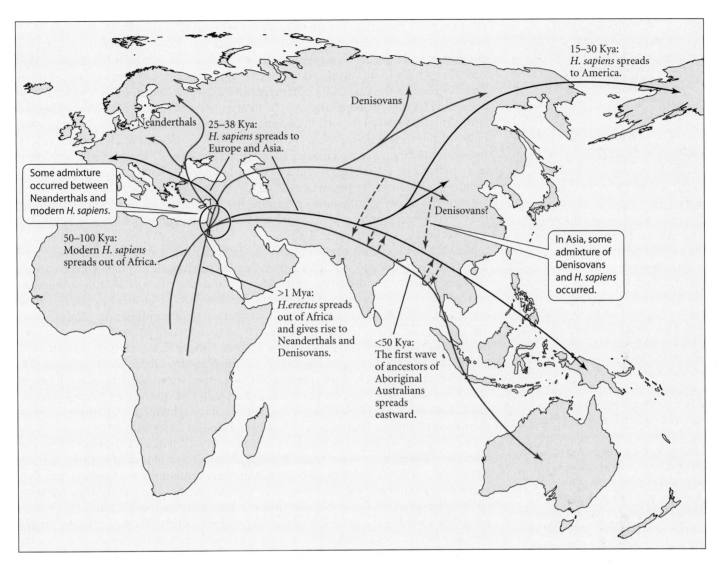

FIGURE 6.21 Spread of hominins from Africa to other parts of the world. More than 1 Mya, *Homo erectus* spread through much of Asia and Europe, giving rise to Neanderthals and Denisovans (blue arrows). Between 50 and 100 Kya, African populations that had evolved into modern *H. sapiens* emerged from Africa (red path). Although they were mostly reproductively isolated from Neanderthals, there was some interbreeding between the two species. Two major migrations of *sapiens* then occurred. The first (green arrow), more than 50 Kya, established the ancestors of Australian Aborigines. In the second migration, 25–38 Kya (black arrows), humans colonized Europe and Asia. From eastern Asia, they spread southeastward to New Guinea and Oceania. Dashed blue arrows indicate that Denisovan-like populations contributed some genes to the first wave of migrants and perhaps to the second wave of migrants as well, and that populations stemming from both migrations exchanged some genes. After about 30 Kya, populations in northeastern Asia expanded over the Bering land bridge into northwestern America. (After Stoneking and Krause 2011 and Rasmussen et al. 2011.)

of several environmental conditions, such as the range of temperatures, the amount of available water, and the availability of suitable food items. That is, both abiotic and biotic aspects of the environment can affect the species' distribution. The highly influential ecologist G. Evelyn Hutchinson (1957) defined the **fundamental ecological niche** of a population as the set of all those environmental conditions in which a species can maintain a stable population size (**Figure 6.22**). A particular locality falls within the species' fundamental niche if all the relevant environmental factors fall within the organism's tolerance limits, but will fall outside the niche if any one variable, such as coldest winter temperature, falls outside these limits. Thus the *potential* niche space is dictated by the overlap of the fundamental niche and the realized (actual) environmental "space," or set of values. Even within the potential niche space, the range of environmental conditions that a species actually occupies—the **realized ecological niche**—may be further restricted if the species is excluded by biotic factors such as competitors or predators.

The physiological tolerance of a species for some factors, such as temperature, can be determined by experiment, but it is

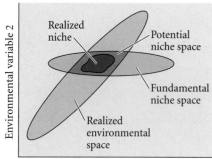

FIGURE 6.22 A conceptualization of ecological niches, showing the combinations of two environmental variables that would permit a particular species population to persist and grow. The fundamental niche space is the set of possible environments in which persistence is possible. The set of environmental combinations that exists in a given region is the realized environmental space. The overlap of the realized environmental space and the fundamental niche space is the potential niche space—the actual set of habitable conditions. However, the species might be excluded from some of these potentially habitable places (perhaps by competing species), so that it is limited to a smaller set of environmental conditions, the realized niche. In reality, many environmental variables (each of which would be represented by another dimension, which we cannot picture) determine population persistence. (After Jackson and Overpeck 2000.)

nevertheless very difficult to identify all the environmental factors that might affect population growth. Furthermore, even if we find a correlation between the geographic range limit and a particular measured variable, such as maximal July temperature, that variable is likely to be correlated with other variables (such as temperature at other times of year, or the variability of temperature, or water deficit, or the availability of a required food item). Nevertheless, the distributions of many species are correlated with climate variables (especially aspects of temperature and rainfall). Some species have shifted their geographic or altitudinal ranges in recent decades, apparently in response to human-caused climate change; for instance, both the northern (or higher) and southern (or lower) range limits of several butterflies have shifted to higher latitudes (or altitudes) (Parmesan et al. 2005). In at least one species, the northern limit is set by the insect's intolerance of extreme low temperatures; in another species, by the effect of temperature on the synchrony between the butterfly's life history and the seasonal development of its food plant.

Related species often have similar ecological requirements, presumably derived from their common ancestor. This pattern, called **phylogenetic niche conservatism**, has been described with respect to both abiotic and biotic factors. A. Towne Peterson and colleagues (1999) predicted the geographic distributions of species of birds, mammals, and butterflies on each side of the Isthmus of Tehuantepec, in Mexico, from the climate characteristics of sites occupied by their sister species (closest relatives) on the other side of the isthmus (**Figure 6.23A**). Almost all the predictions that were based on adequate samples were successful, implying that these animals' ecological requirements had not changed substantially since speciation. Similarly, many lineages of herbivorous insects have remained associated with the same genus or family of food plant (**Figure 6.23B**); some of these associations have remained unchanged for more than 40 million years (Mitter and Farrell 1991; Winkler and Mitter 2008).

Niche conservatism has important consequences (Wiens and Graham 2005). For instance, it contributes to our understanding of the geographic distributions of many clades: oaks (*Quercus*) and dogwoods (*Cornus*) are among the many plant taxa that occur in the temperate regions of eastern North America, Asia, and Europe but have not adapted to warm tropical environments (Donoghue and Smith 2004). Niche conservatism underlies the observation that many species shifted their geographic ranges during the Pleistocene, rather than adapting in situ to changes in climate (see Chapter 5). Clearly, these observations raise important questions about the ability of species to adapt to new environmental conditions.

Range limits: An evolutionary problem

Whether the border of a species' range is set by a climate variable, another species, or an interaction of factors, it poses one of the most puzzling questions in evolutionary biology (Holt 2003): Why doesn't the species adapt to the slightly different conditions just beyond its present range? And if it does succeed in this first small step, what would prevent it from further small adaptive steps that would slowly expand its range indefinitely?

Two principal explanations have been suggested (Bridle and Vines 2007). First, populations may simply lack genetic variation in one or more characteristics necessary for adaptation. For example, populations of two species of rain forest–dwelling *Drosophila* have no detectable genetic variation for desiccation tolerance, which might prevent them from expanding into drier habitats (Kellermann et al. 2006). Second, incursion of genes from populations in favorable environments could prevent recipient populations from adapting to an unfavorable environment at the range margin, since this process of gene flow would counteract natural selection for local adaptation. Several mathematical models have shown that such gene flow could be a powerful limitation on adaptation (Holt and Gaines 1992; Kirkpatrick and Barton 1997a; Lenormand 2002). Chapter 12 presents examples of the conflict between gene flow and natural selection, but this hypothesis on range limits needs more research.

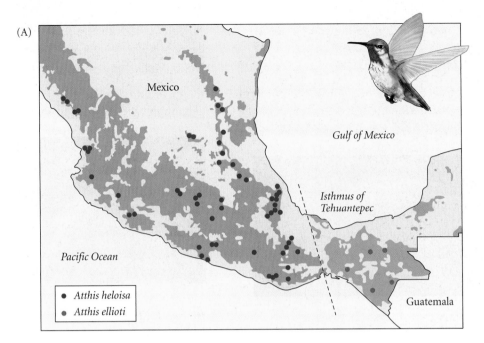

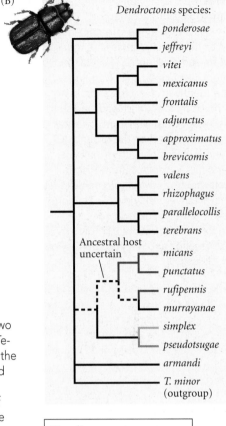

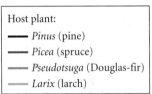

FIGURE 6.23 Two illustrations of phylogenetic niche conservatism. (A) The distribution of two sister species of hummingbirds, *Atthis heloisa* and *A. ellioti*, on either side of the Isthmus of Tehuantepec in Mexico. The climate at localities occupied by each species was used to predict the potential distribution of its sister species, on the assumption that their niches had not evolved since their common ancestor. Green areas west of the dashed line represent the geographic distribution of *A. heloisa*, predicted from that of the eastern species, and brown areas east of the line represent the distribution of *A. ellioti*, predicted from that of the western species. The actual locations where each species was found, indicated by the colored circles, all fall within the predicted range areas. Hence these species have retained similar environmental tolerances or requirements. (B) Many species of herbivorous insects feed only on one or a few kinds of host plants. These associations are often phylogenetically conservative, as shown by bark beetles (*Dendroctonus*), in which lineages of related species tend to feed on the same genus of plant. The colors of the branches of this phylogenetic tree indicate the host plant genus of bark beetles today and the most parsimonious estimate of their ancestors' host associations. (A after Peterson et al. 1999; B after Kelley and Farrell 1998.)

Evolution of Geographic Patterns of Diversity

The field of COMMUNITY ECOLOGY is concerned with explaining the species diversity, species composition, and trophic structure of assemblages of coexisting species (often called COMMUNITIES). These problems are partly biogeographic ones, since the geographic ranges of species determine whether or not they might coexist. Thus historical biogeography bears on community ecology. Until recently, however, many ecologists have attempted to use only ecological theory, based on interactions among species, to predict some aspects of community structure. For example, they propose that competition should tend to prevent the coexistence of species that are too similar in their use of resources. This theory assumes that diversity and other features of communities have reached an equilibrium—which in turn assumes that evolution has provided a supply of potential community members with appropriate characteristics, or that such characteristics will rapidly evolve to suit the situation.

Community convergence

If community structure were predictable from ecological theories, we might expect that if two regions presented a similar array of habitats and resources, species would evolve to use and partition them similarly. In other words, just as individual species often evolve

FIGURE 6.24 Convergent morphologies, or "ecomorphs," of *Anolis* lizards in the Greater Antilles, West Indies. (A) *Anolis lineatopus*, from Jamaica, and (B) *A. strahmi*, from Hispaniola, have both independently evolved the stout head and body, long hind legs, and short tail associated with living on lower tree trunks and on the ground. (C) *Anolis valencienni*, from Jamaica, and (D) *A. insolitus*, from Hispaniola, are both twig-living anoles that have convergently evolved a more slender head and body, shorter legs, and a longer tail. (Photos by K. de Queiroz and R. Glor, courtesy of J. Losos.)

convergent adaptations (see Figures 3.6 and 6.1), sets of species might evolve convergent community structure.

A striking example of community-level convergence has been described in the anoles (*Anolis*) of the West Indies (Williams 1972; Losos 2009). The anoles are a species-rich group of insectivorous, mostly arboreal Neotropical lizards. Different species are known to compete for food, and this competition has influenced the structure of anole communities. Each of the small islands in the Lesser Antilles has either a single (solitary) species or two species. Solitary species are generally moderate in size, whereas larger islands have a small and a large species that can coexist because they take insect prey of different sizes and also occupy different microhabitats. The small species of the various islands are a monophyletic group, and so are the large species. Thus it appears that each island has a pair of species assembled from the small-sized and the large-sized clades.

The large islands of the Greater Antilles (Cuba, Hispaniola, Jamaica, Puerto Rico) harbor greater numbers of anole species. These anoles occupy certain microhabitats, such as tree crown, twig, and trunk, which are filled by different species on each island. The occupants of particular microhabitats, called ECOMORPHS, have consistent, adaptive morphologies (**Figure 6.24**). The species on each of the islands form a monophyletic group that has radiated into species that ecologically and morphologically parallel those on the other islands (**Figure 6.25**). The most reasonable interpretation of this pattern is that as new

FIGURE 6.25 A molecular phylogeny of species of *Anolis* in the Greater Antilles. The letters at the top indicate the island on which each species occurs (C, Cuba; H, Hispaniola; J, Jamaica; P, Puerto Rico). The branch colors indicate the ecomorph classes that occupy different microhabitats. Note that four ecomorphs have evolved independently on all four islands. The islands nevertheless differ in their numbers of anole species, and several ecomorphs are restricted to certain islands. (After Losos et al. 1998.)

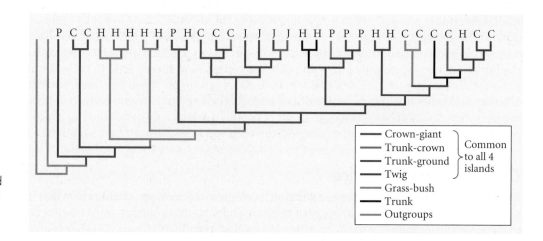

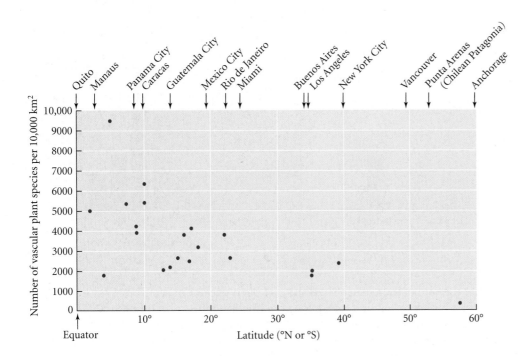

FIGURE 6.26 An example of the latitudinal gradient in species richness. The number of species of vascular plants in various regions of North and South America drops more than tenfold between equatorial and high-latitude regions. The latitudinal locations of some cities are indicated for easy reference. (After Huston 1994.)

species have arisen on each island, they have evolved in similar ways to avoid competition by adapting to the same kinds of previously unused microhabitats.

The degree of convergence shown by the anoles is unusual: ecological niches that are occupied in one region often seem unoccupied in other, climatically similar regions. Blood-feeding (vampire) bats occur in the New World tropics, but not in Africa despite abundant ungulate prey; sea snakes occur in the Indian and Pacific Oceans, but are absent from the Atlantic. Community-level convergence has evolved in some organisms, but not others (Schluter and Ricklefs 1993; Losos 2010; see Chapter 22).

Effects of history on patterns of diversity

There is enormous variation among geographic regions and among environments in their number of species of plants and animals. One of the most dramatic examples is a pattern called the LATITUDINAL DIVERSITY GRADIENT: the numbers of species (and of higher taxa such as genera and families) decline with increasing latitude, both on land and in the ocean (**Figure 6.26**). Most taxa of terrestrial animals and plants are far more diverse in tropical regions, especially in lowlands with abundant rainfall, than in extratropical regions.

Three major hypotheses have been proposed to account for this pattern (**Figure 6.27**). First, ecological factors might enable more tropical species to coexist in a stable community (Figure 6.27A). These factors might include high productivity because of abundant solar energy or fine partitioning of food resources among many species (Willig et al. 2003). Alternatively, the pattern might be explained by evolutionary dynamics over many millions of years (Ricklefs 2004; Wiens and Donoghue 2004). The "diversification rate hypothesis"

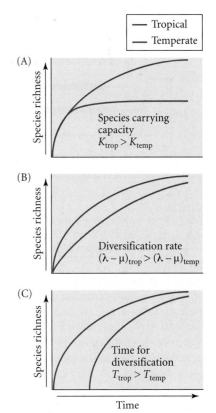

FIGURE 6.27 Three models of species accumulation that have been proposed to account for the latitudinal diversity gradient. (A) Some ecological hypotheses propose that tropical locations can support a higher equilibrium number of species ("carrying capacity," K) than temperate localities. (B) The diversification rate (difference between the speciation rate, λ, and the extinction rate, μ) might be higher in the tropics. Species numbers have not necessarily reached an equilibrium carrying capacity. (C) Lineages diversify at the same rate, but started to diversify more recently in the temperate zone than in the tropics, perhaps because they originated in tropical environments but only recently adapted to the temperate zone. (After Mittelbach et al. 2007.)

FIGURE 6.28 Origin and diversification of the genera of marine bivalves that first occurred in the Miocene. (A) Numbers of genera that originated in tropical regions and outside the tropics. (B) The present-day latitudinal range limits of species in these genera. Many have expanded beyond the tropics (right of the dashed vertical line), while some species in the same genera remain in the tropics. The same pattern has been found for bivalve genera that are first recorded after the Miocene. The tropics seem to be both a "cradle" of new lineages and a "museum" in which lineages persist. (After Jablonski et al. 2006.)

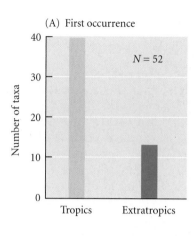

(A) First occurrence

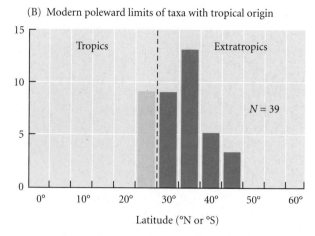

(B) Modern poleward limits of taxa with tropical origin

holds that the rate of increase in diversity has been greater in the tropics for a long time because of a higher speciation rate, a lower extinction rate, or both (Figure 6.27B). David Jablonski and colleagues (2006) determined that new genera of marine bivalves have arisen mostly in tropical areas throughout the last 11 My and have spread from there toward higher latitudes while persisting in tropical regions as well (**Figure 6.28**).

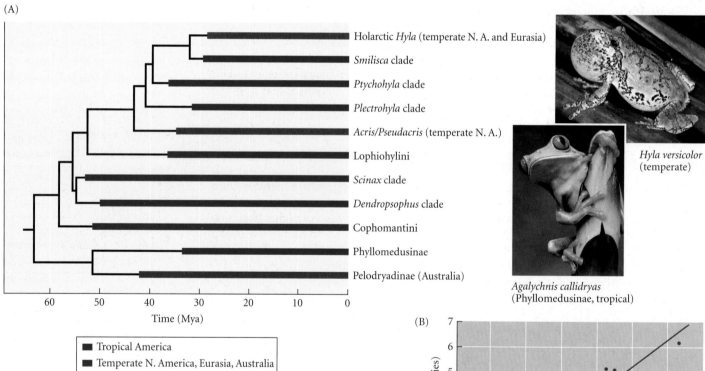

(A)

Holarctic *Hyla* (temperate N. A. and Eurasia)
Smilisca clade
Ptychohyla clade
Plectrohyla clade
Acris/Pseudacris (temperate N. A.)
Lophiohylini
Scinax clade
Dendropsophus clade
Cophomantini
Phyllomedusinae
Pelodryadinae (Australia)

Time (Mya)

■ Tropical America
■ Temperate N. America, Eurasia, Australia

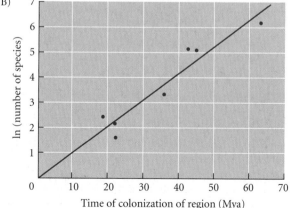

Hyla versicolor (temperate)

Agalychnis callidryas (Phyllomedusinae, tropical)

(B)

FIGURE 6.29 Tree frogs (Hylidae) are much less diverse in the temperate zone than in the tropics because only a few lineages have adapted to the temperate zone, and those only recently. (A) A molecular phylogeny of major lineages of Hylidae shows that this family has radiated in the American tropics and rather recently gave rise to the three lineages (in red) that invaded the temperate zone. (B) The number of tree frog species in a continental region (note log scale) is strongly correlated with the time since the Hylidae first started diversifying in that region, as estimated by applying a molecular clock to the branch points in the phylogeny. (After Wiens et al. 2006.)

Finally, the "time and area hypothesis" (Figure 6.27C) holds that most lineages have been accumulating species for a longer time in tropical than in extratropical environments. During the Cretaceous and the first 40 My of the Cenozoic, Earth was warmer than it is today: more of Earth had a tropical climate then than now. For that reason, most lineages originated in tropical climates, and the relatively few lineages that have evolved adaptations to the stressful temperatures and seasonal fluctuations in food supply that are typical of the temperate zone are younger lineages that have not had time to become as diverse. Thus this hypothesis is based on phylogenetic niche conservatism (Wiens and Donoghue 2004).

John Wiens and colleagues (2006) took an explicitly phylogenetic approach to this question in a massive study of tree frogs (Hylidae). Their molecular phylogeny indicates that all the major lineages of tree frogs and their common ancestors were distributed in tropical America (**Figure 6.29A**). The temperate zone has been invaded by only three lineages. Moreover, the number of species in each region is positively correlated with the time since tree frog clades first inhabited the region, estimated on the basis of clade ages as determined with a molecular clock (**Figure 6.29B**). It appears that the diversification rate has been much the same everywhere, and that tropical regions have simply had more time to accumulate species. In agreement with this hypothesis, Paul Fine and Richard Ree (2006) determined that the number of tree species in tropical, temperate, and boreal ecosystems on each continent is correlated with an index that integrates the area that each of these ecosystems has occupied since the Miocene, Oligocene, or even as far back as the Eocene. A similar model accounts for much of the regional variation in the species richness of vertebrates across the world (Jetz and Fine 2012). Thus tropical environments and vegetation have occupied larger areas for a longer time than other environments, so that is where most genera and species arose.

Go to the
EVOLUTION
Companion Website at
sites.sinauer.com/evolution3e
for quizzes, data analysis and simulation exercises, and other study aids.

Summary

1. Biogeography, the study of organisms' geographic distributions, has both historical and ecological components. Certain distributions are the consequence of long-term evolutionary history; others are the result of current ecological factors.

2. The geographic distributions of organisms provided Darwin and Wallace with some of their strongest evidence for the reality of evolution.

3. The historical processes that affect the distribution of a taxon are extinction, dispersal, and vicariance (fragmentation of a continuous distribution by the emergence of a barrier). These processes may be affected or accompanied by environmental change, adaptation, and speciation.

4. Histories of dispersal or vicariance can often be inferred from phylogenetic data. When a pattern of phylogenetic relationships among species in different areas is repeated for many taxa, a common history of vicariance is likely.

5. Disjunct distributions are attributable in some instances to vicariance and in others to dispersal.

6. Genetic patterns of geographic variation within species can provide information on historical changes in a species' distribution. Studies of this kind are illuminating the origin and spread of human populations.

7. The local distribution of species is affected by ecological factors, including both abiotic aspects of the environment and biotic features such as competitors and predators. Why species do not enlarge their ranges indefinitely, by incrementally adapting to conditions farther and farther away, is a major question in evolutionary biology.

8. In some cases, sets of species in different locations have independently evolved to partition resources in similar ways, suggesting that competition may limit species diversity and may result in different communities with a similar structure. However, convergence of community structure is usually incomplete, suggesting that evolutionary history has had an important effect on ecological assemblages.

9. Geographic patterns in the number and diversity of species may stem partly from current ecological factors, but long-term evolutionary history also may explain them.

Terms and Concepts

biogeographic realm
biogeography
disjunct distribution
dispersal

ecological biogeography
ecological niche (fundamental, potential, realized)

endemic
historical biogeography
phylogenetic niche conservatism

phylogeography
vicariance

Suggestions for Further Reading

M. V. Lomolino, B. R. Riddle, R. J. Whittaker, and J. H. Brown, *Biogeography*, fourth edition (Sinauer Associates, Sunderland, MA, 2010), is the leading textbook of biogeography.

Phylogeography is treated in depth by J. C. Avise in *Phylogeography* (Harvard University Press, Cambridge, MA, 2000), and human phylogeography is included in J. Klein and N. Takahata, *Where Do We Come From? The Molecular Evidence for Human Descent* (Springer-Verlag, New York, 2002).

Problems and Discussion Topics

1. Until recently, the plant family Dipterocarpaceae was thought to be restricted to tropical Asia, where many species are ecologically dominant trees. However, a new species of tree in this family was discovered in the rain forest of Colombia, in northern South America. What hypotheses can account for its presence in South America, and how could you test those hypotheses?

2. The ratites are a very old clade of flightless birds that include the ostrich in Africa, rheas in South America, the emu and cassowaries in Australia, and kiwis and the recently extinguished moas in New Zealand. South American tinamous, which are capable of flight, are closely related to the ratites. The "Gondwanan distribution" of these birds has often been attributed to vicariance, but there is considerable uncertainty about their phylogeny and distributional history. Read several phylogenetic studies of the ratites and discuss how best to explain their distribution: A. Cooper et al., *Nature* 409: 704–707, 2001; O. Haddrath and A. J. Baker, *Proc. Royal Soc. Lond. B* 268: 939–945, 2001; and S. J. Hackett et al., *Science* 320: 1763–1768, 2008.

3. In later chapters, we will see that many characteristics of most species have the genetic variation that is required for those characteristics to evolve, and that many examples of rapid adaptation to human-altered environments have been documented. Discuss whether or not this observation is inconsistent with the claim that many organisms display phylogenetic niche conservatism.

4. Some biogeographers, subscribing to the "cladistic vicariance" school of thought (Humphries and Parenti 1986), hold that vicariance should always be the preferred hypothesis and that dispersal should be invoked only when necessary, because the vicariance hypothesis can be falsified (if it is false), whereas dispersal can account for any pattern and therefore is not falsifiable. What are the pros and cons of this position? (See Endler 1983.)

5. In some cases, it can be shown that species are physiologically incapable of surviving temperatures that prevail beyond the borders of their range. Do such observations prove that cold regions have low species diversity because of their harsh physical conditions?

6. By far the most effective way of saving endangered species is to preserve large areas that include their habitat. For social, political, and economic reasons, the number and distribution of areas that can be allocated as preserves are highly limited. It might be easier to save more species if areas of endemism were correlated among different taxa, such as plants, birds, and mammals. Are they correlated? (See, for example, N. Myers et al., *Nature* 403: 853–858, 2000; J. R. Prendergast et al., *Nature* 365: 335–337, 1993; A. P. Dobson et al., *Science* 275: 550–553, 1993.)

7. Would you expect large numbers of species in a region to have had similar histories of geographic distribution? Why or why not? How could you use phylogeographic analyses, as illustrated in Figure 6.17, to address this question?

8. How might you test the hypothesis that adaptation to conditions beyond the edge of a species' range is prevented by migration and gene flow from populations well within the range?

9. Global climate change, owing largely to human production of greenhouse gases, is occurring faster than at almost any previous time in Earth's history. How do you think the geographic and altitudinal ranges of species will be affected? Will the effects differ among taxa or among geographic regions? Is it likely that many species will adapt to this climate change and maintain their current distribution? What will determine whether or not they will adapt?

The Evolution of Biodiversity

Biological diversity, or biodiversity, poses some of the most interesting questions in biology. Why are there more species of rodents than of primates, of flowering plants than of ferns? Why has the diversity of species changed over evolutionary time? Does diversity increase steadily, or has it reached a limit? Because so many factors can influence the diversity of species, these questions are both intriguing and challenging.

Biodiversity can be studied from the complementary perspectives of ecology and evolutionary history. Ecologists focus primarily on factors that operate over short time scales to influence diversity within local habitats or regions. But factors that operate on longer time scales, such as climate change and evolution, also affect diversity. On a scale of millions of years, extinction, adaptation, speciation, climate change, and geological change create the potential for entirely new assemblages of species.

In this chapter, we will examine long-term patterns of change in diversity, caused by originations and extinctions of taxa on a scale of millions of years. Ecological and evolutionary studies of living species help us to interpret these paleontological patterns. Conversely, understanding factors that have altered biodiversity in the past may help us predict how diversity will be affected by current and future environmental changes, such as the global climate change that is now under way as a result of human use of fossil fuels.

Modern biodiversity
A diverse collection of weevils, scarabs, long-horned and metallic wood-borers, and other beetles. Beetles are by far the largest order of insects, with more than 350,000 described species, and untold numbers yet to be described or discovered. What accounts for their amazing diversity?

Estimating and Modeling Changes in Biological Diversity

In most evolutionary studies, "diversity" refers to the number of taxa, such as genera or species. (The latter is often called SPECIES RICHNESS.) Over long time scales or large areas, diversity is often estimated by compiling records, such as the publications or museum specimens that have been accumulated by many investigators, into faunal or floral lists of species.

Diversity, like almost everything studied by scientists, is estimated from samples. In any higher taxon of organisms, the number of known species is only a sample of those that actually exist—a fairly complete sample for a few groups, such as birds, but a very incomplete sample for most groups. (The number of known species of insects, for example, may be no more than 10 percent of living species.) Thus we may *estimate* diversity, but our picture will be incomplete, and that picture may be systematically biased (misleading) unless the effects of incomplete sampling are considered. Estimates of past diversity are especially incomplete.

Much of this chapter will be concerned with changes in diversity over time, as studied in two ways: by paleontology and by phylogenetic analysis of living species. Both approaches are based on a simple model of change in diversity over time.

Modeling rates of change in diversity

The number of taxa (N) changes over time by origination (as a result of branching of lineages) and extinction. These events are analogous to the births and deaths of individual organisms in a population, so models of population growth have been adapted to describe changes in taxonomic diversity. For each time interval Δt, suppose S new taxa originate per taxon present at the start of the interval, and suppose E is the number of taxa that become extinct, per original taxon, during the interval. Then ΔN, the change in N, equals the number of "births," SN, minus the number of "deaths," EN, and the rate of change in diversity, $\Delta N/\Delta t$, is

$$\frac{\Delta N}{\Delta t} = SN - EN \quad \text{or} \quad \frac{\Delta N}{\Delta t} = DN$$

where $D = (S - E)$ and is the PER CAPITA RATE OF DIVERSIFICATION. The growth in number of taxa is positive if $D > 1$.

Between the beginning (time t_0) and the end (time t_1) of one time interval, the "population" of taxa grows by its original size multiplied by the per capita rate of increase: $N_1 = N_0D$. If the rates S and E remain constant, then after the next interval Δt, the population will be $N_2 = N_1D = N_0D^2$. In general, after t time intervals, the number of taxa will be

$$N_t = N_0D^t$$

as long as the per capita origination and extinction rates remain the same.

This equation describes **exponential growth** of the number of taxa (or of a population of organisms) over discrete time intervals (**Figure 7.1A**). For continuous growth, rather than growth in discrete intervals, the equivalent expressions (using r in place of D) are

$$\frac{dN}{dt} = rN \quad \text{and} \quad N_t = N_0e^{rt}$$

where r is the INSTANTANEOUS PER CAPITA RATE OF DIVERSIFICATION. By taking the logarithm of both sides of this equation, we obtain

$$\log(N_t) = \log(N_0) + rt$$

which is a straight line with slope r. If data on the numbers of taxa at different times fit a straight line, we may infer that the per capita rate of diversification (r) has been constant.

Let us pursue the analogy between the number of taxa in a clade and the number of organisms in a population (in which, as we know, birth and death rates are not constant)

(A) Discontinuous exponential growth

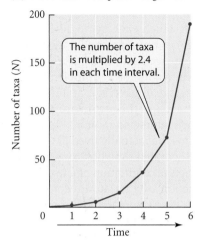

(B) Continuous exponential and logistic growth

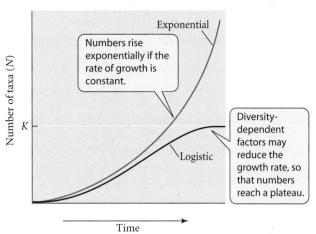

FIGURE 7.1 Models of population growth have been adapted to describe changes in taxonomic diversity. (A) Theoretical increase in the number of taxa (N), according to the equation $N_t = N_0 D^t$, where D is the rate of increase in each discrete time interval (in this example, $D = 2.4$). (B) The increase in the number of taxa when change is continuous and the rate of increase is denoted by r. The number of taxa grows exponentially if r is constant, but levels off if growth is logistic (i.e., if the rate of increase is negatively diversity-dependent).

a bit further. Factors such as severe weather may alter birth and death rates by proportions that are unrelated to the density of the population. (These are termed DENSITY-INDEPENDENT FACTORS.) In contrast, DENSITY-DEPENDENT FACTORS, which include competition for food or space, cause the per capita birth rate to decline, or the death rate to increase, with population density, so that population growth slows down and the population density may reach a stable equilibrium (**Figure 7.1B**). Paleobiologists have suggested that changes in the number of species or higher taxa may be similarly affected by **diversity-dependent factors** that might reduce origination rates or increase extinction rates as the number of species increases. For example, competition among species for resources might limit the possible number of species to some maximal number, denoted by K. Ecologists describe density-dependent population growth by the LOGISTIC EQUATION, which for growth in numbers of species may be expressed as

$$\frac{\Delta N}{\Delta t} = r_0 N \left(\frac{K - N}{K} \right)$$

where r_0 is the per capita rate of increase when the number of species is very low. Then the increase in number, $\Delta N / \Delta t$, declines as N increases and $(K - N)$ goes toward zero. At equilibrium, when $\Delta N / \Delta t = 0$, $N = K$. That is, the number of taxa will remain stable as long as diversity-dependent interactions among species depress the origination rate or increase the extinction rate and thereby reduce the diversification rate to zero.

Of course, this model is a great oversimplification of reality because changes in the environment and in organisms themselves are likely to change rates of origination and extinction, and consequently the rate of diversification, over time.

Diversity in the fossil record

Most paleontological studies of diversity employ counts of higher taxa, such as families and genera, because they generally provide a more complete fossil record than individual species do. Although paleobiologists have used several expressions for rates of origination, extinction, and diversification, the most useful are the numbers per taxon ("per capita") per time interval, as denoted by S, E, and D in the previous section (Foote 2000a).

Paleobiologists continue to study the effects of sampling on their estimates of diversity (e.g., Alroy et al. 2008), and they have developed correction factors that must often be included in data analysis (Raup 1972; Signor 1985; Foote 2000a). For example, rare species are more likely to be included in large samples, that include more individual organisms, than in small samples. If we want to compare the species diversity in two samples that differ in size, we must correct for this problem, perhaps by picking the same number of specimens at random from all the samples.

In addition, the geological or stratigraphic STAGES into which each geological period is divided vary in duration, and more recent geological times are represented by greater volumes and areas of fossil-bearing rock. Therefore it may be necessary to adjust the count of taxa by the amount of time and rock volume represented. An important problem is that records of fossil taxa are often described at the level of the stage (on average, an interval of about 5 to 6 My). Thus the first and last occurrences of a taxon in the record are accurate only to about 5 My, so the estimated duration of the taxon is imprecise.

Because fossils constitute a small sample of the organisms that actually lived at the time they were formed, a taxon is often recorded from several separated time horizons, but not from those in between. We can therefore deduce that the actual origination of a taxon may have occurred before its earliest fossil record, and its extinction after its latest record. It follows that if many taxa actually became extinct in the same time interval, the last recorded occurrences of some are likely to be earlier, so that their *apparent* times of extinction will be spread out over time. Conversely, if many taxa actually originated at the same time, some of them may appear to have originated at later times.

Since our count of living (Recent) species is much more complete than our count of past species, taxa that are still alive today appear to have longer durations and lower extinction rates than they would if they had been recorded only as fossils. That is, we can list a living taxon as present throughout the last 10 My, let us say, even if its only fossil occurrence was 10 My ago. Because the more recently a taxon arose, the more likely it is to still be extant, diversity will seem to increase as we approach the present, a bias called the **pull of the Recent**. This bias can be reduced by counting only fossil occurrences of each living taxon and not listing it for time intervals between its last fossil occurrence and the Recent.

Because of unusually favorable preservation conditions at certain times or other chance events, a taxon may be recorded from only a single geological stage, even though it lived longer than that. Such "singletons" make up a higher proportion of taxa as the completeness of sampling decreases and therefore bias the sample; moreover, they can create a spurious correlation between rates of origination and rates of extinction because they appear to originate and become extinct in the same time interval. Diversity may be more accurately estimated by ignoring such singletons and counting only those taxa that cross the border from one stage to another.

The foregoing discussion illustrates a fundamentally important aspect of every scientific discipline, including evolutionary biology: scientists discuss the ways in which their data could possibly be misinterpreted and lead to false conclusions, and they devise ways of avoiding misinterpretation.

Phylogenetic studies of diversity

Methods have been developed that can infer the rate of increase in the number of species in a clade from a molecular phylogeny of living species, if the ages of branch points in the tree can be estimated from DNA sequence divergence (Nee 2006). For example, **Figure 7.2A** is a phylogeny of the Hawaiian silverswords (see Figure 3.23); notice that most of the branch points are quite recent (Baldwin and Sanderson 1998). The number of lineages increases from 1 (the common ancestor at 15 Mya) to 28 extant species. A **lineage-through-time plot** of the log number of species against time is approximately linear until the recent past, when it curves upward (**Figure 7.2B**). The slope of the linear portion provides an estimate of r, the per capita diversification rate, which is the difference $(S - E)$ between the per capita rates of origination or speciation (S) and extinction (E). We may assume that E has been very low (let us assume zero) in the very recent past because species that arose recently have not yet had time to become extinct. Therefore the slope must be close to S in the very recent past (**Figure 7.2C**). This assumption provides an estimate of S, so we can estimate E from the older (linear) part of the plot. Thus it is possible to

(A)

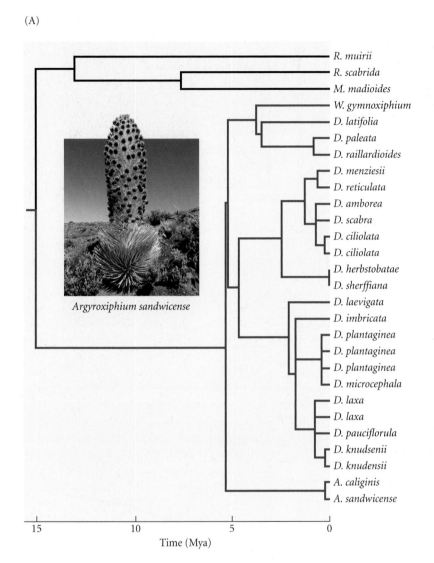

Argyroxiphium sandwicense

R. muirii
R. scabrida
M. madioides
W. gymnoxiphium
D. latifolia
D. paleata
D. raillardioides
D. menziesii
D. reticulata
D. amborea
D. scabra
D. ciliolata
D. ciliolata
D. herbstobatae
D. sherffiana
D. laevigata
D. imbricata
D. plantaginea
D. plantaginea
D. plantaginea
D. microcephala
D. laxa
D. laxa
D. pauciflorula
D. knudsenii
D. knudensii
A. caliginis
A. sandwicense

15   10   5   0
Time (Mya)

(B)

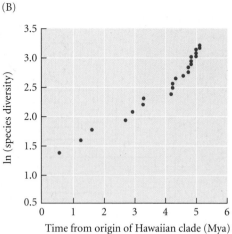

Time from origin of Hawaiian clade (Mya)

(C)

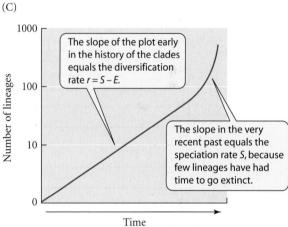

The slope of the plot early in the history of the clades equals the diversification rate $r = S - E$.

The slope in the very recent past equals the speciation rate S, because few lineages have had time to go extinct.

Time

FIGURE 7.2 Growth in the number of lineages over time, estimated from a time-calibrated phylogeny. (A) A phylogeny of Hawaiian silverswords, based on rDNA ITS sequences, with the times of branch points based on a fossil-calibrated estimate of the rate of sequence evolution. The black branches are outgroup taxa. (B) A lineage-through-time plot of the number of lineages of Hawaiian silverswords found by taking "time slices" through the phylogeny in part A. For example, there were three detectable lineages at 5 Mya. (C) A theoretical lineage-through-time plot of the expected cumulative increase in the logarithm of the number of lineages if rates of origination or speciation (S) and extinction (E) of lineages, and consequently the rate of increase r (= $S - E$), are constant. The silversword data resemble this theoretical curve. Notice that the slope is greater in the recent past because few lineages have had enough time to become extinct. (A, B after Baldwin and Sanderson 1998; C after Nee 2006.)

estimate both the speciation rate and the extinction rate from the slopes of a lineage-through-time plot *on the assumption of constant rates of speciation and extinction*. The data on Hawaiian silverswords suggest that the extinction rate was close to zero and that the per capita rate of speciation was 0.56 species per My, a much higher rate than is typical of clades on continents. If a clade continued to diversify indefinitely at a rate of 0.5 per My, it would grow from 1 to about 270,000 species in 25 My (Nee 2006).

In a phylogeny of wood warblers (**Figure 7.3A**), most of the branch points are closer to the base of the tree than in the case of the silverswords, so a lineage-through-time plot shows a decelerating rate of accumulation of species (**Figure 7.3B**). The diversification rate r has declined over time. Such a decline could be caused by a declining rate of speciation (**Figure 7.3C**), an increasing rate of extinction (**Figure 7.3D**), or both. Many clades show such a declining rate of lineage accumulation; relatively few show an increasing diversification rate (McPeek 2008). Considerable debate and research are now addressing the questions of whether or not the effects of changes in S and E can be distinguished and how to account for the effects of unknown extinction events on the inferences drawn from phylogeny (Rabosky 2009, 2010; Liow et al. 2010).

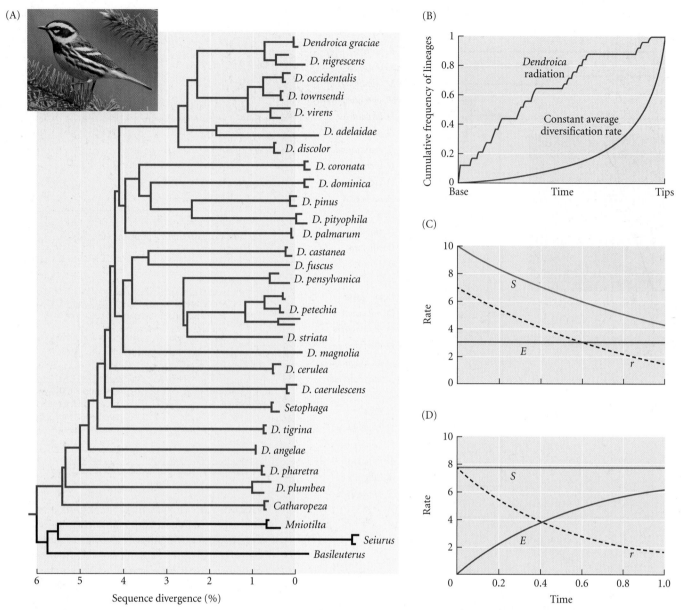

FIGURE 7.3 A phylogeny that shows a decline in the rate of accumulation of species over time. (A) Phylogeny of a clade of American wood warblers, *Dendroica*, and the closely related genus *Catharopeza*, based on mtDNA sequences. The black branches represent outgroup genera in the warbler family. The branch points are plotted according to degree of sequence divergence. (B) A lineage-through-time plot of the *Dendroica* clade, on an arithmetic axis (rather than logarithmic, as in Figure 7.2). Note that the increase slows down and starts to level off. A theoretical curve in which the number of lineages grows at a constant rate is shown for comparison. (C) Theoretical curves showing that the per capita rate of increase *r* (dashed line) declines if the extinction rate *E* (blue line) remains constant but the speciation rate *S* (red line) declines. (D) The rate of increase *r* can also decline if *E* increases while *S* remains constant. (A, B after Lovette and Bermingham 1999; C, D after Rabosky and Lovette 2008.)

Diversity and Disparity through the Phanerozoic

The most complete fossil record has been left by skeletonized marine animals (those with hard parts such as shells or skeletons). Jack Sepkoski (1984, 1993) accomplished the heroic task of compiling data from the paleontological literature on the stratigraphic ranges of more than 4000 skeletonized marine families and 30,000 genera throughout the 542 My of the Phanerozoic. Using this database, he plotted the diversity of families throughout the Phanerozoic, creating one of the most famous graphs in the literature of paleobiology (**Figure 7.4A**). The graph shows a rapid increase in the Cambrian and Ordovician, a plateau throughout the rest of the Paleozoic, and a steady, almost fourfold increase throughout the Mesozoic and Cenozoic. This pattern is interrupted by decreases in diversity caused by mass extinction events. The number of genera

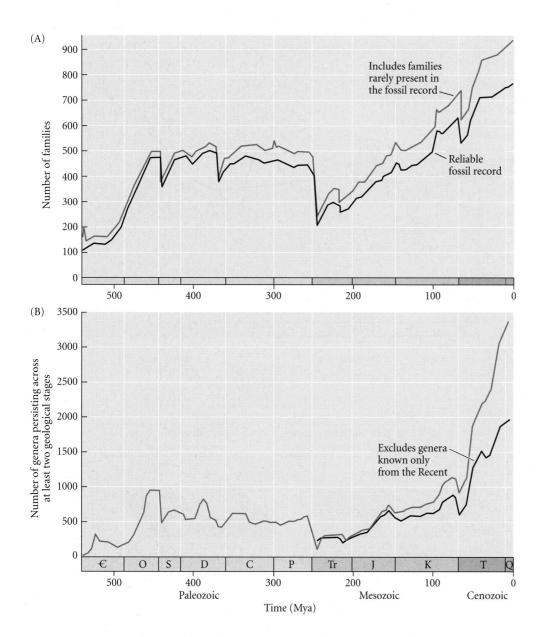

FIGURE 7.4 Taxonomic diversity of skeletonized marine animals during the Phanerozoic. The number of taxa entered for each geological stage includes all those whose known temporal extent includes that stage. (A) Diversity of families. The blue curve includes families that are rarely preserved; the black curve represents only families that have a more reliable fossil record. There are approximately 1900 marine animal families alive today, including those rarely or never preserved as fossils. (B) Diversity of genera, counting only those that cross boundaries between two or more stages. The black curve shows genera only as they are represented in the fossil record, excluding their known occurrences in the Recent. (A after Sepkoski 1984; B after Foote 2000a.)

of skeletonized marine animals shows a similar pattern of change (**Figure 7.4B**). Diversity seems to have increased on land as well, especially after the mid-Cretaceous (Benton 1990).

Sepkoski's plots do not include most of the corrections for sampling errors and biases that we have noted. Recently, paleontologists have reanalyzed the numbers of marine genera, using several procedures to minimize systematic errors (Alroy et al. 2008; Foote 2010). Their results differ from the Sepkoski plot in several ways (**Figure 7.5**): there is a decline in diversity in the Devonian instead of a Paleozoic plateau, an increase in the Permian before the end-Permian extinction, and a much less steep post-Paleozoic increase. This analysis has reopened an earlier debate about whether or not to take the pattern in the raw data, especially the dramatic post-Paleozoic increase, at face value (Miller 2000). The ongoing struggle to understand the history of biodiversity is a fine example of the way science works: data and interpretations are always open to skeptical questioning, and new data or methods of analysis can always modify old interpretations.

Morphological and ecological variety among organisms, termed **disparity**, has increased from the early Paleozoic to the present; among marine animals, for example, the variety of different modes of life and associated adaptations is far greater now than in the Cambrian

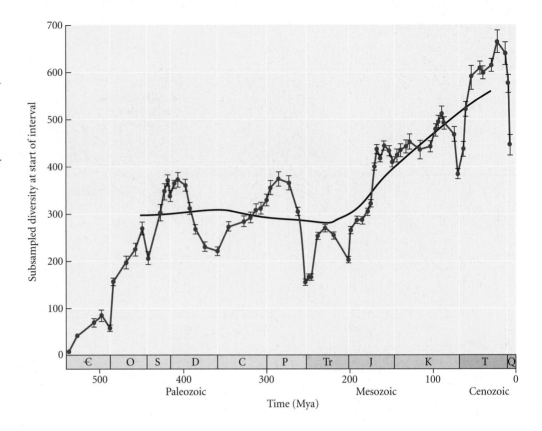

FIGURE 7.5 The numbers of skeletonized marine animal genera over time, corrected for biases such as temporal differences in rock volume and the pull of the Recent. The vertical lines represent the range of values, within which the real number lies. The smooth curve, a running average from the late Ordovician to the mid-Cenozoic, suggests that, aside from mass extinctions and subsequent recoveries, diversity showed a stronger increasing trend in the Mesozoic and Cenozoic than in the Paleozoic. Compare with Figure 7.4B. (After Foote 2010.)

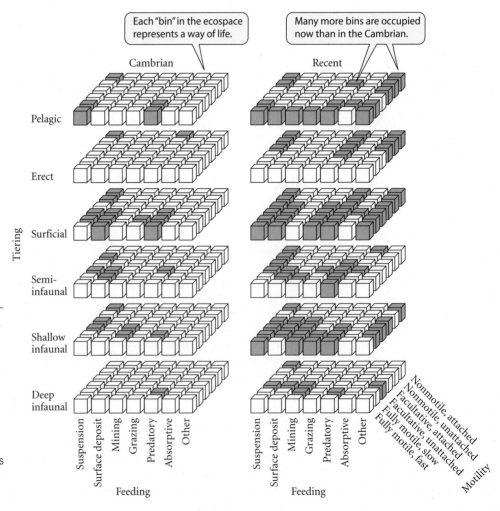

FIGURE 7.6 The use of "ecospace" by marine animals in the Cambrian compared with the Recent (i.e., the present). Each layer represents the vertical space used by animals, from open ocean (pelagic) to deep in the sediment (deep infaunal). In each layer, the "bins" from left to right represent different modes of feeding, and those from front to rear represent different habits of movement (motility). Far more bins, or ways of life, are filled (indicated by green) by animals now than in the Cambrian. (After Bush and Bambach 2011.)

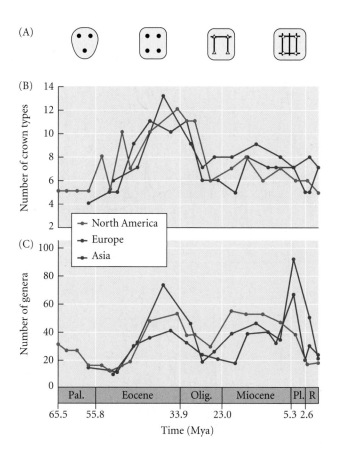

(A)

(B)

(C)

Time (Mya)

FIGURE 7.7 Higher taxa often display their greatest morphological disparity soon after their origin. (A) Diagrams of 4 of the 28 recognized patterns of cheek teeth (premolars and molars) among hoofed mammals. The crown (chewing surface) of the tooth is shown. Dots and triangles indicate cusps, and lines indicate ridges that connect cusps. (B) The disparity, or diversity of form, of cheek teeth increased early in the history of hoofed mammals, and then declined. (C) The number of genera of hoofed mammals also showed an early peak, but also a later peak that was not accompanied by increasing tooth disparity. (After Jernvall et al. 1996.)

(**Figure 7.6**). This increase is largely a consequence of the origin of major new forms of life (higher taxa), for individual higher taxa often display their greatest morphological disparity soon after their origin (**Figure 7.7**), and subsequent speciation is accompanied by relatively slight divergence (Foote 1997).

Rates of origination and extinction

The increase in diversity during the Mesozoic and Cenozoic tells us that on average, the rate of origination of marine animal taxa has been greater than the rate of extinction. However, both rates have fluctuated throughout Phanerozoic history (**Figure 7.8**). Extinction rates, in particular, have varied dramatically (Figure 7.8B). A distinction is often made between episodes during which exceptionally high numbers of taxa become extinct, referred to as **mass extinctions**, and periods of so-called normal or **background extinction**. Five mass extinctions are generally recognized: at the end of the Ordovician, in the late Devonian, at the Permian/Triassic (P/Tr) boundary (the end-Permian extinction), at the end of the Triassic, and at the Cretaceous/Tertiary (K/T) boundary (the K/T extinction). The rate of origination of new genera and families (Figure 7.8A) was highest in the early Paleozoic and immediately after each of the "big five" mass extinctions. Because each mass extinction was immediately followed by a very low extinction rate and a high origination rate of new taxa (Alroy 2008), diversity recovered quickly (in geological terms), usually within "only" 10 My or so.

David Raup and Jack Sepkoski (1982) discovered that the background extinction rate has declined during the Phanerozoic, a conclusion supported by subsequent studies (see Figure 7.8B). The average per capita rate of origination of new taxa has also declined (see Figure 7.8A). What might account for these declines?

Extinctions of species are caused by failure to adapt to environmental changes. Thus it would seem reasonable to expect lineages of organisms to become more resistant to extinction over the course of time as they become better adapted. However, evolutionary

FIGURE 7.8 Per capita rates of (A) origination, (B) extinction, and (C) diversification of genera of marine invertebrates from the late Ordovician through the Quaternary (not including the Recent). Five mass extinctions are seen at the ends of the Ordovician, Devonian, Permian, Triassic, and Cretaceous. The origination rate increases after the mass extinctions. Aside from these major changes, the average (background) extinction rate and origination rate have declined. The net diversification rate does not show a long-term trend. Data from the Cambrian, early Ordovician, and late Cenozoic are not included in order to minimize sampling biases.

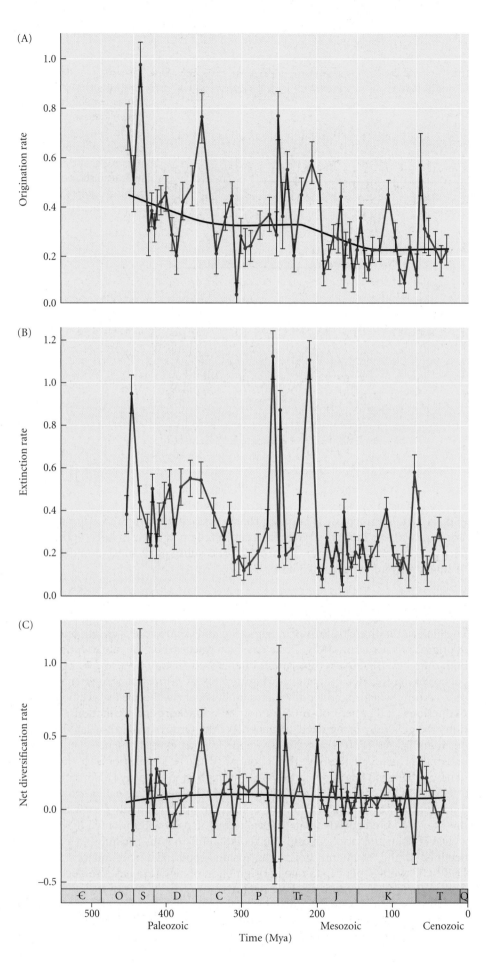

theory does not necessarily predict that species will become more extinction-resistant because natural selection, having no foresight, cannot prepare species for changes in the environment. If the environmental changes that threaten extinction are numerous in kind, we should not expect much carryover of "extinction resistance" from one change to the next. So extinction rates should vary randomly over time if changes in the environment occur at random.

Two hypotheses might account for the decline in extinction rates. The average number of species per genus and per family seems to be greater in the Cenozoic than in earlier eras; this would lower the probability of extinction of higher taxa because a genus or family does not become extinct until all its constituent species are extinct (Alroy 2008). Another explanation is that more volatile clades have been replaced by less volatile clades. Norman Gilinsky (1994) showed that rates of origination and extinction are highly correlated; for example, both rates were higher in ammonoids and trilobites than in gastropods or bivalves. That is, some clades are more VOLATILE than others: they have a higher turnover rate, evolving new families and losing old ones at a higher rate (**Figure 7.9**). These taxa have a shorter life span before they become extinct. The extinction of such taxa leaves the less volatile taxa, those that have longer life spans and lower extinction rates. This process, a form of natural selection among clades (see Chapter 11), results in a decline in the average extinction rate of clades over a long time, as long as highly volatile clades do not evolve anew—which has happened only rarely.

Why are extinction and origination rates correlated? Steven Stanley (1990) suggested that speciation rate may be correlated with extinction rate because both are influenced by certain features of organisms (see also Chapter 18). For these characteristics to determine the extinction rates of families, they must be fairly phylogenetically conservative—they should vary consistently among families. Several characteristics may influence both rates:

1. *Degree of ecological specialization.* Ecologically specialized species are likely to be more vulnerable than generalized species to changes in their environment (Jackson 1974). They may also be more likely to speciate because of their patchier distribution, and newly formed species may be more likely to persist by specializing on different resources and thus avoiding competition with other species. Certain aspects of specialization are phylogenetically conservative in at least some taxa (see Chapter 6).

2. *Geographic range.* Species with broad geographic ranges tend to have a lower risk of extinction because they are not extinguished by local environmental changes (Gaston and Blackburn 2000). They also have lower rates of speciation (**Figure 7.10**), probably because they have a high capacity for dispersal and perhaps broader environmental tolerances; these characteristics could increase their geographic range while lowering the rate at which geographically distant populations become different species (see Chapter 18). Geographic range size is phylogenetically conserved in molluscs, birds, and mammals; that is, there are consistent differences among clades in the average range size of their component species (Jablonski 1987; Brown 1995).

3. *Population dynamics.* Species with low or fluctuating population sizes are especially susceptible to extinction. Some authors believe that speciation is also enhanced by small or fluctuating population sizes, although this hypothesis is controversial.

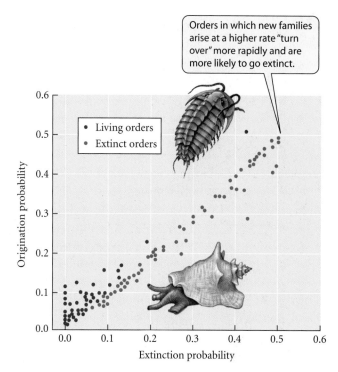

FIGURE 7.9 Groups of marine organisms vary in volatility. The rate (probability) of origination of new families within an order, per time interval, is correlated with the rate (probability) of extinction of families. More volatile orders, with higher rates, have higher turnover, and so are more likely to decrease greatly or become extinct. Most living orders (purple points) have low rates of origination and extinction compared with extinct orders (red points). (After Gilinsky 1994.)

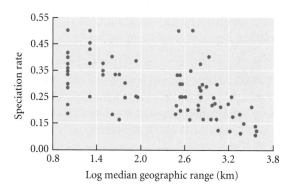

FIGURE 7.10 Speciation rates (the number of new species per lineage per My) are lower in lineages of gastropods (snails) in which the median geographic range is larger. (After Jablonski and Roy 2003.)

(A) Hypothetical curves

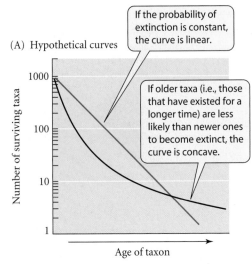

If the probability of extinction is constant, the curve is linear.

If older taxa (i.e., those that have existed for a longer time) are less likely than newer ones to become extinct, the curve is concave.

(B) Data for Ammonoidea

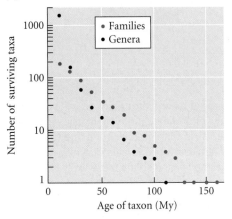

• Families
• Genera

FIGURE 7.11 Taxonomic survivorship curves. Each curve or series of points represents the number of taxa that persisted in the fossil record for a given duration, irrespective of when they originated during geological time. (A) Hypothetical survivorship curves. In a semilogarithmic plot, the curve is linear if the probability of extinction is constant. It is concave if the probability of extinction declines as a taxon ages, as it might if adaptation lowered the long-term probability of extinction. (B) Taxonomic survivorship curves for families and genera of ammonoids. The plot for families suggests an extinction rate that is constant with age, whereas the plot for genera suggests that older survivors have a lower rate of extinction. (B after Van Valen 1973.)

Do extinction rates change as clades age?

Another way of analyzing the rate of extinction of taxa in the fossil record is to plot the fraction of component taxa (e.g., the fraction of genera within a family) that survive for different lengths of time (i.e., their age at extinction). This approach is different from asking whether or not extinction rates have changed over the course of geological time (e.g., whether they were lower in the Jurassic than in the Devonian); instead, we ask whether the rate at which members of a clade become extinct changes over the life span of the clade, irrespective of when the clade originated. If the probability of extinction is independent of clade age, then the proportion of component taxa surviving to increasingly greater ages should decline exponentially (just like the proportion of "surviving" parent atoms in radioactive decay; see Figure 4.2). Plotted logarithmically, the survivorship curve would become a straight line. If taxa become increasingly resistant to causes of extinction as they age, the logarithmic plot should be upwardly concave, with a long tail (**Figure 7.11A**).

When Leigh Van Valen (1973) plotted taxon survivorship in this way, he obtained rather straight curves, and he suggested that the probability of extinction is roughly constant (**Figure 7.11B**). This is what we would expect if organisms are continually assaulted by new environmental changes, each carrying a risk of extinction. One possibility, Van Valen suggested, is that the environment of a taxon is continually deteriorating because of the evolution of other taxa. He proposed the **Red Queen hypothesis**, which states that, like the Red Queen in Lewis Carroll's *Through the Looking-Glass*, each species has to run (i.e., evolve) as fast as possible just to stay in the same place (i.e., survive) because its competitors, predators, and parasites also continue to evolve. There is always a roughly constant chance that it will fail to do so. (Later authors pointed out that some of the survivorship plots are somewhat concave [see Figure 7.4B], and that extinction rates were probably not entirely constant.)

Causes of extinction

Extinction has been the fate of almost all the species that have ever lived, but little is known of its specific causes. Biologists agree that extinction is caused by failure to adapt to changes in the environment. Ecological studies of current populations and species point to habitat destruction as the most frequent cause of extinction by far, and some cases of extinction that are due to introduced predators, diseases, and competitors have been documented.

When a species' environment deteriorates, some populations may become extinct, and the geographic range of the species may contract, unless formerly unsuitable sites become suitable for colonists to establish new populations. Not all environmental changes cause populations to decline, but if they do, the survival of those populations—and perhaps of the entire species—depends on adaptive genetic change. Whether or not this change suffices to prevent extinction depends on how rapidly the environment (and hence the optimal phenotype) changes relative to the rate at which a character evolves. The rate of evolution may depend on the rate at which mutation supplies genetic variation. It may also depend on population size, because smaller populations experience fewer mutations. Thus an environmental change that reduces population size also reduces the chance of adapting to it (Lynch and Lande 1993). Because a change in one environmental factor, such as temperature, may bring about changes in other factors, such as the species composition of a community, the survival of a species may require evolutionary change in several or many features.

It is difficult to identify the cause of extinction of individual taxa during periods of background extinction. Both abiotic and biotic changes have doubtless caused extinctions (Benton 2009). For example, during the Pliocene, the rate of extinction of bivalves and many coral reef inhabitants increased, perhaps due to a decrease in temperature (Stanley 1986; Jackson and Johnson 2000). The role of competition in extinction is discussed later in this chapter.

Mass extinctions

The history of extinction is dominated by the "big five" mass extinctions at the end of the Ordovician, Devonian, Permian, Triassic, and Cretaceous periods (see Figure 7.8B; Bambach 2006). The end-Permian extinction was the most drastic (**Figure 7.12**), eliminating about 54 percent of marine families, 84 percent of genera, and 80 to 90 percent of species (Erwin 2006). On land, major changes in plant assemblages occurred, several orders of insects became extinct, and the dominant amphibians and therapsids were replaced by new groups of therapsids (including the ancestors of mammals) and diapsids (including the ancestors of dinosaurs).

Less severe, but much more famous, was the K/T, or end-Cretaceous, extinction, which marked the demise of many marine and terrestrial plants and animals, including the dinosaurs (except for birds). The K/T extinction is famous because of the truly dramatic hypothesis, first suggested by Walter Alvarez and colleagues (1980), that the nonavian dinosaurs were extinguished by the impact of an extraterrestrial body—an asteroid or large meteorite. Alvarez et al. postulated that this object struck Earth with a force great enough to throw a pall of dust into the atmosphere, darkening the sky and lowering temperatures, and thus reducing photosynthesis. Geologists now agree that such an impact occurred; its site, the Chicxulub crater, has been discovered off the coast of the Yucatán Peninsula of Mexico. As we have seen (in Chapter 5), most paleontologists agree that this impact caused the mass extinction at the K/T boundary, but some argue that the impact was only one of several environmental changes that interacted to cause the K/T extinction. Likewise, the causes of the end-Permian extinction are debated, although there is increasing consensus that gases released by massive volcanic eruptions in Siberia played a dominant role.

Mass extinctions were "selective" in that some taxa were more likely than others to survive. Survival of gastropods (snails and relatives) through the end-Permian extinction was greater for species with wide geographic and ecological distributions and for genera consisting of many species (Erwin 1993). Extinction appears to have been random with respect to other characteristics, such as mode of feeding. The pattern of selectivity was much the same as during periods of background extinction, when gastropods and other taxa with broad geographic distributions had lower rates of extinction than narrowly distributed taxa (Boucot 1975). Patterns of survival through the K/T extinction, however, differed from those during "normal" times (Jablonski 1995). During times of background extinction, survivorship

FIGURE 7.12 Reconstructions of an ancient seabed (A) immediately before and (B) after the end-Permian mass extinction. A rich fauna of burrowing, epifaunal, and swimming organisms was almost completely extinguished. (Artwork © J. Sibbick.)

(A)

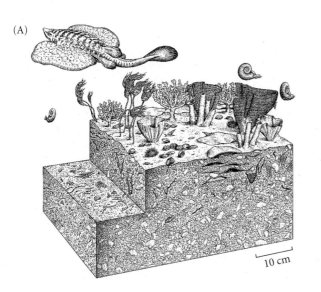

(B)

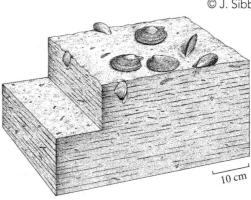

of late-Cretaceous bivalves and gastropods was greater for taxa with PLANKTOTROPHIC larvae (those that feed while being dispersed by currents) and for genera consisting of numerous species, especially if those genera had broad geographic ranges. In contrast, during the end-Cretaceous mass extinction, planktotrophic and nonplanktotrophic taxa had the same extinction rates, and the survival of genera, although enhanced by broad distribution, was not influenced by their species richness. Thus the characteristics that were correlated with survival seem to have differed from those during "normal" times.

During mass extinction events, taxa with otherwise superb adaptive qualities succumbed because they happened not to have some critical feature that might have saved them from extinction under those circumstances. Evolutionary trends initiated in "normal" times were cut off at an early stage. For example, the ability to drill through bivalve shells and feed on the animals inside evolved in a Triassic gastropod lineage, but it was lost when that lineage became extinct in the late-Triassic mass extinction (Fürsich and Jablonski 1984). The same feature evolved again 120 My later, in a different lineage that gave rise to diverse oyster drills. A new adaptation that might have led to a major adaptive radiation in the Triassic was strangled in its cradle, so to speak.

Both abiotic and biotic environmental conditions were probably very different after mass extinctions than before. Perhaps for this reason, many taxa continued to dwindle long after the main extinction events (Jablonski 2002), while others, often members of previously subdominant groups, diversified. Although the diversity of some clades, such as ammonoids, recovered rapidly (**Figure 7.13**), full recovery of other clades took many millions of years. For example, the rate of origination of genera of bivalves (clams and relatives) increased after the K/T extinction and has remained high ever since. New genera have arisen mostly in tropical latitudes, so ongoing recovery from the mass extinction has affected the geographic pattern of diversity that exists today (Krug et al. 2009).

The mass extinction events, especially the end-Permian and K/T extinctions, had an enormous effect on the subsequent history of life because, to a great extent, they wiped the slate clean. Stephen Jay Gould (1985) suggested that there are "tiers" of evolutionary change, each of which must be understood in order to comprehend the full history of evolution. The first tier is microevolutionary change *within populations and species*. The second tier is "species selection," the *differential proliferation and extinction of species* during "normal" geological times, which affects the relative diversities of lineages with different characteristics. The third tier is the *shaping of the biota by mass extinctions*, which can extinguish diverse taxa and reset the stage for new evolutionary radiations, initiating evolutionary histories that are largely decoupled from earlier ones.

Richard Bambach and colleagues (2002) found some support for Gould's idea when they classified Phanerozoic marine animal genera by three functional criteria: whether they were motile or nonmotile, whether they were "buffered" against physiological stress

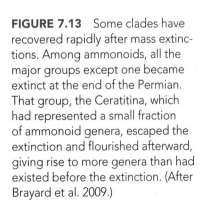

FIGURE 7.13 Some clades have recovered rapidly after mass extinctions. Among ammonoids, all the major groups except one became extinct at the end of the Permian. That group, the Ceratitina, which had represented a small fraction of ammonoid genera, escaped the extinction and flourished afterward, giving rise to more genera than had existed before the extinction. (After Brayard et al. 2009.)

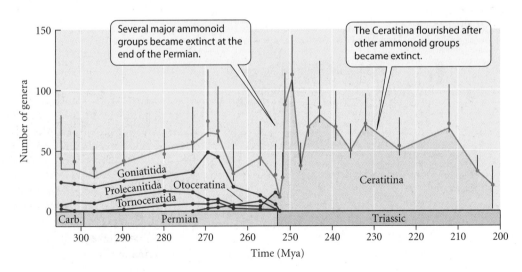

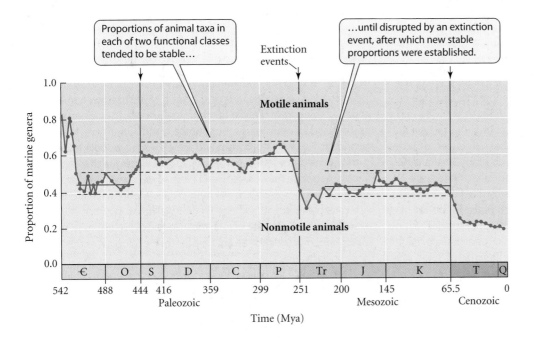

FIGURE 7.14 Changes in the proportions of genera of motile versus nonmotile marine animals during the Phanerozoic. The proportions were roughly stable (dashed horizontal lines) between mass extinctions, but shifted rapidly to a new stable state after mass extinction events at the end of the Ordovician, Permian, and Cretaceous (solid vertical lines). Similar changes (not shown here) occurred in the proportions of predators versus nonpredators and of animals thought to be physiologically buffered versus unbuffered, based on anatomical criteria. (After Bambach et al. 2002.)

(with well-developed gills and circulatory system, such as crustaceans) or not (such as echinoderms), and whether or not they were predatory. With respect to all three kinds of functional groupings, the proportions of taxa with alternative characteristics remained stable over intervals as long as 200 My, even though the total diversity and the taxonomic composition of the marine fauna changed greatly (**Figure 7.14**). However, shifts from one stable configuration to another occurred at the ends of the Ordovician, Permian, and Cretaceous, suggesting that the extinction of long-prevalent (incumbent) taxa permitted the emergence of new community structures.

No truly massive extinction has occurred for 65 My; even the great climate oscillations of the Pleistocene, though they altered geographic distributions and ecological assemblages, had a relatively small impact on the diversity of life. But it is depressingly safe to say that the next mass extinction has begun (**Box 7A**). The course of biodiversity has been altered for the foreseeable future by human domination of Earth, and altered for the worse. Without massive, dedicated action, humanity will suffer profoundly, and much of the glorious variety of the living world will be extinguished almost as quickly as if another asteroid had smashed into the planet and again cast over it a pall of death.

Box 7A The Current Mass Extinction

For the first time in the history of life, a single species has precipitated a mass extinction. Within the next few centuries, the diversity of life will almost certainly plummet at a greater pace than ever before.

The human threat to Earth's biodiversity has accelerated steadily with the advent of ever more powerful technology and the exponential growth of the world's human population, which has surpassed 7 billion. The per capita rate of population growth is greatest in the developing countries, which are chiefly tropical and subtropical, but the per capita impact on the world's environment is greatest in the most highly industrialized countries. An average American, for example, has perhaps 140 times the environmental impact of an average Kenyan, because the United States is so

profligate a consumer of resources harvested throughout the world and of energy (with impacts ranging from strip mines and oil spills to insecticides and production of the "greenhouse gases" that cause global warming).

Some species are threatened by hunting or overfishing and others by species that humans have introduced into new regions. But by far the greatest cause of extinction, now and probably over the course of the twenty-first century, is the destruction of habitat (Sala et al. 2000). It is largely for this reason that 29 percent of North American freshwater fishes are endangered or already extinct, and that about 10 percent of the world's bird species are considered endangered by the International Council for Bird Preservation.

(continued)

Box 7A *(continued)*

The numbers of species likely to be lost are highest in tropical forests, which are being destroyed at a phenomenal and accelerating rate. As E. O. Wilson (1992) said, "in 1989 the surviving rain forests occupied an area about that of the contiguous forty-eight states of the United States, and they were being reduced by an amount equivalent to the size of Florida each year." Several authors have estimated that 10 to 25 percent of tropical rain forest species—accounting for as much as 5 to 10 percent of Earth's species diversity—will become extinct in the next 30 years. To this toll must be added extinctions caused by the destruction of species-rich coral reefs, pollution of other marine habitats, and losses of habitat in areas such as Madagascar and the Cape Province of South Africa, which harbor unusually high numbers of endemic species.

In the long run, an even greater threat to biodiversity may be global warming caused by high and increasing consumption of fossil fuels and production of carbon dioxide and other "greenhouse" gases. Earth's climate has warmed by a global average of 0.6°C during the last century, and the rate of warming is accelerating. Although the effects of climate change will vary among regions (some regions will actually suffer a cooling trend), snow cover, glaciers, and polar ice caps are rapidly shrinking, and some tropical areas are becoming much drier. These changes are happening much faster than most of the climate changes that have occurred in the past. Some species may adapt by genetic change, but there is already evidence that many species will shift their ranges. Such shifts, however, are difficult or impossible for most mountaintop and Arctic species, and for many others whose habitats and the habitat "corridors" along which they might disperse have been destroyed. Computer simulations, based on various scenarios of warming rate and species' capacity for dispersal, suggest that within the next 50 years, between 18 and 35 percent of species will become "committed to extinction"—that is, they will have passed the point of no return (Thomas et al. 2004).

If mass extinctions have happened naturally in the past, why should we be so concerned? Different people have different answers, ranging from utilitarian to aesthetic to spiritual. Some point to the many thousands of species that are used by humans today, ranging from familiar foods to fiber, herbal medicines, and spices used by peoples throughout the world. Others cite the economic value of ecotourism and the enormous popularity of bird-watching in some countries. Biologists will argue that thousands of species may prove useful (as many already have) as pest control agents or as sources of medicinal compounds or industrially valuable materials. Except in a few well-known groups, such as vertebrates and vascular plants, most species have not even been described, much less been studied for their ecological and possible social value.

The rationale for conserving biodiversity is only partly utilitarian, however. Many people (including this author) cannot bear to think that future generations will be deprived of tigers, sea turtles, and macaws. They share with millions of others a deep renewal of spirit in the presence of unspoiled nature. Still others feel that it is in some sense cosmically unjust to extinguish, forever, the species with which we share Earth.

Conservation is an exceedingly complicated topic; it requires not only a concern for other species, but compassion and understanding of the very real needs of people whose lives depend on clearing forests and making other uses of the environment. It requires that we understand not only biology, but also global and local economics, politics, and social issues ranging from the status of women to the reactions of the world's peoples and their governments to what may seem like elitist Western ideas. Anyone who undertakes work in conservation must deal with these complexities. But everyone can play a helpful role, however small. We can try to waste less; influence people about the need to reduce population growth (surely the most pressing problem of all); support conservation organizations; patronize environment-conscious businesses; stay aware of current environmental issues; and communicate our concerns to elected officials at every level of government. Few actions of an enlightened citizen of the world can be more important.

Diversification

We turn now to the questions of what factors have governed the history of diversity and why some lineages have become more diverse than others.

Does species diversity reach equilibrium?

A huge ecological literature is concerned with whether or not the number of coexisting species (of some group such as plants or mammals) tends toward an equilibrium. This question is complex and not entirely resolved, but ecologists agree that some factors tend to limit species diversity. The space that plants compete for and the energy fluxes that organisms depend on are finite, so they can be divided among a limited number of species populations that are still large enough to persist. Phenomena such as competitive exclusion of species from each other's ranges suggest that interactions among species can limit local species diversity.

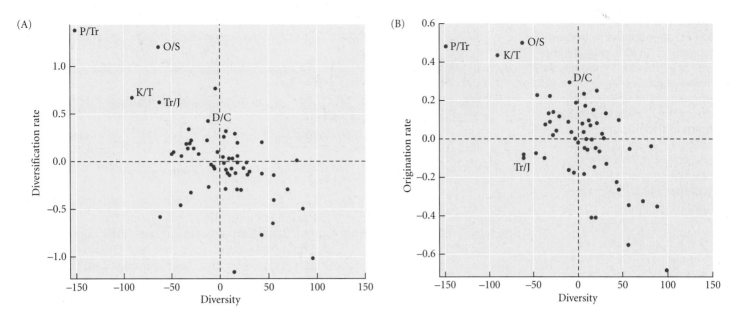

FIGURE 7.15 The per lineage rates of (A) diversification and (B) origination of skeletonized marine invertebrate genera during the Phanerozoic are diversity-dependent. Each point plots the rate of change during a stratigraphic stage against the diversity of taxa at the start of that stage. The higher the diversity, the lower the rate of diversification (in A), and this pattern is attributable to the reduced rate at which new genera arise (in B). The pattern suggests that higher diversity imposes stronger competition and prevents new genera from evolving. For statistical reasons, the points are shown on scales that are centered at zero. Points representing mass extinction events are labeled (O/S, end-Ordovician; D/C, late Devonian; P/Tr, end-Permian; Tr/J, end-Triassic; K/T, Cretaceous/Tertiary boundary). (After Foote 2010.)

The fossil record supports the hypothesis that *the per capita rate of increase in the number of species (or higher taxa) is diversity-dependent*: it decreases as the number grows. For example, Michael Foote (2010), using an updated database of skeletonized marine invertebrates, calculated the per capita rates of origination (S), extinction (E), and diversification (D) of genera from one stratigraphic stage to the next, then correlated these short-term changes with the number of genera present (N) at the beginning of the stage (**Figure 7.15**). Both the diversification rate ($D = S - E$) and the origination rate (S) declined as diversity increased. In a similar analysis, John Alroy (2008) found evidence that extinction rates (E) were higher if diversity at the start of an interval was higher. Moreover, a high extinction rate in one interval was correlated with a high origination rate in the following interval (Alroy 2008). These and other such analyses imply that *the diversity of taxa tends to be stabilized and approach an equilibrium*. The same conclusion may be drawn from phylogenetic studies of the growth in the number of species in a clade, in which the rate of diversification tends to decline as diversity grows (see Figure 7.3).

The most plausible interpretation of this pattern is that, because resources such as food and space are limited, competition among species increases as the number of species grows and reduces the rate at which new species can be added. This hypothesis is supported by several kinds of data.

First, studies of both living and extinct organisms have shown that lineages often have diversified most rapidly when presented with an ecological opportunity: what is often called "ecological space" or "vacant niches" not occupied by other species. In many isolated islands and bodies of water, descendants of just a few original colonizing species have diversified and filled ecological niches that are occupied in other places by unrelated organisms. Such adaptive radiations include the cichlid fishes in the Great Lakes of eastern Africa, the honeycreepers in the Hawaiian Islands, and Darwin's finches in the Galápagos Islands (see Figures 3.22 and 3.24).

Second, the fossil record provides many instances in which the reduction or extinction of one group of organisms has been followed or accompanied by the proliferation of an ecologically similar group. For example, angiosperms (flowering plants) diversified as conifers and other gymnosperms declined, and the orders of placental mammals appear in the fossil record after the late Cretaceous extinction of the nonavian dinosaurs.

Several hypotheses can account for these patterns (Benton 1996; Sepkoski 1996a). Two of these hypotheses involve competition between species in two different clades. On the one hand, the later clade may have *caused* the extinction of the earlier clade by competition, a process called COMPETITIVE DISPLACEMENT (**Figure 7.16A**). On the other hand, an

FIGURE 7.16 Models of competitive displacement and incumbent replacement. In each diagram, the width of a "spindle" represents the number of species. (A) Competitive displacement, in which the increasing diversity of clade 2 causes a decline in clade 1 by direct competitive exclusion. (B) Incumbent replacement, in which the extinction of clade 1 enables clade 2 to diversify. (After Sepkoski 1996a.)

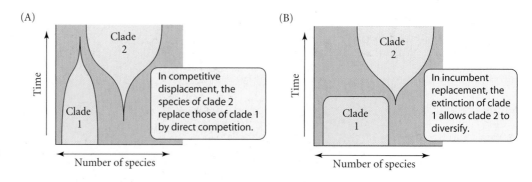

(A)

Clade 2

Clade 1

Time

Number of species

In competitive displacement, the species of clade 2 replace those of clade 1 by direct competition.

(B)

Clade 2

Clade 1

Time

Number of species

In incumbent replacement, the extinction of clade 1 allows clade 2 to diversify.

incumbent taxon may have *prevented* an ecologically similar taxon from diversifying because it already "occupies" resources. Extinction of the incumbent taxon may then have vacated ecological "niche space," *permitting* the second taxon to radiate (**Figure 7.16B**). This process has been called INCUMBENT REPLACEMENT by Rosenzweig and McCord (1991), who argued that even if the second taxon had superior adaptive features, it may not have displaced the earlier taxon by competition.

Sepkoski and colleagues (2000) developed a mathematical model in which two clades increase in diversity, but in which the increase in each clade is inhibited by both its own diversity and that of the other clade. They applied the model to data on the number of genera of two groups of bryozoans ("moss animals"), the cyclostomes and the cheilostomes. These sessile colonial animals spread over rocks or other surfaces by budding. When colonies of these two groups meet, cheilostomes generally overgrow cyclostomes (**Figure 7.17A**). Especially since the end-Cretaceous extinction, the diversity of cheilostomes has

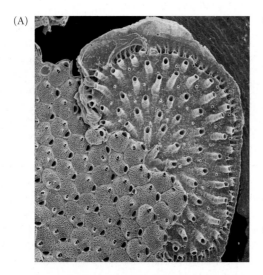

(A)

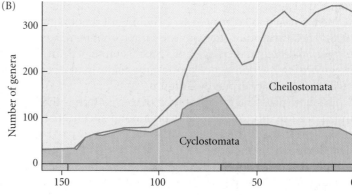

(B)

Number of genera

300

200

100

0

150   100   50   0

Cheilostomata

Cyclostomata

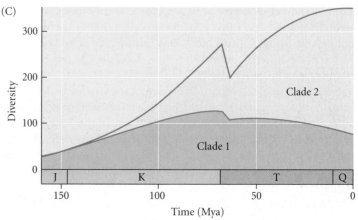

(C)

Diversity

300

200

100

0

Clade 2

Clade 1

J   K   T   Q

150   100   50   0

Time (Mya)

FIGURE 7.17 Competitive displacement among bryozoans, sessile colonial animals that spread over hard surfaces. (A) Colonies of cheilostome bryozoans (left) can overgrow colonies of cyclostome bryozoans (right), causing their death. (B) Cheilostomes appeared in the late Jurassic and soon increased greatly in diversity, whereas the increase of cyclostomes was reversed in the Tertiary. (C) The changes in diversity in a model of competition among species in two clades, assuming that clade 2 species are competitively superior and that the diversity of both clades is reduced by an external perturbation at time 60. (A courtesy of Frank K. McKinney; B, C after Sepkoski et al. 2000.)

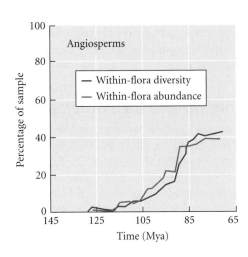

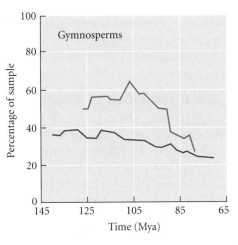

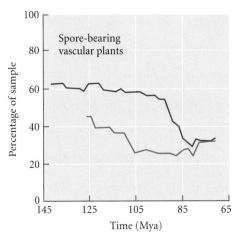

increased, whereas cyclostomes have not recovered (**Figure 7.17B**). When Sepkoski et al. simulated the end-Cretaceous drop in the diversity of both clades, their model rendered a profile of subsequent diversity change that closely matches the data (**Figure 7.17C**). This result does not prove that competition determined the history of bryozoan diversity, but it is consistent with that hypothesis.

A pattern of replacement is consistent with competitive displacement if the earlier and later taxa lived in the same place at the same time, if they used the same resources, if the earlier taxon was not decimated by a mass extinction event, and if the diversity and abundance of the later taxon increased as the earlier one declined (Lupia et al. 1999). Vascular plants, which certainly compete for space and light, showed this pattern during the Cretaceous, when flowering plants increased in diversity and abundance at the expense of nonflowering plants, especially spore-bearing plants such as ferns (**Figure 7.18**).

Incumbent replacement has probably been more common than competitive displacement (Benton 1996). The great radiation of placental mammals in the early Cenozoic is often credited to the K/T extinction of the last nonavian dinosaurs and other large reptiles, which may have suppressed mammals by both competition and predation. This is one of many examples of a long delay between the origin of a clade and its diversification and rise to ecological prominence, a delay that may often be caused by incumbent suppression of the new clade (Jablonski 2008a). The best evidence of incumbency and release is supplied by repeated replacements. For instance, amphichelydians, the "stem group" of turtles, could not retract their heads and necks into their shells (**Figure 7.19**). Two groups of modern turtles, which protect themselves by bending the neck under or within the shell, replaced the amphichelydians in different parts of the world four or five times, especially after the K/T extinction event. The modern

FIGURE 7.18 Changes in relative diversity (blue curves) and abundance (proportion of individuals; red curves) of major groups of vascular plants in fossil samples during the Cretaceous. The increase in both diversity and abundance of flowering plants was mirrored by the decline, in both respects, of the spore-bearing plants (e.g., ferns) and the decline in abundance of gymnosperms (e.g., conifers). This pattern is consistent with competitive displacement. (After Lupia et al. 1999.)

FIGURE 7.19 Pleurodiran and cryptodiran turtles replaced incumbent amphichelydian turtles, which became entirely extinct. (A) Amphichelydians, represented here by the reconstructed skeleton of the earliest known turtle (*Proganochelys quenstedti*, upper Triassic), could not retract their heads for protection. (B) Snakeneck turtles such as *Chelodina longicollis* are pleurodiran turtles, which flex the neck sideways beneath the edge of the carapace. (C) Cryptodiran turtles, represented here by an Eastern box turtle (*Terrapene carolina*), fully retract the head into the shell by flexing the neck vertically. (A courtesy of E. Gaffney, American Museum of Natural History.)

(A)

(B)

(C)

groups evidently could not radiate until the amphichelydians had become extinct. That this replacement occurred in parallel in different places and times makes it a likely example of release from competition (Rosenzweig and McCord 1991).

Why are some kinds of organisms more diverse than others?

Among animal phyla, the Arthropoda include more than 1,000,000 species, and the Acanthocephala (spiny-headed worms) include about 1150. There are about 400,000 described species of beetles (order Coleoptera), but only 550 scorpionflies (Mecoptera). The Rodentia include about 2280 species, and the Tubulidentata only one (the aardvark). There are more than 225,000 species of orchids, but only one species of gingko (see Figure 5.19C). Why do taxa differ so greatly in diversity? Many of the studies addressing this question take a phylogenetic approach.

It has been proposed (see Figure 6.27) that differences in species diversity among different regions might be attributed to differences in the ages of the regions (time for species accumulation), in their rates of increase in diversity, or in their **carrying capacity**—the number of species that can persist there at equilibrium. Exactly the same hypotheses might account for differences in the diversity of different clades (represented by higher taxa). These hypotheses are not mutually exclusive, and they are not always independent. For example, the rate of diversification may decline as a clade approaches its equilibrium diversity, and so may the diversification rate of equally old clades be correlated with their equilibrium numbers.

Much of the variation in species diversity among clades is explained by differences in their ages. Mark McPeek and Jonathan Brown (2007) compiled data from molecular phylogenies of 163 groups of animals in three phyla. They found that, in general, species richness is correlated not with diversification rate, but with the age of the clade, as estimated either by time-calibrated sequence differences (**Figure 7.20A**) or by earliest appearance in the fossil record (**Figure 7.20B**). Thus many taxa have more species simply because they are older and have had more time to grow in number. The correlation is far from perfect, however; indeed, some ancient clades that have few species today (such as the single species of gingko) were once much more diverse.

The relation of species diversity to the per capita rate of diversification (r, the difference between the rate of origination S and the rate of extinction E) has been approached in several ways. One is to estimate the average rate r for many independent clades, using the equation $N_t = N_0 e^{rt}$: N_t is the number of living species and t is estimated from DNA sequence differences among species or from fossil evidence. For example, Susana Magallón and Michael Sanderson (2001) used fossil data to estimate the ages (t) of clades of flowering

FIGURE 7.20 The species richness of clades in relation to their age. (A) Number of extant species in diverse clades of arthropods, molluscs, and vertebrates in relation to the age of the order (as estimated from molecular phylogenies). (B) Number of extant species in orders of insects and vertebrates in relation to the age of the order (defined as first appearance of the crown group in the fossil record). These plots, especially of fossil-based data, show a correlation between species richness and clade age. (After McPeek and Brown 2007.)

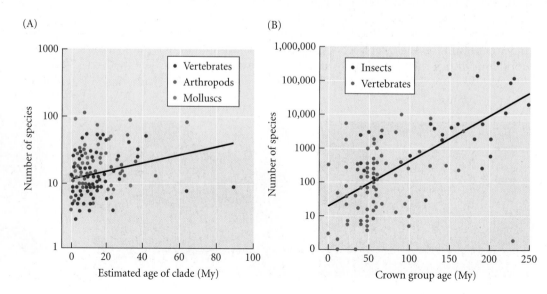

plants, and asked if current diversity of various clades is correlated with r. They determined that although older clades generally have more species, some clades are unusually rich in species, and others unusually poor, considering their age (**Figure 7.21**). That is, some clades have had unusually high or low rates of diversification (r). Clades with very low r have few species for their age (for example, the Nymphaeales, water lilies and relatives, with $r = 0.031$ and 85 species). Almost all clades with exceptionally high r are very species-rich; for example, the Asterales, with $r = 0.272$, has about 26,000 species, including the sunflower family, one of the two largest families of plants. But some large clades, such as poinsettia and its relatives (Euphorbiaceae), owe their species richness to age rather than to a high rate of diversification. In a similar vein, more than 85 percent of living species of vertebrates are members of six clades that have had an exceptionally high rate of diversification: three clades of bony fishes, a clade of lizards and snakes, the birds, and a clade containing most placental mammals (Alfaro et al. 2009). We will shortly explore the possible causes of differences in the rate of increase among clades.

As we have seen, there is both paleontological and phylogenetic evidence that the rate of diversification is diversity-dependent and declines as the number of taxa grows (see Figures 7.3 and 7.15). This observation implies that numbers of taxa approach an equilibrium at which rates of origination and extinction are approximately balanced. The equilibrium diversity is likely to differ among clades because of many factors, such as the variety of different resources that the species in a clade can use. Mark McPeek (2008) has shown by computer simulation that a deceleration in the rate of diversification can result from competition among species: at first, new species can persist if their resource requirements

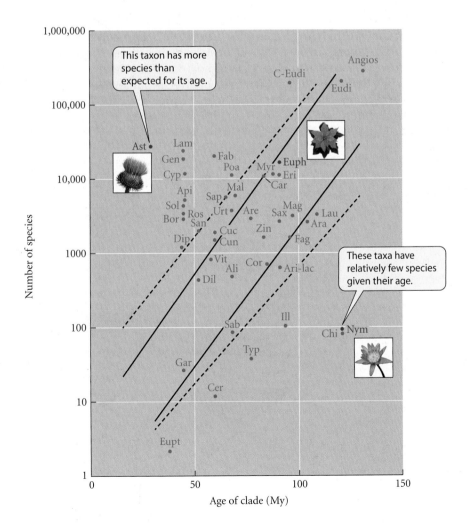

FIGURE 7.21 The diversity of living species in some clades is explained by their rate of diversification. Here, the number of living species in some higher taxa of flowering plants is plotted against the age of the clade. The pairs of parallel lines are the range of diversity values, in relation to age, that would be expected if every clade had diversified at a rate equal to that of angiosperms as a whole, assuming zero extinction (solid lines) or a high extinction rate (dashed lines). Taxa within these bounds are as diverse as expected, given their age. The taxa outside these bounds have exceptionally high (e.g., Ast) or exceptionally low (e.g., Nym) diversity for their age, which must be attributed to their exceptionally high or low rates of diversification. The taxa indicated in green (Ast, Asterales; Euph, Euphorbiaceae; Nym, Nymphaeales) are mentioned in the text. (After Magallón and Sanderson 2001.)

do not overlap too greatly with those of other species, but as the number of species grows, resources are more completely utilized, it is more difficult for new species to fit in, and they are extinguished by competition with existing species, a process called **competitive exclusion**. McPeek also found, though, that if newly arisen species are ecologically identical to existing species, the process of competitive exclusion takes a very long time, and such species can represent a significant fraction of total diversity (see also Hubbell 2006; Rosindell et al. 2010). Moreover, if species evolve to become more specialized and subdivide resources more finely, the equilibrium diversity could increase (see Chapter 19). For several reasons, then, the species diversity in some clades may not have reached equilibrium and may continue to increase. Whether or not diversity is near an equilibrium today is a major unresolved question (Ricklefs 2007; Wiens 2011).

Effects of organisms' features on diversification

Differences in the diversification rates of clades are a consequence of the features that affect the probability of extinction and of speciation. But identifying the feature that has caused one clade to diversify faster than another is difficult because that feature is correlated with other distinctive features of that clade, any of which might have enhanced speciation or reduced the likelihood of extinction. For example, angiosperms may have become more diverse than gymnosperms because of their obvious distinctive feature: flowers that attract animal pollinators. But we cannot exclude the possibility that they diversified because their developing seeds are protected within a chamber (carpel), or because they have double fertilization that forms the triploid endosperm that nourishes the embryo within the seed.

Stronger evidence is provided if the rate of diversification is consistently associated with a particular character that has evolved independently in a number of different clades. Such tests have been applied mostly to living organisms. The diversity of a number of clades with a novel character can be compared with the diversity of their sister groups that retain the ancestral character state. Since sister taxa are equally old, the difference between them in number of species must be due to a difference in rate of diversification, not to age. If the convergently evolved character is consistently associated with high diversity, we have support for the hypothesis that it has caused a higher rate of diversification.

Charles Mitter and colleagues (Mitter et al. 1988; Farrell et al. 1991) applied this method, called REPLICATED SISTER-GROUP COMPARISON, to herbivorous insects and plants. The habit of feeding on the vegetative tissues of green plants has evolved at least 50 times in insects, usually from predatory or detritus-feeding ancestors. Phylogenetic studies have identified the nonherbivorous sister groups of 13 herbivorous clades. In 11 of these cases, the herbivorous lineage has more species than its sister group (**Figure 7.22**). This significant

FIGURE 7.22 Two replicated sister-group comparisons of herbivorous clades of insects with their sister clades that feed on animals, fungi, or detritus. Herbivorous clades are consistently more diverse, demonstrating higher rates of diversification. (Data from Mitter et al. 1988.)

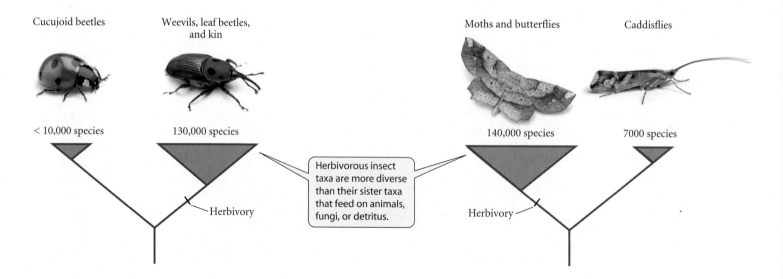

correlation supports the hypothesis that the ability to eat plants has promoted diversification. These researchers then examined the species diversity of 16 clades of plants that have evolved rubbery latex (as in milkweeds) or resin (as in pines), both of which deter attack by herbivorous insects. Thirteen of these clades have more species than their sister clades, which lack latex or resin. These defensive features may have fostered diversification.

Such instances establish a correlation between a trait and the rate at which the number of species increases or decreases. A consistent difference of this kind is termed **species selection**, or **species sorting**, and results in certain characteristics becoming more prevalent than others, among all species taken together. Among gastropods, for example, extinction rates have been lower in clades with highly dispersive planktotrophic larvae than in those with direct-developing, nonplanktotrophic larvae, and plant clades with bilaterally symmetrical flowers have had higher speciation rates than those with radially symmetrical flowers (Jablonski 2008b; Rabosky and McCune 2010). Some traits seem to affect both S and E similarly: molluscs with high dispersal capability have lower extinction rates, but also lower rates of speciation.

Adaptive radiation

Adaptive radiation is the origin of numerous species within a relatively short time and the adaptive modification of these species for different ways of life (see Chapter 3). Some adaptive radiations, such as that of Darwin's finches in the Galápagos Islands (see Figure 3.22), have produced rather few, ecologically diverse species, but some others (such as that of the cichlids of the African Great Lakes; see Figure 3.24) have produced hundreds of species.

Adaptive radiations arise as a multitude of speciation events. Related species often occupy somewhat similar ecological niches, which together are sometimes called an **adaptive zone**. For example, the many species of insectivorous bats and fruit-eating bats, which are nocturnal, occupy two adaptive zones that differ from those of diurnal insectivorous and fruit-eating birds. The term "ecological space" is roughly equivalent to a set of adaptive zones. Adaptive radiation is thought to occur when a lineage has access to ecological opportunity and can diversify to occupy an adaptive zone (Gavrilets and Losos 2009; Glor 2010; Losos 2010). The ecological opportunity must be geographically available; it must be ecologically available, in the sense that the resources are not already fully utilized by competitively superior species; and the progenitor of an adaptive radiation must evolve, or already possess, an adaptation that enables it to use the novel resources or habitat. Such a feature is called a **key adaptation**.

Ecological opportunity is often available to species that colonize islands or lakes, where they may encounter resources that are not already used by other species. Hence many of the best-known adaptive radiations are found in these settings. Resources may also become available when species become extinct; this accounts for the rapid increase in the diversification of many clades that have survived mass extinctions.

Key adaptations have contributed importantly to the increase in diversity over time (Niklas et al. 1983; Bambach 1985). Among the sea urchins (Echinoidea), for example, three orders increased greatly in diversity beginning in the early Mesozoic (**Figure 7.23**). The order Echinacea evolved stronger jaws that enabled its members to use a greater variety of foods. The heart urchins (Atelostomata) and sand dollars (Gnathostomata) became specialized for burrowing in sand, where they feed on fine particles of organic sediment. The key adaptations allowing this major shift of habitat and diet include a flattened form and a variety of highly modified tube feet that can capture fine particles and transfer them to the mouth.

Key adaptations can often be identified by phylogenetic analyses. One such method is replicated sister-group comparison, as described earlier. Other methods identify a novel trait as a key adaptation if it is associated with a burst of speciation within a larger phylogeny (Ree 2005). The nectar spurs on the petals of columbines (see Figure 3.21), for example, are associated with greater diversification, probably because they enabled different species to use morphologically different insects and birds as pollinators (**Figure 7.24**).

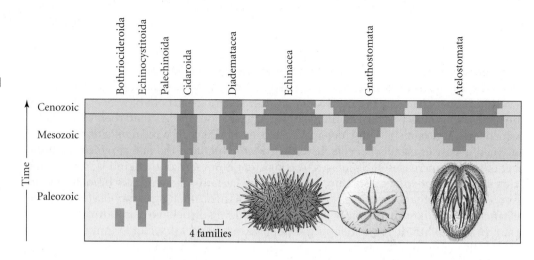

FIGURE 7.23 Changes in the diversity of several groups of echinoid echinoderms over time. The width of the symmetrical profile of each group represents the number of families in that group. The diversity of sea urchins (order Echinacea), sand dollars (Gnathostomata), and heart urchins (Atelostomata) greatly increased during the Mesozoic and Cenozoic, probably because of the key adaptations described in the text. (After Bambach 1985.)

Some adaptive radiations, however, are based on diverse modifications of an ancestral trait, rather than a novel character. Darwin's finches, for example, simply vary in the size and shape of their beaks. Furthermore, the species in some species-rich clades, such as the wood warblers in Figure 7.3, differ only slightly in ecologically important characteristics. In this and many other similar instances, the rate of speciation may be high because of rapidly evolving features that prevent interbreeding but do not confer ecological differences (see Chapter 18). Such cases may not qualify as adaptive radiations.

Computer models of adaptive radiations suggest that they are likely to stem from an "early burst" of increased diversity associated with a burst of ecological and morphological disparity (Gavrilets and Losos 2009). The rate of increase in species diversity and disparity is expected to slow down as the newly added species fill ecological niches and reduce

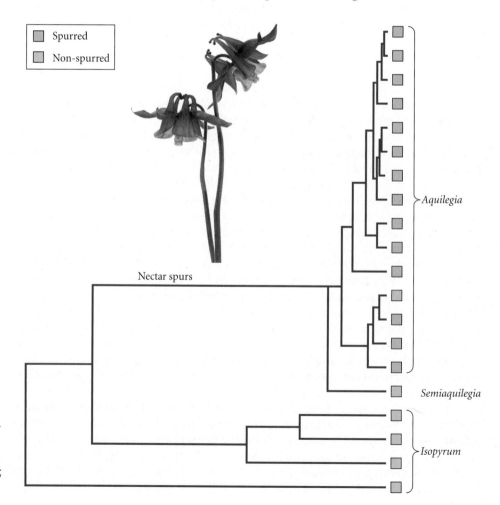

FIGURE 7.24 A phylogeny illustrating an adaptive radiation. The evolution of nectar spurs in the ancestor of columbines (*Aquilegia*) was followed by the origin of numerous species within a short time, as shown by the shortness of the branches between speciation events. The sister group (*Isopyrum*) that lacks spurs did not diversify as abundantly or as quickly. (After Ree 2005; data from S. A. Hodges and M. L. Arnold.)

ecological opportunity. Diversification rates in the fossil record fit this pattern, as we have seen, and the rate of expansion of morphological disparity does as well (e.g., molar tooth patterns in ungulates; see Figure 7.7). As we have seen, bursts of diversification can also be detected in phylogenies as clusters of very short branches between successive branch points (see Figure 7.2A). This pattern sometimes holds not only for species diversity, but also for disparity. In the *Anolis* lizards that have diversified in the Greater Antilles (see Figure 6.24), the rate of increase of disparity in body size and leg length was high at first and became slower over time. It was especially affected by the increase in the number of lineages (Mahler et al. 2010). In many clades of animals, however, evolutionary change in body size and shape continued steadily throughout their history (Harmon et al. 2010).

Other influences on diversity

The degree to which the world's biota is partitioned among geographic regions is called **provinciality**. A faunal or floral province is a region containing high numbers of distinctive, localized taxa (see Chapter 6). The fauna and flora of the present world are divided into more provinces than ever before in the history of life. A trend from a cosmopolitan distribution of taxa to more localized distributions has persisted throughout much of the Mesozoic and Cenozoic and is thought by some paleontologists to be one of the most important causes of the increase in global diversity during this time (Valentine et al. 1978; Signor 1990).

During the Mesozoic, as Pangaea became fragmented first into Laurasia and Gondwana and then into the modern continents, marine animals became distributed among an increasing number of latitudinally arranged provinces in both the Atlantic and Pacific regions. Similarly, among terrestrial vertebrates, a distinct fauna developed on each major land mass during the later Mesozoic and the Cenozoic, and the broad latitudinal distributions of many dinosaurs and other Mesozoic groups gave way to the much narrower latitudinal ranges of today's vertebrates. The spread of land masses over a broader latitudinal span created a stronger latitudinal temperature gradient than ever before (Valentine et al. 1978). Not only did the variety of environments increase, but the fragmentation of land masses also allowed for divergent evolution and prevented interchanges of species that, by competition or predation, might lower diversity.

Although some interactions among species, such as the effects of predators and pathogens, can cause extinction, an increase in the number of species in a clade almost inevitably results, sooner or later, in an increase of affiliated species—especially parasites and mutualists—that use them as resources. The evidence that diversity thereby begets more diversity comes largely from studies of living organisms. For example, tropical forests contain far more species of herbivorous insects than temperate forests do because tropical forests have more species of plants—resources on which the insects are specialized to some extent (Novotny et al. 2006, 2007; Dyer et al. 2007). Using a phylogeny of the largest family of butterflies, Niklas Janz and colleagues (2006) chose pairs of sister taxa that differ in the diversity of the plants that their larvae eat. In 18 of 22 such pairs, the group with the higher diversity of host plants has the higher number of species, as expected if the diversification of plants has contributed to speciation and diversification of the insects. In many such cases, a molecular clock analysis has shown that the insects diversified after the divergence among the plant lineages that they now use (Winkler and Mitter 2008).

Major changes in climate have been associated with increases in extinction as well as with changes in the distribution of habitats and vegetation types that have led to diversification. Changes in the mammalian fauna are associated with changes in climate (Figueirido et al. 2012); for example, in the mid-Eocene, the climate became cooler and drier, subtropical forests were widely replaced by savannahs in much of the temperate zone, and the diversity of primates and other arboreal mammals declined while that of large herbivores increased (Janis 1993). On the whole, however, the importance of climate change in the evolution of diversity, relative to other factors such as key adaptations and biotic interactions, is not yet well understood (Alroy et al. 2000).

Summary

1. The per capita rate of diversification equals the rate of origination (or speciation) per time interval minus the rate of extinction per time interval. Analyses of diversity in the fossil record require procedures to correct for biases caused by the incompleteness of the record. Some inferences about rates of diversification, speciation, and extinction can also be made from time-calibrated phylogenies of living species.

2. The diversity of skeletonized marine animals has increased during the Phanerozoic, but some aspects of this pattern are uncertain. Diversity appears to have increased in the Cambrian to an approximate equilibrium that lasted for almost two-thirds of the Paleozoic; then, after a mass extinction at the end of the Permian, it has increased, with interruptions and at varying rates, since the Mesozoic.

3. The background rate of extinction (in between mass extinctions) has declined during the Phanerozoic, perhaps because higher taxa that were particularly prone to extinction became extinct early.

4. Five major mass extinctions (at the ends of the Ordovician, Devonian, Permian, Triassic, and Cretaceous) are recognized. Although the cause of the incomparably devastating end-Permian extinction is unknown, it may have been the result of a rapid episode of environmental change initiated by the volcanic release of vast quantities of lava. The impact of a large extraterrestrial body at the end of the Cretaceous may have caused the extinction of many taxa, including the last of the nonavian dinosaurs.

5. Periods of high extinction rates have been followed by intervals of rapid origination of new taxa. Their diversification was probably released by the extinction of taxa that had occupied similar ecological space. Newly diversifying groups have sometimes replaced other taxa by direct competitive displacement, but more often they have replaced incumbent taxa after those taxa became extinct.

6. The rates of both extinction and origination of taxa have been diversity-dependent. Such observations imply that diversity tends toward an equilibrium, but how close to equilibrium current diversity may be is an unresolved issue.

7. Differences among taxa in species richness today are attributable to different rates of diversification in some cases, to differences in clade age in others, and probably to differences in equilibrium diversity in still other cases.

8. The increase in diversity over time appears to have been caused mostly by adaptation to vacant or underused adaptive zones ("ecological space"), often as a consequence of the evolution of key adaptations. Diversity has also been affected by increasing provinciality (differentiation of the biota in different geographic regions) owing to plate tectonics and by biological interactions, whereby each new species may be a resource for new species of consumers or mutualists.

Terms and Concepts

adaptive zone
background extinction
carrying capacity
competitive exclusion

disparity
diversity-dependent factors
exponential growth

key adaptation
lineage-through-time plot
mass extinctions
provinciality

pull of the Recent
Red Queen hypothesis
species selection
species sorting

Suggestions for Further Reading

Most of the topics in this chapter are treated clearly in an excellent textbook on paleobiology, *Principles of Paleontology* (third edition), by M. Foote and A. I. Miller (W. H. Freeman, New York, 2007). See also D. Jablonski et al. (eds.), *Evolutionary Paleobiology* (University of Chicago Press, Chicago, 1996).

The end-Permian mass extinction is the subject of a popular book by D. H. Erwin, *Extinction: How Life on Earth Nearly Ended 250 Million Years Ago* (Princeton University Press, Princeton, NJ, 2006). The consequences of mass extinctions are reviewed by R. K. Bambach, "Phanerozoic biodiversity mass extinctions," *Annual Review of Earth and Planetary Sciences* 34: 127–155 (2006) and D. Jablonski, "Mass extinctions and macroevolution," *Paleobiology* 31 (Supplement): 192–210 (2005).

Adaptive radiation is treated by R. E. Glor, "Phylogenetic insights on adaptive radiation," *Annual Review of Ecology, Evolution, and Systematics* 41: 251–270 (2010), and at greater length by D. Schluter in *The Ecology of Adaptive Radiation* (Oxford University Press, Oxford, 2010).

Problems and Discussion Topics

1. Distinguish between the rate of speciation in a higher taxon and its rate of diversification. What are the possible relationships between the present number of species in a taxon, its rate of speciation, and its rate of diversification?

2. What factors might account for differences among taxa in their numbers of extant species? Suggest methods for determining which factor might actually account for an observed difference.

3. Ehrlich and Raven (1964) suggested that coevolution with plants was a major cause of the great diversity of herbivorous insects, and Mitter et al. (1988) presented evidence that the evolution of herbivory was associated with increased rates of insect diversification. However, the increase in the number of insect families in the fossil record was not accelerated by the explosive diversification of flowering plants (Labandeira and Sepkoski 1993). Suggest some hypotheses to account for this apparent conflict and some ways to test them.

4. A factor that might contribute to increasing species numbers over time is the evolution of increased specialization in resource use, whereby more species coexist by more finely partitioning resources. Discuss ways in which, using either fossil or extant organisms, one might test the hypothesis that a clade is composed of increasingly specialized species over the course of evolutionary time.

5. In several phyla of marine invertebrates, lineages classified as new orders appear first in the fossil record in shallow-water environments and are recorded from deep-water environments only later in their history (Jablonski and Bottjer 1990). What might explain this observation? (Note: No one has offered a definitive explanation so far, so use your imagination.)

6. The analysis by McPeek and Brown (2007) suggests that clades with few living species may be very young. Is this necessarily the case? Are there alternative hypotheses? Can you find evidence for any of these hypotheses? What would constitute evidence?

7. The method of replicated sister-group comparison of species richness has been used to implicate certain adaptive characteristics as contributors to higher species richness. Is there any way, conversely, to test hypotheses on what factors may have contributed to the decline or extinction of groups?

8. Scientific debate continues about the history and interpretation of the diversity patterns of many taxa. Analyze such a debate, and decide whether either side has settled the issue. If not, what further research would be needed to do so? An example is whether or not the enormous diversity of leaf beetles (Chrysomelidae) is due to a long history of co-diversification with their host plants. See Farrell 1998; Farrell and Sequeira 2004; and Gómez-Zurita et al. 2007.

The Origin of Genetic Variation

The fundamental process of evolution is a change in the inherited characteristics of a population or species. It is an alteration of the genetic composition of a population. To understand evolution, therefore, it is essential to know the fundamentals of genetics and to understand the several processes that can change organisms' characteristics at a genetic level.

Genetic changes in populations and species begin with changes in the genetic material carried by individual organisms: mutations. Every gene, every variation in DNA, every characteristic of a species, every species itself, owes its existence to processes of mutation. Mutation is not the cause of evolution, any more than fuel in a car's tank is the cause of the car's movement. But it is the sine qua non, the necessary ingredient of evolution, just as fuel is necessary—though not sufficient—for traveling down a highway. The fundamental role of mutation makes it a logical starting point for our analysis of the causes of evolution.

Technological advances in molecular biology have utterly transformed our knowledge of the nature and processes of mutation in the last several decades. Many kinds of alterations of DNA, and their effects on individual genes, have been described, but the study of mutations has now entered a major new phase in which it is possible to describe how various kinds of mutations have shaped entire genomes. One of the landmarks in the history of science was the simultaneous publication in February 2001 of two "drafts" of the complete sequence of the human genome, one by the International Human Genome Sequencing Consortium (2001) and the other by a private company (Venter et al. 2001). Since then, complete genome sequences of other species have been announced almost every week, so it is now possible to compare genetic differences among a multitude of viruses, bacteria, and eukaryotes that range

Ladies of the court In *Las Meninas*, the artist Velásquez portrayed two achondroplasic dwarves (at right) among the ladies attending the daughter of Philip IV of Spain. Achondroplasia is caused by dominant mutations that substitute arginine for glycine at a site in the receptor protein FGF-3, a signal transduction protein. The mutant receptor results in insufficient formation of bone from cartilage. Most mutations that contribute to evolution have less pronounced effects.

from unicellular parasites to rice, sea urchins, and the platypus. Comparisons among these sequences are providing unprecedented volumes of information about the processes and history of evolution (see Chapter 20). Still, as we will see, many questions about the role of mutation in evolution require more than molecular data and are still very imperfectly answered.

Genes and Genomes

We will begin by describing mutation at the molecular level, but doing so requires a short review of the genetic material and its organization.

Except in certain viruses, in which the genetic material is RNA (ribonucleic acid), organisms' genomes consist of DNA (deoxyribonucleic acid), which is made up of a series of nucleotide BASE PAIRS (bp), each consisting of a purine (adenine, A, or guanine, G) and a pyrimidine (thymine, T, or cytosine, C). A haploid (gametic) genome of the fruit fly *Drosophila melanogaster* has about 1.5×10^8 bp, and that of a human about 3.2×10^9 bp (3.2 billion). DNA content varies greatly among organisms: it differs more than a hundredfold, for example, among species of salamanders, some of which have more than a hundred times as much DNA as humans. The genome of the unicellular protist *Amoeba dubia* is 200 times the size of a human's!

Each chromosome consists of a single long, often tightly coiled, DNA molecule together with histone proteins. Some portions of the DNA strand constitute different genes. The term **gene** usually refers to a sequence of DNA that is transcribed into RNA, together with any untranscribed regions that play roles in regulating its transcription. The term **locus** (plural **loci**) technically refers to the chromosome site occupied by a particular gene, but it is often used to refer to the gene itself. Thousands of genes in the human genome encode ribosomal and transfer RNAs (rRNAs and tRNAs) that are not translated into proteins. The number of protein-coding genes is about 26,000 in the small flowering plant *Arabidopsis* (sometimes called mustard weed), 6000 in the fungus (yeast) *Saccharomyces*, 16,000 in *Drosophila*, and 24,000 in mice and humans.

One strand of a protein-coding gene is transcribed into RNA. The transcription process is regulated by **control regions**: untranscribed sequences (**enhancers** and **repressors**) to which regulatory proteins produced by other genes bind. (The control regions are said to be in a *cis* position, adjacent to the coding sequence, whereas the regulatory proteins, or transcription factors, are encoded by genes that are located elsewhere in the genome, and are said to be in a *trans* position, relative to the controlled gene.) A gene may have many different enhancer sequences.

In eukaryotes, the transcribed sequence of a gene consists of coding regions (called **exons**), which in most genes are separated by noncoding regions (called **introns**; **Figure 8.1**). Control regions of a gene (e.g., enhancers) may be located at various distances from the coding regions, often "upstream" of them but sometimes "downstream" or within introns. The average human gene has 1340–1500 bp of coding sequence (encoding 445–500 amino acids) divided into 8.8 exons, which are separated by a total of 3365 bp of introns. After a gene is transcribed, the portions transcribed from exons are processed into a messenger RNA (mRNA) by enzymes that splice out the portions that were transcribed from introns. ALTERNATIVE SPLICING, whereby the mature mRNA corresponds to a variable number of the exons, can result in several proteins being encoded by a single gene. At least 35 percent of human genes appear to be subject to alternative splicing, so the number of possible proteins may greatly exceed the number of genes.

Through the action of ribosomes, enzymes, and tRNAs, mRNA is translated into a polypeptide or protein on the basis of the **genetic code**, whereby a triplet of bases (a **codon**) specifies a particular amino acid in the growing polypeptide chain. The RNA code (**Figure 8.2**), which is complementary to the DNA code, consists of $4^3 = 64$ codons, which, however, encode only the 20 amino acids of which proteins are composed. In other words, the genetic code is redundant: most of the amino acids are encoded by two or more **synonymous codons**. The third position in a codon is the most "degenerate"; for example,

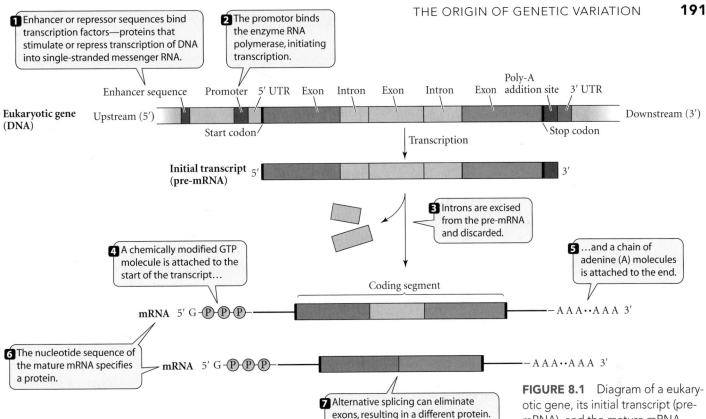

FIGURE 8.1 Diagram of a eukaryotic gene, its initial transcript (pre-mRNA), and the mature mRNA transcript. Proteins that regulate transcription bind to enhancer and repressor sequences, shown here upstream of the coding sequences. Transcription proceeds in the 5′ to 3′ direction. Introns are transcribed, but are spliced out of the pre-mRNA and discarded. The coding segment of the mature mRNA, which is translated to produce a protein, corresponds to the gene's exons. In some genes, alternative splicing may yield several different mRNAs. UTRs (untranslated regions) sometimes include regulatory sequences.

all four CC– codons (CCU, CCC, CCA, CCG) specify proline. The second position is least degenerate—a substitution of one base for another in the second position usually results in an amino acid substitution. Three of the 64 codons are "stop" ("chain end") signals that terminate translation. Substitution in the third positions of five other codons can produce a "stop" codon, which may result in an incomplete, often nonfunctional, protein.

It is a profoundly wonderful fact that the genetic code is nearly universal, from viruses and bacteria to pineapples and mammals. Moreover, the machinery of transcription and translation is remarkably uniform, so that sea urchin DNA or mRNA can be translated into

First nucleotide	Second nucleotide: U	C	A	G	Third nucleotide
U	UUU Phe UUC Phe UUA Leu UUG Leu	UCU Ser UCC Ser UCA Ser UCG Ser	UAU Tyr UAC Tyr UAA Stop UAG Stop	UGU Cys UGC Cys UGA Stop UGG Trp	U C A G
C	CUU Leu CUC Leu CUA Leu CUG Leu	CCU Pro CCC Pro CCA Pro CCG Pro	CAU His CAC His CAA Gln CAG Gln	CGU Arg CGC Arg CGA Arg CGG Arg	U C A G
A	AUU Ile AUC Ile AUA Ile AUG Met	ACU Thr ACC Thr ACA Thr ACG Thr	AAU Asn AAC Asn AAA Lys AAG Lys	AGU Ser AGC Ser AGA Arg AGG Arg	U C A G
G	GUU Val GUC Val GUA Val GUG Val	GCU Ala GCC Ala GCA Ala GCG Ala	GAU Asp GAC Asp GAA Glu GAG Glu	GGU Gly GGC Gly GGA Gly GGG Gly	U C A G

FIGURE 8.2 The genetic code, as expressed in mRNA. Three of the 64 codons are "stop" ("chain end") signals; the other codons encode the 20 amino acids found in proteins. Note that many codons, especially those differing only in the third position, are synonymous. The three-letter abbreviations signify the amino acids.

FIGURE 8.3 The human hemoglobin gene family has two subfamilies, α (green) and β (blue), located on different chromosomes. Each functional gene is indicated by three lines, representing its three exons. Pseudogenes are denoted by ψ. See Figure 3.29 for a history of how these genes have arisen by duplication and divergence. (After Hartwell et al. 2000.)

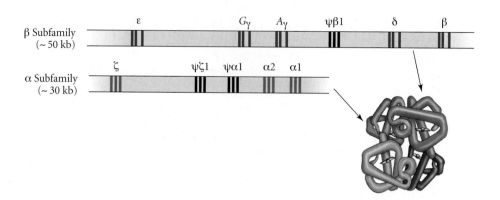

a protein if injected into a frog. This uniformity is the basis of genetic engineering, such as the insertion into crop plants of bacterial genes that encode natural insecticides.

Only about 2 percent of the human genome encodes proteins. As we have seen, noncoding DNA near transcribed genes often plays a role in regulating transcription of those genes, and this is also true of portions of some introns. Until recently, most of the noncoding DNA in eukaryotes was thought to be nonfunctional "junk DNA," but there is evidence that much of this DNA is functionally important. About 80 percent of it is transcribed into RNAs that perform various functions, and many other nontranscribed regions are "control" sequences to which proteins that regulate transcription bind. This DNA's role, however, is imperfectly understood. Mutations in the sequence of base pairs in truly nonfunctional regions of the genome would not affect the organism and so should not be subject to natural selection (see Chapter 10).

Some of the noncoding DNA consists of highly repeated sequences. For instance, at least 45 percent of the human genome consists of as many as 4.3 million repetitive elements that include repeated sequences of a few base pairs each. These very short sequences are sometimes referred to as **microsatellites**, and they often occur as **tandem repeats** that may number more than 2 billion in some species. Other repeated sequences, which may harbor coding DNA, include short interspersed repeats (SINEs) of 100 to 400 base pairs; long interspersed repeats (LINEs) that are more than 5 kilobases (kb) in length; and DNA transposons. These elements either are capable of undergoing or have been formed by the process of **transposition**: the production of copies that become inserted into new positions in the genome. DNA sequences that are capable of transposition are called **transposable elements** (**TEs**).

Many protein-coding genes are members of **gene families**: groups of genes that are similar in sequence and often have related functions. For example, the human hemoglobin gene family includes two subfamilies, α and β, located on different chromosomes (**Figure 8.3**). Different α and β polypeptides are combined to form different types of hemoglobin before and after birth. Such gene families are examples of repeated sequences that have diversified over time. Some gene families, such as that for mammalian olfactory receptors, have more than a thousand functional members. Both the α and β hemoglobin subfamilies also include PSEUDOGENES: sequences that resemble the functional genes but differ from them at several base pair sites and are not transcribed (i.e., they are "silent") because they have internal "stop" codons. **Processed pseudogenes** are pseudogenes that originated by REVERSE TRANSCRIPTION from an mRNA message into a DNA sequence that lacked introns, and later underwent silencing and further base pair changes. The DNA sequence of a processed pseudogene is recognizably similar to the coding portion of the gene from which it arose, but it lacks passages that can be matched with the functional gene's introns.

Mutations: An Overview

The word **mutation** refers both to the process of alteration of a gene or chromosome and to its product, the altered state of a gene or chromosome. It is usually clear from the context which is meant.

Before the development of molecular genetics, a mutation was identified by its effect on a phenotypic character. That is, a mutation was a newly arisen change in morphology, survival, behavior, or some other property that was inherited and could be mapped (at least in principle) to a specific locus on a chromosome. In practice, many mutations are still discovered, characterized, and named by their phenotypic effects. Thus we will frequently use the term "mutation" to refer to an alteration of a gene from one form, or **allele**, to another, the alleles being distinguished by phenotypic effects. In a molecular context, however, a gene mutation is an alteration of a DNA sequence, independent of whether or not it has any phenotypic effect. A particular DNA sequence that differs by one or more mutations from homologous sequences is called a **haplotype**. We will often refer to **genetic markers**, which are detectable mutations that geneticists use to recognize specific regions of chromosomes or genes.

Mutations have evolutionary consequences only if they are transmitted to succeeding generations. Mutations that occur in somatic cells may be inherited in certain animals and plants in which the reproductive structures arise from somatic meristems, but in those animals in which the germ line is segregated from the soma early in development, a mutation is inherited only if it occurs in a germ line cell. Mutations are thought to occur mostly during DNA replication, which usually occurs during cell division. In mammals, *Drosophila*, and many other animals, more new mutations enter the population via sperm than via eggs because more cell divisions have transpired in the germ line before spermatogenesis than before oogenesis in individuals of equal age (Sayres et al. 2011).

DNA is frequently damaged by chemical and physical events, and changes in the base pair sequence can result. Many such changes are repaired by a variety of repair enzymes, but some are not. These mutations are considered by most evolutionary biologists to be errors. That is, *the process of mutation is generally thought to be not an adaptation*, but a consequence of unrepaired damage (see Chapter 15).

A particular mutation occurs in a single cell of a single individual organism. If that cell is in the germ line, it may give rise to a gamete or, quite often, to a number of gametes that carry the mutation, and so may be inherited by several offspring. Initially, the mutation is carried by a very small percentage of individuals in the species population. Because of natural selection or genetic drift, the mutation may ultimately become **fixed**, i.e., carried by the entire population. In that case, it is best referred to as a **substitution**. (This word is also used to describe base pair differences between homologous DNA sequences.) Most mutations do not become fixed. Consequently, *the process of mutation is not equivalent to evolution*. Nor does an organism that carries a new mutation constitute a distinct new species. (The meaning of "species" is discussed in Chapter 17.)

Kinds of mutations

Mutational changes in DNA are many in kind; there is a full spectrum, from changes of a single base pair through alterations that affect longer DNA sequences that may encompass part or all of a gene, to alterations of entire chromosomes or even an entire genome. For convenience, we will arbitrarily distinguish gene mutations from chromosome changes.

GENE MUTATIONS   The simplest mutations are alterations of a single base pair. These mutations include BASE PAIR CHANGES (**Figure 8.4**). In classical genetics, a mutation that maps to a single gene locus is called a **point mutation**; in modern usage, this term is often restricted to single base pair substitutions. A **transition** is a substitution of a purine for a purine (A ↔ G) or a pyrimidine for a pyrimidine (C ↔ T). **Transversions**, of eight possible kinds, are substitutions of purines for pyrimidines, or vice versa (A or G ↔ C or T).

Mutations may have a phenotypic effect if they occur in genes that encode ribosomal and transfer RNA, untranslated regulatory sequences such as enhancers, or protein-coding regions. Because of the redundancy of the genetic code, 24 percent of the possible kinds of base pair mutations in coding regions are **synonymous mutations**, which have no effect on the amino acid sequence of the polypeptide or protein involved. **Nonsynonymous**

Direction of transcription →

Original sequence:	DNA:	AGA	TGA	CGG	TTT	GCA
	RNA:	UCU	ACU	GCC	AAA	CGU
	Protein:	Ser	Thr	Ala	Lys	Arg

Base pair substitutions

Transition (A→G)

GGA	TGA	CGG	TTT	GCA
CCU	ACU	GCC	AAA	CGU
Pro	Thr	Ala	Lys	Arg

Transversion (A→T)

TGA	TGA	CGG	TTT	GCA
ACU	ACU	GCC	AAA	CGU
Thr	Thr	Ala	Lys	Arg

Frameshifts

Insertion (T)

AGT	ATG	ACG	GTT	TGC	A _ _
UCA	UAC	UGC	CAA	ACG	
Ser	Tyr	Cys	Glu	Thr	

Deletion (T)

AGT	ATGA	CGG	TTT	GCA
UCA	UCU	GCC	AAA	CGU
Ser	Ser	Ala	Lys	Arg

FIGURE 8.4 Examples of point mutations and their consequences for mRNA and amino acid sequences. (Only the transcribed, "sense" strand of the DNA is shown.) The boxes at the left show two kinds of base pair substitutions, a transition and a transversion, at the first base position. The boxes at the right show two kinds of frameshift mutations: an insertion and a deletion. Note that a frameshift mutation affects the codons, and thus translation, "downstream" (i.e., to the right). In this example, the deletion, which occurs downstream from the insertion that shifted the reading frame, restores the frame, and thereby the amino acid sequence, downstream.

mutations, in contrast, result in amino acid substitutions; they may have little or no effect on the functional properties of the polypeptide or protein, and thus no effect on the phenotype, or they may have substantial consequences. For example, a change from the (RNA) codon GAA to GUA causes the amino acid valine to be incorporated instead of glutamic acid. This is the mutational event that in humans caused the abnormal β-chain in sickle-cell hemoglobin, which has many phenotypic consequences and is usually lethal in homozygotes.

INSERTIONS or DELETIONS (together called INDELS) of DNA are another common kind of mutation. Indels may involve a single base pair or many. If a single base pair is inserted into or deleted from a coding sequence, the triplet reading frame is shifted by one nucleotide, so that downstream triplets are read as different codons and translated into different amino acids (see Figure 8.4). Thus insertions or deletions often result in **frameshift mutations**. The greatly altered gene product is usually nonfunctional, although exceptions to this generalization are known. Base pair changes occur more frequently than insertions and deletions; for example, they account for more than ten times the number of differences between homologous processed pseudogenes of humans and chimpanzees (Nachman and Crowell 2000).

REPLICATION SLIPPAGE is a process that alters the number of repeats of microsatellites. The growing (3′) end of a DNA strand that is being formed during replication may become dissociated from the template strand and form a loop, so that the next repeat to be copied from the template is one that has already been copied. Thus extra repeats will be formed in the growing strand. Microsatellite "alleles" that differ in copy number arise by replication slippage at a high rate. For instance, the number of tetranucleotide (4 bp) repeats

FIGURE 8.5 Mutational changes in the number of tetranucleotide repeats of a microsatellite in the superb fairy-wren (*Malurus cyaneus*), estimated by comparing repeats in offspring and their parents. Most frequently, only a single repeat was added or deleted. Results are based on alternative assumptions that the progenitor of an altered microsatellite was the largest among the possible parental alleles (red bars), the smallest (blue), or drawn at random (gold). The number of repeats varied from 32 to 93 in this population. (After Beck et al. 2003.)

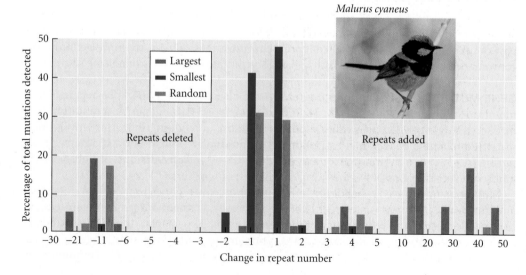

Malurus cyaneus

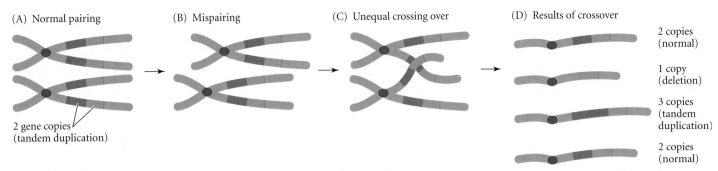

(A) Normal pairing

2 gene copies
(tandem duplication)

(B) Mispairing

(C) Unequal crossing over

(D) Results of crossover

2 copies
(normal)

1 copy
(deletion)

3 copies
(tandem
duplication)

2 copies
(normal)

FIGURE 8.6 Unequal crossing over occurs most commonly when repeated sequences mispair with their homologues. Crossing over then yields a deletion of one of the repeated sequences on one chromatid and a duplication (and hence three copies of that sequence) on the other chromatid. The repeated sequence may be a single base pair, it may encode part of a gene, or it may encode one or more complete genes. In this diagram, the chromosome pair already has two repeats of a sequence when it undergoes unequal exchange that deletes one repeat on one of the chromosomes and adds a third repeat to the other. (After Hartl and Jones 2001.)

at a microsatellite locus was scored in 1615 offspring of 254 females and 288 males in a long-term study of superb fairy-wrens (*Malurus cyaneus*) in Canberra, Australia (**Figure 8.5**). Among the 3230 meioses represented, 45 mutational changes in repeat number were recorded, for a mutation rate of 1.4 percent. This rate is much higher than the rate of point mutation per base pair, as we will see.

SEQUENCE CHANGES ARISING FROM RECOMBINATION Recombination typically begins with precise alignment of the DNA sequences on homologous chromosomes in meiosis. When homologous DNA sequences differ at two or more base pairs, **intragenic recombination** between them can generate new DNA sequences, just as crossing over between genes generates new gene combinations. DNA sequencing has revealed many examples of variant sequences that apparently arose by intragenic recombination.

Unequal crossing over (unequal exchange) can occur between two homologous sequences or chromosomes that are not perfectly aligned. Recombination then results in a TANDEM DUPLICATION on one recombination product and a deletion on the other (**Figure 8.6**). The length of the affected region may range from a single base pair to a large block of loci (a SEGMENTAL DUPLICATION), depending on the amount of displacement of the two misaligned chromosomes. Most unequal crossing over occurs between sequences that already include tandem repeats (e.g., ABBC) because the duplicate regions can pair out of register:

$$\begin{pmatrix} \text{ABBC...} \\ \text{...ABBC} \end{pmatrix}$$

This pairing generates further duplication (ABBBC). Unequal crossing over is one of the processes that has generated the extremely high number of copies of nonfunctional sequences that constitute much of the DNA in most eukaryotes. It has given rise to many gene families, and it has been extremely important in the evolution of greater numbers of functional genes and increases in total DNA (see Chapter 20).

CHANGES CAUSED BY TRANSPOSABLE ELEMENTS Most TEs produce copies that can move to any of many places in the genome, and when they do so, they sometimes carry with them other genes near which they had been located. TEs include genes that encode enzymes that accomplish the transposition (movement). The several kinds of TEs include INSERTION SEQUENCES, which encode only enzymes that cause transposition; TRANSPOSONS, which encode other functional genes as well; and RETROELEMENTS, which carry a gene for the enzyme reverse transcriptase. Retroelements are first transcribed into RNA, which then is reverse-transcribed into a DNA copy (cDNA) that is inserted into the genome. Some retroelements are retroviruses (such as the HIV virus that causes AIDS) whose RNA copies can cross cell boundaries.

TEs often become excised from a site into which they had previously been inserted, but leave behind sequence fragments that tell of their former presence. From such cases and from more direct evidence, TEs are known to have many effects on genomes (Kazazian 2004):

• When inserted into a coding region, they alter, and usually destroy, the function of the protein, often by causing a frameshift or by altering splicing patterns.

FIGURE 8.7 Recombination between copies of a transposable element can result in deletions and inversions. The boxes containing arrows represent TEs, with the polarity of base pair sequence indicated by the arrows. The numerals represent genetic markers. (A) Recombination between two direct repeats (i.e., with the same polarity) excises one repeat and deletes the sequence between the two copies. (B) Recombination between two inverted repeats (with opposite polarity) inverts the sequence between them. (After Lewin 1985.)

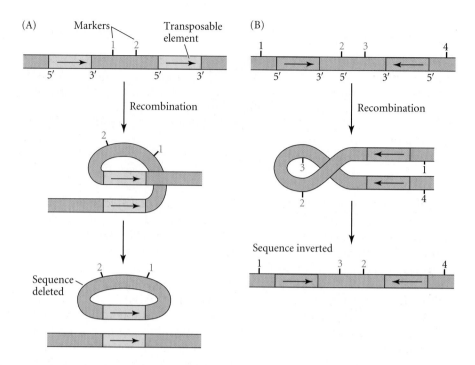

- When inserted into or near control regions, they can alter gene expression (e.g., the timing or amount of transcription), or act as enhancers or other control elements.

- They are known to increase mutation rates of host genes.

- They can cause rearrangements in the host genome, resulting from recombination between two copies of a TE located at different sites (**Figure 8.7**). Just as unequal crossing over between repeated sequences can generate duplications and deletions, so can recombination between copies of a TE located at nonhomologous sites. Recombination between two copies of a TE with the same sequence polarity can *delete* the region between them (Figure 8.7A), whereas recombination between two copies with opposite polarity *inverts* the region between them (Figure 8.7B).

- TEs that encode reverse transcriptase sometimes insert DNA copies (cDNA) not only of their own RNA, but also of RNA transcripts of other genes, into the host genome. These cDNA copies of RNA (RETROSEQUENCES) resemble the exons of an ancestral gene located elsewhere in the genome, but they lack control regions and introns. Most retrosequences are processed pseudogenes, which do not produce functional gene products.

- By transposition and unequal crossing over, TEs can increase or decrease in number, and so change the size of the genome. About 10 percent of the human genome is composed of more than a million copies of a retrotransposable element called *Alu*, which are thought to have built up over the course of about 60 My (Petrov and Wendel 2006).

Most or all of these TE-induced effects have been observed in experimental populations. In laboratory stocks of *Drosophila melanogaster*, for example, transposon insertions cause most of the spontaneous mutations (mutations that have not been deliberately induced by mutagens) that geneticists have identified by their phenotypic effects. The transposition rate of various retroelements ranges from about 10^{-5} to 10^{-3} per copy in inbred lines of *Drosophila*, resulting in an appreciable rate of mutation (Nuzhdin and Mackay 1994). All the kinds of changes engendered by TEs can be found by comparing genes and genomes of organisms of the same or different species. For instance, *L1* retrotransposon insertions are associated with many disease-causing mutations in both mice and humans (Kazazian 2004). Some of the mutations caused by insertion of TEs are advantageous, however. Such mutations have been found in wild populations (González et al. 2008), and some have contributed to major changes during evolution (Lynch et al. 2011).

Examples of mutations

Geneticists have learned an enormous amount about the nature and causes of mutations by studying model organisms such as *Drosophila* and the bacterium *Escherichia coli*. Moreover, many human mutations have been characterized because of their effects on health. Human mutations are usually rather rare variants that can be compared with normal forms of the gene; in some instances, newly arisen mutations have been found that are lacking in both of a patient's parents.

Single base pair substitutions are responsible for conditions such as sickle-cell hemoglobin, described earlier, and for precocious puberty, in which a single amino acid change in the receptor for luteinizing hormone causes a boy to show signs of puberty when about 4 years old. Because many different alterations of a protein can diminish its function, the same phenotypic condition can be caused by many different mutations of a gene. For example, cystic fibrosis, a fatal condition afflicting 1 in 2500 live births in northern Europe, is caused by mutations in the gene encoding a sodium channel protein. The most common such mutation is a 3 bp deletion that deletes a single amino acid from the protein; another converts a codon for arginine into a "stop" codon; another alters splicing so that an exon is missing from the mRNA; and many of the more than 500 other base pair changes recorded in this gene are also thought to cause the disease (Zielenski and Tsui 1995). Mutations in any of the many different genes that contribute to the normal development of some characteristics can also result in similar phenotypes. For example, retinitis pigmentosa, a degeneration of the retina, can be caused by mutations in genes on eight of the twenty-three chromosomes in the (haploid) human genome (Avise 1998).

Hemophilia can be caused by mutations in two different genes that encode blood-clotting proteins. In both genes, many different base pair substitutions, as well as small deletions and duplications that cause frameshifts, are known to cause the disease, and about 20 percent of cases of hemophilia-A are caused by an inversion of a long sequence within one of the genes (Green et al. 1995). Huntington's disease, a fatal neurological disorder that strikes in midlife, is caused by an excessive number of repeats of the sequence CAG: the normal gene has 10 to 30 repeats, the mutant gene more than 75. Unequal crossing over between the two tandemly arranged genes for α-hemoglobin (see Figure 8.3) has given rise to variants with three tandem copies (duplication) and with one copy (deletion). The deletion of one of the loci causes α-thalassemia, a severe anemia. Another case of deletion, which results in high cholesterol levels, is the lack of exon 5 in a low-density lipoprotein gene. This deletion has been attributed to unequal crossing over, facilitated by a short, highly repeated sequence called *Alu* that is located in the introns of this gene (as well as in many other sites in the genome, as mentioned above) (**Figure 8.8**).

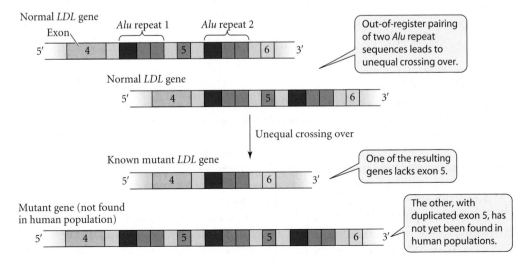

FIGURE 8.8 A mutated low-density lipoprotein (*LDL*) gene in humans lacks exon 5. It is believed to have arisen by unequal crossing over between two normal gene copies as a result of out-of-register pairing between two of the repeated sequences (*Alu*, shown as blue boxes) in the introns. The numbered boxes are exons. (After Hobbs et al. 1986.)

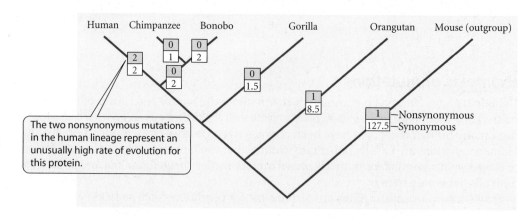

FIGURE 8.9 A phylogeny of the Hominoidea and its divergence from an outgroup, the mouse. Each box shows the number of nonsynonymous (yellow boxes) and synonymous (white boxes) substitutions in the *FOXP2* gene. The two nonsynonymous substitutions in the human lineage represent an unusually high rate of evolution of the FOXP2 protein and they may represent mutations that have been important in the evolution of language and speech. (After Zhang et al. 2002.)

These examples might make it seem as if mutations are nothing but bad news. While this is close to the truth—far more mutations are harmful than helpful—these examples represent a biased sample. Many advantageous mutations have become incorporated into species' genomes (fixed) and thus represent the current **wild-type**, or normal, genes. We will encounter many examples of beneficial mutations, and we will only remark here that they may include not only point mutations, but also some more complex genomic alterations. For example, most sequences that have arisen by reverse transcription from mRNA are nonfunctional pseudogenes, but at least one has been found that is a fully functional member of the human genome. Phosphoglycerate kinase is encoded by two genes. One, on the X chromosome, has a normal structure of 11 exons and 10 introns. The other, on an autosome, which lacks introns, clearly arose from the X-linked gene by reverse transcription. It is expressed only in the testes, a novel pattern of tissue expression that seems to compensate for the lower level of transcription of the X-linked gene in this tissue (McCarrey et al. 1996).

When biologists seek those genes that have been involved in the evolution of a specific characteristic, they often use rare deleterious mutations of the kind just described as indicators of CANDIDATE GENES, those that may be among the genes they seek. For example, a rare mutation in the human *FOXP2* gene (*forkhead box 2*) causes severe speech and language disorders. This gene encodes a transcription factor that regulates genes that are important in the development of parts of the brain. Two research groups independently found that the normal form of this gene has undergone two nonsynonymous (amino acid–changing) substitutions in the human lineage since the divergence of the human and chimpanzee lineages less than 7 Mya (Enard et al. 2002; Zhang et al. 2002). (The same mutations were later found in the *Homo neanderthalensis* gene as well, and so occurred in the common ancestor of *H. neanderthalensis* and *H. sapiens*.) This is a much higher rate of protein evolution than would be expected, considering that only one other such substitution has occurred between these species and the mouse, which diverged almost 90 Mya (**Figure 8.9**). Both research groups proposed that these substitutions are among the important steps in the evolution of human language and speech. The mutations in humans affect the regulation of downstream genes and consequently affect brain regions (Reimers-Kippling et al. 2011). The DNA sequence of *FOXP2* also varies among species of echolocating bats, which use their own vocal signals to orient, find prey, and communicate (Li et al. 2007).

Rates of mutation

Recurrent mutation refers to the repeated origin of a particular mutation. The *rate* at which a particular mutation occurs is typically measured as the number of independent origins per gene copy (e.g., per gamete) per generation or per unit time (e.g., per year). Mutation rates are estimates, not absolutes, and these estimates depend on the method used to detect mutations. In classical genetics, a mutation was detected by its phenotypic effects, such as white versus red eyes in *Drosophila*. Such a mutation, however, might be caused by the alteration of any of many sites within a locus; moreover, many base pair changes have no phenotypic effect. Thus phenotypically detected rates of mutation underestimate the rate at which all mutations occur at a locus. Moreover, mutations may be expressed phenotypically more readily in some genes than in others, because the processes by which phenotypic features develop affect the expression of the genes that contribute to their development (Stern and Orgogozo 2009). With modern molecular methods, mutated DNA sequences can be detected directly, so mutation rates can be expressed per base pair. It is important to distinguish among the mutation rate per base pair, per locus (the sum of detectable mutations at all sites within the locus), and per genome (the sum of detectable mutations at all loci).

Back mutation is mutation of a "mutant" allele back to the allele (usually the wild type) from which it arose. Back mutations are ordinarily detected by their phenotypic effects. They usually occur at a much lower rate than "forward" mutations (from wild type to mutant), presumably because many more changes can impair gene function than can restore it. At the molecular level, most phenotypically detected back mutations are not restorations of the original sequence, but instead result from a second amino acid change, either in the same or a different protein, that restores the function that had been altered by the first mutation. Advantageous mutations arose and compensated for severely deleterious mutations within 200 generations in experimental populations of *E. coli* (Moore et al. 2000).

ESTIMATING MUTATION RATES   Several methods can be used to assess the rate of mutation averaged over different time scales (Drake et al. 1998; Houle and Kondrashov 2006). In the short term, mutations can be screened among offspring of genetically characterized parents or over several generations. This screening is usually done in the laboratory, but occasionally in field populations, as in the case of the study of fairy-wrens cited earlier. On a somewhat longer time scale are MUTATION ACCUMULATION EXPERIMENTS, in which new mutations arising in a laboratory stock (which is usually initially homozygous) are scored after several generations. Because mutation rates per gene are usually very low, reasonably good estimates of the mutation rate require that large numbers of progeny be scored. Thus many of these experiments use very fecund or rapidly reproducing organisms, such as fruit flies or, especially, microorganisms.

There is a risk of underestimating the number of mutations if some cause the death of their carriers and thus do not persist long enough to be counted. This is, of course, natural selection in action, so special efforts are needed to design experiments to minimize or control for the effects of natural selection.

An indirect method that estimates mutation rates averaged over many generations (**Box 8A**) is based on the number of base pair differences between homologous genes in different species, relative to the number of generations that have elapsed since they diverged from their common ancestor. This method depends on the neutral theory of molecular evolution, described in Chapter 10, which specifies that the per-generation rate of origin of selectively neutral mutations, per base pair, equals the proportion of base pairs that differ between two species, divided by twice the number of generations since their common ancestor. SELECTIVELY NEUTRAL mutations are those that do not alter their carriers' survival or reproductive success.

Mutation accumulation experiments with *Drosophila* and the nematode worm *Caenorhabditis elegans* have confirmed that the total mutation rate per gamete is quite high. The first major experiment was performed by Terumi Mukai and colleagues (1972), who screened more than 1.7 million flies in order to estimate the rate at which chromosome 2 accumulates mutations that affect egg-to-adult survival (VIABILITY). They used crosses (see Figure 9.12) in which copies of the wild-type chromosome 2 were carried in a heterozygous condition so that deleterious recessive mutations could persist without being eliminated by natural selection. Every ten generations, they performed crosses that made large numbers of those chromosomes homozygous and measured the proportion of those chromosomes that reduced viability. The mean viability declined, and the variation (variance) in viability among chromosomes increased steadily (**Figure 8.10**). From the changes in the mean and variance, Mukai et al. calculated a mutation rate of about 0.15 per chromosome 2 per gamete. This rate represents the sum, over all loci on the

FIGURE 8.10 Effects of the accumulation of spontaneous mutations on the viability (egg-to-adult survival) of *Drosophila melanogaster*. The mean viability of flies made homozygous for chromosome 2 carrying new recessive mutations decreased, and the variation (variance) in viability among those chromosomes increased. The rate of mutation was estimated from these data. (After Mukai et al. 1972.)

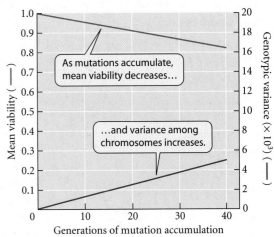

As mutations accumulate, mean viability decreases…

…and variance among chromosomes increases.

Mean viability (——)

Genotypic variance ($\times 10^3$) (——)

Generations of mutation accumulation

Box 8A Estimating Mutation Rates from Comparisons among Species

In Chapter 1, we introduced the neutral theory of molecular evolution. This theory describes the fate of purely neutral mutations—that is, those mutations that neither enhance nor lower fitness. One possible fate for a mutation is that it will become fixed—that is, attain a frequency of 1.0—entirely by chance. The probability that this event will occur equals u, the rate at which neutral mutations arise. In each generation, therefore, the probability is u that a mutation that occurred at some time in the past will become fixed. After the passage of t generations, the fraction of mutations that will have become fixed is therefore ut.

If two species diverged from a common ancestor t generations ago, the expected fraction of fixed mutations in both species is $D = 2ut$, since various mutations have become fixed in both lineages. If the mutations in question are base pair changes, a fraction $D = 2ut$ of the base pairs of a gene should differ between the species, assuming that all base pairs are equally likely to mutate. Thus the average mutation rate per base pair per generation is $u = D/2t$.

Thus we can estimate u if we can measure the fraction of base pairs in a gene that differ between two species (D), and if we can estimate the number of generations since the two species diverged from their common ancestor (t). This method requires an estimate of the length of a generation, information on the absolute time at which the common ancestor existed (based on a fossil record

or perhaps a geologically timed vicariance event, as described in Chapter 6), and an understanding of the phylogenetic relationships among the living and fossilized taxa.

In applying this method to DNA sequence data, it is necessary to assume that most base pair substitutions are neutral and to correct for the possibility that earlier substitutions at some sites in the gene have been replaced by later substitutions ("multiple hits"; see Chapter 2). Uncertainty about the time since divergence from the common ancestor is usually the greatest source of error in estimates obtained by this method.

The best estimates of mutation rates at the molecular level have been obtained from interspecific comparisons of pseudogenes, other untranslated sequences, and fourfold-degenerate third-base positions (those in which all mutations are synonymous), since these are thought to be least subject to natural selection (although probably not entirely free of it). In comparisons among mammal species, the average rate of nucleotide substitution has been about 2.2 per nucleotide site per 10^9 years, for a mutation rate of 2.2×10^{-9} per site per year (Kumar and Subramanian 2002). If the average generation time were 2 years during the history of the lineages studied, the average rate of mutation per site would be about 1.1×10^{-9} per generation.

TABLE 8.1 Spontaneous mutation rates of specific genes, detected by phenotypic effects	
Species and locus	**Mutations per 100,000 cells or gametes**
Escherichia coli	
Streptomycin resistance	0.00004
Resistance to T1 phage	0.003
Arginine independence	0.0004
Salmonella typhimurium	
Tryptophan independence	0.005
Neurospora crassa	
Adenine independence	0.0008–0.029
Drosophila melanogaster	
Yellow body	12.0
Brown eyes	3.0
Eyeless	6.0
Homo sapiens	
Retinoblastinoma	1.2–2.3
Achondroplasia	4.2–14.3
Huntington's disease	0.5

Source: After Dobzhansky 1970.

chromosome, of mutations that affect viability. Because chromosome 2 carries about a third of the *Drosophila* genome, the total mutation rate is about 0.50 per gamete. Thus almost every zygote carries at least one new mutation that reduces viability. Subsequent studies have indicated that the mutation rate for *Drosophila* is at least this high, and that it reduces viability by 1 to 2 percent per generation (Lynch et al. 1999; Haag-Liautard et al. 2007).

Mutation rates vary among genes and even among regions within genes, but on average, as measured by *phenotypic* effects, a locus mutates at a rate of about 10^{-6} to 10^{-5} mutations per gamete per generation (**Table 8.1**). The average mutation rate *per base pair* has been estimated at about 10^{-10} per replication in prokaryotes such as *E. coli* and about 10^{-9} per sexual generation in eukaryotes, based on both direct estimates in experimental *Drosophila* populations (Haag-Liautard et al. 2007) and the indirect method of comparing DNA sequences of different species. For example, Michael Nachman and Susan Crowell (2000) sequenced 18 processed pseudogenes in two humans and one chimpanzee. Because pseudogenes do not encode proteins, they generally lack function, so mutations in pseudogenes are presumably selectively neutral. Among the 16,089 base pairs sequenced in each individual, they found 199 differences, of which 66 percent were transitions, 26 percent were transversions, and 8 percent were insertion-deletion (indel) variants. Averaging over several different estimates of divergence time and length of a generation, Nachman and Crowell estimated the mutation rate to range between 1.3×10^{-8} and 3.4×10^{-8} per site per generation, and suggested a best estimate of 2.5×10^{-8}. The best current estimate, based on this and subsequent research, is 1.1×10^{-8}, implying that a new human carries about 70 new mutations on average (Keightley 2012).

EVOLUTIONARY IMPLICATIONS OF MUTATION RATES With such a low mutation rate per locus, it might seem that mutations occur so rarely that they cannot be important. However, summed over all genes, the input of variation by mutation is considerable. Taking into account the fraction of nucleotides in a mammalian genome that are thought likely to affect fitness if mutated (perhaps 5 percent), Keightley (2012) estimated that a new human zygote might carry 2.2 deleterious mutations on average. The fraction of mutations that might prove advantageous is unknown, but suppose it were 1 percent of the deleterious mutation rate, or 0.02 per genome. Then, in a population of 1 million people, about 20,000 potentially useful mutations would arise in every generation. Even if the fraction of potentially advantageous mutations were much less, the amount of new "raw material" for adaptation would be substantial over the course of thousands or millions of years. Of course, far more deleterious mutations would occur, but many of them would be eliminated by natural selection.

Mutation rates vary among genes and chromosome regions, and they are affected by environmental factors and even by an organism's own genotype (Baer et al. 2007). For example, mutations on chromosome 2 accumulated at a higher rate in lines of *Drosophila* that carried deleterious genes on chromosome 3 (Sharp and Agrawal 2012). MUTAGENS (mutation-causing agents) include ultraviolet light, X-rays, and a great array of chemicals, many of which are environmental pollutants. Mutation rates in birds and mice are elevated in industrial areas, and mice exposed to particulate air pollution in an urban-industrial site showed higher rates of mutation in repetitive elements than mice exposed only to filtered air at the same site or mice placed in a rural location (**Figure 8.11**). Human mutation rates may also be influenced by air pollution (Somers and Cooper 2009). The worst recorded nuclear accident occurred in 1986, when the Chernobyl nuclear power plant in Ukraine (then part of the Soviet Union) exploded. Since then, studies have shown a much higher incidence of mutations in swallows from that region than elsewhere (Ellegren et al. 1997), as well as in the children of fathers who were exposed to the radiation (Dubrova et al. 1996).

As we will describe in Chapters 9 and 13, the variation in most phenotypic characters is POLYGENIC: it is based on several or many different genetic loci. It is difficult to single out any of these loci to study the mutation rate per locus, but it is easy to estimate the **mutational variance** of the character—the increased variation in a population caused by new mutations in each generation. Studies of traits such as bristle number in *Drosophila* have shown that their mutational variance is high enough that an initially homozygous population would take only about 500 generations to achieve the level of genetic variation generally found in a natural population. The magnitude of mutational variance varies somewhat among characters and species (Lynch 1988).

In summary, although any particular mutation is a rare event, the rate of origination of new genetic variation in the genome as a whole, and for individual polygenic characters, is appreciable. However, mutation alone does not cause a character to evolve from one state to another because the rate of mutation is too low. Suppose that alleles A_1 and A_2 determine alternative phenotypes (e.g., red versus purple) of a haploid species, that half the individuals carry A_1, and that the rate of recurrent mutation from A_1 to A_2 is 10^{-5} per gene per generation. In one generation, the proportion of A_2 genes will increase from 0.5 to $0.5 + [(0.5)(10^{-5})] = 0.50000495$. At this rate, it will take about 70,000 generations before A_2 constitutes 75 percent of the population, and another 70,000 generations before it reaches 87 percent. This rate is so slow that, as we will see, factors other than recurrent mutation usually have a much stronger influence on allele frequencies, and thus are responsible for whatever evolutionary change occurs.

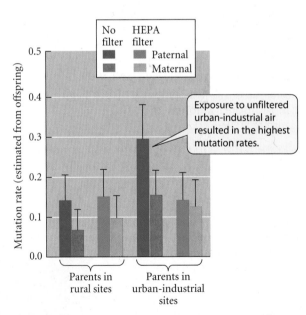

FIGURE 8.11 Mutation rates in mice, estimated from DNA sequences of two loci in their offspring. The mice were placed for 10 weeks in rural sites or in urban-industrial sites near steel mills and a major highway, where they were exposed either directly to the air or to air passed through a HEPA filter. (After Somers et al. 2004.)

(A) Control

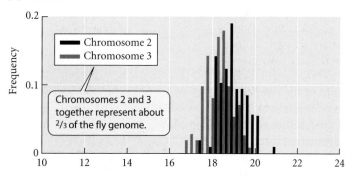

(B) Transposable elements introduced

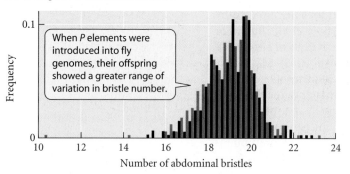

FIGURE 8.12 The frequency distribution of the number of abdominal bristles in (A) 392 homozygous control lines of *Drosophila melanogaster* and (B) 1094 homozygous experimental lines in which researchers used transposable elements (*P* elements) to induce mutations in chromosome 2 or chromosome 3. The mutations both increased and decreased bristle numbers compared with the control lines. (From Lyman et al. 1996.)

Phenotypic effects of mutations

A mutation may alter one or more phenotypic characters, such as size, coloration, or the amount or activity of an enzyme. Alterations in such features may affect survival or reproduction, the major components of FITNESS (see Chapter 11). It is often convenient to distinguish between a mutation's effects on fitness and on other characters, even though they are connected.

The phenotypic effects of mutational changes in DNA sequence range from none to drastic. At one extreme, synonymous base pair changes are expected to have no evident phenotypic effect, and this is also true of some amino acid substitutions that seem not to affect protein function. Little is known about what fraction of mutations in regulatory regions have phenotypic effects. The effects of mutations that contribute to polygenic traits, such as human stature or bristle number in *Drosophila*, range from slight to substantial; in one study, each mutation induced by insertion of transposable elements altered the number of abdominal bristles by about 0.9 bristles on average (**Figure 8.12**).

Even single base pair changes may have novel, beneficial effects. A remarkable example is a mutation prevalent in some populations of the sheep blowfly (*Lucilia cuprina*) that confers resistance to the organophosphate insecticides that have been used to control this serious pest. This mutation substitutes aspartic acid for glycine at one position in the active site of a carboxylesterase, abolishing the enzyme's esterase activity and transforming it into an organophosphate hydrolase (Newcomb et al. 1997).

Among the most fascinating mutations are those in the "master control genes" that regulate the expression of other genes in developmental pathways. (We will discuss these genes in detail in Chapter 21.) HOMEOTIC SELECTOR GENES, for example, determine the basic body plan of an organism, conferring a distinct identity on each segment of the developing body by producing DNA-binding proteins that regulate other genes that determine the features of each such segment. These genes derive their name from HOMEOTIC MUTATIONS in *Drosophila*, which redirect the development of one body segment into another. Mutations in the *Antennapedia* gene, for example, cause legs to develop in place of antennae (**Figure 8.13**). Another master control gene, *Pax6*, switches on about 2500 other genes required for eye development in mammals, insects, and many other animals (Gehring and Ikeo 1999). Mutations in this gene cause malformation or loss of eyes.

FIGURE 8.13 The drastic phenotypic effect of homeotic mutations that switch development from one pathway to another. (A) Frontal view of the head of a wild-type *Drosophila melanogaster*, showing normal antennae and mouthparts. (B) Head of a fly carrying the *Antennapedia* mutation, which converts antennae into legs. The *Antennapedia* gene is part of a large complex of Hox genes that confer identity on segments of the body (see Chapter 21). (Photos courtesy of F. R. Turner.)

(A)

(B)

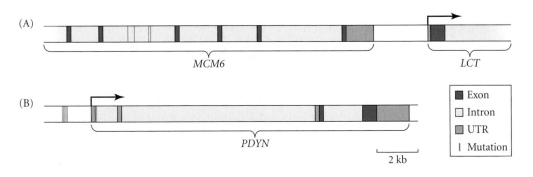

FIGURE 8.14 *Cis*-regulatory mutations in two human genes. The bent arrows mark start sites for transcription. (A) The red bars indicate four mutations in an intron of the *MCM6* locus that maintain transcription of the *LCT* (lactase) locus in adults, a phenotype called lactase persistence. (B) Red bars show six mutations that distinguish human and chimpanzee *PDYN* genes and affect differences in the gene's transcription level. Five are located in an upstream sequence of repeated DNA; one is located in a binding site for a transcription-regulating protein. (After Wray 2007.)

Development of morphological features and many physiological functions depends on regulation of gene expression, usually by altering rates of transcription. Thus changes in gene regulation have been extremely important in evolution. These changes may stem from mutations in the regulatory genes that encode transcription factors or in *cis*-regulatory sequences such as enhancers (Wray 2007; Lynch and Wagner 2008). For example, certain human populations are able to digest lactose in milk as adults because of persistent synthesis of the enzyme lactase, encoded by the gene *LCT*. The elevated transcription of this gene is caused by several mutations in an intron of the neighboring gene *MCM6* (**Figure 8.14A**). In another example, humans and chimpanzees differ in the level of transcription of *PDYN*, which encodes a neuropeptide that affects memory, emotions, and pain. Six mutational differences in a *cis*-regulatory region of the gene appear to account for the difference (**Figure 8.14B**).

Differences in gene expression are often studied by using MICROARRAYS that assay the abundance of the RNA transcripts of thousands of genes in a sample of tissue. This approach has been used to screen for changes in gene expression in mutation accumulation lines of *Drosophila*, *C. elegans*, and the yeast *Saccharomyces cerevisiae* (Landry et al. 2007). For example, Scott Rifkin and colleagues (2005) found mutational variance for level of gene expression in 4658 out of 11,798 genes assayed in 12 initially identical inbred lines of *Drosophila melanogaster* that had accumulated mutations for 200 generations (**Figure 8.15**).

Dominance describes the effect of an allele on a phenotypic character when it is paired with another allele in the heterozygous condition. A fully dominant allele (say, A_1) produces nearly the same phenotype when heterozygous (A_1A_2) as when homozygous (A_1A_1), and its partner allele (A_2) in that instance is fully **recessive**. All degrees of INCOMPLETE DOMINANCE,

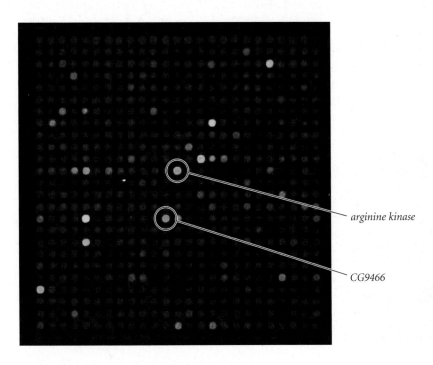

FIGURE 8.15 A microarray shows differences in gene expression that have arisen by spontaneous mutation in two inbred lines of *Drosophila melanogaster*. The figure shows a 25 × 25 block of a much larger array. Each position shows the level of expression, based on mRNA levels, of a single gene. Genes that are more highly expressed in line 71, due to mutation, appear in red (such as the arginine kinase gene in row 12, column 13); those that are more highly expressed in line 51 appear in green (such as *CG9466*, which is involved in mannose metabolism). Genes that are expressed in both lines appear in yellow. (Courtesy of Scott A. Rifkin.)

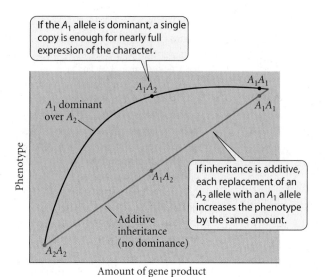

If the A_1 allele is dominant, a single copy is enough for nearly full expression of the character.

A_1 dominant over A_2

Additive inheritance (no dominance)

If inheritance is additive, each replacement of an A_2 allele with an A_1 allele increases the phenotype by the same amount.

Amount of gene product

FIGURE 8.16 Two of the possible relationships between phenotype and genotype at a single locus with two alleles. If inheritance is additive, replacing each A_2 allele with an A_1 allele steadily increases the amount of gene product, and the phenotype changes accordingly. If A_1 is dominant over A_2, the phenotype of A_1A_2 nearly equals that of A_1A_1 because the single dose of A_1 produces enough gene product for full expression of the character.

measured by the degree to which the heterozygote resembles one or the other homozygote, may occur. Inheritance is said to be **additive** if the heterozygote's phenotype is precisely intermediate between those of the homozygotes. For example, A_1A_1, A_1A_2, and A_2A_2 may have phenotypes 3, 2, and 1, respectively; the effects of replacing each A_2 with an A_1 simply add up (**Figure 8.16**). LOSS-OF-FUNCTION mutations, in which the activity of a gene product is reduced, are often at least partly recessive, whereas dominant mutations often result in enhanced gene product activity.

Most genes are **pleiotropic**, meaning that they affect more than one character. Pleiotropy has extremely important consequences for evolution (as we will see in Chapters 11, 12, and 21). Some pleiotropy is attributable to the "promiscuous" activities of many enzymes, which catalyze not only the primary chemical reaction with which they are identified, but also other, sometimes quite different, reactions (Khersonsky and Tawfik 2010). For example, the "native" activity of malonate semialdehyde decarboxylase is decarboxylation, but it also acts in hydration of certain substrates. Some mutational changes in amino acid sequence can enhance the efficiency of a "subsidiary" reaction and confer a new function on the enzyme (as in the case of the blowfly's carboxylesterase mentioned above). In other cases, a gene is pleiotropic because it regulates a number of other genes, or because it participates in a developmental pathway that affects more than one characteristic. This may account for examples such as those illustrated in **Figure 8.17**.

Effects of mutations on fitness

The effects of new mutations on survival and reproduction (i.e., fitness) may range from highly advantageous to highly disadvantageous. Undoubtedly, many mutations are neutral, or nearly so, having very slight effects on fitness (see Chapter 10). The *average*, or net, effect of

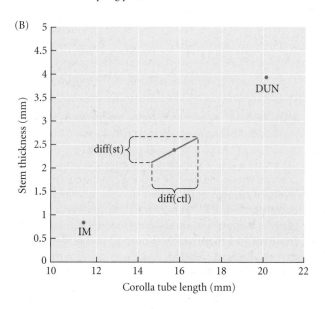

FIGURE 8.17 Two examples of pleiotropy. (A) In different cultures of the nematode *C. elegans*, each represented by a point, mutations accumulated that affected the number of offspring produced at young and at older ages similarly. (B) The mean stem thickness of monkeyflower plants from two geographic populations (DUN and IM) is plotted against the mean length of the corolla (petal) tube. The central point shows the effects of one of the loci that contribute to the character differences. Its position (halfway between the parent means) shows that the locus affects both characters additively. At this locus, the populations have different alleles that cause a difference diff(st) in stem thickness (about 0.5 mm) and diff(ctl) in corolla tube length (about 2 mm). (A after Estes et al. 2005; B after Hall et al. 2006.)

those mutations that do affect fitness, however, is deleterious. Mutations that are identified by their visible phenotypic effects often have deleterious pleiotropic effects. For example, some mutations that affect *Drosophila* bristle number also disrupt the development of the nervous system and reduce the viability of larvae, which do not have bristles (Mackay et al. 1992). The net deleterious effect of mutation is evident in mutation accumulation experiments such as Mukai's *Drosophila* experiment described above, in which mean viability declined (see Figure 8.10). Such experiments have shown that the rate of deleterious mutations in animals and plants generally ranges from 0.01 to more than 1.0 mutations per haploid (gametic) genome (Halligan and Keightley 2009). In yeast, the more characteristics a pleiotropic mutation affects, the more likely it is to be deleterious (**Figure 8.18**).

Among single mutations isolated in experimental cultures of the bacterium *Pseudomonas fluorescens*, a few slightly enhanced fitness, some greatly decreased it, and the majority had small deleterious effects (**Figure 8.19**). The beneficial mutations had an approximately exponential distribution of fitness effects: the majority enhanced fitness only slightly, with fewer showing larger effects. Many experiments that demonstrate advantageous mutations have been done with microorganisms such as phage, bacteria, and yeast (Elena and Lenski 2003; Elena and Sanjuán 2007). Some of these studies have reported a high incidence (e.g., 6 percent) of advantageous mutations (Hall and Joseph 2010).

The proportion of mutations that are advantageous, neutral, and deleterious can also be estimated from patterns of DNA sequence variation within and among species. The methods for doing so are described in Chapter 12, but the basic idea is that *if a particular sequence differs among species very little, compared with the sequence difference in neutral parts of the genome* (such as pseudogenes), *most of the mutations that have occurred in that sequence must have been deleterious*

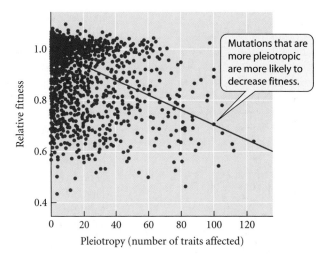

FIGURE 8.18 Mutation of more pleiotropic genes has more harmful effects on fitness. Each point represents a strain of yeast (*Saccharomyces cerevisiae*) in which a single gene has been radically mutated by deleting it. The fitness of the strain, relative to a standard strain, is plotted against the number of phenotypic characteristics that are altered by the mutation. Notice that a single gene can affect more than 100 characteristics. Most of the traits measured are biochemical functions. (After Cooper et al. 2007.)

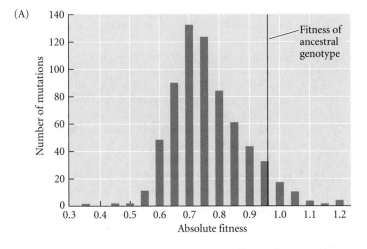
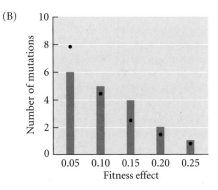

FIGURE 8.19 The distribution of fitness effects of 665 newly arisen mutations in laboratory cultures of the bacterium *Pseudomonas fluorescens*. (A) Fitness effects of all mutations, measured as absolute rate of growth. The solid vertical line marks the fitness of the ancestral genotype. (B) The fitness effects of the 28 beneficial mutations shown to the right of the vertical line in (A), measured as their increase in growth rate relative to the ancestral genotype. The frequency distribution is approximately exponential, fitting expectations from theory (shown by dots) fairly well. (After Kassen and Bataillon 2006.)

and have been expunged by "purifying" natural selection: they have not persisted to differentiate the species. Conversely, if a sequence differs among species much more than expected based on neutral parts of the genome, some of the differences must represent mutations that were advantageous in each species and were fixed by natural selection. Such comparisons among species of *Drosophila* imply that 97 percent of mutations in amino acid–encoding DNA sites have been deleterious (and have been eliminated), and that of the DNA base pairs that differ among the species, about half represent advantageous mutations (Sella et al. 2009). Moreover, as Peter Andolfatto (2005) discovered, more than half of mutations in untranslated DNA regions (including introns and intergenic regions) have been deleterious, for these regions have been highly conserved: they differ very little among species. Moreover, at least 10 percent to 20 percent of the base pairs that do differ in these untranslated sequences have undergone advantageous mutation. Clearly, much of what was once thought to be "junk" DNA is functional, perhaps playing roles in regulating development. This remarkable conclusion also holds for mammalian genomes, including the human genome (Eöry et al. 2010; Halligan et al. 2011).

The role of new beneficial mutations in adaptation has been studied in bacteria, which can be frozen (during which time they undergo no genetic change) and later revived. Samples taken at different times from an evolving population can be stored, and their fitness can later be directly compared. Richard Lenski and colleagues used this method to trace the increase of fitness in laboratory populations of *E. coli* for an astonishing 20,000 generations. Each population was initiated with a single individual and was therefore genetically uniform at the start. Nevertheless, fitness increased substantially—rapidly at first, but at a decelerating rate later (**Figure 8.20A**). In a similar experiment (Bennett et al. 1992), *E. coli* populations adapted rapidly to several different temperatures (**Figure 8.20B–D**).

These *E. coli* populations were grown in a medium in which availability of glucose limited population growth. The medium also contained citrate, a potential energy source that *E. coli* cannot use because it cannot transport citrate into the cell. After 33,000 generations,

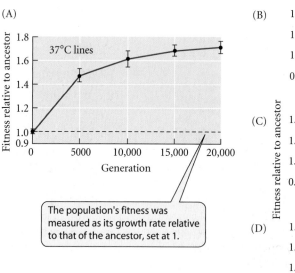

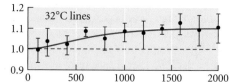

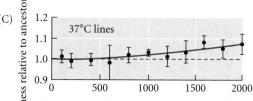

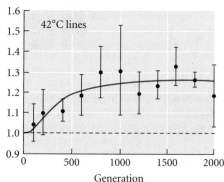

The population's fitness was measured as its growth rate relative to that of the ancestor, set at 1.

FIGURE 8.20 Adaptation in experimental populations of *Escherichia coli*. Vertical bars show a measure of variation (95 percent confidence interval) among replicate populations around the mean fitness. (A) Increase in fitness during 20,000 generations in populations kept at 37°C (the normal temperature of *E. coli*'s habitat). (B–D) Adaptation over a much shorter time (2000 generations) in populations kept at three different temperatures (32°C, 37°C, and 42°C). Because all of these populations initially lacked genetic variation, the increase in adaptation was due to natural selection acting on new advantageous mutations. (A after Cooper and Lenski 2000; B–D after Bennett et al. 1992.)

the researchers observed that one population increased dramatically in density because of a mutation (*Cit⁺*) that enabled it to use citrate (Blount et al. 2008). Long before the 33,000th generation, billions of mutations had transpired in these populations, yet the adaptation had not evolved earlier. Lenski's group showed that another mutation must have occurred in this culture earlier that either facilitated the origin of the *Cit⁺* mutation or made it advantageous. That is, *the evolution of this adaptation was historically contingent on previous genetic change*. This example of the origin of a key adaptation—one that enabled use of a new ecological niche—suggests that evolution of novel characteristics might depend on extremely rare combinations or sequences of mutational events. That is, *the mutational process might limit evolution*.

Bacteria can be screened for mutations that affect their biochemical capacities by placing them on a medium on which that bacterial strain cannot grow, such as a medium that lacks an essential amino acid or other nutrient. Whatever colonies do appear on the medium must have grown from the few cells in which mutations occurred that conferred a new biochemical ability. For example, Barry Hall (1982) studied a strain of *E. coli* that lacks the *lacZ* gene, which encodes β-galactosidase, the enzyme that enables *E. coli* to metabolize the sugar lactose as a source of carbon and energy. Hall screened populations for the ability to grow on lactose and recovered several mutations. A mutation in a different gene (*ebg*) altered an enzyme that normally performs another function so that it could now break down lactose. Another mutation altered regulation of the *ebg* gene, and a third mutation altered the ebg enzyme so that it metabolized lactose into lactulose, which increased the cell's uptake of lactose from the environment. The three mutations together restored the metabolic capacities that had been lost by the deletion of the original *lacZ* gene. Thus *mutation and selection in concert can give rise to complex adaptations*.

The limits of mutation

The great majority of mutations that affect the phenotype alter one or more *pre-existing* traits. Mutations with phenotypic effects alter developmental processes, but they cannot alter developmental foundations that do not exist. We may conceive of winged horses and angels, but no mutant horses or humans will ever sprout wings from their shoulders, for the developmental foundations for such wings are lacking.

The direction of evolution may be constrained if some conceivable mutations are more likely to arise and contribute to evolution than others. In laboratory stocks of the green alga *Volvox carteri*, for example, new mutations affecting the relationship between the size and number of germ cells correspond to the typical states of these characters in other species of *Volvox* (Koufopanou and Bell 1991).

Mutation may not constrain the rate or direction of evolution very much if many different mutations can generate a particular phenotype, as may be the case when several or many loci affect the trait (polygeny). For example, when different copper-tolerant populations of the monkeyflower *Mimulus guttatus* are crossed, the variation in copper tolerance is greater in the F_2 generation than in either parental population, indicating that the populations differ in the loci that confer tolerance (Cohan 1984).

Nevertheless, certain advantageous phenotypes can apparently be produced by mutation at only a few loci, or perhaps only one. In such instances, the supply of rare mutations might limit the capacity of species for adaptation. The rarity of necessary mutations may help to explain why species have not become adapted to a broader range of environments, or why, in general, species are not more adaptable than they are (Bradshaw 1991). For instance, resistance to the insecticide dieldrin in different populations of *Drosophila melanogaster* is based on repeated occurrences of the same mutation, which, moreover, is thought to represent the same gene that confers dieldrin resistance in flies that belong to two other families (ffrench-Constant et al. 1990). This may mean that very few genes—perhaps only this one—undergo mutations that can confer dieldrin resistance, and that such a mutation is a very rare event. In Chapter 3, we saw that mutations of the same gene, *Mc1r*, have affected pigmentation in various species of reptiles, birds, and mammals.

(A)   (B)   (C)

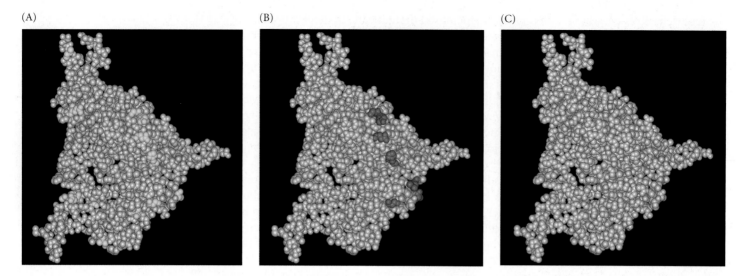

FIGURE 8.21 The surface of the major capsid protein (gpF) of phage strains φX174 and S13, showing parallel amino acid substitutions. (A) Amino acids that underwent substitution in the experimental lines are shown in yellow. (B) Amino acids shown in red are known to affect the fitness of wild phage in either of two bacterial host species. (C) Amino acids that represent differences between the two original strains, φX174 and S13, are shown in blue. Note that all the differences that have naturally evolved between these phage strains also occurred in the experimental lines. (After Wichman et al. 2000.)

Wichman et al. (2000) studied experimental populations of two closely related bacteriophage strains, φX174 and S13, as they adapted to two species of host bacteria at high temperatures. Of the many amino acid substitutions that occurred in the populations, most occurred repeatedly, at a small number of sites, and many of those substitutions matched the variation found in natural populations and even the differences between the two kinds of phage (**Figure 8.21**). This result suggests that the natural evolution of these phage can take only a limited number of pathways, constrained by the possible kinds of advantageous mutations (Wichman et al. 2000).

Mutation as a Random Process

Mutations occur at random. It is extremely important to understand what this statement does and does not mean. It does not mean that all conceivable mutations are equally likely to occur, because, as we have noted, the developmental foundations for some imaginable transformations do not exist. It does not mean that all loci, or all regions within a locus, are equally mutable, for geneticists have described differences in mutation rates, at both the phenotypic and molecular levels, among and within loci. It does not mean that environmental factors cannot influence mutation rates: radiation and chemical mutagens do so.

Mutation *is* random in two senses. First, although we may be able to predict the *probability* that a certain mutation will occur, we cannot predict which of a large number of gene copies will undergo the mutation. The spontaneous process of mutation is stochastic rather than deterministic. Second, mutation is random in the sense that *the chance that a particular mutation will occur is not influenced by whether or not the organism is in an environment in which that mutation would be advantageous*. That is, the environment does not induce adaptive mutations. Indeed, it is hard to imagine how most environmental factors could direct the mutation process by dictating that just the right base pair changes should occur.

The argument that adaptively directed mutation does not occur is one of the fundamental tenets of modern evolutionary theory. If it did occur, it would introduce a Lamarckian element into evolution, for individual organisms would then acquire adaptive hereditary characteristics in response to their environment. Such "neo-Lamarckian" ideas were expunged in the 1940s and 1950s by experiments with bacteria in which spontaneous, random mutation followed by natural selection, rather than mutation directed by the environment, explained adaptation.

One of these experiments was performed by Joshua and Esther Lederberg (1952), who showed that advantageous mutations occur without exposure to the environment in which they would be advantageous to the organism. The Lederbergs used the technique of REPLICA PLATING (**Figure 8.22**). Using a culture of *E. coli* derived from a single cell, the

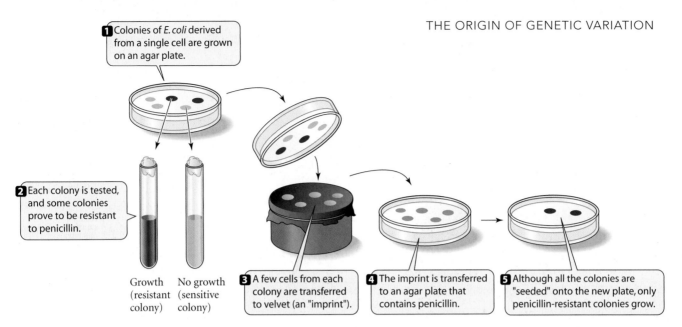

1 Colonies of *E. coli* derived from a single cell are grown on an agar plate.

2 Each colony is tested, and some colonies prove to be resistant to penicillin.

Growth (resistant colony) No growth (sensitive colony)

3 A few cells from each colony are transferred to velvet (an "imprint").

4 The imprint is transferred to an agar plate that contains penicillin.

5 Although all the colonies are "seeded" onto the new plate, only penicillin-resistant colonies grow.

FIGURE 8.22 The method of replica plating, which the Lederbergs used to show that mutations for penicillin resistance arise spontaneously, before exposure to penicillin, rather than being induced by that exposure. They began (step 1) with an agar plate containing numerous colonies of *E. coli*, all derived from a single cell. They tested the colonies on this master plate for penicillin resistance (step 2). They pressed a velvet cloth against the plate (step 3), thereby transferring some cells from each colony to the velvet. They then touched the velvet to a new plate with medium containing the antibiotic penicillin (step 4), transferring bacteria to this replica plate in the same spatial configuration as the colonies on the master plate from which they had been taken. A few colonies appeared on the replica plate (step 5), having grown from penicillin-resistant mutant cells. Only those colonies that had been the source of penicillin-resistant cells on the master plate grew on the replica plate, showing that the mutations had occurred before the bacteria were exposed to penicillin.

Lederbergs spread cells onto a "master" agar plate. Each cell gave rise to a distinct colony. They pressed a velvet cloth against the plate, and then touched the cloth to a new plate with medium containing the antibiotic penicillin, thereby transferring some cells from each colony to the replica plate, in the same spatial configuration as the colonies from which they had been taken. A few colonies appeared on the replica plate, having grown from penicillin-resistant mutant cells. When all the colonies on the master plate were tested for penicillin resistance, those colonies (and only those colonies) that had been the source of penicillin-resistant cells on the replica plate displayed resistance, showing that the mutations had occurred before the bacteria were exposed to penicillin.

Because of such experiments, biologists have generally accepted that mutation is adaptively random rather than directed. There is good evidence that mutation rates throughout the genome may be elevated by certain stressful environments, but no convincing evidence that these environments elicit specific mutations that would be advantageous (Sniegowski and Lenski 1995; Foster 2000).

Alterations of the Karyotype

An organism's **karyotype** is the description of its complement of chromosomes: their number, size, shape, and internal arrangement. In considering alterations of the karyotype, it is important to bear in mind that the loss of a whole chromosome, or a major part of a chromosome, usually reduces the viability of a gamete or an organism because of the loss of genes. Furthermore, a gamete or organism is often inviable or fails to develop properly if it has an **aneuploid**, or "unbalanced," chromosome complement—for example, if a normally diploid organism has three copies of one of its chromosomes. (For instance, humans with three copies of chromosome 21, a condition known as Down syndrome or trisomy-21, have brain and other defects.)

As we have seen, chromosome structure may be altered by duplications and deletions, which change the amount of genetic material (see Figure 8.6). Other alterations of the karyotype are changes in the number of whole sets of chromosomes (**polyploidy**) and rearrangements of one or more chromosomes.

Polyploidy

A DIPLOID organism has two entire sets of homologous chromosomes ($2N$); a POLYPLOID organism has more than two. (In discussing chromosomes, N refers to the number of different chromosomes in the gametic, or HAPLOID, set, and the numeral refers to the number of representatives of each autosome.) Polyploids can be formed in several ways, especially

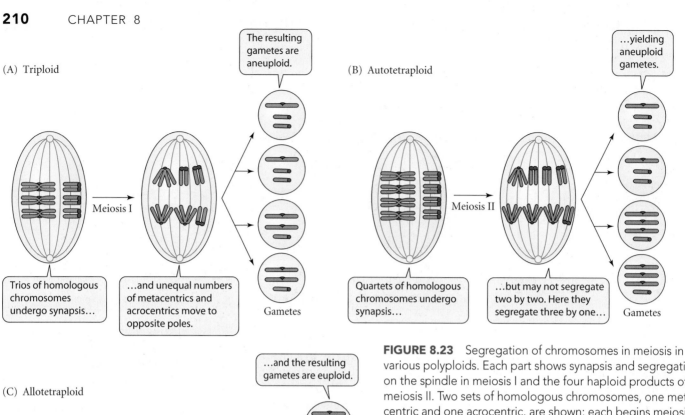

(A) Triploid

The resulting gametes are aneuploid.

Meiosis I

Trios of homologous chromosomes undergo synapsis…

…and unequal numbers of metacentrics and acrocentrics move to opposite poles.

Gametes

(B) Autotetraploid

…yielding aneuploid gametes.

Meiosis II

Quartets of homologous chromosomes undergo synapsis…

…but may not segregate two by two. Here they segregate three by one…

Gametes

(C) Allotetraploid

…and the resulting gametes are euploid.

Meiosis I

Each chromosome pairs with a single homologue from the same parental species.

Segregation is then normal…

Gametes

FIGURE 8.23 Segregation of chromosomes in meiosis in various polyploids. Each part shows synapsis and segregation on the spindle in meiosis I and the four haploid products of meiosis II. Two sets of homologous chromosomes, one metacentric and one acrocentric, are shown; each begins meiosis with two identical sister chromatids. All chromosomes with the same color (red or blue) are derived from one species.

when failure of the reduction division in meiosis produces diploid, or unreduced, gametes (Ramsey and Schemske 1998). The union of an unreduced gamete (with $2N$ chromosomes) and a reduced gamete (with N chromosomes) yields a TRIPLOID ($3N$) zygote. Triploids produce few offspring because most of their gametes have aneuploid chromosome complements. At segregation, each daughter cell may receive one copy of certain chromosomes and two copies of certain others (**Figure 8.23A**). However, tetraploid ($4N$) offspring can be formed if an unreduced ($3N$) gamete of a triploid unites with a normal gamete (N) of a diploid—or if two diploid gametes, whether from triploid or diploid parents, unite. Other such unions can form hexaploids ($6N$), octoploids ($8N$), or genotypes of even higher ploidy. Some polyploids arise by the union of unreduced gametes of the same species; these organisms are known as **autopolyploids**. But the majority are **allopolyploids**, which arise by hybridization between closely related species (see Chapter 18).

Each set of four homologous chromosomes of a tetraploid may be aligned during meiosis into a quartet (quadrivalent), and then may segregate in a balanced (two by two) or unbalanced (one by three) fashion (**Figure 8.23B**). In some such polyploids, aneuploid gametes may result and fertility may be greatly reduced. In other cases, the four chromosomes align not as a quartet, but as two pairs that segregate normally, resulting in balanced (**euploid**), viable gametes, so that fertility is normal, or nearly so (**Figure 8.23C**). This requires that the chromosomes be differentiated so that each can recognize and pair with a single homologue rather than with three others. In allopolyploids, most of the parental species' chromosomes are different enough for the chromosomes from each parent to recognize and pair with each other, so that meiosis in an allotetraploid involves normal segregation of pairs rather than quartets of chromosomes (see Figure 8.23C).

Many species of plants and a few species of trout, tree frogs, and other animals have arisen by polyploidy, and there is evidence that extant vertebrates stem from a polyploid ancestor (see Figure 3.30 and Chapter 18). In flowering plants, 2 to 4 percent of newly formed species are polyploid, and the proportion of angiosperm species that are derived from polyploid ancestors has been estimated at between 30 and 70 percent (Otto and Whitton 2000). Cell size and many other features of newly formed polyploid plants are altered, and such plants can have an immediate advantage in certain environments (Ramsey 2011).

Chromosome rearrangements

Changes in the structure of chromosomes constitute another class of karyotypic alterations. These changes are caused by breaks in chromosomes followed by rejoining of the pieces in new configurations. Some such changes can affect the pattern of segregation in meiosis and therefore affect the proportion of viable gametes. Although most chromosome rearrangements seem not to have direct effects on morphological or other phenotypic features, an alteration of gene sequence sometimes brings certain genes under the influence of the control regions of other genes, and so alters their expression. It is not certain that such "position effects" have contributed to evolutionary change. Individual organisms may be homozygous or heterozygous for a rearranged chromosome and are sometimes referred to as HOMOKARYOTYPES or HETEROKARYOTYPES, respectively.

INVERSIONS Consider a segment of a chromosome in which ABCDE denotes a sequence of markers such as genes. If a loop is formed, and breakage and reunion occur at the point of overlap, a new sequence, such as ADCBE, may be formed. (The inverted sequence is underlined.) Such an **inversion**, with a rearranged gene order, is PERICENTRIC if it includes the centromere and PARACENTRIC if it does not.

During meiotic synapsis in an inversion heterozygote, alignment of the genes on the normal and inverted chromosomes requires the formation of a loop, which can sometimes be observed under the microscope (**Figure 8.24A**). Now suppose that in a paracentric inversion, crossing over occurs between loci such as B and C (**Figure 8.24B**). Two of the four strands are affected. One strand lacks certain gene regions (A), and also lacks a centromere; it will not migrate to either pole, and is lost. The other affected strand not only lacks some genetic material but also has two centromeres, so the chromosome breaks when these centromeres are pulled to opposite poles. The resulting daughter cells lack certain gene regions and will not form viable gametes. Consequently, in inversion heterokaryotypes (but not in homokaryotypes), *fertility is reduced* because many gametes are inviable, and *recombination is effectively suppressed* because gametes carrying the recombinant chromosomes, which lack some genetic material, are inviable.

In *Drosophila* and some other flies (Diptera), however, the incomplete recombinant chromosomes enter the polar bodies during meiosis, so that female

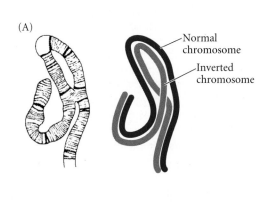

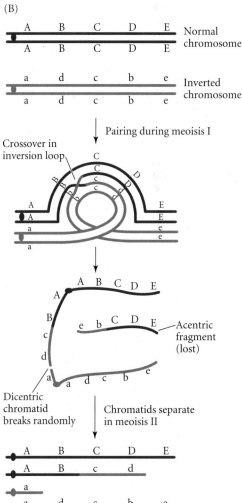

FIGURE 8.24 Chromosome inversions. (A) Synapsed chromosomes in a salivary gland cell of a larval *Drosophila pseudoobscura* heterozygous for *Standard* and *Arrowhead* arrangements. The two homologous chromosomes are so tightly synapsed that they look like a single chromosome. The "bridge" forms the loop shown in the diagram. Similar synapsis occurs in germ line cells undergoing meiosis. (B) Two homologous chromosomes differing by an inversion of the region B–D, and their configuration in synapsis. Crossing over between two chromatids (between B and C) yields products that lack a centromere or substantial blocks of genes. Because these products do not form viable gametes, crossing over appears to be suppressed. Only cells that receive the two chromatids that do not cross over become viable gametes. (After Strickberger 1968.)

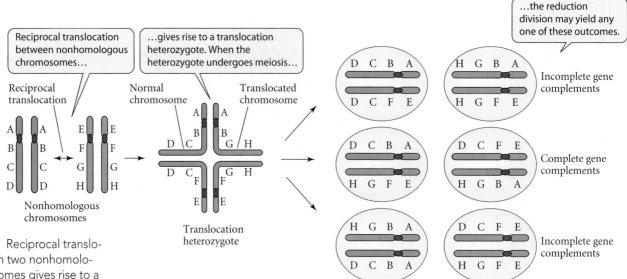

FIGURE 8.25 Reciprocal translocation between two nonhomologous chromosomes gives rise to a translocation heterozygote. When the heterozygote undergoes meiosis, many of the products will have incomplete gene complements.

fecundity is not reduced. It is particularly easy to study inversions in *Drosophila* and some other flies because the larval salivary glands contain giant (polytene) chromosomes that remain in a state of permanent synapsis (so that inversion loops are easily seen), and because these chromosomes display bands, each of which apparently corresponds to a single gene. The banding patterns are as distinct as the bar codes on supermarket products, so an experienced investigator can identify different sequences. INVERSION POLYMORPHISMS are common in *Drosophila*—more than 20 different arrangements of chromosome 3 have been described for *Drosophila pseudoobscura*, for example—and they have been extensively studied from both population genetic and phylogenetic points of view.

TRANSLOCATIONS By breakage and reunion, two nonhomologous chromosomes may exchange segments, resulting in a **reciprocal translocation** (**Figure 8.25**). Meiosis in a translocation heterokaryotype often results in a high proportion of aneuploid gametes, so the fertility of translocation heterokaryotypes is often reduced by 50 percent or more. Consequently, polymorphism for translocations is rare in natural populations. Nevertheless, related species sometimes differ by translocations, which have the effect of moving groups of genes from one chromosome to another. The Y chromosome of the male *Drosophila miranda*, for example, includes a segment that is homologous to part of one of the autosomes of closely related species.

FISSIONS AND FUSIONS It is useful to distinguish ACROCENTRIC chromosomes, in which the centromere is near one end, from METACENTRIC chromosomes, in which the centromere is somewhere in the middle and separates the chromosome into two arms. In the simplest form of chromosome FUSION, two nonhomologous acrocentric chromosomes undergo reciprocal translocation near the centromeres so that they are joined into a metacentric chromosome (**Figure 8.26A**). For example, two chromosomes found in chimpanzees and gorillas have fused to form human chromosome 2, resulting in one less pair of chromosomes in the human genome (de Pontbriand et al. 2002). More rarely, a metacentric chromosome may undergo FISSION. A simple fusion heterokaryotype has a metacentric, which we may refer to as AB, with arms that are homologous to two acrocentrics, A and B. AB, A, and B together synapse as a "trivalent" (**Figure 8.26B**). Viable gametes and zygotes are often formed, but the frequency of aneuploid gametes can be quite high, especially for more complex patterns of fusion. Differences in chromosome number, whether caused by fusion or fission, often distinguish related species or geographic populations of the same species.

CHANGES IN CHROMOSOME NUMBER Polyploidy (especially in plants), translocations, and fusions or fissions of chromosomes are the mutational foundations for the evolution of

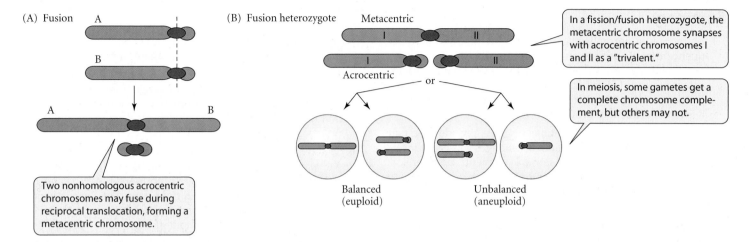

(A) Fusion

Two nonhomologous acrocentric chromosomes may fuse during reciprocal translocation, forming a metacentric chromosome.

(B) Fusion heterozygote

In a fission/fusion heterozygote, the metacentric chromosome synapses with acrocentric chromosomes I and II as a "trivalent."

In meiosis, some gametes get a complete chromosome complement, but others may not.

Balanced (euploid)   Unbalanced (aneuploid)

FIGURE 8.26 Chromosome fusion. (A) A simple fusion of two acrocentric chromosomes to form arms A and B. (B) Segregation in meiosis of a fusion heterozygote can yield euploid (balanced) or aneuploid (unbalanced) complements of genetic material.

chromosome number. For example, the haploid chromosome number varies among mammals from 3 to 42 (Lande 1979), and among insects from 1 in an ant species to about 220 in some butterflies (the highest number known in animals). Related species sometimes differ strikingly in karyotype: in one of the most extreme examples, two very similar species of barking deer, *Muntiacus reevesii* and *M. muntiacus*, have haploid chromosome numbers of 23 and 3 or 4 (in different populations), respectively (**Figure 8.27**).

The spontaneous rate of origin of any given class of chromosome rearrangement (e.g., reciprocal translocation) is quite high: about 10^{-4} to 10^{-3} per gamete (Lande 1979). Only a small fraction of these rearrangements become fixed in populations, however, and the rate at which species undergo karyotype divergence is highly variable. Through DNA sequencing, the relative positions of many genes along chromosomes can be compared across species. On this basis, the number of chromosome breaks per megabase (million base pairs) per million years has been estimated as 2.7×10^{-4} between human and mouse, and ranges from 6.5×10^{-2} to 10.1×10^{-2} for various pairs of *Drosophila* species (Ranz et al. 2001; Bartolomé and Charlesworth 2006). Human and chimpanzee genomes may differ by as many as 1500 inversions (Sharp et al. 2006).

Muntiacus reevesii ($2N = 46$)

Muntiacus muntiacus ($2N = 8$)

FIGURE 8.27 The diploid chromosome complements (taken from micrographs) of two closely related species of muntjacs (barking deer) represent one of the most extreme differences in karyotype known among closely related species. Despite the difference in karyotype, the species are phenotypically very similar.

Summary

1. Mutations of chromosomes or genes are alterations in sequence that are subsequently replicated. They ordinarily do not constitute new species, but instead are variant chromosomes or genes (alleles, haplotypes) within a species. The process of mutation is necessary, but not sufficient, for evolution: it is not equivalent to evolution.

2. At the molecular level, mutations of genes include base pair substitutions, frameshift mutations, changes caused by insertion of various kinds of transposable elements, and duplications and deletions that can range from single base pairs to long segments of chromosomes. New DNA sequences also arise by intragenic recombination.

3. Mutations that result in amino acid substitution in a protein (nonsynonymous mutations) and mutations in regulatory sequences may affect the phenotype and perhaps fitness. Synonymous mutations, those that do not alter the amino acid sequence, may not affect the phenotype or fitness, and so may be selectively neutral. Mutations in much of the noncoding DNA, which dominates eukaryotic genomes, affect fitness, but the function of most noncoding DNA is not well understood.

4. The rate at which any particular mutation arises is quite low: on average, about 10^{-6} to 10^{-5} per gamete for mutations detected by their phenotypic effects, and about 10^{-9} per base pair. The mutation rate, by itself, is too low to cause substantial changes of allele frequencies. However, the total input of genetic variation by mutation, for the genome as a whole or for individual polygenic characters, is appreciable.

5. The magnitude of change in morphological or physiological features caused by a mutation can range from none to drastic. In part because most mutations have pleiotropic effects, the average effect of mutations on fitness is deleterious, but some mutations are advantageous.

6. Mutations alter pre-existing biochemical or developmental pathways, so not all conceivable mutational changes are possible. Some adaptive changes may not be possible without just the right mutation of just the right gene. For these reasons, the rate and direction of evolution may in some instances be affected by the availability of mutations.

7. Mutations appear to be random, in the sense that their probability of occurrence is not directed by the environment in favorable directions, and in the sense that specific mutations cannot be predicted. The likelihood that a mutation will occur does not depend on whether or not it would be advantageous.

8. Mutations of the karyotype (chromosome complement) include polyploidy and structural rearrangements (e.g., inversions, translocations, fissions, fusions). Many such rearrangements reduce fertility in the heterozygous condition. These kinds of chromosome changes are the basis of evolutionary changes in chromosome number.

Terms and Concepts

additive
allele
allopolyploid
aneuploid
autopolyploid
back mutation
cis, trans
codon
control regions (enhancers and repressors)
dominance
euploid

exon
fixed
frameshift mutation
gene
gene family
genetic code
genetic marker
haplotype
intragenic recombination
intron
inversion
karyotype

locus
microsatellite
mutation
mutational variance
nonsynonymous mutation
pleiotropic
point mutation
polyploidy
processed pseudogene
recessive
reciprocal translocation
recurrent mutation

substitution
synonymous codon
synonymous mutation
tandem repeat
transition
transposable element (TE)
transposition
transversion
unequal crossing over
wild type

Suggestions for Further Reading

Evolutionary Genetics: Concepts and Case Studies, edited by C. W. Fox and J. B. Wolf (Oxford University Press, New York, 2006), is an excellent collection of essays by leading authorities on genetic aspects of evolution, written for students. The chapter on mutation, by D. Houle and A. Kondrashov (pp. 32–48), is perhaps the best overview of mutation as an evolutionary process.

D. Grauer and W.-H. Li, *Fundamentals of Molecular Evolution*, second edition (Sinauer Associates, Sunderland, MA, 2000), treats evolutionary aspects of mutation at the molecular level.

Reviews of mutation rates include J. W. Drake et al., "Rates of spontaneous mutation," *Genetics* 148: 1667–1686 (1998), and M. Lynch et al., "Perspective: Spontaneous deleterious mutation," *Evolution* 53: 645–663 (1999).

Problems and Discussion Topics

1. Consider two possible studies. (a) In one, you capture 3000 wild male *Drosophila melanogaster*, mate each with laboratory females heterozygous for the autosomal recessive allele *vg* (*vestigial*, which causes miniature wings when homozygous), and examine each male's offspring. You find that half the offspring of each of three males have miniature wings and have genotype *vgvg*. (b) In another study, you determine the nucleotide sequence of 1000 base pairs for 20 copies of the cytochrome *b* gene, taken from 20 wild mallard ducks. You find that at each of 30 nucleotide sites, one or another gene copy has a different base pair from all others. From these data, can you estimate the rate of mutation from the wild type to the *vg* allele (case a) or from one base pair to another (case b)? Why or why not?

2. From a laboratory stock of *Drosophila* that you believe to be homozygous wild type (++) at the *vestigial* locus, you obtain 10,000 offspring, mate each of them with homozygous *vgvg* flies, and examine a total of 1 million progeny. Two of these are *vgvg*. Estimate the rate of mutation from + to *vg* per gamete. What assumptions must you make?

3. The following DNA sequence represents the beginning of the coding region of the alcohol dehydrogenase (*Adh*) gene of *Drosophila simulans* (Bodmer and Ashburner 1984), arranged into codons:

   CCC ACG ACA GAA CAG TAT TTA AGG AGC TGC GAA GGT

   (a) Find the corresponding mRNA sequence, and use Figure 8.2 to find the amino acid sequence. (b) Again using Figure 8.2, determine for each site how many possible point mutations (changes of individual nucleotides) would cause an amino acid change, and how many would not. For the entire sequence, what proportion of possible mutations are synonymous versus nonsynonymous? What proportion of the possible mutations at first, second, and third base positions within codons are synonymous? (c) What would be the effect on the amino acid sequence of inserting a single base, G, between sites 10 and 11 in the DNA sequence? (d) What would be the effect of deleting nucleotide 16? (e) For the first 15 (or more) sites, classify each possible mutation as a transition or transversion, and determine whether or not the mutation would change the amino acid. Does the proportion of synonymous mutations differ between transitions and transversions?

4. A genus of Antarctic fishes, *Channichthys*, lacks hemoglobin. In its relative *Trematomus*, hemoglobin serves its usual functions. Assuming that the gene encoding hemoglobin in *Channichthys* has no function, and is not transcribed, how might you expect the nucleotide sequence of this gene to differ between these two genera?

5. Ultraviolet light (UV) can induce mutations in organisms such as *Drosophila*. Because it damages DNA, and therefore essential physiological functions, it can also reduce survival. Suppose you expose a large number of *Drosophila* to UV, screen their adult offspring for new mutations, and discover that a few offspring carry mutations that increase the amount of black pigment, which can protect internal organs from UV. The progeny of an equal number of control flies, not exposed to UV, show fewer or no mutations that increase pigmentation. Can you conclude that the process of mutation responds to organisms' need for adaptation to the environment?

6. Researchers have used artificial selection (see Chapter 9) to alter many traits in *Drosophila melanogaster*, such as phototactic behavior and wing length. No one has selected *Drosophila* (about 2 mm long) to be as large (ca. 30 mm) as bumblebees (although I'm not sure if anyone has tried). Do you suppose this could be done? How would you attempt it? If your attempts were unsuccessful, what hypotheses could explain your lack of success? What role might mutation play in your experiment?

7. The text above says that "two chromosomes found in chimpanzees and gorillas have fused to form human chromosome 2, resulting in one less pair of chromosomes in the human genome." What reason is there to think that the difference in chromosome number resulted from fusion rather than fission? In general, can we tell ancestral from mutant (derived) genes?

8. In a study described in this chapter, Nachman and Crowell (2000) estimated the neutral mutation rate from sequence differences between human and chimpanzee in several processed pseudogenes. One methodological detail not mentioned in this chapter was that they also sequenced several kilobases of flanking DNA on each side of the pseudogene. Any base pair differences found in these flanking sequences were not included in the calculation of mutation rates. (a) Why did they choose pseudogenes rather than some other kind of DNA sequence to estimate the neutral mutation rate? (b) Why did they sequence flanking DNA?

9. Most, if not all, attempts to estimate the rate of beneficial mutations are likely to produce underestimates. Why?

Variation:
The Foundation of Evolution

The great changes in organisms that have transpired over time and the differences that have developed among species as they diverged from their common ancestors all originated as genetic variations within species—variation that originated by mutation. Understanding the processes of evolution requires us to understand genetic variation and the ways in which it is transmuted into evolutionary change. Understanding genetic variation, moreover, provides insight into questions—ranging from the significance of "intelligence tests" to the meaning of "race"—that deeply affect human society. Of all the biological sciences, evolutionary biology is most dedicated to analyzing and understanding variation, a profoundly important feature of all biological systems. Genetic variation is the foundation of evolution.

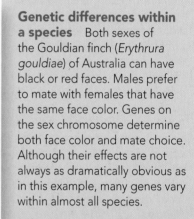

Because genetic variation will be discussed throughout the rest of this book, we need a brief review of its vocabulary:

- **Phenotype** refers to a characteristic in an individual organism, or in a group of individuals that are alike in this respect. All features of an organism except DNA sequences may be called phenotypes, ranging from the level of transcription of a gene to aspects of the organism's morphology, behavior, and life history. Phenotypic variation is largely the result of genetic differences among individuals, but it can have several other sources, as described below, of which the most important are the direct effects of environmental variation on development.

- **Genotype** refers to the genetic constitution of an individual organism, or of a group of organisms that are alike in this respect, at one or more loci singled out for discussion.

Genetic differences within a species Both sexes of the Gouldian finch (*Erythrura gouldiae*) of Australia can have black or red faces. Males prefer to mate with females that have the same face color. Genes on the sex chromosome determine both face color and mate choice. Although their effects are not always as dramatically obvious as in this example, many genes vary within almost all species.

- A LOCUS (plural LOCI) is a site on a chromosome or, more usually, the gene that occupies a site.

- An ALLELE is a particular form of a gene, usually distinguished from other alleles by its effects on the phenotype.

- A HAPLOTYPE is one of the sequences of a gene or DNA segment that can be distinguished from homologous sequences by molecular methods such as DNA sequencing.

- **Gene copy** is the term we will use when counting the number of representatives of a gene.* In a diploid population (such as humans), each individual carries two copies of each autosomal gene, so a sample of 100 individuals represents 200 gene copies. The term "gene copy" refers to all copies without distinguishing allelic or sequence differences among them. To make such distinctions, we will sometimes refer to ALLELE COPIES. Thus a heterozygous individual, A_1A_2, has two copies of the A gene: one copy of allele A_1 and one copy of allele A_2.

The number of alleles or loci that contribute to genetic variation in a phenotypic trait differs from case to case. In the simplest cases, two alleles at a single locus account for most of the variation (e.g., the color forms of the snow goose; **Figure 9.1A**). In other cases, three or more alleles exist within a population. For instance, several forms of the African swallowtail butterfly (*Papilio dardanus*) are palatable mimics of different species of butterflies that birds find unpalatable (**Figure 9.1B**). In this case, the several different color patterns

---

*Gene copies are often referred to as "sequences," but "sequence" has several meanings. In this book, "sequence" refers to a particular array of nucleotide base pairs, which may differ from another array.

FIGURE 9.1 Examples of phenotypic variation caused by genetic differences within a species. In all of these cases, the forms interbreed freely. (A) "Blue" and white forms of the snow goose (*Chen caerulescens*) are caused by two alleles at a single locus. (B) Mimetic variation in the African swallowtail *Papilio dardanus*. Males are *nonmimetic* and all appear similar (top left individual). The other three individuals are females, each of which mimics a distantly related toxic species of butterfly. This female-limited variation in *P. dardanus* is inherited as if it were due to multiple alleles at one locus, but it is actually caused by several closely linked genes. (C) In *Homo sapiens*, alleles at several or many different loci contribute to continuous variation in "quantitative characters" such as skin pigmentation, hair color and texture, and facial features. (B courtesy of Fred Nijhout.)

(A)

(B)

(C)

are inherited as if they were multiple alleles at one locus. Multiple loci contribute to the continuous variation typical of most phenotypic features, such as the familiar variation in human hair and skin color (**Figure 9.1C**).

Sources of Phenotypic Variation

Genetic and environmental sources of variation

The phenotypic variation that is most frequently discussed in evolutionary biology is attributable to genetic differences among individuals (as in Figure 9.1) and to differences in the environment they have experienced. "Environment" in this context means anything external to the organism, including the food it consumes and its interactions with other organisms, including other members of the same species. Some environmental effects are temporary, as in physiological acclimation, enzyme induction, and the expression of many learned behaviors. Other environmental effects are permanent: the sex of turtles, for example, is determined by temperature at a critical stage of development. Some environmental factors act on an embryo before birth or hatching. Consumption of alcohol, tobacco, and many other drugs by pregnant women, for example, increases the risk of nongenetic birth defects in their babies. Thus some CONGENITAL differences among individuals—those present at birth—do not have a genetic basis.

The capacity of an organism, of a given genotype, to express different phenotypes under different environmental conditions is referred to as **phenotypic plasticity** (Pigliucci 2001). Some environmentally determined phenotypic effects are not adaptive and may be unavoidable: for example, most organisms grow more slowly at lower temperatures, and most of us will weigh more if we are too sedentary or ingest too many calories. Other phenotypic effects are responses to the environment that have been shaped by natural selection, and represent adaptive phenotypic plasticity (**Figure 9.2**). Aphids, for example, develop wings if cues such as crowding or poor food quality signal that their environment is deteriorating.

Many characteristics have both genetic and environmental sources of variation (**Figure 9.3**), and in many cases, both kinds of influences affect the same developmental pathway. When variation stems from both genetic and environmental sources, it is pointless to ask whether a characteristic is "genetic" *or* "environmental," as if it had to be exclusively one or the other. Moreover, the relative amounts of genetic and environmental variation can differ with different circumstances, even in the same population. This is especially true if genotypes differ in their **norm of reaction**, the variety of different phenotypic states that can be produced by a single genotype under different environmental conditions. For example, Gupta and Lewontin (1982) measured the mean number of abdominal bristles in

FIGURE 9.2 Examples of phenotypic variation caused by environmental conditions. (A) Many species of aphids develop wings only under certain environmental conditions. In early summer, parthenogenetically reproducing peach aphids (*Myzus persicae*) develop as large, wingless adults (left). Their offspring, however, are winged (right) and capable of dispersing to secondary host plants as summer wanes. (B) The water flea *Daphnia cucullata* develops a protective "helmet" (left) if it is exposed, when young, to chemicals released by certain predators. The two individuals shown here are genetically identical. (B from Agrawal et al. 1999, photo courtesy of A. A. Agrawal.)

(A)

(B) Predator-induced   Typical

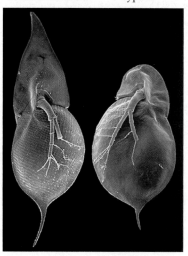

FIGURE 9.3 Both genetic and environmental differences often cause a character to vary. (A) Long-winged form of the planthopper *Sogatella furcifera*. This pest of rice also has a short-winged form. (B) The percentage of planthoppers that develop long wings is greater if they are reared at high population densities—an environmental effect. But compared to control lines (*c*), the percentage that develop long wings is higher (*m*) or lower (*b*) in laboratory lines that were selected for long or short wings for several generations. The differences between the *m*, *c*, and *b* lines are caused by genetic differences. (C) The African snail *Bulinus truncatus* is a self-fertilizing hermaphrodite, but some individuals develop a penis-like phallus and can inseminate other individuals. Phallus formation is more likely in snails that develop at low temperatures. (D) Genetic differences among snails from three local populations cause them to differ in their tendency to develop a phallus when all three are reared in the laboratory at low temperature. (B after Matsumura 1996; C and D after Schrag et al. 1994.)

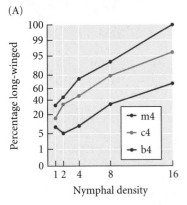

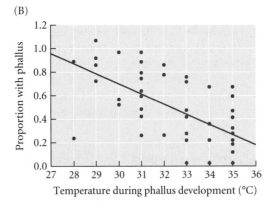

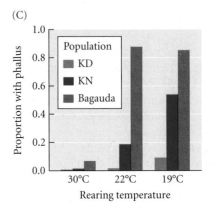

each of ten genetic strains of *Drosophila pseudoobscura*, each of which was derived from a single pair of wild-caught flies. They raised each genotype at three different temperatures. The norms of reaction of the ten genotypes (**Figure 9.4**) show that temperature affected the bristle number of every genotype, and that the genotypes differed in bristle number at any one temperature. A population of flies that had experienced different temperatures would then manifest both genetic variation and environmentally induced variation (often called environmental variance) in this feature. Moreover, the reaction norms of these genotypes are not all parallel: they display **genotype × environment interaction**, meaning that the effect of temperature on phenotype differed among genotypes. The amount of phenotypic variation in a population that is due to genetic differences among individuals thus depends on the particular range of temperatures at which the flies develop.

FIGURE 9.4 An example of genotype × environment interaction and how it can affect the relative contributions of genetic and environmental variation to phenotypic variation. The graph shows the number of bristles on the abdomen of male *Drosophila pseudoobscura* of ten genotypes, each reared at three different temperatures. At any temperature, the genotypes differ in this trait. Temperature affects the development of bristle number differently for each genotype, however, so the phenotype results from an interaction between genotype and environment. (After Gupta and Lewontin 1982.)

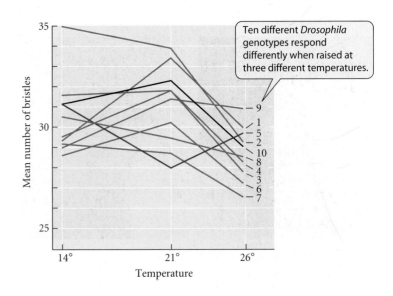

Ten different *Drosophila* genotypes respond differently when raised at three different temperatures.

Because evolution depends on the genetic component of variation, it is often critically important to determine whether variation in a characteristic is genetic, environmental, or both. Several methods are frequently used:

1. Phenotypes can be experimentally crossed to produce F_1, F_2, and backcross progeny. Mendelian ratios among the phenotypes of the progeny (e.g., 3:1 or 1:2:1) are taken as evidence of simple genetic control.

2. Correlation between the average phenotype of offspring and that of their parents, or greater resemblance among siblings than among unrelated individuals, suggests that genetic variation contributes to phenotypic variation. However, we must be sure that siblings (or relatives in general) do not share more similar environments than nonrelatives. For example, human geneticists rely strongly on studies of adopted children to determine whether behavioral or other similarities among siblings are due to shared genes rather than shared environments.

3. The offspring of phenotypically different parents can be reared together in a uniform environment, often referred to as a **common garden**. Differences among offspring from different parents that persist in such circumstances are likely to have a genetic basis. Propagation in the common garden for at least two generations is advisable.

Nongenetic inherited variation

Variation in a character is called GENETIC VARIATION if it is caused by differences in DNA sequence. But there are several other ways in which traits may be inherited (Danchin et al. 2011). The most familiar is CULTURAL INHERITANCE. Humans tend to inherit their parents' language, religion, economic status, and much more. Many other mammals, birds, and some other animals learn certain behavioral traits from adult individuals, so the traits are transmitted within and can diverge among populations. For instance, populations of many species of songbirds have different song dialects as a consequence of learning (**Figure 9.5**). Male birds reared in isolation, that have never heard another male's song, develop a less

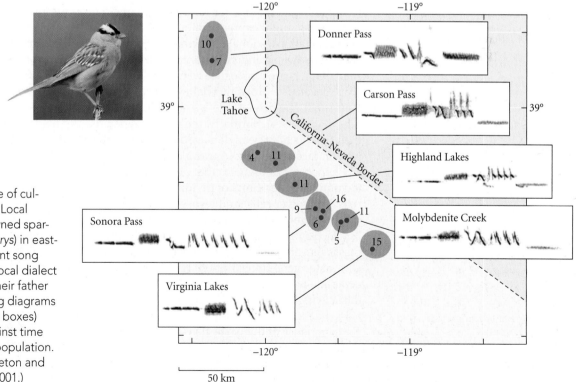

FIGURE 9.5 An example of culturally inherited variation. Local populations of white-crowned sparrows (*Zonotrichia leucophrys*) in eastern California have different song dialects. Males learn the local dialect by hearing the songs of their father and other males. The song diagrams (sonograms; shown inside boxes) plot sound frequency against time of a typical song in each population. (After MacDougall-Shackleton and MacDougall-Shackleton 2001.)

FIGURE 9.6 A maternal effect. The graph shows length of the helmets developed by *Daphnia cucullata* offspring in response to predator chemicals (kairomones) in four experimental treatments: mothers exposed to kairomones before their first brood and offspring exposed after birth; mothers exposed but offspring not exposed; offspring exposed but mothers not exposed; and neither exposed (controls). (After Agrawal et al. 1999.)

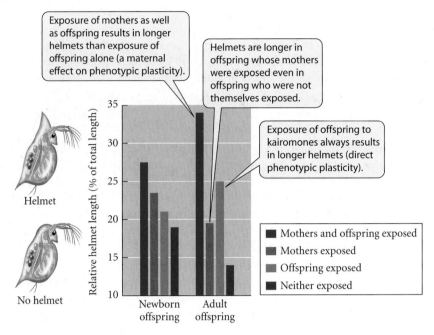

Exposure of mothers as well as offspring results in longer helmets than exposure of offspring alone (a maternal effect on phenotypic plasticity).

Helmets are longer in offspring whose mothers were exposed even in offspring who were not themselves exposed.

Exposure of offspring to kairomones always results in longer helmets (direct phenotypic plasticity).

■ Mothers and offspring exposed
■ Mothers exposed
■ Offspring exposed
■ Neither exposed

structured approximation of the species-typical song. Experimental groups of such birds, descended from isolated fathers, developed more and more species-typical songs over the course of several generations (Fehér et al. 2009). This observation suggests that the song has a genetically based ("innate") aspect as well.

PARENTAL EFFECTS are effects of parents on the phenotype of their offspring that are not caused by the genes they transmitted. The most familiar parental effects are **maternal effects**, resulting from factors such as the amount of yolk or chemicals included in the eggs, the amount and kind of maternal care the mother provides, her hormonal and physiological condition while carrying eggs or embryos, or her environment. Determining whether variation is caused by genes or maternal effects can be difficult. Often investigators will grow descendants of phenotypically different individuals in a common garden for several generations, rather than only one. The assumption is that the first generation of offspring might express maternal effects, but that such effects are likely to dissipate in later generations, so that any remaining variation is likely to have a genetic basis. Maternal effects are diverse and important in many contexts, and they are sometimes adaptive (Rossiter 1996). For example, the water flea *Daphnia cucullata*, a freshwater planktonic crustacean, develops a protective helmet if exposed to the odor of a predaceous insect larva (see Figure 9.2); the helmet size is greater in the offspring of females that were exposed to the odor (**Figure 9.6**). Similarly, the chemical defenses of wild radish plants are enhanced if their parents have been damaged by feeding caterpillars, and caterpillars do not develop as well on these offspring (Agrawal et al. 1999).

Phenotypic differences that are not based on DNA sequence differences are transmitted among generations of dividing cells in multicellular organisms (e.g., mitosis of liver cells yields liver cells) and also, sometimes, from parents to offspring. This phenomenon, called **epigenetic inheritance**, has several molecular causes, including DNA METHYLATION, in which a methyl group is joined to a cytosine (C) in a CG doublet. The methylation process is repeated during DNA replication, and the methylated state, which often reduces or eliminates gene transcription, may be transmitted for a few or many generations. Cross-generational epigenetic inheritance is often manifested as GENOMIC IMPRINTING, whereby the activity of a gene or chromosome—perhaps as many as 1 percent of all genes in mammals—depends on its parental origin. Epigenetic inheritance has been described for many characteristics of many species (Jablonka and Raz 2009) and is responsible for some rare "mutant" phenotypes. For example, one of the five petals of the toadflax *Linaria vulgaris* normally has a long nectar-bearing spur, but in the *peloria* mutant, which was first discovered by Linnaeus, all five petals have this form, transforming the flower's symmetry from

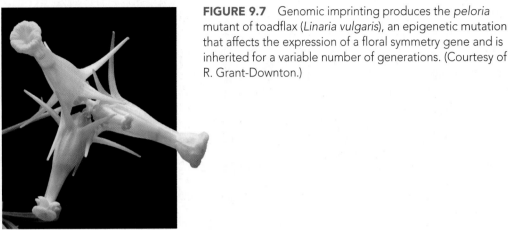

Standard *Linaria vulgaris*   *peloria* form

FIGURE 9.7 Genomic imprinting produces the *peloria* mutant of toadflax (*Linaria vulgaris*), an epigenetic mutation that affects the expression of a floral symmetry gene and is inherited for a variable number of generations. (Courtesy of R. Grant-Downton.)

bilateral to radial—the ancestral condition (**Figure 9.7**). This mutant phenotype can be produced not only by a mutation of the DNA sequence, but also by extensive methylation of a gene that controls flower symmetry, without alteration of the nucleotide sequence (Cubas et al. 1999). There is extensive evidence that genotypes differ in their ability to be epigenetically modified, so the "interpretive machinery" that can affect gene expression is itself able to evolve (Day and Bonduriansky 2011).

In some cases, an environmental factor alters the epigenetic state (Jablonka and Raz 2009). For example, female rats descended from great-grandparents that were treated with a fungicide that reduces androgen levels preferred male mates that lacked a history of exposure, suggesting that epigenetic effects could influence mating patterns in nature (Crews et al. 2007). There is no evidence that epigenetic "mutations" are induced by environments in which the mutation would be adaptive; that is, epigenetic inheritance does not imply Lamarckian evolution (Haig 2007). Although epigenetic variation, such as variation in the methylation state of certain genes, contributes to phenotypic variation of some characteristics in wild populations (Herrera and Bazaga 2011), there has been very little research on the role of epigenetic inheritance in adaptive evolution of natural populations. Theoretical analyses have shown that epigenetic and other forms of nongenetic inheritance could have diverse effects on evolution because phenotypic changes are partly decoupled from genetic changes (Bonduriansky and Day 2009). For example, phenotypic change might continue in the same direction after natural selection stops favoring a more extreme trait; in fact, the direction of selection and the direction of change in the phenotype might differ. The role of epigenetics in evolution is a subject of considerable speculation and research.

Understanding Evolution: Fundamental Principles of Genetic Variation

We embark now on our study of genetic variation and the factors that influence it—that is, the factors that cause evolution within species. *The definitions, concepts, and principles introduced here are absolutely essential for understanding evolutionary theory.* We begin with a short description of these ideas, followed by an explanation of a very important formal model.

At any given gene locus, a population may contain two or more alleles that have arisen over time by mutation. Sometimes one allele is by far the most common (and may be called the WILD TYPE) and the others are very rare; sometimes two or more of the alleles are each quite common. The relative commonness or rarity of an allele—its proportion of all gene copies in the population—is called the **allele frequency** (sometimes imprecisely referred to as the "gene frequency"). In sexually reproducing diploid populations, the alleles, carried in eggs and sperm, become combined into **homozygous** (two copies of the same allele)

FIGURE 9.8 Allele frequencies and genotype frequencies. For a locus with two alleles (A_1, A_2), this diagram shows the frequency of three genotypes among females and males in one generation, the allele frequencies among their eggs and sperm, and the genotype frequencies among the resulting offspring.

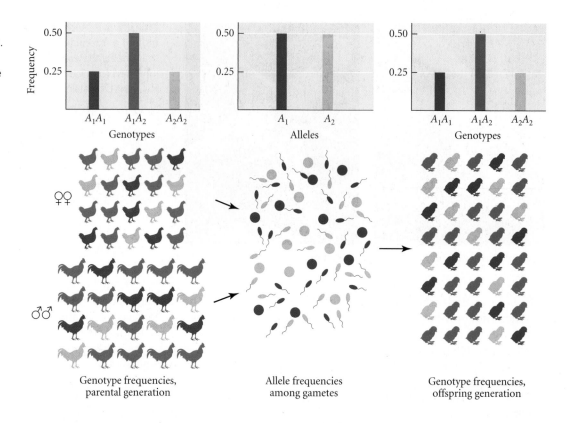

Genotype frequencies, parental generation

Allele frequencies among gametes

Genotype frequencies, offspring generation

and **heterozygous** (one copy each of two different alleles) genotypes. The **genotype frequency** is the proportion of a population that has a certain genotype (**Figure 9.8**). The proportions of the different genotypes are related to the allele frequencies in simple but important ways, as we will soon see. If the genotypes differ in a phenotypic character, the amount of variation in that character will depend not only on how different the genotypes are from one another, but also on the relative abundance (the frequencies) of the genotypes in the population—which in turn depends on the allele frequencies.

An alteration of the genotype frequencies in one generation will usually alter the frequencies of the alleles carried by the population's gametes when reproduction occurs, so the genotype frequencies of the following generation will be altered in turn. *Such alteration, from generation to generation, is the central process of evolutionary change.* However, the genotype and allele frequencies do not change on their own; something has to change them. *The factors that can cause the frequencies to change are the causes of evolution.*

Frequencies of alleles and genotypes: The Hardy-Weinberg principle

Imagine that a diploid population has 1000 individuals, and that for a particular gene locus in that population, there exist only two alleles, A_1 and A_2. Thus there are three possible genotypes for this locus: A_1A_1, A_2A_2 (both homozygous), and A_1A_2 (heterozygous).

Let us say 400 individuals have the genotype A_1A_1, 400 are A_1A_2, and 200 are A_2A_2. Let the frequencies of genotypes A_1A_1, A_1A_2, and A_2A_2 be represented by D, H, and R, respectively, and let the frequencies of the alleles A_1 and A_2 be represented by p and q. Thus the genotype frequencies are $D = 0.4$, $H = 0.4$, and $R = 0.2$, and the allele frequencies are $p = 0.6$ and $q = 0.4$ (**Figure 9.9A,B**). The frequency of allele A_1 is the sum of all the alleles carried by A_1A_1 homozygotes, plus one-half of those carried by heterozygotes. Hence $p = D + H/2 = 0.6$. Likewise, the frequency of allele A_2 is calculated as $q = H/2 + R = 0.4$.

Now let us assume that each genotype is equally represented in females and males, and that *individuals mate entirely at random*. (This is an entirely different situation from

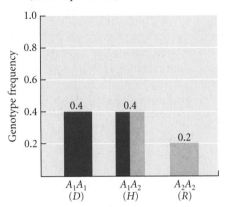

(A) Parental genotype frequencies (not in equilibrium)

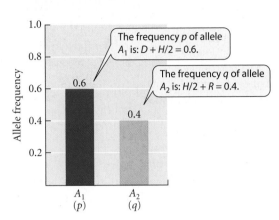

(B) Parental allele frequencies

The frequency p of allele A_1 is: $D + H/2 = 0.6$.

The frequency q of allele A_2 is: $H/2 + R = 0.4$.

FIGURE 9.9 A hypothetical example illustrating attainment of Hardy-Weinberg genotype frequencies after a single generation of random mating. (A) Genotype frequencies in the parental population. (B) Allele frequencies in the parental population. (C) Calculation of genotype frequencies among the offspring. (D) Genotype frequencies among the offspring, if all assumptions of the Hardy-Weinberg principle hold. (E) Allele frequencies among the offspring.

(C)

Offspring	Probability of a given mating producing the genotype		
A_1A_1	$\Pr[A_1 \text{ egg}] \times \Pr[A_1 \text{ sperm}] = p \times p = p^2$	0.6^2	$= 0.36$
A_1A_2	$\begin{cases} \Pr[A_1 \text{ egg}] \times \Pr[A_2 \text{ sperm}] = p \times q = pq & 0.6 \times 0.4 = 0.24 \\ \Pr[A_2 \text{ egg}] \times \Pr[A_1 \text{ sperm}] = q \times p = pq & 0.4 \times 0.6 = 0.24 \end{cases}$		$\Big\} = 0.48$
A_2A_2	$\Pr[A_2 \text{ egg}] \times \Pr[A_2 \text{ sperm}] = q \times q = q^2$	0.4^2	$= 0.16$

(D) Offspring genotype frequencies

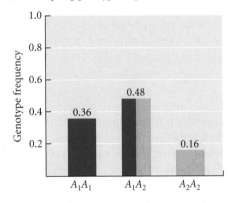

(E) Offspring allele frequencies

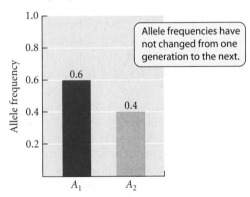

Allele frequencies have not changed from one generation to the next.

the crosses encountered in elementary genetics exercises, in which familiar Mendelian ratios are produced by nonrandomly crossing females of *one* genotype with males of *one* genotype.) Random mating is conceptually the same as mixing eggs and sperm at random, as might occur, for instance, in those aquatic animals that release gametes into the water. A proportion p of eggs carry allele A_1, q carry A_2, and likewise for sperm. The probabilities (Pr) for all possible allele combinations are shown in **Figure 9.9C**, and the resulting offspring genotype and allele frequencies are shown in **Figure 9.9D,E**.

Notice that in the new generation, the allele frequencies are still $p = D + H/2 = 0.6$ and $q = H/2 + R = 0.4$. *The allele frequencies have not changed from one generation to the next*, although the alleles have become distributed among the three genotypes in new proportions. These proportions, denoted as p^2, $2pq$, and q^2, constitute the HARDY-WEINBERG DISTRIBUTION of genotype frequencies.

The principle that genotypes will be formed at these frequencies as a result of random mating is named after G. H. Hardy and W. Weinberg (pronounced "vine-berg"), who independently calculated these results in 1908. The Hardy-Weinberg principle, broadly

Box 9A Derivation of the Hardy-Weinberg Distribution

Let the frequencies of genotypes A_1A_1, A_1A_2, and A_2A_2 be D, H, and R, respectively, and let the frequencies of alleles A_1 and A_2 be p and q. The genotype frequencies sum to 1, as do the allele frequencies. The probability of a mating between a female of any one genotype and a male of any one genotype equals the product of the genotype frequencies. For reciprocal crosses between two different genotypes, we take the sum of the probabilities. The mating frequencies and offspring produced are

Mating	Probability of mating*	Offspring genotype A_1A_1	A_1A_2	A_2A_2
$A_1A_1 \times A_1A_1$	D^2	D^2		
$A_1A_1 \times A_1A_2$	$2DH$	DH	DH	
$A_1A_1 \times A_2A_2$	$2DR$		$2DR$	
$A_1A_2 \times A_1A_2$	H^2	$H^2/4$	$H^2/2$	$H^2/4$
$A_1A_2 \times A_2A_2$	$2HR$		HR	HR
$A_2A_2 \times A_2A_2$	R^2			R^2

*Note that there are a total of 9 mating possibilities: 3 female genotypes × 3 male genotypes. Reciprocal matings between two different genotypes count as 2.

Recalling that $p = D + H/2$ and $q = H/2 + R$, the frequency of each genotype among the offspring is

A_1A_1: $D^2 + DH + H^2/4 = (D + H/2)^2 = p^2$
A_1A_2: $DH + 2DR + H^2/2 + HR = 2[(D + H/2)(H/2 + R)] = 2pq$
A_2A_2: $H^2/4 + HR + R^2 = (H/2 + R)^2 = q^2$

This result may also be obtained by recognizing that if genotypes mate at random, gametes, and therefore genes, also unite at random to form zygotes. Because the probability that an egg carries allele A_1 is p, and the same is true for sperm, the probability of an A_1A_1 offspring is p^2. A Punnett square shows the probability of each gametic union:

		Sperm	
		A_1 (p)	A_2 (q)
Eggs	A_1 (p)	A_1A_1 (p^2)	A_1A_2 (pq)
	A_2 (q)	A_1A_2 (pq)	A_2A_2 (q^2)

The Hardy-Weinberg principle can be extended to more complicated patterns of inheritance, such as multiple alleles. For example, if there are k alleles ($A_1, A_2, \ldots A_k$), the H-W frequency of homozygotes for any allele A_i is p_i^2, and that of heterozygotes for any two alleles A_i and A_j is $2p_ip_j$, where p_i and p_j are the frequencies of the two alleles in question. The total frequency of all heterozygotes combined (H) is sometimes expressed as the complement of the summed frequency of all homozygous genotypes, or $H = 1 - p_i^2$.

For example, if there are three alleles A_1, A_2, and A_3 with frequencies p_1, p_2, and p_3, the H-W frequencies of the three possible homozygotes (A_1A_1, A_2A_2, A_3A_3) are p_1^2, p_2^2, and p_3^2; and the H-W frequencies of the heterozygotes (A_1A_2, A_1A_3, A_2A_3) are $2p_1p_2$, $2p_1p_3$, and $2p_2p_3$. The total frequency of heterozygotes is

$$H = 1 - (p_1^2 + p_2^2 + p_3^2)$$

stated, holds that whatever the initial genotype frequencies for two alleles may be, *after one generation of random mating, the genotype frequencies will be p^2:2pq:q^2$*. Moreover, both these genotype frequencies and the allele frequencies *will remain constant in succeeding generations* unless factors we have not yet considered should change them. When genotypes at a locus have the frequencies predicted by the Hardy-Weinberg principle, the locus is said to be in **Hardy-Weinberg (H-W) equilibrium**. **Box 9A** shows how these results are obtained by tabulating all matings and the proportions of genotypes among the progeny of each.

An example: The human MN *locus*

Among the many variations in proteins on the surface of human red blood cells are those resulting from variation at the *MN* locus. Two alleles (*M, N*) and three genotypes (*MM, MN, NN*) can be distinguished by blood typing. A sample of 320 people in the Sicilian village of Desulo (R. Ceppellini, in Allison 1955) yielded the following numbers of people carrying each genotype:

MM	*MN*	*NN*
187	114	19

We can now estimate the *frequency* of each genotype as the proportion of the total sample (320) that carries that genotype. Thus

Frequency of *MM* = D = 187/320 = 0.584

$$\text{Frequency of } MN = H = 114/320 = 0.356$$

$$\text{Frequency of } NN = R = 19/320 = 0.059$$

Note that these frequencies, or proportions, must sum to 1.

Now we can calculate the allele frequencies. Each person carries two gene copies, so the total sample represents $320 \times 2 = 640$ gene copies. Because *MM* homozygotes each have two *M* alleles and *MN* heterozygotes have one, the number of *M* alleles in the sample is $(187 \times 2) + (114 \times 1) = 488$. The number of *N* alleles is $(19 \times 2) + (114 \times 1) = 152$. Thus

$$\text{Frequency of } M = p = 488/640 = 0.763$$

$$\text{Frequency of } N = q = 152/640 = 0.237$$

As with the genotype frequencies, p and q must sum to 1.

These figures are *estimates* of the true genotype frequencies and allele frequencies in the population because they are based on a sample rather than on a complete census. The larger the sample, the more confident we can be that we are obtaining accurate estimates of the true values. (We assume that the sample is a RANDOM one; that is, that our likelihood of collecting a particular type is equal to the true frequency of that type in the population.)

Our hypothetical example (see Figure 9.9) showed that for a given set of allele frequencies, a population might or might not have H-W genotype frequencies. (The frequencies 0.36, 0.48, and 0.16 in the offspring generation were in H-W equilibrium, but the frequencies 0.40, 0.40, and 0.20 in the parental generation were not.) Our real example shows a close fit to the H-W frequency distribution. The allele frequencies of *M* and *N* were $p = 0.763$ and $q = 0.237$. If we calculate the *expected genotype frequencies* under the H-W principle and multiply them by the sample size (320), we obtain the *expected number* of individuals with each genotype. These values in fact closely fit the *observed* numbers:

	Genotype		
	MM	**MN**	**NN**
	p^2	$2pq$	q^2
Expected H-W frequency	0.582	0.362	0.056
Expected number (H-W frequency × sample size)	186	116	18
Observed number	187	114	19

The significance of the Hardy-Weinberg principle: Factors in evolution

The Hardy-Weinberg principle is the foundation on which almost all of the theory of population genetics of sexually reproducing organisms—which is to say, most of the genetic theory of evolution—rests. We will encounter it in the theory of natural selection and other causes of evolution. It has two important implications: First, genotype frequencies attain their H-W values after a single generation of random mating. If some factor in the past had caused genotype frequencies to deviate from H-W values, a single generation of random mating would erase the imprint of that history. Second, according to the H-W principle, not only genotype frequencies, but also allele frequencies, remain unchanged from generation to generation. A new mutation, for example, will remain at its initial very low allele frequency indefinitely.

Like any mathematical model, the Hardy-Weinberg principle holds only under certain assumptions. Since allele frequencies and genotype frequencies often do change (i.e., evolution occurs), the assumptions of the Hardy-Weinberg formulation must not always hold. Therefore *the study of genetic evolution consists of asking what happens when one or more of the assumptions of the Hardy-Weinberg principle are relaxed.* The most important of these assumptions are the following:

1. *Mating is random.* If a population is not **panmictic**—that is, if members of the population do not mate at random—the genotype frequencies may depart from the ratios $p^2:2pq:q^2$.

2. *The population is infinitely large* (or so large that it can be treated as if it were infinite). The H-W calculations are made in terms of probabilities. If the number of events is finite, the actual outcome is likely to deviate, purely by chance, from the predicted outcome. If we toss an infinite number of unbiased coins, probability theory says that half will come up heads, but if we toss only 100 coins, we are likely not to obtain exactly 50 heads, purely by chance. Similarly, among a finite number of offspring, both the genotype frequencies and the allele frequencies may differ from those in the previous generation, *purely by chance.* Such random changes are called **genetic drift**.

3. *Genes are not added from outside the population.* Immigrants from other populations may carry different frequencies of A_1 and A_2; if they interbreed with residents, this will alter allele frequencies and, consequently, genotype frequencies. Mating among individuals from different populations is termed **gene flow**, or sometimes simply as **migration**. We may restate this assumption as: There is no gene flow.

4. *Genes do not mutate from one allelic state to another.* Mutation, as we saw in Chapter 8, can change allele frequencies, although usually very slowly. The Hardy-Weinberg principle assumes no mutation.

5. *All individuals have equal probabilities of survival and of reproduction.* If these probabilities differ among genotypes (i.e., if there is a consistent difference in the genotypes' rates of survival or reproduction), then the frequencies of alleles and genotypes may be altered from one generation to the next. Thus the Hardy-Weinberg principle assumes that there is *no natural selection* affecting the locus of interest.

Inasmuch as nonrandom mating, chance, gene flow, mutation, and selection can alter the frequencies of alleles and genotypes, these are the major factors that cause evolutionary change within populations.

Certain subsidiary assumptions of the Hardy-Weinberg principle can be important in some contexts. First, the principle, as presented, applies to autosomal loci; it can also be modified for sex-linked loci (which have two copies in one sex and one in the other). Second, the principle assumes that alleles segregate into a heterozygote's gametes in a 1:1 ratio. Deviations from this ratio are known and are called SEGREGATION DISTORTION or MEIOTIC DRIVE.

If the assumptions we have listed hold true for a particular locus, that locus will display Hardy-Weinberg genotype frequencies. But if we observe that a locus displays the Hardy-Weinberg frequency distribution, we cannot conclude that the assumptions hold true! For example, mutation or selection may be occurring, but at such a low rate that we cannot detect a deviation of the genotype frequencies from the expected values. Or, under some forms of natural selection, we might observe deviations from Hardy-Weinberg equilibrium if we measure genotype frequencies at one stage in the life history, but not at other stages.

Frequencies of alleles, genotypes, and phenotypes

At Hardy-Weinberg equilibrium, **heterozygosity** (the frequency of heterozygotes) is greatest when the two alleles have equal frequencies (**Figure 9.10**). When an allele is very rare, almost all its carriers are heterozygous. Because almost all copies of a rare recessive allele are carried by heterozygotes, the allele may not be detected unless large samples are examined. If, for example, the frequency of a recessive allele is $q = 0.01$, only $(0.01)^2 = 0.0001$ of the population displays this recessive allele in its phenotype. Thus populations can carry concealed genetic variation. However, a dominant allele may well be less common than a recessive allele. "Dominance" refers to an allele's phenotypic effect in the heterozygous condition, not to its numerical prevalence. In the British moth *Cleora repandata*, for example, black coloration is inherited as a dominant allele, and "normal" gray coloration is recessive. In a certain forest, 10 percent of the moths were found to be black (Ford 1971).

FIGURE 9.10 Hardy-Weinberg genotype frequencies as a function of allele frequencies at a locus with two alleles. Heterozygotes are the most common genotype in the population if the allele frequencies are between ⅓ and ⅔. (After Hartl and Clark 1989.)

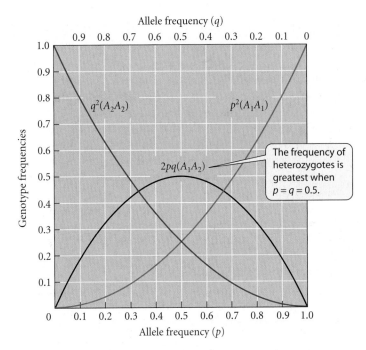

Inbreeding

The Hardy-Weinberg principle assumes that a population is panmictic. What if mating does not occur at random? One form of nonrandom mating is **inbreeding**, which occurs when individuals are more likely to mate with relatives than with nonrelatives, or more generally, when the gene copies in uniting gametes are more likely to be identical by descent than if they joined at random. Gene copies are said to be **identical by descent** if they have descended, by replication, from a common ancestor, relative to other gene copies in the population. The probability that a random pair of gene copies is identical by descent is denoted by F, the **inbreeding coefficient**. **Box 9B** shows that as inbreeding proceeds, the frequency of each homozygous genotype increases, and the frequency of heterozygotes decreases by the same amount. The frequency of heterozygotes is $H = H_0(1 - F)$, where H_0 is the heterozygote frequency expected if the locus were in H-W equilibrium.

The ways in which inbreeding changes genotype frequencies are easily seen if we envision mating of individuals only with their closest relatives—namely, themselves—by **self-fertilization**, or **selfing**. (You may have trouble imagining this if you think about people, but think about plants and you will see that it is not only plausible, but quite common.) The homozygous genotype A_1A_1 can produce only A_1 eggs and A_1 sperm, and thus only A_1A_1 offspring; likewise, A_2A_2 individuals produce only A_2A_2 offspring. Heterozygotes produce A_1 and A_2 eggs in equal proportions, and likewise for sperm. When these eggs and sperm join at random, one-fourth of the offspring are A_1A_1, half are A_1A_2, and one-fourth are A_2A_2. (The two allele copies carried by the homozygous offspring are identical by descent.) Thus the frequency of heterozygotes is halved in each generation and eventually reaches zero. Conversely, F increases as inbreeding continues; in fact, F can be estimated by the deficiency of heterozygotes relative to the H-W equilibrium value (see Box 9B). If CONSANGUINEOUS mating (mating among relatives) is a consistent feature of a population, F will increase over generations at a rate that depends on how closely related mates are, on average.

Self-fertilization occurs in many species of plants and a few animals. The wild oat *Avena fatua*, for example, which reproduces mostly by selfing, displayed a low frequency of heterozygotes at the several loci studied (**Figure 9.11**). The inbreeding coefficients estimated from data on all the loci were nearly equal, which is as it should be because inbreeding affects all loci in the same way.

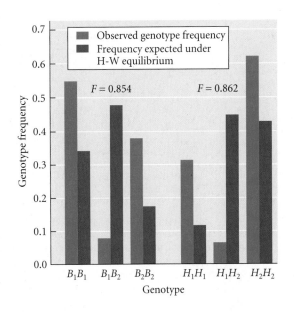

FIGURE 9.11 Genotype frequencies observed at two loci in a population of the self-fertilizing wild oat *Avena fatua* compared with those expected under Hardy-Weinberg equilibrium. Note that heterozygotes are deficient at both loci, and that calculated values of F are nearly the same for the two loci. (Data from Jain and Marshall 1967.)

Box 9B   Change of Genotype Frequencies by Inbreeding

Suppose we could label every gene copy at a locus and trace the descendants of each gene copy through subsequent generations. We would then be tracing alleles that are identical by descent. Some of the gene copies are allele A_1, and others are A_2. If we label one of the A_1 copies A_1^*, then after one generation, a sister and brother may both inherit replicates of A_1^* from a parent (with probability $0.5^2 = 0.25$). Among the progeny of a mating between sister and brother, both heterozygous for A_1^*, one-fourth will be $A_1^*A_1^*$. These individuals carry two gene copies that are not only the same allele (A_1), but are also identical by descent (A_1^*) (Figure 1). The $A_1^*A_1^*$ individuals are said to be not only homozygous, but AUTOZYGOUS. ALLOZYGOUS individuals, by contrast, may be either heterozygous or homozygous (if the two copies of the same allele are not identical by descent).

The inbreeding coefficient, denoted F, is the probability that an individual taken at random from the population will be autozygous. In a population that is not at all inbred, $F = 0$. In a fully inbred population, $F = 1$: all individuals are autozygous.

In a population that is inbred to some extent, F is the fraction of the population that is autozygous, and $1 - F$ is the fraction that is allozygous. If two alleles, A_1 and A_2, have frequencies p and q, the probability that an individual is allozygous and that it is A_1A_1 is $(1 - F) \times p^2$. Likewise, the fraction of the population that is allozygous and heterozygous is $(1 - F) \times 2pq$, and the fraction that is allozygous and A_2A_2 is $(1 - F) \times q^2$.

Turning our attention now to the fraction, F, of the population that is autozygous, we note that none of these individuals is heterozygous, because a heterozygote's alleles are not identical by descent. If an individual is autozygous, the probability that it is autozygous for A_1 is p, the frequency of A_1. Thus the fraction of the population that is autozygous and A_1A_1 is $F \times p$. Likewise, $F \times q$ is the fraction that is autozygous and A_2A_2.

Thus, taking into account the allozygous and autozygous fractions of the population, the genotype frequencies (Figure 2) are

	Allozygous		Autozygous		Genotype frequency
A_1A_1	$p^2(1 - F)$	+	pF	=	$p^2 + Fpq = D$
A_1A_2	$2pq(1 - F)$			=	$2pq(1 - F) = H$
A_2A_2	$q^2(1 - F)$	+	qF	=	$q^2 + Fpq = R$

Therefore the consequence of inbreeding is that the frequency of homozygotes is higher, and the frequency of heterozygotes is lower, than in a population that is in Hardy-Weinberg equilibrium. Note that H, the frequency of heterozygotes in the inbred population, equals $(1 - F)$ multiplied by the frequency of heterozygotes we expect to find in a randomly mating population ($2pq$). Denoting $2pq$ as H_0, we have

$$H = H_0(1 - F) \quad \text{or} \quad F = (H_0 - H)/H_0$$

Thus we can estimate the inbreeding coefficient by two measurable quantities, the observed frequency of heterozygotes, H, and the "expected" frequency, $2pq$, which we can calculate from data on the allele frequencies p and q. In practice, then, the inbreeding coefficient F is a measure of the reduction in heterozygosity compared with a panmictic population with the same allele

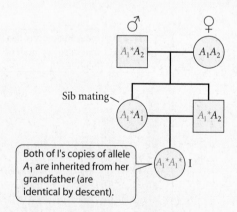

Sib mating

Both of I's copies of allele A_1 are inherited from her grandfather (are identical by descent).

FIGURE 1   A pedigree showing inbreeding due to mating between siblings. Squares represent males, circles females. Copies of an A_1 allele, A_1^* (red type), are traced through three generations. Individual I possesses two copies of A_1^* that are identical by descent (she is autozygous). I's mother is homozygous for A_1, but the two copies are not identical by descent (she is allozygous).

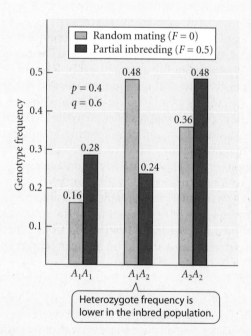

Heterozygote frequency is lower in the inbred population.

FIGURE 2   Genotype frequencies at a locus with allele frequencies $p = 0.4$ and $q = 0.6$ when mating is random ($F = 0$) and when the population is partially inbred ($F = 0.5$).

Genetic Variation in Natural Populations: Individual Genes

Morphology and viability

Genetic **polymorphism** (*poly*, "many"; *morph*, "form") is the presence in a population of two or more variants (alleles or haplotypes). A locus or character that is not polymorphic is **monomorphic**.

Until the 1960s, polymorphisms in natural populations could be detected and studied only by finding variable phenotypic features that showed Mendelian segregation in experimental crosses—which limited study to only a few characteristics and only a few experimentally tractable species. Some of these polymorphisms have been studied extensively (such as the butterfly polymorphism in Figure 9.1B), and some dramatic variations have been shown to be single-locus polymorphisms (see, e.g., Figure 9.1A).

In the 1920s, geneticists found that the same mutations of *Drosophila* that had been studied in laboratory populations—mutations affecting coloration, bristle number, wing shape, and the like—also existed in wild populations. These recessive alleles could be revealed and studied by "extracting" chromosomes—that is, producing flies that are homozygous for a particular chromosome. Chromosome extraction is accomplished by a series of crosses between a wild population and a laboratory stock that carries a dominant marker allele and an inversion that suppresses crossing over (**Figure 9.12**). The inversion causes the entire chromosome to be inherited as a unit, and the genetic marker enables the experimenter to trace its inheritance and to bring two copies of the wild chromosome together in homozygous form. A chromosome that has been made homozygous in this way may carry not only recessive alleles that affect morphological features, but also alleles

FIGURE 9.12 Crossing technique for "extracting" a chromosome from a wild male *Drosophila melanogaster* and making it homozygous in order to detect recessive alleles. In this figure, plus signs denote wild chromosomes with wild-type alleles, and the subscript number identifies a particular chromosome. The process is shown for two wild males (with chromosome pairs $+_1,+_2$ and $+_3,+_4$), each of which is crossed to a laboratory stock that has dominant mutant markers on the homologous chromosomes: *Cy* (curly wing) and *L* (lobed wing) on one and *Pm* (plum eye color) on the other. Each of these chromosomes also has inversions (see Chapter 8) that prevent crossing over. Consequently, one of each male's wild chromosomes (say, $+_2$) is transmitted intact to the F_3 generation. When crosses are performed as shown, the F_3 generation consists, in principle, of equal numbers of four genotypes, including a $+/+$ homozygote and the *Cy L/+* and *Pm/+* heterozygotes. The viability of these genotypes is measured by their proportion in the F_3 generation, relative to the expected 1:1:1:1 ratio. The lowermost family illustrates how flies heterozygous for two wild chromosomes ($+_2$ and $+_3$) can be produced. Their viability is also measured by deviation from the 1:1:1:1 ratio expected in this cross. (After Dobzhansky 1970.)

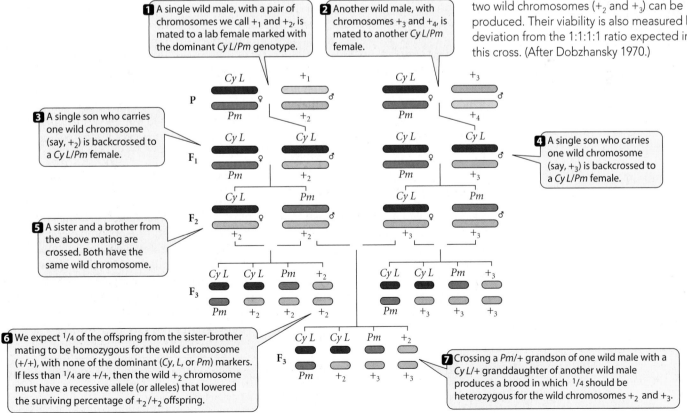

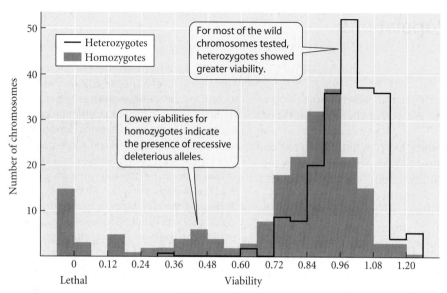

FIGURE 9.13 The frequency distribution of relative viabilities of chromosomes extracted from a wild population of *Drosophila pseudo-obscura* by the method illustrated in Figure 9.12. A viability value of 1.00 indicates conformity to the expected ratios of laboratory and wild-type genotypes in the F₃ generation, as explained in Figure 9.12. The colored distribution shows the viability (relative survival from egg to adult) of homozygotes for 195 wild chromosomes. (After Lewontin 1974.)

that affect traits such as viability (survival from egg to adult). The crosses produce an F₃ generation of flies of which one-fourth are expected to be homozygous for the wild chromosome. If no wild-type adult offspring (i.e., offspring without any of the dominant markers) appear, then the wild chromosome must carry at least one recessive **lethal allele**—that is, one that causes death before the flies reach the adult stage. Performing such crosses with many different wild flies makes it easy to determine what fraction of wild chromosomes cause complete lethality or reduce the likelihood of survival.

The great population geneticist Theodosius Dobzhansky and his collaborators examined hundreds of copies of chromosome 2 sampled from wild *Drosophila pseudoobscura*.* They discovered that about 10 percent of those copies were lethal when made homozygous. About half of the remaining chromosomes reduced survival to at least some extent. The other chromosomes in the genome yielded similar results, leading to the conclusion that almost every wild fly carries at least one chromosome that, if homozygous, substantially reduces the likelihood of survival (**Figure 9.13**). Moreover, many chromosomes cause sterility as well. Using a very different analytical method based on the mortality of children from marriages between relatives, Morton, Crow, and Muller (1956) concluded that humans are just like flies: "the average person carries heterozygously the equivalent of 3–5 recessive lethals [lethal alleles] acting between late fetal and early adult stages."

Such results, which have been confirmed by later work, point to a staggering incidence of life-threatening genetic defects. They imply, moreover, that *natural populations carry an enormous amount of concealed genetic variation* that is manifested only when individuals are homozygous. However, when flies carrying two different lethal chromosomes (i.e., derived from two wild flies) are crossed, the heterozygous progeny generally have nearly normal viability (see Figure 9.13). Hence the two chromosomes must carry recessive lethal alleles at different loci: one lethal homozygote, for example, is *aaBB*, the other is *AAbb*, and the heterozygote, *AaBb*, has normal viability because each recessive lethal is masked by a dominant "normal" allele. From such data, it can be determined that the lethal allele at any one locus is very rare ($q < 0.01$ or so), and that the high proportion of lethal chromosomes is caused by the summation of rare lethal alleles at many loci.

Inbreeding depression

Because populations of humans and other diploid species harbor recessive alleles that have deleterious effects, and because inbreeding increases the proportion of homozygotes, populations in which many matings are consanguineous often manifest a decline in components of fitness, such as survival and fecundity. Such a decline is called **inbreeding depression**. This effect has long been known in human populations (**Figure 9.14**). For example, a very high proportion (27 to 53 percent) of individuals with Tay-Sachs disease, a lethal neurodegenerative disease caused by a recessive allele that is especially prevalent in Ashkenazic Jewish populations, are children of marriages between first cousins (Stern 1973).

*Drosophila melanogaster, D. pseudoobscura, and many other species of *Drosophila* have 4 pairs of chromosomes. denoted X/Y, 2, 3, and 4. Chromosome 4 is very small and has few genes. The X chromosome and the two major autosomes, chromosomes 2 and 3, carry roughly similar numbers of genes.

Inbreeding depression is a well-known problem in the small, captive populations of endangered species that are propagated in zoos in the hope of reestablishing wild populations. Special breeding designs are required to minimize inbreeding in these populations (Frankham et al. 2002). Inbreeding may also increase the risk of extinction of small populations in nature. Thomas Madsen and colleagues (1995, 1999) studied an isolated Swedish population of a small poisonous snake, the adder *Vipera berus*, that consisted of fewer than 40 individuals. The snakes were found to be highly homozygous (we will see shortly how this can be determined), the females had small litter sizes (compared with outbred adders in other populations), and many of the offspring were deformed or stillborn. The authors introduced 20 adult male adders from other populations, left them there for four mating seasons, and then removed them. Soon thereafter, the population increased dramatically (**Figure 9.15**), owing to the improved survival of the outbred offspring.

Genetic variation at the molecular level

Evolution would be very slow if populations were genetically uniform, and if only occasional mutations arose and replaced pre-existing genotypes. In order to know what the potential is for rapid evolutionary change, it would be useful to know how much genetic variation natural populations contain.

Answering this question requires that we know what fraction of the loci in a population are polymorphic, how many alleles are present at each locus, and what their frequencies are. To know this, we need to count both the monomorphic (invariant) and polymorphic loci in a random sample of loci. Ordinary phenotypic characters cannot provide this information because we cannot count how many genes contribute to phenotypically uniform traits.

In 1966, Richard Lewontin and John Hubby, working with *Drosophila pseudoobscura*, addressed this question in a landmark paper. They reasoned that because most loci code for proteins (including enzymes), an invariant enzyme should signal a monomorphic locus, and a variable enzyme should signal a polymorphic locus. At that time, DNA sequencing had not yet been developed, but biochemists could visualize certain proteins

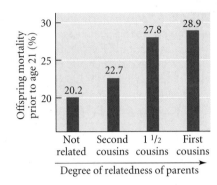

FIGURE 9.14 Inbreeding depression in humans: the more closely related the parents, the higher the mortality rate among their offspring. These data are for offspring up to 21 years of age from marriages registered in 1903–1907 in Italian populations. (After Stern 1973.)

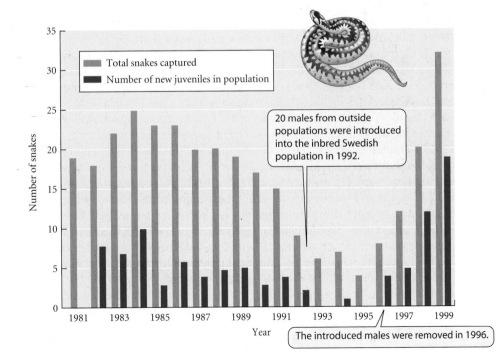

FIGURE 9.15 Population decline and increase in an inbred population of adders in Sweden. The gold bars represent the total number of males found in the population each year; the blue bars show the number of juveniles recruited into the population each year. (After Madsen et al. 1999.)

FIGURE 9.16 An electrophoretic gel, showing variation in the enzyme phosphoglucomutase among 18 individual killifishes (*Fundulus zebrinus*). Five allozymes (representing five different alleles) can be distinguished by differences in mobility. The fastest, at top, is allele 1; the slowest, at bottom, is allele 5. Homozygotes display a single blot, and heterozygotes display two blots. (Courtesy of J. B. Mitton.)

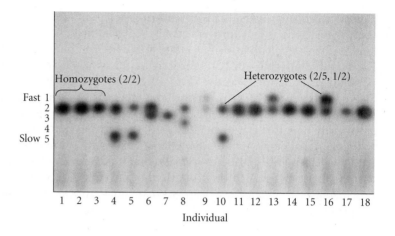

with **electrophoresis**. In this technique, proteins in a tissue extract move slowly through a starch gel or some other medium to which an electrical current is applied. Certain amino acid substitutions alter the net charge of a protein, so some enzyme variants that are encoded by different alleles differ in mobility; their positions can be visualized by reacting the enzyme with a substrate, then making the product visible as a colored blot. The various homozygotes and heterozygotes can then be distinguished (**Figure 9.16**). Electrophoretically distinguishable forms of an enzyme that are encoded by different alleles are called **allozymes**. Electrophoresis often underestimates the amount of genetic variation because not all amino acid substitutions alter electrophoretic mobility and because, of course, this method does not reveal synonymous variation.

Lewontin and Hubby examined 18 loci in populations of *Drosophila pseudoobscura*. In every population, about a third of the loci were polymorphic, represented by two to six different alleles, all of which had quite high frequencies in the population. Heterozygosity is a good measure of how nearly equal in frequency the alleles are (see Figure 9.8 and Box 9A). Assuming Hardy-Weinberg equilibrium, the frequency of heterozygotes (H) at a locus is $1 - \Sigma p_i^2$, where p_i is the frequency of the ith allele and p_i^2 is the frequency of homozygote A_iA_i. Averaged over all 18 loci (including the monomorphic ones), the frequency of heterozygotes was about 0.12 in each of Lewontin and Hubby's populations. This is equivalent to saying that an average individual is heterozygous at 12 percent of its loci. (This calculation is called the AVERAGE HETEROZYGOSITY, $\overline{H}$.) In a similar study of a human population, Harris and Hopkinson (1972) found that 28 percent of 71 loci were polymorphic and that the average heterozygosity ($\overline{H}$) was 0.07. Since these pioneering studies, other investigators have examined hundreds of other species, most of which have proved to have similarly high levels of genetic variation.

Lewontin and Hubby's paper (see also Lewontin 1974) had a great impact on evolutionary biology. Their data, and Harris's, confirmed that *almost every individual in a sexually reproducing species is genetically unique.* (Even with only two alleles each, 3000 polymorphic loci—the estimate for humans—could generate $3^{3000} = 10^{1431}$ genotypes, an unimaginably large number.) Populations are far more genetically diverse than almost anyone had previously imagined. Lewontin and Hubby asked, "Do forces of natural selection maintain this variation, or is it neutral, subject only to the operation of random genetic drift?" This question set in motion a research agenda that has kept population geneticists busy ever since.

This approach resulted in an enormous advance in the study of genetic variation because it revealed, in almost every species, abundant polymorphisms with a clear genetic basis. These polymorphisms could be studied in their own right (e.g., to study natural selection), or they could be used simply as genetic markers to determine, for example, which individuals mate with each other, or how genetically different related species or populations are from each other. Electrophoresis, however, has been superseded by DNA sequencing—which at first was an extraordinarily laborious procedure.

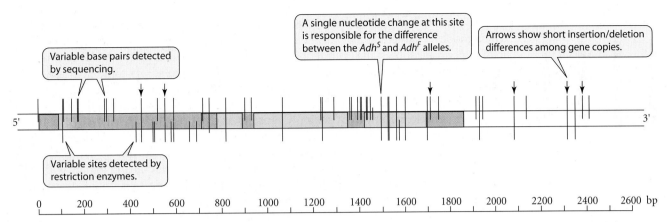

FIGURE 9.17 Nucleotide variation at the *Adh* locus in *Drosophila melanogaster*. Four exons (colored blocks) are separated by introns (gray blocks). The yellow blocks represent the coding regions of the exons. The lines intersecting the diagram from above show the positions of 43 variable base pairs and 6 insertions or deletions (indicated by arrows) detected by sequencing 11 copies of this gene. The lines intersecting the diagram from below show the positions of 27 variable sites detected by another method (restriction enzymes) among 87 gene copies. The scale below shows base pair (bp) position along the sequence. (After Kreitman 1983.)

The first study of genetic variation using complete DNA sequencing was carried out by Martin Kreitman (1983), who studied the locus that encodes alcohol dehydrogenase (ADH) in *Drosophila melanogaster*. Throughout the world, populations of this species are polymorphic for two common electrophoretic alleles, "fast" (*Adh^F*) and "slow" (*Adh^S*). Kreitman sequenced 11 copies of a 2721–base pair region that includes four exons and three introns, as well as noncoding flanking regions on either side (**Figure 9.17**). Among only 11 gene copies, he found 43 variable base pair sites, as well as six insertion/deletion polymorphisms (presence or absence of short runs of base pairs). Fewer sites were variable in the exons (1.8 percent) than in the introns (2.4 percent). All but 1 of the 14 variations in the coding regions were synonymous substitutions, the exception being a single nucleotide change responsible for the single amino acid difference between the *Adh^S* and *Adh^F* alleles.

A measure of sequence variation, analogous to the average heterozygosity ($\overline{H}$) used to quantify allozyme variation, is the average NUCLEOTIDE DIVERSITY (HETEROZYGOSITY) PER SITE (π), the proportion of nucleotide sites at which two gene copies ("sequences") randomly taken from a population differ. For Kreitman's *Adh* sample, $\pi = 0.0065$, although over the genome as a whole, *D. melanogaster* is much more variable ($\pi = 0.05$; Li 1997). In human populations, the average nucleotide diversity per site across loci is $\pi = 0.0015$.

Since Kreitman's pioneering study, DNA sequencing technology has advanced greatly, and the procedure is now relatively inexpensive, so large samples and entire genomes are commonly sequenced. Sequence variation has been found in most of the genes and organisms that have been examined. For example, more than 10 million **single-nucleotide polymorphisms** (**SNPs**), or single-site nucleotide variations, have been documented in the human genome. This variation is the basis of the DNA matching that is used in crime investigations, paternity tests, and other applications, and it is now the basis of a great range of evolutionary studies. DNA sequence variation in natural populations includes all the kinds of mutations described in Chapter 8, and it provides abundant genetic markers for studying mating patterns, population structure (see Chapter 10), phylogeography (see Chapter 6), natural selection (see Chapter 12), and many other topics.

Genetic Variation in Natural Populations: Multiple Loci

All genetic variation owes its origin ultimately to mutation, but in the short term, a great deal of the genetic variation within populations arises through recombination. In sexually reproducing eukaryotes, genetic variation arises from two processes: the union of genetically different gametes, and the formation of gametes with different combinations of alleles, owing to independent segregation of nonhomologous chromosomes and to crossing over between homologous chromosomes.

(A) No linkage; linkage equilibrium

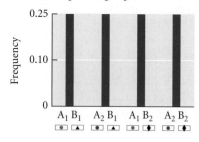

(B) Linkage; linkage equilibrium

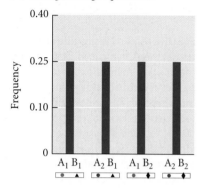

(C) Linkage; linkage disequilibrium

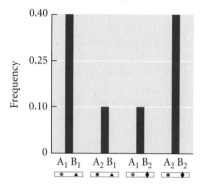

FIGURE 9.18 The distinction between linkage and linkage (dis) equilibrium. The frequencies of alleles A_1 and A_2 are each 0.5, and the same for alleles B_1 and B_2. (A) The A and B loci are unlinked (on different chromosomes) and are in linkage equilibrium: each gene combination has frequency 0.5^2, as expected if alleles are not associated. (B) The loci are linked (on the same chromosome), and are in linkage equilibrium. (C) The linked loci are in linkage disequilibrium: A_1 and B_1 are nonrandomly associated, as are A_2 and B_2.

The potential genetic variation that can be released by recombination is enormous. To cite a modest example, if an individual is heterozygous for only one locus on each of five pairs of chromosomes, independent segregation alone generates $2^5 = 32$ allele combinations among its gametes, and mating between two such individuals could give rise to $3^5 = 243$ genotypes among their progeny. If each locus affects a different feature, this represents a great variety of character combinations. However, recombination can both increase and decrease genetic variation. In sexually reproducing populations, genes are transmitted to the next generation, but genotypes are not: they end with organisms' deaths and are reassembled anew in each generation. Thus an unusual, favorable gene combination may occasionally arise through recombination, but if the individuals bearing it mate with other members of the population, it will be lost immediately by the same process. Consequently, recombination can either enhance or retard adaptation.

Linkage and linkage disequilibrium

Consider two loci (A and B), each with two alleles: A_1 and A_2 have frequencies p_A and q_A, and likewise, B_1 and B_2 have frequencies p_B and q_B. There are nine possible genotypes of zygotes, but haploid gametes have only four possible combinations: A_1B_1, A_2B_1, A_1B_2, and A_2B_2. If, among all the gametes produced by members of a population, the A allele is not correlated with the B allele (i.e., the alleles are not associated), then the expected frequency of each of these gamete types equals the product of the probability of a specified A allele and of a specified B allele. For example, the frequency of the A_1B_2 combination should be $p_A q_B$. It is then easy to predict the frequency of each of the nine zygotes formed by random union of gametes in the "gamete pool." The genotype $A_1A_1B_2B_2$, for example, is formed by union of two A_1B_2 gametes, with probability $(p_A q_B)^2$ (**Figure 9.18A**).

Linkage refers to the physical association of genes on the same chromosome (**Figure 9.18B**). Even if two loci are on the same chromosome, however, the alleles at one locus may segregate randomly with respect to the alleles at the other. In this case and in the previous one, the loci are said to be in a state of **linkage equilibrium** (Figure 9.18A,B). Knowing that an egg carries the A_2 allele would not help you predict which B allele it carries.

But suppose two populations, one consisting only of genotype $A_1A_1B_1B_1$ and the other only of $A_2A_2B_2B_2$, are mixed together and mate at random. In the next generation, there are three genotypes in the population: A_1B_1/A_1B_1, A_1B_1/A_2B_2, and A_2B_2/A_2B_2 (the slash separates the allele combinations an individual receives from its two parents). In this generation, there is a perfect association, or correlation, between specific alleles at the two loci: A_1 with B_1 and A_2 with B_2. The loci are in a state of complete **linkage disequilibrium** (**LD**). The degree of correlation, or LD, between two loci is often incomplete (**Figure 9.18C**). LD is extremely important for analyzing disease inheritance and other problems in human genetics and genomics. (As we will see, linkage disequilibrium is not always due to linkage: alleles at loci on different chromosomes may also be nonrandomly associated, so this term is slightly misleading.)

If alleles at one locus are generally associated with specific alleles at another locus, then any factor (such as natural selection) that causes a change in the frequencies of alleles at one locus should result in a correlated change at the other locus. For instance, if eggs and sperm generally carried either the haploid combination A_1B_1 or A_2B_2, an increase in the frequency of A_1 would automatically cause B_1 to increase (a phenomenon called GENETIC HITCHHIKING). If the loci affected different characteristics, an adaptive change in one feature could then result in a nonadaptive change in another feature. That would not happen if these loci were in linkage equilibrium, even if they were linked. So it is important to know if linkage disequilibrium between alleles is likely to be common.

Recombination during meiosis reduces the association—the level of LD—between specific alleles at two loci and brings the loci toward linkage equilibrium. If there were no recombination in our example, gametes would carry only the allele combinations A_1B_1 and A_2B_2, which, when united, could generate only the same three genotypes as before. However, recombination in the double heterozygote (A_1B_1/A_2B_2) gives rise to some A_1B_2 and A_2B_1 gametes, which, by uniting with A_1B_1 or A_2B_2 gametes, produce genotypes such

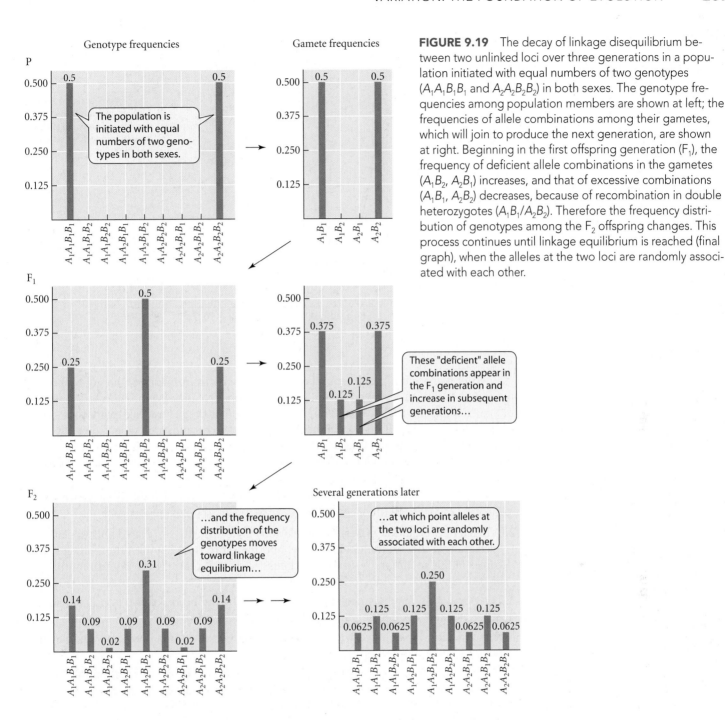

FIGURE 9.19 The decay of linkage disequilibrium between two unlinked loci over three generations in a population initiated with equal numbers of two genotypes ($A_1A_1B_1B_1$ and $A_2A_2B_2B_2$) in both sexes. The genotype frequencies among population members are shown at left; the frequencies of allele combinations among their gametes, which will join to produce the next generation, are shown at right. Beginning in the first offspring generation (F_1), the frequency of deficient allele combinations in the gametes (A_1B_2, A_2B_1) increases, and that of excessive combinations (A_1B_1, A_2B_2) decreases, because of recombination in double heterozygotes (A_1B_1/A_2B_2). Therefore the frequency distribution of genotypes among the F_2 offspring changes. This process continues until linkage equilibrium is reached (final graph), when the alleles at the two loci are randomly associated with each other.

as A_1B_1/A_1B_2, so the degree of LD has been reduced (**Figure 9.19**). Let us denote the frequencies of the gamete types as g_{ij}, where i represents the A allele and j the B allele (for example, g_{12} is the frequency of the A_1B_2 combination). A COEFFICIENT OF LINKAGE DISEQUILIBRIUM may be defined as $D = (g_{11} \times g_{22}) - (g_{12} \times g_{21})$. If $D > 0$, then gametes A_1B_1 and A_2B_2, as well as genotypes formed by their union (such as $A_1A_1B_1B_1$), will be more frequent than expected. As recombination occurs, D declines, because the "deficient" allele combinations (A_1B_2, A_2B_1) and genotypes increase in frequency slowly from generation to generation until the alleles at each locus are randomized with respect to alleles at the other locus and there is no correlation between them. The higher the recombination rate between the loci is, the faster the approach to linkage equilibrium proceeds: it is most rapid for unlinked genes (**Figure 9.20**).

FIGURE 9.20 The decrease in linkage disequilibrium (D) over time, relative to its initial value (D_0), for pairs of loci with different recombination rates (R). $R = 0.50$ if loci are unlinked. (After Hartl and Clark 1989.)

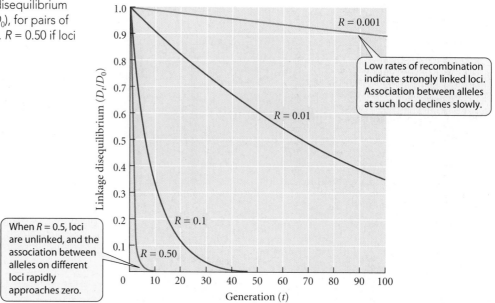

Whether loci are in linkage equilibrium or linkage disequilibrium, the genotype frequencies at each locus, viewed individually, conform to Hardy-Weinberg frequencies. At linkage equilibrium, the loci could be closely linked, but the probability that a copy of the A_1 allele would have a B_1 as its neighbor on the same chromosome would be p_B, its frequency in the population at large.

Why would we ever find loci in linkage disequilibrium if recombination breaks it down? There are several possible reasons.

1. Nonrandom mating can maintain LD. In an extreme case, a sample of organisms that display LD may actually include two reproductively isolated species.

2. When a new mutation arises, the single copy is necessarily associated with specific alleles at other loci on the chromosome, and therefore is in LD with those alleles. The copies of this mutation in subsequent generations will retain this association until it is broken down by recombination.

3. The population may have been formed recently by the union of two populations with different allele frequencies, and LD has not yet decayed.

4. Recombination may be very low or nonexistent. Chromosome inversions and parthenogenesis (asexual reproduction) have this effect.

5. LD may be caused by genetic drift (see Chapter 10). If the recombination rate is very low, the four gamete types in the example above may be thought of as if they were four alleles at one locus. One of these "alleles" may drift to high frequency by chance, creating an excess of that combination relative to others.

6. Natural selection may cause LD if two or more allele combinations are much fitter than recombinant genotypes.

In panmictic sexually reproducing populations, pairs of polymorphic loci are often found to be in linkage equilibrium, or nearly so. There are some interesting exceptions, however. For example, the European primrose *Primula vulgaris* is HETEROSTYLOUS, meaning that plants within a population differ in the lengths of stamens and style (pistil). Almost all plants have either the "pin" phenotype, with long style and short stamens, or the "thrum" phenotype, with short style and long stamens (**Figure 9.21**). Thus the anthers of each morph (at the ends of the stamens) are at the same level as the other morph's stigma. In most experimental crosses, this difference is inherited as if it were due to a single pair of alleles, with *thrum* dominant over *pin*. Rarely, however, "homostylous" progeny are produced, in which the female

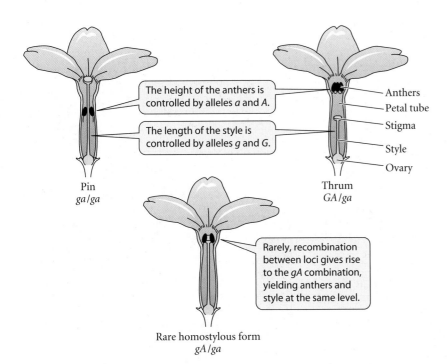

The height of the anthers is controlled by alleles *a* and *A*.

The length of the style is controlled by alleles *g* and *G*.

Pin
ga/ga

Anthers
Petal tube
Stigma
Style
Ovary

Thrum
GA/ga

Rarely, recombination between loci gives rise to the *gA* combination, yielding anthers and style at the same level.

Rare homostylous form
gA/ga

FIGURE 9.21 Heterostyly in the primrose *Primula vulgaris* is an example of linkage disequilibrium. Natural populations consist almost entirely of "pin" and "thrum" plants, which differ at two closely linked loci that determine the positions of stigma and anthers. Rarely, crossing over produces homostylous forms, in which the stigma and anthers are situated at the same level. (After Ford 1971.)

and male structures are equal in length (either short or long). Thus style and stamen length are actually determined by separate, closely linked loci: alleles *G* and *g* determine short and long style, respectively, and alleles *A* and *a* determine long and short stamens, respectively. Thrum plants have genotype *GA/ga*, and pin plants *ga/ga*. The gamete combinations *Ga* and *gA*, which give rise to homostylous plants, are very rare—partly because thrum and pin phenotypes are most successful in cross-pollination, since pollen from one is placed on insect visitors in a position that corresponds to the stigmatic surface of the other.

Linkage disequilibrium is common in asexual populations (see Chapter 15) because they undergo little recombination. *It is also found among very closely situated molecular markers, such as sites within genes.* LD provides important ways of detecting processes such as natural selection from DNA sequence data (see Chapter 12), and researchers in human genetics use molecular markers, in a procedure called LINKAGE DISEQUILIBRIUM MAPPING, to find nearby mutations that cause genetic diseases. A great deal of such research, based on sequence variation in the human genome, is under way, and evolutionary geneticists are playing a major role in developing methods for analyzing such data.

Variation in quantitative traits

Discrete genetic polymorphisms in phenotypic traits, such as pin versus thrum flowers, are much less common than slight differences among individuals, such as variation in the number of bristles on the abdomen of *Drosophila* or in weight or nose shape among humans. Such variation, called **quantitative variation**—or CONTINUOUS or METRIC variation—often approximately fits a **normal distribution** (**Figure 9.22**). The genetic component of such variation is often **polygenic**: that

FIGURE 9.22 The frequency distribution of the number of dermal ridges, summed over all ten fingertips, in a sample of 825 British men. The distribution nearly fits a normal curve (red line). The number of dermal ridges is an additively inherited polygenic character, with a heritability of about 0.95. (After Holt 1955.)

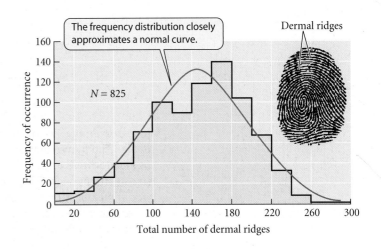

The frequency distribution closely approximates a normal curve.

Dermal ridges

N = 825

Frequency of occurrence

Total number of dermal ridges

FIGURE 9.23 Inheritance of a continuous-ly varying trait, length of the corolla (petals), in the tobacco plant *Nicotiana longiflora*. V_G and V_E represent variation that is due to genes and environment, respectively. The crosses show that the genetic variation is due to multiple genes (polygenic variation) rather than one or two loci. The two parental strains (P) are homozygous genotypes; the F_1 is heterozygous but genetically uniform. The F_2 shows expanded, continuous varia-tion resulting from recombination among the loci that affect the trait. If only one or two loci differed between parental strains, the F_2 would show discrete length catego-ries. Four F_3 families are shown, from par-ents whose means are indicated by arrows. The mean of the offspring in each family is close to that of their parents, indicating that differences among F_2 phenotypes are inher-ited. (After Mather 1949.)

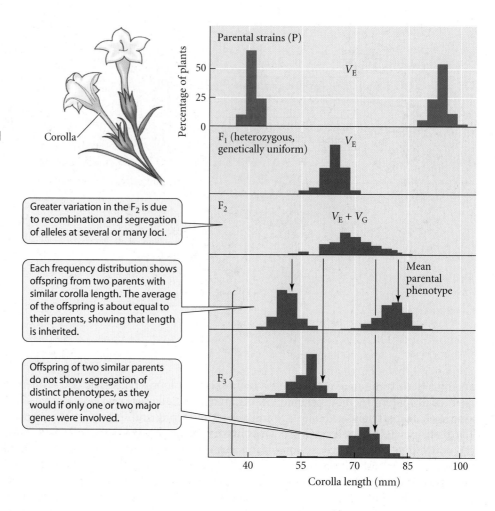

is, it is due to variation at several or many loci, each of which contributes to the variation in phenotype.

Figure 9.23 shows a typical pattern of variation and inheritance when homozygous strains that differ in a polygenic character are crossed. Note that in this case, the average trait value of offspring is about midway between the parents' means, and that variation increases in the F_2 generation because of recombination among many loci, which generates a great variety of genotypes.

A simple model of the relation between genotype and phenotype in a case of quantita-tive variation, in which we envision only two variable loci, might be

	A_1A_1	A_1A_2	A_2A_2
B_1B_1	5	6	7
B_1B_2	7	8	9
B_2B_2	9	10	11

In this example, relative to the genotype $A_1A_1B_1B_1$, each A_2 allele adds one unit, and each B_2 allele adds two units of measurement, on average, to the phenotype. This is a model of purely **additive** allele effects. In Figure 9.23, for example, the "short" and "long" parental strains might be $A_1A_1B_1B_1$ and $A_2A_2B_2B_2$, respectively; the F_1 would be $A_1A_2B_1B_2$, and the F_2 would include all nine genotypes.

Recombination can either increase or decrease variation in a quantitative trait, depend-ing on the initial distribution of genotypes. Imagine that at five loci, + and − alleles add or subtract one unit of measurement, respectively—say, 1 millimeter (mm). If we begin with quintuply heterozygous parents (both +−+−+/−+−+−), both of size 20 mm, recombination can result in offspring ranging in size from 15 mm (−/− at all five loci) to 25 mm (+/+ at all five loci). (Compare the F_1 and F_2 distributions of corolla lengths in Figure 9.23.) However, if we begin with parents that are quintuply homozygous for just + or just − alleles (genotypes +++++/+++++ and −−−−−/−−−−−), with respective sizes of 25 and 15 mm, the F_2 generation will show lower variance than the parental generation, because most offspring inherit various mixtures of + and − alleles. (Compare the P and F_2 generations in Figure 9.23.)

In order to judge how much variation is released by recombination, a team led by Theodosius Dobzhansky studied the effects of chromosomes they had "extracted" (see Figure 9.12) from a wild population of *Drosophila pseudoobscura* (Spassky et al. 1958). Homozygous chromosomes from natural populations of this species show tremendous variation in their effects on survival from egg to adult (see Figure 9.13). However, the Dobzhansky team chose ten homologous chromosomes that conferred almost the same, nearly normal viability when homozygous, and they made all possible crosses between flies bearing those chromosomes. From the F_1 female offspring, in which crossing over had occurred, they then extracted recombinant chromosomes and measured their effect on viability when homozygous. Even though the original ten chromosomes had differed little in their effect on viability, the variance in viability among the recombinant chromosomes was huge, equaling more than 40 percent of the variance found among homozygotes for much larger samples of chromosomes from natural populations. Thus a single episode of recombination among just ten chromosomes generates a large fraction of a wild population's variability. Some of the recombinant chromosomes were "synthetic lethals," meaning that recombination between two chromosomes that yield normal viability produced chromosomes that were lethal when they were made homozygous. This finding implies that each of the original chromosomes carried an allele that did not lower viability on its own, but did cause death when combined with another allele, at another locus, on the other chromosome.

HERITABILITY The description and analysis of quantitative variation are based on statistical measures because until recently, the loci that contribute to quantitative variation could not be easily singled out for study. The amount of genetic variation in a character depends on the number of variable loci that affect it, the genotype frequencies at each locus (**Figure 9.24**), and the phenotypic difference among genotypes.

The most useful statistical measure of variation is the **variance**, which measures the degree to which individuals are spread away from the average; technically, the variance

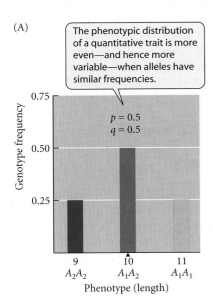

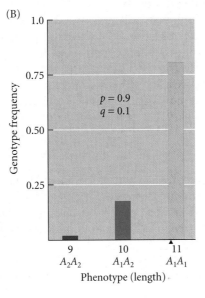

FIGURE 9.24 Variation in a quantitative trait, such as body length, due to alleles with frequencies (A) $p = 0.5$, $q = 0.5$ or (B) $p = 0.9$, $q = 0.1$. The black triangle denotes the mean. The distribution of lengths is more even, hence more variable, when the alleles have the same frequencies. The genetic variance, V_G, equals 0.500 in A and 0.472 in B. Hardy-Weinberg equilibrium is assumed in both cases.

BOX 9C Mean, Variance, and Standard Deviation

Suppose we have measured a character in a number of specimens. The character may vary continuously, such as body length, or discontinuously, such as the number of fin rays in a certain fin of a fish. Let X_i be the value of the variable in the ith specimen (e.g., $X_3 = 10$ cm in fish number 3). If we have measured n specimens, the sum of the values is $X_1 + X_2 + \ldots + X_n$, or

$$\sum_{i=1}^{n} X_i$$

(or simply X). The ARITHMETIC MEAN (or average) is

$$\bar{x} = \frac{\sum X_i}{n}$$

If the variable is discontinuous (e.g., fin rays), we may have n_1 individuals with value X_1, n_2 with value X_2, and so on for k different values. The arithmetic mean is then

$$\bar{x} = \frac{n_1 X_1 + n_2 X_2 + \cdots + n_k X_k}{n_1 + n_2 + \cdots + n_k}$$

The sum of n_i equals n, so we may write this as

$$\bar{x} = \frac{n_1 X_1}{n} + \frac{n_2 X_2}{n} + \cdots + \frac{n_k X_k}{n}$$

If we denote $n_i/n = f_i$, the frequency of individuals with value X_i, this becomes

$$\bar{x} = \sum_{i=1} (f_i X_i)$$

For example, in a sample of $n = 100$ fish, we may have $n_1 = 16$ fish with 9 fin rays ($X_1 = 9$), $n_2 = 48$ fish with 10 rays ($X_2 = 10$), and $n_3 = 36$

fish with 11 rays ($X_3 = 11$). There are thus three phenotypic classes ($k = 3$). The mean is

$$\bar{x} = \sum_{i=1} (f_i X_i) = (0.16)(9) + (0.48)(10) + (0.36)(11) = 10.2$$

Because we have only a sample from the fish population, this sample mean is an *estimate* of the true (PARAMETRIC) mean of the population, which we can know only by measuring every fish in the population.

How shall we measure the amount of variation? We might measure the RANGE (the difference between the two most extreme values), but this measure is very sensitive to sample size. A larger sample might reveal, for example, rare individual fish with 5 or 15 fin rays. These rare individuals do not contribute to our impression of the degree of variation. For this and other reasons, the most commonly used measures of variation are the variance and its close relative, the standard deviation. The true (parametric) variance is estimated by the mean value of the square of an observation's variation from the arithmetic mean:

$$V = \frac{(X_1 - \bar{x})^2 + (X_2 - \bar{x})^2 + \cdots + (X_n - \bar{x})^2}{n_1 - 1}$$

$$= \frac{1}{n-1} \sum_{i=1} n_i (X_i - \bar{x})^2$$

For statistical reasons, the denominator of a sample variance is $n - 1$ rather than n. In our hypothetical data on fin ray counts,

$$V = \frac{16(9 - 10.2)^2 + 48(10 - 10.2)^2 + 36(11 - 10.2)^2}{99} = 0.485$$

is the average squared deviation of observations from the mean (see **Box 9C**). In simple cases, the variance in a phenotypic character (V_P) is the sum of **genetic variance** (V_G) and the **environmental variance** (V_E) caused by the direct effects of environmental differences among individual organisms. That is, $V_P = V_G + V_E$. Oversimplifying, we can imagine that each genotype in a population has an average phenotypic value (of, say, body length), but that individuals with that genotype vary in their phenotypes because of environmental effects. The genetic variance, V_G, measures the amount of variation among the averages of the different genotypes, and the environmental variance, V_E, measures the average amount of variation among individuals with the same genotype (at the relevant loci). The proportion of the phenotypic variance that is genetic variance is the **heritability** of a trait, denoted h^2. Thus

$$h^2 = V_G/(V_G + V_E)$$

One way of detecting a genetic component of variation, and of estimating V_G and h^2, is to measure correlations* between parents and offspring, or between other relatives. For example, suppose that in a population, the mean value of a trait in the members of each brood of offspring were exactly equal to the value of that trait averaged between their two

---

*More properly, the *regression* of offspring mean on the mean of the two parents. The regression coefficient measures the slope of the relationship and is conceptually related to the correlation.

BOX 9C *(continued)*

The variance is a very useful statistical measure, but it is hard to visualize because it is expressed in squared units. It is easier to visualize its square root, the **standard deviation** (see also Figure 4.22):

$$s = \sqrt{V}$$

For our hypothetical data, $S = \sqrt{0.485} = 0.696$. The meaning of this number can perhaps be best understood by contrast with a sample of 1 fish with 9 rays, 18 with 10 rays, and 81 with 11 rays—an intuitively less variable sample. Then $\bar{x} = 10.8$, $V = 0.149$, and $s = 0.387$. V and s are smaller in this sample than in the previous sample because more of the individuals are closer to the mean.

A continuous variable, such as body length, often has a bell-shaped, or normal, frequency distribution (Figure 1). In the mathematically idealized form of this distribution (which many real samples approximate quite well), about 68 percent of the observations fall within one standard deviation on either side of the mean, 96 percent within two standard deviations, and 99.7 percent within three. If, for example, body lengths in a sample of fish are normally distributed, then if $\bar{x} = 100$ cm and $s = 5$ cm (hence $V = 25$ cm²), 68 percent of the fish are expected to range between 95 and 105 cm, and 96 percent to range between 90 and 110 cm. If the standard deviation is greater—say, $s = 10$ cm ($V = 100$ cm²)—then, for the same mean, the range limits embracing 68 percent of the sample are broader: 90 cm and 100 cm.

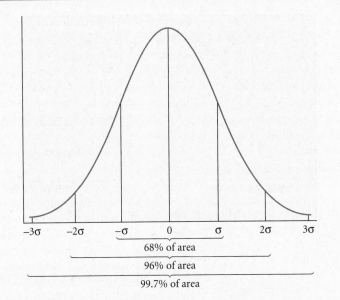

FIGURE 1 The normal distribution curve, with the mean taken as a zero-reference point, showing how the variable represented on the x-axis can be measured in standard deviations (σ). The bracketed areas show the fraction of the area under the curve (that is, the proportion of observations) embraced by one, two, and three standard deviations on either side of the mean. The true (parametric) value of the standard deviation is denoted by σ; the estimate of σ based on a sample is denoted s in this book.

parents (the MIDPARENT MEAN; **Figure 9.25A**). So perfect a correlation clearly would imply a strong genetic basis for the trait. In fact, in this instance, V_G/V_P (i.e., the heritability, h^2) would equal 1.0: all the phenotypic variation would be accounted for by genetic variation. If the correlation were lower, with some of the phenotypic variation perhaps due to environmental variation, the heritability would be lower (**Figure 9.25B**).

In a real-life example, Peter Boag (1983) studied heritability in the medium ground finch (*Geospiza fortis*) of the Galápagos Islands. He kept track of mated pairs of *G. fortis* and their offspring by banding them so that they could be individually recognized. He measured the phenotypic variance of bill depth and several other features, and he calculated the correlation between the phenotypes of parents and of their offspring (**Figure 9.25C**). Boag estimated that the heritability of bill depth was 0.9; that is, about 90 percent of the phenotypic variation among individuals was attributable to genetic differences and 10 percent to environmental differences.

More commonly, genetic variance and the heritability of various traits have been estimated from organisms reared in a greenhouse or laboratory, in which it is easier to set up controlled matings and to keep track of progeny. Most of the characteristics reported, for a great number of species, are genetically variable, with h^2 usually in the range of about 0.1 to about 0.9 (e.g., Mousseau and Roff 1987).

Heritability in human populations has often been estimated by comparing the correlation between dizygotic ("fraternal," or "nonidentical") twins with the correlation between monozygotic ("identical") twins, who are expected to have a higher correlation because they are genetically identical. Physical characteristics such as height, finger length, and head breadth are highly heritable (ca. 0.84–0.94) within populations, and the variation in dermatoglyphic traits (fingerprints) is nearly entirely genetically based (0.96; Lynch and Walsh 1998).

(A)

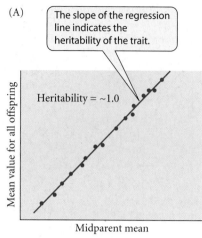

The slope of the regression line indicates the heritability of the trait.

Heritability = ~1.0

Mean value for all offspring

Midparent mean

(B)

Each point represents the mean value of a phenotypic trait in all offspring of two parents plotted against the mean parental value.

Heritability = 0

Mean value for all offspring

Midparent mean

(C)

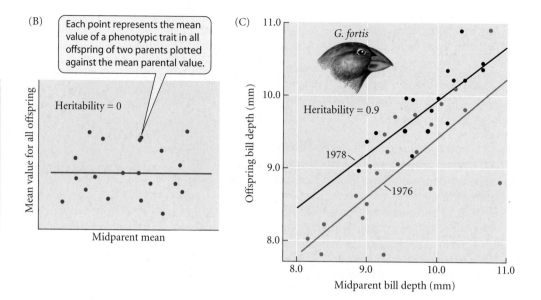

G. fortis

Heritability = 0.9

1978

1976

Offspring bill depth (mm)

Midparent bill depth (mm)

FIGURE 9.25 The relationship between the phenotypes of offspring and parents provides a measure of heritability. Each point represents the mean of a brood of offspring, plotted against the mean of their two parents (midparent mean). (A) A hypothetical case in which offspring means are nearly identical to midparent means. The heritability is nearly 1.00. (B) A hypothetical case in which offspring and midparent means are not correlated. The slope of the relationship and the heritability are approximately 0.00. (C) Bill depth in the ground finch *Geospiza fortis* in 1976 and 1978. Although offspring were larger in 1978, the slope of the relationship between offspring and midparent means was nearly the same in both years. The heritability, estimated from the slope, was 0.9. (C after Grant 1986, based on Boag 1983.)

RESPONSES TO ARTIFICIAL SELECTION Because a character can be altered by selection only if it is genetically variable, **artificial selection** can be used to detect genetic variation in a character. To do this, an investigator (or a plant or animal breeder) breeds only those individuals that possess a particular trait (or combination of traits) of interest. Artificial selection may grade into natural selection, but the conceptual difference is that under artificial selection, the reproductive success of individuals is determined largely by a single characteristic chosen by the investigator, rather than by their overall capacity (based on all characteristics) for survival and reproduction.

Hundreds of such experiments have been performed. For example, Theodosius Dobzhansky and Boris Spassky (1969) bred the offspring of 20 wild female *Drosophila pseudoobscura* flies to form a base population, and from this population drew flies to establish several selected populations, each maintained in a large cage with food. Two of these populations were selected for positive phototaxis (a tendency to move toward light) and two for negative phototaxis (a tendency to move away from light), by placing males and virgin females in a maze in which the flies had to make 15 successive choices between light and dark pathways, ending up in one of 16 tubes (**Figure 9.26A**). The mean and variance of phototactic scores were estimated from the number of flies in each of the 16 tubes. In each generation, 300 flies of each sex, from each population, were released into the maze, and the 25 flies of each sex that had the most extreme high score (in the positively selected population) or low score (in the negatively selected population) were saved to initiate the next generation. This procedure was repeated for 20 generations. Initially, the flies showed no preference, but within a few generations, both the positively and negatively selected populations diverged, in opposite directions, from the initial mean (**Figure 9.26B**). We may therefore infer that variation among flies in their response to light is partly hereditary. From the rapidity of the change, Dobzhansky and Spassky calculated that the heritability of phototaxis is about 0.09.

Experiments of this kind have shown that *Drosophila* species are genetically variable for almost every trait, including features of behavior (e.g., mating speed), morphology, life history (e.g., longevity), physiology (e.g., resistance to insecticides), and even features of the genetic system (e.g., rate of crossing over). Artificial selection has been the major tool of breeders who have produced agricultural varieties of corn, tomatoes, pigs, chickens, and every other domesticated species, which often differ extremely in numerous characteristics. Such evidence has led many evolutionary geneticists to conclude that *species contain genetic variation that could serve as the foundation for the evolution of many of their characteristics,* and that *many or most characters should be able to evolve quite rapidly*—far more quickly than Darwin ever imagined (Barton and Partridge 2000).

(A)

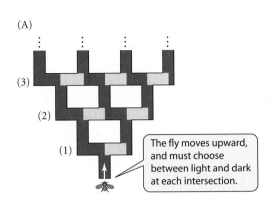

The fly moves upward, and must choose between light and dark at each intersection.

(B)

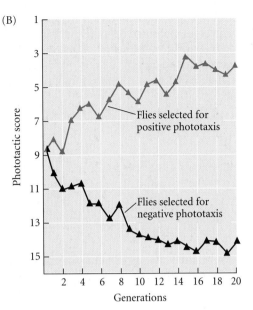

Flies selected for positive phototaxis

Flies selected for negative phototaxis

FIGURE 9.26 Selection for movement in response to light (phototaxis) in *Drosophila pseudoobscura*. (A) Diagram of part of a maze. The maze is oriented vertically below a light source, and flies introduced at the bottom move upward, choosing light or dark at each intersection. Lateral movement across intersections is prevented by barriers. The diagram shows the first 3 of the 15 choices made by the flies in the selection experiment. (B) The mean phototactic scores of flies in the first 20 generations of selection in two populations selected for positive and negative phototaxis. (After Dobzhansky and Spassky 1969.)

Variation among Populations

Almost all species are subdivided into several or many separate populations, with most mating taking place between members of the same population. Such populations of a single species often differ in genetic composition. Studies of differences among populations in different geographic areas, or **geographic variation**, have provided countless insights into the mechanisms of evolution.

Patterns of geographic variation

If distinct forms or populations have overlapping geographic distributions, such that they occupy the same area and can frequently encounter each other, they are said to be **sympatric** (from the Greek *syn*, "together," and *patra*, "fatherland"). Populations with adjacent but nonoverlapping geographic ranges that come into contact are **parapatric** (Greek *para*, "beside"). Populations with separated distributions are **allopatric** (Greek *allos*, "other").

A **subspecies**, or GEOGRAPHIC RACE, in zoological taxonomy means a recognizably distinct population, or group of populations, that occupies a different geographic area from other populations of the same species. (In botanical taxonomy, subspecies names are sometimes given to sympatric, interbreeding forms.) In some instances, subspecies differ in a number of features with concordant patterns of geographic variation. For example, in the northern flicker (*Colaptes auratus*), the subspecies *auratus* and *cafer*, distributed in eastern and western North America, respectively, differ in the color of the underwing, in the crown and mustache marks, in the presence or absence of several other plumage marks, and in size (Short 1965; Moore and Price 1993). Despite these differences, the two populations interbreed in the Great Plains (**Figure 9.27**), forming a wide **hybrid zone** (a region in which genetically distinct parapatric forms interbreed). In some other species, different characters display different geographic patterns.

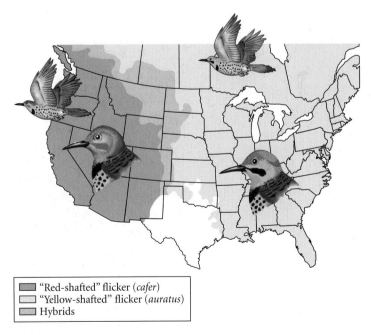

☐ "Red-shafted" flicker (*cafer*)
☐ "Yellow-shafted" flicker (*auratus*)
☐ Hybrids

FIGURE 9.27 Two subspecies of a common North American woodpecker, the northern flicker (*Colaptes auratus*). The eastern ("yellow-shafted") subspecies (*C. a. auratus*) and the western ("red-shafted") subspecies (*C. a. cafer*) form a broad hybrid zone. (After Moore and Price 1993.)

FIGURE 9.28 Human skin color shows clinal variation in the Northern Hemisphere: populations are gradually paler (as measured by reflectance: the proportion of light reflected by the skin) at higher latitudes. The popular notion of discrete "races," based on skin color, is not supported by the data. (After Mielke et al. 2006.)

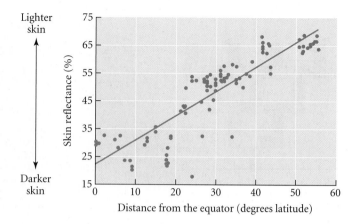

A gradual change in a character or in allele frequencies over geographic distance is called a **cline**. For instance, human populations display a cline in skin pigmentation with latitude (**Figure 9.28**). Many species of mammals and birds have larger body sizes at higher latitudes, a pattern that has been named BERGMANN'S RULE. This pattern, which is too consistent to be attributed to chance, provides important evidence of *adaptive geographic variation* resulting from natural selection. Larger body size is thought to be advantageous for homeothermic animals in colder climates because it reduces the surface area, relative to body mass, over which body heat is lost.

Adaptation is also strongly suggested by the consistency of clines in allele frequencies at the alcohol dehydrogenase locus in *Drosophila melanogaster* (**Figure 9.29**): on several continents, the *Adh^F* allele gradually increases from low to high frequency as one moves toward higher latitudes (Oakeshott et al. 1982). As this example shows, populations of a species often differ in the frequencies of two or more alleles, rather than being fixed for different alleles.

Ecotypes are phenotypes that are associated with a particular habitat, often in a patchy, mosaic pattern. In an important early study, Jens Clausen, David Keck, and William Hiesey

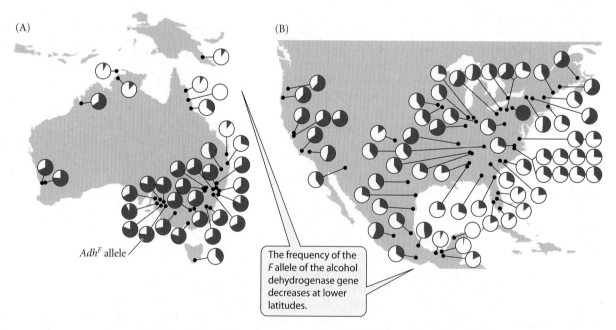

FIGURE 9.29 Clines in the frequency of the *Adh^F* allele at the alcohol dehydrogenase locus of *Drosophila melanogaster* in (A) Australia and (B) North America. The colored area of each "pie" diagram represents the frequency of *Adh^F*, which increases at higher latitudes on both continents. (After Oakeshott et al. 1982.)

(A) *Potentilla glandulosa*

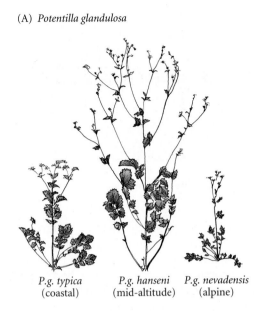

P.g. *typica*
(coastal)

P.g. *hanseni*
(mid-altitude)

P.g. *nevadensis*
(alpine)

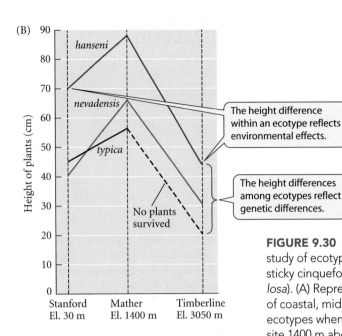

The height difference within an ecotype reflects environmental effects.

The height differences among ecotypes reflect genetic differences.

FIGURE 9.30 A common-garden study of ecotypic variation in the sticky cinquefoil (*Potentilla glandulosa*). (A) Representative specimens of coastal, mid-altitude, and alpine ecotypes when grown together at a site 1400 m above sea level. (B) The mean heights of the three ecotypes when grown in common gardens at three elevations. The differences among ecotypes at the same elevation reflect genetic differences, whereas the differences within each ecotype at different elevations reflect environmental effects. (After Clausen et al. 1940.)

(1940) set out to determine whether the differences among ecotypes are genetically based or are directly caused by differences in the environment. They worked with several plants, including *Potentilla glandulosa*, the sticky cinquefoil, which is distributed in western North America from sea level to above timberline and displays variation among populations that is correlated with altitude. Clausen and colleagues divided, or cloned, each of a number of plants from populations of several different ecotypes growing at sites at different altitudes, then grew them in common gardens at three different altitudes in California. The differences among ecotypes in some features, such as flower color, remained unchanged, irrespective of altitude. Clausen et al. concluded that these features differ genetically and are not substantially affected by the environment. Other features, such as plant height, varied among the three gardens, showing that they were affected by the environment, but the ecotypes nevertheless differed one from another in each garden, implying that these features were also influenced by genetic differences (**Figure 9.30**). In further studies, Clausen et al. showed that the genetic differences among populations in these features are polygenic.

Gene flow

Populations of a species typically exchange genes with one another to a greater or lesser extent. Genes can be carried by moving individuals (such as most animals, as well as seeds and spores) or by moving gametes (such as pollen and the gametes of many marine animals). Migrants that do not succeed in reproducing within their new population do not contribute to gene flow.

Different models of gene flow treat organisms as if they formed either discrete populations (e.g., on islands) or continuously distributed populations (**isolation by distance** models). In an isolation by distance model, each individual is the center of a NEIGHBOR-HOOD, within which the probability of mating declines with distance from the center. The population as a whole consists of overlapping neighborhoods.

Gene flow, if unopposed by other factors, homogenizes the populations of a species—that is, it brings them all to the same allele frequencies unless it is sufficiently counterbalanced by forces of divergence, such as genetic drift or natural selection (see Chapters 10 and 12). For example, if migrants are equally likely to disperse to any of a group of discrete populations, all of equal size, each population will ultimately reach the average allele frequency among the group of populations (**Figure 9.31**). The rate at which this process occurs is proportional to the RATE OF GENE FLOW (m), the proportion of gene copies of the breeding

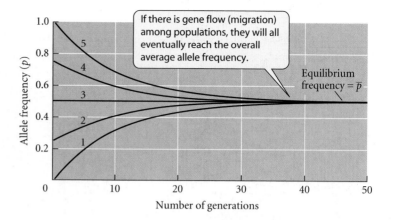

FIGURE 9.31 Gene flow causes populations to converge in their allele frequencies. This model shows the changes over time in the frequency of one allele in five populations that exchange genes equally at the rate $m = 0.1$ per generation. (After Hartl and Clark 1989.)

individuals in each generation that have been carried into that population by immigrants from other populations. This model assumes a fairly constant rate of migration among established populations. In some cases, another kind of gene flow may well be more important (McCauley 1993). If local populations at some sites become extinct, and those sites are then colonized by individuals from several other populations, the allele frequencies in the new colonies are a mixture of those among the source populations.

The characteristics of a species greatly affect its capacity for dispersal and gene flow. For example, animals such as land snails, salamanders, and wingless insects generally move little, and they are divided into relatively small, genetically distinct populations. Gene flow is greater among more mobile organisms, such as far-flying monarch butterflies (*Danaus plexippus*) and the many mussels and other marine invertebrates whose planktonic larvae are carried long distances by currents. However, even seemingly mobile species often display remarkably restricted dispersal. Despite their capacity for long-distance movement, many migratory species of salmon and birds breed near their birthplace, forming genetically distinct populations.

Rates of gene flow among natural populations can be estimated directly by following the dispersal of marked individuals or their gametes. A. J. Bateman (1947) studied gene flow by pollen in corn (maize, *Zea mays*) by counting the number of heterozygous progeny of homozygous recessive plants ("seed parents") situated at various distances from a stand of homozygous dominant plants ("pollen parents"). At only 30 to 50 feet away, less than 1 percent of the progeny were fathered by the dominant plants. Similar studies have shown that gene flow by both pollen and seed dispersal is very restricted in many species of plants (Levin 1984). However, estimates of gene flow based on genetic differences among populations, as we will describe in Chapter 10, are generally more reliable than direct observations of this kind.

Allele frequency differences among populations

Variation in allele frequencies among populations can be quantified in several ways. A commonly used measure, for a locus with two alleles, is the fixation index, F_{ST}:

$$F_{ST} = \frac{V_q}{(\bar{q})(1 - \bar{q})}$$

where $\bar{q}$ is the mean frequency of one of the alleles and V_q is the variance among populations in its frequency. A comparable measure, G_{ST}, can be calculated for a locus with more than two alleles. Both F_{ST} and G_{ST} range from 0 (no variation among populations) to 1 (populations are fixed for different alleles).

Armbruster et al. (1998) estimated G_{ST}, averaged over five allozyme loci, in samples of the mosquito *Wyeomyia smithii*, the larvae of which develop only in the water-holding leaves of pitcher plants. Samples were taken from southern sites in North America and from northern localities that had been covered by the most recent Pleistocene glacier. In both the southern and northern regions, more distant populations differed more in allele frequency (i.e., had a greater G_{ST}), an illustration of isolation by distance (**Figure 9.32**).

However, northern populations were genetically more similar than southern populations that were separated by comparable distances. The most likely interpretation is that there has not been enough time for northern populations to become substantially differentiated since the species colonized the formerly glaciated region.

DNA sequences provide information not only on allele frequency differences among populations, but also on the genealogical (phylogenetic) relationships among alleles, which may cast light on the history of the populations (see Chapters 2 and 6). For example, populations of MacGillivray's warbler (*Oporornis tolmiei*) from temperate western North America all share one common haplotype (*A*) at the mitochondrial cytochrome *b* locus. Rare variants, differing from haplotype *A* by single mutations, have been found in some populations. In contrast, samples from northern Mexico carry a variety of haplotypes, differing from one another by as many as five mutations (**Figure 9.33**). This pattern indicates that the Mexican population has been stable, accumulating genetic diversity for a long time. The temperate North American populations, however, probably stem from recent (postglacial) colonization by a few original founders that carried haplotype *A*. The rare haplotypes have only recently arisen from haplotype *A* by new mutations (Milá et al. 2000).

Almost all MacGillivray's warblers in Mexico have different haplotypes from those in temperate North America, so that almost 90 percent of the total genetic variation is *between* the two regions, and only 10 percent is *within* populations. The situation is the opposite if we compare human populations.

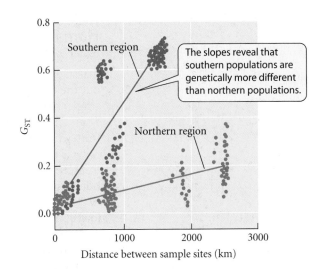

FIGURE 9.32 Genetic differentiation among populations of the North American pitcher-plant mosquito *Wyeomyia smithii*. G_{ST}, a measure of allele frequency differences averaged over five loci, is plotted in relation to the distance between populations. Each point represents a comparison of samples from two southern localities (red points) or two northern localities (green points). Northern populations are genetically more similar to each other than southern populations. (After Armbruster et al. 1998.)

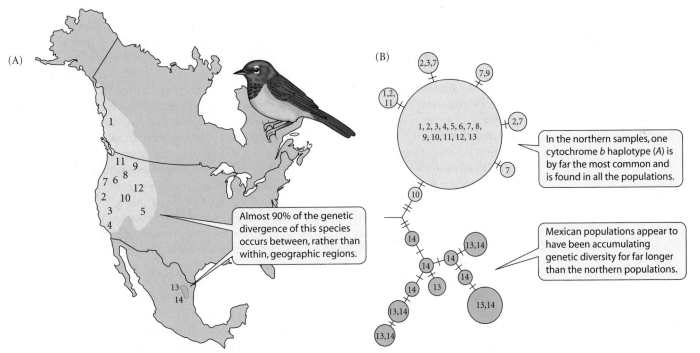

FIGURE 9.33 Geographic variation in mtDNA in MacGillivray's warbler, showing regional differences in genetic diversity. (A) Sample localities in the western United States and Canada (1–12) and in Mexico (13, 14). (B) An evolutionary tree of the 17 cytochrome *b* haplotypes found. Each haplotype is represented by a circle, labeled with the sample localities in which it was found. The size of each circle is proportional to the total frequency of the haplotype it represents. Haplotypes are connected to one another by sequence similarity, with bars across branches indicating single nucleotide changes. Some haplotypes are placed at branch points, implying that they are the ancestors of other haplotypes in the sample. The diversity and the degree of sequence divergence among haplotypes are greater in the Mexican samples than in those from western North America. (After Milá et al. 2000.)

Human genetic variation

Since the nearly complete DNA sequence of the human genome was first reported in 2001, massive efforts have been made to characterize sequence variation in populations throughout the world. The HapMap Project and other research groups, for example, have found single-nucleotide polymorphisms (SNPs) at more than 10 million sites in the genome. There are now more data on genetic variation in *Homo sapiens* than on any other species, from which an increasingly detailed portrait of the structure and history of human populations is being drawn (Novembre and Ramachandran 2011).

More than 60 percent of SNPs are rare, with the frequency of the less common nucleotide less than 0.05. Nonetheless, the average heterozygosity (π) across the genome is 0.001 to 0.002: an average individual is heterozygous at 1 out of every 1000 to 2000 sites (Crawford et al. 2005). The frequencies of these variants vary among human populations, so we may decompose the genetic variation in the entire human species into variation *within populations*, *among populations* within regions (such as continents), and *among regions*. Many such analyses have been performed, and they consistently show that the great majority of variation is within populations (Kittles and Weiss 2003). For example, J. Z. Li and collaborators (2008) analyzed 642,690 autosomal SNPs in 938 individuals from 51 populations throughout the world. They calculated that 89.9 percent of the variation is among individuals in an average population, 2.1 percent is among populations, and 9.0 percent is among major geographic regions. To update a conclusion reached by Lewontin et al. (1984), who performed a similar analysis using allozyme frequencies: "If everyone on earth became extinct except for the Kikuyu of East Africa, about 90 percent of all human variability would still be present in the reconstituted species." Because humans recently dispersed from Africa throughout the rest of the world, most of the genetic variation in non-African populations is a subset of the variation found in African populations, and populations across the globe are genetically very similar.

Human populations are fixed for different alleles at few, if any, loci. Some variations, however, do differ considerably in frequency among human populations, including some that cause genetic diseases. Mutations that cause cystic fibrosis are most prevalent in northern Europe, for example, and the mutation that causes Tay-Sachs disease has its highest frequency in the Ashkenazic Jewish population, originally located in eastern Europe. SNPs also vary in frequency among populations. Some rare SNP alleles are restricted to specific local or regional populations; in other cases, alleles are shared among populations, but differ slightly in frequency. If two populations differ slightly in allele frequencies at enough sites, however, they can be distinguished. Suppose, for example, that at each of n SNP sites, a rare variant has frequency q in population A, but is absent from population B. Perhaps $q = 10^{-3}$. The probability that a randomly chosen member of population A lacks the variant at any specific site is $1 - q$, and the probability that he or she lacks the rare variant at all n sites is $(1 - q)^n$, which becomes a very small number if n is very large. But this is the probability that the individual has the same genotype as a member of population B. Hence the probability of misclassifying an individual can be very low if enough SNPs are assayed, even if populations differ only slightly in SNP frequencies—which means, conversely, that the probability of distinguishing members of the two populations is high.

Taking this approach, Li and collaborators found that clusters of populations in major regions can be distinguished (**Figure 9.34A**). The greatest difference is between sub-Saharan African populations and non-African populations. Populations south of the Sahara have higher SNP heterozygosity (π) and greater frequency differences (F_{ST}) among populations than are seen among indigenous populations in Europe, Asia, or the Americas (Kittles and Weiss 2003). The farther populations are from eastern Africa, the lower their level of SNP variation is (**Figure 9.34B**). All these features are consistent with the "out of Africa" hypothesis (see Chapter 6). Using large numbers of SNPs, Novembre et al. (2008) could even distinguish populations in Europe that are less than 500 miles apart (**Figure 9.34C**). With enough genetic information, a person's ancestry can be traced to a localized region with high probability.

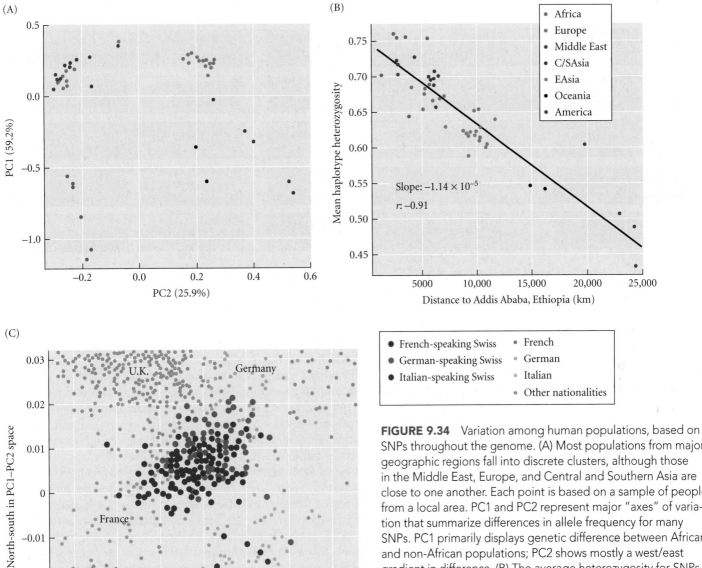

FIGURE 9.34 Variation among human populations, based on SNPs throughout the genome. (A) Most populations from major geographic regions fall into discrete clusters, although those in the Middle East, Europe, and Central and Southern Asia are close to one another. Each point is based on a sample of people from a local area. PC1 and PC2 represent major "axes" of variation that summarize differences in allele frequency for many SNPs. PC1 primarily displays genetic difference between African and non-African populations; PC2 shows mostly a west/east gradient in difference. (B) The average heterozygosity for SNPs, a measure of genetic diversity within a population, decreases in populations that are increasingly distant from eastern Africa, represented by Addis Ababa, Ethiopia. (C) If many SNPs are used, individuals can be assigned even to local regions within Europe. Each symbol represents an individual. Even French-, German-, and Italian-speaking Swiss can be separated to some extent. (A, B after Li et al. 2008; C after Novembre et al. 2008.)

GENE FLOW BETWEEN POPULATIONS *Homo sapiens* is a single biological species. There are no biological barriers to interbreeding among human populations, and even the cultural barriers that do exist often break down. Consequently, there can be interbreeding, or ADMIXTURE, among genetically differentiated groups. For instance, despite racist barriers in the United States, a blood group allele, Fy^a, that is fairly abundant in European populations but virtually absent in African populations was found to have a frequency of 0.11 in the African-American population of Detroit, Michigan, from which it was calculated that 26 percent of that population's genes have been derived from the white population, at an average rate of gene flow (*m*) of about 1 percent per generation (Cavalli-Sforza and Bodmer 1971). By using thousands of SNP markers instead of one gene, it is now possible to estimate the proportion

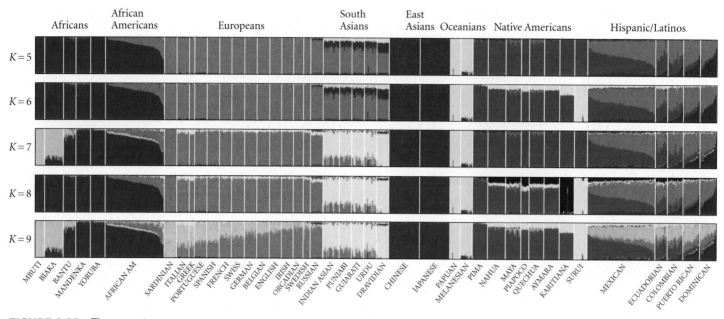

FIGURE 9.35 The genetic population structure of worldwide human populations. Each of 1112 individuals, from 42 different populations, is represented as a narrow vertical bar in each of the five vertically stacked graphs. Each bar is composed of one or more colors, each color representing the proportion of the individual's genome that is estimated to have one or another regional ancestry. For example, green and red represent African and European ancestry, respectively, and some African-Americans show a mixture of these ancestries in their genome. The five graphs differ in the number of ancestral regions assumed; for example, K = 5 (top graph) means that the world was divided into five ancestral regions. Individuals in some populations have more genetic admixture than those in others (e.g., compare Europeans with East Asians in the K = 9 graph). These graphs are based on more than 73,000 SNP markers. (From Henn et al. 2010.)

of a person's genome that has been derived from different populations (**Figure 9.35**). The data show that diverse populations will interbreed, but also that admixture between populations can be curtailed by geographic and cultural barriers (Henn et al. 2010).

Because of interbreeding, and because the expansion from Africa was quite recent, the genetic differences among populations are generally clinal (gradual): there are few sharp geographic breaks in genetic composition. Consequently, humans cannot be clearly divided into the several "races" that were traditionally defined by a few features such as skin color and hair texture. Indeed, various authors have recognized as many as 60 "races." The number of human races that can be recognized is arbitrary, for each supposed racial group can be subdivided into an indefinite number of distinct populations. (Among Africans, for example, Congo pygmies are the shortest of humans, and Maasai are among the tallest.) With enough data, genetic clusters can be described at every level from continents to local rural villages. Recognizing this structure can be useful in health and medicine because some genetic diseases have higher prevalence in some geographic populations than in others. However, many geneticists and anthropologists maintain that the concept of distinct human races has little scientific justification (Barbujani 2005; Koenig et al. 2008). The American Anthropological Association considers "race" to be a human invention that "is about culture, not biology." (See www.aaanet.org and the association's project on race at www.understandingrace.org.) That U.S. President Barack Obama is described as "black," even though his mother was Caucasian, illustrates that "race" is a cultural, not a biological, framework.

VARIATION IN COGNITIVE ABILITIES   No topics in evolutionary biology are more controversial than those concerning the evolution and genetics of human behavioral characteristics, including cognitive abilities described as "intelligence." Although these abilities have evolved, and must therefore have a genetic foundation, much of the variation in their manifestation could well be nongenetic, especially in view of the enormous effects of social conditioning and learning. For example, almost everyone agrees that most of the cultural differences among groups are not genetic, for pronounced cultural differences exist among geographically neighboring peoples who are genetically almost indistinguishable and interbreed.

Determining the heritability of human behavioral traits is difficult because family members typically share not only genes but also environments—and it would be unethical to deliberately rear humans in different experimentally altered environments. For this reason, studies of people adopted as children are critically important. The genetic component of variation is estimated by correlations between twins or other siblings reared apart, or by adoptees' correlations with their biological parents. In a variant of this method, the correlation of adopted children with their adoptive parents is contrasted with that of the biological (nonadopted) children of the same parents, which is expected to be higher if the variation has a genetic component. There is a risk that heritability will be overestimated in such studies, however, because adoption agencies often place children in homes that are similar (in factors such as religion and socioeconomic status) to those of their siblings. Some modern studies try to measure such environmental correlations and take them into account. Twins have played an important role in human genetic studies, since monozygotic ("identical") twins should be more similar than dizygotic ("nonidentical") twins if variation has a genetic component. The genetic correlation between dizygotic twins should be no greater than between non-twin siblings.

It is exceedingly important to bear in mind that a genetic basis for a trait does not mean that the trait is fixed or unalterable (see Figures 9.4 and 9.30). It may be the case, therefore, that a heritability value holds only for the particular population and the particular environment in which it was measured and cannot be reliably extrapolated to other populations. By the same token, high heritability of variation *within* populations does not mean that differences *among* populations have a genetic basis. A character may display high heritability within a population, yet be altered dramatically by change in the environment. Twin studies have suggested that the heritability of human height is 0.8 or more, yet in many industrial nations, mean height has increased considerably within one or two generations as a result of nutritional and other improvements.

To a greater extent than in almost any other context, the "nature versus nurture" debate has centered on variations in the cognitive abilities collectively called "intelligence"—or more properly, on IQ ("intelligence quotient") scores. IQ tests are supposed to be "culture-free," but they have been strongly criticized as favoring white, middle-class individuals. IQ testing has a long, sordid history of social abuse (Gould 1981). Even in the recent past, some individuals have wrongly argued that IQ cannot be improved by compensatory education because it is highly heritable (Jensen 1973), and others have suggested that African-Americans, who on average score lower than European-Americans on IQ tests, have genetically lower intelligence (Herrnstein and Murray 1994). (For counterarguments, see Fraser 1995; Fischer et al. 1996.)

Recent studies of the heritability of cognitive abilities, based on twins reared apart and on adopted compared with nonadopted children, have corrected many of the flaws of earlier studies; for example, they have assessed the similarity of the family environments in which separated twins were reared (Bouchard et al. 1990; McClearn et al. 1997; Plomin et al. 1997). "General cognitive ability" (IQ) appears to be substantially heritable, and, surprisingly, the heritability increases with age, from about 0.40 in childhood to 0.50 in adolescence to about 0.60 in adulthood and old age. The genetic component of specific cognitive abilities, such as verbal comprehension, spatial visualization, perceptual speed, and accuracy, is somewhat lower, and these features are only partly correlated with one another.

Despite the substantial heritability of IQ, there is abundant evidence that education and an enriched environment can substantially increase IQ scores (Fraser 1995). For example, one study found that children who remained in their parents' homes had an average IQ score of 107, those adopted into different homes an average of 116, and those who were returned to their biological mothers after a period of adoption, only 101 (Tizard 1973).

Probably the most incendiary question about IQ is whether genetic differences account for differences in average IQ scores among "racial" or ethnic groups, such as the 15-point difference (one standard deviation) between the average scores of European-Americans and African-Americans. Virtually all the evidence indicates that the average difference between blacks and whites is due to their very different social, economic, and educational environments (Nisbett 1995; Mielke et al. 2006). In studies of black children adopted into white homes or reared in the same residential institution with white children, the two "racial" groups had similar scores. One study correlated individuals' IQ scores with their degree of admixture of alleles for several blood groups that differ in frequency between European and African populations. The correlation between IQ and degree of European ancestry was nearly zero. Finally, the average IQ of German children fathered by white American soldiers during World War II was nearly identical to the IQ of those with black American fathers.

Summary

1. Evolution occurs by the replacement of some genotypes by others. Hence evolution requires genetic variation. The all-important concepts of allele frequency and genotype frequency are central to the Hardy-Weinberg principle, which states that in the absence of perturbing factors, allele and genotype frequencies remain constant over generations. For two alleles with frequencies p and q at an autosomal locus, the Hardy-Weinberg genotype frequencies are in the ratio $p^2:2pq:q^2$.

2. The potential causes of allele frequency changes at a single locus are those factors that can cause deviations from the Hardy-Weinberg equilibrium. These factors are (a) nonrandom mating; (b) finite population size, resulting in random changes in allele frequencies (genetic drift); (c) incursion of genes from other populations (gene flow); (d) mutation; and (e) consistent differences among genes or genotypes in reproductive success (natural selection).

3. Inbreeding occurs when related individuals mate and have offspring. Inbreeding increases the frequency of homozygous genotypes and decreases the frequency of heterozygotes. Most diploid populations contain rare recessive deleterious alleles at many loci, so inbreeding causes a decline in components of fitness (inbreeding depression).

4. Populations of most species contain a great deal of genetic variation. This variation includes rare alleles at many loci, most of which appear to be deleterious. But it also includes many common alleles, so that as many as a third of loci are polymorphic, as revealed by enzyme electrophoresis. Most genes are variable when analyzed at the level of DNA sequence.

5. Alleles at different loci, affecting the same or different traits, are sometimes nonrandomly associated within a population, a condition called linkage disequilibrium, or LD.

6. Many "quantitative," continuously varying phenotypic traits, including morphological, physiological, and behavioral features, exhibit polygenic variation.

7. Variation in most quantitative phenotypic traits includes both a genetic component and a nongenetic ("environmental") component. Variation in some traits may also include components caused by nongenetic maternal effects and epigenetic inheritance. The proportion of the phenotypic variance that is due to genetic variation (genetic variance) is the heritability of the trait. Genetic variance and heritability can be estimated by breeding experiments and by artificial selection. Many characters appear to be so genetically variable that we should expect them to be able to evolve quite rapidly.

8. Genetic differences among different populations of a species take the form of differences in the frequencies of alleles that may also be polymorphic within populations. Unless countered by natural selection or genetic drift, gene flow among populations will cause them to become homogeneous.

9. Patterns of allele frequency differences and the phylogeny of alleles or haplotypes can shed light on the history that gave rise to geographic variation.

Terms and Concepts

additive
allele frequency
allopatric
allozyme
artificial selection
cline
common garden
ecotype
electrophoresis
environmental variance
epigenetic inheritance
gene copy
gene flow
genetic drift

genetic variance
genotype
genotype × environment interaction
genotype frequency
geographic variation
Hardy-Weinberg (H-W) equilibrium
heritability
heterozygosity
heterozygous
homozygous
hybrid zone
identical by descent

inbreeding
inbreeding coefficient
inbreeding depression
isolation by distance
lethal allele
linkage
linkage disequilibrium (LD)
linkage equilibrium
maternal effect
migration
monomorphism
norm of reaction
normal distribution

panmictic
parapatric
phenotype
phenotypic plasticity
polygenic
polymorphism
quantitative variation
self-fertilization (= selfing)
single nucleotide polymorphism (SNP)
standard deviation
subspecies
sympatric
variance

Suggestions for Further Reading

The study of variation builds on a rich history, much of which is summarized in three classic works by major contributors to evolutionary biology. Theodosius Dobzhansky's *Genetics of the Evolutionary Process* (Columbia University Press, New York, 1970) and its predecessors, the several editions of *Genetics and the Origin of Species*, are among the most influential books on evolution from the viewpoint of population genetics. Ernst Mayr's *Animal Species and Evolution* (Harvard University Press, Cambridge, MA, 1963) is a classic work that contains a wealth of information on geographic variation and the nature of species in animals, with important interpretations of speciation and other evolutionary phenomena. Richard C. Lewontin's *The Genetic Basis of Evolutionary Change* (Columbia University Press, New York, 1974) is an insightful analysis of the study of genetic variation by classic methods and by electrophoresis.

Excellent contemporary treatments of variation include *Principles of Population Genetics*, by D. L. Hartl and A. G. Clark, fourth edition (Sinauer Associates, Sunderland, MA, 2007), which includes extensive treatment of much of the material in this and the next four chapters, and *Molecular Markers, Natural History, and Evolution*, by J. C. Avise, second edition (Sinauer Associates, Sunderland, MA, 2004), an outstanding description of the ways in which molecular variation is used to study topics such as mating patterns, kinship and genealogy within species, speciation, hybridization, phylogeny, and conservation. *Human Biological Variation*, by J. H. Mielke, L. W. Konigsberg, and J. H. Relethford (Oxford University Press, 2008) is an excellent introduction to that topic.

Problems and Discussion Topics

1. In an electrophoretic study of enzyme variation in a species of grasshopper, you find 62 A_1A_1, 49 A_1A_2, and 9 A_2A_2 individuals at a particular locus in a sample of 120. Show that $p = 0.72$ and $q = 0.28$ (where p and q are the frequencies of alleles A_1 and A_2), and that the genotype frequencies of A_1A_1, A_1A_2, and A_2A_2 are approximately 0.51, 0.41, and 0.08, respectively. Demonstrate that the genotype frequencies are in Hardy-Weinberg equilibrium.

2. In a sample from a different population of this grasshopper species, you find four alleles at this locus. The frequencies of A_1, A_2, A_3, and A_4 are $p_1 = 0.50$, $p_2 = 0.30$, $p_3 = 0.15$, $p_4 = 0.05$. Assuming Hardy-Weinberg equilibrium, calculate the expected proportion of each of the ten possible genotypes (e.g., that of A_2A_3 should be 0.09). Show that heterozygotes, of all kinds, should constitute 63.9 percent of the population. In a sample of 100 specimens, how many would you expect to be heterozygous for allele A_4? How many would you expect to be homozygous A_4A_4?

3. In the peppered moth (*Biston betularia*), black individuals may be either homozygous (A_1A_1) or heterozygous (A_1A_2), whereas pale gray moths are only homozygous (A_2A_2). Suppose that in a sample of 250 moths from one locality, 108 are black and 142 are gray. (a) Which allele is dominant? (b) Assuming that the locus is in Hardy-Weinberg equilibrium, what are the allele frequencies? (c) Under this assumption, what *proportion* of the sample is heterozygous? What is the *number* of heterozygotes? (d) Under the same assumption, what proportion of black moths is heterozygous? (Answer: approximately 0.85.) (e) Why is it necessary to assume Hardy-Weinberg genotype frequencies in order to answer parts b–d? (f) For a sample from another area consisting of 287 black and 13 gray moths, answer all the preceding questions.

4. In an experimental population of *Drosophila*, a sample of males and virgin females includes 66 A_1A_1, 86 A_1A_2, and 28 A_2A_2 flies. Each genotype is represented equally in both sexes, and each can be distinguished by eye color. Determine the allele and genotype frequencies and whether or not the locus is in Hardy-Weinberg equilibrium. Now suppose you discard half the A_1A_1 flies and breed from the remainder of the sample. Assuming the flies mate at random, what will be the genotype frequencies among their offspring? (*Hint:* The proportion of A_2A_2 should be approximately 0.23.) Now suppose you discarded half of the A_1A_2 flies instead of A_1A_1. What will be the allele and genotype frequencies in the next generation? Why is the outcome so different in this case?

5. In an electrophoretic study of a species of pine, you can distinguish heterozygotes and both homozygotes for each of two genetically variable enzymes, each with two alleles (A_1, A_2 and B_1, B_2). A sample from a natural population yields the following numbers of each genotype: 8 $A_1A_1B_1B_1$, 19 $A_1A_2B_1B_1$, 10 $A_2A_2B_1B_1$, 42 $A_1A_1B_1B_2$, 83 $A_1A_2B_1B_2$, 44 $A_2A_2B_1B_2$, 48 $A_1A_1B_2B_2$, 97 $A_1A_2B_2B_2$, 49 $A_2A_2B_2B_2$. (a) Determine the frequencies of alleles A_1 and A_2 (p_A, q_A) and B_1 and B_2 (p_B, q_B). (b) Determine whether locus A is in Hardy-Weinberg equilibrium. Do the same for locus B. (c) *Assuming* linkage equilibrium, calculate the expected *frequency* of each of the nine genotypes. (*Hint:* The expected frequency of $A_1A_1B_1B_2$ is 0.106.) (d) From the results of part c, calculate the expected *number* of each genotype in the sample and determine whether or not the loci actually are in linkage equilibrium. (e) From these calculations, can you say whether or not the loci are linked?

6. Until a few decades ago, most population geneticists believed that populations are genetically uniform, except for rare deleterious mutations. We now know that most populations are genetically very variable. Contrast the implications of these different views for evolutionary processes.

7. Different characters may vary more or less independently among geographic populations of a species, or they may vary concordantly (as in the flicker example in Figure 9.27). Suggest processes that could produce each pattern.

8. Suppose two alleles additively affect variation in a trait such as finger length. Contrast two populations with the same allele frequencies: mating is random in one ($F = 0$), whereas the other is undergoing inbreeding (perhaps $F = 0.25$). Explain why the variance in finger length is *greater* in the inbreeding population. As a challenge, algebraically determine the ratio of the variance in the inbreeding population to that in the panmictic population.

9. Why would estimates of gene flow from genetic differences among populations be better than those based on directly following the movement of organisms or their gametes?

Genetic Drift: Evolution at Random

O ne of the first and most important lessons a student of science learns is that many words have very different meanings in a scientific context than in everyday speech. The word "chance" is a good example. Many nonscientists think that evolution occurs "by chance." What they mean is that evolution occurs without purpose or goal. But by this token, everything in the natural world—chemical reactions, weather, planetary movements, earthquakes—happens by chance. None of these phenomena has a purpose, for "purpose" implies forethought and goal. In fact, scientists consider purposes or goals to be unique to human thought and do not view any natural phenomena as purposeful. But scientists don't view chemical reactions or planetary movements as chance events, either—because in science, "chance" has a very different meaning.

Although the meaning of "chance" is a complex philosophical issue, scientists use "chance," or RANDOMNESS, to refer to a situation in which physical causes can result in any of several outcomes, but we cannot predict what the outcome will be in any particular case. Nonetheless, we may be able to specify the PROBABILITY, and thus the FREQUENCY, of one or another outcome. Although we cannot predict the sex of some-

one's next child, we can say with considerable certainty that there is a probability of 0.5 that it will be a daughter.

Almost all phenomena are affected simultaneously by both random, or **stochastic** (unpredictable), and non-random, or **deterministic** (predictable), factors. Any of us may experience a car accident resulting from the unpredictable behavior of another driver, but we are predictably more likely to have an accident if we drive after drinking. So it is with evolution. As we will see in the next chapter, natural selection is a deterministic, nonrandom process. But at the same time, there are many important random processes in evolution.

Population size affects genetic variation and evolutionary potential
In small populations of the European adder (*Vipera berus*), genetic drift has reduced genetic variation and fixed harmful alleles. The capacity for evolutionary adaptation is diminished in small populations.

Random Processes in Evolution

We have already discussed some of the ways in which mutation is a random process (in Chapter 8). Another way is the order in which mutations occur, for the fixation of one mutation in a population may affect whether or not certain other subsequent mutations will increase in frequency. Similarly, the sequence of environmental changes affects the evolution of adaptations, some of which may determine which subsequent changes would be advantageous. Asteroid impacts or other causes of the great mass extinctions were random events that had profound effects on life's history (see Chapter 7).

One of the most important random processes in evolution is random fluctuation in the frequencies of alleles or haplotypes owing to "sampling error": the process of random **genetic drift**. Genetic drift and natural selection are the two most important causes of allele substitution—that is, of evolutionary change—in populations. Genetic drift occurs in all natural populations because, unlike ideal populations at Hardy-Weinberg equilibrium, natural populations are finite in size. Random fluctuations in allele frequencies can result in the replacement of old alleles by new ones, resulting in **nonadaptive evolution**. While natural selection results in adaptation, genetic drift does not—so it is not responsible for those anatomical, physiological, and behavioral features of organisms that equip them for reproduction and survival. Genetic drift nevertheless has many important consequences, especially at the molecular genetic level.

Because all populations are finite, alleles at all loci are potentially subject to random genetic drift—but all are not necessarily subject to natural selection. For this reason, and because the expected effects of genetic drift can be mathematically described with some precision, many evolutionary geneticists hold the opinion that genetic drift should be the "null hypothesis" used to explain an evolutionary observation unless there is positive evidence of natural selection or some other factor. This perspective is analogous to the use of the NULL HYPOTHESIS in statistics: the hypothesis that the data do not depart from those expected on the basis of chance alone.* According to this view, we should not assume that a characteristic, or a difference between populations or species, is adaptive or has evolved by natural selection unless there is evidence for that conclusion.

The theory of genetic drift, much of which was developed by the American geneticist Sewall Wright starting in the 1930s and by the Japanese geneticist Motoo Kimura starting in the 1950s, includes some of the most highly refined mathematical models in biology. (But fear not! We shall skirt around almost all the math.) We will first explore the theory and then see how it explains data from real organisms. In our discussion, we will describe random fluctuations in the frequencies (proportions) of two or more kinds of self-reproducing entities that do not differ, *on average* (or differ very little), in reproductive success (fitness). For the purposes of this chapter, those entities are alleles. But the theory applies to any other self-replicating entities, such as chromosomes, asexually reproducing genotypes, or even species.

The Theory of Genetic Drift

Genetic drift as sampling error

That chance should affect allele frequencies is readily understandable. Imagine, for example, that a single mutation, A_2, appears in a large population that is otherwise A_1. If the population size is stable, each mating pair leaves an average of two progeny that survive to reproductive age. From the single mating $A_1A_1 \times A_1A_2$ (for there is only one copy of A_2), the probability that one surviving offspring will be A_1A_1 is ½; therefore the probability that

---

*For example, if we measure height in several samples of people, the null hypothesis is that the observed means differ from one another only because of random sampling and that the parametric means of the populations from which the samples were drawn do not differ. A statistical test, such as a *t*-test or analysis of variance, is designed to show whether or not the null hypothesis can be rejected. It will be rejected if the sample means differ more than would be expected had the samples had been randomly drawn from a single population.

two surviving progeny will both be A_1A_1 is $\frac{1}{2} \times \frac{1}{2} = \frac{1}{4}$—which is the probability that the A_2 allele will be immediately lost from the population. We may assume that mating pairs vary at random, around the mean, in the number of surviving offspring they leave (0, 1, 2, 3, …). In that case, as the pioneering population geneticist Ronald Fisher calculated, the probability that A_2 will be lost, averaged over the population, is 0.368. He went on to calculate that after the passage of 127 generations, the cumulative probability that the allele will be lost is 0.985. This probability, he found, is not greatly different if the new mutation confers a slight advantage: as long as it is rare, it is likely to be lost, just by chance.

In this example, the frequency of an allele can change (in this instance, to zero from a frequency very near zero) because the one or few copies of the A_2 allele may happen not to be included in those gametes that unite into zygotes, or may happen not to be carried by the offspring that survive to reproductive age. The genes included in any generation, whether in newly formed zygotes or in offspring that survive to reproduce, are a *sample* of the genes carried by the previous generation. Any sample is subject to random variation, or **sampling error**. In other words, the proportions of different kinds of items in a sample (in this case, A_1 and A_2 alleles) are likely to differ, by chance, from the proportions in the set of items from which the sample is drawn.

Imagine, for example, a population of land snails in which (for the sake of argument) offspring inherit exactly the brown or the yellow color of their mothers. Suppose 50 snails of each color inhabit a cow pasture. (The proportion of yellow snails is $p = 0.50$.) If two yellow and four brown snails are stepped on by cows, p will change to 0.511. Since it is unlikely that a snail's color affects the chance of its being squashed by cows, the change might just as well have been the reverse, and indeed, it may well be the reverse in another pasture, or in this pasture in the next generation. The chances of increase or decrease in the proportion of yellow snails are equal in each generation, so the proportion will fluctuate. But an increase of, say, 1 percent in one generation need not be compensated by an equal decrease in a later generation—in fact, since this process is random, it is very unlikely that it will be. Therefore the proportion of yellow snails will wander over time, eventually ending up near, and finally at, one of the two possible limits: 0 or 1. It seems reasonable, too, that if the population should start out with, say, 80 percent brown and 20 percent yellow snails, it is more likely that the proportion of yellow snails will wander to zero than to 100 percent. In fact, the probability of yellow being lost from the population is exactly 0.80, which is the same as the probability that brown will reach 100 percent—that is, that it will reach **fixation**.

Coalescence

The concept of random genetic drift is so important that we will take two tacks in developing the idea. **Figure 10.1** shows a hypothetical, but realistic, history of gene lineages. First, imagine the figure as depicting lineages of individual asexual organisms, such as bacteria, rather than genes. We know from our own experience that not all members of our parents' or grandparents' generations had equal numbers of descendants; some had none. Figure 10.1 diagrams this familiar fact. We note that the individuals in generation t (at the right of the figure) are the progeny of only some of those that existed in the previous generation ($t-1$): purely by chance, some individuals in generation $t-1$ failed to leave descendants. Likewise, the population at generation $t-1$ stems from only some of those individuals that existed in generation $t-2$, and similarly back to the original population at time 0.

Now think of the objects in Figure 10.1 as copies of genes at a locus, in either a sexual or an asexual population. Figure 10.1 shows that as time goes on, more and more of the original gene lineages become extinct, so that the population consists of descendants of fewer and fewer of the original gene copies. In fact, if we look backward rather than forward in time, *all the gene copies* in the population ultimately *are descended from a single ancestral gene copy*, because given long enough, all other original gene lineages become extinct. The genealogy of the genes in the present population is said to **coalesce** back to a single common ancestor. Because that ancestor represents only one of the two (or more) original alleles, the population's genes, descended entirely from that ancestral gene copy, must

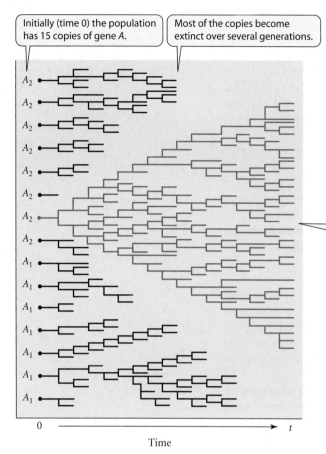

Initially (time 0) the population has 15 copies of gene A.

Most of the copies become extinct over several generations.

By time t, all copies of the gene present in the population are descended from (coalesce to) a single ancestral gene copy.

A_2
A_2
A_2
A_2
A_2
A_2
A_2
A_2
A_1
A_1
A_1
A_1
A_1
A_1
A_1

0 ⟶ t

Time

FIGURE 10.1 A possible history of descent of gene copies in a population that begins (at time 0, at left) with 15 copies, representing two alleles (A_1 and A_2). Each gene copy has 0, 1, or 2 descendants in the next generation. The gene copies present at time t (far right) are all descended from (coalesce to) a single ancestral copy, which happens to be an A_2 allele (the lineage shown in red). Gene lineages descended from all other gene copies have become extinct. If the failure of gene copies to leave descendants is random, then the gene copies at time t could equally likely have descended from any of the original gene copies present at time 0. (After Hartl and Clark 1989.)

eventually become monomorphic: one or another of the original alleles becomes fixed (reaches a frequency of 1). The smaller the population, the more rapidly all gene copies in the current population coalesce back to a single ancestral copy, since it takes longer for many than for few gene lineages to become extinct by chance.

In our example, all gene copies have descended from a copy of an A_2 allele, but because this process is random, A_1 might well have been the "lucky" allele if the sequence of random events had been different. If, in the generation that included the single common ancestor of all of today's gene copies, A_1 and A_2 had been equally frequent ($p = q = 0.5$), then it is equally likely that the ancestral gene copy would have been A_1 or A_2; if, on the other hand, A_1 had had a frequency of 0.9 in that generation, then the probability is 0.9 that an A_1 allele would have been the ancestor of all the later genes at time t. *Our analysis therefore shows that by chance, a population will eventually become monomorphic for one allele or the other, and that the probability that allele A_1 will be fixed, rather than another allele, equals the initial frequency of A_1.*

According to this analysis, for example, all the mitochondria of the entire human population are descended from the mitochondria carried by a single woman, who has been called "mitochondrial Eve," at some time in the past. (Mitochondria are transmitted only through eggs.) This does *not* mean, however, that the population had only one woman at that time: "mitochondrial Eve" happened to be the one among many women to whom all mitochondria trace their ancestry (in a pattern like that seen in Figure 10.1). Similarly, various nuclear genes are descended from single gene copies that were carried in the past by many different members of the ancestral human population.

If genetic drift occurs in a large number of independent, non-interbreeding populations, each with the same initial number of copies of each of two alleles at, say, locus A, then we would expect a fraction p of the populations to become fixed for A_1 and a fraction $1 - p$ to become fixed for A_2. Thus the genetic composition of the populations would diverge by chance. If the original populations had each contained three (or more) different alleles, rather than two, each of those alleles would become fixed in some of the populations, with a probability equal to its initial frequency (say, p_i).

As allele frequencies in a population change by genetic drift, so do the genotype frequencies, which conform to Hardy-Weinberg equilibrium among the new zygotes in each generation. If, for example, the frequencies p and q (that is, p and $1 - p$) of alleles A_1 and A_2 change from 0.5:0.5 to 0.45:0.55, then the frequencies of genotypes A_1A_1, A_1A_2, and A_2A_2 change from 0.25:0.50:0.25 to 0.2025:0.4950:0.3025. As described in Chapter 9, the frequency of heterozygotes, H, declines as one of the allele frequencies shifts closer to 1 (and the other moves toward 0):

$$H = 2p(1 - p)$$

Bear in mind that this model, as developed so far, includes only the effects of random genetic drift. It assumes that other evolutionary processes—namely, mutation, gene flow, and natural selection—do not operate. Thus the model does not describe the evolution of adaptive traits—those that evolve by natural selection. We will incorporate natural selection in the following chapters.

Random fluctuations in allele frequencies

Let us take another, more traditional, approach to the concept of random genetic drift. Assume that the frequencies of alleles A_1 and A_2 are p and q in each of many independent populations, each with N breeding individuals (representing $2N$ gene copies in a diploid species). Small independent populations are sometimes called **demes**, and an ensemble of such populations may be termed a **metapopulation**. As before, we assume that the genotypes do not differ, *on average*, in survival or reproductive success—that is, that the alleles are **neutral** with respect to fitness.

In each generation, the large number of newborn zygotes is reduced to N individuals of reproductive age by mortality that is random with respect to genotype. By sampling error, the proportion of A_1 (p) among the survivors may change. The new p (call it p') could take on any possible value from 0 to 1, just as the proportion of heads among N tossed coins could, in principle, range from no heads to all heads. The probability of each possible value—whether it be the proportion of heads or the proportion of A_1 allele copies—can be calculated from the binomial theorem, generating a PROBABILITY DISTRIBUTION. Among a large number of demes, the new allele frequency (p') will vary, by chance, around a mean—namely, the original frequency, p.

Now if we trace one of the demes, in which p has changed from 0.5 to, say, 0.47, we see that in the following generation, it will change again from 0.47 to some other value, *either higher or lower with equal probability*. This process of random fluctuation continues over time. Since no stabilizing force returns the allele frequency toward 0.5, p will eventually wander (drift) either to 0 or to 1: *the allele is either lost or fixed*. (Once the frequency of an allele has reached either 0 or 1, it cannot change unless another allele is introduced into the population, either by mutation or by gene flow from another population.) The allele frequency describes a **random walk**, analogous to a New Year's Eve reveler staggering along a very long train platform with a railroad track on either side: if he is so drunk that he doesn't compensate for steps toward one side with steps toward the other, he will eventually fall off the edge of the platform onto one of the two tracks (**Figure 10.2**).

Just as an allele's frequency may increase by chance in some demes from one generation to the next, it may decrease in other demes. As a result, allele frequencies may vary among the demes. The variance in allele frequency among the demes continues to increase from generation to generation (**Figure 10.3**). Some demes reach $p = 0$ or $p = 1$ and can no longer change. Among those in which fixation of one or the other allele has not yet occurred, the allele frequencies continue to spread out, with all frequencies between 0 and 1 eventually becoming equally likely (**Figure 10.4**). The number of populations fixed for one or another allele ($p = 0$ or $p = 1$) continues to increase until all demes in the metapopulation have become fixed. Thus *demes that initially are genetically identical evolve by chance to have different genetic constitutions*. (Remember, though, that we are assuming that the alleles have identical effects on fitness—that is, that they are neutral.)

FIGURE 10.2 A "random walk" (or "drunkard's walk"). The reveler eventually falls off the platform if he is too far gone to steer a course toward the middle. The edges of the platform ("0" and "1") represent loss and fixation of an allele, respectively.

Fixation	Loss
← 1	0 →

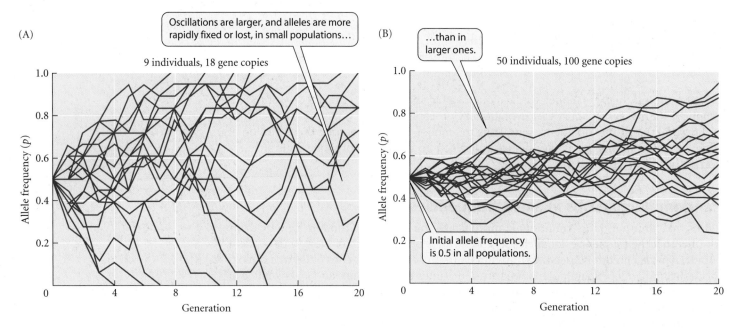

FIGURE 10.3 Computer simulations of random genetic drift in populations of (A) 9 diploid individuals (2N = 18 gene copies) and (B) 50 diploid individuals (2N = 100 gene copies). Each line traces the frequency (p) of one allele for 20 generations. Each panel shows allele frequency changes in 20 replicate populations, all of which begin at p = 0.5 (i.e., half the gene copies are A_1 and half A_2). (After Hartl and Clark 1989.)

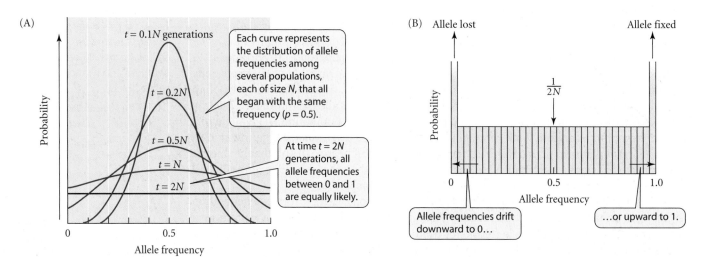

FIGURE 10.4 Changes in the probability that an allele will have various possible frequencies as genetic drift proceeds over time. (A) Each curve shows the probability distribution of allele frequencies between 0 and 1 at different times. The number of generations that elapse (t) is measured in units of the initial population size (N). For example, if the population begins with N = 50 individuals, t = 2N represents the frequency distribution after 100 generations. The probability distribution after t = 0.1N generations is shown by the uppermost curve. This curve may be thought of as the distribution of allele frequencies among several populations, each of size N, that began with the same allele frequency. With the passage of generations, the curve becomes lower and broader as the allele frequencies in all populations drift toward either 0 or 1. At t = 2N generations, all allele frequencies between 0 and 1 are equally likely. (This panel does not show the proportion of populations in which the allele has been fixed or lost.) (B) The proportion of populations with different allele frequencies after t = 2N generations have elapsed, including populations in which the allele has been fixed (p = 1) or lost (p = 0). The proportion of populations in which the allele is lost or fixed increases at the rate of 1/(4N) per generation, and each allele frequency class between 0 and 1 decreases at the rate of 1/(2N) per generation. (A after Kimura 1955; B after Wright 1931.)

Evolution by Genetic Drift

The following points, which follow from the previous discussion, are some of the most important aspects of evolution by genetic drift:

1. Allele (or haplotype) frequencies fluctuate at random within a population, and eventually one or another allele becomes fixed.

2. Therefore the genetic variation at a locus declines and is eventually lost. As the frequency of one allele approaches 1, the frequency of heterozygotes, $H = 2p(1 - p)$, declines. The rate of decline in heterozygosity is often used as a measure of the rate of genetic drift within a population.

3. At any time, an allele's probability of future fixation equals its frequency and is not affected or predicted by its previous history of change in frequency.

4. Therefore populations with the same initial frequency of a particular allele (p) diverge and a proportion p of the populations is expected to become fixed for that allele. A proportion $1 - p$ of the populations becomes fixed for alternative alleles.

5. If an allele has just arisen by mutation and is represented by only one among the $2N$ gene copies in the population, its frequency is

$$p_t = \frac{1}{2N}$$

and this is its likelihood of reaching $p = 1$. Clearly, it is more likely to become fixed in a small than in a large population. Moreover, if the same mutation arises in each of many demes, each of size N, the mutation should eventually be fixed in a proportion $1/(2N)$ of the demes. Similarly, *of all the new mutations (at all loci) that arise in a population, a proportion 1/(2N) should eventually be fixed.*

6. Allele substitution by genetic drift proceeds faster in small than in large populations. In a diploid population, the average time to fixation of a newly arisen neutral allele that does become fixed is $4N$ generations, on average. That is a long time if the population size (N) is large.

7. Among a number of initially identical demes, the average allele frequency ($\bar{p}$) does not change, but since the allele frequency in each deme does change, eventually becoming 0 or 1, the frequency of heterozygotes (H) declines to 0 in each deme and in the metapopulation as a whole.

8. Just as a low-frequency advantageous allele may be lost from a population by genetic drift, a mildly disadvantageous allele may increase in frequency by genetic drift and may even become fixed. Consequently, fitness may decline in small populations. Such declines may be viewed as a form of inbreeding depression (Chapter 9) because matings in small populations are frequently between relatives, whose genes are identical by descent.

9. If we consider two alleles at two linked loci and ignore recombination, the four gamete types (A_1B_1, A_1B_2, A_2B_1, A_2B_2) may be considered as if they were four alleles. Their frequencies will fluctuate by random genetic drift, and one gamete type (A_iB_j) may increase to a frequency greater than that expected (p_ip_j) if it were at linkage equilibrium. Therefore linkage disequilibrium (nonrandom association of alleles) may develop by genetic drift. Recombination breaks down linkage disequilibrium even as genetic drift re-creates it, so the eventual level of linkage disequilibrium depends on the recombination rate and population size.

Effective population size

The theory presented so far assumes highly idealized populations of N breeding adults. If we measure the actual number (N) of adults in real populations, however, the number we count (the CENSUS SIZE) may be greater than the number that actually contribute genes to the next generation. Among elephant seals, for example, a few dominant males mate

FIGURE 10.5 The effective population size among northern elephant seals (*Mirounga angustirostris*) is much lower than the census size because only a few of the large males compete successfully for the smaller females. The winner of contests like the one shown here will father the offspring of an entire "harem" of females.

with all the females in a population, so the alleles those males happen to carry contribute disproportionately to following generations; from a genetic point of view, the unsuccessful subdominant males might as well not exist (**Figure 10.5**). Thus the rate of genetic drift of allele frequencies, and of loss of heterozygosity, will be greater than expected from the population's census size, corresponding to what we would expect of a smaller population. In other words, the population is *effectively* smaller than it seems. The **effective population size** (denoted N_e) of an actual population is the number of individuals in an ideal population (in which every adult reproduces) in which the rate of genetic drift (measured by the rate of decline in heterozygosity) would be the same as it is in the actual population. For instance, if we count 10,000 adults in a population, but only 1000 of them successfully breed, genetic drift proceeds at the same rate as if the population size were 1000, and that is the effective population size, N_e.

The effective population size can be smaller than the census size for several reasons:

1. *Variation in the number of progeny* produced by females, males, or both reduces N_e. The elephant seal represents an extreme example.

2. Similarly, a *sex ratio* different from 1:1 lowers the effective population size.

3. *Natural selection* can lower N_e by increasing variation in progeny number; for instance, if larger individuals have more offspring than smaller ones, the rate of genetic drift may be increased at all neutral loci because smaller individuals contribute fewer gene copies to subsequent generations.

4. If *generations overlap*, offspring may mate with their parents, and since these pairs carry identical copies of the same genes, the effective number of genes propagated is reduced.

5. Perhaps most importantly, *fluctuations in population size* reduce N_e, which is more strongly affected by the smaller than by the larger sizes. For example, if the number of breeding adults in five successive generations is 100, 150, 25, 150, and 125, N_e is approximately 70 (the harmonic mean*) rather than the arithmetic mean, 110.

N_e can be estimated by direct demographic studies of the factors listed, but it is more often estimated from genetic data. The current N_e is often estimated by changes in allele frequencies from generation to generation, since these changes are greater in smaller populations (Waples 1989; Wang and Whitlock 2003). We will soon see that the average N_e over

---

*The HARMONIC MEAN is the reciprocal of the average of a set of reciprocals. If the number of breeding individuals in a series of t generations is $N_0, N_1, ..., N_t$, N_e is calculated from $1/N_e = (1/t)(1/N_0 + 1/N_1 + ... + 1/N_t)$.

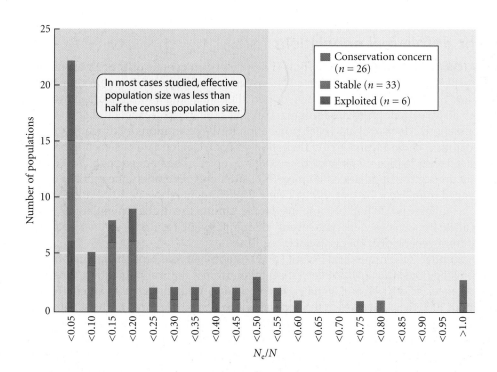

FIGURE 10.6 The ratio of effective population size to census population size (N_e/N) in 65 animal populations. Purple segments represent species of special conservation concern; the red segment represents six exploited fish populations. (After Palstra and Ruzzante 2008.)

In most cases studied, effective population size was less than half the census population size.

■ Conservation concern ($n = 26$)
■ Stable ($n = 33$)
■ Exploited ($n = 6$)

the long term can be estimated from heterozygosity, given an estimate of the neutral mutation rate, because the equilibrium level of genetic variation is set by the balance between the gain of variation attributable to mutation and the loss attributable to genetic drift.

Current N_e values have been estimated for many animal species, especially managed wildlife populations and species of conservation concern. For example, the greater prairie-chicken (*Tympanuchus cupido*) has been largely extirpated from the former prairies of central North America and persists only as fragmented populations. The numbers of microsatellite alleles and mitochondrial DNA (mtDNA) haplotypes have declined over the last 50 years, and the effective sizes of local populations ($N_e = 15$–32 birds in each area) are only a tenth of the census sizes (Johnson et al. 2004). The ratio of the short-term effective population size to the census size (N_e/N) of an important game fish in the Gulf of Mexico, red drum (*Sciaenops ocellatus*), estimated from changes in allele frequencies, is only 0.001, probably because some of the estuaries in which the larvae develop are much less productive than others. The comparable ratio for long-term N_e/N, based on heterozygosity, is an order of magnitude lower (Turner et al. 2002). In general, the ratio N_e/N averages about 0.10 to 0.14 (Frankham 1995; Palstra and Ruzzante 2008; **Figure 10.6**).

Founder effects

Restrictions in size through which populations may pass are called **bottlenecks**. A particularly interesting bottleneck occurs when a new population is established by a small number of colonists, or FOUNDERS—sometimes as few as a single mating pair (or a single inseminated female, as in insects in which females store sperm). Because the founders are a small sample from the source population, allele frequencies in the new population may differ by chance from those in the source population. The resulting random genetic drift is often called a **founder effect**. If the new population rapidly grows to a large size, allele frequencies (and therefore heterozygosity) will probably not be greatly altered from those in the source population, although some rare alleles will not have been carried by the founders. If the colony remains small, however, genetic drift will alter allele frequencies and erode genetic variation. If the colony persists and grows, new mutations eventually restore heterozygosity to higher levels. As is true of genetic drift in general, bottlenecks increase levels of linkage disequilibrium (Charlesworth et al. 2003). The concept of founder effects has been important in evolutionary theory, especially in relation to speciation (see Chapter 18).

Genetic drift in real populations

LABORATORY POPULATIONS   Peter Buri (1956) described genetic drift in an experiment with *Drosophila melanogaster*. He initiated 107 experimental populations of flies, each with 8 males and 8 females, all heterozygous for two alleles (*bw* and *bw*[75]) that affect eye color (by which all three genotypes are recognizable). Thus the initial frequency of *bw*[75] was 0.5 in all populations. He propagated each population for 19 generations by drawing 8 flies of each sex at random and transferring them to a vial of fresh food. (Thus each generation was initiated with 16 flies × 2 gene copies = 32 gene copies.) The frequencies of *bw*[75] rapidly spread out among the populations (**Figure 10.7**); after one generation, the number of *bw*[75] copies ranged from 7 ($q = 7/32 = 0.22$) to 22 ($q = 0.69$). By generation 19, 30 populations had lost the *bw*[75] allele and 28 had become fixed for it; among the unfixed populations, intermediate allele frequencies were quite evenly distributed. The results nicely matched those expected from genetic drift theory (see Figure 10.4).

More recently, McCommas and Bryant (1990) established four replicate laboratory populations, using houseflies (*Musca domestica*) taken from a natural population, at each of three bottleneck sizes: 1, 4, and 16 pairs. Each population rapidly grew to an equilibrium size of about 1000 flies, after which the populations were again reduced to the same bottleneck sizes. This procedure was repeated as many as five times. After each recovery from a bottleneck, the investigators estimated the allele frequencies at four polymorphic enzyme loci for each population, using electrophoresis (see Chapter 9). They found that average heterozygosity ($\overline{H}$) declined steadily after each bottleneck episode, and that the smaller the bottlenecks were, the more rapidly it declined. On the whole, $\overline{H}$ closely matched the values predicted by the mathematical theory of genetic drift.

FIGURE 10.7   Random genetic drift in 107 experimental populations of *Drosophila melanogaster*, each founded with 16 *bw*[75]/*bw* heterozygotes, and each propagated by 16 flies (8 males and 8 females) per generation. The frequency distribution of the number of *bw*[75] copies is read from front to back, and the generations of offspring proceed from left to right. The number of *bw*[75] alleles, which began at 16 copies in the parental populations (i.e., a frequency of 0.5), became more evenly distributed between 0 and 32 copies with the passage of generations, and the *bw*[75] allele was lost (0 copies) or fixed (32 copies) in an increasing number of populations. (After Hartl and Clark 1989.)

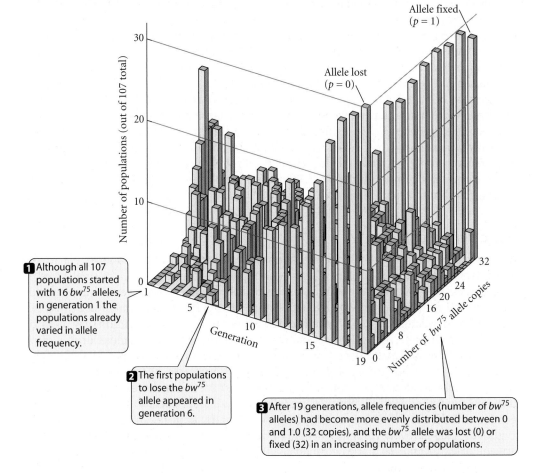

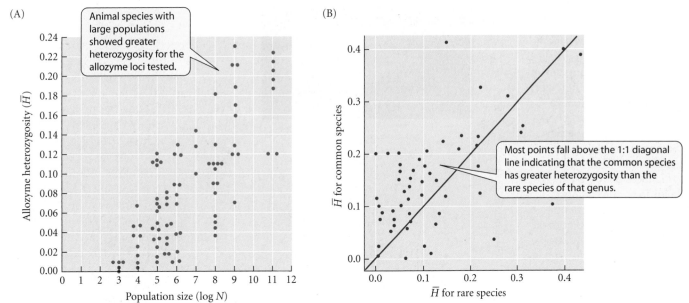

FIGURE 10.8 Genetic variation is reduced in smaller populations. (A) The relation between heterozygosity ($\bar{H}$) for allozymes and estimated population sizes of some species of animals. (B) Each point plots the average heterozygosities ($\bar{H}$) of a pair of species in the same plant genus, one of which is common and the other rare. (A after Soulé 1976; B after Cole 2003.)

NATURAL POPULATIONS   When we describe the genetic features of natural populations, the data are not usually based on experimental manipulations, nor do we usually have detailed information on the populations' histories. We therefore attempt to *infer causes* of evolution (such as genetic drift or natural selection) by *interpreting patterns. Such inferences are possible only on the basis of theories* that tell us what pattern to expect if one or another cause has been most important.

Patterns of molecular genetic variation in natural populations often correspond to what we would expect if the loci were affected by genetic drift. For example, many studies of both animals and plants have shown that allozyme variation is generally lower in rare than in common species (**Figure 10.8**; Frankham et al. 2002).

Occasionally, we can check the validity of our inferences using independent information, such as historical data. For example, a survey of electrophoretic variation in the northern elephant seal (*Mirounga angustirostris*; see Figure 10.5) revealed no variation at any of 24 enzyme-coding loci (Bonnell and Selander 1974)—a highly unusual observation, since most natural populations are highly polymorphic (see Chapter 9). However, although the population of this species now numbers about 30,000, it was reduced by hunting to about 20 animals in the 1890s. Moreover, the effective size was probably even lower, because less than 20 percent of males typically succeed in mating. A later study of mtDNA revealed a severe reduction in sequence diversity compared with DNA from specimens taken before the population was severely reduced (Weber et al. 2000). The hypothesis that genetic drift was responsible for reducing genetic variation is supported by the historical data. European populations of the ruddy duck (*Oxyura jamaicensis*), a North American species, are thought to stem from individuals that escaped from a captive population founded by seven North American birds. Only one haplotype of mtDNA was found in European birds, compared with 23 haplotypes in North America (Muñoz-Fuentes et al. 2006). Ashkenazi Jewish populations, especially in eastern Europe, have a relatively high frequency (up to about 0.03) of mutations that cause Tay-Sachs and other diseases. Patterns of linkage disequilibrium among closely linked genetic markers, as well as other aspects of sequence variation in these genes, correspond to known bottlenecks in the history of the Ashkenazim, such as the diaspora in 70 CE (AD) and the Black Death in 1348 (Risch et al. 2003; Slatkin 2004). Thus bottlenecks can increase the frequency of deleterious mutations, as theory predicts and as has been documented in many natural populations (**Figure 10.9**).

FIGURE 10.9 The failure rate for egg hatching is greater in New Zealand bird species that have suffered a bottleneck in population size. In both graphs, bottleneck size decreases logarithmically to the right. (A) Data for 22 species of native birds, many of which approached extinction after European settlement. Some are still highly endangered, such as no. 17, the kakapo (a large flightless parrot, pictured). (B) Data for 15 species of introduced birds, showing hatching failure in relation to the number of individuals released in New Zealand. Most of these species were introduced from Europe. (After Briskie and Mackintosh 2004.)

Strigops habroptila

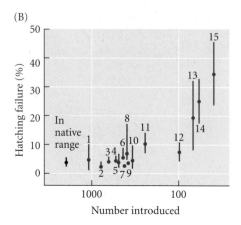

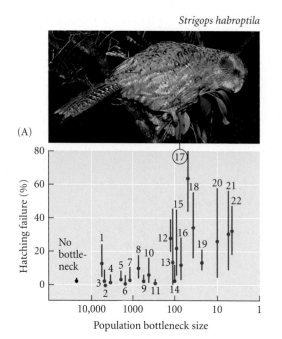

(A)

(B)

The Neutral Theory of Molecular Evolution

Whether or not random genetic drift has played an important role in the evolution of many of the morphological and other phenotypic features of organisms is a subject of considerable debate. There is no question, however, that at the levels of DNA and protein sequences, genetic drift is a significant factor in evolution.

From the evolutionary synthesis of the late 1930s until the mid-1960s, most evolutionary biologists believed that almost all alleles differ in their effects on organisms' fitness, so that their frequencies are affected chiefly by natural selection. This belief was based on studies of genes with morphological or physiological effects. But in the 1960s, the theory of evolution by random genetic drift of selectively neutral alleles became important as two kinds of molecular data became available. In 1966, Lewontin and Hubby showed that a high proportion of enzyme-coding loci are polymorphic (see Chapter 9). They argued that natural selection could not actively maintain so much genetic variation and suggested that much of it might be selectively neutral. At about the same time, Motoo Kimura (1968) calculated the rates of evolution of amino acid sequences of several proteins, using the phylogenetic approach described in Chapter 2 (see Figure 2.20). He concluded that a given protein evolved at a similar rate in different lineages. He argued that such constancy would not be expected to result from natural selection, but would be expected if most evolutionary changes at the molecular level are caused by mutation and genetic drift. These authors and others (King and Jukes 1969) initiated a controversy, known as the "neutralist–selectionist debate," that persists today. Although everyone now agrees that some molecular variation and evolution are neutral (i.e., a result of genetic drift), "selectionists" think a larger fraction of molecular evolutionary changes are due to natural selection than "neutralists" do.

The **neutral theory of molecular evolution**, as developed by Kimura (1983), holds that although a small minority of mutations in DNA or protein sequences are advantageous and are fixed by natural selection, and although many mutations are disadvantageous and are eliminated by natural selection, *the great majority of those mutations that are fixed are effectively neutral with respect to fitness and are fixed by genetic drift*. According to this theory, most genetic variation at the molecular level is selectively neutral and lacks adaptive significance. This theory, moreover, holds that evolutionary substitutions at the molecular level proceed at a roughly constant rate, so that the degree of sequence difference between species can serve as a MOLECULAR CLOCK, enabling us to determine the divergence times of species (see Chapter 2).

The neutral theory plays a major role in contemporary evolutionary biology. It is important to recognize that the neutral theory does *not* hold that the morphological, physiological, and behavioral features of organisms evolve by random genetic drift. Many—perhaps most—such features evolve chiefly by natural selection, and they are based on base pair substitutions that (according to the neutralists) constitute a very small fraction of DNA sequence changes. Furthermore, the neutral theory acknowledges that many mutations are deleterious and are eliminated by natural selection, so that they contribute little to the variation we observe. Thus the neutral theory does not deny the operation of natural selection on *some* base pair or amino acid differences. The theory does postulate, however, that most of the variation we *observe* at the molecular level, both within and among species, has little effect on fitness. For protein-coding sequences, this means that the observed differences in base pair sequence are not translated into amino acid differences, or that most variations in amino acid sequence have little effect on the organism's physiology. [A theory of "nearly neutral" mutations, developed by Tomoko Ohta (1992), will be considered in Chapter 12.]

Principles of the neutral theory

Let us assume that mutations occur in a gene at a constant rate of u_T per gamete per generation, and that because of the great number of mutable sites, every mutation constitutes a new DNA sequence (or allele, or haplotype). Of all such mutations, some fraction (f_0) are effectively neutral, so the neutral mutation rate, $u_0 = f_0 u_T$, is less than the total mutation rate, u_T. By EFFECTIVELY NEUTRAL, we mean that the mutant allele is so similar to other alleles in its effect on survival and reproduction (i.e., fitness) that changes in its frequency are governed by genetic drift alone, not by natural selection. (It is, of course, possible that the mutation does affect fitness slightly. Then, as we will see in Chapter 12, natural selection and genetic drift operate simultaneously, but because genetic drift is stronger in small than in large populations, the changes in the mutant allele's frequency will be governed almost entirely by genetic drift if the population is small enough. Therefore a particular allele may be effectively neutral, relative to another allele, when the population is small, but not when the population is large.)

The rate of origin of effectively neutral alleles at a locus by mutation, u_0, depends on the gene's function. If many of the amino acids in the protein it encodes cannot be altered without seriously affecting an important function—perhaps because they affect the shape of a protein that binds to DNA or to other proteins—then a majority of mutations in the gene will be deleterious rather than neutral, and u_0 will be much lower than the total mutation rate, u_T. Such a locus is said to have many **functional constraints**. On the other hand, if the protein can function well despite any of many amino acid changes (i.e., it is less constrained), u_0 will be higher. Within regions of DNA that encode proteins, we would expect the neutral mutation rate to be highest at third-base positions in codons and lowest at second-base positions, because those positions have the highest and lowest redundancy, respectively (see Figure 8.2). We would expect constraints to be least, or even nonexistent, and the fraction of neutral mutations (or substitutions) to be greatest, for DNA sequences that are not transcribed and have no known function, such as many pseudogenes and probably many introns. (By comparing rates of sequence evolution to predictions of the neutral theory, evolutionary geneticists have found unexpected evidence that much noncoding DNA, including some portions of certain introns and pseudogenes, may actually be functional; see Chapters 8 and 20.)

Now consider a population of effective size N_e in which the rate of neutral mutation at a locus is u_0 per gamete per generation (**Figure 10.10**). The *number* of new mutations is, on average, $u_0 \times 2N_e$, since there are $2N_e$ gene copies that could mutate. From genetic drift theory, we have learned that the probability that a mutation will be fixed by genetic drift is its frequency, p, which equals $1/(2N_e)$ for a newly arisen mutation. Therefore the number of neutral mutations that arise in any generation *and will someday be fixed* is

$$2N_e u_0 \times 1/(2N_e) = u_0$$

FIGURE 10.10 Evolution by genetic drift. Each curve plots the number of copies (*n*) of a neutral mutation in a diploid population of *N* individuals (2*N* gene copies) against time. (A) Most new mutations are lost soon after they arise, but occasionally an allele increases toward fixation by genetic drift. (B) Over a longer time, successive mutations become fixed at this locus. The average time required for such fixation is $\bar{t}$. (C) The average time required for fixation is longer in larger populations. Hence at any time there will be more neutral alleles in a larger population. (After Crow and Kimura 1970.)

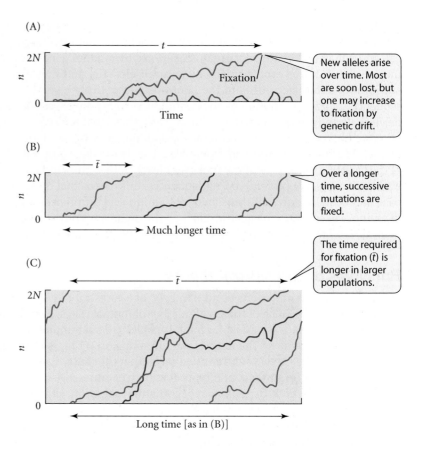

Because u_0 is presumed to be constant in all generations, about the same number of neutral mutations should be fixed in every generation. On average, it will take $4N_e$ generations for such mutations to reach fixation. Therefore *the rate of fixation of mutations is theoretically constant and equals the neutral mutation rate*. This principle is the theoretical basis of the molecular clock. Notice that over the long term, surprisingly, the rate of substitution does not depend on the population size: each mutation drifts toward fixation more slowly if the population is large, but the lower rate is compensated for by the greater number of mutations that arise.

If two species diverged from their common ancestor *t* generations ago, and if each species has experienced u_0 substitutions per generation (relative to the allele in the common ancestor), then the number of base pair differences (*D*) between the two species should be $D = 2u_0t$, because each of the two lineages has accumulated u_0t substitutions. Therefore, if we have an estimate of the number of generations that have passed (see Box 8A), the neutral mutation rate can be estimated as

$$u_0 = D/2t$$

This formula requires qualification, however. We are assuming here that the common ancestor had a single invariant sequence and that all the mutations fixed in the two species occurred since the ancestor was divided into two reproductively isolated populations. And, very importantly, over a sufficiently long time, some sites experience repeated base substitutions (multiple hits): a particular site may undergo substitution from, say, A to C and then from C to T or even back to A. Thus the observed number of differences between species will be less than the number of substitutions that have transpired. As the time since divergence becomes greater, the number of differences begins to plateau, as is evident from the number of differences per base pair between the mtDNA sequences of different mammalian taxa in **Figure 10.11** (see also Figure 2.21). Each point represents a pair of taxa for which the age of the common ancestor has been estimated from the fossil record. The number of base pair differences increases linearly for about 5 to 10 My before it begins to level off because of repeated

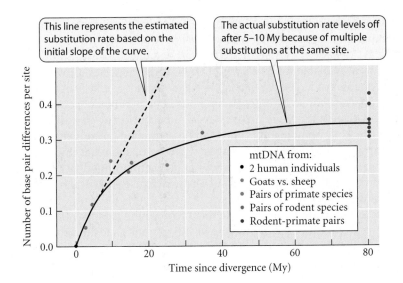

This line represents the estimated substitution rate based on the initial slope of the curve.

The actual substitution rate levels off after 5–10 My because of multiple substitutions at the same site.

mtDNA from:
- 2 human individuals
- Goats vs. sheep
- Pairs of primate species
- Pairs of rodent species
- Rodent-primate pairs

FIGURE 10.11 The number of base pair differences per site between the mtDNA of pairs of mammalian taxa, plotted against the estimated time since their most recent common ancestor. (After Brown et al. 1979.)

substitutions. From the linear part of the curve, the mutation rate can be readily calculated, assuming that all the base pair differences represent neutral substitutions (in Figure 10.11, the rate is about 0.01 mutations per base pair per lineage per My, or about 10^{-8} per year).

Note that if u_0 has been accurately estimated for a lineage, it can be used to estimate the time since related organisms diverged from their common ancestor from the relationship $t = D/2u_0$ (see Chapter 2). It can also be used to estimate the time since gene lineages on a gene tree diverged (see Figures 2.25 and 10.1).

Within a population, there is TURNOVER, or FLUX, of alleles or haplotypes (see Figure 10.10). As one or another allele approaches fixation (about every $4N$ generations, on average), other alleles are lost. But new neutral alleles arise continually by mutation, and although many are immediately lost by genetic drift, others drift to higher frequencies and persist for some time in a polymorphic state before they are lost or fixed. Although the identity of the several or many alleles present in the population changes over time, the level of variation reaches an equilibrium when the rate at which alleles arise by mutation is balanced by the rate at which they are lost by genetic drift. This equilibrium level of variation, represented by the frequency of heterozygotes, $\hat{H}$, is higher in a large population than in a small one. It can be shown mathematically that at equilibrium,

$$\hat{H} = \frac{4N_e u_0}{4N_e u_0 + 1}$$

(Figure 10.12). For example, given the observed mutation rate for allozymes, 10^{-6} per gamete (Voelker et al. 1980), the equilibrium frequency of heterozygotes ($\hat{H}$) would be 0.004 if the effective population size N_e were 1000, but would be 0.50 if N_e were 250,000.

Support for the neutral theory

The major prediction of Kimura's neutral theory is that the rate of evolution should be greater at nucleotide sites or in DNA sequences in which we can safely assume, on biological grounds, that changes would not affect an organism's fitness. In contrast, the evolutionary rate should be lower if changes would be expected to affect fitness because their effects are much more likely to be harmful than beneficial. For example, Kimura cited the peptide hormone insulin, which is formed by the splicing of two segments of a proinsulin chain, the third segment of which (the C peptide) is removed, apparently playing no role other than in the formation of the mature insulin chain. Among mammals,

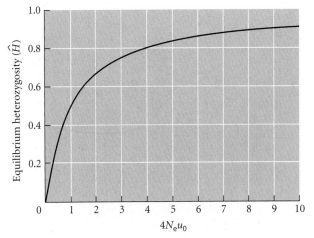

FIGURE 10.12 The equilibrium level of heterozygosity at a locus increases as a function of the product of the effective population size (N_e) and the neutral mutation rate (u_0). (After Hartl and Clark 1989.)

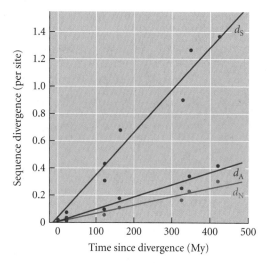

FIGURE 10.13 Average sequence divergence of 4198 nuclear genes between human and nine other vertebrates (six mammals and one each of bird, frog, and bony fish), plotted against estimated time since common ancestor, based on the fossil record. Synonymous nucleotide substitutions (d_S) have accumulated much faster than nonsynonymous nucleotide substitutions (d_N) or amino acid substitutions in the encoded proteins (d_A). This difference provides strong support for the neutral theory. (After Nei et al. 2010.)

the average rate of amino acid substitution has been six times greater in the C peptide locus than in the sites coding for the other portions of proinsulin.

Over the past decade, DNA sequencing has become routine, and DNA sequences are now available for hundreds of thousands of species, providing a wealth of data on rates of molecular evolution. Rates of change can be estimated by comparing species whose divergence time has been estimated from fossils (see Chapter 2). Moreover, the relative rates of substitution at, say, different codon positions within homologous genes can be estimated by comparing the sequences of any two species: more changes might be observed at third-base positions than at second-base positions, even though we may not know the time span over which divergence occurred. All such estimates use statistical methods that correct for biases such as multiple hits.

The most consistent pattern, and the one that provides the most plentiful support for the neutral theory, is the higher rate of synonymous substitutions than nonsynonymous substitutions (**Figure 10.13**; Nei et al. 2010). Amino acid substitutions caused by nonsynonymous mutations are expected to be often deleterious and to be expunged by "purifying" natural selection, whereas synonymous mutations are thought to be much less likely to affect fitness. (However, purifying selection against some synonymous mutations has been detected for certain genes in certain organisms; see Chapter 20.) In one of countless studies that have supported this expectation, Carlos Bustamante et al. (2005) compared 11,624 genes between human and chimpanzee; they found 34,099 fixed base pair differences, of which 1.02 percent were synonymous and 0.24 percent were nonsynonymous, a ratio of 0.24. In early studies of DNA sequences, researchers found that the highest substitution rates were at "fourfold-degenerate" sites (in which all third-base substitutions are synonymous) and in pseudogenes, which are thought generally to lack function (**Figure 10.14**). Although early studies found high substitution rates in introns and flanking regions, this is not always the case, and researchers today suspect that many of these and other noncoding regions are functional.

Molecular clocks, revisited

Although Kimura was inspired to develop the neutral theory by the apparently constant rate of sequence evolution, it has become clear that the rate varies considerably among different groups of organisms (Ellegren 2008; Lanfear et al. 2010). For example, DNA sequences evolve more slowly among Old World primates than among most other groups of mammals, and more slowly in hominoid apes than in monkeys (Steiper et al. 2004). This rate variation among taxa has several likely causes (Bromham 2009). Some of the variation is attributable to generation length: gene frequencies change from one generation to the next, so the speed of substitution per time interval is greater in species with shorter generations. Large-bodied organisms exhibit slower sequence evolution than small species, partly because they generally have longer generations. Furthermore, the rate of mutation varies among species, perhaps partly because of differences in their metabolic

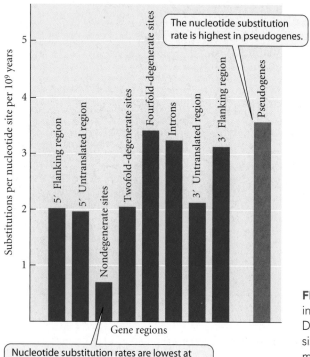

FIGURE 10.14 Average rates of substitution in different parts of genes and in pseudogenes, estimated from comparisons between humans and rodents. Differences in the rate of molecular evolution among these classes are consistent with the neutral theory. Nucleotide positions are nondegenerate if no mutation would be synonymous; twofold-degenerate if one other nucleotide would be synonymous; and fourfold-degenerate if mutation to any of the other three nucleotides would be synonymous (see Figure 8.2). (After Li 1997.)

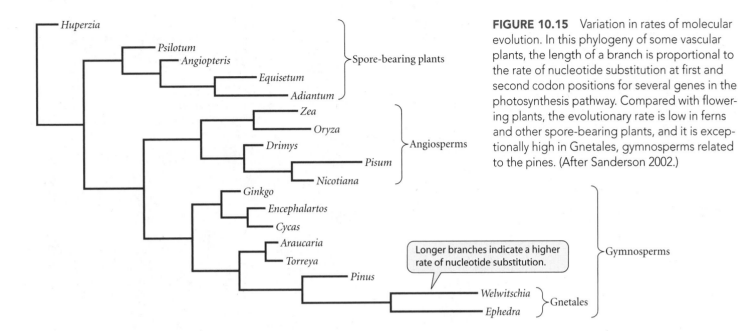

FIGURE 10.15 Variation in rates of molecular evolution. In this phylogeny of some vascular plants, the length of a branch is proportional to the rate of nucleotide substitution at first and second codon positions for several genes in the photosynthesis pathway. Compared with flowering plants, the evolutionary rate is low in ferns and other spore-bearing plants, and it is exceptionally high in Gnetales, gymnosperms related to the pines. (After Sanderson 2002.)

rate. Another important factor is effective population size: slightly deleterious mutations, not just purely neutral ones, can be fixed in small populations, in which genetic drift is more effective relative to natural selection than in large populations (see Chapter 12, p. 330).

Consequently, although closely related species often have similar rates of neutral sequence evolution (as detected by synonymous substitutions), a uniform molecular clock cannot be assumed a priori in studies of large, diverse clades. Instead, various subclades may have "clocks" that tick at different rates (**Figure 10.15**). Several methods have been developed to detect and calibrate such differences in rate (Lanfear et al. 2010).

Gene Flow and Genetic Drift

A measure of the variation in allele frequency among different populations is the fixation index, F_{ST} (see Chapter 9). The rate at which populations drift toward fixation of one allele or another is inversely proportional to the effective population size, N_e (or N, for simplicity). However, the drift toward fixation is counteracted by gene flow from other populations, at rate m. These factors strike a balance, or equilibrium, at which F_{ST} is approximately

$$F_{ST} = \frac{1}{4Nm+1}$$

The quantity Nm is the *number* of immigrants per generation. If $m = 1/(N)$ (i.e., only one breeding individual per population is an immigrant, per generation), then $Nm = 1$, and $F_{ST} = 0.20$. That is, even a little gene flow keeps all the demes fairly similar in allele frequency, and heterozygosity remains high.

Recasting this equation as

$$Nm = \frac{(1/F_{ST})-1}{4}$$

enables us to indirectly estimate average rates of gene flow among natural populations, since F_{ST} can be estimated from the variation of allele frequencies. In fact, such indirect estimates may be better than direct estimates of gene flow, such as those described in Chapter 9, because direct observations might be misleading if they are made when m is unusually high or low and because they are usually insufficient to detect long-distance migration, rare episodes of massive gene flow, and the perhaps rare (but nevertheless important) processes of population extinction and recolonization that can increase gene flow (Slatkin 1985).

FIGURE 10.16 Geographic variation in allele frequencies at two electrophoretic loci in the pocket gopher *Thomomys bottae*. The left half of each circle shows frequencies of up to four alleles of the gene for the enzyme alcohol dehydrogenase (*Adh*); the right half, those of up to three alleles of the lactate dehydrogenase-1 gene (*Ldh1*). (After Patton and Yang 1977.)

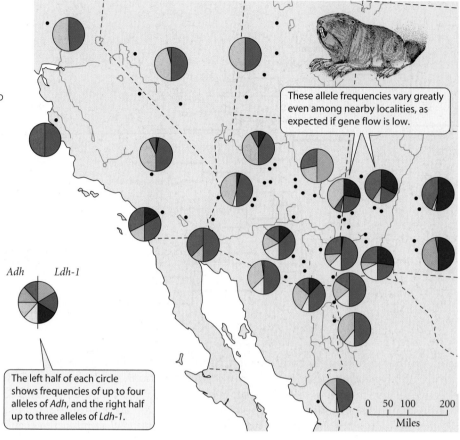

These allele frequencies vary greatly even among nearby localities, as expected if gene flow is low.

Adh   *Ldh-1*

The left half of each circle shows frequencies of up to four alleles of *Adh*, and the right half up to three alleles of *Ldh-1*.

0 50 100 200
Miles

We must assume that the alleles for which we calculate F_{ST} are selectively neutral. (F_{ST} would underestimate gene flow if natural selection favored different alleles in different areas, and it would overestimate gene flow if selection favored the same allele everywhere.) This assumption can be evaluated by the degree of consistency among different loci for which F_{ST} is estimated. Genetic drift and gene flow affect all loci the same way, whereas natural selection affects different loci more or less independently. Therefore, if each of a number of polymorphic loci yields about the same value of F_{ST}, it is likely that selection is not strong. It is also necessary to assume that allele frequencies have reached an equilibrium between gene flow and genetic drift. This might not be the case if, for example, the sampled sites have only recently been colonized and the populations have not yet had time to differentiate by genetic drift. Their genetic similarity would then lead us to overestimate the rate of gene flow.

The pocket gopher *Thomomys bottae* is a burrowing rodent that seldom emerges from the soil. This species is famous for its localized variation in coloration and other morphological features, which has led taxonomists to name more than 150 subspecies. Moreover, local populations differ more in chromosome configuration than in any other known mammalian species. Such geographic variation suggests that gene flow might be relatively low. Indeed, 21 polymorphic enzyme loci in 825 specimens from 50 localities in the southwestern United States and Mexico showed extreme geographic differentiation (**Figure 10.16**). Among all 50 populations, the average F_{ST} was 0.412 (which might imply Nm = 0.36); among localities in Arizona, it was 0.198 (implying Nm = 1.01). The genetically most different populations were most geographically distant or segregated by expanses of unsuitable habitat—both factors that would reduce gene flow (Patton and Yang 1977).

Gene trees and population history

Earlier in this chapter, we introduced the principle of genetic drift by showing that because gene lineages within a population eventually become extinct by chance over the course

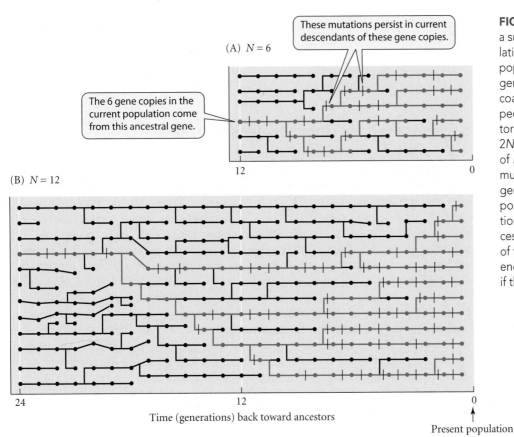

(A) $N = 6$

These mutations persist in current descendants of these gene copies.

The 6 gene copies in the current population come from this ancestral gene.

12   0

(B) $N = 12$

24   12   0

Time (generations) back toward ancestors

Present population

FIGURE 10.17 Coalescence time in (A) a small population and (B) a large population. Gene copies in the present-day population are shown in red, and their genealogy is shown back as far as their coalescence (common ancestry). The expected time back to the common ancestor of all gene copies in the population is 2N generations for a haploid population of N gene copies. Tick marks represent mutations, each at a unique site in the gene. Each gene copy in the present population (at time 0) has all the mutations (tick marks) that occurred in its ancestral gene lineage, back to the left end of the diagram. More mutational differences among gene copies are expected if the coalescence time has been longer.

of time, all gene copies in a population today are descended from a single gene copy that existed at some time in the past. The genealogical history of genes in populations is the basis of COALESCENT THEORY. This theory, applied to DNA sequence data, provides inferences about the structure and the effective size (N_e) of species populations (Hudson 1990). Recall that if the number of breeding individuals changes over time, N_e is approximately equal to the harmonic mean, which is much closer to the smallest number the population has experienced than to the arithmetic mean. If, for example, the population has rapidly grown to its present large size from a historically much smaller size, N_e is close to the earlier size, and this value can be estimated from coalescent theory.

Because the smaller the effective size (N_e) of a population, the more rapidly genetic drift transpires, the existing gene copies in a small population must stem from a more recent common ancestor than the gene copies in a large population (compare **Figures 10.17A and 10.17B**). That is, if we look back in time from the present, it takes longer for the present-day genes in the larger population to coalesce to their common ancestor. Mathematical models show that in a haploid population of N_e individuals, the average time back to the common ancestor of all the gene copies (t_{CA}) is $2N_e$ generations, and in a diploid population, $t_{CA} = 4N_e$ generations. In a diploid population, the common ancestor of a random pair of gene copies occurred $2N_e$ generations ago. (For mitochondrial genes, carried in an effectively haploid state and transmitted only by females, $t_{CA} = N_e$ generations.)

If two randomly sampled gene copies had a common ancestor t generations ago, and each gene lineage experiences, on average, u neutral mutations per generation, then each will have accumulated $u \times t$ mutations since the common ancestor. Therefore the expected number of base pair differences between them (π) will be $2ut$ because there are two gene lineages. Since $t = 2N_e$, $\pi = 4N_e u$. We therefore *expect the average number of base pair differences between gene copies to be greater in large than in small populations.* (This difference is illustrated by the tick marks, representing mutations, on the gene trees in Figure 10.17.) In fact, if we have an estimate of the mutation rate per base pair (u), and if we measure the average proportion of sites that differ between random pairs of gene copies (π), *we can estimate the effective population size* as

$$N_e = \pi/4u$$

(For mitochondrial genes, the estimate of N_e is $N_e = \pi/u$.)

The origin of modern Homo sapiens *revisited*

As we saw in Chapter 6, the gene tree for human mtDNA, sampled from populations throughout the world, is rooted between most African lineages and a clade of lineages that includes some African haplotypes and all non-African haplotypes (see Figure 6.19). Moreover, there are far fewer nucleotide differences among non-African haplotypes, on average, than among African haplotypes. This pattern, which has been found consistently as data have been amassed throughout the nuclear genome, supports the "replacement," or "out-of-Africa," hypothesis, according to which the world's human population outside Africa is descended from a relatively small population that spread from Africa rather recently.

Analyses of genome-wide markers, based on the theory described above, have been used to infer both the time of dispersal from Africa and the effective sizes of populations in Africa and beyond. For example, a study of mtDNA and genes on the X and Y chromosomes concluded that the ancestral Eurasian population was founded only about 40 Kya (Garrigan et al. 2007). Several studies that estimated the historical effective population size concluded that the world's human population is descended from an effective breeding population of only 4600 to 11,200 people (see Hammer 1995; Rogers 1995)! One study of sequence variation in nuclear genes estimated the long-term N_e of humans as 10,400, lower than that of bonobo (12,300), chimpanzee (21,300), or gorilla (25,200; Yu et al. 2004). Another research group conducted a massive analysis of the exon sequences of 800 genes in almost 700 individuals in one African and several European and Asian populations, as well as surveying SNPs throughout 179 human genomes (Gravel et al. 2011). The authors estimated that from an ancestral African population with N_e between 13,000 and 16,000, dispersal into Eurasia occurred about 51,000 years ago (**Figure 10.18**). The relatively few individuals that founded the human population throughout Eurasia had an effective population size of only about 1860, and the effective sizes of the ancestral European and Asian populations, when they separated about 23,000 years ago, were 1000 or less. Contrast these numbers with the current world population, which has passed the 7 billion mark!

As small numbers of colonists dispersed farther and farther from Africa, genetic drift reduced genetic diversity, both for DNA sequences (see Figure 9.34) and for features such as skull measurements (Manica et al. 2007). A higher proportion of SNPs is thought to be associated with potentially damaging alterations of proteins in European-American than in African-American individuals, suggesting that historical bottlenecks may have elevated the frequencies of deleterious mutations in Eurasian populations (Lohmueller et al. 2008). Moreover, genetic drift also increases the average level of linkage disequilibrium (LD) between SNPs because only a sample of gene combinations is transmitted each generation. The farther a population is from Africa, the higher its average LD among SNPs (**Figure 10.19**). When Brenna Henn and coworkers (2011) calculated the correlation between LD in non-African populations and their geographic distance from various African populations, they found the strongest correlation in hunter-gatherer (Bushman) populations in South Africa, which have the highest nucleotide diversity and the lowest levels of linkage disequilibrium of all the human populations that have been studied. In agreement with an earlier study (Tishkoff et al. 2009), they suggest that modern humans originated in southern Africa.

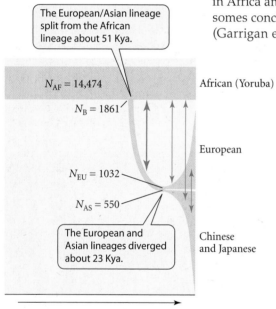

The European/Asian lineage split from the African lineage about 51 Kya.

$N_{AF} = 14,474$

$N_B = 1861$

African (Yoruba)

European

$N_{EU} = 1032$

$N_{AS} = 550$

The European and Asian lineages diverged about 23 Kya.

Chinese and Japanese

Time

FIGURE 10.18 The history of the spread of humans from Africa to Europe and Asia, based on genetic data from African (Yoruba people, Nigeria), European, and Asian (Chinese and Japanese) samples. The width of the colored areas indicates population size. The effective size of the ancestral African population (N_{AF}) was estimated as 14,474. Spread from Africa occurred 51 Kya by a bottlenecked population (B) of effective size $N_B = 1861$. By 23 Kya, humans had spread to Europe and to Asia. In both of these areas, populations grew rapidly in size, but their effective sizes (N_{EU} and N_{AS}) were small because the populations were small at first. During this history, movements of people among populations resulted in some gene flow, indicated by the red arrows. (After Gravel et al. 2011.)

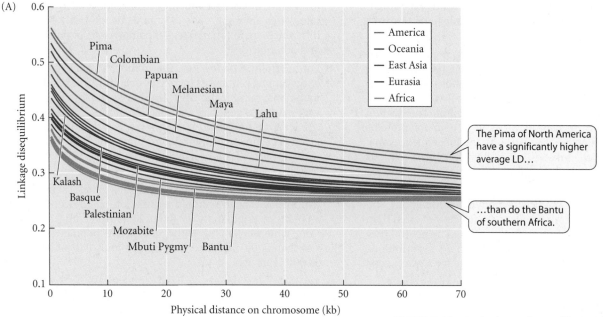

(A)

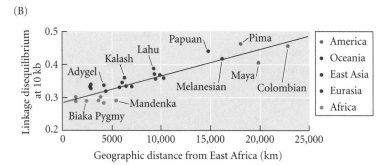

(B)

FIGURE 10.19 Linkage disequilibrium (LD) between genetic markers (SNPs) in human populations. (A) The average LD between many pairs of markers declines with their distance (in kilobases, kb) from each other on a chromosome because of the increasing recombination rate. At any distance between markers, some human populations have higher average LD than others. (B) The average LD is greater the farther these populations are from East Africa, as expected because of successive founder events during colonization of farther regions. Among the populations shown, LD is greatest in Native American (Maya, Pima, and Colombian) populations. (After Jakobsson et al. 2008.)

The history of population growth can be estimated from DNA sequence data using coalescent theory. We have already seen that an estimate of N_e, averaged over many generations, shows that the human population has expanded from a much smaller size. Very recent increases in population size can be detected, partly by a greater number of rare alleles than in a population of constant size: alleles that are rare simply because they arose by very recent mutation. Alon Keinan and Andrew Clark (2012) found exactly this pattern in large samples of human sequences; the genetic data matched the explosive growth that the world's human population has experienced in the last 2000 years or so. DNA sequence variation in European and southeastern Asian populations shows that the rate of population growth has increased fivefold since agriculture was initiated about 12,000 years ago (Gignoux et al. 2011).

Go to the
EVOLUTION
Companion Website at
sites.sinauer.com/evolution3e
for quizzes, data analysis and simulation exercises, and other study aids.

Summary

1. The frequencies of alleles that differ little or not at all in their effect on organisms' fitness (neutral alleles) fluctuate at random because only a sample of the genes in a population at any time is transmitted to the next generation. This process, called random genetic drift, reduces genetic variation and leads eventually to the random fixation of one allele and the loss of others, unless it is countered by other processes, such as gene flow or mutation.

2. Different alleles are fixed by chance in different populations.

3. The probability, at any time, that a particular allele will be fixed in the future equals the frequency of the allele at that time. For example, if a newly arisen mutation is represented by one copy in a diploid population of N individuals, the probability that it will increase to fixation is $1/(2N)$.

4. The smaller the effective size of a population, the more rapidly random genetic drift operates. The effective size is often much smaller than the apparent population size.

5. Patterns of allele frequencies at some loci in both experimental and natural populations conform to predictions from the theory of genetic drift.

6. The theory of genetic drift has been applied especially to variation at the molecular level. The neutral theory of molecular evolution holds that, although many mutations are deleterious, and a few are advantageous, most molecular variation within and among species is selectively neutral. The fraction of mutations that are neutral varies: it is high for proteins that lack strong functional constraints and for sequences that are not transcribed. Likewise, it is higher for synonymous than for nonsynonymous (amino acid–replacing) nucleotide substitutions.

7. As the neutral theory predicts, synonymous mutations and mutations in less constrained genes are fixed more often than those that are more likely to affect function. The neutral theory also predicts that over long spans of time, substitutions will occur at an approximately constant rate for a given gene (providing a basis for the "molecular clock"). However, this rate varies among groups of organisms, probably because of differences in mutation rate, generation time, population size, and other factors.

8. For neutrally evolving loci, the number of nucleotide differences among sequences increases over time as a result of new mutations, but genetic drift, leading to the loss of gene lineages, reduces genetic variation. When these factors balance, the level of sequence variation reaches equilibrium. Thus, given an estimate of the mutation rate, the level of sequence variation provides a basis for estimating the historical effective size (N_e) of a population.

9. These principles, applied to human genes, support the hypothesis that the human population has descended from an African population of about 14,000 breeding members, from which colonists, with an effective population size of less than 2000, migrated into Europe and Asia perhaps only 50,000 years ago.

Terms and Concepts

bottleneck	fixation	metapopulation	random walk
coalescence	founder effect	neutral	sampling error
deme	functional constraint	neutral theory of molecular evolution	stochastic
deterministic	genetic drift (= random genetic drift)	nonadaptive evolution	
effective population size			

Suggestions for Further Reading

Fundamentals of Molecular Evolution, second edition, by D. Graur and W.-H. Li (Sinauer Associates, Sunderland, MA, 2000), is a leading textbook on the topics discussed in this chapter and covers many other topics as well.

The Neutral Theory of Molecular Evolution, by M. Kimura (Cambridge University Press, Cambridge, 1983), is a comprehensive (although very dated) discussion of the neutral theory by its foremost architect.

Human Population Genetics, by J. H. Relethford (Wiley-Blackwell, Hoboken, NJ, 2012), covers major aspects of this fast-moving field.

Problems and Discussion Topics

1. For a diploid species, assume one set of 100 demes, each with a constant size of 50 individuals, and another set of 100 demes, each with 100 individuals. (a) If in each deme the frequencies of neutral alleles A_1 and A_2 are 0.4 and 0.6, respectively, what fraction of demes in each set is likely to become fixed for allele A_1 versus A_2? (b) Assume that a neutral mutation arises in each deme. Calculate the probability that it will become fixed in a population of each size. In what number of demes (approximately) do you expect it to become fixed? (c) If fixation occurs, how many generations do you expect it to take?

2. Assuming that the average rate of neutral mutation is 10^{-9} per base pair per gamete, how many generations would it take, on average, for 20 base pair substitutions to be fixed in a gene with 2000 base pairs? Suppose that the number of base pair differences in this gene between species A and B is 92, between A and C is 49, and between B and C is 91. Assuming that no repeated replacements have occurred at any site in any lineage, draw the phylogenetic tree, estimate the number of fixations that have occurred along each branch, and estimate the number of generations since each of the two speciation events.

3. Some evolutionary biologists have argued that the neutral theory should be taken as the null hypothesis to explain genetic variation within species or populations and genetic differences among them. In this view, adaptation and natural selection should be the preferred explanation only if genetic drift cannot explain the data. Others might argue that since there is so much evidence that natural selection has shaped species' characteristics, selection should be the explanation of choice and that the burden of proof should fall on advocates of the neutral theory. Why might one of these points of view be more convincing than the other?

4. Some critics of Darwin's theory of evolution by natural selection have claimed that the concept of natural selection is tautologous (i.e., it is a "vicious circle"). They say, "Natural selection is the principle of the survival of the fittest. But the fittest

are defined as those that survive, so there is no way to prove or disprove the theory." Argue against this statement, drawing on the contents of this chapter.

5. How can gene trees be used to estimate rates of gene flow among geographic populations of a species? What assumptions would have to be made? (See Slatkin and Maddison 1989.)

6. Suppose that several investigators want to use genetic markers such as SNPs to estimate gene flow (the average number of migrants per generation) among populations of several species. One wants to study movement of howler monkeys among forest patches left by land clearing in Brazil; another plans to study movement among populations of mink frogs in lakes in Ontario, north of Lake Superior; a third intends to study the warbler finch on the 17 major islands of the Galápagos Islands, near the equator. For which of these species is this approach most likely, or least likely, to yield a valid estimate of current gene flow? Why?

7. Could the principles of this chapter be used to estimate when an infectious disease organism first entered the human population? What conditions or characteristics of the organism would help to make this feasible? See, for example, "Timing the ancestor of the HIV-1 pandemic strains," by B. Korber et al. (*Science* 288: 1789–1796, 2000).

Natural Selection and Adaptation

The theory of natural selection is the centerpiece of *The Origin of Species* and of evolutionary theory. It is this theory that accounts for the adaptations of organisms, those innumerable features that so wonderfully equip them for survival and reproduction; it is this theory that accounts for the divergence of species from common ancestors and thus for the endless diversity of life. Natural selection is a simple concept, but it is perhaps the most important idea in biology. It is also one of the most important ideas in the history of human thought—"Darwin's dangerous idea," as the philosopher Daniel Dennett (1995) has called it—for it explains the apparent design of the living world without recourse to a supernatural, omnipotent designer.

An **adaptation** is a characteristic that enhances the survival or reproduction of organisms that bear it relative to alternative character states (especially the ancestral condition in the population in which the adaptation evolved). Natural selection is the only mechanism known to cause the evolution of adaptations, so many biologists would simply define an adaptation as a characteristic that has evolved by natural selection. The word "adaptation" also refers to the process whereby the members of a population become better suited to some feature of their environment through change in a characteristic that affects their survival or reproduction.

Take precautions: Use your head! In several lineages of cavity-dwelling ants, such as this member of the genus *Cephalotes*, the head of some workers is modified to block the entrance and prevent predators from entering the nest. This remarkable adaptation, known as phragmosis, has also evolved in certain tree frogs. In a genus of spiders, the abdomen is modified to serve this function.

Adaptations in Action: Some Examples

We can establish a few important points about adaptations by looking at some striking examples:

• In most terrestrial vertebrates, the skull bones are rather rigidly attached to one another, but in snakes they are loosely joined. Most snakes can swallow prey much larger than their heads, manipulating them with astonishing versatility. The lower jawbones (mandibles) articulate to a long, movable quadrate bone that can be rotated downward so that the mandibles drop away from the skull; the front ends of the two mandibles are not fused (as they are in almost all other vertebrates), but are joined by a stretchable ligament. Thus the mouth opening can be greatly increased (**Figure 11.1A**). Both the mandibles and the tooth-bearing maxillary bones, which are suspended from the skull, independently move forward and backward to pull the prey into the throat. In rattlesnakes and other vipers, the maxilla is short and bears only a long, hollow fang, to which a duct leads from the massive poison gland (a modified salivary gland). The fang lies against the roof of the mouth when the mouth is closed. When the snake opens its mouth, the same lever system that moves the maxilla in nonvenomous snakes rotates the maxilla 90 degrees (**Figure 11.1B**), so that the fang is fully erected. Snakes' skulls, like many anatomical features, are complex mechanisms that *look as if they had been designed* by engineers to perform a specified function. Their features have been achieved by modifications of the same bones that are found in other reptiles.

• Among the 18,000 to 25,000 species of orchids, many have extraordinary modifications of flower structure and astonishing mechanisms of pollination. In pseudocopulatory pollination, for example (**Figure 11.2**), part of the flower looks somewhat like a female insect, and the flower emits a scent that mimics the attractive sex pheromone (scent) of a female bee, fly, or thynnine wasp, depending on the orchid species. As a male insect "mates" with the flower, pollen is deposited precisely on that part of the insect's body that will contact the stigma of the next flower visited. Several aspects of this example are of interest. First, adaptations are found among plants as well as animals. For Darwin, this was an important point, because Lamarck's theory, according to which animals inherit characteristics altered by their parents' behavior, could not explain the adaptations of plants. Second, the floral form and scent are *adaptations to promote reproduction* rather than survival. Third, the plant achieves reproduction by deceiving, or exploiting, another organism. The insect gains nothing from its interaction with the flower; in fact, it would be to the insect's advantage to resist the flower's deceptive allure, since copulating with

FIGURE 11.1 The kinetic skull of snakes. The movable bones of the upper jaw are shown in gold. (A) The skull of a nonvenomous snake with jaws closed (top) and open (bottom). (B) A viper's skull. (C) The head of a red diamondback rattlesnake (*Crotalus ruber*) in strike mode. Abbreviations: ec, ectopterygoid; mx, maxilla; pal, palatine; pt, pterygoid. (A, B after Porter 1972.)

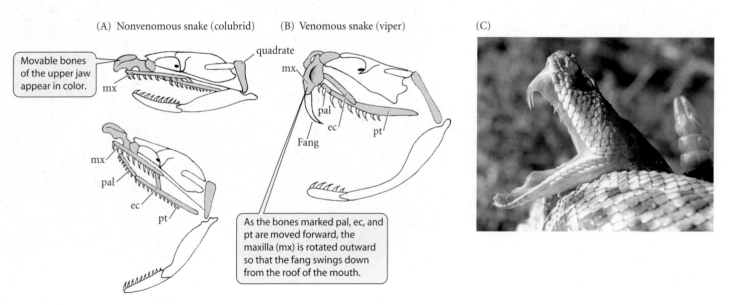

(A) Nonvenomous snake (colubrid)   (B) Venomous snake (viper)   (C)

Movable bones of the upper jaw appear in color.
mx
quadrate
mx
pal
ec
pt
Fang
mx
pal
ec
pt

As the bones marked pal, ec, and pt are moved forward, the maxilla (mx) is rotated outward so that the fang swings down from the roof of the mouth.

(A)

(B)

FIGURE 11.2 Pseudocopulatory pollination. (A) *Ophrys apifera*, one of the "bee orchids," uses phero-mones to attract male bees. It is shaped so that when a bee attempts to copulate with the flower, pollen adheres to the insect's body. (B) A long-horned bee (*Eucera longicornis*) attempts to mate with an *Ophrys scolopax* flower. A yellow pollen mass adheres to the bee's head.

a flower probably reduces its opportunity to find proper mates. So *organisms are not necessarily as well adapted as they could be.*

- After copulation, male redback spiders (*Latrodectus hasselti;* related to the "black widow" spider), often somersault into the female's mouthparts and are eaten (**Figure 11.3A**). This suicidal behavior might be adaptive, because males seldom have the opportunity to mate more than once, and it is possible that a cannibalized male fathers more offspring. Maydi-anne Andrade (1996) tested this hypothesis by presenting females with two males in succession, and using genetic markers to determine the paternity of the females' offspring. She found that females that ate the first male with whom they copulated were less likely to mate a second time, so these cannibalized males fertilized all the eggs. Furthermore, among females that did mate with both males, the percentage of offspring that were fathered by the second male was greater if he was eaten than if he survived. (**Figure 11.3B**). Both outcomes support the hypothesis that sexual suicide enhances males' reproductive success. This example suggests that *prolonged survival is not necessarily advantageous,* and it also illustrates *how hypotheses of adaptation may be formulated and tested.*

- Many species of animals engage in COOPERATIVE BEHAVIOR, but it reaches extremes in some social insects. An ant colony, for example, includes one or more inseminated queens and a number of sterile females, the workers. The arboreal weaver ants (genus *Oecophylla*) in southeastern Asia and Australia construct nests of living leaves by the intricately coordi-

FIGURE 11.3 Prolonged survival is not necessarily advantageous. (A) The small male redback spider somersaults into the larger female's mouthparts after copulation. (B) Among males that were second to copulate with females, cannibalized males fertilized a greater proportion of eggs than noncannibalized males, because they copulated longer, and the proportion of eggs fertilized was correlated with the duration of his copulation. (A after Forster 1992; B after Andrade 1996.)

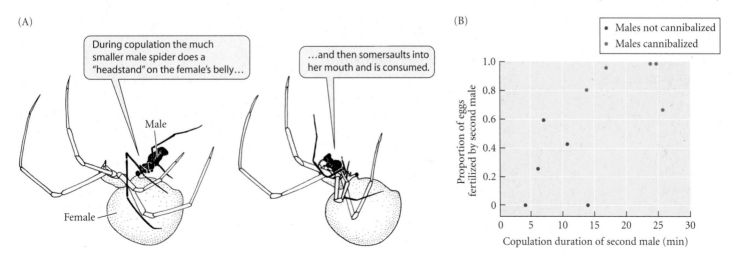

FIGURE 11.4 Weaver ants (*Oecophylla*) constructing a nest. Chains of workers, each seizing another's waist with her mandibles, pull leaves together. (Photo from Hölldobler and Wilson 1983, courtesy of Bert Hölldobler.)

nated action of numerous workers, groups of which draw together the edges of leaves by grasping one leaf in their mandibles while clinging to another. Sometimes several ants form a chain to collectively draw together distant leaf edges (**Figure 11.4**). The leaves are attached to one another by the action of other workers carrying larvae that emit silk from their labial glands. (The adult ants cannot produce silk.) The workers move the larvae back and forth between the leaf edges, forming silk strands that hold the leaves together. In contrast to the larvae of other ants, which spin a silk cocoon in which to pupate, *Oecophylla* larvae produce silk only when used by the workers in this fashion. These genetically determined behaviors are adaptations that enhance the reproductive success not of the worker ants that perform them, since the workers do not reproduce, but rather of their mother, the queen, whose offspring include both workers and reproductive daughters and sons. *In some species, then, individuals have features that benefit other members of the same species.* How such features evolve is a topic of special interest.

The Nature of Natural Selection

Design and mechanism

Most adaptations, such as a snake's skull, are *complex*, and most have the appearance of *design*—that is, they are constructed or arranged so as to accomplish some *function*, such as growth, feeding, or pollination, that appears likely to promote survival or reproduction. In inanimate nature, we see nothing comparable—we would not be inclined to think of erosion, for example, as a process designed to shape mountains.

The complexity and evident function of organisms' adaptations cannot conceivably arise from the random action of physical forces. For hundreds of years, it seemed that adaptive design could be explained only by an intelligent designer; in fact, this "argument from design" was considered one of the strongest proofs of the existence of God. For example, the Reverend William Paley wrote in *Natural Theology* (1802) that, just as the intricacy of a watch implies an intelligent, purposeful watchmaker, so every aspect of living nature, such as the human eye, displays "every indication of contrivance, every manifestation of design, which exists in the watch," and must, likewise, have had a Designer.

Supernatural processes cannot be the subject of science, so when Darwin offered a purely natural, materialistic alternative to the argument from design, he not only shook the foundations of theology and philosophy, but *brought every aspect of the study of life into the realm of science*. His alternative to intelligent design was design by the completely mindless process of natural selection, according to which organisms possessing variations that enhance survival or reproduction replace those less suitably endowed, which therefore survive or reproduce in lesser degree. This process cannot have a goal, any more than erosion has the goal of forming canyons, for *the future cannot cause material events in the present*. Thus the concepts of goals or purposes have no place in biology (or in any of the other natural sciences), except in studies of human behavior. According to Darwin and contemporary evolutionary theory, the weaver ants' behavior has the appearance of design because among many random genetic variations (mutations) affecting the behavior of an ancestral ant species, those displayed by *Oecophylla* enhanced survival and reproduction under its particular ecological circumstances.

Adaptive biological processes *appear* to have goals: weaver ants act as if they have the goal of constructing a nest; an orchid's flower develops toward a suitable shape and stops developing when that shape is attained. We may loosely describe such features by TELEO-LOGICAL statements, which express goals (e.g., "She studied *in order* to pass the exam"). But no conscious anticipation of the future resides in the cell divisions that shape a flower or, as far as we can tell, in the behavior of weaver ants. Rather, the apparent goal-directedness is caused by the operation of a program—coded or prearranged information, residing in DNA sequences—that controls a process (Mayr 1988a). A program likewise resides in a computer chip, but whereas that program has been shaped by an intelligent designer, the information in DNA has been shaped by a historical process of natural selection. Modern biology views the development, physiology, and behavior of organisms as the results of purely mechanical processes, resulting from interactions between programmed instructions and environmental conditions or triggers.

Definitions of natural selection

It is important to recognize that *"natural selection" is not synonymous with "evolution."* Evolution can occur by processes other than natural selection, especially genetic drift. And natural selection can occur without any evolutionary change, as when natural selection maintains the status quo by eliminating deviants from the optimal phenotype.

Evolution by natural selection occurs if three conditions hold (Lewontin 1970): (1) Individuals in a population of reproducing entities (e.g., organisms) differ in one or more features (phenotypic variation). (2) Different phenotypes differ in rates of survival and reproduction (differential fitness). (3) There is a correlation between parents and offspring in their contribution to future generations (fitness is heritable). Among the many proposed definitions of natural selection (Endler 1986), we will therefore choose the following one: natural selection is *any consistent difference in fitness among phenotypically different classes of biological entities*. Let us explore this definition in more detail.

In most discourse about evolution, the "entities" that differ in survival and reproduction are individual organisms. However, differences in fitness also exist below the organismal level, among genes and cell lineages, and above the organismal level, among populations and species. In other words, different kinds of biological entities may vary in fitness, resulting in different **levels of selection**. Of these, selection among individual organisms (**individual selection**) and among genes (**genic selection**) are the most important.

The **fitness**—often called the **reproductive success**—of a biological entity is its average per capita rate of increase in numbers. When we speak of natural selection among different genotypes or phenotypes, the components of fitness generally consist of (1) the probability of survival to reproductive age, (2) the average number of offspring (e.g., eggs, seeds) produced via female function, and (3) the average number of offspring produced via male function. "Reproductive success" has the same components as "fitness," since survival is a prerequisite for reproduction.

Variation in the number of offspring produced as a consequence of competition for mates is often referred to as **sexual selection**, which some authors distinguish from natural selection. We will follow the more common practice of regarding sexual selection as a kind of natural selection.

Because the *probability* of survival and the *average* number of offspring enter into the definition of fitness, and because these concepts apply only to *groups* of events or objects, fitness is best defined for a *set* of like entities, such as all the individuals with a particular genotype. That is, *natural selection exists if there is an average (i.e., statistically consistent) difference in reproductive success*. It is not useful to refer to the fitness of a single individual, since its history of reproduction and survival may have been affected by chance to an unknown degree, as we will see shortly.

Natural selection can exist only if different classes of entities differ in one or more features, or traits, that affect fitness. Evolutionary biologists differ on whether or not the definition of natural selection should require that these differences be inherited. We will adopt the position taken by those (e.g., Lande and Arnold 1983) who define selection among individual organisms as a consistent difference in fitness among phenotypes. Whether or not this variation in fitness results in evolution (i.e., alteration of the frequencies of genotypes in subsequent generations) depends on whether and how the phenotypes are inherited—but that determines the *response to selection*, not the process of selection itself. Although we adopt the phenotypic perspective, we will almost always discuss natural selection among heritable phenotypes because selection seldom has a lasting evolutionary effect unless there is inheritance. Most of our discussion will assume that inheritance of a trait is based on genes. However, many of the principles of evolution by natural selection also apply if inheritance is epigenetic (based on, for example, differences in DNA methylation; see Chapter 9) or is based on cultural transmission, especially from parents to offspring. CULTURE has been defined as "information capable of affecting individuals' behavior that they acquire from other members of their species through teaching, imitation, and other forms of social transmission" (Richerson and Boyd 2005, p. 5).

Notice that according to our definition, natural selection exists whenever there is variation in fitness. *Natural selection is not an external force or agent, and certainly not a purposeful one. It is a name for statistical differences in reproductive success among genes, organisms, or populations, and nothing more.*

Natural selection and chance

If one neutral allele replaces another in a population by random genetic drift (see Chapter 10), then the bearers of the first allele have had a greater rate of increase than the bearers of the other. Natural selection has not occurred, however, because the genotypes do not differ *consistently* in fitness: the alternative allele could just as well have been the one to increase. There is no *average* difference between the alleles, no *bias* toward the increase of one relative to the other. Fitness differences, in contrast, are *average* differences, *biases*, differences in the *probability* of reproductive success. This does not mean that every individual of a fitter genotype (or phenotype) survives and reproduces prolifically while every individual of an inferior genotype perishes; some variation in survival and reproduction occurs independently of—that is, at random with respect to—phenotypic differences. But natural selection resides in the difference in rates of increase among biological entities that is *not* due to chance. *Natural selection is the antithesis of chance.*

If fitness and natural selection are defined by consistent, or average, differences, then we cannot tell whether a difference in reproductive success between two *individuals* is due to chance or to a difference in fitness. We cannot say that one identical twin had lower fitness than the other because she was struck by lightning (Sober 1984), or that the genotype of the Russian composer Tchaikovsky, who had no children, was less fit than the genotype of Johann Sebastian Bach, who had twenty. We can ascribe genetic changes to natural selection rather than random genetic drift only if we observe consistent, nonrandom changes in

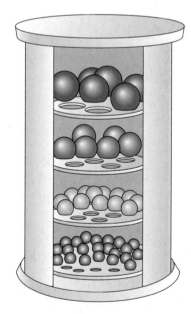

FIGURE 11.5 A child's toy that selects small balls, which drop through smaller and smaller holes from top to bottom. In this case there is selection *of* red balls, which happen to be the smallest, but selection is *for* small size. (After Sober 1984.)

replicate populations, or if we measure numerous individuals of each phenotype and find an average difference in reproductive success.

Selection of and selection for

In the child's "selection toy" pictured in **Figure 11.5**, the holes in each partition are smaller than in the one above. Balls of several sizes, when placed in the top compartment, fall through the holes in the partitions If the smallest balls in the toy are all red, and the larger ones are all other colors, the toy will select the small, red balls. Thus we must distinguish *selection of objects* from *selection for properties* (Sober 1984). Balls are selected *for* the property of small size—that is, *because of* their small size. They are not selected for their color, or because of their color; nonetheless, here there is selection *of* red balls. Natural selection may similarly be considered a sieve that selects *for* a certain body size, mating behavior, or other feature. There may be incidental selection *of* other features that are correlated with those features.

The importance of this semantic point is that when we speak of the **function** of a feature, we imply that there has been natural selection *of* organisms with that feature and *of* genes that program it, but selection *for* the feature itself. We suppose that the feature *caused* its bearers to have higher fitness. The feature may have other **effects**, or consequences, that were not its function, and *for* which there was no selection. For instance, there was selection for an opposable thumb and digital dexterity in early hominins, with the incidental effect, millions of years later, that we can play the piano. Similarly, a fish species may be selected for coloration that makes it less conspicuous to predators. The *function* of the coloration, then, is predator avoidance. An *effect* of this evolutionary change might well be a lower likelihood that the population will become extinct, but *avoidance of extinction is not a cause of evolution* of the coloration.

The analogy of natural selection to the toy emphasizes that *selection is a process of sorting* of phenotypes over time, from one generation to another, by some phenotypic trait. Sorting can also occur in space. In a species that is expanding its geographic range, new populations are founded at the edges of the range by colonists that may be a nonrandom sample from the source population—namely, those most capable of dispersal (Shine et al. 2011). For example, in newly founded populations of several species of bush crickets that are expanding northward due to global warming, the proportion of long-winged, highly dispersive individuals is higher than in older populations to the south (**Figure 11.6A**). In Australia, the introduced South American cane toad (*Rhinella marina*) is expanding at an accelerating rate because there is automatic selection in new populations of longer-legged, more active individuals (**Figure 11.6B,C**). These fast-dispersing individuals are also prone to spinal arthritis, however, showing that the selection process can have harmful side effects at the individual level (Brown et al. 2007). Evolutionary changes at the expanding range edge are short-lived because the features selected by the dispersal process tend to be disadvantageous in stable populations.

The effective environment depends on the organism

The environmental factors that impose natural selection on a species are greatly influenced by the characteristics of the species itself: the evolutionary history of a species affects its relationship to the environment (Lewontin 2000). The branching structure of trees in a forest is important for many tree-nesting birds such as orioles, but almost irrelevant for ground-nesting species such as partridges; the viscosity of water, which varies with temperature, is much more important for a ciliate than for a fish. To some extent,

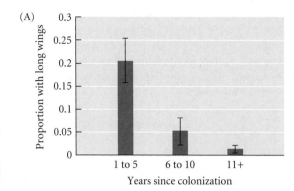

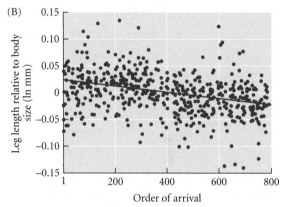

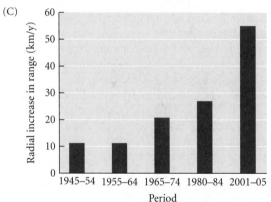

FIGURE 11.6 Expansion of a species into new territory automatically selects for characteristics that enhance dispersal. (A) Areas newly colonized by bush crickets (*Metrioptera roeselii*) have a higher proportion of long-winged individuals. (B) The earliest cane toads to arrive at a research site in the path of the species' advance have the longest legs; those that arrive later have shorter legs. (C) Because of the sorting process, the advancing front of the cane toad invasion is populated by faster and faster toads, so the rate of advance has increased since the toad was introduced in 1935. (A after Simmons and Thomas 2004; B, C after Phillips et al. 2006).

FIGURE 11.7 The evolutionary histories of some animals have made them less reliant on vision, so selection for visual acuity has been relaxed. (A) Army ants (genus *Eciton*) rely almost entirely on chemical information. They have highly reduced compound eyes, consisting of a single unit (ommatidium) rather than the many that compose most insects' eyes. (B) Similarly, burrowing blind snakes (Typhlopidae) have highly reduced eyes that perceive light, but cannot form an image.

(A)

(B)

FIGURE 11.8 Natural selection on mutations in the β-galactosidase gene of *Escherichia coli* in laboratory populations maintained on lactose. In each case, a strain bearing a mutation competed with a control strain bearing the wild-type allele. Populations were initiated with equal numbers of cells of each genotype; that is, with log (ratio of mutant/control) initially equal to zero. Without selection, no change in the log ratio would be expected. (A) One mutant strain decreased in frequency, showing a selective disadvantage. (B) Another mutant strain increased in frequency, demonstrating its selective (adaptive) advantage. (After Dean et al. 1986.)

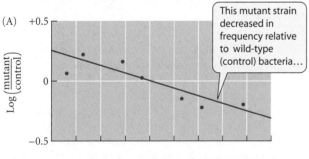

This mutant strain decreased in frequency relative to wild-type (control) bacteria...

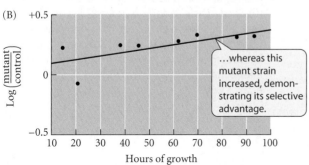

...whereas this mutant strain increased, demonstrating its selective advantage.

Hours of growth

organisms "construct" their ecological niche (Odling-Smee et al. 2003), literally (as does a beaver) or more metaphorically. Organisms "screen off" some aspects of their environment, which may then cease to exert natural selection. Many species of ants, rodents, and other animals have become so reliant on chemical signals that they experience no selection for visual acuity (**Figure 11.7**), whereas humans have lost the function of many olfactory receptor genes, having become so much more reliant on vision than smell.

Examples of Natural Selection

We can illustrate important principles of natural selection by several examples, some of which show how natural selection can be studied.

Experimental evolution

Because of their very rapid population growth, bacteria and other microbes are useful for "experimental evolution" studies, in which evolutionary changes occur in experimental populations as a consequence of conditions that are imposed by the experimenter (Kawecki et al. 2012). Anthony Dean and colleagues (1986) studied competition between a wild-type strain of *Escherichia coli* and each of several strains that differed from the wild type only by mutations of the gene that codes for β-galactosidase, the enzyme that breaks down lactose. Pairs of genotypes, each consisting of the wild type and one mutant strain, were cultured together in vessels with lactose as their sole source of energy. The populations were so large that changes in genotype frequencies attributable to genetic drift alone would have been almost undetectably slow. Among the mutant strains, one decreased in frequency, and so had lower fitness than the wild type, apparently because of its lower enzyme activity. Another mutant strain, with higher enzyme activity, increased in frequency, displaying a greater rate of increase than the wild type (**Figure 11.8**).

This experiment conveys the essence of natural selection: it is a completely mindless process without forethought or goal. Adaptation—the evolution of a bacterial population with a higher average ability to metabolize lactose—resulted from a difference in the rates of reproduction of different genotypes caused by a phenotypic difference (enzyme activity).

Another experiment with bacteria illustrates *the distinction between "selection of" and "selection for."* In *E. coli*, the wild-type allele *his*⁺ codes for an enzyme that synthesizes histidine, an essential amino acid, whereas *his*⁻ alleles are nonfunctional. The *his*⁻ alleles are selectively neutral, however, if histidine is supplied to the bacteria so that cells with the mutant allele can grow. Atwood and colleagues (1951) observed, to their surprise, that every few hundred generations, the allele frequencies changed rapidly and drastically in experimental cultures that were supplied with histidine (**Figure 11.9**). The experimenters showed that the *his* alleles were **hitchhiking** with advantageous mutations at other loci—a phenomenon readily observed in bacteria because their rate of recombination is extremely low. Occasionally, a genotype (say, *his*⁻) would increase rapidly in frequency because of linkage to an advantageous mutation that had occurred at another locus. Subsequently, the alternative allele (*his*⁺) might increase because of linkage to a new advantageous mutation at another locus altogether. Thus there was selection *for* new advantageous mutations in these bacterial populations, and selection *of* neutral alleles at the linked *his* locus.

Male reproductive success

The courting males of many species of animals have elaborate morphological features and engage in conspicuous displays; roosters provide a familiar example. Darwin posited that some such features evolved through female choice of males with conspicuous features, which therefore enjoy higher reproductive success than less elaborate males (see Chapter 15). Malte Andersson (1982) tested this hypothesis with male long-tailed widowbirds (*Euplectes progne*), which have extremely long tail feathers. He shortened the tail feathers of some wild males and attached the clippings to the tail feathers of others, thus elongating them well beyond the natural length. He observed that, as predicted by Darwin's hypothesis, the number of new nests per territory was greatest for males with artificially elongated tails, evidence of their enhanced mating success (**Figure 11.10**).

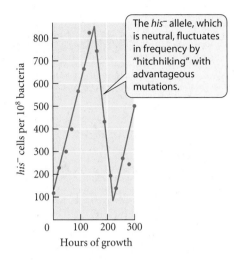

The *his*⁻ allele, which is neutral, fluctuates in frequency by "hitchhiking" with advantageous mutations.

FIGURE 11.9 Allele frequency fluctuates because of hitchhiking in a laboratory population of *Escherichia coli*. The *y*-axis represents the frequency of the selectively neutral *his*⁻ allele compared with that of the wild-type *his*⁺ allele. The frequency of the *his*⁻ allele increases if a cell bearing it experiences an advantageous mutation at another locus, then decreases if a different, more advantageous mutation occurs in a *his*⁺ cell. (After Nestmann and Hill 1973.)

FIGURE 11.10 An example of sexual selection. (A) A male long-tailed widowbird (*Euplectes progne*) in territorial flight. (B) Effects of experimental alterations of tail length on males' mating success, measured by the number of new nests in each male's territory. Nine birds were chosen for each of four treatments: shortening or elongation of the tail feathers, or controls of two types: one in which the tail feathers were cut and repasted, and one in which the tail was not manipulated. (B after Andersson 1982.)

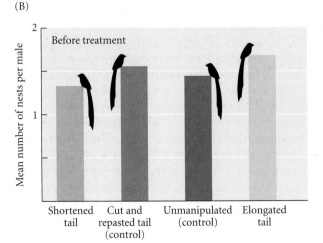

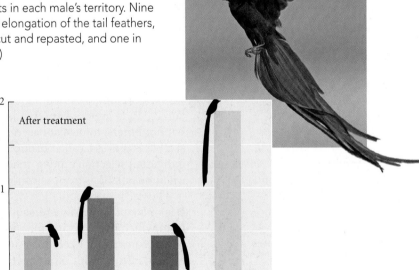

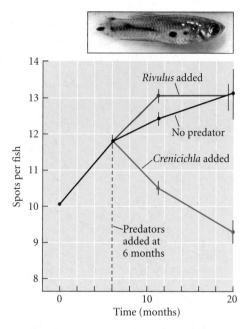

FIGURE 11.11 Evolution of male color pattern in experimental populations of guppies. Six months after the populations were established, some were exposed to a major predator of adult guppies (*Crenicichla*), some to a less effective predator that feeds mainly on juvenile guppies (*Rivulus*), and some were left free of predators (controls). Numbers of spots were counted after 4 and 10 generations. The vertical bars indicate the variation among males. (After Endler 1980; photo courtesy of Anne Houde.)

Male guppies (*Poecilia reticulata*) have a highly variable pattern of colorful spots. In Trinidad, males have smaller, less contrasting spots in streams inhabited by their major predator, the fish *Crenicichla*, than in streams without this predator. John Endler (1980) moved 200 guppies from a *Crenicichla*-inhabited stream to a site that lacked the predator. About 2 years (15 generations) later, he found that the newly established population had larger spots and a greater diversity of color patterns, so that the population now resembled those that naturally inhabit *Crenicichla*-free streams. Endler also set up populations in large artificial ponds in a greenhouse. After 6 months of population growth, he introduced *Crenicichla* into four ponds, released a less dangerous predatory fish (*Rivulus*) into four others, and maintained two control populations free of predators. In censuses after 4 and 10 generations, the number and brightness of spots per fish had increased in the ponds without *Crenicichla* and had declined in those with it (**Figure 11.11**). Males with more and brighter spots have greater mating success, but they are also more susceptible to being seen and captured by *Crenicichla*.

These experiments show that natural selection may consist of differences in reproductive rate owing to differences in mating success—what Darwin called sexual selection. The guppy experiments also show that a feature may be subjected to *conflicting selection pressures* (such as sexual selection and predation), and that the direction of evolution may then depend on which is stronger. Many advantageous characters, in fact, carry corresponding disadvantages, often called COSTS or TRADE-OFFS.

Group selection

The small beetle *Tribolium castaneum* breeds in stored grains and can be reared in containers of flour. Larvae and adults feed on flour but also eat (cannibalize) eggs and pupae. Michael Wade (1977, 1979) set up 48 experimental populations under each of three treatments (**Figure 11.12A**). Each population was propagated from 16 adult beetles each generation. In treatment A, Wade deliberately selected for large population size by initiating each generation's 48 populations with sets of 16 beetles taken only from those few populations (out of the 48) in which the largest number of beetles had developed to adulthood. In treatment B, small population size was selected for in the same way, by propagating beetles only from the smallest populations. The control populations (treatment C) were propagated simply by moving beetles to a new vial of flour: each population in one generation gave rise to one population in the next.

Over the course of nine generations, the average population size declined in all three treatments, most markedly in treatment B and least in treatment A (**Figure 11.12B**). The net reproductive rate also declined. In treatment C, these declines must have been due to evolution *within* each population, with natural selection acting on the genotypes of individual beetles within a population. This process is individual selection, of the same kind we have assumed to operate on, say, the color patterns of guppies. But in treatments A and B, Wade imposed another level of selection by allowing some populations, or groups, but not others, to persist based on a phenotypic characteristic of each *group*—namely, its size. This process, called **group selection**, or **interdemic selection**, operates *in addition to* individual selection among genotypes within populations. We must distinguish selection *within* populations from selection *among* populations.

The decline of population size in the control (C) populations seems like the very antithesis of adaptation. Wade discovered, however, that compared with the foundation stock from which the experimental populations had been derived, adults in the C populations had become more likely to cannibalize pupae, and females were prone to lay fewer eggs when confined with other beetles. For an individual beetle, cannibalism is an advantageous way of obtaining protein, and it may be advantageous for a female not to lay eggs if she perceives the presence of other beetles that may eat them. But although these features are advantageous to the individual, they are disadvantageous to the population, whose growth rate declines.

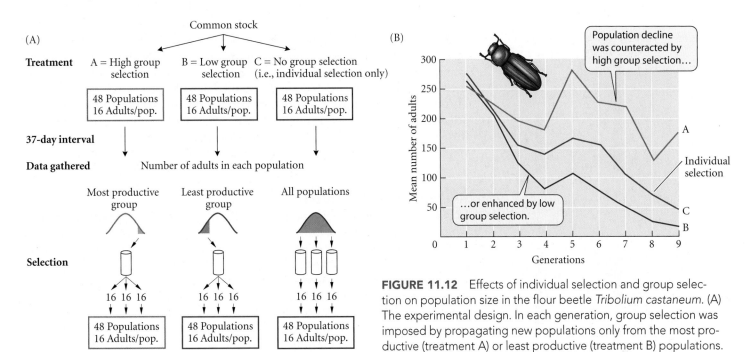

FIGURE 11.12 Effects of individual selection and group selection on population size in the flour beetle *Tribolium castaneum*. (A) The experimental design. In each generation, group selection was imposed by propagating new populations only from the most productive (treatment A) or least productive (treatment B) populations. In treatment C (controls), all populations were propagated, so there was no group selection; any changes were due to natural selection among individuals within populations. (B) Changes in the mean number of adult beetles in the three treatments. (After Wade 1977.)

By selecting groups for small population size (treatment B), Wade *reinforced* these same tendencies. In treatment A, however, selection at the group level for large population size *opposed* the consequences of individual selection within populations. Compared with the C populations, beetles from treatment A had higher fecundity in the presence of other beetles, and they were less likely to cannibalize eggs and pupae. Thus selection among groups had affected the course of evolution.

This experiment shows that the size or growth rate of a population may decline as a result of natural selection even as individual organisms become fitter. *It also illustrates that selection might operate at two levels*: among individuals and among populations. In this case, selection at the group level was imposed by the investigator, so the experiment shows that it is possible, but whether or not it is important in nature is an open question.

Kin selection

To continue the cannibal theme, the aquatic larvae of the tiger salamander (*Ambystoma tigrinum*) often develop into a distinct phenotype that eats smaller larvae. Most, although not all, cannibals tend to avoid eating close relatives, such as siblings. One hypothesis for the evolution of such kin discrimination is that an allele that influences its bearer to spare its siblings' lives should increase in frequency because the siblings are likely to carry other copies of that same allele, which are identical by descent. (This is the concept of kin selection, to which we will soon return.) There are alternative hypotheses, however, of which the most likely is that cannibals are at risk of contracting infectious diseases, especially if they share with their relatives a genetic susceptibility to certain pathogens—in which case it would be advantageous not to eat kin.

David Pfennig and colleagues (1999) tested these and other hypotheses. They reasoned that if the disease hypothesis were correct, cannibals ought to avoid eating diseased larvae, and that nonkin prey would be less likely to transmit disease to a cannibal than would kin prey. But when they offered cannibals a choice between diseased and healthy prey, or exposed them to diseased kin versus diseased nonkin, neither of those predictions was

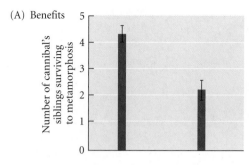

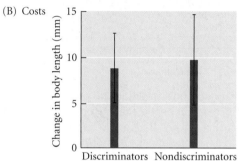

FIGURE 11.13 Genetic benefit and cost of kin discrimination by cannibalistic tiger salamanders (*Ambystoma tigrinum*). (A) Discriminators benefit by sparing their siblings' lives, since siblings share copies of their genes. (B) At least as indicated by the rate of their growth in length, discriminators suffer no reduction in fitness, or cost. (After Pfennig et al. 1999.)

confirmed. However, when Pfennig et al. presented several related and unrelated prey to discriminating versus nondiscriminating cannibals, they found that discriminators suffered no reduction in growth rate (i.e., no evident cost) by sparing their siblings' lives, whereas doing so resulted in a high genetic benefit: survival of copies of the cannibal's own genes (**Figure 11.13**). They concluded that kin selection explains the discriminatory behavior. The concept of natural selection invoked here is that *a gene may change in frequency because of its effect on the survival of copies of itself*, even if these copies are carried by other individuals of the species.

Selfish genetic elements

In many species of animals and plants, there exist **"selfish" genetic elements**, which are transmitted at a higher rate than the rest of an individual's genome and are detrimental (or at least not advantageous) to the organism (Hurst and Werren 2001; Burt and Trivers 2006). Many of these elements exhibit **segregation distortion**, a form of **gametic selection**. Some cases of segregation distortion are caused by **meiotic drive**, which in the strict sense means that an allele is more likely to segregate into the oocyte than into a polar body.

An example of a selfish genetic element that exhibits segregation distortion is the *t* locus of the house mouse (*Mus musculus*). In a male heterozygous for a *t* allele and for the normal allele *T*, more than 90 percent of the sperm carry the *t* allele because it kills sperm that carry the normal allele. In the homozygous condition, certain *t* alleles cause death of the embryo, and others cause males to be sterile. Despite these disadvantages to the individual, segregation distortion is so great that the disadvantageous *t* alleles can reach a high frequency in the population. Another selfish element is a small chromosome called *psr* (which stands for "paternal sex ratio") in the parasitic wasp *Nasonia vitripennis*. It is transmitted mostly through sperm, and not through eggs. When an egg is fertilized by a sperm containing this genetic element, it causes the destruction of all the other paternal chromosomes, leaving only the maternal set. In *Nasonia*, as in all Hymenoptera, diploid eggs become females and haploid eggs become males. The *psr* element thus converts female eggs into male eggs, thereby ensuring its own future propagation through sperm, even though this could possibly so skew the sex ratio of a population as to threaten its survival.

Selfish genetic elements forcefully illustrate the nature of natural selection: it is nothing more than differential reproductive success (of genes in this case), which need not result in adaptation or improvement in any sense. These elements also exemplify different levels of selection: in these cases, genic selection acts in opposition to individual selection. Natural selection among genes may not only be harmful to individual organisms, but might also cause the extinction of populations or species.

Levels of Selection

The last three examples introduced the idea of different levels of selection, corresponding to levels of biological organization. The philosopher of science Samir Okasha (2006) refers to higher-level units as COLLECTIVES and the lower-level units they include as PARTICLES. Thus collective/particle pairs might be clade/species, species/population, population/individual organism, or organismal genotype/gene. If the particles (e.g., genotypes) in a collective (e.g., population) vary in some character, then collectives will differ in their mean (average) for that character. Sometimes this is a straightforward relationship: the frequency distribution of body sizes among individuals determines the mean body size in a population, which is therefore an AGGREGATE property of its members. But there may also be features of a population (e.g., its abundance or geographic distribution) that cannot be measured on an individual, even if they are the consequences of individual organisms' properties. Such features are called EMERGENT or RELATIONAL characteristics (Damuth and

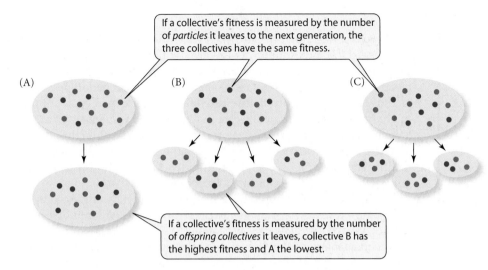

If a collective's fitness is measured by the number of *particles* it leaves to the next generation, the three collectives have the same fitness.

If a collective's fitness is measured by the number of *offspring collectives* it leaves, collective B has the highest fitness and A the lowest.

FIGURE 11.14 Concepts of the fitness of particles and of "collectives." The small red and blue circles represent individual particles, such as individual asexual organisms of two different genotypes. Each of the blue ovals represents a collective of individuals. The fitness (per capita rate of increase) of blue particles is greater than the fitness of red particles in collective A, lower in collective B, and equal in collective C. Collectives B and C both produce "offspring collectives" by colonization, but B has more offspring collectives than C, and has higher fitness. (After Okasha 2006.)

Heisler 1988). Both aggregate and emergent characteristics can affect the rates at which populations become extinct or give rise to new populations. Note, then, that we can distinguish two measures of the fitness of a collective: the mean reproductive success of its constituent members, or the rate at which it produces "offspring collectives" (**Figure 11.14**).

Conceptualizing the level at which selection acts is a philosophically complex issue on which a great deal has been written. If, for instance, alleles *A* and *a* determine dark versus pale color in an insect, and populations of dark insects have a lower extinction rate than populations of pale insects, does this represent selection at the level of populations, of individual insects, or of genes? Okasha suggests that we can take a "gene's-eye view," in which allele *A* increases as a consequence of both the survival of populations and the reproductive success of individuals. This "view," though, is not the same as genic selection, which requires that the cause of gene frequency change operate at the level of the gene, not at the level of the individual genotype (such as greater susceptibility of pale insects to predators) or of the population. In contrast, a selfish "outlaw gene," such as the *t* allele in mice, increases because of the activity of the gene itself, not of the collective (the genotype of the mouse) of which the gene is a part, and so is subject to genic selection.

Evolutionary biologists have extensively discussed selection at the level of gene, genotype, population, and species (Sober 1984; Okasha 2006). Selection can also occur at other levels, such as among cell lineages within a multicellular organism, which is the basis of cancer. Evolutionary biologists have studied this topic as well (Pepper et al. 2009).

Selection of organisms and groups

By "natural selection," both Darwin and contemporary evolutionary biologists usually mean consistent differences in fitness among phenotypically and genetically different organisms within populations. However, people (including biologists) have been known to express statements to the effect that oysters have a high reproductive rate "to ensure the survival of the species," or that antelopes with sharp horns refrain from physical combat because combat would lead to the species' extinction. These statements betray a misunderstanding of natural selection as it is usually conceived. If traits evolve by individual selection—by the replacement of less fit by more fit individuals, generation by generation—then the possibility of future extinction cannot possibly affect the course of evolution. Moreover, an **altruistic trait**—a feature that reduces the fitness of an individual that bears it for the benefit of the population or species—cannot evolve by individual selection. An altruistic genotype amid other genotypes that were not altruistic would necessarily decline in frequency, simply because it would leave fewer offspring per capita than the others. Conversely, if a population were to consist of altruistic genotypes, a selfish mutant—a "cheater"—would increase to fixation, even if a population of such selfish organisms had a higher risk of extinction (**Figure 11.15**).

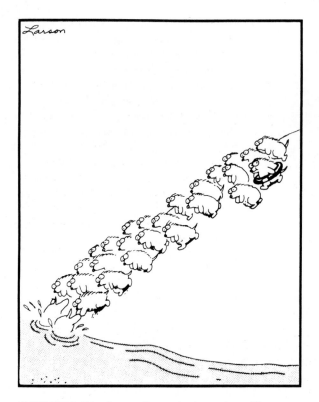

FIGURE 11.15 A popular myth about the self-sacrificial behavior of lemmings holds that they rush en masse into the sea to prevent overpopulation. Cartoonist Gary Larson, in *The Far Side*, illustrates the "cheater" principle and shows why such altruistic behavior would not be expected to evolve. (Reprinted with permission of Chronicle Features, San Francisco.)

There is a way, however, in which traits that benefit the population at a cost to the individual might evolve, which is by group selection: differential production or survival of groups that differ in genetic composition. For instance, populations made up of selfish genotypes, such as those with high reproductive rates that exhaust their food supply, might have a higher extinction rate than populations made up of altruistic genotypes that have lower reproductive rates. If so, then the species as a whole might evolve altruism through the greater survival of groups of altruistic individuals, even though individual selection within each group would act in the opposite direction (**Figure 11.16A**).

The hypothesis of group selection, most prominently advanced by V. C. Wynne-Edwards (1962), was criticized by George Williams (1966) in his influential book *Adaptation and Natural Selection*. Williams argued that supposed adaptations that benefit the population or species, rather than the individual, do not exist: either the feature in question is not an adaptation at all, or it can be plausibly explained by benefit to the individual or the individual's genes. For example, females of many species lay fewer eggs when population densities are high, but not to ensure a sufficient food supply for the good of the species. At high densities, when food is scarce, a female simply cannot form as many eggs, so her reduced fecundity may be a physiological necessity, not an adaptation. Moreover, an individual female may indeed be more fit if she forms fewer eggs in these circumstances and allocates the energy to surviving until food becomes more abundant, when she may reproduce more prolifically.

Williams based his opposition to group selection on a simple argument. Individual organisms are much more numerous than the populations into which they are aggregated, and they TURN OVER—are born and die—much more rapidly than populations, which are born (formed by colonization) and die (become extinct) at relatively low rates. Selection at either level requires differences—among individuals or among populations—in rates of birth or death. Thus the rate of replacement of less fit by more fit individuals is potentially much greater than the rate of replacement of less fit by more fit populations, so individual selection will generally prevail over group selection (**Figure 11.16B**). Among evolutionary biologists, the majority view is that *few characteristics have evolved because they benefit the population or species*. We will consider an alternative position, that group selection is important in evolution, in Chapter 16.

If adaptations that benefit the population are so rare, how do we explain worker ants that labor for the colony and do not reproduce, or birds that emit a warning cry when they see a predator approaching the flock? Among the possible explanations is one posited by William Hamilton (1964): such seemingly altruistic behaviors have evolved by **kin selection**, which is best understood from the "viewpoint" of a gene (see Chapter 16). An allele for altruistic behavior can increase in frequency in a population if the beneficiaries of the behavior are usually related to the individual performing it (see Figure 11.13). Since the altruist's relatives are more likely to carry copies of the altruistic allele than are members of the population at large, when the altruist enhances the fitness of its relatives, even at some cost to its own fitness, it can increase the frequency of the allele. We may therefore define kin selection as a form of selection in which alleles differ in fitness by influencing the effect of their bearers on the reproductive success of individuals (kin) who carry the same allele by common descent.

Species selection

Selection among groups of organisms is called **species selection** when the groups involved are species and there is a correlation between some characteristic and the rate of speciation or extinction (Gould 2002; Jablonski 2008b; Rabosky and McCune 2010). The consequence

(A)

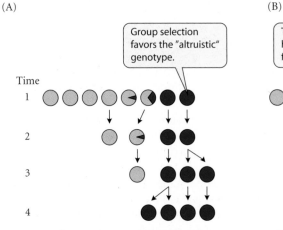

Group selection favors the "altruistic" genotype.

(B)

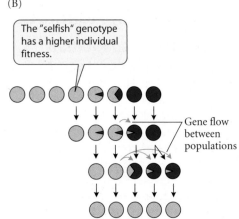

The "selfish" genotype has a higher individual fitness.

Gene flow between populations

Wynne-Edwards: Altruistic behavior will evolve because group selection favors it (i.e., more "selfish" populations go extinct.)

Williams: Within-population selection favors the "selfish" allele and increases it more rapidly than whole-population selection can act, so the "selfish" allele will become fixed.

FIGURE 11.16 Conflict between group and individual selection. Each circle represents a population of a species, traced through four time intervals. Some new populations are founded by colonists from established populations, and some populations become extinct. The proportions of blue and pink in each circle represent the proportions of an "altruistic" and a "self-ish" genotype in the population. Individuals with the selfish genotype are assumed to have higher fitness. Lateral arrows indicate gene flow between populations. (A) An altruistic trait may evolve by group selection if the rate of extinction of populations of the selfish genotype is very high. (B) Williams's argument: Because individual selection operates so much more rapidly than group selection, the selfish genotype increases rapidly within populations and may spread by gene flow into populations of altruists. Thus the selfish genotype becomes fixed, even if it increases the rate of population extinction.

of species selection is that the proportion of species that have one character state rather than another changes over time (**Figure 11.17**). Species selection does not shape adaptations of organisms, but it does affect the DISPARITY—the diversity of biological characteristics—of the world's organisms.

Some authors use "species selection" to refer to both "aggregate" and "emergent" collective features, whereas others restrict it to the few "emergent" characteristics of a species. The geological duration of species of Cretaceous molluscs is correlated with the size of their geographic range, which may be considered an emergent property; moreover, range size is similar among related species and therefore is "heritable" at the species level (Jablonski 1987; Jablonski and Hunt 2006). The combination of species selection and species-level heritability might have resulted in a long-term increase in the average range size of species, but the K-T mass extinction event cut short any possible trend, and range size did not affect the chances of species' survival at that time.

Many aggregate traits appear to affect rates of diversification and thus exemplify species selection. For example, most lineages of insects that evolved the habit of feeding on living plants are more diverse than their nonherbivorous sister groups (see Figure 7.22). Their high rate of diversification may explain why they account for about half of known insect species and about one-fourth of the described species of animals and plants. Another likely example of the effects of species selection is the prevalence of sexual species compared with closely related asexual forms. Many groups of plants and animals have given rise to asexually reproducing lineages, but, with some interesting exceptions, asexual lineages tend to be young, as indicated by their close genetic similarity to sexual forms. This observation implies that asexual forms have a higher rate of extinction than sexual populations, since most asexual forms that arose long ago have not persisted (Normark et al. 2003; see Figure 15.2).

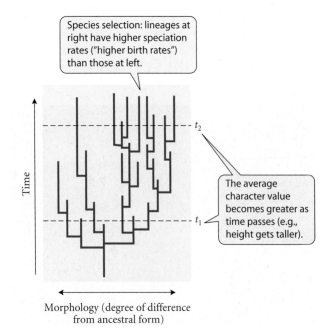

Species selection: lineages at right have higher speciation rates ("higher birth rates") than those at left.

The average character value becomes greater as time passes (e.g., height gets taller).

Time

Morphology (degree of difference from ancestral form)

FIGURE 11.17 Species selection caused by a correlation between speciation rate and a morphological character, such as body size (x-axis). The higher speciation rate of lineages with large body sizes is analogous to a higher birth rate of individual organisms in individual selection. The character value, averaged across species, is greater at time t_2 (upper dashed line) than at time t_1 (lower dashed line). (After Gould 1982.)

The Nature of Adaptations

Definitions of adaptation

All biologists agree that an adaptive trait is one that enhances fitness compared with at least some alternative traits. However, some authors include a historical perspective in their definition of adaptations and others do not.

An ahistorical definition was provided by Kern Reeve and Paul Sherman (1993): "An adaptation is a phenotypic variant that results in the highest fitness among a specified set of variants in a given environment." This definition refers only to the current effects of the trait on fitness compared with those of other variants. At the other extreme, Paul Harvey and Mark Pagel (1991) hold that "for a character to be regarded as an adaptation, it must be a derived character that evolved in response to a specific selective agent." This history-based definition requires that we compare a character's effects on fitness with those of a specific variant; namely, the ancestral character state from which it evolved. Phylogenetic or paleontological data may provide information about the ancestral state.

One reason for this emphasis on history is that a character state may be a simple consequence of phylogenetic history, rather than an adaptation. Darwin saw clearly that a feature might be beneficial, yet not have evolved for the function it serves today, or for any function at all: "The sutures in the skulls of young mammals have been advanced as a beautiful adaptation for aiding parturition [birth], and no doubt they facilitate, or may be indispensable for this act; but as sutures occur in the skulls of young birds and reptiles, which have only to escape from a broken egg, we may infer that this structure has arisen from the laws of growth, and has been taken advantage of in the parturition of the higher animals" (*The Origin of Species*, chapter 6). Whether or not we should postulate that a trait is an adaptation depends on such insights. For example, we know that pigmentation varies in many species of birds (see Figure 9.1A), so it makes sense to ask whether there is an adaptive reason for color differences among closely related species. But it is not sensible to ask whether it is adaptive for a goose to have four toes rather than five, because the ancestor of birds lost the fifth toe and it has never been regained in any bird since. Five toes are probably not an option for birds because of genetic developmental constraints. Thus, if we ask why a species has one feature rather than another, the answer may be adaptation, or it may be phylogenetic history.

A **preadaptation** is a feature that fortuitously serves a new function. For instance, parrots have strong, sharp beaks, used for feeding on fruits and seeds. When domesticated sheep were introduced into New Zealand, some were attacked by an indigenous parrot, the kea (*Nestor notabilis*), which pierced the skin and fed on the sheep's fat. The kea's beak was fortuitously suitable for a new function and may be viewed as a preadaptation for slicing skin.

FIGURE 11.18 Exaptation and adaptation. (A) The wing might be called an exaptation for underwater "flight" in members of the auk family, such as this puffin (*Fratercula arctica*). (B) The modifications of the wing for efficient underwater locomotion in penguins (these are Humboldt penguins, *Spheniscus humboldti*) may be considered adaptations.

(A)

(B)

Preadaptations that have been co-opted to serve a new function have been termed **exaptations** (Gould and Vrba 1982). For example, the wings of alcids (birds in the auk family) may be considered exaptations for swimming: these birds "fly" under water as well as in air (**Figure 11.18A**). An exaptation may be further modified by selection so that the modifications are adaptations for the feature's new function: the wings of penguins have been modified into flippers that enhance swimming but cannot support flight in air (**Figure 11.18B**). Some proteins have been "exapted" to serve new functions, and some play dual roles (Piatigorsky 2007). For instance, the diverse crystallin proteins of animal eye lenses have been co-opted from several phylogenetically widespread proteins, including stress proteins that stabilize cellular function and enzymes such as lactate dehydrogenase (**Figure 11.19**). In some cases, exactly the same protein serves both its ancestral and new roles, such as the τ-crystallin of reptiles and birds, which doubles as α-enolase; in other cases, the ancestral gene was duplicated, and the crystallin encoded by one of the duplicates has undergone some amino acid substitutions. Exaptation is a very common early stage in the evolution of new adaptations.

Recognizing adaptations

Not all traits are adaptations. There are several other possible explanations of organisms' characteristics. First, a trait may be a necessary consequence of physics or chemistry. Hemoglobin gives blood a red color, but there is no reason to think that redness is an adaptation; it is a by-product of the structure of hemoglobin. Second, the trait may have evolved by random genetic drift rather than by natural selection. Third, the feature may have evolved not because it conferred an adaptive advantage, but because it was correlated with another feature that did. (Genetic hitchhiking, as exemplified in the bacterial experiment by Atwood et al. described earlier in this chapter, is one cause of such correlation; pleiotropy—the phenotypic effect of a gene on multiple characters—is another.) Fourth, as we saw in the previous section, a character state may be a consequence of phylogenetic history. It may be an ancestral character state, as Darwin recognized in his analysis of skull sutures.

Because there are so many alternative hypotheses, many authors believe that we should not assume that a feature is an adaptation unless the evidence favors this interpretation

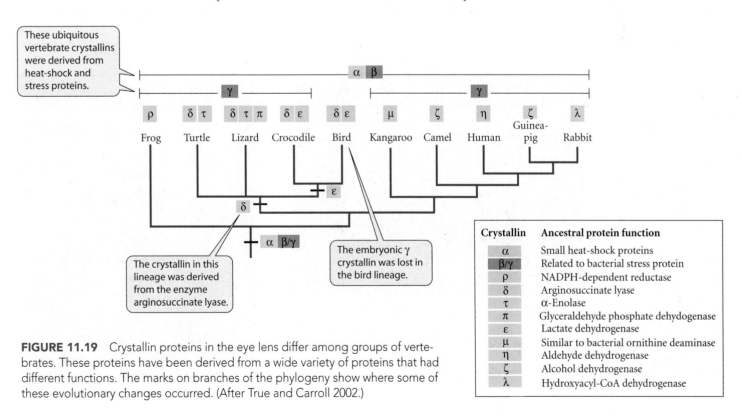

FIGURE 11.19 Crystallin proteins in the eye lens differ among groups of vertebrates. These proteins have been derived from a wide variety of proteins that had different functions. The marks on branches of the phylogeny show where some of these evolutionary changes occurred. (After True and Carroll 2002.)

Crystallin	Ancestral protein function
α	Small heat-shock proteins
β/γ	Related to bacterial stress protein
ρ	NADPH-dependent reductase
δ	Arginosuccinate lyase
τ	α-Enolase
π	Glyceraldehyde phosphate dehydrogenase
ε	Lactate dehydrogenase
μ	Similar to bacterial ornithine deaminase
η	Aldehyde dehydrogenase
ζ	Alcohol dehydrogenase
λ	Hydroxyacyl-CoA dehydrogenase

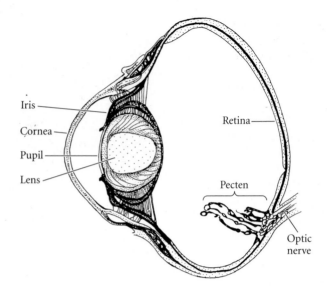

FIGURE 11.20 Sagittal section of a bird's eye, showing the pecten. Among the 30 or so hypotheses that have been proposed for the function of the pecten, the most likely is that it supplies oxygen to the retina. (After Gill 1995.)

(Williams 1966). This is not to deny that a great many of an organism's features, perhaps the majority, are adaptations. Several methods are used to infer that a feature is an adaptation for some particular function. We shall note these methods only briefly and incompletely at this point, exemplifying them more extensively in later chapters. The approaches described here apply to phenotypic characters; in the next chapter, we will describe how selection can be inferred from DNA sequence data.

COMPLEXITY   Even if we cannot immediately guess the function of a feature, *we often suspect it has an adaptive function if it is complex*, for complexity cannot evolve except by natural selection. For example, a peculiar, highly vascularized structure called a pecten projects in front of the retina in the eyes of birds (**Figure 11.20**). Only recently has evidence been developed to show that the pecten supplies oxygen to the retina, but it has always been assumed to play some important functional role because of its complexity and because it is ubiquitous among bird species.

DESIGN   The function of a character is often inferred from its *correspondence with the design* an engineer might use to accomplish some task, or with the *predictions of a model* about its function. For instance, many plants that grow in hot environments have leaves that are finely divided into leaflets, or which tear along fracture lines (**Figure 11.21**). These features conform to a model in which the thin, hot "boundary layer" of air at the surface of a leaf is more readily dissipated by wind passing over a small than a large surface, so that a divided leaf's temperature is more effectively reduced. The fields of functional morphology and ecological physiology are concerned with analyses of this kind.

EXPERIMENTS   Experiments may show that a feature enhances survival or reproduction, or enhances performance (e.g., locomotion or defense) in a way that is likely to increase fitness, relative to individuals with other features. For example, several floral characters have

(A)

(B)

FIGURE 11.21 Functional morphological analyses have shown that small surfaces shed the hot "boundary layer" of air that forms around them more readily than large surfaces. Many tropical and desert-dwelling plants have large leaves that are broken up into leaflets, as in *Acacia karroo* (A), or split into small sections, as in the banana (B). The form of these leaves is therefore believed to be an adaptation for reducing leaf temperature.

(A)

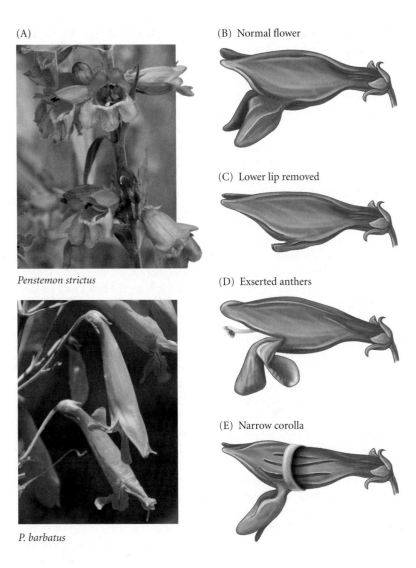

Penstemon strictus

P. barbatus

(B) Normal flower

(C) Lower lip removed

(D) Exserted anthers

(E) Narrow corolla

FIGURE 11.22 Experimental test of a hypothesis of adaptation. (A) Bee-pollinated *Penstemon strictus* (top) and hummingbird-pollinated *P. barbatus* (bottom). (B–E) Experimental modifications of flowers of a bee-pollinated species (*Penstemon strictus*) to mimic features of hummingbird-pollinated species of *Penstemon*. (B) The normal flower of *P. strictus*. Modifications included (C) removal of the lower lip "landing platform," (D) reattaching stamens so that the anthers project from the flower, and (E) constriction to form a narrower corolla tube. (B–E after Castellanos et al. 2004.)

evolved convergently in the many plant lineages that have shifted from insect pollination to bird pollination (**Figure 11.22A**). Maria Castellanos and coworkers (2004) tested the hypothesis that the advantage of some of these features is that they facilitate bird pollination, and that the advantage of others is that they discourage bees, which comb much of the pollen into a mass that they feed to their larvae and so are less effective pollinators. The researchers surgically altered several features of flowers on a bee-pollinated plant to resemble those of related hummingbird-pollinated species (**Figure 11.22B–E**). They then measured pollen transfer from the altered flowers by bumblebees and hummingbirds. The researchers concluded that the lower "lip" typical of bee-pollinated flowers, which bees use as a landing platform (Figure 11.22B), has been reduced or lost in some bird-pollinated species because its absence discourages bees (Figure 11.22C). The exserted anthers of bird-pollinated plants also seem to be an "anti-bee" adaptation (Figure 11.22D), and the narrowly constricted corolla tube (Figure 11.22E) is both "pro-bird" and "anti-bee": it forces hummingbirds to remove more pollen, but prevents bees from easily obtaining nectar.

THE COMPARATIVE METHOD A powerful means of inferring the adaptive significance of a feature is the **comparative method**, which consists of *comparing sets of species to pose or test hypotheses on adaptation and other evolutionary phenomena*. This method takes advantage of "natural evolutionary experiments" provided by convergent evolution. If a feature evolves independently in many lineages because of a similar selection pressure, we can often infer the function of that feature by determining the ecological or other selective factor with which it is correlated.

For instance, a long, slender beak has evolved in at least six lineages of birds that feed on nectar (see Figure 3.8). Human digestion of milk provides another example. Most adult humans are sickened by lactose, the principal carbohydrate in milk, and cannot digest it because once humans mature, their production of the digestive enzyme lactase-phloridzin hydrolase (lactase, for short) is regulated at low levels. However, people in several populations around the world have persistently high lactase levels in adulthood, especially in northern Europe and in certain populations in Africa. Milk and milk products have traditionally been an important part of the diet of all these populations. Adult lactose digestion has evolved at least three times in different populations (Holden and Mace 1997; Tishkoff et al. 2007), in which it is based on three different mutations, marked by different SNPs, in a regulatory DNA sequence upstream of the gene for lactase. DNA sequence evidence suggests that these evolutionary changes occurred after dairying was adopted. That a similar feature arose independently several times in a similar ecological context—a diet of dairy products—strongly suggests that it is an adaptation to that context.

Biologists often predict such correlations by postulating, perhaps on the basis of a model, the adaptive features we would expect to evolve repeatedly in response to a given selective factor. For example, in species in which a female mates with multiple males, the several males' sperm compete to fertilize eggs. Males that produce more abundant sperm should therefore have a reproductive advantage. In primates, the quantity of sperm produced is correlated with the size of the testes, so large testes should be expected to provide a greater reproductive advantage in polygamous than in monogamous species. Paul Harvey and collaborators compiled data from prior publications on the mating behavior and testes size of various primates. They confirmed that, as predicted, the weight of the testes, relative to body weight, is significantly higher among polygamous than monogamous taxa (**Figure 11.23**).

This example raises several important points. First, although all the data needed to test this hypothesis already existed, the relationship between the two variables was not known until Harvey and collaborators compiled the data, because no one had had any reason to do so until an adaptive hypothesis had been formulated. Hypotheses about adaptation can be fruitful because they suggest investigations that would not otherwise occur to us.

Second, because the consistent relationship between testes size and mating system was not known a priori, the hypothesis generated a prediction. The predictions made by evolutionary theory, like those in many other scientific disciplines, are usually predictions of what we will find when we collect data. Prediction in evolutionary theory does *not* usually mean that we predict the future course of evolution of a species. Predictions of what we will find, deduced from hypotheses, constitute the **hypothetico-deductive method**, of which Darwin was one of the first effective exponents (Ghiselin 1969; Ruse 1979).

Third, the investigators found support for their hypothesis by demonstrating that the average testes sizes of polygamous and monogamous taxa show a STATISTICALLY SIGNIFICANT difference. To do this, it is necessary to have a sufficient number of data points—that is, a large enough sample size. Furthermore, for a statistical test to be valid, each data point must be INDEPENDENT of all others. Harvey et al. could have had a larger sample size if they had included, say, 30 species of marmosets and tamarins (Callithricidae) as separate data points, rather than using only one. All the members of this family are monogamous. That observation suggests that monogamy evolved only once and has been retained by all callithricids for unknown reasons: perhaps monogamy is advantageous for all the species, or perhaps an internal constraint of some kind prevents the evolution of polygamy even if it would be adaptive. Because our hypothesis is that testes size *evolves* in response to the mating system, we must suspect that the different species

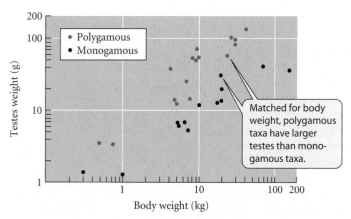

FIGURE 11.23 The relationship between the weight of the testes and body weight among polygamous and monogamous primate taxa. (After Harvey and Pagel 1991.)

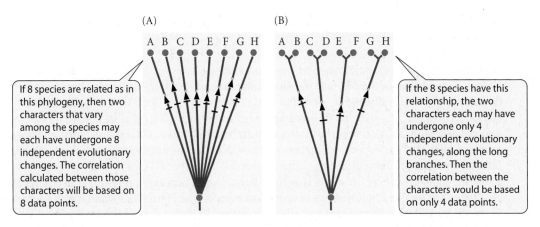

If 8 species are related as in this phylogeny, then two characters that vary among the species may each have undergone 8 independent evolutionary changes. The correlation calculated between those characters will be based on 8 data points.

If the 8 species have this relationship, the two characters each may have undergone only 4 independent evolutionary changes, along the long branches. Then the correlation between the characters would be based on only 4 data points.

FIGURE 11.24 The problem of phylogenetic correlation in employing the comparative method. Suppose we test a hypothesis about adaptation by calculating the correlation between two characters, such as testes size (arrowheads) and mating system (ticks), in eight species (A–H). (A) If the species are related as shown in this phylogenetic tree, the character states in each species have evolved independently, as shown by ticks and arrowheads, and we have a sample of eight. (B) If the species are related as shown in this phylogenetic tree, the states of both characters may be similar in each pair of closely related species because of their common ancestry, rather than independent adaptive evolution. Some authors maintain that the two species in each pair are not independent tests of the hypothesis; we would have a sample of four in this case. (After Felsenstein 1985.)

of callithricids represent only one evolutionary change, and so provide only one data point (**Figure 11.24**). Therefore, if we use convergent evolution (i.e., the comparative method) to test hypotheses of adaptation, we should count the *number of independent convergent evolutionary events* by which a character state evolved in the presence of one selective factor versus another (Felsenstein 1985; Harvey and Pagel 1991). Consequently, methods that employ PHYLOGENETICALLY INDEPENDENT CONTRASTS (Felsenstein 1985) are usually used in comparative studies of adaptation.

Adaptive Evolution Observed

The course of adaptive evolution has been documented with historical data on many morphological, physiological, and behavioral traits, especially in species that have been introduced into new regions or subjected to human alterations of their environment (Hendry and Kinnison 1999; Palumbi 2001; Smith and Bernatchez 2008). For instance, several species of insects, such as the soapberry bug (*Jadera haematoloma*; Carroll and Boyd 1992; Carroll et al. 1997), have adapted rapidly to new food plants. The soapberry bug feeds on seeds of plants in the soapberry family (Sapindaceae) by piercing the enveloping seed pod with its slender beak (**Figure 11.25**). In Texas, its natural host plant—the soapberry tree—has a small pod. Its major contemporary host plant, however, is the Asian golden rain tree, which has become common only within the last 20 to 50 years. The golden rain tree has a larger pod than the soapberry tree, and the beak length of local populations of the bug that feed on it has evolved to be 8 percent longer. In Florida, conversely, the beak has become 25 percent shorter in soapberry bug populations that feed on the introduced flat-podded rain tree,

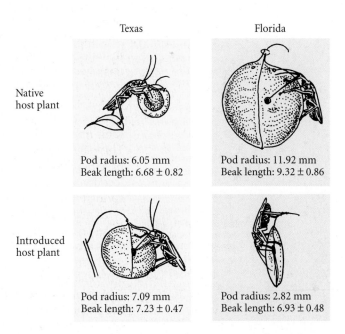

FIGURE 11.25 Soapberry bugs (*Jadera haematoloma*) and their native and introduced host plants in Texas and Florida, drawn to scale. The bug's beak is the needlelike organ projecting from the head at right angles to the body. The average pod radius of each host species and the average (with standard deviation) beak length of associated *Jadera* populations are given. Beak length has evolved rapidly as an adaptation to the new host plants. (After Carroll and Boyd 1992.)

also from Asia, which has a flatter, smaller seed pod than the native soapberry trees in that region. Adaptation to an altered environment—to new food plants—has resulted in large, rapid changes in morphology.

In many organisms, resistance to novel poisons evolves very rapidly (Palumbi 2001). Within 3 years after penicillin was put to widespread use, penicillin-resistant bacteria were described. Each new antibiotic, developed in response to the declining efficacy of earlier ones, has evoked the evolution of bacterial resistance in turn. Methicillin-resistant *Staphylococcus aureus* (MRSA), dangerously frequent in hospitals, has evolved resistance not only to methicillin and the other beta-lactam antibiotics used against most bacterial infections, but also to tetracyclines, sulfa drugs, and others; it is now treated mostly with vancomycin, but some vancomycin-resistant strains have been found. Similarly, resistance to chemical pesticides has evolved in hundreds of species of insects (see Figure 12.9), and many species of weeds have evolved resistance to herbicides within 10 to 20 years of field exposure. Copper, zinc, and other heavy metals are toxic to plants, but in several species of grasses and other plants, metal-tolerant populations have evolved where soils have been contaminated by mine works that range from more than 700 to less than 100 years old. In some cases, tolerance has evolved within decades on a microgeographic scale, such as in the vicinity of a zinc fence. Tolerance is based on a variable number of genes, depending on the species and population. When tolerant and nontolerant genotypes of a species are grown in competition with other plant species in the absence of the metal, the relative fitness of the tolerant genotypes is often much lower than that of the nontolerant genotypes, implying a cost of adaptation (Antonovics et al. 1971; Macnair 1981).

Commercial overexploitation has severely depleted populations of many species of fish and has resulted in evolutionary changes as well (Kuparinen and Merilä 2007). In many species, there has been a trend toward earlier sexual maturation at a smaller size, as would be predicted if larger age classes are more subject to predation (see Chapter 14). In some species, such as Atlantic cod (*Gadus morhua*), these changes clearly have a genetic basis (**Figure 11.26A**). Similarly, trophy hunting for bighorn sheep (*Ovis canadensis*) with the largest horns has resulted in the evolution of smaller horns (**Figure 11.26B**). In both instances, the very quality that adds value to the resource has been diminished by the response to selection.

Our final examples illustrate likely adaptation to the ongoing climate change, caused by human production of CO_2 and other greenhouse gases. Since the 1950s, German populations of the blackcap (*Sylvia atricapilla*), a European songbird, have migrated not only to their traditional winter range in the western Mediterranean region, but also to Britain. This new direction of migration, to the northwest rather than the southwest, is genetically based and has evolved in a few decades (Berthold et al. 1992). The change in wintering location is probably advantageous because of warming winters in Britain and an earlier spring return to Germany from Britain than from the Mediterranean.

In many insects, the cue for entering diapause, a state of low metabolic activity that is necessary for surviving the winter, is a critical photoperiod (day length). Northern populations are typically genetically programmed to enter diapause at a longer day length than southern populations because winter arrives at northern latitudes sooner, when days are still relatively long. William Bradshaw and Christina Holzapfel (2001) determined the critical photoperiod in a wide range of populations, from southern Canada to Florida, of the pitcher-plant mosquito (*Wyeomyia smithii*), sampling the same populations four times between 1972 and 1996. They found that the critical photoperiod has become shorter during this time: the insects are now programmed to enter diapause later in autumn (**Figure 11.27**). The change has been greatest in the most northerly populations, as expected because the increase in temperature has been greater at higher latitudes. The speed of evolution has been amazing, having taken as little as 5 years.

These and many other examples show that species can often evolve at rates far greater than the average evolutionary rates documented in the fossil record—although over a much shorter span of time. The rapidity of these changes is a consequence of the presence of genetic variation in many characteristics (described in Chapter 9) and the often

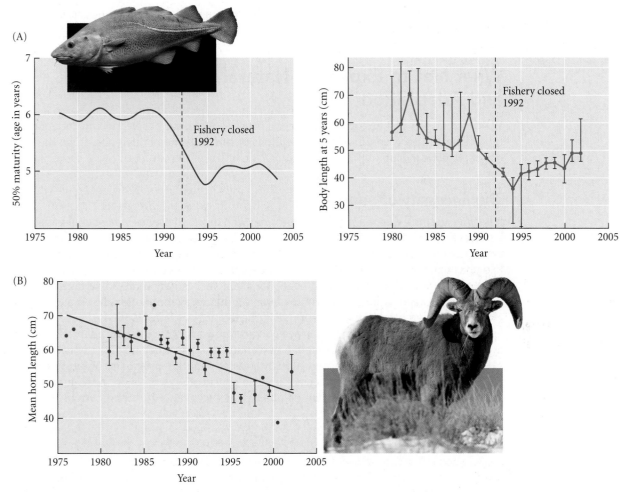

FIGURE 11.26 Evolutionary changes caused by human harvesting. (A) The age at which 50 percent of Atlantic cod reached maturity (left graph) declined until 1992, when the fishery was closed because of overfishing. Body length at 5 years of age (right graph) also declined until 1992. (B) Mean horn length of 4-year-old bighorn sheep rams declined because of selection imposed by hunting. (A after Olsen et al. 2004; B after Coltman et al. 2003.)

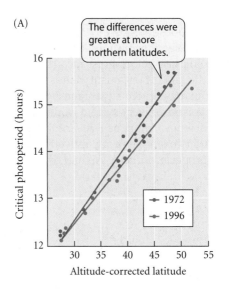
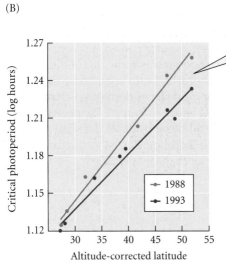

FIGURE 11.27 In North America, the critical photoperiod for entering diapause, in relation to latitude, in the pitcher-plant mosquito has decreased between 1972 and 1996, especially in more northern populations. (A) Comparison of 1972 and 1996 data. (B) Comparison of 1988 and 1993 data. The comparisons between years are shown in two separate graphs because somewhat different methods of measuring critical day length were used in 1972 and 1996 and in 1988 and 1993. (After Bradshaw and Holzapfel 2001.)

stringent natural selection that populations experience in altered environments, especially those altered by human activities (see Chapters 12 and 13).

What Not to Expect of Natural Selection and Adaptation

We conclude this discussion of the general properties of natural selection and adaptation by considering a few common misconceptions of, and misguided inferences from, the theory of adaptive evolution.

The necessity of adaptation

It is naïve to think that if a species' environment changes, the species must either adapt or become extinct. Not all environmental changes reduce population size. Nonetheless, an environmental change that does not threaten extinction may set up selection for change in some characteristics, because they enhance survival or reproduction, *compared to* other character states. White fur in polar bears, for example, is surely more advantageous than brown fur because white bears may sneak up on their prey more often, but that does not imply that brown bears could not survive in the Arctic (Williams 1966). Just as a changed environment need not set in motion selection for new adaptations, new adaptations may evolve in an unchanging environment if new mutations arise that are superior to any pre-existing genetic variations. We have already stressed that the probability of future extinction of a population or species does not in itself constitute selection on individual organisms, and so cannot cause the evolution of adaptations.

Perfection

Darwin noted that "natural selection will not produce absolute perfection, nor do we always meet, as far as we can judge, with this high standard in nature" (*The Origin of Species*, chapter 6). Selection can fix only those genetic variants with a higher fitness than other genetic variants in a particular population at a particular time. It cannot fix the best of all conceivable variants if they do not arise, or have not yet arisen, and the best possible variants often fall short of perfection because of various constraints. For example, with a fixed amount of available energy or nutrients, a plant might evolve higher seed numbers, but only by reducing the size of its seeds or some other part of its structure (see Chapter 14).

Progress

Whether or not evolution is "progressive" is a complicated question (Ruse 1996). The word "progress" has the connotation of a goal, and as we have seen, evolution does not have goals. But even if we strip away this connotation and hold only that progress means "betterment," the possible criteria for "better" depend on the kind of organism. Better learning ability or greater brain complexity has no more evident adaptive advantage for most animals—for example, rattlesnakes—than an effective poison delivery system would have for humans. Measurements of "improvement" or "efficiency" must be relevant to each species' special niche or tasks. There are, of course, many examples of adaptive trends, each of which might be viewed as progressive within its special context. We will consider this topic in depth in Chapter 22.

Harmony and the balance of nature

As we have seen, selection at the level of genes and individual organisms is inherently "selfish": the gene or genotype with the highest rate of increase increases at the expense of others. The variety of selfish behaviors that organisms inflict on conspecific individuals, ranging from territory defense to parasitism and infanticide, is truly stunning. Indeed, cooperation among organisms requires special explanations. For example, a parent that forages for food for her offspring, at the risk of exposing herself to predators, is cooperative, but for an obvious reason: her own genes, including those coding for this parental behavior, are carried by her offspring, and the genes of individuals that do not forage for their offspring are less likely to survive than the genes of individuals that do. The evolution of this behavior is an example of kin selection, an important basis for the evolution of cooperation within species (see Chapter 16).

Because the principle of kin selection cannot operate across species, "natural selection cannot possibly produce any modification in a species exclusively for the good of another species" (*The Origin of Species*, chapter 6). If a species exhibits behavior that benefits another species, either the behavior is profitable to the individuals performing it (as in bees that obtain food from the flowers they pollinate), or they have been duped or manipulated by the species that profits (as are insects that copulate with orchids). Most mutualistic interactions between species consist of reciprocal exploitation (see Chapter 19).

The equilibrium we may observe in ecological communities—the so-called balance of nature—likewise does not reflect any striving for harmony. We observe coexistence of predators and prey not because predators restrain themselves, but because prey species are well enough defended to persist, or because the abundance of predators is limited by some factor other than food supply. Nitrogen and mineral nutrients are rapidly and "efficiently" recycled within tropical rain forests not because ecosystems are selected for or strive for efficiency, but because under competition for sparse nutrients, microorganisms have evolved to decompose litter rapidly, while plants have evolved to capture the nutrients released by decomposition. Selection of individual organisms for their ability to capture nutrients has the *effect*, in aggregate, of producing a dynamic that we measure as ecosystem "efficiency." There is no scientific foundation for the notion that ecosystems evolve toward harmony and balance (Williams 1992a).

Morality and ethics

Natural selection is just a name for differences among organisms or genes in reproductive success. Therefore it cannot be described as moral or immoral, just or unjust, kind or cruel, any more than wind, erosion, or entropy can. Hence it cannot be used as a justification or model for human morality or ethics. Nevertheless, evolutionary theory has often been misused in just this way. Darwin expressed distress over an article "showing that I have proved 'might is right,' and therefore that Napoleon is right, and every cheating tradesman is also right." In the late nineteenth and early twentieth century, Social Darwinism, a philosophy promulgated by Herbert Spencer, considered natural selection to be a beneficent law of nature that would produce social progress as a result of untrammeled struggle among individuals, races, and nations. Evolutionary theory has likewise been used to justify eugenics and racism, most perniciously by the Nazis. But neither evolutionary theory nor any other field of science can speak of or find evidence of morality or immorality. These precepts do not exist in nonhuman nature, and science describes only what *is*, not what *ought to be*. The **naturalistic fallacy**, the supposition that what is "natural" is necessarily "good," has no legitimate philosophical foundation.

Summary

1. A feature is an adaptation for a particular function if it has evolved by natural selection for that function by enhancing the relative rate of increase—the fitness—of biological entities with that feature.

2. Natural selection is a consistent difference in fitness among phenotypically different biological entities. It is the antithesis of chance. Natural selection may occur at different levels, such as genes, individual organisms, populations, and species.

3. Selection at the level of genes or organisms is likely to be the most important because the numbers and turnover rates of these entities are greater than those of populations or species. Therefore most features are unlikely to have evolved by group selection, the one form of selection that could in theory promote the evolution of features that benefit the species even though they are disadvantageous to the individual organism.

4. Not all features are adaptations. Methods for identifying and elucidating adaptations include studies of function and design, experimental studies of the correspondence between fitness and variation within species, and correlations between the traits of species and environmental or other features (the comparative method). Phylogenetic information may be necessary for proper use of the comparative method.

5. Because many characteristics are genetically variable in natural populations, they may evolve rapidly if selection pressures change. Especially because humans drastically alter environments and move species into new environments, many historical examples of rapid adaptive evolution have been documented, often within much less than a century.

6. Natural selection does not necessarily produce anything that we can justly call evolutionary progress. It need not promote harmony or balance in nature, and, utterly lacking any moral content, it provides no foundation for morality or ethics in human behavior.

Terms and Concepts

adaptation
altruistic trait
comparative method
exaptation
fitness
function (vs. effect)

gametic selection
genic selection
group selection
 (= interdemic selection)
hitchhiking
hypothetico-deductive
 method

individual selection
kin selection
levels of selection
meiotic drive
 (= segregation
 distortion)
naturalistic fallacy

preadaptation
reproductive success
selfish genetic elements
sexual selection
species selection

Suggestions for Further Reading

Adaptation and Natural Selection, by G. C. Williams (Princeton University Press, Princeton, NJ, 1966), is a classic: a clear, insightful, and influential essay on the nature of individual and group selection. See also the same author's *Natural Selection: Domains, Levels, and Challenges* (Oxford University Press, New York, 1992).

Two books by R. Dawkins, *The Selfish Gene* (Oxford University Press, Oxford, 1989) and *The Blind Watchmaker* (W. W. Norton, New York, 1986), explore the nature of natural selection in depth, as well as treating many other topics in a vivid style for general audiences.

Levels of selection and related issues are treated in *The Nature of Selection: Evolutionary Theory in Philosophical Focus*, by E. Sober (MIT Press, Cambridge, MA, 1984); *Evolution and the Levels of Selection*, by S. Okasha (Oxford University Press, Oxford, 2006); and *Levels of Selection in Evolution*, edited by L. Keller (Princeton University Press, Princeton, NJ, 1999).

The Comparative Method in Evolutionary Biology, by P. H. Harvey and M. D. Pagel (Oxford University Press, Oxford, 1991), treats the use and phylogenetic foundations of the comparative method.

Experimental evolution, used extensively in elucidating evolutionary processes and in medical and technological applications, is reviewed in *Experimental Evolution*, edited by T. Garland, Jr., and M. R. Rose (University of California Press, Berkeley, 2009), and in "Experimental evolution," by T. J. Kawecki, et al. (*Trends Ecol. Evol.* 27: 547–560, 2012).

Problems and Discussion Topics

1. Discuss criteria or measurements by which you might conclude that a population is better adapted after a certain evolutionary change than before.

2. Consider the first copy of an allele for insecticide resistance that arises by mutation in a population of insects exposed to an insecticide. Is this mutation an adaptation? If, after some generations, we find that most of the population is resistant, is the resistance an adaptation? If we discover genetic variation for insecticide resistance in a population that has had no experience of insecticides, is that variation an adaptation? If an insect population is polymorphic for two alleles, each of which confers resistance against one of two pesticides that are alternately applied, is that variation an adaptation? Or is each of the two resistance traits an adaptation?

3. It is often proposed that a feature that is advantageous to individual organisms is the reason for the great number of species

in certain clades. For example, wings have been postulated to be a cause of the great diversity of winged insects compared with the few species of primitively wingless insects. How could an individually advantageous feature cause greater species diversity? How can one test a hypothesis that a certain feature has caused the great diversity of certain groups of organisms?

4. Provide an adaptive and a nonadaptive hypothesis for the evolutionary loss of useless organs, such as eyes in many cave-dwelling animals. How might these hypotheses be tested?

5. Could natural selection, at any level of organization, ever cause the extinction of a population or species?

6. If natural selection has no foresight, how can it explain features that seem to prepare organisms for future events? For example, deciduous trees at high latitudes drop their leaves before winter arrives, male birds establish territories before females arrive in the spring, and animals such as squirrels and jays store food as winter approaches.

7. List the possible criteria by which evolution by natural selection might be supposed to result in "progress," and search the biological literature for evidence bearing on one or more of these criteria.

The Genetic Theory of Natural Selection

Natural selection is the most important concept in the theory of evolutionary processes. It is surely the explanation for most of the characteristics of organisms that we find most interesting, ranging from the origin of DNA to the complexities of the human brain. Natural selection—differential reproductive success—is fundamentally a very simple concept. But its explanatory power is much greater if we take into account the many ways in which it can act and the ways in which its outcome is affected not only by the environment, but also by genetic factors such as recombination and the relationship between phenotype and genotype. When we take into account these complexities, we can begin to address a great variety of questions: Why are some characteristics more variable than others within species? How great a difference can we expect to see among different populations of a species, such as our own? Are different populations of a species likely to evolve the same adaptation to a particular environmental challenge? How do cooperative and selfish behaviors evolve? Why do some species reproduce sexually and others asexually? How can we explain the extraordinary display feathers of the peacock, the immense fecundity of mushrooms and oysters, the brevity of a mayfly's life, the pregnancy of the male sea horse, the abundance of transposable elements in our own genome?

Darwin fully realized that a truly complete theory of evolutionary change would require an understanding of the mechanism of inheritance. That understanding began to develop only in 1900, when Mendel's work was rediscovered. Modern evolutionary theory started

Human adaptation to a cultural environment The Maasai are one of several human populations that have evolved the capacity to digest lactose (milk sugar) as adults, and whose diet is largely based on milk. Analyses of the DNA sequence of the gene that confers this ability have revealed its evolutionary history.

to develop as the growing understanding of Mendelian genetics was synthesized with Darwin's theory of selection (as described in Chapter 1). The "genetical theory of natural selection" (as the pioneering population geneticist R. A. Fisher titled his seminal 1930 book) is the keystone of contemporary evolutionary theory, on which our understanding of adaptive evolution depends.

As we delve into the genetical theory of natural selection, it is important to keep in mind the following important points about natural selection:

- *Natural selection is not the same as evolution.* Evolution is a two-step process: the origin of genetic variation by mutation or recombination, followed by changes in the frequencies of alleles and genotypes, caused chiefly by genetic drift or natural selection. Neither natural selection nor genetic drift accounts for the origin of variation.

- *Natural selection is different from evolution by natural selection.* In some instances, selection occurs—that is, in each generation, genotypes differ in survival or fecundity—yet the proportions of genotypes and alleles stay the same from one generation to another; that is, there is no evolutionary change.

- Although natural selection may be said to exist whenever different phenotypes vary in average reproductive success, *natural selection can have no evolutionary effect unless phenotypic differences are inherited.* For instance, selection among genetically identical members of a clone, even though they may differ in phenotype, can have no evolutionary consequences. Therefore it is useful to describe the reproductive success, or fitness, of genotypes, even though genotypes differ in fitness only because of differences in phenotype.

- *A feature cannot evolve by natural selection unless that feature affects reproduction or survival.* The long-haired tail of a horse, used as a fly-switch, could not have evolved merely because it increases horses' comfort; it must have resulted in increased reproductive success, perhaps by lowering mortality caused by fly-borne diseases. Similarly, there can be no natural selection to reduce the pain associated with wounds, such as evisceration, that result in certain death.

Unlike genetic drift, inbreeding, and gene flow, which act at the same rate on all loci in a genome, the allele frequency changes caused by natural selection in a sexually reproducing species proceed largely independently at different loci. Moreover, different characteristics of a species evolve at different rates (mosaic evolution), as we would expect if natural selection brings about changes in certain features while holding others constant (see Chapter 3). Thus we are justified in beginning our analysis of natural selection with its effect on a single variable locus that alters a phenotypic character.

Fitness

Unless otherwise specified, the following discussion of natural selection concerns selection at the level of individual organisms in sexually reproducing populations. The consequences of natural selection depend on (1) the relationship between phenotype and fitness and (2) the relationship between phenotype and genotype. These relationships, then, yield (3) a relationship between fitness and genotype, which determines (4) whether or not evolutionary change occurs.

Modes of selection

The different ways in which fitness varies among different genotypes at a locus are the MODES of selection, which have different evolutionary outcomes. Let us denote the fitness of a genotype as w. Supposing there are two alleles (A_1, A_2) at a locus (**Figure 12.1A**), the fitnesses of the three genotypes A_1A_1, A_1A_2, and A_2A_2 are then w_{11}, w_{12}, and w_{22}. Selection is said to be **directional** if $w_{11} \geq w_{12} > w_{22}$. If the heterozygote has the highest fitness ($w_{11} < w_{12} > w_{22}$), the locus is said to be OVERDOMINANT, and if it has the lowest fitness ($w_{11} > w_{12} < w_{22}$), the locus is UNDERDOMINANT. In the case of directional selection, the disadvantageous

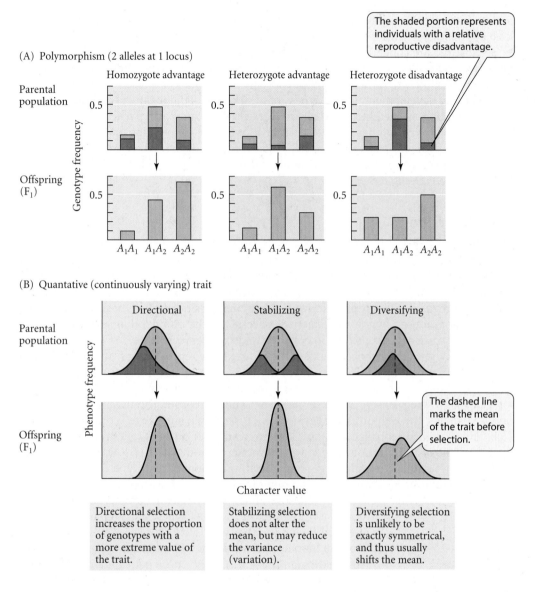

(A) Polymorphism (2 alleles at 1 locus)

The shaded portion represents individuals with a relative reproductive disadvantage.

Homozygote advantage   Heterozygote advantage   Heterozygote disadvantage

Parental population

Offspring (F₁)

A_1A_1 A_1A_2 A_2A_2 A_1A_1 A_1A_2 A_2A_2 A_1A_1 A_1A_2 A_2A_2

(B) Quantative (continuously varying) trait

Directional   Stabilizing   Diversifying

Parental population

Offspring (F₁)

The dashed line marks the mean of the trait before selection.

Character value

Directional selection increases the proportion of genotypes with a more extreme value of the trait.

Stabilizing selection does not alter the mean, but may reduce the variance (variation).

Diversifying selection is unlikely to be exactly symmetrical, and thus usually shifts the mean.

FIGURE 12.1 Modes of selection on (A) a polymorphism consisting of two alleles at one locus and (B) a heritable quantitative (continuously varying) character. The upper graphs in both (A) and (B) show the frequency distribution in the parental generation, before selection occurs. The shaded portions represent individuals with a relative disadvantage (lower fitness). The dashed line in (B) represents the mean character value in the parental generation. The lower graphs in both (A) and (B) show the frequency distribution in the F₁ generation, after selection has occurred. (After Endler 1986.)

allele (A_2 in the example given) is lowered in frequency and may be entirely eliminated, and selection on this allele is sometimes said to be **purifying**. Some literature uses POSITIVE and NEGATIVE selection to refer to whether a new mutation is selected for or against. For a quantitative (continuously varying) trait, such as size, selection is directional if phenotypes greater (or less) than the mean (average) have higher fitness than the mean phenotype, **stabilizing** (NORMALIZING) if the mean phenotype in the population has highest fitness, or **diversifying** (**disruptive**) if two or more phenotypes are fitter than the intermediates between them (**Figure 12.1B**).

Which *genotype* has the highest fitness under a given selection regime depends on the relationship between phenotype and genotype. For example, under directional selection for large size, genotype A_1A_1 would be fittest if it were largest, but A_1A_2 would be favored if it were larger than either homozygote. As we will soon see, this difference would have important evolutionary results: the population would become fixed for the largest phenotype if the homozygote were largest, but not if the heterozygote were largest.

The relationship between phenotype and fitness can depend on the environment, since the OPTIMAL (fittest) phenotype often differs under different environmental conditions. It also depends on how the mean of a character and its variation are distributed relative to the fitness/phenotype relationship. Thus, if the mean body size is below the optimum, body size will be directionally selected until it corresponds to the optimum (at least approximately); after that, it will be subject to stabilizing selection.

Because we are concerned with only those effects of selection that depend on inheritance, we will use models in which an average fitness value is assigned to each genotype. A genotype is likely to have different phenotypic expressions as a result of environmental influences on development, so the fitness of a genotype is the mean of the fitnesses of its several phenotypes, weighted by their frequencies. For example, a particular genotype of *Drosophila pseudoobscura* has a variable number of bristles, depending on the temperature at which the fly develops (see Figure 9.4). Thus, if fitness depended on bristle number, the fitness of a given genotype would depend on the proportions of flies with that genotype that developed at each temperature.

Defining fitness

The **fitness** of a genotype is the average per capita lifetime contribution of individuals of that genotype to the population after one or more generations (measured at the same stage in the life history). Often it suffices to measure fitness as the average number of eggs or offspring one generation hence that are descended from an average egg or offspring born. This measure, also called **reproductive success**, includes not simply the average number of offspring produced by the reproductive process, but the number that survive to reproductive age, since organisms that do not survive do not reproduce.

Fitness is most easily conceptualized for an asexually reproducing (PARTHENOGENETIC) population in which all adults reproduce only once, all at the same time (nonoverlapping generations), and then die, as in some parthenogenetic weevils and other insects that live for a single growing season. Suppose that in a population of such an organism, consisting only of females, the proportion of eggs of genotype *A* that survive to reproductive age is 0.05, and that each reproductive adult lays an average of 60 eggs (her FECUNDITY). Then the fitness of *A* is (proportion surviving) × (average fecundity) = 0.05 × 60 = 3. This value is the number of offspring an average newborn individual of genotype *A* contributes to the next generation. A population made up of this genotype would grow by a factor of 3 per generation. Thus this value is the genotype's or population's PER CAPITA GROWTH RATE (or per capita replacement rate), denoted *R*—or, in other words, its fitness. Likewise, genotype *B* might have a survival rate of 0.10 and an average fecundity of 40, yielding a fitness of 4. With these per capita growth rates, both genotypes will increase in number, but the *proportion* of the genotype with higher *R* will increase rapidly (**Figure 12.2**).

The per capita growth rate, R_i, of each genotype *i* is that genotype's **absolute fitness**. The **relative fitness** of a genotype, w_i, is its value of *R* relative to that of some reference genotype. By convention, the reference genotype, often the one with highest *R*, is assigned a relative fitness of 1.0. Thus, in our example in Figure 12.2, $w_A = \frac{3}{4} = 0.75$ and $w_B = 1.0$. The **mean fitness**, $\bar{w}$, is then *the average fitness of individuals in a population relative to the fittest genotype*. In our example, if the frequencies of genotypes *A* and *B* were 0.2 and 0.8, respectively, the mean fitness would be $\bar{w} = (0.2)(0.75) + (0.8)(1.0) = 0.95$. The mean fitness does not indicate whether or not the population is growing because it is a relative measure.

Another important term is the **coefficient of selection**, usually denoted *s*, which is the amount by which the fitness of a genotype differs from the reference genotype. In our example, $w_A = 0.75$, so $s = 0.25$. The coefficient of selection measures the **selective advantage** of the fitter genotype, or the intensity of selection against the less fit genotype.

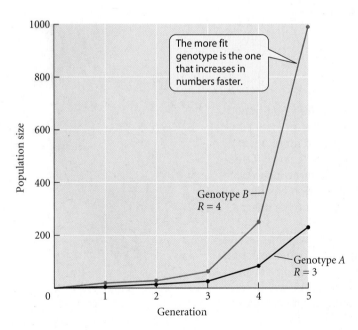

The more fit genotype is the one that increases in numbers faster.

Genotype *B*
R = 4

Genotype *A*
R = 3

FIGURE 12.2 The growth of two genotypes with different per capita growth rates in an asexually reproducing population with nonoverlapping generations. The growth rate per capita, *R*, is a measure of fitness.

It is easy to show mathematically that *the rate of genetic change under selection depends on the relative*, not the absolute, *fitnesses of genotypes*. The rate at which a population would become dominated by genotype *B* in our hypothetical example would be the same whether genotypes *A* and *B* had *R* values of 6 versus 8, or 15 versus 20, or 300 versus 400.

Components of fitness

Survival and fecundity are only two of the possible components of fitness. Fitness may be more complex if a species reproduces sexually and if individuals reproduce repeatedly during their lifetime. Differences among males in reproductive success often contribute to differences in fitness. When generations overlap, as in humans and many other species that reproduce repeatedly, the absolute fitness of a genotype may be measured in large part by its per capita rate of increase per time interval, *r* (see Chapter 14). (The difference between *R* and *r* is that they express population growth rate per generation versus per unit time.) This rate of increase depends on the proportion of individuals surviving to each age class and on the fecundity of each age class. Moreover, *r* is strongly affected by the age at which females have offspring, not just by their number; that is, genotypes may differ in the length of a generation. If females of genotypes *A* and *B* have the same number of offspring, but *A* reproduces at age 6 months and *B* reproduces at age 12 months, the rate of increase (the fitness) of *A* is about twice that of *B* because *A* will produce twice as many generations of descendants per time interval.

In sexually reproducing species, genotypes do not simply make copies of themselves; instead, they transmit haploid gametes. Therefore genotype frequencies depend on the allele frequencies among uniting gametes (see Figure 9.8). These allele frequencies are affected by several components of selection at the zygotic (organismal) stage, and sometimes by selection at the gametic stage as well (**Figure 12.3**). **Table 12.1** summarizes the components of selection in a sexual species.

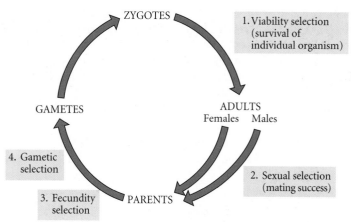

FIGURE 12.3 Components of natural selection that may affect the fitness of a sexually reproducing organism over the life cycle. Beginning with newly formed zygotes, (1) genotypes may differ in survival to adulthood; (2) they may differ in the numbers of mates they obtain, especially males; (3) those that become parents may differ in fecundity (number of gametes produced, especially eggs); and (4) selection may occur among the haploid genotypes of gametes, as in meiotic drive, differential gamete viability, or fertilization success. (After Christiansen 1984.)

TABLE 12.1 Components of selection in sexually reproducing organisms

I. Zygotic selection

A. *Viability*. The probability of survival of the genotype up to and through each of the ages at which reproduction can occur. After the age of last reproduction, the length or probability of survival does not usually affect the genotype's contribution to subsequent generations, and so does not usually affect fitness.

B. *Mating success*. The number of mates obtained by an individual. Mating success is a component of fitness if the number of mates affects the individual's number of progeny, as is often the case for males, but less often for females, all of whose eggs may be fertilized by a single male. Variation in mating success is the basis of sexual selection.

C. *Fecundity*. The average number of viable offspring per female. In species with repeated reproduction, the contribution of each offspring to fitness depends on the age at which it is produced (see Chapter 14). The fertility of a mating may depend only on the maternal genotype (e.g., number of eggs or ovules), or it may depend on the genotypes of both mates (e.g., if they display some reproductive incompatibility).

II. Gametic selection

D. *Segregation advantage* (meiotic drive or segregation distortion). An allele has an advantage if it segregates into more than half the gametes of a heterozygote.

E. *Gamete viability*. Dependence of a gamete's viability on the allele it carries.

F. *Fertilization success*. An allele may affect the gamete's ability to fertilize an ovum (e.g., if there is variation in the rate at which a pollen tube grows down a style).

Box 12A Selection Models with Constant Fitnesses

We first present a general model of allele frequency change under natural selection (Hartl and Clark 1997) and then modify it for specific cases. Suppose three genotypes at a locus with two alleles differ in relative fitness due to differences in survival:

	A_1A_1	A_1A_2	A_2A_2
Frequency at birth	p^2	$2pq$	q^2
Relative fitness	w_{11}	w_{12}	w_{22}

The ratio of A_1A_1: A_1A_2: A_2A_2 among surviving adults is

$$p^2w_{11}:2pqw_{12}:q^2w_{22}$$

and the ratio of the alleles (A_1: A_2) among their gametes is

$$[p^2w_{11} + \tfrac{1}{2}(2pqw_{12})]:[\tfrac{1}{2}(2pqw_{12}) + q^2w_{22}]$$

which simplifies to

$$p(pw_{11} + qw_{12}):q(pw_{12} + qw_{22})$$

The gamete frequencies, which are the allele frequencies among the next generation of offspring, are found by dividing each term by the sum of the gametes, which is

$$p(pw_{11} + qw_{12}) + q(pw_{12} + qw_{22})$$
$$= p_2w_{11} + 2pqw_{12} + q_2w_{22}$$
$$= \overline{w}$$

Thus the allele frequencies after selection (p', q' are the gamete frequencies, or

$$p' = \frac{p(pw_{11} + qw_{12})}{\overline{w}}$$

$$q' = \frac{q(pw_{12} + qw_{22})}{\overline{w}}$$

The change in allele frequency between generations is $\Delta p = p' - p$, or

$$\Delta p = \frac{p(pw_{11} + qw_{12}) - p\overline{w}}{\overline{w}}$$

Substituting for $\overline{w}$ and doing the algebra yields

$$\Delta p = \frac{pq[p(w_{11} - w_{12}) + q(w_{12} - w_{22})]}{\overline{w}} \quad \text{(A1)}$$

We can analyze various cases of selection by entering explicit fitness values for the w's. A few important cases are the following:

1. Advantageous dominant allele, disadvantageous recessive allele ($w_{11} = w_{12} > w_{22}$).

For w_{11}, w_{12}, and w_{22}, substitute 1, 1, and $1 - s$ respectively in Equation A1. The mean fitness is $p^2(1) + 2pq(1) + q^2(1 - s) = 1 - sq^2$ (bearing in mind that $p^2 + 2pq + q^2 = 1$). The equation for allele frequency change is

$$\Delta p = \frac{spq^2}{1 - sq^2}$$

or, equivalently,

$$\Delta q = \frac{-spq^2}{1 - sq^2} \quad \text{(A2)}$$

2. Advantageous allele partially dominant, disadvantageous allele partially recessive ($w_{11} > w_{12} > w_{22}$).

Let h, lying between 0 and 1, measure the degree of dominance for fitness, and substitute 1, $1 - hs$, and $1 - s$ for w_{11}, w_{12}, and w_{22}. (If $h = 0$, allele A_2 is fully recessive.) After sufficient algebra, we find that

$$\Delta p = \frac{spq[h(1 - 2q) + sq]}{1 - 2pqhs - sq^2} \quad \text{(A3)}$$

which is positive for all $q > 0$, so allele A_1 increases to fixation. If $h = \frac{1}{2}$, Equation A3 reduces to

$$\Delta p = \frac{spq}{[2(1 - sq)]}$$

3. Fitness of heterozygote is greater than that of either homozygote ($w_{11} < w_{12} > w_{22}$).

Using s and t as selection coefficients, let the fitnesses of A_1A_1, A_1A_2, and A_2A_2 be $1 - s$, 1, and $1 - t$ respectively. Substituting these in Equation A1, we obtain

$$\Delta p = \frac{pq(-sp + tq)}{1 - sp^2 - tq^2} \quad \text{(A4)}$$

There is a stable "internal equilibrium" that can be found by setting $\Delta p = 0$; then $sp = tq$. Substituting $1 - p$ for q, the equilibrium frequency p is $t/(s + t)$. Thus the frequency of A_1 is proportional to the relative strength of selection against A_2A_2.

4. Fitness of heterozygote is less than that of either homozygote (that is, $w_{11} > w_{12} < w_{22}$).

As this is the reverse of the preceding case, let $1 + s$, 1, and $1 + t$ be the fitnesses of A_1A_1, A_1A_2, and A_2A_2. The equation for allele frequency change is

$$\Delta p = \frac{pq(sp - tq)}{1 + sp^2 + tq^2} \quad \text{(A5)}$$

Δp is positive if $sp > tq$, and negative if $sp < tq$. Setting $\Delta p = 0$ and solving for p, we find an internal equilibrium at $p = t/(s + t)$, but this is an unstable equilibrium. For example, if $s = t$, the unstable equilibrium is $p = 0.5$, but then Δp is positive if $p > q$ (i.e., if $p > 0.5$), and negative if $p < q$. The allele frequency therefore arrives at either of two stable equilibria, $p = 1$, or $p = 0$.

Evolution by natural selection depends on the way in which changes in allele frequencies are determined by the components of fitness of each zygotic and each gametic genotype. These components of fitness are combined (usually by multiplying them) into the overall fitness of each genotype. For instance, the fitness of the genotypes in the simple example above was found by multiplying the survival rate and fecundity of each genotype. In that example, one genotype had superior fecundity and the other had a superior survival rate: a genotype may be superior to another in certain components of fitness and inferior in others, but its *overall fitness determines the outcome of natural selection*.

Models of Selection

In the following discussion, we make the simplifying assumptions that the population is very large, so genetic drift may be ignored; that mating occurs at random; that mutation and gene flow do not occur; and that selection at other loci does not affect the locus we are considering. We will later consider the consequences of changing these unrealistic assumptions. We also assume, for the sake of simplicity, that selection acts through differential survival among genotypes in a species with discrete generations. The principles are much the same for other components of selection and for species with overlapping generations.

If a locus has two alleles (A_1, A_2) with frequencies p and q, respectively, the change in the frequency of A_1 from one generation to the next is expressed by Δp, which is positive if the allele is increasing in frequency, negative if it is decreasing, and zero if it is at equilibrium (similarly, Δq refers to a change in the frequency of A_2). In any model of selection, the change in allele frequencies depends on the relative fitnesses of the different genotypes and on the allele frequencies themselves. **Box 12A** provides a mathematical framework for several models of selection.

Directional selection

THEORY   The replacement of relatively disadvantageous alleles by more advantageous alleles is the fundamental basis of adaptive evolution. This replacement occurs when the homozygote for an advantageous allele has a fitness equal to or greater than that of the heterozygote or of any other genotype in the population (Box 12A, Cases 1, 2).

An advantageous allele may initially be fairly common if under previous environmental circumstances it was selectively neutral or was maintained by one of several forms of balancing selection (described later in this chapter). However, an advantageous allele is likely to be initially very rare if it is a newly arisen mutation or if it was disadvantageous before an environmental change made it advantageous.

An advantageous allele that increases from a very low frequency is often said to INVADE a population. *Unless an allele can increase in frequency when it is very rare, it is unlikely to become fixed in the population.* According to this principle, some conceivable adaptations are unlikely to evolve because they could not increase if they were initially very rare. For instance, venomous coral snakes (*Micrurus*) are brilliantly patterned in red, yellow, and black (**Figure 12.4**). This pattern is presumed to be APOSEMATIC (warning) coloration, which is beneficial because predators associate the colors with danger and avoid attacking such snakes. How this coloration initially evolved has long been a puzzle, however, since the first few mutant snakes with brilliant colors would presumably have been easily seen and killed by naïve predators. Given that all coral snakes are aposematically colored, it is understandable that predators might evolve an aversion to them (and, indeed,

FIGURE 12.4 Warning (aposematic) coloration in a western coral snake (*Micrurus euryxanthus*). If a population of dangerous or unpalatable organisms has a high frequency of such a color pattern, predators may rapidly learn to avoid the aposematically colored organisms, or may evolve to avoid them. It is less obvious how a new, rare mutation for such coloration can increase in frequency if it makes the organisms conspicuous to naïve predators.

some predatory birds seem to have an innate aversion to coral snakes)—but how the snake's adaptation "got off the ground" is uncertain. (One possibility is that predators generalized from other brilliantly colored unpalatable or dangerous organisms, such as wasps, and avoided aposematic snakes from the beginning.)

A simple example of directional selection occurs if the fitness of the heterozygote is precisely intermediate between that of the two homozygotes (i.e., neither allele is dominant with respect to fitness). The frequencies and fitnesses of the three genotypes may be denoted as follows:

Genotype	A_1A_1	A_1A_2	A_2A_2
Frequency	p^2	$2pq$	q^2
Fitness	1	$1 - (s/2)$	$1 - s$

These fitnesses may be entered into Equation A1 in Box 12A, which, when solved, shows that the advantageous allele A_1 increases in frequency, per generation, by the amount

$$\Delta p = \frac{\frac{1}{2}spq}{1 - sq} \tag{12.1}$$

where $(1 - sq)$ equals the mean fitness, $\overline{w}$.

Equation 12.1 tells us that Δp is positive whenever p and q are greater than zero. Therefore allele A_1 increases to fixation ($p = 1$), and $p = 1$ is a STABLE EQUILIBRIUM. The rate of increase (the magnitude of Δp) is proportional to both the coefficient of selection s and the allele frequencies p and q, which appear in the numerator. Therefore the rate of evolutionary change increases as the variation at the locus increases. (It is approximately proportional to $2pq$, the frequency of heterozygotes, when selection is weak.) Conversely, the rate is low when one allele is much more frequent than the other.

Another important aspect of Equation 12.1 is that Δp is positive as long as s is greater than zero, even if it is very small. Therefore, as long as no other evolutionary factors intervene, *a character state with even a minuscule advantage will be fixed by natural selection.* Hence even very slight differences among species, in seemingly trivial characters such as the distribution of hairs on a fly or veins on a leaf, could conceivably have evolved as adaptations. This principle explains the extraordinary apparent "perfection" of some features. Some katydids, for example, resemble dead leaves to an astonishing degree, with transparent "windows" in the wings that resemble holes and blotches that resemble spots of fungi or algae (**Figure 12.5**). One might suppose that a less detailed resemblance would provide sufficient protection against predators, and some species are indeed less elaborately cryptic; but if an extra blotch increases the likelihood of survival by even the slightest amount, it may be fixed by selection (providing, again, that no other factors intervene).

The same equations that describe the increase of an advantageous allele describe the fate of a disadvantageous allele: If A_1 and A_2 are advantageous and disadvantageous alleles, respectively, with frequencies p and q, and if $p + q = 1$, then $\Delta p = -\Delta q$. That is, purifying selection reduces the frequency of a deleterious mutation or eliminates it altogether.

The number of generations required for an advantageous allele to replace one that is disadvantageous depends on the initial allele frequencies, the selection coefficient, and the degree of dominance (**Figure 12.6**). An advantageous allele can increase from low frequency more rapidly if it is dominant than if it is recessive because it is expressed in the heterozygous state, and until it reaches a fairly high frequency, it is carried almost entirely by heterozygotes. After a dominant

FIGURE12.5 This tropical katydid (*Mimetica mortuifolia*) is well named: it resembles, in great detail, a leaf that has been chewed and turned partly brown by insects such as caterpillars or beetles. Such examples of extraordinary cryptic form and coloration, which protect the animal against predation, are marvels of adaptation by natural selection.

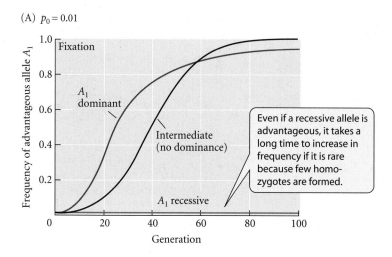

(A) $p_0 = 0.01$

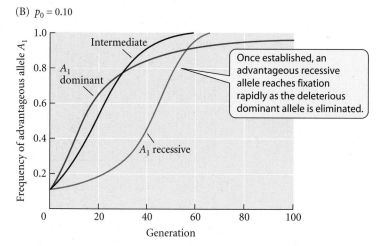

(B) $p_0 = 0.10$

FIGURE 12.6 Increase of an advantageous allele (A_1) from initial frequencies of (A) $p_0 = 0.01$ and (B) $p_0 = 0.10$. The three curves in each graph show the increase of a fully dominant allele (green), an allele with intermediate dominance (black), and a recessive allele (red). For the advantageous dominant A_1, the fitnesses of genotypes A_1A_1, A_1A_2, and A_2A_2 are 1.0, 1.0, and 0.8 respectively; for the "intermediate" case, they are 1.0, 0.9, and 0.8; and for the advantageous recessive A_1, they are 1.0, 0.8, and 0.8.

advantageous allele attains high frequency, however, the deleterious recessive allele is eliminated very slowly, because a rare recessive allele occurs mostly in heterozygous form and is thus shielded from selection.

The denominator in Equation 12.1 is the mean fitness of individuals in the population, $\overline{w}$, which increases as the frequency (q) of the deleterious allele decreases. The mean fitness therefore increases as natural selection proceeds. In a graphical representation of this relationship (**Figure 12.7**), we may think of the population as climbing up a "hillside" of increasing mean fitness until it arrives at the summit. (The arrowheads on the curves in Figure 12.7 show the direction of change in p for four different cases.)

Equation 12.1, finally, can be used to draw an interesting inference from data. If we have data on the frequencies of genotypes at a locus (and therefore also have estimates of the frequencies of alleles, p and q), and if we also have data on allele frequencies in successive generations (i.e., an estimate of Δp), we can solve for s in the equation. This is one way of estimating the strength of natural selection. Several other methods are used to estimate selection coefficients, such as estimating the survival rates (or other components of fitness) of different genotypes in natural populations (Endler 1986).

EXAMPLES OF DIRECTIONAL SELECTION If a locus has experienced consistent directional selection for a long time, the advantageous allele should be near equilibrium—that is, near fixation. Thus the dynamics of directional selection are best studied in recently altered environments, such as those altered by human activities. Many examples of rapid evolution under such circumstances have been observed. Many are changes in polygenic traits, as described in the next chapter.

FIGURE 12.7 Plots of mean fitness ($\overline{w}$) against allele frequency (p) for one locus with two alleles when genotypes differ in survival. Each of these plots represents an "adaptive landscape," and may be thought of as a surface, or hillside, over which the population moves. From any given frequency of allele A_1 (p), the allele frequency moves in a direction that increases mean fitness ($\overline{w}$). The arrowheads show the direction of allele frequency change. (A) Directional selection. Here, A_1A_1 is the favored genotype. The equilibrium ($\hat{p} = 1$) is stable: the allele frequency returns to $p = 1$ if displaced. (B) Directional selection, in which the relative fitnesses are reversed compared with (A), perhaps because of changed environmental conditions. A_2A_2 is now the favored genotype. (C) Overdominance (heterozygote advantage). From any starting point, the population arrives at a stable polymorphic equilibrium ($\hat{p}$). (D) Underdominance (heterozygote disadvantage). The interior equilibrium ($\hat{p} = 0.4$ in this example) is unstable because even a slight displacement initiates a change in allele frequency toward one of two stable equilibria: ($\hat{p} = 0$ (loss of A_1) or $\hat{p} = 1$ (fixation of A_1). Therefore this curve represents an adaptive landscape with two peaks. (After Hartl and Clark 1989.)

(A) Directional selection
$w_{11} = 1$, $w_{12} = 0.8$, $w_{22} = 0.2$

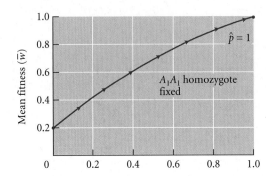

(B) Directional selection
$w_{11} = 0.2$, $w_{12} = 0.8$, $w_{22} = 1$

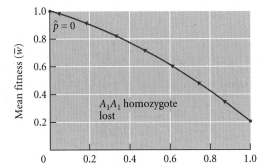

Frequency (p) of allele A_1

(C) Overdominance (heterozygote advantage)
$w_{11} = 0.6$, $w_{12} = 1$, $w_{22} = 0.2$

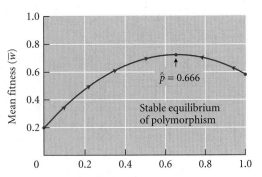

(D) Underdominance (heterozygote disadvantage)
$w_{11} = 1$, $w_{12} = 0.4$, $w_{22} = 0.8$

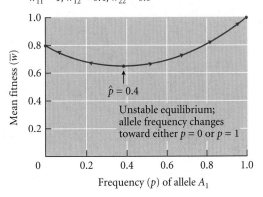

Frequency (p) of allele A_1

An example of rapid evolution of a single-locus trait is the case of warfarin resistance in brown rats (*Rattus norvegicus*; Bishop 1981). Warfarin is an anticoagulant: it inhibits an enzyme responsible for the regeneration of vitamin K, a necessary cofactor in the production of blood-clotting factors. Susceptible rats poisoned with warfarin often bleed to death from slight wounds. A mutation confers resistance by altering the enzyme to a form that is less sensitive to warfarin, but also less efficient in regenerating vitamin K, so that a higher dietary intake of the vitamin is necessary.

Starting in 1953, warfarin was used as a rat poison in Britain, and by 1958 resistance was reported in certain rat populations. Under exposure to warfarin, resistant rats (with allele R_1) have a strong survival advantage, and the frequency of the mutation has been known to increase rapidly to nearly 1.0 (**Figure 12.8**). Resistant rats suffer a strong disadvantage compared with susceptible rats (genotype R_2R_2), however, because of their greater need for vitamin K, and the frequency of the resistance allele drops rapidly if the poison is not administered. The decline in the frequency of resistance portrayed in Figure 12.8 is best explained by a model in which the relative fitnesses of the genotypes are 1.00 (R_1R_1), 0.75 (R_1R_2), and 0.46 (R_2R_2). The genotypes clearly differed greatly in fitness.

Resistance to insecticides appeared in many species of insects in the 1940s, when synthetic pesticides came into wide use. By 1990, populations of more than 500 species were known to be resistant to one or more insecticides (**Figure 12.9**). Some species, such as the Colorado potato beetle (*Leptinotarsa decemlineata*), have evolved resistance to all the major classes of pesticides. The evolution of resistance adds immensely to the cost of agriculture

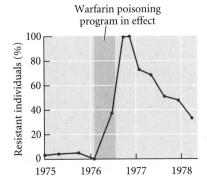

FIGURE 12.8 Proportion of warfarin-resistant individuals in a population of rats in Wales. The proportion increased when a warfarin poisoning program was initiated in 1976, but decreased after the program ended. (After Bishop 1981.)

and is a major obstacle in the fight against insect-borne diseases such as malaria. For these reasons, as well as the devastating toxic effects of many pesticides on natural ecosystems and on human health, supplementary or alternative methods of pest control are a major topic of research in entomology.

Insecticide resistance in natural populations of insects is often based primarily on single mutations of large effect (Roush and Tabashnik 1990). The resistance allele (R) is usually partially or fully dominant over the allele for susceptibility. R alleles generally have very low frequencies in populations that have not been exposed to insecticides, but they often increase nearly to fixation within 2 or 3 years after a pesticide is applied because the mortality of susceptible genotypes is extremely high. In the absence of the insecticide, however, resistant genotypes are about 5 to 10 percent less fit than susceptible genotypes, and they decline in frequency. Like warfarin resistance, resistance to insecticides illustrates a COST OF ADAPTATION, or TRADE-OFF: advantageous traits often have "side effects" that are disadvantageous, at least in some environments.

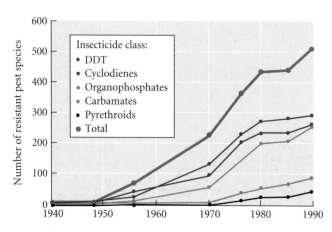

FIGURE 12.9 Cumulative numbers of arthropod pest species known to have evolved resistance to five classes of insecticides. The upper curve shows the total number of insecticide-resistant species. (After Metcalf and Luckmann 1994.)

Deleterious alleles in natural populations

Although the most advantageous allele at a locus should, in principle, be fixed by directional selection, deleterious alleles often persist because they are repeatedly reintroduced, either by recurrent mutation or by gene flow from other populations in which they are favored by a different environment. In either case, the frequency of the deleterious allele moves toward a *stable equilibrium* that is a *balance between the rate at which it is eliminated by selection and the rate at which it is introduced by mutation or gene flow*.

SELECTION AND MUTATION Suppose that a deleterious recessive allele A_2 with frequency q arises at a mutation rate u from other alleles that have a collective frequency of $p = 1 - q$. The increase in frequency of A_2 that is due to mutation in each generation is up, whereas the decrease in its frequency that is due to selection (from Equation A2 in Box 12A) is $-spq^2/\overline{w}$. At equilibrium, the rate of increase equals the rate of decrease:

$$up = \frac{spq^2}{\overline{w}}$$

where $\overline{w} = 1 - sq^2$.

Let us assume that A_2 is rare, so that $\overline{w}$ is approximately equal to 1, and solve for q. The result is the equilibrium frequency, denoted $\hat{q}$. We find that $\hat{q}^2 = u/s$, and therefore that

$$\hat{q} = \sqrt{\frac{u}{s}}$$

The equilibrium frequency of a deleterious recessive allele, therefore, is directly proportional to the mutation rate and inversely proportional to the strength of selection. Thus, if s is much greater than u, the allele will be very rare. For example, if $s = 1$ (i.e., A_2 is a recessive lethal allele) and the mutation rate is 10^{-6}, the equilibrium frequency will equal 0.001, but almost all the zygotes with this allele will be heterozygous. If the deleterious allele is partly or entirely dominant, its equilibrium frequency will be even lower because of selection against both homozygous and heterozygous carriers.

Such MUTATION-SELECTION BALANCE explains why many chromosomes in *Drosophila* populations (as well as in humans and many other species) carry rare mutations that slightly reduce fitness in heterozygous condition and are strongly deleterious or even lethal when homozygous (Crow 1993; see Figure 9.13). In humans, for example, abnormal neck vertebrae bearing ribs characterize about half of European fetuses that die before birth, but less than 2 percent of the adult population (Galis et al. 2006). Selection against this abnormal phenotype is very strong (*s* is approximately 0.81), probably because the many mutations that cause cervical ribs have pleiotropic effects: these fetuses commonly have many other abnormalities. These are just a few of the many deleterious mutations that arise each generation (about 2.2 per human zygote; see Chapter 8), which undoubtedly contribute to the high rate of spontaneous abortion. [As many as half of all human embryos may die within 12 weeks of implantation, mostly without the mother's knowledge (Benagiano et al. 2010).] These prenatal deaths provide scope, or opportunity, for so-called purifying selection.

SELECTION AND GENE FLOW   Very often, different environmental conditions favor different alleles in different populations of a species. Thus, in the absence of gene flow, the frequency *q* of an allele A_2 will evolve to be 1 in certain populations and 0 in others. Gene flow among populations can introduce each allele into populations in which it is deleterious, and the allele frequency in each population thus arrives at an equilibrium ($\hat{q}$) set by the balance between the incursion of alleles by gene flow and their elimination by selection. Thus gene flow can contribute to genetic variation within populations and can aid adaptation. However, gene flow can also reduce adaptation to local conditions. For example, if gene flow is much greater than selection, a population inhabiting a patch of a distinctive environment will not become genetically differentiated from surrounding populations (Lenormand 2002). If local populations are distributed along an environmental gradient over which the fitness of different genotypes changes (**Figure 12.10A**), then even if the environment changes gradually, we should expect an abrupt shift in allele frequencies (a STEP CLINE) in the absence of gene flow among populations along the gradient. This is true as long as one homozygote or the other has highest fitness in each population. However, a *gradual cline in allele frequencies may be established if there is gene flow* among populations along the gradient (**Figure 12.10B**). The width of the cline (the distance over which *q* changes from, say, 0.2 to 0.8) is proportional to *V*/*s*, where *V* measures the distance that genes disperse and *s* is the strength of selection against them (the coefficient of selection). Thus, if selection is strong relative to gene flow, steep clines in allele frequencies will result, so that populations are strongly differentiated.

The dark and pale forms of the pocket mouse *Chaetodipus intermedius*, illustrated in Figure 3.6, display steep clines between dark lava beds and the light rock in the surrounding desert (**Figure 12.11A**). The difference in coloration is based on a single allele difference at the *Mc1r* locus. Previous studies of other species with this pattern showed high mortality of mice mismatched with their environment due to predation by owls. In *Chaetodipus*, selection for the dark (*D*) allele on lava beds and for the pale allele (*d*) on light rocks is opposed by gene flow, which accounts for low frequencies of the "wrong" allele in both lava bed– and light rock–inhabiting populations (**Figure 12.11B**).

Hopi Hoekstra and colleagues (2004) estimated gene flow (*m*) between lava beds and light rock areas, using presumably neutral sequence variation in mitochondrial genes (mtDNA). All the mouse populations have the same common mtDNA haplotype, indicating high gene flow (**Figure 12.11C**). Recall from Chapter 10 (p. 271) that the average frequency of heterozygotes at neutral DNA sites is proportional to $N_e u$ at equilibrium. The neutral mutation rate of mitochondrial genes has been estimated as 10^{-6} to 10^{-7} for

FIGURE 12.10 Geographic variation arising from selection and gene flow when genotype fitnesses differ gradually along an environmental gradient. (A) Relative fitnesses w_{11}, w_{12}, and w_{22} of genotypes A_1A_1, A_1A_2, and A_2A_2 are plotted for populations along an east–west gradient. A_1A_1 has higher fitness than A_2A_2 in the west, and vice versa in the east. (B) Frequency of the A_2 allele in an array of populations along the gradient. Each curve represents a different level of gene flow (*g*), ranging from 0 to 100 percent (*g* = 1). The lower the level of gene flow, the steeper the cline in allele frequency. (After Endler 1973.)

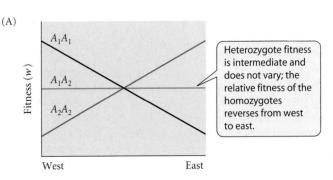

(A)

Heterozygote fitness is intermediate and does not vary; the relative fitness of the homozygotes reverses from west to east.

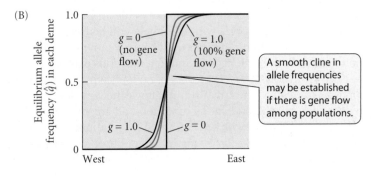

(B)

A smooth cline in allele frequencies may be established if there is gene flow among populations.

FIGURE 12.11 Opposing effects of natural selection and gene flow on three characteristics in pocket mice. The mice inhabit dark lava beds in the three central sampling locations and pale rock outcrops in the flanking localities. The frequencies of (A) the melanic (dark) phenotype and (B) the D allele at the Mc1r locus are higher in the populations on lava beds, and lower in the populations on light rocks, but gene flow has prevented complete fixation of the alternative alleles in the locations where they are favored by divergent selection. (C) The frequency of a common mtDNA variant is almost the same among all of these populations, showing that gene flow homogenizes populations when it is not opposed by selection. (After Hoekstra et al. 2004.)

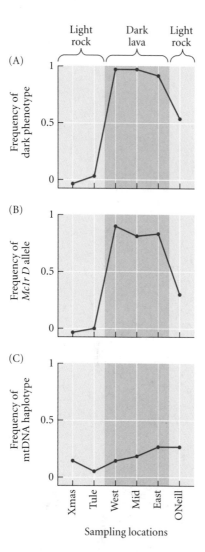

mammals, so Hoekstra et al. could estimate N_e, the effective population size. Recall also that F_{ST}, the variation in mtDNA haplotype frequencies among populations, equals $1/(4N_e m + 1)$ at equilibrium (see p. 273 in Chapter 10). The rate of gene flow between mouse populations (m) could therefore be estimated, given the measured F_{ST} and the estimate of N_e. Hoekstra et al. estimated m to be about 4×10^{-3} to 2.5×10^{-4} between the lava bed populations and the nearest light rock populations. Having estimated gene flow, which is the same for all genes, they could estimate s, the selection coefficient against a deleterious Mc1r allele, assuming that the allele frequency is balanced by the opposing forces of selection and gene flow. They concluded that selection against light mice on dark lava beds is very strong (s perhaps as great as 0.4), and that there is weaker, but nevertheless substantial, selection against dark mice on a light rock background.

Gene flow can reduce the level of adaptation of populations to their local environment. For example, Jon Bridle and coworkers (2009) studied cold resistance in samples of the Australian rain forest fly *Drosophila birchii* collected along altitudinal transects on two mountain ranges, both extending from 100 meters to 1000 meters above sea level. One transect was much steeper, extending over 4 kilometers, than the other, which was 10 kilometers long. High-altitude flies were more cold-resistant than low-altitude flies on the shallow transect, but not on the steep transect (**Figure 12.12**). Because the altitudinally separated populations

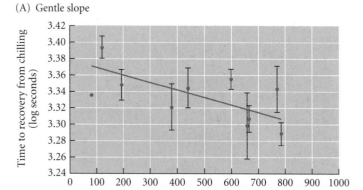

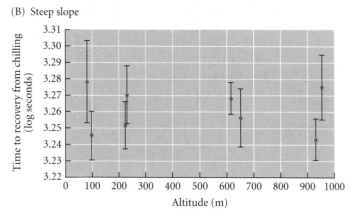

FIGURE 12.12 Evidence that gene flow can prevent populations from adapting to their local environment. Different populations of *Drosophila birchii* along an altitudinal transect, tested for recovery from cold, differed more, showing greater local adaptation, along a transect up a gentle mountain slope (A) than a steep one (B). Populations at different altitudes are closer to each other along the steep slope, so there is a higher rate of dispersal and gene flow among them. (After Bridle et al. 2009.)

along the steeper slope are separated by shorter distances, gene flow among them is thought to be higher, preventing them from adapting to different thermal environments.

Polymorphism Maintained by Balancing Selection

Until the 1940s, the prevalent, or classic, view had been that at each locus, a best allele (the "wild type") should be nearly fixed by natural selection, so that the only variation should consist of rare deleterious alleles, recently arisen by mutation and fated to be eliminated by purifying selection. As we saw in Chapter 9, studies of natural populations revealed instead a wealth of variation. The factors that might be responsible for this variation are (1) recurrent mutation producing deleterious alleles subject to only weak selection; (2) gene flow of locally deleterious alleles from other populations in which they are favored by selection; (3) selective neutrality (i.e., genetic drift); and (4) maintenance of polymorphism by natural selection. The last of these hypotheses was championed by British ecological geneticists led by E. B. Ford and American population geneticists influenced by Theodosius Dobzhansky. They represented the BALANCE SCHOOL, which holds that a great deal of genetic variation is maintained by **balancing selection** (which is simply selection that maintains polymorphism).

There is still uncertainty about how much of the extensive genetic variation in natural populations fits the models represented by these contrasting points of view. Several models of natural selection can account for persistent, stable polymorphisms, but we do not know the extent to which each accounts for the observed genetic variation within populations.

Heterozygote advantage

If the heterozygote has higher fitness than either homozygote, both alleles will be propagated in successive generations, in which union of gametes will yield all three genotypes among the zygotes. Such **heterozygote advantage** is also termed **overdominance** or SINGLE-LOCUS HETEROSIS. It results in a stable equilibrium at which the allele frequencies depend on the balance between the fitness values (hence the selection coefficients) of the two homozygotes (**Figure 12.13A**; see also Case 3 in Box 12A and Figure 12.7C).

(A)

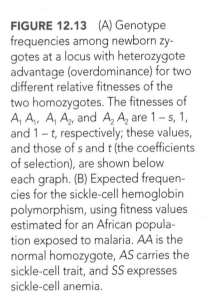

If the two homozygotes are both less fit than the heterozygote but have unequal fitness ($s \neq t$), the equilibrium genotype frequencies will be skewed.

	A_1A_1	A_1A_2	A_2A_2
$w =$	0.90	1.0	0.95

$s = 0.10, t = 0.05$

If both homozygotes are less fit than the heterozygote to the same extent ($s = t$), they both will have a frequency of 0.25 at equilibrium.

	A_1A_1	A_1A_2	A_2A_2
$w =$	0.90	1.0	0.90

$s = 0.10, t = 0.10$

(B)

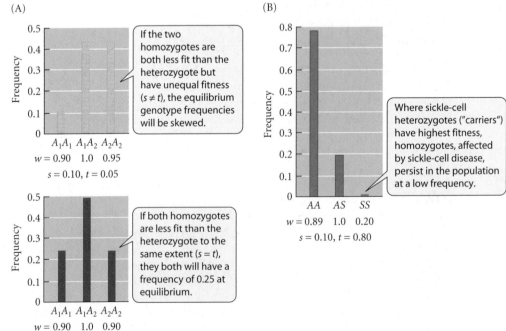

Where sickle-cell heterozygotes ("carriers") have highest fitness, homozygotes, affected by sickle-cell disease, persist in the population at a low frequency.

	AA	AS	SS
$w =$	0.89	1.0	0.20

$s = 0.10, t = 0.80$

FIGURE 12.13 (A) Genotype frequencies among newborn zygotes at a locus with heterozygote advantage (overdominance) for two different relative fitnesses of the two homozygotes. The fitnesses of $A_1 A_1$, $A_1 A_2$, and $A_2 A_2$ are $1 - s$, 1, and $1 - t$, respectively; these values, and those of s and t (the coefficients of selection), are shown below each graph. (B) Expected frequencies for the sickle-cell hemoglobin polymorphism, using fitness values estimated for an African population exposed to malaria. *AA* is the normal homozygote, *AS* carries the sickle-cell trait, and *SS* expresses sickle-cell anemia.

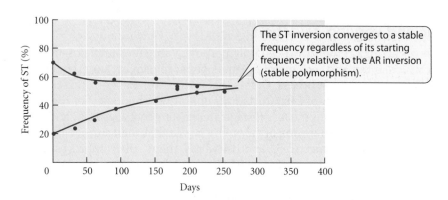

The ST inversion converges to a stable frequency regardless of its starting frequency relative to the AR inversion (stable polymorphism).

FIGURE 12.14 Demonstration of a stable equilibrium. *Drosophila pseudoobscura* flies that are heterozygous for the chromosome inversions ST and AR have higher fitness than either homozygote. Different laboratory populations that started with very different frequencies of the ST inversion converged to the same equilibrium frequency of ST, as theory predicts. (After Dobzhansky 1948.)

A stable equilibrium of this kind is illustrated in experiments by Theodosius Dobzhansky and his collaborators. Natural populations of *Drosophila pseudoobscura* are highly polymorphic for inversions—that is, differences in the arrangement of genes along a chromosome resulting from 180-degree reversals in the orientation of chromosome segments (see Chapter 8). Dobzhansky suspected that these inversions affected fitness when he observed that the frequencies of several inversions in natural populations displayed a regular seasonal cycle. In the laboratory, populations initiated with very different frequencies of two inversions rapidly converged to the same equilibrium frequency (**Figure 12.14**).

Few cases of overdominance for fitness have been well documented, even though genotypes that are heterozygous at several or many loci often appear to be fitter than more homozygous genotypes. For example, inbreeding depression is commonly observed when organisms become more homozygous under inbreeding (see Chapter 9). However, this phenomenon can be explained by dominance rather than overdominance: inbred lines become homozygous for deleterious recessive alleles. Likewise, it can be very difficult to show that the heterozygote (*Aa*) at a particular locus has higher fitness than both homozygotes (*AA, aa*). Suppose most of the chromosomes in a population are *Ab* and *aB*, where *A* and *B* are favorable dominant alleles and *a* and *b* are unfavorable recessive alleles at two closely linked loci. The homozygotes *Ab/Ab* and *aB/aB*, each expressing an unfavorable recessive, would have lower fitness than the heterozygote *Ab/aB*. If we were aware of the phenotypic effect of only one of the loci, it would appear to display overdominance for fitness. Such apparent but spurious superiority of the heterozygote at the observed locus is called ASSOCIATIVE OVERDOMINANCE.

The best-understood case of heterozygote advantage is the β-hemoglobin locus in some African and Mediterranean human populations (Cavalli-Sforza and Bodmer 1971). One allele at this locus encodes sickle-cell hemoglobin (*S*), which is distinguished by a single amino acid substitution from normal hemoglobin (encoded by the *A* allele). At low oxygen concentrations, *S* hemoglobin forms elongate crystals, which carry oxygen less effectively, causing the red blood cells to adopt a sickle shape and to be broken down more rapidly. Heterozygotes (*AS*) suffer slight anemia; homozygotes (*SS*) suffer severe anemia (sickle-cell disease) and usually die before reproducing. "Normal" homozygotes (*AA*) suffer higher mortality from malaria than heterozygotes (*AS*), however, because the organism that causes the most serious form of malaria (*Plasmodium falciparum*) develops in red blood cells. Since the red blood cells of heterozygotes are broken down more rapidly, the growth of *Plasmodium* is curtailed. Thus heterozygotes survive at a higher rate than either homozygote; their relative fitnesses have been estimated as $w_{AA} = 0.89$, $w_{AS} = 1.0$, $w_{SS} = 0.20$ (**Figure 12.13B**). Consequently, the frequency of the *S* allele is quite high (about $q = 0.13$) in some parts of Africa with a high incidence of malaria. The heterozygote advantage therefore arises from a balance of OPPOSING SELECTIVE FACTORS: anemia and malaria. In the absence of malaria, balancing selection gives way to directional selection, because then the *AA* genotype has the highest fitness. In the African-American population, which is not subject to malaria, the frequency of *S* is about 0.05, and it is presumably declining—unfortunately, as a result of mortality.

Antagonistic and varying selection

The opposing forces acting on the sickle-cell polymorphism are an example of **antagonistic selection**, which in this instance maintains the polymorphism because the heterozygote happens to have the highest fitness. Unless that is the case, antagonistic selection usually does not maintain polymorphism (Curtsinger et al. 1994). Suppose, for example, that the survival rates of an insect in the larval stage are 0.5 for genotypes A_1A_1 and A_1A_2 and 0.4 for A_2A_2, whereas the proportions of surviving larvae that then survive through the pupal stage are 0.6 and 0.9, respectively. In other words, the A_1 allele improves larval survival, but reduces pupal survival, compared with the A_2 allele. For genotypes A_1A_1 and A_1A_2, the proportion surviving to reproductive adulthood is $(0.5)(0.6) = 0.30$, and for A_2A_2 it is $(0.4)(0.9) = 0.36$. Thus A_2A_2 has a net selective advantage, and allele A_2 will be fixed.

Within a single breeding population, a fluctuating environment may favor different genotypes in different generations (TEMPORAL FLUCTUATION), or different genotypes may be best adapted to different microhabitats or resources (SPATIAL VARIATION). This form of antagonistic selection may maintain allelic variation at a locus, but does not necessarily do so (Hedrick 1986).

Temporal fluctuation in the environment may slow down the rate at which one or another allele approaches fixation, but it usually does not preserve multiple alleles indefinitely, because the decrease of each allele's frequency when it is disadvantageous is unlikely to be exactly balanced by its increase when it is advantageous—and unless the balance is exact, one of the alleles will eventually zigzag its way to fixation. Spatial variation in the environment, on the other hand, can often maintain polymorphism if different homozygotes in a single population are best adapted to different microhabitats or resources—that is, if they have different "niches." (This phenomenon is sometimes called **multiple-niche polymorphism**.) A stable multiple-niche polymorphism is more likely if each individual organism usually experiences only one of the environments. It is also more likely if selection is "soft" rather than "hard." SOFT SELECTION occurs when the number of survivors in a patch of a particular microenvironment is determined by competition for a limiting factor, such as space or food, and the *relatively* superior genotype has a higher probability of survival. Then selection determines the genotype frequencies among the surviving adults, but not their total number. HARD SELECTION occurs when the likelihood of survival of an individual in a microenvironment depends on its *absolute* fitness, not on the density of competitors. Selection then determines not only the genotype frequencies, but also the total number, of survivors. (For example, selection for insecticide resistance may be hard, since the survival of an individual insect may depend only on whether it has a resistant or susceptible genotype. In contrast, selection for long mouthparts that enable a bee to obtain nectar from flowers more rapidly than shorter-tongued bees would be soft selection if the shorter-tongued bees could obtain nectar, and survive and reproduce prolifically, in the absence of superior longer-tongued competitors.)

The black-bellied seedcracker (*Pyrenestes ostrinus*) provides a nice example of multiple-niche polymorphism (Smith 1993; **Figure 12.15**). Populations of this African finch have a bimodal distribution of bill width. The difference between wide and narrow bills seems to be due to a single allele difference. The two morphs differ in the efficiency with which they process seeds of different species of sedges, their major food: wide-billed birds process hard seeds, and narrow-billed birds soft seeds, more efficiently. Thomas B. Smith banded more than 2700 juvenile birds and found that survival to adulthood was lower for birds with intermediate-sized bills than for either wide- or narrow-billed birds. Thus diversifying selection, arising from the superior fitness of different genotypes on different resources,

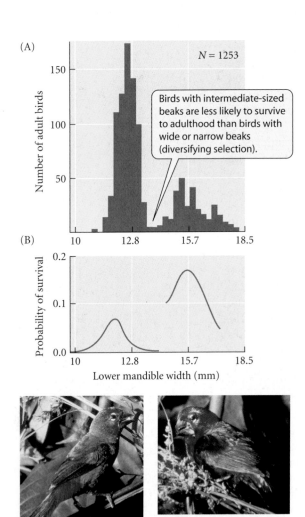

FIGURE 12.15 Multiple-niche polymorphism in the black-bellied seedcracker. (A) Probability of survival to adulthood of banded juvenile birds in relation to their lower mandible width, a measure of bill size. (B) The distribution of lower mandible width among adults is bimodal. The small- and large-billed morphs are shown at left and right, respectively. (After Smith 1993; photos courtesy of Thomas B. Smith.)

appears to maintain the polymorphism. Populations of many species are composed of individuals that differ in their use of resources (Bolnick et al. 2003).

The two sexes may be thought of as different "environments" in which different alleles may confer different fitness. In particular, an allele that enhances fitness in one sex may diminish fitness, relative to an alternative allele, in the other sex. Mathematical models show that such INTRALOCUS SEXUAL CONFLICT can often maintain polymorphism. Antagonistic selection between the sexes has several interesting consequences (Bonduriansky and Chenoweth 2009) and appears to be common (see Chapter 13).

Frequency-dependent selection

In the models we have considered so far, the fitness of each genotype is assumed to be constant within a given environment. Very often, however, *the fitness of a genotype depends on its own frequency relative to the frequencies of other genotypes in the population*. This phenomenon is referred to as **frequency-dependent selection**. This kind of balancing selection appears to maintain many polymorphisms, and it has many other important consequences, especially for the evolution of animal behavior, as we will see in Chapter 16.

INVERSE FREQUENCY-DEPENDENT SELECTION   In **inverse frequency-dependent selection**, the rarer a phenotype is in the population, the greater its fitness (**Figure 12.16A**). For example, the per capita rate of survival and reproduction of a dominant phenotype (with genotype A_1A_1 or A_1A_2) may be greatest when it is very rare and may decrease as it becomes more common; the same may be true of the recessive phenotype (with genotype

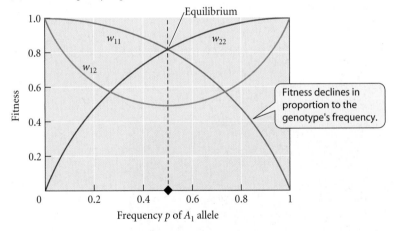

(A) Inverse frequency-dependent selection

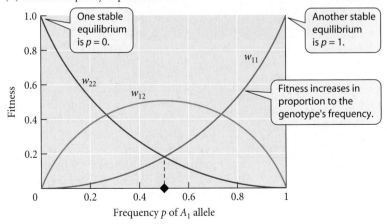

(B) Positive frequency-dependent selection

FIGURE 12.16 Two forms of frequency-dependent selection, in which the fitness of each genotype depends on its frequency in a population. (A) Inverse frequency-dependent selection. A genotype's fitness declines with its frequency, which depends on allele frequencies (expressed here as p, the frequency of A_1). The fitnesses of A_1A_1, A_1A_2, and A_2A_2 are $w_{11} = 1 - sp^2$, $w_{12} = 1 - spq$, and $w_{22} = 1 - sq^2$ in this model; the curves are calculated for $s = 1$. A stable equilibrium exists at $p = q = 0.5$. (B) Positive frequency-dependent selection, in which a genotype's fitness increases with its frequency. The fitnesses of A_1A_1, A_1A_2, and A_2A_2 are $1 + sp^2$, $1 + 2spq$, and $1 + sq^2$. (After Hartl and Clark 1989.)

A_2A_2). Thus when A_2 is at high frequency, it declines because A_2A_2 has lower fitness than A_1A_1 and A_1A_2, and likewise for A_1. Whatever the initial allele frequencies may be, they shift toward a stable equilibrium value, which in this case occurs when the frequencies of the two phenotypes are equal (i.e., when $q^2 = 0.5$). At this point, the mean fitnesses of both phenotypes are the same: neither has an advantage over the other.

Many biological phenomena can give rise to inverse frequency-dependent selection. The self-incompatibility alleles of many plant species are a striking example. These alleles enforce outcrossing because pollen carrying any one allele cannot effectively grow down the stigma of the plant that produced it or of any other plant that carries the same allele. Thus, if there are three alleles, S_1, S_2, and S_3, in a plant population, pollen of type S_1 can grow on stigmas, and fertilize ovules, of genotype S_2S_3, but not on S_1S_2 or S_1S_3; S_2 pollen can fertilize only S_1S_3 plants, and S_3 pollen can fertilize only S_1S_2 plants. (Notice that plants cannot be homozygous at this locus.) A new S allele that arises by mutation can fertilize almost all plants because it is rare at first, and so can increase in frequency until it reaches a frequency of about $1/k$, where k is the number of S alleles in the population. Therefore we should expect to find a large number of self-incompatibility alleles, all about equal in frequency, at equilibrium. In some plant species, indeed, hundreds of such alleles exist.

European populations of the alpine elderflower orchid (*Dactylorhiza sambucina*) consist of about half purple-flowered and half yellow-flowered plants. Luc Gigord and coworkers (2001) placed arrays of 50 plants, with varying proportions of the two morphs, in ten locations. They found that pollinating bumblebees visited flowers of the minority morph more frequently, removing their pollen masses more often—an indicator of the plant's success in the male role (**Figure 12.17A**). The rarer morph also had higher fitness in its female role, receiving pollen and developing fruit more often (**Figure 12.17B**). Like many orchids, *Dactylorhiza* does not provide pollinators with a "reward": it lacks nectar, and the pollen forms a solid mass that bees do not use to feed their larvae. The authors suggest that bees learn to avoid nonrewarding flowers and that they learn faster if a flower type (color) is common.

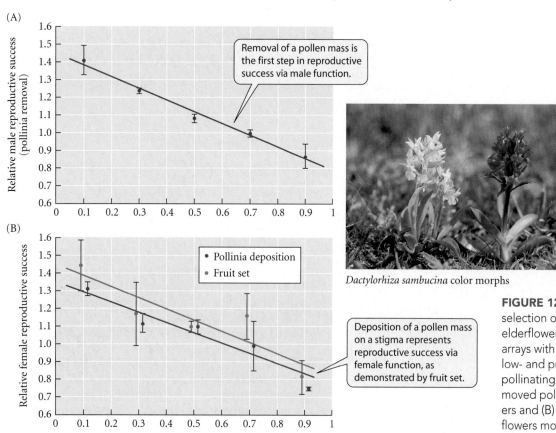

(A)

Relative male reproductive success (pollinia removal)

Removal of a pollen mass is the first step in reproductive success via male function.

(B)

Relative female reproductive success

• Pollinia deposition
• Fruit set

Deposition of a pollen mass on a stigma represents reproductive success via female function, as demonstrated by fruit set.

Frequency of yellow morph

Dactylorhiza sambucina color morphs

FIGURE 12.17 Frequency-dependent selection of flower color in the alpine elderflower orchid. In experimental arrays with different proportions of yellow- and purple-flowered individuals, pollinating bumblebees both (A) removed pollen masses from yellow flowers and (B) deposited them on yellow flowers most often when they were the minority type. (After Gigord et al. 2001.)

Selection is often frequency-dependent when genotypes compete for limiting resources, as in the model of soft selection described above. Suppose that two genetically determined phenotypes, P_1 and P_2, can survive on either resource 1 or resource 2, but that P_1 is a superior competitor for resource 1 and P_2 is a superior competitor for resource 2. If P_1 is rare, each P_1 individual will compete for resource 1 chiefly with inferior P_2 competitors, and so will have a higher per capita rate of increase than P_2 individuals. As P_1 increases in frequency, however, each P_1 individual competes with more individuals of the same genotype, so its per capita advantage, relative to P_2, declines. The same pattern applies to phenotype P_2.

This effect is illustrated in an experiment on the grass *Anthoxanthum odoratum*, which can be propagated asexually by vegetative cuttings. In a natural environment, Norman Ellstrand and Janis Antonovics (1984) planted small "focal" cuttings and surrounded each with competitors. The surrounding plants were either the same genotype as the focal individual, taken as vegetative shoots from the same parent plant, or different genotypes, obtained from sexually produced seed. After a season's growth, focal individuals on average were larger and produced more seed (i.e., had higher fitness) if surrounded by different genotypes than if surrounded by the same genotype. This pattern is just what we would expect if different genotypes use somewhat different resources. (Perhaps certain mineral nutrients are partitioned differently by adjacent plants that differ in genotype.) The more intense competition among individuals of the same genotype would then impose inverse frequency-dependent selection. In the same vein, agricultural researchers have sometimes found that fields planted with a mixture of crop varieties yield more than fields of a single variety.

EVOLUTION OF THE SEX RATIO BY FREQUENCY-DEPENDENT SELECTION Why is the sex ratio about even (1:1) in so many species of animals? This is quite a puzzle, because from a group-selectionist perspective, we might expect that a female-biased sex ratio (i.e., production of more females than males) would be advantageous because such a population could grow more rapidly. If sex ratio evolves by individual selection, however, and if all females have the same number of progeny, why should a genotype producing an even sex ratio have an advantage over any other?

The solution to this puzzle was provided by the great population geneticist R. A. Fisher (1930), who realized that because every individual has both a mother and a father, females and males must contribute equally to the ancestry of subsequent generations. Therefore individuals that vary in the sex ratio of their progeny can differ in the number of their grandchildren (and later descendants), and thus can differ in fitness if it is measured over two or more generations.

To see why this is so, we must distinguish the sex ratio in a population (the POPULATION SEX RATIO, S) from that in the progeny of an individual female (an INDIVIDUAL SEX RATIO, s). The sex ratio is defined as the proportion of males. In a large, randomly mating population, the fitness of a genotype that produces a given individual sex ratio depends on the population sex ratio, which in turn depends on the frequencies of the various individual-sex-ratio genotypes. Selection favors genotypes whose individual sex ratios are biased toward that sex that is in the minority in the population as a whole because the average per capita reproductive success of the minority sex is greater than that of the majority sex.

For example, suppose a population's sex ratio is 0.25 (1 male to 3 females) because it consists of a genotype with this individual sex ratio. Suppose each female has 4 offspring. The average number of progeny per female is 4, but since every offspring has a father, the average number of progeny sired by a male is 12 (since each male mates with 3 females on average). Thus a female has 4 grandchildren through each of her daughters and 12 through each son, for a total of $(3 \times 4) + (1 \times 12) = 24$ grandchildren. Now suppose a rare genotype with an individual sex ratio of 0.50 (2 daughters and 2 sons) enters the population. Each such individual has $2 \times 4 = 8$ grandchildren through her daughters and $2 \times 12 = 24$ grandchildren through her sons, for a total of 32. Since this is a greater number of descendants than the mean number per individual of the prevalent female-biased genotype, any allele that causes a more male-biased progeny in this female-biased population will increase in frequency. Likewise, if the population sex ratio were male-biased, an allele for female-biased

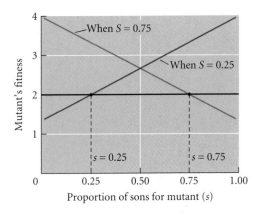

FIGURE 12.18 Frequency-dependent selection on sex ratio. Mutants may arise that vary in individual sex ratio (*s*, the proportion of sons among a female's progeny). The fitness of each such mutant, based on its average number of grandchildren, depends on the sex ratio in the population (*S*). The average fitness of individuals in the population equals 2. When *S* = 0.25—i.e., when 25 percent of the population is male—the fitness of a mutant is directly proportional to the proportion of sons among its offspring, and it is greater than the mean fitness of the prevalent genotype if the mutant's *s* exceeds 0.25. Any such mutant will therefore increase in frequency. Conversely, if *S* = 0.75, the fitness of a mutant is inversely proportional to its individual sex ratio, and its fitness increases in frequency if its *s* is less than 0.75. (After Charnov 1982.)

individual sex ratios would spread. By this reasoning, a genotype that produces an even sex ratio (0.5) has the highest fitness and cannot be replaced by any other genotype (**Figure 12.18**). (A genotype that produces a sex ratio of 0.5 represents an EVOLUTIONARILY STABLE STRATEGY, or ESS, as we will see in Chapter 14.)

Alexandra Basolo (1994) tested this theory using the platyfish *Xiphophorus maculatus*, which has three kinds of sex chromosomes, W, X, and Y. Females are XX, WX, or WY, and males are XY or YY. Of the six possible crosses, four yield a sex ratio of 0.5, but XX mated with YY yields all sons, and WX mated with XY yields 0.25 sons, so there is variation among genotypes in individual sex ratios. Basolo set up experimental populations with different frequencies of these chromosomes, which carry different color-pattern alleles so that chromosome frequencies can be followed. Two populations were initiated with sex ratios (proportion of males) of 0.25 percent and 0.78 percent, respectively. Within only two generations, the sex ratio in both populations evolved nearly to 0.5, as Fisher's theory predicts (**Figure 12.19**).

Multiple Outcomes of Evolutionary Change

One of the most important principles in evolution is that initial genetic conditions often determine which of several paths, or trajectories, of genetic change a population will follow. Thus the evolution of a population often depends on its previous evolutionary history. Positive frequency-dependent selection and heterozygote disadvantage are two important factors that can give rise to multiple outcomes—**multiple stable equilibria**.

Positive frequency-dependent selection

In **positive frequency-dependent selection**, the fitness of a genotype is greater the more frequent it is in a population. As a result, *whichever allele is initially more frequent will be fixed* (Figure 12.16B).

For example, the unpalatable tropical butterfly *Heliconius erato* has many different geographic races that differ markedly in their warning coloration patterns (**Figure 12.20**). Each race is monomorphic. Adjacent geographic races interbreed at zones only a few kilometers wide. (The geographic variation in color pattern in this species is closely paralleled by that in another unpalatable species of *Heliconius*—a spectacular example of Müllerian mimicry, as described in Chapter 3.) James Mallet and Nicholas Barton (1989) showed that within *Heliconius erato*, gene flow from one geographic race to another is countered by positive frequency-dependent selection:

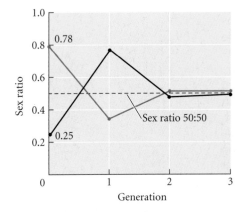

FIGURE 12.19 Changes in sex ratio (proportion of males) in two experimental populations of platyfish, initiated with sex ratios (proportions of males) of 0.25 and 0.78. In both populations, the sex ratio evolved nearly to 0.5 within only two generations. (After Basolo 1994.)

immigrant butterflies that deviate from the locally prevalent color pattern are selected against because predators have not learned to avoid attacking butterflies with those unusual color patterns. On either side of a contact zone between two geographic races in Peru, Mallet released *H. erato* of the other race and, as a control, butterflies of the same race from a different locality. The butterflies, marked so they could be recognized, were repeatedly recaptured for some time thereafter. Compared with control butterflies with the same color pattern as the population into which they were released, far fewer of those with the allopatric color pattern were captured. From bill marks left on the wings of butterflies that had escaped from birds, the authors concluded that the missing butterflies were lost to bird predation. They calculated an average selection coefficient of 0.52 against the "wrong" color pattern in either population, which amounts to a selection coefficient of about $s = 0.17$ at each of the three major loci that control the differences in color pattern between the races. Individuals that conform to the locally prevalent color pattern are strongly favored over those that deviate from it.

Heterozygote disadvantage

The case in which the heterozygote has lower fitness than either homozygote is called **heterozygote disadvantage** or **underdominance**. If a population is initially monomorphic for A_1A_1, and A_2 then enters at low frequency by mutation or gene flow, almost all A_2 alleles are carried by heterozygotes (A_1A_2); since their fitness is lower than that of A_1A_1, selection reduces the frequency (q) of A_2 to zero. Likewise, A_1 is eliminated if it enters a monomorphic population of A_2A_2, which also has a greater fitness than A_1A_2. Monomorphism for either A_1A_1 or A_2A_2 is therefore a stable equilibrium: the alternative allele cannot increase in frequency because it is carried mostly by less fit heterozygotes (see Case 4 in Box 12A). This model applies to some chromosome rearrangements, such as inversions and translocations, which may reduce the fertility of heterozygotes as a result of improper segregation in meiosis (see Figures 8.24, 8.25, and the next section of this chapter).

FIGURE 12.20 An extraordinary case of geographic variation and Müllerian mimicry. The butterflies *Heliconius melpomene* and *H. erato* both vary geographically in coloration, but vary precisely in parallel. In each pair of butterflies shown here, the upper is *H. melpomene* and the lower is *H. erato*. Both species are unpalatable to predators. Races of *H. erato* interbreed with one another to some extent where they meet, but the hybrid zone between them is very narrow; the same is true of *H. melpomene*. (Photos courtesy of Andrew Brower.)

If the two homozygotes' fitnesses are different, but both are greater than that of the heterozygote, the mean fitness in a population that has become fixed for the less fit homozygote is less than if it were fixed for the other homozygote, but *selection cannot move the population from the less fit to the more fit condition*. Thus *a population is not necessarily driven by natural selection to the most adaptive possible genetic constitution.*

Adaptive landscapes

Recall that we can calculate the mean fitness ($\overline{w}$) of individuals in a population with any conceivable allele frequency (p) and plot a curve showing $\overline{w}$ as a function of p (see Figure 12.7). When fitnesses are constant, natural selection changes allele frequencies in such a way that mean fitness ($\overline{w}$) increases, so that the population moves up the slope of this curve. The current location of the population on this slope is then a simple guide to how allele frequencies will change under selection: simply see which direction of allele frequency change will increase $\bar{w}$. For an underdominant locus, the curve dips in the middle and slopes upward to $p = 0$ and $p = 1$ (see Figure 12.7D and Figure 12.21). Thus natural selection decreases or increases p depending on whether a population begins to the left or the right of the minimum of the $\bar{w}$ curve.

Curves such as those in Figures 12.7 and 12.21 are often referred to as **adaptive landscapes**, or ADAPTIVE TOPOGRAPHIES. The curve in these figures represents two **adaptive peaks** separated by an **adaptive valley**. This metaphor, introduced by Sewall Wright, is widely used in evolutionary biology, and we will refer to it in Chapter 18. Each point on the curve (the adaptive landscape) represents the *average* fitness of individuals in a hypothetical population made up of (in this case) three genotypes with frequencies p^2, $2pq$, and q^2. Each possible value of p—each possible hypothetical population—yields a different value on the *x*-axis, and therefore exists at a different point on the landscape. When, as in Figure 12.7D, the relationship of $\bar{w}$ to p has two or more maxima, two genetically different populations (e.g., with $p = 0$ or $p = 1$ in this case) can (but need not) have the same mean fitness ($\bar{w}$) *under the same environmental conditions*. Different environments, which might alter the fitnesses of the genotypes relative to one another, are represented not by different points on any one landscape, but rather by *different landscapes*—different relationships between $\bar{w}$ and p.

Interaction of Selection and Genetic Drift

In developing the theory of selection so far, we have assumed an effectively infinite population size. *In a finite population*, however, *allele frequencies are simultaneously affected by both selection and chance*. Just as the movement of an airborne dust particle is affected both by the deterministic force of gravity and by random collisions with gas molecules (Brownian motion), the effective size (N_e) of a population and the strength of selection (s) both affect changes in allele frequencies. The effect of random genetic drift is negligible if selection on a locus is strong relative to the population size—that is, if s is much greater than $1/(4N_e)$ (i.e., if $4N_e s > 1$). Conversely, if s is much less than $1/(4N_e)$ (i.e., if $4N_e s < 1$), selection is so weak that the allele frequencies change mostly by genetic drift: the alleles are *nearly neutral*. The critical value is $4N_e s$: genetic drift predominates if selection is weak or the effective population size is small.

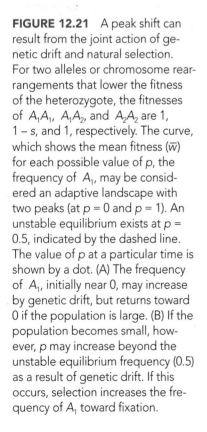

FIGURE 12.21 A peak shift can result from the joint action of genetic drift and natural selection. For two alleles or chromosome rearrangements that lower the fitness of the heterozygote, the fitnesses of A_1A_1, A_1A_2, and A_2A_2 are 1, $1 - s$, and 1, respectively. The curve, which shows the mean fitness ($\bar{w}$) for each possible value of p, the frequency of A_1, may be considered an adaptive landscape with two peaks (at $p = 0$ and $p = 1$). An unstable equilibrium exists at $p = 0.5$, indicated by the dashed line. The value of p at a particular time is shown by a dot. (A) The frequency of A_1, initially near 0, may increase by genetic drift, but returns toward 0 if the population is large. (B) If the population becomes small, however, p may increase beyond the unstable equilibrium frequency (0.5) as a result of genetic drift. If this occurs, selection increases the frequency of A_1 toward fixation.

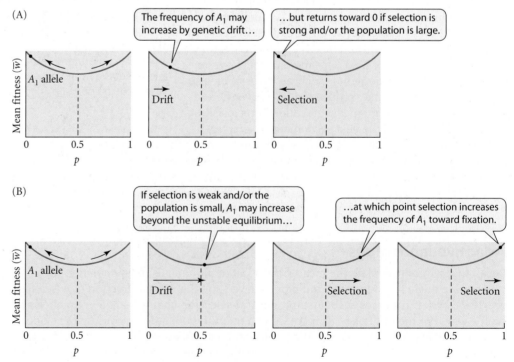

This principle may be important if heterozygotes are inferior in fitness, so that the adaptive landscape has two peaks (see Figure 12.7D). Selection alone cannot move a population down the slope of one peak and across a valley to the slope of another peak, even if the second peak is higher: a population does not first become poorly adapted so that it can then become better adapted (**Figure 12.21A**). But during episodes of very low population size, allele frequencies may fluctuate so far by genetic drift that they cross the adaptive valley—after which selection can move the population "uphill" to the other peak (**Figure 12.21B**). The probability that such a **peak shift** will occur (Barton and Charlesworth 1984) depends on the population size and on the difference in height (mean fitness) between the valley and the initially occupied peak. Thus, when there are multiple stable equilibria, *genetic drift and selection may act in concert to accomplish what selection alone cannot*, moving a population from one adaptive peak to another.

This theory explains how populations can come to differ in underdominant chromosome rearrangements such as translocations and pericentric inversions. Some chromosome rearrangements are thought to conform to the model of underdominance because heterozygotes have lower fertility than either homozygote (see Chapter 8). For example, local populations of the Australian grasshopper *Vandiemenella viatica* are monomorphic for different chromosome fusions and pericentric inversions. Heterozygotes for these chromosomes have many aneuploid gametes. Any such chromosome, introduced by gene flow into a population monomorphic for a different arrangement, is reduced in frequency by natural selection, so no two "chromosome races" are sympatric; instead, they meet in "tension zones" only 200 to 300 meters wide (**Figure 12.22**; White 1978). Because these grasshoppers are flightless and quite sedentary, local populations are small, providing the opportunity for genetic drift to occasionally initiate a peak shift whereby a new chromosome arrangement is fixed.

The effect of population size on the efficacy of selection has several important consequences. First, a population may not attain exactly the equilibrium allele frequency predicted from its genotypes' fitnesses; instead, it is likely to wander by genetic drift in the vicinity of the equilibrium frequency. Second, the probability of fixation of new mutations

FIGURE 12.22 The chromosome configurations and geographic distributions of three chromosome races of the flightless grasshopper *Vandiemenella viatica* on Kangaroo Island, off the coast of South Australia. The dots on the map indicate sampling sites. The diagrams below the map show the chromosome configurations of the three races; each chromosome is represented by a vertical bar. Only one of the two sets of chromosomes is shown for each race. The (probably) ancestral configuration, found in race *viatica*$_{19}$, has 10 pairs of chromosomes. Compared with race *viatica*$_{19}$, chromosomes B (blue) and 6 (purple) are fused in the other two races. In race P24 (XY), a chromosome 1 (red) has become translocated to the X chromosome (green). The other member of the chromosome 1 pair is therefore left unpaired, and has the status of a Y chromosome. A translocated or fused chromosome may have become established in a geographic area by a "peak shift" in a small population. (After White 1978.)

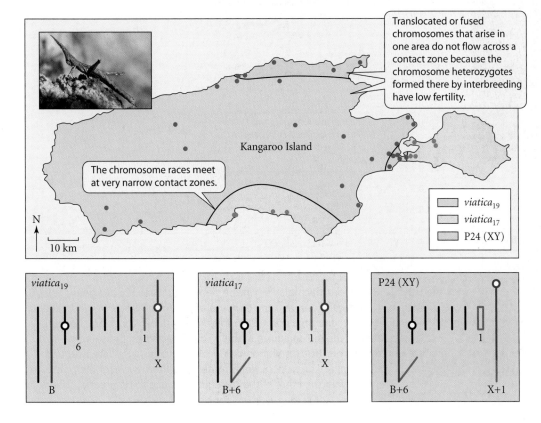

Translocated or fused chromosomes that arise in one area do not flow across a contact zone because the chromosome heterozygotes formed there by interbreeding have low fertility.

The chromosome races meet at very narrow contact zones.

Kangaroo Island

N

10 km

viatica$_{19}$
viatica$_{17}$
P24 (XY)

viatica$_{19}$ 6   1   X   B

viatica$_{17}$ 1   X   B+6

P24 (XY)   1   B+6   X+1

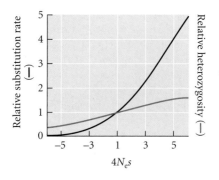

FIGURE 12.23 The substitution rate of selected relative to neutral mutations (black line) and the relative frequency of heterozygotes (red line) both increase as a function of $4N_es$, the product of effective population size and the selection coefficient. (After Fay and Wu 2003.)

is affected by both genetic drift and selection, a subject developed at length by Tomoko Ohta (1992) in her NEARLY NEUTRAL THEORY of molecular evolution. *Slightly deleterious mutations can become fixed by genetic drift*, especially if the effective population size is small, as it may be if population bottlenecks occur. Most importantly, *a slightly advantageous mutation is less likely to be fixed by selection if the population is small than if it is large, because it is more likely to be lost simply by chance*. Thus, if mutations with a selective advantage *s* arise repeatedly, both the rate of fixation of advantageous mutations compared with neutral mutations and the relative levels of heterozygosity for the two types of mutations increase with increasing $4N_es$ (**Figure 12.23**; Fay and Wu 2003).

The Strength of Natural Selection

Until the 1930s, most evolutionary biologists followed Darwin in assuming that the intensity of natural selection is usually very slight. By the 1930s, however, examples of very strong selection came to light. One of the first examples was INDUSTRIAL MELANISM in the peppered moth, *Biston betularia* (Majerus 1998; Grant 2012). In some parts of England, after the onset of the Industrial Revolution in the mid-nineteenth century had darkened the trunks of trees, a black form of this moth carrying a dominant allele increased in frequency relative to the pale gray form that had been typical before that time (**Figure 12.24A**). Museum specimens acquired continuously since the mid-nineteenth century showed that in less than a century, the black (melanic) form increased from about 1 percent to more than 90 percent of the moths in some areas. This rate of change is so great that it implies a very substantial selective advantage, possibly as high as 50 percent, for the melanic form (Haldane 1932). As air pollution became regulated and conditions reverted to their preindustrial state during the late twentieth century, the frequency of the melanic form declined rapidly, and the pale gray genotype increased (**Figure 12.24B**; Cook 2003).

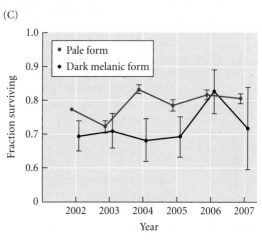

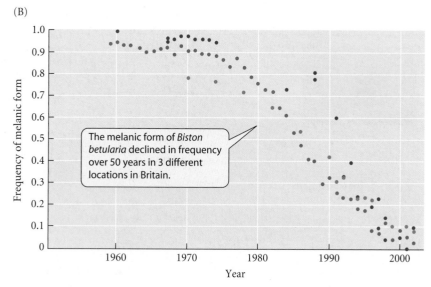

The melanic form of *Biston betularia* declined in frequency over 50 years in 3 different locations in Britain.

FIGURE 12.24 Industrial melanism in the peppered moth (*Biston betularia*). (A) The pale gray "typical" form and the dark melanic form on a tree trunk darkened by air pollution (left) and on a normal, nonblackened trunk (right). (B) The decline in the frequency of the melanic form in three British localities, indicated by symbols of different colors, as air pollution was mitigated during the late twentieth century. (C) In a locality with nonblackened tree bark, the survival of the pale form was consistently higher than that of the melanic form due to differential predation by birds. (B after Cook 2003; C after Cook et al. 2012.)

(A) Survival differences

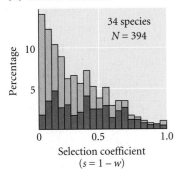

(B) Reproductive differences

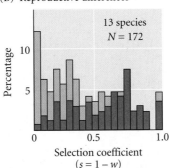

FIGURE 12.25 A compilation of selection coefficients (s) reported in the literature for discrete, genetically polymorphic traits in natural populations of various species. The total height of each bar represents the percentage of all reported values in each interval, and the red portion represents the percentage shown by statistical analysis to be significantly different from zero. N is the number of values reported, based on one or more traits from each of several species. (A) Selection based on differences in survival. (B) Selection based on differences in reproduction (fecundity, fertility, and sexual selection). (After Endler 1986.)

In the 1950s, British researcher H. B. D. Kettlewell showed that birds attacked a greater proportion of pale than melanic moths where tree trunks were blackened by air pollution. From 2002 to 2007, Michael Majerus showed that among 4864 moths placed on normal, nonblackened trees, the survival rate of melanic moths was only 91 percent that of the pale form (**Figure 12.24C**). Majerus directly observed that birds selectively captured melanic moths. Thus strong selection by differential predation accounts for both the increase and the subsequent decline of the melanic allele (Cook et al. 2012).

In the last few decades, components of fitness have been quantified for polymorphic traits in many species. The selection coefficients estimated in natural populations range from low to very high (**Figure 12.25**). Moreover, selection acting through both survival and reproduction can be very strong. Natural selection is a powerful factor in evolution and is often far stronger than Darwin ever would have imagined.

Molecular Evidence for Natural Selection

Until recently, studies of natural selection were performed mostly on field or laboratory populations and focused on phenotypic characteristics. Today such studies have been joined by a very different approach: inferring selection from variation in DNA sequences. The advantages of this approach are many: in almost any species, a great many genes can be studied in a short time without lengthy and often difficult observations; and the genes themselves, rather than inherited phenotypes, are studied directly. Without further biological or ecological information, however, this approach does not reveal the causes of selection or, often, even the function of the gene under study. Furthermore, the methods depend on assumptions that may not hold in particular cases, and it can be difficult to distinguish the effects of selection from those of other factors, such as population structure and growth. Nevertheless, these powerful approaches are providing deep insights into selection in humans and other species.

Detecting selection from geographic variation

Biologists have long used patterns of variation among populations of a species, such as correlations between organisms' characteristics and environmental factors, to infer that a characteristic is likely to be adaptive (see Chapter 9). Consistent correlations between allele frequency and an ecological variable can reveal genes that are probably responding to selection. For example, a genome-wide scan of SNPs among human populations revealed several loci at which different alleles are more common in populations that subsist on foraging than in agricultural or pastoral populations (**Figure 12.26**).

When acting on selectively neutral variation, genetic drift causes divergence among populations, but gene flow tends to homogenize them (see Chapter 10). These factors affect all loci in the genome to the same extent, so all loci should arrive at much the same level of variation among populations. The variation among populations in allele frequency is measured by the fixation index, F_{ST} (see Chapter 10). F_{ST} values will vary among different

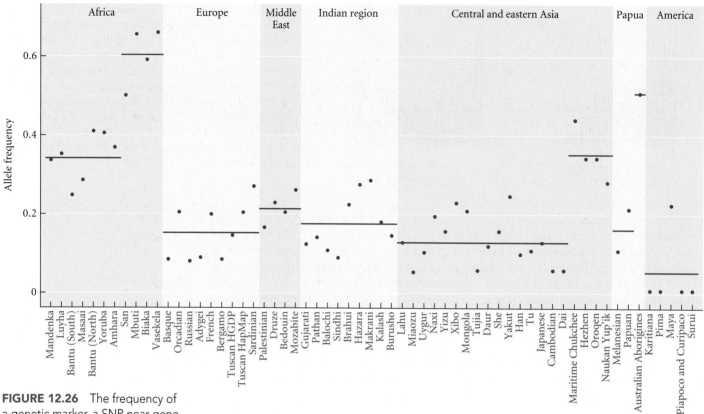

FIGURE 12.26 The frequency of a genetic marker, a SNP near gene region IL22, in human samples from throughout the world. Populations that subsist by foraging are shown in brown, and all others are shown in blue; horizontal bars show mean frequency in sets of foraging and nonforaging populations. The consistently higher frequency of the marker in foraging populations implies that a nearby gene confers an advantage to people with that way of life. The gene and its function have not yet been identified. (After Hancock et al. 2010.)

neutral loci in the genome, by chance, but the amount of such variation can be specified by mathematical modeling.

If different alleles at a locus are selected in different populations, and if selection is strong relative to gene flow, their frequencies should differ more among populations than those at neutral loci, and the locus should have an unusually high F_{ST}. Thus loci that have exceptionally high F_{ST} values, compared with the majority of loci screened by molecular methods, are likely to be affected by selection (Beaumont and Nichols 1996; Storz 2005). Among a number of allozyme polymorphisms studied in the deer mouse *Peromyscus maniculatus*, for example, the albumin locus revealed unusually great allele frequency differences among populations, and the degree of difference was greater for pairs of populations at extremely different altitudes (**Figure 12.27**). These observations provide strong evidence that the genetic difference at this locus (or possibly a closely linked locus) is adaptive, probably in relation to an altitudinal variable such as oxygen concentration. The F_{ST} test has also

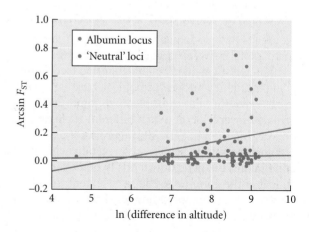

FIGURE 12.27 F_{ST}, the variance in allele frequency among samples of the mouse *Peromyscus maniculatus* from different locations, is plotted for several presumptively neutral enzyme-encoding loci and for the albumin locus. Each point is F_{ST} for a pair of localities, plotted against the difference in their altitude. The exceptionally high F_{ST} values for the albumin locus suggest that different alleles confer greater fitness at different altitudes. (After Storz 2005).

attested to natural selection on human genes. For example, the Duffy blood group locus shows a much higher F_{ST} between African and non-African populations than do other loci; some African populations are nearly fixed for an allele that confers resistance to malaria (Hamblin et al. 2002). F_{ST} is exceptionally high for at least six genes that affect skin pigmentation, one of several lines of evidence that differences in this trait among populations are adaptive (Norton et al. 2007). Dark skin is thought to confer protection against harmful effects of ultraviolet (UV) light, which is strongest at lower latitudes, whereas reduced melanization enhances light-dependent synthesis of previtamin D, which may be limited at higher latitudes (Jablonski and Chaplin 2000).

A test for selection: Variation within and among species

According to the neutral theory of molecular evolution, the rate of allele substitution over time and the equilibrium level of heterozygosity are both proportional to the neutral mutation rate, u_0. If, because of differences in constraints or other factors, various kinds of DNA sequences or nucleotide sites differ in their rate of neutral mutation, those sequences or sites that differ more *among* related species should also display greater levels of variation *within* species. That is, *there should be a positive correlation between the heterozygosity at a locus and its rate of evolutionary change in its DNA sequence.* If one set of nucleotide sites is twice as heterozygous as another within a species, it should also display twice the number of substitutions between that species and a related species—if all the variation is selectively neutral.

John McDonald and Martin Kreitman (1991) applied this principle in an analysis of DNA sequence variation in the coding region of the *Adh* (alcohol dehydrogenase) gene (see Figure 9.17) in three closely related species of *Drosophila*. Polymorphic sites (differences within species) and substitutions (differences among species) were classified as either synonymous or nonsynonymous (amino acid–replacing). If the neutral mutation rate is u_R for replacement (nonsynonymous) changes and u_S for synonymous changes, then according to the neutral theory, the ratio of replacement to synonymous differences should be the same—$u_R:u_S$—for both polymorphisms and substitutions, if indeed the replacement changes are subject only to genetic drift. The data (**Table 12.2**) showed, however, that only 5 percent of the polymorphisms, but fully 29 percent of the substitutions that distinguish species, were replacement changes.

Because the ratio of replacement to synonymous substitutions differed between polymorphisms and between-species substitutions, and thus did not fit the prediction of the neutral theory, McDonald and Kreitman inferred that the evolution of amino acid–replacing substitutions is an adaptive process governed by natural selection. (Their analysis, adopted by many other researchers since then, is called the **McDonald-Kreitman [MK] test**). Their logic was that if most replacement substitutions are advantageous rather than neutral, they will increase in frequency and be fixed more rapidly than by genetic drift alone. They will therefore spend less time in a polymorphic state than selectively neutral synonymous changes do, and will thus contribute less to polymorphic variation within species. Many of the methods used to find evidence of natural selection in DNA sequences are variations of the McDonald-Kreitman test. By applying the MK test to great numbers

TABLE 12.2 Replacement (nonsynonymous) and synonymous substitutions and polymorphisms within and among three *Drosophila* species[a]

	Polymorphisms	Substitutions
Replacement	2	7
Synonymous	42	17
Percentage replacement	4.5	29.2

Source: Data from McDonald and Kreitman 1991.

[a]*D. melanogaster, D. simulans,* and *D. yakuba.*

of genes, several researchers have concluded that 40 to 50 percent of replacement (non-synonymous) substitutions among species of *Drosophila* are adaptive (Sella et al. 2009). A similar analysis suggested that 10 to 20 percent of nonsynonymous differences between chimpanzee and human genomes were propelled by natural selection (Boyko et al. 2008).

In a conceptually related approach, many authors have sought indication of selection in the ratio of nonsynonymous differences (d_n) to synonymous differences (d_s) between species, on the assumption that synonymous substitutions are mostly neutral. This ratio is expected to equal 1 if the nonsynonymous differences are selectively neutral. A ratio $d_n/d_s < 1$ implies that many possible mutations are deleterious and have been purged by purifying selection, whereas $d_n/d_s > 1$ implies a preponderance of fixed advantageous mutations. This test is very conservative because it reveals positive selection only if enough advantageous mutations have occurred to stand out against the neutral mutations. One interesting result of applying this test is that reproductive proteins, such as gamete surface proteins and seminal fluid proteins, evolve in amino acid sequence much faster than the neutral expectation in a wide variety of animals (Swanson and Vacquier 2002), including humans and other apes (Wyckoff et al. 2000).

Detecting selection from DNA sequences: Theoretical expectations

Selection on DNA sequences may be inferred if the pattern of variation differs from patterns expected under the neutral theory of molecular evolution. We saw in Chapter 10, for example, that at equilibrium between mutation and genetic drift, the expected amount of neutral variation in a diploid population, as expressed by the frequency of heterozygotes per nucleotide site, is

$$\frac{4N_e u_0}{4N_e u_0 + 1}$$

where N_e is the effective population size and u_0 is the rate of neutral mutation. Furthermore, the nucleotides at different variable sites in a DNA sequence should not be correlated with one another at equilibrium because recombination, even between closely linked sites, should eventually lead to linkage equilibrium (see Chapter 9).

Now suppose that an advantageous mutation occurs at a particular nucleotide site in a gene, and consider the effects of positive selection of this mutation on *neutral variation at sites that are closely linked to the selected site*. As the frequency of the new mutation increases, all of the gene copies with that mutation will have descended from the single copy in which the mutation first occurred. As John Maynard Smith and J. Haigh (1974) pointed out long before DNA sequencing was feasible, the neutral variant nucleotides in that ancestral gene copy, linked to the mutation, will also increase, by hitchhiking. Thus not only the advantageous mutation, but also the DNA sequence that flanks it for some distance on both sides, becomes fixed, and *all neutral variation in the gene is eliminated* by a **selective sweep** (**Figure 12.28**). Therefore POSITIVE DIRECTIONAL SELECTION (selection for an advantageous mutation) reduces variation at closely linked sites, and variation is only slowly reconstituted as new neutral mutations occur among the copies of the advantageous gene (Nielsen 2005). If, furthermore, we observe the population when the advantageous mutation has reached intermediate frequency, but before it is fixed, we will observe *linkage disequilibrium (LD) among various neutral polymorphic sites*, because the advantageous mutation has remained associated with particular nucleotides at linked sites, which are therefore correlated with one another.

How far do these effects extend along the DNA sequence from the selected site? The recombination rate (r) is greater between the selected site and sites more distant from it. As r increases, the initial hitchhiking effect is weaker, and the LD between a given site and the advantageous mutation breaks down more rapidly (**Figure 12.29**). However, the higher the selective advantage (s) of the mutation, the more rapidly the selected haplotype (the mutation plus its linked neutral variants) increases in frequency, so that closer sites have

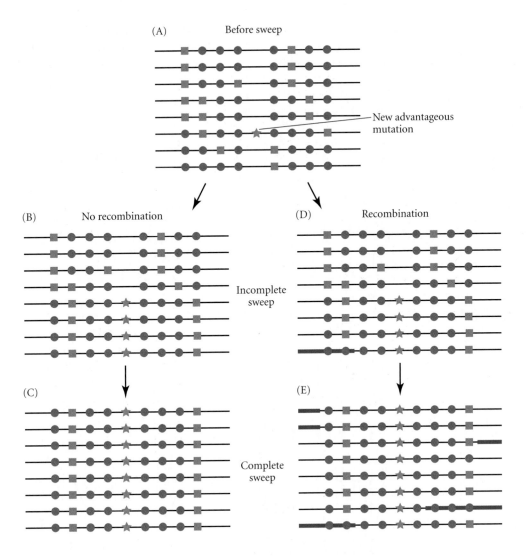

(A) Before sweep
(B) No recombination
(D) Recombination
Incomplete sweep
(C)
(E)
Complete sweep
— New advantageous mutation

FIGURE 12.28 Selective sweeps. Each panel shows a sample of eight chromosomes from a population. Eight neutral polymorphisms, such as SNPs, are indicated, each with two states (green square or red circle). (A) A new advantageous mutation (star) arises on one chromosome. (B) If there is little or no recombination, the mutation increases in frequency (lower four chromosomes), causing a partial sweep that reduces the amount of linked variation. Note that the SNP variants linked to the mutation are perfectly correlated with one another, forming a long haplotype that has intermediate frequency. The SNPs on chromosomes that lack the mutation show little or no linkage disequilibrium. (C) The advantageous mutation has been fixed and has completely swept away linked variation. (D) If there is recombination, the increase in frequency of the advantageous mutation has much the same effect as in (B), but note that recombination has partly broken down linkage disequilibrium of SNPs with the mutation. The colored segment of the lowermost chromosome has been acquired from another chromosome by crossing over. (E) Fixation of the advantageous allele results in considerable linkage disequilibrium and a high frequency of long haplotypes, but to a lesser extent than if there were no recombination. (After Nielsen et al. 2007.)

less opportunity to recombine with other haplotypes in the population. Thus the higher *s* is and the lower *r* is, the longer the haplotype. Because LD declines due to recombination (see Figure 9.20), one might use the length of the LD segment to estimate when an advantageous mutation arose (if one knows *r* from genetic data and can assume or estimate a value of *s*).

The effects of various forms of selection are evident in gene genealogies. Consider two unlinked loci, one that has been evolving solely by genetic drift (**Figure 12.30A**) and one that has experienced a selective sweep (**Figure 12.30B**). Compared with the neutrally evolving gene, the copies of the gene that was fixed by selection are descended from a more recent common ancestor (the one in which the favorable mutation occurred); they have had less time to accumulate different

FIGURE 12.29 A selective sweep causes an excess of the most common allele and a deficiency of intermediate-frequency alleles. Variability (*S*), as measured by the number of variable nucleotide sites in a sample of sequences, is lowered near the site of an advantageous mutation, as is Tajima's *D*, an index of the departure of the frequency distribution of different sequences (alleles) from that expected under the neutral theory. Linkage disequilibrium is greatest near the selected site. (After Nielsen 2005.)

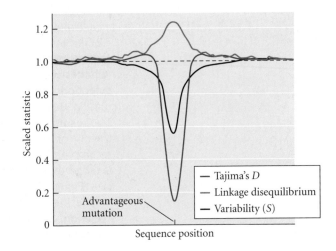

Advantageous mutation

Sequence position

— Tajima's *D*
— Linkage disequilibrium
— Variability (*S*)

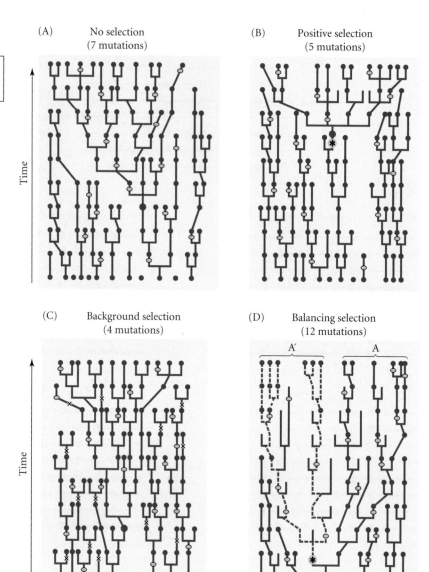

| Neutral mutation |
| Advantageous mutation |
| Deleterious mutation |

FIGURE 12.30 Schematic genealogies of gene copies in a population, showing the effects of three modes of selection on nucleotide diversity compared with the neutral model (compare with Figure 10.17). The current population (represented by 12 gene copies) is at the top of each diagram, and the ancestry of current gene copies is marked by the red gene trees. In each diagram, some gene lineages (blue lines) become extinct by chance. These diagrams assume that no recombination occurs in the gene. Ovals (yellow) represent selectively neutral mutations, each at a different site in the gene. (A) Neutral mutations only. Contemporary gene copies vary by seven mutations. (B) Positive selection of an advantageous mutation, marked by an asterisk. In a selective sweep, this gene lineage replaces all others, so contemporary copies differ only in the five mutations that have occurred since the advantageous mutation. (C) Background selection. Deleterious mutations, marked by ×, eliminate some gene copies, hence reducing the number (four) of surviving neutral mutations. (D) Balancing selection of alleles *A* and *A'*, the latter lineage (dashed lines) arising from the mutation marked with an asterisk. The two gene lineages persist for a long time, and so accumulate more mutations (twelve) than in the neutral case.

neutral mutations, and so are more similar in sequence. *A selective sweep resembles a bottleneck in population size in that it reduces variation and increases the genealogical relatedness among gene copies, but a bottleneck will affect the entire genome, not just the portion surrounding an advantageous mutation.*

Like positive directional selection, purifying selection against deleterious mutations reduces neutral polymorphism at closely linked sites. Brian Charlesworth and colleagues (Charlesworth et al. 1993; Charlesworth 1994a), who have termed this effect **background selection**, pointed out that when a copy of a deleterious mutation is eliminated from a population, selectively neutral mutations linked to it are eliminated as well (**Figure 12.30C**). The resulting reduction in the level of heterozygosity for neutral mutations is greatest if the rate of deleterious mutation is high, if the mutations are strongly deleterious, and if the recombination rate is very low. As predicted by this theory, the level of nucleotide variation is reduced in chromosome regions with low recombination, in *Drosophila* and in other species (**Figure 12.31**).

The effects of balancing selection (e.g., heterozygote advantage or frequency-dependent selection) are opposite those of positive directional selection. Suppose that two variants are maintained at a polymorphic site, and that recombination is low in the vicinity of that

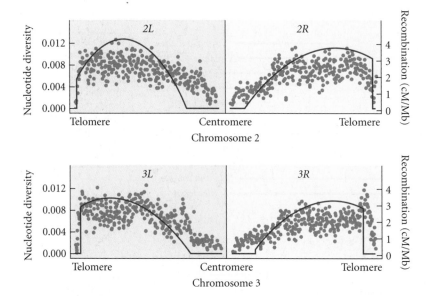

FIGURE 12.31 Nucleotide diversity (indicated by points) plotted along the two arms (L and R) of chromosomes 2 and 3 of *Drosophila melanogaster*. The solid curves show the recombination rate at sites along the length of a chromosome arm. As predicted by theory, nucleotide diversity is lower in regions of low recombination (e.g., near the centromeres), conforming to the effects of background selection or positive selection. The unit cM (centimorgans) refers to recombination distance along a chromosome. (After Mackay et al. 2012.)

site. The gene copies in the population are all descended from two ancestral copies (bearing the original, selectively advantageous alternative nucleotides), each of which was the progenitor of a lineage of genes that have accumulated neutral mutations in the vicinity of the selected site (**Figure 12.30D**). Thus, compared with a gene with solely neutral variation, a gene subjected to balancing selection will display elevated variation in the vicinity of the selected site (Strobeck 1983). In a genealogy of sequences sampled from a population, the common ancestor of all the sequences may be older than if they had been evolving solely by genetic drift because selection has maintained two gene lineages longer. A study of 13,400 genes in two human populations revealed 60 genes that had exceptionally high sequence variation and are thought to reflect balancing selection. Many of these genes encode keratins and defensive proteins such as immunoglobulins (Andrés et al. 2009).

In some cases, balancing selection has maintained polymorphism for so long that speciation has occurred in the interim. In that case, both lineages of genes may have been inherited by two (or more) species, and some gene copies in each species may be genealogically more closely related to genes in the other species than to other genes in the same species. This pattern has been found at the self-incompatibility locus in the family Solanaceae, in which many of the alleles in different genera of plants that diverged more than 30 Mya (e.g., petunia and tobacco) are genealogically more closely related to one another than to other alleles in the same species (**Figure 12.32A**). Similarly, allele lineages of certain of the major histocompatibility (MHC) genes are older than the divergence between humans and chimpanzees (**Figure 12.32B**). The proteins encoded by these genes bind foreign peptides (antigens), a key step in the immune response, and variant MHC proteins may be specific for different antigens.

Molecular signatures of selection

The imprints, or SIGNATURES, of selection are the subject of a great deal of research in evolutionary genomics (described in Chapter 20), the study of variation in genome sequences within and among species—especially *Homo sapiens* (Sabeti et al. 2006, 2007; Nielsen et al. 2007; Novembre and Di Rienzo 2009). The statistical methods used in practice are more complicated than the following descriptions, which convey only the reasoning behind them.

As we have seen, a selective sweep has two important features: (1) variation at nucleotide sites near the selected (advantageous) site is reduced—that is, homozygosity increases—and (2) the selected site is associated, by LD, with distant genetic markers, thus forming a LONG HAPLOTYPE. A higher frequency of such a haplotype generally implies greater age since it started to increase by selection, but the longer the haplotype, the younger it is likely to be, because recombination dissociates sites as time passes. The

FIGURE 12.32 Because of long-lasting balancing selection, a polymorphism may be inherited by two or more species from their common ancestor, and certain haplotypes in each species may be most closely related to haplotypes in the other species. These trees show relationships among gene sequences (haplotypes) from two or more species; each species is indicated by a different color. (A) Alleles at the self-incompatibility locus in the potato family, Solanaceae. The common ancestor of the six alleles sequenced in the tobacco *Nicotiana alata* is older than the common ancestor of the three genera *Nicotiana*, *Petunia*, and *Solanum*. (B) The phylogenetic relationships among six alleles in humans and four alleles in chimpanzees at the major histocompatibility (MHC) loci A and B. Both species have loci A and B, which form monophyletic clusters, indicating that the two loci arose by gene duplication before speciation gave rise to the human and chimpanzee lineages. At both loci, each chimpanzee allele is more closely related (and has more similar nucleotide sequence) to a human allele than to other chimpanzee alleles. Thus polymorphism at each locus in the common ancestor has been carried over into both descendant species. These polymorphisms are at least 5 My old. (A after Ioerger et al. 1990; B after Nei and Hughes 1991.)

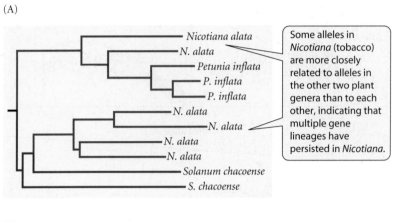

(A)

Some alleles in *Nicotiana* (tobacco) are more closely related to alleles in the other two plant genera than to each other, indicating that multiple gene lineages have persisted in *Nicotiana*.

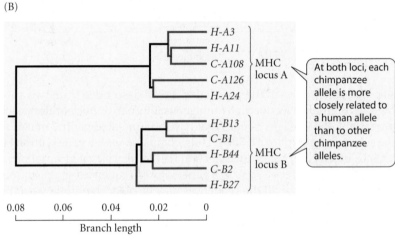

(B)

At both loci, each chimpanzee allele is more closely related to a human allele than to other chimpanzee alleles.

strength of selection can be inferred from the frequency and length of the longest haplotypes, and the age of a "core" haplotype (a short region of interest) can be gauged by the pattern of decay of LD with markers at various distances (Sabeti et al. 2002, 2006).

For example, the persistence of adult humans' ability to digest lactose in milk is based on mutations of the gene *LCT*, which encodes the enzyme lactase (lactase-phloridzin hydrolase). As we saw in Chapter 11, lactase persistence is common only in populations in which dairy products are an important part of the diet, such as some populations in sub-Saharan Africa and northern Europe. Different mutations of *LCT* confer lactase persistence in Eurasia and in Africa, implying that this trait evolved independently in those regions. Todd Bersaglieri and colleagues (2004) found that the lactase persistence mutation in Eurasia, denoted *T-13910*, resides in a haplotype that has high frequency (0.77) and extends with little breakdown for about 1000 kilobases (1 megabase, Mb). In contrast, most chromosomes with the nonpersistence allele (*C-13910*) have much shorter distances of LD (**Figure 12.33A**). The investigators estimated the strength of selection to be *s* = 0.014–0.150, and they proposed, on the basis of the decay of LD at increasing distances from the core region, that the selective sweep occurred between 2200 and 20,600 years ago. This estimate is consistent with the time of origin of dairy farming in northern Europe, about 9000 years ago. In pastoral populations in eastern Africa, Sarah Tishkoff and colleagues (2007) found that lactase persistence is based on two independent mutations. The *C-14010* allele, which has a frequency as high as 0.46 in some tribal populations, resides in an extended haplotype of more than 2 Mb, far longer than that of any nonpersistence alleles (**Figure 12.33B**). Homozygosity is much greater for haplotypes containing the advantageous mutation than for those containing the ancestral allele, as theory predicts (**Figure 12.33C**). Tishkoff et al. postulate that *s* is between 0.035 and 0.097, and that the allele started to increase 6000 to 7000 years ago, consistent with the origin of pastoralism in Egypt less than 9000 years ago and its spread

(A) Eurasian C/T-13910

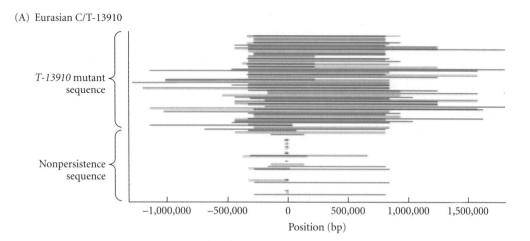

(B) African G/C-14010

(C)

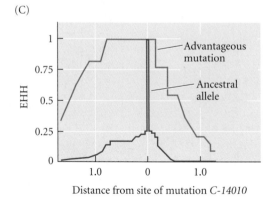

FIGURE 12.33 Linkage disequilibrium (LD) marks selective sweeps at the *LCT* (lactase persistence) locus in human populations. Each horizontal line indicates the length (in base pairs) of a long haplotype, with markers in LD. Some haplotypes are shorter because they have descended from longer haplotypes that underwent recombination, reducing the length of LD. (A) Haplotypes in European and Asian populations are longer in sequences with the lactase persistence mutation *T-13910* (green; at position 0) than in sequences with the nonpersistence allele (orange), as expected if the lactase persistence allele has recently increased in a partial selective sweep. (B) The same pattern is seen in a comparison of sequences with the lactase persistence mutation *C-14010* (red) with nonpersistence sequences (blue) in Kenya and Tanzania. (C) EHH is a measure of the relative frequency with which pairs of randomly chosen sequences in a population are homozygous for SNP markers at various distances from the site of the advantageous lactase persistence mutation *C-14010* (at position 0) in a Kenyan population. EHH is thus an inverse measure of variation. The farther a site is from 0, the lower the EHH, indicating that variation is greater and that longer haplotypes are less frequent than shorter ones because recombination has broken down LD. The drop-off is much more gradual for sequences with the derived, advantageous mutation than for sequences with the ancestral (nonpersistence) allele. (After Tishkoff et al. 2007; A based on data of Bersaglieri et al. 2004.)

to eastern sub-Saharan Africa within the last 4500 years. Because milk is a source of water as well as nutrients, and because milk consumption causes diarrhea in lactose-intolerant people, the ability to digest milk could have had a large selective advantage.

Adaptive evolution across the human genome

Tests for selection are now being applied to thousands of human genes, using increasingly large databases on human sequence variation, as well as complete genome sequences for chimpanzee, house mouse, and other species (including other primates), that enable us to distinguish derived from ancestral alleles and to describe large-scale patterns of genome evolution (see Chapter 20).

Comparisons between human and chimpanzee genes test for selection on a time scale of millions of years. One research group (Clark et al. 2003; Bustamante et al. 2005) used d_n/d_s as a test for selection, estimating rates of substitution in both lineages with reference to a distantly related outgroup, the mouse. They found small numbers of loci that had been subject to both positive directional selection and purifying selection (**Figure 12.34A**). Among the classes of genes in which evidence for positive selection was prominent were those involved in immune system defenses, olfaction, and spermatogenesis, as well as some genes that affect the cell cycle and which may be related to cancer. Genes that show their maximal expression (transcription) in testes are prone to positive selection, but (perhaps surprisingly) genes that are expressed in the brain are among the most conservative genes of all. An interesting discovery was a strong correlation between the strength of purifying selection on a gene, as judged from the human-chimpanzee comparison, and the likelihood that the gene is known to display disease-causing mutations in humans (**Figure 12.34B**).

A few years later, another research group compared human and chimpanzee genome sequences with those of a more closely related outgroup, the rhesus macaque (Bakewell et al. 2007). Most of the genes that showed evidence of positive selection had functions, such as ion transport, that are not obviously related to the anatomical differences (e.g., in the brain) that most distinguish humans from other apes. Perhaps surprisingly, Bakewell et al. found far more selected genes in the chimpanzee than in the human lineage.

Selection within and among human populations is a focus of extensive research using all the methods described in this chapter. For example, Benjamin Voight and colleagues (2006) detected many instances of long haplotypes, defined by linkage disequilibrium, in a scan of more than 800,000 genome-wide SNPs in samples from sub-Saharan Africa (the Yoruba of Nigeria), eastern Asia, and Europe. (The assembly of these samples is called the HapMap project.) They found a considerable number of selective sweeps, most of which have occurred in the European and Asian population groups (**Figure 12.35A**). Most sweeps

FIGURE 12.34 Comparison of homologous human and chimpanzee genes indicates selection throughout the human genome. (A) The numbers of genes showing negative and positive selection of different strengths, based on differences in d_n/d_s between human polymorphisms and fixed differences. Red and blue regions of the histogram represent selection significantly different from zero (neutrality). (B) A low d_n (proportion of fixed nonsynonymous differences between human and chimpanzee) indicates that most mutations in a gene are deleterious and have been eliminated by natural selection. Such genes were predicted to cause human disease if mutated, and indeed, had a greater probability of having been recorded as a "disease locus." (After Bustamante et al. 2005.)

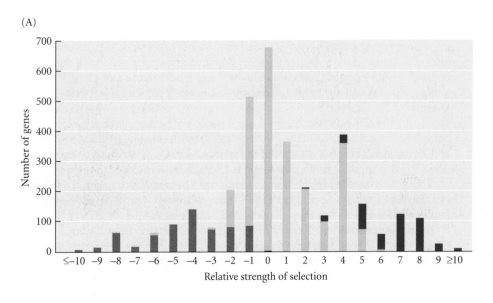

(A)

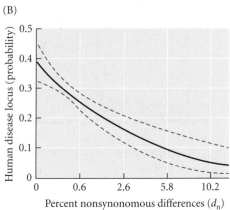

(B)

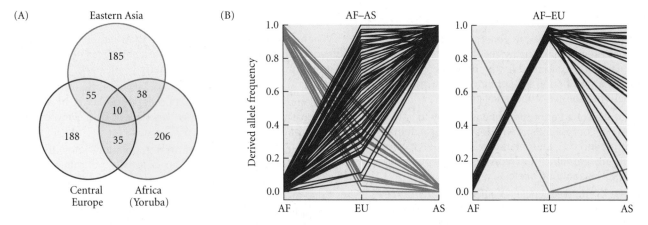

FIGURE 12.35   Recent selective sweeps in the human genome. (A) The numbers of loci exhibiting selective sweeps in the Yoruba people of Africa and in eastern Asian and central European populations. The numbers in the overlapping sectors, which represent sweeps that are shared among populations, are relatively low, showing that most sweeps have been specific to a region. (B) The frequency of the derived allele at loci in which different alleles are nearly fixed between the African (AF) versus Asian (AS) populations (AF–AS), and between the African versus the European (EU) populations (AF–EU). Each line indicates the frequency of a derived allele in all three populations. Far more of the derived alleles have high frequency in Asia and Europe (purple lines) than in Africa (brown lines). (A after Voight et al. 2006; B after Coop et al. 2009.)

have been more recent in the non-African populations than in the Yoruba, and the great majority represent derived alleles (**Figure 12.35B**). (Homologous chimpanzee sequences were used to distinguish ancestral from derived alleles.) In these and other studies, signatures of selection have been detected especially in genes related to neurogenesis, olfaction, immunity and defense, and steroid metabolism, as well as diet-related functions such as starch digestion and protein metabolism (Haygood et al. 2010; Babbitt et al. 2011). These observations may reflect both the early divergence of the human diet from that of other apes and cultural diet norms, including cooking food. Haygood et al. (2010) reported that neural function may have adapted mostly by changes in gene expression, whereas protein-coding changes account for adaptive evolution of immunity, male reproduction, and other functions. A study of 55 human populations revealed variation in about 100 genes that was strongly correlated with the local diversity of pathogens—perhaps the most important source of selection in the recent past, and possibly still today (Fumagalli et al. 2011).

Adaptation based on new versus standing variation

A selective sweep is expected if a newly arisen mutation increases fitness. But if, as much of the theory presented earlier in this chapter presumes, adaptation to a changed environment is based on formerly rare alleles (perhaps maintained by recurrent mutation), any signature of a selective sweep may be faint or nonexistent. The formerly rare allele would have been present in several or many copies that may all have contributed adapted descendants to the population, and each gene copy in the ancestral population would have been associated with different linked variations. In other words, the nucleotide base pair that makes the allele advantageous in the new environment would have been in linkage equilibrium with nearby variant sites, and there would exist no unique, adaptive long haplotype. Some cases of adaptation certainly have been based on rare alleles that contributed to the "standing" genetic variation that most traits display (Barrett and Schluter 2008). As we will see in the next chapter, most characteristics have a polygenic basis and may well have evolved from standing variation without producing selective sweeps. Many human adaptations may have evolved this way (Pritchard et al. 2010). It is not yet clear whether or not adaptation most often requires new mutations.

Go to the
EVOLUTION
Companion Website at
sites.sinauer.com/evolution3e
for quizzes, data analysis and simulation exercises, and other study aids.

Summary

1. Even at a single locus, the diverse genetic effects of natural selection cannot be summarized by the slogan "survival of the fittest." Selection may indeed fix the fittest genotype, or it may maintain a population in a state of stable polymorphism, in which inferior genotypes may persist.

2. The absolute fitness of a genotype is measured by its rate of increase, the major components of which are survival, female and male mating success, and fecundity. In sexual species, differences among gametic (haploid) genotypes may also contribute to selection among alleles.

3. Rates of change in the frequencies of alleles and genotypes are determined by differences in their relative fitness and are also affected by genotype frequencies and the degree of dominance at a locus.

4. Much of adaptive evolution by natural selection consists of replacement of previously prevalent genotypes by a superior homozygote (directional, or positive, selection). However, genetic variation at a locus often persists at a stable equilibrium owing to a balance between selection and recurrent mutation, between selection and gene flow, or because of any of several forms of balancing selection.

5. The kinds of balancing selection that maintain polymorphism include heterozygote advantage, inverse frequency-dependent selection, and variable selection arising from variation in the environment.

6. Often the final equilibrium state to which selection brings a population depends on its initial genetic constitution: there may be multiple possible outcomes, even under the same environmental conditions. This is especially likely if the genotypes' fitnesses depend on their frequencies, or if two homozygotes both have higher fitness than the heterozygote.

When the heterozygote is less fit than either homozygote, genetic drift is necessary to initiate a shift from one homozygous equilibrium state to the other.

7. When genotypes differ in fitness, selection determines the outcome of evolution if the population is large; in a sufficiently small population, however, genetic drift is more powerful than selection.

8. Studies of variable loci in natural populations show that the strength of natural selection varies greatly, but that selection is often strong, and is thus a powerful force of evolution.

9. Selection can often be inferred if the alleles at a locus display unusually great differences in frequency among various populations of a species. Variation in DNA sequences can also provide evidence of natural selection. Compared with the level of variation expected under neutral mutation and genetic drift alone, positive selection (of an advantageous mutation) causes "selective sweeps" that reduce the level of neutral variation at closely linked sites and create linkage disequilibrium among neutral variants in the region of the advantageous mutation. Purifying (background) selection against deleterious mutations also reduces linked neutral variation. Balancing selection results in higher levels of linked variation than expected under the neutral theory. Studies of DNA sequence variation in humans and other species have provided evidence of extensive recent positive selection.

10. Selection can also be inferred from DNA sequence comparisons among different species, as indicated by the incidence of nonsynonymous versus synonymous nucleotide substitutions and their relationship to levels of polymorphism with the species.

Terms and Concepts

absolute fitness
adaptive landscape
adaptive peak/valley
antagonistic selection
background selection
balancing selection
coefficient of selection
directional selection

diversifying (disruptive) selection
fitness
frequency-dependent selection (inverse and positive)
heterozygote advantage/disadvantage

McDonald-Kreitman (MK) test
mean fitness
multiple-niche polymorphism
multiple stable equilibria
overdominance

peak shift
purifying selection
relative fitness
reproductive success
selective advantage
selective sweep
stabilizing selection
underdominance

Suggestions for Further Reading

Natural Selection in the Wild by J. A. Endler (Princeton University Press, Princeton, NJ, 1986) analyzes methods of detecting and measuring natural selection and reviews studies of selection in natural populations.

Textbooks of population genetics, such as D. L. Hartl and A. G. Clark's *Principles of Population Genetics* (fourth edition, Sinauer Associates, Sunderland, MA, 2007) and P. W. Hedrick's *Genetics of Populations* (Jones and Bartlett, Sudbury,

MA, 2000), present the mathematical theory of selection in depth.

R. Nielsen, "Molecular signatures of natural selection," *Annual Review of Genetics* 39: 197–218 (2005), reviews methods of studying natural selection at the DNA sequence level, and applications of these approaches to human genetic variation are reviewed by P. C. Sabeti et al., "Positive natural selection in the human lineage," *Science* 312: 1614–1620 (2006).

Problems and Discussion Topics

1. If a recessive lethal allele has a frequency of 0.050 in newly formed zygotes in one generation, and the locus is in Hardy-Weinberg equilibrium, what will be the allele frequency and the genotype frequencies at this locus at the beginning of the next generation? (Check your understanding: the answer is $q = 0.048$; $p^2 = 0.9071$, $2pq = 0.0907$, $q^2 = 0.0023$.) Calculate these values for the succeeding generation. If the lethal allele arises by mutation at a rate of 10^{-6} per gamete, what will be its frequency at equilibrium?

2. Suppose the egg-to-adult survival of A_1A_1 is 80 percent as great as that of A_1A_2, and the survival of A_2A_2 is 95 percent as great. What is the frequency (p) of A_1 at equilibrium? What are the genotype frequencies among zygotes at equilibrium? Now suppose the population has reached this equilibrium, but that the environment then changes so that the relative survival rates of A_1A_1, A_1A_2, and A_2A_2 become 1.0, 0.95, and 0.90. What will the frequency of A_1 be after one generation in the new environment? (Answer: 0.208.)

3. If the egg-to-adult survival rates of genotypes A_1A_1, A_1A_2, and A_2A_2 are 90, 85, and 75 percent, respectively, and their fecundity values are 50, 55, and 70 eggs per female, what are the approximate absolute fitnesses (R) and relative fitnesses of these genotypes? What are the allele frequencies at equilibrium? Suppose the species has two generations per year, that the genotypes do not differ in survival, and that the fecundity values are 50, 55, and 70 in the spring generation and 70, 65, and 55 in the fall generation. Will polymorphism persist, or will one allele become fixed? What if the fecundity values are 55, 65, 75 in the spring and 75, 65, 55 in the fall?

4. In pines, mussels, and other organisms, investigators have often found that components of fitness such as growth rate and survival are positively correlated with the number of allozyme loci at which individuals are heterozygous rather than homozygous (Mitton and Grant 1984; Zouros 1987). The interpretation of such data has been controversial (see references in Avise 2004). Provide two hypotheses to explain the data and discuss how those hypotheses might be distinguished.

5. Considering the principles of mutation, genetic drift, and natural selection discussed in this and previous chapters, do you expect *adaptive* evolution to occur more rapidly in small or in large populations? Why? Answer the same question with respect to nonadaptive (*neutral*) evolution.

6. Discuss whether or not natural selection would be expected to (a) increase the abundance (population size) of populations or species; (b) increase the rate at which new species evolve from ancestral species, thus increasing the number of species.

7. What would be the difference in the rapidity of a population's adaptation to an environmental change if an advantageous allele were already present at low frequency (say, 10^{-4} to 10^{-3}), compared with adaptation based on a newly arisen mutation at that locus? Would both of these events be accompanied by a selective sweep? Would they both be detectable by studying DNA sequence variation?

8. Suppose that in an asexually reproducing population of bacteria, a single advantageous mutation occurs that enhances the ability to metabolize a carbohydrate and obtain energy, and that at about the same time, another single advantageous mutation at another locus reduces thermal stress. Both result in an increased rate of cell division (asexual reproduction). Will evolution at either locus be affected by the mutation at the other locus? (*Hint:* These two mutations occur in different members of the population, since the chance that both occur in the same cell are low.) Now suppose the same two mutations occur in a sexually reproducing species, at different genes in a chromosome region that has a low recombination rate. How will these mutations affect each other's increase, if at all?

9. Describe the conditions under which evolution does not occur even though natural selection on a genetically variable character is occurring.

10. What might account for the observation that the chimpanzee genome apparently has more genes that have experienced adaptive evolution than the human genome?

Phenotypic Evolution

An adaptation based on many genes Variation in the size of the bill in species of Darwin's finches is gradual: the differences among individuals and species are based on small contributions from many genes. The massive bill of the large ground finch (*Geospiza magnirostris*) is adapted for feeding on large, hard seeds.

W hen we seek to understand how a morphological, physiological, or other phenotypic feature of a species evolves by natural selection, the one-locus models described in the previous chapter provide an indispensable framework. However, there are many characteristics they do not adequately describe, and many questions they do not adequately address. The variation in most features is based on allelic variation at several or many loci as well as direct effects of environment. Many characteristics are complex, with functions that depend on the coaction of several component features. Furthermore, some characteristics are correlated with one another. From these considerations, a host of questions arise: How does evolution of a feature proceed if it is based on many genes? Do the genes that contribute to phenotypic evolution have small or large effects? What determines the speed of evolution of a trait? Are there limits to how much it can change? How do correlated characters affect one another's evolution? Does nongenetic variation play any evolutionary role? Considering that the vast majority of species that have ever lived are extinct, what limits the ability of species to adapt to changing environments?

Box 13A Two methods of QTL mapping

QTL mapping uses correspondence between genetic markers (such as SNPs) and phenotypic differences to estimate the number and locations of loci that contribute to variation in a character. One common method, used to analyze variation both within species and between related species that are not fully intersterile, scores variation in F_2 offspring. Suppose, as a simple hypothetical example, that we have two inbred strains of a fruit fly species that has two pairs of chromosomes, that the strains differ in wing length (indicated by the numbers denoted L), and that they also carry different alleles at genetic marker sites X, Y, and Z on two chromosomes (**Figure 1**). Strain 1 is homozygous for alleles X_1, Y_1, and Z_1, while strain 2 is homozygous for alleles X_2, Y_2, and Z_2. A backcross between the F_1 generation and strain 2 ($X_1X_2Y_1Y_2Z_1Z_2 \times X_2X_2Y_2Z_2Z_2$) yields F_2 progeny that provide the information we seek because they have various combinations of markers, owing to recombination in their F_1 parents. As illustrated by progeny genotypes (1) and (2) in Figure 1A, some progeny, such as $\mathbf{X_1}X_2Y_1Y_2Z_1Z_2$ and $\mathbf{X_2}X_2Y_1Y_2Z_1Z_2$, differ only with respect to chromosome 1. If they also differ in wing length, then chromosome 1 must carry at least one locus, linked to locus X, that contributes to variation. Genotypes 1 and 3 in Figure 1A differ at both the Y and Z markers; since they too differ in wing length, chromosome 2 must also carry at least one gene that affects this character. Now compare genotypes

2 and 4 in Figure 1A ($X_2X_2\mathbf{Y_1}Y_2Z_1Z_2$ and $X_2X_2\mathbf{Y_2}Y_2Z_1Z_2$), which differ only at the Y marker. They differ in wing length, so the region of chromosome 2 marked by Y contributes to wing length variation. But the difference in wing length between genotypes 2 and 4 is not as great as the difference between genotypes 1 and 3, so there are probably genes elsewhere on chromosome 2, not closely linked to the marker locus Y, that affect wing length. A comparison of

FIGURE 1 Illustration of the principle underlying estimation of the number of loci affecting variation in a quantitative trait. The mean value of the trait (L) is shown for each genotype. Strains 1 and 2 are homozygous for three marker loci (X, Y, Z). (A) Backcrossing the F_1 to strain 2 yields offspring genotypes 1–4 (among others). Differences in L among genotypes are correlated with differences in the markers they carry, providing an estimate of the minimal number of genes contributing to variation in the trait. For example, contrasting genotype 1 with both 2 and 3 shows that there must be at least one locus on each of the two chromosomes that contributes to the character difference between strains 1 and 2. (B) Recombinant inbred lines (RILs) result from pairing F_2 offspring of two strains that differ in a character (L), and propagating the descendants of each pair separately by brother × sister mating for many generations. This inbreeding results in a uniform line with a single homozygous genotype that differs from other RILs. Among these lines, associations between genetic markers and character values indicate loci (QTL) that contribute to the variation in the trait.

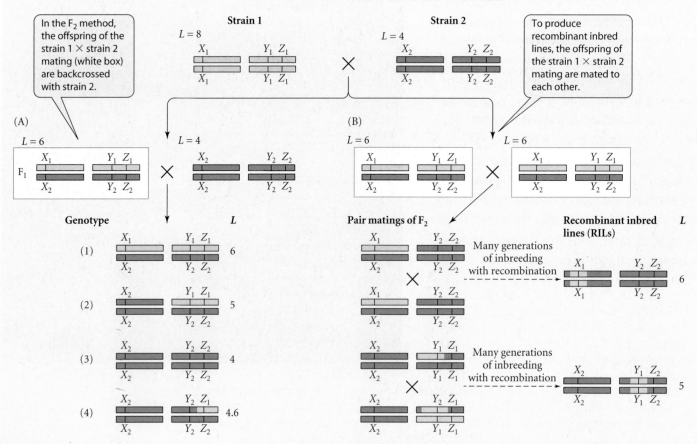

Box 13A *(continued)*

genotypes that differ only at locus Z (not shown) might confirm this suspicion. The more markers there are, distributed along all the organism's chromosomes, the greater the number of chromosome regions that can be compared in this way, and the greater the potential number of trait-affecting loci that can be detected.

An alternative to this analysis of F_2 individuals is to start with two strains that differ in phenotype and cross their F_1 offspring to yield a great variety of F_2 genotypes (Figure 1B). Many (typically hundreds of) male-female pairs of F_2 offspring are then crossed,

and each such line is propagated by sister × brother mating for 10 to 25 generations. Because inbreeding results in homozygosity (see Chapter 9), this process produces RECOMBINANT INBRED LINES (RILs), each of which is expected to be homozygous for a different mixture of chromosomal segments, with different genetic markers, from the two parental strains. The phenotypic variation among these lines can be related to the genetic markers—that is, the QTL can be mapped—in the same way as in the F_2 method.

Genetic Architecture of Phenotypic Traits

The term GENETIC ARCHITECTURE refers to the genetic basis of a trait and its relationship to other traits (Hansen 2006). The components of genetic architecture are the *number of loci* that contribute to a trait's development, its variation within species, and sometimes the differences between species or populations; the *magnitude of the effect* on the trait of different alleles at each locus (allelic effects); the pattern of *additive effects* and *dominance* at the various loci; the phenotypic effect of *synergistic interactions* among loci (**epistasis**); and the *pleiotropic effects* of the loci—the extent to which individual genes affect more than one characteristic. The variation in some traits is simple, based on just one allelic difference, but more often, as we have noted, variation is continuous (quantitative) because of the effects of several genes and the environment. The field of **quantitative genetics** was developed to analyze quantitative (continuously varying) characters, and its methods are used by biologists who study the evolution of morphology, life history characteristics, behavior, and other continuous phenotypic traits. The main traditional tool of quantitative genetics is statistical analysis of variation among relatives, an approach that has now been supplemented by the use of molecular genetic markers to find some of the genes that contribute to the variation.

The number of loci that contribute to variation in a character may be less than the number that actually contribute to its development. Only variable loci can be detected, however—and detecting those that have small phenotypic effects is not easy. The principal method of detection relies on associating phenotypic differences with detectable genetic markers with known positions on the chromosomes (**Box 13A**). This method is called **QTL mapping**. ("QTL" stands for "quantitative trait locus" or "loci.") The more markers are available, and the more widely they are distributed along all the organism's chromosomes, the greater the number of chromosome regions that can be compared in this way, and the greater the potential number of trait-affecting loci that can be detected. (Of course, some markers may not be associated with any difference in the trait studied, suggesting that no closely linked genes affect the character, or at least none with effects large enough to detect.) Moreover, as the number of markers along a chromosome increases, the position of each detected QTL can be localized to a shorter region of the chromosome. QTL mapping is usually carried out with controlled crosses, as described in Box 13A, but QTL can be also be detected and mapped in samples from a natural population, without experimental crossing, if (and only if) a QTL and a genetic marker are in linkage disequilibrium in the population (i.e., if they are actually associated with each other). This approach, called LINKAGE DISEQUILIBRIUM MAPPING, is the basis of GENOME-WIDE ASSOCIATION STUDIES (GWAS) of variation in human genomes, which are used especially to study inherited diseases.

A QTL is not necessarily a single gene, but rather a chromosome interval that may include several or many genes, of which one or more affect the character under study. In order to detect and locate many potential loci by the use of many molecular markers, very large numbers of progeny must be compared and the data subjected to special statistical analyses.

The genetic architecture of bristle number in *Drosophila melanogaster,* studied by Trudy Mackay and her colleagues (e.g., Nuzhdin et al. 1999; Dilda and Mackay 2002), appears to be typical of most quantitative characteristics. The bristles they studied, on the abdomen and the sternopleuron (part of the side of the thorax), have a sensory function and are part of the peripheral nervous system. The investigators selected for high and low bristle numbers in laboratory populations founded by large numbers of wild flies, and they used transposable elements on the X chromosome and chromosome 3 as markers to analyze the genetic differences between the high- and low-selected strains. They detected 53 QTL, of which 33 affected sternopleural bristle number, 31 affected abdominal bristle number, and 11 affected both characters (**Figure 13.1A**). The 53 QTL detected are certainly only a moderate fraction of the loci that can affect these features, since the two chromosomes scanned represent only 60 percent of the fly genome. Moreover, when mutations were induced by inserting transposable elements at random throughout the genome, 262 mutated loci were estimated to have affected bristle number (Mackay and Lyman 2005). Some QTL alleles had rather large effects, adding or subtracting several bristles, but many had smaller effects (**Figure 13.1B**). Many QTL displayed additive effects on the phenotype (i.e., alleles were neither dominant nor recessive; see below). Some had strong epistatic interactions, meaning that the joint effect of some pairs of loci differed from the sum of their individual effects. The investigators also found that the effects of some QTL depended on the temperature at which the flies were raised (that is, those QTL displayed genotype × environment interactions, as in Figure 9.4). Most of the QTL mapped to chromosome sites where CANDIDATE LOCI were located: genes known from past studies to affect bristle morphology or development of the peripheral nervous system. DNA sequence variation at some of these loci has been shown to correlate with bristle variation, and alleles of certain loci distinguished by DNA sequence segregate at high frequencies in natural populations of *Drosophila* (Lai et al. 1994). The high frequency of these alleles suggests that they are either selectively neutral or are maintained by balancing selection (see Chapter 12). However, many of the QTL that affect bristle number have been shown to reduce fitness (Mackay 2001a).

(A) QTL for sternopleural bristles: Chromosome 3

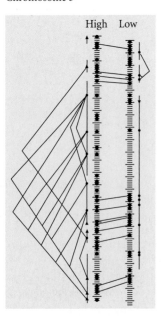

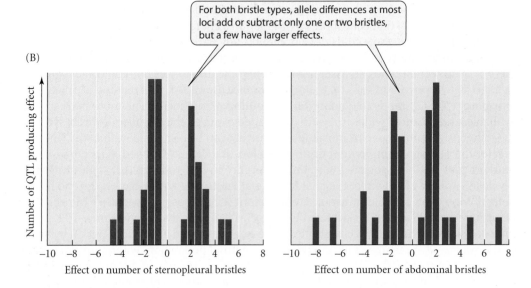

FIGURE 13.1 (A) The locations of QTL on chromosome 3 that contribute to differences in the number of sternopleural bristles between populations of *Drosophila melanogaster* selected for more ("High") or fewer ("Low") bristles. Lines between chromosomes join markers found in both the "High" and "Low" populations. Angled lines join QTL that interact to affect bristle number. (B) Frequency distribution of allelic effects on sternopleural and abdominal bristles. (A from Dilda and Mackay 2002; B after Mackay 2001b.)

Many quantitative characters, in diverse species, show similar results (Flint and Mackay 2009). For example, a genome-wide association study of human height, in which 2,834,208 SNP markers were screened in 133,653 individuals of European descent, detected at least 180 loci that contribute to variation (Allen et al. 2010). These and other such studies have important implications for the study of quantitative traits (Mackay 2001a; Flint and Mackay 2009). Statistical models of quantitative traits generally assume that variation in a trait is based on many loci, each with alleles that have small, nearly equal effects. However, some of the variation is usually due to loci with large effects. Furthermore, the models have often assumed that variation stems from the sum of additively acting alleles at functionally interchangeable loci. Variation in bristle number, however, is due to loci that are not interchangeable: the loci have different developmental roles, and they interact with one another in producing the phenotype. Models that take into account such complex gene interactions are important for researchers in this field, but are beyond the scope of this book.

Components of Phenotypic Variation

In order to discuss the evolution of quantitative traits, we must recall and expand on some concepts introduced in Chapter 9. An important measure of the variation within a sample is the VARIANCE (V), defined as the average of the squared deviations of observations from the arithmetic mean of the sample. The square root of the variance is the STANDARD DEVIATION ($s = \sqrt{V}$), measured in the same units as the observations. If a variable has a normal (bell-shaped) frequency distribution, about 68 percent of the observations lie within one standard deviation on either side of the mean, 96 percent within two standard deviations, and 99.7 percent within three (see Box 9C).

The **phenotypic variance (V_P)** in a phenotypic trait is the sum of the variance that is due to differences among genotypes (the **genetic variance, V_G**) and the variance that is due to direct effects of the environment and developmental noise (the **environmental variance, V_E**). Thus $V_P = V_G + V_E$. Considering for the moment the effect of only one locus on the phenotype, we may take the midpoint between the two homozygotes' means as a point of reference (**Figure 13.2**). Then the mean phenotype of A_1A_1 individuals deviates from that midpoint by $+a$, and that of A_2A_2 by $-a$. The quantity a, the ADDITIVE EFFECT of an allele, measures how greatly the phenotype is affected by the genotype at this locus. If the inheritance of the phenotype is entirely additive—that is, if neither allele is dominant or recessive—the heterozygote's phenotype is exactly in between that of the two homozygotes. Such additive effects are responsible for a critically important component of genetic variance, the **additive genetic variance**, denoted **V_A**. (The genetic variance V_G may also include nonadditive components, resulting from dominance and epistatic gene interactions, but we will not concern ourselves with those components in this chapter. In the subsequent discussion, we will assume that the genetic variance is entirely additive, so $V_G = V_A$.)

The additive genetic variance depends both on the magnitude of the additive effects of alleles on the phenotype and on the genotype frequencies. If one genotype is by far the most common, most individuals will be close to the average phenotype, so the variance will be lower than if the several genotypes had more equitable frequencies, as they would if the allele frequencies were more nearly equal. When, as in Figure 13.2, two alleles (with frequencies p and q) have purely additive effects, V_A at a single locus is

$$V_A = 2pqa^2$$

When several loci contribute additively to the phenotype, the average phenotype of any particular genotype is the sum of the phenotypic values of each of the loci. Likewise, V_A is the sum of the additive genetic variance contributed by each of the loci: $V_A = 2 \Sigma p_i q_i a_i^2$. The values of p, q, and a may differ for each locus, i. The expression given here for V_A would be modified if there were dominance at any of the loci.

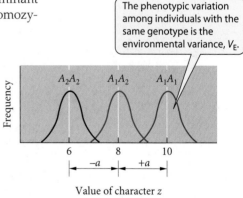

FIGURE 13.2 Additive effects of two alleles at one locus on the value of a character. When inheritance is purely additive, the heterozygote lies at the midpoint between the homozygotes. The additive effect of substituting an A_1 or an A_2 allele in the genotype is a (in this case, $a = 2$). The magnitude of a affects that of the additive genetic variance, V_A. The phenotypic variation within each genotype represents the environmental variance, V_E.

Additive genetic variance plays a key role in evolutionary theory because the additive effects of alleles are responsible for the degree of similarity between parents and offspring and are therefore the basis for response to selection within populations. When alleles have additive effects, the expected average phenotype of a brood of offspring equals the average of their parents' phenotypes.* Evolution by natural selection requires that selection among phenotypically different parents be reflected in the mean phenotype of the next generation. Therefore V_A enables a **response to selection**—a change in the mean character state of one generation as a result of selection in the previous generation.

The proportion of phenotypic variance that is due to additive genetic differences among individuals is referred to as the character's **heritability** in the narrow sense, h_N^2. (We will not be concerned with "broad-sense" heritability, which includes nonadditive components of genetic variance.) The heritability is determined by the additive genetic variance (V_A), which depends on allele frequencies, and by the environmental variance (V_E), which depends in part on how variable the environmental factors are that affect the development or expression of the character. That is,

$$h_N^2 = \frac{V_A}{V_P}$$

where $V_P = V_A + V_E$. Because allele frequencies and environmental conditions may vary among populations, an estimate of heritability is strictly valid only for the population in which it was measured, and only in the environment in which it was measured. Moreover, it is wrong to think that if a character has a heritability of 0.75, the feature is ¾ "genetic" and ¼ "environmental," as if a character were formed by mixing genes and environment the way one would mix paints to achieve a desired color. It is the *variation* in the character that is statistically partitioned, not the condition of an individual organism. Although heritability is often used in evolutionary studies, additive genetic variance is more informative because it does not include V_E and therefore provides more information about the potential ability of a character to evolve (its EVOLVABILITY) (Houle 1992).

The additive genetic variance of a character can be estimated from resemblances among relatives. In order to achieve accurate estimates, it is important that relatives not develop in more similar environments than nonrelatives; otherwise, it may not be possible to distinguish similarity that is due to shared genes from similarity that is due to shared environments.

The narrow-sense heritability (h_N^2, or simply h^2) of a trait equals the slope of the regression of offspring phenotypes (y) on the average phenotype of the two parents of each brood of offspring (x) (**Figure 13.3**; see also Figure 9.25). The slope of this relationship is calculated so as to minimize the sum of the squared deviations of all the values of y from the line. As the correspondence between offspring and parents is reduced, the slope of the regression declines (compare Figures 13.3A and 13.3B). Therefore anything that reduces the similarity between parents and their offspring (such as environmental effects on phenotype, or dominance) reduces heritability. There are also ways of estimating V_A and other components of genetic variance from similarities among other relatives, such as siblings ("full sibs") or half sibs (individuals that have only one parent in common).

Heritable variation has been reported for the great majority of traits in which it has been sought, in diverse species (Lynch and Walsh 1998). Commonly, h^2 is in the range of about 0.4 to 0.6, but not all characters are equally variable. For example, characters strongly correlated with fitness (such as fecundity) tend to have lower heritability than characters that seem unlikely to affect fitness as strongly (Mousseau and Roff 1987). However, the low h^2 of fitness components arises from the greater magnitude of other variance components, especially V_E, in the denominator of the expression $h^2 = V_A/V_P$. The additive genetic variance (V_A) of components of fitness is actually higher than that of morphological and other traits, probably because many physiological and morphological characteristics affect fitness (Houle et al. 1996).

---

*The correlation between parent and offspring phenotypes is lower if there is dominance at the locus.

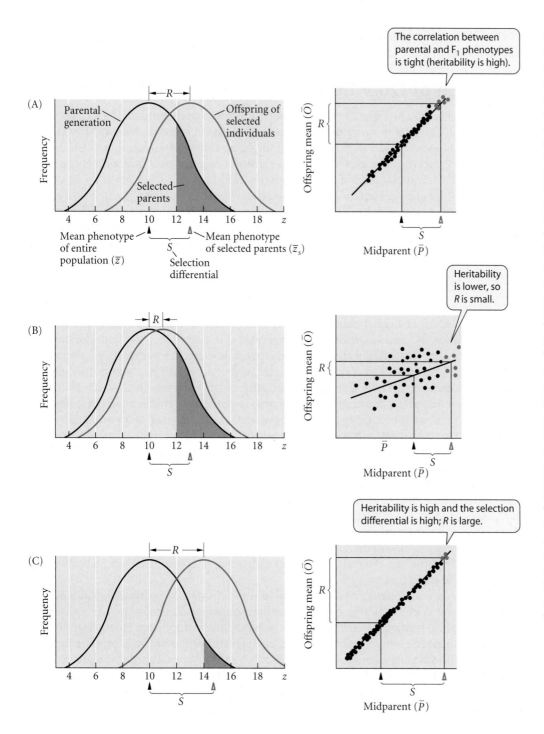

FIGURE 13.3 The response of a quantitative character to selection depends on the heritability of the character and the selection differential (S). (A) The black curve shows a normally distributed trait, z, with an initial mean of 10. Truncation selection is imposed, such that individuals with $z > 12$ reproduce. The graph at the right shows a strong correspondence between the mean phenotype of pairs of selected parents (the midparent mean, $\bar{P}$) and that of their broods of offspring ($\bar{O}$), i.e., high heritability. Brown circles represent the selected parents and offspring, black circles the rest of the population (were it to have bred). Because heritability is high, the selection differential S results in a large response to selection (R). The brown curve is the distribution of z in the next generation, whose mean lies R units to the right of the mean of the parental generation. (B) The circumstances are the same, but the relationship between phenotypes of parents and offspring is more variable and has a lower slope in the graph at the right, i.e., lower heritability. Consequently, the selection differential S translates into a smaller response (R). The frequency distribution of offspring (brown curve) shifts only slightly to the right. (C) Here heritability is very high, as in (A), but the selected parents have $z > 14$. Thus the selection differential S is larger, resulting in a greater response to selection (R).

In some cases, though, traits do not appear to be genetically variable at all. The paucity of genetic variation would then be a genetic constraint that could affect the direction of evolution (for example, an insect might adapt to some species of plants rather than to others) or prevent adaptation altogether. For instance, in sites near mines where the soil concentration of copper or zinc is high, a few species of grasses have evolved tolerance for these toxic metals, but most species of plants have not. Populations of various species of grasses growing on normal soils were screened for copper tolerance by sowing large numbers of seeds in copper-impregnated soil. Small numbers of tolerant seedlings were found in those species that had evolved tolerance in other locations, but no tolerant seedlings were found in most of the species that had not formed tolerant populations (Bradshaw

1991). Genetic variation that might enable the evolution of copper tolerance seems to be rare or absent in those species.

Evolution of Quantitative Traits by Genetic Drift

Quantitative characters will evolve by genetic drift if alleles at the underlying loci are selectively neutral. Selection might be inferred if data do not fit a model of neutral evolution. Bear in mind that variation in the trait itself might be selectively neutral, yet variation at the underlying loci might nevertheless affect fitness because of pleiotropic effects on other traits, so that selection would indirectly affect the trait.

First, consider changes in a trait over a relatively short time, during which new mutations have little effect. Changes in the frequencies of existing alleles at variable loci will cause fluctuations in the mean trait value, which will eventually drift away from its original value. In this "short-term" model, the smaller the population, the faster the change in the mean will be, since genetic drift changes allele frequencies more rapidly in smaller populations (see Chapter 10).

Alternatively (Turelli et al. 1988), consider changes over a very long time, during which the variation and evolution of the character are affected only by new mutations (which increase variation) and genetic drift (which erodes it). The variance that arises per generation by mutation, V_m, is proportional to the number of mutating loci, the mutation rate per locus, and the average phenotypic effect of a mutation (see Chapter 8). V_m is often about 0.001 times V_E (Lynch 1988). Over the long term, mutation will be balanced by genetic drift, so the genetic variance should, theoretically, reach a stable value (which is $2N_eV_m$). As mutations that affect the character arise and are fixed by genetic drift, the mean will fluctuate at random. If a number of isolated populations are derived from an initially uniform ancestral population, mutation and genetic drift can cause genetic divergence among them in a polygenic character, just as they do at a single locus (see Chapter 10). In this "long-term" model, the rate of divergence among the population means equals V_m per generation: it depends only on the mutation rate, not on the population size. (Recall that this is also true of the rate at which populations diverge by neutral substitutions at a single locus; see Chapter 10.) If a trait has diverged among populations or species much faster than the expected neutral rate, it is likely to have evolved by directional selection. If it has changed much more slowly than the neutral rate, one may suspect that stabilizing selection has operated—either on the trait itself or on other characteristics that are pleiotropically affected by the same genes.

Sonya Clegg and colleagues (2002) applied these models to morphological features, such as bill and leg length, of a small Australian songbird, *Zosterops lateralis* (the silvereye), and its close relative *Z. tephropleurus*, which inhabits distant Lord Howe Island. DNA sequence differences between the two species suggest that they have been separated for hundreds of thousands of years. Clegg and colleagues also analyzed populations of *Z. lateralis* formed by colonization of New Zealand and its coastal islands within the last 170 years. Applying the short-term model to differences between the recently formed New Zealand populations and Australian birds, Clegg et al. concluded that the rate of evolution was so high that evolution by genetic drift would have required effective population sizes much smaller than historical records indicate. They argued, therefore, that directional selection has contributed to evolution. In contrast, Clegg et al. used the long-term model to analyze the differences between the Australian and Lord Howe Island species, assuming that V_m is about 0.001. They concluded that the rate of change has been slow—perhaps even slower than predicted by the mutation rate. Therefore stabilizing selection may have maintained the similarity between these two species.

In the same vein, Michael Lynch (1990) studied the rates of evolution of skeletal features of several groups of mammals, using data from fossils (e.g., horse and hominid lineages) as well as on differences between living species (such as lion and cheetah) for which there are good estimates of the time since the common ancestor. Lynch also assumed that the

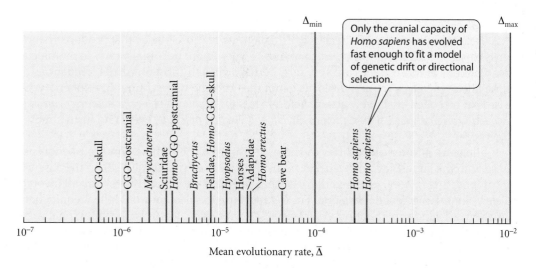

FIGURE 13.4 The mean evolutionary rates ($\bar{\Delta}$) of morphological characters in several mammals. The values Δ_{min} and Δ_{max} are the minimal and maximal rates that fit a model of mutation and genetic drift. Most characters have evolved so slowly over the long term that stabilizing or fluctuating selection is likely to have prevailed. "CGO" refers to comparisons among chimpanzee, gorilla, and orangutan. (After Lynch 1990.)

mutation rate (V_m) for such features is about 0.001, as has been estimated for skeletal features of mice and other living species of mammals. He found that almost all the features had evolved at much lower rates than expected under mutation and genetic drift (**Figure 13.4**). Only the cranial capacity of *Homo sapiens* has evolved at a rate equal to or greater than expected from the neutral model. The very low rate at which most characters seem to have evolved suggests that stabilizing selection—or perhaps rapidly fluctuating selection—has maintained them at roughly constant values for long periods. Fossils show that features of many lineages fluctuate rather rapidly, but show little net change (see Figure 4.23), so that the rate of evolution over the long term is much less than it is over shorter intervals of time.

Selection on Quantitative Traits

Response to directional selection

Suppose an experimenter imposes selection for greater tail length by breeding only those rats in a captive population with tails longer than a certain cutoff value. This form of selection is called TRUNCATION SELECTION, and it can be used as an approximation of the way directional natural selection acts. The mean tail length of the selected parents ($\bar{z}_S$) differs from that of the population from which they were taken ($\bar{z}$) by an amount S, the **selection differential** (see Figure 13.3A). The average tail length among the offspring of the selected parents ($\bar{z}'$) differs from that of the parental generation as a whole by an amount R, the response to selection (Figure 13.3A, right-hand graph). The magnitude of R is proportional to the heritability of the trait (compare Figures 13.3A and 13.3B) and to the selection differential S (compare Figures 13.3A and 13.3C). In fact, the change in mean phenotype of offspring, R, that is due to selection of parents, S, can be read directly from the regression line as

$$R = h^2 S$$

Since this equation* can be rearranged as $h^2 = R/S$, heritability can be estimated from a selection experiment in which S (which is under the experimenter's control) and R are measured. Such an estimate of h^2 is called the **realized heritability**. This is how (as you undoubtedly recall from Chapter 9) Dobzhansky and Spassky estimated that the heritability of phototaxis in *Drosophila pseudoobscura* was about 0.09 (see Figure 9.26).

---

*The simple equation presented here is the "breeders' equation," used for predicting responses to artificial selection in domesticated species. A conceptually related equation is used more often by evolutionary biologists: $\Delta\bar{z} = V_A/\bar{w} \times d\bar{w}/d\bar{z}$, where z is a character state, V_A is the additive genetic variance, and w is fitness. Thus the rate of evolution of the character mean ($\Delta\bar{z}$) is proportional to V_A and to the derivative of fitness with respect to the character value (Lande 1976). Because of its direct role in determining the rate of evolution, V_A is better than the heritability, h^2, as a measure of "evolvability" (Houle 1992).

As selection proceeds, it increases the frequencies of those alleles that produce phenotypes closer to the optimal value. Each of those alleles might be initially rare, so multilocus combinations of those alleles are very rare, perhaps virtually absent. As the frequencies of those alleles increase, these multilocus genotypes become more common, so phenotypes arise that had been effectively absent before. Thus *the mean of a polygenic character can shift beyond the original range of variation as directional selection proceeds*, even if no further mutations occur.

If alleles at different loci differ in the magnitude of their effects on the phenotype, those with the largest favorable effects are likely to be fixed first (Orr 1998). In the absence of complicating factors, prolonged directional selection should ultimately fix all favored alleles, eliminating genetic variation. Further response to selection would then require new variation, arising from mutation. As we have seen, for many features, the mutational variance, V_m, is on the order of $10^{-3} \times V_E$ per generation, or about 0.001 times the environmental variance. Thus a fully homozygous population could, by mutation, attain a heritability value $V_A/(V_A + V_E) = h^2 = 0.5$, for an entirely additively inherited trait, in about 1000 generations. This observation suggests that the ongoing origin of new mutations would enable further adaptation to proceed at a slow but steady pace. If, however, many of these mutations have harmful pleiotropic effects and have a net selective disadvantage, the "usable" mutational variance may be much lower (Hansen and Houle 2004).

Responses to artificial selection

Animal and plant breeders have used artificial selection to alter domesticated species in extraordinary ways. Darwin opened *The Origin of Species* with an analysis of such changes, and evolutionary biologists have drawn useful inferences about evolution from artificial selection ever since then. Artificial selection differs from natural selection because the human experimenter focuses on one trait, rather than on the organism's overall fitness. Nevertheless, natural selection often operates much like artificial selection.

Responses to artificial selection over just a few generations generally are rather close to those predicted from estimates of heritability based on correlations among relatives, such as parents and offspring. These heritability estimates seldom predict accurately the change in a trait over many generations of artificial selection, however, because of changes in linkage disequilibrium and genetic variance, input of new genetic variation by mutation, and the action of natural selection, which often opposes artificial selection (Hill and Caballero 1992). Such effects were found, for example, in an experiment by B. H. Yoo (1980), who selected for increased numbers of abdominal bristles in three lines of *Drosophila melanogaster* taken from a single laboratory population. For each of 86 generations, Yoo scored bristle numbers on 250 flies, then bred the next generation from the top 50 flies of each sex. In the base population from which the selection lines were drawn, the mean bristle number was 9.35 in females, and more than 99 percent of females had fewer than 14 bristles (i.e., three standard deviations above the mean). After 86 generations, mean bristle numbers in the experimental populations had increased to 35 to 45 (**Figure 13.5**). This change represents an average increase of 316 percent, or 12 to 19 phenotypic standard deviations—far beyond the original range of variation. In a very short time, selection had accomplished an enormous evolutionary change, at a rate far higher than is usually observed in the fossil record.

This progress was not constant, however. Some of the experimental populations showed periods of little change followed by episodes of rapid increase. Such irregularities in response are partly due to the origin and fixation of new mutations with rather large effects (Mackay et al. 1994). Several lines eventually stopped responding: they reached a **selection plateau**. This cessation of response to selection was not caused by loss of genetic variation, because when Yoo terminated ("relaxed") selection after 86 generations, mean bristle number declined, proving that genetic variation was still present.

A selection plateau and a decline when selection is relaxed are commonly observed in selection experiments. These patterns are caused by natural selection, which opposes

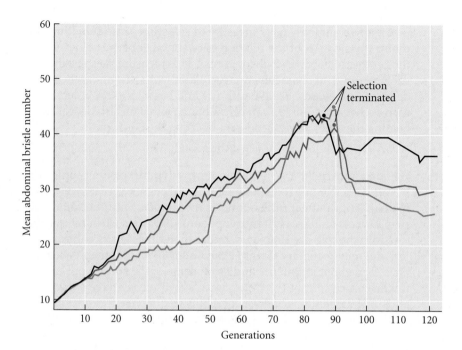

FIGURE 13.5 Responses to artificial selection for increased numbers of abdominal bristles in three laboratory populations of *Drosophila melanogaster*. After about 86 generations, the means had increased greatly. Selection was terminated at the points indicated by colored circles, and bristle number declined thereafter, indicating that genotypes with fewer bristles had higher fitness. (After Yoo 1980.)

artificial selection: genotypes with extreme values of the selected trait have low fitness. The changes in fitness are due to both hitchhiking of deleterious alleles (linkage disequilibrium) and to pleiotropy. Yoo found, for example, that lethal alleles had increased in frequency in the selected populations because they were closely linked to alleles that increased bristle number. Other investigators have shown that some alleles affecting bristle number have pleiotropic effects that reduce viability (Kearsey and Barnes 1970; Mackay et al. 1992).

Several investigators have found that artificially selected traits change faster in large than in small experimental populations (e.g., Weber and Diggins 1990; López and López-Fanjul 1993). This result has several explanations: more genetic variation is introduced by mutation in large than in small populations, large populations lose variation by genetic drift more slowly, and selection is more efficient in large populations. (Recall from Chapter 12 that allele frequency change is affected more by genetic drift than by selection if the effective population size is very low, i.e., if $N_e < 1/s$.) In some of these experiments, however, the progress of artificial selection has not been impeded by counteracting natural selection. For example, after more than 100 generations of selection for faster flight, experimental populations of *Drosophila melanogaster* showed no evidence of diminished fitness (Weber 1996), suggesting that in this case, the underlying alleles did not have deleterious pleiotropic effects.

Directional selection in natural populations

In studies of natural populations, several measures of the strength of natural selection on quantitative traits have been used. The simplest indices of selection can be used if the mean ($\bar{z}$) and variance (V) of a trait are measured in a single generation before ($\bar{z}_b$, V_b) and then again after ($\bar{z}_a$, V_a) selection has occurred. (For instance, these measurements may be made on juveniles and then on those individuals that successfully survive to adulthood and reproduce.) Then, if selection is directional, an index of the INTENSITY OF SELECTION is

$$i = \frac{\bar{z}_a - \bar{z}_b}{\sqrt{V_P}}$$

where V_P is the phenotypic variance.

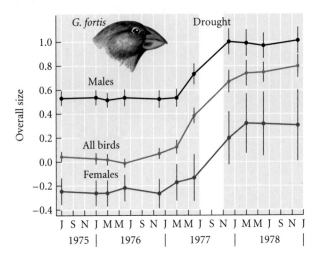

FIGURE 13.6 Changes in the mean size of the medium ground finch (*Geospiza fortis*) on Daphne Island resulting from mortality during a drought in 1977. Changes occurring in 1977 and 1978 were all the result of mortality; no reproduction occurred during this period. "Overall size" is a composite of measurements of several characters. (After Grant 1986.)

A more commonly used measure of selection is the **selection gradient**: the slope, *b*, of the relation between phenotype values (*z*) and the fitnesses (*w*) of those phenotypes.* This measure is especially useful when several characters are correlated with one another to some degree, such as, say, beak length (z_1) and body size (z_2). Selection on each such feature (say, z_1) can be estimated while (in a statistical sense) holding the other (say, z_2) constant by using the equation

$$w = a + b_1 z_1 + b_2 z_2$$

(Lande and Arnold 1983). The slopes (partial regression coefficients) b_1 and b_2 enable one to estimate, for instance, how greatly variations in beak length affect fitness among individuals with the same body size. In most studies, certain components of fitness (such as juvenile-to-adult survival) are estimated, rather than fitness in its entirety.

Peter and Rosemary Grant (1989; Grant 1986) and their colleagues have carried out long-term studies on some of the species of Darwin's finches on certain of the Galápagos Islands. They have shown that birds with larger (especially deeper) bills feed more efficiently on large, hard seeds, whereas there is some evidence that small, soft seeds are more efficiently used by birds with smaller bills. When the islands suffered a severe drought in 1977, seeds, especially small ones, became sparse, medium ground finches (*Geospiza fortis*) did not reproduce, and their population size declined greatly as a result of mortality. Compared with the pre-drought population, the survivors were larger and had larger bills (**Figure 13.6**). From the differences in morphology between the survivors ($\bar{z}_a$) and the pre-drought population ($\bar{z}_b$), the intensity of selection *i* and the selection gradient *b* were calculated for three characters:

Character	*i*	*b*
Weight	0.28	0.23
Bill length	0.21	–0.17
Bill depth	0.30	0.43

The values of *i* show that each character increased by about 0.2 to 0.3 standard deviations, a very considerable change to have occurred in one generation. The values of *b* show the strength of the relationship between survival and each character while holding the other characters constant. Selection strongly favored birds that were larger and had deeper bills because they could more effectively feed on large, hard seeds, virtually the only available food. The negative *b* value shows that selection favored shorter bills. Nevertheless, bill length increased, in opposition to the direction of selection, because bill length is correlated with bill depth. Thus a feature can evolve in a direction opposite to the direction of selection if it is strongly correlated with another trait that is more strongly selected. (We will soon return to this theme.)

Why don't these finches evolve ever larger bills? The Grants found that during normal years, birds with smaller bills survive better in their first year of life, probably because they feed more efficiently on abundant small seeds; in addition, small females tend to breed earlier in life than large ones. Thus conflicting selection pressures, averaged over time, create stabilizing selection that favors an intermediate bill size. Many studies have found evidence of fluctuations in the direction of selection (Kingsolver and Diamond 2011).

---

*The slope is the derivative of fitness with respect to character state and is closely related to $d\bar{w}/d\bar{z}$, one of the determinants of the rate of character evolution in Lande's equation (see the previous footnote, p. 355).

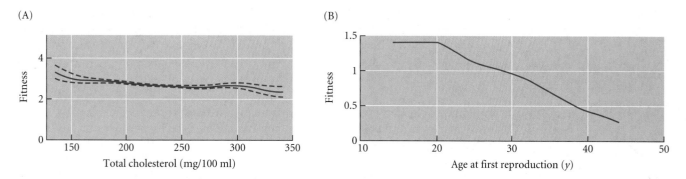

(A)

(B)

FIGURE 13.7 Examples of directional selection in contemporary human populations. Fitness was measured by women's lifetime reproductive success. (A) Selection for lower total blood cholesterol in the northeastern United States. The dashed lines refer to statistical error around the estimate shown by the solid line. (B) Selection for earlier age at first reproduction, based on a sample of Australian twins. (A after Byars et al. 2010; B after Kirk et al. 2001.)

Are humans still evolving by directional selection? For at least a few traits, the answer seems to be yes. One research group (Byars et al. 2010; Stearns et al. 2010) applied the Lande and Arnold method to data on survival and reproduction in a 40-year record of more than 5000 people and their children in Massachusetts. They found evidence of directional selection on six heritable traits, including selection against high cholesterol level and blood pressure, and for shorter stature, earlier reproduction, and later menopause (**Figure 13.7A**). In a study of 1200 pairs of female twins in the Australian Twin Registry, Katherine Kirk and collaborators (2001) found statistically significant selection for earlier age at first reproduction (**Figure 13.7B**). Comparison of monozygotic ("identical") and dizygotic twins suggested a heritability of 0.23 for this trait, so ongoing evolution is likely. Directional selection seems to have caused actual evolution of an earlier age at first reproduction in an isolated French-Canadian population (Milot et al. 2011). Among women who married between 1799 and 1940, the average age at which they bore their first child declined from 26 to 22 years. The authors found strong indications that the change has a largely genetic basis.

Stabilizing and disruptive selection

If selection is stabilizing or diversifying, the change within a generation in the phenotypic variance provides an index of the intensity of SELECTION ON VARIANCE:

$$j = \frac{V_a - V_b}{V_b}$$

(Endler 1986). This index is negative if selection is stabilizing, positive if it is diversifying.

Many traits are subject to stabilizing selection, so that the mean changes little, if at all. For example, human infants have lower rates of mortality if they are near the population mean for birth weight than if they are lighter or heavier (**Figure 13.8**; Karn and Penrose 1951). In birds, a clutch of either too few or too many eggs is disadvantageous. Birds with too low a reproductive rate obviously make a small contribution to population growth. Experimental alterations of the numbers of eggs in nests of blue tits (*Cyanistes caeruleus*) in England showed that fledglings from nests with more than 12 eggs had higher mortality, and so did the females that had cared for such large broods (Nur 1984). Consequently, a clutch of 12 eggs maximized fitness (**Figure 13.9**). Stabilizing selection often occurs because of trade-offs or antagonistic

FIGURE 13.8 Stabilizing selection for birth weight in humans. The rate of infant mortality is shown by the red points and the line fitting them. The histogram shows the distribution of birth weights in the population. (After Cavalli-Sforza and Bodmer 1971.)

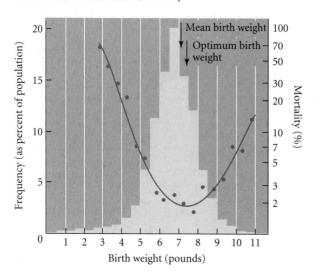

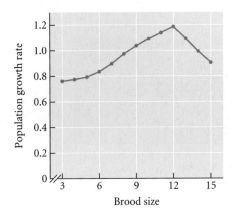

FIGURE 13.9 Blue tits that lay clutches of 12 eggs have higher fitness than those that lay fewer or more, as measured by the growth rate of a population if all its members had the clutch size indicated. Higher fecundity enhances fitness up a point, beyond which the survival of both parents and offspring is lowered enough to diminish fitness. (After Nur 1984.)

agents of selection (Travis 1989). For example, sexually antagonistic selection, favoring different character states in females and males, is common (Bonduriansky and Chenoweth 2009). Although the size and weight of collared flycatchers (*Ficedula albicollis*) does not differ between the sexes when they leave the nest, mortality results in smaller males and larger females by the time the birds reproduce in the next year (**Figure 13.10**). It is possible that smaller males survive better because they require less food when they arrive at the breeding sites, at a season when food is still scarce (Merilä et al. 1997).

Correlated Evolution of Quantitative Traits

Evolutionary change in one character is often correlated with change in other features. For example, species of animals that differ in body size differ predictably in many individual features, such as the length of their legs or intestines. Correlated evolution can have two causes: correlated selection and genetic correlation.

Correlated selection

In **correlated selection**, there is independent genetic variation in two or more characters, but selection favors some combination of character states over others, usually because the characters are functionally related.

Edmund Brodie (1992) found evidence of correlated selection on color pattern and escape behavior in the garter snake (*Thamnophis ordinoides*). This snake can have a uniform color, spots, or lengthwise stripes. Brodie found that when chased down an experimental runway, some snakes fled in a straight line, while others repeatedly reversed their course. Resemblance among siblings indicated that both coloration and escape behavior are highly heritable. Other investigators had noted that among different species of snakes, spotted species have irregular escape behavior or tend to be sedentary, relying on their cryptic patterns to avoid predation. Striped species tend to flee rapidly in a single direction, probably because predators that hunt visually find it difficult to judge the speed and position of a moving stripe.

Brodie scored color pattern and propensity to reverse course in 646 newborn snakes born to 126 pregnant females. He marked each snake so it could be individually recognized, released the snakes in a suitable habitat, and periodically sought them thereafter. Brodie had reason to believe that most of the snakes that were not recaptured were eaten by crows and other predators. He found that the survival rate was greatest for those that had both strong striping and a low reversal propensity, and for those that had both a nonstriped pattern and a high reversal propensity (**Figure 13.11**). Other combinations, such as striped snakes that reversed course when chased, had lower survival rates. Thus there was correlated selection on color pattern and escape behavior in the direction that had been predicted from comparisons among species of snakes and from the theory of visual perception.

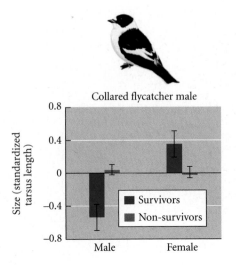

FIGURE 13.10 Sexually antagonistic selection favors different character states in females and males. The length of the tarsal segment of the leg is a measure of size in birds. In collared flycatchers, males with shorter tarsi had higher survival rates, while the reverse was true of females. The four averages shown are standardized in relation to the population mean, which is arbitrarily set at zero. (After Merilä et al. 1997.)

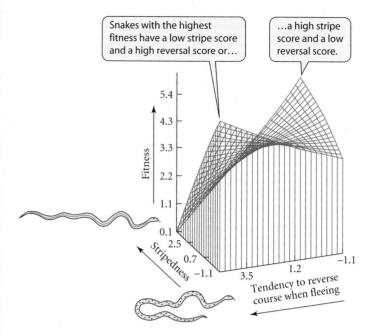

Snakes with the highest fitness have a low stripe score and a high reversal score or...

...a high stripe score and a low reversal score.

FIGURE 13.11 An example of correlated selection. The fitness surface shows fitnesses (relative survival rates) for combinations of two traits in the garter snake (*Thamnophis ordinoides*), based on individuals' survival in the field. The height of a point on the surface represents the fitness of individuals with particular values of stripedness and reversal (tendency to reverse course when fleeing). (After Brodie 1992.)

Genetic correlation

We often observe that characters are correlated within a species: tall people, for example, tend to have long arms. The magnitude of a correlation between two characters—the degree to which they vary in concert—is expressed by the CORRELATION COEFFICIENT (*r*), which ranges from +1.0 (for a perfect correlation in which both features increase or decrease together) to –1.0 (when one feature decreases exactly in proportion to the other's increase). For uncorrelated characters, $r = 0$.

The **phenotypic correlation**, r_P, between, say, body size and fecundity is simply what we measure in a random sample from a population. Just as the phenotypic variance may have both genetic and environmental components, so too may the phenotypic correlation. Two features of individuals with the same genotype may vary together because both are affected by environmental factors, such as nutrition. Such features display **environmental correlation**, r_E. In genetically variable populations, correlated variation may also be caused by genetic differences that affect both characters, causing **genetic correlation**, r_G.

Genetic correlations can have two causes. One cause is *linkage disequilibrium*, nonrandom association among the genes that independently affect each character. The other cause is *pleiotropy* the influence of the same genes on different characters. A genetic correlation that is due to pleiotropy will be perfect ($r_G = 1.0$ or -1.0) if all the alleles that increase one character increase (or all decrease) the other character as well. If some genes affect only one of the characters, or if some alter both characters in the same direction (+,+ or –,–) while others have opposite effects on the two characters (+,– or –,+), the genetic correlation will be imperfect.

Genetic correlations need not be constant, and they may evolve (Turelli 1988). A genetic correlation caused by linkage disequilibrium, such as the correlation between style (pistil) length and stamen height in the primrose *Primula vulgaris* (see Figure 9.21), will decline as a result of recombination unless selection for the adaptive gene combinations maintains it. Correlations caused by pleiotropy may also change, although more slowly than those caused by linkage disequilibrium. Some alleles may become fixed, so that the loci contribute no variation and therefore no correlation, whereas other loci, perhaps affecting only one character, remain variable—and changes in allele frequency at these loci will also affect r_G.

Genetic correlation between traits can be estimated from correlations among relatives in the same way that genetic variance can. For example, Juan Fornoni and collaborators (2003) examined the relationship between resistance to herbivorous insects and tolerance of insect damage in jimsonweed (*Datura stramonium*) in central Mexico. Fornoni and colleagues placed pollen from each of 25 plants (male, or "pollen parents") on flowers of

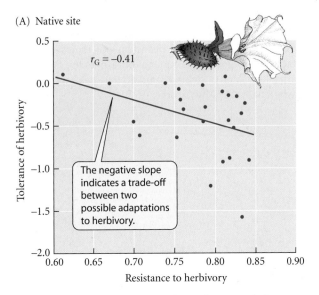

(A) Native site

$r_G = -0.41$

> The negative slope indicates a trade-off between two possible adaptations to herbivory.

Tolerance of herbivory

Resistance to herbivory

FIGURE 13.12 Genetic correlation between tolerance of and resistance to insect herbivory in jimsonweed (*Datura stramonium*). Each point represents the joint mean of the two characters in the offspring of a single pollen parent. The negative correlation implies a trade-off between these two possible adaptations to herbivory. (After Fornoni et al. 2003.)

two other plants (female, or "seed parents"), then planted several offspring from each seed parent in a common garden in a natural environment. They measured resistance—the capacity to repel herbivores by chemical or physical defenses—as the proportion of leaf area that was not eaten by insects. They measured tolerance—a plant's ability to sustain high fitness even if damaged—by the number of seeds a plant produced: the higher the number of seeds, relative to the amount of leaf tissue that had been consumed by insects, the higher the plant's tolerance. The variation among family means for different pollen parents represents the additive genetic variance in each trait, and the correlation between the means for the two characters estimates the genetic correlation. (Calculations based on pollen-parent means are free of whatever nongenetic maternal effects might exist.) The results in one such common garden (**Figure 13.12**) revealed a negative genetic correlation: more resistant genotypes proved to be less tolerant of insect damage. This result is consistent with the hypothesis that the plant allocates energy or resources to each of these traits at the expense of the other.

How genetic correlation affects evolution

Genetic correlations among characters can cause them to evolve in concert. They can also either enhance or retard the rate of adaptive evolution, depending on the circumstances. In some cases, they may severely constrain adaptation.

If two characters, z_1 and z_2, are genetically correlated, the rate and direction of evolution of z_1 depend on both direct selection (on z_1 itself) and selection on z_2. If selection on z_2 is much stronger than on z_1, z_1 may change mostly as a result of its correlation with z_2, and the change may not be in an adaptive direction. For example, Stevan Arnold (1981) estimated a genetic correlation of 0.89 between the feeding reaction of newborn garter snakes to slugs (z_1) and to leeches (z_2; **Figure 13.13**). If both slugs and leeches are present in the

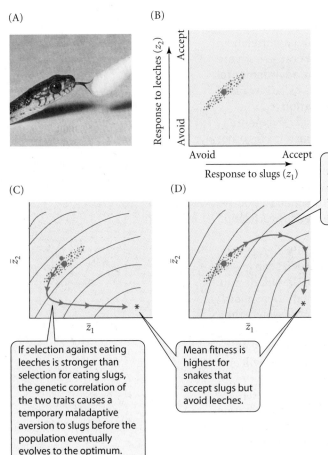

(A)

(B)

Response to leeches (z_2)

Accept

Avoid

Avoid — Accept

Response to slugs (z_1)

> If selection for eating slugs is stronger than selection against eating leeches, the propensity to eat leeches may temporarily increase due to the positive genetic correlation.

(C)

$\bar{z}_2$

$\bar{z}_1$

> If selection against eating leeches is stronger than selection for eating slugs, the genetic correlation of the two traits causes a temporary maladaptive aversion to slugs before the population eventually evolves to the optimum.

(D)

$\bar{z}_2$

$\bar{z}_1$

> Mean fitness is highest for snakes that accept slugs but avoid leeches.

FIGURE 13.13 Possible evolutionary implications of a genetic correlation between feeding responses in garter snakes. (A) A newborn garter snake investigating the odor of a potential prey item on a cotton swab. Snakes show a higher response to favored prey types by flicking the tongue more often. (B) The positive genetic correlation (r_G) between response to slugs (z_1) and response to leeches (z_2). The "cloud" of small points represents individual genotypes; the large point is the population mean for both traits. (C, D) Possible fitness landscapes for a snake population in places where slugs and leeches both occur. Mean fitness, shown by contours for various mean responses to slugs and leeches, is highest (*) for populations that accept slugs but avoid leeches. If the traits were not genetically correlated, they would evolve directly to the optimum (*). Because of the genetic correlation, the joint evolution of these two traits takes a curved path, which may result in temporary maladaptation in one trait. (Photo courtesy of S. J. Arnold.)

environment, there may be direct selection in favor of slug feeding (increased z_1), but stronger selection against leech feeding (decreased z_2), since a leech can kill a snake by biting its digestive tract after being swallowed. If adaptive aversion to leeches evolves (decreased z_2), maladaptive aversion to slugs (decreased z_1) may also evolve, at least temporarily, as a correlated effect (Figure 13.13C). (We assume that other kinds of food are available, so that the population can persist even if it evolves aversion to both leeches and slugs.) Conversely, if leeches are rare and slugs are the most abundant food, selection for feeding on slugs may be stronger, the population may evolve a slug-feeding habit, and it may also evolve the maladaptive habit of occasionally eating leeches (Figure 13.13D). After the more strongly selected trait approaches its optimum (e.g., z_2, strong aversion to leeches, in Figure 13.13C), the more weakly selected trait can evolve to its optimum (z_1, positive feeding response to slugs). That change is based mostly on those genes that affect only z_1.

A conflict may therefore exist between the genetic correlation of characters and directional selection on those characters. When such a conflict exists, the two characters may evolve to their optimal states only slowly, and they may even evolve temporarily in a maladaptive direction. (We have already seen that selection for a deeper bill in the Galápagos finch *Geospiza fortis* caused average bill length to increase, even though selection favored a shorter bill.) In some cases, a genetic correlation may be so strong that one or both traits cannot reach their optimal states. For example, the genetic correlation between male and female body size in the collared flycatcher is approximately 1.0 (Merilä et al. 1998). The optimal size appears to differ between the sexes (see Figure 13.10), but such a strong genetic correlation would prevent the sexes from evolving to their different optimal sizes.

A common constraint is the trade-off between the number and the size of eggs (or seeds) that an organism can produce because the resources that it can allocate to reproduction are limited. This trade-off creates a negative genetic correlation, with some genotypes producing more but smaller eggs and others producing fewer but larger ones (see Chapter 14). Although selection might favor both more and larger eggs, increasing both is possible to only a very limited extent. Thus genetic correlations, owing in some cases to trade-offs of this kind, can sometimes act as genetic constraints on evolution. Whether or not a genetic correlation acts as a long-term constraint depends on several factors, such as how readily the genetic correlation changes.

In other cases, genetic correlations may enhance adaptive evolution, rather than constrain it (Wagner 1988). Multiple characters may evolve as an integrated ensemble more rapidly if they are genetically correlated, as when they are subject to the same developmental controls. This is especially true if the characters are functionally related. For instance, the size of each organ (e.g., lungs, gut, bones) must match the overall body size if an animal is to function properly. Body size would evolve much more slowly in response to selection if every organ had to undergo independent genetic change than if there were coordinated increases or decreases in the sizes of the various organs. During development, in fact, the various organs grow in concert, and alleles that change body size have correlated effects on most body parts (see the discussion of allometric growth in Chapter 3). However, not all functionally related features are so strongly integrated.

Can Genetics Predict Long-Term Evolution?

If the response to selection in natural populations were never limited by the availability of genetic variation in single characters or combinations of characters, the rate and direction of adaptive evolution would depend only on the strength and direction of natural selection. There is reason, however, to think that in some

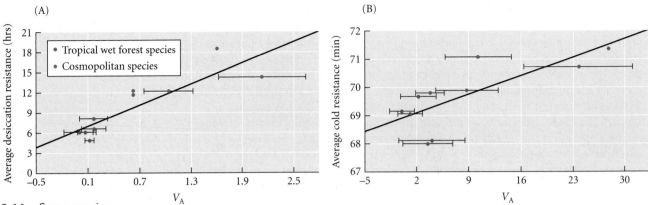

FIGURE 13.14 Some species may lack genetic variation for some characters. The mean trait value is plotted against the additive genetic variance (V_A) for two characters, (A) resistance to desiccation and (B) resistance to cold, in ten species of *Drosophila*. These flies include five species that inhabit only tropical wet forest and five species that are cosmopolitan and are found in many habitats, often associated with humans. The lower V_A in the rain forest species suggests that they would not adapt readily to dry or cold conditions that might be encountered outside their rain forest habitat. (After Kellermann et al. 2009.)

instances, evolution may proceed along "genetic lines of least resistance" (Stebbins 1974; Schluter 1996). Some characters may have very little genetic variation, and therefore may constrain, or at least bias, the direction of evolution. For example, when species of host-specific herbivorous beetles were screened for their propensity to feed on novel plants that they do not normally eat, genetic variation was found in the feeding responses to only certain plants, especially those most closely related to the insects' normal host plants. This pattern suggested that these beetles might adapt more readily to feed on closely related plants than on distantly related plants—which is exactly what has occurred in the evolution of these beetles and many other groups of herbivorous insects (Futuyma et al. 1995). Species of *Drosophila* that inhabit tropical rain forest may also lack genetic variation for some traits: they exhibit much less genetic variance for resistance to cold and dry conditions than more widely distributed species do (**Figure 13.14**).

If two characters are genetically correlated, the greatest genetic variation lies along the long axis of the ellipse formed by a plot of genotypes' joint character values (**Figure 13.15A**). Dolph Schluter (2000) referred to this axis as the "maximal genetic variation" (g_{max}) and predicted that, at least in the short term, adaptive evolution should be greatest along this axis because of the constraining effect of genetic correlation between characters under directional selection. Over time, however, g_{max} should have less of an effect, because characters that are not perfectly genetically correlated can eventually evolve to their optimal values (as described earlier). (The same principle holds for multiple characters. A table of values of V_A for each character and the genetic covariance—a value closely related to the correlation—between each pair of characters is called the GENETIC VARIANCE-COVARIANCE MATRIX, usually referred to simply as the G MATRIX, or G.)

For several adaptive morphological characters in stickleback fishes, sparrows, and other vertebrates, Schluter determined the direction of divergence of a species from a closely related species and compared it with g_{max}, the direction of evolution predicted by the genetic correlation between characters (**Figure 13.15B**). As predicted, the deviation between these two directions was least for species that had diverged only recently, and it increased with time since common ancestry. The rather close initial correspondence between the actual and genetically predicted directions suggests that patterns of genetic variation and correlation can persist for long periods and can influence the direction of evolution. Schluter estimated that this influence may last for up to 4 million years.

The pattern Schluter describes suggests that genetic correlations among characters may remain consistent for a long time. Whether or not this is generally true is one of the most important, and most poorly understood, aspects of phenotypic evolution. Some investigators have found that the genetic correlations among certain characters are fairly similar among geographic populations of a species or among closely related species, but other studies have found lower similarities, suggesting that the strength of genetic correlations can evolve rather rapidly (Steppan et al. 2002). Mattieu Bégin and Derek Roff (2004) found that the additive genetic variances and genetic covariances (a statistic related to

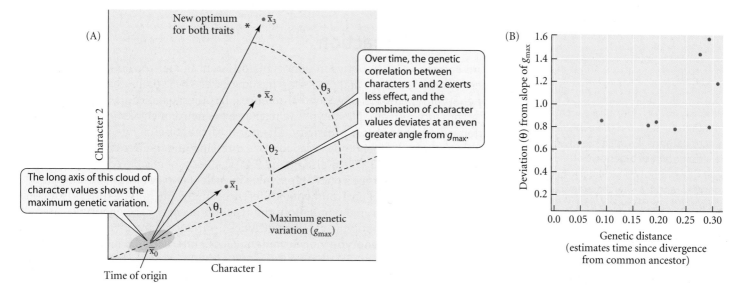

FIGURE 13.15 Evolution along genetic lines of least resistance. (A) The ellipse shows the distribution of values of the genetically correlated characters 1 and 2. The maximal genetic variation (g_{max}) is for combinations of trait values along the long axis of the ellipse. If the ancestral population mean is at point $\bar{x}_0$ and it experiences directional selection toward a new optimum (*), the genetic correlation will cause evolution toward $\bar{x}_1$ at first. The direction of evolution (arrow) deviates by angle θ_1 from g_{max}. Over time, the genetic correlation will exert a lesser effect, so that the line between $\bar{x}_0$ and $\bar{x}_2$ deviates more from g_{max} (angle θ_2). The deviation is still greater (θ_3) at a later time, when the population approaches the optimum for both traits. (B) Genetic correlations among several adaptive traits (e.g., bill dimensions) were estimated in song sparrows (*Melospiza melodia*) and used to determine g_{max}. Points show the deviation (angle θ in panel A) between g_{max} and the difference in these traits between the song sparrow and nine other species of sparrows, in relation to the molecular genetic distance between these species and the song sparrow. The lower deviation of more closely related species (those at lower genetic distance) from g_{max} shows that g_{max} in the song sparrow predicts the initial direction of evolution fairly well. The theory of genetic correlation predicts that the traits of more distantly related species should display a larger deviation from g_{max}, as is observed in these data. (After Schluter 1996.)

correlation) of several morphological measurements are quite similar among species of field crickets and, furthermore, that the pattern of correlation of characters among species closely matched the average pattern of genetic correlation within species (**Figure 13.16**). They suggested that this correspondence indicates genetic constraint. In some studies of other species, however, no correspondence has been found between divergence and the "genetic lines of least resistance," indicating that evolution by natural selection has not been genetically constrained.

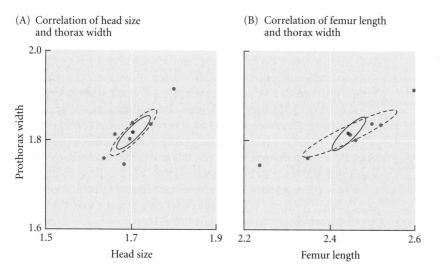

FIGURE 13.16 Orientation of the genetic correlation between pairs of morphological characters within a species of cricket (solid ellipses) compared with the correlation among species for the same pairs of characters (dashed ellipses). The joint mean values for each of the seven species are indicated by red circles; the blue circles show the character means of the species in which the genetic correlation was estimated. (A) The axis of the genetic correlation between head size and prothorax width is nearly identical to the axis of variation among species, suggesting that divergence among them has followed the genetic line of least resistance. (B) Variation among species in the relation between femur length and prothorax width does not match the within-species genetic correlation as closely. (After Bégin and Roff 2004.)

Norms of Reaction

The correspondence between genotypic differences and phenotypic differences depends on developmental processes. In some cases, these processes may reduce the phenotypic expression of genetic differences. In other instances, a single genotype may produce different phenotypes in response to environmental stimuli, a phenomenon called **phenotypic plasticity**. The **norm of reaction** of a genotype is the set of phenotypes that genotype is capable of expressing under different environmental conditions (**Figure 13.17**; see also Figure 9.4). The norm of reaction can be visualized by plotting the genotype's phenotypic value in each of two or more environments.

When the effect of environmental differences on the phenotype differs from one genotype to another in a population, the reaction norms of the genotypes are not parallel, and the phenotypic variance includes a variance component (referred to as $V_{G \times E}$) that is due to **genotype × environment (G × E) interaction** (Figure 13.17B). If all the genotypes have parallel reaction norms (Figure 13.17A), there is no G × E interaction ($V_{G \times E} = 0$). With genotype × environment interaction, the reaction norm may be capable of evolving.

Canalization

For many characteristics, the most adaptive norm of reaction may be a constant phenotype, buffered against alteration by the environment (Figure 13.17C). It may be advantageous, for example, for an animal to attain a fixed body size at maturity or metamorphosis, despite variations in nutrition or temperature that affect the rate of growth. The developmental system underlying the character may then evolve so that it resists environmental influences on the phenotype (Scharloo 1991). This principle was articulated independently by a leading Russian biologist, Ivan Schmalhausen (1986; Russian publication 1947) and by British developmental biologist Conrad Waddington, who referred to this phenomenon as **canalization**.

Waddington (1953) used the concept of canalization to interpret some curious experimental results. A crossvein in the wing of *Drosophila* sometimes fails to develop if the fly is subjected to heat shock as a pupa. By selecting and propagating flies that developed a crossveinless condition in response to heat shock, Waddington bred a population in which most individuals were crossveinless when treated with heat. But after further selection, a considerable portion of the population was crossveinless even without heat shock, and the crossveinless condition was heritable. A character state that initially developed in response to the environment had become genetically determined, a phenomenon that Waddington called **genetic assimilation**.

Although this result is reminiscent of the discredited theory of inheritance of acquired characteristics, it has a simple genetic interpretation. Genotypes of flies differ in their susceptibility to the influence of the environment (in this case, temperature)—that is, they differ in their degree of canalization, so that some are more easily deflected into an aberrant developmental pattern. Selection for this developmental pattern favors alleles that canalize development into the newly favored pathway. As such alleles accumulate, less environmental stimulus is required to produce the new phenotype. The finding that genetic assimilation does not occur in inbred populations that lack genetic variation supports this interpretation (Scharloo 1991).

More recently, some genes that confer canalization have been identified. One is *heat shock protein 90* (*Hsp90*), the protein

FIGURE 13.17 Genotype × environment interaction and the evolution of reaction norms. Each line represents the reaction norm of a genotype—its expression of a phenotypic character in environments E_1 and E_2. The character states expressed are labeled z_1 and z_2. The arrows indicate the adaptively optimal phenotype in each environment. (A) The genotypes do not differ in the effect of environment on phenotype; there is no G × E interaction. The optimal norm of reaction cannot evolve in this case because no genotype matches the arrows. (B) The effect of environment on phenotype differs among genotypes; G × E interaction exists. The genotype with the norm of reaction closest to the optima in E_1 and E_2 (red line) will be fixed. New mutations that bring the phenotype closer to the optimum for each environment may be fixed thereafter. (C) Selection may favor a constant phenotype, irrespective of environment; thus a genotype with a horizontal reaction norm (red line) may be optimal.

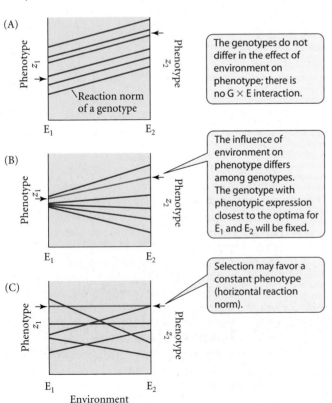

The genotypes do not differ in the effect of environment on phenotype; there is no G × E interaction.

The influence of environment on phenotype differs among genotypes. The genotype with phenotypic expression closest to the optima for E_1 and E_2 will be fixed.

Selection may favor a constant phenotype (horizontal reaction norm).

product of which stabilizes a wide variety of signal-transducing proteins in eukaryotes. When *Hsp90* function is impaired by mutation or chemical treatment of *Drosophila* or the wild mustard *Arabidopsis*, morphological abnormalities of all kinds appear, revealing cryptic genetic variation (Rutherford and Lindquist 1998; Queitsch et al. 2002).

Phenotypic plasticity

In many species, adaptive phenotypic plasticity has evolved: a genotype has the capacity to produce different phenotypes, suitable for different environmental conditions (Pigliucci 2001; West-Eberhard 2003). Phenotypic plasticity includes rapidly reversible changes in morphology, physiology, and behavior as well as "developmental switches" that cannot be reversed during the organism's lifetime (**Figure 13.18A**). In some semiaquatic plants, for instance, the form of a leaf depends on whether it develops below, above, or on the surface of water (**Figure 13.18B**). Adaptive phenotypic plasticity of this kind has evolved by natural selection for those genotypes with norms of reaction that most nearly yield the optimal phenotype for the various environments the organism commonly encounters (Schlichting and Pigliucci 1998).

The evolution of an optimal degree of plasticity, assuming it is not limited by genetic variation in reaction norm, depends on its advantages relative to the costs of plasticity (DeWitt et al. 1998; Auld et al. 2010). Several kinds of costs of plasticity have been identified. For example, costs may be incurred to obtain information about environmental conditions. Most studies of costs have focused on "maintenance costs," which refer not to the cost of producing a specific phenotype (such as one of the leaf types illustrated in Figure

(A)

Catkins

Caterpillars

(B)

Submerged   Air-water interface   Aerial

FIGURE 13.18 Examples of phenotypic plasticity. (A) Larvae of the geometrid moth *Nemoria arizonaria* that hatch in the spring (left) resemble the oak flowers (catkins) on which they feed. Those that hatch in the summer (right) feed on oak leaves and resemble twigs. (B) The form of a leaf of the water-crowfoot *Ranunculus aquatilis* depends on whether it is submerged, aerial, or situated at the air-water interface during development. (A, photos courtesy of Erick Greene; B from Cook 1968.)

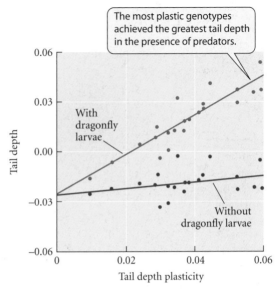

The most plastic genotypes achieved the greatest tail depth in the presence of predators.

FIGURE 13.19 Genetic variation in phenotypic plasticity. The mean tail depth and the plasticity of tail depth (measured as the difference between tail depths in the two environments) are shown for each of 21 half-sib families of wood frog tadpoles. Each family is represented by two points, which show its mean tail depth when reared without dragonfly larvae (green circles) and with larvae (red circles). The spread along the x-axis shows genetic variation for degree of plasticity. (After Relyea 2002.)

13.18B), but rather to the energy and materials a plastic genotype requires to maintain the ability to detect and respond to environmental conditions, compared with a nonplastic genotype that does not make such an investment. For example, tadpoles of the wood frog (*Rana sylvatica*) develop several features, including a deeper tail fin, when they perceive the presence of dragonfly larvae (dangerous predators of tadpoles). Rick Relyea (2002) examined genetic variation in plasticity by rearing 21 families of half-sibs, each family divided between containers with and without dragonfly larvae. He found that families differed in their degree of plasticity, measured as the difference in tail depth between the two environments (**Figure 13.19**). In the presence of dragonfly larvae, tadpoles with deeper tails had higher survival rates, as previous experiments had also shown. In the absence of dragonfly larvae, however, those families that exhibited higher plasticity had lower survival, even though the families did not differ in tail depth in that environment. Thus the capacity for plastic development carries a cost, the exact cause of which is unknown.

Evolution of variability

Whereas "variation" refers to the differences actually present within a sample or a species, the word "variability," in the strict sense, refers to the ability, or potential, to vary (Wagner et al. 1997). For example, in insects, the number of compound eyes (two, or in a few species, none) seems able to vary much less than the number of units (ommatidia) that compose each compound eye. In mammals, because of the developmental correlation between the size of the body and the size of the brain and intestines, some conceivable variations—large bodies with tiny brains, for instance—are seldom or never seen. Developmental processes therefore affect **variability**, the extent to which genetic variation can be expressed as phenotypic variation. Does variability depend solely on immutable "laws" of development, or does it evolve by natural selection? This question applies to both variability in individual characters and correlations among characters.

The variability of individual characters is affected by the evolution of canalization. A character that is insensitive to alteration by environmental factors is ENVIRONMENTALLY CANALIZED. A character may also become GENETICALLY CANALIZED; that is, it may acquire low sensitivity to the effects of mutations. In such instances, the phenotype may remain unchanged even if the genes underlying its development vary considerably.

Threshold traits, for example, are expressed as discrete alternatives but are controlled by polygenic variation rather than by single loci. The polygenic variation is not expressed phenotypically unless development is perturbed substantially (beyond a threshold) by a large enough genetic or environmental change. For example, in natural populations of *Drosophila melanogaster* and related species, there is almost no variation in the number of bristles (four) on the scutellum (part of the thorax). In homozygotes for the *scute* (*sc*) mutation, however, bristle number is variable because of the expression of polygenic variation at other loci (**Figure 13.20A**). Thus this mutation breaks down canalization, and conversely, the normal allele at the *scute* locus may be considered to exert genetic canalization. Using an experimental population that was homozygous for *scute* and was therefore variable in bristle number, James Rendel and colleagues (1966) imposed selection against variation by breeding from the least variable families (i.e., those that most consistently had two bristles). Within about 30 generations, the phenotypic variation became greatly reduced (**Figure 13.20B**). The investigators apparently had selected for genes that canalized development into a new pathway.

Can natural selection produce the same result? According to mathematical models, alleles for environmental canalization should increase if there is prolonged stabilizing selection against deviations from an optimal phenotype. Genetic canalization, however, evolves only if stabilizing selection is not too strong. Strong stabilizing selection eliminates new mutations so fast that few individuals deviate from the optimum, so there is little

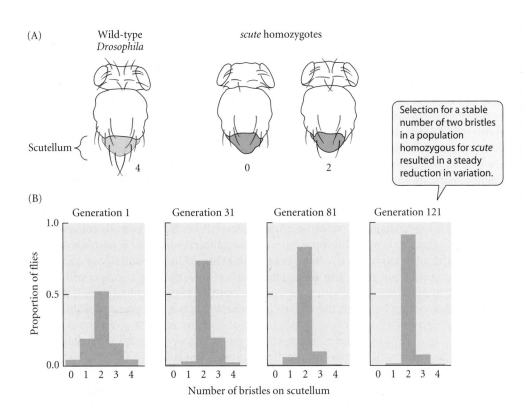

FIGURE 13.20 Canalization by artificial selection. (A) The top of the head and thorax of *Drosophila melanogaster*. The posterior part of the thorax, the scutellum, bears four bristles in wild-type flies, but a variable number (e.g., zero or two) in *scute* homozygotes. (B) Selection for a stable number of two bristles in a population homozygous for *scute* resulted in a steady reduction of variation, shown here at 1, 31, 81, and 121 generations of selection. (Data from Rendel et al. 1966.)

Selection for a stable number of two bristles in a population homozygous for *scute* resulted in a steady reduction in variation.

selection for alleles that prevent the phenotypic expression of the mutations (Wagner et al. 1997; Kawecki 2000). Some patterns of variation within species imply that selection for canalization has been effective. For example, the floral structures of some animal-pollinated plants, which are thought to be strongly selected for successful pollination, are less variable than their leaves (**Figure 13.21**).

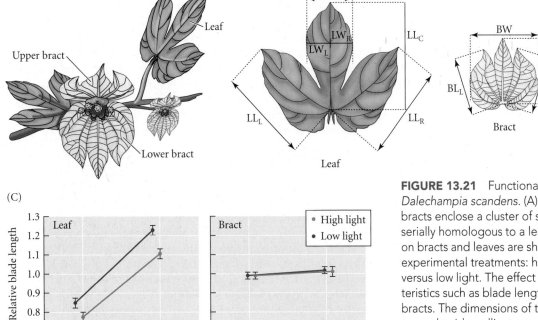

FIGURE 13.21 Functional integration in a tropical vine, *Dalechampia scandens*. (A) Upper and lower pale green bracts enclose a cluster of small flowers. (B) A bract (right) is serially homologous to a leaf (left). The measurements taken on bracts and leaves are shown. (C) Plants were grown in four experimental treatments: high versus low nutrition and high versus low light. The effect of these treatments on characteristics such as blade length was stronger on leaves than on bracts. The dimensions of the bracts, which protect the flowers and guide pollinators, are thought to be under stronger canalizing selection. (After Pélabon et al. 2011.)

The theory of canalization may explain why some characters—such as some synapomorphies of higher taxa—have remained unchanged over vast periods of time (see Chapter 22). For example, the earliest known Devonian tetrapods had a variable number of about eight or nine toes (see Chapter 4). Soon afterward, however, tetrapods "settled on" five-digit (pentadactyl) limbs, and almost no tetrapod vertebrates since then have had more than five digits. Therefore there was no inescapable rule that feet could have no more than five toes, but the developmental processes evolved so that the maximum digit number came to be constrained.

The hypothesis of morphological integration (Olson and Miller 1958), or more generally, **phenotypic integration** (Pigliucci and Preston 2004), holds that functionally related characteristics should be genetically correlated with one another. Günter Wagner and Lee Altenberg (1996) have shown theoretically that prolonged directional selection favors modifier alleles that enhance a pleiotropic correlation between functionally related traits along an axis pointing toward the optimum for the characters (marked with an asterisk in Figure 13.15A). For example, if it were functionally important for the upper and lower mandibles of a bird's bill to be the same length, then selection for a longer bill would include selection for alleles that coordinate the development of the two mandibles, creating a pleiotropic correlation between them. This hypothesis implies that pleiotropic effects of genes can evolve, which could occur by alteration of a gene's expression in the development of one or both traits, either by changes in its *cis*-regulatory regions or by changes in another gene that affects its expression (Pavlicev and Wagner 2012). The potential for such evolution has been shown in experiments such as those performed by Lynda Delph and collaborators (2011) on *Silene latifolia* (bladder campion), which has female and male plants. Although the genetic correlation between the size of female and male flowers is 1.0, Delph et al. were able to reduce r_G within five generations by artificial selection.

Just how prevalent adaptive phenotypic integration is remains to be seen, but the hypothesis is supported by considerable evidence. Among the most intriguing is an analysis of correlations within and between the forelimbs and hindlimbs of monkeys and apes, including humans (Young et al. 2010). Forelimbs and hindlimbs differ in function, and may be considered different FUNCTIONAL MODULES, but they are serially homologous (see p. 63 in Chapter 3), and the same Hox genes are expressed during development of the limbs' corresponding segments. The segments may be considered different DEVELOPMENTAL MODULES (**Figure 13.22A**). Monkeys are quadrupedal; their forelimbs and hindlimbs are roughly equally long and have somewhat similar functions. Humans are drastically different, with much longer hindlimbs, which are an adaptation to bipedal locomotion, and modifications of the arms and hands that provide enhanced manual dexterity. Nonhuman apes have relatively long arms, which they (especially gibbons and orangutans) use for specialized locomotion in trees. Young and collaborators found that forelimb lengths and hindlimb length are correlated within all the primate species they examined. The correlation was

FIGURE 13.22 Developmental and functional modules in the limbs of humans and their relatives. (A) Five Hox genes (9–13) are differentially expressed in the development of the three major sections of the limbs, marking three developmental modules. Expression of each Hox gene is indicated by the presence of a bar in the diagram at left; the darker bars indicate higher expression. Note that the genes are expressed in corresponding parts of both forelimbs and hindlimbs, which are distinct functional modules. H, R, and MC are humerus, radius, and metacarpals in the forelimb; F, T, and MT are femur, tibia, and metatarsals in the hindlimb. (B) For each species in the phylogeny, the diagram at the top shows correlations (numbers) between lettered boxes that represent bones in the forelimb (left boxes) and the hindlimb (right boxes). Each number is the correlation between lengths of the adjacent bones. Correlations between forelimb and hindlimb are stronger (indicated by darker gray) in the monkeys (the four species at right) than in the hominoid apes (the four species at left). The diagrams just above the phylogeny show that the overall modularity (the average degree of correlation among limb segments) is lower in apes. (After Young et al. 2010.)

weaker in apes and humans than in monkeys, presumably mirroring the evolution of greater functional differentiation in apes; but they remain correlated nevertheless. The most interesting observation was that serially homologous segments (e.g., humerus and femur, or radius and tibia) that are strongly correlated in the monkeys have become less strongly correlated in apes and humans (**Figure 13.22B**). Within limbs, adjacent segments of the hindlimb are strongly correlated in humans, reflecting its new function. Overall, the

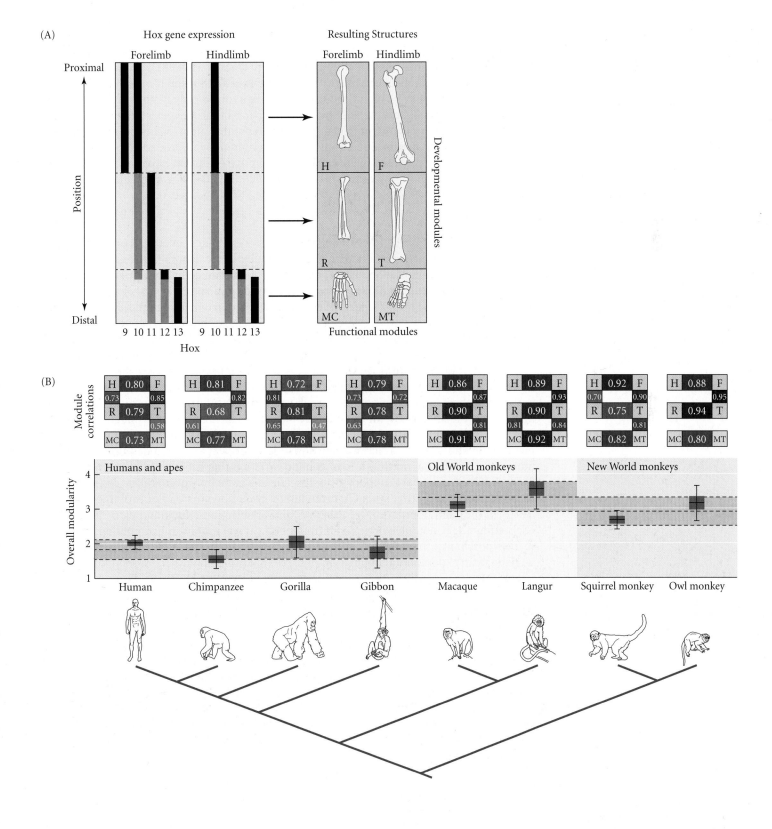

degree of correlation among limb segments is lower in apes and humans than in monkeys, perhaps indicating the evolution of reduced constraint that facilitated the evolution of very different locomotion and other functions. Correlations among humans' leg segments suggest that a new pattern of phenotypic integration then evolved.

Adaptation and Constraint

The variety of finely attuned, often astonishing adaptations of organisms has evolved by the action of natural selection on genetic variation. But failures of adaptation also require explanation. For example, relatively few species of plants have adapted to saline or other stressful environments, even if they grow nearby (Bradshaw 1991); there appear to be "empty niches" for kinds of organisms that have evolved less often than ecological opportunity might suggest (e.g., blood-feeding bats have evolved only once and are limited to the American tropics), and the vast majority of species that have existed became extinct because of their failure to adapt to changes in their environment. Many other phenomena also suggest that adaptation must have been prevented, or restricted to some channels rather than others. The possible constraints on adaptive evolution have become an important focus of contemporary research (Hansen and Houle 2004; Blows and Hoffmann 2005; Futuyma 2010).

Genetic constraints on adaptation

Some constraints on adaptation arise in part from extrinsic factors (Barton and Partridge 2000). For example, adaptation of a population may be inhibited by gene flow from populations that are adapted to a different environment (see Figure 12.12). Other constraints on adaptation must arise from the kind of genetic variation on which natural selection can act. Evolution may be constrained by elements of the G matrix: the amount of genetic variation (V_A) in individual characters and genetic correlations (r_G) among characters (Blows and Hoffmann 2005).

As we have seen, genetic variation appears to be slight or absent for some traits, such as desiccation resistance in rain forest *Drosophila*, feeding responses to certain plants by herbivorous beetles, and metal tolerance in some species of grasses. Because many of the alleles that contribute to V_A for quantitative characters have deleterious pleiotropic effects, they may not contribute to adaptation, so the "usable" V_A may be far less than meets the eye (Houle et al. 1996). Adaptation in some cases appears to have been based on rare mutations rather than on the store of genetic variation that most populations harbor. For example, resistance to organophosphate insecticides in the mosquito *Culex pipiens* is based on a single mutation that spread worldwide by gene flow, instead of evolving independently in different populations exposed to similar selection (Raymond et al. 2001). Thus adaptation might not have occurred if this specific novel mutation had not happened.

This possibility, that adaptation may depend on an improbable historical accident, was realized in an experimental population of *E. coli* that was maintained for more than 30,000 generations in a medium that contained citrate (Blount et al. 2008). This species normally is citrate-negative: that is, it cannot use citrate as a carbon source because it cannot transport it into the cell. In one of many replicate populations, a high frequency of citrate-using (citrate-positive) cells was found after 31,500 generations. Moreover, replicate lines founded by citrate-negative cells from this same population rapidly evolved to be citrate-positive. During the previous 30,000 generations, every possible base pair mutation must have occurred in each population, yet the genetic capacity for citrate use had not evolved. What enabled this population to evolve citrate use, repeatedly, was the previous origin of a mutation at a different locus (locus *A*) that enabled later mutations at a second locus (*B*) to confer citrate transport. That is, the advantageous function of one mutated gene depended on its interaction with a previous mutation that "set the stage." This fortuitous combination of mutations was very improbable.

(A) Armored plates

(B) Population phylogeny

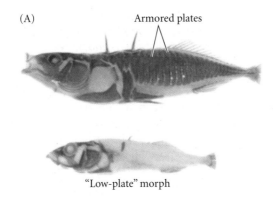

"Low-plate" morph

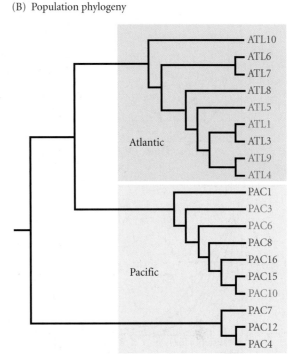

Such examples raise the question of how often adaptation is based on a supply of new mutations (and may therefore be limited by that supply) and how often it is based on plentiful standing genetic variation (Barrett and Schluter 2008). Evidence of selective sweeps (see Chapter 12) suggests that some mutations must have been beneficial as soon as they occurred, because the copies of an advantageous allele that had persisted in the population at low frequency for a long time before it became advantageous would have been separated by recombination from specific linked SNP variants and would show little evidence of a selective sweep (Przeworski et al. 2005). In contrast, some adaptations certainly evolved from selection on low-frequency alleles in the pool of standing variation. For example, reduced body armor has evolved independently in many freshwater populations of three-spined sticklebacks (*Gasterosteus aculeatus*), but it is based on the same allele at the *Ectodysplasin* (*Eda*) locus in all of those populations (**Figure 13.23**). This allele arose more than 2 Mya, based on its sequence divergence from the ancestral allele that characterizes fully armored populations, and has been found at low frequency in those populations (Colosimo et al. 2005). Much of the genetic divergence between marine and freshwater populations of this species shows a similar pattern (Jones et al. 2012).

The genetic correlations that result from pleiotropy create more genetic variation for some combinations of trait values than for others, and evolution sometimes occurs mostly along the directions of greatest genetic variation (see Figures 13.15 and 13.16). It is quite possible for each of several characteristics to display V_A individually, yet for there to exist no genetic variation at all for certain character combinations (Walsh and Blows 2009). For example, Mark Blows and colleagues (2004) found that the concentration of each of the nine cuticular hydrocarbons (CHCs) on the body surface of male *Drosophila serrata*, which affect females' choice of mates, is genetically variable. However, the CHC combinations most preferred by females varied hardly at all, and it appears that the CHCs are unlikely to evolve in response to sexual selection. In a study of QTL affecting 70 skeletal measurements that vary among inbred strains of mice, Günter Wagner and colleagues (2008) estimated that the average locus affected 7.8 traits, and that some affected as many as 30. Such extensive pleiotropy can result in strong genetic correlations among sets of measurements, so that the phenotype consists of a number of "modules," each composed of some strongly correlated elements. In this case, the "effective number" of independent characters may be far less than we at first may suppose (Kirkpatrick 2009; Walsh and Blows 2009), and the variety of paths that evolution might take will be restricted.

Can adaptation rescue species from extinction?

Although every extant species owes its existence to the many adaptive changes its ancestors underwent in the past, the great majority of past species are extinct, so adaptation to environmental changes must sometimes be insufficient. The question of what determines

FIGURE 13.23 Evolution based on standing variation in the stickleback *Gasterosteus aculeatus*. (A) Cleared and stained specimens of (left) the ancestral, completely armored morph, found in marine and some freshwater populations, and (right) the "low-plate" morph, found in many freshwater populations in northern Eurasia and North America. The low-plate phenotype is caused by an allele of the *Ectodysplasin* (*Eda*) gene, which encodes a signaling protein that is required for differentiation of ectodermal features. (B) A phylogeny of the populations based on SNPs at 25 random loci shows that several low-plated populations (blue) have evolved independently. The *Eda* sequences in all the low-plated populations form a single branch of the gene tree (not shown). Therefore the *Eda* sequences in these populations are descended from a single mutated allele that has been maintained in the species and has increased in frequency independently in different populations. (After Colosimo et al. 2005.)

(A)

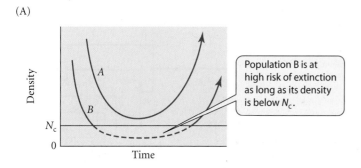

Population B is at high risk of extinction as long as its density is below N_c.

(B)

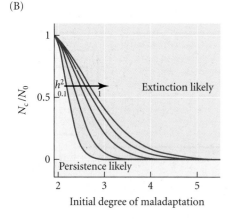

(C)

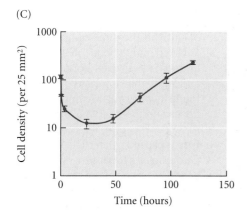

FIGURE 13.24 Evolution may avert extinction of a population in an abruptly changed environment. (A) In Gomulkiewicz and Holt's model, two populations, A and B, decline in abundance because their growth rate in the altered environment is negative. Their density rebounds as adapted genotypes increase in frequency. Population A is not at high risk of extinction because it adapts before its density drops below a critical level (N_c). Population B risks extinction because its density drops below N_c, where random fluctuations may extinguish it. If it persists long enough to adapt, it may grow to a safe density, above N_c. (B) Combinations of parameters that determine whether a population is likely to persist or become extinct. The y-axis is the critical density N_c, below which extinction is likely, relative to the initial density N_0. The x-axis is the degree of maladaptation before genetic adaptation occurs; it is directly proportional to the difference between the actual and the optimal mean trait values. The several curves show that higher heritability (h^2) increases the zone of likely persistence. (C) The average change in population density in cultures of yeast subjected to lethal concentrations of salt. The populations contained some genetic variation for resistance and reversed their decline by adapting. Compare this curve with the theoretical curves in panel A. (A, B after Gomulkiewicz and Holt 1995; C after Bell and Gonzalez 2009.)

whether or not evolution will rescue a population or species from rapid environmental changes is one of the most important questions in evolutionary biology, because humans are changing the environment in many ways, ranging from pollution and habitat destruction to global climate change.

If the per capita growth rate (r) of a population falls below zero, the population will eventually become extinct. This trajectory may be reversed if adaptive evolution occurs rapidly enough: the fate of the population depends on a race between demographic and evolutionary processes. Evolutionary theorists have modeled these processes when the environment changes abruptly (e.g., sudden pollution of a river by mine waste) and when it changes gradually (e.g., the ongoing increase in global temperature). Most of the models treat a quantitative, polygenic trait that affects r.

Richard Gomulkiewicz and Robert Holt (1995) modeled an abrupt environmental change in which a population declines to a critically low density (N_c), below which it is destined for extinction if evolutionary adaptation is not fast enough (**Figure 13.24A**). The rate of adaptation depends on the genetic variance in the adaptive trait. The degree of maladaptation is greater, the greater the difference between the initial trait mean and the new optimal value of the trait. The probability of adaptive "rescue" is greater if the population had initially high density, if the degree of maladaptation is low, and if the heritability is high (**Figure 13.24B**). In an experiment in which hundreds of laboratory yeast populations were subjected to lethal salt concentrations, Graham Bell and Andrew Gonzalez (2009) found a close match to the model (**Figure 13.24C**).

Michael Lynch and Russell Lande (1993) developed the first of several models of adaptation to a steadily changing environment, in which the optimal trait changes in the same direction at rate k per unit time. There must be a critical rate of change, above which adaptive evolution cannot keep pace, so that the difference between the actual trait mean and the optimum becomes steadily greater, and the population declines to extinction. Because the genetic variation initially present in the population will be "used up" by directional selection, ongoing evolution of the trait depends on the mutational variance (V_m), a continuous supply of new mutations that replenish V_A (see p. 201). Because V_m is proportional to effective population size, larger populations can adapt to higher rates of environmental change (**Figure 13.25**).

FIGURE 13.25 Persistence versus extinction in a model of a directionally changing environment. Above a critical rate of environmental change (i.e., a critical rate of change of the optimal character state), a population cannot evolve fast enough and is likely to become extinct. The rate of adaptation, hence the chance to keeping up with the environment, is greater for populations with larger effective sizes because of the higher genetic variance supplied by new mutations. The risk of extinction is somewhat greater (marked by the brown and green lines) for higher σ_θ^2, the degree of fluctuation in the environment around its long-term trend. (After Lynch and Lande 1993.)

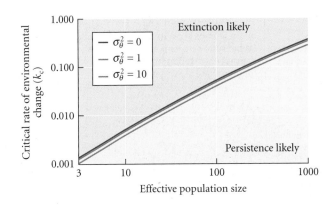

These models make many simplifying assumptions, some of which have been explored by other researchers. For example, phenotypic plasticity can increase the chance of successful adaptation because it may enhance survival in the altered environment (Chevin et al. 2010). If the environment (and hence the optimal phenotype) fluctuates around a long-term trend, adaptation suffers setbacks, and extinction is more likely. Perhaps most importantly, successful adaptation may require evolution of multiple characteristics, especially if the changing environment imposes multiple stresses. For example, climate change alters ecological interactions among species, some of which may be better suited to the new climate and inhibit adaptation by other species (de Mazancourt et al. 2008). A population's persistence may then depend on evolutionary changes both in temperature tolerance and in resistance to competitors or predators.

Because most phenotypic traits are genetically variable, and because rapid adaptation has been demonstrated in many species (see Chapter 11), it is likely that evolution will rescue some populations from ongoing and future human alterations of the environment—but not all (Kinnison and Hairston 2007). Recall (from Chapter 6) that during the Pleistocene, when temperatures changed drastically but much more slowly than they are changing now, many species shifted their geographic ranges and became extinct in previously inhabited regions (where they were replaced by other species). Genetic variation in an apparently key trait does not guarantee that a population will adapt to a changing environment. For example, the fitness of the European winter moth (*Operophtera brumata*) depends on a close match between the time of egg hatch and the time of spring bud break (leafing out) of the oaks on which the larvae often feed. If the larvae hatch before bud break, they starve; if they hatch too late, they must feed on tougher, chemically defended mature leaves, and they become small adults, with low fecundity. The timing of egg hatch has been advancing faster than the advance of bud break of oak trees as temperature has risen during the last two decades (van Asch et al. 2007). The relationship of egg hatch to spring temperature is genetically variable (h^2 ranges from 0.63 to 0.94), but for unknown reasons, hatching time seems not to be evolving to match the time of bud break. If the moth population declines, there will probably be a severe impact on populations of birds that depend on the larvae for feeding nestlings.

Genetic correlations among traits might also retard the rate of adaptation. Julie Etterson and Ruth Shaw (2001; Etterson 2004) grew seedlings of an annual prairie plant, partridge-pea (*Chamaecrista fasciculata*), from northern (Minnesota), mid-latitude (Kansas), and southern (Oklahoma) populations at all three latitudes. In this way, they simulated the responses of populations to global climate change: by 2025–2035, according to climate change models, the Minnesota population will experience temperature and drought conditions that match those of Kansas today. Etterson and Shaw estimated V_A and genetic correlations (r_G) for several traits in each population grown in each environment, and they estimated the direction and intensity of selection on each trait by its correlation with seed production. Most features displayed heritable variation, but the expected rate of evolution of each trait toward the future optimum was significantly reduced by genetic correlations among traits that were antagonistic to the direction of selection (**Figure 13.26A**). The

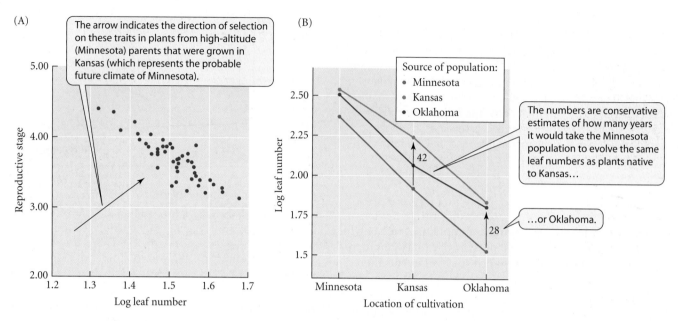

FIGURE 13.26 Why the partridge-pea might not adapt to climate change. (A) The genetic correlation between two traits (leaf number and reproductive stage, representing rate of development) is shown by the points. There is little genetic variation along the axis of selection (arrow). (B) Average leaf number of plants derived from populations in Minnesota, Oklahoma, and Kansas, grown in each of those locations. (After Etterson and Shaw 2001.)

Go to the

EVOLUTION
Companion Website at
sites.sinauer.com/evolution3e
for quizzes, data analysis and
simulation exercises, and other
study aids.

investigators concluded that in the Minnesota population, some key traits would not evolve to match the mean of today's Kansas population by the time the Minnesota population experiences a Kansas-like climate (**Figure 13.26B**). They suspect that adaptation will take even longer than calculated from the genetic parameters because seed production by the Minnesota population in the hot, dry Kansas environment is so greatly diminished that the population will probably dwindle and lose genetic variation by genetic drift. Because of CO_2 produced by human combustion of fossil fuels, climates are changing at rates more than 100 times greater than during the Pleistocene, and it seems likely that thousands of species, like the partridge-pea, will fail to adapt fast enough to avert extinction.

Summary

1. Quantitative trait loci (QTL) can be mapped using molecular or other markers. The variation in many traits is caused by variation at several or many loci, some with large and others with small effects. Some of the genes involved in certain characters have been identified and their functions are known.

2. Variation (variance) in a phenotypic trait (V_P) may include genetic variance (V_G) and variance due to the environment (V_E). Genetic variance may include both additive genetic variance (V_A) due to the additive effects of alleles and nonadditive genetic variance due to dominance and epistasis. Only the additive variance creates a correlation between parents and offspring (and it can be measured by this correlation). Thus only V_A enables response to selection.

3. If alleles that contribute to variation in a polygenic trait are selectively neutral and change in frequency by genetic drift, the short-term rate of evolution depends on the effective population size, but the long-term rate is proportional to the rate of polygenic neutral mutation. Evolutionary rates are often lower than the neutral model predicts, implying that stabilizing selection or purifying selection has acted.

4. The ratio V_A/V_P is the heritability (h_N^2, or simply h^2) of a trait. Heritability is not fixed, but depends on allele frequencies and on the amount of phenotypic variation induced by environmental variation. The short-term effect of selection ("response" to selection) on a character can be predicted from the heritability and the strength of selection. The additive genetic variance itself, however, is a better indicator of the ability to evolve.

5. Most, although not all, characters show substantial genetic variance in natural populations and may therefore evolve rapidly if selection pressures change. Artificial selection experiments show that traits can often be made to evolve far beyond the initial range of variation. The response to

selection is based on both genetic variation in the original population and new mutations that occur during the experiments. Directional selection has been detected in many populations, including humans.

6. Some traits exhibit stabilizing selection, either because the character is nearly at its optimum or because conflicting selection pressures or negative pleiotropic effects prevent further change. Sex differences in the optimal value of a trait can contribute to antagonistic selection.

7. Linkage disequilibrium and especially pleiotropy cause genetic correlations among characters, which, together with correlations caused by environmental factors, give rise to phenotypic correlations. The evolution of a trait is governed both by selection on that trait and by selection on other traits with which it is genetically correlated. The effect of a genetic correlation depends on its strength and degree of permanence. Genetic correlations can enhance the rate of adaptation (if functionally interdependent features show adaptive correlation), can cause a trait to evolve in a maladaptive direction (if selection on a correlated trait is strong enough), or may reduce the rate at which characters evolve toward their optimal states. It is not certain whether genetic correlations are especially strong among characters that are functionally integrated (the hypothesis of phenotypic integration).

8. The norm of reaction—the expression of the phenotype under different environmental conditions—can evolve if genotypes vary in the degree to which the phenotype is altered by the environment in which an individual develops. Some characters exhibit adaptive phenotypic plasticity, whereas selection in other cases favors constancy of phenotype despite differences in environment.

9. Canalization is the buffering of development against alteration by environmental or genetic variation. Canalized characters include threshold characters, in which underlying polygenic variation is not phenotypically expressed unless a drastic mutation or environmental perturbation breaks down canalization. Canalization can evolve under some circumstances. The evolution of canalization may explain the constancy of some characters over vast periods of evolutionary time.

10. Although many characters are genetically variable and can evolve rapidly, evolution appears often to be constrained, partly because of limitations on genetic variation, including limitations arising from genetic correlation among characters. Understanding genetic constraints is the key to understanding phenomena that range from phylogenetically conservative characters to extinction, both in the past and in the near future.

11. The likelihood that adaptive evolution will rescue a population from extinction in an abruptly altered or continuously changing environment depends on many factors, especially the amount of genetic variation. When the environment is gradually but steadily changing, the supply of new variation by mutation, which is proportional to population size, is critical. Some, but by no means all, natural populations are likely to survive the many great changes in environment that humans are causing.

Terms and Concepts

additive genetic variance (V_A)

canalization

correlated selection

environmental correlation (r_E)

environmental variance (V_E)

epistasis

genetic assimilation

genetic correlation (r_G)

genetic variance (V_G)

genotype × environment (G × E) interaction

heritability

norm of reaction

phenotypic correlation (r_P)

phenotypic integration

phenotypic plasticity

phenotypic variance (V_P)

QTL mapping

quantitative genetics

realized heritability

response to selection

selection differential

selection gradient

selection plateau

threshold trait

variability

Suggestions for Further Reading

Introduction to Quantitative Genetics, by D. S. Falconer and T. F. C. Mackay, fourth edition (Longman Group Ltd., Harlow, UK, 1996) is a widely read, clear introduction to the subject. An advanced treatment is *Genetics and Analysis of Quantitative Traits*, by M. Lynch and J. B. Walsh (Sinauer Associates, Sunderland, MA, 1998).

Phenotypic plasticity, canalization, and related topics are the subject of C. D. Schlichting and M. Pigliucci's *Phenotypic Evolution: A Reaction Norm Perspective* (Sinauer Associates, Sunderland, MA, 1998).

Useful overviews of genetic constraints on adaptation are "A reassessment of genetic limits to evolutionary change," by M. W. Blows and A. A. Hoffmann, *Ecology* 86: 1371–1384 (2005), and "Evolutionary constraints and ecological consequences," by D. J. Futuyma, *Evolution* 64: 1865–1884 (2010).

Problems and Discussion Topics

1. Under artificial selection for increased body weight, what will be the response to selection (R), after one generation, for the following values of phenotypic variance (V_P), additive genetic variance (V_A), environmental variance (V_E), and selection differential (S)? (a) $V_P = 2.0$ grams^2, $V_A = 1.25$ g^2, $V_E = 0.75$ g^2, $S = 1.33$ g; (b) $V_P = 2.0$ g^2, $V_A = 0.95$ g^2, $V_E = 1.05$ g^2, $S = 1.33$ g; (c) $V_P = 2.0$ g^2, $V_A = 1.25$ g^2, $V_E = 0.75$ g^2, $S = 2.67$ g. (Answer for part a: Mean weight will increase by about 0.83 g.) If the parameters remain the same for successive generations of selection, and the initial mean weight is 10 g, what is the expected mean after two generations of selection in each case?

2. If most quantitative genetic variation within populations were maintained by a balance between the origin of new mutations and selection against them, then most mutations might be eliminated before environments could change and favor them. If the "residence times" of most mutations were short enough, the alleles that distinguish different populations or species would generally not be those found segregating within populations (Houle et al. 1996). How might one determine whether the alleles that contribute to among-population differences in the means of quantitative traits are also polymorphic within populations?

3. It has been suggested that genetic correlation between the expressions of a trait in the two sexes is responsible for some apparently nonadaptive traits, such as nipples in men and the muted presence in many female birds of the bright colors used by males in their displays (Lande 1980). Suggest ways of testing this hypothesis. What other traits might have evolved because they are genetically correlated with adaptive traits, rather than being adaptive themselves?

4. Debate the proposition that paucity of genetic variation and genetic correlations does not generally constrain the rate or direction of evolution.

5. Consider a character that is typical of the species in a higher taxon and may indeed be an important synapomorphic character for the clade. (For example, the number of petals is such a character in many genera and families of plants.) How might you decide whether this consistency is a result of intrinsic, unchangeable developmental "rules" or of a history of selection for canalization?

6. Traditional quantitative genetics is based on a theory of multiple anonymous loci, the functional roles of which are unknown. Using QTL mapping, together with "candidate loci," the sequences and functions of some of these loci are now being discovered. In what ways is this important for understanding the evolution of phenotypic traits?

7. How can adaptive phenotypic plasticity be distinguished from nonadaptive phenotypic plasticity?

8. Consider the number of characters a gene pleiotropically affects and the magnitude of the effects of different alleles on each of those characters. How would these variables affect the rate of adaptation to a novel environment? (See Orr 2000 and Wagner et al. 2008.)

9. Many ecologists are studying the likely effects of global climate change on species and communities. Do they generally assume that species will adapt, or will not adapt, to changing climate? (You might scan papers in the journal *Global Change Biology*.)

10. Debate the proposition that most species will successfully evolve adaptations to climate change.

The Evolution of Life Histories

Much of the richness of biology lies in the diverse, often astonishing, and sometimes bizarre adaptations of organisms—features that have evolved because they increased fitness relative to ancestral characteristics. "Fitness" appears as an abstract variable in much of the theory of natural selection as we have considered it thus far, but as we have seen, fitness has several major components. These components include survival, female fecundity, and male mating success, which in turn are the results of real anatomical, physiological, and cellular features. When we turn our attention to the components of fitness, we encounter variations and even paradoxes that cry out for explanation.

Consider some differences among species in various aspects of their life cycles. Sea anemones and corals can live for close to a century, bristlecone pine trees (*Pinus aristata*) have survived for 4600 years, and vegetatively propagating clones of quaking aspen (*Populus tremuloides*) can live for more than 10,000 years. In contrast, annual plants die less than a year after germinating, and many small animals, such as some rotifers, live for at most a few weeks. Fecundity likewise varies. Many bivalve molluscs and other marine invertebrates release thousands or millions of tiny eggs in each spawning, whereas a blue whale (*Balaenoptera musculus*) gives birth to a single offspring that weighs as much as an adult elephant, and a kiwi (*Apteryx*) lays a single egg that weighs 25 percent as much as its mother (**Figure 14.1**). Some birds, such as cranes and parrots, form

Showing its age Bristlecone pines, surviving in the punishing environment of desert mountaintops in California, are among the oldest known individual organisms. Evolutionary theory seeks to understand why some species live so much longer than others.

FIGURE 14.1 This X-ray of a kiwi (*Apteryx*) shows the bird's enormous egg, which weighs 25 percent of the female's body weight. (Photo courtesy of Otorohanga Zoological Society.)

lifelong pair-bonds, while in other birds, such as many grouse and hummingbirds, a male may mate with many females and form a pair-bond with none (**Figure 14.2**). In some such species, males have baroque ornaments that may be necessary for mating success. Many species, such as humans, reproduce repeatedly, whereas others, such as century plants (*Agave*) and some species of salmon (*Oncorhynchus*), reproduce only once and then die. Reproductive age may be reached rapidly or slowly. A newly laid egg of *Drosophila melanogaster* may be a reproducing adult 10 days later, and a parthenogenetic aphid may carry an embryo even before she herself is born. In contrast, periodical cicadas feed underground for 13 or 17 years before they emerge, reproduce, and then die within a month (**Figure 14.3**). Many animals and plants have separate sexes, but earthworms and many plants are SIMULTANEOUS HERMAPHRODITES that have both sexual functions at the same time, whereas certain sea bass, squash, and other SEQUENTIAL HERMAPHRODITES develop first as one sex and later switch to the other sex. The process of reproduction often involves sex, accompanied by genetic recombination, but many organisms reproduce by parthenogenesis—development from an unfertilized egg.

What accounts for such extraordinary variation in species' survival and reproduction—the very features that we would expect to be most intimately related to their fitness? There are apparently many ways to achieve high fitness. To make sense of all this diversity, evolutionary ecologists and evolutionary geneticists have developed theories of the evolution of **life histories**: the age-specific probabilities of survival and reproduction characteristic of a species.

(A)

(B)

FIGURE 14.2 Two bird species with different mating strategies. (A) Inflatable sacs, white breast feathers, and spiky tail feathers are displayed by a male greater sage grouse (*Centrocercus urophasianus*). In this species, males compete to mate with as many females as possible and do not form a pair-bond or care for the offspring. (B) Atlantic puffins (*Fratercula arctica*) are sexually monomorphic seabirds that form pair-bonds. Both parents care for the offspring.

(A)

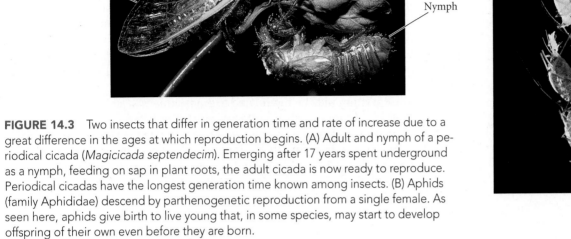

Adult

Nymph

(B)

FIGURE 14.3 Two insects that differ in generation time and rate of increase due to a great difference in the ages at which reproduction begins. (A) Adult and nymph of a periodical cicada (*Magicicada septendecim*). Emerging after 17 years spent underground as a nymph, feeding on sap in plant roots, the adult cicada is now ready to reproduce. Periodical cicadas have the longest generation time known among insects. (B) Aphids (family Aphididae) descend by parthenogenetic reproduction from a single female. As seen here, aphids give birth to live young that, in some species, may start to develop offspring of their own even before they are born.

Individual Selection and Group Selection

Why do codfish produce hundreds of thousands of eggs? Is it to compensate for the high mortality of both eggs and juveniles, thus helping to ensure the survival of the species? Why do people die of "old age"? Is it to make room for the vigorous new generation that will propagate the species? Why do so many species reproduce sexually? Is it because species need sex in order to adapt to a changing environment?

Even some professional biologists have been known to answer "yes" to these questions. But either they have assumed that these characteristics did not evolve by Darwinian natural selection (selection among individuals), or they have not realized that the good of the species does not affect the course of selection among individuals. That is, they may have not fully understood the meaning of natural selection.

Because they are components of fitness, differences in fecundity and life span must have evolved at least partly by natural selection. Selection among populations—the only possible cause of evolution of a trait that is harmful to the individual but beneficial to the population or species—is generally a weaker force than selection among individuals, as we saw in Chapter 11. This must be especially true for life history traits, which are components of individuals' fitness.

The possibility of a species' future extinction as a result of excessive population growth or inadequate reproduction is irrelevant to, and cannot affect, the course of natural selection among individuals. A mutation that increased the fecundity of humans (or any other species) would increase individual fitness (if it had no other effects) and would therefore become fixed—even if overpopulation and mass starvation were to ensue. Instead of supposing that a species' fecundity evolves to balance mortality, *we should consider the level of mortality to be the ecological consequence of the level of fecundity*, since the size of most populations is limited by food and other resources (Williams 1966). The more offspring are born, compared with the number that resources can support, the more will die.

At first surmise, then, we should expect any species to evolve ever greater fecundity and an ever longer life span. The problem, therefore, is to understand what advantage low fecundity or a short life span—or the genes that underlie them—might provide to individual organisms, rather than to entire populations or species. By the same token, we must beware of supposing that sexual reproduction has evolved because it provides species with the ability to adapt to future conditions.

Modeling Optimal Phenotypes

Optimality theory, also called OPTIMIZATION THEORY, is an important approach to understanding some adaptations, including life history traits. This approach consists of specifying, often on the basis of mathematical models, which state of some character, among a specified set of plausible states (often called STRATEGIES), would maximize individual fitness, subject to specified constraints (Parker and Maynard Smith 1990). (In some cases, INCLUSIVE FITNESS, as described in Chapter 16, is used instead of individual fitness.) Often the criterion of optimization in a model is a variable, such as rate of food acquisition, that is assumed to be correlated with fitness. The "character" to be optimized may be a reaction norm rather than a single state. This is the case for many aspects of behavior, in which the optimum may be modulated to environmental variables; for instance, the optimal time an animal spends foraging in a patch of habitat may depend on the travel time between suitable patches. Optimality theory is used extensively in the study of animal behavior (the field of BEHAVIORAL ECOLOGY) and is often applied to the ideal expression of a flexible behavior that is modulated by experience and environmental conditions, including the behavior of conspecific individuals. For many features, however, the optimum may be envisioned as a single genetically determined state.

Optimality theory rests on the assumption that sufficient genetic variation has been available for natural selection to shape a feature, so it ignores the history of genetic change and examines only the expected outcome of selection. Optimality theory has been criticized on the grounds that constraints such as limited genetic variation or genetic correlations may prevent attainment of an optimum (e.g., Gould and Lewontin 1979). Defenders of optimality theory (e.g., Parker and Maynard Smith 1990) reply that the approach does not actually assume that organisms are perfectly adapted, but instead aims to understand specific examples of adaptation.

"General" optimal models provide qualitative insights into the kinds of broad "solutions" to adaptive problems that might be expected to evolve, and they often predict the direction of difference between populations that face different problems. "Specific" models may use data on particular species to make quantitative predictions. For instance, the relation of surface area (A) to volume (V), $A \propto V^{2/3}$, dictates that gases, water, and heat are exchanged at higher rates between small objects and the environment than between large objects and the environment. Consequently, we might expect, and do observe, that birds and mammals in cold environments frequently have larger body sizes and shorter appendages than closely related forms in hot environments, and that many plant species in dry environments have thick leaves or none at all (**Figure 14.4**). The fields of functional morphology and physiology often use this and other principles of physics to calculate expected values of morphological features.

Simple optimal models are frequency-independent. A special class of optimal models,

FIGURE 14.4 Adaptations based on surface/volume relationships. (A) Appendages such as ears and legs are relatively shorter, reducing the surface area over which heat may be lost, in the Arctic hare (*Lepus arcticus*, left) than in the antelope jackrabbit (*L. alleni*, right), which occupies hot deserts. (B) Water loss is reduced by the growth forms of plants such as the "living stone" (*Aloinopsis schooneesii*, left), in which masses of small, thick, water-storing succulent leaves grow close to the ground. Similarly, in members of the cactus family such as the golden barrel (*Echinocactus grusonii*, right), globular stems and the lack of leaves minimize the surface/volume ratio and thus minimize water loss.

(A)

(B)

however, treats situations in which the optimal feature for an individual depends on other individuals with which it interacts; fitness is then frequency-dependent (see Chapter 12). These models, based on the mathematical theory of games, use the concept of an **evolutionarily stable strategy** (**ESS**), which John Maynard Smith (1982) defined as *"a strategy such that, if all the members of a population adopt it, then no mutant strategy could invade under the influence of natural selection."* That is, an ESS is a phenotype that cannot be replaced by any other phenotype under the prevailing conditions. A strategy may be PURE, meaning that an individual always has the same phenotype, or MIXED, meaning that an individual's phenotype varies over time, as is often the case with behavior. ESS models frequently describe interactions between two individuals, each of which has one of two or more phenotypes. For each possible pairwise combination of strategies, an individual has a different payoff—the increment or decrement of fitness that it receives. The payoff depends not only on the individual's own phenotype, but also on that of the individual with which it interacts. We encountered an example of an ESS model when we considered the optimal sex ratio among a female's offspring, which depends on the sex ratio in the population at large (see Figure 12.18). **Box 14A** provides examples of both simple and ESS approaches. The ESS example shows, incidentally, that evolution under frequency-dependent selection may not result in maximal population fitness.

Box 14A Optimal Models: Examples

A simple model: Optimal diet choice

If an individual predator encounters two kinds of prey (1 and 2) that differ in energy content (E_1 and E_2), would the predator do best to attack one or both, assuming that fitness would be increased by maximizing the rate of energy gain? The classic optimal foraging model (Krebs and McCleery 1984) assumes that prey types 1 and 2 are encountered at rates L_1 and L_2, and that if captured they require handling times H_1 and H_2. The model assumes that handling time, which may depend on various aspects of the predator's behavior and morphology, is a fixed constraint (i.e., it does not evolve); that handling and searching cannot be done simultaneously; that prey are recognized immediately and without error; and that prey are encountered at random.

A nonselective predator (a GENERALIST) foraging for time T_s will obtain $E = L_1 T_s E_1 + L_2 T_s E_2$ calories if it does not reject any prey, and the total time expended will be $T = (T_s) + (L_1 T_s H_1 + L_2 T_s H_2)$, where the terms in parentheses represent total search time and total handling time, respectively. Thus the average rate of intake will be

$$\frac{E}{T} = \frac{L_1 E_1 + L_2 E_2}{1 + L_1 H_1 + L_2 H_2}$$

Now if prey type 1 yields more calories per unit handling time ($E_1/H_1 > E_2/H_2$), the optimal diet should be only type 1 if

$$\frac{L_1 E_1}{1 + L_1 H_1} > \frac{L_1 E_1 + L_2 E_2}{1 + L_1 H_1 + L_2 H_2}$$

that is, if the caloric intake is greater for a specialist on type 1 than for a generalist, which expends time handling less profitable prey. With a bit of algebra, this reduces to

$$\frac{L_1 E_1}{1 + L_1 H_1} > \frac{E_2}{H_2}$$

So specialization is favored if the specialist's average rate of intake is greater than the calories gained per unit time spent handling a less profitable prey item. The abundance of the more profitable prey (L_1) affects whether or not a predator should specialize on it, whereas the abundance of a less profitable prey (L_2) has dropped out of the inequality, and so has no bearing on the optimal diet.

An ESS model: Tree height

Hanna Kokko (2007) provides a simple example of an ESS model, which we present here in barest outline. Suppose that the vegetative biomass of a plant can be apportioned between stem (h, ranging from 0 to 1) and leaves (f, which also ranges from 0 to 1), that leaves have photosynthetic rate g, and that fitness is proportional to fg, the amount of carbon fixed (which may then be used for reproduction). If $h = 1$, fitness is 0, because the plant has no photosynthesis. Because taller stems must be thicker to support the added weight without buckling, f must decline with h more rapidly at larger values of h. Now suppose that plant A has a neighbor, plant B, which reduces A's rate of photosynthesis, g_A, if it is taller and casts shade, and suppose further that g_A declines as the difference in height ($h_B - h_A$) increases. Two plants of the same height are assumed to reduce each other's fg to some extent. Is there a strategy—a height h—such that a population of plants with that strategy would not be replaced by a genotype with a different ("mutant") strategy?

The answer can be found mathematically, but Kokko provides a numerical example that illustrates the reasoning. She envisions a "payoff matrix" (below) in which fg values, based on the above considerations, are specified for two competing plants A and B when they have heights h_A and h_B, respectively, and there are four possible heights. In each cell, the left and right numbers are the payoffs, fg, to plants A and B, respectively.

(continued)

Box 14A *(continued)*

Height of A (h_A)	Height of B (h_B)			
	0	**⅓**	**⅔**	**1**
0	0.625; 0.625	0.369; 0.848	0.276; 0.686	0.255; 0
⅓	0.848; 0.369	0.602; 0.602	0.356; 0.620	0.266; 0
⅔	0.685; 0.276	0.620; 0.356	0.440; 0.440	0.256; 0
1	0; 0.255	0; 0.266	0; 0.260	0; 0

Let us move through this matrix to an equilibrium, if one exists. If both plants have strategy $h = 0$ (upper left cell of the table), they have equal payoffs (0.625), less than maximal because they shade each other equally. If B has strategy $h = ⅓$ (one cell to the right), it gains ($fg = 0.848$) and A's payoff is reduced to 0.369, so $h = ⅓$ is a superior strategy. But if the entire population (i.e., both plants) has

$h = ⅓$ (down one cell), both have a payoff of 0.602. If B "mutates" to $h = ⅔$, its fg rises to 0.620 and A's payoff drops to 0.356. So $h = ⅔$ is a superior strategy. If both A and B (the entire population) have this strategy, they both drop to a payoff of 0.440 as a result of mutual shading. As seen in **Figure 1A**, this is the ESS because the only possible further change, to $h = 1$, reduces fitness to 0. The same principle illustrated here by two individuals holds if a population of plants is modeled, and if h is a continuous instead of a discrete variable (**Figure 1B**). The ESS is found at the intersection of the "best response" curves.

Notice that in this model, the total amount of carbon a plant fixes (fg) at equilibrium, a measure of average fitness, is lower (0.440) than the maximum possible (0.625, when $h = 0$). Evolution in response to competition has led to lower average fitness. This is a common result in models of frequency-dependent selection.

(A) Discrete

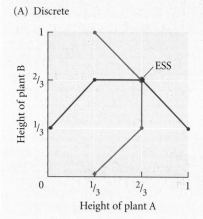

(B) Continuous

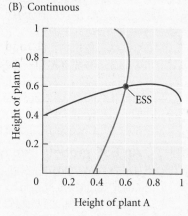

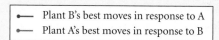
- Plant B's best moves in response to A
- Plant A's best moves in response to B

FIGURE 1 A graphical depiction of the "best responses" of competing plants to each other's height, according to the game-theoretical approach. (A) Here the plants display one or the other of four possible heights, ranging from 0 to 1. (B) The corresponding curves when height is a continuous variable.

Life History Traits as Components of Fitness

Among the most important components of fitness are survival to and through the reproductive ages, the number of offspring (fecundity) of females, and the number of offspring a male fathers. Here we will focus on the evolution of life history traits, especially the potential life span, the ages at which reproduction begins and ends, and female fecundity at each age (Stearns 1992; Charlesworth 1994b; Roff 2002). These traits affect the growth rates of populations.

Fecundity, semelparity, and iteroparity

In Chapter 12, we defined the fitness of a genotype for the simple case in which females reproduce once and then die (a **semelparous** life history). In that case, reproductive success—the number of descendants of an average female after one generation—is R, the product of the probability of a female's survival to reproductive age (L) and the average number of offspring per survivor (M):

$$R = LM$$

For **iteroparous** species—those in which females reproduce more than once—the calculation of fitness is more complex. The average number of offspring per female is the sum

(A) Survivorship

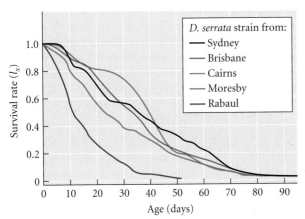

(B) Fecundity

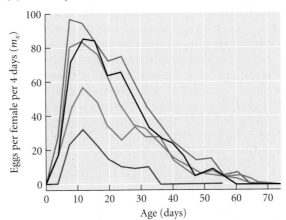

FIGURE 14.5 Genetic variation in life history characteristics. The graphs show (A) age-specific probability of survival, l_x, and (B) fecundity, m_x, in strains of the Australian fruit fly *Drosophila serrata* from five localities when raised in the laboratory. The survival curves show the fraction of newborns that survive to each age, and the fecundity curves show the average egg production per female at each age. Egg production peaks a few days after the flies transform from the pupal to the adult stage. These curves show only the adult (post-pupal) stage of the life history. (After Birch et al. 1963.)

of the offspring an average female produces at each age, weighted by the probability that a female survives to that age. We use x to denote age, l_x to denote the probability of survival to age x (i.e., the proportion of eggs or newborns that survive to age x), and m_x to denote the average fecundity (number of eggs or newborns produced) at age x. (**Figure 14.5** illustrates l_x and m_x for several populations of a species of *Drosophila*.)

Suppose that at ages 1, 2, 3, and 4 years, females lay 0, 4, 8, and 0 eggs, respectively, and that their chances of surviving to those ages are 0.75, 0.50, 0.25, and 0.10, respectively. We can use these data to write a simple LIFE TABLE:

x	l_x	m_x	$l_x m_x$
0	1.00	0	0
1	0.75	0	0
2	0.50	4	2
3	0.25	8	2
4	0.10	0	0
5	0.00	0	0
$\Sigma = R$			4

R is calculated as

$$R = \Sigma\, l_x m_x$$

Thus each female is replaced, on average, by $R = 4$ offspring. (By convention, ecologists generally count only daughters in such analyses and assume that sons are produced in equal numbers.) This sum is also the growth rate, per generation, of the genotype: if the population starts with N_0 individuals, its size after g generations will be

$$N = N_0 R^g$$

A genotype with higher R would increase faster in numbers, and would thus have higher fitness.

If genotypes differ in the length of a generation, their fitness can be compared only by their increase per unit time, not per generation. The per capita rate of increase per unit time is denoted r, which is related to R by $R = e^r$. If the population starts with N_0 individuals, its size after t time units will be

$$N = N_0\, e^{rt}$$

Like R, r depends on the probability of survival and the fecundity at each age. Under most circumstances, r, which is often termed the INSTANTANEOUS RATE OF INCREASE, is a suitable measure of a genotype's fitness.

All else being equal, increasing l_x—survival to any age x—up to and including the reproductive ages will increase R (or r) and therefore increase fitness. If, as for human females (or in the hypothetical life table above), there is a postreproductive life span (when $m_x = 0$), changing the probability of survival to advanced postreproductive ages does not alter R. Therefore *natural selection usually does not favor postreproductive survival.* (Postreproductive survival may be advantageous, however, if postreproductive parents care for their offspring, as in humans.) If, however, the reproductive period is extended into older ages, then increasing survival to those ages does increase R. Similarly, increasing m_x (fecundity at any age x) increases fitness, all else being equal.

Age structure and reproductive success

Offspring produced at an earlier age increase fitness more because they contribute more to population growth (r) than the same number of offspring produced at a later age. That is, they have greater "value" in terms of fitness. For instance, suppose females of two different genotypes reproduce at age 2 and age 3, respectively, but have the same fecundity. The 2-year-old females will contribute more to the future population size than the 3-year-olds. Because fewer individuals will survive to age 3 than to age 2, 2-year-old reproducers, collectively, will leave more offspring. Moreover, population growth is like compound interest. Just as your bank account grows faster if you make a deposit now than if you wait, the offspring of 2-year-olds will themselves contribute offspring (i.e., "gain interest") before the offspring of 3-year-olds do so. Thus a genotype that reproduces earlier in life has a shorter generation time, and higher fitness (as measured by r), than a genotype that delays reproduction until a later age.

Because reproduction early in life contributes more to the rate of population growth than an equal production of offspring later in life, the SENSITIVITY of fitness to small changes in life history traits—the effect of a given magnitude of change in fecundity or survival on

TABLE 14.1 A hypothetical example of a life table and of the sensitivity of fitness (r) to changes in age-specific survival and fecundity[a]

Age class	Number of survivors	Fraction of survivors	Average fecundity	Survival × fecundity			Sensitivity of r to m_x	Sensitivity of r to l_x
x		l_x	m_x	$l_x m_x$	e^{-rx}	$e^{-rx}l_x m_x$	$S_m(x)$	$S_s(x)$
0	1000	1.000	0.00	0.00	1.000	0.000	0.335	0.334
1	750	0.750	0.00	0.000	0.796	0.000	0.200	0.334
2	600	0.600	1.20	0.720	0.634	0.456	0.128	0.182
3	480	0.480	1.40	0.672	0.505	0.339	0.081	0.068
4	360	0.360	1.03	0.396	0.402	0.159	0.049	0.018
5	180	0.180	0.96	0.144	0.320	0.046	0.019	0.018
6	100	0.100	0.00	0.000	0.255	0.000	0.011	—
Sums:				1.932 = R				1.000

Source: After Stearns 1992.

[a]The instantaneous rate of increase, r, is calculated by "trial and error" from the equation $1 = \sum_{x=a}^{x=z} e^{-rx} l_m m_x$ and is found to be 0.228.

The sensitivity coefficients $S_m(x)$ and $S_s(x)$ indicate the effect on r of a small change in fecundity (m_x) or survival (l_x), respectively, at age x. They are calculated, respectively, as $S_m(x) = \dfrac{e^{-rx} l_m}{T}$ and $S_s(x) = \sum_{y=x}^{y=z} \dfrac{e^{-ry} l_y m_y}{T}$.

fitness (r)—depends on the age at which the changes are expressed (Charlesworth 1994b). For the hypothetical data in **Table 14.1**, for example, an increase in survival to the first age class (l_1) would obviously increase fitness. On the other hand, increasing survival from age 5 to age 6 (l_5) would not alter fitness at all because this species does not reproduce beyond age 5 ($m_6 = 0$).

Furthermore, the selective advantage of a slight increase in survival or fecundity at an advanced age increases fitness (r) less than an equal increase at an early age because the contribution of a cohort to population growth declines with age.* (Note the decline in the "sensitivity values" $S_t(x)$ and $S_m(x)$ and $S_s(x)$ with age in Table 14.1.) A simple reason is that older females are less likely to be alive. Another reason, as we have seen, is that offspring born to older females contribute less to the rate of population growth than do those born to young females. The consequence of this theory is that, at least in growing populations, *the optimal reproductive strategy is to reproduce as much as possible early in life, the earlier the better.*

Trade-Offs

Because characters evolve so as to maximize fitness, we might naïvely expect organisms to evolve ever greater fecundity, ever longer life, and ever earlier maturation. That all organisms are in fact limited in these respects may be attributed to various constraints.

PHYLOGENETIC CONSTRAINTS arise from the history of evolution, which has bequeathed to each lineage certain features that constrain the evolution of life history traits and other characters. For instance, although many insects feed as adults, and so obtain energy and protein that enable them to form successive clutches of eggs, adult silkworm moths and some other insect groups lack functional mouthparts, so their fecundity is limited by the resources they stored when they fed as larvae. Most such insects lay only one batch of eggs and then die. In most groups of birds, the number of eggs per clutch varies within and among species, but all species in the order Procellariiformes (albatrosses, petrels, and relatives) lay only a single egg.

Other constraints on evolution, termed PHYSIOLOGICAL or GENETIC CONSTRAINTS, are less well understood, but may be detected by comparisons among different genotypes or phenotypes. Some such constraints constitute **trade-offs**, whereby the advantage of a change in one character is correlated with a disadvantage in other respects. For example, the reproductive activities of animals often increase their risk of predation, so there is a trade-off between reproduction and survival.

Trade-offs can influence the evolution of virtually all characteristics. They are a major component of most optimal models and of many genetic models as well. For instance, selection for different alleles in different environments occurs if there are trade-offs in fitness among genotypes across environments.

One of the most important trade-offs is between allocation of energy and nutrients to reproduction (often called **reproductive effort**) versus an individual's own growth or maintenance. Genotypes that allocate more to reproduction and less to themselves may display decreased subsequent survival or growth, an example of a **cost of reproduction**. This ALLOCATION TRADE-OFF would be manifested as a *negative genetic correlation* between reproduction and survival. If there were also genetic variation in the amount of resources individuals acquired from the environment, however, that variation could give rise to a *positive genetic correlation* between reproduction and survival (**Figure 14.6**; Bell and Koufopanou 1986; van Noordwijk and deJong 1986). The allocation trade-off might still constrain evolution, but the trade-off might be difficult to detect in this case.

There are several ways to detect trade-offs (Reznick 1985):

1. Correlations between the means of two or more traits in different populations or species can strongly suggest a trade-off, although such correlations might result from other,

---

*Members of a population that were born at the same time constitute a cohort.

FIGURE 14.6 Factors giving rise to positive or negative genetic correlations between life history traits such as survival (or growth) and reproduction. (A) Variation at many loci, such as locus A, affects the amount of energy or other resources that an individual acquires from the environment. Variation at many other loci, such as locus B, affects allocation of resources to functions such as growth or self-maintenance versus reproduction, in proportions x and 1 − x. (B) Genotypes that differ at "acquisition loci" are represented by green circles, those that differ at "allocation loci" by red circles. The overall genetic correlation between survival and reproduction depends on the relative magnitudes of variation in resource acquisition versus resource allocation.

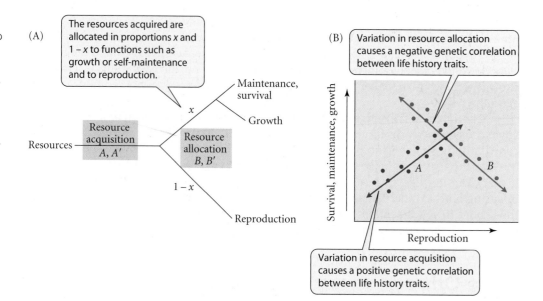

(A) The resources acquired are allocated in proportions x and 1 − x to functions such as growth or self-maintenance and to reproduction.

Resources → Resource acquisition A, A′ → Resource allocation B, B′

x — Maintenance, survival — Growth

1 − x — Reproduction

(B) Variation in resource allocation causes a negative genetic correlation between life history traits.

Variation in resource acquisition causes a positive genetic correlation between life history traits.

FIGURE 14.7 The relationship between number and weight of propagules (seeds) among species of goldenrods (*Solidago*) suggests an allocation trade-off. The colonizing species, which grow in old-field habitats, tend to produce more and smaller seeds than those species that grow in more stable prairie habitats, where competition may be intense and favor larger offspring. (After Werner and Platt 1976.)

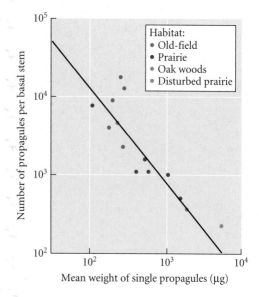

Habitat:
- Old-field
- Prairie
- Oak woods
- Disturbed prairie

unknown differences among the populations. For example, we would expect, and can often document, an allocation trade-off between many small versus fewer large offspring if parents must allocate limited resources (**Figure 14.7**).

2. Phenotypic or, better, genetic correlations between traits within populations can be useful indicators of the extent to which enhancement of one component of fitness is immediately accompanied by reduction of another (see Chapter 13). For instance, Law and colleagues (1979) grew individuals of each of many families of meadow grass (*Poa annua*) in a randomized array. They found that families that, on average, produced more inflorescences in their first season produced fewer in the second, and also achieved less vegetative growth. This experiment demonstrated a genetic basis for a cost of reproduction.

3. Correlated responses to artificial or natural selection provide some of the most consistent evidence of trade-offs (Reznick 1985; Stearns 1992). Linda Partridge and colleagues (1999) set up ten selection lines of *Drosophila melanogaster* from the same base population. They selected five "young" populations by rearing offspring from eggs laid by females less than 1 week old, and five "old" populations by propagating from eggs laid when females were 3 to 4 weeks old. After 19 generations, the mean life span of the "young" populations did not differ from that of the base population, but the longevity of the "old" populations had increased—as we would expect, since only flies that lived at least 3 weeks could contribute genes to subsequent generations (**Figure 14.8A**). However, the fecundity of 1-week-old females in the "old" populations was lower than that of the base or "young" populations (**Figure 14.8B**). Thus survival to greater age seems to have been achieved at the expense of reproduction early in life—an important result, as we will see shortly.

4. Experimental manipulation of one trait and observation of the effect on other traits often reveals trade-offs. For instance, Sgrò and Partridge (1999) followed up on the selection experiment on *Drosophila* longevity by experimentally sterilizing females from both "young" and "old" populations, either by gamma radiation or by inheritance of a dominant allele that causes female sterility. In both experimental treatments, the difference in longevity between the "young" and "old" populations disappeared, proving that longevity is affected by a physiological cost of reproduction (**Figure 14.8C**). These results are consistent with other evidence that in *Drosophila* and many other insects, mating activity and egg production reduce the longevity of both sexes, and that virgins live longer than nonvirgins (Bell and Koufopanou 1986; Fowler and Partridge 1989).

(A)

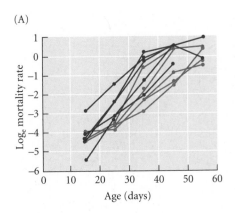

(B)

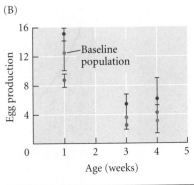

- Selected for reproduction at "young" age
- Selected for reproduction at "old" age

(C)

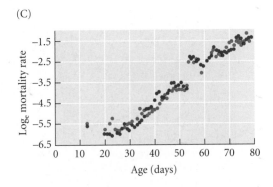

FIGURE 14.8 Results of selection of laboratory populations of *Drosophila* for age at reproduction. (A) The mortality rate, per 10-day interval, is lower at older ages for populations that were selected for reproduction at "old" ages than for populations that were selected for reproduction at "young" ages. (B) Relative to the "young" populations and the baseline population, the "old" populations had lower egg production when young. (C) The difference in mortality rate between "old" and "young" populations disappeared when a gene that prevents female reproduction was crossed into the populations, suggesting that the costs of reproduction increase mortality. (After Partridge et al. 1999 and Sgrò and Partridge 1999.)

The Evolution of Life History Traits

Life span and senescence

Most organisms in which germ cells are distinct from somatic tissues undergo physiological degeneration with age, a process called SENESCENCE. Why does this occur? Many physiological processes may contribute to senescence; for example, it has been suggested that a contributing factor may be the shortening of TELO-MERES, the DNA caps at the ends of chromosomes, with age. Telomeres shorten more slowly in long-lived than in short-lived species, and even lengthen in at least one long-lived bird species (Haussmann et al. 2003). However, senescence and life span, and their immediate physiological causes, differ among species, and they clearly evolve. Thus telomere loss, or any other single physiological cause of senescence, does not explain why species have the life span they have, since they could have evolved a different life span. These explanations of senescence provide a clear example of the difference between a PROXIMATE (mechanistic) explanation and an ULTIMATE (evolutionary) explanation of a biological phenomenon.

Two related hypotheses on the evolution of senescence and limited life spans have been proposed (Rose 1991). Both rest on the principle that *the selective advantage of an enhanced probability of survival declines with age.* Peter Medawar (1952) proposed that deleterious mutations that affect older age classes accumulate in populations at a higher frequency than those that affect younger age classes because selection against them is weak. If there are many loci affected by such mutations, then the causes of senescence should vary among individuals. This hypothesis predicts that genetic variation in fitness-related traits (such as those affecting survival) should be greater in old than in young age classes. The other hypothesis, proposed by George Williams (1957), postulates a genetic trade-off based on **antagonistic pleiotropy**: an allele that has a beneficial effect on one trait may have a deleterious effect on another trait. Because of the greater contribution of younger age classes to fitness, an allele that increases reproductive effort early in life has a selective advantage even if it is deleterious later in life (perhaps by reducing allocation of energy and materials to maintenance, repair, and defense). That is, *senescence may be one of the costs of reproduction.*

The evidence for Medawar's mutation accumulation hypothesis is limited, but Williams's hypothesis of antagonistic pleiotropy is strongly supported by selection experiments that provide evidence of a negative relationship between early reproduction and both longevity and later reproduction (Partridge 2001; see Figure 14.8). These experiments are among the most striking confirmations of evolutionary hypotheses that had been posed long before the experiments took place.

(A)

(B)

FIGURE 14.9 Some semelparous plants reproduce once, after many years, and then die. (A) The cabbage palm *Corypha utan*, of southeastern Asia, may produce up to a million flowers. (B) Bamboos engage in highly synchronous reproduction and then die, resulting in years of great food scarcity for animals that eat the shoots (such as specialized insects and giant pandas) and great food abundance for those that eat the seeds (such as finches). (Photos by the author.)

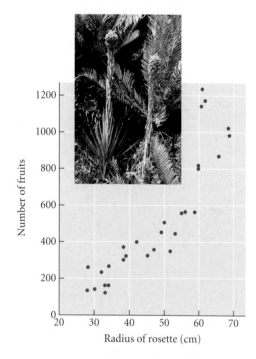

FIGURE 14.10 The number of fruits produced by the terrestrial bromeliad *Puya dasylirioides* in relation to the radius of the rosette of leaves (a measure of plant size). Note the exponential relationship: larger (and older) plants produce disproportionately more seeds. This species grows in montane bogs in Central America. (After Augspurger 1984; photo by the author.)

Age schedules of reproduction

If survival contributes to fitness only as long as reproduction continues, why don't organisms reproduce indefinitely? The answer is that all else being equal, there is always an advantage to reproducing earlier in life. Since early reproduction is correlated with lowered subsequent reproduction, we might expect organisms to be semelparous, allocating all their resources to a single early burst of reproduction rather than to maintaining themselves. Annual plants and many short-lived animals fit this expectation. However, two common life history patterns do not. Perennial herbs, most trees, most familiar vertebrates, and many other species are iteroparous. Some of them, such as albatrosses and humans, have delayed rather than early reproduction. And some species are semelparous, but reproduce at an advanced age (**Figure 14.9**). How have these life histories evolved?

Because of costs of reproduction, reproducing at an early age may increase the risk of death, decrease growth, or decrease subsequent fecundity, so that r is lower than it would be if reproduction were deferred. (By analogy, if you think a stock's value will increase for some time and then level off, you might do well to wait until it approaches its maximum expected value before selling it.) Fecundity, for example, is often correlated with body mass in species that grow throughout life, such as many plants and fishes. In such species, allocating resources to growth, self-maintenance, and self-defense rather than to immediate reproduction is an investment in the much greater fecundity that may be attained later in life. The question then is whether, once reproduction begins, it is optimal to reproduce only once or repeatedly.

In theory, a semelparous "big bang" life history may be optimal if (1) the rate of growth of body mass declines as an individual grows larger, (2) the probability of survival increases with body mass, and (3) there is an exponential relationship between body mass and reproductive output (Schaffer 1974; Metcalf et al. 2003). These conditions have been documented in many species of semelparous plants (Metcalf et al. 2003; **Figure 14.10**). Bamboos, agaves, cabbage palms, and other species that reproduce only after many years, and then die, produce massive numbers of seeds.

Iteroparity can also be advantageous if greater fecundity can be achieved by deferring reproduction to older age classes. Mathematical models show that *repeated reproduction is more likely to evolve if adults have high survival rates from one age class to the next and if the rate of population increase is low.* Another factor that makes repeated reproduction advantageous is environmental variation. If reproduction is likely to fail in some years and be successful in others, it is advantageous to spread the risk by reproducing repeatedly and allocating some energy each time to survival until the next reproductive opportunity.

The same factors also favor later, rather than earlier, reproductive maturity in an iteroparous species and can favor the evolution of a genetically longer life span. Since some resources are used for growth, maintenance, and defense, the effort devoted by an iteroparous species to reproduction will be less at each reproductive age than the effort a semelparous species devotes to reproduction in its single "big bang" reproductive episode. As individuals age, however, the benefit of withholding energy from reproduction declines because the prospects of subsequent reproduction decline. Therefore we would expect that at some point in life, *the proportion of energy or other resources devoted to reproduction by iteroparous species should increase with age* (Williams 1966; Charlesworth 1994b).

These theoretical predictions have been supported by many comparative studies. Comparisons among species within several taxa support the prediction that reproductive effort, in each reproductive episode, should be lower in iteroparous than in semelparous organisms (Roff 2002). For example, inflorescences make up a lower proportion of plant weight in perennial than in annual species of grasses (**Figure 14.11**; Wilson and Thompson 1989). Likewise, the prediction that high adult survival rates favor delayed maturation and high reproductive effort later in life is upheld by studies of mammals, fishes, lizards and snakes, and other groups: species that have longer life spans in nature also mature at a later age (**Figure 14.12**; Promislow and Harvey 1991; Shine and Charnov 1992). Species with larger body sizes tend to have higher adult survival rates and longer life spans (**Figure 14.13A**). As theory predicts, they also tend to have delayed reproduction (**Figure 14.13B**) and a lower number of offspring in each reproductive episode (**Figure 14.13C**).

Some experiments have also provided strong support for life history theory. David Reznick and colleagues have studied guppies in Trinidad (e.g., Reznick et al. 1990; Reznick and Travis 2002). In some streams, the cichlid fish *Crenicichla* preys heavily on large (mature) guppies (see Chapter 11). In other streams, or above waterfalls, *Crenicichla* is absent, and there is much less predation. Predation by *Crenicichla* should favor the evolution of maturity and reproduction early in life, and guppies from *Crenicichla*-dominated streams do indeed mature faster and at smaller sizes, reproduce more frequently, have higher reproductive effort (measured as weight of embryos relative to weight of mother), and have more and smaller offspring than guppies from low-predation streams. In two streams, Reznick and colleagues moved guppies from below a waterfall, where they were preyed on by *Crenicichla*, to sites above the waterfall, where guppies and *Crenicichla* were absent. After several generations, the researchers took guppies from the sites of origin and introduction and reared their offspring in a common laboratory environment. As predicted by life history theory, the populations relieved of predation on large adults had evolved delayed maturation

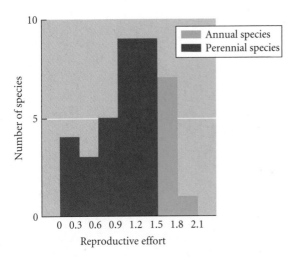

FIGURE 14.11 Reproductive effort—an index of the proportion of biomass allocated to inflorescences—in annual (semelparous) and perennial (iteroparous) species of British grasses. Allocation to reproduction is greater in the semelparous annual species. (After Wilson and Thompson 1989.)

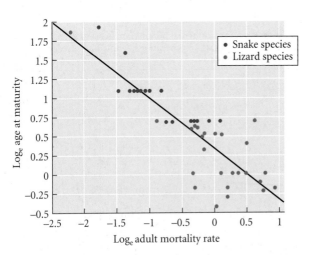

FIGURE 14.12 Among species of snakes and lizards, the lower the annual mortality rate of adults, the later they reach reproductive maturity. This pattern conforms to the prediction that delayed onset of reproduction is most likely to evolve in species with high rates of adult survival. (After Shine and Charnov 1992.)

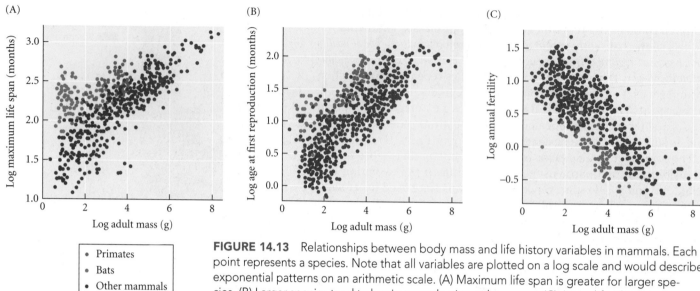

● Primates
● Bats
● Other mammals

FIGURE 14.13 Relationships between body mass and life history variables in mammals. Each point represents a species. Note that all variables are plotted on a log scale and would describe exponential patterns on an arithmetic scale. (A) Maximum life span is greater for larger species. (B) Larger species tend to begin reproduction at later ages. (C) Annual fertility (number of offspring), is lower in larger species. These patterns conform to the theoretical relationships expected for iteroparous species. (After Jones 2011.)

and larger adult size, and they tended toward fewer, larger offspring and lower reproductive effort (**Figure 14.14**). Recall that the opposite effects—earlier maturation at smaller sizes—have been seen in fish populations that have experienced increased human predation (overfishing) on adults (see Figure 11.26A).

▲ Contrasts the means of two natural populations in high-predation streams and two in low-predation streams.

● Contrasts an experimental population, isolated from predators for 7 generations, with a downstream control population that experiences high predation.

● Contrasts an experimental population, isolated from predators for about 18 generations, with a high-predation downstream control.

FIGURE 14.14 Differences between guppy (*Poecilia reticulata*) populations from high-predation and low-predation environments. These differences were assayed in common-garden comparisons of second-generation, laboratory-reared offspring of wild females. Asterisks indicate statistically significant differences. Low-predation populations tended to evolve (A) a later age at maturity in males and (B) in females; (C) a larger size at maturity in females; and (D) a larger offspring size at birth. (Data from Reznick and Travis 2002.)

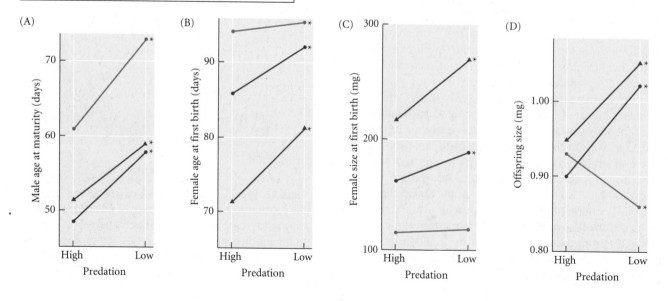

Number and size of offspring

All else being equal, a genotype with higher fecundity has higher fitness than one with lower fecundity. Why, then, do some species, such as humans, albatrosses, and kiwis, have so few offspring?

OPTIMAL CLUTCH SIZE   The British ecologist David Lack (1954) proposed that the OPTI-MAL CLUTCH SIZE for a bird is the number of eggs that yields the greatest number of surviving offspring. The number of surviving offspring from larger broods may be less than the number from more modest clutches because parents are unable to feed larger broods adequately. This decrease in offspring survival has proved to be one of several costs of large clutch size in birds; excessively large clutches may also reduce the parents' subsequent clutch size and survival (Stearns 1992).

Joost Tinbergen and Serge Daan (1990) analyzed data from a long-term study, performed in Holland, of survival and reproduction in great tits, which involved a 5-year program of decreasing and increasing brood size by moving hatchlings among nests as well as observations of unmanipulated nests. They estimated the effects of these treatments on the reproductive value of both the parents and the brood of young. The REPRODUCTIVE VALUE (V) of individuals of a certain age is their expected future contribution to population growth, taking into account their expected future age-specific survival and reproduction (which are assumed to be the l_x and m_x values estimated from the long-term study of the population). Tinbergen and Daan found that artificially increasing brood size decreased V because it lowered survival in the nest, survival from fledging to the next breeding season, and the probability that the parents would lay a second clutch of eggs in the same year. Decreasing brood size also reduced V, simply because the nests produced fewer fledglings (**Figure 14.15A**). Tinbergen and Daan calculated that the optimal clutch size was 8.9 eggs, close to the natural mean, 9.2. Curiously, the birds that naturally had highest V were those

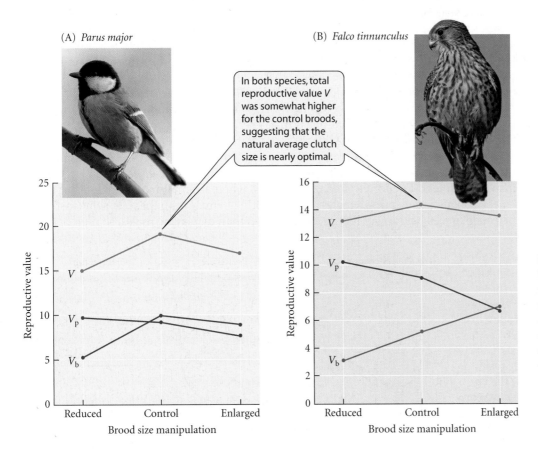

(A) *Parus major*

(B) *Falco tinnunculus*

In both species, total reproductive value V was somewhat higher for the control broods, suggesting that the natural average clutch size is nearly optimal.

FIGURE 14.15   Mean clutch sizes in birds may be close to their optimal values. The graphs show the reproductive value—a measure of expected future reproductive success—of broods (V_b), their parents (V_p), and the sum (V) for (A) great tits and (B) Eurasian kestrels in which brood sizes were experimentally reduced, enlarged, or left unchanged (controls). (A after Tinbergen and Daan 1990; B after Daan et al. 1990.)

few that produced clutches of 15 eggs. The authors suggested that individual pairs may adjust their clutch size to what would be best for them individually, based on such factors as their physiological condition and the quality of their territory.

The same research group (Daan et al. 1990) performed a similar study of a small falcon, the Eurasian kestrel. Artificially increased clutch sizes reduced the subsequent survival (and *V*) of parents, but did not diminish that of offspring. Total reproductive value was greatest for unmanipulated clutches, again suggesting that clutch size is close to the optimum (**Figure 14.15B**).

PARENTAL INVESTMENT   When parents can provide only a limited amount of yolk, endosperm, nourishment, or other forms of parental care (collectively referred to as **parental investment**), there may be a negative correlation between number and size of offspring. Greater initial offspring size usually enhances survival and growth rate (for example, human twins have a lower birth weight, on average, than single-born infants, and infant survival of human quadruplets and quintuplets is notoriously precarious). Barry Sinervo (1990) experimentally manipulated hatchling size in the lizard *Sceloporus occidentalis* by removing yolk from some eggs with a hypodermic needle. Smaller hatchlings ran more slowly, which probably would reduce their survival in the wild. In such cases, females' reproductive success—the number of *surviving* offspring they leave to the next generation—should be maximized by producing a more modest number of offspring that are larger and better equipped for survival. This is the most plausible explanation of the very low fecundity of species such as humans, whales, and elephants.

Life Histories and Mating Strategies

Males as well as females are subject to costs of reproduction, and that fact underlies some interesting variations in life histories. For example, as we saw at the opening of this chapter, some plants, annelid worms, fishes, and other organisms change sex over the course of the life span (a phenomenon called sequential hermaphroditism). In species that grow in size throughout reproductive life, a sex change can be advantageous if reproductive success increases with size to a greater extent in one sex than in the other (**Figure 14.16**). In the bluehead wrasse, for instance, some individuals start life as females and later become brightly colored "terminal-phase males" that defend territories. Other individuals begin life as "initial-phase males," which resemble females and spawn in groups (**Figure 14.17**). A female usually does not produce as many eggs as a large terminal-phase male typically fertilizes. Both females and initial-phase males become terminal-phase males at about the size at which this form achieves superior reproductive success (Warner 1984).

The reproductive behavior of initial-phase and terminal-phase males is an example of the ALTERNATIVE MATING STRATEGIES that are observed in many species. Often large males display or defend territories to attract females, whereas small males do not, but rather "sneak" about, intercepting females and attempting to mate with them. In some instances, "sneaker" males have lower reproductive success than displaying males, so their behavior

FIGURE 14.16   A model for the evolution of sequential hermaphroditism. (A) When reproductive success increases equally with body size in both sexes, there is no selection for sex change. (B) A switch from female to male (protogyny) is optimal if male reproductive success increases more steeply with size than female reproductive success. (C) The opposite relationship favors the evolution of protandry, in which males become females when they grow to a large size. (After Warner 1984.)

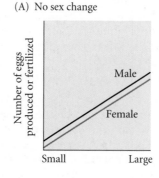
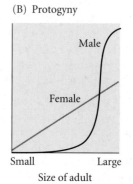
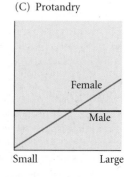

(A)

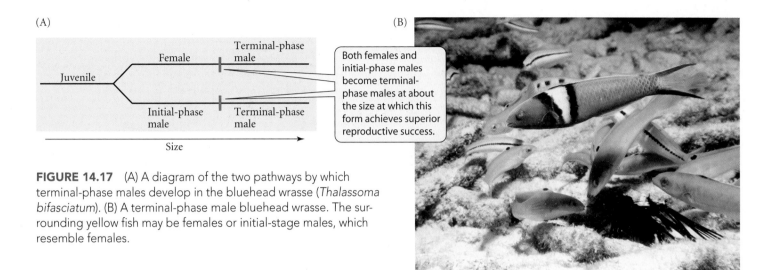

FIGURE 14.17   (A) A diagram of the two pathways by which terminal-phase males develop in the bluehead wrasse (*Thalassoma bifasciatum*). (B) A terminal-phase male bluehead wrasse. The surrounding yellow fish may be females or initial-stage males, which resemble females.

is probably not an adaptation; they are probably making the best of a bad situation, as they are unable to compete successfully. In other instances, however, the sneaker strategy appears to be an alternative adaptation, yielding the same fitness as the display strategy. For example, in the Pacific coho salmon (*Oncorhynchus kisutch*), large, red "hooknose" males develop hooked jaws and enlarged teeth, and fight over females, whereas "jack" males are smaller, resemble females, do not fight, and breed when they are only about a third as old as the hooknose males. Based on data on survival to breeding age and frequency of mating, the fitnesses of these two types of males appear to be nearly equal (Gross 1984).

Evolution of the Rate of Increase

Because the per capita rate of increase (r) of a genotype is the measure of its fitness, we might suppose that species would always evolve higher rates of increase. We have seen, however, that a shorter life span, lower fecundity, and delayed maturation, all of which lower r, can each be advantageous. Thus the potential rate of population growth can evolve—and certainly has evolved—to lower levels in many species. One simple reason is that in many species, the actual instantaneous rate of increase of a population, r, is usually lower than the **intrinsic** (potential) **rate of increase** (r_m, the maximal rate of increase that is expressed at low population densities), because density-dependent factors such as resource limitation or predation are reducing birth rates or increasing death rates. Different genotypes are likely to have higher r when the population density is high than when it is low (when r_m is realized), as illustrated in **Figure 14.18**. The genotypes that have higher r when the population is dense often have lower fecundity and later maturation, because of trade-offs between these traits and the ability to compete for resources or resist predation.

As the population density approaches equilibrium, a more competitive genotype may sustain positive population growth while inferior competitors decline in density (have negative r), and the more competitive genotype is therefore likely to achieve a

FIGURE 14.18   A model of density-dependent selection of rates of increase. Assume that a population contains two different genotypes, *A* and *B*. The instantaneous per capita rate of increase, *r*, declines for both *A* and *B* as population density (*N*) increases. The intrinsic rate of increase (r_m)—the population growth rate at very low density—is lower for genotype *B*, but this genotype has a selective advantage at high density and attains a higher equilibrium density (*K*). (After Roughgarden 1971.)

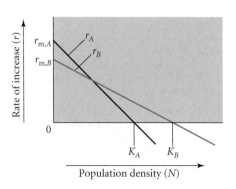

(A)

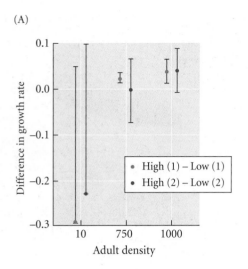

(B)

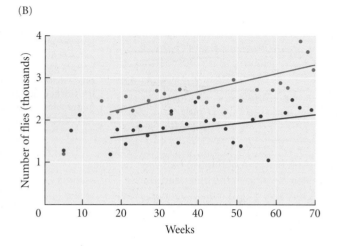

FIGURE 14.19 Experimental evolution of population growth rate and density in *Drosophila* species. (A) Each point shows the difference in per capita population growth rate between two sets of *Drosophila melanogaster* populations [Low (1) and Low (2)] that were maintained at low density for 200 generations and two sets of populations [High (1) and High (2)] derived from the first sets but maintained at high density for 25 generations. The per capita growth rates were measured at several adult population densities. The negative differences in growth rate at densities of 10 show that the Low strains had higher growth rates than the High strains when tested at this low density. The positive differences in growth rate at higher densities show that High strains had higher growths rate when tested at high densities. Thus selection at high densities for 25 generations resulted in evolution similar to the difference between the hypothetical genotypes *A* and *B* in Figure 14.18. (B) The densities of two experimental populations (red and blue points) of *Drosophila serrata* increased over many generations, implying adaptation to high densities and improved conversion of food (supplied at a constant rate) into flies. The potential rate of increase of this species is so great that the population would have reached carrying capacity in less than 10 weeks. (A after Mueller et al. 1991; B after Ayala 1968.)

higher equilibrium density (*K*). Precisely this difference has been found between laboratory populations of *Drosophila* that were maintained at low versus high densities for 200 and 25 generations, respectively, and then assayed for population growth rate (*r*) and productivity at each of several densities (**Figure 14.19A**). These populations evolved so that the high-density-selected populations had higher productivity under crowded conditions, but a lower growth rate under uncrowded conditions, than the low-density-selected populations. Notice, further, that Figure 14.18 implies that genotypes that can attain a higher equilibrium population density (*K*) at high densities have higher fitness, so we might expect selection to have the effect of increasing abundance. This prediction has been verified in experimental *Drosophila* populations that displayed genetically based trends toward higher population densities (**Figure 14.19B**).

In populations that are regulated by density-dependent factors (see Chapter 7) and occupy relatively stable environments, predation or competition for resources often causes heavier mortality among juveniles than adults. The mortality of seedling trees in a mature forest, for example, is exceedingly high, but if a tree does survive beyond the sapling stage—perhaps because a treefall has opened a light gap—it is likely to have a long life. As we have seen, these conditions favor the evolution of iteroparous reproduction late in life, and thus favor the evolution of long life spans. Moreover, under competitive conditions, juvenile survival can be enhanced by large size, so producing large eggs or offspring—and fewer of them—may be advantageous. Thus many authors have concluded that traits associated with a low intrinsic rate of increase—delayed maturation, production of few, large offspring, a long life span—are likely to evolve in species that occupy stable, competitive, or resource-poor environments. For example, species of beetles, fish, and other animals that inhabit caves generally develop very slowly and produce large eggs at an extraordinarily low rate (Culver 1982).

Summary

1. Life history features such as reproductive rates and longevity do not evolve to perpetuate the species. They can best be understood from the perspective of individual selection. Life history traits are components of the fitness of individual genotypes, the basis for natural selection.

2. Models of the evolution of adaptive characteristics include both population genetic models and optimal models, which attempt to determine which character states might be expected to evolve under specified conditions and under specified constraints. Optimal models include those that find the ESS (evolutionarily stable strategy) when fitness depends on the frequencies of different phenotypes among interacting individuals.

3. The major components of fitness (the per capita rate of increase of a genotype, r) are the age-specific values of survival, female fecundity, and male mating success. Natural selection on morphological and other phenotypic characters results chiefly from the effects of those characters on these life history traits.

4. Constraints, especially trade-offs between reproduction and survival and among the several components of reproduction, such as the number and size of offspring, prevent organisms from evolving indefinitely long life spans and infinite fecundity.

5. The effect of changes in survival (l_x) or fecundity (m_x) on fitness depends on the age at which such changes are expressed and declines with age. Hence selection for reproduction and survival at advanced ages is weak.

6. Consequently, senescence (physiological aging) evolves. Senescence appears to be a result, in part, of the negative pleiotropic effects on later age classes of genes that have advantageous effects on earlier age classes.

7. If reproduction is very costly (in terms of growth or survival), repeated (iteroparous) or delayed reproduction may evolve, provided that reproductive success at later ages more than compensates for the loss of fitness incurred by not reproducing earlier. Otherwise a semelparous life history, in which all of the organism's resources are allocated to a single reproductive effort, is optimal. Iteroparity is especially likely to evolve if juvenile mortality is high relative to adult mortality and if population density is stable, or if reproductive success is highly variable because of environmental fluctuation.

8. Because lower fecundity and delayed reproduction can evolve, the intrinsic rate of population increase, the maximal rate of increase that is expressed at low population densities, may evolve to be lower. In such populations, however, the rate of population growth is often close to zero because of density-dependent limitations.

Terms and Concepts

antagonistic pleiotropy

cost of reproduction

evolutionarily stable strategy (ESS)

intrinsic rate of increase

iteroparous

life history

optimality theory

parental investment

reproductive effort

semelparous

trade-off

Suggestions for Further Reading

The Evolution of Life Histories, by S. C. Stearns (Oxford University Press, Oxford, 1992), and *Life History Evolution*, by D. A. Roff (Sinauer Associates, Sunderland, MA, 2002) are comprehensive treatments of the topics discussed in this chapter.

Problems and Discussion Topics

1. Female parasitoid wasps search for insect hosts in which to lay eggs, and they can often discriminate among individual hosts that are more or less suitable for their offspring. Behavioral ecologists have asked whether or not the wasps' willingness to lay eggs in less suitable hosts varies with the female's age. On the basis of life history theory, what pattern of change would you predict? Does life history theory make any other predictions about animal behavior?

2. Suppose that a mutation in a species of annual plant increases allocation to chemical defenses against herbivores, but decreases production of flowers and seeds (i.e., there is an allocation trade-off). What would you have to measure in a field study in order to predict whether or not the frequency of the mutation will increase?

3. In many species of birds and mammals, clutch size is larger in populations at high latitudes than in populations at low latitudes. Species of lizards and snakes at high latitudes often have smaller clutches, however, and are more frequently viviparous (bear live young rather than laying eggs), than low-latitude species (see references in Stearns 1992). What selective factors might be responsible for these patterns?

4. An important life history characteristic, not discussed in this chapter, is dispersal between hatching and reproductive age. The extent of dispersal varies considerably among different organisms. What are the advantages versus disadvantages of dispersing? How might the evolution of dispersal be affected by group selection versus individual selection? (See Olivieri et al. 1995 and references therein.)

5. Another consequence of dispersal might be selection among genotypes that differ in characteristics that affect dispersal. What might be some effects of dispersal on life history or other characters? (See the July 2008 supplement in *American Naturalist*, vol. 172.)

6. Figure 14.13 illustrates that, compared with most other mammals, primates and bats have lower fecundity and a later age at first reproduction. Why might that be? (See Jones 2011.)

Sex and Reproductive Success

I n the previous chapter, we treated those components of fitness—age-specific schedules of survival and female reproduction—that affect the increase
or decrease in numbers of individual organisms with different genotypes.
Changes in the relative numbers of variant alleles, however, stem not only from
those fitness components but from several others as well (see Chapter 11). The
most obvious of these is variation in mating success, or sexual selection, which is
usually more pronounced in males than in females. But why are there
males, anyway? Many species do not have separate sexes, but consist of hermaphroditic, or cosexual, members. Why should some
species have separate sexes? Moreover, many organisms
can reproduce asexually. Why does sexual reproduction
exist at all?

Organisms vary greatly in what is sometimes called their
GENETIC SYSTEM: whether they reproduce sexually or asexually, self-fertilize or outcross, are hermaphroditic or have
separate sexes. Discovering why and how each of these
characters evolved poses some of the most challenging
problems in evolutionary biology, and these problems are
the subject of some of the most creative contemporary
research on evolution.

The genetic system affects genetic variation, which of
course is necessary for the long-term survival of species.
This fact has been cited for more than a century as the reason for the existence of recombination and sexual reproduction. But as we have seen, arguments that invoke long-
term benefits to the species must be carefully examined
because they often rely on group selection, which many
biologists consider a weak agent of evolution. The question, then, is whether or not natural selection *within* populations can account for features of the genetic system.

**A big investment in repro-
duction**   Like other wind-
pollinated plants, the lodgepole
pine (*Pinus contorta*) releases
huge numbers of pollen grains.
Modern evolutionary theory at-
tributes this to Darwin's concept
of sexual selection: males that
produce more pollen than others
are likely to fertilize more eggs,
have more offspring, and pass on
more genes.

The Evolution of Mutation Rates

How can natural selection account for genetic systems, if not by group selection and the long-term survival of species? We can first address this question by thinking about the evolution of mutation rates. Two hypotheses have been proposed: either the mutation rate has evolved to some optimal level, or it has evolved to the lowest possible level. Variation in factors such as the efficacy of DNA repair provides potential genetic variation in the genomic mutation rate. According to the hypothesis of optimal mutation rate, selection has favored somewhat inefficient DNA repair. According to the hypothesis of minimal mutation rate, mutation exists only because the repair system is as efficient as it can be, or because selection is not strong enough to favor investment of energy in a more efficient repair system. According to this hypothesis, the process of mutation is not an adaptation.

An optimal (greater than zero) mutation rate could be favored because populations, or lineages within populations, that experience beneficial mutations—and thus adapt faster to changing environments—might persist longer than populations or lineages that do not experience mutation. We do not know how rapidly this process of long-term selection would occur, because the faster the environment changes, the higher the mutation rate must be to avert extinction (Lynch and Lande 1993; see Chapter 13).

Alternatively, we can ask how evolution within populations affects the mutation rate, by postulating a "mutator locus" that affects the mutation rate of other genes. Let us assume that mutator alleles that increase the mutation rate affect the fitness of their bearers only indirectly, via the mutations they cause. It turns out that the fate of such a mutator allele depends on the level of recombination. In an asexual population, the mutator allele is likely to decline in frequency because copies of the allele are permanently associated with the mutations they cause, and far more mutations reduce than increase fitness. But occasionally a mutator allele causes a beneficial mutation. It will then increase in frequency by hitchhiking with the mutation it has caused.

In a sexual population, however, recombination will soon separate the mutator allele from a beneficial mutation it has caused, so it will not hitchhike to high frequency. Because deleterious mutations occur at so many loci, the mutator allele will usually be associated with one or another of them at first, and will therefore decline in frequency. Therefore *natural selection in sexual populations will tend to eliminate any allele that increases mutation rates, and mutation rates should evolve toward the minimal achievable level*—even though mutation is necessary for the long-term survival of a species. Consequently, most evolutionary biologists believe that the existence of mutation is not an adaptation in most organisms, but is rather a by-product of imperfect DNA replication that either cannot be improved or can be improved only at too great a cost in fitness (Leigh 1973; Sniegowski et al. 2000). Greater fidelity of replication does seem to be costly: in one experiment, genetic changes in an RNA virus that reduced the mutation rate also reduced the rate of virus proliferation (Elena and Sanjuán 2007).

Mutator alleles actually occur at considerable frequencies in some natural populations of the bacterium *Escherichia coli*. Because *E. coli* reproduces mostly asexually, it is understandable that mutator alleles sometimes increase to fixation in experimental populations (**Figure 15.1**). It has been shown that these increases are caused by hitchhiking with new beneficial mutations (Shaver et al. 2002).

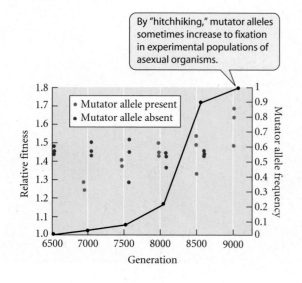

> By "hitchhiking," mutator alleles sometimes increase to fixation in experimental populations of asexual organisms.

FIGURE 15.1 Mutator alleles may increase to fixation in asexually reproducing populations. In an experimental population of *Escherichia coli*, an allele that increases the mutation rate throughout the genome increased in frequency (solid line), and eventually became fixed, by hitchhiking with one or more advantageous mutations. Samples of clones carrying the mutator allele showed an increase in fitness over time, whereas the fitness of clones that lacked the allele did not change. (After Shaver et al. 2002.)

Sexual and Asexual Reproduction

Sexual reproduction, or SEX, usually refers to the union (SYNGAMY) of two genomes, usually carried by gametes, followed at some later time by REDUCTION, ordinarily by the process of meiosis and gametogenesis. Sex often involves outcrossing between two individuals, but it can occur by self-fertilization in many plants and some other organisms. Sex almost always includes SEGREGATION of alleles and RECOMBINATION among loci, although the extent of recombination varies greatly. Most sexually reproducing species have distinct female and male functions, which are defined by a difference in the size of the gametes they produce (ANISOGAMY). In ISOGAMOUS organisms, such as *Chlamydomonas* and many other unicellular eukaryotes, the uniting cells are the same size; such species have MATING TYPES but not distinct sexes. Species in which individuals are either female or male, such as willow trees and mammals, are termed **dioecious** or GONOCHORISTIC; species such as roses and earthworms, in which an individual can produce both kinds of gametes, are **hermaphroditic** or COSEXUAL.

Asexual reproduction may be carried out by **vegetative propagation**, in which an offspring arises from a group of cells, as in plants that spread by runners or stolons, or by **parthenogenesis**, in which an offspring develops from a single cell. The most common kind of parthenogenesis is **apomixis**, whereby meiosis is suppressed and an offspring develops from an unfertilized egg. The offspring is genetically identical to its mother, except for whatever new mutations may have arisen in the cell lineage from which the egg arose. A lineage of asexually produced, and thus genetically nearly identical, individuals may be called a **clone**.

In some taxa, recombination and the mode of reproduction can evolve rather rapidly. Using artificial selection in laboratory populations of *Drosophila*, investigators have altered the rate of crossing over between particular pairs of loci, and have even developed parthenogenetic strains from sexual ancestors (Carson 1967; Brooks 1988). Asexual lineages have often evolved from sexual ancestors, and asexual populations are known in many otherwise sexual species of plants and animals. However, many of the features required for sexual reproduction seem to degenerate rapidly in populations that have evolved obligate asexual reproduction, so reversal from asexual to sexual reproduction becomes very unlikely (Normark et al. 2003). Some species are facultatively sexual: they reproduce asexually except at a certain season or under certain stressful conditions, when they engage in sexual reproduction.

The Paradox of Sex

Benefits and costs of recombination and sex

The traditional explanation of the existence of recombination and sex is that they increase genetic variation (i.e., the variety of genetic combinations), thereby facilitating the adaptive evolution of a species and reducing the risk of extinction. That there is indeed a long-term advantage of sex is indicated by phylogenetic evidence that most asexual lineages of eukaryotes have arisen quite recently from sexual ancestors, and by the observation that such lineages often retain structures that once had a sexual function. A typical example is the dandelion *Taraxacum officinale*, which is completely apomictic but is very similar to sexual species of *Taraxacum*, retaining nonfunctional stamens and the brightly colored petal-like structures that in its sexual relatives serve to attract cross-pollinating insects (**Figure 15.2A**). If a parthenogenetic lineage were able to persist for many millions of years, it should have diverged greatly from its sexual relatives and given rise to a morphologically and ecologically diverse clade. Among eukaryotes, such diversity, betokening persistence since an ancient origin of asexuality, is rare, although it is found in some unicellular clades and in a very few multicellular groups, such as the bdelloid rotifers (**Figure 15.2B**; Normark et al. 2003). Most asexual lineages that arose a very long time ago must have become extinct, since so few ancient asexual forms exist today.

FIGURE 15.2 Most asexual eukaryotes have arisen recently from sexual ancestors. (A) The dandelion *Taraxacum officinale* reproduces entirely asexually, by apomixis. The brightly colored flower, which evolved as an adaptation for attracting pollinating insects, suggests that asexual reproduction in this species evolved only recently. In fact, some species of the genus *Taraxacum* reproduce sexually. (B) Scanning electron micrograph of a bdelloid rotifer, *Rotaria tardigrada*. This group of rotifers is unusual among metazoans because it has apparently been parthenogenetic for a very long time. (B from Diego et al. 2007.)

(A)

(B)
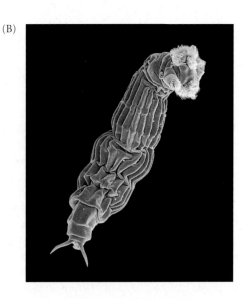

The recency of most parthenogenetic lineages suggests that sex reduces the risk of extinction. If this were the reason for its prevalence, sex might be one of the few characteristics of organisms that has evolved by group selection. But recombination and sex also have serious disadvantages, so serious that the existence of sex is a true enigma (Otto 2009). First, recombination can reduce, rather than expand, genetic variation. For instance, recombination among the many loci that contribute to variation in a quantitative character results in an approximately normal frequency distribution, in which extreme phenotypes are rare (see Figure 9.22). In contrast, the frequency distribution in an asexual population can take any form, such as abundant phenotypes at both extremes. A related, and important, disadvantage of recombination is that *it destroys adaptive combinations of alleles*, both within and between loci. If a heterozygous combination of alleles happens to be most fit, an asexual mother will have entirely heterozygous progeny, but sexual reproduction will produce some less fit homozygotes (e.g., sickle-cell hemoglobin). For a two-locus example, recall that in the primrose *Primula vulgaris*, plants with a short style and long stamens carry a chromosome with the gene combination *GA*, whereas those with a long style and short stamens are homozygous for the combination *ga* (see Figure 9.21). These combinations are adaptive because each can successfully cross with the other. Recombination, however, produces combinations such as *Ga*, which has both short stamens and a short style, and is less successful in pollination. In general, asexual reproduction preserves adaptive combinations of genes, whereas sexual reproduction breaks them down and reduces linkage disequilibrium between them. In a constant, uniform environment, an allele that promotes recombination will decline in frequency because it will often be associated with gene combinations that reduce fitness (**Figure 15.3**).

Sexual reproduction has additional disadvantages that make its existence one of the most difficult puzzles in biology (Lehtonen et al. 2012). Its ecological costs include the time, effort, and risks involved in seeking, competing for, and choosing mates, as well as the risk of infectious disease. In isogamous organisms that multiply by cell division, meiosis takes 5 to 100 times longer than mitosis and thus reduces the reproductive rate. Anisogamous species bear a special burden, the *cost of producing males*, often referred to simply as the **cost of sex**. Imagine two genotypes of females, one sexual and one asexual. In many sexual species, only half of all offspring are female. However, all the offspring of an asexual female are female (because they inherit their mother's sex-determining genes). If sexual and asexual females have the same fecundity, then a sexual female will have only half as many grandchildren as

FIGURE 15.3 Recombination destroys adaptive combinations of alleles. An allele (*r*) that promotes recombination will be selected against if genotypes *Ab* and *aB* have lower fitness than *AB* and *ab*. The parental generation has two copies of *R* and two copies of *r*. Allele *R*, which suppresses sex and recombination, increases in frequency because of its association with the favored genotypes *AB* and *ab*.

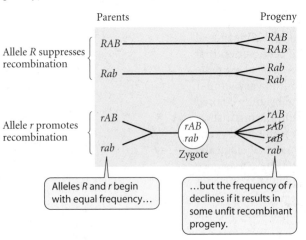

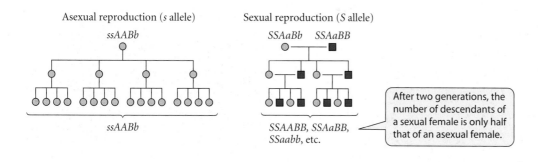

FIGURE 15.4 The cost of sex. The diagram shows the disadvantage of an allele *S*, which codes for sexual reproduction, compared with an allele *s*, which codes for asexual reproduction. Circles represent females, squares males. Each female produces four equally fit offspring. In a population with equal numbers of both kinds of females, the frequency of the *S* allele drops rapidly in each generation, because half of the progeny are males.

an asexual female (**Figure 15.4**). Therefore the rate of increase of an asexual genotype is approximately twice that of a sexual genotype, so an asexual mutant allele would very rapidly be fixed if it occurred in a sexual population. [The cost of sex is lower if the sex ratio is female-biased, or if males enhance females' reproductive success, for example, by sharing parental care (Lehtonen et al. 2012).] In the long term, of course, the evolution of asexuality might doom a population to extinction, but given this twofold advantage of asexuality, it is doubtful that extinction occurs frequently enough to offset the replacement of sexual with asexual genotypes within populations.

The problem, therefore, is to discover whether there are any *short-term advantages* of sex that can overcome its short-term disadvantages (Williams 1975; Maynard Smith 1978; Otto 2009). The problem is to envision conditions under which individuals carrying an allele for sexual reproduction have more descendants within a few generations than asexual individuals.

Hypotheses for the advantage of sex and recombination

Among the many proposed explanations for the prevalence of recombination and sex, a common theme is that recombination breaks down linkage disequilibrium, so that combinations of deleterious alleles on the one hand, and of advantageous alleles on the other, arise and thus increase the variance in fitness (**Figure 15.5**). Alleles that cause sexual reproduction might then increase because they are carried by fitter genotypes.

ADAPTATION UNDER DIRECTIONAL SELECTION The most apparent advantage of sex is that it enhances the rate of adaptation to new environmental conditions by combining new mutations or rare alleles to create new genotypes (**Figure 15.6**). In an asexual population, beneficial mutations *A* and *B* would be combined only when a second mutation (*B*) occurs in a growing lineage that has already experienced mutation *A* (or vice versa). Furthermore, a clone with one advantageous mutation might begin to increase, but then dwindle when a slightly more advantageous mutation arises in a different clone (note replacement of the genotype AC by AB in population 1 in Figure

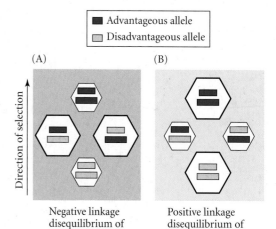

FIGURE 15.5 How linkage disequilibrium (LD) affects the response to selection. Long and short bars represent two different loci, each with an advantageous (green) and a disadvantageous (gray) allele. Hexagons are gene combinations, and their size is proportional to the frequency of the gene combination in a population. (A) LD is negative (i.e., an advantageous allele is associated with a disadvantageous allele), as indicated by the position of the large hexagons. Most of the genetic variation is aligned at right angles to the direction of selection, so adaptive evolution is slow. (B) Positive LD (association between two advantageous alleles), shown by the position of the large hexagons, maximizes the genetic variation along the direction of selection. A possible outcome is rapid fixation of the advantageous two-locus genotype. (After Poon and Chao 2004.)

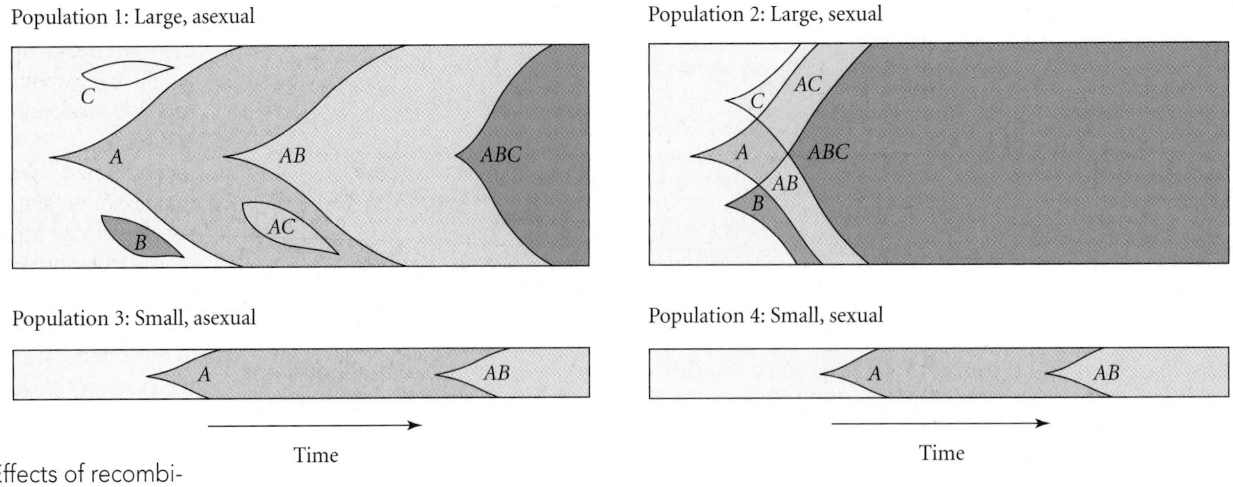

Population 1: Large, asexual

Population 2: Large, sexual

Population 3: Small, asexual

Population 4: Small, sexual

Time

Time

FIGURE 15.6 Effects of recombination on the rate of evolution. *A*, *B*, and *C* are new mutations that are advantageous in concert. In asexual populations (1 and 3), combination *AB* (or *ABC*) is not formed until a second mutation (such as *B*) occurs in a lineage that already bears the first mutation (*A*). In a large sexual population (2), independent mutations can be brought together in a lineage more rapidly by recombination, so adaptation is more rapidly achieved. In a small sexual population (4), however, the interval between the occurrences of favorable mutations is so long that the population does not adapt more rapidly than an asexual population. Notice an example of clonal interference in population 1: the advantageous genotypes *B*, *C*, and *AC* are eliminated by the increase of the more fit genotypes *A* and *AB*. (After Crow and Kimura 1965.)

15.6). This process of CLONAL INTERFERENCE (Gerrish and Lenski 1998) has been described in experimental populations of viruses and bacteria (e.g., Miralles et al. 1999). In a sexual population, in contrast, mutations that have occurred in different lineages can be combined more rapidly. However, this difference in evolutionary rate depends on population size. In small populations, mutations are so few that the first mutation (*A*) is likely to be fixed by selection before the second (*B*) arises, whether the population is asexual or sexual. Exactly this pattern was observed in laboratory populations of the unicellular green alga *Chlamydomonas reinhardtii*: sexual populations evolved higher fitness more rapidly than asexual populations, but this difference occurred only in large populations (Colegrave 2002). Nevertheless, directional selection is unlikely to explain the prevalence of sexual reproduction because it is not frequent enough to counteract the great short-term advantage of asexual reproduction.

ADAPTATION TO VARYING ENVIRONMENTS Suppose a polygenic character is subject to stabilizing selection, but the optimal character state fluctuates because of a fluctuating environment (Maynard Smith 1980). Let us assume that alleles *A*, *B*, … increase a trait such as body size, and alleles *a*, *b*,… decrease it. Stabilizing selection for intermediate size reduces the variance and creates linkage disequilibrium between alleles with opposite effects, so that combinations such as *Ab* and *aB* are present in excess (see Chapter 13). If the environment changes so that larger size is favored, combinations such as *AB* may not exist in an asexual population, but they can arise in a sexual population as recombination breaks down linkage disequilibrium. This capacity can provide not only a long-term advantage for sex (a higher rate of adaptation), but a short-term advantage as well, because sexual parents are likely to leave more surviving offspring than asexual parents. For this hypothesis to work, the selection regime must fluctuate rather frequently, and some factor must maintain genetic variation, because a long-term regime of stabilizing selection for a constant optimal phenotype would fix a homozygous genotype (such as *AAbb*).

One popular possibility is that genetic variation is maintained, and sex is favored, by continuing coevolution with parasites (see Chapter 19). As a resistant host genotype (e.g., *ABCD*) increased in frequency, a parasite might evolve to attack it. Parasite genotypes that could attack less common host genotypes (such as *abCD*) would then become rare, so the uncommon host genotypes would acquire higher fitness and increase in frequency. Continuing cycles of coevolution between host and parasite might select for sex, which would continually regenerate rare combinations of alleles.

Although this hypothesis may require very strong selection and may not provide a general explanation for the advantage of sex (Otto and Nuismer 2004), it is appealing and has considerable support. Curtis Lively and collaborators have studied populations of the freshwater snail *Potamopyrgus antipodarum* that include both sexual and asexual genotypes

(A)

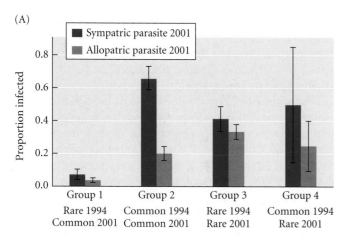

(B)

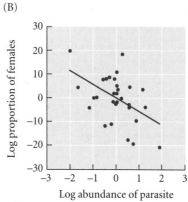

FIGURE 15.7 An evolving parasite causes evolutionary change in a host population and selects for sexual reproduction. (A) The proportions of snails in four groups of closely related clonal genotypes that became infected when experimentally inoculated with sympatric trematodes (from the same lake as the snails) or allopatric trematodes (from a different lake) in 2001. Groups 2 and 4, which were common in 1994, were susceptible to infection, especially by sympatric trematodes, whereas group 1 genotypes, which had been rare, were resistant and had become common by 2001. (B) The proportion of sexual genotypes was greater in snail populations that were exposed to greater abundances of the parasite, as shown by the decreased proportion of females in such populations. (A after Jokela et al. 2009; B after Jokela and Lively 1995.)

and which are infected by trematodes (Lively and Dybdahl 2000; Jokela et al. 2009). Over the course of 7 years, common asexual clones became rare and rare ones common. At the end of this interval, experimental tests showed that previously abundant clones were more susceptible to trematode infection than previously rare (now more common) clones because the parasite had adapted to the formerly prevalent clones (**Figure 15.7A**). Despite the changes in selection by an evolving parasite population, the abundance of sexual snails remained much the same throughout this period. Moreover, the frequency of sexual genotypes was greater in populations where the trematode was more abundant (**Figure 15.7B**).

Levi Morran and coworkers (2011) directly demonstrated the effect of host/parasite coevolution in laboratory populations of the nematode *Caenorhabditis elegans* that were composed of males and hermaphrodites of three types. One type reproduced only by (obligate) self-fertilization, one was an obligate outcrosser (mating with males), and one, the wild type, could do both. Populations of all three types were reared for 30 generations without a pathogen (the controls), with a single, non-evolving strain of a lethal pathogenic bacterium, or with a strain that was allowed to coevolve with the nematodes it infected. The rate of outcrossing increased greatly in the nematode populations maintained with the coevolving pathogen, but not in the other treatments (**Figure 15.8**).

SEPARATING BENEFICIAL FROM HARMFUL MUTATIONS   Because of genetic drift in finite populations, recombination can increase fitness by separating beneficial alleles from deleterious mutations at other loci. This idea was first introduced by Herman Muller (1964), who won a Nobel Prize for his role in discovering the mutagenic effect of radiation. His

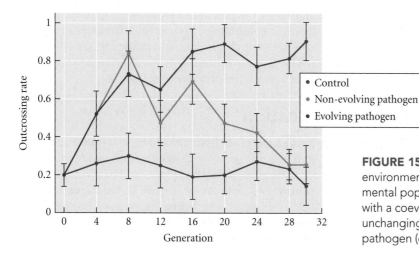

FIGURE 15.8 Recombination may be advantageous in a changing environment. The frequency of outcrossing increased rapidly in experimental populations of the nematode *C. elegans* that were maintained with a coevolving pathogen, but not in populations exposed to an unchanging pathogen genotype or which were not exposed to the pathogen (controls). (After Morran et al. 2011.)

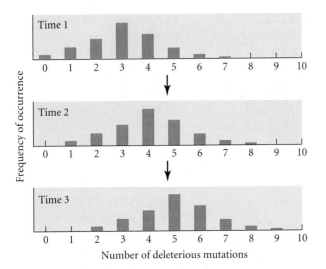

FIGURE 15.9   Muller's ratchet. The frequencies of individuals with different numbers of deleterious mutations (0–10) are shown for an asexual population at three successive times. The class with the lowest mutation load (0 in top graph, 1 in middle graph) is lost over time, both by genetic drift and by its acquisition of new mutations. In a sexual population, class 0 can be reconstituted, since recombination between genomes in class 1 that bear different mutations can generate progeny with none. (After Maynard Smith 1988.)

hypothesis has been named MULLER'S RATCHET (**Figure 15.9**). In an asexual population, deleterious mutations at various loci create a spectrum of genotypes carrying 0, 1, 2, …, *m* mutations. Individuals can carry more mutations (a greater MUTATIONAL LOAD) than their ancestors did (because of new mutations), but not fewer. Thus the zero-mutation class declines over time because its members experience new deleterious mutations. Moreover, because of genetic drift, the zero-mutation class may be lost by chance, despite its superior fitness. (The smaller the population, the more likely this is to happen.) Thus all remaining genotypes have at least one deleterious mutation. Sooner or later, by the same process of drift, the one-mutation class is lost, and all remaining individuals carry at least two mutations. The continuing accidental loss of superior genotypes is an irreversible process—a ratchet. The resulting reduction of fitness is likely to lower population size, and this, in turn, increases the rate at which the least mutation-laden genotypes are lost by genetic drift. Thus there may be an accelerated decline of fitness—a "mutational meltdown"—leading to extinction (Lynch et al. 1993). In contrast, *recombination in a sexual population reconstitutes the least mutation-laden classes of genotypes* by generating progeny with new combinations of favorable alleles. Nancy Moran (1996) has suggested that Muller's ratchet explains several features of the genome of the asexual bacterium *Buchnera*, an endosymbiont of aphids, such as accelerated rates of nonsynonymous mutations but not of synonymous mutations.

Sarah Otto has proposed a related hypothesis: sex and recombination are advantageous because they liberate advantageous alleles from their association with less beneficial ones (Otto 2009; Hartfield and Otto 2011). Owing to genetic drift and recurrent mutation in finite populations, linkage disequilibrium develops between alleles that enhance fitness and neighboring sites that lower fitness, and this association retards or prevents the increase of the beneficial alleles. This phenomenon is known as the HILL-ROBERTSON EFFECT (Hill and Robertson 1966). Individuals that carry a gene that increases recombination will have more descendants because their beneficial alleles will be freed from their undesirable associates, and their descendants will inherit the recombination allele, which will therefore increase in frequency. This model seems to explain the results of an experiment in which populations of *E. coli* bacteria were constructed with and without an allele enabling recombination (*rec⁺* and *rec⁻*), and with and without an allele that increased the mutation rate throughout the genome (*mut⁺* and *mut⁻*) (Cooper 2007). Cultures with all four possible combinations of these alleles were grown for 1000 generations. Recombination caused a greater increase of fitness in the high-mutation than in the low-mutation lines, as this hypothesis predicts (**Figure 15.10**).

The advantage of recombination in this model can be strong enough to overcome the twofold cost of sex. It may well be the most common and powerful advantage of recombination because it operates under both directional selection and purifying selection, and because it applies to genes throughout the genome (Otto 2009). Nonetheless, it might well be that a combination of factors will best explain the prevalence of sex.

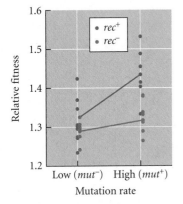

FIGURE 15.10   Recombination can increase fitness by separating deleterious mutations from advantageous parts of the genome. Strains of *E. coli* with a recombination allele (*rec⁺*) had higher fitness than strains without recombination (*rec⁻*), especially when the mutation rate was increased by a mutator allele (*mut⁺*). (After Cooper 2007.)

Sex Ratios and Sex Allocation

We turn now to the question of why some species are hermaphroditic and others are dioecious, and what accounts for variation in sex ratio among dioecious species. The theory of sex allocation has been developed to explain such variation (Charnov 1982; West 2009).

The **sex ratio** is defined as the proportion of males. As in Chapter 12, we distinguish the sex ratio in a population (the POPULATION SEX RATIO) from that in the progeny of an individual female (an INDIVIDUAL SEX RATIO). In Chapter 12, we saw that in a large, randomly mating population, a genotype with a male-biased individual sex ratio has an advantage if the population sex ratio is female-biased, and vice versa, because each individual of the minority sex has a higher number of offspring, on average, than each individual of the majority sex. A genotype with an individual sex ratio of 0.5 is an ESS (evolutionarily stable strategy) because it has, per capita, the greatest number of grandchildren (see Figure 12.18).

In some species, however, mating occurs not randomly among members of a large population, but instead within small local groups descended from one or a few founders. After one or a few generations, progeny emerge into the population at large, then colonize patches of habitat and repeat the cycle. In many species of parasitoid wasps, for example, the progeny of one or a few females emerge from a single host and almost immediately mate with each other; the daughters then disperse in search of new hosts. Such species often have female-biased sex ratios.

William Hamilton (1967) explained such "extraordinary sex ratios" by what he termed LOCAL MATE COMPETITION (LMC) (Antolin 1993). Whereas in a large population, a female's sons compete for mates with many other females' sons, they compete only with one another in a local group founded by their mother. Thus the founding female's genes can be propagated most prolifically by producing mostly daughters, with only enough sons to inseminate all of them. Additional sons would be redundant, since they all carry their mother's genes. Another way of viewing this situation is to recognize that groups founded by genotypes whose individual sex ratio is biased in favor of females contribute more individuals (and genes) to the population as a whole than groups founded by unbiased genotypes. The difference among local groups in their production of females increases the frequency, in the population as a whole, of female-biasing alleles (**Figure 15.11**; Wilson and Colwell 1981). The smaller the number of founders of a group, the less even the optimal sex ratio will be. According to this hypothesis, sex ratio evolves by a form of selection among small groups, called TRAIT GROUPS, that differ because of random variation in allele frequencies (D. S. Wilson 1975; Eldakar and Wilson 2011).

Hamilton's hypothesis is a fine example of an evolutionary theory that has received strong support long after it was proposed. For example, individual sex ratio is adaptively plastic in the parasitoid wasp *Nasonia vitripennis*. The progeny of one or more females develop in a fly pupa and mate with each other immediately after emerging. Based on the theory described above, we would expect females to produce more daughters than sons, but we would expect the sex ratio to increase with the number of families developing in a host. Moreover, if the second wasp that lays eggs in a host can detect previous

FIGURE 15.11 A model of the evolution of a female-biased sex ratio in a population that is structured into local groups, but periodically forms a single pool of dispersers. The frequency of A_1, an allele that biases the individual sex ratio toward daughters, is indicated by the dark portion of each circle. The size of each circle represents the size of a group or population. From a genetically variable pool, groups are established by one or a few founders. The frequency of A_1 varies among those groups by chance. Although A_1 declines in frequency within each group over the course of several generations, the growth in group size is greater the higher the frequency of A_1 (because of the greater production of daughters). When individuals emerge from the groups, they form a pool of dispersers, in which A_1 has increased in frequency since the previous dispersal episode. (After Wilson and Colwell 1981.)

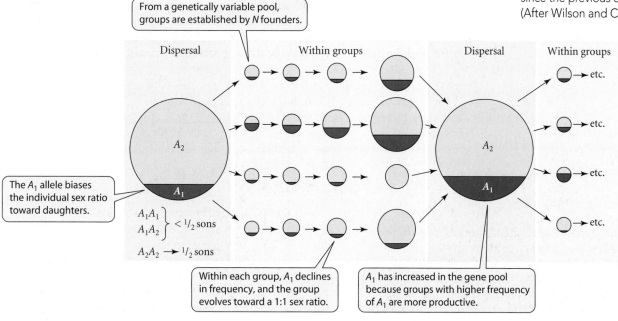

From a genetically variable pool, groups are established by N founders.

The A_1 allele biases the individual sex ratio toward daughters.

$$\left.\begin{array}{l} A_1A_1 \\ A_1A_2 \end{array}\right\} < {}^1/_2 \text{ sons}$$

$$A_2A_2 \rightarrow {}^1/_2 \text{ sons}$$

Within each group, A_1 declines in frequency, and the group evolves toward a 1:1 sex ratio.

A_1 has increased in the gene pool because groups with higher frequency of A_1 are more productive.

FIGURE 15.12 Adaptive adjustment of individual sex ratio by a parasitoid wasp (*Nasonia vitripennis*) in which females usually mate with males that emerge from the same host. The points show the relationship between the proportion of sons among the offspring of a second female (i.e., one that lays eggs in a fly pupa in which another female has already laid eggs) and the proportion of her offspring in the host. The curved line is the theoretically predicted individual sex ratio in second females' broods, as a function of the sex ratio in the first female's brood and the relative number of the two females' offspring. If the second female's offspring make up only a small fraction of the total, that female's optimal "strategy" should be to produce mostly sons, which could potentially inseminate many female offspring of the first female. As predicted, the fewer the progeny of the second female, the more of them are sons. (After Werren 1980.)

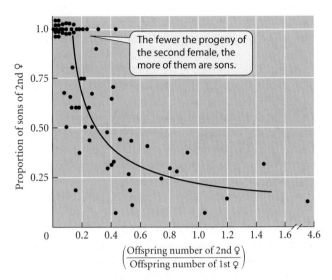

parasitization, we would expect her to adjust her individual sex ratio toward a higher proportion of sons than the first female. John Werren (1980) calculated the theoretically optimal individual sex ratio of a second female, then measured the individual sex ratios of second females by exposing fly pupae successively to two different strains of *Nasonia* that were distinguishable by an eye-color mutation. On the whole, his data fit the theoretical prediction very well (**Figure 15.12**). In another study, spider mites (*Tetranychus urticae*), in which individual sex ratio is genetically variable, were reared for 54 generations under two regimes: one foundress (fertilized female) per patch and 10 foundresses per patch. A third regime, with more than 100 foundresses per patch, ran for 14 generations. At the end of the study, the individual sex ratio had evolved to be low in the single-foundress lines, and nearly even (0.5) in those with more foundresses, just as Hamilton's hypothesis predicts (**Figure 15.13**).

The principle that explains the even sex ratio in many species—namely, that fitness through female and male functions must be equal—also helps to explain why some species have separate sexes and others do not (Charnov 1982). Flowering plants vary greatly in their reproductive systems (Barrett 2002, 2010): most are hermaphroditic, some are dioecious (composed of unisexual individuals), and some are GYNODIOECIOUS, having mixed populations of hermaphrodites and females. (A few species have both hermaphrodites and males.) The fitness of a hermaphrodite is the sum of its reproductive success via male function (successful pollination) and via female function (successful seed production). Dioecious species arise from hermaphroditic ancestors by mutations for male sterility

FIGURE 15.13 Sex ratios in experimental populations of spider mites evolved nearly to the optimum predicted by the local mate competition hypothesis. Mating occurred among offspring of 1 (LMC+), 10 (LMC–), or more than 100 founding females (panmixia). The gold and purple lines show the predicted sex ratios for the LMC– and panmixia lines. (After Macke et al. 2011.)

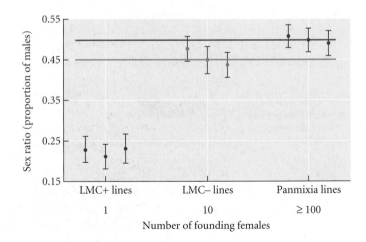

(A)

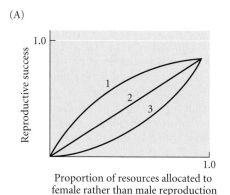

Proportion of resources allocated to
female rather than male reproduction

(B)

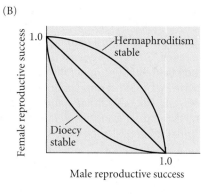

Male reproductive success

FIGURE 15.14 The theory of sex allocation. (A) The reproductive success of an individual as a function of the fraction of resources allocated to one sexual function (here, female) rather than the other. Increasing allocation to that sexual function may yield decelerating (1), linear (2), or accelerating (3) gains in reproductive success. (B) Reproductive success gained through female function plotted against that gained through male function. Because resources are allocated between these functions, there is a trade-off between the reproductive success an individual achieves through either sexual function. If reproductive success is linearly related to allocation (curve 2 in A), this trade-off is linear, and the sum of female reproductive success and male reproductive success equals 1.0 at each point on the trade-off curve. If the gain in reproductive success is a decelerating function of allocation to one sex or the other (curve 1 in A), then the fitness of a hermaphrodite exceeds that of a unisexual individual (with reproductive success = 1.0 for one sexual function and 0 for the other). If reproductive success is an accelerating function of allocation (curve 3 in A), then the trade-off curve is concave, and dioecy (separate sexes) is stable—that is, a hermaphrodite's fitness is lower than that of either dioecious type. (After Thomson and Brunet 1990.)

(producing females) and for female sterility (producing males), mutations that abolish one of the pathways of reproductive success. A unisexual mutant (e.g., a male-sterile female), however, may increase in a population for two combined reasons. First, females, which can only outcross, may have greater fitness than hermaphrodites if self-fertilization is common and inbreeding depression is severe. Second, assume that there is a trade-off in how a potentially hermaphroditic individual allocates energy and resources to female functions (e.g., seeds) and to male functions (e.g., pollen). Individuals that allocate more toward one reproductive function (e.g., female) may have higher fitness than hermaphrodites if their reproductive success is an accelerating function of allocation (as in curve 3 in **Figure 15.14**). Models show that female mutants can increase if they produce more than twice as many seeds as hermaphrodites, compensating for the loss of male reproductive function. These conditions appear to hold true for a gynodioecious Hawaiian shrub, *Schiedea adamantis*, in which nearly 70 percent of offspring result from self-fertilization, inbreeding depression is more than 60 percent, and females produce about 2.3 times as many seeds as hermaphroditic individuals (Sakai et al. 1997).

Inbreeding and Outcrossing

Characteristics that promote outcrossing include dioecy (separate sexes), asynchronous male and female function (e.g., maturation and dispersal of pollen either before or after the stigma of the same flower is receptive), and several kinds of self-incompatibility (Matton et al. 1994). The DNA sequences of self-incompatibility alleles, which prevent pollen from fertilizing plants that carry the pollen's allele, show that these polymorphisms are stable and very old, and that they have been maintained by balancing selection (see Figure 12.32A). These features are thought to have evolved largely to avoid inbreeding depression (Charlesworth and Charlesworth 1978; Lloyd 1992). Nevertheless, self-fertilization (selfing), which results in inbreeding, has originated more than a thousand times among plants and occurs in some hermaphroditic animals (such as certain snails) as well. Many plants can both self-fertilize and outcross, but some, such as wheat, have evolved a strong tendency toward self-fertilization within flowers. Many such species produce only a little pollen and have small, inconspicuous flowers that lack the scent and markings to which pollinators are attracted. In some cases, flowers remain budlike and do not open at all.

In animals, the major mechanisms for reducing inbreeding are prereproductive dispersal and INCEST AVOIDANCE: avoidance of mating with relatives (Thornhill 1993), which inevitably brings to mind the well-known "incest taboo," often codified in religious and civil law, in human societies. Incest is a highly controversial topic with respect to both its actual incidence and the interpretation of the social taboo. Societies vary as to which kin matings are prohibited (Ralls et al. 1986). Moreover, the incidence of closely incestuous sexual activity is evidently far higher than society has generally been ready to recognize,

Outcrossing   Selfing

FIGURE 15.15 Inbreeding may purge deleterious alleles. The mean number of flowers produced by *Eichhornia paniculata* plants from (A) a naturally outcrossing Brazilian population and (B) a naturally inbred Jamaican population in the first outcrossed generation (O1), in five generations of selfing (S1–S5), and in outcrosses between selfed plants after five generations (O5). The naturally outcrossing Brazilian population displayed inbreeding depression (compare O1 through S5) and heterosis in outcrossed plants (O5), but the naturally inbred Jamaican population did not. The data suggest that the mutation load has been purged by inbreeding in the Jamaican population. (After Barrett and Charlesworth 1991; photo courtesy of S. C. H Barrett.)

(A) Brazil

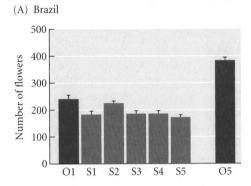

(B) Jamaica

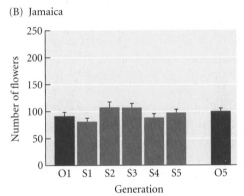

especially in the form of sexual attention forced upon young women by fathers and uncles. These observations raise doubt about whether a strong, genetically based aversion to incest has evolved in our species (at least in males). Some anthropologists hold that outbreeding is a social device (not an evolved genetic trait) to establish coalitions between families or larger groups in order to reap economic and other benefits of cooperation.

Inbreeding increases homozygosity, and it is often accompanied by inbreeding depression, which is usually caused by homozygosity for deleterious recessive (or nearly recessive) alleles (see Chapter 9). Conversely, crosses between inbred parents often show HYBRID VIGOR, or HETEROSIS, because the recessive alleles in each genotype are masked in the hybrid. Because inbreeding exposes the recessive alleles to natural selection, they should be "purged" from a population as inbreeding continues, so genetic variation should decline and mean fitness should increase. The mean fitness of a long-inbred population might therefore equal or even exceed that of the initial outbreeding population (Lande and Schemske 1985). Spencer Barrett and Deborah Charlesworth (1991) provided some evidence for this hypothesis, using a highly outcrossing Brazilian population and an almost exclusively self-fertilizing Jamaican population of the aquatic plant *Eichhornia paniculata*. For five generations, Barrett and Charlesworth self-pollinated plants in each population, then cross-pollinated the inbred lines. In the naturally outcrossing Brazilian lines, flower number declined as inbreeding proceeded (**Figure 15.15A**), but it increased dramatically in crosses between the inbred lines. In contrast, the naturally selfing Jamaican lines conformed to the purging theory: they showed neither inbreeding depression when self-pollinated nor heterosis when crossed with each other (**Figure 15.15B**).

Because outcrossing populations show reduced fitness when inbred, the evolution of obligate or predominant self-fertilization must have overcome the obstacle of inbreeding depression as well as the loss of reproductive success through outcross pollen. Several possible advantages of selfing might outweigh these disadvantages. For example, even if selfing is disadvantageous on average, it may occasionally produce a highly fit homozygous genotype that sweeps to fixation, carrying with it the alleles that increase the selfing rate (Holsinger 1991). A related possibility is that selfing may "protect" locally adapted genotypes from OUTBREEDING DEPRESSION as a result of gene flow and recombination. Populations of the grass species *Anthoxanthum odoratum* in the vicinity of mines have evolved tolerance for heavy metals within the last several centuries (see Chapter 13). Although they have diverged from neighboring nontolerant populations on uncontaminated soil, they receive gene flow from those populations via wind-borne pollen. Several of these tolerant populations self-pollinate more frequently than nontolerant populations do, having become more self-compatible (**Figure 15.16**).

Perhaps the most important advantage of selfing is REPRODUCTIVE ASSURANCE: a plant is almost certain to produce some seeds by selfing, even if scarcity of pollinators, low population density, or other adverse environmental conditions prevent cross-pollination. There is abundant support for this hypothesis (Wyatt 1988; Jarne and Charlesworth 1993). For example, adaptations for self-fertilization are especially common in plants that grow

FIGURE 15.16 Self-fertilization evolved in a population of the grass species *Anthoxanthum odoratum* that had adapted to soil contaminated with lead and zinc near a mine. (A) Sample sites and distances along a transect from the contaminated area (blue shading) to a pasture with normal (uncontaminated) soil. (B) Seed set by plants whose flowers were bagged to prevent cross-pollination and which therefore could only self-fertilize. The experiment, performed in two successive years, showed a much higher frequency of self-fertilization by the plants growing on contaminated soil. Selfing was advantageous to these plants because it prevented pollination by nonadapted pasture plants, which would produce offspring less adapted to the metals than their parents. (After Antonovics 1968; McNeilly and Antonovics 1968.)

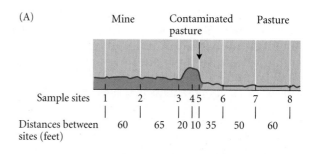

(A)

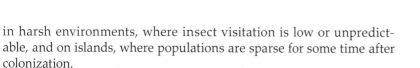

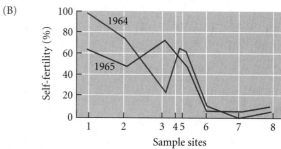

(B)

in harsh environments, where insect visitation is low or unpredictable, and on islands, where populations are sparse for some time after colonization.

The Concept of Sexual Selection

Mating success is a major component of fitness in sexually reproducing species (see Figure 12.3). Darwin introduced the concept of **sexual selection** to describe differences among individuals of a sex in the number of mates they obtain or those mates' reproductive capacity, although the concept may be extended to include competition among individuals for access to the gametes of the opposite sex (Andersson 1994; Kokko et al. 2006). Darwin focused mostly on males' reproductive success, which may depend on the number of their mates, the fecundity of their mates, and the proportion of their mates' eggs that they fertilize when they mate. Sexual selection was Darwin's solution to the problem of why conspicuous male traits such as bright colors, horns, and elaborate displays have evolved. He proposed two forms of sexual selection: contests between males for access to females, and female choice (or "preference") of some male phenotypes over others. Several other mechanisms for sexual selection have also been recognized (**Table 15.1**).

Sexual selection of males exists because females produce relatively few, large gametes (eggs) and males produce many small gametes (sperm). *This difference creates an automatic conflict* between the reproductive strategies of the sexes: a male can mate with many females, and often suffers little reduction in fitness if he mates with an inappropriate

TABLE 15.1 Mechanisms of sexual selection and traits likely to be favored

Mechanism	Traits favored
Same-sex contests	Traits improving success in confrontation (e.g., large size, strength, weapons, threat signals); avoidance of contests with superior rivals
Mate preference	Attractive and stimulatory features; offering of food, territory, by opposite sex or other resources that improve mate's reproductive success
Scrambles	Early search and rapid location of mates; well-developed sensory and locomotory organs
Endurance rivalry	Ability to remain reproductively active during much of season
Sperm competition	Ability to displace rival sperm; production of abundant sperm; mate guarding or other ways of preventing rivals from copulating with mate
Coercion	Adaptations for forced copulation and other coercive behavior
Infanticide	Similar traits as for same-sex contests
Antagonistic	Ability to counteract the other sex's resistance to mating (by, e.g., coevolution hyperstimulation); egg's resistance to sperm entry

Source: After Andersson 1994; Andersson and Iwasa 1996.

FIGURE 15.17 Variation in mating success. Among elephant seals, a few dominant males defend harems of females against other males (see Figure 10.5). A plot of the number of offspring produced by each of 140 elephant seals over the course of several breeding seasons shows much greater variation in reproductive success among males than among females. (After Gould and Gould 1989, based on data of B. J. LeBoeuf and J. Reiter.)

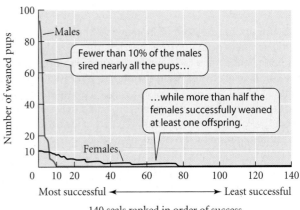

female, whereas all of a female's eggs can potentially be fertilized by a single male, and her fitness can be significantly lowered by inappropriate matings. Thus encounters between males and females often entail conflicts of reproductive interest (Trivers 1972; Parker 1979). Furthermore, the OPERATIONAL SEX RATIO—the relative numbers of males and females in the mating pool at any time—is often male-biased because males mate more frequently and because the cost of breeding is greater for females, which may have little opportunity to mate and reproduce repeatedly during a breeding season because of energetic and time constraints (Kokko et al. 2006). Commonly, then, *females are a limiting resource for males*, which compete for mates, but males are less commonly a limiting resource for females. Thus *variation in mating success is generally greater among males than among females* (**Figure 15.17**), and indeed, is a measure of the intensity of sexual selection (Wade and Arnold 1980). Likewise, eggs are a limiting resource for sperm, engendering competition among sperm of different males, or among pollen from different plants.

Contests between Males and between Sperm

Male animals often compete for mating opportunities. The males of some species fight outright and possess weapons, such as horns or tusks, that can inflict injury (**Figure 15.18**; Emlen 2008). Others compete through visual displays of bright colors or other ornaments. Plumage patterns and songs are used by many birds to establish dominance. The male

FIGURE 15.18 Hornlike structures have evolved independently in the males of diverse animals, which use them to compete for access to females. Such structures are illustrated here for 13 species, including several that are extinct (marked with an asterisk): 1, narwhal (*Monodon monoceros*); 2, chameleon [*Chamaeleo (Trioceros) montium*]; 3, trilobite (*Morocconites malladoides**); 4, unicorn fish (*Naso annulatus*); 5, ceratopsid dinosaur (*Styracosaurus albertensis**); 6, horned pig (*Kubanochoerus gigas**); 7, protoceratid ungulate (*Synthetoceras* sp.*); 8, dung beetle (*Onthophagus raffrayi*); 9, brontothere (*Brontops robustus**); 10, rhinoceros beetle [*Allomyrina (Trypoxylus) dichotomus*]; 11, isopod (*Ceratocephalus grayanus*); 12, horned rodent (*Epigaulus* sp.*); 13, giant rhinoceros (*Elasmotherium sibiricum**). (Illustration courtesy of Douglas Emlen.)

red-winged blackbird (*Agelaius phoeniceus*), for example, sings a distinctive song and has a bright red shoulder patch (Andersson 1994). Male blackbirds that were experimentally silenced were likely to lose their territories to intruders; however, intruders were deterred by tape recordings of male song. When territorial males were removed and replaced with stuffed and mounted specimens whose shoulder patches had been painted over to varying extents, the specimens with the largest red patches were avoided by potential trespassers.

In sexual selection by male contest, directional selection for greater size, weaponry, or display features can cause an "arms race" that results in evolution of ever more extreme traits. Among antelopes and their relatives (Bovidae), the size of males' horns is greater in species in which socially dominant males guard larger "harems" of females. Sexual selection is stronger in these species, in which the operational sex ratio is more male-biased (Bro-Jorgensen 2007). Such "escalation" becomes limited by opposing ecological selection (i.e., selection imposed by ecological factors) if the cost of larger size or weaponry becomes sufficiently great (West-Eberhard 1983). As Darwin noted, the duller coloration and lack of exaggerated display features in females and nonbreeding males of many species imply that those features of breeding males are ecologically disadvantageous.

Closely related to contests among males for mating opportunities are the numerous ways in which males reduce the likelihood that other males' sperm will fertilize a female's eggs (Thornhill and Alcock 1983; Birkhead and Møller 1992; Simmons 2001). Males of many species of birds defend territories, keeping other males away from their mates (although studies of DNA markers show that females of many such species nevertheless engage in high rates of extra-pair copulation). The males of many frogs, crustaceans, and insects clasp a female, guarding her against other males for as long as she produces fertilizable eggs. In some species of *Drosophila*, snakes, and other animals with internal fertilization, the seminal fluid of a mating male forms a copulatory plug in the vagina, reduces the sexual attractiveness of the female to other males, or reduces her receptivity to further mating (Partridge and Hurst 1998).

Sexual competition between males may continue during and after copulation. In many damselflies, the males' genitalia are adapted to remove the sperm of previous mates from the female's reproductive tract (**Figure 15.19**). In many animals, **sperm competition** occurs when the sperm of two or more males have the opportunity to fertilize a female's eggs (Parker 1970; Birkhead et al. 2009). In some such cases, a male can achieve greater reproductive success than other males simply by producing more sperm. This explains why polygamous species of primates tend to have larger testes than monogamous species (see Figure 11.23). SPERM PRECEDENCE, whereby most of a female's eggs are fertilized by the sperm of only one of the males with which she has mated, occurs in many insects

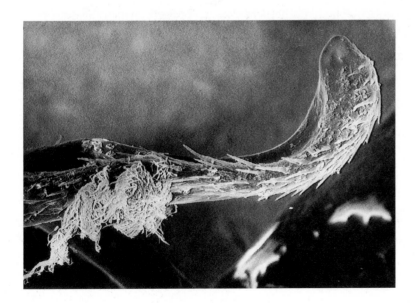

FIGURE 15.19 An elaborate mechanism for improving a male's likelihood of paternity. This micrograph of a structure near the end of the penis of the black-winged damselfly *Calopteryx* showing its spinelike hairs and a clump of a rival's sperm. (Courtesy of J. Waage.)

and other species. In *Drosophila melanogaster,* the degree of sperm precedence is affected by genetic variation among females and by genetic variation among males in the ability of their sperm to displace other males' sperm and to resist displacement (Clark and Begun 1998). Competition among sperm (as well as male-female interactions, as we will see) may explain variation among species in sperm morphology, such as the giant sperm of some insects. *Drosophila bifurca,* with a body length less than 3 mm, has the longest known sperm of any animal, at 58.3 mm!

(A)

Sexual Selection by Mate Choice

In many species of animals, individuals of one sex (usually the male) compete to be chosen by the other. The evolution of sexually selected traits by mate choice presents some of the most intriguing problems in evolutionary biology and is the subject of intense research (Andersson 1994; Johnstone 1995; Andersson and Simmons 2006).

Females of many species of animals mate preferentially with males that have larger, more intense, or more exaggerated characters such as color patterns, ornaments, vocalizations, or display behaviors. The extraordinarily long, ornamented train of back feathers of the peacock is one of many examples (**Figure 15.20**; see also Figure 11.10). The preferred male characters are often ecologically disadvantageous. For example, males of the cricket *Teleogryllus oceanicus* strike their wings together to produce a calling song that attracts females, but on the Hawaiian island Kauai, a recently introduced fly (*Ormia ochracea*) uses the cricket's calling song to find its prey. Marlene Zuk and colleagues (2006) have found that over the course of fewer than 20 generations, the cricket population has become almost entirely silent: most males have modified wings, probably due to a single mutation, that do not produce sound. Phylogenetic studies have revealed many examples in which sexually selected ornaments have been lost in evolution (Wiens 2001).

Subject to limits imposed by ecological selection, male traits will obviously evolve to exaggerated states if they enhance mating success. But why should females have a preference for these traits, especially for features that seem so arbitrary—and even dangerous to the males that bear them? Several hypotheses have been proposed.

Direct benefits of mate choice

The least controversial hypothesis applies to species in which the male provides a direct benefit to the female or her offspring, such as nutrition, a superior territory with resources for rearing offspring, or parental care. Under these circumstances there is selection pressure on females to recognize males that are superior providers by means of some feature that is correlated with their

(B)

(C)

FIGURE 15.20 Some features of male birds used in sexual display to females. (A) The elongated, ornamented train of back feathers (not a tail!) of the Indian peacock (*Pavo cristatus*). (B) Males of many species of hummingbirds have striking display features, such as the modified tail feathers and the white leg puffs of the booted racket-tail (*Ocreatus underwoodii*), found high in the Andes. (C) In western New Guinea, the male Vogelkop bowerbird (*Amblyornis inornata*) is a very plain brown bird that builds an extraordinary structure, a bower, that is more than 1.5 meters wide and provisioned with a display of contrastingly colored fruits, fungi, beetle wing covers, and other objects. Females mate preferentially with males that have more elaborate, diverse displays. The bower is not a nest, and the male does not help the female or his offspring. (C courtesy of the author.)

(A)

(B)

FIGURE 15.21 Males that provide direct benefits to females. (A) This female Mormon cricket (*Anabrus simplex*) will eat the large white spermatophore her mate has placed in her genital opening as a "nuptial gift." (B) The male ornate moth (*Utetheisa ornatrix*) transfers pyrrholizidine alkaloids to the female during mating. (A courtesy of John Alcock.)

ability to provide. Once this capacity has evolved in females, their preference selects for males with the distinctive, correlated feature. For example, the males of many insects provide "nuptial gifts" to females, often in the form of nutrients and chemical compounds that they transfer to females together with sperm (Gwynne 2008; **Figure 15.21**). In the ornate moth (*Utetheisa ornatrix*), these compounds include alkaloids that the insect sequestered from its food plant as a larva, which are toxic and repellent to predators (but not to the moth). Females provide their eggs with both alkaloids that they themselves sequestered and alkaloids that they acquire by mating. Females prefer to mate with males that release more of a sex pheromone that the male produces from metabolized plant alkaloids, and larger males release more of this pheromone. Vikram Iyengar and Thomas Eisner (1999) demonstrated experimentally that females preferred larger males when given a choice, and showed that fewer of the eggs from matings with preferred males were eaten by ladybird beetles than eggs from "forced" matings with nonpreferred males, presumably because they contained more alkaloids. In some species of birds, more vocal or brightly colored males provide more care for the female's offspring; for example, male house finches (*Carpodacus mexicanus*) with more extensive red coloration bring food to nestlings at a higher rate (Hill 1991).

Mate choice without direct benefits

The most difficult problem in accounting for the evolution of female preferences is presented by species in which the male provides no direct benefit to either the female or her offspring, but contributes only his genes. In this case, alleles affecting female mate choice increase or decrease in frequency depending on the fitness of the females' offspring. Thus females may benefit indirectly from their choice of mates (Kirkpatrick and Barton 1997; Kokko et al. 2002). The evolution of choice due to indirect benefits may be aided, in some cases, by an apparently nonadaptive phenomenon, sensory bias.

SENSORY BIAS Researchers in animal behavior discovered long ago that animals frequently show greater responses to SUPERNORMAL STIMULI—those that are outside the usual range of stimulus intensity—than to normal stimuli. In the context of mate choice, this inherent tendency to react to exaggerated stimuli has been termed **sensory bias**. Phylogenetic studies have shown that female preferences sometimes evolve before the preferred male trait, simply because of the organization of the sensory system (**Figure 15.22A**; Ryan 1998). For instance, in some species of the fish genus *Xiphophorus* (swordtails), part of the male's tail is elongated into a "sword" (**Figure 15.22B**). Alexandra Basolo (1995, 1998) found that females preferred males fitted with plastic swords not only in sword-bearing species of

FIGURE 15.22 Evidence supporting sexual selection as a result of sensory bias in female mate choice. (A) A hypothetical phylogeny of three species that differ in presence (+) or absence (–) of a male trait (T) and in female preference (P) for that trait. Both T and P are absent in the outgroup species, evidence that their presence is derived. That P is present in both species 1 and 2, whereas T is present in 2 only, is evidence that a female preference, or sensory bias, for the trait T evolved before the trait did. (B) Such evidence has been found in poeciliid fishes that lack (top) or possess (bottom) a swordlike tail ornament in the male. (A after Ryan 1998; B photos courtesy of Alexandra Basolo.)

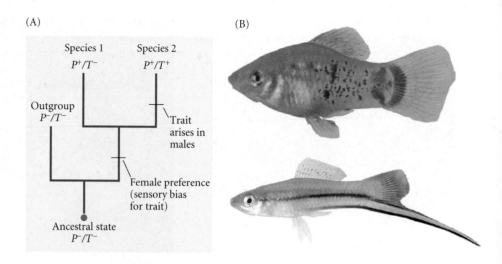

Xiphophorus, but also in a species of *Xiphophorus* that lacks the sword and in the swordless genus *Priapella*, the sister group of *Xiphophorus*. In fact, female *Priapella* had a stronger preference for swords than females of a sword-bearing species of *Xiphophorus* did. The female preference evidently evolved in the common ancestor of these genera, and thus provided selection for a male sword when the mutations for this feature arose.

A sensory bias of this kind could initiate a process of sexual selection that results in both a more exaggerated male trait and a more extreme female preference. Two major models of this process have been advanced.

RUNAWAY SEXUAL SELECTION   In the **runaway sexual selection** model (also called the "sexy son" hypothesis) proposed by R. A. Fisher (1930), the evolution of a male trait and a female preference, once initiated, becomes a self-reinforcing, snowballing, or "runaway" process (Lande 1981; Kirkpatrick 1982; Pomiankowski and Iwasa 1998). According to this hypothesis, female choice among males does not benefit the survival of offspring, but only increases the mating success of sons: the sons of females that choose a male trait have improved mating success because they inherit the trait that made their fathers appealing to their mothers.

In the simplest form of this model (**Figure 15.23**), males of genotypes T_1 and T_2 have frequencies of t_1 and t_2, respectively. T_2 codes for a more exaggerated trait, such as a longer tail, that carries an ecological disadvantage, such as increasing the risk of predation. Females with a rare mutation P_2 (with frequency p_2) prefer males of type T_2 (perhaps because of a sensory bias), whereas P_1 females exhibit little preference (or prefer T_1). It is assumed that alleles P_1 and P_2 do not affect survival or fecundity, and thus are selectively neutral.

Although the expression of genes P and T is sex-limited, both sexes carry both genes and transmit them to offspring. (In most species, including humans, each sex carries almost all the genes that underlie the development of the features of the opposite sex.) Because P_2 females and T_2 males tend to mate with each other, linkage disequilibrium will develop: their offspring of both sexes tend to inherit both the P_2 and T_2 alleles. *The male trait and the female preference thus become genetically correlated* (see Chapter 13), so that *any increase in the frequency of the male trait is accompanied by an increase in the frequency of the female preference through hitchhiking.*

Perhaps T_2 males have a slight mating advantage over T_1 males because they are both acceptable to P_1 females and preferred by the still rare P_2 females. Whether for this or another reason, suppose t_2 increases slightly. Because of the genetic correlation between the loci, an increase in t_2 is accompanied by an increase in p_2. That is, T_2 males have more progeny, and their daughters tend to inherit the P_2 allele, so P_2 also increases in frequency. As P_2 increases, T_2 males have a still greater mating advantage because they are preferred by more females; thus the association of the P_2 allele with the increasing T_2 allele can

FIGURE 15.23 A model of runaway sexual selection by female choice. (A) T_1 and T_2 represent male genotypes that differ in some trait, such as tail length. P_1 and P_2 females have different preferences for T_1 versus T_2 males. (B) The resulting pattern of mating creates a correlation in the next generation between the tail length of sons and the mate preference of daughters. Thus a genetic correlation is established in the population, in which alleles P_2 and T_2 are associated to some extent. Any change in the frequency (t_2) of the allele T_2 thus causes a corresponding change in p_2. (C) Each point in the space represents a possible population with some pair of frequencies p_2 and t_2, and the vectors show the direction of evolution. When p_2 is low, t_2 declines as a result of ecological selection, so p_2 declines through hitchhiking. When p_2 is high, sexual selection for T_2 males outweighs ecological selection, so t_2 increases, and p_2 also increases through hitchhiking. Along the solid line, allele frequencies are not changed by selection, but may change by genetic drift. (C after Pomiankowski 1988.)

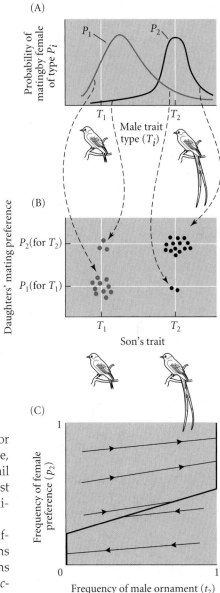

increase p_2 still further. In this way, both T_2 and P_2, hence both the exaggerated trait and the female preference for it, may increase to fixation.

If both the male trait and the female preference are polygenically inherited, the same principle holds, but new genetic variation resulting from mutation can enable continuing evolution of both characteristics, and if the genetic correlation between them is strong enough, the evolutionary process "runs away" toward more extreme preference for an ever more extreme male trait.

Because this model "automatically" generates female choice and does not include ecological selection, Richard Prum (2010) has urged that it be considered the "null model" for female choice, just as mutation and genetic drift are the null model in much of evolutionary genetics. These models are default explanations because they are simpler than models that also include ecological agents of natural selection.

As we have seen, many exaggerated sexually selected traits carry ecological costs for the males that bear them. Female choice may also carry an ecological cost; for instance, rejecting a male and searching for a more acceptable one may delay reproduction or entail other risks. Such costs may prevent the runaway process from occurring, although the cost of female choice may be offset by having attractive, reproductively successful sons (Pomiankowski et al. 1991).

If females have genetically variable responses to each of several or many male traits, different traits or combinations of traits may evolve, depending on initial genetic conditions (Pomiankowski and Iwasa 1998). Thus runaway sexual selection can follow different paths in different populations, so that *populations may diverge in mate choice and become reproductively isolated.* Sexual selection is therefore a powerful potential cause of speciation (see Chapter 18). Runaway sexual selection of this kind could explain the extraordinary variety of male ornaments among different species of hummingbirds and many other kinds of animals.

INDICATORS OF GENETIC QUALITY The major alternatives to the runaway model are GOOD GENES MODELS, or INDICATOR MODELS, in which the preferred male trait indicates high viability, which is inherited by the offspring of females who choose such males. According to these models, females should evolve to choose males with high genetic quality, so that their offspring will inherit "good genes" and thus have a superior prospect of survival and reproduction. Any male trait that is correlated with genetic quality—any INDICATOR of "good genes"—could be used by females as a guide to advantageous matings, so selection would favor a genetic propensity in females to choose mates on this basis. Alleles affecting the female preference would increase because they would be carried by offspring with high survival rates. (That is, an association between preference alleles and male trait alleles would develop, just as in the runaway model.)

Female preference should be most likely to evolve if the trait is a **condition-dependent indicator** of fitness. In this case, males develop the indicator trait to a fuller extent if they are in good physiological condition because they also carry "good genes," such as allele B. Suppose males with allele CDT_2 at a "condition-dependent trait locus" can develop a more conspicuous, more reliable indicator trait than those with allele CDT_1. Because females with a "preference allele" (P_2) are more likely to mate with males that carry both CDT_2 and B, their offspring inherit all three alleles. Thus linkage disequilibrium develops among the three loci, and both P_2 and CDT_2 increase in frequency because of their association with the advantageous allele B.

Good physiological condition is likely not to be based on a single allele ("B"), but to be polygenic instead. Locke Rowe and David Houle (1996) developed a model of *"capture" of genetic variance by condition-dependent sexually selected traits*. In this model, a male's condition, and thus the size of the display ornament he can develop, is inversely proportional to the number of deleterious mutations he carries (**Figure 15.24**). Continuing occurrence of mutations will enable long-sustained evolution of both more extreme ornaments and more extreme female preferences. Other authors have suggested that continual change in the genetic composition of parasite populations may maintain genetic variation in resistance traits in populations of their hosts, and that sexually selected male traits may be indicators of resistance to parasites (Hamilton and Zuk 1982).

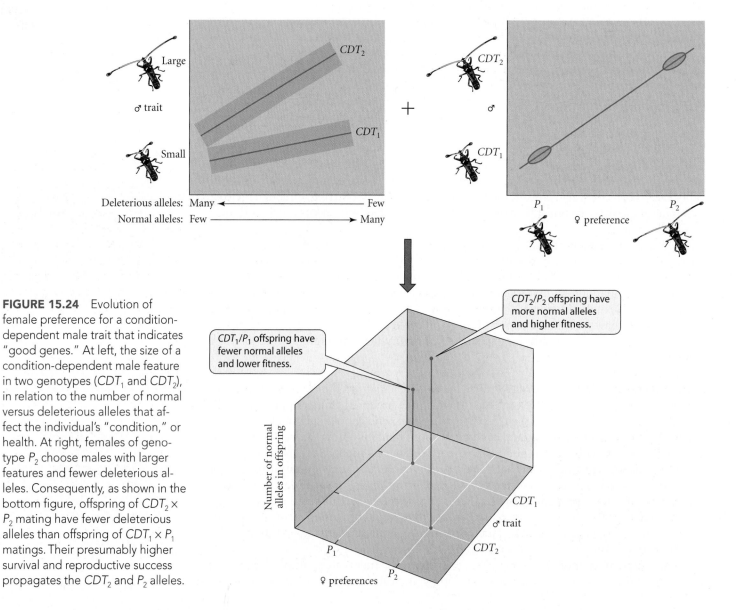

FIGURE 15.24 Evolution of female preference for a condition-dependent male trait that indicates "good genes." At left, the size of a condition-dependent male feature in two genotypes (CDT_1 and CDT_2), in relation to the number of normal versus deleterious alleles that affect the individual's "condition," or health. At right, females of genotype P_2 choose males with larger features and fewer deleterious alleles. Consequently, as shown in the bottom figure, offspring of $CDT_2 \times P_2$ mating have fewer deleterious alleles than offspring of $CDT_1 \times P_1$ matings. Their presumably higher survival and reproductive success propagates the CDT_2 and P_2 alleles.

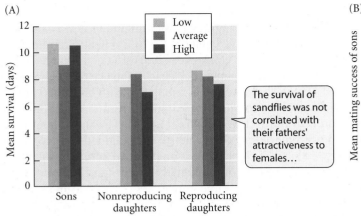

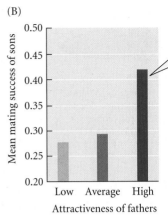

The survival of sandflies was not correlated with their fathers' attractiveness to females…

…but the sons of "attractive" males had higher mating success.

FIGURE 15.25 Evidence for Fisher's "sexy son" hypothesis. (A) Survival is shown for the sons and for both the reproducing and nonreproducing daughters of three classes of sandfly fathers that differed in their attractiveness to females. There is no indication that attractive fathers contribute "good genes" to their offspring. (B) The *sons* of highly attractive male sandflies, however, show strikingly high mating success when females choose between them and the sons of less attractive fathers. Thus the indirect benefit to a female of mating with an attractive male is that her sons will enjoy greater reproductive success. (After Jones et al. 1998.)

EVIDENCE FOR THE ROLE OF INDIRECT EFFECTS The authors of a recent synthesis of many experimental studies concluded that the Fisherian runaway process ("sexy son" hypothesis) may be the most important explanation of female choice: the male traits that females prefer are more strongly correlated with the mating success of their sons than with any other component of their offspring's fitness (Prokop et al. 2012). An example is the sand fly *Lutzomyia longipalpis*, males of which aggregate in LEKS (areas where males gather and compete for females that visit those areas in order to mate). Females typically reject several males before choosing one with whom to mate. T. M. Jones and coworkers (1998), using experimental groups of males, determined which individuals were most often, and which least often, chosen by females. Females who mated with the most "attractive" versus the least "attractive" males did not differ in survival or in the number of eggs they laid; that is, there was no evidence of a direct benefit of mating with attractive males. Nor was the father's attractiveness correlated with his offspring's survival or with his daughters' fecundity; that is, there was no evidence that attractive fathers transmitted good genes for survival or egg production (**Figure 15.25A**). However, when the investigators provided virgin females with trios of males—one the son of a highly attractive father, one the son of an "average" father, and one the son of an unattractive father—the sons of attractive fathers were by far the most successful in mating, as predicted by the Fisherian "sexy son" hypothesis (**Figure 15.25B**).

There is also evidence for the good genes hypothesis. First, the size or intensity of male display traits in some species is correlated with physiological condition or health, as the theory predicts. In the extraordinary stalk-eyed flies, for instance, the length of the eyestalks varies with the animal's physiological condition, and is more affected by nutritional or other stresses than other characteristics are (**Figure 15.26**). Second, in some species,

FIGURE 15.26 A condition-dependent male trait. (A) In stalk-eyed flies (*Cyrtodiopsis*) the sides of the head are greatly elongated, especially in males, and bear the eyes and antennae at the tips. Females mate preferentially with males with wider eye spans. (B) Eye span is affected by diet in the larval stage to a greater extent than other features such as wing length, and so is more condition-dependent. (After Cotton et al. 2004.)

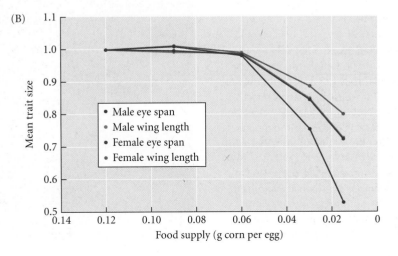

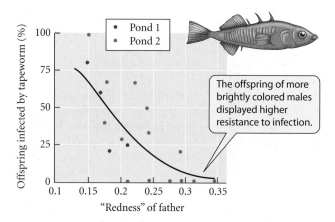

The offspring of more brightly colored males displayed higher resistance to infection.

FIGURE 15.27 Evidence for the "good genes" hypothesis of female choice. The percentage of young sticklebacks (*Gasterosteus aculeatus*) that became infected when exposed to tapeworm larvae was inversely correlated with the intensity of their fathers' red coloration. (After Barber et al. 2001.)

males that have more exaggerated ornaments or which are most preferred by females have been shown to sire healthier offspring. For instance, Barber and colleagues (2001) studied a character preferred by female three-spined sticklebacks: intense red coloration of the male's belly. They found that young fish with bright red fathers were more resistant to infection by tapeworms than their half-sibs who had dull red fathers (**Figure 15.27A**). The red coloration is based on carotenoid pigments, which are obtained from food and appear to enhance development of an effective immune system.

In one of the most comprehensive studies of the good genes hypothesis, Megan Head and coworkers (2005) confined female house crickets with males that had been scored as attractive or unattractive, based on the responses of other females. The investigators scored the lifetime survival of those females, the lifetime survival and egg production of their daughters, and the survival and mating success of their sons, then used those data to estimate the intrinsic rate of increase of each female and her descendants (a measure of overall fitness). They found that mating with attractive males had a substantial cost in terms of female survival, but that this cost was outweighed by the higher rate of increase of descendants of attractive males, whose daughters were more fecund and whose sons were more successful in attracting females. This experiment provided evidence both for direct costs of mating with attractive males and for substantial indirect benefits.

Variations on the theme of sexual selection

REVERSED AND MUTUAL SEXUAL SELECTION In cases of so-called SEX ROLE REVERSAL, in which males care for the young, as in phalaropes and sea horses (**Figure 15.28**), males can tend fewer offspring than a female can produce. For this or other reasons, such as a sex difference in survival, the operational sex ratio (the ratio of sexually active males to receptive females) in some populations may be low, and females in such species often compete for males (Clutton-Brock 2007; Kraaijeveld et al. 2007).

In many species, such as the puffins in Figure 14.2B, both sexes have ornaments or bright colors. According to the indicator model of sexual selection, mutual sexual selection is likely when both parents are socially monogamous and care for the offspring, for then both sexes profit from selecting a highly fit mate. A small seabird that fits this pattern is the crested

FIGURE 15.28 Sex role reversal. (A) Two female red phalaropes (*Phalaropus fulicarius*) fight over the smaller, duller-plumaged male on their breeding grounds in Alaskan tundra. In contrast to most birds, female phalaropes court males, which care for the eggs and young. (B) A male Australian sea horse (*Hippocampus breviceps*) giving birth. Males of this species carry and nurture developing young in their pouch, and are courted by females.

(A)

(B)

FIGURE 15.29 Crested auklets, *Aethia cristatella*. Both sexes care for the young and have similar ornaments that appear to be the object of mutual sexual selection.

auklet (**Figure 15.29**), a relative of the puffins, in which both sexes display greater sexual responsiveness to mounted specimens with longer crests (Jones and Hunter 1999).

Female display traits may also evolve by SOCIAL SELECTION, which Mary Jane West-Eberhard (1979) defined as "differential reproductive success … due to differential success in social competition, whatever the resource at stake." For males, the resource is often mates; for females, it is often territory or other resources. Sexual and social selection may act together; in the crested auklet, long-crested individuals of both sexes are dominant in aggressive same-sex encounters.

SEXUAL CONFLICT AND SEXUAL SELECTION Although the sexes must cooperate in order to produce offspring, conflict between the sexes is also pervasive (Parker 1979, 2006; Arnqvist and Rowe 2005). It will often be optimal for males to mate with previously mated females, since there is some chance that they will father some offspring, but for females to mate only once or a few times, since mating can entail injury, disease, or costs in energy or time. Moreover, a male profits if he can cause a female to produce as many of his offspring as possible, even if this reduces her subsequent ability to survive and reproduce. Because of such costs of mating with too many males, females may be selected to evolve resistance to the male displays or other features that induce females to mate (**Figure 15.30**). In this model, female preference (i.e., acceptance only of males with extreme traits) evolves not because of the benefits a male provides, but as an escape from harmful matings. As the female resistance threshold evolves, selection may favor more extreme male traits that overcome that threshold. There may ensue a prolonged "arms race" in which change in a male character is parried or neutralized by evolution of a female character, which in turn selects for change in the male character, and so on in a chain reaction (Parker 1979; Gavrilets 2000; Gavrilets et al. 2001) that has been termed CHASE-AWAY SEXUAL SELECTION (Holland and Rice 1998).

In many species, males inflict harm on their mates. Forced copulation by groups of male mallard ducks has been known to drown females, and the genitalia of many male insects have spines that injure the female reproductive tract and reduce survival. Female bedbugs suffer reduced survival and reproduction

(A)

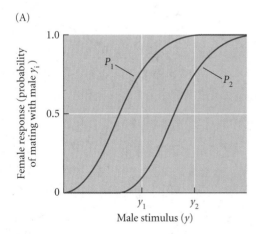

(B)

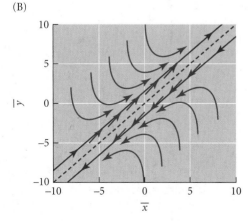

FIGURE 15.30 A model of the evolution of female mate choice caused by sexual conflict. (A) The curves P_1 and P_2 show the probability that a female will accept a male with a stimulating trait of value y. The larger the trait, the higher the likelihood of mating. Female type P_2 is more resistant to the male stimulus and requires a larger value of y to mate: she is much less likely than P_1 to accept male phenotype y_1, but is fairly likely to mate with y_2. The position of the P curve is a measure of female resistance. If mating is costly or harmful, P_2 will have higher fitness than P_1, because she accepts fewer males. Hence the preference curve should evolve toward the right, which in turn selects for males with higher y value, resulting in a coevolutionary "arms race." (B) The resulting coevolution is portrayed by the movement (shown by an arrow) of a point that represents the average male stimulus ($\bar{y}$) and the average female resistance ($\bar{x}$) at any time. The vectors show that if females suffer a strong cost of multiple mating, there is runaway coevolution toward either higher or lower female resistance and male stimulus. (After Gavrilets 2000; Gavrilets et al. 2001.)

FIGURE 15.31 Experimental evidence of genetic conflict between the sexes. (A) Measures of fitness of males from two experimental populations in which only males could evolve, relative to that of males from control populations. The measures are net fitness (number of sons per male), the rate at which the males re-mated with previously mated females, and sperm defense (the ability of a male's sperm to fertilize eggs of a female that was given the opportunity to mate with a second male). (B) When males from these experimental populations were mated with females from another stock, they caused higher female mortality than did males from four control populations. (After Rice 1996.)

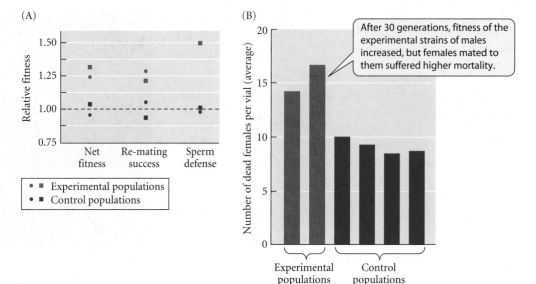

from repeated "traumatic insemination:" the male pierces his mate's abdominal wall with his genital structure, rather than inserting it into her genital opening (Stutt and Siva-Jothy 2001). In most cases, the harm males inflict is probably a side effect of characteristics selected to enhance success in mating and fertilization, rather than a direct advantage to the male (Parker 2006).

The seminal fluid of *Drosophila melanogaster* contains toxic proteins that reduce female survival, and males whose sperm best resist displacement by a second male's sperm cause the greatest reduction of their mates' life span (Civetta and Clark 2000). In a series of clever experiments, William Rice and Brett Holland have shown that male and female *Drosophila* are locked in a coevolutionary "arms race" (Rice 1992, 1996; Holland and Rice 1999). In one experiment, they constructed experimental populations of flies in which males could evolve but females could not. (In each generation, the males were mated with females from a separate, non-evolving stock with chromosome rearrangements that prevented recombination with the males' chromosomes, and only sons were saved for the next generation.) After 30 generations, the fitness of these males had increased, compared with control populations, but females that mated with these males suffered greater mortality, probably because of enhanced semen toxicity (**Figure 15.31**). This experiment suggests that males and females are continually evolving, but in balance, so that we cannot see the change unless evolution in one sex is prevented.

Sexually antagonistic coevolution has been demonstrated in other species as well, such as water striders, in which males force females to copulate. The evolution of features that enable males to grip females has been accompanied by the evolution of a spine and other features that enable females to break the males' grip (**Figure 15.32**). In the water strider *Gerris incognitus*, male mating success is enhanced by the match of the male's abdomen to the female's deterrent spine, and these features vary in parallel among geographic populations (Perry and Rowe 2012).

Sexual conflict probably accounts for many aspects of gamete evolution (Parker 1979; Birkhead et al. 2009). Sperm compete to fertilize eggs, so selection on sperm always favors a greater ability to penetrate eggs rapidly. But selection on eggs should favor features that slow sperm entry, or else POLYSPERMY (entry by multiple sperm) may result. Polyspermy disrupts development, and eggs have elaborate mechanisms to prevent it. The conflict between the interests of egg and sperm may result in continual antagonistic coevolution of greater penetration ability by sperm and countermeasures by eggs. Such coevolution might be the cause of the extraordinarily rapid evolution of the amino acid sequence of the sperm lysin protein of abalones (*Haliotis*), a protein released by sperm that bores a hole through

(A)

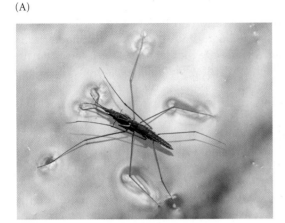

(B) *G. incognitus* ♂

(C) *G. incognitus* ♀

(D) *G. thoracicus* ♂

(E) *G. thoracicus* ♀

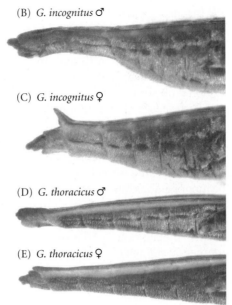

FIGURE 15.32 Among water striders (genus *Gerris*), males forcibly copulate with females. (A) A mated pair of *G. lacustris*. (B, C) In *G. incognitus*, the abdomen of the male has exaggerated grasping adaptations (extended, flattened genital segments), and that of the female has features (prolonged, erect abdominal spines; curved abdomen tip) that obstruct the male's grip during premating struggles. (D, E) The closely related species *G. thoracicus* shows the ancestral condition, in which these adaptations have not evolved. (B–E from Arnqvist and Rowe 2002.)

the vitelline envelope surrounding the egg. Nonsynonymous differences between the lysin genes of different species of abalones have evolved much faster than synonymous differences, a sure sign of natural selection. Gamete surface proteins and other reproductive proteins of animals have generally evolved at extraordinarily high rates (Swanson and Vacquier 2002; Vacquier and Swanson 2011). Similarly, sperm morphology evolves rapidly, and varies among species more than that of any other cell type, probably because of both competition among sperm and antagonistic coevolution between sperm and the female reproductive tract. Among species of *Drosophila* and many other taxa, sperm length is correlated with the length of the female structure in which sperm is stored after mating (**Figure 15.33**). The diversity and evolution of interactions between male and female reproductive systems is one of the most intriguing and least understood subjects in evolutionary biology.

ADAPTIVE EFFECTS OF SEXUAL SELECTION For a long time, biologists have speculated that sexual selection might decrease the mean fitness of populations. For instance, if male characteristics such as the peacock's train are ecologically very disadvantageous, but nevertheless evolve by a runaway process, they could theoretically cause the extinction of the species (Lande 1980). Likewise, sexually antagonistic selection may lower mean fitness if alleles advantageous to male mating are harmful to females. But there also exists

FIGURE 15.33 Joint evolution of sperm length and female reproductive tract among species of *Drosophila*. (A) The female reproductive tract of *Drosophila pseudoobscura* (left) and *D. bifurca* (right), showing the paired ovaries (o), vagina (v), and the seminal receptacle (SR) in which sperm are stored after mating. In *D. bifurca*, the SR is 82 mm long, corresponding to the male's extraordinary 58-mm-long sperm. (B) This image, compiled from scanning electron micrographs, illustrates that male *Drosophila bifurca* produce about six extraordinarily long sperm for each egg that females produce—the lowest sperm:egg ratio known. The sperm tail forms a highly convoluted ball in the seminal vesicle. (C) The correlation between length of sperm and SR among 46 species of *Drosophila*. (A from Patterson, 1943; B courtesy of Adam Bjork and Romano Dallai; C after Pitnick et al. 1999.)

(A)

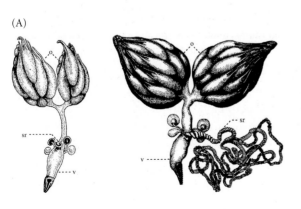

(B)

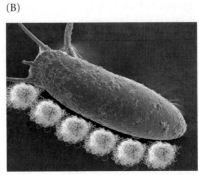

(C)

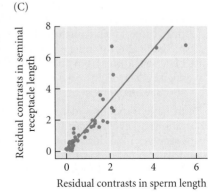

the intriguing possibility that sexual selection might help to purge deleterious mutations from populations. This possibility is a direct implication of the "capture of genetic variance" model, in which females evolve to choose males with condition-dependent traits that indicate good health and physiological condition: that is, males with fewer harmful alleles in their genetic makeup (Whitlock and Agrawal 2009). This model has been explored by experiments in which fitness has been compared between populations of *Drosophila melanogaster* maintained for generations under enforced monogamy (each female is given a single male) and other populations in which females can choose among males. In some such studies, females from the polygamous lines (with sexual selection allowed) had higher fitness, as the theory would predict. In another study, small laboratory populations of a mite became extinct, due to inbreeding depression, more often if they were forced to be monogamous and were thus deprived of the genetic benefits of sexual selection (Jarzebowska and Radwan 2010). Similarly, the opportunity for sexual selection enhanced the rate at which experimental populations of a seed beetle adapted to a novel host plant, supporting the hypothesis that sexual selection may sometimes increase adaptation to environmental change (Fricke and Arnqvist 2007).

Go to the

EVOLUTION
Companion Website at
sites.sinauer.com/evolution3e

for quizzes, data analysis and simulation exercises, and other study aids.

Summary

1. Alleles that increase mutation rates are generally selected against because they are associated with the deleterious mutations they cause. Therefore we would expect mutation rates to evolve to the minimal achievable level, even if this reduces genetic variation and increases the possibility of a species' extinction.

2. Asexual populations have a high extinction rate, so sex has a group-level advantage in the long term. But this advantage is unlikely to offset the short-term advantage of asexual reproduction.

3. In a constant environment, alleles that decrease the recombination rate are advantageous because they lower the proportion of offspring with unfit recombinant genotypes. In addition, asexual reproduction has approximately a twofold advantage over sexual reproduction because only half of the offspring of sexuals (i.e., the females) contribute to population growth, whereas all of the (all-female) offspring of asexuals do so. Therefore the prevalence of recombination and sex requires explanation.

4. Among the several hypotheses for the short-term advantage of sex are the following: (a) the rate of adaptation, by fixation of combinations of advantageous mutations, may be higher in sexual than in asexual populations, if the populations are large; (b) recombination enables the mean of a polygenic character to evolve to new, changing optima in a fluctuating environment; (c) in asexual populations, fitness declines because genotypes with few deleterious mutations, if lost by genetic drift, cannot be reconstituted, as they are in populations with recombination (Muller's ratchet); (d) recombination separates beneficial alleles from linked deleterious alleles, enabling beneficial alleles to increase in frequency. This last advantage of sex may be the most important.

5. In large, randomly mating populations, a 1:1 individual sex ratio is an evolutionarily stable strategy, because if the population sex ratio deviates from 1:1, a genotype that produces a greater proportion of the minority sex has higher fitness. If, however, populations are characteristically subdivided into small local groups whose offspring then colonize patches of habitat anew, a female-biased sex ratio can evolve because female-biased groups contribute a greater proportion of offspring to the population as a whole.

6. Outcrossing can be advantageous because it prevents inbreeding depression in an individual's progeny. Conversely, self-fertilization may evolve if it economizes the energy or resources required to develop reproductive structures, if an allele for selfing becomes associated with advantageous homozygous genotypes, or if selfing ensures reproduction despite low population density or scarcity of pollinators.

7. Differences between the sexes in the size and number of gametes give rise to conflicts of reproductive interest and to sexual selection, in which individuals of one sex compete for mates. The several forms of sexual selection include direct competition between males, or between their sperm, and female choice among male phenotypes.

8. Females may prefer certain male phenotypes because of sensory bias, direct contributions of the male to the fitness of the female or her offspring, or indirect contributions to female or offspring fitness. Indirect benefits may include fathering offspring that are genetically superior with respect to mating success (the runaway sexual selection model) or with respect to components of viability (good genes models). For example, a male display feature may be a condition-dependent indicator of the individual's physiological health, which is depressed if he carries many deleterious mutations. In species in which only males care for offspring, sexual selection may cause the evolution of display characters in females. In some species with biparental care, both sexes may select mates with characters that indicate that they can provide superior care of offspring, resulting in sexually monomorphic ornaments and displays.

9. A conflict of evolutionary interest between the sexes is common, and mating can consequently be harmful, especially for females. This can result in antagonistic coevolution, in which females evolve to resist mating and males evolve stimuli better able to overcome female resistance.

Terms and Concepts

apomixis	cost of sex	runaway sexual selection	sexual reproduction (sex)
clone	dioecious	sensory bias	sexual selection
condition-dependent indicator	hermaphroditic	sex ratio	sperm competition
	parthenogenesis		vegetative propagation

Suggestions for Further Reading

The evolution of sex and recombination is reviewed by S. P. Otto, "The evolutionary enigma of sex," *American Naturalist* 174 (suppl.): S1–S14 (2009).

Sexual Selection, by M. Andersson (Princeton University Press, Princeton, NJ, 1994) is a dated but comprehensive review of sexual selection that is a good starting point in this area. U. Candolin and J. Heuschele, "Is sexual selection beneficial during adaptation to climate change?," *Trends in Ecology and Evolution* 23: 446–452 (2008), reviews the effects of sexual selection on adaptation.

Sexual Conflict, by G. Arnqvist and L. Rowe (Princeton University Press, Princeton, NJ, 2005) is a comprehensive analy-

sis of that subject, which is briefly but comprehensively discussed by G. A. Parker, "Sexual conflict over mating and fertilization: An overview," *Philosophical Transactions of the Royal Society* B 361: 235–259 (2006).

An informative and provocative book for a general audience by a prominent evolutionary biologist is J. Roughgarden's *Evolution's Rainbow: Diversity, Gender, and Sexuality in Nature and People* (University of California Press, Berkeley, 2004). A controversial subject is the role of sexual selection in the evolution of human behavior, the subject of *The Mating Mind: How Sexual Choice Shaped the Evolution of Human Nature*, by G. F. Miller (Doubleday, New York, 2000).

Problems and Discussion Topics

1. Populations of some species of fish, insects, and crustaceans consist of both sexually and obligately asexually reproducing individuals. Would you expect such populations to become entirely asexual or sexual? What factors might maintain both reproductive modes? How might studies shed light on the factors that maintain sexual reproduction?

2. Some mites, scale insects, and gall midges display "paternal genome loss" (or *pseudoarrhenotoky*, a fine word for parlor games). Males develop from diploid (fertilized) eggs, but the chromosomes inherited from the father become heterochromatinized and nonfunctional early in development, so that males are functionally haploid. How might this peculiar genetic system have evolved?

3. Many parthenogenetic "species" of plants and animals are known to be genetically highly variable. What processes might account for this variation?

4. The text says that "the recency of most parthenogenetic lineages suggests that sex reduces the risk of extinction." Explain why.

5. The number of chromosome pairs in the genome ranges from one (in an ant species) to several hundred (in some butterflies and ferns). In a single genus of butterflies (*Lysandra*), the haploid number varies from 24 to about 220 (White 1978). Is it likely that chromosome number evolves by natural selection? Is there evidence for or against this hypothesis?

6. Would you expect sexual selection to increase or decrease adaptation of a population to its environment? Might this depend on the form of sexual selection (e.g., male conflict, female mate choice)? Do the several hypotheses for the evolution of female choice (e.g., runaway selection versus good

genes hypotheses) differ in their implications for adaptation to the environment?

7. In many socially monogamous species of parrots and other birds, both sexes are brilliantly colored or highly ornamented. Is sexual selection likely to be responsible for the coloration and ornaments of both sexes? How can there be sexual selection in pair-bonding species with a 1:1 sex ratio, since every individual presumably obtains a mate? Which of the models of sexual selection described in this chapter might account for these species' characteristics?

8. A possible consequence of "chase-away" selection arising from sexual conflict is that males may possess features that no longer are highly stimulatory to females, but which are nevertheless necessary for acceptance by females. That is, the feature may be necessary, but not sufficient, for female acceptance. What evidence might test this hypothesis? Have you already encountered a possible example in this book?

9. There is an enormous and very controversial literature on whether behavioral features in humans have a genetic basis and have evolved for reasons postulated in evolutionary models, or whether they are culturally formed. Analyze reasons for and against the proposition that men are more aggressive than women because of sexual selection for competitiveness. (See, for example, Daly and Wilson 1983 versus Kitcher 1985.)

10. In the same vein as problem 9, discuss aspects of human behavior, physiology, and morphology that might be explained by sexual selection. (A book by Miller [2000] is one of many treatments of this topic.) What are the alternative hypotheses, and how might we tell which is (or are) correct?

Conflict and Cooperation

Parental care The costs versus benefits in fitness determine whether both sexes, one sex, or neither cares for offspring. Parental care in frogs, when present, is the male's task. This male poison dart frog (*Epipedobates trivittatus*) is transporting recently hatched tadpoles to water.

Darwin first conceived of natural selection when he read the economist Thomas Malthus's theory that human population growth must inevitably cause competition for food and other resources. Malthus's *Essay on Population* inspired Darwin's realization that in all species, only a fraction of those born survive to reproduce, and that the survivors must often be those best equipped to compete for limiting resources. Thus conflict has been inherent in the idea of natural selection from the start. Darwin soon realized, however, that not all of natural selection stems from overt struggle among members of a species. We also observe cooperation in nature: among the cells and organs of an individual organism, between mates in the act of sexual reproduction, among individuals in many social species of animals, and even between some mutualistic pairs of species. Yet conflict is so ubiquitous, even inevitable, that the question of how evolutionary theory can account for cooperation has occupied evolutionary biologists ever since Darwin.

Almost all instances of cooperation include at least the possibility of conflict as well. Conflict and cooperation pervade social interactions among individuals, including interactions between mates and between parents and their offspring. They are found not only in animals, but in plants and microbes as well; they are the basis of evolutionary battles even between different genes within individual organisms. An understanding of conflict and cooperation is fundamental for understanding the evolution of a vast variety of biological phenomena, ranging from mating displays to sterile social insects, from cannibalism to some peculiar features of human pregnancy.

Modelling Conflict

Many of the world's fisheries, such as the cod fishery in the North Atlantic, have been devastated by overfishing. It profits each boat to harvest as many fish as possible, but the outcome of such individually rational behavior is the destruction of the resource on which all participants depend. This outcome has been called the "tragedy of the commons," in reference to communal grazing lands that are similarly destroyed by overgrazing. Similar problems arise in the natural world, and they pose the question of how restraint can evolve, especially when a "cheater" genotype that maximizes its personal gain has higher fitness, at least in the short run (as mentioned in Chapter 11). Explaining how cooperation can evolve, when selfish genotypes would be expected to have higher fitness, is an extensive field of both theoretical and empirical inquiry.

Conflict and selfishness are ubiquitous. For example, more than 150 species of birds are known to practice "conspecific brood parasitism," wherein a female lays eggs in other females' nests, in addition to those she herself incubates (Lyon and Eadie 2008). In the American coot (*Fulica americana*), this behavior greatly increases the reproductive output of parasitic females compared with that of nonparasitic females (**Figure 16.1A**). The host female's fitness is reduced because her chicks compete with the parasitic female's chicks for limited food. But just as the reproductive success of parasitic females causes an increase in the frequency of a mutation that disposes females toward this behavior, selection favors mutations that enable parasitized females to protect their offspring. As one might expect, coots have evolved the capacity to detect and evict foreign eggs from their nests, based on variation in egg color (**Figure 16.1B,C**).

The evolution of characteristics that enhance fitness in competitive and cooperative interactions can be understood by means of population genetic models (as described in

FIGURE 16.1 Intraspecific conflict: brood parasitism in the American coot (*Fulica americana*). (A) The total fecundity of parasitic females (number of eggs in own nest plus eggs laid in other females' nests) is much greater than that of nonparasitic females. The notations along the *x*-axis indicate data from four wetlands (R, J, S, and K) and different years. (B) Eggs from a single nest show variation in color, which can be ranked from 1 (lightest) through 7 (darkest); subtracting the lower number from the higher gives the "color rank difference" between eggs. (C) Brood-parasitic eggs rejected by host females differed more from the host's own eggs than those that were accepted by the host. (A after Lyon 1993; C after Lyon 2003.)

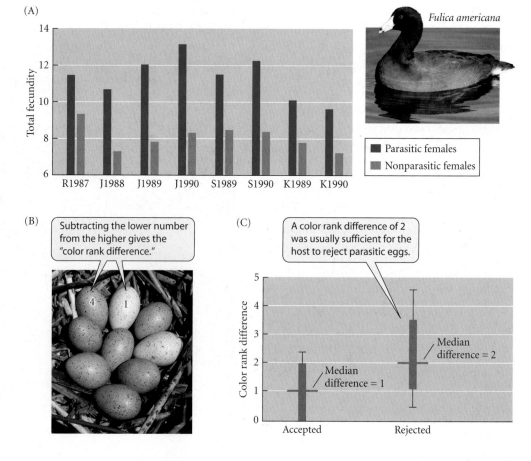

Chapters 12 and 13) or by phenotypic models, in which the optimal phenotype, given specified constraints and assumptions about costs and benefits that are measured in terms of fitness, is discovered. The evolution of fitness of contending genotypes in social interactions is often modeled using approaches based on game theory to calculate the EVOLUTIONARILY STABLE STRATEGY, or ESS: a phenotype that cannot be replaced by any other phenotype under the specified conditions (see Chapter 14). For example, ESS theory can be applied to conflict among animals, such as males competing for a territory or a mate.

In many species, aggressive encounters consist largely of displays, which seldom escalate into physical conflicts that could cause injury or death. The question is whether it is individually advantageous not to escalate conflicts. John Maynard Smith (1982), one of the first to apply game theory to evolution, postulated two strategies: a "Hawk" escalates the conflict either until it is injured or until its opponent retreats, whereas a "Dove" retreats as soon as its opponent escalates the conflict. Is it better to be a Hawk, more likely to win the prize, or a Dove, more likely to escape injury and live to compete another day? The Dove strategy is never an ESS, because a Hawk genotype can always increase in a population of Doves. The Hawk strategy is an ESS if the fitness gained in a successful contest (V), even with another Hawk, is greater than the cost of injury (C). If, however, $C > V$, ESS analysis shows that a pure Hawk phenotype can be replaced by a "mixed strategy," such that an individual will adopt Hawk behavior with probability $p = V/C$. In this case, the model predicts that the optimal behavior is variable and is contingent on conditions such as the value of the resource.

We can also postulate an Assessor strategy, in which the individual escalates the conflict if it judges the opponent to be smaller or weaker, but retreats if it judges the opponent to be larger or stronger. In most theoretical cases, the Assessor strategy is an ESS. In accord with these models, many animals react aggressively or not depending on their opponent's size or correlated features. In some species of *Anolis* lizards, for example, the size of a territorial male's dewlap, a brightly colored throat ornament that is displayed in aggressive encounters, is correlated with the force of the male's bite (**Figure 16.2**; Vanhooydonck et al. 2005).

Characters such as the lizard's dewlap are HONEST SIGNALS of the individual's fighting ability or resource-holding potential. Theoretically, deceptive signals, indicating greater fighting ability than the individual actually has, should be unstable over evolutionary time because selection would favor competitors that ignore the signals, which, having then lost their utility, would be lost in subsequent evolution. Thus many existing signals of resource-holding potential are probably honest assessment signals (Johnstone and Norris 1993). However, dishonest signals do exist (Bradbury and Vehrencamp 1998). The greatly enlarged claw of a male fiddler crab, for example, is used for display and fighting with other males. In *Uca annulipes*, males that lose their large claw regenerate one that is almost the same size, but has less muscle and is much weaker. Although such males lose physical fights with intact males, they

FIGURE 16.2 The threat displays of male *Anolis* lizards use an honest signal. (A) A male *Anolis grahami* displays a brightly colored dewlap. (B) Dewlap size is correlated with the force of the male's bite. Because both dewlap size and bite force are correlated with body size, the points in this graph have been statistically adjusted for variation in body size; some values appear negative because of this adjustment. (A courtesy of J. Losos; B after Vanhooydonck et al. 2005.)

(A)

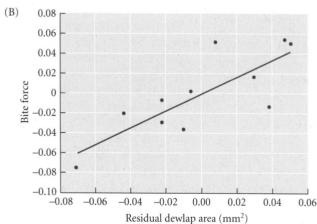

(B)

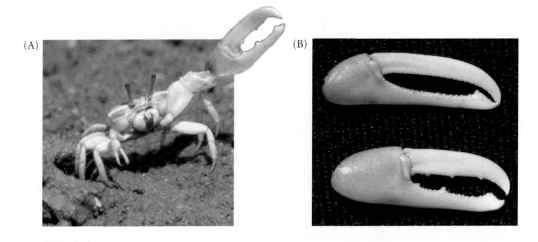

FIGURE 16.3 A dishonest signal. (A) A male fiddler crab (*Uca* sp.) signaling with its greatly enlarged claw. (B) A slender regenerated claw (top) and a robust original claw of *Uca annulipes*. The regenerated claw is an effective bluff, even though it does not have the strength needed to win a fight with an intact male. (B from Backwell et al. 2000; photos courtesy of Patricia Backwell.)

can effectively bluff and deter potential opponents, and they are apparently just as successful as intact males in attracting females (**Figure 16.3**; Backwell et al. 2000).

Social Interactions and Cooperation

Helping activities are of two kinds: cooperation and altruism (Lehmann and Keller 2006). **Cooperation** is activity that provides a benefit to other individuals and to the actor as well; **altruism** is activity that enhances the fitness of other individuals but lowers the fitness of the actor. DEFECTORS or CHEATERS receive a fitness benefit, but do not provide any.

In this chapter, we will often be concerned with entities at one level that interact to form higher-level entities. For instance, individual cells may form a multicellular organism; female and male organisms may form mated pairs, at least briefly; individual organisms may be organized into flocks or colonies. Both cooperation and conflict may be, and usually are, inherent in such interactions. The reduction of conflict sometimes results in the emergence of higher-level entities with distinct identity, such as multicellular organisms (Maynard Smith and Szathmáry 1995; Michod 1999).

Individuals that engage in social interactions usually form groups, which may be short-lived or long-lasting. There has been extensive debate among biologists about whether fitness at the level of the individual organism or at the level of the group best explains the evolution of cooperation and altruism (see Chapter 11). Most evolutionary biologists seek explanations of the evolution of cooperation and conflict based on selection at the level of the individual organism or the gene. This approach has been very fruitful and has deepened our understanding of many otherwise puzzling phenomena.

Even very "simple" organisms can engage in social interactions that illustrate the conflict between individual and group benefit. The social amoeba *Dictyostelium discoideum* and the myxobacterium *Myxococcus xanthus*, although distantly related, have similar, very odd life histories (Dao et al. 2000). When food is scarce, individual cells in the soil aggregate and form a single moving organism (a "slug"), which settles and forms an erect stalk with a cap within which spores develop. The spore-forming cells reproduce, but the stalk-forming cells die without reproducing, apparently sacrificing themselves for the good of the colony. In both species, "cheater" mutations have been described that preferentially make spores (**Figure 16.4A**). In laboratory culture, the frequency of a "cheater" allele increased over the course of several life cycles (**Figure 16.4B**). However, cheaters have not taken over natural populations, for several reasons (Strassmann and Queller 2011). For example, "slugs" with a high proportion of cheater cells produce fewer spores. Another reason appears on p. 435.

Many models of the evolution of cooperation and conflict have been developed and elaborated. Most of them fit into several major categories, as described by Sachs and colleagues (2004), Nowak (2006), and Lehmann and Keller (2006). An important distinction is whether or not interacting individuals share greater than random genetic identity. We first describe interactions among unrelated individuals.

(A)

Wild-type cells aggregate to pre-vegetative (stalk) region of the slug.

chtA mutant cells migrate to prepro-ductive (spores) region of the slug.

(B)

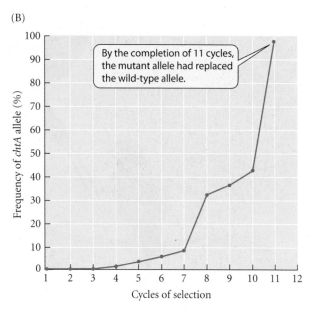

By the completion of 11 cycles, the mutant allele had replaced the wild-type allele.

FIGURE 16.4 Effects of the selfish mutation of the gene *chtA* in the slime mold *Dictyostelium*. (A) In a "slug" made up of a mixture of mutant cells and wild-type cells (which appear darker because of a genetic label), wild-type cells moved to the upper cap of the slug, a region that becomes the stalk of the fruiting body rather than the reproductive spores. (B) Over the course of 11 growth and development cycles, the frequency of the selfish mutant *chtA* allele increased in laboratory culture. (A courtesy of H. L. Ennis; B after Dao et al. 2000.)

Cooperation among unrelated individuals

DIRECT BENEFITS OF COOPERATION   Individuals may form associations or groups entirely out of self-interest. An individual that joins a flock or herd, for instance, lowers its risk of predation simply by finding safety in numbers. It is often advantageous to each individual to be as close as possible to the center of the group, thus using other group members as shields against approaching predators. The effect of this behavior is to increase the compactness and cohesion of the group (Hamilton 1971). Other active interactions among group members provide more direct, immediate benefits. Predators that hunt in groups, such as wolves, share prey that a single individual would be unlikely to capture (Clutton-Brock 2009).

A tropical American cuckoo, the greater ani, illustrates the direct benefits that can flow from cooperating. As many as four breeding pairs of unrelated birds build a communal nest, and all contribute to incubating eggs and feeding the offspring without discriminating among them. Each female, before laying her first egg, evicts the eggs already in the nest, but she no longer does so after she has laid an egg. Earlier-laying females therefore suffer more egg loss, and lay more eggs to compensate, than later-laying females (**Figure 16.5A**). Christina Riehl (2011) discovered that nests attended by three pairs of anis were less likely to suffer egg predation by mammals and snakes than nests with two pairs. (She saw only two solitary pairs, neither of which reproduced successfully.) By marking

(A)

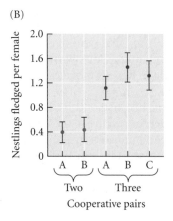

(B)

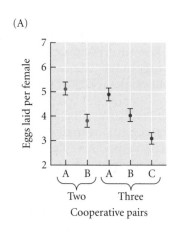

FIGURE 16.5 Reproductive success of female greater anis (*Crotophaga major*) in nests with two versus three cooperative pairs. A, B, and C indicate first, second, and third females in the egg-laying sequence. (A) Earlier-laying females lay more eggs, to compensate for those removed by later-laying females. (B) The number of successfully fledged offspring is about the same for all the females within each group, but is greater in three-pair than in two-pair nests. (After Riehl 2011.)

FIGURE 16.6 Two, or occasionally more, male long-tailed manakins (*Chiroxiphia linearis*) perform a joint leapfrogging courtship display to females. The dominant male obtains all, or almost all, copulations. (From Alcock 2013.)

individuals and tracing parent-offspring relationships using variable microsatellite markers, Riehl found that the average number of fledging offspring was greater for each female in three-pair nests than in two-pair nests (**Figure 16.5B**), showing that the benefit of cooperating in a larger group exceeds the cost.

In some cases, cooperating seems to offer little advantage, but is nevertheless more likely to result in reproductive success than the solitary, individualistic alternative. For example, a subordinate male long-tailed manakin (**Figure 16.6**) forms a long-lasting association with an unrelated dominant (alpha) male. The team performs a coordinated "leapfrogging" courtship display. Females strongly prefer male teams with highly coordinated displays, and they almost always mate with the alpha male (McDonald and Potts 1994). Subordinate males rarely obtain copulations, but they succeed the alpha male when he dies—although they may have to wait as long as 13 years. The benefit of joining a dominant male is delayed and uncertain, but a subordinate male must do so if he is to have any chance at all of reproductive success. Similarly, among unrelated female paper wasps (*Polistes dominulus*) that build a nest together, one aggressively achieves dominance over the others and prevents them from reproducing, but a subordinate female takes over the nest and reproduces if the dominant female dies (Queller et al. 2000).

RECIPROCITY Robert Trivers (1971) first suggested that cooperation could evolve if an individual X profits from benefiting another individual Y, as long as Y is highly likely to provide reciprocal aid in the future. This kind of cooperation, known as **reciprocity**, can evolve only if there are repeated interactions between individuals, individuals recognize and remember each other, and the benefit is great enough to outweigh the cost of helping (Lehmann and Keller 2006).

Reciprocity is a complex topic because of the obvious possibility that Y will defect and X will have paid the cost of helping without recompense. The conditions under which reciprocity can evolve have been analyzed by game theory, especially a game called the "prisoner's dilemma" (Axelrod and Hamilton 1981). This game is named for the situation in which two imprisoned suspects can cooperate by agreeing not to give evidence against each other (cooperation that would benefit both), but each is rewarded by the police if he or she selfishly acts as an informant. The simplest winning strategy is "tit for tat," in which

(A)

(B)

FIGURE 16.7 Mammal species that may display reciprocity between individuals. (A) Vampire bats (*Desmodus rotundus*) form roosting groups in which members that have fed successfully sometimes feed regurgitated blood to other members of the group. (B) Among primates, such as these chacma baboons (*Papio hamadryas ursinus*), social alliances between individuals are reinforced by activities such as grooming. (B photo by Anne Engh, courtesy of Dorothy Cheney.)

each "player" starts with cooperation and then does whatever the other has done in the previous round. However, this strategy does not allow for accidental mistakes. Among various alternatives, a perhaps more stable strategy that maintains cooperation is one in which each individual repeats its previous move whenever it is doing well, and changes otherwise (Nowak 2006). (Bear in mind that the term "strategy" does not imply conscious reasoning or planning; a strategy is simply a phenotypic character, such as a particular behavior, of a certain genotype.)

Cooperation of this kind, based on "partner choice," appears to be uncommon, perhaps because it requires rather complex information processing to distinguish different individuals and remember their past behavior (Hammerstein 2003; Clutton-Brock 2009). Even in apparent cases of reciprocity, it is difficult to rule out alternative explanations (Clutton-Brock 2009). For example, vampire bats, which feed on mammalian blood, form roosting groups in which members that have fed successfully sometimes feed regurgitated blood to other group members, which reciprocate at other times (Wilkinson 1988; **Figure 16.7A**). However, many of the interacting individuals are related, so kin selection (described below) might play a role in this interaction. Reciprocal grooming (**Figure 16.7B**), a major feature of many primate societies, may well constitute real reciprocity (Clutton-Brock 2009). However, even though chimpanzees are capable of delaying gratification, detecting inequities, and choosing previous collaborators over non-collaborators, experiments provide only weak evidence that they are capable of true reciprocity. Human children, on the other hand, understand and practice reciprocity at an early age. Marc Hauser and colleagues (2009) suggest that only humans have evolved the ability to integrate the cognitive functions needed for reciprocity.

The theory of reciprocity is part of a more general TRANSACTIONAL MODEL OF REPRODUCTIVE SKEW, first developed by Sandra Vehrencamp (1983). The basic idea is that dominant individuals gain from the assistance of subordinate helpers and "pay" those helpers by allowing them to reproduce just a little more than they could if they left the group and reproduced on their own (Keller and Reeve 1994, 1999). For example, a dominant dwarf mongoose (*Helogale parvula*), a weasel-like African carnivore, suppresses reproduction by older subordinates less than it does reproduction by younger subordinates. As subordinates grow older, they are better able to disperse and breed in another group, so the model predicts that dominants should offer them a greater share of the group's reproductive output as an incentive to stay (Creel and Waser 1991).

PARTNER FIDELITY FEEDBACK Reciprocity can be favored by PARTNER FIDELITY FEEDBACK (Sachs et al. 2004), in which the association between individuals is so long-lasting that the benefits each partner provides to the other feed back to the individual's own benefit. In *The Defiant Ones* (1958), Tony Curtis and Sidney Poitier portray two escaped convicts who,

FIGURE 16.8 A model of the evolution of nitrogen (N_2) fixation in legume-associated rhizobia. The blue trace shows the rate of N_2 fixation that would be evolutionarily stable if the average coefficient of relationship among rhizobia coinfecting a plant were r. The red trace shows the ESS level of N_2 fixation when plant sanctions are modeled (i.e., when the plant supplies carbon to a nodule in proportion to the rate of N_2 fixation by the rhizobia in that nodule). (After West et al. 2002.)

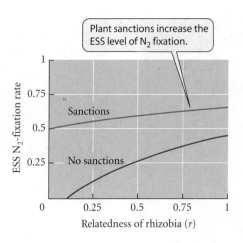

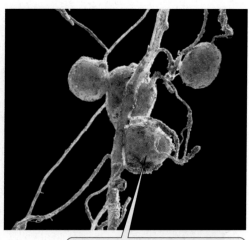

Rhizobia invade the roots of legumes and induce formation of an enveloping nodule, within which they fix N_2.

because they are chained to each other, must cooperate even though they dislike each other. When the fitness of each member of a group depends on the fitness of the other members—of the group as a whole—cooperation is clearly in every individual's interest. This principle is important in many biological contexts; perhaps the simplest examples are cooperation among the genes in a cell and among the cells in an organism. If the cell dies, so do all the included genes; if the organism dies, so do its cells. Selection at the higher level—cell or organism—thus eliminates outlaw genes or renegade cells that selfishly diminish the survival of the group. It also favors mechanisms that suppress or destroy such outlaws when they arise. We will see clear examples of this principle later in this chapter.

ENFORCING COOPERATION   In many situations, cooperation is enhanced if one partner in the interaction punishes selfish noncooperators: punishment alters the ratio of benefit to cost (Frank 2003). The punishing partner often imposes "sanctions," terminating the relationship by withholding benefits from the noncooperator. For example, Stuart West and collaborators (2002) have modeled the interaction between legumes (such as beans) and the mutualistic bacteria—called rhizobia—that invade legume roots, induce formation of an enveloping nodule, and fix nitrogen (N_2), which benefits the plant. The rhizobia obtain photosynthate (carbon) from the plant. This interaction is potentially unstable because a rhizobial genotype that cheats, and receives carbon without expending energy on N_2 fixation, would have an advantage. West and colleagues showed that the advantage of N_2 fixation depends partly on kin selection (described in the next section), based on a high degree of relationship (r) among bacteria. But the level of N_2 fixation that evolves is greater if the plant preferentially supplies carbon to nodules that fix more N_2 or withholds it from nodules that fix less (**Figure 16.8**). Experiments have shown that legumes do impose "sanctions" against poorly performing nodules (Kiers et al. 2003).

The evolution of altruism by means of shared genes

If we think of selection at the level of the gene, we will recognize that an allele replaces another allele in a population if it leaves more copies of itself in successive generations by means of whatever effect it may have. Commonly, the effect is on the individual organism that carries the allele, but an allele may also leave more copies of itself by enhancing the fitness of other individual organisms that carry copies of that same allele. For instance, an allele for maternal care of offspring increases because it increases the chance of survival of copies of the mother's allele that are carried by her offspring. In a pathbreaking paper, William D. Hamilton (1964) introduced the concept of the **inclusive fitness** of an allele:

its effect on both the fitness of the individual bearing it (**direct fitness**) and the fitness of other individuals that carry copies of the same allele (**indirect fitness**). An individual organism, likewise, has inclusive fitness, with both direct and indirect components. Inclusive fitness theory has been important in understanding not only cooperation, but also parent-offspring conflict, spite, sex ratios, dispersal, cannibalism, genomic imprinting, and many other phenomena. The great majority of researchers in the field agree that inclusive fitness theory is the leading explanation for the evolution of social behavior (Lehmann and Keller 2006; Abbot et al. 2011; Bourke 2011).

An allele may accrue indirect fitness in two ways. First, we can imagine a gene (or perhaps a group of tightly linked genes) that codes for a phenotypic trait (say, a green beard), enables its carrier to recognize other individuals with the same trait, and inclines its carrier to help those individuals (at a cost to its own fitness). In this so-called GREEN BEARD MODEL of the evolution of altruism (Dawkins 1989), the cooperating individuals need share only this gene and no others. A real-life example is the *csA* gene of the social amoeba *Dictyostelium discoideum*, in which cells aggregate into a "slug" that differentiates into reproductive spores and a stalk made up of cells that altruistically die (see Figure 16.4). The *csA* gene promotes formation of the slug because it encodes a cell adhesion protein that binds to the same protein in the membrane of other cells. When David Queller and collaborators (2003) mixed wild-type cells and cells in which *csA* had been knocked out, the knockout cells were excluded from the slug, and few of them became spores. Wild-type *csA* genes promote their own reproduction by forming colonies of cells in which some *csA* bearers sacrifice themselves for the reproductive benefit of others.

Second, individuals may share genes for cooperation because they are related: they are kin. In this, the much more commonly discussed case, selection based on inclusive fitness is called **kin selection**. Let us suppose that an individual performs an act that benefits another individual, but incurs a cost to itself: a reduction in its own (direct) fitness. The fundamental principle of kin selection is that an allele for such an altruistic trait can increase in frequency only if the number of extra copies of the allele passed on by the altruist's beneficiary (or beneficiaries) to the next generation as a result of the altruistic interaction is greater, on average, than the number of allele copies lost by the altruist. This principle is formalized in **Hamilton's rule**, which states that *an altruistic trait can increase in frequency if the fitness benefit* (b) *received by the donor's relatives, weighted by their relationship* (r) *to the donor, exceeds the cost* (c) *of the trait to the donor's fitness*. That is, altruism spreads if $rb > c$.

The **coefficient of relationship**, r, is the fraction of the donor's genes that are identical by descent to any of the recipient's genes (Grafen 1991). ("Identical by descent" means derived from common ancestral genes; see Chapter 9.) For example, suppose that at an autosomal locus in a diploid species, an offspring inherits one of its mother's two gene copies, so $r = 0.5$ for mother and offspring (**Figure 16.9**). For two full siblings, $r = 0.5$ also, because the probability is 0.25 that both siblings inherit copies of the same gene from their mother, and likewise from their father.

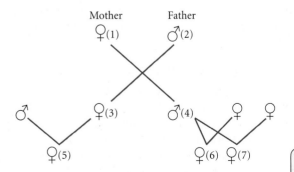

Donor, Recipient	r
Mother (1), offspring (3 or 4)	0.5
Father (2), offspring (3 or 4)	0.5
Full siblings (3, 4)	0.5
Half-siblings (6, 7)	0.25
Aunt (3), niece (6)	0.25
Full cousins (5, 6)	0.125

The coefficient of relationship (*r*) between donor and the recipient of the action is the average proportion of the donor's genes that are present and identical by descent in the recipient.

FIGURE 16.9 Coefficients of relationship (*r*) among some relatives in a diploid species. Individuals 3 and 4 are full siblings, the offspring of female 1 and male 2. Individuals 3 and 4 mate with unrelated individuals, producing offspring 5, 6, and 7. The table gives the values of *r* between a donor who provides a benefit and the recipient of that benefit.

The simplest example of a trait that has evolved by kin selection is parental care. If females with allele *A* enhance the survival of their offspring by caring for them, whereas females lacking this allele do not, then if parental care results in more than two extra surviving offspring, *A* will increase in frequency, even if parental care should cost the mother her life. If *c* = 1 (death of mother) and *b* = 1 (survival of an extra offspring), then since a mother's *A* allele has a probability of *r* = 0.5 of being carried by her offspring, Hamilton's rule is satisfied by the survival of more than two extra offspring, relative to those of noncaring mothers.

Interactions among other ("collateral") relatives also follow Hamilton's rule. For instance, the relationship between an individual and her niece or nephew is *r* = 0.25, so alleles that cause aunts to care for nieces and nephews will spread only if they increase the fitness benefit by more than fourfold the cost of care. The more distantly related the beneficiaries are to the donor, the greater the benefit to them must be for the allele for an altruistic trait to spread.

Parental care illustrates why indiscriminate altruism cannot evolve by individual selection. If allele *A* caused females to dispense care to young individuals in the population at random, it could not increase in frequency because, on average, the fitness of all genotypes in the population, whether they carried *A* or not, would be equally enhanced. Thus the only difference in fitness among genotypes would be the reduction in fitness associated with dispensing care.

CONDITIONS FOR THE OPERATION OF KIN SELECTION Kin selection can operate only if individuals are more likely to help kin than nonkin. This can be achieved in two ways. First, kin selection may operate if individuals are usually associated with kin, at least during the period in their life history, or under environmental circumstances, when helping behavior is expressed. Such a population structure requires that individuals not become randomly mixed before the time of dispersal. For example, local colonies and troops of many primates, prairie dogs, and other mammals are composed largely of relatives (Manno et al. 2007). In such cases, kin recognition would not be necessary.

Alternatively, individuals must be able to distinguish related from unrelated individuals, perhaps by assessing their similarity with respect to one or more characteristics that are highly variable in the population (Sherman et al. 1997). This variation might be genetically based, or it might be caused by a shared environmental imprint; for example, individual colonies of many ants and other social insects have a distinctive "colony odor" that is apparently derived from food or other environmental factors (Wilson 1971). An example of a kin-selected characteristic, based on kin recognition, is provided by tadpoles of the spadefoot toads *Spea bombifrons* and *S. multiplicata*, which develop into detritus- and plant-feeding omnivores if they eat those materials early in life, or into cannibalistic carnivores, with large horny beaks and large mouth cavities, if they eat animal prey (Pfennig and Frankino 1997). Tadpoles are less likely to develop the cannibalistic phenotype if they develop in the company of full siblings than if they develop alone or with a mixture of related and nonrelated individuals. Moreover, omnivores associate more with their siblings than with nonrelatives, whereas carnivores do the opposite, and carnivores eat siblings much less frequently than they eat unrelated individuals. Even plants may be able to recognize kin: sea rockets (*Cakile edentula*) allocated more growth to roots (thereby increasing their competitive ability) when they were surrounded by unrelated individuals than when surrounded by siblings (Dudley and File 2007).

EXAMPLES OF COOPERATION AMONG KIN Kin selection often accounts for cooperation among both organisms and their cells, as two examples illustrate.

In some animals, sperm cells form cooperative groups that have increased swimming velocity and gain an advantage in competition with other sperm. Heidi Fisher and Hopi Hoekstra (2010) discovered sperm aggregations in mice of the genus *Peromyscus*. They predicted, and verified, that in the deer mouse, *P. maniculatus*, a sexually promiscuous species in which sperm of different males may compete within a female's reproductive tract, sperm aggregate more frequently with other sperm from the same male than with sperm from unrelated males. Sperm from a monogamous species, *P. polionotus*, did not evince

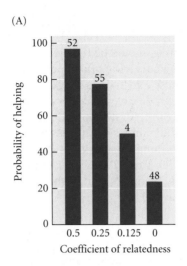

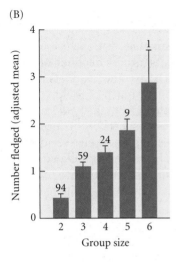

FIGURE 16.10 Cooperative breeding in white-fronted bee-eaters (*Merops bullockoides*). (A) Potential helpers had a higher likelihood of becoming actual helpers, the more closely related they were to the nestlings they might rear. (B) The number of successfully fledged offspring increased with the number of helpers at a nest. A group size of 2 indicates a mated pair without helpers. In both graphs, the numbers above the columns indicate sample sizes. (After Emlen and Wrege 1991.)

discrimination. Fisher and Hoekstra suggest that although only one of the sperm in an aggregate that contacts an egg will fertilize it, its teammates gain inclusive fitness by kinship.

Unlike the greater ani mentioned earlier, most of the many species of birds that engage in cooperative breeding live in kin-based groups, in which some individuals help their parents or other relatives rear offspring (Clutton-Brock 2002; Cockburn 2006; Hatchwell 2009). One such species is the African white-fronted bee-eater. During a 5-year study of individually marked birds, Stephen Emlen and Peter Wrege (1988, 1991, 1992) discovered that nonbreeding helpers were generally related, with an average *r* of 0.33, to the offspring they helped, and that they were more likely to help, the closer their relationship to the offspring they help to raise (**Figure 16.10A**). In 94 percent of the cases in which helpers had a choice between two broods that differed in degree of relatedness to themselves, they chose to help the more closely related young. A high proportion of nestlings die of starvation, but helpers increased the number of survivors by about 0.5 per helper (**Figure 16.10B**). The average inclusive fitness of a helper was *rN*, where *N* is the number of extra offspring reared, or $0.33 \times 0.5 = 0.165$ (and was 0.25 for those that helped their parents). The average inclusive fitness for an unaided pair was 0.23. The similarity of these values suggests that helping may sometimes be a better path to fitness than attempting to breed on one's own.

SPITE SPITEFUL traits are harmful to both the actor and the recipient. Surprising as it may seem, inclusive fitness theory predicts that such traits might evolve, under conditions precisely the opposite of those that favor altruism. It requires that the actor be less closely related to the recipient than to an average member of the population, and that harming the recipient enhance the fitness of other individuals more closely related to the actor (West and Gardner 2010). Among the best examples of spite is the production, by many bacteria, of bacteriocins, toxins that are secreted into the cell's environment and kill susceptible bacterial cells (Riley and Wertz 2002). Bacteriocin-producing genotypes are resistant to the toxin because of a gene that is very tightly linked to the gene for toxin production. Producing bacteriocin is costly because it reduces growth (and in *E. coli* is released by cell death). However, producer genotypes increase in laboratory cultures because by killing nearby susceptible cells, they enhance the growth and reproduction of related cells that carry the producer gene (Inglis et al. 2009; **Figure 16.11**).

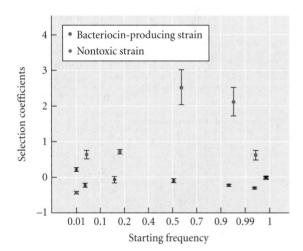

FIGURE 16.11 The coefficient of selection for a strain of *Pseudomonas* that produces a bacteriocin that kills nearby sensitive cells, in relation to the initial frequency of the strain, when competing against a standard genotype. The advantage of the toxic strain is greatest at intermediate frequencies, because it then kills off the greatest number of competitors. The corresponding values for a control, nontoxic strain, also competing against the standard strain, were close to zero. (After Inglis et al. 2009.)

An Arena for Cooperation and Conflict: The Family

At first surmise, it would seem that relationships within families should be the epitome of cooperation, since the parents' fitness depends on producing surviving offspring. However, evolutionary biologists have come to understand that these interactions are pervaded with potential conflict, and that much of the diversity of reproductive behavior and life histories among organisms stems from the balance between conflict and cooperation. Evolutionary hypotheses for many aspects of the formation and dynamics of families, based on ecology (which may affect whether relatives disperse or remain together), kin selection, and reproductive skew, have been formulated (Emlen 1995; Mock 2004). (Incidentally, the ways in which some animal species behave toward family members starkly show that natural selection utterly lacks morality, as pointed out in Chapter 11.)

Mating systems and parental care

Whether or not one or both parents care for offspring varies greatly among animal species and partly determines the MATING SYSTEM: the pattern of how many mates individuals have and whether or not they form pair-bonds (Clutton-Brock 1991; Davies 1991). Providing care (such as guarding eggs against predators, or feeding offspring) increases offspring survival, which enhances the fitness of both parents (and of the offspring). But parental care is also likely to have a cost. It entails risk, and it requires the expenditure of time and energy that the parent might instead allocate to further reproduction: a female might lay more eggs, and a male might find more mates.

In most animals, neither parent provides care to the eggs or offspring after eggs are laid, and one or both sexes, especially in long-lived species, may mate with multiple partners (PROMISCUOUS MATING). In many birds and mammals, females provide care but males do not, and males may mate with multiple females (POLYGYNY). In some fishes and frogs and a few species of birds, only males guard eggs or care for offspring, and in some such species, the female may mate with multiple males (POLYANDRY). (These terms stem from the Greek *polys*, "many"; *gyne*, "woman"; *andros*, "man.") In many species of birds, some mammals, and a few insects such as dung beetles, a female and male form a "socially monogamous" pair-bond, and both provide parental care of offspring. However, many such birds engage in frequent "extra-pair copulation" and are not sexually monogamous. As we have seen, females may also increase their reproductive success by laying eggs in the nests of unwitting foster parents. Females in some pair-bonding species of birds appear to favor highly ornamented males with characteristics that serve as honest signals of paternal caregiving (**Figure 16.12**). This is as some theory predicts (Kokko 1998), but it has also been argued that such ornaments are not necessarily honest signals: highly ornamented males might devote less time to offspring care and more to extra-pair copulation, in which they may be especially likely to succeed (Houston et al. 2005).

Each parent in a mated pair will maximize her or his fitness by some combination of investing care in current offspring and attempting to produce still more offspring. In a species with biparental care, each parent should profit by leaving as much care as possible to the other partner, as long as any resulting loss of her (or his) fitness owing to the death of her (or his) current offspring

FIGURE 16.12 Examples of variations in paternal care that are correlated with signal traits in songbirds. (A) Paternal care is correlated with the brightness of the male's breast in the northern cardinal. (B) Among sedge warblers, males that sing a larger repertoire of songs also bring food to offspring more frequently. (A after Linville et al. 1998; B after Buchanan and Catchpole 2000.)

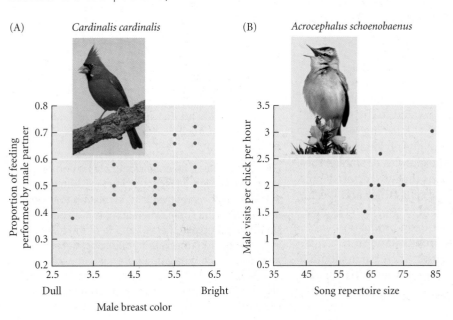

(A) *Cardinalis cardinalis*

Proportion of feeding performed by male partner

Male breast color — Dull / Bright

(B) *Acrocephalus schoenobaenus*

Male visits per chick per hour

Song repertoire size

were more than compensated by the offspring she (or he) would have from additional matings. If offspring survival were almost as great with uniparental care as with biparental care, selection would favor females that defected, abandoning the brood to the care of the male—or vice versa (**Figure 16.13**). Thus a conflict between mates arises as to which will evolve a promiscuous mating habit and which will care for the eggs or young. Selection favors defection more strongly in the sex for which parental care is more costly (in terms of lost opportunities for further reproduction). However, if the survival of offspring depends strongly on care, both parents may maximize their fitness by forming a socially monogamous pair-bond and cooperating in offspring care.

This theory may explain why in birds and mammals, parental care is generally provided by females or by both mates, whereas in fishes and frogs it is usually provided by males (Clutton-Brock 1991). Fishes and frogs guard the eggs or young, but do not feed them. Males can often mate with multiple females and guard all their eggs in a single nest (**Figure 16.14A**); thus they pay a smaller cost than females, whose subsequent reproduction depends on replenishing the massive resources they expend in egg production. In birds and mammals that must feed their young, parental care is more costly for males than for females, since males could potentially obtain many matings in the time they must spend rearing one brood exclusively (**Figure 16.14B**).

The strength of natural selection for parental care is greater if the opportunity for reproduction outside the pair-bond is restricted, if offspring survival greatly depends on maximal care, and if individuals are likely to be actually caring for their own offspring rather than someone else's. Male parental care, for example, is expected to be proportional to a male's CONFIDENCE OF PATERNITY—which may be quite low in species in which extra-pair copulation is frequent

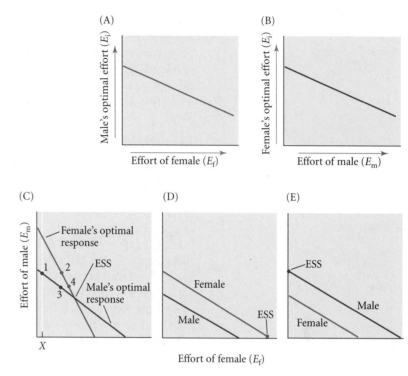

FIGURE 16.13 An ESS model of parental care. (A, B) The optimal parental effort expended by each sex declines, the more effort its partner expends. (C) Curves for males and females plotted together. Their intersection marks the ESS, the evolutionarily stable strategy. If, for example, the population starts with female effort (E_f) equal to X, male effort (E_m) evolves to point 1; but then the optimal E_f is at point 2 on the female's optimality line. When E_f evolves to point 2, E_m evolves to point 3; but then E_f evolves to point 4. Eventually, E_m and E_f evolve to the intersection (the ESS), no matter what the initial conditions are. (D, E) Conditions can be envisioned in which the optimal curves for the sexes do not intersect and the ESS is care by only the female (D) or the male (E). (After Clutton-Brock and Godfray 1991.)

FIGURE 16.14 Parental care. (A) A male stickleback (*Gasterosteus aculeatus*) builds and cares for a nest containing egg clutches fathered by him. Such nest-brooding activity can attract additional females to mate with the male, and the resulting eggs are added to his nest. (B) Great crested grebes (*Podiceps cristatus*) exemplify the many bird species in which both parents care for the young.

(Whittingham et al. 1992; Westneat and Stewart 2003). The dunnock (*Prunella modularis*) is a songbird with a variable mating system: some individuals form monogamous pairs, some polygynous trios (two females and one male), and some polyandrous trios (two males and one female). Polyandrous males each provide paternal care to the female's offspring only in proportion to the amount of time they spent with her during her fertile period, presumably using this as an estimate of their paternity (Davies 1992).

Infanticide, abortion, sibling competition, and siblicide

There are a number of circumstances in which an individual's fitness may be enhanced by killing young members of their species (Hausfater and Hrdy 1984; Clutton-Brock 1991). Infanticide, for example, may be sexually selected. In many species of mammals, including lions and some primates and rodents, a male that replaces a mated male kills the existing offspring of his new mate (**Figure 16.15**). The infanticidal male can then father his own offspring faster, because females become fertile and sexually receptive sooner if they are not nursing young. DNA analyses of wild langur monkeys (*Presbytis entellus*) showed that males that killed infants were the fathers of the infants that were subsequently born to the same mothers (Borries et al. 1999).

Parents sometimes kill offspring as a way of adaptively regulating brood size (Mock 2004). A parent's fitness is proportional (all else being equal) to the number of surviving offspring, which equals (size of egg clutch or brood) × (per capita probability of survival). In species with parental care, the probability of survival may decrease as brood or clutch size increases because of competition among the offspring for food, and the parental care expended on an excessively large brood can reduce the parent's survival and subsequent reproduction, reducing its lifetime fitness (see Chapter 14). Thus it can be adaptive for a parent to reduce the offspring to an optimal number (possibly even zero). For example, females of some species of mice kill some or all of their young if food is scarce or the number born is too high. Likewise, shortage of food or disturbance by predators can induce many birds to desert their eggs or young, especially early in the breeding season (when the parents have the opportunity to renest elsewhere). In the same vein, plants typically abort the development of many of their offspring (seeds), allocating limited resources to fewer but larger seeds with a greater chance of survival.

FIGURE 16.15 Infanticide. This male lion has killed a cub after displacing the cub's father and other adult males in the group he has joined. (Photograph courtesy of G. Schaller.)

Individual A foregoes some meals in favor of its sibling B.

While B grows strong, A is weakened by its own actions...

...and siblings C and D, who now must compete for food with the more vigorous B, may also be adversely affected.

FIGURE 16.16 Competition among kin can counteract the kin-selected advantage of altruism, which may reduce the fitness of some kin more than it increases the benefit to others. (Illustration © John Norton.)

Aggregation can increase competition among kin for food and other resources, counteracting kin selection for cooperation because the benefits accrued by the recipients of altruistic help are balanced by the reduced fitness of the relatives with which the recipients compete (**Figure 16.16**; Queller 1994; Frank 1998). When the offspring in a brood (sibs) share an environment, such as young birds in a nest or larval parasitic wasps within a caterpillar, they may actively fight for resources, and larger individuals may kill smaller siblings (SIBLICIDE). Siblicide is the norm in some species of eagles and boobies, in which the female lays two eggs but one of the nestlings always kills the other (**Figure 16.17**). (The second egg may be the female's "insurance" in case the first is inviable.) The young of some shark species eat their siblings while they share their mother's uterus.

Parent-offspring conflict

Parents and offspring typically differ with respect to the optimal level of parental care (Trivers 1974; Godfray 1999). A parent's investment of energy in caring for one offspring may reduce her production of other (e.g., future) offspring. Production of other offspring increases the inclusive fitness of both the parent and each current offspring. However, the parent is equally related to all her offspring ($r = 0.5$), whereas each offspring values parental investment in itself twice as much as investment in a full sib ($r = 0.5$) and four times as much as investment in a half sib ($r = 0.25$). Therefore offspring are expected to try to obtain more resources from a parent than it is optimal for the parent to give, resulting in **parent-offspring conflict**, as is seen when young mammals persist in suckling while their mothers try to wean them. Animals may evolve features that enable offspring to extract extra resources from their parents; for example, baby American coots have unusual reddish plumes that stimulate their parents to feed them (Lyon et al. 1994). Parent-offspring conflict can take other forms as well. In their study of white-fronted bee-eaters, for example, Emlen and Wrege (1992) observed many instances in which older breeding males harassed another pair and prevented it from breeding. In more than one-third of these cases, the harassed male joined the breeding effort of the harasser as a helper, and in more than 60 per cent of such cases, the harassed helper was the son of the older male. The young helper increased his parents' fitness by rearing younger siblings, and his own indirect fitness was nearly the same as if he had nested, unaided, on his own.

FIGURE 16.17 Siblicide in the brown booby (*Sula leucogaster*). The parent is sheltering a large chick that has forced its sibling out of the nest. The parent ignores the dying chick (foreground). (Photo by John Alcock.)

Eusociality

The most extreme altruism is found in EUSOCIAL animals: those in which nearly or completely sterile individuals (workers) rear the offspring of reproductive individuals, usually their parent (or parents). Eusociality is known in one species of mammal (the naked mole-rat; **Figure 16.18A**), in all species of termites (Isoptera; **Figure 16.18B**), in many members of the order Hymenoptera (**Figure 16.18C,D**), and in a few other kinds of insects (Wilson 1971; Keller 1993; Bourke and Franks 1995; Crozier and Pamilo 1996).

Eusociality has evolved independently many times among the wasps, bees, and ants in the order Hymenoptera. Phylogenetic studies show that all of these lineages have arisen from solitary, stinging, wasplike insects in which females provide parental care to their young, usually by constructing a burrow or nest and provisioning it with food (e.g., paralyzed insects) for their larvae. The ancestral mating system in all these lineages was monogamy: females mated with a single male (Hughes et al. 2008). In all eusocial Hymenoptera, the workers are females that do not mate. The workers and one or more egg-laying females (queens) constitute a colony that may persist for a single season or for many years, depending on the species. Most of the queen's eggs develop into workers, but at certain times, eggs are produced that develop into reproductive females and males. In most species, whether a female develops into a queen or a worker depends on her diet, which is often controlled by the workers. It appears likely that the ancestors of social Hymenoptera were preadapted (see Chapter 11) for social life because they practiced parental care of defenseless, almost immobile larvae and had stings that were "preadapted" for nest defense (Strassmann and Queller 2007). Sociality may have first evolved because daughters that stayed at the natal nest would increase the chance of their younger siblings' survival if their mother perished while foraging.

Hymenoptera are HAPLODIPLOID: females develop from fertilized eggs and are diploid, but males develop from unfertilized eggs and are haploid. Consequently, coefficients of relationship (r) among relatives differ from those in

(A)

(B)

(C)

(D)

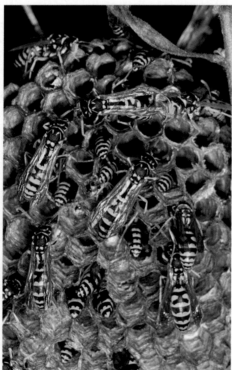

FIGURE 16.18 Some eusocial animals. (A) Naked mole-rats (*Heterocephalus glaber*) are the only mammalian species known to be eusocial. (B) This queen termite is attended by small, sterile workers and large-headed, sterile soldiers. (C) Australian honeypot ants (*Camponotus inflatus*), engorged with nectar, hang from roof of their nest's larder. These "repletes" regurgitate nectar on demand to their worker nestmates. (D) Paper wasps (*Polistes gallicus*) at their nest.

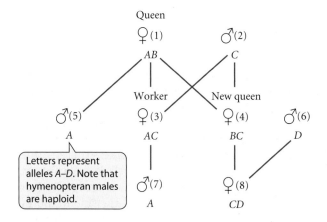

Letters represent alleles *A–D*. Note that hymenopteran males are haploid.

Donor, Recipient	*r*
Female (1), Daughter (3, 4)	0.5
Female (1 or 3), Son (5 or 7)	0.5
Male (2), Daughter (3, 4)	1.0
Male (2), Mate's son (5)	0.0
Sister (3), Sister (4)	0.75
Sister (3), Brother (5)	0.25
Female (3), Niece (8)	0.375
Female (4), Nephew (7)	0.375

FIGURE 16.19 Coefficients of relationship among relatives in hymenopteran species, assuming a single, singly mated queen per colony. Individuals in the pedigree are numbered as in the table, and one of the possible genotypes of each is labeled with letters representing alleles *A–D*. The coefficient of relationship (*r*) is calculated by finding the proportion (1.0 or 0.5) of a donor's genes that are inherited from one of its parents, multiplying by the probability that a copy of the same gene has also been inherited by the recipient, and summing these products over the donor's one or two parents. Notice that in these haplodiploid species, sisters are more closely related to one another than mother to daughter or mother to son. Compare with the analogous situation for diploid species shown in Figure 16.9. (After Bourke and Franks 1995.)

diploid species (**Figure 16.19**). Whereas *r* in diploid species is 0.5 both between parent and offspring and between full siblings, a female hymenopteran is more closely related to her sisters (*r* = 0.75) than to her daughters (*r* = 0.5) and is less closely related to her brothers (*r* = 0.25). Kin selection is a powerful explanation of many aspects of cooperation and conflict in eusocial insects (Foster et al. 2006; Ratnieks et al. 2006; Strassmann and Queller 2007; Bourke 2011).

Much of the evidence for the role of kin selection is based on the consequences of conflict within colonies. In most eusocial Hymenoptera, workers can lay haploid eggs that develop into males. A worker is more closely related to her own sons (*r* = 0.5) than to her mother's (the queen's) sons (*r* = 0.25). Therefore workers gain direct fitness by laying eggs, and they are altruistic if they do not. Because the queen is more closely related to her own sons (*r* = 0.5) than to her daughters' sons (*r* = 0.25), and because male offspring do not contribute to colony growth and welfare, queens practice "policing" in many species by destroying workers' eggs—and in some species, workers destroy the eggs of other workers (Ratnieks et al. 2006; Wenseleers and Ratnieks 2006). The level of reproduction by workers is lower in species in which a high degree of policing has evolved. Policing usually does not occur if the queen dies or is removed, and under these conditions, fewer workers lay eggs—more workers retain their altruistic behavior—in those species in which the level of relatedness among workers is higher (**Figure 16.20**).

In most social insects, workers can control the sex ratio among the larvae destined to be males or queens by withholding care from male larvae and by altering the proportion of female larvae that develop into queens versus workers. In a colony with a single queen, a worker's inclusive fitness is maximized by rearing young queens and males in a 3:1 ratio, because a worker is related by *r* = 0.75 to her queen sisters, but only by *r* = 0.25 to her

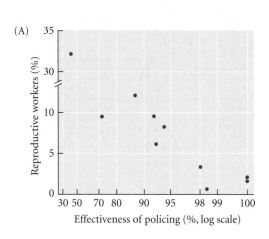

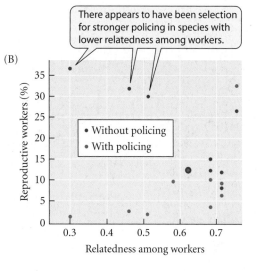

There appears to have been selection for stronger policing in species with lower relatedness among workers.

FIGURE 16.20 Policing in eusocial insects, illustrated by comparisons of nine wasp species and the honeybee. (A) In species in which more worker-laid eggs are destroyed by nestmates, fewer workers attempt to reproduce. This pattern implies that policing is effective. (B) In queenless colonies, in which policing does not occur, fewer workers lay eggs in species in which workers are more closely related to one another. That is, workers are more altruistic, as predicted by Hamilton's rule. However, when queens are present and policing occurs, the relationship is reversed, probably because there has been selection for stronger policing in species with lower relatedness among workers. (After Ratnieks and Wenseleers 2007.)

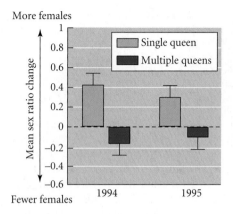

More females

Fewer females

1994   1995

FIGURE 16.21 Evidence that workers in hymenopteran colonies manipulate the sex ratio to increase their inclusive fitness, as predicted by kin selection theory. The bars show the changes in the proportion of females between the egg and pupal stage in colonies of wood ants with single queens and in colonies with multiple queens, in which each worker is unrelated to many of the offspring. (After Sundström et al. 1996.)

brothers. In colonies with multiple queens, however, workers would have a coefficient of relationship lower than 0.75 to many of the female offspring they rear (since they are not full sisters), so the sex ratio should be closer to 0.5 (1:1). A resident queen's inclusive fitness is maximized by a 1:1 sex ratio, since she is equally related to her daughters and sons.

Many data support these predictions (Crozier and Pamilo 1996). The average sex ratio is closer to 3:1 in single-queen than in multiple-queen species of ants, and it is almost exactly 1:1 in slave-making species of ants, in which the workers that rear the slave-maker's brood are members of other species, captured by the slave-maker's own workers. The slaves have no genetic interest whatever in the captor's reproductive success, and so are not expected to alter the sex ratio by preferential treatment. Moreover, there is direct evidence that wood ant (*Formica exsecta*) workers manipulate the sex ratio as kin selection theory predicts (Sundström et al. 1996). Although all colonies have about the same sex ratio among eggs, the sex ratio among pupae becomes shifted toward females in single-queen colonies, but toward males in multiple-queen colonies, which would be advantageous for the queens (**Figure 16.21**). This observation provides some of the most convincing evidence of the role of kin selection in social animals.

Kin Selection or Group Selection?

So far in this chapter, we have described explanations, based on direct or indirect benefits to the individual, for the increase and persistence of genes for cooperation. In Chapter 11, we saw that V. C. Wynne-Edwards (1962) interpreted cooperative behaviors of animals as adaptations for the good of the species and proposed that they evolved by a higher rate of extinction of populations composed of noncooperating genotypes. George Williams, John Maynard Smith, and others convinced most biologists that this model of group selection is very unlikely to explain adaptations.

In 1975, David Sloan Wilson proposed a new model of group selection, based on the principle that individuals often interact within groups (which he referred to as TRAIT GROUPS) that last for a less than a generation or so and that may differ in the frequencies of alleles that determine social behaviors (such as cooperation or defection). He proposed that trait groups with many cooperators may produce more offspring than trait groups with fewer. When the groups dissolve and random mating takes place before the establishment of new groups, the productive groups of cooperators contribute more genes to the total gene pool, so the allele for cooperation increases in frequency (**Figure 16.22**; see also Figure 15.11). Wilson's proposal is a model of selection at multiple levels (see Chapter 11), in which selection among temporary groups is

FIGURE 16.22 "Old" and "new" conceptions of group selection. (A) In the "old" notion of group selection, discarded by evolutionary biologists in the 1960s, groups are discrete for generations and differ in rate of extinction. Selfish genotypes increase within groups. (B) In the "new" concept, groups are less well defined and soon break down into a common reproductive pool before new temporary groups are established. Groups with cooperators contribute more to the pool. These groups are often groups of kin. (After West et al. 2007.)

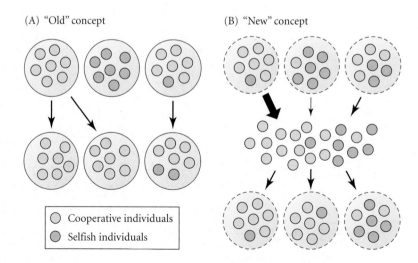

stronger than selection within those groups. Some authors have argued that this process vindicates a role for group selection in evolution (Wilson 1997; Nowak et al. 2010; Eldakar and Wilson 2011). Some other authors, however, have pointed out that trait groups are made up of individuals that, with respect to genes for cooperation, are more closely related to one another than expected by chance—usually because they are kin. Within kin groups, altruism is disfavored by individual selection (see Figure 16.16), but kin groups that include altruistic genotypes may be more productive than kin groups that do not. Consequently, they argue that kin selection and trait group selection are two equivalent ways of describing the same process (Frank 1998; Foster et al. 2006; West et al. 2007). For many reasons, however, the kin selection approach has been more useful to biologists who study cooperation among real organisms (West et al. 2007; Lion et al. 2011). For instance, groups of organisms are often more difficult to define and measure than individuals or genes, and the kin selection approach makes it easier to distinguish direct from indirect benefits.

Genetic Conflict

Conflict can exist not only between individual organisms, but also between different genes (or, more broadly, genetic elements) within individual organisms (**Figure 16.23**). Understanding such conflicts and their evolutionary effects requires that we think in terms of selection at the level of the gene (see p. 292 in Chapter 11). This perspective informs the concept of kin selection, in which the fitness of an allele depends on its effects both on its bearer and on other individuals that possess the same allele. Here we consider selection of selfish genetic elements, which enhance their own transmission relative to that of other elements in an individual's genome. The enhanced transmission of a selfish genetic element reduces the transmission of other genes, creating a **genetic conflict** that may result in selection for genes that suppress the selfish "outlaws" (Burt and Trivers 2006; Werren 2011). Genetic conflict can be generated by cytoplasmic inheritance, meiotic drive, post-segregation distorters, and transposable elements.

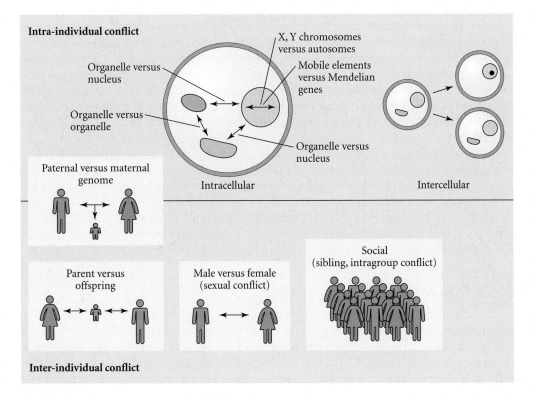

FIGURE 16.23 Types of genetic conflicts. Conflicts *within* individual organisms are diagrammed in the upper half, and conflicts *between* individuals are shown in the lower half, of the figure. Within-individual conflicts can be intercellular (e.g., between normal and cancerous cells) or intracellular (i.e., between genetic elements within the cell). The box referring to conflicts between paternal and maternal genomic elements straddles the intra- and intercellular divide, since it has elements of both. (After Werren 2011.)

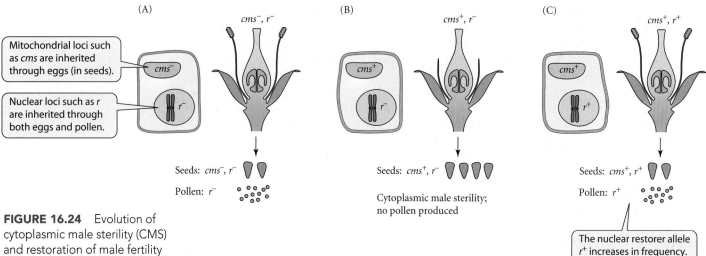

FIGURE 16.24 Evolution of cytoplasmic male sterility (CMS) and restoration of male fertility exemplifies a genetic conflict. The diagrams portray a flower and a cell of each genotype. (The nucleus is depicted as haploid for simplicity.) (A) Mitochondrial loci such as *cms* are inherited only through eggs (i.e., within seeds), and nuclear loci such as *r* are inherited through both eggs and pollen. (B) A *cms*⁺ mutation eliminates pollen production, and it may increase seed number due to the plant's reallocation of energy and resources from pollen to eggs. Therefore *cms*⁺ increases in frequency. (C) Because the fitness of nuclear genes is increased by transmission through both eggs and pollen, selection favors the restorer allele *r*⁺, which counteracts the effects of *cms*⁺. The population may become fixed for antagonistic alleles *cms*⁺ and *r*⁺, but may be phenotypically indistinguishable from the ancestral population, with *cms*⁻ and *r*⁻.

Cytoplasmic inheritance

Some thyme plants (*Thymus vulgaris*) have hermaphroditic flowers, with both male (stamens) and female (pistil) parts. Many individual thyme plants, however, lack anthers; they are "male-sterile," or female. Male sterility is caused by a mitochondrial allele. The interesting point is that all thyme plants carry this cytoplasmic male sterility (CMS) factor. Hermaphrodites have male function only because they carry a "restorer" allele, inherited on a nuclear chromosome, that counteracts CMS. Remarkably, there exist hermaphroditic plant species in which all individuals carry both a CMS factor and a nuclear restorer (**Figure 16.24**). Why should a gene exist, only to be counteracted by another gene?

Because mitochondria are maternally inherited (a type of CYTOPLASMIC INHERITANCE), any mitochondrial allele that can increase the number of female relative to male offspring will increase in frequency relative to an allele that does not alter the sex ratio from 1:1. Because they divert protein and other resources from pollen to seed production, male-sterile plants produce more seeds than hermaphroditic plants. Thus mitochondrially inherited *cms*⁺ alleles may increase in frequency. In contrast, recall (from Chapter 12) that because every individual has a mother and a father, nuclear alleles that produce an even sex ratio are advantageous. Thus production of excess females is disadvantageous to nuclear genes, since it reduces their transmission through pollen. A mutation of a nuclear gene that nullifies the effect of a *cms*⁺ allele may therefore increase in frequency. The function of such a gene, then, is to counteract the effect of a selfish gene at another locus.

Mitochondrial and other cytoplasmically inherited genes commonly conflict with nuclear genes, as cytoplasmic male sterility illustrates. Moreover, cytoplasmically inherited genes that have harmful effects on males are shielded from purifying selection because males do not transmit them anyway. For this reason, humans, fruit flies, and other species have a high incidence of mitochondrial genetic variants that cause male-specific deleterious effects and that even affect the male-specific expression of the nuclear genes with which mitochondrial genes interact (Gemmell et al. 2004; Innocenti et al. 2011).

Meiotic drive

In "fair" meiosis, two homologous chromosomes have equal chances of being included in functional eggs or sperm; in cases of segregation distortion or MEIOTIC DRIVE, one of the two is more likely to be included. This can occur either because of preferential segregation or because of a "gamete killer system," in which a "driving gene" causes the death or disruption of other gametes that lack it. In many species, the sex ratio among the progeny of a single mating depends on the proportion of eggs fertilized by X-bearing versus Y-bearing

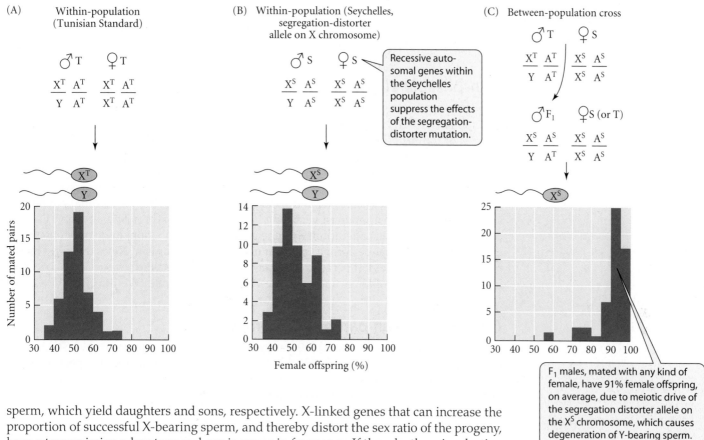

sperm, which yield daughters and sons, respectively. X-linked genes that can increase the proportion of successful X-bearing sperm, and thereby distort the sex ratio of the progeny, have a transmission advantage and can increase in frequency. If they do, there is selection for autosomal genes that achieve highest fitness by restoring the 1:1 sex ratio. The restoration conceals the segregation distorter allele, but its existence is sometimes revealed by crossing a population that bears it with other populations. For example, in matings of *Drosophila simulans* from the Seychelles, the progeny sex ratio is about 1:1, but the F_1 males from a cross between Seychelles females and males from elsewhere in the world produce a great excess of female offspring (**Figure 16.25**). This skewed sex ratio is caused by a segregation distorter allele on the X chromosomes of Seychelles flies that causes degeneration of the male's Y-bearing sperm. The effect of this allele is not ordinarily seen because the Seychelles population also has recessive autosomal genes that suppress the distortion. Other populations of *D. simulans* lack both the X-linked distorter and the suppressor alleles, so the effects of the distorter become evident when it is crossed into their genetic background. In the Seychelles and in other areas where the distorter allele has increased in frequency, there has been selection for autosomal suppression of its effects (Atlan et al. 1997).

Post-segregation distorters

POST-SEGREGATION DISTORTERS (PSD) often act by reducing the survival of offspring that have lost the distorter gene. Examples of PSDs are "killer plasmids" in yeast and bacteria, which encode a toxin that is transmitted to daughter cells, which die unless they inherit the plasmid that "rescues" them. Similarly, offspring of female flour beetles (*Tribolium*) that carry the "Medea" element (named for the character in Greek mythology who, according to Euripides, was abandoned by her husband and took revenge by murdering their children) survive only if they inherit the element. Another kind of PSD (already mentioned in Chapter 11) has been found in certain parasitic wasps, which, like all Hymenoptera, are haplodiploid. Some males carry an extra chromosome, called *psr* ("paternal sex ratio"). Offspring of these males begin life as diploid (female) zygotes, but then *psr* causes all of the father's chromosomes (except itself) to degenerate, turning the

FIGURE 16.25 Effects of segregation distortion on sex ratio in *Drosophila simulans*. Each graph shows the numbers of male-female pairs producing offspring with a certain sex ratio (percentage of females). Results are shown for (A) crosses within a "standard" population from Tunis (T), (B) crosses within the Seychelles population (S), and (C) offspring of F_1 males from crosses between these two populations. The parents' genotypes are symbolized by X, Y, and A for X chromosomes, Y chromosomes, and autosomes, respectively. The highly female-skewed sex ratio in the cross between populations (C) reveals that the Seychelles population has a segregation distorter allele whose effect is suppressed by a restorer allele that is absent in the Tunisian population. (After Atlan et al. 1997.)

zygote into a haploid male that carries only the maternal chromosomes (which are destined to be destroyed in the next generation) and *psr*. This bizarre case is the most extreme selfish genetic element known.

Transposable elements

TRANSPOSABLE ELEMENTS (TEs) are the epitome of selfish elements. They proliferate in the genome not unlike viruses or other parasites, harming the "host" organism because of the cost of transcription and translation and the deleterious mutations and chromosome rearrangements that they cause (see Chapter 8). Mechanisms to suppress TE activity, such as DNA methylation and interference RNA, have been described in animals, plants, and fungi (Johnson 2007). These "defense" mechanisms appear to be successful, because TEs seldom show gene trees across related species (see Chapter 2), the way ordinary genes do, showing that they do not persist very long before they are extinguished. Some TEs have produced advantageous alterations of gene structure and function, and some have even evolved into genes that perform a function useful to the organism. Nevertheless, TEs are maintained in genomes not because they may occasionally become beneficial, but because, like parasites, they proliferate and infect new hosts at a high rate (Werren 2011).

Conflict between parental genomes

Our final example of genetic conflict may be a surprise: it concerns pregnancy in humans and other mammals (Haig 1993, 2002). Here the cooperation one might expect between mother, father, and offspring is compromised by parent-offspring conflict and a *conflict between parents' genes* that is played out in the offspring.

We have seen that an offspring is expected to try to get more resources from its mother than the mother should willingly give, since she profits more than the offspring does from reserving some resources for subsequent offspring. Some of the interactions between the human fetus and its mother follow this prediction. For example, mothers increase their production of insulin (which causes cells to remove glucose from the blood) during pregnancy—the very time you might expect them to decrease insulin levels in order to provide sugar to the fetus. However, the fetus produces extremely high levels of a hormone (hPL) that counteracts insulin, so that the net result is no alteration of the mother's blood glucose concentration. This escalation of opposing hormones seems to serve no purpose, but is just what one might expect of a parent-offspring conflict.

Even more remarkable is the expression of certain genes that are IMPRINTED so that whether or not they are transcribed in the embryo depends on whether they were inherited from the mother or the father. [For example, the maternal copy of a gene may be methylated, in which case it is not expressed in the offspring (see Chapter 9).] In terms of genic selection, the fitness of paternally inherited genes will be greatest if the embryo survives, since copies of those genes will not be carried by the mother's subsequent offspring if she mates with a different male. In contrast, maternally transmitted genes will also be carried by the mother's subsequent offspring, so their inclusive fitness is greatest if they prevent the mother from nurturing the current embryo exclusively, to the detriment of later offspring. Therefore paternally inherited genes should enhance the embryo's ability to obtain nourishment from the mother, whereas maternally inherited genes should temper the embryo's ability to do so. In mice, a pair of conflicting genes shows just this pattern. Only the paternal copy of the gene for IGF-2 (insulin-like growth factor 2), which promotes the early embryo's ability to obtain nutrition from the uterus, is expressed in the embryo. IGF-2 is degraded by another protein (IGF-2R, insulin-like growth factor 2 receptor); only the maternal allele of the gene for this protein is expressed. The level of IGF-2 is thus balanced by the opposing effects of paternal and maternal genes. Several human pathologies also seem to be caused by an imbalance in expression of one parent's allele relative to the other's (Frank and Crespi 2011).

Parasites, Mutualists, Individuals, and Levels of Organization

The concept of genetic conflict bears on many topics in evolutionary biology (Hurst et al. 1996; Burt and Trivers 2006). For example, the same principles that underlie the evolution of transposable elements apply to any element that can replicate faster than the rest of the genome with which it is associated. One such element is a bacterium that lives within a host organism's cells. (Such an organism is called an ENDOSYMBIONT.) If the population of endosymbionts within a single host is genetically variable, selection within that population favors (by definition) the genotype that increases in numbers faster than others. Because the growth and reproduction of the endosymbionts depend on resources obtained from the host, excessive growth in the number of symbionts may reduce the host's fitness.

The fitness of an endosymbiont genotype—its growth in numbers—is measured by the number of new hosts it infects per generation. If the endosymbionts exhibit **horizontal transmission**—that is, if they are transmitted laterally among members of the host population (**Figure 16.26A**)—the number of new hosts infected may be proportional to the number of symbiont progeny released from each old host. If symbionts escape to new hosts before the old host dies, their fitness does not depend strongly on the reproductive success of the individual host in which the parent symbionts reside. Therefore selection favors symbiont genotypes with a high reproductive rate, even if this causes the host's death (after the symbionts have spread to other hosts). In other words, selection favors evolution of a symbiont that is parasitic and which may be highly virulent.

Suppose, in contrast, that symbionts exhibit **vertical transmission**, from a mother host to daughter hosts (**Figure 16.26B**). Symbiont and host are now chained to each other, and the reproductive success of the symbionts now depends entirely on the fitness of the host. Selection for high proliferation *within* the symbiont population occupying each host is opposed by selection *among* the populations of symbionts that occupy different hosts. On balance, selection at the group level favors genotypes with restrained reproduction—those that do not extract so many resources from the host as to cause its death before it can transmit the endosymbionts to its progeny. Selection may even favor alleles in the symbiont that enhance the host's fitness, or that cooperate with host genes, since those strategies also enhance the fitness of the symbionts. If the symbiont does not evolve sufficient restraint, selection will

FIGURE 16.26 Selection pressures on endosymbionts vary with their mode of transmission. (A) Horizontal transmission of endosymbiotic elements (v) from one host to unrelated hosts selects for a high level of virulence. (B) Vertical transmission of endosymbiotic elements from a host to its descendants favors relatively benign endosymbionts (b) with a lower reproductive rate. (C) In the extreme case, a vertically transmitted symbiont may become an integral part of the host. These intracellular bacteria (*Buchnera*) in specialized cells (bacteriocytes) of an aphid supply essential amino acids to the host. (C, photo courtesy of N. Moran and J. White.)

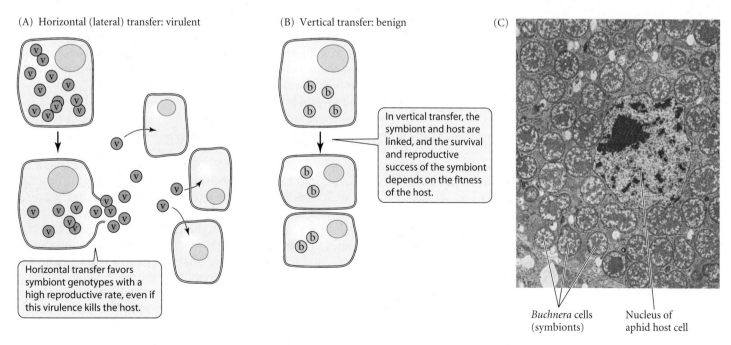

(A) Horizontal (lateral) transfer: virulent

Horizontal transfer favors symbiont genotypes with a high reproductive rate, even if this virulence kills the host.

(B) Vertical transfer: benign

In vertical transfer, the symbiont and host are linked, and the survival and reproductive success of the symbiont depends on the fitness of the host.

(C)

Buchnera cells (symbionts)   Nucleus of aphid host cell

favor host alleles that restrain its growth. Evolution in both the symbiont and its host may therefore result in **mutualism**: an interaction in which two genetic entities enhance each other's fitness. (We will return to the evolution of parasitism and mutualism in Chapter 19.)

In the extreme case, the symbiont may become an integral, essential part of the host. Many eukaryotes harbor vertically transmitted intracellular bacteria that play indispensable biochemical roles, such as those that reside in special cells in aphids and synthesize essential amino acids (**Figure 16.26C**). Mitochondria and chloroplasts are considered organelles in eukaryotic cells, but they were originally endosymbiotic bacteria (see Chapter 5). In these cases, the host and the symbiont have highly correlated reproductive interests: any advantage to one party provides an advantage to the other. The common interest of associated endosymbiont and host genomes resulted in the evolution of a new kind of collective entity, the eukaryotic cell: a higher LEVEL OF ORGANIZATION.

The strong correlation between the reproductive interests of associated elements—such as the organelle and nuclear genomes of eukaryotes—leads us to understand why the most familiar level of organization, the individual multicellular organism, exists at all (L. W. Buss 1987; Maynard Smith and Szathmáry 1995; Frank 1998; Michod 1999). An organism is more than a group of cells; for example, dividing bacteria that remain loosely attached to, but physiologically independent of, one another do not constitute an organism. The cells of a multicellular organism cooperate and typically play different roles—which include the distinction between cells that give rise to gametes (germ cells) and those that do not (the soma). Why should unicellular ancestors, in which each cell had a prospect of reproduction, have given rise to multicellular descendants in which some cells sacrifice this prospect?

The fundamental answer is kin selection: if the cell lineages in a multicellular organism arise by mitosis from a unicellular egg or zygote, the genes of cooperative cells that sacrifice reproduction for the good of the cell "colony" are propagated by closely related reproductive cells. However, the coefficient of relationship is reduced if genetically different cells invade the colony, or if mutational differences arise among cells. A mutation that increases the rate of cell division has a selective advantage *within* the colony, but unregulated cell division—as in cancer—usually harms the organism. Selection at the level of whole colonies of cells—organisms—therefore opposes selection among cells within colonies. It has favored mechanisms of "policing" that regulate cell division and prevent renegade cell genotypes from disrupting the integrated function of the organism. In animals, it has resulted in the evolution of a germ line that is segregated from the soma early in development, thereby excluding most disruptive mutations from the gametes. Selection for organismal integration may be responsible for the familiar but remarkable fact that almost all organisms begin life as a single cell, rather than as a group of cells. This feature increases the kinship among all the cells of the developing organism, reducing genetic variation and competition within the organism and increasing the heritability of fitness. The result, then, has been the emergence of the "individual," and with it, the level of biological organization at which much of natural selection and evolution take place.

Human Behavior and Human Societies

Few subjects are more controversial than evolutionary approaches to human behavior, on which thousands of articles and books have been written. These topics are controversial partly because the questions are intrinsically difficult to answer, concerning as they do a species with extraordinarily complex behavior that cannot be studied in the same experimental ways as in other species; partly because it can be difficult to sort out the well-informed hypotheses and studies from a great mass of poorly informed speculation and poorly analyzed data; and partly because many people (including many social scientists) are skeptical of and often hostile to biological interpretations of human behavior. This reaction is easy to understand because there is a long, ugly history of citing biology to justify human prejudice and oppression of non-Western peoples, "racial" groups, women, homosexuals, and others. Male dominance over women, prejudice against gay people, and imperialistic domination of some cultures by others did not start with Darwinism, but after 1859, in an abuse of evolutionary theory, they were justified by reference to so-called human nature. Racism has long been justified by arguments that its targets were genetically inferior, and the horrendous extremes to which this ideology was taken by the Nazis caused a revulsion against the idea of genetic determinism that has had a lasting effect.

Nevertheless, it is possible and useful to study human evolutionary biology, including its implications for understanding human behavior, if it is done rigorously and with careful explication of the inferences that can and cannot be legitimately drawn from data. All the behaviors of any species, including humans, are biological phenomena at some level, since they originate in a physical, biological structure, the brain. Many behaviors are likely to be expressions of genetic reaction norms with greater or lesser phenotypic plasticity, depending on the trait. For example, the capacity for language clearly has a genetic foundation that has evolved since the human lineage separated from our common ancestor with chimpanzees. Although captive chimpanzees, bonobos, and gorillas can learn and use sign language and other representational modes of communication, with an apparent command of syntax and representational meaning, they do not begin to match the abilities of even very young human children. So human language ability has evolved, but its expression—the actual language an individual speaks—is completely a matter of learning.

Evolutionary biology, applied judiciously, is among the disciplines that can lead us to greater understanding of ourselves. Finding a biological basis for a behavior or any other trait does not justify ill treatment of anyone. We do not abandon our ideals or lessen our respect for people by diagnosing the biological basis of a physical or mental disability; likewise, we need not abandon progressive ideals by finding a biological component in a widespread behavioral characteristic.

Evolutionary approaches to human behavior

Other primates, especially other apes, have been extensively studied in order to discern possible evolutionary origins of human behavioral capacities. Primate studies include both field observations and experiments with captives, including tests much like those devised by psychologists and behavioral economists to study human responses. The capacity for altruism, which is far more highly developed in humans than in other mammals, is a particularly interesting topic. Altruism is commonly seen in social species of primates, in which it is directed almost exclusively toward kin and reciprocating social partners and almost never toward members of other social groups (Silk and House 2011).

Pre-state human cultures, such as hunter-gatherer communities whose mode of life may be adapted to conditions rather like those of early humans, show both similarities to and differences from social groups of other primates. For example, cooperation in the African Hadza people is correlated with degree of kinship and reciprocity, but also with factors such as physical strength, perhaps because strength is useful in cooperative tasks (Apicella et al. 2012). In the Turkana, a nomadic society in which warfare reaps benefits (cattle) for very large social groups, individuals who shirk combat are punished by groups of warriors, a kind of "third party" enforcement of cooperation unknown in other primates (Mathew and Boyd 2011).

Some primates seem to have a "theory of mind," meaning that they apparently attribute mental states, such as intentions, to other individuals (Cheney 2011). Chimpanzees and very young children both help others achieve simple goals, but chimpanzees are less likely to share except with a few social partners (Warneken and Tomasello 2009). Chimpanzees seem to provide information only in order to achieve their own goals, whereas human infants spontaneously share information with anyone. Children are initially generous and informative, although they soon develop a concern for reciprocity and a preference for specific partners, possibly reflecting an evolutionary heritage. With time, they shape their altruistic behavior to conform to culturally received social norms. Both as children and as adults, then, humans show important cognitive, emotional, and social differences from other apes.

Most evolutionary studies of human behavior fall into several subdisciplines (Brown et al. 2011). HUMAN SOCIOBIOLOGY was initiated by E. O. Wilson (1975), an authority on social insects, who interpreted many aspects of human behavior as characteristics that had evolved by natural selection. Wilson recognized that many human behaviors are affected by social and other environmental variables, but nonetheless emphasized their underlying genetic basis. Three major approaches to applying evolutionary theory to human behavior have developed since the 1970s: HUMAN BEHAVIORAL ECOLOGY, EVOLUTIONARY PSYCHOLOGY, and CULTURAL EVOLUTION (see also Chapter 23). These fields differ partly in their subject matter and partly in their assumptions about the role of genes and environment in determining human behaviors. An additional approach, HUMAN EVOLUTIONARY GENETICS, is well suited for determining the genetic basis of features that vary among individuals within populations. A leading question in these fields is why the behavior exists: what is its adaptive function, if any?

HUMAN EVOLUTIONARY GENETICS   An example of a variable behavioral trait that has prompted considerable evolutionary speculation is the continuum from exclusive heterosexual through degrees of bisexual to exclusive homosexual orientation. Many environmental psychological hypotheses have been advanced to account for homosexuality (although seldom for heterosexuality), such as the supposed influence of a distant father or a doting mother, but none of these has been supported by convincing evidence (Wilson and Rahman 2005). In contrast, there is growing evidence, from studies of twins and other relatives (see p. 243 in Chapter 9), that the sexual orientation of both women and men is partly inherited, with heritability of about 0.2 to 0.4 (Iemmola and Camperio-Ciani 2009; Langstrom et al. 2010). Factors on the X chromosome appear to contribute to male sexual orientation. Although a heritability of less than 1.0 implies that environmental factors also contribute to variation, sexual orientation clearly begins to develop early in life; no contemporary researchers think that sexual orientation (as a self-identity) is chosen (Wilson and Rahman 2005). At least a small percentage of the male population in almost all cultures studied, throughout the world and throughout history, has expressed homosexuality (although not necessarily exclusively so), and in some cultures this has been the norm (Duberman et al. 1989). It seems likely, then, that sexual orientation has been variable since before our species spread throughout the world. Homosexual behavior has been recorded in diverse animals, including more than a hundred species of mammals, and is a subject of increasing attention by evolutionary biologists (Bailey and Zuk 2009; Poiani 2010).

To the extent that homosexual orientation has a genetic basis, homosexuality, like sterile worker insects, seems to present an evolutionary enigma, since homosexuals are generally supposed to reproduce less than heterosexuals. This assumption is actually a weak one, since it is quite possible that throughout most of human history, as in some societies both today and in the past (e.g., ancient Greece), homosexuals were accepted or tolerated as long as they played a socially expected procreative role, and thus may have been as reproductively prolific as heterosexuals. If so, alleles affecting sexual orientation today might have been nearly selectively neutral. Assuming, however, that male homosexuals have a low average reproductive rate, several population genetic models could account for a stable polymorphism in sexual orientation (Gavrilets and Rice 2006; Camperio-Ciani et al. 2008). The models that qualitatively match the data best ascribe homosexual inclination to at least two loci, including at least one on the X chromosome, with effects such that the reproductive disadvantage of male homosexuality is balanced by the increased fecundity of females with the same X-linked allele. This situation would be an example of a polymorphism maintained by intralocus sexual conflict, also called sexually antagonistic selection (see Chapter 12). In northern Italy, gay male subjects reported a significantly higher proportion of homosexual maternal relatives (5 percent) than paternal relatives (2 percent), consistent with inheritance on the X chromosome. Moreover, both the mothers and maternal aunts of gay men had significantly more children than those of heterosexual men (**Figure 16.27**; Camperio-Ciani et al. 2004; Iemmola and Camperio-Ciani 2009). It appears increasingly likely that genetic factors inclining men toward homosexual orientation provide a reproductive advantage to women. [Very recently, evolutionary geneticist William Rice and collaborators (2012) have

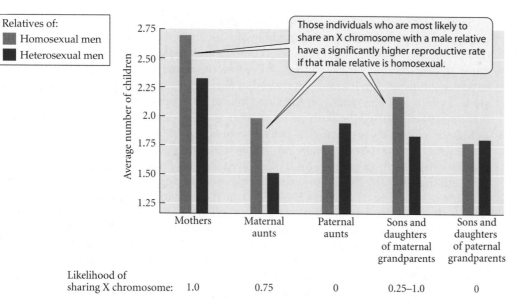

FIGURE 16.27 Does male homosexuality reflect a polymorphism maintained by intralocus sexual conflict? The graphs show the average number of children of relatives of homosexual and heterosexual men in northern Italy. (Data from Camperio-Ciani et al. 2004.)

suggested that data on the inheritance of sexual orientation and the higher fitness of homosexuals' relatives are best explained by inherited epigenetic modification of genes that affect the sensitivity of fetal brain tissue to androgen hormones.]

HUMAN BEHAVIORAL ECOLOGY   Human behavioral ecologists adopt optimality approaches to evolutionary biology, such as life history theory (see Chapter 14) and ESS models of interactions (see Chapter 12), to address questions such as whether humans hunt in optimal group sizes and whether marriage patterns appear to maximize reproductive success. For example, Monique Borgerhoff Mulder (1990) determined that women in a certain African tribe prefer to marry wealthy men, even if they already have wives. She compared her data with a model that describes some species of birds, in which a female's fitness is maximized by joining an already-mated male if his resources will enable her to raise more offspring than if she were to join a bachelor male who controls fewer resources. Human behavioral ecologists do not necessarily assume that a specific behavior is genetically determined; they may assume, instead, that humans have evolved the cognitive abilities to respond adaptively to various environmental circumstances. The response may have been learned, perhaps from a cultural norm, but this norm is itself adaptive.

EVOLUTIONARY PSYCHOLOGY   Some behavioral features are said to be universal among human populations. These features include body adornment, cooperation, crying, death rituals, division of labor, various facial expressions, gossip, language, males being more politically dominating and more aggressive, marriage, music, reciprocity, and many others (Pinker 2002). Whereas many or most social scientists have strongly favored the view that most universal social behaviors are culturally determined, evolutionary psychologists, like sociobiologists, assume that many such behaviors have an innate, genetically determined foundation, the expression of which is often modulated by cultural or other environmental influences. Evolutionary psychologists employ a model of the brain in which genetically determined "modules," or neural circuits, are designed to solve certain classes of problems, such as interacting with kin, detecting cheaters, or learning a language (Tooby and Cosmides 1992; Cartwright 2000; Pinker 2002). These psychological mechanisms are assumed (as in psychology generally: Norenzayan and Heine 2005) to underlie universal human capabilities that have evolved by natural selection. Some of these cognitive capacities are highly flexible, enabling people to adjust in novel ways to novel circumstances. Evolutionary psychologists do not deny culture, but they distinguish universal traits from cultural traits that can change rapidly.

Some research in evolutionary psychology uses psychological methods to test hypotheses about the postulated adaptive modules. For example, Leda Cosmides and John Tooby (1992) proposed that in social exchanges among unrelated individuals (reciprocity), cheating is an ever-present threat, so that evolved mechanisms for detecting cheaters should have design features that are not activated in nonsocial contexts. They presented college students with problems that had the same logical form but different content, and found that the students solved the problem more often if it described cheating than if it did not.

Many evolutionary psychologists use a cross-cultural approach, on the supposition that a postulated behavioral adaptation that is much the same among culturally very different populations may be an evolved trait. An example in a particularly controversial area is a cross-cultural study by David Buss of mate preferences among 37 diverse cultures in different parts of the world (D. M. Buss 1989, 1998). Buss aimed to test several hypotheses based on expectations from sexual selection theory: that females, more than males, should prefer mates with indicators (such as ambition and earning potential) of high potential for guarding and care of both mate and offspring; that men should place higher value than women on physical attractiveness in potential mates (a supposed indicator of age-specific reproductive potential); and that males should value chastity in their mates more than women do (since extra-pair copulation places males at greater risk than females of unknowingly caring for another's offspring). Buss reported that a large majority of cultures conformed to every prediction. For example, women are said to value earning potential more than men do in 97 percent of the cultures, men prefer women younger than themselves in 100 percent, and men value chastity in their mates more than women do in 62 percent. Similarly, many of the physical features that are considered appealing by the opposite sex are consistent across very different cultures, and some seem to be indicators of reproductive fitness (Gallup and Frederick 2010).

Adaptive evolutionary hypotheses have been proposed for many human behavioral and social features, from murder and jealousy to music-making and religious belief and ritual (see, e.g., Pinker 2002; Wilson 2002). It remains difficult, in many cases, to determine whether these characteristics are best explained by evolutionary (genetic) adaptation or by cultural adaptation, which might be expected to produce much the same phenotype, but by different means.

Cultural evolution and gene-culture coevolution

Differences in language, technologies, social customs, and way of life among the world's many different cultural groups clearly reflect a history of divergent change (**Figure 16.28**). Of the many scientists who have endeavored to understand how cultures change, Peter Richerson and Robert Boyd, of the University of California, have perhaps been the most inspired by models of genetic evolution to develop a judiciously analogous model of **cultural evolution**, based on processes within and among populations (Richerson and Boyd 2005; Boyd and Richerson 2009). The cognitive capacities envisioned by evolutionary psychologists may well exist, but they are not enough to enable an unschooled individual to make a boat or an arrow, much less a light bulb or even a nail. It is human culture that enables adaptive behaviors to accumulate over the course of many generations, so that societies acquire knowledge and capabilities far beyond what any one individual could invent or learn de novo. Only the human species has evolved this ability.

Just as individual genes mutate, Richerson and Boyd say, *nongenetic variations* of cultural traits are produced by individual people in a population. Some such changes are random, such as a mispronunciation of a word or misremembering of a cultural item. Others are nonrandom, perhaps transformed during learning or intentionally altered by an individual, possibly in an effort to improve a method. The frequencies of cultural variants may change by random fluctuation, or "cultural drift" (e.g., the few practitioners of a special craft in a small population may die before having trained apprentices), or by BIASED or NONRANDOM TRANSMISSION. A cultural variant increases in frequency if it is copied or learned by other individuals— that is, if it is transmitted to them. Individuals may adopt or imitate a cultural trait

(A)

(B)

(C)

FIGURE 16.28 Variant cultural traits. Wardrobe and technology are some of the cultural differences among contemporary human populations such as (A) the Huli of montane Papua New Guinea, (B) the Inuit of northern Canada, and (C) urban professionals in the United States and western Europe. (A, photo by the author.)

because of its content (e.g., it is perceived to be advantageous, or is simply easy to remember), because of features of the individuals who already exhibit the trait (e.g., copying prestigious or successful persons), or because the trait is already common and people perceive a reason to conform to the norm. Copying others' behavior is often advantageous in the absence of any other information; or the society may actively enforce group norms by coercion.

Cultural traits are also subject to *natural selection* if they affect people's behavior in ways that increase the probability of the trait's transmission to other people. (Recall that natural selection was defined in Chapter 11 as a consistent difference in fitness among phenotypically different classes of biological entities. Here the entities are cultural traits, which are biological in the sense that they are produced by brains.) Selection may occur within populations if, for example, people acquire traits from their parents and those traits affect the parents' survival or reproduction. Hygienic or food-gathering traits can affect survival, for instance; religious beliefs and doctrines can affect reproductive traits, even in modern societies. Traits that are acquired by imitating celebrities, teachers, and other nonparental figures may be selected if they affect the likelihood of becoming such a figure.

Richerson and Boyd emphasize that cultural differences among groups, such as tribes, can evolve by group selection as a result of competition among groups. Even if groups are very large, they suggest, group selection can be more effective in cultural than in genetic evolution because cultural traits are often very homogeneous within groups, since group norms may be forcibly maintained. A culture of Christianity, for example, has been maintained in many populations not only by inheritance of parents' beliefs, but also by policing that ranges from the Inquisitions of the past to the social exclusion of nonconformists in some societies today. The ways in which traits may be subjected to group selection include differential population growth based on agricultural practice and superiority in warfare. Darwin envisioned such group selection in *The Descent of Man* (1874), writing that "although a high standard of morality gives but a slight or no advantage to each individual man and his children over other men of the same tribe, … an increase in the number of well-endowed men and an advancement in the standard of morality will certainly give an immense advantage to one tribe over another."

Many traits, however, will spread by biased transmission rather than selection, and although they may well be advantageous in a particular ecological or social environment, they need not be. Maladaptive traits can spread if they have a conformist bias (e.g., belief in witchcraft) or a prestige bias (e.g., smoking among teenagers), and most people at most times in history have not had any information that would enable them to tell whether or not a belief (e.g., in witches) or cultural practice (e.g., sacrificing to the gods) was true or beneficial.

Many cultural traits concern marriage rules, cooperation, conflict resolution, and other social interactions that are the subject of adaptive hypotheses in human behavioral ecology and evolutionary psychology. Richerson and Boyd suggest that genes and culture may have coevolved. They note that large brains and cumulative cultural adaptation both seem to have developed about 500,000 years ago, so cultural environments could have imposed selection on genes for about 20,000 generations. (For example, mutations that enable adults to digest lactose were selected in dairying cultures; see Chapter 12).

Language may epitomize gene-culture coevolution. Humans have not only an innate capacity for language that is far greater than in other apes, but also physical adaptations such as the position of the larynx, which is lower in the throat than in apes, enabling a greater variety of sounds. These features must have evolved because there was selection for greater linguistic versatility, probably based on its advantages in social interactions. This selection could have occurred only in a population that already was using rudimentary language of some kind. The subsequent changes in larynx position and in language centers of the brain, in turn, enabled languages to develop greater complexity by cultural change.

Similarly, there may have been selection for genotypes with a greater predisposition to adopt certain kinds of cultural traits or for characteristics that influence the processes of cultural change. If cultural traits that promote cooperation were advantageous to survival or reproduction (of individuals or of groups), genotypes more inclined to cooperate would increase in frequency. If conforming to group beliefs generally enhanced survival, a genetic tendency toward conformity could evolve. To some extent, then, some cognitive features such as those postulated by evolutionary psychologists might have evolved, but in cultural contexts that themselves could change. Richerson and Boyd suggest that we might have two sets of innate predispositions that evolved genetically. The older set, shared with many of our primate relatives, evolved by kin selection and reciprocity; it concerns interactions among family members, mate preferences, and relatively simple social interactions. The younger set of such predispositions, which evolved during the last few hundred thousand years or less, may involve features, such as true empathy and altruism, that are advantageous in the context of large tribes and societies, in which simple reciprocity is ineffective. But they are predispositions only, and they allow for an immense range of cultural expressions and individual potentialities. No one should fear that such evolved biological foundations of our psychology and social constructions dictate what we can, much less what we should, do with our lives.

Go to the
EVOLUTION
Companion Website at
sites.sinauer.com/evolution3e
for quizzes, data analysis and simulation exercises, and other study aids.

Summary

1. Many biological phenomena result from conflict or cooperation among organisms or among genes. The evolution of most interactions can be explained best by selection at the level of individual organisms or genes.

2. Characteristics that contribute to conflict and cooperation often evolve by frequency-dependent selection. Such features can sometimes be modeled by calculating the evolutionarily stable strategy (ESS): the phenotype that, once established, cannot be replaced by mutant phenotypes. In animals, interactions are often mediated by signals that may or may not honestly indicate the individual's strength, potential parental caregiving ability, or other relevant phenotypic variable.

3. Altruism benefits other individuals and reduces the fitness of the actor, whereas cooperative behavior need not reduce the

actor's fitness. Cooperation can evolve because it is directly beneficial to the actor, although the benefit may be delayed, or by reciprocity, based on repeated interactions between individuals or on long-lasting associations in which the fitness interests of the associates are aligned. Cooperative interactions, both within and between species, can be maintained in part by "policing," or punishment of cheaters.

4. Altruism generally evolves by kin selection, which is based on differences among alleles or genotypes in their inclusive fitness: the combination of an allele's direct effects (on its carrier's fitness) and its indirect effects (on other copies of the allele, borne by the carrier's kin). Hamilton's rule describes the conditions for increase of an allele for an altruistic trait in terms of the benefit to the beneficiary, the cost to the donor,

and the coefficient of relationship between them. Short-lived groups containing altruistic genotypes may contribute disproportionately to the entire population, but these groups generally consist of kin. Thus this form of group selection is generally the same as kin selection.

5. Conflict and kin selection together affect the evolution of many interactions among family members. The genetic benefit of caring for offspring is an increase in the number of current offspring that survive. The genetic cost is the number of additional offspring that the parent could expect to have if she/he abandoned the offspring and reproduced again. Parental care is expected to evolve only if its genetic benefit exceeds its genetic cost. Whether or not one or both parents evolve to provide care can depend on the male's confidence of paternity and on the relative cost/benefit ratio for each parent.

6. Conflicts between parents and offspring may arise because a parent's fitness may be increased by allocating some resources to its own survival and future reproduction, thus providing fewer resources to current offspring than would be optimal from the offspring's point of view. This principle may be one of several reasons why in many species, parents may reduce their brood size by aborting some embryos or killing some offspring.

7. The most extreme examples of cooperation and altruism are in eusocial insects and humans. In eusocial insects, in which unmated workers rear other workers and reproductive individuals (queens and males), many interactions are governed by kin selection and by policing.

8. Conflicts may exist among different genes in a species' genome. For example, a particular gene may spread at a faster rate than other parts of the genome, engendering selection for other genes to prevent it from doing so. At loci that are transmitted through only one sex, genic selection favors alleles that alter the sex ratio in favor of that sex. Such alteration creates selection at other loci for suppressors that restore the 1:1 sex ratio.

9. Other phenomena explained by genetic conflicts include genomic imprinting, which affects the expression of maternally and paternally derived alleles in mammalian embryos, and the evolution of integration among cells—the very essence of multicellular organisms.

10. The extent to which human behaviors, including social behaviors, have an evolved genetic foundation is a highly controversial topic. It is most readily addressed in studies of traits that vary within populations. Human behavioral ecology and evolutionary psychology are two of the contemporary approaches to understanding the evolution and adaptive significance of human behaviors.

11. Human traits that are claimed to be universal, although usually variable in expression (e.g., language, body adornment), are likely to have some genetic foundation that has evolved since our common ancestor with other apes. How trait-specific the genetic bases may be is unknown. Evolutionary psychologists propose that the brain includes many rather problem-specific psychological "modules." Another view is that we have broad genetic predispositions, some of which are held in common with other apes (e.g., altruism toward close kin) and others of which evolved more recently by natural selection stemming from cultural environments (e.g., capacity for empathy with strangers in large societies). Cultures evolve nongenetically by processes that are partly analogous to genetic evolution, and cultural and genetic changes may influence each other. Such genetic predispositions as may constitute so-called human nature, however, do not evidently constrain or limit any groups of people more than others.

Terms and Concepts

altruism
coefficient of
    relationship
cooperation
cultural evolution

direct fitness
genetic conflict
Hamilton's rule
horizontal
    transmission

inclusive fitness
indirect fitness
kin selection
mutualism

parent-offspring
    conflict
reciprocity
vertical transmission

Suggestions for Further Reading

An outstanding, easily readable introduction to the study of behavior, emphasizing evolutionary explanations, is *Animal Behavior: An Evolutionary Approach*, by J. Alcock (tenth edition, Sinauer Associates, Sunderland, MA, 2013). *Evolutionary Behavioral Ecology*, edited by D. Westneat and C. W. Fox (Oxford University Press, Oxford, 2010), includes chapters by authorities on all the major topics in this field. The evolution of social behavior and its implications for major transitions in evolution are comprehensively treated by A. F. G. Bourke in *Principles of Social Evolution* (Oxford University Press, Oxford, 2011).

An excellent set of essays on many aspects of cooperation and conflict is *Levels of Selection*, edited by L. Keller (Princeton University Press, Princeton, NJ, 1999). Genetic conflict is reviewed in A. R. Burt and R. Trivers, *Genes in Conflict: The Biology of Selfish Genetic Elements* (Harvard University Press, Cambridge, MA, 2006).

A few of the many books on evolution and human behavior are J. Cartwright, *Evolution and Human Behavior* (MIT Press, Cambridge, MA, 2000); R. I. M. Dunbar and L. Barrett (eds.), *Oxford Handbook of Evolutionary Psychology* (Oxford University Press, Oxford, 2009); D. M. Buss, *Evolutionary Psychology: The New Science of the Mind*, fourth edition (Pearson Allyn & Bacon, Boston, 2012); S. Pinker, *The Blank Slate: The Modern Denial of Human Nature* (Penguin, New York, 2002); and P. J. Richerson and R. Boyd, *Not by* *Genes Alone: How Culture Transformed Human Evolution* (University of Chicago Press, Chicago, 2005).

Problems and Discussion Topics

1. Describe the contexts in which animals might evolve signals used in interactions among members of the same species or in interactions with different species. Would you expect each kind of signal to be honest or dishonest? Why?

2. Speculate about why some genetic elements (e.g., mitochondria and chloroplasts) are usually inherited through only one parent rather than biparentally, as are most nuclear genes. (See Hurst et al. 1996.)

3. Kin selection explains why organisms may provide benefits to relatives. Is there a conflict between the principle of kin selection and the evolution of siblicide and abortion?

4. Explain why we may expect mitochondrial genes to have exceptionally high frequencies of mutations that are harmful to males.

5. Analyze arguments for and against the proposition that men are more aggressive than women because of sexual selection for competitiveness. (See, for example, Daly and Wilson 1983 versus Kitcher 1985.)

6. Standardized for body weight, testes are larger in polygamous than in monogamous species of primates and several other groups of animals (see Figure 11.23). This observation has been used to infer what the predominant mating system of *Homo sapiens* has been for most of its evolutionary history (see, e.g., Harvey and May 1989). What has been the conclusion, and how might we evaluate its validity?

7. Why is it more difficult to test the hypothesis that a species-typical trait (such as the supposedly universal human behavioral traits discussed in this chapter) has an evolved genetic foundation than to do so for a variable trait? Is it sufficient to conclude that a trait has a genetic foundation if it differs consistently between two closely related species, such as human and chimpanzee?

8. Many cultural traits that do not have a genetic basis, such as various methods of obtaining and preparing food and shelter, are adaptive. Is the propensity to believe in powerful immaterial entities, such as spirits and gods, a cultural adaptation? How might it have (culturally or biologically) evolved, and how might the hypotheses be tested? (See Wilson 2002 and Dennett 2006.)

9. What are the similarities and differences between the likely processes of biological evolution and cultural evolution?

10. Traditional group selection explained cooperative behaviors as the consequence of the enhanced survival of populations of cooperative or altruistic genotypes compared with populations of selfish genotypes. Describe at least three kinds of observations that are more consistent with kin selection, reciprocity, or direct benefit than with traditional group selection.

Species

Speciation forms the bridge between the evolution of populations and the evolution of taxonomic diversity. The diversity of organisms is the consequence of CLADOGENESIS, the branching or multiplication of lineages, each of which then evolves (by ANAGENESIS, or evolution within species) along its own path. Each branching point in the great phylogenetic tree of life marks a speciation event: the origin of two species from one. In speciation lies the origin of diversity, and the study of speciation bridges microevolution and macroevolution. The most important consequence of speciation is that different species undergo largely independent divergence, maintaining separate identities, evolutionary tendencies, and fates (Wiley 1978). It is also possible that speciation facilitates the evolution of new morphological and other phenotypic characters.

Many events in the history of evolution are revealed to us only by virtue of speciation. If a single lineage evolves great changes but does not branch, the record of all steps toward its present form is erased, unless they can be found in the fossil record. But if the lineage branches frequently, and if intermediate stages of a character are retained in some branches that survive to the present, then the history of evolution of the feature may be represented, at least in part, among living species. This fact is used routinely to infer phylogenetic relationships among living taxa and to trace the evolution of characteristics on phylogenetic trees (see Figure 3.1).

Two species that can hybridize Although the indigo bunting (*Passerina cyanea*) and the lazuli bunting (*Passerina amoena*) differ in plumage, they hybridize where their ranges overlap.

Some steps toward speciation may occur fast enough for us to study directly, but the full history of the process is usually too prolonged for one generation, or even a few generations, of scientists to observe. Conversely, speciation is often too fast to be fully documented in the fossil record, and even an ideal fossil record could not document some of the genetic processes in speciation that are still inadequately understood. Thus the study of speciation is based largely on inferences from living species.

What Are Species?

Many definitions of "species"—which is Latin for "kind"—have been proposed. It is important to bear in mind that a definition is not true or false, because the definition of a word is a convention. Still, if a conventional definition of a word has been established, one can apply it in error. Although a rose by any other name would smell as sweet, we would be wrong to call a rose a skunk cabbage. A definition can be more or less useful, and it can be more or less successful in accurately characterizing a concept or an object of discussion. Probably no definition of "species" suffices for all the contexts in which a species-like concept is used.

Linnaeus and other early taxonomists held what Ernst Mayr (1942, 1963) has called a TYPOLOGICAL or ESSENTIALIST notion of species. Individuals were members of a given species if they sufficiently conformed to that "type," or ideal, in certain morphological characters that were "essential" fixed properties of the species—a concept descended from Plato's "ideas" (see Chapter 1). Thus a bird specimen was a member of the species *Corvus corone*, the carrion crow, if it looked like a carrion crow, and it was a common raven (*Corvus corax*) if it had certain different features (**Figure 17.1A,B**). But typological notions of species were challenged by variation. Carrion crows, which are entirely black, are readily distinguished from hooded crows (*Corvus cornix*), which are black and gray (**Figure 17.1C**)—except in a narrow region in central Europe, where crows have various amounts of gray (**Figure 17.1D**). Are there two species or one?

Almost all definitions of species share the notion that species are independently evolving lineages (de Queiroz 2007). They differ, however, as to the properties that define those lineages. Among the many proposed definitions of species (**Table 17.1**), those most commonly advocated at this time are the **biological species concept** (**BSC**) and several variations on the **phylogenetic species concept** (**PSC**). These two types of definitions differ chiefly in that phylogenetic species concepts emphasize species as the outcome of evolution—the *products* of a history of evolutionary divergence—whereas the biological species concept emphasizes the *process* by which species arise and takes a prospective view of the future status of populations (Harrison 1998). The biological species concept and closely related definitions have been and continue to be most frequently used by evolutionary biologists who are concerned with processes of evolution. No matter which species concept is adopted, however, some populations of organisms will not be unambiguously assigned

FIGURE 17.1 Three closely related birds that illustrate the problem of defining species. (A) Common ravens (*Corvus corax*) are distinguished from crows by their larger size, heavier bill, shaggy throat feathers, more pointed tail (when viewed from above or below), and voice. Even though (B) carrion crows (*Corvus corone*) and (C) hooded crows (*C. cornix*) are superficially more different from each other than from the common raven, hybrids showing various mixtures of their plumage patterns are found in a hybrid zone in central Europe (D). *C. corone* and *C. cornix* have been classified by some taxonomists as subspecies of a single species, but because they appear to exchange genes only to a very limited extent, they might better be considered species. (D after Mayr 1963.)

(A) *Corvus corax*

(B) *Corvus corone*

(C) *Corvus cornix*

(D)

Hybrid zone

corone

cornix

TABLE 17.1 Some species concepts

Biological species concept   Species are groups of actually or potentially interbreeding natural populations that are reproductively isolated from other such groups (Mayr 1942).

Evolutionary species concept   A species is a single lineage (an ancestor–descendant sequence) of populations or organisms that maintains an identity separate from other such lineages and which has its own evolutionary tendencies and historical fate (Wiley 1978).

Phylogenetic species concepts   (1) A phylogenetic species is an irreducible (basal) cluster of organisms that is diagnosably distinct from other such clusters, and within which there is a parental pattern of ancestry and descent (Cracraft 1989). (2) A species is the smallest monophyletic group of common ancestry (de Queiroz and Donoghue 1990).

Genealogical species concept   Species are "exclusive" groups of organisms, where an exclusive group is one whose members are all more closely related to one another than to any organism outside the group (Baum and Shaw 1995).

Recognition species concept   A species is the most inclusive population of individual biparental organisms that share a common fertilization system (Paterson 1985).

Cohesion species concept   A species is the most inclusive population of individuals having the potential for phenotypic cohesion through intrinsic cohesion mechanisms (Templeton 1989).

---

to one species or another. There are always borderline cases, because the species properties advanced in every definition evolve gradually.

Phylogenetic species concepts

Phylogenetic species concepts, which are accepted by many systematists, emphasize the phylogenetic history of organisms. In one widely accepted definition, lineages are different species if they can be distinguished (diagnosed): a phylogenetic species is "an irreducible (basal) cluster of organisms diagnosably different from other such clusters, and within which there is a parental pattern of ancestry and descent" (Cracraft 1989). This definition would presumably apply to both sexual and asexual organisms. According to this definition, speciation would occur whenever a population undergoes fixation of a genetic difference—even a single DNA base pair—that distinguishes it from related populations. This would be the case if all the gene copies in each of the species were to coalesce (i.e., be traceable to different, individual ancestral gene copies that differ by one or more mutations; see the discussion of coalescence in Chapter 10 and later in this chapter.) Because of incomplete lineage sorting, however, two reproductively isolated populations may undergo substantial genetic divergence (i.e., allele frequency differentiation at many loci) before complete coalescence occurs at any one locus. Thus a species definition based on fixed genetic differences may not recognize some recently formed species (Hudson and Coyne 2002). For this reason, some biologists recognize species by evidence, based on several or many loci or genetic markers, that populations have diverged enough to be considered separate lineages (Knowles and Carstens 2007; Fujita et al. 2012).

The biological species concept

The biological species concept used throughout this book was defined by Ernst Mayr (1942) as follows: "*Species are groups of actually or potentially interbreeding populations, which are reproductively isolated from other such groups.*" **Reproductive isolation** means that any of several biological differences between the populations greatly reduce gene exchange between them, even when they are not geographically separated. These differences may or may not result in mortality or sterility of hybrids between the two forms. Mayr (1942) and other advocates of the BSC have not insisted that populations must be 100 percent reproductively isolated in order to qualify as species; they recognize that, as in the case of the crows in Figure 17.1, there can be a little genetic "leakage" between some species.

Although genetic or phenotypic differences do not *define* species according to the BSC, such differences enable us to *recognize* and distinguish them.

The roots of the biological species concept are very old, for it has always been recognized that morphologically very different organisms (e.g., different sexes) might be born of the same parents and thus be **conspecific** (members of the same species). Nevertheless, the BSC arose from studies of variation that showed that morphological similarity and difference do not suffice to define species. Several critical observations contributed to its development:

1. *Variation within populations.* Characteristics vary among the members of a single population of interbreeding individuals. The white and blue forms of the snow goose (see Figure 9.1A), known to be born to the same mother, represent a genetic polymorphism, not different species. A mutation that causes a fruit fly to have four wings rather than two is just that: a mutation, not a new species.

2. *Geographic variation.* Populations of a species differ; there exists a spectrum from slight to great difference; and intermediate forms, providing evidence of interbreeding, are often found where such populations meet. Human populations are a conspicuous example.

3. *Sibling species.* **Sibling species** are reproductively isolated populations that are difficult or impossible to distinguish by morphological features, but which are often recognized by differences in ecology, behavior, chromosomes, or genetic markers. The discovery that the European mosquito *Anopheles "maculipennis"* was actually a cluster of six sibling species had great practical importance because some transmit human malaria and others do not. (**Box 17A** provides definitions of sibling species and some other terms related to the biological species concept; Box 17B describes the diagnosis of a sibling species.)

Box 17A Some Terms Encountered in the Literature on Species

Some of the following terms are used frequently in the literature on species, others infrequently. These definitions conform to usage by adherents to the biological species concept (e.g., Mayr 1963).

Geographic isolation Reduction or prevention of gene flow between populations by an extrinsic barrier to movement, such as topographic features or unfavorable habitat.

Reproductive isolation Reduction or prevention of gene flow between populations by genetically determined differences between them.

Allopatric populations Populations occupying separated geographic areas.

Parapatric populations Populations occupying adjacent geographic areas, meeting at the border.

Sympatric populations Populations occupying the same geographic area and capable of encountering one another.

Hybrid zone A region where genetically distinct populations meet and interbreed to some extent, resulting in some individuals of mixed ancestry ("hybrids").

Introgression The movement, or incorporation, of genes from one genetically distinct population (usually considered a species or semispecies) into another.

Sibling species Reproductively isolated species that are difficult to distinguish by morphological characteristics.

Sister species Species that are thought, on the basis of phylogenetic analysis, to be each other's closest relatives, derived from an immediate common ancestor. (Cf. sister groups in phylogenetic systematics.)

Chronospecies Phenotypically distinguishable forms in an ancestor-descendant series in the fossil record that are given different names.

Subspecies Populations of a species that are distinguishable by one or more characteristics and are given subspecific names (see the subspecies of flickers, Figure 9.27). In zoology, subspecies have different (allopatric or parapatric) geographic distributions and are equivalent to "geographic races." In botany, they may be sympatric forms.

Race A vague term, sometimes equivalent to subspecies and sometimes to polymorphic genetic forms within a population.

Semispecies Usually, one of two or more parapatric, genetically differentiated groups of populations that are thought to be partially, but not fully, reproductively isolated; nearly, but not quite, different species.

Ecotype Used mostly in botany to designate a phenotypic variant of a species that is associated with a particular type of habitat; may be designated a subspecies.

Domain and application of the biological species concept

All concepts have limitations. A concept may have a limited *domain of application* (for example, "matter" is an ambiguous concept on a subatomic scale). A concept may inadequately describe *borderline cases*. (The concept of an "individual organism," for instance, is ambiguous in a grove of aspen trees that have grown by vegetative propagation from a single seed.) There may also be *practical limitations* in applying the concept. (The "human population of the world" is a presumably valid concept, but technological and economic limitations prevent us from counting the population accurately.)

DOMAIN The domain of the BSC is restricted to sexual, outcrossing organisms. It is also limited to short intervals of time, since it is meaningless to ask whether an ancestral population could have interbred with its descendants a million years later. To be sure, asexually reproducing organisms are given binomial names, such as *Escherichia coli*, and putative ancestral and descendant populations in the fossil record may be distinguished by such names, such as the mid-Pleistocene *Homo erectus* and the later *Homo sapiens*. These cases illustrate that the word "species" has two overlapping but distinct meanings in biology. One meaning is embodied in the BSC. The other meaning is *a taxonomic category*, just like "genus" and "family." Some organisms that bear binomial names (such as *Escherichia coli*) are taxa in the species category, but are not biological species.

BORDERLINE CASES Interbreeding versus reproductive isolation is not an either/or, all-or-none distinction. There exist graded levels of gene exchange among adjacent (parapatric) populations and sometimes between sympatric, more or less distinct populations. Several such situations are encountered frequently.

Narrow **hybrid zones** exist where genetically distinct populations meet and interbreed to a limited extent, but in which there exist partial barriers to gene exchange (see Figure 17.1D). The hybridizing entities are often recognized as species but may be called **semispecies**.

Partial, but not completely free, gene exchange sometimes occurs between populations that are rather broadly sympatric. Such SYMPATRIC HYBRIDIZATION is not uncommon in plants (**Figure 17.2**), although it is concentrated in a small minority of plant families and genera (Ellstrand et al. 1996). Molecular markers have revealed that in many animals as well, parts of the genome are exchanged between related species by low levels of interbreeding (Schwenk et al. 2008).

PRACTICAL DIFFICULTIES The greatest practical limitation of the BSC lies in the difficulty of determining whether or not geographically segregated (allopatric) populations belong to the same species, for applying the BSC requires us to assess whether or not they would *potentially* interbreed if they should someday encounter each other. In many instances, range extension or colonization could well bring presently separate populations

FIGURE 17.2 An example of sympatric hybridization. The gray oak (*Quercus grisea*) and Gambel's oak (*Q. gambelii*) have broadly overlapping ranges in the southwestern United States, including much of Texas, New Mexico, Arizona, and Colorado. Hybrids showing variation in leaf shape and other features are found in many of these places. Hybridization occurs among many species of oaks, willows, and some other plant taxa. (Photos courtesy of M. Cain.)

Quercus grisea (gray oak)

Hybrids

Quercus gambelii (Gambel's oak)

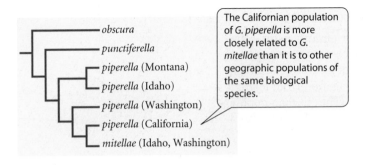

The Californian population of *G. piperella* is more closely related to *G. mitellae* than it is to other geographic populations of the same biological species.

FIGURE 17.3 The phylogeny of some species and populations in the moth genus *Greya*, based on mitochondrial DNA sequence data. This pattern would be expected if a local population of ancestral *G. piperella* evolved reproductive isolation and became the distinct biological species *G. mitellae*. According to the biological species concept, *G. mitellae* and *G. piperella* would be recognized as two species, one of which (*G. piperella*) is paraphyletic. According to the phylogenetic species concept, if *G. mitellae* is recognized as a species, the various populations of *G. piperella* should also be called species. (After Harrison 1998.)

into contact, so the evolutionary future of the populations depends on whether or not they have evolved reproductive isolation. Humans have inadvertently or purposely introduced many species into new areas, some of which have hybridized with native populations (Abbott 1992), and many native species have extended their ranges considerably over the course of decades. After the Pleistocene glaciations, disjunct populations of a great many species expanded their ranges, met one another, and in many instances now interbreed. Future changes in climate will undoubtedly affect some currently disjunct populations in similar ways.

In principle, one can test for reproductive isolation between allopatric populations, and such tests have been performed with *Drosophila* and many other organisms by bringing them together in a laboratory or garden. For many organisms, however, such tests are not feasible (although the mere impracticality of such studies does not invalidate the concept of reproductive isolation). In practice, therefore, the classification (i.e., naming) of allopatric populations is often somewhat arbitrary. Commonly, allopatric populations have been classified as species if their differences in phenotype or in DNA sequence are as great as those usually displayed by sympatric species in the same group (Tobias et al. 2010b).

WHEN SPECIES CONCEPTS CONFLICT   Advocates of the BSC and the PSC are most likely to classify populations differently in two circumstances. First, allopatric populations that can be distinguished by fixed characters are species according to the PSC, but if the diagnostic differences are slight, advocates of the BSC may recognize the populations as geographic variants of a single species.

Second, in some cases a local population of a widespread species evolves reproductive isolation from other populations, which remain reproductively compatible with one another (Funk and Omland 2003). Phylogenetic study may then show that the "new" species is more closely related to some populations of the "old" species than some of the "old" populations are to one another. Under the BSC, two species would be recognized, one of which is paraphyletic (**Figure 17.3**). (Recall from Chapter 2 that a paraphyletic taxon lacks one or more of the descendants of the common ancestor of the members of that taxon, and that paraphyletic taxa are not acceptable under cladistic rules of classification.) Under the PSC, the various distinguishable populations of the paraphyletic group might be named as distinct species.

Barriers to Gene Flow

Gene flow between biological species is largely or entirely prevented by biological differences that have often been called ISOLATING MECHANISMS, but which we will term **isolating barriers**, or BARRIERS TO GENE FLOW. Under the BSC, therefore, speciation—the origin of two species from a common ancestral species—consists of the evolution of biological barriers to gene flow. As we noted above, mere physical isolation does not define populations as different species, although isolation by topographic or other barriers is considered to be instrumental in the formation of species.

It is an error to think that sterility of hybrids is the criterion of species as conceived in the BSC. There are many kinds of isolating barriers (**Table 17.2**). The most important

---

TABLE 17.2 A classification of isolating barriers

I. Premating barriers: Features that impede transfer of gametes to members of other species

  A. Ecological isolation: Potential mates (although sympatric) do not meet

    1. Temporal isolation (populations breed at different seasons or times of day)

    2. Habitat isolation (populations have propensities to breed in different habitats in the same general area, and so are spatially segregated)

    3. Immigrant inviability (immigrants between populations do not survive long enough to interbreed)

  B. Potential mates meet but do not mate

    1. Sexual (behavioral or ethological) isolation (in animals, differences prevent populations from mating)

    2. Pollinator isolation (in plants, pollen is transferred in different populations by different animal species or on different body parts of a single pollinator; may also be classified as ecological isolation)

II. Postmating, prezygotic barriers: Mating or gamete transfer occurs, but zygotes are not formed

  A. Mechanical isolation (copulation occurs, but no transfer of male gametes takes place because of failure of mechanical fit of reproductive structures)

  B. Copulatory behavioral isolation (failure of fertilization because of behavior during copulation or because genitalia fail to stimulate properly)

  C. Gametic isolation [failure of proper transfer of gametes or of fertilization, either due to competition between conspecific and heterospecific gametes (conspecific sperm precedence or pollen tube precedence) or intrinsic incompatibility]

  D. Immigrant inviability after mating occurs in "alien" habitat

III. Postzygotic barriers: Hybrid zygotes are formed but have reduced fitness

  A. Extrinsic (hybrid fitness depends on context)

    1. Ecological inviability (hybrids do not have an ecological niche in which they are competitively equal to parent species)

    2. Behavioral sterility (hybrids are less successful than parent species in obtaining mates)

  B. Intrinsic (hybrid fitness is low because of problems that are relatively independent of environmental context)

    1. Hybrid inviability (developmental problems cause reduced survival)

    2. Hybrid sterility (usually due to reduced ability to produce viable gametes; also "behavioral sterility," neurological incapacity to perform normal courtship)

Source: After Coyne and Orr 2004.

---

distinction is between **prezygotic barriers** and **postzygotic barriers**. Most prezygotic barriers are premating barriers, although some species are isolated by postmating, prezygotic barriers. Any of these barriers may be incomplete; for example, interspecific mating may occur at a low rate, or hybrid offspring may have reduced fertility but not be entirely sterile. Especially if the chief barriers to gene exchange are prezygotic and the species' environment is altered, reproductive isolation can sometimes break down and speciation may be reversed (Seehausen 2006b). For example, pollution (eutrophication) of lakes has caused increased interbreeding among closely related species of fishes such as whitefishes and cichlids (Seehausen et al. 1997; Vonlanthen et al. 2012). In the cichlids, pollution altered the light spectrum so that individuals could no longer perceive differences in coloration that had served as mating signals.

Premating barriers

Premating barriers prevent (or reduce the likelihood of) transfer of gametes to members of other species.

ECOLOGICAL ISOLATION **Ecological isolation** includes mechanisms that reduce the opportunities for closely related sympatric species to meet and mate. Many species breed at different times of year (seasonal isolation). Two closely related field crickets (*Gryllus*

FIGURE 17.4 Oscillograms of the songs of three morphologically indistinguishable species of green lacewings (*Chrysoperla*). Each oscillogram displays a plot of amplitude (intensity of sound) against time. Two of the species, *C. adamsi* and *C. johnsoni*, were distinguished and named only after studies of songs and DNA differences made it clear that they are distinct species. (After Martínez Wells and Henry 1992a.)

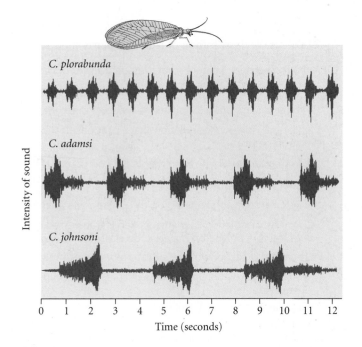

pennsylvanicus and *G. veletis*), for example, reach reproductive age in fall and spring, respectively, in the northeastern United States (Harrison 1979). Some species are isolated by habitat, so potential mates seldom meet. For instance, two Japanese species of herbivorous ladybird beetles (*Henosepilachna nipponica* and *H. yasutomii*) feed on thistles (*Cirsium*) and blue cohosh (*Caulophyllum*), respectively. Each species mates exclusively on its own host plant, and this ecological segregation appears to be the only barrier to gene exchange (Katakura and Hosogai 1994). Disturbance of habitat sometimes results in a breakdown of ecological isolation; for example, the wild irises *Iris fulva* and *I. hexagona* hybridize in Louisiana where their habitats (bayous and marshes, respectively) have been disturbed (Nason et al. 1992). An important form of ecological isolation is IMMIGRANT INVIABILITY, which occurs if individuals of one species or population fail to survive when they disperse into the habitat of the other species or population and consequently fail to mate with the residents (Nosil et al. 2005).

SEXUAL ISOLATION In animals, **sexual isolation** (also called BEHAVIORAL ISOLATION or ETHOLOGICAL ISOLATION) is an important barrier to gene flow among sympatric species that frequently encounter each other, but simply do not mate. Commonly, one sex (often the female) will not respond to inappropriate male display signals. In three morphologically indistinguishable species of green lacewings (*Chrysoperla*), for example, a male and a female engage in a duet, initiated by the male, of low-frequency songs produced by vibrating the abdomen (Martínez Wells and Henry 1992b). Mating does not occur unless the female sings back to the male. The three species produce very different songs (**Figure 17.4**), and females respond much more often to recordings of their own species than to those of other species. Moreover, they discriminate against the intermediate songs produced by hybrids.

Sexual isolation in many animals (e.g., many mammals and insects) is based on differences in chemical mating signals (sex pheromones). Many other groups (e.g., many birds, fishes, jumping spiders) use visual signals, sometimes accompanied by acoustic or chemical signals, in their courtship displays (**Figure 17.5**). Differences in such signals often underlie sexual isolation. In many organisms, the courtship signals have not been identified, but it is nevertheless possible to measure sexual isolation by comparing the frequency of conspecific and heterospecific matings in experimental settings.

In plants, the nearest equivalent of sexual isolation is pollination of different species by different pollinating animals that respond to differences in the color, form, or scent of

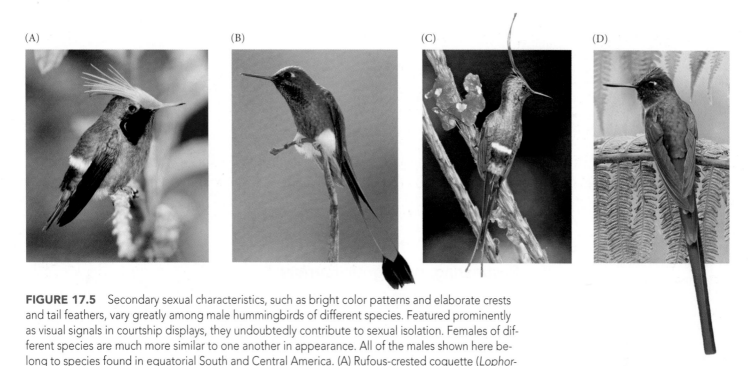

FIGURE 17.5 Secondary sexual characteristics, such as bright color patterns and elaborate crests and tail feathers, vary greatly among male hummingbirds of different species. Featured prominently as visual signals in courtship displays, they undoubtedly contribute to sexual isolation. Females of different species are much more similar to one another in appearance. All of the males shown here belong to species found in equatorial South and Central America. (A) Rufous-crested coquette (*Lophornis delattrei*). (B) Booted racket-tail (*Ocreatus underwoodii*). (C) Wire-crested thorntail (*Discosura popelairii*). (D) Violet-tailed sylph (*Aglaiocercus coelestis*).

flowers (Grant 1981). For example, the monkeyflower *Mimulus lewisii*, which, like most members of its genus, is pollinated by bees, has pink flowers with a wide corolla (**Figure 17.6A**). Its close relative *M. cardinalis* has a narrow, red, tubular corolla and is pollinated by hummingbirds (Schemske and Bradshaw 1999; **Figure 17.6C**).

FIGURE 17.6 Pollinator isolation in monkeyflowers. (A) *Mimulus lewisii* has the broadly splayed petals characteristic of many bee-pollinated flowers. (B) An F$_1$ hybrid between *M. lewisii* and *M. cardinalis*. (C) *M. cardinalis* has the red coloration and narrow, tubular form that have evolved independently in many bird-pollinated flowers. (D–F) Some F$_2$ hybrids, showing the variation that Schemske and Bradshaw used to analyze the genetic basis of differences between these two species. (From Schemske and Bradshaw 1999.)

(A)

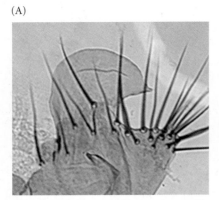

(B)

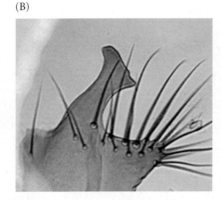

(C)

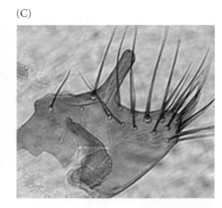

FIGURE 17.7 The posterior lobe of the genital arch in males of three closely related species of *Drosophila*: (A) *D. simulans*, (B) *D. sechellia*, and (C) *D. mauritiana*. This morphological feature is almost the only one by which these species differ. Differences in genitalia can contribute to reproductive isolation between species if copulation between them occurs. (Photos courtesy of J. R. True.)

Postmating, prezygotic barriers

On the border between premating and postzygotic barriers are features that prevent successful formation of hybrid zygotes even if mating takes place. In many groups of insects and some other taxa, the genitalia of related species differ in morphology, and it was suggested long ago that each species' male genitalia are a special "key" that can open only a conspecific female's "lock." Only a few studies support this hypothesis, but there is good reason to believe that females terminate mating, and prevent transfer of sperm, if a male's genitalia do not provide suitable tactile stimulation (Eberhard 1996; **Figure 17.7**).

Transfer of sperm is no guarantee that those sperm will fertilize a female's eggs. In some insects, such as ground crickets (*Nemobius*), fertilization by a heterospecific male's sperm will occur if the female mates only with that male, but if she also mates with a conspecific male, only the conspecific sperm will be successful in fertilizing her eggs. This phenomenon is known as CONSPECIFIC SPERM PRECEDENCE (Howard 1999). A similar effect is seen in some plants, in which heterospecific pollen cannot compete well against conspecific pollen in growing down through the style to reach the ovules.

GAMETIC ISOLATION occurs when gametes of different species fail to unite. This barrier is important in many externally fertilizing species of marine invertebrates that release eggs and sperm into the water. Because cell surface proteins determine whether or not sperm can adhere to and penetrate an egg, divergence in such proteins can result in gametic isolation (Palumbi 1998). Among species of abalones (large gastropods), the sperm protein lysin dissolves the vitelline envelope of only conspecific eggs. The failure of heterospecific eggs and sperm to unite is related to the high rate of divergence among abalone species in the amino acid sequences of both lysin and the vitelline envelope protein with which it interacts (Galindo et al. 2003; see Chapter 20).

Postzygotic barriers

Postzygotic barriers consist of reduced survival or reproductive rates of hybrid zygotes that would otherwise backcross to the parent populations, introducing genes from each to the other. These barriers are sometimes classified as either extrinsic or intrinsic, depending on whether or not their effect depends on the environment.

HYBRID INVIABILITY   Hybrids between species often (though by no means always) have lower survival rates than nonhybrids. Quite often mortality is INTRINSIC, occurring in the embryonic stages because of failure of proper development, irrespective of the environment. The causes of inviability and the genes involved are probably diverse. For example, nucleoporin genes that encode channels in the nuclear membrane interact with other genes to cause the death of hybrids between *Drosophila melanogaster* and *D. simulans*; in other cases, in both *Drosophila* and mice, segregation distorters that contribute to genetic conflict (see Chapter 16) contribute to hybrid inviability (Presgraves 2010). Hybrid inviability is EXTRINSIC if hybrids have reduced survival in some environments but not others. For example, some plant hybrids have higher survival in intermediate or disturbed habitats than in those occupied by the parent species (Anderson 1949; Cruzan and Arnold 1993).

HYBRID STERILITY   The reduced fertility of many hybrids is an intrinsic barrier that can be caused by two types of differences between the parent species. *Structural differences between the chromosomes* inherited from the two parents can cause segregation of some ANEUPLOID gametes during meiosis (i.e., gametes with an unbalanced complement of chromosomes). Hybrid sterility can also be caused by *disharmonious interactions between the different genes* inherited from the two parents (such interactions exemplify EPISTASIS; see Chapter 13). These two causes are difficult to distinguish, and it is not clear how often differences in chromosome structure reduce fertility (King 1993; Rieseberg 2001).

As is sometimes true of hybrid inviability, hybrid sterility is often limited to the heterogametic sex. (The HETEROGAMETIC sex is the one with two different sex chromosomes, or with only one sex chromosome; the HOMOGAMETIC sex has two sex chromosomes of the same type. The male is heterogametic in mammals and most insects; the female is heterogametic in birds and butterflies.) This generalization is called **Haldane's rule**.

Hybrid sterility and inviability may be manifested not only in F_1 hybrids, but also in F_2 and backcross offspring. This phenomenon has been observed not only in crosses between species, but also in crosses among different geographic populations of the same species. For example, survival of F_2 larvae in a cross between *Drosophila pseudoobscura* from California and from Utah was lower than that in either "pure" population. This observation was interpreted to mean that recombination in the F_1 generated various combinations of alleles that were "disharmonious." In contrast, alleles at different loci within the same population have presumably been selected to form harmonious combinations. They are said to be **coadapted**, and each population is said to have a COADAPTED GENE POOL (Dobzhansky 1955).

Multiple isolating barriers

Commonly, related species differ by several kinds of isolating barriers. These barriers are likely to come into play in a temporal sequence, so that a potentially strong barrier may be inconsequential in nature. For example, Justin Ramsey and colleagues (2003) used distributional data, field observations of pollinator movements between plants, and experimental data from greenhouse-grown plants to estimate the strength of several barriers to gene exchange between the two species of monkeyflowers shown in Figure 17.6. These species are about 58 percent separated by geography or habitat. In sympatric populations, 97.6 percent of pollinator transitions are between conspecific plants. As a result of conspecific pollen precedence, 70 to 95 percent of ovules are fertilized by conspecific pollen, even in cross-pollinated plants. F_1 hybrids have somewhat reduced seed germination, and the viability of their pollen is reduced by more than 60 percent. Despite these multiple barriers to gene exchange, pollinator fidelity accounts for more than 97 percent of the total reproductive isolation between sympatric populations of the two species, so the later-acting factors have little effect.

How Species Are Diagnosed

Biological species are defined as reproductively isolated populations, but diagnosing species—distinguishing them in practice—is seldom done by directly testing their propensity to interbreed or their ability to produce fertile offspring. Morphological and other phenotypic characters are the usual evidence used for diagnosing species (**Figure 17.8**), even though species are not defined by their degree of phenotypic difference. Morphological and other phenotypic characters, judiciously interpreted, serve as *markers* for reproductive isolation among sympatric populations. If a sample of sympatric organisms falls into two discrete clusters that differ in two or more characters, it is likely to represent two species. Likewise, genetic markers can reveal the existence of two or more sympatric species.

FIGURE 17.8 An example of species distinguished by morphological characters. These seven species of horned lizards (*Phrynosoma*) from western North America can be distinguished by differences in the number, size, and arrangement of horns and scales as well as body size and proportions, color pattern, and habitat. (From Stebbins 1954.)

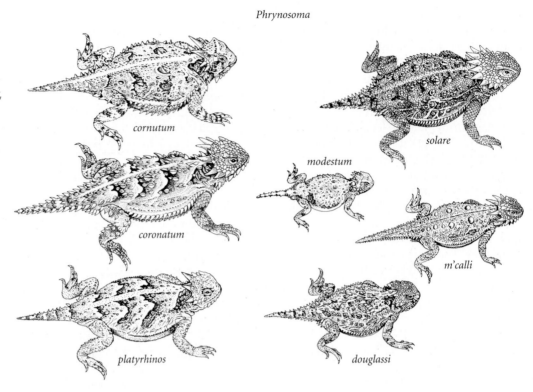

Phrynosoma

cornutum

solare

modestum

coronatum

m'calli

platyrhinos

douglassi

How can phenotypic differences indicate a barrier to gene exchange? From elementary population genetics (see Chapter 9), we know that a locus in a single population with random mating should conform fairly closely to Hardy-Weinberg genotype frequencies. Furthermore, two or more loci should be nearly at linkage equilibrium unless very strong selection or suppression of recombination exists. If these loci affect a more or less additively inherited character, its variation will have a single-peaked, more or less normal distribution. If they affect different characters, variation in those characters is likely not to be strongly correlated. Conversely, if a sample of organisms includes two (or more) reproductively isolated populations with different allele frequencies, then a single locus should show a deficiency of heterozygotes compared with the Hardy-Weinberg expectation; variation in a polygenic character may have a bimodal distribution; and variation in different, genetically independent characters may be strongly correlated. (**Box 17B** describes an application of these principles.) Applying these same principles to molecular markers at two or more loci can provide an even clearer indication of reproductive isolation than can phenotypic traits. (Criteria for considering allopatric populations to be different species are discussed below.)

Differences among Species

Some of the differences among species include those that are responsible for reproductive isolation. Others are adaptive differences related to ecological factors, such as temperature tolerance and habitat use. Still others are presumably neutral differences that have arisen by mutation and genetic drift. Any such character difference may have evolved partly in geographically segregated populations before they became different species, partly during the process of speciation, and partly after the reproductive barriers evolved. Genetic differences between species, estimated from DNA markers, are generally greater than those among conspecific populations, but there is a continuum of the degree of differentiation, not a sharp break between populations and species (Hey and Pinho 2012).

The genetic distance (D) between populations, measured by the degree of difference in their allozyme allele frequencies or in their DNA sequences, can be used as an approximate

Box 17B Diagnosis of a New Species

Each species in the leaf beetle genus *Ophraella* feeds on one species or a few related species of plants. *O. notulata*, for example, has been recorded only from two species of *Iva* along the eastern coast of the United States. This species is most readily distinguished from other species of *Ophraella* by the number and pattern of dark stripes on each wing cover.

Some leaf beetles found in Florida closely resembled *O. notulata*, but were collected on ragweed, *Ambrosia artemisiifolia*. This host association suggested the possibility that these beetles were a different species. In a broader study of the genus, I collected samples of beetles from both *Ambrosia* and *Iva* throughout Florida and examined them by enzyme electrophoresis (Futuyma 1991). I found consistent differences in allele frequencies between samples from *Iva* and from *Ambrosia* at three loci, even in samples from both plants in the same locality. In the most extreme case, one allele had an overall frequency of 0.968 in *Ambrosia*-derived specimens, but was absent in *Iva*-derived specimens, in which a different allele had a frequency of 0.989. No specimens had heterozygous allele profiles that would suggest hybridization. Thus these genetic markers were evidence of two reproductively isolated gene pools.

A careful examination then revealed average differences between *Ambrosia*- and *Iva*-associated beetles in a few morphological characters, such as the shape of one of the mouthparts and the relative length of the legs. None of these morphological differences distinguished all individuals of one species from all individuals of the other. Later studies showed that adults and newly hatched larvae strongly prefer their natural host plant (*Ambrosia* or *Iva*) when given a choice, and that the beetles mate preferentially with their own species. In laboratory crosses, viable eggs were obtained by mating female *Ambrosia* beetles with males from *Iva*, but not the reverse. Few of the hybrid larvae survived to adulthood, and none laid viable eggs. Based on all of this evidence, I concluded that the *Ambrosia*-associated form, which I named *Ophraella slobodkini* in honor of the ecologist Lawrence Slobodkin, is a sibling species of *O. notulata*.

Ophraella slobodkini

"molecular clock" to estimate the relative times of divergence of various pairs of populations or species. Coyne and Orr (1989, 1997) used such information to plot the temporal pattern of the evolution of reproductive isolation. They compiled data on reproductive isolation from reports, published over the previous 60 years, of experiments on numerous combinations of populations or species of *Drosophila*. They arrived at several important conclusions:

1. The strength of both prezygotic and postzygotic isolation increases gradually with the time since the separation of the populations (**Figure 17.9**). That is, speciation is a gradual process.

2. The time required for full reproductive isolation to evolve is highly variable, but on average, it is achieved when D is about 0.30–0.53, which (based on a molecular clock calibrated from the few fossils of *Drosophila*) corresponds to about 1.5–3.5 My. However, a considerable number of fully distinct species appear to have evolved in less than 1 My.

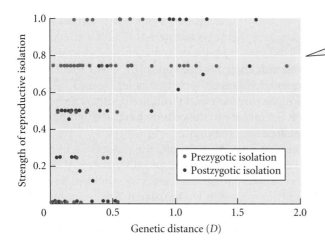

The strength of both prezygotic and postzygotic isolation increases gradually with the time since the separation of the populations, measured here as the genetic distance.

FIGURE 17.9 The strength of prezygotic or postzygotic reproductive isolation between pairs of populations and species of *Drosophila*, plotted against genetic distance (*D*; time since divergence can be inferred from genetic distance). The strength of prezygotic isolation was measured by observing mating versus failure to mate between flies from different populations when confined together in the laboratory. The strength of postzygotic isolation was measured by survival and fertility of hybrid individuals. (After Coyne and Orr 1997.)

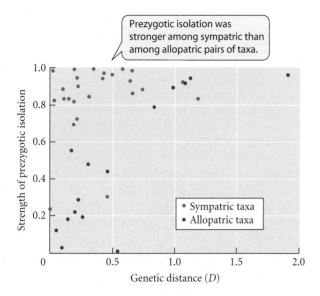

FIGURE 17.10 The strength of prezygotic isolation among allopatric and sympatric pairs of *Drosophila* populations, plotted against genetic distance (*D*). At low genetic distances—interpreted as recent divergence—prezygotic isolation is stronger among sympatric than among allopatric populations. (After Coyne and Orr 1997.)

3. Among recently diverged populations or species, premating isolation is, overall, a stronger barrier to gene exchange than postzygotic isolation (hybrid sterility or inviability). However, in Coyne and Orr's results, this effect was entirely due to sympatric taxa. Prezygotic isolation was stronger among sympatric than among allopatric pairs of taxa (**Figure 17.10**). This finding bears on a controversy about whether or not sexual isolation evolves to prevent hybridization, as we will see in the next chapter.

4. In the early stages of speciation, hybrid sterility or inviability is almost always seen in males only; female sterility or inviability appears only when taxa are older (an example of Haldane's rule). Thus postzygotic isolation evolves more rapidly in males than in females.

Similar analyses of several other groups of organisms have yielded similar patterns (see Coyne and Orr 2004). Because genetic differences continue to accumulate long after two species achieve complete reproductive isolation, some of the genes, and even some of the traits, that now confer reproductive isolation may not have been instrumental in forming the species in the first place. Thus it can be very difficult to tell which of several demonstrable isolating barriers, or which of many gene differences that may confer hybrid sterility, was the cause of speciation. Such information can be obtained by studying populations that have achieved reproductive isolation only very recently.

The Genetic Basis of Reproductive Isolation

In analyzing barriers to gene exchange, we wish to know whether the genetic differences required for speciation consist of few or many genes and how those genes act. Because some genetic differences accrue after speciation has occurred, we must compare populations that have speciated very recently, or are still in the process of doing so, in order to answer these questions.

Chromosome differences and postzygotic isolation

Chromosome differences among species include differences in chromosome structure and differences in the number of chromosome sets (polyploidy, treated in Chapter 18). The role of alterations of chromosome structure in postzygotic isolation and speciation is controversial (King 1993; Rieseberg 2001; Coyne and Orr 2004). An important question is whether heterozygosity for chromosome rearrangements causes reduced fertility (postzygotic isolation) in hybrids as a result of segregation of aneuploid gametes in meiosis (see Chapter 8).

A RECIPROCAL TRANSLOCATION is an exchange between two nonhomologous chromosomes. Suppose, for example, that 1.2 and 3.4 represent two metacentric chromosomes in one population, with 1 and 2 representing the arms of one and 3 and 4 the arms of the other. A second population that is fixed for a translocation might have chromosomes 1.4 and 3.2. The F_1 hybrid would have all four chromosome types (1.2, 3.4, 1.4, 3.2). Only if the two parental combinations segregated (1.2 and 3.4 to one pole, 1.4 and 3.2 to the other) would balanced (euploid) gametes be formed (see Figure 8.25). Other patterns of segregation (e.g., 1.2 and 3.2 to one pole, 3.4 and 1.4 to the other) would yield unbalanced (aneuploid) gametes that lack considerable genetic material.

Some species differ by multiple translocations. For example, the jimsonweeds *Datura stramonium* and *D. discolor* each have twelve pairs of chromosomes (Dobzhansky 1951). In the F_1 hybrid, seven pairs form normal synapsed pairs in meiosis. The other five pairs differ by multiple translocations. If these chromosomes in *D. stramonium* are designated 1.2, 3.4, 5.6, 7.8, and 9.10, those of *D. discolor* represent 1.3, 2.7, 4.10, 5.9, and 6.8. In synapsis, a ring of ten chromosomes is formed, with the two arms of each *stramonium* chromosome

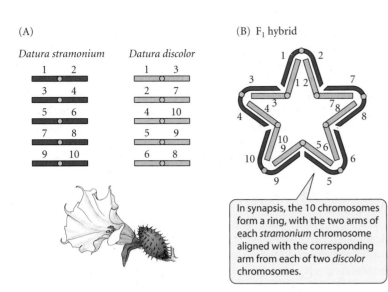

FIGURE 17.11 (A) Five chromosomes of the jimsonweeds *Datura stramonium* and *D. discolor*, which differ by five reciprocal translocations. Homologous chromosome arms are correspondingly numbered. Only one member of each chromosome pair is shown. (B) A diagram of the possible arrangement of these chromosomes in synapsis in an F₁ hybrid.

In synapsis, the 10 chromosomes form a ring, with the two arms of each *stramonium* chromosome aligned with the corresponding arm from each of two *discolor* chromosomes.

aligned with an arm of each of two *discolor* chromosomes (**Figure 17.11**). The opportunities for aneuploid segregation are numerous.

Perhaps for this reason, such chromosome rearrangements are seldom found as polymorphisms within populations. More often, they are nearly or entirely monomorphic in different populations, except where those populations meet in narrow hybrid zones. For example, parapatric races of the burrowing mole-rat *Spalax ehrenbergi* in Israel differ in their number of chromosome pairs due to chromosome fusions. Hybrids between these races are found in zones that range from 2.8 kilometers to only 0.3 kilometers wide (**Figure 17.12**).

This pattern is expected if chromosomal heterozygotes have lower fitness than homozygotes (are UNDERDOMINANT), perhaps owing to reduced fertility caused by aneuploidy. If so, then a chromosome introduced by gene flow from one population into another should seldom increase in frequency because its initial frequency would be low, it would occur mostly in heterozygous condition, and it would probably be eliminated by selection (see Chapter 11). It is very difficult to tell, however, whether the reduced fitness of hybrids is caused by the structural differences between the chromosomes or by differences between the genes of the parent populations. Certainly, chromosome rearrangements reduce gene exchange between populations. For example, the sunflowers *Helianthus annuus* and *H. petiolaris* differ by inversions and translocations that affect some chromosomes but not others (Rieseberg et al. 1999). The rearranged chromosomes show a more abrupt transition in a hybrid zone between these species than do the chromosomes lacking rearrangements. But as we will see in our discussion of hybrid zones below, this pattern may be due to genic incompatibility, not to production of aneuploid gametes. According to Coyne and Orr (2004), the evidence, so far, that chromosome heterozygotes have reduced fertility because of meiotic irregularities is convincing for some plants and mammals, but not for other organisms.

Genes affecting reproductive isolation

The most extensive information on the genetics of reproductive isolation has been obtained for certain *Drosophila* species because so many genetic markers—the sine qua non of any genetic analysis—are available for those well-studied species. Theodosius Dobzhansky (1936, 1937), one of the most influential figures of the evolutionary synthesis, pioneered the use of genetic markers to study hybrid sterility. Until recently, morphological mutations

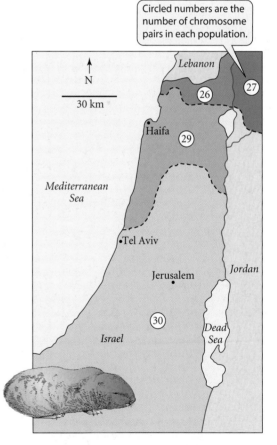

Circled numbers are the number of chromosome pairs in each population.

FIGURE 17.12 The distribution of four "races" of the mole-rat *Spalax ehrenbergi* with different chromosome numbers. Pairs of races meet at very narrow hybrid zones, indicated by the dashed lines. (After Nevo 1991.)

were used as markers; today, molecular markers are used for QTL (quantitative trait locus) mapping (see Chapter 13). In both cases, the markers are used to indicate linked genes that differ between the species and contribute to their reproductive isolation.

The principle is well illustrated by Jerry Coyne's (1984) analysis of hybrid sterility in crosses between *D. simulans* and its close relative *D. mauritiana*. F_1 hybrid males are sterile, but hybrid females are fertile (Haldane's rule again!). Coyne used a *D. simulans* stock with recessive visible mutations (which served as genetic markers) on the X chromosome and on each arm of both of the autosomes. Female *D. simulans* were crossed to *D. mauritiana* males, and the fertile F_1 hybrid females were backcrossed to *D. simulans* males. Among the resulting male progeny, a recessive mutant phenotype showed which *D. simulans* chromosome arms were carried in homozygous condition. Coyne scored males for motile versus immotile sperm (immotility is correlated with male sterility). Sperm motility differed for each pair of genotypes distinguished by one or more recessive markers (**Figure 17.13**). Therefore each chromosome arm carries at least one gene difference between *D. simulans* and *D. mauritiana* that contributes to male hybrid sterility.

Coyne used only five genetic markers in this experiment, so he could detect no more than five linked genetic factors contributing to hybrid sterility. Subsequent researchers have used multiple molecular markers. For example, Chung-I Wu and his coworkers backcrossed many small marked segments of the X chromosomes of *Drosophila mauritiana* and *D. sechellia* into a genome otherwise derived from *D. simulans*. Males had lowered fertility whenever they carried two or more such segments. By extrapolation from these short segments of chromosomes, Wu and his coworkers suggested that as many as 40 gene differences on the X chromosome and 120 in the genome as a whole might cause hybrid male sterility among these closely related species (Wu and Hollocher 1998). A similar study of *Drosophila simulans* and *D. melanogaster* suggested that about 200 genes can contribute to

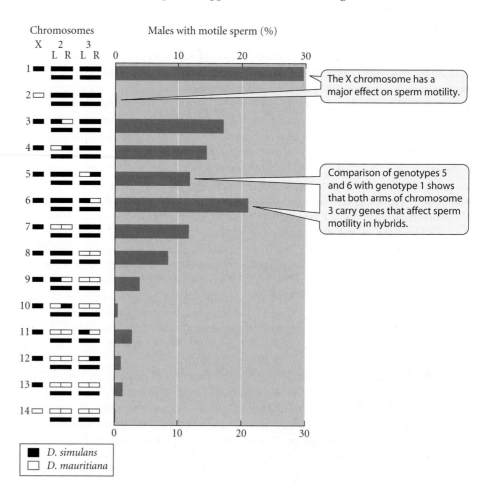

FIGURE 17.13 The proportion of males with motile sperm in nonhybrid *Drosophila simulans* (genotype 1) and in backcross hybrids with various combinations of chromosome arms from *D. simulans* and *D. mauritiana* (genotypes 2–14). All genotypes have a *D. simulans* Y chromosome (not shown). Note that every chromosome arm derived from *D. mauritiana* reduced sperm motility compared with that of the standard *D. simulans* genotype. (After Coyne 1984.)

hybrid inviability (Presgraves 2003). However, far fewer gene differences suffice to confer postzygotic isolation. Male sterility of F₁ hybrids between *Drosophila pseudoobscura* and *D. bogotana*, which are thought to have diverged less than 200,000 years ago, appears to be based on only about five gene regions, of which four are required for any sterility at all (Orr and Irving 2001). It therefore appears that the early acquisition of hybrid sterility requires few gene differences, but many other sterility-causing mutations accumulate after speciation has been completed.

Genes that contribute to hybrid sterility or inviability do not have these effects in nonhybrid individuals; these effects must therefore stem from interactions between genes in the two different species. That is, *epistatic interactions* contribute to postzygotic isolation (**Figure 17.14A**). A pair of such genes causes inviability of F₁ hybrids between *Drosophila melanogaster* and *D. simulans* (**Figure 17.14B**). Relative to the common ancestor, mutations have been fixed in the *Hmr* gene of *D. melanogaster* and the *Lhr* gene of *D. simulans* that together cause the death of hybrid males. The *Hmr* ("hybrid male rescue") and *Lhr* ("lethal hybrid rescue") genes were first found because mutations in either gene enable hybrid males to survive.

Studies of this kind generally show that the X chromosome has a greater effect than any of the autosomes. This result sheds some light on the more rapid evolution of sterility in male than in female *Drosophila* hybrids, an instance of Haldane's rule. It has been proposed that X-linked genes diverge faster than autosomal genes because favorable X-linked recessive alleles are most exposed to natural selection (since males carry only one X). In addition, autosomal genes affecting male sterility have diverged faster than those affecting female sterility, possibly because of sexual selection.

Like postzygotic isolation, premating isolation is frequently based on polygenic traits, although in some cases only a few genes are involved. In a massive QTL study of F₂ progeny from crosses between the bee-pollinated monkeyflower *Mimulus lewisii* and its hummingbird-pollinated relative *M. cardinalis* (see Figure 17.6), Bradshaw and colleagues (1998) found that each of twelve characters that distinguish the species' flowers differed by between one and six loci. For most of these features, one locus accounted for at least 25 percent of the difference between the species. Schemske and Bradshaw (1999) then placed a full array of F₂ hybrids in an area where the two species are sympatric and observed pollination. At least four of the twelve floral characters affected the relative frequency of visitation by bees versus hummingbirds. Two QTL underlying these characters could be shown to have a significant effect on pollinator isolation. Alleles at one locus that strongly affect flower color were then backcrossed from each species into the other, resulting in flowers that were normal in every respect except for color (Bradshaw and Schemske 2003). The normally hummingbird-pollinated *M. cardinalis* then received 74 times more bee visits than the wild type, and the normally bee-pollinated *M. lewisii* received 68 times as many hummingbird visits as the wild type. These observations suggest that a single mutation can greatly alter pollination and contribute substantially to reproductive isolation.

An important question is whether or not genes affecting the mating display of one sex (e.g., male) are closely linked to those that affect the mating preferences of the other sex, because a tight association of the preference with the preferred trait should strengthen the barrier to gene exchange between divergent populations. In some cases, such as production of and response to a sex pheromone in some moths, there appears to be no genetic linkage (Roelofs et al. 1987), but in some animals, the genes are tightly linked. Kerry Shaw and her colleagues backcrossed five QTL that affect the pulse rate of the male song of the cricket *Laupala paranigra* into its sister species *L. kohalensis*. In fourth-generation backcross hybrids, in which 97 percent of the genome is derived from *kohalensis*, four of five *paranigra* QTL produced not only a *paranigra*–like pulse rate in males, but also female preference for that pulse rate (Wiley et al. 2012). Thus it is possible that the same genes affect both the male signal and the female response.

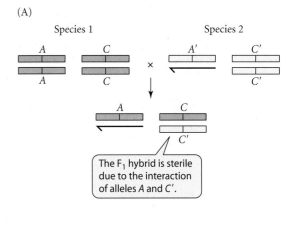

The F₁ hybrid is sterile due to the interaction of alleles *A* and *C'*.

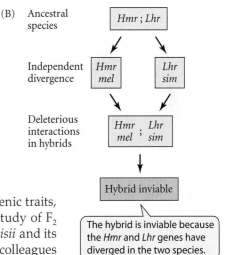

The hybrid is inviable because the *Hmr* and *Lhr* genes have diverged in the two species.

FIGURE 17.14 Gene interactions that cause sterility or inviability in hybrids between *Drosophila* species. (A) In a cross between a female of species 1 and a male of species 2, the sex chromosomes and one pair of autosomes of the two species are shown above the F₁ genotype. The interaction between loci *A* and *C* causes sterility or inviability. Such an interaction between loci is called a Dobzhansky-Muller incompatibility (see p. 476). (B) The genes *Hmr* and *Lhr* have undergone changes, because of natural selection, in *Drosophila melanogaster* and *D. simulans*, respectively. Together, these changes cause inviability of the male hybrid. (B after Brideau et al. 2006.)

The significance of genetic studies of reproductive isolation

Both prezygotic and postzygotic isolation are generally due to differences between populations at several gene loci. A mismatch between genes affecting sexual communication, such as a male courtship signal and a female response, creates reproductive isolation. Similarly, a functional mismatch between genes gives rise to hybrid sterility or inviability. In both cases, reproductive isolation requires that populations diverge by at least two allele substitutions. Thus an $A_1A_1B_1B_1$ ancestral population may give rise to populations with genotypes $A_1A_1B_2B_2$ and $A_2A_2B_1B_1$, in which the incompatibility between A_2 and B_2 is the cause of reproductive isolation in the F$_1$ hybrid ($A_1A_2B_1B_2$). This kind of epistatic incompatibility between loci is called a **Dobzhansky-Muller (DM) incompatibility** because Theodosius Dobzhansky, in 1934, and Hermann Muller, in 1940, postulated that such interactions underlie postzygotic isolation. (Some writers refer to BDM, rather than DM, incompatibility because the idea was foreshadowed by William Bateson in 1909.) The interaction between the *Hmr* and *Lhr* genes that caused inviability in male hybrids between *Drosophila melanogaster* and *D. simulans* (see Figure 17.14B) is exactly what these authors envisioned. Similarly, two wild strains of the mustard *Arabidopsis thaliana* displayed a DM incompatibility caused by mutational silencing of two duplicate genes that both encode an enzyme required to synthesize the essential amino acid histidine (Bikard et al. 2009). If we denote these loci *a* and *b*, one strain was homozygous for genotype a^+b^-, and the other for genotype a^-b^+ (where the "minus" superscript indicates a nonfunctional gene). Some recombinant progeny, with genotype a^-b^-, were inviable. That silencing of duplicate genes might cause hybrid incompatibility had been postulated many years before this and other examples came to light (Werth and Windham 1991).

Genetic studies have shown that differences among species, including those that confer reproductive isolation, have the same kinds of genetic foundations as variation within species. Thus there is no foundation for the opinion, held by some earlier biologists such as the geneticist Richard Goldschmidt (1940), that species and higher taxa arise through qualitatively new kinds of genetic and developmental repatterning. Moreover, reproductive isolation, like divergence in any other character, usually evolves in relatively small steps, by the successive substitution of alleles in populations.

Genetic Divergence and Exchange

So far, we have considered genes that contribute to reproductive isolation and which have different alleles fixed in related species. We now expand our view to genes throughout the genome. Populations that are in the process of speciation, or which have recently achieved species status, are likely to share variation at some loci, for two major reasons: they may retain variation that was segregating in their common ancestor, or they may continue to exchange some genes through interbreeding.

Ancestral variation and coalescence

The pattern of molecular variation within and between **sister species**—two species with an immediate common ancestor—can shed light on their history of divergence. Two populations (or species) that become isolated from each other will at first share many of the same gene lineages, inherited from their polymorphic common ancestor (**Figure 17.15**). Each population is at first polyphyletic with respect to gene lineages (i.e., its gene copies are derived from several ancestral gene copies). Thus, with respect to some loci, individuals in each population are genealogically less closely related to one another than they are to some individuals in the other population (Figure 17.15, time t_1).

According to COALESCENT THEORY, described in Chapter 10, genetic drift in each species eventually results in the loss of all the ancestral lineages of DNA sequence variants except one; that is, coalescence to a common ancestral gene copy occurs in each species. (This process can also be caused by directional selection for a favorable mutation.) Gene lineages are

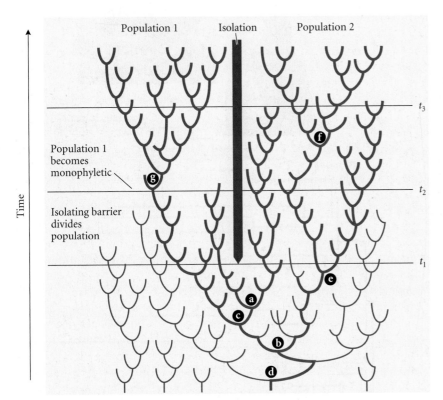

FIGURE 17.15 The transition from genetic polyphyly to paraphyly to monophyly in speciation. The gene tree shows lineages of haplotypes at a single locus in a population that becomes divided into two populations at time t_1. As a result of genetic drift and natural selection, all ancestral haplotype lineages except one are eventually lost, so that all the gene copies in the current populations (at the top) coalesce to ancestral gene copy *b*. Population 1 loses haplotype lineages more rapidly than population 2, perhaps because it is smaller. Between times t_1 and t_2, both populations are polyphyletic for haplotype lineages, since some gene copies in each population, such as those derived from gene copy *a*, are more closely related to some gene copies in the other population than to some other gene copies (e.g., those derived from *b*) in their own population. Population 1 becomes monophyletic (coalescing to gene copy *g*) sooner—at time t_2—than population 2, perhaps because of its smaller size. Between times t_2 and t_3, population 2 is genetically paraphyletic, since its gene copies derived from ancestral gene copy *c* are more closely related to population 1's gene copies than they are to gene copies in population 2 that have been derived from ancestral gene copy *e*. At time t_3, population 2 also becomes monophyletic, with all gene copies derived from *f*. (After Avise and Ball 1990.)

lost by genetic drift at a rate inversely proportional to the effective population size. At some point, one population (population 1 in Figure 17.15) will become monophyletic for a single gene lineage, while the other population, if it is larger, retains both this and other gene lineages (population 2 in Figure 17.15, time t_2). At this time, the more genetically diverse population will be paraphyletic with respect to this gene, and some gene copies sampled from population 2 will be more closely related to population 1's gene copies than they are to other copies from population 2. Thus the phylogenetic relationships among genes from organisms in both populations will not correspond to the relationships among the individual organisms or the populations. Eventually, however, both populations will become monophyletic for gene lineages (Figure 17.15, time t_3), and the relationships among genes will reflect the relationships among populations. This process of sorting of gene lineages into species, or **lineage sorting**, is faster if effective population sizes (N_e) are small (Neigel and Avise 1986).

Closely related species often share ancestral polymorphisms; in other words, lineage sorting is incomplete (Funk and Omland 2003). For example, several haplotype lineages of the nuclear α-enolase gene are shared between the crested auklet (*Aethia cristatella*) and the least auklet (*A. pusilla*), two small seabirds that split from a common ancestor at least 2.8 Mya (**Figure 17.16**). This pattern is expected if species have arisen very recently, or if they have had large population sizes, so that genetic drift, resulting in coalescence, has proceeded slowly (see Chapter 10). Moreover, shared polymorphisms can persist for a very long time if natural selection maintains the variation in both species (see Figure 12.32). For example, humans and chimpanzees, which are each other's closest relatives, share several gene lineages at two loci in the major histocompatibility complex (MHC) that have been retained since they diverged from their common ancestor more than 5 Mya. The polymorphism is thought to have been retained because of the role that MHC proteins play in defense against pathogens.

Because the process of lineage sorting may take longer than the interval between successive speciation events, the phylogeny of a particular gene (the GENE TREE) among three

FIGURE 17.16 Incomplete lineage sorting results in a poly-phyletic gene tree for the α-enolase locus in two closely related seabird species. Blue (C) and red (L) branches mark haplotype lineages found in the crested auklet (*Aethia cristatella*) and the least auklet (*A. pusilla*), respectively. (After Walsh et al. 2005; *A. cristatella* photo courtesy of Art Sowls, U.S. Fish and Wildlife Service.)

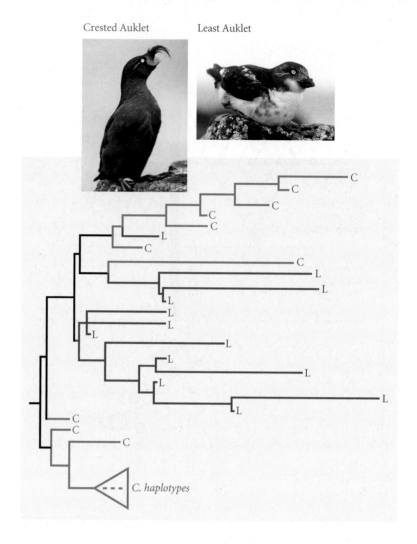

Crested Auklet   Least Auklet

C. haplotypes

or more species may differ from the sequence of population branching (i.e., the phylogeny of the species, or SPECIES TREE; see Figure 2.25). Moreover, the gene trees for different loci may differ from the species tree in different ways. By applying models of coalescence to several or many sequenced loci, researchers can determine whether or not different populations represent genetically distinct lineages and determine the phylogenetic relationships among those lineages (Knowles and Carstens 2007; Yang and Rannala 2010; Fujita et al. 2012). Allopatric populations that are shown by this approach to be distinct lineages are often considered to be species.

Gene flow and hybridization

Hybridization occurs when offspring are produced by interbreeding between genetically distinct populations (Harrison 1990). Hybridization in nature interests evolutionary biologists because the hybridizing populations sometimes represent intermediate stages in the process of speciation. In some cases, hybridization may be the source of new adaptations or even of new species (Arnold 1997; Mallet 2007).

A hybrid zone is a region where genetically distinct populations meet and mate, resulting in at least some offspring of mixed ancestry (Harrison 1990, 1993). A character or locus that changes across a hybrid zone exhibits a CLINE that may be quite steep; for example, the alleles for the gray hood in hybridizing hooded and carrion crows display a steep cline in frequency across the zone of hybridization between these forms (see Figure 17.1D). Hybrid zones are thought to be caused by two processes. **Primary hybrid zones** originate in situ as geographic differences in natural selection alter allele frequencies in a series of more or less continuously distributed populations. Thus the position of the zone

(A) Primary hybrid zone

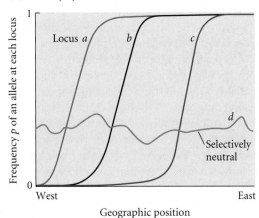

(B) Secondary hybrid zone

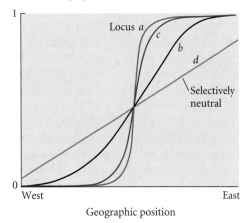

FIGURE 17.17 Expected patterns of variation in the frequency (p) of alleles or characters across primary and secondary hybrid zones. Clines are shown for four loci. (A) A primary hybrid zone originates by divergent selection along an environmental gradient. Loci a, b, and c are affected differently by that gradient, so their clines have different geographic positions. Allelic variation at locus d (orange line) is nearly selectively neutral, so the variation at that locus does not show a cline. (B) In a secondary hybrid zone, clines for all loci, including that for the nearly neutral alleles at locus d, are expected to have about the same locations. The steepness of each cline depends on the strength of selection, relative to gene flow, at the locus or at closely linked loci. After a long enough time, the cline for locus d will come to resemble that in (A) because of gene flow.

is likely to correspond to a sharp change in one or more environmental factors. **Secondary hybrid zones** are formed when two formerly allopatric populations that have become genetically differentiated expand so that they meet and interbreed (**secondary contact**). In both cases, we can expect that allele frequencies will have diverged at certain loci because of natural selection, including loci that contribute to any reproductive isolation that might exist between the differentiated populations. Neutral variants may also have diverged in frequency, perhaps to a lesser extent. Both neutral and selected alleles might be expected to show more COINCIDENT (i.e., located in the same place) frequency clines in secondary than in primary hybrid zones (**Figure 17.17**) because they share the same history of separation and rejoining.

Dispersal, selection, and linkage all affect the distribution of alleles and phenotypic characters in hybrid zones. Let us consider the case in which hybrids have low fitness because of epistatic incompatibility or heterozygote disadvantage at certain loci (Barton and Gale 1993). Suppose populations (semispecies) 1 and 2 have come into contact, that they are fixed for alleles A_1 and A_2, respectively, and that the fitness of A_1A_2 is lower than that of either homozygote. Dispersal of individuals of each semispecies into the range of the other, followed by random mating, constitutes gene flow that tends to make the cline in allele frequency broader and shallower. Gene flow continues if F$_1$ hybrids backcross with parental (nonhybrid) genotypes. However, allele A_1 cannot increase in frequency within population 2, nor can A_2 within population 1, because of the heterozygote's disadvantage (see Chapter 12). The lower the fitness of the F$_1$ hybrid, the fewer A_1 or A_2 allele copies will be introduced into the parent populations, and the stronger the barrier to gene flow will be. The steepness of the cline at the A locus depends on the strength of selection against hybrids relative to the frequency of dispersal (the level of gene flow).

Now consider another locus, with alleles B_1 and B_2 fixed in populations 1 and 2. If there is selection against B_1B_2 heterozygotes, the cline at this locus will parallel the cline at the A locus; in fact, the two clines will reinforce each other, since the two loci both reduce the F$_1$ hybrid's fitness. Suppose, though, that the B_1 and B_2 alleles are selectively neutral. If they are very closely linked to the A locus, then they will hitchhike with the A alleles and form a cline coincident with that of the A locus in position and steepness. In fact, they will mark the existence and location of the locus (A) that contributes to low hybrid fitness. If the chromosomal location of B is distant from A or any other divergently selected locus, however, the B_1 and B_2 alleles will diffuse through the hybrid zone and spread into the other semispecies, forming a cline that becomes more shallow over time (see Figure 17.17B). The diffusion occurs because if F$_1$ hybrids reproduce at all, some of the offspring of the backcross $A_1A_2B_1B_2$ × $A_1A_1B_1B_1$ (i.e., F$_1$ × population 1) will be $A_1A_1B_1B_2$. Thus some copies of the B_2 allele, dissociated by recombination from the deleterious heterozygous allele combination A_1A_2, will be introduced into population 1. Backcrosses to population 2 will similarly spread B_1 allele copies into that population. Thus some alleles or characters can spread from each semispecies into the other (a process called **introgression**), whereas others cannot.

A lowered fitness of hybrids at one locus therefore reduces the flow of neutral (or advantageous) alleles at other loci between populations, but the reduction is greater for alleles at closely linked loci than for those at loosely linked or unlinked loci. Any factor that reduces recombination, such as a chromosome rearrangement, will strengthen the association between neutral sites and selected loci, reducing the diffusion of neutral alleles between the hybridizing populations. If a selected locus lies within a chromosome rearrangement, such as an inversion, that reduces crossing over (see Chapter 9), then neutral alleles at all loci within the rearrangement, as well as the rearrangement itself, will display the same steep cline as the selected locus. In the hybrid zone between species of sunflowers described earlier in this chapter, rearranged chromosomes have steeper clines than structurally similar chromosomes, and they include genes that reduce hybrid fertility (Rieseberg 2001).

Because some neutral genetic markers will be tightly linked to a locus that reduces hybrid fitness, whereas others may be unlinked, the number of markers with steep clines provides an estimate of the number of loci contributing to the low fitness of hybrids (i.e., to postzygotic reproductive isolation). For example, the eastern European fire-bellied toad (*Bombina bombina*) and the western European yellow-bellied toad (*B. variegata*) meet in a long hybrid zone that is only about 6 kilometers wide. The two species (or semispecies) differ in allozymes and in several morphological features (**Figure 17.18**), and hybrids have lower rates of survival than nonhybrids because of epistatic incompatibility. These taxa are thought to have arisen during the Pliocene, and to have formed a secondary hybrid zone after spreading from different refuges in southeastern and southwestern Europe that they

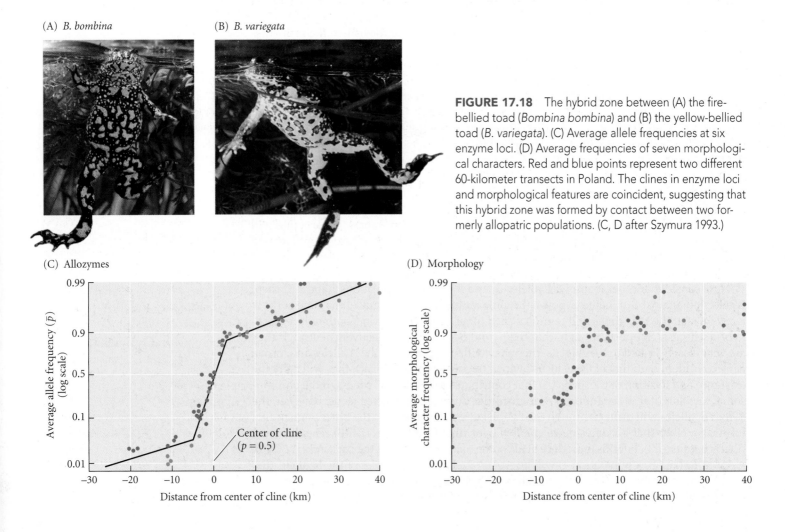

(A) *B. bombina*

(B) *B. variegata*

FIGURE 17.18 The hybrid zone between (A) the fire-bellied toad (*Bombina bombina*) and (B) the yellow-bellied toad (*B. variegata*). (C) Average allele frequencies at six enzyme loci. (D) Average frequencies of seven morphological characters. Red and blue points represent two different 60-kilometer transects in Poland. The clines in enzyme loci and morphological features are coincident, suggesting that this hybrid zone was formed by contact between two formerly allopatric populations. (C, D after Szymura 1993.)

(C) Allozymes

Average allele frequency ($\bar{p}$) (log scale)

Center of cline ($p = 0.5$)

Distance from center of cline (km)

(D) Morphology

Average morphological character frequency (log scale)

Distance from center of cline (km)

occupied during the Pleistocene glacial periods (Szymura 1993). Allozyme clines in this hybrid zone have been used to estimate that about 55 loci contribute to the reduction in the fitness of hybrids between these species of toads (Barton and Gale 1993; Szymura 1993).

Today many studies of speciation and hybridization use DNA sequencing to screen great numbers of genetic markers, such as SNPs, throughout the genomes of divergent populations and species. Although the interpretation of such data is based on the same principles of population genetics that are used in more classical studies, genome-level studies provide far more information about genetic differences bewteeen species, and about processes of speciation. We will consider the "genomics of speciation" in Chapter 18.

The fate of hybrid zones

Hybrid zones may have several fates:

1. A hybrid zone may persist indefinitely, with selection maintaining steep clines at some loci even while the clines in neutral alleles dissipate because of introgression.

2. Natural selection may favor alleles that enhance prezygotic isolation, resulting ultimately in full reproductive isolation.

3. Alleles that improve the fitness of hybrids may increase in frequency. In the extreme case, the postzygotic barrier to gene exchange may break down, and the semispecies may merge into one species.

4. Some hybrids may become reproductively isolated from the parent forms and become a third species.

These possibilities will be discussed in Chapter 18 as we examine the processes of speciation.

Summary

1. Many definitions of "species" have been proposed. The biological species concept (or variants of it) is the one most widely used by evolutionary biologists. It defines species by reproductive discontinuity based on differences between populations, not by phenotypic differences (although phenotypic differences may be indicators of reproductive discontinuity). Among other species concepts, the most widely adopted is the phylogenetic species concept, according to which species are sets of populations with character states that distinguish them.

2. The biological species concept (BSC) has a restricted domain; moreover, some populations cannot be readily classified as species because the evolution of reproductive discontinuity is a gradual process. Although the BSC applies to allopatric populations, it is often difficult to determine whether or not such populations are distinct species.

3. The biological differences that constitute barriers to gene exchange are many in kind, the chief distinction being between prezygotic barriers (e.g., ecological or sexual isolation) and postzygotic barriers (hybrid inviability or sterility). Some species are also isolated by postmating, prezygotic barriers (e.g., gametic isolation).

4. Among prezygotic barriers to gene exchange, sexual (behavioral) isolation is important in animals. It entails a breakdown in communication between the courting and the courted sexes, and therefore, usually, genetic divergence in both the signal and the response.

5. Postzygotic isolation is usually caused by differences in nuclear genes and sometimes by structural differences in chromosomes. Genic differences that yield hybrid sterility or inviability consist of differences at two or more (usually considerably more) loci that interact disharmoniously in the hybrid. The role of chromosome differences in hybrid sterility is not well understood.

6. Reproductive isolation evolves gradually. Hybrid sterility or inviability of the heterogametic sex usually evolves before its manifestation in the homogametic sex; this is known as Haldane's rule.

7. The degree of molecular divergence between closely related species varies greatly, and some recently formed species cannot be distinguished by molecular markers. Some species share ancestral molecular polymorphisms. In some cases, some gene copies in one species are more closely related to gene copies in another species than to other gene copies in the same species. In such cases, the gene tree may not match the phylogeny of the species that carry the genes.

8. Species (or semispecies) sometimes hybridize, often in hybrid zones that result from contact between formerly allopatric populations. Alleles at some loci, but not others, may spread between hybridizing populations by gene flow, forming allele frequency clines of varying steepness. The steepness of such a cline depends on the rate of dispersal between populations, the strength of selection, and linkage to selected loci. Divergently selected loci, and closely linked neutral variants, show steeper clines than loosely linked regions of the genome.

Terms and Concepts

biological species
    concept (BSC)
coadapted
conspecific
Dobzhansky-Muller
    (DM) incompatibility
ecological isolation

Haldane's rule
hybrid zone (primary,
    secondary)
hybridization
introgression
isolating barriers

lineage sorting
phylogenetic species
    concept (PSC)
postzygotic barriers
premating barriers
prezygotic barriers

reproductive isolation
secondary contact
semispecies
sexual isolation
sibling species
sister species

Suggestions for Further Reading

Speciation, by J. A. Coyne and H. A. Orr (Sinauer Associates, Sunderland, MA, 2004), is the most comprehensive recent book about speciation. The authors analyze hypotheses and data on speciation carefully and summarize a great amount of relevant literature. They provide an extensive discussion of species concepts and a justification of the biological species concept in particular.

Three historically important books by Ernst Mayr—*Animal Species and Evolution* (Harvard University Press, Cambridge,

MA, 1963), its abridged successor, *Populations, Species, and Evolution* (Harvard University Press, Cambridge, MA, 1970), and its predecessor, *Systematics and the Origin of Species* (Columbia University Press, New York, 1942)—are the classic works on the nature of animal species and speciation. For plants, an equally foundational work is *Variation and Evolution in Plants*, by G. L. Stebbins (Columbia University Press, New York, 1950). This topic was also treated by V. Grant in *Plant Speciation* (Columbia University Press, New York, 1981).

Problems and Discussion Topics

1. Some degree of genetic exchange occurs in bacteria, which reproduce mostly asexually. What evolutionary factors should be considered in debating whether or not the biological species concept can be applied to bacteria?

2. Suppose the phylogenetic species concept were adopted in place of the biological species concept. What would be the implications for (a) evolutionary discourse on the mechanisms of speciation; (b) studies of species diversity in ecological communities; (c) estimates of species diversity on a worldwide basis; (d) conservation practices under such legal frameworks as the U.S. Endangered Species Act?

3. How might the fate of two hybridizing populations (i.e., whether or not they persist as distinct populations) depend on the kind of isolating barriers that reduce gene exchange between them?

4. Identify two species of plants in the same genus that grow in your area. Propose a research program that would enable you to (a) judge whether or not there is any gene flow between the species and (b) determine what the isolating barriers are that maintain the differences between them.

5. Studies of hybrid zones have shown that mitochondrial and chloroplast DNA markers frequently introgress farther, and have higher frequencies, than nuclear gene markers (see Avise 2004, pp. 367–374). Thus, far from the hybrid zone, individuals that have no phenotypic indications of hybrid

ancestry may have mitochondrial or chloroplast genomes of the other species (or semispecies). How would you account for this pattern?

6. Suppose you are interested in a poorly studied genus of leaf beetles that occurs in a distant land you can't afford to visit. A biologist friend who will be doing research at a field station in that region volunteers to collect some specimens and preserve them in alcohol for you. You can then study both their morphology and DNA sequences. How will you decide how many species are in the collection your friend brings back? Suppose your friend extracts mitochondrial DNA from each specimen and brings it back safely, but loses the specimens en route. How can you decide how many species are represented—or can you?

7. Suppose that two or more related taxa are polyphyletic for gene lineages, as in the grasshoppers in Figure 2.26. What evidence might enable you to decide whether this pattern is due to incomplete sorting (i.e., lack of coalescence) of ancestral polymorphism or to introgressive hybridization?

8. Postzygotic isolation between populations or species is sometimes based on interactions between nuclear genes and cytoplasmic genetic elements such as mitochondria or endosymbiotic bacteria. How might such incompatibility act and how might it arise? Refer to work on *Wolbachia*, an endosymbiotic bacterium in insects, as a particularly well-studied example (Werren 1998).

Speciation

In this chapter, we turn to the question of how species arise. If we considered species to be merely populations with distinguishing characteristics, the question of how they originate would be easily answered: natural selection or genetic drift can fix novel alleles or characteristics (see Chapters 10–13). But if the permanence of these distinctions depends on reproductive isolation, and if we consider reproductive isolation a defining feature of species, then the central question about speciation must be how genetically based barriers to gene exchange arise. Our description of the forms of reproductive isolation between species must now be complemented by understanding how they evolve.

The difficulty this question poses is most readily seen if we consider the postzygotic reproductive barriers discussed in Chapter 17, such as hybrid inviability or sterility. If two populations are fixed for genotypes A_1A_1 and A_2A_2, but the heterozygote A_1A_2 has lower viability or fertility than either homozygous genotype, how could these populations have diverged? Whatever allele the ancestral population may have carried (say, A_1), the low fitness of A_1A_2 would have prevented the alternative allele (A_2) from increasing in frequency and thus forming a reproductively incompatible population.

Suppose, instead, that reproductive isolation between the populations is based on more than one locus. The problem then is that *recombination generates intermediates*. If several loci govern, for example, time of breeding, $A_1A_1B_1B_1C_1C_1$ and $A_2A_2B_2B_2C_2C_2$ might breed

Sexual selection: A major cause of speciation?
Speciation has been most prolific in many groups of animals, such as birds of paradise, in which sexual selection appears to be intense. Here, a male Raggiana bird of paradise (*Paradisaea raggiana*) displays his bounteous plumes to a female. In many such species, only a few males mate with most of the females.

early and late in the season, respectively, and so be reproductively isolated. But if mutations A_2, B_2, and C_2 occur in an initially $A_1A_1B_1B_1C_1C_1$ population and increase in frequency, many genotypes with intermediate breeding seasons, such as $A_1A_2B_1B_1C_1C_2$, are formed by recombination, and these genotypes constitute a "bridge" for the flow of genes between the two extreme genotypes.

The problem of speciation, then, is how two different populations can be formed without intermediates. This problem holds, whatever the character that confers prezygotic or postzygotic isolation may be. The many conceivable solutions to this problem are referred to as MODES OF SPECIATION.

Modes of Speciation

The modes of speciation that have been hypothesized can be classified by several criteria (**Table 18.1**), including the geographic origin of barriers to gene exchange, the genetic bases of those barriers, and the causes of evolution of those barriers. These criteria are independent of one another; so, for example, two species may conceivably form by geographic separation (allopatry) of populations, in which reproductive isolation then evolves by either natural selection or genetic drift, which results in few or many genetic differences.

Speciation may occur in three kinds of geographic settings. **Allopatric speciation** is the evolution of reproductive barriers in populations that are prevented by a geographic barrier from exchanging genes at more than a negligible rate. A distinction is often made between allopatric speciation by **vicariance** (divergence of two large populations; **Figure 18.1A**) and **peripatric speciation** (divergence of a small population from a widely distributed ancestral form; **Figure 18.1B**). In **parapatric speciation**, adjacent, spatially distinct populations, between which there is some gene flow, diverge and become reproductively isolated (**Figure 18.1C**). **Sympatric speciation** is the evolution of reproductive barriers within a single, initially randomly mating (panmictic) population (**Figure 18.1D**).

Allopatric, parapatric, and sympatric speciation form a continuum, differing in the initial level of gene flow (m) between diverging populations. Strictly defined, $m = 0$ if speciation is allopatric, and it is maximal ($m = 0.5$) if speciation is sympatric; intermediate cases ($0 < m < 0.5$) represent parapatric speciation (Fitzpatrick et al. 2008). The initial reduction of gene exchange is accomplished by a physical barrier extrinsic to the organisms in allopatric speciation, but by evolutionary change in the biological characteristics of the organisms themselves in sympatric speciation. Allopatric or "nearly allopatric" (with low initial m) speciation is widely acknowledged to be a common mode of speciation; the incidence of sympatric or "nearly sympatric" (with high m) speciation is debated.

From a genetic point of view, the reproductive barriers that arise may be based on genetic divergence (allele differences at, usually, several or many loci), cytoplasmic incompatibility, or cytological divergence (polyploidy or structural rearrangement of chromosomes). We will devote most of this chapter to speciation by genetic divergence.

The causes of the evolution of reproductive barriers, as of any characters, are genetic drift and natural selection of genetic alterations that have arisen by mutation. Peripatric speciation, a hypothetical form of speciation that is also referred to as TRANSILIENCE or FOUNDER EFFECT SPECIATION, requires both genetic drift and natural selection. Both sexual selection and ecological causes of natural selection may result in speciation. In some cases, there may be *selection for reproductive isolation*—that is, to

---

TABLE 18.1   Modes of speciation

I. Classified by geographic origin of reproductive barriers
   A. Allopatric speciation
      1. Vicariance
      2. Peripatric speciation
   B. Parapatric speciation
   C. Sympatric speciation
II. Classified by genetic and causal bases[a]
   A. Genetic divergence (allele substitutions)
      1. Genetic drift
      2. Peak shift (genetic drift + natural selection)
      3. Natural selection
         a. Ecological selection
            i. Ecological trait causes reproductive isolation
            ii. Pleiotropic genes correlate ecological difference and reproductive isolation
         b. Sexual selection
   B. Cytoplasmic incompatibility
   C. Cytological divergence
      1. Polyploidy
      2. Chromosome rearrangement
   D. Recombinational speciation

[a]Most of the genetic and causal bases might act in an allopatric, parapatric, or sympatric geographic context, and some of the causal bases listed under "Genetic divergence" also apply to cytoplasmic incompatibility, cytological divergence, and recombinational speciation.

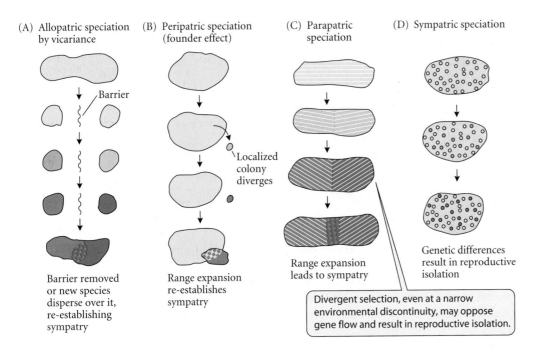

FIGURE 18.1 Schematic diagrams showing the successive stages in models of speciation that differ in their geographic setting. (A) The vicariance model of allopatric speciation. (B) The peripatric, or founder effect, model of allopatric speciation. (C) Parapatric speciation. (D) Sympatric speciation.

(A) Allopatric speciation by vicariance

Barrier

Barrier removed or new species disperse over it, re-establishing sympatry

(B) Peripatric speciation (founder effect)

Localized colony diverges

Range expansion re-establishes sympatry

(C) Parapatric speciation

Range expansion leads to sympatry

Divergent selection, even at a narrow environmental discontinuity, may oppose gene flow and result in reproductive isolation.

(D) Sympatric speciation

Genetic differences result in reproductive isolation

prevent hybridization. (Recall the distinction between *selection for* and *selection of* traits, discussed in Chapter 11.) Alternatively, reproductive isolation may arise as a *by-product of genetic changes* that occur for other reasons (Muller 1940; Mayr 1963). In this case, there may be ADAPTIVE DIVERGENCE of the isolating character itself (e.g., climate factors may favor breeding in two different seasons, with the effect that the populations do not interbreed), or the reproductive barrier may arise as a pleiotropic by-product of genes that are selected for their other functions.

Allopatric Speciation

Allopatric speciation is *the evolution of genetic reproductive barriers between populations that are geographically separated* by a physical barrier such as a topographic feature, water (or land), or unfavorable habitat. The physical barrier reduces gene flow enough for sufficient genetic differences to evolve to prevent gene exchange between the populations should they later come into contact (see Figure 18.1A). Although some authors' definitions of allopatric speciation require zero gene flow ($m = 0$) between the populations, in this discussion, we will assume only that m is so low that divergence by very weak selection or even genetic drift is possible. Allopatry is defined by a severe reduction of movement of individuals or their gametes, not by geographic distance. Thus in species that disperse little or are faithful to a particular habitat, populations may be "microgeographically" isolated (e.g., among patches of a favored habitat along a lakeshore). All evolutionary biologists agree that allopatric speciation occurs, and many hold that it is the prevalent mode of speciation, at least in animals (Mayr 1963; Coyne and Orr 2004).

From paleontological and genetic studies (see Chapters 5 and 6), we know that species' geographic ranges change over time, and that populations may become separated and later rejoined. (Consider, for example, the postglacial range expansions portrayed in Figure 6.16.) Thus allopatric populations may expand their range and come into contact. If sufficiently strong isolating barriers have evolved during the period of allopatry, the populations may become sympatric without exchanging genes. If incomplete reproductive isolation has evolved, they will form a hybrid zone (see Chapter 17, where we described the possible fates of hybrid zones). Sympatric sister species that we observe today may well have speciated allopatrically and then expanded their ranges; *current sympatry, in itself, is not evidence that speciation occurred sympatrically.*

FIGURE 18.2 The strength of sexual isolation between populations of the salamander *Desmognathus ochrophaeus* is correlated with (A) the geographic distance between the populations as well as (B) their genetic distance (Nei's *D*, which measures the difference in allozyme frequencies at several loci). (After Tilley et al. 1990.)

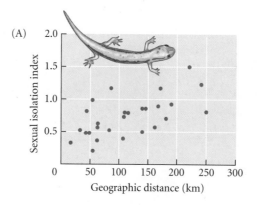
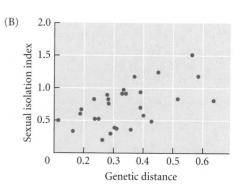

Evidence for allopatric speciation

Because both natural selection and genetic drift cause populations to diverge in genetic composition, it is probably inevitable that if separated long enough, geographically separated populations will become different species. Many species show incipient prezygotic or postzygotic reproductive isolation among geographic populations. For example, Stephen Tilley and colleagues (1990) examined sexual isolation among dusky salamanders (*Desmognathus ochrophaeus*) from various localities in the southern Appalachian Mountains of the eastern United States. They brought males and females from different populations (heterotypic pairs) and from the same population (homotypic pairs) together and scored the proportion of the pairs that mated. Among the various pairs of populations, an index of the strength of sexual isolation varied continuously, from almost no isolation to almost complete failure to mate. The more geographically distant the populations, the more genetically different they were, and the less likely they were to mate (**Figure 18.2**).

Speciation can often be related to the geological history of barriers. For example, the emergence of the Isthmus of Panama in the Pliocene divided many marine organisms into Pacific and Caribbean populations, some of which have diverged into distinct species. Among seven such species pairs of snapping shrimp, only about 1 percent of interspecific matings in the laboratory produced viable offspring (Knowlton et al. 1993).

In some cases, CONTACT ZONES between differentiated forms mark the meeting of formerly allopatric populations. For example, Eldredge Bermingham and John Avise (1986; Avise 1994) analyzed the genealogy of mitochondrial DNA in samples of six fish species from rivers throughout the coastal plain of the southeastern United States. In all six species, DNA sequences form two distinct clades characterizing eastern and western populations, and the two clades make contact in the same region of western Florida (**Figure 18.3**). This pattern implies that gene flow between east and west was reduced at some

FIGURE 18.3 Evidence for allopatric genetic divergence followed by range expansion and secondary contact. In each of six freshwater fish species of the southeastern United States, mitochondrial DNA sequences fall into two clades, one with a western and one with an eastern distribution. Three families of fishes are represented here: Centrarchidae (sunfishes), Amiidae (bowfin), and Poeciliidae (mosquitofish). (After Avise 1994.)

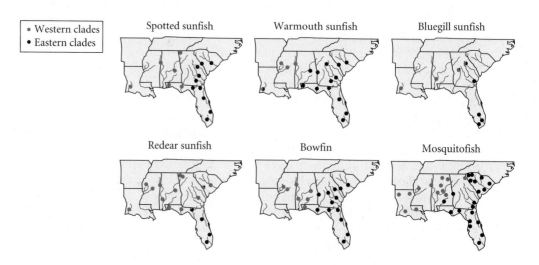

FIGURE 18.4 The degree of overlap in the geographic ranges of pairs of closely related species, plotted against the genetic divergence between them, which is an index of time since speciation. Overlap increases with time in fairy wrens and swordtail fish, as expected if speciation was allopatric in these groups. There is no correlation between overlap and time since divergence in tiger beetles or *Rhagoletis* fruit flies, a pattern consistent with sympatric speciation. (After Barraclough and Vogler 2000.)

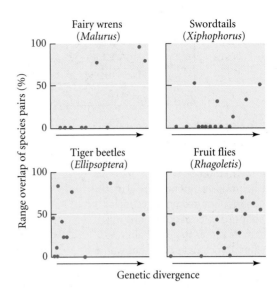

time in the past. The amount of sequence divergence between the two clades suggests that isolation occurred 3 to 4 Mya. At that time, sea level was much higher than it is at present, forming a barrier to dispersal by freshwater fishes.

Tim Barraclough and Alfried Vogler (2000) reasoned that over time, the amount of overlap between the geographic ranges of species that have formed by allopatric speciation can only increase from zero, whereas overlap between species that originated by sympatric speciation should stay the same or decrease. For several clades of closely related birds, insects, and fishes, they plotted degree of range overlap against degree of molecular difference between species, which they used as an index of time since gene exchange was curtailed. Several groups showed increasing overlap with time, as expected in the case of allopatric speciation, whereas two groups of insects displayed a pattern consistent with the possibility of sympatric speciation (**Figure 18.4**).

Species on islands have provided abundant evidence of allopatric speciation. Where two or more closely related species of birds occur together on an island, other islands or a continent can be identified as a source of invading species, and in all cases there is evidence that the ancestors of the several species invaded the island at separate times. For example, many of the islands in the Galápagos archipelago harbor two or more species of Darwin's finches, which evolved on different islands and later became sympatric. But Cocos Island, isolated far to the northeast of the Galápagos, has only one species of finch, which occupies several of the ecological niches that its relatives in the Galápagos Islands fill (Werner and Sherry 1987; see Figure 3.22). In contrast to archipelagoes, no pairs of sister species of birds occur together on any isolated island smaller than 10,000 square kilometers in area. This observation implies that speciation in birds does not occur on land masses that are too small to provide geographic isolation between populations (Coyne and Price 2000). A similar pattern is found in many other taxa (Kisel and Barraclough 2010). Moreover, taxa in which dispersal, and therefore gene flow, over long distances is high (such as bats) have speciated only on much larger islands than taxa (such as snails) in which gene flow is very limited (**Figure 18.5**). This pattern is as expected, because gene flow opposes the genetic divergence required for speciation.

FIGURE 18.5 Speciation is more likely on larger islands. (A) The proportion of lizard lineages that have undergone speciation within isolated oceanic islands is higher on larger islands. (B) For various taxa, the minimum island size allowing speciation is larger in taxa that have higher rates of gene flow (at left). Taxa with low rates of gene flow, such as snails, can speciate within much smaller islands. Gene flow is inversely related to the fixation index (F_{ST}), a measure of the variation in allele frequencies among local populations of a species. (After Kisel and Barraclough 2010.)

(B)

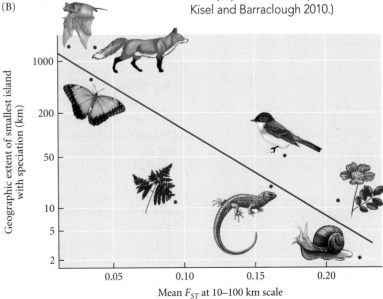

(A)

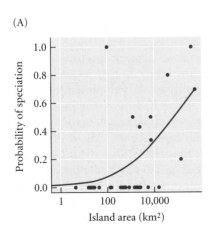

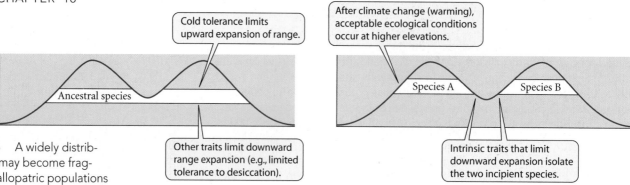

Cold tolerance limits upward expansion of range.

After climate change (warming), acceptable ecological conditions occur at higher elevations.

Ancestral species

Species A   Species B

Other traits limit downward range expansion (e.g., limited tolerance to desiccation).

Intrinsic traits that limit downward expansion isolate the two incipient species.

FIGURE 18.6 A widely distributed species may become fragmented into allopatric populations if the habitat on which it depends becomes fragmented by climate change. For example, a species adapted to cool conditions may shift its range to higher altitudes when the climate becomes warmer, and the populations, isolated on different mountains, may become different species. Phylogenetic niche conservatism can therefore contribute to speciation. (After Wiens 2004.)

The role of barriers such as the Isthmus of Panama in curtailing gene flow between populations is obvious, but what kinds of barriers could have produced the great numbers of species, in many taxa, that are found on continents? An important consideration is phylogenetic niche conservatism (see Chapter 6). Geographic distributions may be fragmented if populations maintain dependence on specific environmental conditions, such as climate regimes or habitats. For example, a species that is widely distributed at low elevations in a mountain range when the climate is cool may move upward and form separate populations on different mountains when the climate becomes warmer (**Figure 18.6**). Exactly this pattern has been found for allopatric sister species of salamanders, which are found in locations with similar climate conditions and are absent from intervening regions with different climate conditions (Kozak and Wiens 2006). Populations of a species may also become separated if they become adapted to different habitats that are geographically segregated (Sobel et al. 2010).

Mechanisms of vicariant allopatric speciation

Models of vicariant allopatric speciation based on genetic drift, natural selection, and a combination of these two factors have been proposed. The combination of genetic drift and selection is discussed later, in relation to peripatric speciation.

THE ORIGIN OF INCOMPATIBILITY How can failure to interbreed, or inability of hybrids to reproduce, arise if they imply fixation of alleles that lower reproductive success? The increase of such alleles to fixation, of course, would be counter to natural selection. Theodosius Dobzhansky (1936) and Hermann Muller (1940) provided a theoretical solution to this problem that does not envision increasing an allele's frequency in opposition to selection. It requires that the reproductive barrier be based on differences at two or more loci that have complementary effects on fitness. In other words, fitness depends on the combined action of the "right" alleles at both loci. The "wrong" combinations of alleles result in Dobzhansky-Muller (DM) incompatibility, as illustrated by the *Drosophila* genes described in Figure 17.14.

Suppose the ancestral genotype in two allopatric populations is $A_2A_2B_2B_2$ (**Figure 18.7**). For some reason, A_1 replaces A_2 in population 1 and B_1 replaces B_2 in population 2, yielding populations monomorphic for $A_1A_1B_2B_2$ and $A_2A_2B_1B_1$, respectively. Both A_1A_2 and A_1A_1 have fitness equal to or greater than A_2A_2 in population 1, as long as the genetic background is B_2B_2; likewise, B_1B_2 and B_1B_1 are equal or superior to B_2B_2, as long as the genetic background is A_2A_2. Therefore these allele substitutions can occur by natural selection (if the fitnesses differ) or by genetic drift (if they do not). However, an epistatic interaction between A_1 and B_1 causes incompatibility, so that either the hybrid $A_1A_2B_1B_2$ has lowered viability or fertility, or $A_1A_1B_2B_2$ and $A_2A_2B_1B_1$ are isolated by a prezygotic barrier, such as different sexual behavior. *The important feature of this model is that neither population has passed through a stage in which inferior heterozygotes existed.* Neither of the incompatible alleles has ever been "tested" against the other within the same population.

This model is supported by genetic data showing that reproductive isolation is based on epistatic interactions among several or many loci (see Chapter 17). It is theoretically possible that the allele substitutions could be caused by either genetic drift or natural selection. However, no convincing examples have been described in which speciation can be

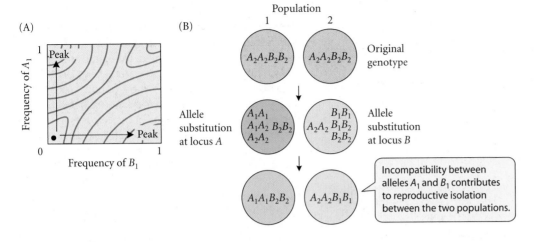

FIGURE 18.7 The Dobzhansky-Muller theory explains how allele substitution can lead to reproductive isolation. We begin with two populations, both initially composed of genotype $A_2A_2B_2B_2$. (A) The adaptive landscape, in which contour lines represent mean fitness as a function of allele frequencies at both loci, shows how the two populations may move uphill toward different adaptive peaks. (B) Each population undergoes an allele substitution at a different locus (substituting either A_1 or B_1). The hybrid combination $A_1A_2B_1B_2$ has low fitness (as indicated by the "valley" in the center of the landscape) because of prezygotic or postzygotic incompatibility between A_1 and B_1.

attributed entirely to genetic drift (Coyne and Orr 2004). In contrast, natural selection may contribute to the origin of species in several ways.

THE ROLE OF NATURAL SELECTION   The most widely held view of vicariant allopatric speciation is that it is caused by *natural selection, which causes the evolution of genetic differences that create prezygotic or postzygotic isolation.* Some—perhaps most—of this reproductive isolation evolves while the populations are allopatric, so that a substantial or complete barrier to gene exchange exists when the populations meet again if their ranges expand (Mayr 1963). Thus speciation is usually an *effect*—a by-product—of natural selection that occurred during allopatry. That selection may be either ecological selection or sexual selection.

The other possibility is that natural selection favors prezygotic (e.g., sexual) reproductive barriers *because of* their isolating function—because they prevent their bearers from having unfit hybrid progeny. Selection would then result in *reinforcement of reproductive isolation.* This reinforcement would occur only when the genetically different populations come into contact and have the opportunity to hybridize. In this scenario, some degree of postzygotic isolation (low hybrid fitness) evolves while the populations are allopatric, but speciation is completed when the incipient species come into contact.

Ecological selection and speciation

Ecological selection might cause speciation in two ways, which have been termed mutation-order speciation and ecological speciation (Schluter 2009). In MUTATION-ORDER SPECIATION, mutations at different genes occur in each population, are selected for the same reason (e.g., they provide adaptation to the same selective factor), and confer DM incompatibility (e.g., mutations A_1 and B_1 in Figure 18.7). The most likely examples of mutation-order speciation described so far are based not on ecological selection, but on genetic conflict, such as that arising from meiotic drive (see Chapter 16). **Ecological speciation** refers to the evolution of barriers to gene flow caused by divergent ecologically based selection (Rundle and Nosil 2005; Schluter 2009; Nosil 2012).

Allopatric populations and species undergo both adaptive divergence and evolution of reproductive isolation, but showing that reproductive isolation is a result of adaptive divergence requires evidence that the two processes are genetically and causally related to each other. The most direct evidence comes from laboratory studies of *Drosophila* and houseflies, in which investigators tested for reproductive isolation among subpopulations drawn from a single base population and subjected to divergent selection for various morphological, behavioral, or physiological characteristics (Rice and Hostert 1993). In many of these studies, partial sexual isolation or postzygotic isolation developed, demonstrating that substantial progress toward speciation can be observed in the laboratory, and that it can be caused by divergent selection. That is, reproductive isolation in these studies was due to pleiotropic effects of genes for the divergently selected character, or closely linked genes.

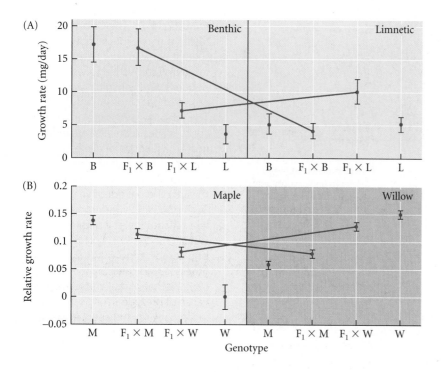

FIGURE 18.8 Postzygotic reproductive isolation, manifested as reduced fitness of hybrids, may depend on environmental context. (A) Blue circles show growth rates of limnetic (open water; L) and benthic (lake bottom; B) forms of the three-spined stickleback when caged in benthic and limnetic habitat. Red circles show growth rates of backcrosses to both parent forms ($F_1 \times B$ and $F_1 \times L$). Each backcross type has high fitness in the habitat of its backcross parent and low fitness in the habitat of the other parent type. (B) The same experimental design applied to populations of the leaf beetle *Neochlamisus bebbianae* that are adapted to maple (M; *Acer rubrum*) and willow (W; *Salix bebbiana*). Each backcross hybrid ($F_1 \times M$; $F_1 \times W$) shows highest fitness on the host plant of its "pure" parent type. (A after Rundle 2002; B after Egan and Funk 2009).

In some cases, reproductive isolation is clearly the direct result of ecologically selected character differences. For example, the species of monkeyflowers (*Mimulus*) in Figure 17.6 avoid interbreeding almost entirely by attracting the different pollinators (bees versus hummingbirds) to which their different flowers are adapted. Many incipient species are reproductively isolated by immigrant inviability (Nosil et al. 2005), in which populations are genetically adapted to different environments and have low fitness in each other's environments (see Table 17.2). Examples include stickleback fish that are adapted for foraging in limnetic (open water) versus benthic (lake bottom) habitats (**Figure 18.8A**) and "host races" of insects adapted to different host plants (**Figure 18.8B**). Characters that confer both ecological adaptation and reproductive isolation have been termed "magic traits" by Sergei Gavrilets (2004). Such traits make speciation relatively "easy," and they appear to be fairly common (Servedio et al. 2011).

Alternatively, reproductive isolation can be a by-product of ecological adaptation, presumably because of pleiotropic effects of alleles that affect adaptation. For example, three-spined sticklebacks (*Gasterosteus aculeatus*) have undergone PARALLEL SPECIATION in several Canadian lakes, where a limnetic (open-water) ECOMORPH coexists with a benthic (bottom-feeding) ecomorph that is smaller and differs in shape. These ecomorphs, which are sexually isolated, have evolved independently in each lake; that is, speciation has occurred in parallel (**Figure 18.9A**). Parallel ecological divergence implies that ecological selection has shaped the differences between the ecomorphs. In laboratory trials, fish of the same ecomorph from different lakes mate almost as readily as those from the same lake, but different ecomorphs mate much less frequently (**Figure 18.9B**). Thus features associated with ecological divergence affect sexual isolation (Rundle et al. 2000). Exactly why they do so is not known.

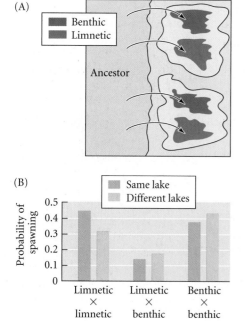

FIGURE 18.9 Parallel speciation in the three-spined stickleback (*Gasterosteus*). (A) Pairs of open-water (limnetic) and bottom-feeding (benthic) ecomorphs have arisen independently in different lakes. (B) Females mate preferentially with males on the basis of their morphology, whether they are from the same or different lakes. This isolating character is evidently adaptive, since it has evolved repeatedly in the same way. (A after Schluter and Nagel 1995; B after Rundle et al. 2000.)

(A)

(B)

(C)

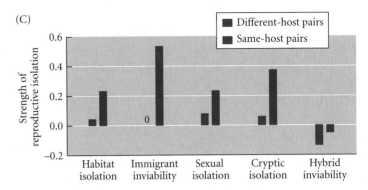

FIGURE 18.10 Ecomorphs of the stick insect *Timema cristinae* found on the shrubs *Adenostoma fasciculatum* (A) and *Ceanothus spinosus* (B) differ in body form and color pattern, closely matching the foliage on which they feed. (C) Pairs of populations from different hosts are more strongly reproductively isolated in several respects than are pairs from the same host. The negative values for hybrid inviability indicate enhanced viability of hybrids, but here these values are not statistically significant. Cryptic isolation refers to reduced fecundity in interpopulation matings. (Photos courtesy of Patrik Nosil; C after Nosil 2007.)

Two ecomorphs of the stick insect *Timema cristinae* are associated with different host plants in the chaparral vegetation of California, and they differ in several morphological features that make each ecomorph better camouflaged on its own host plant than on the other. Patrik Nosil (2007) has studied several components of reproductive isolation between multiple pairs of different-host populations of this species (**Figure 18.10**). He concluded that ecological divergence directly reduces gene flow in two ways. First, habitat isolation reduces gene flow because large patches of chaparral vegetation are dominated by one or the other plant. Second, immigrant inviability results from the high mortality the insects suffer if they disperse to the "wrong" host, where they are easy targets for birds. Indirect effects of ecological divergence include sexual isolation, which is greater between different-host than same-host populations, and reduced fertility in cross-matings: females that mate with the other ecomorph lay fewer eggs.

Daniel Funk and colleagues (2006) provided evidence that divergent ecological adaptation commonly contributes to speciation by compiling data from the literature on reproductive isolation, indicators of ecological divergence among species, and genetic distances among species in several groups of plants, insects, fishes, frogs, and birds. Genetic distance was used as an index of time since pairs of species had diverged from their common ancestor. By statistically controlling for time, the investigators showed that the level of reproductive isolation achieved at any time is correlated with the degree of ecological divergence between species (**Figure 18.11**).

FIGURE 18.11 A method for testing the hypothesis that reproductive isolation evolves as a by-product of ecological divergence. (A) For several pairs of species, experimental estimates of reproductive isolation are plotted against both the time since common ancestry (estimated by genetic distance, as in Figure 17.9) and a measure of ecological difference. (B) The amount of "residual" reproductive isolation that is not accounted for by time since common ancestry is plotted against ecological divergence and tested for correlation. (C) An example of real data, showing that postzygotic isolation between pairs of flowering plant species is significantly correlated with difference in habitat use. (After Funk et al. 2006.)

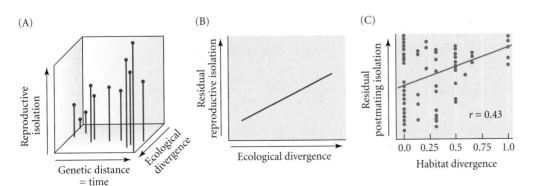

(A)   (B)   (C)

Sexual selection and speciation

Closely related species of animals are often sexually isolated by female preferences for features of conspecific males. In fact, many authors consider sexual isolation to be the most important reproductive barrier, although this view is controversial (Coyne and Orr 2004). One hypothesis proposed to explain the differences between species in these characteristics is that they enable individuals to recognize conspecific mates and avoid hybridization, which would be disadvantageous if hybrid offspring have low fitness. An alternative hypothesis is that divergent sexual selection in different geographic populations of a species results in different male display traits and female preferences (Fisher 1930; West-Eberhard 1983). This hypothesis has been supported by mathematical models (Lande 1981; Pomiankowski and Iwasa 1998; Turelli et al. 2001).

It is very likely that sexual selection has been an important cause of speciation, especially in highly diverse groups such as Hawaiian *Drosophila*, birds of paradise, and hummingbirds (see Figure 17.5), in which males are often highly (and diversely) colored or ornamented (Panhuis et al. 2001). The male color patterns of some closely related African lake cichlids act as reproductive barriers between species, and it is likely that sexual selection has contributed to the extraordinarily high species diversity of these fishes (Seehausen et al. 1999). Some comparisons of the species diversity of sister groups of birds suggest that sexual selection has enhanced diversity (**Figure 18.12**). Groups of birds with promiscuous mating systems have higher diversity than sister clades in which pair-bonds are formed and the variance in male mating success is presumably lower—resulting in weaker sexual selection (Mitra et al. 1996).

The role of sexual selection in speciation has been extensively studied in orthopteran insects (e.g., crickets) and frogs, in which males produce fairly simple acoustic mating signals that can be simulated and varied electronically. For example, two sister species of crickets (*Gryllus texensis* and *G. rubens*) in the southern United States differ in the pulse rate of the male song, and females of both species show a much stronger response to

FIGURE 18.12 In sister clades of birds that differ in their mating system, those clades that mate promiscuously and do not form a pair-bond (A, C) tend to have more species than nonpromiscuous clades that do form pair-bonds (B, D). The promiscuously mating clades are thought to experience stronger sexual selection. (A) A promiscuous male magnificent bird of paradise (*Cicinnurus magnificus*) and (B) a nonpromiscuous manucode (*Manucodia comrii*). (C) A male violet sabrewing hummingbird (*Campylopterus hemileucurus*) and (D) a common swift (*Apus apus*), a member of a sister clade that forms pair-bonds.

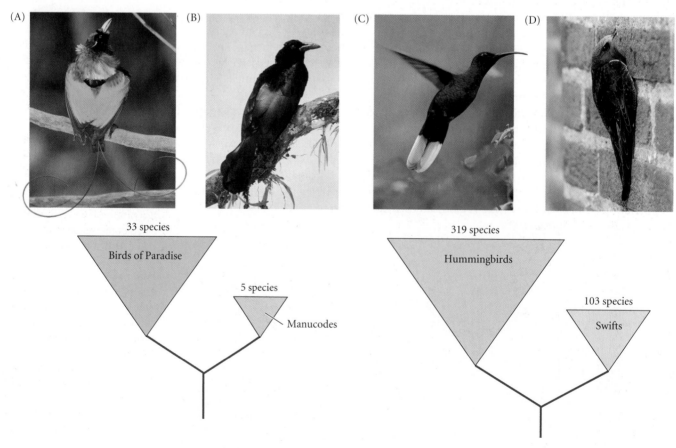

synthesized songs with pulse rates characteristic of their own species (Gray and Cade 2000). Hybrids produced in the laboratory are viable and fully fertile, and there is no evidence that the differences in male song or female preference have evolved to reduce hybridization, because distant allopatric populations of the two species, which have no opportunity to hybridize, are just as different as sympatric populations. Among populations of a Hawaiian cricket (*Laupala cerasina*), male calls also vary in pulse rate, and female preferences are strongly correlated with the rates of conspecific males (**Figure 18.13**; Grace and Shaw 2011). In crosses between two other species of *Laupala*, male pulse rate and female pulse preference were genetically correlated and possibly controlled by the same genes. Such correlation that would facilitate divergence by sexual selection (see Chapter 17).

There has been little research on why sexual selection varies among populations, leading to divergence and sexual isolation. In some cases, visual signals (e.g., coloration) and acoustic signals have been shaped in part by selection for more effective transmission and reception, which can be affected by the environment (Endler and Basolo 1998). For example, the songs of birds that live in the undergrowth of Amazonian forests have higher frequencies ("pitch") than those of close relatives that inhabit stands of bamboo, a difference that corresponds to the most effectively transmitted signal in each environment (Tobias et al. 2010a). It is also likely that the course of runaway sexual selection, or of selection for condition-dependent indicators of fitness, comes to differ between populations, but whether natural selection or genetic drift causes such changes in course is a problem for future research. Sexual conflict (see Chapter 15) can easily lead to the evolution of reproductive isolation caused by different male features that reduce female fitness, and different female countermeasures, in different populations (Gavrilets 2000). Göran Arnqvist et al. (2000) found that among 25 pairs of sister clades, species richness was greater in those with polygamous females, in which sperm of multiple males may compete, than in those with monogamous females.

Sexual selection also probably plays a role in the evolution of gametic isolation, a major barrier to gene flow between species of broadcast-spawning marine invertebrates (**Figure 18.14A**; Palumbi 2009). In some cases, as in abalones and related snails, both a sperm-surface protein and the egg protein with which it interacts during fertilization have diverged rapidly by natural selection (**Figure**

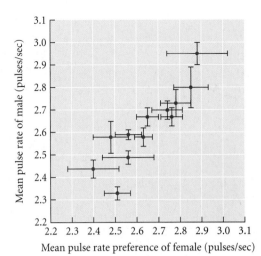

FIGURE 18.13 The pulse rate of the mating call of male crickets (*Laupala cerasina*) and the pulse rate preferred by females both vary among local populations. These differences are genetically based. The confidence intervals around each point show that females of the most widely different populations would not readily mate with males at the other extreme. (After Grace and Shaw 2011.)

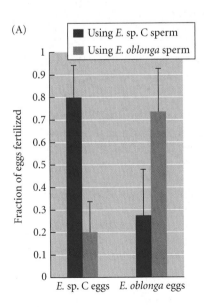

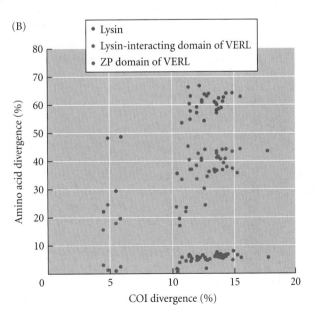

FIGURE 18.14 Gametic isolation based on protein differences. (A) When eggs from two closely related sea urchin species, *Echinometra oblonga* and *E.* species C, are exposed to a mixture of sperm of both species, conspecific sperm are much more successful in fertilization. (B) Fertilization in the marine snail genus *Tegula* is partly based on the ability of the sperm protein lysin to dissolve the vitelline envelope, a species-specific reaction that depends on the binding of lysin to an envelope protein called VERL (vitelline envelope receptor for lysin). Among species of *Tegula*, sequence divergence of lysin and the domain of VERL with which lysin interacts is high and has increased with time since speciation, as measured by divergence of a mitochondrial gene (COI). The ZP domain of VERL, which does not interact with lysin, has diverged more slowly. (A after Palumbi 2009; B after Hellberg et al. 2012.)

18.14B), resulting in a block to crossing between species. How natural selection has caused this divergence is unclear. One possibility is that egg surface proteins evolve to prevent infection by pathogens, and that sperm proteins must adjust. Sexual conflict is perhaps a more likely answer: changes in the egg surface that slow down sperm entry are advantageous because fertilization by more than one sperm kills the egg. Any such changes in the egg will impose selection for sperm that can beat their competitors by penetrating more quickly.

Reinforcement of reproductive isolation

We have seen that reproductive isolation can arise as a *side effect* of genetic divergence due to natural selection. However, many biologists have supposed that reproductive isolation evolves, at least in part, as an *adaptation to prevent the production of unfit hybrids*. The champion of this viewpoint was Theodosius Dobzhansky, who expressed the hypothesis this way:

> Assume that incipient species, A and B, are in contact in a certain territory. Mutations arise in either or in both species which make their carriers less likely to mate with the other species. The nonmutant individuals of A which cross to B will produce a progeny which is adaptively inferior to the pure species. Since the mutants breed only or mostly within the species, their progeny will be adaptively superior to that of the nonmutants. Consequently, natural selection will favor the spread and establishment of the mutant condition. (Dobzhansky 1951, p. 208)

Dobzhansky introduced the term "isolating mechanisms" to designate reproductive barriers, which he believed were indeed mechanisms designed to isolate. In contrast, Ernst Mayr (1963), among others, held that although natural selection might enhance reproductive isolation, reproductive barriers arise mostly as side effects of allopatric divergence, whatever its cause may be. Mayr cited several lines of evidence: sexual isolation exists among fully allopatric forms; it has failed to evolve in some hybrid zones that are thought to be thousands of years old; features that promote sexual isolation between species are usually not limited to regions where the species are sympatric and face the "threat" of hybridization. It is now generally agreed that natural selection can enhance prezygotic reproductive isolation between hybridizing populations, but how often this process plays a role in speciation is not known (Servedio and Noor 2003; Butlin et al. 2012).

The enhancement of prezygotic barriers that Dobzhansky envisioned is often called **reinforcement** of prezygotic isolation.

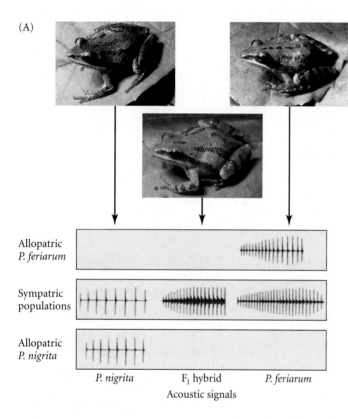

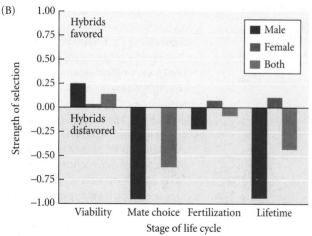

FIGURE 18.15 Character displacement and reinforcement of sexual isolation. (A) The range of the chorus frog *Pseudacris feriarum* in the eastern United States partly overlaps that of the more southern *P. nigrita*. The oscillograms (displaying amplitude plotted against time) show that male calls differ more in pulse rate between sympatric populations of the two species than between allopatric populations. The call of the F₁ hybrid is also shown. (B) The geographic pattern of character displacement in these frogs was probably caused by reinforcement of prezygotic isolation. Hybrid males have reduced fitness both because they are less preferred by nonhybrid females (mate choice) and because they have lower average success in fertilizing eggs. (After Lemmon and Lemmon 2010; photos courtesy of Emily Lemmon.)

Reinforcement has been cited as a cause of **reproductive character displacement**, meaning a *pattern* whereby characters differ more where two taxa are sympatric than where they are allopatric (Brown and Wilson 1956). Population genetic models have shown that reinforcement can evolve: alleles that reduce the likelihood of interbreeding can increase in frequency because they are more likely to be inherited by viable nonhybrid offspring than are alleles that permit random mating—which will decline in frequency if they are inherited by unfit hybrids (Servedio and Noor 2003).

Reinforcement of prezygotic isolation appears to occur fairly often (Noor 1999). For example, the pulse rate and pulse number of the male mating call of the chorus frog *Pseudacris feriarum*, of eastern North America, are higher in populations that are sympatric with a more southern species, *Pseudacris nigrita*, and females have likewise shifted their preference for male calls (**Figure 18.15A**). Male hybrids have lower fertility than nonhybrids, and female *P. feriarum* discriminate against hybrid male calls (**Figure 18.15B**). This pattern of character displacement is the expected consequence of reinforcement. Similarly, *Drosophila serrata* and its close relative *D. birchii* are sympatric in northern Australia, but *D. serrata* extends much farther south than *D. birchii* does. Females of both species choose males based on their relative proportions of several hydrocarbon compounds in the cuticle (CHCs). There is an abrupt difference in male CHCs between allopatric and sympatric populations of *D. serrata* (**Figure 18.16A**). When Megan Higgie and Mark Blows (2007) confined allopatric *D. serrata* with *D. birchii* in laboratory populations, the composition of male CHCs evolved within nine generations toward the composition found in sympatric *D. serrata* (**Figure 18.16B**), showing that reinforcement was not only predictable, but could occur very rapidly. In another experiment, Higgie and Blows (2008) found that in laboratory populations formed by crossing sympatric and allopatric populations of *D. serrata*, male CHCs evolved toward the natural allopatric composition, and female preference evolved toward the CHC composition found in allopatric populations. Sexual selection alone, then, favors a different CHC composition than selection for reinforcement.

When Jerry Coyne and Allen Orr (1989) compiled experimental data on reproductive isolation in *Drosophila*, they found that sexual (but not postzygotic) isolation was stronger between sympatric pairs of species (or populations) than between allopatric populations

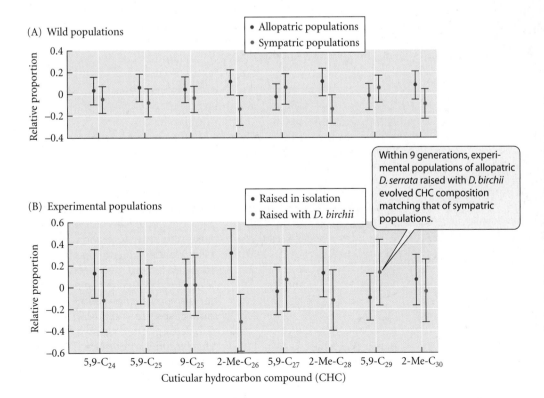

(A) Wild populations

• Allopatric populations
• Sympatric populations

(B) Experimental populations

• Raised in isolation
• Raised with *D. birchii*

Within 9 generations, experimental populations of allopatric *D. serrata* raised with *D. birchii* evolved CHC composition matching that of sympatric populations.

Cuticular hydrocarbon compound (CHC)

FIGURE 18.16 Evidence of reinforcement of differences in hydrocarbon composition of the male cuticle (CHC) in two species of *Drosophila*. Female mate choice is mediated by the relative proportions of cuticular hydrocarbons. (A) The relative proportions of eight CHCs in populations of *D. serrata* that are allopatric and sympatric with *D. birchii*. (B) The relative proportions of these CHCs in *D. serrata* from allopatric populations differed between stocks kept in the same cages as *D. birchii* for nine generations and those kept isolated from *D. birchii*. (After Higgie and Blows 2007.)

of the same estimated age (see Figure 17.10). They suggested that this pattern was a consequence of reinforcement. The evidence for this interpretation has grown. For example, Roman Yukilevich (2012), in an analysis of a similar but larger data set, compared asymmetry in both sexual isolation and postzygotic isolation in reciprocal crosses between species or populations of *Drosophila*. That is, the hybrid offspring of the cross female A × male B may have lower viability or fertility (indicating a stronger postzygotic barrier) than the offspring of female B × male A. Yukilevich found that for almost every sympatric pair, the cross that produces lower hybrid fitness also shows stronger sexual isolation than the reciprocal cross. Allopatric pairs, in contrast, showed no correlation between sexual and postzygotic isolation. This pattern is predicted by the hypothesis that selection for reinforcement of sexual isolation is stronger if the fitness penalty for cross-mating is greater.

Peripatric speciation

THE PERIPATRIC SPECIATION HYPOTHESIS   One of Ernst Mayr's most influential and controversial hypotheses was founder effect speciation (1954), which he later termed peripatric speciation (1982b). He based this hypothesis on the observation that, in many birds and other animals, isolated populations with restricted distributions, in locations peripheral to the distribution of a probable "parent" species, are often highly divergent from those parent species, to the point of being classified as different species or even genera. For example, the paradise-kingfisher varies little throughout the large island of New Guinea, but has differentiated into several distinctly different forms on small islands along its coast (**Figure 18.17**).

Mayr proposed that genetic change could be very rapid in localized populations founded by a few individuals and cut off from gene exchange with the main body of the species. He reasoned that in such populations, allele frequencies at some loci would differ from those in the parent population because of accidents of sampling (i.e., genetic drift), simply because a small number of colonists would carry only some of the alleles from the source population, and at different frequencies. (He termed this initial alteration of allele frequencies the FOUNDER EFFECT; see Chapter 10.) Because epistatic interactions among genes affect fitness, this initial change in allele frequencies at some loci would alter the selective value of genotypes at other, interacting loci. Hence selection would alter allele frequencies at these loci, and this in turn might select for

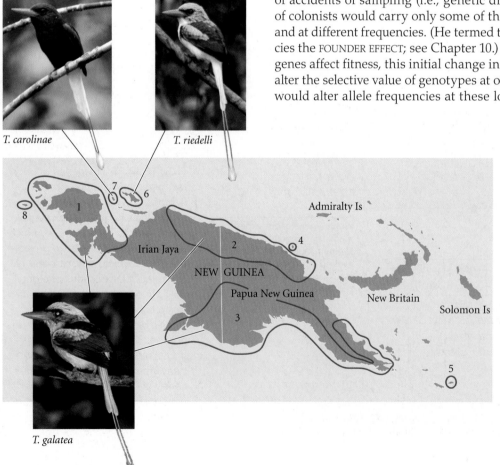

T. carolinae

T. riedelli

T. galatea

FIGURE 18.17 Variation among paradise-kingfishers in New Guinea. *Tanysiptera galatea* is distributed throughout the New Guinea lowlands (regions 1, 2, 3) and some satellite islands (4, 5), whereas the very localized forms *T. riedelii* on Biak Island (6) and *T. carolinae* on Numfor Island (7) are now recognized as distinct species. (After Mayr 1954; *T. galatea* photo courtesy of Rob Hutchinson/Birdtour Asia; *T. riedelii* and *T. carolinae* courtesy of Mehd Halaouate.)

changes at still other epistatically interacting loci. The "snowballing" genetic change that might result would incidentally yield reproductive isolation.

As Mayr (1954) pointed out, this hypothesis implies that substantial evolution would occur so rapidly, and on so localized a geographic scale, that it would probably not be documented in the fossil record. If such a new species expanded its range, it would appear suddenly in the fossil record, without evidence of the intermediate phenotypic changes that had occurred. Thus this hypothesis, he said, might help to explain the rarity of fossilized transitional forms among species and genera. Mayr thus anticipated, and provided the theoretical foundation for, the idea of PUNCTUATED EQUILIBRIUM (see Chapters 4 and 22) advanced by Niles Eldredge and Stephen Jay Gould (1972). Hampton Carson (1975) and Alan Templeton (1980) later advanced a related hypothesis, which they called **founder-flush speciation**, in which they emphasized that during the rapid initial population growth ("flush") of such a colony, advantageous combinations of rare alleles at different loci might be more likely to arise and be fixed than in a stable population. These genetic changes might affect characteristics that contribute to reproductive isolation.

One interpretation of these hypotheses employs the metaphor of an adaptive landscape (see Figure 12.21). According to this interpretation, the colony undergoes a PEAK SHIFT from one adaptive (coadapted) combination of genes (that of the parent population) through a less adaptive genetic constitution (an adaptive valley) to a new adaptive equilibrium (**Figure 18.18A**). The process begins when genetic drift in the small, newly founded population shifts allele frequencies from the vicinity of one adaptive peak (with high frequency of, say, genotype $A_1A_1B_1B_1$) to the slope of another adaptive peak (at which genotype $A_2A_2B_2B_2$ has high frequency). This stage can be accomplished by genetic drift, but not by natural selection, since selection cannot reduce mean fitness. However, selection can move the allele frequencies up the slope away from the valley toward the new peak. Some population geneticists consider peak shifts unlikely because genetic drift is unlikely to move a population's genetic composition across an adaptive valley in opposition to natural selection (Charlesworth and Rouhani 1988; Turelli et al. 2001). However, speciation can occur by genetic drift if an isolated population moves along an "adaptive ridge" to the other side of an adaptive valley from the parent population (**Figure 18.18B**; Gavrilets 2004). Moreover, as Montgomery Slatkin (1996) pointed out, genetic drift is actually weak during rapid population growth because the average number of offspring per capita is greater than in a stable population. Therefore rare combinations of advantageous alleles are more likely to increase in frequency, enabling a population to ascend a new fitness peak. This theory, together with evidence that epistatic interactions among genes are very common (Phillips 2008; Zwarts et al. 2011), makes founder effect speciation a plausible possibility.

EVIDENCE FOR PERIPATRIC SPECIATION Several investigators have passed laboratory populations through repeated bottlenecks to see whether reproductive isolation can evolve in this way. Some investigators interpret these experiments as providing little evidence of reproductive isolation (Rundle 2003; Coyne and Orr 2004); others argue that although incipient reproductive isolation may have evolved in only a fraction of experimental populations, it happens often enough to support the founder-flush hypothesis (Templeton 2008). For example, Agustí Galiana and colleagues (1993) passed 45 laboratory populations of *Drosophila pseudoobscura* through repeated phases of rapid population growth and bottlenecks of a few (1–9) pairs, then tested the populations for sexual isolation from one another. Almost half the experimental populations, especially those passed through more severe bottlenecks, displayed some evidence of sexual isolation.

(A) Peak shift across an adaptive valley

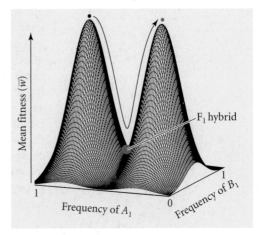

(B) Genetic drift along an adaptive ridge

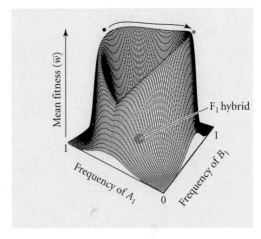

FIGURE 18.18 Two adaptive landscapes show how peripatric speciation might occur. The height of a point on the three-dimensional landscape represents the mean fitness of a population ($\overline{w}$). The mean fitness is a function of allele frequencies at loci *A* and *B*. (A) In a peak shift, a population evolves from one adaptive peak to another by moving downhill (lowering fitness) and then uphill. The F_1 hybrid of a cross between populations on the two peaks lies in the valley; that is, it has low fitness, which causes some reproductive isolation between the populations. (B) Genetic drift along an adaptive ridge, in which genetic constitutions with the same fitness connect the beginning and end states of a population. The F_1 hybrid between populations with these genetic constitutions lies inside the crater. (After Gavrilets and Hastings 1996.)

Many species do originate, as Mayr said, as localized "buds" from a widespread parent species. This is shown by gene trees such as that of the *Greya* moths in Figure 17.3, in which a geographically localized species is more closely related to certain populations of a more widespread species than the populations of the widespread species are to one another (Avise 1994). But this pattern, in itself, does not tell us whether the population experienced a bottleneck that might have triggered the evolution of reproductive isolation. Evidence on this last point can be provided by the pattern of DNA sequence variation. We saw in Chapter 10 (p. 275) that the effective size (N_e) of a population—the measure of a population's size that is most affected by bottlenecks in its history—can be estimated from the proportion of neutral polymorphic sites in its DNA sequences. Extension of the same coalescent theory makes it possible to estimate N_e for the ancestral population that split into two populations or species (Wakeley and Hey 1997). High levels of shared polymorphism (indicating incomplete lineage sorting; see Figure 2.25) indicate that the populations have not suffered drastic reduction in size since they became separated.

Several investigators have used this approach to judge whether or not divergent, peripheral populations have undergone severe bottlenecks that could have induced founder-flush speciation. The web-toed salamander *Hydromantes brunus*, which occupies a small geographic range on the periphery of that of the more broadly distributed *H. platycephalus*, clearly evolved as a localized population of their common ancestor (**Figure 18.19A**). The lower level of sequence variation in *H. brunus* implies that it may have been founded by about 1500 individuals (**Figure 18.19B**)—a small number relative to the ancestor of *H. platycephalus*, but probably too large to have enabled speciation by the founder-flush mechanism (Rovito 2010). A similar analysis has been applied to the zebra finch (*Taeniopygia guttata*), which consists of morphologically and behaviorally different populations in Australia, where it is very abundant, and in the Lesser Sunda Islands to the northwest (Balakrishnan and Edwards 2009). The Lesser Sunda

(A)

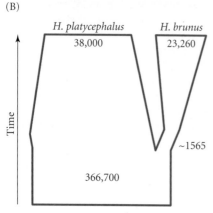

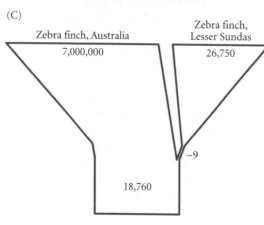

(B)

(C)

FIGURE 18.19 Tests for a population bottleneck in recent or incipient speciation. (A) A gene tree shows that the salamander *Hydromantes brunus* is nested within the more widespread species *H. platycephalus*. (B) Estimates of the effective population sizes of these species at present, in their common ancestor, and during the origin of *H. brunus*. (C) A similar estimate of effective population sizes for the zebra finch *Taeniopygia guttata*. Unlike the salamander *H. brunus*, the Lesser Sunda population of the zebra finch, which appears to be an incipient species, has undergone a strong reduction of population size, suggesting that it is a candidate for founder effect or founder-flush speciation. (A after Rovito 2010; B, data from Rovito 2010; C, data from Balakrishnan and Edwards 2009.)

subspecies may have been founded by as few as 9 individuals, from which it increased to a current N_e estimated at more than 26,000 (**Figure 18.19C**). In most analyses of recently derived sister species (see Figure 17.16), however, incomplete sorting of shared gene lineages provides evidence that both species have had large effective population sizes and have not experienced severe bottlenecks.

Alternatives to Allopatric Speciation: Speciation with Gene Flow

Allopatric, parapatric, and sympatric speciation form a continuum, from little to more to much gene flow between the diverging groups that eventually evolve biological barriers to gene exchange. The higher the rate of gene flow between two populations, the stronger divergent selection must be for their allele frequencies to differ (see Figure 12.10). Even in allopatric speciation, there may be some gene flow between populations, but it is very low compared with the divergent action of natural selection or genetic drift. Parapatric speciation is essentially the same process, but since the rate of gene flow is higher, the force of selection must be correspondingly stronger to engender genetic differences that create reproductive isolation, and it must be stronger still to produce the genetic discontinuities required for sympatric speciation.

SPECIATION WITH GENE FLOW occurs if reproductive isolation evolves while the incipient species are exchanging genes (Pinho and Hey 2010). This term includes parapatric and sympatric speciation, as well as cases in which incipient reproductive isolation between formerly allopatric populations is reinforced (as described above). An important feature of speciation with gene flow is the existence of substantial genetic difference in genomic regions that harbor divergently selected loci, but little genetic difference in regions that are not divergently selected. In these latter regions, gene flow between the populations opposes differentiation (see Figure 17.17). Thus there may exist ISLANDS OF DIVERGENCE between otherwise undifferentiated genomes, a pattern that is being revealed in many closely related species pairs by studies of multiple genetic markers.

We have seen that reproductive isolation between populations is almost always based on a combination of allele differences at two or more loci (the Dobzhansky-Muller model). In Figure 18.7, for example, the gene combinations A_1B_2 and A_2B_1 are incompatible (they display at least partial pre- or postzygotic isolation). Thus, for any substantial reproductive isolation to occur, *the alleles at these loci must be in strong linkage disequilibrium* (i.e., there must be a strong association between A_1 and B_2, and between A_2 and B_1). Recombination tends to break down associations between alleles (see Figure 9.20). Unless the reproductive incompatibility between carriers of these gene combinations is very strong, they will produce some hybrid offspring (A_1B_2/A_2B_1), in which recombination will give rise to the other allele combinations, reducing linkage disequilibrium. Importantly, recombination strongly opposes the evolution of a new incompatible gene combination in an initially randomly mating population. In order for two partly incompatible subpopulations to form when there is an initially predominant genotype (say, A_2B_2, as in Figure 18.7), the rare mutations A_1 and B_1 must increase in frequency and become associated with their complementary partners (B_2 and A_2, respectively); but recombination continually breaks down these initially uncommon gene combinations. Recombination opposes the formation of reproductively isolated gene combinations whenever there is gene flow. In allopatric populations, in contrast, the substitution of different alleles in each population, resulting in incompatible combinations (as in Figure 18.7), is not opposed by the recombination that stems from gene flow. *The breakdown of different combinations of genes in linkage disequilibrium, due to gene flow and recombination, is the most powerful obstacle to speciation with gene flow* (Felsenstein 1981; Gavrilets 2004). The more genes are needed to establish reproductive isolation, the more serious this obstacle is. We now consider how speciation with gene flow might occur despite this obstacle.

Genomic studies of speciation with gene flow

Recent investigations have used GENOME SCANS, employing large numbers of markers such as SNPs, to characterize the extent and pattern of genetic difference during speciation (Nosil and Feder 2012). Frequency differences (calculated as indexes such as F_{ST}; see Chapter 10) between the populations are calculated for all such markers. The theory of selection versus gene flow leads us to expect islands of strong genetic divergence near selected loci, standing above the "sea" of slight or modest differentiation at neutral loci (**Figure 18.20A**). When many nearby sites have been selected, larger islands or "continents" of divergence might be observed (**Figure 18.20B**). These will be indicated by neutral genetic markers (such as SNPs) in genomic regions that have undergone selective sweeps (see Chapter 12, p. 336). Many of the differences between populations and species, however, including those that underlie DM incompatibility and other components of reproductive isolation, are genetically variable within the populations, so evolution based on standing variation is likely (Cutter 2012). For this reason, among others, the selective sweeps around some selected differences may be too short to include any genetic markers and may not be detected (Strasburg et al. 2012). Hence the number of genomic islands is likely to underestimate the number of selected differences between populations.

The number of genomic islands detected between recently diverged populations and species varies among the organisms studied so far (Nosil and Feder 2012). In some cases, genomic divergence has been greater in genome regions with reduced recombination rates, as theory predicts. For example, the genomic difference between the closely related species *Drosophila pseudoobscura* and *D. persimilis* is greater within than outside the several chromosome inversions that distinguish these species (McGaugh and Noor 2012). (Recall that recombination is greatly reduced between chromosomes that differ by an inversion; see Chapter 9). Genome studies have shown that some gene flow often occurs even between some readily distinguished species, despite considerable reproductive isolation. For example, even though the sunflower species *Helianthus annuus* and *H. petiolaris* are strongly reproductively isolated, they are differentiated only by small genomic islands, the

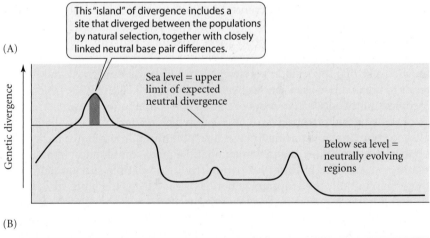

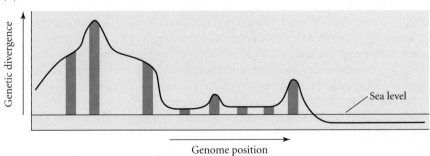

FIGURE 18.20 Hypothetical "islands" and "continents" of genetic divergence between populations. The amount of divergence (DNA sequence difference) is plotted along a region of the genome. (A) Above the horizontal line ("sea level"), the degree of divergence is statistically distinguishable (has "outlier status") from the expected level of divergence by random genetic drift. (B) If many selected loci have diverged, they and their closely linked neutral variants may form divergent genomic "continents." (After Nosil and Feder 2012.)

nearby genes having been largely homogenized by gene flow and recombination (Sambatti et al. 2012). In some cases, islands of divergence include genes known to contribute to reproductive isolation. For example, DNA sequence divergence among hybridizing populations of *Heliconius* butterflies (see Figure 12.20) was greatest at the loci that determine differences in the pattern of warning coloration (Nadeau et al. 2012). These color patterns are subject to strong positive frequency-dependent selection (see Figure 12.16B), and also contribute to behavioral isolation between some species.

Parapatric speciation

Parapatric speciation can theoretically occur if gene flow between populations that occupy adjacent regions with different selective pressures is weaker than divergent selection for different gene combinations (Endler 1977; Gavrilets 2004). Hybrids may have low fitness, and individuals with the "wrong" genotype or phenotype that migrate across the border may fail to survive and reproduce (Nosil et al. 2005). Consequently, clines at various loci may tend to develop at the same location, resulting in a primary hybrid zone that has developed in situ, but may look like a secondary hybrid zone (Endler 1977; Barton and Hewitt 1985). Steady genetic divergence may eventually result in complete reproductive isolation.

Another possibility is that populations separated by distance can evolve reproductive incompatibility, even though the species is distributed throughout the intervening region. Divergent features that arise at widely separated sites in the species' distribution may spread, supplanting ancestral features as they travel and preventing gene exchange when they eventually meet. Russell Lande (1982) has theorized that prezygotic isolation could arise in this way as a result of divergent sexual selection.

Parapatric speciation undoubtedly occurs and may even be common, but it has been difficult to demonstrate that it provides a better explanation than allopatric speciation for real cases (Coyne and Orr 2004). An example of parapatric divergence is provided by three-spined sticklebacks (*Gasterosteus aculeatus*) in four lakes, each with an outlet stream, on Vancouver Island in western Canada (Roesti et al. 2012). Even though there is no external barrier to movement between lake and stream subpopulations, the subpopulations differ adaptively in several morphological features. (Studies of other populations of sticklebacks have shown that morphologically different forms are often sexually isolated to some degree; see Figure 18.9.) A genome scan revealed that genetic differences between lake and stream populations were most pronounced in central chromosome regions where the recombination rate is low, and that, even taking this into account, the lake/stream pairs were highly divergent at many genomic sites (**Figure 18.21**). Thus there exists extensive genetic divergence, despite gene flow, caused by selection at many sites. It is difficult to determine whether the divergence occurred in the face of gene flow, or in allopatric populations that subsequently met and interbred.

FIGURE 18.21 Genomic differentiation among parapatric and allopatric populations of sticklebacks. Each of the four panels shows differentiation between two samples along all 21 chromosomes. The vertical marks indicate values of F_{ST}, a measure of the difference in frequency of SNPs, at each variable site found on a chromosome. (These values were standardized to correct for differences in F_{ST} caused by location within a chromosome.) In each panel, marks that rise above the horizontal red line are sites with significantly different frequencies between the samples. The upper three panels show genomic differences between parapatric pairs of populations in three lakes and their connected streams. The lowest panel contrasts allopatric populations, in two lakes. Despite gene flow, the parapatric populations show about as many genetic differences as the allopatric populations. (After Roesti et al. 2012.)

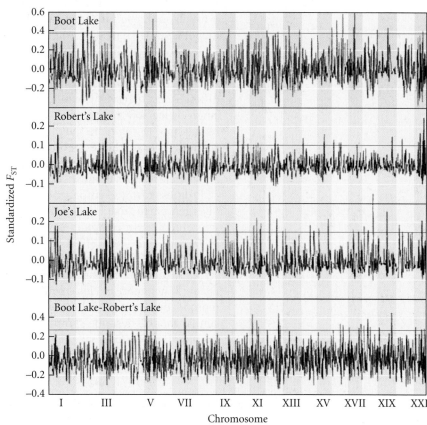

Sympatric speciation

Sympatric speciation is a highly controversial subject. Speciation would be sympatric if a biological barrier to gene exchange arose within an initially randomly mating population without any spatial segregation of the incipient species—that is, if speciation occurred despite high initial gene flow. The difficulty any model of sympatric speciation must overcome is how to establish linkage disequilibrium among a set of genes that together confer reproductive isolation: in other words, how to reduce the frequency of the intermediate genotypes that would act as a conduit of gene exchange between the incipient species. Ernst Mayr (1942, 1963), the most vigorous and influential critic of the sympatric speciation hypothesis, showed that many supposed cases are unconvincing and that the hypothesis must overcome severe theoretical difficulties. Under certain special circumstances, however, these difficulties are not all that severe (Turelli et al. 2001; Gavrilets 2004; Bolnick and Fitzpatrick 2007).

MODELS OF SYMPATRIC SPECIATION Most models of sympatric speciation postulate disruptive (diversifying) ecological selection (see Chapter 12) at one or several loci at which different alleles confer adaptation to two distinct resources. For example, different homozygous genotypes (say, A_1A_1 and A_2A_2) might have high fitness on one or the other of two resources, and intermediate (heterozygous A_1A_2) phenotypes might have lower fitness because they are not as well adapted to either resource. In some cases (Servedio et al. 2011), the selected ecological trait is a "magic trait" (Gavrilets 2004) that also creates reproductive isolation. For example, differences in the color or form of flowers that serve to attract different pollinators (as in the monkeyflowers shown in Figure 17.6) are adaptations that may also cause reproductive isolation. Similarly, if a species of herbivorous insect has the habit of finding mates on its preferred host plant, different genotypes that prefer different host plants will tend to be reproductively isolated (Bush 1969).

Alternatively, selection may favor not only ecologically adaptive alleles (such as A_1 and A_2), but also alleles, at different loci, that tend to make their carriers mate nonrandomly, if this would result in their having fewer poorly adapted heterozygous offspring. (This model is somewhat similar to reinforcement.) If such alleles increased in frequency, and if an association (linkage disequilibrium) between alleles at the ecological trait loci and the mating locus (say, M_1 and M_2) were to develop, the result would be two partially reproductively isolated populations (say, $A_1A_1M_1M_1$ and $A_2A_2M_2M_2$) that are adapted to different resources. This would be a step toward speciation (but only a step, because the single M locus would provide only partial premating isolation, and more such loci would be required to complete the process.) Two models of nonrandom mating have been considered: (1) **assortative mating**, in which individuals prefer mates that match their own phenotype, and (2) trait-preference, in which different genes control a female preference and a male trait (as in most sexual selection models; see Chapter 15).

Progress toward sympatric speciation is theoretically somewhat more likely if there is divergence in a "magic trait" that affects ecological divergence and automatically restricts interbreeding. In computer simulations of an insect in which some loci affect host plant preference and others affect physiological adaptation to the different plants, the frequencies of both kinds of alleles may rapidly increase (**Figure 18.22A**), and gene flow may be strongly reduced, so that the population divides into two host-associated, ecologically isolated incipient species. However, if host preference is a continuous, polygenic trait, reproductive isolation will not evolve unless selection is strong (**Figure 18.22B**). Somewhat similar models describe sympatric speciation by adaptation to a continuously distributed resource, such as prey size (Dieckmann and Doebeli 1999; Kondrashov and Kondrashov 1999). Some authors have questioned how realistic these models are (Coyne and Orr 2004; Gavrilets 2004).

EVIDENCE FOR SYMPATRIC SPECIATION The conditions required for sympatric speciation to occur are theoretically more limited than those for allopatric speciation, and

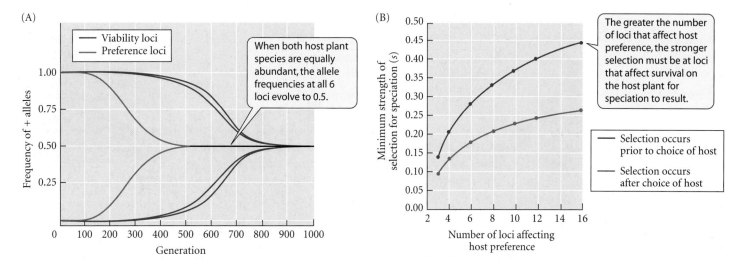

FIGURE 18.22 Some results of a computer simulation of sympatric speciation in an insect that mates on its host plant. (A) Alleles that enhance survival on or preference for one host plant species are referred to as "+ alleles"; those that have the complementary effect are called "– alleles." The simulation shows changes in the frequency of the + allele at four loci that affect survival (viability loci) and two loci that affect host preference (preference loci) when the + allele at each locus begins with a frequency near 0 or 1. Eventually, half the population prefers and survives better on one host and half on the other, representing progress toward reproductive isolation. (B) These curves show how strong selection at viability loci has to be to result in speciation when preference is controlled by multiple loci. The strength of selection at each viability locus is *s* (the coefficient of selection; see Chapter 12). The two curves model life histories in which selection occurs before and after the choice of host occurs. (After Fry 2003.)

sympatric speciation does not occur easily (Gavrilets 2004). Ecological and genetic research can help to determine whether or not conditions are favorable for sympatric speciation in specific groups of organisms (Bolnick and Fitzpatrick 2007). However, because there is so much evidence for allopatric speciation, sympatric speciation must be demonstrated, rather than assumed, for most groups of organisms. Demonstrating sympatric speciation is quite difficult because evidence must show that a past allopatric phase of genetic differentiation is very unlikely (Coyne and Orr 2004). However, many possible examples, supported by varying degrees of evidence, have been proposed.

Just as allopatric speciation results from division in space, ALLOCHRONIC speciation might result from division in time (i.e., if the breeding season of a population becomes divided). For example, in Japanese sites with mild winters, the moth *Inurois punctigera* breeds throughout the winter, and moths breeding from November to March show little difference in genetic composition. In contrast, populations in sites with very cold winters consist of genetically different subpopulations that mate in early winter and in late winter (Yamamoto and Sota 2009).

"Host races" of specialized herbivorous insects—partially reproductively isolated subpopulations that feed on different host plants—have often been proposed to represent sympatric speciation in progress, although the evidence for this is limited (Futuyma 2008). The most renowned case, studied first by Guy Bush (1969) and later by Jeffrey Feder and colleagues (2005), is that of the apple maggot fly (*Rhagoletis pomonella*). The larvae develop in ripe fruits and overwinter in the ground as pupae; adult flies emerge in July and August and mate on the host plant. The major ancestral host plants throughout eastern North America were hawthorns (*Crataegus*). About 150 years ago, *R. pomonella* was first recorded in the northeastern United States as a pest of cultivated apples (*Malus*), and infestation of apples spread from there. Allele frequencies at many loci differ significantly between apple- and hawthorn-derived flies, showing that gene exchange between them is limited

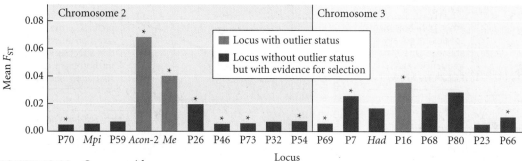

FIGURE 18.23 Genome differences in the apple maggot fly (*Rhagoletis pomonella*), in which populations associated with different host plants have diverged by natural selection. The difference in allele frequency between the populations is measured as F_{ST} for several loci on each of two chromosomes. Similar patterns were found for loci on the other chromosomes. Loci that were statistically distinguishable from neutral differences are shown in green. The purple bars with asterisks below them represent marked regions of the genome that differ less in frequency, but which were shown in two experiments to affect survival in environments corresponding to the different host plants. Bars without asterisks are loci that have significantly different allele frequencies in the two host species. (After Michel et al. 2010.)

(**Figure 18.23**). Gene exchange is reduced (to about 2 percent) by several factors, including a difference in host preference (for apples versus hawthorns) that appears to be based on differences at about four loci (Feder and Forbes 2008), and a difference of about 3 weeks between breeding activity on apple and on hawthorn, corresponding to a difference in the fruiting time of these plants. Because of this timing difference, larvae and pupae in apples experience higher temperatures and a longer pupal period than hawthorn flies. Experiments showed that these differences impose strong divergent selection at several loci (Filchak et al. 2000). Although divergence in host preference presumably occurred in sympatry, the earlier development time evolved in hawthorn-feeding populations in Mexico and was fortuitously advantageous for development on apples (Michel et al. 2007). Thus incipient speciation in this case has had both geographic and sympatric components.

Probably the best-documented examples of sympatric speciation are sister species that inhabit small isolated islands where there has been no opportunity for spatial separation. Several genera of plants conform to this criterion on Lord Howe Island, east of Australia (**Figure 18.24**; Savolainen et al. 2006; Papadopulos et al. 2011). A similar example has been described in two groups of cichlid fishes that are confined to two small crater lakes (Schliewen et al. 1994). Mitochondrial DNA sequence data indicate that the cichlid species in each lake are monophyletic, suggesting that speciation has occurred within the crater lakes. The lakes lie in simple conical basins that lack habitat heterogeneity and opportunity for spatial isolation. It has often been suggested that the enormous diversity of cichlid fishes in the African Great Lakes (see Figure 3.24) arose by sympatric speciation, but there are plentiful opportunities for allopatric speciation within each lake because these sedentary

FIGURE 18.24 Sister species of palms on Lord Howe Island, which is less than 12 square kilometers in area. (A) *Howea forsteriana* (kentia palm) has straight leaves with drooping leaflets. (B) *H. belmoreana* (curly palm) has curved leaves with ascending leaflets. (Photos courtesy of W. J. Baker, Royal Botanic Gardens, Kew.)

(A)

(B)

species are restricted to distinct, discontinuously distributed habitats along the lake periphery. Conspecific populations of these cichlids differ genetically, even over short distances (Rico et al. 2003), suggesting that spatial separation has played a role in their speciation.

Polyploidy and Recombinational Speciation

Polyploidy

A POLYPLOID is an organism with more than two complements of chromosomes (see Chapter 8). A tetraploid, for example, has four chromosome complements in its somatic cells; a hexaploid has six. Polyploid populations are reproductively isolated by postzygotic barriers from their diploid (or other) progenitors, and are therefore distinct biological species. Speciation by polyploidy is the only known mode of instantaneous speciation by a single genetic event.

For reasons that are not well understood, polyploid species are rare among sexually reproducing animals, although many parthenogenetic polyploid animals have been described. Polyploidy is very common in plants. Perhaps 30 to 70 percent of plant species are descended from polyploid ancestors (Otto and Whitton 2000), and polyploidy accompanies about 15 percent of speciation events in flowering plants and 31 percent in ferns (Wood et al. 2009). Natural polyploids span a continuum between two extremes, called autopolyploidy and allopolyploidy. An AUTOPOLYPLOID is formed by the union of unreduced gametes from genetically and chromosomally compatible individuals that may be thought of as belonging to the same species. The cultivated potato (*Solanum tuberosum*), for example, is an autotetraploid of a South American diploid species. An ALLOPOLYPLOID is a polyploid derivative of a diploid hybrid between two species.

SPECIATION BY POLYPLOIDY   Polyploidy usually occurs because of a failure of the reduction division in meiosis (Ramsey and Schemske 1998). For example, the union of an unreduced ($2n$) gamete with a haploid (n) gamete yields a triploid ($3n$) individual; a tetraploid is then formed if an unreduced $3n$ gamete unites with a reduced (n) gamete. Plants with odd-numbered ploidy (e.g., triploids, $3n$, and pentaploids, $5n$) are generally nearly sterile because most of their gametes are aneuploid. Because the hybrid between a tetraploid and its diploid ancestor would be triploid, the tetraploid is reproductively isolated and is therefore a distinct biological species (and the same is true at higher ploidy levels).

A milestone in the study of speciation was the experimental production of a natural polyploid species by Arne Müntzing in 1930. Müntzing suspected that the mint *Galeopsis tetrahit*, with $2n = 32$ chromosomes, might be an allotetraploid derived from the diploid ($2n = 16$) ancestors *G. pubescens* and *G. speciosa*. By crossing these species and selecting among their hybrid descendants, Müntzing obtained tetraploid offspring that closely resembled *G. tetrahit* in morphology, were highly fertile, and were reproductively isolated from the diploid species, but yielded fertile progeny when crossed with wild *G. tetrahit*.

In this and some other experimental studies of allopolyploids, diploid hybrids between species are mostly sterile and form few bivalents in meiosis, whereas the tetraploid offspring of these hybrids are highly fertile and have normal, bivalent chromosome pairing. In these cases, *the sterility of the diploid hybrid is not due to functional interactions between the genes of the two parent species, but instead to the mechanisms that inhibit chromosome pairing*. The diploid and tetraploid hybrids have the same genes in the same proportions, so genic differences cannot account for the sterility of the one but not the other (Darlington 1939; Stebbins 1950).

Molecular studies have cast additional light on the origin of polyploid species. For example, three diploid European species of goatsbeards (Asteraceae), *Tragopogon dubius, T. porrifolius,* and *T. pratensis,* have become broadly distributed in North America. F_1 hybrids between them have low fertility. In 1950, Marion Ownbey described two fertile tetraploid species, *T. mirus* and *T. miscellus,* and postulated that *T. mirus* was a recent tetraploid hybrid of *T. dubius* and *T. porrifolius,* and that *T. miscellus* was likewise derived from *T. dubius* × *T.*

(A)

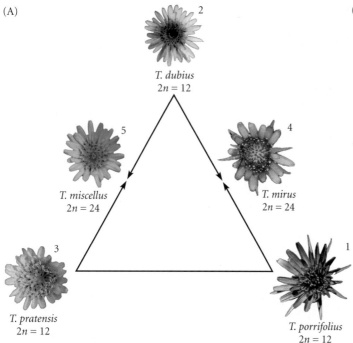

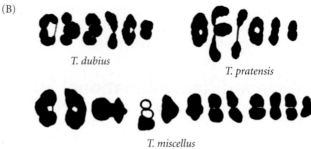

FIGURE 18.25 Allotetraploid species of goatsbeards (*Tragopogon*). (A) The flower heads of the diploid species *T. porrifolius* (1), *T. dubius* (2), and *T. pratensis* (3) and of the fertile tetraploid species *T. mirus* (4, from 1 × 2) and *T. miscellus* (5, from 2 × 3). (B) Drawings of the chromosomes of the diploids *T. dubius* and *T. pratensis* and of their tetraploid hybrid derivative, *T. miscellus*. The tetrapoid has twice as many chromosomes as the diploid species. (A from Pires et al. 2004; B from Ownbey 1950.)

pratensis (**Figure 18.25**). Decades later, Douglas Soltis, Pamela Soltis, and their collaborators (2004) found that the tetraploid species have exactly the combinations of DNA markers from the diploid species that are predicted by Ownbey's hypothesis. In *Tragopogon* and other plants, allopolyploid species have typically arisen independently several times by hybridization between their diploid parents (Pires et al. 2004).

ESTABLISHMENT AND FATE OF POLYPLOID POPULATIONS How polyploid species become established is not fully understood. If a newly arisen tetraploid within a diploid population crosses at random, its reproductive success is expected to be lower than that of the diploids because many of its offspring will be inviable or sterile triploids, formed by backcrossing with the surrounding diploids. A study of mixed experimental populations of diploid and tetraploid fireweeds (*Chamerion angustifolium*) showed that the lower the frequency of tetraploids in the population, the lower their seed production, as a result of increased pollination from the diploids (Husband 2000).

Several conditions—self-fertilization, vegetative propagation, higher fitness than the diploid, or ecological and habitat segregation from the diploid—might enable a new polyploid to increase and form a viable population (Fowler and Levin 1984; Rodríguez 1996). Indeed, many polyploid taxa reproduce by selfing or vegetative propagation, and most differ from their diploid progenitors in habitat and distribution, and so would be segregated from them. Increases in ploidy alter cell size, water content, rate of development, and many other physiological properties (Levin 1983; Otto and Whitton

FIGURE 18.26 Differences between a newly formed polyploid and its ancestor may confer ecological differences that could reduce the opportunity for crossing between them. Survival (A) and flowering time (B) of a newly originated hexaploid (neo-6*n*) yarrow (*Achillea*), planted in a dry dune, were intermediate between its tetraploid parent (4*n*) and an existing hexaploid (6*n*) species. (After Ramsey 2011.)

(A) Survival

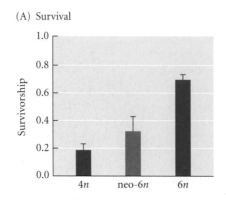

(B) Flowering time

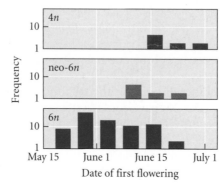

2000). But do polyploids display such differences from diploids when they are first formed? If the yarrow *Achillea borealis* is representative, the answer is yes. In California, tetraploids and hexaploids grow in wetter and drier habitats, respectively. Justin Ramsey (2011) planted seedlings of both forms, as well as "neohexaploids" that had originated de novo from tetraploid parents, in dry dunes. The neohexaploids survived better and flowered earlier than the tetraploids (**Figure 18.26**), showing that they would be partly isolated from the tetraploids, by habitat and flowering time, immediately upon their origin.

Because it apparently confers new physiological and ecological capabilities, polyploidy may play an important role in plant evolution; moreover, the increased number of genes in polyploids may enhance their adaptability (Otto and Whitton 2000). However, polyploidy does not confer major new morphological characteristics, such as differences in the structure of flowers or fruits, and it seems unlikely to cause the evolution of new genera or other higher taxa (Stebbins 1950).

Recombinational speciation

Hybridization sometimes gives rise not only to polyploid species, but also to distinct species with the same ploidy as their parents. Among the great variety of recombinant offspring produced by F_1 hybrids between two species, certain genotypes may be fertile but reproductively isolated from the parent species. These genotypes may then increase in frequency, forming a distinct population (Rieseberg 1997; Gross and Rieseberg 2005). This process has been called **recombinational speciation** or HYBRID SPECIATION (Grant 1981).

Hybrid speciation seems to be more common in plants than in animals (Rieseberg 1997; Mallet 2007). In a molecular phylogenetic analysis of part of the sunflower genus, *Helianthus*, Loren Rieseberg and coworkers found that hybridization between *Helianthus annuus* and *H. petiolaris* has given rise to three other distinct species (*H. anomalus, H. paradoxus,* and *H. deserticola*; **Figure 18.27**). Although F_1 hybrids between the parent species have low fertility, the derivative species are fully fertile and are genetically isolated from the parent species by postzygotic incompatibility. The recombinant species grow in very different (drier or saltier) habitats than either parent species, flower later, and have unique morphological and chemical features. *H. anomalus*, for example, has thicker, more succulent leaves and smaller flower heads than either parent species. Such "extreme" traits transgress the range of variation between the two parent species. Rieseberg and coworkers (2003) crossed *H. annuus* and *H. petiolaris*, the parent species, and grew the backcross progeny in a greenhouse for several generations. Using genetic markers on all the chromosomes, they found

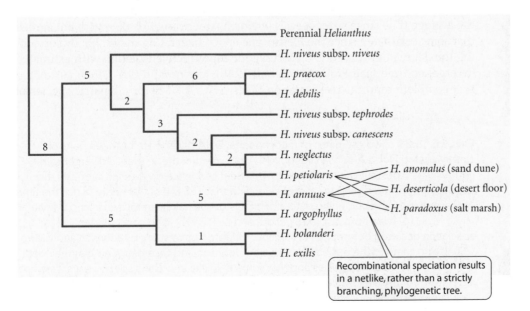

FIGURE 18.27 The hybrid origin of some diploid species of sunflowers. The phylogeny, based on sequences of chloroplast DNA and nuclear ribosomal DNA, shows that *Helianthus anomalus, H. paradoxus,* and *H. deserticola* have arisen from hybrids between *H. annuus* and *H. petiolaris*. The numbers of synapomorphic base pair substitutions are shown along the branches of the phylogenetic tree. (After Gross and Rieseberg 2005.)

that the experimental hybrids had combinations of *annuus* and *petiolaris* chromosome segments that matched those found in the three hybrid species—confirming that those species indeed arose from hybridization. Almost all the extreme, "transgressive" traits of *H. anomalus* and the other two hybrid species, such as small flower heads, occurred among the experimentally produced hybrids. To a considerable degree, then, the hybridization experiment replayed the origin of these species.

How Fast Is Speciation?

The phrase "rate of speciation" has several meanings (Coyne and Orr 2004). One is the duration of the process, or **time for speciation** (**TFS**)—the time required for (nearly) complete reproductive isolation to evolve once the process has started. Another is the BIOLOGICAL SPECIATION INTERVAL (BSI), the average time between the origin of a new species and when that species branches (speciates) again. The BSI includes not only the TFS, but also the "waiting time" before the process of speciation begins again. For example, in a clade that speciates by polyploidy, a new polyploid species may originate rarely (i.e., the waiting time, and therefore the BSI, is long), but when it does, reproductive isolation is achieved within one or two generations (the TFS is very short).

The diversification rate, R, or increase in species number per unit time, equals the difference between the rates of speciation (S) and extinction (E). R can be estimated for a monophyletic group if the age of the group (t) can be estimated and if we assume that the number of species (N) has increased exponentially according to the equation

$$N_t = e^{Rt}$$

(We encountered this approach in Chapter 7 when we considered long-term rates of diversification in the fossil record.) The average time between branching events on the phylogeny is $1/R$, the reciprocal of the diversification rate. This number estimates the BSI, the average time between speciation events, if we assume there has been no extinction ($E = 0$). According to estimates made using this approach, BSI in animals ranges from less than 0.3 My (in the phenomenal adaptive radiation of cichlid fishes in the African Great Lakes) to more than 10 My in various groups of molluscs. When estimates of E from the fossil record are taken into account, BSI is about 3 My for horses and is still very long (6–11 My) for bivalve molluscs.

An upper bound on TFS can be estimated if geological evidence or calibrated DNA sequence divergence enables us to judge when young pairs of sister species were formed. For example, endemic species of *Drosophila* have evolved on the "big island" of Hawaii, which is less than 800,000 years old. Two research groups that used the sequence divergence of mitochondrial DNA to measure TFS disagreed on the average time since sister species of North American birds diverged, but agreed that some originated since the Pleistocene epoch began (**Figure 18.28**). By correlating the degree of prezygotic or postzygotic reproductive isolation with estimated divergence time (see Figures 17.9 and 17.10), Coyne and Orr (1997) estimated that complete reproductive isolation takes 1.1 to 2.7 My for allopatric species of

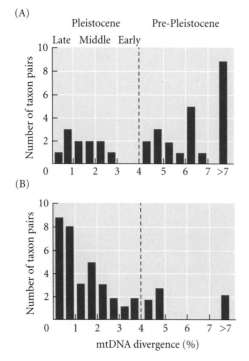

FIGURE 18.28 Two estimates of mitochondrial DNA divergence between pairs of closely related North American songbirds. Assuming that the sequence divergence rate of mtDNA is 2 percent per My, less than 4 percent sequence divergence suggests that speciation occurred during the Pleistocene, and that TFS is less than 2 My. (Since the time of these studies, the beginning date of the Pleistocene has been revised from 1.8 Mya to 2.6 Mya.) (A) Data from a study by Klicka and Zink (1997), who concluded that most speciation occurred before the Pleistocene. (B) Data from a study by Johnson and Cicero (2004), who came to the opposite conclusion. Johnson and Cicero argued that their data are based only on sister species and that some of Klicka and Zink's species pairs were not sister taxa. Johnson and Cicero also classified some forms as species that were classified as subspecies by other authors. (After Lovette 2005.)

Drosophila, but only 0.08 to 0.20 My for sympatric species. (They attributed this difference to reinforcement of prezygotic isolation between sympatric forms.) A similar approach suggested that frogs take 1.5 My, on average, to complete speciation (Sasa et al. 1998).

Clearly, time for speciation varies greatly—as we would expect from theories of speciation. We expect the process of speciation to be excruciatingly slow if it proceeds by mutation and drift of neutral alleles; we expect it to be faster if it is driven by ecological or sexual selection, and to be accelerated if reinforcement plays a role. Allopatric speciation could be slow or very rapid, depending on the strength of divergent selection and on genetic variation in relevant traits. Some possible modes of speciation, such as polyploidy, recombinational speciation, and sympatric speciation, should be very rapid when they occur— although they may occur rarely, resulting in long intervals (BSI) between speciation events. As we have already seen, ecological speciation can be rapid (Hendry et al. 2007): substantial reproductive isolation apparently evolved within about a century in the apple maggot fly and the hybrid sunflower species *Helianthus anomalus*. On the other hand, some sister taxa of snapping shrimp (*Alpheus*) on opposite sides of the Isthmus of Panama have not achieved full reproductive incompatibility in the 3.5 My since the isthmus arose (Knowlton et al. 1993).

What characteristics favor high rates of speciation (low BSI)? The best way to approach this question is to compare the species diversity of replicated sister groups that differ in the characteristics of interest (see Figure 18.12), although it is often hard to tell whether those characteristics enhance the speciation rate or diminish the extinction rate. Among the characteristics studied so far, those that seem most likely to have increased the speciation rate as such seem to be animal (rather than wind) pollination in plants and features that indicate intense sexual selection in animals (Coyne and Orr 2004). These observations suggest that diversification in some groups of animals owes more to the simple evolution of reproductive isolation (due to sexual selection) than to ecological diversification. This conclusion may call into question the hypothesis that ecological divergence is the main engine of evolutionary radiation (Schluter 2000).

Consequences of Speciation

The most important consequence of speciation is diversity. For sexually reproducing organisms, every branch in the great phylogenetic tree of life represents a speciation event, in which populations became reproductively isolated and therefore capable of independent, divergent evolution, including, eventually, the acquisition of those differences that mark genera, families, and still higher taxa. Speciation, then, stands at the border between MICROEVOLUTION—the genetic changes within and among populations—and MACROEVOLUTION—the evolution of the higher taxa in all their glorious diversity.

In their hypothesis of punctuated equilibrium, Eldredge and Gould (1972; see also Stanley 1979; Gould and Eldredge 1993) proposed that speciation may be required for morphological evolution to occur at all (see Figure 4.19). They suggested, based on Mayr's (1954) proposal that founder events trigger rapid evolution from one genetic equilibrium to another, that most evolutionary changes in morphology are triggered by and associated with peripatric speciation. However, population geneticists argued (Charlesworth et al. 1982), and Gould (2002) himself conceded, that there is no compelling reason to think that speciation (acquisition of reproductive isolation) triggers morphological evolution. Morphological characters vary among populations of a species, just as they do among reproductively isolated species, and there is little evidence that founder-flush effects account for most speciation.

Nevertheless, evolutionary change often appears to be correlated with speciation. Speciation seems to be associated with morphological evolution in the great majority of lineages of unicellular foraminiferans (see Figure 4.3), which have a detailed enough fossil record to distinguish cladogenesis from anagenesis (Strotz and Allen 2013). Moreover, the rate of evolution of a lineage, and therefore the amount of evolutionary change from the root of a phylogenetic tree to any extant species (PATH LENGTH), is expected to increase

(A) Phyletic gradualism

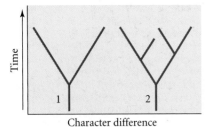

(B) Punctuated equilibria

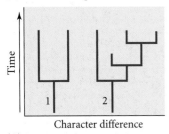

(C)

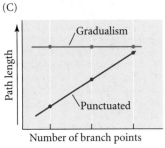

FIGURE 18.29 Models of (A) phyletic gradualism and (B) punctuated equilibria suggest how phylogenetic data might be used to determine whether speciation is associated with enhanced evolution of molecular or morphological characters. In both models, lineages 1 and 2 differ in the number of speciation events (branch points). With phyletic gradualism, the variation among living species (character difference) and the amount of evolutionary change from the root to any living species (path length) are not affected by the number of speciation events. The correlation between path length and number of branch points in the phylogeny is expected to be zero, as indicated by the horizontal line in (C). In the punctuated equilibria model, character evolution occurs only at speciation, so the variation among living species, and thus the path length, is expected to be correlated with the number of branch points. Note that if some species have become extinct, as illustrated, the number of branch points in the phylogeny of extant species will underestimate the number of speciation events. (C after Pagel et al. 2006.)

with the number of speciation events in the punctuated equilibrium model, but not in the phyletic gradualism model (**Figure 18.29**). Mark Pagel and colleagues (2006) found that in 27 percent of the phylogenies of animals, fungi, and plants, path lengths, as measured by numbers of nucleotide substitutions, were significantly correlated with the number of species, as predicted if speciation accelerates evolution. In a related approach, Tiina Mattila and Folmer Bokma (2008) concluded that speciation explains more than two-thirds of the variance in body mass among species of mammals, and that gradual evolution accounts for little variation. Perhaps as a consequence, the body sizes of terrestrial vertebrates show almost no accumulated change in less than a million years, even though size is genetically variable and can evolve rapidly over very short intervals (**Figure 18.30**; Uyeda et al. 2011). Evidently, most such short-term changes are impermanent fluctuations.

What might cause these patterns? Morphological change might be associated with speciation because reproductive isolation enables morphological differences between populations to persist in the long term (Futuyma 1987; Eldredge et al. 2005). Although different local populations may diverge rapidly as a result of selection, local populations are ephemeral: as climate and other ecological circumstances change, divergent populations move about and come into contact sooner or later. Much of the divergence that has occurred may then be lost by interbreeding—unless reproductive isolation has

FIGURE 18.30 The amount of evolutionary change in body size in various terrestrial vertebrates, in relation to the elapsed time between samples. Using the change in the logarithm of size standardizes for absolute size (enabling comparison of evolutionary changes in mice with those in elephants). The key shows the kind of data used (field study of recent and contemporary changes; the fossil record; and divergence inferred from time-calibrated phylogenies of living species). Even though rates of change are high on short time scales (see Figure 4.23), little change accumulates before about a million years. (From Uyeda et al. 2011.)

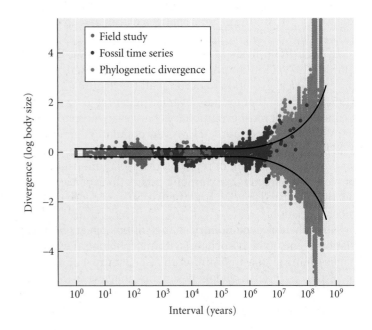

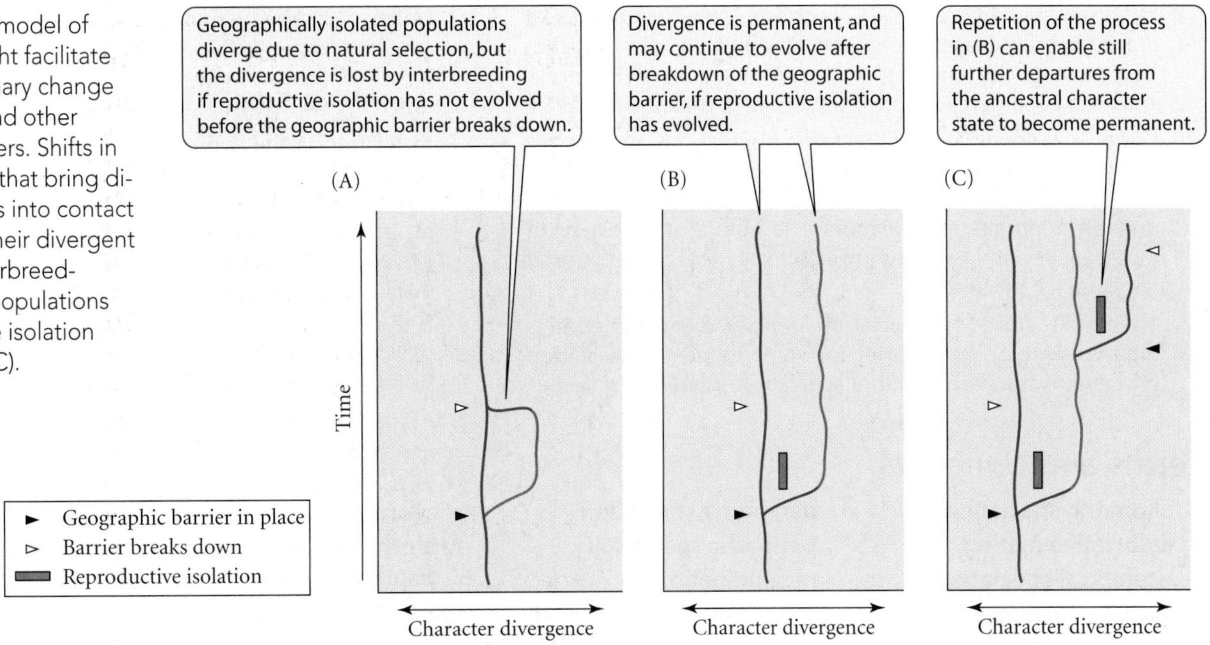

FIGURE 18.31 A model of how speciation might facilitate long-term evolutionary change in morphological and other phenotypic characters. Shifts in geographic ranges that bring divergent populations into contact may cause loss of their divergent features due to interbreeding (A), unless the populations evolve reproductive isolation while allopatric (B, C).

Geographically isolated populations diverge due to natural selection, but the divergence is lost by interbreeding if reproductive isolation has not evolved before the geographic barrier breaks down.

Divergence is permanent, and may continue to evolve after breakdown of the geographic barrier, if reproductive isolation has evolved.

Repetition of the process in (B) can enable still further departures from the ancestral character state to become permanent.

(A)   (B)   (C)

Time

► Geographic barrier in place
▷ Barrier breaks down
▬ Reproductive isolation

Character divergence   Character divergence   Character divergence

evolved (**Figure 18.31**). A succession of speciation events, each "capturing" further change in a character, may result in a long-term trend. Perhaps, as Ernst Mayr (1963, p. 621) wrote, "Speciation…is the method by which evolution advances. Without speciation, there would be no diversification of the organic world, no adaptive radiation, and very little evolutionary progress. The species, then, is the keystone of evolution."

Go to the
EVOLUTION
Companion Website at
sites.sinauer.com/evolution3e
for quizzes, data analysis and simulation exercises, and other study aids.

Summary

1. Probably the most common mode of speciation is allopatric speciation, in which gene flow between populations is reduced by geographic or habitat barriers, allowing genetic divergence by natural selection and/or genetic drift.

2. In vicariant allopatric speciation, a widespread species becomes sundered by a geographic barrier, and one or both populations diverge from the ancestral state.

3. In a simple model of the evolution of reproductive isolation, complementary allele substitutions that do not reduce the fitness of heterozygotes occur at two or more loci in one or both populations. Epistatic interactions between alleles fixed in the two populations may reduce the fitness of hybrids formed when the populations meet. Likewise, genetic divergence may result in prezygotic isolation.

4. Reproductive isolation between allopatric populations appears to evolve as a side effect of divergent ecological or sexual selection. Both processes require further study before their relative importance can be assessed. There is no evidence that reproductive isolation evolves by random genetic drift.

5. Prezygotic isolation evolves mostly while populations are allopatric, but may be reinforced when the populations become parapatric or sympatric.

6. Peripatric speciation, or founder effect speciation, is a hypothetical form of allopatric speciation in which genetic drift in a small peripheral population initiates rapid evolution, and reproductive isolation is a by-product of that evolutionary change. It may be especially likely if colonization is followed by rapid population growth, during which genetic drift is reduced and new adaptive gene combinations may be fixed. The likelihood of this form of speciation differs depending on the mathematical model used, and although some possible examples have been described, there is little evidence that this form of speciation is common.

7. Speciation with gene flow, i.e. parapatric or sympatric speciation, may occur if divergent selection is stronger than gene flow. Sympatric speciation—the origin of reproductive isolation within an initially randomly mating population—is controversial. The sympatric evolution of sexual isolation is opposed by recombination among loci affecting mating and loci affecting the disruptively selected character. Sympatric speciation may occur, however, if recombination does not oppose selection. For example, if disruptive selection favors preference for different habitats and if mating occurs within those habitats, prezygotic isolation may result. How often this occurs is debated.

8. Instantaneous speciation by polyploidy is common in plants. Allopolyploid species arise from hybrids between genetically divergent populations. Establishment of a polyploid population probably requires ecological or spatial segregation from the diploid ancestors because backcross offspring have low reproductive success. Polyploid species can have multiple origins.

9. In recombinational (hybrid) speciation, some genotypes of diploid hybrids are fertile and are reproductively isolated from the parent species, and so give rise to new species. This process has been documented more often in plants than in animals.

10. Genomic comparisons of diverging populations and species can be used to locate some of the genetic differences that have been caused by natural selection, including those that underlie reproductive isolation. The number and extent of divergent genomic regions are affected by several factors and vary among species.

11. The time required for speciation to proceed to completion is highly variable. It is shorter for some modes of speciation (polyploidy, recombinational speciation) than others (espe-cially speciation by mutation and drift of neutral alleles that confer incompatibility). The process of speciation may require 2 to 3 My, on average, for some groups of organisms; it is much longer in some cases and very much shorter in others.

12. Speciation is the source of the diversity of sexually repro-ducing organisms, and it is the event responsible for every branch in their phylogeny. It probably does not stimulate evolutionary change in morphological characters, as sug-gested by the hypothesis of punctuated equilibria. Rates of evolutionary change may nevertheless be correlated with speciation, perhaps because speciation prevents interbreed-ing between populations from undoing the changes wrought by natural selection or genetic drift.

Terms and Concepts

allopatric speciation
assortative mating
ecological speciation
founder-flush speciation

parapatric speciation
peripatric speciation
recombinational speciation

reinforcement
reproductive character displacement
sympatric speciation

time for speciation (TFS)
vicariance

Suggestions for Further Reading

As noted in Chapter 17, *Speciation*, by J. A. Coyne and H. A. Orr (Sinauer Associates, Sunderland, MA, 2004), is the most comprehensive recent work on the subject. Those with a mathematical bent will enjoy the wide-ranging treatment of models of speciation by S. Gavrilets in *Fitness Land-scapes and the Origin of Species* (Princeton University Press, Princeton, NJ, 2004). An essay by the Marie Curie Specia-tion Network, "What do we need to know about specia-tion?" [Butlin et al., *Trends in Ecology and Evolution* 27: 27–39 (2012)], succinctly reviews present knowledge and direc-tions of future research in speciation.

Problems and Discussion Topics

1. Why is it difficult to demonstrate that speciation has occurred parapatrically or sympatrically?

2. Coyne and Orr (1997) found that sexual isolation is more pro-nounced between sympatric populations than between allo-patric populations of the same apparent age, and took this find-ing as evidence for reinforcement of sexual isolation. It might be argued, though, that any pairs of sympatric populations that were not strongly sexually isolated would have merged, and so would have been unavailable for study. Thus the degree of sexual isolation in sympatric compared with allopatric popula-tions might be biased. How might one rule out this possible bias? (Read Coyne and Orr after suggesting an answer.)

3. Can postzygotic isolation (low hybrid fertility or viability) be reinforced (i.e., accentuated) by natural selection in hybrid zones? Is this a way in which natural selection can reduce mix-ing between gene pools? See Grant (1966) or Coyne (1974).

4. Suppose that full reproductive isolation between two popula-tions has evolved. Can speciation in this case be reversed, so that the two forms merge into a single species? Under what conditions is this probable or improbable?

5. Referring to the discussion of parallel speciation in stickle-backs, can a single biological species arise more than once (i.e., polyphyletically)? How might this possibility depend on the nature of the reproductive barrier between such a species and its closest relative?

6. The heritability of an animal's preference for different habitats or host plants might be high or low. How might heritability affect the likelihood of sympatric speciation by divergence in habitat or host preference?

7. Biological species of sexually reproducing organisms usually differ in morphological or other phenotypic traits. The same is often true of taxonomic species of asexual organisms such as bacteria and apomictic plants. What factors might cause dis-crete phenotypic "clusters" of organisms in each case?

8. In many groups of plants, low levels of hybridization between related species are not uncommon, yet only a few cases of the origin of "hybrid species" by recombinational speciation have been documented. What factors make recombinational spe-ciation likely versus unlikely?

9. If speciation occurs by divergent pathways of sexual selection in different populations, what might cause the nature of sex-ual selection to differ?

10. Genetic drift and natural selection give rise to geographic variation among populations of a species. How do we account, then, for the features that are uniform among all populations of a species? (See Morjan and Rieseberg 2004.)

11. Choose a topic from this chapter and discuss how its treat-ment would be altered if one adopted a phylogenetic species concept rather than the biological species concept.

The Evolution of Interactions among Species

Nearly 20 years before he published *On the Origin of Species*, Charles Darwin started to study orchids, intrigued by the extraordinary features of their flowers. He roamed the British landscape, making observations and experiments, and he grew tropical species, solicited from horticulturists, in his greenhouse. His field studies, experiments, and dissections culminated in the first book he wrote after *On the Origin of Species*, in 1862, *On the Various Contrivances by which British and Foreign Orchids are Fertilised by Insects, and on the Good Effects of Intercrossing*. It is a landmark work, in which Darwin put into practice his principles of descent with modification and natural selection. Parting with the prevailing theological interpretation, that flowers were shaped by God to inspire us with beauty, Darwin showed that the astonishingly diverse and peculiar features of orchids were modified from the structures of their more prosaic relatives, and that each modification is useful to the plant, enhancing the likelihood that it will attract insects and deposit pollen on them in so precise a way as to ensure cross-pollination. Among these remarkable plants was a species from Madagascar, *Angraecum sesquipedale*, with a nectar-bearing tube up to 30 centimeters long. Other plants with much shorter nectar spurs are visited by insects with tongues long enough to reach the nectar, so Darwin predicted that there must exist in Madagascar a moth with a similarly long proboscis. One reviewer ridiculed this idea, and indeed the very idea that the features of flowers are useful, but in 1903, a sphinx moth with a proboscis up to 30 centimeters long was described from Madagascar, and was fittingly named *Xanthopan morganii praedicta*.

Angraecum and its moth demonstrated perfectly Darwin's speculation, in *On the Origin of Species*, that both a flower and a pollinating insect "might slowly become, either simultaneously or one after the other, modified and adapted in the most perfect manner,

A coevolved interaction The orchid *Angraecum sesquipedale* bears nectar in an exceedingly long spur, and is pollinated by the long-tongued sphinx moth *Xanthopan morganii praedicta*. The moth was discovered about 40 years after Darwin predicted its existence. Each of the species in this mutualism is adapted to obtain something from the other.

by the continued preservation of individuals presenting mutual and slightly favourable deviations of structure." With these words, Darwin became the first to describe the evolution of interactions among species: what we today call **coevolution**.

Interactions among Species

Most of the species with which an individual might interact can be classified as RESOURCES (used as nutrition or habitat), COMPETITORS (for resources such as food, space, or habitat), ENEMIES (species for which the focal species is a consumable resource), or COMMENSALS (species that profit from, but have no effect on, the focal species). In MUTUALISTIC interactions (such as the relationship between the *Angraecum* orchid and its pollinator), each species obtains a benefit from the other. Some interactions are more complex, often because they involve multiple species. For example, leaf-cutter ants (*Atta* and *Acromyrmex*) in tropical America eat only a fungus (Lepiotaceae) that they grow underground on fragments of leaves (see photo on cover). The fungus, which exists nowhere else, is threatened by other fungi, in the genus *Escovopsis*, that can attack and devastate the ants' fungal garden. The ants counter this threat by harboring actinomycete bacteria, *Pseudonocardia*, that produce an antibiotic that inhibits *Escovopsis* growth. The mutualistic bacteria are housed in many small pits on the ants' exoskeletons and are apparently nourished by secretions of glands that are unique to attine ants. These four organisms are thus bound to one another in a web of antagonistic and mutualistic interactions, and they have been adapting to one another since the origin of the leaf-cutter ants, about 50 Mya (Currie et al. 2003, 2006).

The nature and strength of an interaction between two species may vary depending on genotype, environmental conditions, and other species with which those species interact. There is genetic variation, for example, in virulence within species of parasites and in resistance within species of hosts. Some mycorrhizal fungi, associated with plant roots, enhance plant growth in infertile soil, but depress it in fertile soil. In the limber pine, populations in areas where squirrels eat the seeds have cones that reduce squirrel depredation, but are also less favorable for the Clark's nutcracker, a bird that the pine depends on for seed dispersal (**Figure 19.1**). Thus the selection that species exert on one another may

(A)

Nucifraga columbiana

Squirrels absent

Tamiasciurus hudsonicus

Squirrels present

(B)

- Sierra Nevada with squirrels
- Rocky Mountains with squirrels
- Great Basin without squirrels

Axis 2: longer cones, with more scales and seeds

Axis 1: wider and heavier cones, with thicker scales and peduncles, fewer seeds, and thicker seed coats

FIGURE 19.1 A geographic mosaic of interactions. (A) Typical cones of limber pine (*Pinus flexilis*) populations that (at right) are adapted to resist seed-eating squirrels or (at left) are adapted for seed dispersal by Clark's nutcracker where squirrels are absent. (B) A plot of two variables, each of which combines several measurements of cones and seeds, shows that pines in an area without squirrels differ from those in two areas with squirrels. Each point represents one tree. (After Siepelski and Benkman 2007; pine cone photos courtesy of C. Benkman.)

differ among populations, resulting in a "geographic mosaic" of coevolution that differs from one place to another (Thompson 2009).

Coevolution

In many cases, only one species has adapted to an interaction; for instance, the orchids that mimic an insect sex pheromone are adapted to attract insect species that have not adapted to the deception (see Figure 11.2). However, the possibility that an evolutionary change in one species will evoke a reciprocal change in another species distinguishes selection in interspecific interactions from selection imposed by the physical environment. Reciprocal genetic change in interacting species, owing to natural selection imposed by each on the other, is coevolution in the narrow sense.

The term "coevolution" includes several concepts (Futuyma and Slatkin 1983; Thompson 1994). In its simplest form—SPECIFIC COEVOLUTION—two species evolve in response to each other (**Figure 19.2A**). For example, Darwin envisioned predatory mammals, such as wolves, and their prey, such as deer, evolving ever greater fleetness, each improvement in one causing selection for compensating improvement in the other, in an "evolutionary arms race" between prey and predator. GUILD COEVOLUTION, sometimes called DIFFUSE COEVOLUTION, occurs when several species are involved and their effects are not independent (**Figure 19.2B**). For example, genetic variation in the resistance of a host to two different species of parasites might be correlated (Hougen-Eitzman and Rausher 1994). In ESCAPE-AND-RADIATE COEVOLUTION, a species evolves a defense against enemies and is thereby enabled to radiate into a diverse clade (**Figure 19.2C**). For example, Paul Ehrlich and Peter Raven (1964) proposed that plant species that evolved effective chemical defenses were freed from predation by most herbivorous insects, and thus diversified, evolving into a chemically diverse array of food sources to which different insects later adapted and then diversified in turn.

Phylogenetic aspects of species associations

The term "coevolution" has also been applied to histories of parallel diversification, as revealed by concordant phylogenies, of associated organisms such as hosts and their parasites or endosymbionts. For example, aphids have endosymbiotic bacteria (*Buchnera*) that live in special cells and supply the essential amino acid tryptophan to their hosts (see Figure 16.26C). The phylogeny of these bacteria is completely concordant with that of their aphid hosts (**Figure 19.3**). The simplest interpretation of this pattern is that the association between *Buchnera* and aphids dates from the origin of this insect family, that there has been little if any cross-infection between aphid lineages, and that the bacteria have diverged in concert with speciation in their hosts.

Differences among host species may prevent symbionts from switching to distantly related species. For example, the phylogeny of feather lice in the genus *Columbicola* is partly congruent with the phylogeny of their hosts, pigeons and doves, suggesting that the lice have cospeciated with their hosts rather frequently. Dale Clayton and colleagues (2003) found that the body size of a louse species is correlated with its host's body size,

(A) Specific coevolution

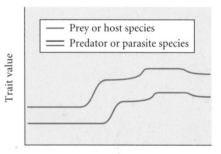

(B) Guild (diffuse) coevolution

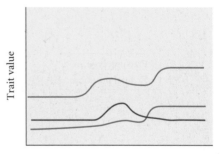

(C) Escape-and-radiate coevolution

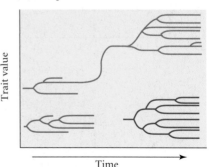

FIGURE 19.2 Three kinds of coevolution. In each graph, the horizontal axis represents evolutionary time, and the vertical axis shows the state of a character in a species of prey or host and one or more species of predators or parasites. (A) Specific coevolution. (B) Guild, or diffuse, coevolution, in which a prey species interacts with two or more predators, can take many paths. In this case, a prey species becomes better defended against two predators, only one of which (blue trace) becomes better able to capture the prey. (C) Escape-and-radiate coevolution. A prey or host species (red trace) evolves a major new defense, escapes association with a predator or parasite (blue trace), and diversifies. Later, a different predator or parasite (green trace) adapts to the host clade and diversifies.

FIGURE 19.3 The phylogeny of endosymbiotic bacteria included under the name *Buchnera aphidicola* is perfectly congruent with that of their aphid hosts. Several related bacteria (names in red) were included as outgroups in this analysis. Names of the aphid hosts of the *Buchnera* lineages are given in blue. The estimated ages of the aphid lineages are based on fossils and biogeography. (After Moran and Baumann 1994.)

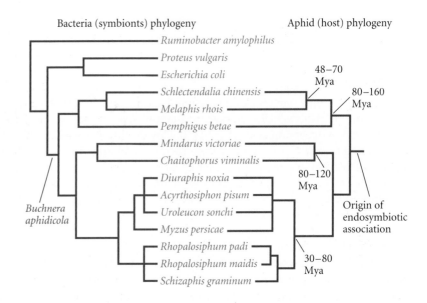

which in turn is correlated with the distance between the adjacent units (barbs) that make up a feather. Therefore lice fit better into the feathers of their normal host than into the feathers of larger or smaller birds. Lice transferred to smaller species of columbids had reduced survival rates because the birds could easily remove them by preening their feathers. On birds that could not preen effectively because their bills were fitted with bits, the lice survived quite well (**Figure 19.4**). The investigators concluded that lice have seldom switched between bird species that differ in size because colonists would have had low fitness.

Associations between hosts and their parasites or symbionts rarely show such evidence of cospeciation, however. Instead, there are often mismatches, caused by horizontal transfer, or HOST SWITCHING, and by several other factors, such as extinction of parasite lineages (Page 2002). Host switching may be likely if parasites disperse from one host to another through the environment, as plant-feeding insects do. Among host-specific herbivorous insects, related species often feed on plants in the same family, and these associations may be very old. Nevertheless, the phylogeny of the insect species seldom matches that of the host plants very closely, and the species of insects have often originated long after their host plant lineages diverged (Futuyma et al. 1995; Winkler and Mitter 2008). Flea beetles in the genus *Blepharida* are associated with the plant family Burseraceae in both Mexico and Africa, and this association is more than 100 My old. The beetles are adapted to the terpene-containing resins that appear to be the plants' main defense against nonadapted insects. Judith Becerra (1997) found that in Mexico, closely related host-specialized species of *Blepharida* are more likely to feed on chemically similar species

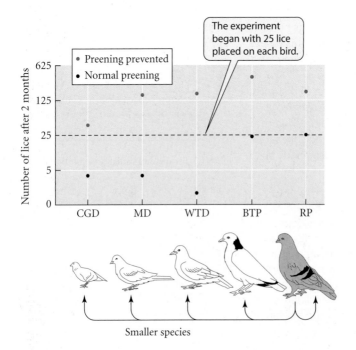

FIGURE 19.4 The difficulty of host switching by feather lice may contribute to the match between their phylogeny and that of their hosts. A starting population of 25 lice transferred from rock pigeons (shaded) to other individuals of the same and four different species increased on all of those species when a bit placed on the birds' bills prevented them from preening. Louse populations on birds that could preen normally decreased on the three smaller species. RP, rock pigeon (*Columba livia*); BTP, band-tailed pigeon (*Patagioenas fasciata*); WTD, white-tipped dove (*Leptotila verreauxi*); MD, mourning dove (*Zenaida macroura*); CGD, common ground dove (*Columbina passerina*). (After Clayton et al. 2003.)

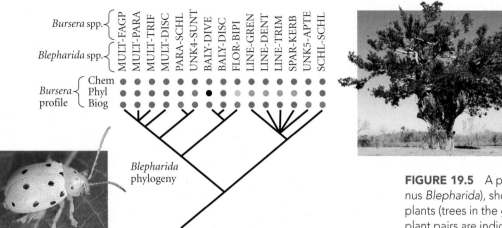

of *Bursera* than on phylogenetically closely related species (**Figure 19.5**). This pattern suggests that newly arisen beetle species have adapted to plant lineages that had already evolved, and furthermore, that these insects adapt more readily to new hosts that are chemically similar to the beetles' ancestral host than to chemically different plants.

Coevolution of Enemies and Victims

In considering the processes of evolutionary change in interacting species, we will begin with interactions between enemies and victims: predators and their prey, parasites and their hosts, herbivores and their host plants. Predators and parasites have evolved some extraordinary adaptations for capturing, subduing, or infecting their victims (**Figure 19.6**). Defenses against predation and parasitism can be equally impressive, ranging from cryptic patterning (see Figure 12.5) to the most versatile of all defenses: the vertebrate immune system, which can generate antibodies against thousands of foreign compounds. Many such adaptations appear to be directed at a variety of different enemies or prey species, so although it is easy to demonstrate adaptations in a predator or a prey species, it is usually difficult to show how any one species has coevolved with another.

Theoretically, the coevolution of predator and prey might take any of several courses (Abrams 2000): it might continue indefinitely in an unending escalation of an evolutionary arms race (Dawkins and Krebs 1979); it might result in a stable genetic equilibrium; it might cause

FIGURE 19.5 A phylogeny of some leaf beetles (genus *Blepharida*), showing features of each species' host plants (trees in the genus *Bursera*); associated beetle-plant pairs are indicated by four-letter abbreviations. The host plants fall into categories, indicated by colored dots, with respect to their chemical profile (Chem), phylogenetic clade (Phyl), and geographic area (Biog). The diagram shows that closely related species of beetles tend to feed on plants that are chemically similar, but not necessarily closely related or sharing similar geographic distribution. For example, *Blepharida balyi* (BALY) and *B. flohri* (FLOR) feed on *Bursera diversifolia* (DIVE), *B. discolor* (DISC), and *B. bipinnata* (BIPI), which are chemically similar (all red dots), but phylogenetically diverse (black, blue, and yellow dots) and occupy different geographic regions (red, blue, and purple dots). The pattern suggests that these insects adapt to chemically similar plants more often than to chemically different plants. (After Becerra and Venable 1999; *Blepharida* photo courtesy of Judith Becerra.)

FIGURE 19.6 Predators and parasites have evolved many extraordinary adaptations to capture prey or infect hosts. (A) The dorsal fin spine of a deep-sea anglerfish (*Himantolophus*) is situated above the mouth and modified into a luminescent fishing lure. (B) The larva of a parasitic trematode (*Leucochloridium*) migrates to the eyestalk of its intermediate host, a land snail, and turns it a bright color to make the snail more visible to the next host in the parasite's life cycle, a snail-eating bird such as a thrush. (B, photo by P. Lewis, courtesy of J. Moore.)

continual cycles (or irregular fluctuations) in the genetic composition of both species; or it might even lead to the extinction of one or both species.

An unending arms race is unlikely because adaptations that increase the offensive capacity of the predator or the defensive capacity of the prey entail allocations of energy and other costs that at some point outweigh their benefits. Consequently, a stable equilibrium may occur when costs equal benefits. Many examples of such costs have been demonstrated. For example, production of the toxic secondary compounds that plants use as defenses against herbivores, such as the tannins of oaks and the terpenes of pines and of *Bursera*, can account for more than 10 percent of a plant's energy budget. Genetic lines of wild parsnip (*Pastinaca sativa*) containing high concentrations of toxic furanocoumarins suffered less damage from webworms, and matured more seeds, than lines with lower levels when grown outdoors. In the greenhouse, however, where the plants were free from insect attack, the lines with higher concentrations of furanocoumarins had lower seed production (Berenbaum and Zangerl 1988). Levels of chemical defenses are higher in plant species that inhabit nutrient-poor environments than in species that inhabit richer environments, where plants can more easily replace tissues that are consumed by herbivores, so that the benefit of chemical defenses is lower relative to their cost (Coley et al. 1985; Fine et al. 2006).

Another kind of cost arises if a defense against one enemy makes the prey more vulnerable to another enemy. For example, terpenoid compounds called cucurbitacins enhance the resistance of cucumber plants (*Cucumis sativus*) to spider mites, but they attract certain cucumber-feeding leaf beetles (Dacosta and Jones 1971).

Models of enemy-victim coevolution

QUANTITATIVE TRAIT MODELS   Coevolutionary models of a defensive polygenic character (y) in a prey species and a corresponding polygenic character (x) in a predator are mathematically complex and include many variables that can affect the outcome (Abrams 2000; Nuismer et al. 2007). An important distinction is whether the capture rate of the prey by the predator decreases with the difference $|x - y|$ (e.g., if it depends on a close match between the size of the prey and the size of the predator's mouth) or increases (e.g., when the predator's speed is greater than the prey's). In the first case, in which the capture rate depends on a close match between x and y, suppose that deviation too great in either direction increases the cost of x (or y), and that $\bar{x} = \bar{y}$. Then either increasing or decreasing y will improve prey survival. In this case, y will evolve in one or the other direction, and x will evolve to track y. Eventually, y may evolve in the opposite direction as its cost becomes too great, and x will evolve likewise. The result may be continuing cycles of change in the characteristics of both species, and these changes may contribute to cycles in population density (**Figure 19.7**). If,

FIGURE 19.7 Computer simulation of coevolution between prey and predator in which the optimal predator phenotype (e.g., mouth size) matches a prey phenotype (e.g., body size). (A) Evolution of character state means. As a character state diverges from a reference value, its fitness cost prevents it from evolving indefinitely in either direction. The evolution of the predator's character state lags behind the prey's. (B) Changes in character state means may be paralleled by cycles in population density, arising partly from changes in the match between the predator's character and the prey's. (After Abrams and Matsuda 1997.)

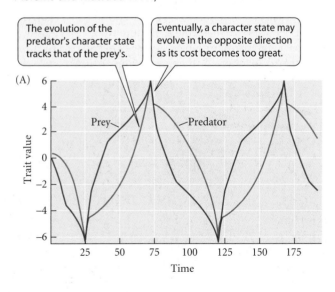

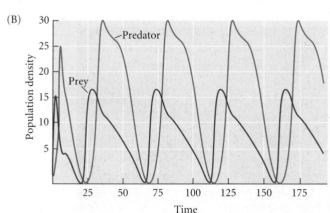

FIGURE 19.8 Fitness of four parasite genotypes on each host genotype in two genetic models of coevolution. (A) Gene-for-gene model, in which the host's alleles A_2 and B_2 confer resistance, and the corresponding parasite alleles α_2 and β_2 confer infectivity. A parasite can infect a host with resistance allele A_2 or B_2 only if it has the corresponding virulence allele α_2 or β_2, respectively. (B) Matching alleles model, in which the host's two resistance alleles are "locks" that can be "opened" by matching "keys" of the parasite. (After Agrawal and Lively 2002.)

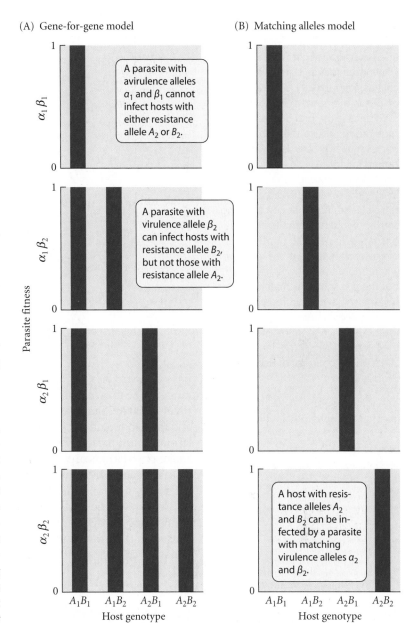

(A) Gene-for-gene model

A parasite with avirulence alleles α_1 and β_1 cannot infect hosts with either resistance allele A_2 or B_2.

A parasite with virulence allele β_2 can infect hosts with resistance allele B_2, but not those with resistance allele A_2.

(B) Matching alleles model

A host with resistance alleles A_2 and B_2 can be infected by a parasite with matching virulence alleles α_2 and β_2.

on the other hand, the capture rate increases with the difference $|x - y|$, cycles may occur if selection is strong enough, but both species will often evolve in the same direction (e.g., toward greater speed), arriving at an equilibrium point that is determined by physiological limits or excessive investment costs.

GENE FREQUENCY MODELS Models of interactions between parasites (including pathogens) and their hosts are based on empirical studies that suggest that the ability of a parasite genotype to infect a host genotype depends on one or a few gene loci in each species (Agrawal and Lively 2002). GENE-FOR-GENE MODELS (**Figure 19.8A**) are based on interactions between plants, such as cultivated flax (*Linum usitatissimum*), and fungal pathogens, such as flax rust (*Melampsora lini*). Similar systems have been described or inferred in several dozen other pairs of plants and fungi, as well as in the interaction between cultivated wheat (*Triticum*) and a major pest, the Hessian fly (*Mayetiola destructor*). In these cases, the host has several loci at which a dominant allele (A_1 or B_1 in Figure 19.8A) confers resistance to the parasite. At each of several corresponding loci in the parasite, a recessive allele (α_1 or β_1) confers INFECTIVITY—the ability to infect and grow—in a host with a particular resistance allele. The most broadly resistant host genotype and the most broadly infective pathogen genotype will be fixed, unless resistance has a cost. In that case, any particular resistance allele will decline in frequency when the parasite's corresponding infectivity allele has high frequency, because its cost will exceed its benefit. As a different resistance allele increases in frequency in the host population, the corresponding infectivity allele increases in the parasite population. Such frequency-dependent selection can cause cycles or irregular fluctuations in allele frequencies.

Cycles are even more likely in the MATCHING ALLELES MODEL (**Figure 19.8B**), which assumes that a pathogen can infect a host only if it has a protein that matches a cell surface receptor protein of the host, like a key and a lock.

Examples of predator-prey evolution

The rough-skinned newt (*Taricha granulosa*) of northwestern North America has one of the most potent known defenses against predation: the neurotoxin tetrodotoxin (TTX). Most populations have high levels of TTX in the skin (one newt has enough to kill 25,000 laboratory mice), but a few populations, such as the one on Vancouver Island, British Columbia,

FIGURE 19.9 Toxicity of rough-skinned newts (*Taricha granulosa*) and resistance of garter snakes (*Thamnophis sirtalis*) in several localities. Prey toxicity is the amount of TTX in the newt; predator resistance is the oral dose of TTX required to reduce the speed of a garter snake by 50%. Below the 85% boundary, snakes can consume co-occurring newts with no reduction in speed; above the 15% boundary, toxicity is so high that co-occurring snakes would be completely incapacitated. In general, populations of garter snakes are more resistant where more toxic newts are found, but there is some mismatch: almost half the snake populations fall below the 85% boundary, and are therefore much more resistant than they need to be. (After Hanifin et al. 2008; photo courtesy of Edmund D. Brodie, Jr.)

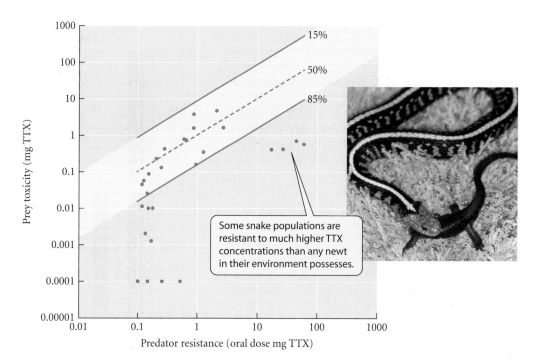

Some snake populations are resistant to much higher TTX concentrations than any newt in their environment possesses.

have almost none. Populations of the garter snake *Thamnophis sirtalis* from outside the range of the toxic newt species have almost no resistance to TTX (Brodie et al. 2002; Hanifin et al. 2008). But snake populations that are sympatric with toxic newts feed on them, and those populations are resistant to TTX. The average level of resistance in snake populations is not perfectly matched with the average toxicity of sympatric newts, for some snake populations are resistant to much higher TTX concentrations than any newt possesses (**Figure 19.9**). There is no selection for increased resistance in these populations, for the snakes suffer no variation in survival from eating the highly toxic newts.

Brood-parasitic birds, such as cowbirds and some species of cuckoos, lay eggs only in the nests of certain other bird species. Cuckoo nestlings hatch first and eject their host's eggs from the nest, so the host ends up rearing only the parasite (**Figure 19.10A**). Adults of host species do not treat parasite nestlings any differently from their own young, but

FIGURE 19.10 Adaptations for and against brood parasitism. (A) A fledgling common cuckoo (*Cuculus canorus*) being fed by its foster parent, a much smaller reed warbler (*Acrocephalus scirpaceus*). (B) Mimetic egg polymorphism in the common cuckoo. The left column shows eggs of six species parasitized by the cuckoo (from top: European robin, pied wagtail, dunnock, reed warbler, meadow pipit, great reed warbler). The middle column shows a cuckoo egg laid in the corresponding host's nest. The match is quite close except in the dunnock nest. The right column shows artificial eggs used by researchers to test rejection responses. (B, photo by M. Brooke, courtesy of N. B. Davies.)

(A)

(B)

some host species do recognize parasite eggs and either eject them or desert the nest and start a new nest and clutch. Many brood parasites have counter-adapted by laying mimetic eggs (Rothstein and Robinson 1998). Each population of the common cuckoo (*Cuculus canorus*) contains several different genotypes, which prefer different hosts and lay eggs closely resembling those of their preferred hosts (**Figure 19.10B**). Some other individuals lay nonmimetic eggs. By tracing the fate of artificial cuckoo eggs placed in the nests of various host species, Nick Davies and Michael Brooke (1998) found that bird species that are not parasitized by cuckoos (because of unsuitable nest sites or feeding habits) tend not to eject cuckoo eggs, whereas among the cuckoos' preferred hosts, those species whose eggs are mimicked by cuckoos rejected artificial eggs more often than those whose eggs are not mimicked. These species have evidently adapted to brood parasitism. Moreover, populations of two host species in Iceland, where cuckoos are absent, accepted artificial cuckoo eggs, whereas in Britain, where those two species are favored hosts, they rejected such eggs. This study provides one of the clearest examples of reciprocal responses to selection among enemies and their victims.

Aposematism and mimicry

In diverse animals, aposematic, or warning, coloration provides a defense against predators, which learn to associate the color pattern with unpalatability or danger. The warning pattern is subject to positive frequency-dependent selection because individuals that deviate from the pattern that predators have learned are likely to be attacked (see p. 328 in Chapter 12). How new aposematic phenotypes evolve is therefore a puzzle, yet related species, and different populations of many species, often differ in coloration (e.g., the geographic races of the unpalatable butterflies *Heliconius erato* and *H. melpomene*; see Figure 12.20). It is possible that genetic drift in places or at times when selection by predators is relaxed may enable a peak shift (see p. 330 in Chapter 12) from one adaptive phenotype to another (Mallet 2010). A possible example is provided by a transition zone between geographic populations of a highly toxic poison dart frog (*Ranitomeya imitator*) in Peru. In this zone, where the frog's coloration is variable, Matieu Chouteau and Bernard Angers (2012) showed, by scoring the rate of bird attacks on clay models of frogs with different color patterns, that the rate of predation is lower and that selection is weaker than in the widespread monomorphic populations outside the transition zone (**Figure 19.11**).

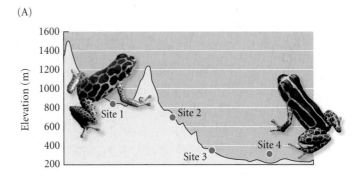

(A)

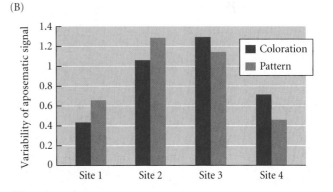

(B)

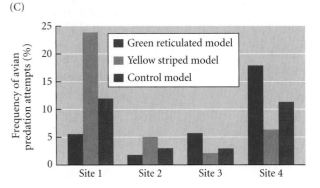

(C)

FIGURE 19.11 Where selection by predators is relaxed, genetic drift may enable new aposematic phenotypes to become established. (A) Transect through a region of Peru that is home to two forms of the poison dart frog *Ranitomeya imitator*. Site 1 is home to a form with a green, reticulated color pattern (left), whereas site 4 is home to a form with a yellow, striped color pattern (right). Sites 2 and 3 constitute a zone of transition between the two forms. (B) Variation in color and pattern is highest in sites 2 and 3. (C) Within the relatively "pure" populations at sites 1 and 4, bird attacks on colored clay models placed in the habitat (revealed by beak impressions in the clay) were lower on the locally abundant color pattern than on the other color pattern or on a brown "control" model. In the zone of transition (sites 2 and 3), predation attempts were less frequent and apparently less selective. (After Chouteau and Angers 2012.)

FIGURE 19.12 A mimicry ring. *Heliconius erato* and *H. melpomene* have a very different color pattern along the upper Huallaga River, in eastern Peru, than in the lower Huallaga basin, where they join a mimicry ring with a "rayed" pattern. This ring of unpalatable species includes four other species of *Heliconius*, three other genera of unpalatable butterflies (the top three species in the center column), and an unpalatable day-flying moth (family Arctiidae, subfamily Pericopinae: center column, bottom). All these species are Müllerian mimics. (Courtesy of J. Mallet.)

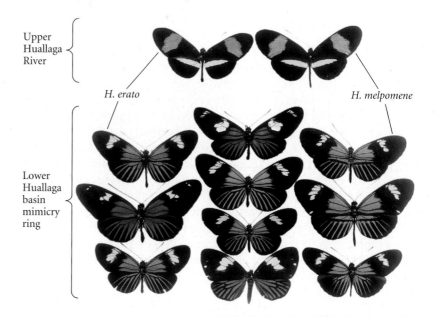

Defensive mimicry is usually based on the aposematic coloration of other species (Mallet and Joron 1999). Two common forms of defensive mimicry, which we described in Chapter 3, are named for the naturalists who first recognized them. In BATESIAN MIMICRY, a palatable species (a mimic) resembles an unpalatable species (a model). Selection on a mimetic phenotype can depend on both its density, relative to that of a model species, and the degree of unpalatability of the model. A predator is more likely to avoid eating a butterfly that looks like an unpalatable model if it has had a recent reinforcing experience (e.g., swallowing a butterfly with that pattern and then vomiting). If it has recently swallowed a tasty butterfly, however, it will be more, not less, inclined to eat the next butterfly with the same phenotype. Thus the rarer a palatable Batesian mimic is relative to an unpalatable model, the more likely predators are to associate its color pattern with unpalatability, and so the greater the advantage of resembling the model will be. If a rare new phenotype arises that mimics a different model species, it will have higher fitness, and so a mimetic polymorphism can be maintained by frequency-dependent selection, as is seen in the African swallowtail *Papilio dardanus* (see Figure 9.1B).

The other major form of mimicry is MÜLLERIAN MIMICRY, in which two or more unpalatable species are co-mimics (or co-models) that jointly reinforce aversion learning by predators. Groups of species that benefit from defensive mimicry are known as MIMICRY RINGS (**Figure 19.12**). In many cases, mimicry rings include both unpalatable Müllerian mimic species and palatable Batesian mimics.

Plants and herbivores

Almost all plants synthesize a variety of SECONDARY COMPOUNDS (so called because they play little or no role in basic metabolism). Thousands of such compounds have been described, including many that humans have found useful as drugs (e.g., salicylic acid, the active ingredient of aspirin), stimulants (caffeine), condiments (capsaicin, the "hot" element in chili peppers), and in other applications (e.g., cannabinol, in marijuana). Higher taxa of plants are often characterized by particular groups of similar compounds, such as cardiac glycosides in milkweeds (Apocynaceae) and glucosinolates in mustards (Brassicaceae). Some of these compounds are known to be toxic to animals, and some to be repellent, so the hypothesis was posed that these features evolved as defenses against herbivores, especially insects (e.g., Fraenkel 1959).

Paul Ehrlich and Peter Raven (1964) proposed a scenario of escape-and-radiate coevolution (see Figure 19.2C), in which a plant species that evolves a new and highly effective chemical defense may escape many of its associated herbivores and give rise to a clade of species that share the novel defense. Eventually, though, some insect species from other

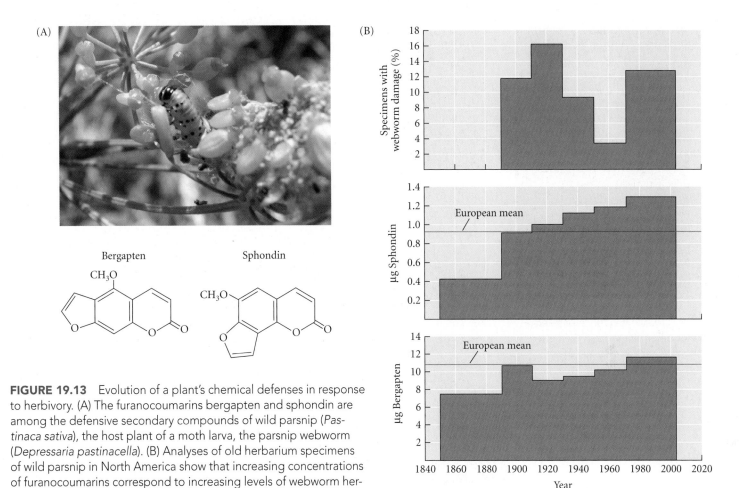

FIGURE 19.13 Evolution of a plant's chemical defenses in response to herbivory. (A) The furanocoumarins bergapten and sphondin are among the defensive secondary compounds of wild parsnip (*Pastinaca sativa*), the host plant of a moth larva, the parsnip webworm (*Depressaria pastinacella*). (B) Analyses of old herbarium specimens of wild parsnip in North America show that increasing concentrations of furanocoumarins correspond to increasing levels of webworm herbivory. The furanocoumarin levels in North American plants reached or surpassed those in native European populations of the plants, indicated by the horizontal lines. (A, photo courtesy of May Berenbaum; B after Zangerl and Berenbaum 2005.)

hosts shift to these plants, adapt to their defense, and give rise to a clade of adapted herbivores. The association of the diverse species of *Blepharida* leaf beetles with the many species of *Bursera* (see Figure 19.5) is one of many examples of adaptive radiations of plants and associated insects that may have arisen in this way.

These hypotheses have been largely supported by subsequent research (Futuyma and Agrawal 2009). Insects and other herbivores certainly impose selection for chemical and other defenses in plants. For example, May Berenbaum and her colleagues (1986) found that variation in the resistance of wild parsnip to its major herbivore, the seed-eating parsnip webworm (*Depressaria pastinacella*), is mostly attributable to the concentrations of two furanocoumarin compounds in the seeds (**Figure 19.13A**). Wild parsnip was introduced in the early 1600s from Europe into North America, where the webworm, also native to Europe, was not recorded before 1869. Analysis of preserved herbarium specimens revealed that North American plants in the 1800s had lower furanocoumarin concentrations than European plants, but that concentrations increased thereafter, in concert with damage inflicted by webworms (Zangerl and Berenbaum 2005; **Figure 19.13B**).

Anurag Agrawal (2005) explicitly analyzed selection on the chemical defenses of the common milkweed (*Asclepias syriaca*) by planting multiple families in a common garden and measuring seed production. He found protective effects of cardenolide compounds, which reduced insect growth (**Figure 19.14A**), and latex (the gummy white fluid for which the plant is named), which reduced the abundance and impact of insects on a plant (**Figure**

FIGURE 19.14 Evidence of selection for defensive traits in the common milkweed (*Asclepias syriaca*). (After Agrawal 2005.)

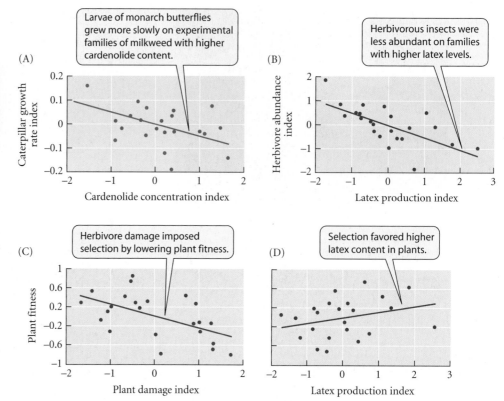

19.14B). Because plant fitness is significantly lowered by insect damage (**Figure 19.14C**), higher levels of latex and other defenses improve plant fitness (**Figure 19.14D**).

There is some evidence that, as we might expect, defenses against an herbivore are stronger in plant species that have a long history of exposure to it. The viburnum leaf beetle (*Pyrrhalta viburni*), an Old World species accidentally introduced into North America, is devastating American species of *Viburnum* shrubs. *Viburnum* species mount a wound response that crushes many of the beetle's eggs. This response is much stronger and more effective in Old World species than in any of the three clades of American species that have evolved independently from Old World ancestors (Desurmont et al. 2011). In this case, as in many others, specialized herbivorous insects can feed on novel plants that are closely related to their natural hosts. This observation is an important consideration in the control of weeds, most of which are plants that have been transported by humans to new regions, unaccompanied by their natural enemies. Host-specialized insects or pathogens that attack the weed in its region of origin are sometimes introduced into the region of weed introduction as biological controls. Such introductions pose a possible risk to crops or native plants, but the risk usually seems to be limited to native plant species that are closely related to the target weeds (Pemberton 2000).

Not surprisingly, herbivores often adapt to plant defenses (Bernays and Chapman 1994; Tilmon 2008). For example, among populations of parsnip webworms, variations in resistance to the toxic effects of furanocoumarins are correlated with differences in the concentrations of these compounds among populations of the host plant (Zangerl and Berenbaum 2003). The seeds of the Japanese camellia (*Camellia japonica*) are enclosed by a woody fruit wall (pericarp) that is much thicker in southern than in northern populations (**Figure 19.15A**). A high proportion of seeds are consumed by larvae of the camellia weevil (*Curculio camelliae*), which inserts eggs into the seed chamber through a hole she bores with her mandibles, located at the end of her long snout, or rostrum (**Figure 19.15B**). Hirokazu Toju and Teiji Sota (2006) showed that the weevils' success in boring through to the seed chamber depends on their rostrum length, relative to the thickness of a fruit's

(A)

(B)

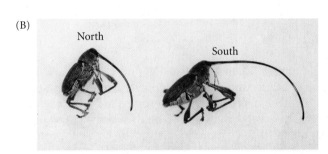

(C)

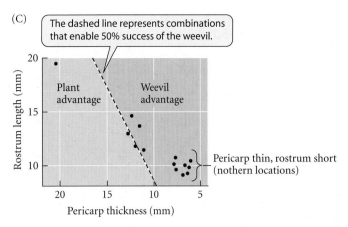

FIGURE 19.15 Imbalance in a coevolutionary conflict. (A) Fruits of the Japanese camellia (*Camellia japonica*) have a much thicker pericarp in the south than in the north of Japan. (B) The rostrum of the camellia weevil (*Curculio camelliae*) is much longer in southern than in northern populations. (C) Average rostrum length and pericarp thickness of associated populations of weevil and camellia in various localities in Japan. The dashed line represents combinations of these variables that enable 50 percent success of the weevil in boring through the pericarp to the seed chamber. In populations to the right of the line, the weevil is favored because its rostrum is long relative to the plant's pericarp thickness. In the population to the left of the line, the plant has the advantage. (After Toju and Sota 2006; photos courtesy of Hiro Toju.)

pericarp. Although southern weevil populations have a much longer rostrum, the southern plant population is "ahead" in this conflict, with pericarps thick enough to reduce the weevils' success to less than 50 percent (**Figure 19.15C**). In the north, weevil populations are "ahead"—their rostra are long enough to ensure a success rate well over 50 percent. These species may be engaged in a coevolutionary "arms race."

Parasite-host interactions and infectious disease

The two greatest challenges a parasite faces are moving itself or its progeny from one host to another (TRANSMISSION) and overcoming the host's defenses. Some parasites are transmitted vertically, from a host parent to her offspring, as in the case of *Wolbachia* bacteria, which are transmitted in insects' eggs. Other parasites are transmitted horizontally among hosts via the external environment (e.g., tapeworms, fleas, and human rhinoviruses, the cause of the common cold, which are discharged by sneezing), via contact between hosts (e.g., the human pubic louse, the gonorrhea bacterium, other venereal disease agents), or via VECTORS (carriers such as the mosquitoes that transmit the yellow fever virus and the malaria-causing protist *Plasmodium*). The effects of parasites on their hosts vary greatly. Those that reduce the survival or reproduction of their hosts are considered VIRULENT.

VIRULENCE AND RESISTANCE IN NATURAL POPULATIONS   According to the gene-for-gene and matching alleles models described earlier, the outcomes of coevolution between a host and a parasite can vary greatly. If new mutations that enhance host resistance or parasite infectivity do not have strong negative side effects (costs), an "arms race" of ever greater defense and offense may occur. In other cases, we should expect frequency-dependent selection to favor rare genotypes in both species, perhaps resulting in prolonged cycles of genetic change. Evidence from natural populations supports both scenarios. For example, *Daphnia magna*, a planktonic freshwater crustacean, can produce eggs that remain dormant in pond sediments for many years. The eggs may harbor resting spores

FIGURE 19.16 Parasite-host coevolution. (A) When *Daphnia* hatched from eggs from several layers of pond sediment, dating from different years, were experimentally exposed to *Pasteuria* from the same ("contemporary"), previous ("past"), or following ("future") sediment layers, the bacteria were generally most successful in infecting "contemporary" *Daphnia*. Each line presents results for *Daphnia* from a particular sediment layer; the starred line is the mean infectivity of all trials, some of which, for simplicity, are not shown here. (B) Reduction of *Daphnia* fecundity by *Pasteuria* taken from different sediment layers show that virulence increased over time. (After Decaestecker et al. 2007.)

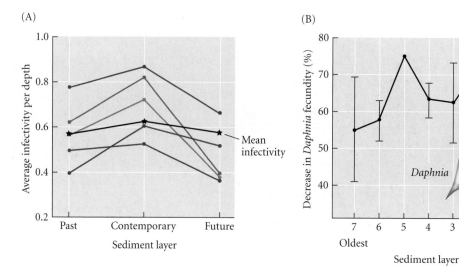

of a parasite, the bacterium *Pasteuria ramosa*. Ellen Decaestecker and colleagues (2007) revived eggs and bacteria from layers of lake sediment deposited over a number of years, then experimentally cross-infected *Daphnia* from several different years with bacteria from the same ("contemporary") year, the preceding ("past") year, and the subsequent ("future") year. They discovered that the hosts were more frequently infected by "contemporary" than by "past" or "future" bacteria (**Figure 19.16A**). These observations suggest that the *Daphnia* population underwent genetic change from year to year and that the bacteria evolved in concert, as in gene-for-gene models of coevolution. Even though both host and parasite underwent continual coevolution, the average virulence of the parasite increased over time, suggesting that virulence alleles were fixed at some loci even while allele frequencies were cycling at other loci (**Figure 19.16B**).

The selective advantage of rare alleles for host resistance can theoretically maintain polymorphisms over the long term. Such polymorphisms have been described for many proteins that defend hosts against pathogens (Woolhouse et al. 2002; Brown and Tellier 2011). The pattern of variation in the resistance genes of many plants indicates that polymorphisms have been maintained by selection for thousands or millions of years. For example, alleles at resistance loci in *Arabidopsis*, a member of the mustard family (Brassicaceae), have a high proportion of nonsynonymous nucleotide differences and are highly divergent in sequence, an indicator that that polymorphism has been maintained by balancing selection, as the frequency-dependent gene-for-gene model predicts (Bergelson et al. 2001; Bakker et al. 2006). Polymorphism at loci in the major histocompatibility complex (MHC) has been shared by humans and chimpanzees since they diverged from a common ancestor more than 5 Mya (see Figure 12.32B). MHC proteins, which present antigens to the immune system, vary in specificity for different pathogen antigens.

We should expect different local populations of a host and its associated parasite to undergo more or less independent turnover of alleles, resulting in geographic variation. This pattern has been described for several associations, such as that between Australian flax (*Linum marginale*) and a host-specific rust fungus (Thrall and Burdon 2003). Another example is the crustacean *Daphnia magna*, which is parasitized by a microsporidian protist (*Pleistophora intestinalis*) that reproduces in the gut epithelium and releases daughter spores in the host's feces. In experimental pairs of infected and uninfected *Daphnia*, the greater the number of parasites in the infected individual, the more likely the other was to become infected. Moreover, the parasites produced more spores, and caused greater mortality, when they infected *Daphnia* from their own or nearby populations than when they infected hosts from distant populations (**Figure 19.17**). Thus populations of this parasite are best adapted to their local host population, and their more virulent effect on sympatric than

FIGURE 19.17 The fitnesses of three strains of a microsporidian parasite and their effects on various populations of their host species, the water flea *Daphnia magna*. Each strain, represented by a different color, was tested in hosts from its locality of origin (solid symbols) and from localities at various distances away (open symbols). (A) The number of parasite spores produced per host (spore load) was greatest when the parasite infected individuals from its own locality, showing that parasites are best adapted to local host populations. (B) Host mortality was greatest in the parasite's own or nearby host populations, showing that the parasite is most virulent in the host population with which it has coevolved. (After Ebert 1994.)

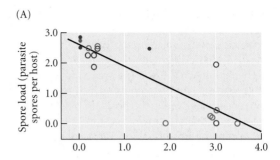

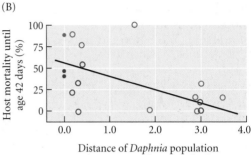

Distance of *Daphnia* population
from source of parasite (km)

on allopatric host populations contradicts the naïve hypothesis that parasites always evolve to be more benign.

THE EVOLUTION OF VIRULENCE   Many people imagine that parasites generally evolve to be benign (AVIRULENT) because the parasite's survival depends on that of the host population. However, a parasite may evolve to be more benign or more virulent depending on many factors (May and Anderson 1983; Bull 1994; Frank 1996). This topic has immense medical implications because the evolution of virulence can be rapid in pathogenic "microparasites" such as viruses and bacteria (Bull 1994; Ewald 1994). The level of virulence depends on the evolution of both host and parasite. For example (Fenner and Ratcliffe 1965), after the introduced European rabbit (*Oryctolagus cuniculus*) became a severe rangeland pest in Australia, the myxoma virus, taken from a South American rabbit, was introduced to control it. Periodically after the introduction, wild rabbits were tested for resistance to a standard strain of the virus, and virus samples from wild rabbits were tested for virulence in a standard laboratory strain of rabbits. Over time, the rabbits evolved greater resistance to the virus, and the virus evolved a lower level of virulence—although it never became completely avirulent (**Figure 19.18**).

Virulence—weakening or killing of the host—is seldom directly beneficial to a parasite, and indeed, it reduces the parasite's fitness if the host dies before the parasite's progeny are transmitted. But the survival and reproduction of an individual parasite depends on its extracting resources from its host, which causes greater damage the more numerous the parasites are. The fitness of a parasite genotype is proportional to the number of hosts its progeny infect, and it is likely to be correlated with the genotype's fecundity. Moreover, competition among parasite genotypes within a host can heighten selection for rapid, prolific reproduction, resulting in higher virulence. Jacobus de Roode and colleagues (2005) found exactly this effect when they experimentally infected mice with pairs of strains of rodent malaria (**Figure 19.19**). Strains that achieved higher densities by reproducing more within the host were also more virulent, causing greater anemia in the mice. The same strains were carried at higher frequencies by mosquitoes that fed on the mice, and so had a higher rate of transmission.

Among many factors that may affect the level of virulence that evolves, three merit special mention. First, selection may occur at two levels. Each host may be viewed as containing a temporary population (deme) of parasites.

FIGURE 19.18 An experimental measure of the mortality ("index of virulence") caused by the myxoma virus in Australian rabbits declined due to changes in both the host ("rabbit component") and the virus ("virus component"). (After Woolhouse et al. 2002.)

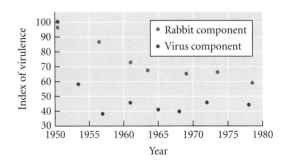

FIGURE 19.19 Virulence increases the fitness of the rodent malarial parasite *Plasmodium chabaudi* when multiple genotypes co-infect mice. (A) *Plasmodium* genotypes that caused higher anemia in the host (i.e., were more virulent) were also more competitive (i.e., attained higher densities within the host). (B) Mosquitoes that fed on the infected mice carried a higher proportion of those *Plasmodium* genotypes that were more abundant in the mice. The more virulent genotypes would therefore have an advantage in transmission to new hosts. (After de Roode et al. 2005.)

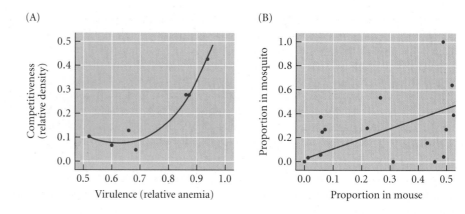

Demes that kill their host before transmission contribute less to the total parasite population than more benign demes, so interdemic selection (group selection) favors low virulence. If a host typically becomes infected by only one individual parasite, or by closely related individuals, the demes are kin groups, so interdemic selection is then tantamount to kin selection, and low virulence may evolve. But as the malaria experiment shows, selection among multiple, unrelated genotypes within demes favors genotypes with high reproductive rates and higher virulence (Frank 1996).

A second factor affecting the level of virulence is whether parasites are transmitted horizontally or vertically (see Figure 16.26A,B). The transmission (and thus fitness) of horizontally transmitted parasites does not depend on the reproduction of their host (or, therefore, on its long-term survival). In contrast, the progeny of a vertically transmitted parasite are "inherited" directly and therefore depend on the host's reproductive success. Hence we may expect evolution toward a less virulent state in vertically transmitted parasites, compared to horizontally transmitted parasites. This hypothesis was supported by an experiment with bacteriophage, in which a phage genotype that reduces its host's growth declined in frequency, and a more "benevolent" genotype increased, when horizontal transmission was prevented (Bull et al. 1991).

Third, if vertebrate hosts rapidly become immune to the parasite, selection favors rapid reproduction—that is, outrunning the host's immune system—by the parasite. Because this may entail greater virulence, an effective immune system (or a drug that rapidly kills the parasite) may sometimes induce the evolution of higher virulence.

EVOLUTION AND EPIDEMICS   The genetics and evolution of parasite-host interactions are highly relevant to human health, as well as that of other species of concern. The examples we have considered suggest that genetic diversity in host populations may be important for maintaining resistance to pathogens. Conversely, populations that are inbred or have low genetic diversity may be at risk of infection. For example, in 1970, 85 percent of the hybrid seed corn planted in the United States carried a cytoplasmic genetic factor for male sterility that was considered useful for preventing unintended cross-pollination. Unfortunately, this genetic factor also caused susceptibility to the southern corn leaf blight (*Helminthosporium maydis*), and about 30 percent of the country's corn crop—and up to 100 percent in some places—was lost to this fungus (Ullstrup 1972). Widely planting a genetically uniform crop is a prescription for disaster.

Among the greatest threats to human health are "emerging pathogens," many of which enter the human population from other species. In some cases, evolutionary change in the pathogen plays a role in its transition to humans (Woolhouse et al. 2005). Phylogenetic analyses are routinely used to trace the origins of new pathogens, such as the evolution of human immunodeficiency viruses (HIV-1 and HIV-2) from chimpanzees and African monkeys (see Figure 3.2). Phylogenetic studies show that some families of viruses have shifted to new host species much more frequently than others have (Jackson and Charleston 2004).

When the origins of a new pathogen can be discovered, it may be possible to determine the genetic basis of the pathogen's adaptation to its new host. For instance, canine parvovirus arose and became pandemic in dogs throughout the world in 1978. Phylogenetic analysis showed that it arose from a virus that infects cats and several other carnivores. Six amino acid changes in the capsid protein of the virus enable it to infect dog cells by specifically binding the canine transferrin receptor. After the virus first entered the dog population, several additional evolutionary changes made it more effective at binding the dog receptor and unable to bind that of its original feline host (Hueffer et al. 2003).

Influenza A viruses are major threats to human health: the most devastating of many influenza A pandemics, the "Spanish flu" in 1918, killed at least 40 million people worldwide. Influenza A viruses are widespread in birds and sometimes enter the human population directly from birds, but usually do so through intermediate vehicles such as pigs (Hay et al. 2001). The shift from birds to mammals usually seems to be based, at least in part, on major genetic changes caused by recombination among fairly different virus strains. Recombination probably played a role in the emergence of the Spanish flu (Gibbs et al. 2001), and certainly in the origin of a strain of the H5N1 avian flu that caused thousands of deaths worldwide in 2009 and 2010. There is concern that H5N1 could evolve further and become pandemic (Longini et al. 2005), either by increased virulence or increased transmissibility. Highly controversial experiments with a strain called H1N1 in captive ferrets showed that only four or five mutational changes enabled the virus to spread by airborne respiratory droplets (e.g., by sneezing). Several of these mutations have been detected in wild populations of the virus. Extensive mathematical simulations, taking into account natural variation, mutation rates, functional effects of amino acid substitutions, possible forms of natural selection, and other factors, showed that the evolution of airborne transmissibility is a potentially serious threat (Russell et al. 2012).

Mutualisms

Mutualisms are interactions between species that benefit individuals of both species. In *The Origin of Species*, Darwin challenged his readers to find an instance of a species having been modified solely for the benefit of another species, "for such could not have been produced through natural selection." No one has met Darwin's challenge. Mutualisms exemplify not altruism, but reciprocal exploitation, in which each species obtains something from the other. Some mutualisms, in fact, have arisen from parasitic or other exploitative relationships. Yuccas (*Yucca*, Agavaceae), for example, are pollinated only by female yucca moths (*Tegeticula* and *Parategeticula*), which carefully pollinate a yucca flower and then lay eggs in it (**Figure 19.20A**). The larvae consume some of the many seeds that develop. Some of the closest relatives of *Tegeticula* simply feed on developing seeds, and one of these species incidentally pollinates the flowers in which it lays its eggs, illustrating what may have been a transitional step from seed predation to mutualism (**Figure 19.20B**).

The theoretical bases for the evolution of mutualism between species are similar to those governing the evolution of reciprocity between conspecific individuals (see Chapter 16). As in the case of intraspecific cooperation, there is always the potential for conflict within mutualisms because a genotype that "cheats" by exploiting its partner without paying the cost of providing a benefit in exchange is likely to have a selective advantage. Thus selection will always favor protective mechanisms, including punishment of cheaters, to prevent overexploitation (Bull and Rice 1991). Moreover, selection will favor "honest" genotypes if the individual's genetic self-interest depends on the fitness of its host or partner (Herre et al. 1999). Thus the factors that should favor evolutionary stability of mutualisms include vertical transmission of endosymbionts from parents to offspring, repeated or lifelong association with the same individual host or partner, and restricted opportunities to switch to other partners or to use other resources. Some mutualisms, although not all, indeed appear to conform to these principles. For example, the *Buchnera* bacteria that live in the cells of aphids and are vertically transmitted are all mutualistic, as far as is known.

(A)

FIGURE 19.20 Mutualisms may result in extreme adaptations. (A) Yucca moths of the genus *Tegeticula* not only lay eggs in yucca flowers, but also use specialized mouthparts to actively pollinate the flowers in which they oviposit. (B) A phylogeny of the yucca moth family, showing major evolutionary changes. The genera other than the "habitual pollinators" *Parategeticula* and *Tegeticula* are seed predators, some species of which (in *Greya*) incidentally pollinate the flowers in which they lay eggs. Intimate mutualism evolved in the ancestor of *Tegeticula* and *Parategeticula*, and "cheating" later evolved twice in *Tegeticula*. (A courtesy of O. Pellmyr; B after Pellmyr and Leebens-Mack 1999.)

(B)

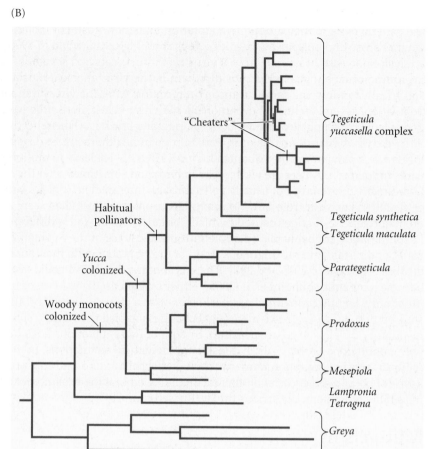

An example of how evolutionary stability can be achieved by punishment is provided by the interaction between yucca species and the moths that are their sole pollinators (Pellmyr and Huth 1994). Typically, a moth lays only a few eggs in each flower, so that only a few of the flower's many developing seeds are consumed by her larvae. Why doesn't she lay more eggs per flower? The answer lies, in part, in the fact that the plant has only enough resources to mature about 15 percent of its many flowers into fruits, and aborts the rest. Pellmyr and Huth found more moth eggs, on average, in aborted than in maturing fruits, suggesting that the plant is more likely to abort a fruit if many eggs have been laid in it. Fruit abortion imposes strong selection on moths that lay too many eggs in a flower because all of the larvae in an aborted flower or fruit perish. Thus the moth has evolved restraint by individual selection and self-interest.

Mutualisms are not always stable over evolutionary time, however: many species cheat. For instance, many orchids secrete no nectar for their pollinators; some of them, as we have seen, deceive male insects that accomplish pollination while "copulating" with the flower (see Figure 11.2). Two lineages of yucca moths that have evolved from mutualistic ancestors do not pollinate, and they lay so many eggs that the larvae consume most or all of the yucca seeds (see Figure 19.20B). These "cheaters" circumvent the plant's abortion response to high numbers of eggs by laying their eggs after the critical period when fruit abortion occurs (Pellmyr and Leebens-Mack 1999).

Partly because of genome studies, mutualism is increasingly recognized as an important basis for adaptation and the evolution of biochemical complexity (Moran 2007). The best-known examples are the evolution of mitochondria from purple bacteria and chloroplasts from cyanobacteria (see Figure 5.4). When a new, "compound" organism is formed from an intimate symbiosis, the subsequent evolution of both genomes is affected. For example, chloroplasts have fewer than 10 percent as many genes as free-living cyanobacteria, but many of the original cyanobacterial genes have been transferred to the plant nuclear genome. These genes may account for as many as 18 percent of the protein-coding genes of *Arabidopsis* (Martin et al. 2002).

Some mutualistic symbioses provide one or both partners with new capabilities (Moran 2007). For example, many features of bacteria are encoded by phage-borne genes. Bacteria and other microbes have formed intimate mutualisms with diverse multicellular organisms, especially animals, which lack the ability to synthesize essential amino acids and vitamins. Some extreme such associations are in sap-sucking homopteran insects (aphids, leafhoppers, cicadas, and relatives). Plant sap lacks many nutrients, including essential amino acids, which in all homopterans are supplied by bacteria that are harbored in specialized cells (bacteriocytes). Some individual insects carry up to six types of symbionts. Almost all plants and animals, however, harbor many kinds of symbionts, whose effects are largely unknown but are the subject of increasing research. Hundreds of species of fungi live within leaves, and thousands of species of bacteria—the "'human microbiome"—inhabit the human body and have diverse beneficial and harmful effects. A case can be made that an "individual" plant or animal should be understood to include its diverse symbionts (Gilbert et al. 2012).

The Evolution of Competitive Interactions

The population densities of many species are limited, at least at times, by resources such as food, space, or nesting sites. Consequently, competition for resources occurs within many species (INTRASPECIFIC COMPETITION) and between different species if they use some or all of the same resources (INTERSPECIFIC COMPETITION). Interspecific competition has two major effects. First, the **competitive exclusion principle** holds that two (or more) competing species that use exactly the same resources cannot coexist indefinitely: one will be driven to extinction. Second, competition imposes selection for divergence in resource use. This effect was first postulated by Darwin, who viewed it as a major reason for divergence of species. There is now considerable evidence that evolution in response to competition is one of the major causes of adaptive radiation (Schluter 2000).

Some species compete for NONSUBSTITUTABLE RESOURCES (resources that cannot be substituted for each other). For example, all plants require both nitrogen (N) and phosphorus (P), which they take up from soil. According to a theory developed by Peter Abrams (1987; see also Fox and Vasseur 2008), competition between species for these nonsubstitutable resources will tend to cause them to evolve more similar uptake ratios (N:P) than if either were growing alone. The reason is that the optimal uptake ratio of N and P equals the ratio in which the organism requires these resources. If uptake N:P is less than the required N:P ratio, growth (and presumably fitness) is limited by the uptake of N, so selection should favor a mechanism that increases the rate at which N is acquired. (The same reasoning holds, in reverse, if the uptake ratio is greater than the requirement ratio.) If a low supply of N lowers the uptake ratio, selection for more efficient uptake of N will occur. Exactly the same will happen if a competing species, with more efficient N uptake, invades and lowers the supply of N. The first species will be selected to increase its efficiency of N uptake, *converging* toward the competing species. Unfortunately, no one has rigorously tested this theory so far.

If an organism can compensate for lower consumption of one resource by increasing consumption of another, the resources are SUBSTITUTABLE. For instance, a predator might eat several small prey in place of a single large one. In this case, a genotype that consumes a resource that is not fully exploited by other genotypes has higher fitness and may increase in frequency until its numbers have depleted the resource. This form of frequency-dependent selection maintains genetic and phenotypic variation within populations of many species (Smith and Skulason 1996; Bolnick et al. 2003), such as the black-bellied seedcracker, in which birds that differ in bill size feed on different kinds of seeds (see Figure 12.15).

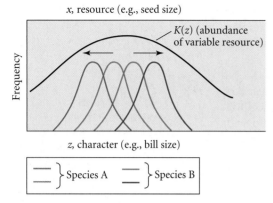

x, resource (e.g., seed size)

z, character (e.g., bill size)

FIGURE 19.21 A model of evolutionary divergence in response to competition. The *x*-axis represents a quantitative phenotypic character (*z*), such as bill size, that is closely correlated with some quality of a resource (*x*), such as the average size of the food items eaten by that phenotype. The curve *K*(*z*) represents the frequency distribution of food items that vary in size, and consequently the density that pure populations of various phenotypes, *z*, could attain. Two variable species (orange and green) initially overlap greatly in *z*, and therefore in the food items they depend on. Those phenotypes in each species that overlap with the fewest members of the other species experience less competition, and so may have higher fitness. Divergent selection on the two species is expected to shift their character distributions (to red and dark green) so that they overlap less.

Suppose that two species of seed-eating birds both vary in bill size, and that the frequency distributions of the two species overlap greatly, so that most individuals compete both with members of their own species and with the other species (**Figure 19.21**). Then, as long as there is a broad range of resource types, the individuals with the most extreme phenotypes (e.g., extremely small or large bills) will experience less intraspecific competition than more "central" phenotypes because they are less abundant, and they will experience less interspecific competition because they tend not to use the same resources as the other species. Therefore the most extreme genotypes will have higher fitness. Such density-dependent diversifying selection can result in the two species *evolving less overlap* in their use of resources and in a shift of their phenotype distributions away from each other (Slatkin 1980; Taper and Case 1992). Divergence in response to competition between species is often called **ecological character displacement**, a term coined by William L. Brown and Edward O. Wilson (1956) to describe a pattern of geographic variation wherein sympatric populations of two species differ more in a characteristic than do allopatric populations. The term is also used to mean the process of divergence that is due to competition.

Observations of the kind of geographic pattern that Brown and Wilson described have provided strong evidence for evolutionary divergence in response to competition (Taper and Case 1992; Schluter 2000). In northwestern North America, for example, several lakes contain reproductively isolated open-water and bottom-dwelling ecomorphs of the three-spined stickleback (*Gasterosteus aculeatus*; see Figure 18.9). These fish differ in body shape (an adaptation to their habitat) as well as in mouth morphology and the number and length of the gill rakers (adaptations to feeding on different prey). Other lakes have only a single form of this stickleback, with intermediate morphology (Schluter and McPhail 1992).

Among seed-eating Galápagos ground finches, bill size is correlated with the efficiency with which the birds process seeds that differ in size and hardness. Illustrating a famous case of character displacement, the Galápagos ground finches *Geospiza fortis* and *G. fuliginosa* differ more in bill size where they coexist than where they occur singly (**Figure 19.22**). In a long-term study, Peter Grant and Rosemary Grant (2006) observed the process of

FIGURE 19.22 Character displacement in bill size in seed-eating ground finches of the Galápagos Islands. Bill depth is correlated with the size and hardness of the seeds most used by each population; arrowheads show average bill depths. (A) Only *Geospiza fuliginosa* occurs on Los Hermanos, and only *G. fortis* occurs on Daphne Major. (B) The two species coexist on Santa Cruz, where they differ more in bill depth. (After Grant 1986; photos courtesy of Peter R. Grant.)

(A) Separate populations

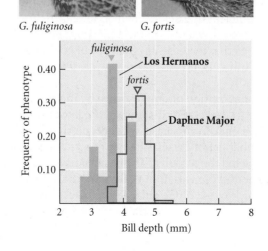

(B) Coexisting populations

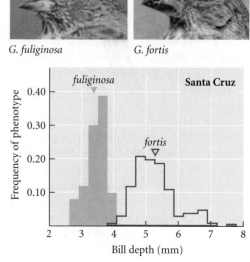

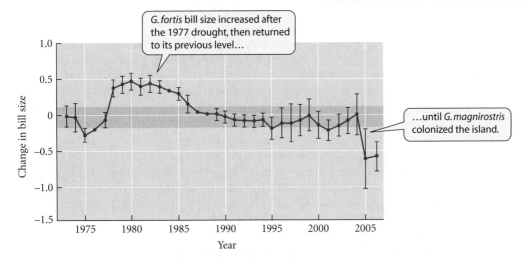

FIGURE 19.23 The history of change in mean bill size in the ground finch *Geospiza fortis* on the island of Daphne Major. Bill size increased after a 1977 drought, then slowly returned to its previous level until the drought of 2004, when it decreased due to character displacement. At that time, *G. magnirostris*, which has a larger bill, invaded the island and depleted the supply of large seeds. The darker shaded bar marks the range of mean bill size in 1973; significant deviations from that range appear above and below that range. (After Grant and Grant 2006.)

character displacement in the population of *G. fortis* on the island of Daphne Major, where this species faced no competitors until *G. magnirostris*, a larger-billed species that can feed on the largest seeds, colonized and built up a large population (**Figure 19.23**). When a drought occurred in 2004, *G. magnirostris* consumed most of the large seeds, and the *G. fortis* population evolved a smaller bill, for survival depended on eating the smallest available seeds, which *G. magnirostris* did not eat. This change in bill size was heritable: offspring of the small-billed survivors, measured in 2006, had small bills.

The concept of character displacement extends to greater numbers of species that might evolve to partition resources (NICHE PARTITIONING). Using mathematical models, the pioneering evolutionary ecologists Robert MacArthur and Richard Levins (1967) predicted that coexisting species might be expected to be evenly spaced with respect to their average resource use, such as the sizes of seeds or prey items. Morphological adaptations might reflect these ecological differences. Many examples fit this prediction (Schluter 2000). For example, among sympatric cats in Israel, differences between sexes and species in the size of the canine teeth, which are used to kill prey, are more nearly equal than expected by chance, and are thought to reflect differences in the average sizes of prey taken by these cats (**Figure 19.24**; Dayan et al. 1990).

Ecological release is another pattern of geographic variation, wherein a species or population exhibits greater variation in resource use, and in associated phenotypic characters, if it occurs alone than if it coexists with competing species. Ecological release is most often characteristic of island populations. For example, in the sole finch species on Cocos Island (in the Pacific Ocean, northeast of the Galápagos Islands), individuals differ in diet and mode of foraging, so that the species population has a much broader diet than any of its relatives in the Galápagos Islands, where there are many more finch species (Werner and Sherry 1987). Similarly, the only species of woodpecker on the Caribbean island of Hispaniola exhibits greater sexual dimorphism in the length of the bill and tongue than do related continental species that coexist with other woodpeckers, and the sexes differ in

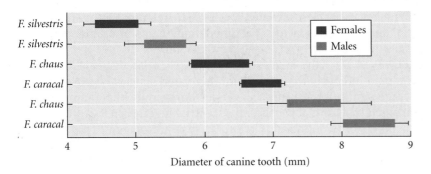

FIGURE 19.24 A nonrandom pattern of equal spacing among multiple species of related predators may have evolved by character displacement, owing to competition for food. The size of the canine teeth differs among the three sympatric species of cats in Israel, and between the sexes in each species. These differences are thought to be adaptations to feeding on prey that differ in average size. The species are the wildcat (*Felis silvestris*), jungle cat (*F. chaus*), and caracal (*F. caracal*). (After Dayan et al. 1990.)

where and how they forage. Alleviation of competition, resulting in ecological release, may enhance rates of speciation and diversification. The fossil record of diversity suggests that when incumbent taxa have become extinct, ecological opportunity has sometimes allowed diversification of other clades on a macroevolutionary scale (see Figure 7.19).

Some species compete not only by depleting resources, but also by INTERFERENCE COMPETITION, whereby individuals suppress competitors or exclude them from defendable resources by aggressive behavior, as in many animals, or by other means, such as poisoning them (as do some plants, fungi, and bacteria). There may then exist selection for competitive ability (Gill 1974). For example, populations of brook sticklebacks (*Culaea inconstans*) have evolved increased aggressiveness where they are sympatric with nine-spined sticklebacks (*Pungitius pungitius*; Peiman and Robinson 2007). Interference competition can result in the exclusion or extinction of some species (Mack et al. 2000).

Evolution and Community Structure

What determines which species occur together in a local community, such as a forest or lake? Ecologists have developed extensive theory, such as elaborations of the competitive exclusion principle, that describes how species interactions affect the ability of species to invade and coexist, but it has become clear that a full understanding of community structure requires an evolutionary perspective as well (Cavender-Bares et al. 2009). First, members of a local community are a subset of the potential members: the species in the larger region (e.g., southeastern North America). These species belong to clades that, perhaps in the remote past, originated in the region or dispersed into the region, and those clades have increased or decreased in species number and in the disparity (variety) of their characteristics over time. Thus, biogeographic history and past patterns of speciation, adaptation, and extinction have left their mark on the present (Webb et al. 2002; see Chapters 6 and 7).

The potential members of a local community are "filtered," first, by environmental factors such as temperature or availability of required resources (e.g., specific host plants for an herbivorous insect), and second, by their ability to coexist with other resident species. The traits that affect whether or not a species gets through these filters have an evolutionary history. Some of these traits are phylogenetically conservative (see p. 154 in Chapter 6): for example, cacti (Cactaceae) generally grow best in dry environments, and *Blepharida* leaf beetles eat *Bursera* foliage (see Figure 19.5). For this reason, members of some clades, but not others, are likely to occur in any habitat or local area (**Figure 19.25**). For example, the chaparral vegetation of coastal California has largely been assembled from clades of plants that originated elsewhere and had features (e.g., small, thick leaves) that "preadapted" them to the dry California summers (Ackerly 2004). Almost every species of leaf beetle in New York State has relatives (in the same genus) that feed on the same families of host plants in other biogeographic regions, such as tropical America. It is likely that they would not have colonized New York if plants related to their ancestral hosts had not been available (Futuyma and Mitter 1996). Campbell Webb (2000) found that the trees that occurred together in small plots in a forest in Borneo were more closely related than a random sample of the trees in the entire forest would have been. These results suggest that closely related species share features that are favored by the particular environmental factors that differ among plots. Thus environmental filtering results in PHYLOGENETIC CLUSTERING by habitat.

Recall, however, Darwin's suggestion that closely related species are more likely to compete intensely than are distantly related species because of their greater average similarity. Jeannine Cavender-Bares and collaborators (2004) found that oaks (*Quercus*) that consistently occurred together in local communities were members of two different clades. These results suggest that although environmental filtering acts on major clades that differ substantially in phenotype, closely related species are too similar to coexist (see Figure 19.25). The resulting

FIGURE 19.25 Factors that may affect phylogenetic relationships among the members of an ecological community. The phylogeny depicts a hypothetical plant clade, the ancestor of which was adapted to moist soils with intermediate pH. Among its contemporary descendants, one clade (a) has become adapted to, and is now found in, acidic soils, and another (c) is associated with dry soils. In a landscape that includes wetter and drier sites but lacks acidic soils, clade a is not found. Drier sites will generally have species in clade c, and wetter sites species in clade b; the environment acts as a filter, resulting in phylogenetic clustering by habitat. But variation among species in each clade leads to stronger competition and exclusion between the closest relatives, so the closest neighbors are phylogenetically overdispersed: they are not as closely related as might be found in random samples of the species.

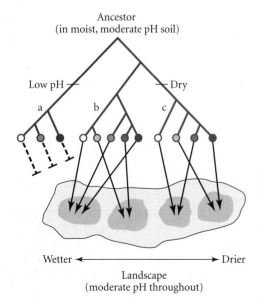

pattern is called PHYLOGENETIC OVERDISPERSION. Such species might be competing for the same resources, or they could be interacting via common enemies. For example, Webb et al. (2006) found in the Bornean forest that the survival of seedlings is greater, the more distantly related they are to the trees in their immediate neighborhood. The likely reason is that a seedling is more likely to be attacked by specialized fungi or herbivores carried by nearby plants if it is more closely related to those plants. This hypothesis has been proposed (and supported by other evidence) to account for the great number of tree species that coexist in many rain forests (Janzen 1970; Connell 1971). In the same vein, phylogenetically diverse plant communities are less easily invaded by non-native plants than are less diverse communities (Gerhold et al. 2011), and a non-native plant has a greater chance of invading if it is more distantly related to the resident species (Strauss et al. 2006).

Both rapid, ongoing evolutionary changes in species interactions and phylogenetic legacies from the distant evolutionary past influence the distribution, dynamics, and interactions of species and help to explain the patterns that are the subject of ecology. Neither evolution nor ecology can be fully understood without reference to the other.

> Go to the
> # EVOLUTION
> Companion Website at
> **sites.sinauer.com/evolution3e**
> for quizzes, data analysis and simulation exercises, and other study aids.

Summary

1. Coevolution is reciprocal evolutionary change in two or more species resulting from the interaction among them. Species also display many adaptations to interspecific interactions that appear one-sided rather than reciprocal.

2. Phylogenetic studies can provide information on the age of associations among species and on whether or not they have co-diversified or acquired adaptations to each other. The phylogenies of certain symbionts and parasites are congruent with the phylogenies of their hosts, implying cospeciation, but in other cases such phylogenies are incongruent and imply shifts between host lineages.

3. Coevolution in predator-prey and parasite-host interactions can theoretically result in an ongoing evolutionary arms race, a stable genetic equilibrium, indefinite fluctuations in genetic composition, or even extinction. Among the many interesting adaptations in predator-prey interactions are aposematism and mimicry.

4. Parasites (including pathogenic microorganisms) may evolve to be more or less virulent depending on the correlation between virulence and the parasite's reproductive rate, the parasite's mode of transmission between hosts (vertical versus horizontal), infection of hosts by single versus multiple parasite genotypes, and other factors. Parasites do not necessarily evolve to be benign.

5. Mutualism is best viewed as reciprocal exploitation. Selection favors genotypes that provide benefits to another species if this action yields benefits to the individual in return. Thus the conditions that favor low virulence in parasites, such as vertical transmission, can also favor the evolution of mutualisms. Mutualisms may be unstable if "cheating" is advantageous, or stable if it is individually advantageous for each partner to provide a benefit to the other.

6. Evolutionary responses to competition among species may lead to divergence in resource use and sometimes in morphology (character displacement). However, competition for nonsubstitutable resources may result in convergent evolution, and selection for greater ability to compete for defendable resources can result in greater aggression and competitive exclusion of less competitive species. Competition is a cause of ecological diversification.

7. Both ongoing evolution and phylogenetic legacies can influence which species coexist in local ecological communities. Phylogenetically conservative characters may be subject to environmental filtering, so that the species in a habitat are phylogenetically clustered; conversely, very closely related species are likely to be phylogenetically overdispersed.

Terms and Concepts

coevolution

competitive exclusion principle

ecological character displacement

ecological release

Suggestions for Further Reading

J. N. Thompson, in *The Geographic Mosaic of Coevolution* (University of Chicago Press, Chicago, 2004), reviews many aspects of coevolution and provides numerous examples. Plant-animal interactions are the focus of essays by prominent researchers in *Plant-Animal Interactions: An Evolutionary Approach*, edited by C. M. Herrera and O. Pellmyr (Blackwell Science, Oxford, 2002).

M. E. J. Woolhouse and colleagues provide an outstanding overview of parasite-host coevolution in "Biological and biomedical implications of the co-evolution of pathogens and their hosts," *Nature Genetics* 32: 569–577 (2002). "Models of parasite virulence," by S. A. Frank, *Quarterly Review of Biology* 71: 37–78 (1996), is an excellent entry into this subject.

The Ecology of Adaptive Radiation, by D. Schluter (Oxford University Press, Oxford, 2000), includes extensive treatment of the evolution of ecological interactions and their role in diversification. "The merging of community ecology and

phylogenetic biology," by J. Cavender-Bares et al., *Ecology Letters* 12: 693–715 (2009), is an excellent overview of the subject.

Problems and Discussion Topics

1. How might coevolution between a specialized parasite and a host be affected by the occurrence of other species of parasites?

2. How might phylogenetic analyses of predators and prey, or of parasites and hosts, help to determine whether or not there has been a coevolutionary arms race?

3. The generation time of a tree species is likely to be 50 to 100 times longer than that of many species of herbivorous insects and parasitic fungi, so its potential rate of evolution should be slower. Why have trees, or other organisms with long generation times, not become extinct as a result of the potentially more rapid evolution of their natural enemies?

4. Design experiments to determine whether greater virulence is advantageous in a horizontally transmitted parasite and in a vertically transmitted parasite.

5. Do you expect that an infectious pathogen such as the bacterium *Staphylococcus aureus* or the HIV virus that causes AIDS will evolve to become more or less virulent? What do you need to know in order to make your best projection? You may want to read about the biology of one such pathogen in order to arrive at an answer.

6. Some authors have suggested that selection by predators may have favored host specialization in herbivorous insects (e.g., Bernays and Graham 1988). How might this occur? Compare the pattern of niche differences among species that might diverge as a result of predation with the pattern that might evolve because of competition for resources.

7. Provide a hypothesis to account for the extremely long nectar spur of the orchid *Angraecum sesquipedale* (chapter opening photo) and the long proboscis of its pollinator. Although it may be fairly obvious why the pollinator's proboscis should match the nectar spur's length, why should both have evolved to be so very long? How would you test your hypothesis?

8. It seems surprising that certain orchids successfully deceive insects into "copulating" with their flowers. Have these species of insects, evidently failing to perceive the difference between a flower and a female of their own species, failed to adapt? If so, what might account for this failure?

9. In simple ecological models, two resource-limited species cannot coexist stably if they use the same resources. Hence coexisting species are expected to differ in resource use because of the extinction or exclusion, by competition, of species that are too similar. Therefore coexisting species could differ either because of this purely ecological process of "sorting" or because of evolutionary divergence in response to competition. How might one distinguish which process has caused an observed pattern? (See Losos 1992.)

10. Suppose that, among related host species that carry related symbionts, the relationship is mutualistic in some pairs and parasitic in others. How would you (a) tell which is which, (b) determine what the direction of evolutionary change has been, and (c) determine whether the change from one to the other kind of interaction has been a result of evolutionary change in the symbiont, in the host, or both?

Evolution of Genes and Genomes

From the time of Aristotle in the fourth century BC to Darwin, Huxley, and Owen in the nineteenth century, the study of the diversity and history of life focused on morphology and, to a lesser extent, behavior. Only in the second half of the twentieth century did it become possible to compare the genes, molecules, and whole genomes of different species in order to understand their evolutionary relationships in the context of population processes such as gene flow and genetic drift. The impact of molecular biology on evolutionary biology has been so profound that it is hard to imagine that evolutionary biology will experience further methodological and conceptual shifts of similar magnitude. Yet the tools of genomics—a suite of biotechnologies that can be described as molecular biology writ large—are having just such an impact. Genomics is to twenty-first-century evolutionary biology what protein electrophoresis and DNA sequencing were to the field in the twentieth century, and it likewise promises to open up as many questions as answers.

Today it is routine for the complete nucleotide sequence of all the genes of an organism to be determined, whether it be an archaean with a 2 megabase (million base pair, Mb) genome or a vertebrate with a genome of 2 or 3 gigabases (billion base pairs, Gb). As this chapter is being written, the list of microbes whose genomes have been completely sequenced contains several thousand species, and complete genome sequences exist for taxa sprinkled across the tree of life: many eukaryotic pathogens, yeasts, nematodes such as *C. elegans*, a sea urchin,

An elephant-nose fish (*Gnathonemus petersii*), a type of electric fish Recent genomic studies reveal convergent evolution in key genes underlying the ability of multiple lineages of electric fish to emit electric signals to detect prey. These studies also reveal the molecular details underlying the evolution of the electric organ from skeletal muscle.

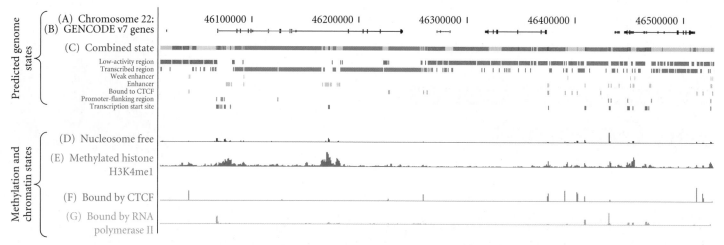

FIGURE 20.1 The ENCODE project. Shown here is a 507,000-base-pair region on human chromosome 22 of a specific cell type called GM12878, and the various states in which that part of the genome occurs. (For simplicity, only a subset of the states tracked by the ENCODE project are shown here.) Each colored row corresponds to a different type of genome state. These states include the presence of various types of genome modification, binding sites for transcription factors, methylation sites, transcription start sites, enhancer regions, and DNA regions that are tightly bound chromatin. At the top (A) are the genome coordinates on chromosome 22. The next row (B) shows the genes, with boxes indicating exons and lines indicating introns. The next row (C) shows the composite status of the genome at a given location, based on the combined distributions of genome states in the next seven rows. In blue (D) are regions available for transcription, as measured by the lack of nucleosomes in those regions. Tightly bound chromatin is less available for binding of transcription factors, and therefore is less likely to be expressed. In red (E) are genome states corresponding to methylation of an associated histone (a modification that correlates with expression status of genes). In dark green (F) are regions enriched for sites bound by a specific type of transcription factor called CTCF; this protein regulates many genomic processes, including transcription of genes and chromatin structure. At the bottom (light green, row G) are regions bound by RNA polymerase II, the key enzyme catalyzing transcription of RNA from the genome. (After the ENCODE Consortium 2012.)

a tunicate, many insects (especially species of *Drosophila*), many plant species, including rice, banana, and the model organism *Arabidopsis*, and several mammals, fishes, and birds. At the time of this writing, the complete genome of a species of Darwin's finch—a member of the group of birds that so inspired Darwin and his theory of natural selection—has been placed online for free download and analysis by the scientific community.

What a bonanza! Biologists can now study in detail the population genetics of hundreds of genes simultaneously and compare the evolutionary trajectories of genes on the same or different chromosomes. Yet we are learning that the mere sequence of a genome is just the beginning of our understanding and use of it. The hierarchical complexity of the genome, in the form of genic, chromosomal, and whole-genome duplications, is slowly being unraveled. The first installment of a major initiative called the Encyclopedia of DNA Elements, or the ENCODE project, was published in 2012 (ENCODE Consortium 2012) and consists of a nearly complete catalogue of the transcriptional, regulatory, and epigenetic activity of the entire human genome (**Figure 20.1**). A vast suite of experimental approaches is yielding orders of magnitude more evidence on the actual functioning of the genome, especially for noncoding regions. Such experimental evidence is a rich source of information for both biomedical science and evolutionary biology, and these additional genomic details are allowing researchers to get closer to the myriad links between genome and organismal evolution.

Comparative genomics—the comparative study of whole genomes—began less than 20 years ago, and for complex eukaryotes, only about 10 years ago. The reach of comparative genomics has already extended beyond model species, such as mice and *Drosophila*, and species of immediate health concern, such as the malaria parasite *Plasmodium falciparum* and its mosquito host *Anopheles gambiae*, to species suited for addressing diverse evolutionary questions. It has extended even to extinct species and pathogens, such as the plague bacterium (*Yersinia pestis*) that afflicted Europeans in the fourteenth century during the Black Death (Bos et al. 2011). Technologies for assaying RNA transcripts, by which the expression status of thousands of genes can be monitored for an organism in different physiological or behavioral states, enable us to study the evolution of gene expression (see Figure 8.15). Comparative genomics not only promises to refine in ever greater detail the tree of life; it also enables us to link genome function with phenotype and to trace the travels of genes as they evolve within species and move between species.

Diverse Players and Evolutionary Processes in Genomes

The advent of rapidly changing technologies and the discovery of novel molecular processes has revealed a host of new molecules and molecular processes that influence the evolution of genomes and their deployment during development. For example, RNA EDITING is a recently discovered process by which the sequence of a gene's mRNA transcript can be altered, often in ways that compensate for premature stop codons or other defects in the gene. Horizontal gene transfer is now known to be much more frequent among microbial species, and even among vertebrates, than previously thought. Whole new classes of molecules, such as the many types of short RNA molecules that we now know are crucial to development and phenotypic variation, have been discovered that are radically changing traditional views of the functioning of gene regulation and expression. Newly discovered molecules called MICRORNAS (RNA sequences of about 22 base pairs that bind to RNA transcripts and repress their translation into proteins) represent one such new class of genomic molecule. These diverse sequences, some represented by as many as 1000 different types and up to 50,000 copies per genome, are now known to perform important functions in gene regulation and development (Bartel 2004). But microRNAs are now drawing the attention of evolutionary biologists as well. Wheeler and colleagues, for example, studied the diversity of microRNAs across the Metazoa. By collecting sequences of microRNAs from many lineages of metazoans, they were able to confirm a pattern that was only dimly evident through the comparison of genome sequences from model organisms: there is a strong signal of progressive accumulation of microRNAs from ancestral to derived lineages of metazoans, and microRNAs tend to persist in metazoan genomes once they have evolved and become incorporated into the gene regulatory network (**Figure 20.2**).

FIGURE 20.2 Continual acquisition of microRNA families throughout metazoan evolution. The numbers beside the branches indicate the addition (black) or loss (red) of identifiable microRNA families. MicroRNA families tend to persist in genomes once they are acquired and integrated into gene regulatory pathways. Thus losses of microRNA families are less common than gains. (After Wheeler et al. 2009.)

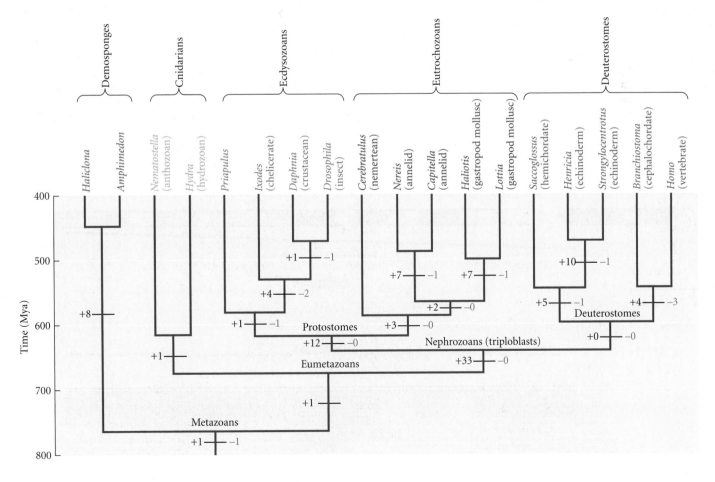

With the discovery of the incredible diversity of gene structures, patterns of duplication, gene splicing patterns, and regulatory interactions, even the definition of the gene itself has undergone radical revision. Some biologists now view the gene as a "computational module," rather than favoring the traditional definition of the gene as a template for an eventual amino acid sequence, which emphasizes its information content.

A key question for scientists has been what fraction of the human genome, and other genomes, is functional. Eukaryotic genomes contain far more noncoding DNA than prokaryotic genomes do. Only about 1.5 percent of the human genome, for example, is composed of protein-coding sequences. Much of a typical human gene (up to 95 percent) consists of introns. The vast regions of vertebrate genomes that lie outside of coding regions of genes were for many decades considered "junk DNA" because no proteins appeared to be derived from those regions. Much of the genome appeared to be derived from repeated DNA such as transposable elements (TEs), which usually wrecked, rather than enhanced, gene function. In general, scientists could ascribe no clear function to the many "gene deserts" of the human genome.

Several lines of evidence have caused scientists to change this view. For example, we now suspect that many regions outside of genes are functional because of their high level of conservation among species. **Ultraconserved elements**—noncoding regions that are conserved, sometimes completely, across species as distant as fish and humans—often occur far from any genes, yet are presumed to perform regulatory functions for one or more genes; however, the function of most ultraconserved elements is unknown (**Figure 20.3**). Scientists estimate that 3–5 percent of the human genome evolves slowly enough to be considered functionally important. Although still a minute fraction of the genome, this is a higher percentage of the genome than is contained in coding regions. By contrast, the ENCODE project has established that 80 percent of the human genome, including much of the non-protein-coding DNA, is transcribed into RNA. These contrasting measures of genome evolution provide contrasting views of the fraction of the genome that is functional. Clearly we cannot equate the functional portion of the genome solely with those regions that are strongly conserved between species, because many adaptively evolving genome regions, such as the genes of the immune system, evolve quickly and hence lose conservation rapidly. But neither can we equate the functional portion of the genome solely with those regions that are transcribed into RNA, because some transcribed regions probably represent simple transcriptional noise. Several transcribed but untranslated regions have been shown to have functions in cell regulation and in transcription and deployment of genes, so the argument that such regions are nonfunctional is becoming less compelling. Still, the precise fraction of the genome that is functionally important is still unclear—it could be anywhere between 3 percent (the minimum fraction conserved)

FIGURE 20.3 Screenshot from the UCSC Genome Browser, showing an analysis of about 4 kbp (kilobase pairs) of human chromosome 15. The track "RefSeqGenes" indicates regions that encode proteins; it is evident that the region shown here is not near any coding regions. At the bottom are the segments of DNA that are conserved (indicated by vertical black lines) between humans and the various species listed at left. The track "Mammal Cons" indicates the level of conservation across the 4 kbp region relative to a neutral model; this region is not exceptionally conserved, except for the 500 bp segment labeled LCNS (Long Conserved Noncoding Sequence). (After Janes et al. 2010.)

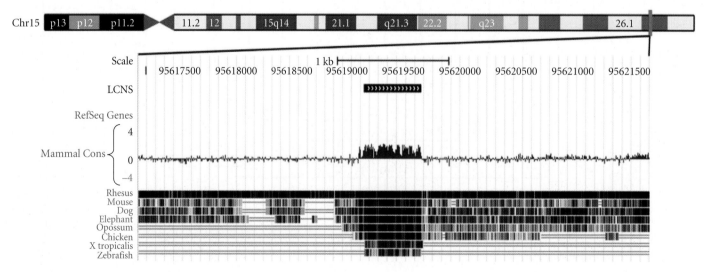

and 80 percent (the fraction transcribed). This broad range represents a remarkable degree of uncertainty; furthermore, it is likely that the functional fraction of the genome is turning over constantly (Ponting et al. 2011). RNA is proving to have myriad roles, not just as a template for translation of proteins, but in vastly complex regulatory networks that control many aspects of the phenotype. Thus the genome appears increasingly fluid and dynamic: many regions formerly considered "junk" turn out to have important functions, and the categories of gene, exon, protein, and RNA transcript are becoming less distinct.

Nonadaptive Processes in Genome Evolution

The roles and prevalence of natural selection versus neutral processes such as genetic drift and mutation in the evolution of genomes is hotly debated. Building on the "nearly neutral" theory of Tomoko Ohta, a Japanese population geneticist who worked closely with Motoo Kimura, Michael Lynch (Lynch 2007) has championed a neutral theory of genome evolution that emphasizes the ability of genetic drift to cause slightly deleterious and nonadaptive mutations to become fixed in evolving lineages (see Chapter 12). Lynch suggests that a variety of genomic features that initially may have had little fitness advantage for organisms—large genome size, introns, transposable elements, large tracts of noncoding DNA, many amino acid substitutions—have become fixed in lineages not because of adaptive processes, but rather by their opposite: the inability of adaptive evolution to work effectively, and the increased ease of fixation of deleterious mutations, in small populations due to strong genetic drift. Lynch suggests, for example, that genetic drift can explain the large diversity of genome sizes—ranging from a few kilobases in the smallest viruses to many megabases in plants and other eukaryotes. Because viruses and bacteria have extremely large population sizes, natural selection can readily fix slightly advantageous mutations, such as mutations that delete nonessential DNA. By contrast, larger organisms with smaller effective population sizes tend to have larger genomes rich in transposable elements or their remnants. Such genomic detritus accumulates in lineages with small populations because natural selection is unable to eliminate deleterious and recurrent mutations in small populations.

Lynch's theory is important because it is an attempt to provide an overarching explanation for the many features of genomes, large and small, that we observe. When applying the theory, it is important to consider the types of mutations that are likely to be deleterious, because it is these mutations that will be more easily fixed in smaller populations and will result in the large-scale trends that we observe. For example, Kuo and colleagues (2009) found that genome sizes were smaller in lineages of bacteria that were estimated to have smaller effective population sizes on the basis of population genetic data (**Figure 20.4**). They suggested that bacteria have an inherent bias toward

FIGURE 20.4 The relationship between genome size and effective population size in bacteria. Each point represents a bacterial species of one of three types: free-living, facultatively parasitic, or obligately parasitic. The logarithm of genome size, in millions of base pairs (Mb) is plotted against an index of effective population size, namely, the ratio of nonsynonymous to synonymous substitutions in protein-coding regions (d_n/d_s). This ratio is sometimes used as a measure of effective population size under the assumption that, in general, nonsynonymous substitutions in DNA are slightly deleterious. If this is the case, then they are most likely to become fixed in populations not by natural selection, but by strong genetic drift, which can sometimes cause slightly deleterious substitutions to accumulate. Thus smaller population sizes occur toward the right on the x-axis. At first glance, the trend of the data, with genome size decreasing with increasing population size, appears contrary to Lynch's theory. But if deletions in DNA represent the major class of spontaneous mutations, and if deletions are slightly deleterious, then the trends are consistent with Lynch's theory, at least across groups. (After Kuo et al. 2009.)

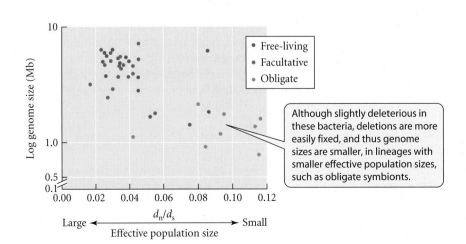

Although slightly deleterious in these bacteria, deletions are more easily fixed, and thus genome sizes are smaller, in lineages with smaller effective population sizes, such as obligate symbionts.

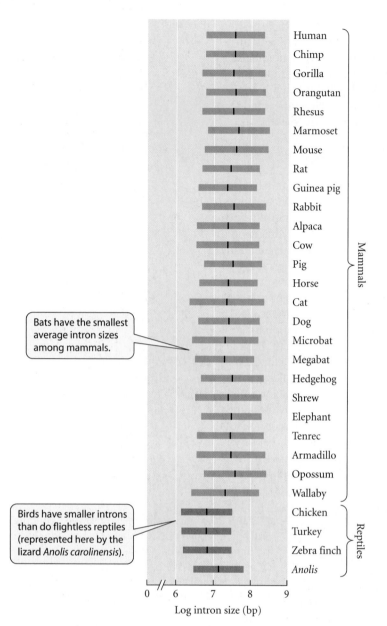

Bats have the smallest average intron sizes among mammals.

Birds have smaller introns than do flightless reptiles (represented here by the lizard *Anolis carolinensis*).

Log intron size (bp)

FIGURE 20.5 Relationship between flight and average intron size in mammals (blue) and reptiles (red). The average intron size for each species is indicated by a solid black vertical line; confidence limits are indicated by the colored bars. (After Zhang and Edwards 2012.)

deletions in their genome, and that these deletions get fixed more frequently in lineages of bacteria with smaller effective population sizes, such as bacteria that are obligate symbionts. This conclusion further suggests that genome reductions, rather than increases, are slightly deleterious in bacteria and are fixed more rapidly by drift, resulting in a genome size trend different from that seen across the domains of life.

Genome size variation in amniotes is less easily explained by nonadaptive consequences of population size. Using whole-genome scans, Zhang and Edwards (2012) found that introns in amniote lineages with powered flight—namely, birds and bats—are shorter than those in lineages without flight (**Figure 20.5**). Following on earlier speculations, the authors proposed that intron size has been reduced in these lineages not by selection for small intron or genome size per se, but by selection for small cell volume, which facilitates the high metabolic rates required for activities such as flight. Selection for small cell volume, in turn, is thought to drive down genome size. Genetic drift is a less likely explanation of these data because birds and bats do not systematically have smaller populations than other reptiles and mammals.

Lynch's theory has still not been tested extensively, but is being studied from many different angles, from gene regulation and expression to the pattern and rate of mutation. Whatever the outcome may be, Lynch's theory, like Kimura's neutral theory of molecular evolution, provides an important starting point for investigating the origins of genome structure and diversity, and it has impressed on genome scientists the importance of population genetic theory for understanding genome evolution.

Rates and Patterns of Protein Evolution

We saw in Chapter 10 that the neutral theory predicts lower rates of evolution for genes under stronger functional constraint. With the availability of multiple types of genomic data for certain organisms, including **transcriptomes**, consisting of the entirety of the expressed genome of an organism or tissue, scientists have been able to study in detail the patterns and evolutionary rates of proteins. This endeavor is made easier by the relatively simple genetic code linking genome sequence and protein sequence.

Codon bias

Although efforts like the ENCODE project are just beginning to unravel the underlying "code" of the vast noncoding regions of the genome, we have understood the code of protein-coding genes since the 1960s (see Figure 8.2). Protein-coding genes are diverse, and their evolution has been modulated in important ways by the structure of the genetic code. For example, many proteins evolve under stabilizing or purifying selection, in which

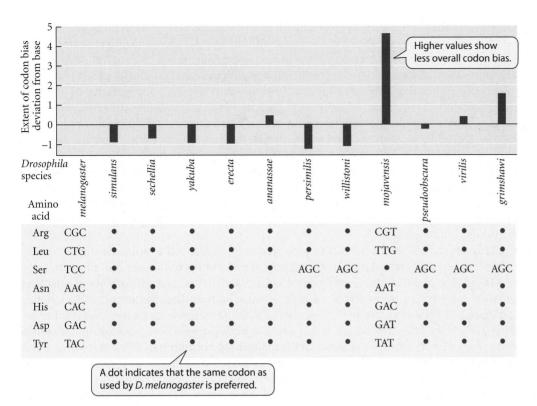

FIGURE 20.6 Extent of codon bias in 12 *Drosophila* species. The histogram shows the deviation in extent of codon bias of each species from *D. melanogaster*, which has been assigned a value of 0. Higher values show less overall codon bias. The chart below shows the most numerous ("preferred") codons in each species for the amino acids given at the left. (After Clark et al. 2007.)

the number of synonymous substitutions exceeds the number of nonsynonymous substitutions ($d_n/d_s < 1$; see Chapter 12). Protein-coding genes also exhibit **codon bias**, which is a measure of the departure from equality of the frequencies of the different synonymous codons specifying a given amino acid. For example, arginine is specified by two codons, AGA and AGG. The null hypothesis is that all codons for a particular amino acid should be used with equal frequency, but codon usage is usually highly uneven, and one codon often dominates over others.

Which codon is used to encode a particular amino acid is correlated with the overall base composition of the gene. Thus in both mitochondrial DNA and the nuclear genome, genes that are generally high in nucleotides G or C tend to use codons that end in G or C (Li 1997). Correlations between transfer RNA abundance and codon usage have been found in several organisms and are thought to underlie some cases of codon bias. Each tRNA is specific for a particular codon. If the tRNA for AGA is more abundant than that for AGG, the expression of a gene will be faster and more efficient if it uses AGA codons for arginine rather than AGG. Typically, genes with very high codon bias tend to be highly expressed and are thought to use those codons that correspond to the anticodons of the most abundant tRNAs. Recent comparisons among whole genomes of *Drosophila* species have shown, surprisingly, that codon bias can vary substantially among species, differing in a consistent way across hundreds of genes in the genome (**Figure 20.6**; Clark et al. 2007).

Gene dispensability and selection for translational robustness

The rate of evolution—a key question even in Kimura's day—is still an active area of research, fueled most recently by the remarkable increase in the scale and type of genomic data that can be generated. Baker's yeast (*Saccharomyces cerevisiae*) was the first eukaryote to have its entire genome sequenced. We now have not only the complete genome sequence for baker's yeast, but also the complete inventory of expression levels for all of its genes in many physiological states, as well as multiple measurements of protein abundance for virtually all of its genes. In addition, for every gene in the yeast genome, we have a measure of **gene dispensability**; that is, the degree to which the fitness of yeast cells is sustained if the gene is physically deleted from the genome. Surprisingly, the correlation between the cell's ability to survive without a particular gene and the rate of evolution of that gene (predicted to be low for indispensable genes) is relatively weak. This low correlation might result from the fact that fitness in yeast is typically measured in the laboratory rather than under natural conditions, providing a poor measure of a gene's dispensability. (Many mouse "knockout" strains, which are missing a specific gene, also seem to fare well without the gene, again because fitness usually is measured only in the laboratory.) Thus we don't yet know the precise relationship between a gene's dispensability and its rate of evolution. In contrast, the rate of evolution is significantly inversely correlated with expression level or protein abundance. Drummond and Wilke (2008) suggest that the rate of protein evolution can be largely explained by selection to prevent and withstand missense errors (those resulting in nonsynonymous codons) during translation, which should impose the strongest selection pressure on highly expressed genes that encode abundant proteins.

Errors in translation of mRNA to protein can be prevented to some degree by codon usage that increases translational accuracy, as may occur if a gene uses codons at the relative frequency at which their corresponding tRNAs are available. This type of selection pressure mainly constrains the nucleotide sequence rather than the amino acid sequence it encodes. However, an amino acid sequence can also experience its own selection pressure. The ability of an amino acid sequence to maintain proper protein folding in the face of ongoing mutation is referred to as its **translational robustness**. Protein folding is the process by which most protein sequences attain a three-dimensional shape that possesses biological activity, such as the capacity of hemoglobin to bind oxygen. Because attaining this functional shape depends critically on interactions between amino acid residues within a protein, even a single missense error in translation can have devastating effects on the folding process. Natural selection to maintain reliable folding is strongest on abundant proteins because they produce the largest amount of waste and the largest cost to the cell and organism if folding fails. Consistent with the selection pressure imposed on such a basic cellular process, the inverse correlation between rate of evolution and level of gene expression holds across multiple species, from *E. coli* to fruit fly to human—organisms with different cellular structures, body sizes, metabolic rates, and ecological niches.

Protein interactions and rates of evolution

A full accounting of the rate of protein evolution involves not only gene expression and protein folding, but also the other proteins with which a focal protein interacts in a cell. Recent work by Qian and colleagues has shown that the correlation between expression and evolutionary rate persists even for specific amino acid sites that are not responsible for protein folding (usually sites near the surface of a folded protein; Qian et al. 2011). This correlation suggests that the stringency of protein folding cannot fully explain variation among proteins in evolutionary rates. Selection against protein misinteractions—the tendency for some proteins to interact with proteins other than their correct protein partners—is considered by some researchers to be an additional and possibly more important factor than protein folding in explaining a protein's evolutionary rate. In cells, hydrophobic amino acids in proteins tend to stick to other proteins by chance more often than do hydrophilic amino acids. This observation leads to a prediction that, because highly expressed proteins are more likely to interact nonspecifically with other proteins than are proteins expressed at a low level, we should expect to see fewer hydrophobic amino acids on the surface of such highly expressed proteins. Additionally, we should expect that "non-sticky" (i.e., hydrophilic) amino acids on the surface of proteins, will change to hydrophobic amino acids only rarely. A combination of computer simulation and analysis of yeast genomic data have confirmed both of these predictions (Qian et al. 2011). By amplifying a relatively weak signal in comparative genomic data, these researchers supported earlier work that found that the rate of protein evolution is inversely, but only weakly, related to the number of other proteins with which a protein interacts.

Developmental biology and rates of protein evolution

Yet another approach to understanding the variation in rates of evolution among proteins comes from developmental biology. For over 150 years, from the time of Haeckel and von Baer, developmental and comparative biologists have noted that during development, organisms pass through a so-called phylotypic stage, in which embryos of diverse evolutionary lineages look similar to one another. Before and especially after this phylotypic stage, embryos diverge, assuming lineage-specific morphologies, tissue types, and developmental stages. The availability of whole-genome data, as well as expression data from the transcriptome of each developmental stage of embryos of several species, has allowed biologists to discover relationships between a protein's long-term rate of evolution and the timing of its expression during development. Early analyses of this relationship, such as that performed by Davis et al. (2005) on data from *Drosophila*, revealed that, when compared across species, proteins expressed earlier in development tended to evolve more slowly than proteins expressed late in development. This and other studies found that different developmental stages are characterized by the expression of distinct transcriptomes, each consisting of hundreds of RNAs encoding proteins with diverse functions. HOUSEKEEPING GENES—genes that act in a wide range of tissue types and that have core (often ancient) metabolic functions—tend to be expressed early in development, whereas later in development, closer to the adult stage, transcriptomes are dominated by transcripts of genes with more specific functions. Housekeeping genes tend to evolve slowly, as measured by d_n, the rate of nonsynonymous substitution, whereas proteins expressed in later stages tend to evolve more quickly.

More detailed examination suggested that mid-to-late embryonic development in *Drosophila*, which corresponds to the phylotypic stage and occurs roughly between 8 and 10 hours after hatching, is characterized by a high rate of expression of slowly evolving genes, such as the Hox genes *wingless*, *Antennapedia*, and *abdominal A* (see Chapter 21 for more on Hox genes). The low rate of evolution of proteins expressed during the phylotypic stage might reflect strong stabilizing selection on proteins at this morphologically conservative stage. In 2010, researchers showed that the phylogenetic age of genes expressed during the phylotypic stage of diverse metazoans, such as *Drosophila*, zebrafish, and *C. elegans*, was greater than that of genes expressed at other developmental stages, insofar as these genes can be found in diverse metazoan lineages and hence are likely to have arisen in their common ancestor (**Figure 20.7**; Domazet-Loso and Tautz 2010). This pattern of gene and expression conservation is

FIGURE 20.7 Hourglass model of transcriptome evolution in *Drosophila*. The transcriptome age index—a measure of the relative age (time of origin as measured on a phylogeny) of genes expressed in a given transcriptome—varies over the course of *Drosophila* development. Note the dip in the transcriptome age index during the mid-to-late embryo stage (segmentation and pharyngula), indicating that genes expressed during this phylotypic stage are phylogenetically old. (After Domazet-Loso and Tautz 2010.)

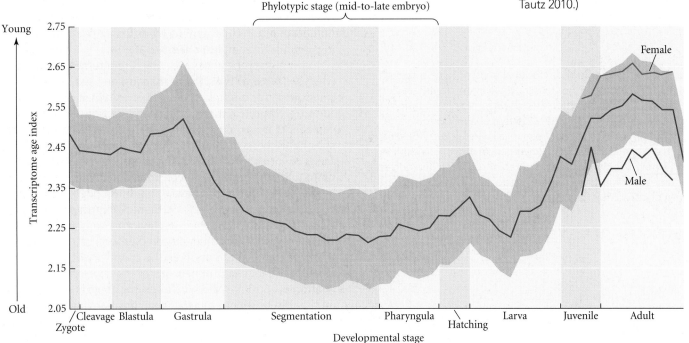

(A)

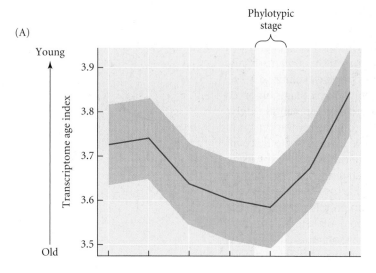

(B)

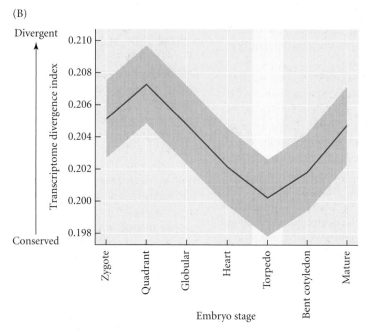

FIGURE 20.8 The hourglass model of transcriptome evolution in plants. (A) Variation in the transcriptome age index (measured as in Figure 20.7) during embryonic development in *Arabidopsis thaliana*. Note the similarity in the pattern of transcriptome age variation seen here to that of *Drosophila* in Figure 20.7. (B) Variation in the average rate of evolution (divergence) of proteins expressed during embryonic development in *Arabidopsis*. Note the decrease in average rate during the torpedo stage, which is the phylotypic stage in plants. (After Quint et al. 2012.)

called the HOURGLASS MODEL because the diversity and rates of evolution of genes expressed at the phylotypic stage are narrower, or more conservative, than those of genes expressed earlier and later in development, mirroring the pattern of morphological conservatism at that stage. More recently, researchers confirmed the hourglass model for transcriptome evolution in plants, using *Arabidopsis* as a test case (**Figure 20.8**; Quint et al. 2012). Together these results bolster the idea that the phylotypic stage of both animals and plants is characterized by expression of ancient regulatory genes, and that genes expressed during earlier and later embryonic stages are generally younger, reflecting the increased need for specialization at these stages and the evolution of lineage-specific genes. The plant results also suggest that the hourglass pattern arose independently in plants and animals, since multicellularity in these two groups evolved convergently. These studies provide a dramatic example of the new insights that can be gained by combining genome information with data on gene expression, protein interactions, and protein structure to develop a comprehensive set of predictions for the rate of protein evolution.

Genome Diversity and Evolution

Diversity of genome structure

The structure of genomes differs widely across the major branches of life. Whereas viral and bacterial genomes are models of efficiency, maximizing the speed of genome replication and minimizing unnecessary genes, eukaryotic genomes—particularly those of mammals, amphibians, and some plants—are very large by comparison, harboring vast regions of noncoding and repeated DNA with unknown functions. Introns are a key feature contributing to the larger genomes of eukaryotes, where they comprise up to 25% of the DNA, relative to bacteria, where they are absent. Indeed, the number of introns in a genome tends to increase with increasing organismal complexity (**Figure 20.9**).

Many eukaryotic genes are subject to **alternative splicing**, whereby many mRNAs, rather than just one, are encoded by a single gene. For example, the *CD44* gene, which encodes a cell surface glycoprotein that modulates interactions between cells, contains 21 exons, at least 12 of which can undergo alternative splicing (Roberts and Smith 2002), potentially yielding thousands of variant proteins. The gene encoding the protein DS-CAM, which is involved in controlling synaptic connections between neurons in the developing brain of *Drosophila*, has 95 exons (Zipursky et al. 2006). Remarkably, this gene is predicted to produce more than 38,000 forms of protein based on alternative splicing of different combinations of its exons. This diversity is thought to underlie the specificity of connections among the thousands of neurons in the developing fly brain (Hattori et al. 2007). Alternative splicing appears to be a major mechanism by which metazoans can increase functional diversity with a limited set of genes. For example, over one-third of alternative splicing events in human cells are distinct from

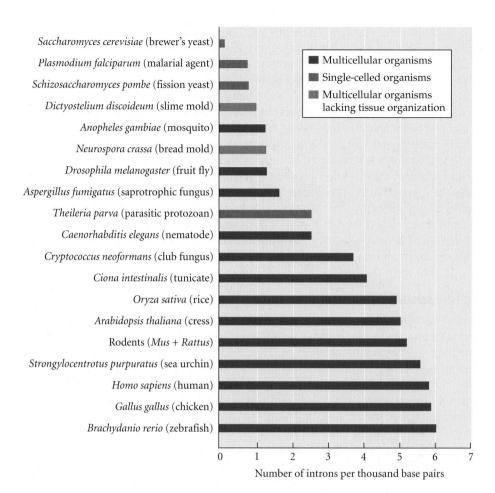

FIGURE 20.9 Phylogenetic distribution of introns, showing the number of introns per kilobase of genomic DNA in 19 eukaryotes. Note the increasing number of introns in moving from unicellular eukaryotes such as *Saccharomyces cerevisiae* to more complex ones. (Compare with comparable data for gene number, shown in Figure 3.27.) (After Carmel et al. 2007.)

those occurring for the same genes in mice (Mudge et al. 2011), and recent research shows that the pattern of alternative splicing can evolve surprisingly quickly (Barbarosa-Morais et al. 2012). Research on *Arabidopsis* suggests that paralogous genes derived from whole-genome duplication (described below) have often diverged in their pattern of alternative splicing. Additionally, alternative splicing occurs in an environment- and tissue-dependent manner in *Arabidopsis*, suggesting a link between alternative splicing and functional diversification (Zhang et al. 2010). Alternative splicing appears to be as important as amino acid substitution in the functional diversification of proteins. Another finding with implications for molecular evolution is that alternatively spliced exons have a faster rate of nonsynonymous substitution (d_n) than do exons that are always included in the mature protein (Chen et al. 2006). This faster rate of nonsynonymous substitution may indicate that the functional constraints on alternatively spliced exons are weaker than those on constant exons, perhaps because the former are less likely to encode important protein domains.

Viral and microbial genomes: The smallest genomes

According to life history theory (see Chapter 14), rapid growth and early reproduction are advantageous in organisms whose populations are frequently expanding to new niches or hosts. A small genome can be replicated more quickly than a large one and thus facilitates rapid reproduction in single-celled organisms. Indeed, many viruses and bacteria have streamlined their genomes by doing away with many genes. Some of them, by exploiting the genomes of their hosts, make do with extremely small and focused genomes. Many viruses, for example, carry genes only for replication of their genome, for construction of their outer coat (capsid), and for integration into the host genome. The RNA genome of HIV, for instance, has only 9.8 kb, encoding nine open reading frames (genes). The bacterium *Mycoplasma genitalium*, a parasite of the genitalia and respiratory tracts of primates (including humans, in which it causes sexually transmitted diseases), has a genome that is only about 580 kb, containing 468 protein-coding genes. These genes primarily govern basic molecular and metabolic functions as well as adaptations for parasitic life, such as variable surface proteins that evade the host's immune system (Razin 1997).

As we saw in Chapter 3, *Buchnera*, which has been an intracellular symbiont of aphids for the past 200 My, has lost more than 2000 genes relative to the common ancestor it shares with its nonparasitic relative *E. coli*. Intriguingly, the gene order has remained constant among *Buchnera* strains, even though they have been diverging from one another for about 150 My. This constancy may be a consequence of the loss of many ribosomal RNA genes and transposable elements that in other bacterial lineages can facilitate recombination between nonhomologous regions of the genome and rearrangement of genes (see Figure 8.7). A recent comparison of seven closely related strains of *Buchnera* revealed a high rate of spontaneous point mutations and deletions, many of which are fixed by genetic drift in the small populations of this symbiont and go on to disrupt formerly functional genes (Moran et al. 2009).

Repetitive sequences and transposable elements

A substantial fraction of the genome of a typical multicellular eukaryote—nearly half the human genome and about 34 percent of the *Drosophila* genome—consists of repeated DNA sequences that are referred to as low repetitive, middle repetitive, and highly repetitive DNA, depending on copy number. These sequences are also termed SATELLITE DNA because of their location in a chemical gradient in a centrifuge. A major source of repeated DNA in human and other mammalian genomes is transposable elements (TEs), sequences that can copy and transpose themselves, or "jump," to other regions of the genome (see Chapter 8). Most TEs are retroelements that produce an RNA transcript, which is then reverse-transcribed and integrated into the host genome as DNA. Many TEs do not contribute to the development or function of the host organism; rather, they encode only proteins essential for their own replication and transposition and are thus examples of SELFISH GENETIC ELEMENTS, or "selfish genes." An increasing number of such elements, however, are now known to have resulted in novel functions for the host and have thus been a potent source of genomic innovation. For example, David Haussler and colleagues discovered a short interspersed nuclear element (SINE) that was active and proliferating in the ancient lineage of lobe-finned fishes represented by the coelacanth (Bejerano et al. 2006; see Figure 22.7B). Copies of this SINE were retained throughout vertebrate evolution and have been co-opted as a functional enhancer regulating the expression of the mammalian developmental gene *ISL1* (**Figure 20.10**). This SINE is also conserved in some human exons and is alternatively spliced in some cases. This example shows that some anciently proliferated TEs, many of which lie quiescent in the genome, can later be co-opted to perform important functions.

The dynamics of the evolution of TEs can be inferred from their presence in different clades. For example, *Alu* elements—a type of SINE—are abundant in hominoid primates, to which they are restricted. Since they are not found in other Old World primates, they must have originated in an ancestor of the hominoid lineage after its divergence from those primates, about 25 Mya. There are more than 500,000 copies in the human genome, making up over 10 percent of human DNA. The number of point mutations observed among hundreds of *Alu* elements scattered across the human genome implies that they underwent an ancient proliferation about 40 to 50 Mya and have since slowed down in their rate of transposition (**Figure 20.11**).

Variation among genomes in the number of repetitive elements goes a long way toward explaining the **C-value paradox**: the finding that organismal complexity and total genome size do not display a tight correlation, at least within eukaryotes (see Figure 3.26 and Chapter 22). The number of TEs, and total genome size, for a species such as corn (maize, *Zea mays*) vastly exceeds that for humans, even though humans are generally considered more complex organisms than corn.

Many transposition events are deleterious because they affect the function of genes near or within which they are inserted. TEs tend to be found in regions between genes

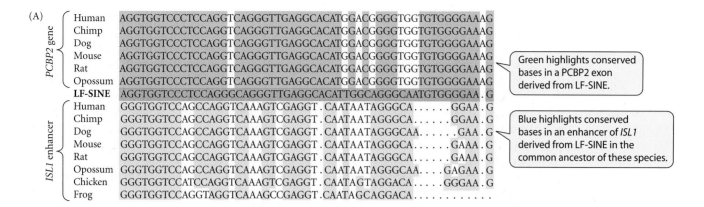

(A)

Green highlights conserved bases in a PCBP2 exon derived from LF-SINE.

Blue highlights conserved bases in an enhancer of *ISL1* derived from LF-SINE in the common ancestor of these species.

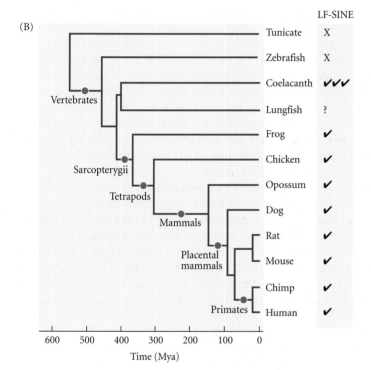

(B)

FIGURE 20.10 Conservation and evolution of a novel SINE in vertebrates. Many copies of the LF-SINE (lobe-fin SINE) are found in the genome of the Indonesian coelacanth (*Latimeria menadoensis*), a lobe-finned fish. (A) Conservation of the LF-SINE sequence in the genomes of various vertebrates. *PCBP2* is an RNA-processing gene; the green sequences indicate conserved bases in an exon of *PCBP2*. *ISL1* is a gene encoding a homeobox transcription factor promoting neuronal development; the blue sequences are conserved bases within a putative enhancer of *ISL1* in a noncoding genomic region of diverse vertebrates. (B) Phylogenetic distribution of LF-SINE. Check marks indicate the relative abundance of the element in various chordate genomes; "X" indicates that the element is not present. (After Bejerano et al. 2006.)

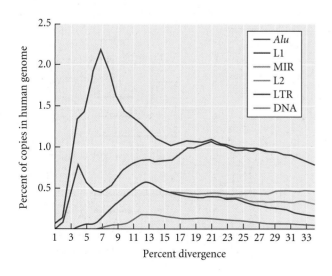

FIGURE 20.11 Age distribution of retroelements in the human genome. The x-axis shows the percent divergence between two sampled copies of a retroelement; the y-axis shows the percentage of copies with that degree of divergence. Distributions are shown for six different retroelements: *Alu* elements, L1 and L2 long interspersed nuclear elements (LINEs), mammalian interspersed repeats (MIRs), long terminal repeats (LTRs), and DNA transposons (DNA). If degree of divergence can be used as a rough proxy for time, we can conclude that many *Alu* elements underwent a burst of proliferation about 40 to 50 Mya (showing 7 percent divergence), but had lower rates of proliferation at earlier and later dates. By contrast, L1 LINEs seem to have had a protracted proliferative phase a very long time ago, with a second spike in numbers corresponding to roughly 4 percent divergence (about 25 Mya). MIRs and L2 LINEs are uniformly ancient and on their way to degradation and loss from the genome. (After Deininger and Batzer 2002.)

and in introns, probably because those that occur in coding regions often cause deleterious mutations and are eliminated by purifying selection. Transpositions can have at least two kinds of genetic effects. First, they can cause mutations: 10 percent of all genetic mutations in mice are thought to result from retroelement transpositions, many of them into coding regions or control regions. Second, the repeated copies of a TE in different parts of the genome can provide templates for nonhomologous recombination, resulting in chromosome or gene rearrangements that often cause deletions of some genetic material (see Figure 8.7). Approximately 0.3 percent of all human genetic disorders, including many hereditary leukemias, are thought to arise in this way (Deininger and Batzer 2002). The reduction of host fitness that is due to deleterious mutations and chromosome rearrangements is thought to be the chief reason that TEs are not even more abundant in host genomes than they are (Charlesworth and Langley 1989).

Natural Selection Across the Genome

As described in Chapter 12, the rapid rate of fixation of advantageous mutations is key evidence of positive selection on DNA sequences. For example, whereas synonymous substitutions accumulate at a rate governed by the mutation rate, nonsynonymous substitutions that are advantageous can accumulate much faster. If the ratio of nonsynonymous substitutions to synonymous substitutions, denoted as d_n/d_s, or ω, exceeds 1 for a protein-coding gene, we can conclude that the number of advantageous substitutions has exceeded the number of neutral substitutions, resulting in a net trajectory marked by positive Darwinian selection. By contrast, if ω is less than 1, purifying selection against novel amino acid substitutions is inferred.

As recently as a few years ago, it was feasible to measure positive selection in only one or, at best, a few genes at a time. Now, with the ability to compare multiple genomes, we can measure the rates of synonymous and nonsynonymous substitutions across thousands of genes and across multiple species. For example, researchers recently sequenced the complete genomes of 12 species of *Drosophila*, which allowed them to study patterns of adaptive evolution of genes in the context of a phylogeny of species as well as across classes of genes with different biological functions. The *Drosophila* 12 Genomes Consortium (Clark et al. 2007) showed that most gene classes among the 12 species of *Drosophila* were under stabilizing selection, with values of ω of less than 1. Several gene classes, however, including genes involved in immune defense and proteolysis, and a class containing genes of unknown function, had higher average values of ω and contained some genes with ω values greater than 1.

The 12 *Drosophila* genome sequences enabled researchers to study the evolution of coding regions in exquisite detail. For example, the consortium was able to measure the

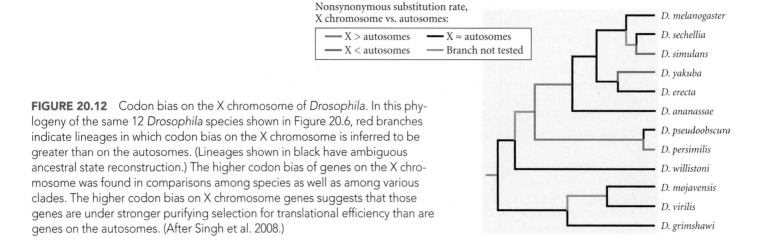

FIGURE 20.12 Codon bias on the X chromosome of *Drosophila*. In this phylogeny of the same 12 *Drosophila* species shown in Figure 20.6, red branches indicate lineages in which codon bias on the X chromosome is inferred to be greater than on the autosomes. (Lineages shown in black have ambiguous ancestral state reconstruction.) The higher codon bias of genes on the X chromosome was found in comparisons among species as well as among various clades. The higher codon bias on X chromosome genes suggests that those genes are under stronger purifying selection for translational efficiency than are genes on the autosomes. (After Singh et al. 2008.)

extent of both positive and purifying selection on genes on the X chromosome and on autosomes in various ways. Theory predicts that, because male fruit flies have only a single ("hemizygous") X chromosome, novel adaptive or deleterious mutations on the X chromosome in males will not be masked by another allele because often the Y chromosome does not have a corresponding orthologue. As a result, rates of positively selected mutations should be higher, and rates of deleterious substitutions lower, on the X chromosome than on the autosomes.

Researchers have shown that codon bias is a sensitive indicator of the extent of pervasive but weak purifying selection because it indicates that genes have been under selection pressure to maximize their translational efficiency. The *Drosophila* consortium found an overwhelming signal of greater codon bias for genes on the X chromosome than for genes on the autosomes, consistent with the idea that purifying selection (for more efficient translation) is stronger on the X chromosome (Clark et al. 2007; **Figure 20.12**). In general, however, genome-scale data such as that in *Drosophila* highlight the fact that the mechanisms promoting adaptive evolution are complex and that the fixation of various types of substitutions involves a complex interplay of demographic and selective forces.

Molecular convergence as evidence for natural selection

The many examples of convergence among organismal traits are a testament to the pervasive action of natural selection on morphology. One might expect convergence at the molecular level to be common as well, but evidence for convergence of specific amino acid sites in proteins has been surprisingly sparse, perhaps because of the many redundancies at the level of the genome. Such redundancies include that of the genetic code, as well as the redundancy that can arise when a gene is duplicated, resulting in two genes, or when different genes share similar functions. The first solid evidence for convergent evolution of functionally important amino acid sites came in the early 1990s for the gene encoding lysozyme, an enzyme that digests bacterial cell walls. In animals with foreguts, lysozyme helps regulate the abundance of foregut bacteria that help break down the cell walls of ingested plant material. In leaf-eating langur monkeys and cattle, two lineages that evolved foregut fermentation independently, the lysozyme gene has undergone convergent amino acid substitutions. The lysozyme gene in the hoatzin, a leaf-eating bird found in South America, has also experienced substitutions convergent with those found in cattle and langurs. In recent years, other cases of convergence at the molecular level have been found, each time associated with a highly specialized behavior or trait. These examples provide insight into how natural selection acts at the molecular level, and they highlight the similarities and differences between convergence at the molecular and organismal levels.

Harold Zakon, David Hillis, and colleagues have investigated convergent evolution of voltage-gated sodium ion channels in several diverse clades. These proteins facilitate electrical activity in the nervous system and muscles of animals, creating channels comprised of four domains (DI, DII, DIII, and DIV) that reside in cell membranes and form a pore through which sodium passes when the channel is open. Sodium ion channels themselves evolved convergently from the same calcium-permeable channel at the origin of the nervous system in metazoans (Liebeskind et al. 2011). Bilaterian animals and jellyfish each sustained the same amino acid substitution (glutamate to lysine) that confers sodium permeability through the pore. However, the substitution occurred in domain DII in the jellyfish lineage and in domain DIII in the bilaterians (**Figure. 20.13A**), indicating that convergence in function occurred via surprisingly similar but not identical means.

In electric fishes, sodium channels have undergone further convergent evolution in both sequence and expression (Arnegard et al. 2010). Electric fishes have an electric organ located in the tail that emits weak electrical signals used for communication and detecting prey. Phylogenetic analysis shows that electric organs evolved convergently from muscle in two divergent groups of electric fishes, the gymnotiforms of South America and the mormyriforms of Africa. One type of sodium ion channel, $Na_v1.4$, is expressed in muscle

(A)

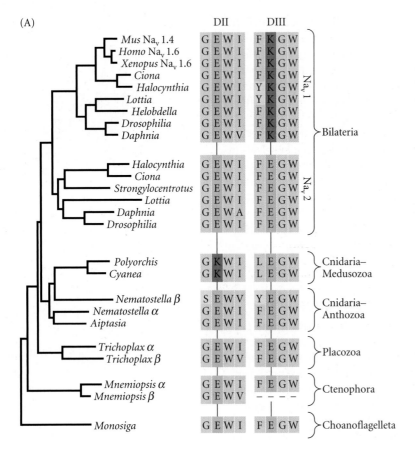

(C)

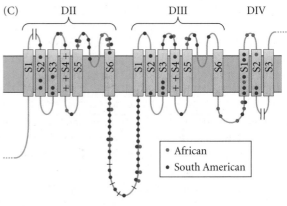

(B)

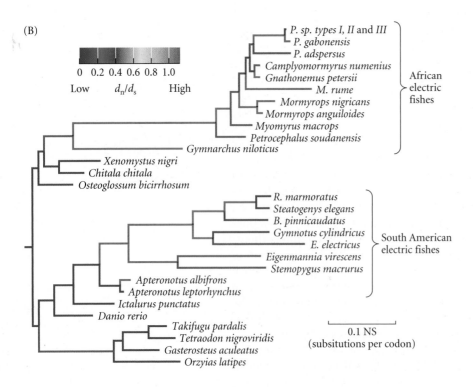

FIGURE 20.13 Convergence in sodium ion channels of metazoans (A) and electric fishes (B, C). (A) A phylogeny of sodium ion channels from diverse metazoans, including choanoflagellates and ctenophores as outgroups. Paralogous gene copies (α and β) are present for some taxa; other paralogues are designated as sodium ion channels 1 and 2 (Na$_v$1 and Na$_v$2), and taxa bearing these copies appear twice in the tree. On the right are amino acid alignments of small portions of domains DII and DIII, showing the convergent lysines (K; highlighted in red) in Medusozoa (jellyfishes) and Bilateria. These substitutions increase the channel's selectivity for sodium. (B) The Na$_v$1.4a sodium ion channel genes in African (mormyriform) and South American (gymnotiform) electric fishes have higher numbers of nonsynonymous substitutions than those in nonelectric fishes. Colors along the branches indicate the ratios of nonsynonymous to synonymous mutations (ω = d_n/d_s) in each lineage according to the scale at top, with red indicating the highest ratios. Brackets at right indicate genes that are expressed exclusively in the electric organ. (C) Positions of amino acid substitutions in the Na$_v$1.4a sodium ion channel expressed in the electric organs of electric fishes. In this diagram, membrane-spanning segments are depicted as numbered rectangles and extra- and intracellular loops as lines. Domains DII and DIII—each consisting of six transmembrane segments, S1–S6—form part of the channel pore. Domain DI and the last three transmembrane segments of domain DIV are not shown. Amino acid substitutions are indicated by red (African electric fishes) and blue (South American electric fishes) dots. Most of the substitutions occur in non-membrane-spanning segments. (A after Liebeskind et al. 2011; B after Arnegard et al. 2010; C after Zakon et al. 2006.)

across all vertebrates. The gene for this channel duplicated at the origin of bony fishes, giving rise to the paralogues $Na_v1.4a$ and $Na_v1.4b$. Both genes are still expressed in muscle in most extant bony fishes. In the two groups of electric fishes, however, $Na_v1.4a$ expression independently became restricted to the newly evolved electric organs, where the channel took on the new function of generating electrical signals. $Na_v1.4a$ also accumulated numerous amino acid substitutions in both lineages of electric fish independently (**Figure 20.13B**). These substitutions are not identical in the two lineages but nonetheless are in both lineages concentrated in regions of the channel protein that loop out of the cell membrane and serve to rapidly terminate the flow of sodium ions (**Figure 20.13C**). Variation in these substitutions lends species specificity to the electrical discharges emitted.

Molecular evolution in the human lineage

Genes that exhibit adaptive evolution on the branch leading exclusively to humans could be responsible for some of the traits that distinguish humans from other primates—such as human speech. We saw in Chapter 8 that the rate of evolution of the *FOXP2* gene, which has been associated with the capacity for language, has accelerated in the human lineage since it separated from chimpanzees. Another gene, *sarcomeric myosin heavy chain (MYH)*, is highly expressed in the chewing muscles of chimpanzees and is responsible for the large size of those muscles. In humans, this gene has been inactivated by a frameshift mutation that occurred about 2.4 Mya, after humans separated from chimpanzees (Stedman et al. 2004). The loss of function of *MYH* is thought to be responsible for the reduced size of chewing muscles in humans. In this example, ω increased from less than 0.1 to approximately 1, as expected if constraints were relaxed following loss of function and if the gene were evolving neutrally in the recent past.

Olfactory perception plays an important role in many mammalian behaviors, but primates, especially humans, possess a much less sensitive sense of smell than most other mammals. The olfactory receptor (OR) genes present a molecular example of the tendency in organisms for "structures of little use" to degenerate, as Darwin noted in *The Origin of Species*. There are approximately 1200 functional OR genes in the mouse genome, but only about 550 in humans. Apparently, selection pressure to maintain OR gene functionality has been relaxed in humans, and mutations that reduce gene function and cause pseudogene formation have been fixed (Gilad et al. 2003). Zhang and Webb (2003) found the same tendency in five genes that are normally expressed in the vomeronasal organ in primates. Mammals use this organ to perceive pheromones, which are used extensively in social interactions. Two distinct families of these genes are functional and conserved in several lineages of Old World primates (catarrhines), but these gene families have degenerated dramatically in hominoids, having accumulated numerous stop codons since the separation of hominoids from other catarrhines some 23 Mya.

Many researchers are specifically interested in that part of the human genome that does not encode proteins. Like the ultraconserved regions of the genome discussed earlier in this chapter, such regions may be rich in regulatory sequences that have influenced phenotypic evolution in humans and other species, but we know much less about such regions than we do about protein-coding regions. To search for regions of the human genome whose regulatory roles may have given rise to human-specific traits, McLean et al. (2011) scanned the human genome for regions that lie outside protein-coding genes and are highly conserved across mammals and other vertebrates, yet have been lost from the human genome. McLean and colleagues reasoned that deletion in humans of regions that have been conserved and are presumably functional in other vertebrates may have had an important effect on human evolution. They identified 583 such regions, which together account for 0.14 percent of the human genome. Many of these regions lie near genes involved in steroid hormone receptor signaling and neurological function. One such region was found to control a gene that is expressed in the human forebrain. The loss of an enhancer of this gene is strongly implicated in increased proliferation of cells in specific regions of the telencephalon. Such studies show that deletion of noncoding genomic regions can have an important role in the evolution of human-specific traits. They also show how information from other species can facilitate our understanding of modern humans.

Origin of New Genes

The approximately 24,000 different functional genes in mammalian genomes must have evolved from a much smaller number in the earliest ancestor of organisms. Presumably, all genes in the human genome ultimately descend from a single gene or set of genes that provided the first programs for life on Earth. Moreover, as we have seen, the number of functional genes differs greatly among major groups of organisms. How do new genes arise?

Evolutionary biologists have described several mechanisms by which the genes in a species' genome may have originated, either from pre-existing genes in the same genome or from the genome of a different species. These mechanisms include horizontal gene transfer, exon shuffling, gene chimerism, retrotransposition, domain multiplication, and gene duplication (Ding et al. 2012). New genes are now also known to arise directly from regions that were inferred to be noncoding in common ancestors.

Horizontal gene transfer

In the 1970s, Carl Woese (1928-2012) analyzed 16S rRNA sequences and found that the tree of life is divided into three major "empires" or "domains": the Bacteria, Archaea, and Eucarya (see Figure 2.1). In Woese's analysis, the Archaea and Eucarya appear to be sister clades. This phylogeny is supported by many gene sequences, but some gene sequences seem to indicate a sister-group relationship between Eucarya and Bacteria instead. As we saw in Chapter 2 (see Figure 2.24), when different genes provide strong support for different phylogenies, **horizontal gene transfer** (**HGT**, also called **lateral gene transfer**, or **LGT**) between lineages is a likely hypothesis. It is now thought that genetic material was transferred many times across quite different lineages early in the history of life, although most authors believe that, despite extensive HGT, an overarching phylogenetic tree for bacteria can be reconstructed. HGT also occurs frequently between the organellar and nuclear genomes of eukaryotes. Plant mitochondrial DNA exhibits extensive diversity in structure and gene content, much of it modulated by HGT to the nucleus. For example, the mitochondrial genomes of some plant species are relatively small (about 200 kb) and have experienced massive transfer of genes to the nucleus (Adams and Palmer 2003), whereas other plant mitochondrial genomes, such as that of cucumber, reach several megabases in length. These mitochondrial genomes, which are larger than the genomes of most bacteria, have retained most of their genes and experienced extensive proliferation of TEs as well (Sloan et al. 2012).

Horizontal gene transfer has also occurred more recently; in fact, HGT among microbial lineages is thought to be an important source of gene functions that are novel for the recipient organism. For example, the eukaryotic protist *Entamoeba histolytica*, which causes more than 50 million cases of amoebic dysentery annually, can live anaerobically in the human colon and in tissue abscesses because it produces fermentation enzymes that most other eukaryotes lack. A phylogenetic analysis showed that several of this protist's fermentation genes were obtained by HGT from archaea (Field et al. 2000). Genome sequence data suggest that HGT may be quite frequent among prokaryotes and that novel adaptive mechanisms, often borne on chromosomally distinct plasmids, are especially likely to spread by HGT (Ochman et al. 2000). In addition, endosymbiotic associations, such as that of *Wolbachia* bacteria with many insect species, provide a context conducive to HGT. Massive HGTs, involving segments ranging from fewer than 500 base pairs to nearly the entire 1 Mb *Wolbachia* genome, have occurred between *Wolbachia* and the nuclear genomes of four insect and four nematode hosts (**Figure 20.14**; Hotopp et al. 2007).

HGT also appears to occur with detectable frequency between microbes and vertebrates, and even among vertebrates. Perhaps 40 to 50 human genes are thought to have

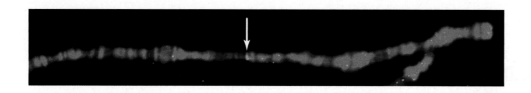

FIGURE 20.14 A polytene chromosome of *Drosophila ananassae* (red) with evidence of integration of a horizontally transferred gene from the intracellular symbiont *Wolbachia* (green). To produce this image, a segment of the *Wolbachia* genome was labeled with a green fluorescent dye and hybridized to the chromosomes of *D. ananassae*. The presence of *Wolbachia* sequences in the *Drosophila* chromosome is direct evidence of horizontal transfer of DNA from the symbiont to the host genome. (From Hotopp et al. 2007.)

their origins in bacteria (Salzberg et al. 2001). The recent sequencing of the genome of a lizard, *Anolis carolinensis*, revealed extensive horizontal transfer of transposons between diverse vertebrates, including primates, bats, and marsupials. These transposons showed low sequence divergence between species, and they differed primarily at synonymous sites. These observations suggest recent interspecific transfers, presumably mediated by viruses (Novick et al. 2010).

Exon shuffling, protein domain evolution, and chimerism

Horizontal gene transfer adds genes to the genome of a particular lineage, but those genes already exist in the source species. Several other mechanisms produce truly new genes. One of these mechanisms is **exon shuffling**. As we have seen, eukaryotic genes include introns and exons. There is often a close correspondence between the division of a gene into exons and the division of the protein it encodes into domains. A **domain** (or "module") is a small (about 100 amino acids) segment of a protein that can fold into a specific three-dimensional structure independently of other domains in the protein. Domains frequently have specific functions, although they usually cannot perform these functions completely in the absence of the other domains that together make up a mature protein. For example, antibodies function primarily through their immunoglobulin domains, two of which together form the major cleft into which foreign proteins fit during an immune response.

The approximate correspondence between exons and domains in some proteins has led to the hypothesis of exon shuffling, which states that much of the diversity of genes has evolved as new combinations of exons have been produced by nonhomologous recombination between genes. Manyuan Long and colleagues (2003) estimate that 19 percent of the exons in eukaryotic genomes have arisen from pre-existing exons via exon duplication and subsequent shuffling. The resulting genes can be considered **chimeric genes**, in that they consist of pieces derived from two or more different ancestral genes.

The first example of a chimeric gene was the *jingwei* gene (**Figure 20.15**), which is found only in *Drosophila teissieri* and *D. yakuba* and consists of four exons. The first three exons, which are homologous to exons in the *yellow-emperor* (*Ymp*) gene found in both these and other *Drosophila* species, have a length typical for *Drosophila* exons, whereas the fourth exon is as long as an entire typical gene. The sequence of this fourth exon is more than 90 percent similar to the entire coding sequence of the well-studied alcohol dehydrogenase (*Adh*) gene. But although *Adh* contains multiple introns, the fourth "exon" of *jingwei* lacks introns. Clearly, a retrotransposed copy of *Adh*—a copy that was first transcribed into RNA, had its introns removed, and was then inserted back into the genome as DNA—landed in the middle of the third intron of a duplicate copy of the *Ymp* gene in the ancestor of *D. teissieri*

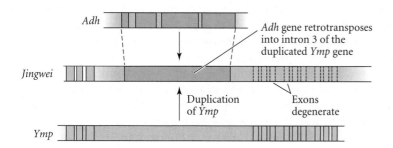

FIGURE 20.15 Origin of a new *Drosophila* gene, *jingwei*, via retrotransposition of a pre-existing gene into an intron of *Ymp* (*yellow-emperor*). Two copies of *Ymp* arose by gene duplication. An ancestral *Adh* gene was retrotransposed into the third intron of one of these copies of *Ymp* approximately 2.5 Mya. After the retrotransposition event, the exons downstream of the novel exon of *Ymp* degenerated because the *Adh* transcript provided a new stop codon. (After Long et al. 2003.)

(A)

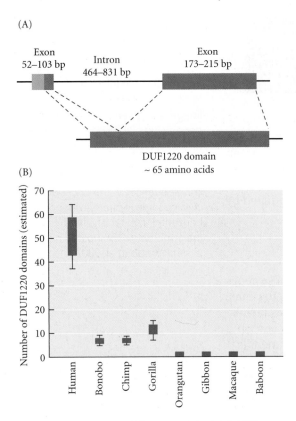

(B)

FIGURE 20.16 Amplification of the DUF1220 domain in the human lineage. (A) The relationship of exons to the DUF1220 domain in proteins. The domain is encoded by two exons in this example. (B) The number of DUF1220 domains in various primates, estimated by direct analysis of levels of *DUF1220* transcript. The domain is estimated to be 4 to 5 times more numerous in the human genome than in the genomes of the great apes and other Old World primates. (After Popesco et al. 2006.)

and *D. yakuba*. The exons derived from *Ymp* modified the expression pattern of the retrotransposed *Adh* gene, because *jingwei* exhibits the precise testis-specific expression pattern shown by *Ymp*. Thus *jingwei* illustrates not only gene chimerism, but also the important role that retrotransposition sometimes plays in the origin of new genes.

Many such retrotransposition events result in new, functional genes, such as *jingwei*, whose function is related to the metabolism of hormones and pheromones (Zhang et al. 2004). About 1 percent of the functional genes in the human and mouse genomes show signatures of retrotransposition, such as a lack of introns compared with their progenitor genes. Genes originating from retrotransposition are often actively transcribed, sometimes in novel tissues and contexts. But most retrotransposition events result in nonfunctional sequences called **processed pseudogenes**. Although their DNA sequences may resemble those of related genes (from which they were originally copied), processed pseudogenes commonly have deletions that destroy the reading frame of the gene and stop codons that occur before the correct termination point. Processed pseudogenes are common in the genomes of humans and other eukaryotes; in humans, there are at least 8000 processed pseudogenes.

Other genes have undergone multiplication of protein domains within genes. A striking example of domain multiplication may have contributed to the higher cognitive functions that distinguish humans from other great apes. A computational survey conducted by Magdalena Popesco and colleagues (2006) revealed 134 genes that appear to have undergone extensive within-gene domain multiplication in humans after their divergence from chimpanzees and gorillas. About 50 copies of a conspicuous protein domain, called DUF1220, were found across 34 of these human genes, whereas other primate species have fewer than 12 copies (**Figure 20.16**). These copies are marked by high rates of nonsynonymous substitution, implicating adaptive evolution. Furthermore, RNA transcripts containing the DUF1220 domain are expressed extensively throughout regions of the human brain associated with higher cognitive functions, such as the hippocampus and neocortex. The function of this domain is unknown at this time.

Gene Duplication

One of the most common ways in which new genes originate is by GENE DUPLICATION, in which new genes arise as copies of pre-existing genes. Many genes are members of larger groups of genes—called **multigene families** (or simply gene families)—that are related to one another by clear ancestry and descent. The members of multigene families often have diverse functions that nonetheless have a common theme. For example, globin genes (see Figure 8.3) all encode proteins that have a heme-binding domain and can bind oxygen. In mammals, the ε-, ζ-, and γ-globin chains have the highest oxygen affinities and are expressed in embryonic tissues, whereas the α- and β-globins function in the adult. The molecular mechanisms of gene duplication are still poorly understood, although we do know that it occurs at the DNA level (since the members of gene families usually have similar intron-exon structures), and unequal crossing over (see Figure 8.6) is known to change copy number in multigene families.

The relationships among members of a multigene family can be analyzed phylogenetically, both among and within species. These two kinds of relationships among genes represent two forms of homology: orthology and paralogy (see Figure 3.28). ORTHOLOGOUS genes are found in different organisms (e.g., different species) and have diverged from a common ancestral gene by phylogenetic splitting at the organismal level. In contrast, the members of multigene families are PARALOGOUS genes that have originated from a common ancestral gene by gene duplication. Multigene families vary greatly in size, from just 2 members (the most common number in both yeast and human genomes; Gu et al. 2002; Dujon et al. 2004) to 800 (e.g., the immunoglobulin superfamily) or more than 1000 (e.g., human ribosomal RNA genes). The sizes of multigene families often coincide with specific adaptations or loss of adaptations. For example, mammals, as we have seen, have hundreds of genes that encode olfactory receptor (OR) proteins, each of which binds one or a few odorant chemicals. In contrast, birds, most of which have a less acute sense of smell than the most sensitive mammals (e.g., rodents and dogs), have relatively few OR genes (Hillier et al. 2004); likewise, primates have fewer OR genes than do rodents.

The fates of duplicate genes

Duplicate genes can have several possible fates: (1) They may diverge in sequence and (usually) in function, in several different ways that we will describe shortly. (2) One copy may remain functional while the other becomes a nonfunctional pseudogene. (3) It is also possible for a locus to be deleted. Thus genes may undergo "birth" and "death," resulting in turnover of the membership of a gene family. (4) Finally, duplicate genes may undergo **gene conversion**, which occurs when sequence information from one locus is transferred unidirectionally to other members of the gene family, so that some or all family members acquire essentially the same sequence.

The molecular process of gene conversion is poorly understood in most organisms, but its consequence, **concerted evolution** of the gene family, is well known. It results in production of the same gene product from multiple loci, which can be adaptive if large quantities of the product are needed. Ribosomal RNA, a major component of ribosomes, is an example of a product encoded by a large multigene family. It would be deleterious if different transcripts of rRNA had different sequences. In another example, in abalone, the vitelline receptor for lysin (VERL) is expressed on the vitelline envelope of the egg and plays an important role in fertilization, helping to maintain species-specific fertilization (e.g., in *Tegula* marine snails; see Figure 18.14B). VERL, which consists of a series of amino acid motifs that have undergone extensive duplication but are still very similar in sequence within each species, is another example of concerted evolution (**Figure 20.17**), which in this case helps maintain species specificity of fertilization.

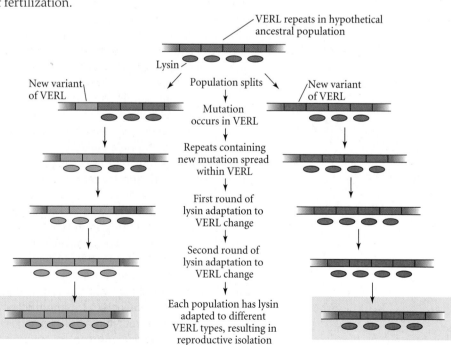

FIGURE 20.17 Concerted evolution of species-specific differences in VERL proteins. During fertilization, lysin, which is initially expressed on the acrosome of sperm, is released and then binds to VERL protein, in the vitelline envelope of the egg. Binding of VERL by lysin occurs in a species-specific manner, necessitating sequence changes in lysin when the conspecific VERL evolves. The several VERL repeats acquire the same new sequence by a process known as gene conversion. Lysin molecules adapt to VERL repeats until they match their species' VERL array. If enough divergence has occurred, lysin of one population will not interact properly with VERL of the other population, resulting in reproductive isolation and consequent speciation. (After Swanson and Vacquier 2002.)

Ohno's dilemma, molecular promiscuity, and the selective fates of recently duplicated loci

Paralogous genes are initially redundant: when gene duplication occurs, two identical copies of a gene suddenly exist in the genome, opening the possibility for functional diversification if one of the copies does not become a pseudogene. As we have just seen, selection for increased abundance of the product of a specific gene is one way in which gene duplicates can be selectively maintained before they acquire new functions.

In 1970, Susumo Ohno (1928–2000) first articulated the classic model of **neofunctionalization**, whereby one of the duplicates retains its original function and the other acquires a new function by the fixation of new mutations. Evidence of this process is the rapid accumulation of nonsynonymous substitutions in only one of the recently duplicated copies. This process will occur only if advantageous new mutations occur before one of the duplicates loses function and becomes a pseudogene, which happens if disabling mutations are fixed by genetic drift. Two genes in the ribonuclease gene family in primates, those encoding eosinophil cationic protein (ECP) and eosinophil-derived neurotoxin (EDN), provide an example of neofunctionalization. Jianzhi Zhang and colleagues (1998) found that the *ECP* gene experienced a large number of nonsynonymous substitutions after duplication from the *EDN* gene in the ancestor of primates Moreover, ECP possesses an antipathogen function that EDN lacks, suggesting that functional diversity was acquired after duplication by rapid accumulation of amino acid substitutions.

The requirement that advantageous mutations occur before one of the recently duplicated genes loses function or is deleted is stringent. On the other hand, maintenance of the original function of the recently duplicated gene by stabilizing selection would not allow for the accumulation of the novel amino acid substitutions required to achieve a new function. This perceived difficulty of gene duplicates ever achieving new functions via neofunctionalization is known as Ohno's dilemma. Although the neofunctionalization model was popular for many decades, in recent years scientists have realized that additional details are required to make it work.

One possible solution to Ohno's dilemma is **subfunctionalization**, whereby each gene duplicate becomes specialized for a subset of the functions originally performed by the ancestral single-copy gene. A model of this process, called the DDC model by Force and colleagues (1999; for duplication-degeneration-complementation), hypothesizes that an ancestral gene had two or more functions, and that in each paralogue, complementary mutations are fixed that reduce or eliminate a different one of these functions. The paralogues are therefore no longer redundant, so both are preserved by natural selection, and they may later undergo further functional specialization and evolutionary change. The DDC model differs from the classic neofunctionalization model in that *both* duplicate loci are expected to undergo changes in sequence and function compared with the ancestral gene. It also predicts a higher rate of retention of gene duplicates. In polyploid plants, more than 15 percent of gene duplicates tend to be functionally retained, a much higher rate than predicted by the classic model (Prince and Pickett 2002).

There is good evidence for a role for subfunctionalization in genome evolution. In the zebrafish, the single *Hoxb1* gene found in other vertebrates, such as mice, has been duplicated into *Hoxb1a* and *Hoxb1b*. Whereas mouse *Hoxb1* is expressed continuously in the developing hindbrain, zebrafish *Hoxb1a* and *b* are expressed sequentially, with *Hoxb1a* expression terminating about 10 hours after fertilization, at which point *Hoxb1b* takes over (Prince 2002; Prince and Pickett 2002). *Hoxb1a* and *b* have each lost single regulatory sequences upstream and downstream of the gene, both of which are present and functional in mouse *Hoxb1*. The complementary expression profiles and complementary degenerate mutations of the two zebrafish genes exemplify subfunctionalization (**Figure 20.18**).

Subfunctionalization does not explain the origin of *new* functions in gene duplicates, however, because the descendant gene duplicates have simply partitioned the functions of the ancestral gene. Selection for amplification of a gene product was also offered as a way out of Ohno's dilemma: if more of a specific gene product were advantageous, both identical gene duplicates would confer an advantage and would thus be retained, even if temporarily.

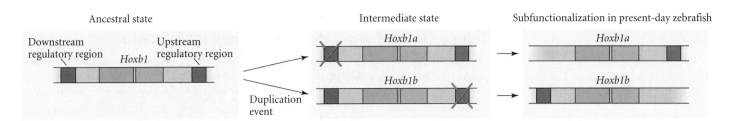

FIGURE 20.18 The DDC model of subfunctionalization as illustrated by Hox genes in zebrafish. In an ancestor of zebrafish, the *Hoxb1* gene had dual functions and was regulated by dual upstream and downstream regulatory regions. In an intermediate derived species, this gene underwent a duplication event, followed by loss of one regulatory region or the other and a change of expression pattern in each of the two novel paralogues (center). The derived state of the duplicated genes in the zebrafish (right) consists of two paralogues, each with a single function. (After Prince and Pickett 2002.)

Another factor, molecular promiscuity, may commonly help to retain recently duplicated genes. MOLECULAR PROMISCUITY refers to the minor, often unrelated, functions that a single gene product often has in addition to its primary function (Khersonsky and Tawfik 2010). Microbiologists have known for many years that many knockout strains of *E. coli*, in which a particular gene has been entirely deleted, are nonetheless viable. In many such cases, an unrelated enzyme has taken over the role of the knocked-out gene product, even while maintaining its primary function. For example, hydrogenases play an important role in metabolism by catalyzing the oxidation of molecular hydrogen (H_2). Mutants missing the usual hydrogenases can survive, however, because of the phosphite-dependent hydrogenase activity of the enzyme alkaline phosphatase, whose primary function is to remove phosphate groups from diverse molecules. An enzyme can also be promiscuous by binding multiple substrates in its active site, by assuming multiple three-dimensional structures in different cellular environments, or by expanding its specificity when novel hydrogen bonds, mediated by water, are formed.

Bergthorsson et al. (2007) proposed a model for the maintenance of duplicated genes, called the IAD MODEL (for innovation, amplification, and divergence), based on the growing evidence for widespread molecular promiscuity. In the IAD model, gene duplication provides an advantage to the organism through the product's promiscuous function (function B in **Figure 20.19**), rather than its primary function (function A), often in response to the organism's colonization of a new ecological niche. Duplicated gene copies are retained because more of function B is provided by the gene duplicates. However, when one of the new gene duplicates improves and begins to specialize in function B, selection on other duplicate copies of the gene (those with only a minor function B) is relaxed, and those copies may be deleted from the genome without consequences for fitness. The model predicts that duplication of identical genes can provide novel functions due to amplification of the minor gene function. It also predicts that selection acts continuously throughout the amplification and duplication process, resulting in high rates of amino acid substitution, such as that seen in the primate ribonucleases and the electric fish sodium ion channels mentioned earlier.

FIGURE 20.19 An example of molecular promiscuity. Each blue shape represents a particular conformation of a protein. The colored shapes nestled in these conformations are ligands (other molecules bound by the folded protein). The primary ligand is the main ligand bound most of the time by the protein (representing its major function). The promiscuous ligand is a minor ligand that is bound rarely (representing the protein's minor function). Arrows indicate transitions to different conformational states that bind alternative ligands.

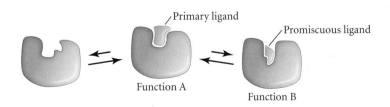

Multigene families and the origin of key innovations

Genomes record in remarkable detail the molecular underpinnings of some of the major transitions of life, and they also allow biologists to immediately grasp the molecular basis of novel adaptations. By sequencing the genomes of species that represent major lineages in the tree of life, biologists can trace the increasing complexity of life at the molecular level, and they have usually found that that complexity has its foundation in the expansion of specific multigene families. This is the case in the origins of multicellularity, a key event in several groups, such as animals, plants, and fungi.

In multicellular organisms, cells adhere to one another and exchange molecules that are critical for conveying information. A lineage that provides important insight into the origin of multicellularity is the choanoflagellates (**Figure 20.20A**; also see Figure 5.11). All of these simple organisms have a unicellular stage in their development, but some species can also occur as colonies in the adult stage. When feeding as single cells, choanoflagellates closely resemble the feeding cells of sponges, among the most primitive of metazoans. But choanoflagellates are not metazoans; they fall outside the sponges and cnidarians at the base of the metazoan tree. Thus the genome of choanoflagellates was predicted to shed light on the earliest stages of multicellularity.

(A)

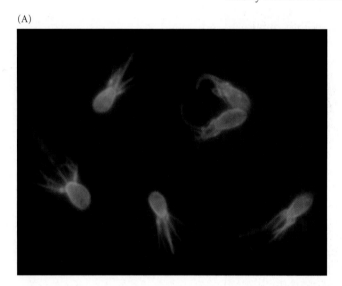

FIGURE 20.20 Ancient origin of cadherin genes as revealed in the genome of the choanoflagellate *Monosiga brevicollis*. (A) This image is a fluorescence micrograph of *M. brevicollis* cells that have been labeled with fluorescent dyes that bind specifically to different cellular components. (B) The structure of a fat-related cadherin protein in choanoflagellates and in various metazoans is shown next to a phylogenetic tree for these groups. This cadherin contains epidermal growth factor and laminin G domains (green boxes) as well as domains linked to extracellular cadherin repeats (in purple). (A, photo courtesy of N. King; B after Abedin and King 2008.)

(B)

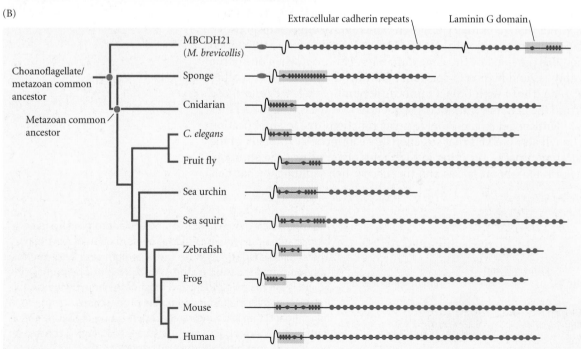

Cadherins are a class of molecules that play important roles in cell adhesion, movement, and communication. Before the first choanoflagellate genome was sequenced, cadherins were known only from metazoans. When Nicole King and colleagues (King et al. 2008) sequenced the first genome from a choanoflagellate (*Monosiga brevicollis*), they were surprised to discover 23 cadherin genes (Abedin and King 2008). The multifunctional nature of the cadherin multigene family was evident in the many types of domains found in the cadherin genes, including domains for growth factors and immunoglobulins (**Figure 20.20B**). These diverse types of cadherin genes encoded a variety of proteins that act as signaling molecules, transmitting information about changes in the environment. *Monosiga brevicollis* is not known to engage in any contacts between cells, so the molecules that enable communication between cells in metazoans must have originally had different functions, with later functionality added to an initial, simpler toolkit. Thus genomes reveal a pattern in which a trait with a set of current functions originally had different functions—a pattern frequently encountered in the history of life (see Chapters 3 and 22).

As new genomes are sequenced, particularly those representing clades that have not yet been studied at the genomic level, scientists learn more about the major environmental and evolutionary events that have shaped those clades. Often the genomic complexity of a particular species is unexpected. For example, the genome of the purple sea urchin (*Strongylocentrotus purpuratus*), a model organism for the study of embryogenesis and gene regulation, was sequenced in 2006 (Sodergren et al. 2006). A major surprise from this genome was the unexpected complexity of the urchin's immune system, as revealed by the number of genes with immune functions (Rast et al. 2006). It had been thought that among the deuterostome phyla, the immune systems of vertebrates differed from those of echinoderms (such as sea urchins) primarily in possessing genes for adaptive immunity—the arm of the immune system that is molecularly diverse, evolves rapidly, and sometimes undergoes somatic mutation in response to novel pathogens. The innate immune system, by contrast, is shared between vertebrates and other deuterostomes. The genome of the sea urchin was found to contain unexpected complexity in the innate immune system as well as the telltale beginnings of an adaptive immune system. For example, purple sea urchins posses 222 genes encoding proteins known as Toll-like receptors (TLRs), which exhibit little polymorphism but are important for combating pathogenic bacteria. By contrast, humans posses only about 10 of these genes (**Figure 20.21**). In addition, the sea urchin possesses recombination-activating genes *rag1* and *rag2*, hitherto unknown from invertebrates, which play an important role in coordinating the recombination and somatic diversification of T cell receptor genes, a crucial component of the adaptive immune system.

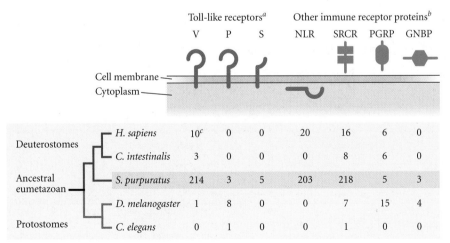

		Toll-like receptors[a]			Other immune receptor proteins[b]			
		V	P	S	NLR	SRCR	PGRP	GNBP
Deuterostomes	*H. sapiens*	10[c]	0	0	20	16	6	0
	C. intestinalis	3	0	0	0	8	6	0
Ancestral eumetazoan	*S. purpuratus*	214	3	5	203	218	5	3
	D. melanogaster	1	8	0	0	7	15	4
Protostomes	*C. elegans*	0	1	0	0	1	0	0

[a] Toll-like receptors include vertebrate type (V), protostome type (P), and short type (S).
[b] NLR = NACHT domain/LRR proteins; SRCR = scavenger receptor cysteine-rich proteins; PGRP = peptidoglycan (bacterial cell wall) recognition proteins; GNBP = Gram-negative bacterial binding proteins.
[c] Plus 1 pseudogene.

FIGURE 20.21 Genes encoding several types of immune receptor proteins have been found in the genome of the purple sea urchin (*Strongylocentrotus purpuratus*). The diversity of these genes rivals or exceeds that in the genomes of other metazoans, indicating unexpected complexity of innate immune defenses in sea urchins. (After Rast et al. 2006.)

Although T cell receptors have not been discovered in sea urchins, the *rag* genes suggest that sea urchins may harbor some capacity to engage in adaptive immunity like their vertebrate relatives.

Genome and Chromosome Duplication

Many genes get duplicated as parts of large chromosomal blocks, or even as parts of whole genomes (polyploidy; see Chapter 18). **Copy number variants** are an important source of variation within species, including humans. Such variants are also called SEGMENTAL DUPLICATIONS because in these cases, a segment of the genome, often containing many genes, is duplicated. Some such variants are implicated in human cancers and diseases. About 1500 regions, spanning about 12 percent of the human genome, consist of duplicated chromosome segments that have diverged in sequence from their paralogues. Because sequence polymorphisms constitute only about 1 in 1000 sites in the human genome, copy number variants are a principle source of variation in humans. The Neanderthal genome sequence revealed 111 likely Neanderthal-specific segmental duplications distinct from those known in modern humans.

Segmental duplication is apparent when comparisons of DNA sequences within the same genome show that many genes seem to have been duplicated at approximately the same time. Whether an entire genome was duplicated in a "big bang" or was duplicated piecemeal over a long time can be determined by examining the age distribution of divergence between many duplicated paralogues, making judicious use of molecular clocks. For example, as mentioned in Chapter 3, it has been postulated that the entire genome was duplicated in the common ancestor of jawed vertebrates and then again in the ancestor of fishes (Prince and Pickett 2002). The distribution of the divergence times of 1739 gene duplication events inferred from a phylogenetic analysis of 749 different gene families in the human genome provided evidence for a "big bang" (a major peak in the frequency distribution at about 500 My ago, before the origin of fishes) (Gu et al. 2002; **Figure 20.22**). There was also a continuous distribution of later divergence times, including a peak near the basal radiation of mammals. Thus the large-scale structure of mammalian genomes may result from both a large-scale and many small-scale duplications.

FIGURE 20.22 Age distribution of gene duplication events in the human genome. The x-axis shows the estimated age of divergence of gene duplicates based on analysis of sequence differences. The y-axis shows the number of gene duplicates in each age class. The figure shows three waves of duplication, I, II, and III. Wave I corresponds to gene duplications that occurred during the radiation of mammals and consists of certain large gene families, such as the immunoglobulin family in mammals. Wave II corresponds to duplications that took place during early vertebrate evolution and consists of tissue-specific isoforms and other developmental loci. Wave III corresponds to duplications that took place very early in metazoan evolution, when several novelties in signal transduction pathways are thought to have originated. (After Gu et al. 2002.)

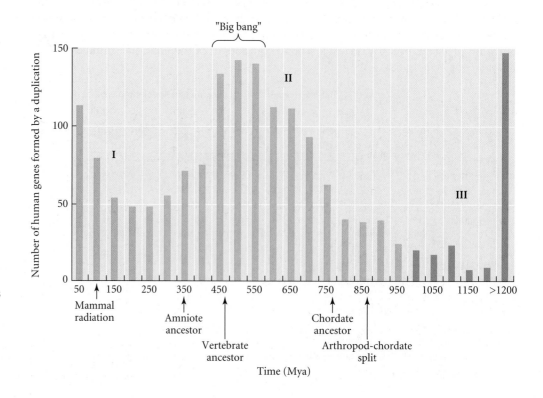

FIGURE 20.23 Complex patterns of genome duplication in banana and other monocots. (A) Circle diagram showing paralogous relationships within the banana genome. These relationships were generated by a whole-genome duplication event that gave rise to several chromosomal "blocks," including block 2 (red) and block 8 (green), shown here. Each segment around the circle represents a different banana chromosome (Chr1 to Chr11). (B) Orthologous relationships between banana genes in blocks 2 and 8 and genes in the rice genome. Three blocks of rice genes are shown (ρ2, ρ5, and σ6). Note that each segment of banana chromosome in this circle is linked to multiple lines leading to different segments of the rice genome. This indicates that paralogous regions resulting from a whole-genome duplication event in rice (or its ancestors) were retained in the banana genome. (After D'Hont et al. 2012.)

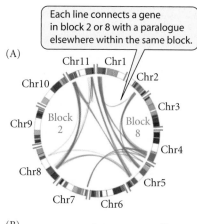

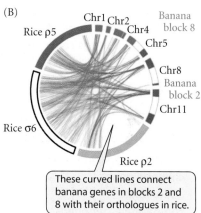

Genome duplication and reduction have played important roles in the evolution of the genes and genomes of plants, in which polyploidy has arisen very frequently (see Chapter 18). The genome of the black cottonwood (*Populus trichocarpa*), an important model species in the genetic study of woody plants, contains more than 45,000 protein-coding genes, about twice the number estimated for humans, even though its size is only about 500 Mb, which is only one-sixth the size of the human genome (Tuskan et al. 2006). About 8000 pairs of genes resulting from a whole-genome duplication continue to survive in the cottonwood genome, for reasons that are still unclear. For about half of the genes in *Arabidopsis*, the cottonwood has an extra paralogue that arose by duplication. In contrast to the cottonwood, *Arabidopsis* lost many genes after its genome underwent a duplication in an ancestral species. The recently sequenced genome of the banana (D'Hont et al. 2012) revealed extensive evidence of gene duplication as revealed in the phylogeny of multigene families. Paralogous regions of the banana genome show clear similarity to other regions of the banana genome as well as to chromosomal regions from other species (e.g., rice), indicating ancient duplications (**Figure 20.23**).

Copy number variants illustrate the complex interplay between recombination, selection, and gene and chromosome duplication in modulating evolution across the genome. This discovery, like others emerging in this new era of comparative genomics, not only illustrates the relevance of decades-old population genetic theory to genome-wide studies, but also shows that new types of molecular players, such as microRNAs and regulatory regions, are posing new explanatory challenges. With the flood of new data coming from genome sequencing projects, and with the genomes of nontraditional study species being sequenced in ever greater numbers, we can expect an increasingly detailed view of genome evolution in the years to come.

Summary

1. New technologies that have advanced the field of genomics have enabled biologists to study genomes on hitherto unprecedented scales, and they have revealed several new elements of genomes, such as diverse classes of untranslated RNA sequences. Patterns of ancestry and evolution can be detected from comparisons among the structure of thousands of genes and entire genomes. Vast new genomic data sets are informed and interpreted not only by decades-old population genetic theory, but also by newer coalescent theory and new ways of detecting natural selection.

2. A major hypothesis that has emerged in recent years suggests that many features of genomes, including their size, abundance of transposable elements, and intron lengths, are the result of nonadaptive processes brought to fixation by genetic drift rather than natural selection. This hypothesis predicts that in lineages with large population sizes, selection is able to act more efficiently and therefore override the effects of

drift that predominate in lineages with small population sizes. Although the hypothesis remains to be tested more extensively, and although it contains several contradictions, it underscores the importance of population genetic thinking for understanding broad-scale patterns of genome evolution.

3. Among the factors that influence the rate of protein evolution are purifying selection against misfolding of proteins, selection against misinteraction with other proteins, and the timing of gene expression during an organism's development.

4. Introns are prominent and often recently evolved components of eukaryotic genes. Alternative splicing provides a powerful source of variation in protein structure within and between species.

5. Genome size varies by several orders of magnitude among organisms. The C-value paradox refers to the discrepancy between genome size and organismal complexity in eukaryotes. It was resolved by noting that the coding portion of genomes

may increase with organismal complexity, whereas the noncoding portion, comprising highly repetitive DNA, transposable elements, and other types of "selfish DNA," varies with factors other than complexity.

6. Molecular convergence is a potent sign of strong natural selection on DNA and amino acid sequences of proteins. Molecular convergence often underlies intriguing adaptations of organisms and novel behaviors.

7. A species may acquire new genes through a variety of mechanisms. Horizontal gene transfer occurs when a gene is transferred between unrelated genomes, presumably by viruses or other genomic vehicles. Novel genes can arise from pre-existing genes by shuffling of exons, which can result in chimeric genes with novel functions. Novel adaptations do not require novel coding regions, but can sometimes arise from deletion of phylogenetically conserved noncoding regions.

8. Genes can also arise by retrotransposition, a process that can generate intronless genes and often produces nonfunctional processed pseudogenes. Such retrotransposed genes can sometimes produce new exons and regulatory regions up- and downstream, thereby gaining functions and expression patterns that differ from those of the progenitor gene.

9. Gene duplication is the major mode of growth of multigene families, which, along with alternative splicing, increases coding region diversity in genomes and hs enabled some of the major innovations in the history of life, such as the origin of multicellularity and immunity. Multigene families diversify in function via the processes of neofunctionalization or subfunctionalization.

10. Genes can be duplicated individually or as parts of large chromosomal regions, and sometimes as part of whole-genome duplications. Gene duplication is frequent in eukaryotic and prokaryotic genomes and provides an opportunity for the generation of novel genes with derived functions. New gene duplicates may be retained in genomes in part by molecular promiscuity, whereby a formerly secondary function of an ancestral gene becomes selected for in one of the descendant gene duplicates. Gene duplication can also occur as a by-product of whole-genome duplication (particularly in plants) as well as chromosome and segmental duplication, resulting in copy number variation of groups of genes.

Terms and Concepts

alternative splicing
chimeric gene
codon bias
comparative genomics
concerted evolution
copy number variants

C-value paradox
domain
exon shuffling
gene conversion
gene dispensability

horizontal (lateral) gene transfer
multigene family
neofunctionalization
processed pseudogenes

subfunctionalization
transcriptome
translational robustness
ultraconserved elements

Suggestions for Further Reading

The Origin of Genome Architecture, by Michael Lynch (Sinauer Associates, Sunderland, MA, 2007), provides a forceful argument for the importance of genetic drift and nonadaptive processes in the evolution of genomes.

Fundamentals of Molecular Evolution, by D. Graur and W.-H. Li (Sinauer Associates, Sunderland, MA, 2000), is a highly readable introduction to many aspects of sequence evolution. Many of these topics are treated in greater depth by A. L. Hughes in *Adaptive Evolution of Genes and Genomes* (Oxford University Press, Oxford, 2000).

In *A Primer of Genome Science*, third edition (Sinauer Associates, Sunderland, MA, 2009), G. Gibson and S. V. Muse provide an introduction to the techniques and promise of this important new field.

Problems and Discussion Topics

1. Codon bias is a phenomenon whereby certain codons are substituted more often in phylogenetic lineages than other (synonymous) codons that encode the same amino acid. What might account for codon bias?

2. Why are highly expressed genes often characterized by a low rate of evolution?

3. What hypotheses could account for differences in the abundance of introns among species?

4. What are the major differences between viral, bacterial, and eukaryotic genomes? What demographic differences among species might account for some of the differences we observe at the level of whole genomes?

5. Distinguish the phylogenetic consequences of different scenarios involving gene duplication, speciation, and concerted evolution. How do these three processes interact to produce different types of phylogenies of multigene families?

6. Describe the difference between neofunctionalization and subfunctionalization of a newly duplicated gene.

7. Give an example of an exaptation—a trait whose current function is different from its original function, or a trait that had no original function but has been co-opted to serve a functional role—at the level of the genome.

8. Give some examples of recently sequenced genomes that have revealed surprising complexity in species that were assumed to be simpler than mammals, for example.

9. Suppose you are discussing evidence for evolution with someone who does not believe in evolution. Describe three kinds of molecular data that provide evidence that different lineages (e.g., humans and chimpanzees) have evolved from a common ancestor.

Evolution and Development

The great morphological complexity and diversity of multicellular organisms is produced by developmental processes, occurring throughout the life history of individuals, that have evolved in response to natural selection. But how do these developmental processes evolve? The evolutionary forces and genetic mechanisms promoting biodiversity, such as the rich panoply of form and color in angiosperm flowers, have mystified biologists for more than a century. As we now know, flower structures are arranged in concentric "whorls" whose underlying development depends on the localized expression, either singly or in overlapping pairs, of specific sets of developmental genes. Clues to the existence of these genes came initially from studies of flower mutants in unrelated groups of plants. Comparisons of the morphology and development of mutants with that of wild-type strains have been highly informative pursuits in both classical developmental biology and modern evolutionary developmental biology. Examining the effects of similar flower mutations in multiple species raises several questions: What are the developmental processes that enable complexity to arise, and how do those processes evolve to result in diverse forms of a structure such as a flower? What happens when these processes evolve independently in different lineages? These questions are closely interwoven with the selective processes underlying the evolution, conservation, and diversification of biological form, themes that recur throughout this text.

The field of **evolutionary developmental biology**, or **EDB** (often called "Evo-Devo"), seeks to understand the mechanisms by which development has evolved, both in terms of developmental processes (for example,

Flower structures evolved from leaves The underlying homology between leaves and flower parts was first proposed in 1790 by the renowned German poet, statesman, and scientist Johann Wolfgang von Goethe. Angiosperm flowers, such as these petunias, have become important models for uncovering how diversification of a shared developmental patterning system leads to spectacular morphological diversity.

what novel cell or tissue interactions are responsible for novel morphologies in certain taxa) and in terms of evolutionary processes (for example, what selection pressures promoted the evolution of novel morphologies). Two of the main questions or themes that concern evolutionary developmental biologists are, first, *what role has developmental evolution played in the history of life on Earth?* and second, *do the developmental trajectories that produce phenotypes bias the production of variation or constrain trajectories of evolutionary change?* Natural selection acts on phenotypes produced by development, but ultimately we want to understand how the developmental processes that create those phenotypes affect evolutionary potentials and trajectories. By deciphering the evolution of these developmental processes, we hope to understand how the key morphological traits that define higher taxa first arose. Thus EDB provides a long-sought avenue for unlocking the secrets of macroevolution.

Hox Genes and the Dawn of Modern EDB

Biologists dating back to Geoffroy Saint-Hilaire (1772–1844), Karl Ernst von Baer (1792–1876), and Darwin himself were fascinated by the patterns of similarity and divergence in development among species. Until recently, the fields of evolutionary biology and developmental biology proceeded along mostly separate paths, with seemingly distinct research programs and methods (Gould 1977; Depew and Weber 1995; Wilkins 2002). In the past three decades, however, burgeoning information about the genetic mechanisms of morphogenesis in model organisms and the genome sequences of many additional species, as well as the molecular genetic techniques developed to obtain that information, have been integrated with many strands of evolutionary research to form the highly interdisciplinary field of EDB.

The discovery and characterization of the Hox genes in animals in the 1970s and 1980s marked the dawn of modern EDB (Wilkins 2002). The **Hox genes** are the best-known class of **homeotic selector genes**, which control the patterning of specific body structures, as we saw in Chapter 8. Hox genes control the identity of segments along the anterior-posterior body axis of all metazoans. Mutations in Hox genes often cause transformations of one type of segment into another. In *Drosophila melanogaster*, for example, a mutation of the *Ultrabithorax* (*Ubx*) gene transforms the third thoracic segment (T3), which normally bears tiny structures called halteres (the *Drosophila* homologue of the hindwing of four-winged insects), into a second thoracic segment (T2), which bears wings (**Figure 21.1**). A mutation in the Hox gene *Antennapedia* (*Antp*) causes the misexpression of Antp protein in the cells that normally give rise to antennae, resulting in the replacement of antennae with legs (see Figure 8.13).

In *D. melanogaster*, the Hox genes occur in two complexes (clusters) on chromosome 3, termed the Antennapedia complex and the bithorax complex. The pioneering genetic work on the bithorax complex was done between the 1940s and the 1970s by E. B. Lewis, and that on the Antennapedia complex in the 1970s and 1980s by Thomas Kaufman and his colleagues. These investigators found that the genes in both complexes control the anterior-posterior identity of segments corresponding to their order on the chromosome (**Figure 21.2**). They also discovered that the eight *Drosophila* Hox genes are members of a single gene family, and that the proteins these genes encode share a particular amino acid sequence that binds DNA. This sequence was subsequently named the HOMEOBOX (in the gene) or the HOMEODOMAIN (in the protein). These findings supported Lewis's idea, proposed in the 1960s, that the Hox genes regulate the transcription of other genes. Other researchers were stunned to discover that *all* animal phyla appear to possess Hox genes. These genes have homeobox sequences similar to those of their homologues in *Drosophila*; in addition, they have the same gene order and orientation as in *Drosophila*. Hox genes form a single complex in most animals. Mammals, however, have four paralogous Hox gene complexes (denoted *Hoxa*, *Hoxb*, *Hoxc*, and *Hoxd*) in different parts of the genome, and a total of thirteen different Hox genes (as opposed to only eight in *Drosophila*).

In *Drosophila*, staining for Hox proteins or mRNAs (using methods described in **Box 21A**) showed that the anterior-posterior expression of the Hox genes corresponds to their mutant

FIGURE 21.1 Effects of homeotic mutations. (A) A wild-type *Drosophila melanogaster* has a single pair of wings, borne on the second thoracic segment, and a pair of small winglike structures called halteres, borne on the third thoracic segment. (B) This mutant fly was experimentally produced by combining several mutations in the regulatory region of the *Ultrabithorax* (*Ubx*) gene. The third thoracic segment has been transformed into another second thoracic segment, bearing wings instead of halteres. (Photos courtesy of E. B. Lewis.)

(A)

Haltere

(B)

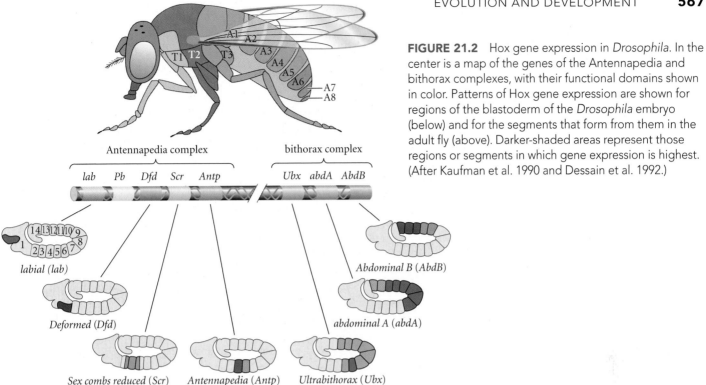

FIGURE 21.2 Hox gene expression in *Drosophila*. In the center is a map of the genes of the Antennapedia and bithorax complexes, with their functional domains shown in color. Patterns of Hox gene expression are shown for regions of the blastoderm of the *Drosophila* embryo (below) and for the segments that form from them in the adult fly (above). Darker-shaded areas represent those regions or segments in which gene expression is highest. (After Kaufman et al. 1990 and Dessain et al. 1992.)

Box 21A Characterizing Gene and Protein Expression during Development

Gene transcription occurs in SPATIO-TEMPORAL patterns defined by both spatial components (specific cells, tissues, segments, or structures) and temporal components (specific developmental stages). The three most common methods of visualizing transcription levels require different tools.

In IN SITU HYBRIDIZATION, tissues or whole specimens are subjected to a chemical process designed to stabilize mRNA molecules in the cells in which they are produced. Then a species-specific, single-stranded RNA or DNA PROBE corresponding to the gene of interest is applied to the specimen, where it hybridizes by base pairing with the mRNA of interest. The probe is either chemically modified so that it can be detected by a staining procedure or labeled with a radioisotope so that it can be detected by autoradiography (**Figure 1A**). Alternatively, extracts of mRNA from different tissues or developmental stages are run in separate lanes on an electrophoretic gel (see Chapter 9). The mRNA from the gel is then blotted onto a membrane, which is exposed to the same sort of chemically or radioactively labeled probe used in the in situ approach. This procedure is known as a NORTHERN BLOT.

A more easily quantified but less spatially detailed method of assaying gene transcription involves collecting mRNA from tissues or stages of interest and converting it into COMPLEMENTARY DNA (cDNA) using reverse transcriptase, then using a version of the polymerase chain reaction (PCR) technique in which

(A)   (B)   (C)

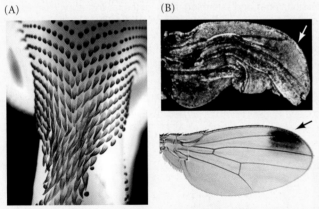

(C) Green fluorescent protein (GFP) expression (bright green) by a transgenic reporter construct containing *cis*-regulatory DNA from the gene *myo-2*, which directs expression in the pharynx of the nematode *Caenorhabditis briggsae*. (A, photo by Matthew Harris; B, photos by John True; C, photo by Eric Haag, courtesy of Takao Inoue and Eric Haag.)

FIGURE 1 Methods of visualizing gene expression patterns in developing animal tissues. (A) In situ hybridization reveals *sonic hedgehog* mRNA (dark blue areas) in developing feathers on the neck region of a chicken. (B) Fluorescent antibody staining of the proteins Yellow (in green) and Ebony (in pink) in the pupal wing of a male *Drosophila biarmipes* (upper photo). The *ebony* gene is expressed where the pigmented spot is located in the fully developed wing (lower photo).

(continued)

Box 21A *(continued)*

fluorescent reagents enable the measurement of accumulated reaction products in each amplification cycle. This method is called REVERSE TRANSCRIPTASE PCR (or REAL-TIME PCR; rtPCR).

Finally, fluorescently labeled samples from different tissues, individuals, species, or stages can be hybridized to MICROARRAYS, tiny rectangular chips onto which cDNA LIBRARIES from known samples are blotted by robots. Differences in fluorescence of specific locations on the arrays, quantified using digital imaging, represent differences in gene expression between samples.

Levels of protein expression can be analyzed using antibodies. (The mRNA and protein expression patterns of a given gene may not be identical due to translational regulation.) Antibodies are produced by injecting a mammal (e.g., a rat) with the protein of interest (the ANTIGEN). The animal produces antibodies (immunoglobulin molecules) that bind specifically to that protein. Tissues or specimens are prepared in a similar way as for in situ hybridization and incubated with the primary antibody. A secondary antibody, an immunoglobulin that specifically binds to the primary antibody, is then applied to the specimen. The secondary antibody is modified so that it can be detected either by an enzymatic reaction that produces a colored product or by fluorescence (**Figure 1B**). An alternative to this in situ technique is to prepare protein extracts from different tissues or stages and run each extract as a separate lane on an electrophoretic gel. The protein from the gel is then blotted onto a membrane, as in the Northern blotting technique described above, and that membrane is incubated with primary and secondary antibodies. This procedure, using proteins rather than mRNA, is known as a WESTERN BLOT.

Finally, the transcription patterns of particular fragments of putative *cis*-regulatory DNA can be studied using REPORTER CONSTRUCTS inserted into cultured cells or transgenic (genetically engineered) individuals. Reporter constructs consist of the regulatory DNA of interest, spliced upstream of a REPORTER GENE that encodes a protein whose expression can be easily visualized under the microscope. One such protein is β-galactosidase, a bacterial enzyme that processes a particular sugar into a blue product. Another is a protein from jellyfish (green fluorescent protein, GFP) that fluoresces bright green when irradiated with light of a particular wavelength. Because reporter construct analysis requires the use of gene transfer technology, it can be undertaken only in certain well-studied model species, such as *Drosophila*, *Caenorhabditis elegans*, *Arabidopsis*, and mice. **Figure 1C** shows the nematode *C. briggsae* expressing a GFP reporter construct containing *cis*-regulatory DNA from the *myo-2* gene, which directs the reporter gene's expression in the pharynx.

In genetic model species, the integration of genomic with genetic and developmental data provides a powerful base of knowledge for studies in evolutionary genetics and evolutionary developmental biology. Public online databases such as FlyBase (http://flybase.org) house many of these data, which are frequently updated and augmented. Among the many forms of data available on FlyBase for most of the genes in the *Drosophila melanogaster* genome are compendia of quantitative gene expression data, measured using microarrays (see above) across various tissues and developmental stages. **Figure 2** shows an example of a temporal expression profile for the *ebony* gene, which is required for both pigment development and eye function. As data from more species are added, many investigations connecting these gene expression data with the evolution of species differences will become possible.

FIGURE 2 Time course of *ebony* expression in *Drosophila melanogaster* as determined by microarray analysis. The developmental stage is indicated by the *x*-axis. Levels of transcription are indicated on the *y*-axis. Red bars indicate low expression and green bars indicate high expression, measured as the logarithm of the ratio of *ebony* expression to the average expression of a pool of all mRNAs in all stages of development. In adult stages at the right of the graph, the blue line indicates male expression, the purple line, female expression. (After http://flybase.org/; data from Arbeitman et al. 2002.)

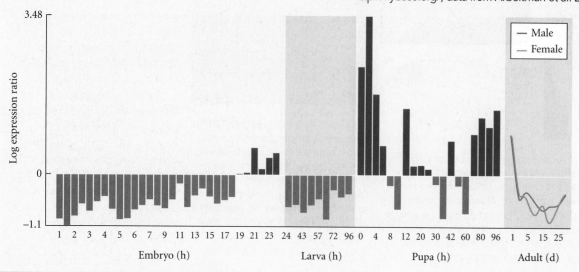

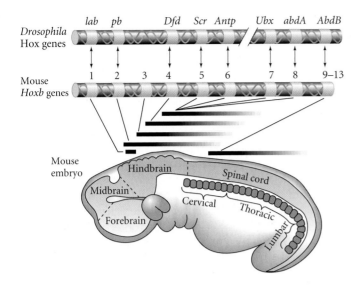

FIGURE 21.3 Segment-specific patterning functions of Hox genes in the vertebrate hindbrain. In this schematic diagram of a mouse embryo, the horizontal bars indicate segmental patterns of *Hoxb* gene expression in the hindbrain and spinal cord, with darker color corresponding to areas of relatively high gene expression. The double-headed arrows connect the genes in the *Hoxb* cluster to the homologous Hox genes in *Drosophila*. (After McGinnis and Krumlauf 1992.)

phenotypes. For example, as predicted, *Ubx* is expressed in the T3 segment (as well as the anterior abdomen), where it is required for normal segment identity (see Figure 21.2). Vertebrate Hox genes are also generally expressed in specific anterior-posterior patterns (**Figure 21.3**), although those expression patterns are more complex.

Mapping the presence and absence of Hox genes on the metazoan phylogenetic tree reveals their evolutionary history (**Figure 21.4**). At least four Hox genes are present in the radially symmetrical Cnidaria (jellyfishes and corals; Figure 21.4 shows the two best known of these genes), which are the sister group of the Bilateria. Several novel Hox genes arose in the lineage leading to all Bilateria, representing new Hox

FIGURE 21.4 Probable evolution of the metazoan Hox gene complex. Vertical white lines delineate currently accepted groups of orthologous Hox genes. Important gene duplication events are indicated by the labeled tick marks. Solid outlines indicate that the complete homeobox sequence for a gene is known; dashed outlines indicate that only partial homeobox sequences are known. Two genes in the Hox complex that are not considered Hox genes in *Drosophila*, *bcd* and *ftz*, are also shown. The orthologues of these two genes are Hox genes in other animals. (After Carroll et al. 2005.)

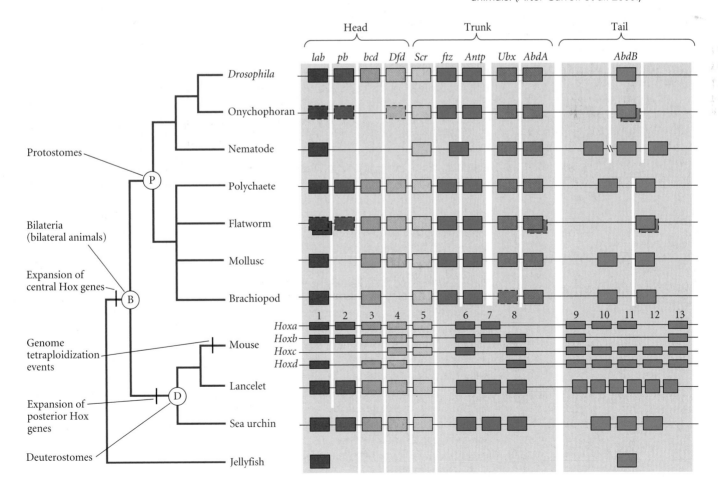

classes (as evidenced by their homeodomain sequences) that presumably can define increasing degrees of anterior-posterior segment identity. The presence of an extra set of Hox gene clusters in teleost fishes (which have seven to eight clusters, compared with most vertebrates, which have only four; see Figure 3.30) has also been proposed as an important causal factor enabling the great morphological diversification of the teleosts (Crow et al. 2006).

We can hardly overstate the importance of the Hox gene discoveries for our understanding of how animal diversity evolved. For the first time, a common developmental genetic framework unified the ontogeny of all metazoans; before then, few biologists imagined that vertebrates and invertebrates would share such fundamental developmental genetic underpinnings. These discoveries set the stage for further investigations into the potential commonalities among animals in other aspects of development (e.g., in dorsal-ventral patterning and in limb development), which have turned out to be plentiful (Wilkins 2002; Carroll et al. 2005). Ironically, such interest in morphological evolution was precipitated by the discovery of *conserved*, as opposed to *evolving*, features.

Evolution of Hox gene expression and function

The discovery of fundamental underlying similarities in the developmental patterning systems of many different taxa raised two new questions: What is the basis of body plan *differences* among taxa, and how can such seemingly conserved genetic factors play a role in those differences? These questions are among the most actively investigated in EDB today. For example, the *Ubx* gene was used to test an early hypothesis that major structural differences, such as hindwing differences between dipteran flies and butterflies, might result from simply turning transcription of a particular Hox gene on or off. Sean Carroll and his colleagues showed that the differences between the tiny *Drosophila* haltere and the large butterfly hindwing were, in fact, *not* due to differences in Hox gene expression; *Ubx* is expressed in both (Warren et al. 1994). Therefore divergence among taxa in hindwing morphology must be due to differences in the expression of other genes.

The Hox genes encode TRANSCRIPTIONAL REGULATORY PROTEINS (also called **transcription factors**) that regulate transcription by binding to the DNA control regions (called **enhancers**, **promoters**, or **cis-regulatory elements**) of "downstream" **target genes** (see Box 21B). Therefore it is likely that morphological divergence is caused by changes in the expression of genes that the Hox genes regulate. In fact, *Drosophila* and butterfly hindwings differ in the expression of several *Ubx* target genes (Carroll et al. 1995; Weatherbee et al. 1998). Further comparisons of Hox gene function in insects have revealed even deeper evolutionary insights. In the order Diptera, which contains *Drosophila*, the development of a single pair of wings on T2 requires no expression of *Antp*, the Hox gene whose expression defines that segment in other insects. The ancestral state of winged insects is to have two pairs of wings, one on T2 and one on T3, as in dragonflies. The key difference between dragonflies and dipterans is that in the latter, the Ubx protein represses wing development in T3, as noted above. This observation led to the general prediction that Hox genes in insects only modify segments toward non-wing development from an ancestral winged state. However, in the order Coleoptera (beetles), wings develop on T3, and elytra, which are modified wings that act as thick, hard coverings over the flight wings, develop on T2. Ubx is required to promote wing identity in the flour beetle *Tribolium castaneum* (Tomoyasu et al. 2005, 2009). When *Ubx* function was experimentally removed, elytra developed on T3. This observation indicates that Hox genes have been very flexible during the evolution of insects in the morphogenetic pathways that they are able to repress and recruit (wing repression and haltere development in flies, elytra suppression and wing development in beetles).

A second type of investigation has shown that differences in regions of Hox expression are strikingly correlated with the evolution of animal body plans. For example, in all the groups of crustaceans except the basal branchiopods, which have only one type of thoracic appendage, the anterior margin of expression of *Ubx* and *abdA* corresponds to the boundary between thoracic segments bearing maxillipeds (small, leglike appendages specialized for feeding) and those bearing thoracic limbs (see Figure 3.5). This observation suggests

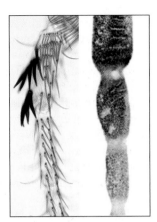

D. ficusphila   *D. nikananu*   *D. biarmipes*

FIGURE 21.5 *Sex combs reduced* (*Scr*) expression patterns are correlated with species differences in sex comb size in *Drosophila*. The photos show the structure of the mature leg (at left) and the distribution of the Scr protein in the developing leg (at right; dark areas, detected by antibody staining) for each species. *D. ficusphila* and *D. biarmipes* have sex combs on both the first and second tarsal segments, whereas *D. nikananu* has a sex comb only on the first tarsal segment. (Photos courtesy of Artyom Kopp.)

that change in the spatial expression of these Hox genes has enabled the segments and their appendages to become different (to become individualized; see Chapter 3, p. 63). Many correlations of this kind have been found throughout the Bilateria (Carroll et al. 2005), implying that evolutionary change in the expression patterns of Hox genes may underlie key body plan adaptations both within and among phyla.

Investigations of differences among more closely related species have also implicated genetic differences at Hox loci. For example, *Sex combs reduced* (*Scr*), which controls the identity of the first thoracic segment (T1) in insects, plays a role in some species in determining differences in a T1-specific bristle structure called the sex comb, which is found only in males and is thought to be used to stimulate females during copulation (Ng and Kopp 2008). From the early embryo stage in all flies, *Scr* is expressed throughout the developing T1 segment. But toward the end of development in the pupal stage in one lineage of the *Drosophila* genus, high levels of Scr are found only in the small region of the developing T1 leg that is destined to give rise to the sex comb (Barmina and Kopp 2007; Randsholt and Santamaria 2008). The pattern of *Scr* expression in each species reflects the size of the sex comb (**Figure 21.5**). The finding that sex comb development depends on the presence of Scr indicates that regulation of this gene has evolved quickly, resulting in morphological diversity among closely related species (Barmina and Kopp 2007). We will say much more about gene regulation later in this chapter.

New concepts of homology

As the Hox gene work illustrates, establishing homologies between structures in different taxa is a critical initial step in investigating the role of development in their evolution. The phylogenetic concept of homology, which is fundamental to all of comparative biology and systematics, states that homologous features are those that have been inherited, with more or less modification, from a common ancestor. In other words, homologous structures are synapomorphies (G. P. Wagner 1989b; Donoghue 1992) (see Chapter 2). We would therefore expect homologous structures to have similar genetic and developmental bases. Many observations conflict with these expectations, however, leading several evolutionary biologists to propose an additional concept, biological homology (Roth 1988, 1991; G. P. Wagner 1989a; P. J. Wagner 1996). The **biological homology** concept considers shared developmental genetic underpinnings to be an essential determinant of the homology of morphological structures in different taxa. Conflicts between phylogenetic and biological homology occur when phylogenetically homologous traits have different developmental and genetic foundations. For example, digits differentiate sequentially from back (postaxial) to front (preaxial) in all tetrapods except salamanders, whose digits differentiate in the reverse order. As another example, all vertebrate eye lenses contain various crystallin proteins, but the lens crystallins of different vertebrate lineages have evolved from a wide variety of different ancestral proteins (see Figure 11.19). No one doubts that all tetrapod digits and all vertebrate eyes are phylogenetically homologous, but because their developmental underpinnings have diverged in different lineages, they may not be fully biologically homologous between two given lineages. Conversely, developmentally and functionally similar structures in different taxa may not be phylogenetically homologous. In perhaps the best example, animal eyes evolved independently in very diverse taxa, but in all of those taxa, a highly conserved transcription factor, Pax6, controls eye development. We will examine this example in more detail later in this chapter.

The concept of biological homology suggests that a feature may be homologous among species at one level of organization (e.g., phenotypic), but not at another level (e.g., genetic or developmental). This observation implies evolutionary change in the building blocks of phenotypic traits, in a manner analogous to what occurs in a cathedral that has stood for centuries, but most pieces of which have been replaced over the years. It also suggests the idea that multicellular organisms are constructed from a set of more or less conserved tools. This **genetic toolkit** (Carroll et al. 2005) consists of individual genes and proteins as well as developmental pathways consisting of multiple proteins. Morphological evolution has consisted, in large part, of "tinkering" with this toolkit (Jacob 1977). SERIALLY HOMOL-OGOUS structures, such as different pairs of legs on an insect, share much of the same developmental genetic machinery in their ontogeny, but they are clearly not historically homologous within a species. In the rest of this chapter, we will see how this framework of gene regulatory evolution can help us understand how organismal form evolves.

Evidence of Developmental Evolution Underlying Morphological Evolution

In EDB, patterns of gene expression (revealed by techniques like those described in Box 21A) are frequently used together with morphological, comparative embryological, and phylogenetic data to infer the developmental genetic origins and histories of morphological characters. Developmental genetic data, such as information from mutant phenotypes or from individuals that have been genetically manipulated to under- or overexpress a gene or protein of interest, can definitively show that a particular gene is required for the development of a tissue or structure.

Recent work on the evolution of flowers among the angiosperms illustrates how information from mutant phenotypes can be used to form evolutionary hypotheses. In the flowers of most angiosperms, three classes of transcription factors encoded by the MADS-box gene family (named after a conserved sequence motif that all members bear)—designated A, B, and C—are required to pattern the four whorls (concentric rings) of distinct structures that make up a flower: sepals, petals, stamens, and carpels (**Figure 21.6A**; Coen and Meyerowitz 1991; Ma and DePamphilis 2000). The sepals are determined by A gene expression alone, the petals by overlapping A and B gene expression, the stamens by overlapping B and C gene expression, and the carpels by C gene expression alone. Loss of expression of a particular gene class during flower development results in conversion of one structure into another. For example, in *Arabidopsis thaliana*, mutations in the B-class gene *APETALA3* (*AP3*) convert petals to sepals and stamens to carpels. Similar mutant phenotypes and expression patterns of homologues of the ABC genes in a very distantly related angiosperm, maize (corn), indicate that the floral patterning system is ancient and have helped to confirm long-supposed homologies between floral structures in different plant taxa (Ambrose et al. 2000). Subsequently, expression of an additional class of MADS-box transcription factors, the E-class, was found to be required in order for the A, B, and C genes to function (**Figure 21.6B**), giving rise to the ABCE model.

Figure 21.6C shows a model, based on comparisons of genomes and gene expression from many plant taxa, that was recently proposed to explain the evolution of the ABCE system (Chanderbali et al. 2010). Gymnosperms, such as pines, employ a system involving orthologues of the B and C genes to determine the sex of cones (Figure 21.6C, bottom). B and C genes are expressed in the male cone and C genes alone are expressed in the female

(A)

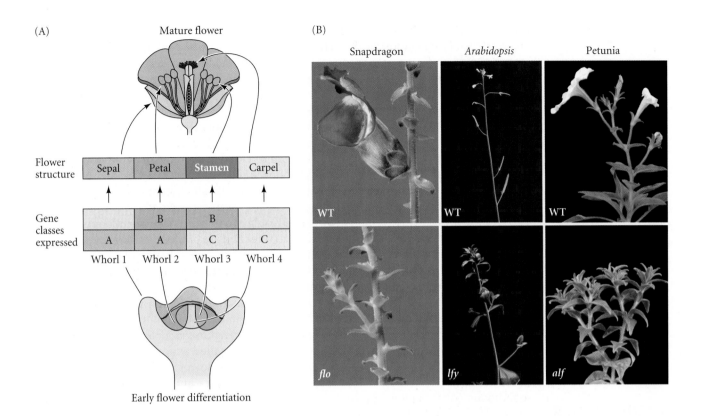

(B)

FIGURE 21.6 Flower patterning genes and flower development in angiosperms and gymnosperms. (A) The ABC system of flower development found in most angiosperms. The pattern of expression of three classes of flower patterning genes in early floral organ differentiation determines the organ identity of the four whorls in the mature flower. (B) Mutations in E-class flower patterning genes—*flo* in snapdragon, *lfy* in *Arabidopsis*, and *alf* in petunia—prevent the development of flowers. Wild-type inflorescences are shown in the top row and those of mutants in the bottom row. These observations are consistent with the hypothesis that E-class gene expression is required to differentiate flowers from tissue that otherwise would develop into leaf-bearing branches. (C) A model for the evolution of the ABCE system of floral organ determination. Derived angiosperms, such as *Arabidopsis* and *Eschscholzia* (poppy), have the fully developed system. Basal angiosperms, such as *Persea* and *Nuphar* (Nymphaeaceae), follow the "fading borders" system of flower patterning, involving overlapping expression patterns of A, B, and C gene orthologues. The BC system is found in gymnosperms and the putative common ancestor of both gymnosperms and angiosperms. (B from Moyroud et al. 2010; C after Chanderbali et al. 2010).

(C)

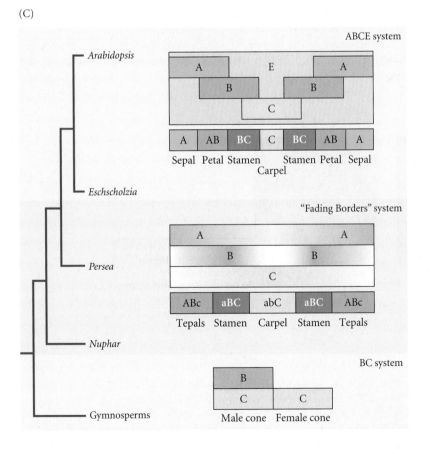

cone. This system reflects the putative ancestral state of both the gymnosperms and the angiosperms. Basal angiosperms such as water lilies (family Nymphaeaceae) and avocado (genus *Persea*) employ a patterning system called the "fading borders" system, involving orthologues of the A, B, and C genes (Figure 21.6C, middle). The carpel at the center of the flower exhibits strong C gene expression, whereas the intermediate whorls (such as stamens) are defined by high, overlapping expression of B and C genes, and the outer whorl is defined by high, overlapping expression of A and B genes. Finally, more derived angiosperms employ the full ABCE system (Figure 21.6C, top). Yet another class of MADS-box genes, termed the D-class, was found to be required for development of ovules (not shown; reviewed by Erbar 2007). Hence the model is now known as the ABCDE model.

Nucleotide sequence data suggest that much of the ABCDE system of flower patterning originated before the divergence of angiosperms and gymnosperms 300 Mya (Purugganan 1997; Rutledge et al. 1998; Winter et al. 1999). Going even deeper into evolutionary history, MADS-box family genes are present in all eukaryotes and are proposed to have been derived from part of a topoisomerase gene present in the ancestor of all eukaryotes (Gramzow et al. 2010). Topoisomerases are proteins that function in winding and unwinding DNA so that it can be accessed by the replication and transcriptional machinery of the cell.

Another group of transcription factors, called the class II TCP family, has played recurrent roles in the bilateral symmetry that has evolved in diverse plant lineages. The ancestral angiosperm flower is thought to have been radially symmetrical, as in the flower of the water lily (**Figure 21.7A**). Bilaterally symmetrical (also called zygomorphic) flowers have evolved independently many times in eudicots, and this flower morphology is known to be attractive to certain pollinators (e.g., bees). In the order Lamiales, which includes the snapdragon (*Antirrhinum*; **Figure 21.7B**), the class II TCP genes *CYCLOIDEA* (*CYC*) and *DICHOTOMOA* (*DICH*) are expressed in developing dorsal regions, causing cell proliferation to decrease in comparison to the rest of the floral meristem and promoting the differentiation of the dorsal petals (Hileman and Cubas 2009). In the related Lamiales genus *Aragoa* (**Figure 21.7C**), expression of the gene *RADIALIS* (*RAD*), which was ancestrally activated by CYC, was lost, presumably through regulatory mutations, leading to a reversal back to radial symmetry.

(A)

(B)

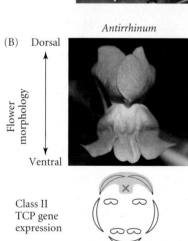

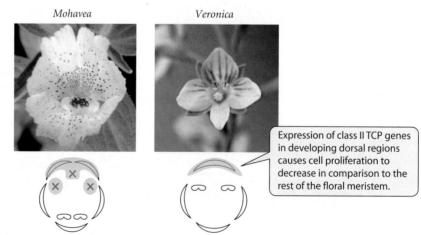

Expression of class II TCP genes in developing dorsal regions causes cell proliferation to decrease in comparison to the rest of the floral meristem.

(C)

FIGURE 21.7 Class II TCP genes and the evolution of floral asymmetry. (A) The radially symmetrical flower of the water lily (genus *Nymphaea*) represents the ancestral state in angiosperms. (B) Schematic illustration of floral meristem expression patterns of class II TCP genes in three genera of the order Lamiales. (C) Flower of *Aragoa abietina*, showing secondarily evolved radial symmetry. (B after Hileman and Cubas 2009, photos courtesy of Lena Hileman; C courtesy of Jill Preston.)

Evolutionarily Conserved Developmental Pathways

The genes that regulate morphogenesis function in hierarchies or networks termed **developmental pathways** or **developmental circuits** (**Box 21B**). These pathways involve extracellular signaling molecules, which relay molecular signals between cells; receptor proteins, which detect extracellular signals; intracellular signaling molecules, which relay signals from receptors to the nucleus; and transcription factors, which respond to these signals by increasing (up-regulating) or decreasing (down-regulating) transcription at target genes. Several developmental pathways that control the formation of major organs or appendages seem to be largely controlled by highly conserved transcription factors (reviewed in Carroll et al. 2005).

A famous example of such a gene is *eyeless*, originally discovered in a classic *Drosophila* mutant with greatly reduced or missing eyes. Mutations in the mammalian homologue of *eyeless*, which is called *Pax6*, also cause reduction of the eyes. *Pax6/eyeless* activates the transcription of a hierarchy of regulatory proteins that control the development and differentiation of the eye. Expression of *Pax6/eyeless* is localized to the developing eye in embryos (**Figure 21.8A–D**). Amazingly, when researchers induced expression of *eyeless* in parts of the developing *Drosophila* body where it is not normally found (referred to as ECTOPIC expression), they discovered that expression of this gene was sufficient to induce

Box 21B Components of Developmental Pathways

Hox genes are examples of homeotic selector genes, which control cascades of gene expression (i.e., transcription) during the patterning and development of particular tissues, organs, or regions of the body. If a homeotic selector gene such as *Ultrabithorax* (*Ubx*) is not expressed properly, specific tissues or organs may not develop at all or may be transformed into inappropriate structures (see Figure 21.1).

Transcription factors, which include the products of most homeotic selector genes, control the expression of many other genes, including the "structural" genes that encode the proteins, such as enzymes and cell structural components, that actually do the work of morphogenesis. The actions of transcription factors are often regulated in part by cell signaling pathways (**Figure 1**). These pathways rely on receptor proteins in the cell membrane that respond to extracellular signals such as hormones or short-range signaling proteins called MORPHOGENS. The receptor proteins relay these signals to the nucleus, resulting in changes in gene transcription levels. Seven such cell signaling pathways (each named for a constituent protein, such as Hedgehog or Notch) have been found in animals; others are known in plants. All of these pathways are conserved between *Drosophila* and mammals, and all are involved in many aspects of morphogenesis and pattern formation throughout the developing body, having evolved many specific functions in particular animal lineages, often through gene duplication. Most cell signaling pathways are used multiple times during development, suggesting that morphogenetic novelty may often evolve through redeployment of these pathways in different tissues and at different developmental stages.

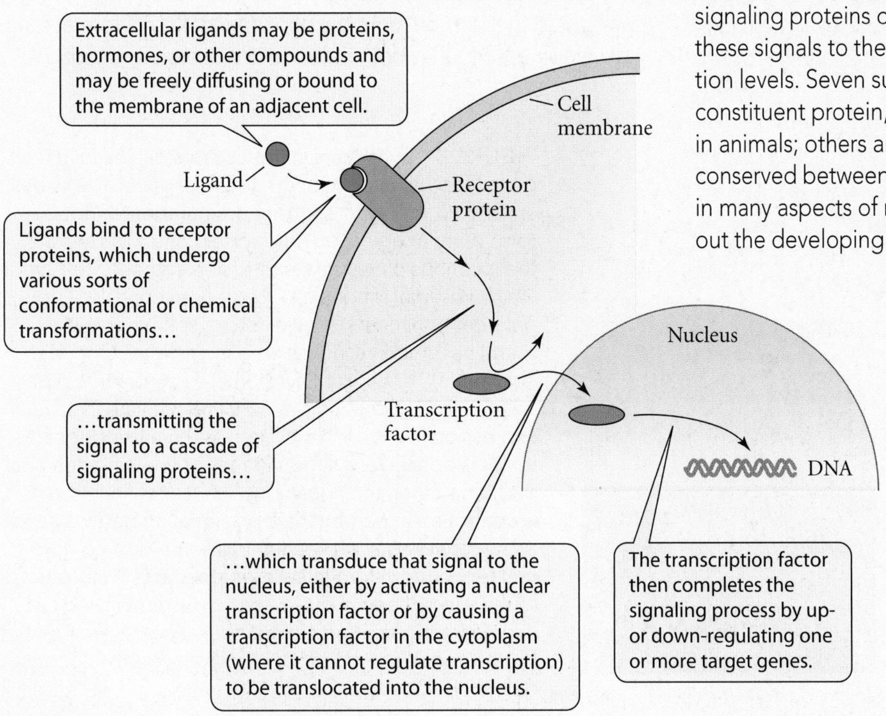

Extracellular ligands may be proteins, hormones, or other compounds and may be freely diffusing or bound to the membrane of an adjacent cell.

Ligand

Ligands bind to receptor proteins, which undergo various sorts of conformational or chemical transformations…

…transmitting the signal to a cascade of signaling proteins…

Cell membrane

Receptor protein

Transcription factor

Nucleus

DNA

…which transduce that signal to the nucleus, either by activating a nuclear transcription factor or by causing a transcription factor in the cytoplasm (where it cannot regulate transcription) to be translocated into the nucleus.

The transcription factor then completes the signaling process by up- or down-regulating one or more target genes.

FIGURE 1 Transduction of intercellular signals through a cell signaling pathway.

(continued)

Box 21B *(continued)*

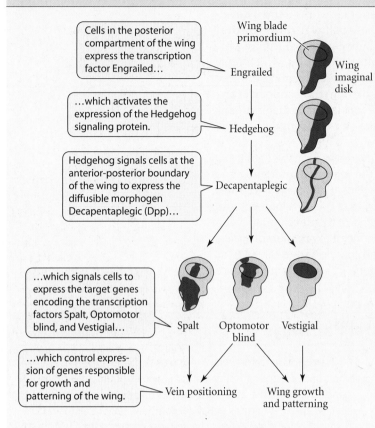

Cells in the posterior compartment of the wing express the transcription factor Engrailed…

…which activates the expression of the Hedgehog signaling protein.

Hedgehog signals cells at the anterior-posterior boundary of the wing to express the diffusible morphogen Decapentaplegic (Dpp)…

…which signals cells to express the target genes encoding the transcription factors Spalt, Optomotor blind, and Vestigial…

…which control expression of genes responsible for growth and patterning of the wing.

Wing blade primordium

Engrailed

Hedgehog

Decapentaplegic

Spalt Optomotor blind Vestigial

Vein positioning Wing growth and patterning

Wing imaginal disk

FIGURE 2 A developmental pathway consisting of cell signaling and transcriptional activation events in the developing *Drosophila* wing imaginal disc. The gene expression patterns are shown in dark brown. Imaginal discs are sacs of epidermal cells that are set aside early in the larval development of some insects. These discs undergo growth and patterning throughout the larval stages; during the pupal stage, they develop into external adult structures, such as wings.

Cell signaling pathways and transcription factors are linked into developmental pathways. The end results of developmental pathways are patterns of gene expression that guide the development of a structure such as the *Drosophila* wing (**Figure 2**). Variation in the expression of several signaling proteins has been shown to be involved in the adaptive evolution of beak morphology in Darwin's finches (see Chapter 22).

The binding of transcription factors to a target gene's enhancer sequences in specific cells and at specific developmental stages determines whether transcription of the gene is increased (up-regulated) or decreased (down-regulated) in those cells and stages. Enhancers therefore act as genetic switches by controlling transcription of the gene in specific spatio-temporal patterns determined by the transcription factors. Some transcription factors act exclusively to increase or to decrease gene expression, whereas others may have either effect in particular contexts, depending on the other proteins they interact with and on the specific enhancer sequence they bind to. **Figure 3** shows two enhancers in the *Drosophila vestigial* gene that affect *vestigial* expression in the developing wing differently. Transcription factors in the Notch signaling pathway bind to an enhancer that directs *vestigial* expression at the anterior-posterior and dorsal-ventral boundaries of the wing field. Transcription factors in the Dpp pathway bind to an enhancer that directs expression in the four quadrants of the wing field that complement the boundary pattern. Normal wing development requires both patterns of expression of *vestigial*.

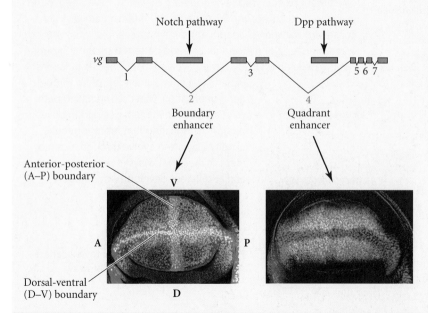

Notch pathway   Dpp pathway

vg

1   2   3   4   5 6 7

Boundary enhancer   Quadrant enhancer

Anterior-posterior (A–P) boundary

V

A   P

Dorsal-ventral (D–V) boundary

D

FIGURE 3 Enhancers in the *Drosophila vestigial* (*vg*) gene. Top: Schematic of the *vestigial* gene. The exons are shown in gray; the introns are numbered. Bottom: Wing imaginal disc. Reporter constructs show the locations of *vestigial* expression in green; cells in which *vestigial* is not expressed appear purple. Two different enhancers (shown in blue in the gene diagram) have been characterized in *vestigial*. One, the "boundary" enhancer (in intron 2), is controlled in part by the Notch signaling pathway and activates *vestigial* expression in a crosslike pattern at the anterior-posterior and dorsal-ventral boundaries of the wing imaginal disc. The other, the "quadrant" enhancer (in intron 4), is controlled in part by the Dpp signaling pathway and activates *vestigial* expression in a somewhat complementary pattern (green cells in upper and lower parts of the wing imaginal disc with slightly diminished expression on the vertical boundary line). A = anterior, P = posterior, D = dorsal, V = ventral. (Photos courtesy of Sean B. Carroll.)

(A)

(B)

(C)

(E)

(D)

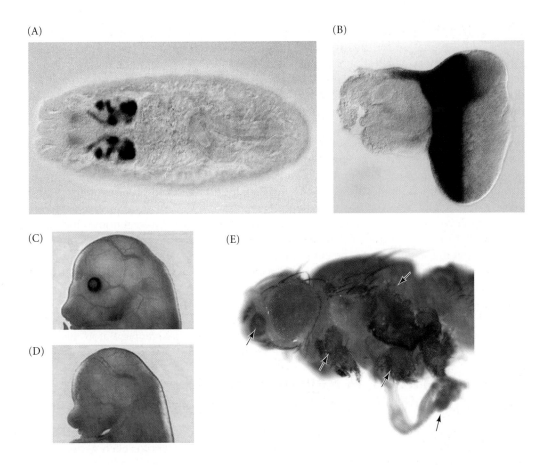

FIGURE 21.8 *Pax6/eyeless* genes in animal eye development. (A, B) Expression of *eyeless* in *Drosophila*, visualized by expression of a β-galactosidase reporter gene (dark blue areas). (A) Expression of *eyeless* in the larval eye precursors and other nearby tissues. (B) Close-up of eye imaginal disc (large lobe on right), a sac of undifferentiated larval cells that develops into the adult eye. The *eyeless* gene is expressed in the cells anterior to a boundary-like feature called the morphogenetic furrow, which moves from posterior to anterior during eye development and delineates the boundary between differentiated and undifferentiated cells. (C, D) The role of *Pax6* in mouse eye development. (C) The developing mouse eye can be seen as a circle of pigment. (D) The mouse mutant *Small eye* (*Sey*) lacks *Pax6* expression and has small or missing eyes. (E) Ectopic expression of the human *PAX6* gene (driven by enhancers for the gene *dpp*) in *Drosophila* causes eyes to form in many locations (arrows). (A, B courtesy of Georg Halder; C, D courtesy of Robert Hill; E courtesy of Nadean Brown.)

the development of ectopic eyes at these positions. Even more astounding is the functional conservation of the *Pax6/eyeless* gene between vertebrates and invertebrates: *Pax6* genes from other animal species, including humans, can induce recognizably eyelike structures (called "ectopic eyes" but lacking the size and full structure of the normal eye) when expressed in *Drosophila* (**Figure 21.8E**). This ancient functional conservation encompasses all metazoans: a transcription factor from the jellyfish *Cladonema radiatum*, *Pax-A*, a gene closely related to *Pax6* but in a different subfamily (the *Pax6* subfamily is only distinguished in bilaterians), was also able to induce ectopic eyes in *Drosophila* (Suga et al. 2010).

Subsequent studies found that at least two genes regulated by *Pax6* have conserved functions in *Drosophila* and mammalian eye development (Oliver et al. 1995; Xu et al. 1997), which helps to explain how *Pax6* homologues can function when placed in the genomes of such divergent animal species. A remarkable degree of development occurs in the resulting ectopic eyes. Those that form in the antenna, an organ that develops from tissues adjacent to the eye in *Drosophila*, develop projections of axons into the central nervous system. If these projections are able to contact the optic lobes, the visual centers of the fly brain, they can form what appear to be fully functional synapses that are able to process information from light hitting the ectopic eyes (Clements et al. 2008). This observation suggests that the early processes of eye development involving the photoreceptor cells are able to induce a large portion of the morphogenesis and differentiation of tissues involved in producing a functioning eye.

The question of how an organ as complex as the vertebrate eye evolved is one in which Darwin himself, and many theorists after him, invested a great deal of thought. Darwin suggested that various intermediate stages, capable of photoreception, may have had adaptive value, leading to the evolution of more complex eyes. Consistent with this idea, basal animal phyla such as Cnidaria have very simple "eyespots," consisting of a few cells containing photoreceptive pigments (rhodopsins), and many types of eyes, varying in complexity, are found among other invertebrates (see Figure 22.13). There is still controversy over whether the common ancestor of the Bilateria possessed functional eyes. Nevertheless, an explanation for the diversity of animal eyes that has gained wide acceptance among evolutionary developmental biologists is that *Pax6* has a very ancient function in regulating the expression of universal components of photoreceptor organs, such as rhodopsin pigments. Thus, in various lineages of animals, different morphological features of eyes evolved independently, possibly because of the actions of different genes that became regulated by the *Pax6* pathway (see Wilkins 2002, pp. 148–155). Therefore the capacity for photoreception may be homologous, but the structural features that carry out this function—eyes—have evolved independently in many different lineages. A genomic study of gene expression in the developing eyes of human and octopus, which are structurally similar but independently evolved (see Figure 3.4), found that of the 1052 protein-coding genes expressed in the octopus eye, 729 (69.3 percent) were also expressed in the developing human eye (Ogura et al. 2004). Phylogenetic analysis of metazoan genomes indicated that the vast majority of the 1052 genes expressed in the octopus eye were likely to have been present in the genome of the common ancestor of all bilaterians. Therefore, as with the photoreceptor cells, the genomic tools for building an eye were present well before the great diversity of animal eyes evolved. Other cases of organogenesis, such as vertebrate heart development, which appears to involve regulation by the transcription factor NKX2-5/tinman across all bilaterians, may also involve convergent recruitment of "downstream" cascades of pre-existing genes by a highly conserved "master regulator" protein.

Gene Regulation: A Keystone of Developmental Evolution

The regulation of gene expression, whereby genes are activated or repressed at particular times and in particular tissues, is achieved by discrete enhancers for each gene. These DNA sequences bind particular sets of transcription factors in specific cells or at specific developmental stages (see Box 21B, Figure 3). For example, several genes expressed in the developing *Drosophila* wing are regulated by the transcription factors Scalloped and Vestigial, which activate genes required for wing development (Guss et al. 2001). The noncoding DNA (introns) of these genes contains one to several binding sites, 8 to 9 nucleotides long, for Scalloped and Vestigial. An enhancer consists of one or more such sites, which bind one or more transcription factors. Enhancers can be studied using **reporter constructs**, in which the DNA sequence containing the putative enhancer is spliced to a REPORTER GENE that encodes a protein whose expression can be easily visualized, and then placing this DNA construct in cultured cells or transgenic (genetically engineered) individuals (see Box 21A).

A particular gene often has a number of different enhancers, each of which may control its expression in particular tissues at particular times during life history. This **regulatory modularity** is thought to enable evolutionary changes in the development of specific tissues and body structures. That is, changes in the enhancers, rather than changes in the amino acid sequences of proteins, are thought to be responsible for many phenotypic adaptations. Several examples of such changes have been well characterized in *Drosophila*.

The *Drosophila* gene *yellow*, which encodes an extracellular protein required for dark melanin pigmentation, is expressed in flies wherever there is dark pigmentation on the body. This pigmentation is often expressed in rapidly evolving sexually dimorphic patterns, such as the male-specific wing spots that occur in various *Drosophila* species. In spot-bearing species such as *D. elegans* (**Figure 21.9A**), the pattern of expression of *yellow* in wing spots, which occurs during the late pupal stage, is directed by an enhancer upstream of the *yellow* coding region (**Figure 21.9B**; Prud'homme et al. 2006). This enhancer resides within a larger enhancer that was previously demonstrated to direct expression of *yellow* in the wing blade (i.e., all of the wing cells except the veins) in *D. melanogaster* and other *Drosophila* species. Interestingly, *D. tristis*, which evolved similar male wing spots independently, also expresses *yellow* in the developing spot region (**Figure 21.9C**). *D. tristis* also

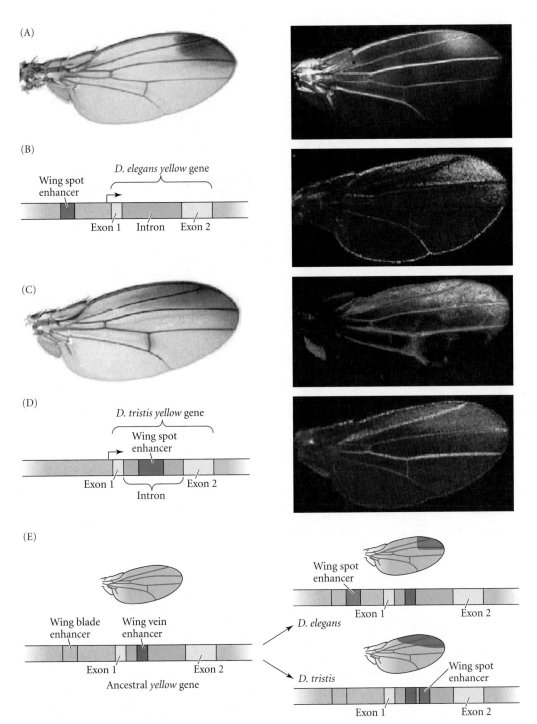

FIGURE 21.9 Independent evolution of *Drosophila* wing spot enhancers in the *yellow* gene. (A) *D. elegans* male wing spot (left) and Yellow protein expression as detected using fluorescent antibody staining (right). (B) Location of wing spot enhancer in the *D. elegans yellow* gene (left) and wing spot reporter expression (right). Arrow indicates transcription start site and direction. (C) *D. tristis* male wing spot (left) and Yellow protein expression as detected using fluorescent antibody staining (right). (D) Location of wing spot enhancer in the *D. trystis yellow* gene (left) and wing spot reporter expression (right). (E) Model of convergent *yellow* wing spot evolution in *Drosophila*. Wing blade enhancer and expression are shown in light blue. Wing vein enhancer and expression are shown in green. Exons of the *yellow* gene are shown in yellow. (From Prud'homme et al. 2006; photos and staining images courtesy of Sean Carroll.)

possesses a "spot enhancer" in the *yellow* gene, but this enhancer is not in its "wing blade" enhancer, but rather in the intron of the *yellow* gene (**Figure 21.9D**). Thus convergent evolution has involved surprisingly different regulatory components of the same gene (**Figure 21.9E**).

Many genes such as *yellow*, originally found as mutations in a model organism such as *D. melanogaster*, are considered "candidate genes" for phenotypic evolution because their known functions are consistent with causation of a difference between species. Confirming their involvement in a specific case of phenotypic evolution requires experimental data. Another example in which a candidate gene has been implicated in a species difference is

FIGURE 21.10 Abdominal pigmentation differences between *Drosophila yakuba* and *D. santomea* are due in large part to *cis*-regulatory changes in the *tan* gene, which is required for melanin synthesis. (A) Abdominal cuticle preparations (lateral view) of *D. yakuba* male, which has dark abdominal pigmentation in the fifth and sixth abdominal segments (A5 and A6), and *D. santomea* male, which has very light abdominal pigmentation. (B) The pigmentation difference between *D. yakuba* and *D. santomea* resembles that between wild-type and *tan* mutant *D. melanogaster*, suggesting *tan* as a candidate gene for pigmentation differences between species. (C) Abdominal expression of *tan* is directed by a *cis*-regulatory element residing between two genes (with pink exons) located 5′ of *tan*, as determined by reporter assays in transgenic *D. melanogaster*. Blue boxes indicate *tan* exons. Green bars below the sequence indicate fragments of the sequence that cause abdominal expression of *tan* when fused to a reporter gene. Red bars indicate fragments that do not cause such expression. The black arrows indicate start sites and direction of transcription for *tan* and for the two 5′ genes. (D) The loss of abdominal expression of *tan* in one naturally occurring *D. santomea tan* allele is due predominantly to two derived nucleotide substitutions (in red) within the enhancer for abdominal expression of *tan* shown in (C), as shown in this phylogeny of species closely related to *D. yakuba* and *D. santomea*. This expression difference is presumably due to changes in transcription factor binding sites, although the transcription factors involved have not yet been identified. Other *D. santomea tan* alleles have different changes in this region, which also cause greatly decreased expression of *tan* in the abdomen. (Photos courtesy of Sean Carroll; C, D after Jeong et al. 2008.)

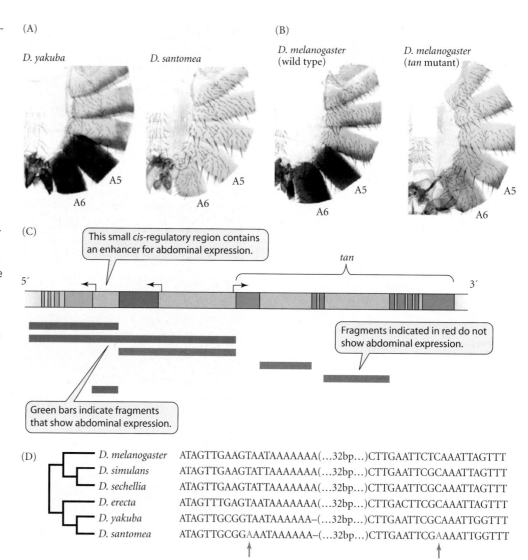

found in *D. santomea*, which has much lighter abdominal melanin pigmentation than its sibling species *D. yakuba*. The light pigmentation of *D. santomea* resembles that of several *D. melanogaster* pigmentation mutants, including a mutant called *tan*. QTL mapping (see Chapter 13) determined that the genomic region around *tan* contributed strongly to the pigmentation difference between these species (Carbone et al. 2005). Molecular analyses demonstrated that differences in *tan* expression explained most of the abdominal pigmentation divergence between *D. yakuba* and *D. santomea* (Jeong et al. 2008). The divergence was shown to be due to just two nucleotide substitutions in a small region of *cis*-regulatory DNA containing an enhancer for abdominal expression of *tan* (**Figure 21.10**). In *D. santomea*, expression of *tan* in the abdomen is essentially turned off.

An evolutionary change in the expression of a gene can be caused either by the alteration of a closely linked (*cis*) regulatory sequence (such as an enhancer), as in the previous example, or by a difference in **trans-regulation**, involving changes in genes whose products regulate the transcription of the focal gene by interacting with its *cis*-regulatory sequences. It was recently shown that of 231 new mutations causing changes in the expression of a fluorescent reporter gene in yeast (*Saccharomyces cerevisiae*), 7 percent were nonsynonymous changes in the coding region of the reporter gene; 2 percent were mutations in the *cis*-regulatory region, which was similar in length (around 600 base pairs) to the coding region; and 81 percent were *trans*-acting mutations (Gruber et al. 2012). This study represents one of the first detailed characterizations of a possible mutation spectrum available to natural selection. It suggests that mutations acting on *trans*-regulatory sequences to alter expression of a particular gene may be much more common than *cis*-acting mutations. Nevertheless, *cis*-acting mutations, even in a small target region of DNA, were still readily detected and thus are likely to be present for all genes in natural populations. During speciation, many variants affecting gene expression become fixed differentially in sibling species. In a study of 29 genes showing differences in expression between the sibling species *D. melanogaster* and *D. simulans*, about half the genes displayed only *cis*-regulatory divergence, while the other half showed signs of both *cis*- and *trans*-regulatory divergence (Wittkopp et al. 2004). These findings suggest that changes in *cis*-regulatory sequences are likely to account for many evolutionary changes in gene expression.

An important pursuit in recent EDB studies is the comparison of gene regulatory changes that contribute to morphological differences both between and within species. The three-spined stickleback (*Gasterosteus aculeatus*), described in Chapter 13, has invaded freshwater habitats from its ancestral marine habitat many times in the last 10,000 years. During these invasions, the same morphological adaptations for living in fresh water have repeatedly and independently evolved. The best characterized of these adaptations are reductions in the pelvic spines and armor plates, which are needed for defense against predators in the ocean, but have been lost in many freshwater habitats in which predation is not a strong selection pressure (see Figure 13.23). Since all of these populations of *G. aculeatus* are interfertile, genetic methods such as QTL mapping can readily determine the genetic basis of these adaptations.

Developmental genetic data from zebrafish and other vertebrate model species have been useful in identifying candidate genes. Shapiro and colleagues (2004) found that among five QTL for the pelvic armor reduction phenotype in the stickleback, the one with the predominant effect maps to the *Pituitary homeobox transcription factor 1* (*Pitx1*) gene, which in mice is required for pelvic limb development. *Pitx1* is expressed in many parts of the body during development in vertebrates, but most importantly, it is expressed in the developing pelvis in marine fishes. The pelvic pattern of expression has been lost in the freshwater stickleback study population, which has reduced pelvic armor (**Figure 21.11**). Since no coding-region differences between marine and freshwater genotypes were found, Shapiro and colleagues concluded that a regulatory change reducing *Pitx1* expression is responsible for the freshwater adaptation. They demonstrated that other populations of *G. aculeatus* with reduced pelvises also carry *Pitx1* alleles with reduced expression in the

(A) Marine

Pitx1 expression (ventral view)

Dorsal spines

Pelvic spine

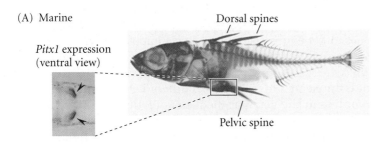

(B) Freshwater

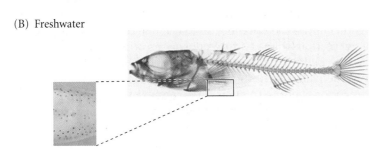

FIGURE 21.11 Loss of expression of *Pitx1* in the pelvis is associated with pelvic armor reduction in freshwater populations of the three-spined stickleback (*Gasterosteus aculeatus*). Adult specimens from (A) marine and (B) freshwater populations. Bony plates and pelvic spines characterize the marine population, but not the freshwater population. In the magnified ventral views of embryos (enlargements), in situ hybridization reveals much greater *Pitx1* expression (purple) in the pelvic area in the marine population than in the freshwater population. The pinkish-brown dots are pigment cells (melanocytes). (After Shapiro et al. 2004; photos courtesy of D. M. Kingsley.)

pelvis. These results suggest that the ancestral marine population harbors these recessive alleles at a low frequency. The enhancer responsible for pelvic expression of *Pitx1* exhibits evidence of positive selection in populations with reduced pelvises (Chan et al. 2010). One interesting feature resulting from the loss of *Pitx1* function in several vertebrate lineages has provided strong evidence that this gene has been involved much more broadly in adaptations of the vertebrate skeleton. In both three-spined (*G. aculeatus*) and nine-spined sticklebacks (*Pungitius pungitius*) with pelvic armor reduction, as well as in mouse mutants lacking *Pitx1* expression, the remaining pelvic and hindlimb elements show directional asymmetry, with more elements on the left than on the right side (reviewed by Shapiro et al. 2006). This asymmetry is thought to be the result of left-biased embryonic expression of the closely related gene *Pitx2*, which may overlap with *Pitx1* in some functions and thus be able to partially compensate for loss of *Pitx1* on the left side. Manatees (*Trichechus manatus*), which have lost their hindlimbs and bear only vestigial elements of the pelvis, also show a pronounced left bias in the remaining pelvic elements (Shapiro et al. 2006), suggesting that loss of *Pitx1* may have been involved in the profound evolutionary modification of the manatee skeleton.

Reductions in body armor plates in different freshwater populations of *G. aculeatus* are caused by regulatory changes in the gene encoding ectodysplasin (*Eda*; see Figure 13.23; Colosimo et al. 2005), a signaling protein required for the development of teeth, hair, and other ectodermal structures in mammals. Among five additional genes in the Eda signaling pathway, only the gene encoding the Eda receptor (*EdaR*) was found to harbor natural variation affecting the size of the armor plates (Knecht et al. 2007). This observation suggests that only particular genes in the armor developmental pathway contribute to adaptive differences in nature.

Protein-Coding Sequences and Phenotypic Evolution

Changes in the regulation of genes are thought by many researchers to be more likely causes of phenotypic trait evolution than are changes in amino acid sequences (Stern and Orgogozo 2009). Genes often have more than one function during development (that is, they are pleiotropic). Changes in the protein-coding sequence of a gene are expected to cause functional alterations in the protein, some or all of which are likely to be disadvantageous, whereas changes in the regulation of the gene at particular stages, or in particular cells or tissues, are expected to affect the gene's function only in those specific stages or tissues. It has been argued, however, that many proteins—transcription factors in particular—are modular, such that their individual domains can evolve to affect small subsets of their functions (e.g., to promote or inhibit their binding with a specific protein cofactor present in only one cell type), thus avoiding strong pleiotropic effects (Wagner and Lynch 2008). The current evidence is rather equivocal on the relative importance of changes in regulatory versus protein-coding sequences in adaptive evolution (Hoekstra and Coyne 2007; Wray 2007).

A surprisingly recurrent example of a morphological difference caused by protein-coding sequence variation occurs in the gene for the melanocortin-1 receptor (*Mc1r*), which was discussed in Chapter 3. The receptor protein encoded by *Mc1r* is expressed on the surfaces of melanocytes (pigment-producing cells), where it receives signals from the signaling peptides (hormones) melanocortin and Agouti. Binding of melanocortin to the receptor signals the melanocyte to produce eumelanin, resulting in dark pigmentation, whereas binding of Agouti signals the melanocyte to switch to pheomelanin production, resulting in lighter pigmentation. In many different species of birds and mammals, naturally occurring amino acid substitutions cause the Mc1r receptor to "behave" as if it is constitutively (i.e., constantly) activated by melanocortin. The result is genetically dominant melanism (dark pigmentation) because melanocytes cannot respond to the Agouti signal, and thus dark pigment is produced throughout the development of hair or feathers (Mundy 2005). Mutations in Mc1r, which is also a receptor for α-, β- and γ-melanocyte-stimulating hormones and adrenocorticotropic hormone, are highly pleiotropic, with many effects on behavior, stress responses, metabolism, and the immune system. Many traits are significantly positively correlated with dark pigmentation in various vertebrates (Ducrest et al. 2008). Such traits include increased sexual behavior, aggressiveness, and resistance to environmental stresses, and larger body size. The frequency of such correlations strongly suggests that adaptations of vertebrate pigmentation have a high propensity to be involved in complex, multi-trait syndromes that, depending on the ecological context, may either constrain or accelerate those adaptations.

The Molecular Genetic Basis of Gene Regulatory Evolution

Most morphological variation within species, and most novel trait differences that arise between species, are thought to be polygenic. QTL mapping can be used to uncover the genetic architecture of novel trait differences between populations and closely related species that can be interbred (see Box 13A). The kinds of genes and alleles involved in these differences (i.e., *cis*-regulatory vs. protein-coding sequences) are particularly interesting to EDB researchers. The few QTL studies that implicate specific genes suggest that developmental regulatory genes are commonly involved in morphological differences between species, and that these genes can include even major developmental regulatory genes such as *Ubx* (reviewed by Simpson 2002). Thus the many crucial developmental functions of these genes do not preclude their involvement in short-term evolutionary change in particular characteristics. As we described above for the Hox gene *Scr* (see Figure 21.5) and will see later in the case of butterfly wing patterns (see Figure 21.15B,C), the once-surprising role of "master regulatory" genes in the fine-scale morphology of specific body segments is due to the continuing roles of these genes in transcriptional regulation late in development. Changes in the regulatory regions of these genes have been implicated more often than changes in their protein-coding sequences, confirming the importance of variation in gene regulation in morphological evolution. In *Drosophila melanogaster*, for example, naturally occurring variation in bristle number is caused partly by nucleotide changes in the enhancers of genes encoding both cell signaling proteins and transcription factors (Lai et al. 1994; Long et al. 1998, 2000).

Molecular evolutionary studies in closely related species show that even when enhancer function is conserved, the binding sites within an enhancer can change in surprising ways. In a classic set of experiments, Michael Ludwig, Martin Kreitman, and colleagues (Ludwig and Kreitman 1995; Ludwig et al. 1998, 2000) examined sequence and functional differences among *Drosophila* species in a specific *cis*-regulatory sequence from the gene *even-skipped* (*eve*), a transcription factor required for segmentation. In *Drosophila melanogaster*, this enhancer, called *eve stripe-2*, is a 670-base-pair segment of DNA that is necessary and sufficient to direct expression of *even-skipped* in the second of the seven stripes in which the Even-skipped protein is expressed in wild-type embryos (**Figure 21.12A**). This enhancer contains 17 binding sites for transcription factors encoded by four different regulatory genes that are expressed in anterior-posterior gradients and which, by activation and repression, combine to determine the precise limits of the *even-skipped* stripes (**Figure 21.12B**). Kreitman's group sequenced *eve stripe-2* from five other *Drosophila* species and examined the number and positions of these binding sites, as well as the function of the five species' enhancers when placed in transgenic *D. melanogaster*. Surprisingly, even though the sequences from all five species directed the correct stripe of expression in *D. melanogaster*, the nucleotide sequences of only 3 of the 17 binding sites were perfectly conserved across all five species (**Figure 21.12C**). In fact, two of the binding sites were not found at all in the two species most distantly related to *D. melanogaster*.

Ludwig and Kreitman wondered whether the function of the *eve stripe-2* enhancer was conserved despite these sequence differences because each individual part of the enhancer maintained its independent role, or because changes in some parts were accompanied by compensatory changes in other parts (i.e., concerted evolution of the entire element). They

(A)

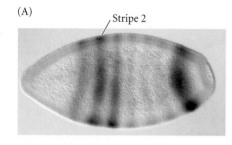

Stripe 2

FIGURE 21.12 Evolution of the *eve stripe-2* enhancer of the *even-skipped* (*eve*) gene in *Drosophila*. (A) Embryo in the blastoderm cellularization stage stained for expression of an *eve stripe-2* reporter construct (dark blue staining) and the Even-skipped protein (brown staining). The *even-skipped* (*eve*) gene is a "pair-rule" gene that is expressed in seven stripes, one in every other segment, in the embryo. The *eve stripe-2* construct is expressed in stripes 3 and 7 as well as in stripe 2. (B) Schematic of the *eve stripe-2* enhancer, showing binding sites for four transcription factors: the maternal effect proteins Bicoid (Bcd) and Hunchback (Hb) and the "gap" proteins Giant (Gt) and Krüppel (Kr). (C) Conservation and evolution of transcription factor binding sites in the *eve stripe-2* enhancer among *Drosophila* species. Binding sites are numbered according to a convention established for the *D. melanogaster eve stripe-2* sequence, which has all 17 sites. The column at right indicates the estimated time of divergence of each species from *D. melanogaster*. Species more distantly related to *D. melanogaster* have more absent or weakly conserved binding sites, even though the overall expression pattern is conserved in all species. (After Carroll et al. 2005.)

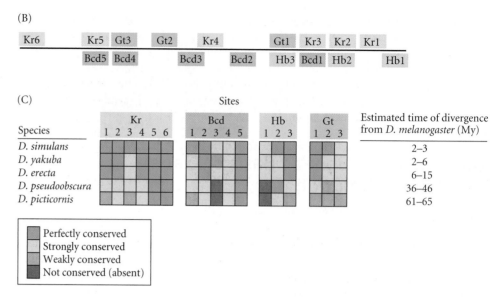

constructed enhancers made up of DNA from two different *Drosophila* species. When these chimeric enhancers were placed in transgenic *D. melanogaster*, the discrete stripe of *even-skipped* expression was usually not seen, showing that the chimeric enhancers could not promote the normal pattern of transcriptional activation. Therefore it appears that within each species, the entire enhancer has evolved in a concerted fashion to maintain its function. If the fixation (by selection or drift) of a nucleotide substitution in one position in the enhancer caused a minor alteration in stripe expression, that change might provide selection pressure for a compensatory substitution elsewhere in the enhancer in order to maintain the stripe expression. Many key interactions of the master regulatory genes involved in morphogenesis have been conserved over very long evolutionary periods (Rebeiz et al. 2012). Thus it is becoming clear that the developmental genetic toolkit involves an ancient framework of key regulatory interactions, often between highly pleiotropic master regulatory proteins that regulate programs of transcription. These interactions occur in a background of binding sites for other transcription factors, which may control finer-scale aspects of gene expression that evolve over shorter time spans.

Protein sequence evolution can also be an important source of regulatory novelty in evolution. Amino acid substitutions in transcription factors, affecting either their interactions with other regulatory proteins or their binding to *cis*-regulatory elements, represent an important potential source of novelty because these changes can potentially affect the expression of many downstream genes during development (Lynch and Wagner 2008). Lynch and colleagues recently demonstrated a strikingly novel function of a transcription factor called CCAAT/enhancer binding protein-β (C/EBP-β) that arose during the evolution of mammals (Lynch et al. 2011). C/EBP-β interacts with another transcription factor, FOXO1A, to turn on expression of the prolactin (*PRL*) gene during the development of endometrial stromal cells. The endometrium—the inner surface of the uterus—plays a critical role in the development of the placenta during the initiation of pregnancy in eutherian (placental) mammals. The eutherian placenta is adapted to enable a long gestation time, which is not seen in marsupial mammals and was probably absent in the common ancestor of eutherians and marsupials. Lynch and colleagues used phylogenetic reconstruction to determine the C/EBP-β and FOXO1A protein sequences most likely present in the most recent common ancestor of all eutherians and in the most recent common ancestor of eutherians and marsupials. Then, in a clever set of experiments, they engineered gene constructs to express the ancestral proteins and the proteins from several extant species in cultured human cells. They tested the interaction of each C/EBP-β protein with the corresponding (same-species) FOXO1A protein, as well as the ability of each protein pair to turn on expression of the *PRL* gene in the human cells. They found that C/EBP-β and FOXO1A proteins from a eutherian (human), a marsupial (opossum), both reconstructed ancestors, and an outgroup species (chicken) were able to bind to each other. However, only the protein pairs of the extant eutherian (human) and the ancestral eutherian showed a "cooperative" or "synergistic" interaction, in which expression of the *PRL* gene was severalfold higher than when either of the activating proteins was expressed by itself. Lynch et al. further showed that this ability of the two proteins to interact cooperatively was due to three amino acid substitutions in the eutherian C/EBP-β protein. Thus it is clear that even a small number of changes in a regulatory protein can lead to important macroevolutionary novelties.

Modularity in morphological evolution

The body plans of multicellular organisms are composed of discernible units, such as body segments, appendages, or the sepals, petals, stamens, and carpels of angiosperm flowers. In animals, many serially homologous structures have quite different morphologies (e.g., different arthropod appendages). Much of the spectacular diversity that we see in nature is the result of such changes in individual segments or structures. Thus the individual parts of the body plan show INDIVIDUALIZATION—that is, they can develop and evolve in an independent manner. This phenomenon is an example of MOSAIC EVOLUTION (see Chapter 3).

The degree to which the development of different body structures is independent is referred to as **modularity**, and the individual structures or units can be thought of as morphological MODULES. How modularity is achieved by developmental pathways is a central question in EDB. It is clear that the EFFECTOR genes—those whose products, such as enzymes and structural components, act to produce the final form of each morphological module—are expressed differentially through the action of regulatory genes, such as those encoding transcription factors. The Hox genes discussed above provide some of the best-known examples of how animals achieve their modular body plans. Because distinct Hox proteins are expressed in different body segments, the effector genes are able to be expressed in very different ways in the different segments. The mechanisms by which Hox genes regulate this segment specificity are likely to be different in each segment. For example, development of sternopleural (Sp) bristles in *Drosophila*, which normally are found only in the second thoracic segment (T2), is repressed differently in the first (T1) and third (T3) segments (**Figure 21.13**; Tsubota et al. 2008). In T1, the Hox protein Scr represses expression of the Hox gene *Antp* and the gene *spineless* (*ss*), which are jointly required for Sp bristle development in T2. Ectopic *ss* expression (in genetically engineered flies) is sufficient

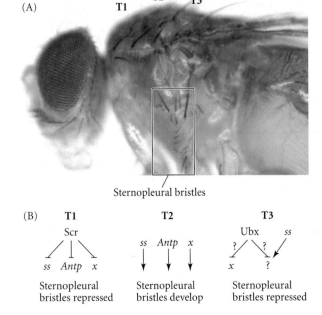

FIGURE 21.13 Hox proteins repress sternopleural (Sp) bristle development differently in the first (T1) and in the third (T3) thoracic segments of *Drosophila melanogaster*. (A) Sp bristles normally develop only in the second thoracic segment (T2). (B) Gene expression in segments T1–T3. Arrows indicate activation of gene expression; lines ending in horizontal lines indicate repression of gene expression. Center: In T2, the *spineless* gene (*ss*), along with the Hox gene *Antennapedia* (*Antp*) and possible unknown genes (indicated by *x*), promotes development of Sp bristles. Left: The Hox protein Scr represses *ss* and *Antp* expression in T1. Right: In T3, Sp bristle development must be repressed through a different mechanism than in T1, because ectopic *ss* expression in T3 does not promote Sp bristle development, as it does in T1. Instead, the Hox protein Ubx may act by inhibiting expression of one or more genes downstream of *ss* or one or more genes required in parallel with *ss* (question marks). (Photo by J. True.)

to cause the development of ectopic Sp bristles in T1. In T3, however, ectopic *ss* expression does not cause the formation of Sp bristles. Instead, the Hox protein Ubx represses development of these bristles by acting at a different stage, perhaps downstream of *ss*. It is not yet clear whether the Sp bristles first evolved only in T2, or whether they evolved in multiple thoracic segments and then disappeared from T1 and T3. In any case, the observed differences in how different Hox genes repress expression of the same genes and structures in different segments seem to reflect the mosaic evolutionary history of these segments.

Macroevolution and the evolution of novel characters

Now that we have seen several illustrations of the molecular and developmental mechanisms underlying evolutionary novelty, we can see how the field of EDB is beginning to address the question of how macroevolutionary change occurs. The idea that "macroevolution is microevolution writ large" is often attributed to Darwin and thus precedes genetic knowledge. Advances in both genomics and developmental genetics are now enabling EDB researchers to truly test this notion, and they have found strong indications in many studies that changing uses of the genetic toolkit are one of the main themes of macroevolution. But general answers to many questions have yet to emerge. One of these, which has been controversial ever since the evolutionary synthesis, is whether evolution is typically gradual, as Darwin postulated, or is more often discontinuous, involving large phenotypic leaps (see Chapter 3).

Genetic analyses of morphological evolution, such as those discussed in this chapter, reveal that a variety of genetic architectures are involved, from single genes with large effects to cases in which many genes each contribute a small fraction to a species difference. In the former case, mutations of major effect (e.g., Hox gene mutations) can certainly cause discontinuous jumps in the morphology of a body structure, giving rise to the possibility that closely related species might differ radically in their development. One of the most spectacular examples of such a difference concerns the development and life history of two congeneric species of sea urchins (see Figure 3.9). In *Helicocidaris tuberculata*, development is indirect: the embryo first develops into a pluteus larva, which later metamorphoses into an adult. In *H. erythrogramma*, on the other hand, development is direct: the embryo initiates the program for adult morphogenesis without an intervening pluteus stage. The observation that these two species are still able to form hybrids (Raff et al. 1999) suggests not only rapid evolution, but also the ability of the developmental system of one life history stage to change radically without disrupting other stages.

The occurrence of such major developmental shifts over short evolutionary time spans carries the implication that they have simple genetic bases consisting of one or a few genes. Although this hypothesis has yet to be tested in the *Helicocidaris* case above, the observation that many genes have multiple roles in development strongly implies that over the course of evolution, many genes and pathways have been redeployed to serve new functions. For example, the transcription factor Distal-less is required to organize the development of legs, wings, and antennae in all insects, but in some butterflies, it is also expressed later in development in specific positions on the developing wing, where it is involved in setting up eyespot color patterns.

Change in the function of pre-existing features in adaptive evolution has been known ever since Darwin. Stephen Jay Gould and Elisabeth Vrba (1982) coined the term EXAPTATION to refer to a novel use of a pre-existing morphological trait (see Chapter 11). Developmental biologists have used the terms **recruitment** (Wilkins 2002) and **co-option** (reviewed by True and Carroll 2002) to refer to the evolution of novel functions for pre-existing genes and developmental pathways.

There are two ways in which a new structure might evolve by co-option of a developmental pathway. First, expression of a protein involved in morphogenesis might persist after the developmental stage in which it was ancestrally used (see the discussion of heterochrony in Chapter 3). The involvement of the *yellow* gene in late pupal development of *Drosophila* wing spots after its earlier use throughout the wing or wing veins, as described above, is an example of this process. If the co-opted gene encodes a transcription factor, then expression of one or more of its target genes may be recruited, resulting in the evolution of a new morphogenetic process later in development. Alternatively, a developmental pathway originally expressed in one region of the embryo might become expressed in a different region, leading to a duplication of that structure in the new region. In the genus *Drosophila*, changes in gene expression resulting in differences among species appear to be common, but truly new expression patterns are quite rare (Rebeiz et al. 2011). One such rare case was that of the gene *Neprilysin-1* (*Nep1*), which exhibits a novel pattern of expression in optic lobe neuroblasts in *D. santomea*. Rebeiz and colleagues found that the enhancer directing this expression had, through a small number of nucleotide changes, evolved the capacity to be activated in the optic lobe by the binding of transcription factors that already bound nearby enhancers with other activities during development. This co-option of "latent activity" of transcriptional activators seems likely to become a common theme in the evolution of novel patterns of gene expression.

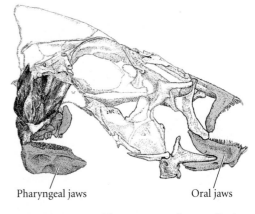

FIGURE 21.14   The two sets of jaws of Lake Malawi cichlids. (From Fraser et al. 2009.)

Pharyngeal jaws   Oral jaws

When genes and developmental pathways are regulated as modules, they can be co-opted independently. An excellent example of such independence comes from a study of the evolution of teeth on the oral jaw of fishes (Fraser et al. 2009). The oral jaw of gnathostomes (jawed vertebrates) evolved from the first pharyngeal arch of agnathans (jawless fishes such as lampreys). Teeth are thought to have evolved on the agnathan pharyngeal arch prior to the evolution of the jaw. Some fish lineages have both oral jaws and pharyngeal jaws (which evolved from a more posterior pharyngeal arch), either or both of which may be toothed (**Figure 21.14**). Fraser et al. (2009) found that tooth number and positioning were highly correlated between the oral and pharyngeal jaws of Lake Malawi cichlids, one of the teleost groups possessing pharyngeal teeth, suggesting that the two sets of teeth share the same regulatory network. However, the upstream regulators determining oral versus pharyngeal jaws are very different: development of the pharyngeal arches requires Hox gene input, whereas development of the oral jaw does not. In fact, loss of Hox gene expression is thought to be one of the major genetic factors involved in the evolutionary appearance of the oral jaw. Therefore it appears that development of the first teeth in agnathans was regulated by Hox genes in the pharynx, but that this network came under different control during the evolution of teeth on the gnathostome oral jaw.

Redeployment of genes controlling various morphogenetic pathways has played fundamental and dynamic roles in evolution. Patterning of the tetrapod limb, for example, involves the expression of Hox genes in nested anterior-posterior patterns, similar to the way in which they are initially expressed during patterning of the anterior-posterior body axis (**Figure 21.15A**). Another example involves the eyespots on the wings of the Nymphalidae, a large family of butterflies. These striking concentric rings of pigmentation resemble eyes and are thought to function in deflecting predator strikes toward the wings and away from the body. The eyespots are among the last morphological features to develop. Prior to the appearance of pigments, several regulatory proteins are expressed in the eyespot primordia (cells destined to constitute the eyespot). The same proteins are also required earlier in development for segmentation and for early wing patterning, such as the differentiation between anterior and posterior compartments of the wing (**Figure 21.15B**). A recent phylogenetic analysis of eyespot gene expression in the Nymphalidae found that recruitment of four different transcription factors in the eyespot focus—the tissue that signals to surrounding cells to form the eyespot—has been very dynamic (**Figure 21.15C**; Shirai et al. 2012). Only one of the four proteins showed unambiguous evidence for a single origin of eyespot expression, whereas the others showed evidence for independent recruitment in different lineages. Another interesting observation was that different combinations of these proteins are expressed in eyespots with very similar morphology, and that similar combinations of expression occur in eyespots with very different morphologies.

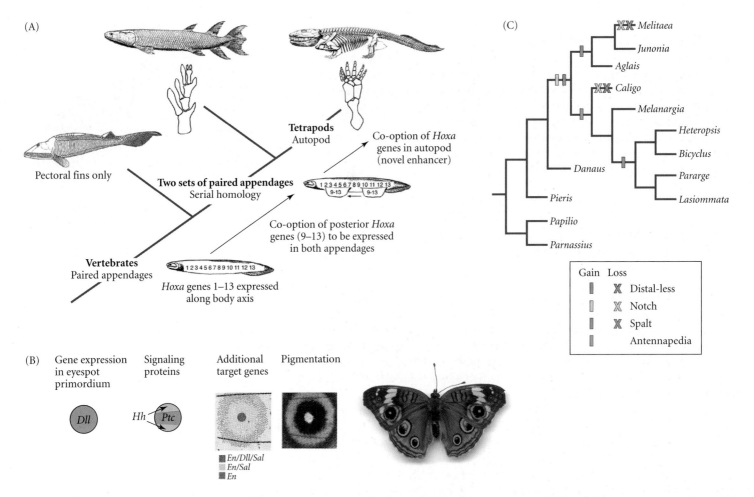

FIGURE 21.15 Co-option of developmental pathways in the evolution of novelties. (A) Co-option of the vertebrate *Hoxa* genes during evolution of the tetrapod appendage. Ancestrally, Hox genes were expressed only along the anterior-posterior axis of the developing body. The evolution of paired forelimbs and hindlimbs involved novel gene expression, presumably using novel enhancer sequences of *Hoxa* genes 9–13. The evolution of hands and feet (autopods) involved further novel expression patterns of the *Hoxa* genes. (B) Butterfly eyespots are the developmental products not only of genes for pigmentation, but also of many co-opted proteins and pathways that play important roles in establishing the body plan. These proteins include the signaling proteins Hedgehog (Hh), Notch, and Patched (Ptc), and the transcription factors Distal-less (Dll), Spalt (Sal), Antennapedia, and Engrailed (En). (C) Phylogenetic reconstruction of gains and losses of transcription factor gene expression during eyespot development in butterflies. (From True and Carroll 2002; C after Shirai et al. 2012.)

In plants, a developmental pathway has been recruited in the evolution of divided, or compound, leaves, which consist of more or less separate leaflets that develop along the main axis of the leaf (reviewed by Bharathan and Sinha 2001; Byrne et al. 2001). *KNOX1*, a member of a homeobox protein family, is expressed in the apical meristem of most plants. *KNOX1* expression must be deactivated in the leaf primordium to enable normal leaf development. In the developing compound leaves of the tomato plant, however, leaflet primordia form as bulges on the main leaf primordium, and *KNOX1* expression reappears in those regions. A growth control pathway involving protein-protein interaction among *KNOX1* family genes has been co-opted in the evolution of compound leaves, which has occurred many times in angiosperms (Goliber et al. 1999; Bharathan and Sinha 2001). Even among closely related species, there is great diversity in leaf shape. In several wild tomato species that Charles Darwin collected in the Galápagos Islands, high expression of the *KNOX1* family gene *Petroselinum* (*Pts*) is associated with complex leaf shapes (**Figure 21.16A,B**). This complex leaf phenotype is also seen in the domesticated tomato (*Solanum lycopersicum*) mutants *Pts* and *bipinnata* (*bip*) (**Figure 21.16C,D**). In both of these mutants, KNOX1 is prevented from entering the nucleus to regulate its target genes during leaf development (Kimura et al. 2008), resulting in alterations in leaf shape. It has been proposed that such variations in the amount of KNOX1 reaching the nucleus may play a major role in leaf shape evolution in many different plant lineages.

(A) *S. lycopersicum*

(B) *S. galapagense*

(C) *S. lycopersicum* mutant *Pts*

(D) *S. lycopersicum* mutant *bip*

FIGURE 21.16 Variation in compound leaf morphology in the genus *Solanum*. (A) Compound leaf of wild-type cultivated tomato (*S. lycopersicum*). 1° indicates primary leaflet, 2° indicates secondary leaflet. Scale bar = 6 cm. (B) Compound leaf from the Galápagos tomato *S. galapagense*, which has a very high degree of dissection (division into parts) of the leaves compared with wild-type *S. lycopersicum*. Scale bar = 2.5 cm. (C) Compound leaf from *S. lycopersicum* mutant *Pts*, showing primary, secondary, and tertiary (3°) leaflets. Scale bar = 6 cm. (D) Compound leaf from *S. lycopersicum* mutant *bip*, which shows a morphology similar to that of *Pts*. Scale bar = 3 cm. (Photos from Kimura et al. 2008.)

The Evolution of Morphological Form

The shapes of organismal bodies and appendages, and their allometries (see Chapter 3), represent some of the most fascinating dimensions of biodiversity. Much of the EDB research that we have highlighted so far in this chapter has been about uncovering the roles of genetic change in morphological evolution, such as which genes are involved and how their functions have changed at the molecular level. However, the other side of the mystery of the evolution of form is how the underlying cellular processes and characteristics have evolved, in terms of cell proliferation and death, cell size, and cell shape.

In *Drosophila*, the size of the wings relative to the size of the body is generally larger in colder environments, reflecting either an adaptation for better flight performance in these environments or a correlated response to selection for larger body size, which is also generally found in these environments. Latitudinal gradients of wing size in *Drosophila subobscura* have been studied in Europe, North America, and South America. The clines on the latter two continents have evolved recently and rapidly (Gilchrist et al. 2001). Surprisingly, the North American cline is largely caused by differences in wing cell size (measured as cell area on the surface of the wing, which is a very flat bilayer of epidermal cells), whereas the South American and European clines are largely caused by differences in wing cell number (Calboli et al. 2003). This observation indicates that the underlying cellular bases of morphological evolution can be diverse, reflecting abundant variation in evolutionary processes. *Drosophila* wing shape also varies among species and between sexes within species (Gidaszewski et al. 2009). A recent analysis of *D. melanogaster* mutations (Carreira et

FIGURE 21.17 The *unpaired-like* (*upd-like*) locus plays a major role in wing shape differences between two species of *Nasonia* wasps. (A) *N. vitripennis* and *N. giraulti* adults, showing differences in wing size and shape. (B) Dissected adult wings from *N. vitripennis* and a hybrid strain (wdw-135k) in which a chromosomal segment from *N. giraulti* containing the *upd-like* locus has been genetically introgressed into *N. vitripennis*. (C) Prepupal wings, showing mRNA expression of the *upd-like* gene in *N. vitripennis* and in the wdw-135k introgression strain. Arrowheads indicate the ends of the expression domains; the domain is much larger in wdw-135k than in *N. vitripennis*. Scale bar = 200 μm. (Photos from Loehlin and Werren 2012, courtesy of Jack Werren.)

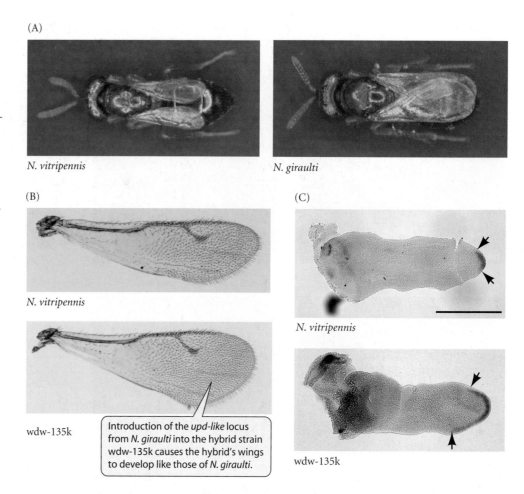

(A)

N. vitripennis   *N. giraulti*

(B)

N. vitripennis

wdw-135k

Introduction of the *upd-like* locus from *N. giraulti* into the hybrid strain wdw-135k causes the hybrid's wings to develop like those of *N. giraulti*.

(C)

N. vitripennis

wdw-135k

al. 2011) found that a majority of transposon-induced mutations had effects on wing shape and that about a third of those mutations affected the two sexes differently. This observation suggests that wing shape variation and sexual dimorphism may have a complex and polygenic basis. This conclusion is consistent with the results of a QTL analysis of a large set of genetically variable inbred lines derived from a wild California population of *D. melanogaster*, which found at least 34 QTL with significant effects on wing shape (Mezey et al. 2005). This study used genetic complementation tests, in which lines at extreme ends of the phenotypic range were crossed to laboratory mutants, and it found significant interactions of a handful of the QTL with mutations in two of the signaling pathways (Hedgehog and Decapentaplegic; see Box 21B) that are known to affect wing shape. These findings indicate that variation in cell signaling during wing morphogenesis may play a significant role in natural variation in wing shape.

Rapid advances in genome sequencing and genotyping methods are leading to insights in many non-model species as well. Loehlin and Werren (2012) used an interspecific cross with molecular marker genotyping to determine that regulatory change in a gene called *unpaired-like* is the principal genetic determinant of wing size and shape differences between two species of wasps, *N. vitripennis* and *N. giraulti* (**Figure 21.17**). The sequence of this gene is very similar to that of the *Drosophila* gene *unpaired*, which functions in cell signaling and proliferation during growth.

The developmental genetics of heterochrony

Much of morphological evolution has entailed HETEROCHRONY: evolutionary changes in the timing of development (see Chapter 3), either at the level of the whole organism or of

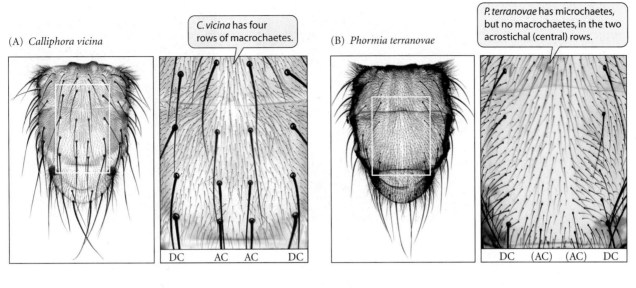

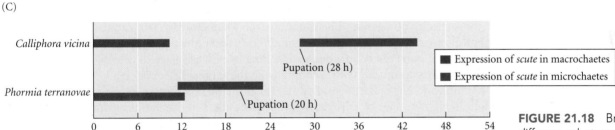

FIGURE 21.18 Bristle pattern differences between the blowflies (A) *Calliphora vicina* and (B) *Phormia terranovae* are correlated with differences in the timing of gene expression. The left photos show the dorsal thorax of each species; rectangles indicate the area of the enlargements on the right. DC = dorsocentral bristle row, AC = acrostichal bristle row. (C) Time course of *scute* expression during development of the puparium (pupal shell) and pupa of *C. vicina* and *P. terranovae*. (A, B from Skaer et al. 2002; C after Skaer et al. 2002.)

individual parts. Studying the evolution of heterochrony necessarily involves comparisons among species, and its developmental genetic basis has not yet been well studied, even in model organisms.

Modularity of regulation provides for great possible variation in the timing of activation and repression of morphogenetic pathways during development. In blowflies (family Calliphoridae), the timing of expression of the gene *scute*, which encodes a transcription factor whose expression pattern determines the patterns of the bristles on the top of the thorax, is significantly correlated with bristle pattern differences between two species, *Calliphora vicina* (**Figure 21.18A**) and its close relative *Phormia terranovae* (**Figure 21.18B**; Skaer et al. 2002). In *C. vicina*, there are four longitudinal rows of large bristles (macrochaetes) on the thorax, which are believed to reflect the condition of the common ancestor of these two species. *P. terranovae* lacks the two central rows. Both species show the same spatial pattern of *scute* expression in the wing imaginal disc during the early pupal stage, which is required in two temporal phases for bristle formation: an early phase, when macrochaetes are determined, and a late phase, when small bristles (microchaetes) are determined (**Figure 21.18C**). In *C. vicina*, these two phases of *scute* expression are separated by about 18 hours, but in *P. terranovae*, the two phases overlap, and the second phase occurs much earlier than it does in *C. vicina*. Because of this overlap, the last precursors of the macrochaetes, which in *C. vicina* are the two central rows, arise at the same time as the first precursors of the microchaetes. This causes them to develop as microchaetes and results in the absence of macrochaetes from the central region of the thorax.

One of the most common types of heterochronic evolution is PAEDOMORPHOSIS (see Chapter 3), in which juvenile traits of an ancestral species are retained in the adults of a derived species. One of the best-studied examples of a paedomorphic trait is the webbed feet of some salamanders (**Figure 21.19**), which result from the suppression of apoptosis,

FIGURE 21.19 Variation in a pae-domorphic trait. (A) Webbed foot of *Hydromantes shastae*. (B) Foot of *H. brunus*, which exhibits less webbing.

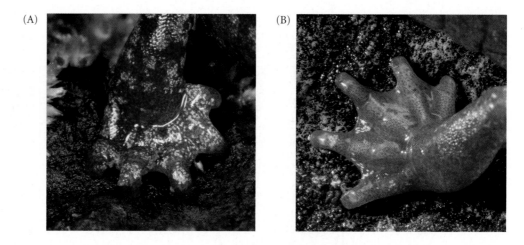

or cell death, between the developing digits, and which have evolved repeatedly in independent lineages. In both arboreal and cave salamanders, foot webbing is thought to be an important adaptation for climbing. Webbing is not the only result of paedomorphosis in most species possessing this trait, however. There are also truncations of limb development (in species that have evolved small body size) and changes in other skeletal traits. Thus it is an open question whether webbing itself was selected, or whether other paedomorphic traits correlated with webbing were the target of selection.

Adams and Nistri (2010) studied the relationship between webbing and limb development in eight species of European cave salamanders in the genus *Hydromantes*, in which foot webbing has evolved multiple times. Three of the eight species showed increases in webbing relative to other limb and digit traits during growth, whereas the other five did not. Therefore, at least for these three species, Adams and Nistri concluded that webbing itself was probably adaptive. In addition, there was more interspecific variation among species at juvenile stages than at adult stages, suggesting that each species has evolved distinct developmental trajectories all converging on a roughly uniform final adult morphology.

Salamanders have also been the subject of genetic studies of paedomorphosis (Shaffer and Voss 1996; Voss et al. 2003). In tiger salamanders (*Ambystoma tigrinum*), the tail fin and external gills are lost during metamorphosis from aquatic larva to terrestrial adult. Several related species, including the axolotl (*A. mexicanum*), attain reproductive maturity while remaining fully aquatic, retaining the gills and other larval features (see Figure 3.15). Some *Ambystoma* species are capable of facultatively transforming into the typical terrestrial adult form if their habitats dry up, whereas others have lost this ability completely.

In salamanders, thyrotropin-releasing hormone (TRH) stimulates the pituitary gland to release another hormone, thyroid-stimulating hormone (TSH), which stimulates the thyroid gland to release a third hormone, thyroxin. Thyroxin triggers metamorphosis by inducing morphogenetic events in several different tissues. In the paedomorphic salamander *A. mexicanum*, the TRH cascade does not take place, but metamorphosis can be induced by injecting TRH into the animal (Shaffer and Voss 1996). This observation suggests that the evolution of paedomorphosis in *A. mexicanum* involved inactivation of the TRH cascade. Genetic analysis of hybrids between a metamorphic strain of *A. mexicanum* (referred to as *Att*) and two different non-metamorphic strains of this species (called *lab* and *wild*) revealed that the *Att* allele of a single locus, called *met*, is necessary for metamorphosis. Recent genome-level studies of transcription suggest that the evolution of paedomorphosis in *A. mexicanum* largely involved turning off gene pathways promoting metamorphosis (Page et al. 2008, 2010).

As we saw in the blowfly case above, evolutionary changes in the timing of developmental gene expression in certain tissues of the embryo can result in substantial morphological divergence. An excellent elaboration of this process was uncovered by comparing vertebral development across diverse species (Gomez et al. 2008). The repeated segments

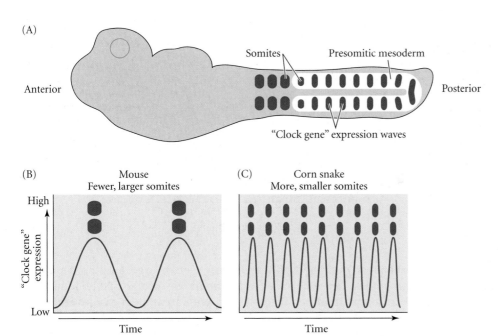

FIGURE 21.20 The segmentation clock in vertebrates. (A) Somites, which develop into individual vertebral segments and their associated structures, develop sequentially from the anterior presomitic mesoderm as a result of regular waves of expression of a battery of patterning genes called "clock genes." (B) In the mouse and other vertebrates with relatively few vertebral segments, the "clock" runs slowly. (C) The "clock" runs approximately four times faster in the corn snake, resulting in the formation of many more vertebral segments. (After Vonk and Richardson 2008.)

of the vertebrate body develop from segments of the developing embryo called SOMITES. Gomez and colleagues compared somite differentiation in zebrafish, chickens, mice, and corn snakes (*Elaphe guttata*). They found that all of these species employ a "clock-and-wavefront" mechanism whereby the mesoderm that eventually gets divided into somites (the presomitic mesoderm, or PSM) first proliferates and then buds off somite segments from its anterior end, which form the vertebral segments. Regular waves of expression of certain patterning genes cause somites to form at regular time intervals, like a ticking clock. Since there is a limited amount of starting tissue in the PSM, there is a trade-off between somite number and somite size. When there is no more PSM to bud off, somite differentiation ceases. Snakes are able to develop many more segments than most other vertebrates because their somite differentiation clock genes "tick" much faster (**Figure 21.20**; Gomez et al. 2008). However, the snake PSM produces small segments compared with other species. It is clear from the evolution of this timing mechanism that a continuous and wide range of final vertebral numbers is possible.

The evolution of allometry

ALLOMETRY refers to the differential growth rates of different parts of the body (see Chapter 3). Allometric relationships between different body parts can evolve quite rapidly, as in the dung beetle *Onthophagus taurus*. Males of this beetle exhibit a conspicuous POLYPHENISM: a non–genetically based polymorphism in which environmental differences, such as seasonal or nutritional factors, cause development to shift between alternate states. In *O. taurus*, small males develop short, stubby horns, whereas those that attain a threshold body size develop long horns that are used in male-male combat. In the mid-twentieth century, this beetle was introduced from Europe into North America and Australia in order to reduce dung accumulation in pastures. The threshold size for horn development has diverged from the ancestral condition in both introduced populations, even though variation of this magnitude was not evident in the founding populations (**Figure 21.21**).

As in the cases of heterochrony discussed above, these evolutionary changes are believed to be caused by changes in the responses of specific tissues (i.e., the cells destined to give rise to the horns) to global hormonal signals that appear to be responses to environmental factors, such as food supply. Douglas Emlen and H. Frederick Nijhout (1999) found that females and small males of *O. taurus*, which normally do not produce long horns, release a small pulse of the hormone ecdysone during the second of the five

(A)

Horned male

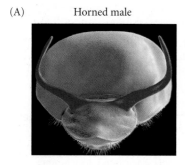

Hornless male

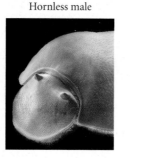

(B)

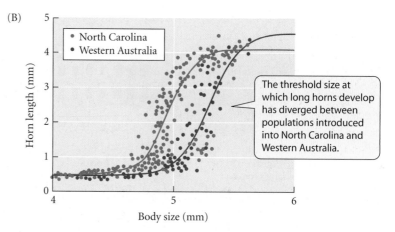

The threshold size at which long horns develop has diverged between populations introduced into North Carolina and Western Australia.

FIGURE 21.21 Rapid evolution of an allometric threshold in the dung beetle *Onthophagus taurus*. (A) Morphology of horned and hornless males at the same developmental stage; horns have been artificially emphasized with blue. (B) The allometric relationship between horn length and body size has diverged between populations of this beetle introduced into different new environments. (A, photos courtesy of Doug Emlen; B after Moczek et al. 2002.)

larval instars (stages between molting events) that is not seen in large, long-horn-producing males. In addition, males whose horns would normally be small could be induced to develop large horns if the investigators injected them with an analogue of juvenile hormone (JH) at a critical stage of the fourth larval instar. Emlen and Nijhout hypothesized that these two hormones "read" the nutritional and sexual states to switch on horn growth only in well-fed males.

Surprisingly, the role of JH is not the same among beetle species. Shelby and coworkers (2007) compared the effects of JH analogue injection in females of three *Onthophagus* species with sexual dimorphisms in both head and pronotal (anterior thoracic) horns. In two species with the standard sexual dimorphism in which males, but not females, grow large horns in both locations, JH application exaggerated the dimorphism by causing injected females to grow even smaller than normal pronotal horns. In a third species with the reverse type of dimorphism, in which females, but not males, grow large horns in both locations, JH injection had no effect on the pronotal horns, but caused the head horns to increase in size in females. Finally, one of the two species with the typical sexual dimorphism showed an alteration of horn shape toward a more male-like form in response to the JH treatment. These results show that although the hormones controlling the development of sexually dimorphic horns in beetles are conserved, there has been considerable evolutionary change in the specific actions of those hormones. One possible explanation might be that these species have diverged in the target genes that are under transcriptional regulation by the JH pathway as a result of *cis*-regulatory evolution of those target genes.

Developmental Constraints and Morphological Evolution

Evolutionary theory generally explains how natural selection, genetic drift, and gene flow, acting on the raw material of genetic variation, have produced the astonishing variety of organisms. Many evolutionary biologists have also asked what the **constraints** on evolution might be. Aside from the constraints experienced by all organisms living on Earth, such as having to develop in the presence of gravity, several more biologically interesting types of constraints on evolution have been distinguished. PHYLOGENETIC CONSTRAINTS, which affect only particular groups of related organisms, are limitations set by the genome and genetic variation of their most recent common ancestor. PHYSICAL CONSTRAINTS, such as the properties of biological materials (e.g., bones, epidermis, DNA), prevent certain morphologies from evolving. In insects, for example, respiratory gas exchange is conducted by diffusion through narrow tubes, or tracheae, that branch throughout the body. Limits on diffusion rates are thought to set an upper limit on insect body size. FUNCTIONAL (or SELECTIVE) CONSTRAINTS prevent the evolution of traits that are always disadvantageous or may interfere with the function of an existing trait. GENETIC CONSTRAINTS prevent certain evolutionary trajectories from occurring because the genetic variation enabling those trajectories is not present (see Chapter 13). John Maynard Smith and colleagues (1985) defined DEVELOPMENTAL CONSTRAINT as "a bias on the production of various phenotypes caused by the structure, character, composition, or dynamics of the developmental system." The two most common phenomena attributed to developmental constraint are absence or paucity of phenotypic variation, including the absence of morphogenetic capacity (i.e., lack of cells, proteins, or genes required for the development of a structure), and strong correlations among characters, which may result from interactions between tissues during development. Developmental pathways may also have different capacities for variation due to pleiotropy or epistasis (gene interactions) of their components.

Distinguishing among functional, genetic, and developmental constraints is difficult and generally requires extensive experimental work. One of the most thorough research programs in this area involves the evolution of eyespots on the wings of *Bicyclus* butterflies. Butterfly eyespots provide excellent evidence of the differing strengths of genetic correlations. In species bearing more than one eyespot on their wings, there are often strong correlations between the sizes and color patterns of the different eyespots. This is to be expected, because the different eyespots are likely to use similar sets of genes for their development. Artificial selection experiments have been useful in testing whether correlations among eyespots on the wings of *Bicyclus anynana* could be decoupled. Patricia Beldade and coworkers (2002) showed that the sizes of two phenotypically correlated eyespots were readily decoupled after only 11 generations of selection. Cerisse Allen and coworkers (2008) confirmed this finding (**Figure 21.22A**), but also found that unlike eyespot size, correlations in the color composition

FIGURE 21.22 Artificial selection experiments showing different levels of constraint on different eyespot traits in *Bicyclus anynana*. (A) Trajectories of selection, starting from center of plot, on dorsal forewing eyespots, with axes in units of phenotypic standard deviations. The graph shows the results of selection for both eyespots to be large or small (Large-large and Small-small, green trajectories) and for one to be large and the other to be small (Large-small and Small-large, purple trajectories). Each point represents one generation. (B) Results of similar experiments on the fourth and sixth eyespots of the ventral hindwing (Black-black and Gold-gold, blue trajectories; Black-gold and Gold-black, red trajectories) Arrows in B indicate the two eyespots that were selected to be uncoupled. The colors of the outer two concentric rings show the most noticeable differences among treatments. Note that the red trajectories in B do not extend far from the starting point at the center, indicating that selection for uncoupling the colors of the two eyespots was not successful. This is also reflected by the lack of differences in the color of these eyespots (compare upper left and lower right in B). (Courtesy of Cerisse Allen and Paul Brakefield.)

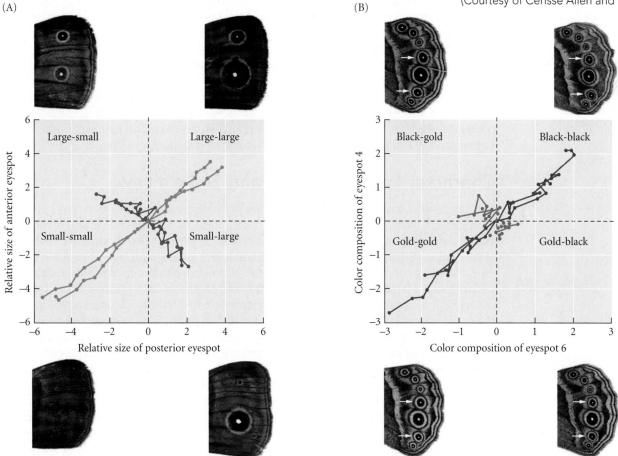

(A)

(B)

FIGURE 21.23 Evidence for developmental constraints. (A) X-ray of the right hind foot of an axolotl salamander (*Ambystoma mexicanum*), showing the normal five-toed condition. (B) The left hind foot of the same individual, which was treated with an inhibitor of mitosis during the limb bud stage. The foot lacks the postaxial toe and some toe segments, and is smaller than the control foot. (C) A normal left hind foot of the four-toed salamander (*Hemidactylium scutatum*) has the same features as the experimentally treated foot of the axolotl. (From Alberch and Gale 1985; photos courtesy of P. Alberch.)

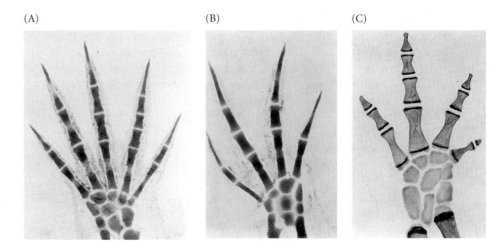

(A)   (B)   (C)

of the eyespots could not be decoupled (**Figure 21.22B**). Their results suggest an absence of genetic variation in the regulation of pigmentation genes that would allow them to respond differently to selection on different regions of the wing.

Developmental constraints can also be revealed by embryological manipulations in the laboratory. In a classic experiment, Pere Alberch and Emily Gale (1985) used the mitosis-inhibiting chemical colchicine to inhibit digit development in the limb buds of frogs (*Xenopus*) and salamanders (*Ambystoma*; **Figure 21.23A,B**). The treatment consistently caused specific digits to be missing in each species, and the missing digits were the preaxial ones in frogs and the postaxial ones in salamanders. These results reflected the different order of digit differentiation in the two taxa: the last digits to form tended to be the most sensitive to the colchicine treatment. Furthermore, the results strongly reflected evolutionary trends: salamanders have often lost postaxial digits (**Figure 21.23C**), and frogs have repeatedly experienced preaxial digit reduction, during evolution. Although the digit number variation in this study was produced artificially, the results suggest that naturally occurring variation in developmental systems may be constrained by intrinsic, species-specific developmental programs.

Although in practice it is very difficult to rule out functional constraints, developmental or genetic constraints might explain some common evolutionary patterns, such as morphological stasis (see Chapter 22), the absence of features in certain lineages, parallel evolution, and the relative reversibility of morphological changes (see Chapter 3).

Character Loss, Reversal, and Dollo's Law

One of the most puzzling features of evolutionary trajectories revealed by recent molecular phylogenetic analyses has been the apparent disappearance and reappearance of morphological characters. A long-standing notion in biology is DOLLO'S LAW (Simpson 1953), which states that evolutionary loss of complex characters is virtually always irreversible. However, recent comparative and developmental studies have found many cases that appear to run counter to this generalization (Collin and Miglietta 2008; Cronk 2011). Detailed ontogenetic analyses of rapidly evolving characters in a variety of species indicate that the development of structures can be dynamically altered by natural selection (as we have already seen in the case of foot webbing in salamanders and bristle patterns in flies). Evolutionary loss of characters may often be the result of **developmental arrest** of a character that developed fully in an ancestral species (e.g., as in paedomorphic characters). Ehab Abouheif and Gregory Wray (2002) showed that in the development of wingless castes of ants, the tissues that give rise to adult wings go through some of the patterning events of ancestral wing development, but the process is arrested. After arrest, the partially developed tissue is resorbed. The sexually dimorphic morphology of the horns of *Onthophagus* beetles has also been shown to result from the differential timing and extent

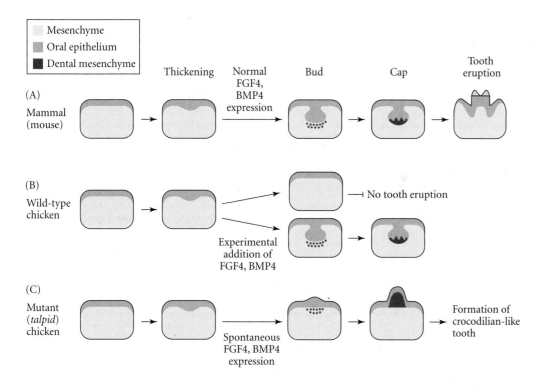

FIGURE 21.24 Early tooth development in a mouse, wild-type chicken, and *talpid* mutant chicken. (A) At the thickening stage, a battery of genes—two of which encode the signaling proteins BMP4 and FGF4—are expressed in the mouse (and other mammals), resulting in tooth formation. (B) In wild-type chickens, these genes are not expressed, resulting in developmental arrest after the thickening stage and no tooth formation. However, when BMP4 and FGF4 proteins are added to experimental (in vitro) cultures, dental mesenchyme forms, and tooth development proceeds to the cap stage. (C) In the *talpid* mutant chicken, BMP4 and FGF4 expression occurs spontaneously, and dental mesenchyme forms, resulting in an outgrowth that eventually forms a structure similar to a crocodilian tooth. The process by which this structure forms lacks the later stages typical of mammalian tooth development. (After Collin and Miglietta 2008.)

of developmental arrest and tissue resorption in male and female larvae (Moczek 2006; Kijimoto et al. 2010).

The retention of some or all of the developmental capacity to produce a structure, even though it is absent in the adult form, provides a mechanistic basis for character reversals in evolution. Evidence of the developmental potential to "re-evolve" a character has been found in modern birds, which lack teeth, but are descended from dinosaurs, which possessed teeth. A mutant strain of chicken, called *talpid*, forms toothlike structures in early development that are indistinguishable from early-stage crocodilian teeth (Harris et al. 2006). Teeth arise from a ridge of oral epithelial cells that overlies mesenchyme tissue derived from embryonic neural crest cells. Thickening of this dental mesenchyme then leads to the formation of a tooth bud, which eventually develops into a tooth. These events require specific sets of genes to be expressed in the oral epithelium and dental mesenchyme (**Figure 21.24A**). Very early in chicken development, this process begins, but then stops. The chicken tissue is capable of continuing further in tooth development when two proteins, BMP4 and FGF4, are added (**Figure 21.24B**; Collin and Miglietta 2008). In *talpid* mutants, these additions are not needed because the genes of the tooth development pathway are expressed, resulting in the formation of early-stage tooth structures (**Figure 21.24C**; Harris et al. 2006).

Ecological Developmental Biology

Biologists have long known that the environment in which development occurs is as critical to its outcome as the genetic pathways that direct it. The phenomena of phenotypic plasticity and polyphenism (see Chapter 13) indicate that the capture of reliable signals from the environment has resulted in morphological and life history adaptations. Research on such adaptations has occurred for over a century, but recently, with significant inspiration from EDB, the field has coalesced into the discipline of ECOLOGICAL DEVELOPMENTAL BIOLOGY (Gilbert and Epel 2009).

As we saw in Chapter 13, the reaction norm of a trait is a representation of its phenotypic plasticity in relation to an environmental variable. If the reaction norm encompasses the phenotypic optimum for one or more new environments, then no new selection on the genotype may be needed, and adaptive evolution of the trait will not occur in the new environment. Mary Jane West-Eberhard (2003) and others have proposed that new environments may actually induce the development of phenotypes that fortuitously turn out to be adaptive. Subsequently, such phenotypes may be subject to genetic assimilation (see Chapter 13), whereby alleles contributing to the tendency for the new advantageous trait to develop will be selectively favored and fixed. West-Eberhard (2003) points out many examples of closely related species pairs in which the reaction norm of one species includes a phenotype that is invariant in the other species. For example, larvae of the moth *Manduca quinquemaculata* develop black pigmentation at low temperatures and green pigmentation at higher temperatures, whereas larvae in the related species *M. sexta* develop green pigmentation at all temperatures (Suzuki and Nijhout 2006). In another possible example, in the lizard *Anolis sagrei*, individuals reared in the presence of narrow perching sticks tended to develop slightly shorter hindlimbs than those reared in the presence of broader sticks. This difference correlates with differences among *Anolis* species that tend to perch on narrower versus thicker twigs (Losos et al. 2000).

Since new environments are often stressful, they may disrupt or alter development in ways that are not adaptive. However, some alterations of development in response to a change in environment may happen to move the phenotype in the direction of the new phenotypic optimum (**Figure 21.25A**; Ghalambor et al. 2007). Until a species experiences a new environment, however, the capacity for development to change in a potential new environment can be considered a neutral trait (**Figure 21.25B**).

One example of potentially maladaptive phenotypic plasticity occurs in vertebrates living at low altitudes, including humans. Acclimatizing to higher altitudes entails constriction of the blood vessels throughout the lungs, which causes increased blood flow away from the surfaces of gas exchange (Storz et el. 2010). This response causes pulmonary hypertension, which has potentially health-threatening consequences. Therefore, in order for adaptation to higher altitudes to occur, natural selection needs to overcome this plasticity such that individuals having the lowest magnitude of this response become the most fit.

In many organisms, specific cues from the environment are sensed during development, resulting in outcomes that are clearly adaptive. If adaptive phenotypic plasticity has enabled rapid adaptive evolution in one species, its ancestral species, or a close relative with the ancestral trait, can be used as a model to test for the presence of genetic variation in plasticity. This idea was recently tested in spadefoot toads (*Spea*; Ledon-Rettig et al. 2010). Most anuran tadpoles, including the ancestor of the genus *Spea*, feed on detritus, but when *Spea* tadpoles develop in the presence of shrimp, they often switch to feeding on shrimp and other animal prey, such as other tadpoles (Pfennig and Murphy 2000). This shift involves development of a shorter gut and a larger head and mouth to accommodate larger prey. Ledon-Rettig

FIGURE 21.25 Evolution of adaptive phenotypic plasticity. (A) Consequences of phenotypic plasticity in a population when placed in a novel environment, represented by concentric circles of increasing fitness in a two dimensional environmental space; the optimal phenotype (X_1) for the new environment is within the center circle. The phenotype is implicitly understood to be a composite of several traits. Each vector represents a genotype with a different reaction norm between the two environments, a starting phenotype in the current environment (X) and different phenotypes in the new environments (X_1, X_2, and X_3). Vector A represents a genotype with perfect phenotypic plasticity. It develops the optimal phenotype, and no evolutionary change ensues. Vector B represents a genotype whose plasticity does not encompass the optimum. Selection may favor enhanced and somewhat reoriented plasticity. Vector C is a genotype with nonadaptive plasticity that moves the phenotype away from the optimum and thus does not facilitate adaptation to the novel environment. (B) Trajectories of phenotypic plasticity of three different genotypes within and outside the normal range of environments encountered by a species. In a population occupying a novel environment, plasticity outside the normal range is unpredictable and is expected to evolve as a neutral trait, as it is never exposed to selection. (After Ghalambor et al. 2007.)

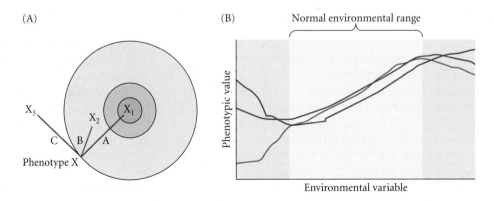

(A)

(B)

et al. found that tadpoles of *Scaphiopus*—the sister genus of *Spea*, which has the ancestral diet—developed more slowly and had shorter guts if raised on shrimp or exposed to the hormone corticosterone than if fed detritus and not exposed to the hormone. (Corticosterone is known to be elevated in tadpoles fed a carnivorous diet.) Moreover, the heritability of these traits also increased in the experimental food and hormone treatments compared with the detritus treatment. This finding demonstrates that cues from novel environments can uncover previously cryptic genetic variation, which can then be selected.

As more potential examples of the role of the environment in adaptive developmental evolution are elucidated, the next step in understanding this process will be to identify the molecular genetic basis of these adaptations. Alison Scoville and Michael Pfrender (2010) studied the molecular basis of an apparent case of adaptive plasticity in the water flea *Daphnia melanica*, which inhabits freshwater lakes at various altitudes. High-altitude lakes are exposed to higher levels of ultraviolet radiation (UV) than are lakes at lower altitudes. In high-altitude lakes, *D. melanica* develops dark pigmentation as a protection against UV (**Figure 21.26A**). However, this dark pigmentation makes them more visible to fish predators than individuals with light pigmentation, which is the usual phenotype of this species at lower altitudes (**Figure 21.26B**). Scoville and Pfrender showed that two ancestral populations in lakes where fish predators were absent were phenotypically plastic, expressing higher melanization (darker pigmentation) when placed in high-UV environments, whereas two independently derived populations in lakes into which predatory trout had been introduced by humans showed constitutive expression of the low-melanin (light-pigmentation) phenotype (**Figure 21.26C**). The occurrence of this change

(A)  (B)

FIGURE 21.26 Adaptive phenotypic plasticity in the pigmentation of *Daphnia melanica*. (A) The dark morph. (B) The light morph. (C–E) Reaction norms of *D. melanica* from lakes with predators versus those from lakes without predators, in response to varying intensities of UV. (C) Melanin concentration. (D) Expression of *Ddc*. (E) Expression of *ebony*. The genes *Ddc* and *ebony* interact to influence pigment production. (Photos courtesy of Michael Pfrender; C–E after Scoville and Pfrender 2010.)

▲ ⎫ Ancestral populations   ▲ ⎫ Derived populations
● ⎭ (predators absent)   ● ⎭ (predators present)

(C)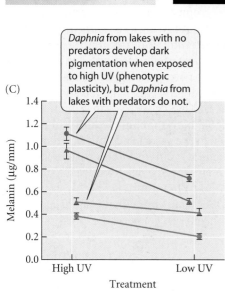

Daphnia from lakes with no predators develop dark pigmentation when exposed to high UV (phenotypic plasticity), but *Daphnia* from lakes with predators do not.

(D)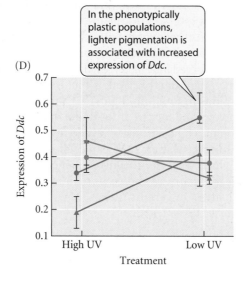

In the phenotypically plastic populations, lighter pigmentation is associated with increased expression of *Ddc*.

(E)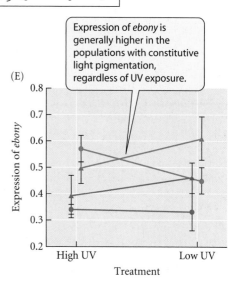

Expression of *ebony* is generally higher in the populations with constitutive light pigmentation, regardless of UV exposure.

in reaction norms independently in two different lineages is a strong indicator that the change is adaptive.

Using candidate gene information from insects, Scoville and Pfrender then tested the involvement of the genes *Ddc* (*Dopa decarboxylase*), which is required for production of precursors for both melanized and non-melanized cuticle, and *ebony*, which is required for development of non-melanized cuticle. *Ddc* expression was plastic in the ancestral populations, and its expression was higher in low-UV environments (**Figure 21.26D**), but this plasticity was lost in the derived populations. This loss of plastic *Ddc* expression was associated with a shift toward constitutive (nonplastic) high *ebony* expression (**Figure 21.26E**), which promotes the light-pigmentation phenotype that is adaptive in the presence of predators. This study represents one of the first examples in which genetic assimilation in nature has been traced to regulatory changes in specific genes.

Evolution of Human Development

As the field of evolutionary developmental biology unfolds, one of its most important and fascinating endeavors will be to elucidate the developmental genetic and evolutionary mechanisms involved in the appearance of traits unique to humans, such as our large brain size, craniofacial morphology, vertebral, limb, and digit innovations, reduced hair cover, and, of course, our complex behavioral and cultural traits (reviewed by Carroll 2003). Based on studies in model organisms, we can expect that many of the innovations that evolved in the human lineage involved several or many genes. Comparative genomic data indicate that many or most of the DNA-level changes responsible caused alterations in the regulation of developmental and structural proteins that we share with our primate and other mammalian relatives. Identifying the functional roles (if any) of nucleotide differences in noncoding DNA will be difficult, but these potential regulatory regions have long been thought to be of primary importance in differences between humans and chimpanzees (King and Wilson 1975). Comparative genomics is providing powerful new approaches to this task.

One clue that a particular regulatory region might have an important function is that its sequence is highly conserved among species. Prabhaker and colleagues (2008) scanned for such regions by comparing the genome sequences of humans and other primates. They assayed the gene expression directed by one such region, called *human-accelerated conserved noncoding sequence-1*, or *HACNS1*, which showed an intriguing cluster of human-specific nucleotide substitutions at positions that were otherwise conserved among non-human vertebrates (**Figure 21.27A**). The expression assays were done by engineering reporter constructs of this sequence from humans, chimpanzees, and rhesus monkeys, and inserting these into transgenic mice. Constructs from all three species were expressed in the developing eye, ear, and pharyngeal arch, but interestingly, only the human construct was expressed in the anterior region of the forelimb and hindlimb buds (**Figure 21.27B**). In further experiments, these expression differences were attributed to a cluster of 13 nucleotide substitutions in humans, found in an 81-base-pair region that is otherwise highly conserved among terrestrial vertebrates (see Figure 21.27A). An interesting hypothesis emerging from this work is that the genes near *HACNS1* play roles in human-specific innovations in forelimb digit structure, which enable high manual dexterity, and in the morphology and relative inflexibility of the toes, which are associated with bipedalism.

Genomic comparisons have also revealed deletions in humans of otherwise conserved DNA. In a recent study (McLean et al. 2011), 510 human-specific deletions (hDELs), almost all of which were noncoding sequences, were found to be otherwise highly conserved among chimpanzees and other mammals. One hDEL, near a gene that is associated with growth arrest in various tissues, removed an enhancer with conserved expression in the

(A)

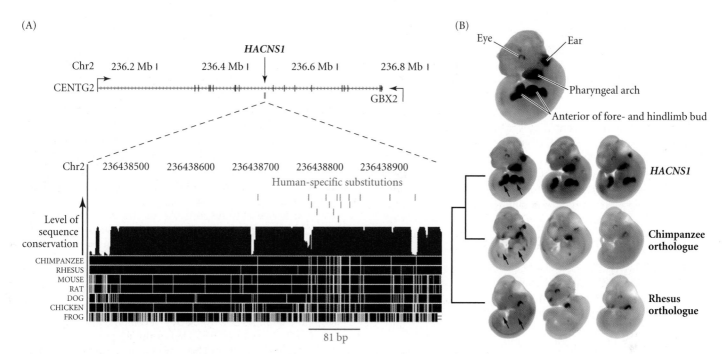

(B)

FIGURE 21.27 Gain of function in a developmentally expressed enhancer in the human lineage. (A) The *human-accelerated conserved noncoding sequence-1* (*HACNS1*) region is highly conserved among vertebrates. The top diagram represents the human genome region surrounding *HACNS1*; vertical bars represent known transcribed genes. The bottom diagram shows conservation of the *HACNS1* region among vertebrates. The dark blue area indicates the degree of conservation. Black vertical bars below the dark blue area indicate conserved regions among taxa. Red vertical bars indicate human-specific nucleotide substitutions. Gray vertical bars in nonhuman taxa indicate differences from the human sequence. Yellow vertical bars indicate conservation among nonhuman vertebrates, but not humans. (B) Evolution of *HACNS1* expression in humans as assayed by reporter construct expression (dark blue) in transgenic mice during embryonic development (three independent transgenic lines are shown for each construct). The top figure and top row show human *HACNS1* expression with primordia of structures indicated. The second and third rows show expression of chimpanzee and rhesus monkey orthologues, respectively. Arrows indicate the forelimb and hindlimb buds. The absence of expression of the chimpanzee and rhesus HACNS1 orthologues in the limb buds reflects the ancestral state. (From Prabhakar et al. 2008. © AAAS. Used by permission.)

ventral telencephalon and diencephalon of the forebrain in early mouse and chimpanzee development (human and chimpanzee expression were assayed in mice using reporter constructs). The chimpanzee and mouse sequences showed expression in cultured human cells, indicating that the enhancer would probably function in humans if it were present. Cell proliferation in these areas of the brain is associated with the great expansion of the primate and human neocortex, suggesting that the deletion of this tissue-specific enhancer, which ancestrally functioned in suppressing growth, might have had a significant evolutionary role.

As these recent examples illustrate, large-scale identification of the genomic changes responsible for developmental changes in the human lineage usually requires experimental work. Many clues to the genetic bases of human traits will also come from studying human variation, including genetic disorders, and development in many different model species. The examples cited in this chapter illustrate new opportunities to elucidate the history and dynamics of the molecular genetic variation contributing to phenotypic evolution at scales ranging from variation within populations to macroevolutionary differences among higher taxa.

Summary

1. Evolutionary developmental biology (EDB) seeks to integrate data from comparative embryology and developmental genetics with theory and data on morphological evolution and population genetics.

2. Phylogenetic homology can differ from biological homology, which reflects many recent discoveries that independently evolved traits can have similar genetic and developmental bases.

3. Many of the genes and developmental pathways underlying morphogenesis in multicellular organisms are conserved across wide phyletic ranges, showing that the vast diversity of multicellular eukaryotes is largely due to diverse uses of a highly conserved "toolkit" of genes and developmental pathways.

4. Some genes may be involved in phenotypic evolution more often than others in the same pathway (e.g., *yellow* in *Drosophila* species and *Mc1r* in vertebrates), suggesting that not all components of developmental genetic pathways are readily available for natural selection to act upon, possibly because of differential pleiotropy among genes.

5. Noncoding DNA sequences called enhancers or *cis*-regulatory elements independently control expression of each gene by binding different sets of transcription factors that are present in different areas of the developing body. The evolution of differences in gene expression is largely due to changes in these enhancers.

6. Developmental pathways include signaling proteins, transcription factors, and structural genes. Evolutionary change in the regulatory connections among signaling pathways and transcription factors, and between transcription factors and their targets, is believed to underlie much of the phenotypic diversity seen in nature. Morphological variation within and among species may be caused by changes in either regulatory or protein-coding sequences. The relative importance of these two fundamental types of genetic change to phenotypic evolution is currently under debate.

7. Modularity among body parts is achieved by patterning mechanisms whose regulation is often specific to certain structures, segments, and life history stages. Modularity has been important in enabling different parts of the body to develop divergent morphologies. Differences in morphology among segments of bilaterian animals, regulated by Hox proteins, provide many classic examples of mosaic evolution enabled by modular developmental pathways that can be deployed differently in each segment.

8. During evolution, genes and developmental pathways have often been co-opted, or recruited, for new functions, a process that is probably responsible for the evolution of many novel morphological traits. This process results from evolutionary changes in gene regulation and function. The proliferation of many gene families, especially those encoding transcription factors, is associated with the evolution of novel gene functions and expression patterns during development.

9. Many differences among species are due to heterochronic or allometric changes in the relative developmental rates of different body parts or in the rates or durations of different life history stages. The modularity of morphogenesis in different body parts and in different developmental stages facilitates such changes.

10. Several kinds of constraints on evolution may determine that certain evolutionary trajectories are followed and others are not. Some correlations of traits in natural populations can be easily broken using artificial selection, whereas others cannot. This observation suggests that some pairs of traits are constrained to vary in limited ways due to limitations in the function and regulation of the underlying genes.

11. Arrest of developing structures can underlie morphological differences between closely related species and plays a significant role in the evolutionary loss of complex morphological traits and their reappearance in some lineages.

12. Ecological developmental biology is a field of growing interest that studies the interactions between developmental processes and environmental cues and how these interactions have been involved in adaptation.

Terms and Concepts

biological homology
cis-regulatory elements
constraints
co-option
developmental arrest

developmental pathways (= developmental circuits)
enhancers
evolutionary developmental biology (EDB)

genetic toolkit
homeotic selector genes
Hox genes
modularity
promoters
recruitment

regulatory modularity
reporter constructs
target genes
transcription factors
trans-regulation

Suggestions for Further Reading

Evolutionary biologists' interest in development was rekindled in the 1970s largely by Stephen Jay Gould, who also portrayed the early history of the subject in *Ontogeny and Phylogeny* (Harvard University Press, Cambridge, MA, 1977). An excellent, very readable introduction to contemporary evolutionary developmental biology, emphasizing regulation of gene expression, is *From DNA to Diversity: Molecular Genetics and the Evolution of Animal Design*, by S. B. Carroll, J. K. Grenier, and S. D. Weatherbee, second edition (Blackwell Science, Malden, MA, 2005). A more extended treatment is *The Evolution of Developmental Pathways*, by A. S. Wilkins (Sinauer Associates, Sunderland, MA, 2002).

For the more general science reader, Sean B. Carroll describes the development of EDB in *Endless Forms Most Beautiful: The New Science of Evo-Devo* (W.W. Norton, New York, 2006) and connects EDB to both the fossil record and the fundamentals of natural selection in *The Making of the Fittest:*

DNA and the Ultimate Forensic Record of Evolution (W.W. Norton, New York, 2007).

Rudolph A. Raff, one of the founders of EDB, portrayed the field somewhat earlier in his influential book *The Shape of Life: Genes, Development, and the Evolution of Animal Form* (University of Chicago Press, Chicago, 1996). Eric H. Davidson, a developmental biologist whose insights helped to shape the field, does much the same in *The Regulatory Genome: Gene Regulatory Networks in Development and Evolution* (Academic Press, London, 2006).

For a recent review of EDB from a plant perspective, see N. D. Pires and L. Dolan, "Morphological evolution in land plants: New designs with old genes," *Philosophical Transactions of the Royal Society* B 367: 508–518 (2012).

Defining one of the most rapidly growing research areas in EDB, Scott Gilbert and David Epel have written a comprehensive text, *Ecological Developmental Biology* (Sinauer Associates, Sunderland, MA 2009.)

Problems and Discussion Topics

1. If two allometrically related traits show a strong correlation both within and among species, what kinds of experiments would you use to test whether these correlations are due to natural selection or to developmental genetic constraints? If some correlations, but not others, are found to be constrained, what does this suggest about the underlying genes?

2. How might differential expression of and regulation by Hox genes contribute to mosaic evolution in which different segments of an animal body plan evolve different morphology?

3. If mutations such as those of the *Ubx* gene can drastically change morphology in a single step, why should most evolutionary biologists maintain that evolution has generally proceeded by successive small steps?

4. Describe how convergent evolution of similar morphology in two different lineages might involve DNA sequence evolution in different parts of the same developmental gene.

5. In various groups of organisms, such as insects and flowering plants, morphological traits have been lost in some species due to the evolution of developmental arrest, followed by remodeling or resorption of the tissue originally destined to develop into the ancestral structure. What does this observation suggest about the nature of mutations and adaptive evolution in these cases?

6. How might the regulatory DNA sequences underlying a heterochronic change (e.g., an increase in the developmental rate of the larval stage of an insect, resulting in a shorter larval period) differ from those responsible for an evolutionary novelty in a particular body segment (e.g., a novel wing pigmentation spot)? What spatio-temporal components of the developmental system would you expect to be acting in these two cases, and what sorts of genes (encoding transcription factors, signaling proteins, hormones) would you look for as candidate genes underlying these two types of evolutionary change?

7. Development of a morphological structure involves many different types of gene products, including transcription factors, signaling proteins, and "effectors" such as enzymes. When a morphological change occurs in a single mutational step, which of these types might be more or less likely to be involved? Within a gene, would such single-step events be more likely to involve coding or noncoding sequences, and what characteristics of the gene's function might affect this likelihood?

Macroevolution: Evolution above the Species Level

T he phenomena of evolution are often divided into **microevolution** (meaning, mostly, processes that occur within species) and **macroevolution**, which is often defined as "evolution above the species level." "Macroevolution" has slightly different meanings to different authors. To Stephen Jay Gould (2002, p. 38), it meant "evolutionary phenomenology from the origin of species on up." These macroevolutionary phenomena include patterns of origination, extinction, and diversification of higher taxa, the subject of Chapter 7. To other authors, macroevolution is restricted to the evolution of great phenotypic changes, or the origin of characteristics that diagnose higher taxa (e.g., Levinton 2001). The subject matter of macroevolutionary studies, however defined, includes patterns that have developed over great periods of evolutionary time—patterns that are usually revealed by paleontological or comparative phylogenetic studies, even if their explanation lies in genetic and ecological processes that can be studied in living organisms. Thus we want to know how fast evolution happens and what determines its pace, whether the great differences that distinguish higher taxa have arisen gradually or discontinuously, what the mechanisms are by which novel features have come into existence, and whether or not there are grand trends, or progress of any kind, in the history of life.

Much of the modern study of macroevolution stems from themes and principles developed by the paleontologist George Gaylord Simpson (1947, 1953), who focused on rates

A major change in form and function The tusk of the male narwhal (*Monodon monoceros*), used in establishing dominance hierarchies, is a highly modified upper left canine tooth. Characteristics that are novel in evolution have almost always been modifed from ancestral features that differed in function.

and directions of evolution perceived in the fossil record, and by Bernhard Rensch (1959), a zoologist who inferred patterns of evolution from comparative morphology. Contemporary macroevolutionary studies draw on the fossil record, on phylogenetic patterns of evolutionary change, on evolutionary developmental biology, and on our understanding of genetic and ecological processes.

Rates of Evolution

As we noted in Chapters 4 and 7, rates of evolution vary greatly. Simpson (1953), who pioneered the study of evolutionary rates, distinguished rates at which single characters or complexes of characters evolve (which he called PHYLOGENETIC RATES) from TAXONOMIC RATES, the rates at which taxa with different characteristics originate, become extinct, and replace one another. Steven Stanley (1979) analyzed taxonomic rates of evolution in several clades that have increased exponentially in species diversity during the Cenozoic. Among the most rapidly radiating groups are murid rodents (mice and rats) and colubrid snakes, diverse groups that arose in the Miocene and took about 1.98 My and 1.24 My, respectively, to double in species number. These rates mean that, without extinction, each rodent species would speciate, on average, within about 2 My (assuming that each species bifurcates into two "daughter" species). This interval is roughly the same as, or even greater than, the time required for speciation that has been estimated from genetic differences between sister species of living organisms (see Chapter 18). Thus a duration of 1 or 2 million years per speciation event is more than enough to account for the evolution of great diversity, even in the most species-rich groups.

Individual characters evolve at rates that differ greatly (see Chapters 4 and 13). Many features in fossil lineages show the pattern that Niles Eldredge and Stephen Jay Gould (1972) called **punctuated equilibrium**: long periods of little change (**stasis**) interrupted by brief episodes of much more rapid change (**Figure 22.1A**). During these brief periods (of hundreds of thousands of years), the rate of change per generation is about the same as, or less than, the evolutionary rates of characteristics that have been altered by novel selection

FIGURE 22.1 Averaged over long periods, the rate of evolution may be low, even though there are episodes of rapid evolution. (A) A punctuated pattern of shift from one rather static character mean to another. If we had fossils only from times 1 and 3, we would not know there had been periods of slow and rapid evolution. (B) A pattern of rapid fluctuations in the character mean, but with little net change over a long period. Given only fossils from times 1 and 8, we might think that little evolution had occurred.

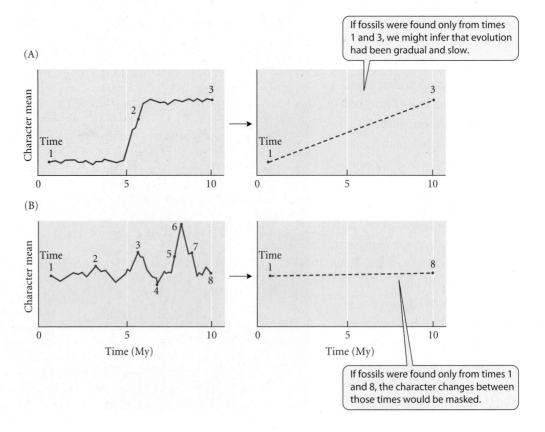

pressures within the last few centuries (see Chapter 13). Over time intervals of several million years, the average rate of evolution of most characters is much lower, because the long-term average rate masks both "punctuational" episodes of rapid evolution and fluctuations around a long-term mean (**Figure 22.1B**).

In their theoretical model of punctuated equilibrium, Eldredge and Gould (1972), applying Ernst Mayr's model of founder effect speciation (see Chapter 18) to the fossil record, proposed that stasis may be caused by genetic constraints, which are loosened in small, local populations that become new species. Although paleontologists have described examples in which morphological change has been associated with true speciation (e.g., the bryozoan genus *Metrarhabdotos*; see Figure 4.22), the punctuated equilibrium hypothesis requires that morphological evolution be almost inevitably accompanied by speciation. Luke Strotz and Andrew Allen (2013) provided evidence that this is the case in Cenozoic planktonic Foraminifera (single-celled protists), but whether or not the hypothesis holds for other groups is uncertain. Geographic variation within species and the rapid adaptive evolution of populations exposed to new selection pressures show that speciation is not required for adaptive phenotypic change. Moreover, there is rather little evidence for peripatric or founder-flush speciation, which some population geneticists consider unlikely (see Chapter 18). Eldredge (1989) and Gould (2002, p. 796) themselves came to agree that speciation is not a necessary trigger for adaptive, directional morphological evolution.

The controversy about punctuated equilibrium had the healthy effect of drawing attention to many interesting questions about macroevolution. Perhaps most importantly, it established "stasis" as an important and puzzling phenomenon (**Figure 22.2**). Rapid evolution is not a problem for evolutionary biology to explain, because rapid rates in the fossil record are fully consistent with information on mutation, genetic variation, short-term rates of evolution in the recent past (see Chapter 13), and divergence among populations and among closely related species (see Chapter 17). The problem, rather, is to explain why evolution is often so slow. For example, Stanley and Yang (1987) measured 24 shell characters in large samples of 19 lineages of bivalves. They compared their measurements for early Pliocene (4 Mya) fossils of each lineage with those for their nearest living relatives (most of which bear the same species names as the fossils). They compared these differences, in turn, with variation among geographic populations of 8 of the living species. With few exceptions, the difference over the span of 4 My was no greater than the difference among contemporary conspecific populations (**Figure 22.3**). Similarly, in phylogenetic analyses, the body size of terrestrial vertebrates seems to evolve hardly at all over time spans of less than a million years, even though it is known to be capable of evolving rapidly (Uyeda et al. 2011; see Figure 18.30).

Three major hypotheses have been proposed to account for stasis within species lineages. First, Eldredge and Gould (1972) proposed that stasis is caused by internal *genetic or developmental constraints*, which would be manifested by lack of genetic variation or by genetic correlations too strong to permit characters to evolve independently to new optima. But although such constraints may indeed play a role in evolution, they cannot explain the constancy of size and shape of many quantitative characters, which are almost always genetically variable and only imperfectly correlated with one another (see Chapter 13).

The second, most commonly suggested explanation of stasis is *stabilizing selection* for a constant optimum phenotype (Charlesworth et al. 1982). It may seem unlikely that natural selection could favor the same character state over millions of years, during which both physical and biotic environmental factors would almost inevitably change. A dramatic example of such change is the series of glacial and interglacial episodes of the Pleistocene, during which climates, geographic ranges of species, and associations among species changed drastically and repeatedly (see Chapters 5 and 6). However, a species' "effective environment" may be far more constant over time than we might expect because of **habitat tracking** (Eldredge 1989): the shifting of the geographic distributions of species in concert with the distribution of their typical habitat. The distribution of cold-climate plants such as hemlock (*Tsuga canadensis*), for example, has shifted southward during glacial times and northward during interglacial times (see Figure 5.29). Many animals, by actively choosing a favored habitat (**habitat selection**), are buffered against some environmental changes.

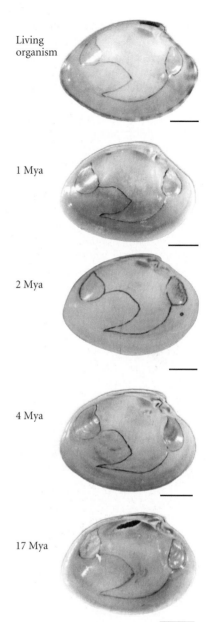

Living organism

1 Mya

2 Mya

4 Mya

17 Mya

FIGURE 22.2 An example of stasis: specimens of the bivalve *Macrocallista maculata* from a living population and from fossil deposits dated at 1, 2, 4, and 17 Mya. All are from Florida. Scale bars = 1 cm. (Photos courtesy of Steven M. Stanley.)

FIGURE 22.3 A quantitative expression of stasis in shell characters of bivalves in the fossil record. Each graph plots the number of characters, in several species, that show a certain percentage difference in the mean between samples of (A) different living geographic populations and (B) living and Pliocene populations. The two upper graphs show measurements pertaining to the outline of the shell; the lower graphs show interior measurements. The variation between Pliocene and living populations is about the same, overall, as that between different geographic populations of living species. (After Stanley and Yang 1987.)

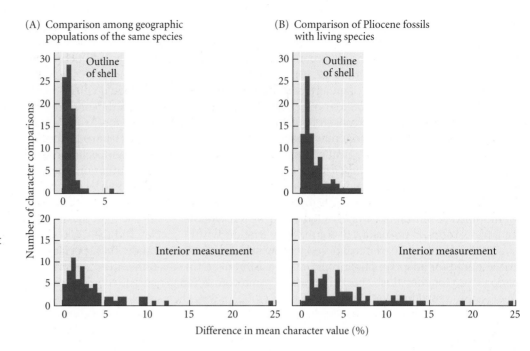

Third is the possibility that *most evolutionary changes in local populations do not contribute to long-term change* because they remain spatially localized and are temporally evanescent. Niles Eldredge, John Thompson, and coauthors (2005) point out that if an adaptive change arises in a geographically localized population of a species, it will become a feature of the entire species (and perhaps of descendant species as well) only if it is spread to other populations by gene flow and is advantageous in those populations. They note that many, perhaps most, advantageous variants are advantageous in only part of a species' geographic range because of geographic variation in selection pressures. For example, interactions with other species, such as predators, parasites, and competitors, vary from place to place in a "geographic mosaic of coevolution," and so will select for different and perhaps everchanging characteristics in different populations of a species (Thompson 2004; see Chapter 19). In a similar vein, Futuyma (1987) proposed that an adaptation—perhaps to a different resource—that arises in one local population may not easily spread to other places where it would be advantageous if migrants interbreed with intervening populations that are adapted to the ancestral resource. Moreover, if the geographic distribution of patches with the new resource shifts (perhaps because of climate change), the divergent character state may be lost as a result of interbreeding with the (probably widespread) ancestral phenotype as individuals disperse and colonize newly favorable sites, and as the populations from which they dispersed become extinct. Therefore the existence of the new phenotype might be as brief as the intervals between climate-induced shifts in the geographic distribution of resources and species, and for this reason might not be registered in the fossil record. Speciation, however, could preserve the new adaptation (see Figure 18.31). Thus, although speciation may not cause anagenetic adaptive change, it may confer long life on such changes, leading to a possible association between speciation and morphological evolution (i.e., the *pattern* of punctuated equilibrium). Whether for this reason or others, there is some phylogenetic evidence that morphological change is accelerated by speciation (Mattila and Bokma 2008; Monroe and Bokma 2010). This scenario implies *that long-term evolutionary change may be more likely to occur in relatively stable than in frequently changing environments*. Evidence of sustained evolution in stable environments has led some paleontologists to the same conclusion, and there is considerable evidence that the drastic climate fluctuations of the Pleistocene inhibited both speciation and persistent adaptive phenotypic change (Jansson and Dynesius 2002).

Gradualism and Saltation

Darwin proposed that evolution proceeds gradually, by small steps. In *The Origin of Species*, he wrote that "if it could be demonstrated that any complex organ existed, which could not possibly have been formed by numerous, successive, slight modifications, my theory would absolutely break down." His ardent supporter Thomas Henry Huxley, however, cautioned that Darwin's theory of evolution would be just as valid even if evolution proceeded by leaps (sometimes called **saltations**). Some later evolutionary biologists, such as the paleontologist Otto Schindewolf (1950) and the accomplished geneticist Richard Goldschmidt, proposed just this. Goldschmidt argued in *The Material Basis of Evolution* (1940) that species and higher taxa arise not from the genetic variation that resides within species, but instead "in single evolutionary steps as completely new genetic systems." He postulated that major changes of the chromosomal material, or "systemic mutations," would give rise to highly altered creatures. Most would have little chance of survival, but some few would be "hopeful monsters" adapted to new ways of life. Goldschmidt's genetic system hypothesis has been completely repudiated, but the possibility of evolution by more modest jumps remains one of the most enduring controversies in evolutionary theory. Quite different species are often connected by intermediate forms, so that it becomes arbitrary whether the complex is classified as two genera (or subfamilies, or families) or as one (see Chapter 3). Nonetheless, there exist many conspicuous gaps, especially among higher taxa such as orders and classes. No living species bridge the gap between cetaceans (dolphins and whales) and other mammals, for example.

The most obvious explanation of phenotypic gaps among living species is extinction of intermediate forms that once existed—as the cetaceans themselves illustrate. Of course, the common ancestor of two quite different forms need not have appeared precisely intermediate between them, because the two phyletic lines may have undergone quite different modifications (**Figure 22.4**). For instance, DNA sequences imply that, among living animals, whales are most closely related to hippopotamuses, but early fossil cetaceans do not have a hippo-like appearance in the slightest (see Figure 4.11). Although the fossil record provides many examples of the gradual evolution of higher taxa (see Chapter 4), many higher taxa appear in the fossil record without intermediate antecedents. The punctuated equilibrium hypothesis, proposed to explain the sudden appearance of only slightly different phenotypes, allows that changes in morphology might (or might not) have been continuous, passing through many intermediate stages, but so rapid and so geographically localized that the fossil record presents the appearance of a discontinuous change. The saltation hypothesis, in contrast, holds that intermediates never existed—that mutant individuals differed drastically from their parents.

In judging gradualism and saltationism, we must distinguish between the evolution of *taxa* and of their *characters*. Gradualists hold that many characters of higher taxa evolved independently and sequentially (MOSAIC EVOLUTION). Both comparisons of living species and the fossil record provide abundant evidence of mosaic evolution (see Chapters 3 and 4), although developmentally integrated features can sometimes evolve as a unit (see Chapter 13). There is some disagreement, however, on whether each of the distinguishing *characters* of a higher taxon—the reduction and the fusion of birds' tail vertebrae, for example—might have evolved discontinuously.

Certainly many mutations arise that have discontinuous, large, even drastic effects on the phenotype. Many of these, however, have such important pleiotropic effects that they greatly reduce viability. A mutation of the *Ultrabithorax* gene (*Ubx*) in *Drosophila*, for example, converts the halteres into wings (see Figure 21.1). It may be tempting to think that this mutation reverses evolution, and that a mutation in this gene caused the evolutionary transformation of the second pair of insect wings

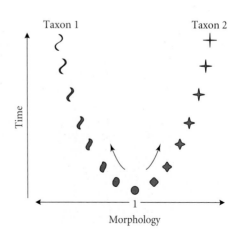

FIGURE 22.4 Two very different taxa may have evolved gradually from a common ancestor, even though no form precisely intermediate between them ever existed.

into halteres in the ancestor of the Diptera. The *Ubx* mutation does not restore "real" hind-wings, however; it transforms the third thoracic segment into a replicated second segment, including a replicated set of forewings. The ancestors of flies did not have identical second and third segments, and the hindwings differ from forewings in all four-winged insects. Mutations that reduce the function of this master control gene interfere with a complex developmental pathway, and development is routed into a "default" pathway that produces the features of the second thoracic segment (including wings). The whole system can be shut down in a single step by turning a master switch, but that does not mean the system came into existence by a single step.

What, then, explains the observation that major morphological differences between higher taxa, such as the presence or absence of flower parts (see Figure 21.6), often appear to be largely caused by a major developmental switch gene? In some cases, the gene merely extends or truncates a developmental trajectory without causing harmful side effects. For example, the few genetic changes that determine the heterochronic difference between metamorphosing tiger salamanders (*Ambystoma tigrinum*) and their paedomorphic relative the axolotl (*A. mexicanum*; see Figure 3.15) do not engender an entirely new complex morphology, but merely truncate a complex, integrated pathway of development that presumably evolved by many small steps. In other cases, as David Stern and Virginie Orgogozo (2008, 2009) propose, there has been stepwise, gradual change by the accumulation of successive mutations—but these mutations have occurred mostly in a single HOTSPOT GENE that controls a key point in a developmental pathway. The important feature of such a gene is that it has few pleiotropic effects on other characters, so mutations are less likely to have deleterious side effects that would prevent them from increasing by natural selection. For example, larvae of *Drosophila sechellia* lack the dorsal trichomes (hairlike extensions of cell cuticle) possessed by its relatives, such as *D. melanogaster* (**Figure 22.5A**). The absence of trichomes (a derived trait) is caused by mutations in three different *cis*-regulatory regions of a single gene, *shavenbaby*, that regulates expression of three downstream batteries of genes that are necessary for trichome development, and which itself is regulated by an array of upstream genes that determine where *shavenbaby* is expressed (**Figure 22.5B**). Mutation of any single downstream gene would not suffice to alter trichome development, and mutation of the upstream genes, all of which have multiple functions, would be likely

FIGURE 22.5 How a morphological difference between species may be caused by repeated evolution in a single gene. (A) Larval *Drosophila sechellia* (right) lack the dorsal trichomes that are present in *D. melanogaster* (left) and other relatives. (B) The absence of trichomes is caused by several mutations in the *cis*-regulatory control region of the *shavenbaby* gene. This gene is regulated by several developmental patterning genes, and in turn regulates the expression of downstream genes that determine actin distribution, membrane matrix, and cuticle, which together transform an epidermal cell into a trichome. (From Stern and Orgogozo 2009.)

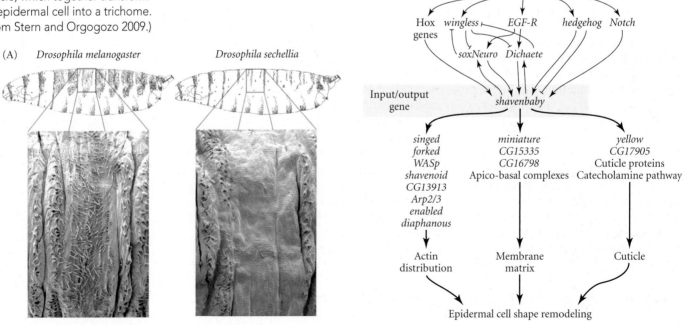

to alter other organs as well. This model suggests that the key genes that control the development of very divergent homologous organs may have accumulated their divergent effects gradually, perhaps in part by successively recruiting different downstream genes and pathways.

Neo-Darwinians have always accepted that characters can evolve by minor jumps—that is, by mutations that have fairly large (but not huge) effects. For example, variation both within and among species in characters such as bristle number in *Drosophila* is often caused by a mixture of quantitative trait loci with both small and large effects (Orr and Coyne 1992; see Chapters 13 and 18). Alleles with large effects contribute importantly to mimetic phenotypes in butterflies such as *Heliconius* (see Figure 19.12), in which a species has converged toward the phenotype of another unpalatable species. Were phenotypes to arise that deviated only slightly from one mimetic pattern toward another, they would lack protective resemblance to either unpalatable model and presumably would suffer a disadvantage. Thus it is likely that the evolution of one mimetic pattern from another was initiated by a mutation of large enough effect to provide substantial resemblance to a different model species, followed by selection of alleles with smaller effects that "fine-tuned" the phenotype (**Figure 22.6**). Genetic analysis of the color patterns in *Heliconius* supports this hypothesis (Baxter et al. 2009).

FIGURE 22.6 A model of the evolution of color pattern in Müllerian mimics such as *Heliconius* butterflies. The red curves represent the degree of protection against predators for two species, A and B, that differ in phenotype and abundance. The wavy lines indicate the distribution of phenotypes within each species. Selection favors convergence of the less abundant species B toward the pattern of the more abundant species A because predators more often learn to avoid the more abundant species. A mutation of small effect that slightly alters the phenotype of species B will be selectively disadvantageous. However, a mutation of large effect that causes members of species B to acquire a phenotype with a modest resemblance to species A (e.g., just left of phenotype *y*) will be selectively advantageous. Subsequently, allele substitutions with small effects that bring B closer to the peak for A will be advantageous. (After Charlesworth 1990.)

> The less abundant species B can converge toward the pattern of the more abundant species A only if a mutation changes its phenotype to the left of *y*, because only such a mutation will increase protection.

Phylogenetic Conservatism and Change

Given the great diversity of phenotypes among even closely related species, and given the rapidity with which evolution can occur, biologists are challenged to understand the existence of "living fossils"—organisms such as the ginkgo (see Figure 5.19B,C), the tadpole shrimp *Triops granarius* (**Figure 22.7A**), and the coelacanth (**Figure 22.7B**) that have changed so little over many millions of years that they closely resemble their Mesozoic or even Paleozoic relatives. The synapomorphies (shared derived characters) of large clades also represent conservatism: almost all mammals, no matter how long or short their necks, have seven neck vertebrae, and almost no tetrapods have had more than five digits per limb. (The earliest tetrapods had more, but soon settled on five.) The puzzle of stasis in individual lineages is magnified in such instances, in which characters have remained little changed despite extensive speciation, and despite diversification in other characteristics.

(A)

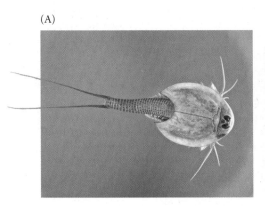

(B)

FIGURE 22.7 Two "living fossils." (A) The tadpole shrimp *Triops granarius*, found in temporary pools in arid regions of Eurasia and northern Africa, has undergone no evident morphological evolution since the Triassic. (B) Coelacanths are lobe-finned fishes that originated in the Devonian and were thought to have become extinct in the Cretaceous, until this living species, *Latimeria chalumnae*, was discovered in 1938.

The hypotheses proposed to explain phylogenetic conservatism include stabilizing selection and internal constraints. These explanations may be related to each other, and perhaps amount to much the same thing.

One important reason why the optimal condition of a character may remain unchanged is **phylogenetic niche conservatism**: long-continued dependence of related species on much the same resources and environmental conditions (Holt 1996; Wiens and Graham 2005). For example, closely related species of plants have climatically similar geographic ranges in Asia and in North America (Ricklefs and Latham 1992); the larvae of all species of the butterfly tribe Heliconiini feed on plants in the passionflower family (Passifloraceae), and apparently have done so since the tribe originated in the Oligocene. By occupying one niche (e.g., host plant group, climate zone) rather than another, a species subjects itself to some selection pressures and screens off others; it may even be said to "construct" or determine its own niche, and therefore many aspects of its potential evolutionary future (Lewontin 2000; Odling-Smee et al. 2003). Niches remain conservative for two major, sometimes interacting, reasons: First, other species, often by acting as competitors, may prevent a species from shifting or expanding its niche; conversely, lineages may evolve and radiate when competition is alleviated (see Chapters 3, 4, and 7). Second, and more generally, if there is gene exchange among individuals that inhabit the ancestral niche (e.g., microhabitat) and those that inhabit a novel niche, and if there is a fitness trade-off between character states that improve fitness in the two environments, then selection will generally favor the ancestral character state (i.e., stabilizing selection will prevail) simply because most of the population occupies the ancestral environment (Holt 1996).

As the degree of adaptation to any one environment increases, the differential in fitness between that environment and a new environment also increases, so that adaptation to an alternative environment may become steadily less likely. In some cases, a species may lose the ability to vary in features that would be necessary for a substantial ecological shift. For example, blue and purple pigments, products of a biosynthetic pathway that uses the substrate cyanidin, are ancestral in the morning glory genus *Ipomoea*, but are replaced in a bird-pollinated clade by red pigments that are synthesized from pelargonidin (**Figure 22.8**). Rebecca Zufall and Mark Rauscher (2004) showed that in a red-flowered species, the F3'H enzyme, which initiates the cyanidin pathway, has undergone both regulatory and functional inactivation, so that only the pelargonidin pathway is active. Perhaps as a consequence of the inactivation of the cyanidin pathway, another enzyme, DFR, has lost the ability to metabolize its substrate in the cyanidin pathway (although it is still active in the pelargonidin pathway). The authors point out that it would require at least two or three restorative changes to evolve blue pigments anew. In general, unused genes acquire disabling mutations such as stop codons and become pseudogenes, as shown by such examples as the reduced number of functional olfactory receptors in primates and the degeneration of many genes and phenotypic functions in parasites.

(A)

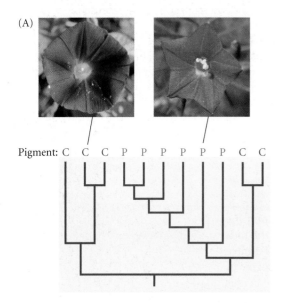

Pigment: C   C   C   P   P   P   P   P   P   C   C

(B)

3 Malonyl-Co-A + Coumaroyl-CoA

↓ CHS

Chalcone

↓ CHI

Naringenin

↓ F3H

Dihydrokaempferol (DHK)

↓ F3'H

Dihydroquercetin (DHQ)

↓ DFR

Leucocyanidin   Leucopelargonidin

↓ ANS   ↓ ANS

Cyanidin   Pelargonidin

↓ UF3GT   ↓ UF3GT

Cyanidin-3-glucoside   Pelargonidin-3-glucoside

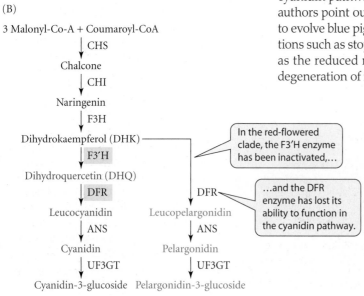

In the red-flowered clade, the F3'H enzyme has been inactivated,…

…and the DFR enzyme has lost its ability to function in the cyanidin pathway.

FIGURE 22.8 Adaptive genetic changes may restrict subsequent evolutionary potential. (A) A molecular phylogeny of species of *Ipomoea* shows that a red-flowered clade, with pelargonidin-derived pigment (P), is derived from blue-flowered ancestors with cyanidin-derived pigment (C). (B) The anthocyanin biosynthetic pathway, with branches to cyanidin (blue) and pelargonidin (red) pigments. In the red-flowered clade, evolutionary changes in two enzymes in the pathway to cyanidin make it unlikely that blue coloration could re-evolve. (After Zufall and Rauscher 2004; photos courtesy of Mark D. Rauscher.)

FIGURE 22.9 Complex structures, if lost, are generally not regained, but their function may be. (A) *Hesperornis*, a marine bird of the late Cretaceous, had teeth that enabled it to grip fish. (B) A typical living fish-eating bird, the grey heron (*Ardea cinerea*), lacks teeth and has a smooth-edged bill. (C) The bill of a red-breasted merganser (*Mergus serrator*), a fish-eating duck, has a substitute for teeth: serrations that enable it to grip fish. (A courtesy of Larry Martin, in Feduccia 1999.)

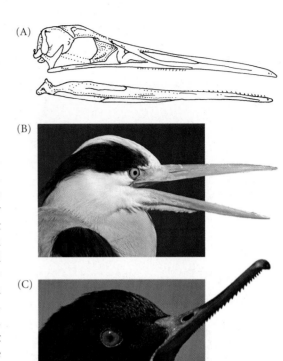

The loss of gene function is one likely basis of DOLLO'S LAW, the generalization that a complex character, if lost, is seldom regained in its original form. Although exceptions to Dollo's law are known (see Figure 3.7; Chapter 21), the loss of ancestral developmental and biosynthetic pathways is highlighted by cases in which other features have been modified to serve the function of a lost character. For example, tooth development is initiated in birds, but stops at a very early stage unless two proteins are added that are not normally expressed (see Figure 21.24). A mutation is known that enables tooth development to proceed further (Collin and Miglietta 2008); nevertheless, birds have not had teeth since the end of the Cretaceous. "Substitute teeth," serrations of the bill margin, have evolved in mergansers, enabling these ducks to catch fish (**Figure 22.9**). Similarly, an adaptive function is often performed not by the structure we might expect, but by another structure that has been modified instead. Moles and the giant panda, for example, have independently evolved a sixth apparent finger, which enlarges the mole's hand for digging and enables the panda to manipulate the bamboo on which it feeds. In both cases, the outermost "finger" (or "thumb"), is not a true digit, but a sesamoid bone that develops from cartilage (**Figure 22.10**). The development of the mole's sesamoid is associated with the expression of two genes that are important in development of the true digits

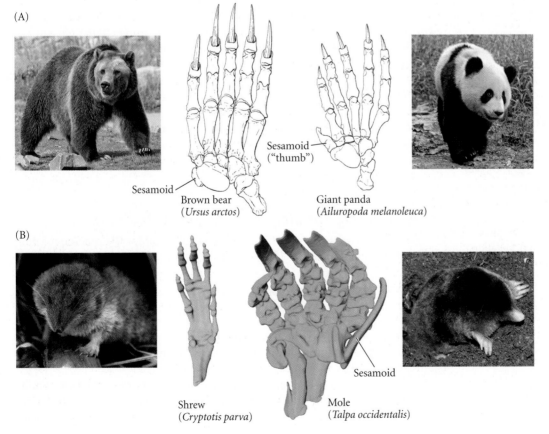

FIGURE 22.10 False fingers ("thumbs") have evolved from sesamoid bones in the giant panda (A) and in moles (B). The unmodified hands of these animals' close relatives, a bear (in A) and a shrew (in B), are shown for comparison. Developmental constraints probably prevented the evolution of a sixth true digit. These structures exemplify what has been called "tinkering" by natural selection: evolution of adaptations from whatever variable characters a lineage happens to already have. (After Davis 1964 and Mitgutsch et al. 2011; shrew and mole hand photos courtesy of Marcelo Sánchez-Villagra.)

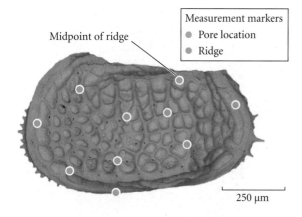

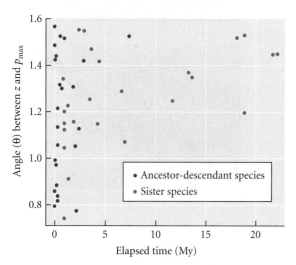

FIGURE 22.11 The relationship between elapsed time and the propensity for evolutionary changes to be in directions close to the axis of greatest phenotypic variation in fossilized lineages of the marine ostracode crustacean *Poseidonamicus*. For multiple measures of the carapace (above), p_{max} expresses the character combination with maximal variation within populations, and z is the direction of change between ancestor-descendant transitions or between sister species derived from a common ancestor. Smaller angles represent a direction of evolution closer to p_{max} (see Figure 13.16). Although there is much variation in the direction of evolution over short intervals, it tends to depart more over longer time spans from the direction that the variation within populations would predict. (After Hunt 2007b; photo courtesy of Gene Hunt.)

(Mitgutsch et al. 2011), but the genes that are necessary for true digit structure (e.g., the phalanges and claw) evidently are not expressed in that region of the developing hand.

From a genetic point of view, there are many possible kinds of constraints. Because of genetic correlations stemming from pleiotropy, genetic variation may exist for certain combinations of characters, but not others, even if each character individually is variable (Blows and Hoffmann 2005). If genes have consistent patterns of pleiotropy, such character correlations may be very long lasting, and they could be an important determinant, in the long run, of the pattern of genetic variation that is available for evolutionary change (Jones et al. 2007). Perhaps for this reason, researchers are increasingly finding that the directions of evolutionary differences among species are correlated with the "genetic lines of least resistance" that have been estimated from genetic or phenotypic correlations in contemporary populations (see Figure 13.15). For example, the evolution of carapace shape characteristics in multiple lineages of the early Tertiary ostracode crustacean *Poseidonamicus* proceeded largely in the direction of those character combinations that showed the greatest variation within populations (Hunt 2007b; **Figure 22.11**). Similarly, the pattern of divergence in wing shape among various clades of *Drosophila* is broadly similar to the genetic variance-covariance matrix within *D. melanogaster*, suggesting that the pattern of genetic variation has been fairly consistent for more than 50 My (Hansen and Houle 2008). On the other hand, experiments have shown that pleiotropic correlations can be reduced by selection at other loci that compensate for pleiotropic effects (see Chapter 13; Pavlicev and Wagner 2012). Just how long genetic constraints may persist is not yet known.

The Evolution of Novelty

How do major changes in characters evolve, and how do new features originate? These questions have two distinct meanings. First, we can ask what the genetic and developmental bases of such changes are—the subject of Chapter 21. Second, we can ask what role natural selection plays in their evolution. For instance, we may well ask whether each step, from the slightest initial alteration of a feature to the full complexity of form displayed by later descendants, could have been guided by selection. What functional advantage can there be, skeptics ask, in an incompletely developed eye? And we can ask how complex characters could have evolved if their proper function depends on the mutually adjusted form of each of their many components.

Accounting for incipient and novel features

Several pathways of evolutionary change account for the macroevolution of phenotypic characters (see Mayr 1960; Nitecki 1990; Müller and Wagner 1991; Galis 1996).

A feature may originate as a *new structure* or as a *novel modification of an existing structure*. For example, sesamoid bones often develop in connective tissue. Such bones are the origin of novel skeletal elements, such as the extra "finger" of the giant panda and mole (see Figure 22.10) and the patella (kneecap) in mammals, which is lacking in reptiles (Müller and Wagner 1991).

A feature may be a *by-product* of other adaptive features, perhaps initially nonadaptive but later recruited or modified to serve an adaptive function. For instance, by excreting nitrogenous wastes as crystalline uric acid, insects

FIGURE 22.12 A lungless bolitoglossine salamander (*Hydromantes supramontis*) captures prey with its extraordinarily long tongue. The rapid tongue extension is accomplished with a modified hyobranchial apparatus, which in other families of salamanders plays an important role in ventilating the lungs. (From Deban et al. 1997, courtesy of S. Deban.)

lose less water than if they excreted ammonia or urea. Excreting uric acid is surely an adaptation, but the white color of uric acid is not. However, pierine butterflies such as the cabbage white butterfly (*Pieris rapae*) sequester uric acid in their wing scales, imparting to the wings a white color that plays a role in thermoregulation and probably in other functions.

Decoupling the multiple functions of an ancestral feature frees it from functional constraints and may lead to its elaboration. For example, the locomotory muscles of many reptiles insert on the ribs, so that these animals cannot breathe effectively while running. In birds and mammals, the muscle insertions have shifted to processes on the vertebrae, so that breathing and running (or flying) are decoupled, and many of these lineages have evolved features associated with rapid locomotion (Galis 1996). David Wake (1982) has proposed that the loss of lungs in the largest family of salamanders (Plethodontidae) has relieved a functional constraint on the evolution of the tongue. In other salamanders, the bones that support the tongue are also used for moving air into and out of the lungs. In plethodontids, this hyobranchial skeleton, no longer used for ventilating the lungs, has been modified into a set of long elements that can be greatly extended from a folded configuration. This modification enables plethodontids to catch prey by projecting the tongue, in some species to extraordinary lengths at extraordinary speed (**Figure 22.12**).

Duplication with divergence gives rise to diversity at the morphological level, as it does at the level of genes and proteins (see Chapter 20). The diversity of a mammal's multiple teeth, for example, provides enormous functional possibilities: molars can slice or grind, while canines stab. Some teeth—such as the tusks of elephants and the upper left canine tooth of the male narwhal (see p. 605)—are used more for social interactions than for eating.

A *change in the function* of a feature alters the selective regime, leading to its modification. This principle, already recognized by Darwin, is one of the most important in macroevolution (Mayr 1960), and every group of organisms presents numerous examples. A bee's sting is a modified ovipositor, or egg-laying device. The wings of auks and several other aquatic birds are used in the same way in both air and water; in penguins, the wings have become entirely modified for underwater flight (see Figure 11.18). Many proteins have been co-opted or modified for new functions, such as lactate dehydrogenase and the many other enzymes that, with little or no modification, form the crystallin lens in vertebrate eyes (see Figure 11.19).

Complex characteristics

FUNCTIONAL INTERMEDIATES A common argument against Darwinian evolution is based on what is sometimes termed "irreducible complexity": the proposition that a complex organismal feature cannot function effectively except by the coordinated action of all its components, so that the feature must have required all of its components from the beginning. Since they could not have all arisen in a single mutational step, the feature could (it is claimed) not have evolved.

Needless to say, the first person to recognize this potential problem was Darwin himself, in *The Origin of Species*: "That the eye, with all its inimitable contrivances for adjusting the focus to different distances, for admitting different amounts of light, and for the correction of spherical and chromatic aberration, could have been formed by natural selection seems, I freely confess, absurd in the highest possible degree." But he then proceeded to supply examples of animals' eyes as evidence that "if numerous gradations from a perfect and complex eye to one very imperfect and simple, each grade being useful to its possessor, can be shown to exist; if further, the eye does vary ever so slightly, and the variations be inherited, which is certainly the case; and if any variation or modification in the organ be ever useful to an animal under changing conditions of life, then the difficulty of believing that a perfect and complex eye could be formed by natural selection, though insuperable by our imagination, can hardly be considered real."

Darwin's claim has been fully supported by later research (e.g., Nilsson and Pelger 1994; Osorio 1994). The eyes of various animals range from small groups of merely light-sensitive cells (in some flatworms, annelid worms, and others), to cuplike or "pinhole camera" eyes (in cnidarians, molluscs, and others), to the "closed" eyes, capable of registering precise images, that have evolved independently in cnidarians, snails, bivalves, polychaete worms, arthropods, and vertebrates (**Figure 22.13**). The evolution of eyes is apparently not so improbable! Each of the many grades of photoreceptors, from the simplest to the most complex, serves an adaptive function. Simple epidermal photoreceptors and cups are most common in slowly moving or burrowing animals; highly elaborated structures are typical of more mobile animals. The molecular basis of vision has also evolved by comprehensible steps. The visual pigments, or opsins, evolved from a G protein-coupled receptor protein of the kind that is ubiquitous in animals and fungi (Oakley and Pankey 2008), and we have seen that lens crystallin proteins have evolved from a variety of other proteins. Neither at the morphological nor the molecular level is the notion of "irreducible complexity" a barrier to evolution.

The antievolutionary argument also fails to recognize that a component of a functional complex that was initially merely superior can become indispensable because other characters evolve to become functionally integrated with it. Although the eyes of many

FIGURE 22.13 Intermediate stages in the evolution of complex eyes. (A) Schematic diagrams of stages of eye complexity in various animals, from a simple photosensitive epithelium, through the deepening of the eye cup (providing progressively more information on the direction of the light source), through gradual evolution toward a "pinhole camera" eye, eventually including a refractive lens and a pigmented iris for more perfect focusing. (B) Most of these stages can be found among various gastropod species, as shown in these drawings. (A after Osorio 1994; B after Salvini-Plawen and Mayr 1977.)

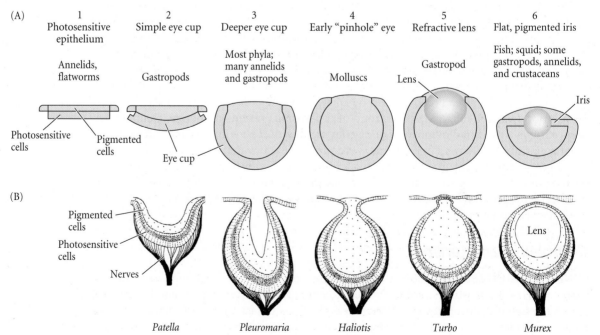

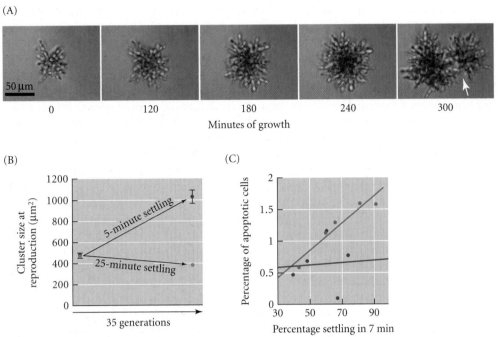

(A)

50 μm

0   120   180   240   300

Minutes of growth

(B)

Cluster size at reproduction (μm²)

5-minute settling

25-minute settling

35 generations

(C)

Percentage of apoptotic cells

30   50   70   90

Percentage settling in 7 min

FIGURE 22.14 Incipient multicellularity evolved in laboratory cultures of yeast under artificial selection. (A) Growth and division (reproduction) of a multicellular cluster of cells. The arrow at 300 minutes shows where a single cluster is dividing into two. (B) Over 35 generations, populations selected for more rapid settling (5 minutes) evolved to delay reproduction until they reached a larger cluster size. Under relaxed selection (25-minute settling time), there was a slight decline in cluster size. (C) After 60 generations (red), the rate of settling was correlated with the proportion of apoptotic cells, which by dying enabled cluster division. In earlier generations (blue) there was no such correlation. (After Ratcliff et al. 2012; photos courtesy of W. C. Ratcliff.)

animals do not have a lens, those animals do quite well without the visual acuity that a lens can provide. But a lens is indispensable for eagles, since their way of hunting prey has been acquired and made possible only by such acuity. Eagles and mammals have *acquired* dependence on the elements of a complex eye. Such dependence, indeed, is often lost: many burrowing and cave-dwelling vertebrates have degenerate eyes.

The first steps in the evolution of multicellularity, surely an epitome of complexity, have been observed in laboratory cultures of yeast (*Saccharomyces cerevisiae*). Clusters of yeast cells settle more rapidly than single cells. William Ratcliff and colleagues (2012) used gravity to select rapidly settling mutants. Within 60 generations, all of their replicate selected populations were dominated by "snowflake-like" clusters of genetically identical cells that adhered instead of separating after mitosis (**Figure 22.14**). Selection for more rapid settling resulted in larger clusters that developed longer before they fragmented, and they produced more fragments, of smaller size. Clusters broke into fragments by the death (apoptosis) of individual cells that linked parts of the cluster—an example of altruism attributable to kin selection, made possible by the genetic identity of the cluster's cells. The evolution of higher rates of apoptosis and larger clusters implies that selection at the level of clusters was stronger than selection among individual cells.

DEVELOPMENTAL BIOLOGY AND COMPLEXITY Throughout the sciences, understanding of complex phenomena has been advanced by "reductionist" investigation: dissecting a system into low-level components, analyzing them, and then reconstructing the elements of the larger system. Today biologists' understanding of how complex organismal phenotypes evolve is being greatly enhanced by advances in molecular, cell, and developmental biology. We now know that animal development is heavily based on a GENETIC TOOLKIT consisting largely of genes that encode transcription factors and components of signaling pathways, and that many of them have been conserved since the origin of animals, or even earlier (Carroll et al. 2001; see Chapter 21). No fewer than 79 percent of mouse genes, for example, are homologous across animal phyla, and thus date from Precambrian time (Gerhart and Kirschner 2007). Many of these genes have retained similar functions across phyla, such as *Pax6*, which initiates the development of photoreceptors, and *Distal-less*, which controls the development of outgrowths such as limbs. But we also know that the location and timing of gene expression can be altered by changes in the *cis*-regulatory elements of the genes that transcription factors regulate. Furthermore, changes in *cis*-regulatory elements can alter functional relationships between controlling and controlled genes (Prud'homme et al. 2007). *Distal-less*, for example, controls not only limb development, but also spot patterns in butterfly wings (see Figure 21.15). *Ultrabithorax* (*Ubx*), a Hox gene instrumental in specifying the identity of body regions, represses the development of hindwings in *Drosophila*, so that they develop into halteres, whereas in the beetle *Tribolium*, *Ubx* is necessary for normal development of the hindwings, and it activates different genes than in *Drosophila* (Tomoyasu et al. 2005; Prud'homme et al. 2007). In such cases, changes in transcription factors evoke different expression of entire downstream pathways: a single genetic change may trigger coherent changes in the activity of many genes.

These revelations of evolutionary developmental biology contribute to the theory of FACILITATED VARIATION espoused by Marc Kirschner and John Gerhart (2005). These authors argue that the "core processes" of protein activity and cell and organ development have properties of *robustness* and *adaptability* that, together with the establishment of new regulatory interactions among ancient, widely shared genes, cause variation to arise in ways that facilitate evolution. For example, developing muscles, nerves, and blood vessels in a limb respond to signals from developing bone and dermis, and so grow into their proper positions. Thus genetic changes in the limb skeleton result in altered but functional limbs, without the necessity for independent genetic changes in musculature and vasculature. Similarly, differences in beak size and shape among the Galápagos finches of the genus *Geospiza* are based on growth differences in the prenasal cartilage (pnc) and later in the premaxillary bone (pmx). These differences correspond to different patterns of expression of the *Bmp* (bone morphogenetic protein) gene in the pnc and of three other genes in the pmx (**Figure 22.15**). When the level of expression of these genes was experimentally altered in chicken embryos, the genes produced the same effects seen in the finch species; enhancing *Bmp* expression, for example, increased all dimensions of the beak, whereas the genes expressed in the pmx increased depth and length, but not width (Abzhanov et al. 2004; Mallarino et al. 2011). These genes encode transcription factors and components of signaling pathways with well-known functions in craniofacial development. Changes in their expression produce altered but well-formed beaks because, Kirschner and Gerhart suggest, developmental and other organismal processes have been selected to be coordinated and resistant to perturbations—they have been canalized (see Chapter 13). Consequently, genetic and environmental variation in the resulting characters is more likely to be viable than we might otherwise suppose, facilitating the evolution of complex characters.

The discovery of aspects of facilitated variation such as recruitment of genes and signaling pathways for new functions is helping biologists understand the origin of novelties. For example, an entire developmental pathway may be triggered heterotopically in a different part of the body. A Mexican plant, *Lacandonia schismatica*, has perfectly formed stamens in the center of the flower, surrounded by carpels, the reverse of the usual arrangement (Kramer and Jaramillo 2005). Similarly, the most anterior digit of a bird's hand is morphologically equivalent to digit 1 in the hand of related dinosaurs, and it expresses the genes characteristic of a first digit (or "thumb"). However, it develops in digit position 2 and is phylogenetically homologous to the dinosaur's second digit. During the evolution of

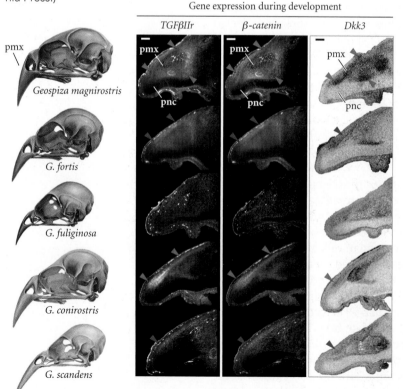

FIGURE 22.15 Among species of Darwin's finches (*Geospiza*), differences in the depth and length of the premaxilla (pmx) are determined largely by differences in the expression of three genes at a critical stage in development. The colored arrows indicate differences in gene expression between species with similar body size but different beak morphology [e.g., *G. fortis* is compared with *G. scandens* (blue arrows)]. (After Mallarino et al. 2011; photos courtesy of R. Mallarino and A. Abzhanov; skull images originally in Bowman 1961, reproduced with permission of University of California Press.)

Gene expression during development

TGFβIIr   *β-catenin*   *Dkk3*

pmx — *Geospiza magnirostris* — *G. fortis* — *G. fuliginosa* — *G. conirostris* — *G. scandens*

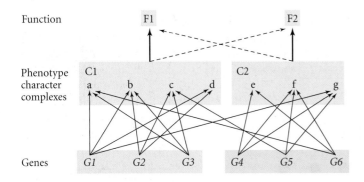

FIGURE 22.16 A schematic representation of the modular organization of a phenotype and its genetic basis. The phenotype is organized into complex characters, such as C1 and C2, each of which consists of component characters (a–d in C1, e–g in C2). Genes (*G1–G6*) have pleiotropic effects, indicated by arrows from each gene to multiple characters. The phenotype is modular to the extent that each gene's pleiotropic effects are mostly on component characters within one complex (C1 or C2), as shown. The more modular a character, the stronger the genetic correlations among the component characters, compared with genetic correlations among characters that are components of different complexes. The diagram suggests that each complex character has a primary function (F1 for C1, F2 for C2) and weakly affects the other's function. (After Wagner and Altenberg 1996.)

birds from nonavian dinosaurs, the thumb was lost and the second digit underwent a shift in developmental identity, taking on the features of a thumb (Wang et al. 2011). Alessandro Minelli (2000, 2003) has suggested that the segmented appendages of arthropods arose as evaginations of the body wall under the influence of the *Distal-less* pathway that underlies evaginations throughout the animal phyla. These evaginations became segmented, he suggests, by recruiting the developmental program, based on Hox genes, that had already established the main, segmented body axis. A similar history of recruitment would account for vertebrate limbs (see Figure 21.15A).

EVOLVABILITY The idea that the production of genetic and phenotypic variation is structured in a way that makes adaptive evolution more likely is one of several concepts that have been termed **evolvability**, a term that Günter Wagner and Lee Altenberg (1996) use to mean "the genome's ability to produce adaptive variants when acted upon by the genetic system." Wagner and others have proposed that evolvability itself can evolve (Hansen et al. 2006; Draghi and Wagner 2008; Wagner 2010). For example, in asexual populations, mutation rates can theoretically evolve to an optimum through selection on "mutator" alleles; similarly, recombination rates and sexual versus asexual reproduction certainly affect the production of genetic variation, and they clearly evolve (see Chapter 15). It is also possible that evolvability can evolve through changes in the relationship between genotype and phenotype, largely by evolution of the pleiotropic correlations among characters (Wagner and Zhang 2011). The traits of an organism are bundled into MODULES to the extent that genes act pleiotropically on traits that are part of the same module, but not on traits in different modules (**Figure 22.16**). To the extent that genes pleiotropically affect different developmental pathways, the resulting traits will be genetically correlated (see Chapter 13).

Genetic correlations among traits can enhance adaptation if the characters are subject to correlational selection—that is, if selection favors specific combinations of characteristics (see Figure 13.15)—or they can constrain evolution if selection on the several traits is antagonistic (e.g., Figure 13.13C,D). If many traits are pleiotropically correlated (i.e., if there is little modularity), directional selection on one trait is likely to be opposed by stabilizing selection on correlated traits, a COST OF COMPLEXITY that reduces the rate of adaptation (Orr 2000). If pleiotropic connections among traits can evolve, then selection may favor PARCELLATION of a highly integrated network of characters into different modules, each consisting of a few functionally related characters that develop in a coordinated manner, and which may themselves evolve to be more highly correlated with one another (INTEGRATION) so that they are functionally matched even if they vary (**Figure 22.17A**).

Pleiotropy can evolve if one locus affects the action of another locus on two or more different characters (**Figure 22.17B**). For example, a mutation at the *ms* locus in the butterfly *Bicyclus anynana* (see Figure 21.22) reduces the eyespots only on the hindwing if the insect has the normal allele at the *sp* locus, but has this effect on both the hindwing and the forewing in the presence of a mutant *sp* allele (Monteiro et al. 2007). Artificial selection experiments have shown that it is often possible to change two or more traits in different

FIGURE 22.17 Modularity of genes and characters can evolve by means of changes in pleiotropic effects. (A) In parcellation, the pleiotropic effects of genes on different phenotypic characters are reduced. Such changes increase the possibility of independent character evolution. In integration, the pleiotropic effects of genes are increased, molding initially independent characters into modular complexes. (B) How a "relationship gene" (rQTL) can change the pleiotropy of another gene. Gene *B* affects only trait T1 if the genotype at the rQTL is *AA* (left). Gene *B* is pleiotropic, however, affecting both T1 and T2, if the rQTL genotype is *Aa* (right). (After G. P. Wagner 1996.)

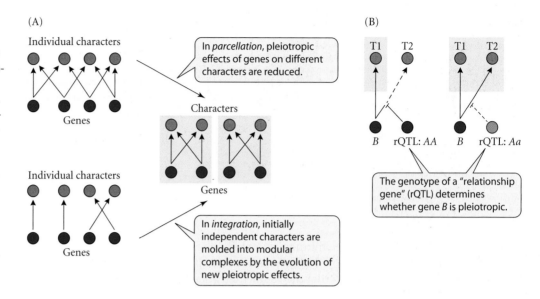

directions even if they are highly genetically correlated and to reduce the genetic correlation in the process (Conner et al. 2011; see Figure 21.22A). Drawing on recent data on the functions of most genes in the genomes of the yeast *Saccharomyces cerevisiae*, the nematode *Caenorhabditis elegans*, and the house mouse *Mus musculus*, Zhi Wang et al. (2010) found that most pleiotropic genes affect only a few morphological or physiological traits, and that the pleiotropy is organized into strongly modular sets of genes affecting the same small groups of characteristics (**Figure 22.18**). That the relationship between genotype and phenotype may evolve in an adaptive manner seems likely, based on the still limited evidence.

Trends, Predictability, and Progress

For many decades after the publication of *The Origin of Species*, many of those who accepted the historical reality of evolution viewed it as a cosmic history of progress. As humanity had been the highest earthly link in the pre-evolutionary Great Chain of Being, just below the angels (see Chapter 1), so humans were seen as the supreme achievement of the evolutionary process (and Western Europeans as the pinnacle of human evolution). Darwin distinguished himself from his contemporaries by denying the necessity of progress or improvement in evolution (Fisher 1986), but almost everyone else viewed progress as an intrinsic, even defining, property of evolution.

In this section, we will examine the nature and possible causes of trends in evolution and ask whether the concept of evolutionary progress is meaningful. A **trend** may be described objectively as a directional shift over time. PROGRESS, in contrast, implies betterment, which requires a value judgment of what "better" might mean.

FIGURE 22.18 Frequency distribution of the degree of pleiotropy in genes affecting physiological traits of yeast (A) and morphological and physiological traits of mouse (B). The degree of pleiotropy is the number (*n*) of traits affected by a gene. The columns show the proportions (frequency) of genes with various degrees of pleiotropy. The yeast data are based on 774 genes and 22 traits, and the mouse data on 4915 genes and 308 traits. (After Wang et al. 2010.)

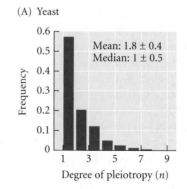

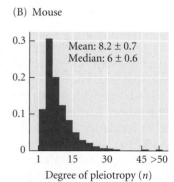

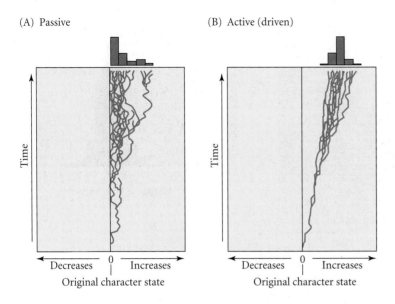

(A) Passive   (B) Active (driven)

FIGURE 22.19 Computer simulations of the diversification of a clade. (A) A passive trend. A character shift in either direction is equally likely, but the character value cannot go beyond the boundary at left. The mean increases, but many lineages retain the original character value. (B) A driven trend. The entire distribution of character values is shifted by a bias in the direction of change. (After McShea 1994.)

Trends: Kinds and causes

A trend is a persistent, directional change in the average value of a feature, or perhaps its maximal (or minimal) value, in a clade over the course of time. A phylogenetically LOCAL TREND applies to an individual clade, whereas a GLOBAL TREND characterizes all of life. Trends can also be classified as passive or driven (McShea 1994). In a **passive trend**, lineages in the clade evolve in both directions with equal probability, but if there is an impassable boundary in one direction (e.g., a minimal possible body size), the variation among lineages can expand only in the other direction. Because the variance expands, so do the mean and the maximum. Although the mean increases, some lineages may remain near the ancestral character value (**Figure 22.19A**). In a **driven**, or **active**, **trend**, changes within lineages in one direction are more likely than changes in the other (i.e., there is a "bias" in direction), so both the maximal and the minimal character values change along with the mean (**Figure 22.19B**).

Both driven and passive trends could have any of several causes. Neutral evolution by *mutation and genetic drift* results in increasing variance among lineages (see Chapter 12) and could produce a passive trend if variation were bounded as in Figure 22.19A. *Individual selection* could be responsible for all the changes within lineages and could result in either a passive or a driven trend, depending on whether or not ancestral character states remained advantageous for some lineages. The mean character state among species in a clade could also change as a result of a correlation with speciation or extinction rates (see Figure 11.17). The character might cause a rate difference (SPECIES SELECTION in the broad sense; see Chapter 11), or it might simply be correlated with another character that causes a rate difference. The latter process has been called **species hitchhiking** (Levinton 2001) by analogy with the hitchhiking of linked genes in populations.

Examples of trends

Surely the most common pattern in morphological evolution is divergence among species in a clade as they become adapted to different resources, microhabitats, or other species with which they interact. In these adaptive radiations (see Chapters 3 and 7), characters often evolve in different directions (smaller/larger, shorter/longer) in different lineages. Nevertheless, trends can sometimes be discerned in individual features. Increases in the body size of mammals may represent a

FIGURE 22.20 A passive trend: Cope's rule in late Cretaceous and Cenozoic North American mammals. (A) Each of 1534 species is plotted as a line showing its temporal extension and its body mass (estimated from tooth size). Although small mammals persist throughout the Cenozoic, there is an increasing number of large species over the course of time. (B) Change in body mass (negative or positive) plotted for 779 pairs of older and younger species in the same genera (likely ancestor-descendant pairs). There are significantly more positive than negative changes, implicating natural selection as the cause of the trend toward increased size. (After Alroy 1998.)

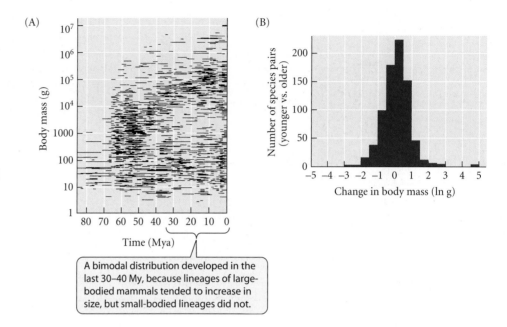

A bimodal distribution developed in the last 30–40 My, because lineages of large-bodied mammals tended to increase in size, but small-bodied lineages did not.

passive trend. Paleontologists noticed long ago that the maximal body size in many animal groups has tended to increase over time, a trend dubbed COPE'S RULE. A plot of the body sizes of 1534 species of North American late Cretaceous and Cenozoic mammals (**Figure 22.20A**) against their dates in the fossil record shows such a passive trend (Alroy 1998). Mammals were small before the K/T mass extinction, and the lower size limit has remained nearly the same ever since. However, mean and maximal sizes have increased, especially since the K/T extinction, when the explosive diversification of mammals began. Changes in body size between 779 matched pairs of older and younger species in the same genera (likely ancestor-descendant pairs) occurred in both directions, but were significantly biased toward increases (**Figure 22.20B**). This nonrandomness strongly suggests that the trend was caused by natural selection rather than genetic drift.

Whereas data for all mammals taken together suggest a passive trend, body size in the horse family (Equidae) conforms to the pattern of a driven trend (MacFadden 1986; McShea 1994). Not only the maximal and the mean, but also the minimal, size increased during the Cenozoic (**Figure 22.21**). Ancestor-descendant pairs showed an increase in size much more often than a decrease. These trends in mammals and in the horse family can be attributed to individual selection, and studies of selection in contemporary populations, especially of birds, insects, and plants, show that directional selection generally favors larger size (Kingsolver and Pfennig 2004).

In plants such as the Solanaceae (tomato family), self-compatibility has evolved in a great many self-incompatible lineages (see Chapter 15), apparently by individual selection, but evolution in the opposite direction never occurs. This asymmetry should eventually result in an entirely self-compatible clade of species. However, more than half the species in the Solanaceae are self-incompatible, because of species selection: self-incompatible species have a lower extinction rate than self-compatible species (Goldberg et al. 2010). An actual trend caused by species selection is the increase in the ratio of nonplanktotrophic to planktotrophic species in several clades of Cenozoic gastropods (**Figure 22.22**). Species that lack a planktotrophic dispersal stage are more susceptible to extinction than are planktotrophic species (species that feed as planktonic larvae). However, the nonplanktotrophic species

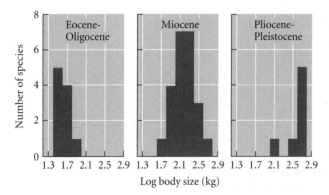

FIGURE 22.21 A driven trend: Cope's rule in the horse family, Equidae. The entire distribution of body sizes in the family shifted toward larger sizes during the Cenozoic. (After McShea 1994.)

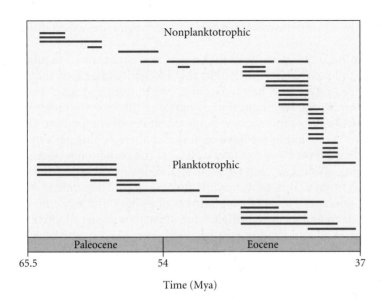

FIGURE 22.22 A trend caused by species selection. The bars show the stratigraphic distributions of fossil species of volutid snails. Although nonplanktotrophic species had shorter durations, they arose by speciation at a higher rate, so the ratio of nonplanktotrophic to planktotrophic species increased over time. (After Hansen 1980.)

more than compensate by their higher rate of speciation, probably because their lower rate of dispersal reduces the rate of gene flow among populations (Hansen 1980; Jablonski and Lutz 1983).

Trends that are due to species hitchhiking are probably very common because if any one character causes one clade to become richer in species than other clades as a result of its effect on the rate of speciation or extinction, then all the other features of that clade will also tend toward greater frequencies. For example, coiled, sucking adult mouthparts may have become more prevalent among insects because they are a feature of the extremely diverse Lepidoptera (moths and butterflies). Lepidopteran larvae are herbivorous, and the herbivorous habit has consistently been associated with a high diversification rate in insects (Mitter et al. 1988; see Figure 7.22). So the increased frequency of sucking mouthparts may be a result of hitchhiking with herbivory.

The boundaries that enforce passive trends may be either functional or developmental genetic constraints. For example, the smallest birds and mammals may have reached the lower functional limit of body size because a smaller animal might be unable to maintain its body temperature because of its greater surface/volume ratio. In addition, developmental pathways may evolve that act as RATCHETS—mechanisms that make evolutionary reversal unlikely, such as losses of complex features, which in some cases may be irreversible (Bull and Charnov 1985).

Are there major trends in the history of life?

Do any trends or directions characterize the entire evolutionary history of life? Although many have been postulated, it is probably safe to say that no uniform driven trends can be discerned, because exceptions can be cited for every proposed trend. Still, one might ask if there is any feature that, on the whole, has evolved with enough consistency of direction that one would be able to tell, from snapshots of life at different times in the past, which was taken earlier and which later. Let us consider a few possibilities (see McShea 1998; Knoll and Bambach 2000).

EFFICIENCY AND ADAPTEDNESS Innumerable examples exist of improvements in the form of features that serve a specific function. The mammal-like reptiles, for example, show trends in feeding and locomotory structures associated with higher metabolism and activity levels, culminating in the typical body plan of mammals (see Chapter 4). There may well be a global trend toward greater efficiency (Ghiselin 1995). However, efficiency and effectiveness are difficult to measure, and they must always be defined relative to the task set by the context—by the organism's environment and way of life, which differ from species to species.

If efficiency of design has increased, does that mean that organisms are more highly adapted than in the past? Darwin thought this likely; he imagined that if long-extinct species were revived and had to compete with today's species, they would lose the competition badly. We might find evidence of the evolution of competitive superiority if the fossil record provided many examples of competitive *displacement* of early by later taxa—but as we have seen (in Chapter 7), this pattern is much less common than *replacement* by later taxa well after the earlier ones became extinct. And even though natural selection within populations increases mean fitness (specifically, *relative* fitness), fitness values are always context-dependent. We cannot meaningfully compare the level of adaptedness of a shark and a falcon, or even of a bird-hunting falcon and a bat-hunting falcon, since they are as adapted to different tasks as are flat-head and Phillips-head screwdrivers. And it may be difficult, even in principle, to compare the adaptedness of a species with that of its long-extinct ancestor, since they may have experienced quite different selective regimes.

We might suppose that species longevity would be a measure of increase in adaptedness, but this need not be so. The consequence of natural selection is the adaptation of a population to the currently prevailing environment, not to future environments, so selection does not imbue a species with insurance against environmental change. We have seen that within many clades, the age of a genus or family does not influence its probability of extinction, implying that a lineage does not become more extinction-resistant over time (see Figure 7.11). Extinction rates were generally higher in the Paleozoic than afterward (see Figure 7.8B), perhaps because major extinction-prone clades did not survive—but this clade-level hypothesis does not require us to postulate that extinction resistance evolves within lineages.

COMPLEXITY   It is difficult to define, measure, or compare complexity among very different organisms, although anatomical complexity may be considered to be proportional to the number of different kinds of parts that comprise an organism and to the irregularity of their arrangement (McShea 1991).

One of the most striking aspects of the history of life has been an increase in the maximal level of HIERARCHICAL ORGANIZATION, whereby entities have emerged that consist of functionally integrated associations of lower-level individuals (Maynard Smith and Szathmáry 1995; McShea 2001). The first cells arose from compartments of replicating molecules; the eukaryotic cell evolved from an association of prokaryotic cells; multicellular organisms with differentiated cell types evolved from aggregations of unicellular ancestors; and the highly integrated colonies of a few kinds of multicellular organisms, among them invertebrates such as corals, social insects, certain social mammals, and humans (**Figure 22.23**) are the "pinnacles of social evolution" (Wilson 1975). The difficulty to be overcome in all these transitions was that selection at the level of the component units (e.g., individual cells) could threaten the integrity of the larger unit (e.g., multicellular organism; see Chapter 16). In general, such conflict has been suppressed by development through a stage (e.g., the unicellular egg) that establishes high relatedness (and thus the power of kin selection) among the component units (e.g., the genetic identity of the cells of a multicellular organism; Maynard Smith and Szathmáry 1995; Michod 1999). Richard Grosberg and Richard Strathmann (2007) describe the many other defenses that multicellular organisms have against disruption by

FIGURE 22.23 Two "pinnacles of social evolution" and their technology. African termites of the genus *Macrotermes* cooperate to build mounds in which a constant temperature is maintained by air conditioning: cool air flows into the base of the mound, passes through vertical tunnels as it warms, and flows out at the top. The author, shown for scale, has been transported from New York to Ethiopia by, and in every way depends on, the technological products that social cooperation has made possible. (Photo by the author's camera.)

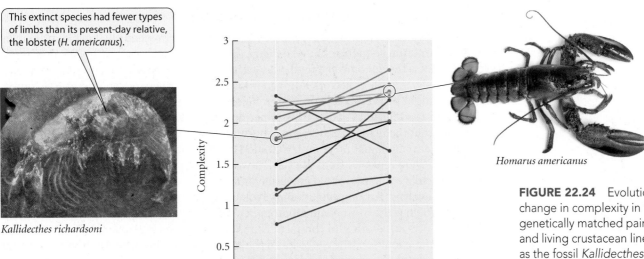

This extinct species had fewer types of limbs than its present-day relative, the lobster (*H. americanus*).

Kallidecthes richardsoni

Homarus americanus

FIGURE 22.24 Evolutionary change in complexity in 12 phylogenetically matched pairs of fossil and living crustacean lineages, such as the fossil *Kallidecthes richardsoni* (left) and the living *Homarus americanus* (right). Complexity was measured by an index based on the number of morphologically different limb types, which also differ in function. (After Adamowicz et al. 2008; photo of *Kallidecthes richardsoni* courtesy of James St. John.)

rogue cells, and they argue that the evolution of multicellularity was rather easy because many of the prerequisites, such as cell adhesion mechanisms, had already evolved in unicellular forms (see Figure 22.14 and Newman et al. 2006).

The major changes in hierarchical organization represent only a few evolutionary events, in which the great majority of lineages did not participate. In contrast, the anatomical complexity of Cambrian animals was arguably as great as that of living forms, and many characteristics have evolved toward simplification or loss in innumerable clades (see Chapter 3). Certainly, complexity has increased in some clades; for example, the number of types of appendages has increased in many lineages of crustaceans (**Figure 22.24**). However, phylogenetic evidence seems to indicate that less pronounced changes in hierarchical complexity, such as the difference between green algae with one cell type versus two, have been almost equally divided between increases and decreases, with a bias toward decreased complexity in several clades (Marcot and McShea 2007). This is true of both morphology and behavior. For example, eusociality has evolved several times in the Hymenoptera, but has frequently been lost (Danforth et al. 2003). The advantage of eusociality, or probably the capacity for any other complex behavior, must depend on the environment, and there is no guarantee that it will always increase.

Genome size, meaning the amount of DNA in the cell nucleus (C-value; see Chapter 3), varies by five orders of magnitude among eukaryotes (Oliver et al. 2007). Although genome size has decreased in some lineages, such as *Arabidopsis*, increases have been more prevalent because of mechanisms such as polyploidy and the proliferation of transposable elements. However, the number of coding sequences does not appear to be correlated with our traditional (although superficial and perhaps erroneous) impressions of phenotypic complexity; we like to think that mammals such as humans, with 24,000 genes, are more complex than water fleas (*Daphnia*) with 31,000, or rice, with 60,000 (see Figure 3.27). The number of functional genes has not yet been determined for enough organisms to permit a phylogenetic analysis of changes in their number, but we may expect that just as for DNA content, both increases and decreases will ultimately be revealed. We know that in parasites and in endosymbiotic bacteria, the number of genes is often much lower than in free-living relatives (McCutcheon and Moran 2012; see Chapter 19). We do not yet know if there has been a trend in the information content of genomes throughout evolution; however, the number of genes may not be the best measure of information content because the variety of functions that reside in a genome may be greatly amplified by alternative splicing of genes, multiple binding sites for different transcription factors, and other processes.

Predictability and contingency in evolution

A course of events (a history) is said to be PREDICTABLE if it proceeds by lawlike principles that determine the sequence of events (e.g., A, B, …, E), each caused by the preceding event. Physics is the epitome of a predictive science, in which physical laws are used to predict events (e.g., missile trajectories), given information on relevant conditions. (Of course, the events might be predictable only in principle, not in practice, if our knowledge of the principles or the relevant conditions is imperfect.) For many people, predictability has the implication that the realized course of events was inevitable from the start. In this view, as the evolutionary biologist Stephen Jay Gould (1989) expressed it, if the tape (or movie) of life on Earth were replayed from the start, it would be very similar to the actual history of life. In particular, humans, or a comparably intelligent life form (a "humanoid," perhaps) would have evolved. Among evolutionary biologists, paleontologist Simon Conway Morris (2003, also 2008) is the most eloquent proponent of this viewpoint (see also Dennett 1995).

HISTORICAL CONTINGENCY, in contrast, means that although each event (e.g., E) is caused by a preceding event (D), the outcome of the history would be different (say E′, not E) if any of many antecedent events had been different (C′ rather than C, or B′ rather than B). Perhaps, in principle, we could know the series of events that caused B′ and C′, rather than B and C, and therefore E′ rather than E, to occur, but realistically, we can never know all of the incredibly vast number of possible causal chains. In this view, which Gould championed, the "tape of life" would be very different if it were to start again from any point in life's long history. The concept of contingency has been familiar in human history at least since the seventeenth-century philosopher Blaise Pascal mused about the effect of Cleopatra's nose: if Mark Antony, one of the triumvirate that ruled Rome from 43 to 33 BC, had not been smitten by Cleopatra's beauty and become her lover, Octavian would not have battled and vanquished him, so ending the triumvirate's rule and in turn the Roman empire. On a slightly less grand scale, most of us can think of "chance" events that changed the course of our own lives.

There is certainly some predictability in evolution, and it is the basis of a great deal of the evolutionary theory presented in this book. Some of the selection equations in population genetics, for example, deterministically predict allele frequency changes in large populations (see Chapter 12). Organisms conform to physical principles, so massive terrestrial vertebrates like elephants have, predictably, disproportionately thicker leg bones. Many features of organisms are more or less successfully predicted by "optimality" theories of life history evolution, behavioral ecology, and functional morphology (see Chapters 14–16). These theories are successful largely because of the high incidence of convergent evolution: the many instances in which similar adaptations to similar environmental selection pressures have evolved independently. Conway Morris (2003, 2008) depends on convergent evolution to support his argument that humans, or humanoids, were an inevitable outcome of evolution: crows, parrots, dolphins, and other animals show signs of "intelligence," so if intelligence had not evolved in our primate lineage, it would have evolved in some other animal lineage. By the same argument, many people, including many physical scientists, are convinced that there must exist intelligent humanoids elsewhere in the universe; this conviction is the basis of SETI (Search for Extraterrestrial Intelligence) and similar projects.

Some—probably most—evolutionary biologists reject this position and argue for a strong role of contingency in the history of life. In some selection equations (e.g., when heterozygotes have lowest fitness), different alleles may be fixed depending on the slightest difference in their initial frequency (see Chapter 12)—and this is even without considering the role of random genetic drift, which is more important than selection when $4N_e s < 1$ (see Chapter 12). The course of evolution, moreover, can depend on rare mutations or combinations of interacting mutations: in one of twelve experimental populations of *E. coli*, the ability to metabolize citrate (based on a combination of two mutations) evolved only after 30,000 generations, during which billions of mutations had occurred (Blount et al. 2008; see Chapter 8). Thus adaptive changes may depend on rare genetic events—and also on environmental events, for the history of adaptation in a lineage depends on the sequence of environment, both physical and biological, that it experiences. Consequently, convergence between closely related lineages, which have similar genetic and developmental potentialities and may inhabit similar environments, may be very close, but convergence between remotely related lineages is usually more superficial. For example, the adaptive radiation of *Anolis* lizards in the American tropics, which is largely a consequence of toe pads that enable them to climb trees, is very imperfectly mirrored by the diurnal arboreal geckos in the Indian Ocean region, which are less diverse in form and use a very different set of microhabitats (Losos 2010). Convergence, moreover, can be difficult to define; for instance, biologists may disagree on whether or not humans and fungus-growing leafcutter ants both practice "agriculture" and whether or not "tool use" is convergent between humans and certain wasps that close their burrows with pebbles.

Examples of convergence abound, but so do unique events in the history of life. As far as we know, life originated only once, as did flowers, vertebrates, terrestrial vertebrates, the amnion, the feather, the mammalian diaphragm, the elephant's trunk, and countless other examples. A single origin is very close to no origin at all, and any biologist (or science fiction writer) can imagine creatures that have never existed, such as cellulose-digesting or photosynthetic mammals (**Figure 22.25**). Species that play important ecological roles in one part of the world often have no equivalents in regions with a different evolutionary history (see Chapter 6); in much of the world, woodpeckers make holes in which many other animals live—but not in Australia, where even surrogate woodpeckers are lacking. Moreover, extinction, including mass extinction events, has cut short the possible evolutionary future of the vast majority of lineages that have ever lived, just as accidents have tragically befallen countless young people whose success or failure can only be imagined. A clam-drilling predatory snail that evolved in the Triassic became extinct later in that period, and no other lineage evolved to fill that role for the next 120 million years (Fürsich and Jablonski 1984; see Chapter 7). No equivalents of trilobites, ammonoids, dinosaurs, and many other extinct groups have ever replaced them. As Gould (1989) emphasized, if any species in our long line of ancestors, back to the first vertebrate or even beyond, had become extinct, intelligent hominids would never have evolved. Among the billion or more lineages (species) of organisms in Earth's history, only one evolved human intelligence, and this happened only after at least 3 billion years of cellular life (see Chapter 5). We have no reason to suppose that any human equivalent would have evolved in our stead. For these reasons, George Gaylord Simpson (1964) and Ernst Mayr (1988b), two of the most influential biologists of the twentieth century, argued that the probability that there exists another life form in

(A)

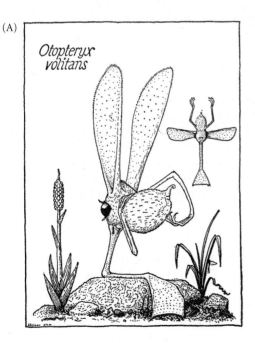

(B)

FIGURE 22.25 In a work both amusing and well informed by evolutionary principles, Harald Stümpke imagined an adaptive radiation of "Rhinogradentia," or "snouters," mammals with noses elaborated for diverse functions. *Otopteryx* flies backward, using its ears as wings and its nose as rudder. *Orchidiopsis* feeds on insects attracted to its petal-like nose and ears. Stümpke's fantasy illustrates some of the many conceivable phenotypes that have never evolved. (From Stümpke 1957.)

the universe that we have the faintest hope of detecting, much less communicating with, is, for all intents and purposes, zero, and that our own evolutionary history was far from inevitable.

The question of progress

Many people who accept evolution still conceive it as a purposive, progressive process, culminating in the emergence of consciousness and intellect. Even some evolutionary biologists have taken some of the qualities we most prize in ourselves, such as intellect or empathy, as criteria of progress, and quite ignoring the innumerable lineages that have not evolved at all in these directions, have seen in evolution a history of progress toward the emergence of humankind (see Ruse 1996).

The word "progress" usually implies movement toward a goal, as well as improvement or betterment. But the processes of evolution, such as mutation and natural selection, cannot imbue evolution with a goal. Moreover, progress in the sense of betterment implies a value judgment, and here we must guard against the parochial, anthropocentric view that human features are better than those of other species. Were a rattlesnake or a knifefish capable of conscious reflection, it doubtless would measure evolutionary progress by the elegance of an animal's venom delivery system or its ability to communicate by electrical signals. Certainly these features became more effective during evolution, but only because more effective venom delivery or electrical signaling is advantageous *in the context* of the environment and lifestyle of these lineages. The great majority of animal lineages—to say nothing of plants and fungi—show no evolutionary trend toward greater "intelligence" (however it might be defined and measured), which must be seen as a special adaptation appropriate to some ways of life, but not others. It is difficult, if not impossible, to specify a universal criterion by which to measure "improvement" that is not laden with our human-centered values.

Many evolutionary biologists have therefore concluded that we cannot objectively find progress in evolutionary history, except in the sense of context-dependent adaptive improvements (Ruse 1996). As we have seen, it is hard to document even objective trends, especially driven trends, in any feature such as complexity or adaptedness. The most characteristic feature of the history of evolution, rather, is the unceasing proliferation of new forms of life, of new ways of living, of seemingly boundless, exquisite diversity. The majesty of this history inspired Darwin to end *The Origin of Species* by reflecting on the "grandeur in this view of life," that "whilst this planet has gone cycling on according to the fixed law of gravity, from so simple a beginning endless forms most beautiful and most wonderful have been, and are being, evolved."

Go to the

EVOLUTION

Companion Website at
sites.sinauer.com/evolution3e

for quizzes, data analysis and
simulation exercises, and other
study aids.

Summary

1. The average rate of evolution of most characters is very low because long periods of little change (stasis) are averaged with short periods of rapid evolution, or the character mean fluctuates without long-term directional change. The highest rates of character evolution in the fossil record are comparable to rates observed in contemporary populations and can readily be explained by known processes such as mutation, genetic drift, natural selection, and speciation. The time required for speciation is short enough to account for the observed rates of increase in species diversity.

2. The fossil record provides examples of both gradual change and the pattern called punctuated equilibrium: a rapid shift from one static phenotype to another. The hypothesis that such shifts require the occurrence of peripatric speciation is not widely accepted because responses to selection do not depend on speciation.

3. Stasis can be explained by genetic constraints, stabilizing selection (owing largely to habitat tracking), or the erasure of divergence by episodic massive gene flow among populations with ancestral and derived character states.

4. Higher taxa arise not in single steps, by macromutational jumps (saltation), but by multiple changes in genetically independent characters (mosaic evolution). Most such characters evolve gradually, through intermediate stages, but some characters evolve discontinuously as a result of mutations with moderately large effects.

5. Characters may be phylogenetically conservative because of limits on the origin of variation (genetic and developmental constraints) or because of phylogenetic niche conservatism (resulting in stabilizing selection).

6. New features are often advantageous even at their inception. They often evolve by modification of pre-existing characters to serve accentuated or new functions, or sometimes as by-products of the development of other structures. Evolutionary novelties often result when two or more functions of a structure are decoupled, or when structures are duplicated and diverge in structure and function.

7. Complex structures such as eyes evolve by rather small, individually advantageous steps. They may acquire functional integration with other structures so that they become indispensable.

8. Although novel features, by definition, are not simply modifications of pre-existing features, they have arisen, in at least some cases, by the recruitment of integrated genetic and developmental pathways in new contexts or combinations. That is, the processes by which they arose were not altogether new.

9. Some fundamental characteristics of developmental processes and organismal integration, such as modularity, may enhance evolvability, the capacity of a genome to produce variants that are potentially adaptive. Some aspects of evolvability, such as patterns of pleiotropy, can probably evolve by natural selection.

10. Long-term trends may result from individual selection, species selection, or species hitchhiking, the phylogenetic association of a character with other characters that affect speciation or extinction rates. Driven trends, whereby the entire frequency distribution of a character among species in a clade shifts in a consistent direction over time, are less common than passive trends, in which variation among species (and therefore the mean of the clade) expands from an ancestral state that is located near a boundary (e.g., the clade may begin near a minimal body size).

11. Probably no feature exhibits a trend common to all living things. Features such as genome size and structural complexity display passive trends, in that the maximum has increased since very early in evolutionary history, but such changes have been inconsistent among lineages. There is no clear evidence of trends in measures of adaptedness such as the longevity of species or higher taxa in geological time.

12. Certain aspects of evolution are predictable, especially in the short term, and may be manifested by convergent evolution. However, considerable evidence supports the view that long evolutionary histories are contingent: that is, particular evolutionary events would have differed, or would not have occurred, if any of a great many previous events had been different. Unique events such as the emergence of human intelligence may have been highly contingent and improbable.

13. If "progress" implies a goal, then there can be no progress in evolution. If it implies betterment or improvement, we still cannot identify objective criteria by which the history of evolution can be shown to be one of "improvement." Characters improve in their capacity to serve certain functions, but those functions are specific to the ecological context of each species.

Terms and Concepts

driven trend (= active trend)
evolvability
habitat tracking

habitat selection
macroevolution
microevolution
passive trend

phylogenetic niche conservatism
punctuated equilibrium
saltation

species hitchhiking
stasis
trend

Suggestions for Further Reading

Tempo and Mode in Evolution, by G. G. Simpson (Columbia University Press, New York, 1944), and *Evolution above the Species Level*, by B. Rensch (Columbia University Press, New York, 1959), are classic works of the evolutionary synthesis, in which the authors reconcile macroevolutionary phenomena with neo-Darwinian theory. *Punctuated Equilibrium*, by Stephen Jay Gould (Harvard University Press, Cambridge, MA, 2007), is the posthumously published central chapter of his magnum opus, *The Structure of Evolutionary Theory*.

J. S. Levinton, in *Genetics, Paleontology, and Macroevolution*, second edition (Cambridge University Press, Cambridge, 2001), discusses many of the topics of this chapter, taking a strong stand on controversial issues. D. W. McShea provides a good introduction to contemporary research on evolutionary trends in "Possible largest-scale trends in organismal evolution: Eight 'live hypotheses'," *Annual Review of Ecology and Systematics* 29: 293–318 (1998). Evolvability is treated by G. P. Wagner in *Evolution Since Darwin: The First 150 Years*, M. A. Bell et al. (eds.), (Sinauer Associates, Sunderland, MA, 2010), pp. 197–213.

The evolution of eyes, and why it is not a mystery, is the subject of a special issue of the journal *Evolution: Education and Outreach* (volume 1, issue 4, October 2008).

Predictability in evolution is defended by S. Conway Morris in *Inevitable Humans in a Lonely Universe* (Cambridge University Press, Cambridge, 2003). Outstanding arguments for contingency, well worth reading today, are "The nonprevalence of humanoids," by G. G. Simpson, *Science* 143: 769–775 (1964), and "The probability of extraterrestrial intelligent life," by E. Mayr, reprinted as pp. 67–74 in *Toward a New Philosophy of Biology* (Harvard University Press, Cambridge, MA, 1988). Michael Ruse, in *Monad to Man: The Concept of Progress in Evolutionary Biology* (Harvard University Press, Cambridge, MA, 1996), surveys the history of thinking on this topic.

Problems and Discussion Topics

1. Snapdragons (*Antirrhinum*) and their relatives in the traditionally recognized family Scrophulariaceae have bilaterally symmetrical flowers, derived from the radially symmetrical condition of their ancestors. A mutation in the *cycloidea* gene makes snapdragon flowers radially symmetrical. Should we conclude that evolution from radial to bilateral symmetry was caused by change in this one gene? Would that be a saltation? Are there other possible evolutionary histories that are not saltational but which are compatible with this observation?

2. Despite the close relationship between humans and chimpanzees, and despite the genetic variation in most of their morphological characteristics, there are no records of humans giving birth to babies with chimpanzee morphology. Would you expect this to occur if humans have evolved gradually from apelike ancestors? If not, why not?

3. Find and evaluate the evolutionary literature that describes how wings could have evolved gradually, through advantageous intermediate steps. (See Dudley et al. 2007.)

4. How might one test the three hypotheses proposed to explain stasis that are mentioned in this chapter?

5. How can we tell whether a trend was caused by species selection or individual selection? How might we tell whether a passive trend was due to natural selection or genetic drift? How can we distinguish a passive from an driven trend?

6. Paleontologists and some biologists commonly infer function, and even behavior, from anatomical details. Skeletal features, for example, are often used to infer that an extinct mammal (such as an early hominin) was highly, somewhat, or not at all arboreal. This inference assumes a good fit of form to function (i.e., optimal form). Can that assumption be justified?

7. Would you expect "living fossils," such as the tadpole shrimp, to differ from other species in amount of genetic variation, genetic correlations among characters, canalization, or any other feature that might affect "evolvability"?

8. Are you surprised that the number of distinct genes in the human genome seems not to differ greatly from that in many other animals? Why or why not?

9. When the stress protein hsp90 is rendered less functional by mutation in *Drosophila* or *Arabidopsis*, individuals display various morphological abnormalities. Does this observation suggest that the *hsp90* gene acts as a regulator of variation and is therefore a mechanism of evolvability?

10. Explain how the relationship of phenotype to genotype might evolve so that genetic variation might be more likely to include potentially adaptive variants. How is this theory different from Lamarckian inheritance or from supposing that natural selection has foresight?

Evolutionary Science, Creationism, and Society

In all scientific disciplines, research continues because many questions are still imperfectly answered. In biology, likewise, we cannot pretend to know more than a small fraction of the evolutionary history of life or to understand fully all the mechanisms of evolution, and there are many differences of opinion among evolutionary biologists about the details of both history and mechanisms. However, the historical reality of evolution—the descent, with modification, of all organisms from common ancestors—has not been in question among scientists for well over a century. It is as much a scientific fact as the atomic constitution of matter or the revolution of Earth around the Sun.

Nevertheless, many people do not accept evolution. More than 40 percent of people in the United States believe that the human species was created directly by God, rather than having evolved from a common ancestor shared with other primates (and from a much more remote common ancestor shared with all other species). People who hold this belief are often referred to as **creationists**. In contrast, a great majority of people in Europe do not question the reality of evolution (even in countries such as Italy that have an officially established religion), and they are often astonished that antiscientific attitudes on evolution flourish in the technologically and scientifically most prominent country in the world. Creationists are becoming more vocal in western Europe, however, and an Islamic creationist movement is very active in Turkey. Among 34 Western countries, Turkey ranks lowest in public acceptance of evolution, and the United States second lowest. Creationist pressure has greatly weakened science education in the United States, for even teachers who accept evolution often compromise their teaching, or minimize their coverage, in order to avoid controversy (Berkman and Plutzer 2011).

Too complex? The eye of a red-eyed tree frog (*Agalychnis callidryas*), veiled by a partly transparent eyelid that makes the eye, and therefore the frog, less conspicuous to predators. Overwhelming evidence and the principles of evolution refute the creationist claim that features such as the vertebrate eye are too complex to have evolved by natural processes.

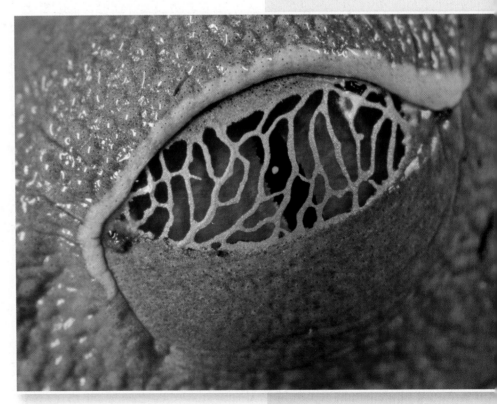

In this chapter, we will look at the beliefs of creationists and their arguments against evolution, and we will review the evidence for evolution presented throughout this book. We will also look at the many ways in which human society depends on an understanding of evolutionary science.

Creationists and Other Skeptics

Most disbelievers in evolution reject the idea because they think it conflicts with their religious beliefs. Many, perhaps most, are Christian, Muslim, or (rarely) Jewish fundamentalists who adhere to a literal, or almost literal, reading of sacred texts. For Christian and Jewish fundamentalists, evolution conflicts with their interpretation of the Bible, especially the first chapters of Genesis, which portray God's creation of the heavens, Earth, plants, animals, and humans in six days. However, many Western religions understand these biblical descriptions to contain symbolic truths, not literal or scientific ones. Many deeply religious people accept evolution, viewing it as the natural mechanism by which God has enabled creation to proceed. Some scientists, including some researchers in evolutionary biology and some of the most impassioned opponents of creationism, subscribe to this view (see Miller 1999, 2008). Some religious leaders have made clear that they accept evolution. (See an array of such statements under "Voices for Evolution" at www.ncseweb.org, the website of the National Center for Science Education.) For example, Pope John Paul II affirmed the validity of evolution in 1996, although he reserved a divine origin for the human soul. [The text of his letter was reprinted in the *Quarterly Review of Biology* 72: 381–396 (December 1997)]. The pope's position was close to the argument generally known as **theistic evolution**, which holds that God established natural laws (such as natural selection) and then let the universe run on its own, without further supernatural intervention.

The beliefs of creationists vary considerably (**Figure 23.1**; Numbers 2006). The most extreme creationists interpret every statement in the Bible literally. They include "young Earth" creationists who believe in **special creation** (the doctrine that each species, living or extinct, was created independently by God, essentially in its present form) and in a

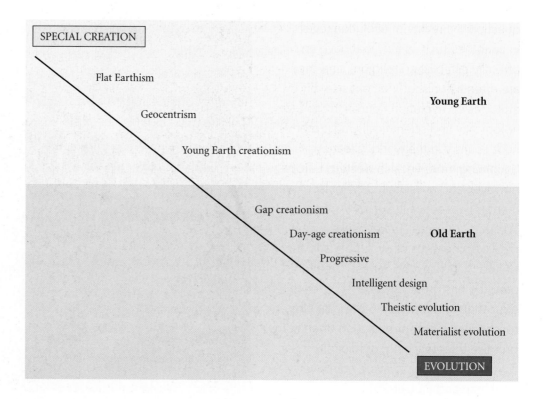

FIGURE 23.1 The special creation–evolution continuum. "Young Earth" proponents affirm that Earth is young (thousands of years old), not old (billions of years old). Adherents of special creation tend to consider the Bible to be literally true. (After Scott 1997.)

young universe and Earth (less than 10,000 years old), a deluge that drowned Earth, and an ark in which Noah preserved a pair of every living species. They must therefore deny not only evolution, but also most of geology and physics (including radiometric and astronomical evidence of the great age of the universe). Some other creationists share many or most of the biblical literalists' beliefs, including special creation, but grant the antiquity of Earth and the rest of the universe. (The "day-age" position holds that each of the six days of creation cited in Genesis was millions of years long.) Still other creationists allow that mutation and natural selection can occur, and even that very similar species can arise from a common ancestor. However, they deny that higher taxa (genera, families, etc.) have evolved from common ancestors, and most of them vigorously assert that the human species was specially created by God, in His image—which is their major reason for caring about evolution at all.

Most of the efforts of activist creationists are devoted to suppressing the teaching of evolution in schools, or at least insisting on "equal time" for their views. In the United States, however, the constitutional prohibition of state sponsorship of religion has been interpreted by the courts to mean that any explicitly religious version of the origin of life's diversity cannot be promulgated in public schools. The activists have therefore adopted several forms of camouflage. One is SCIENTIFIC CREATIONISM, which consists of attacks on, and supposedly scientific disproofs of, evolution. These arguments have not succeeded, chiefly because they do not have, and cannot have, any scientific content. Scientific creationism has been succeeded by **intelligent design** (**ID**) theory. ID proponents generally do not publicly invoke special creation by God. Some of them even accept certain aspects of evolution, such as development of different species from common ancestors. They argue, however, that many biological phenomena are too complicated to have arisen by natural processes and can therefore be explained only by an intelligent designer. They claim that intelligent design is a scientific, not a religious, concept. The designer they envision, however, is a supernatural rather than a material being (i.e., a being equivalent to God). (See Pennock 2003 for an analysis of ID.) Some ID advocates have explicitly stated that their aim is much broader than evolutionary biology: they propose to defeat "scientific materialism" and to replace materialistic explanations with the understanding that nature and human beings are created by God (Forrest and Gross 2004). These creationists' targets include geology, physics, and cosmology (Kaiser 2007).

Bills requiring public schools to give "creation science" equal time have been introduced in state legislatures, but the U.S. Supreme Court in 1987 found one such law unconstitutional because it "endorses religion by advocating the religious belief that a supernatural being created humankind," and because it was written "to restructure the science curriculum to conform with a particular religious viewpoint." As a result, local and state legislators and school officials use a variety of subterfuges to get creationism into school science curricula. In a prominent recent case, parents in the town of Dover, Pennsylvania, filed a suit against local school board officials who read a statement to science classes, warning them that evolution is a theory rather than fact and suggesting that students read a particular book in the school library that advocates intelligent design. In December 2005, the judge ruled for the parents in a decision that is well worth reading as a model statement of what science is and why intelligent design is not science, but rather thinly disguised religious doctrine. (See www.ncseweb.org or www.pamd.uscourts.gov/kitzmiller/kitzmiller_342.pdf.)

Since the Dover decision, the creationist strategy in the United States has been to urge that students be encouraged to examine the evidence for both sides of "controversial scientific issues"—which sounds marvelously enlightened, except that evolution and one or two other topics, such as global warming, are usually singled out as "controversial," with the clear implication that the scientific theories on these topics are particularly doubtful. Bear in mind that the fundamental principles of evolution are not controversial within science. Nevertheless, many bills purporting to support "critical analysis" or "academic freedom" to question these scientific topics have been introduced into state legislatures every year (Branch et al. 2010.)

Science, Belief, and Education

The nature of science

Those who argue against including any form of creationism in science curricula are not against free speech, and they are not trying to extinguish religious belief. They simply hold that although it might be acceptable to teach about creation stories in classes on, say, history, literature, or contemporary society, such beliefs are not valid scientific hypotheses and have no place in science classes. Unfortunately, most people have little understanding of what science is and how it works, even if they have had science courses—and this understanding is critically important to the evolution-versus-creationism debate. This debate has far greater implications than public acceptance of evolution: it is fundamentally about the acceptance and use of science and reason.

Science is not a collection of facts, contrary to popular belief, but rather a *process* of acquiring an understanding of natural phenomena. This process consists largely of posing hypotheses and testing them with observational or experimental evidence. Despite loose talk about "proving" hypotheses, most scientists agree that the hypothesis that currently best explains the data is *provisionally* accepted, with the understanding that it may be altered, expanded, or rejected if subsequent evidence warrants doing so, or if a better hypothesis, not yet imagined, is devised. Sometimes, indeed, a radically new "paradigm" replaces an old one; for example, plate tectonics revolutionized geology in the 1950s, replacing the conviction that continents are fixed in position. More often, old hypotheses are incrementally modified and expanded over time. For instance, Mendel's laws of assortment and independent segregation, which initiated modern genetics, were modified when phenomena such as linkage and meiotic drive were discovered, but Mendel's underlying principle of inheritance based on "particles" (genes) holds true today.

This process reflects one of the most important and valuable features of science: even if individual scientists may be committed to a hypothesis, scientists as a group are not irrevocably committed to any belief and do not maintain it in the face of convincing contrary evidence. They must, and do, change their minds if the evidence so warrants. Indeed, much of science consists of seeking chinks in the armor of established ideas, and few successes will burnish a scientist's reputation more than showing that an important orthodox hypothesis is inadequate or flawed. Thus science, as a social process, is tentative; it questions belief and authority; it continuously tests its views against evidence. Scientific claims, in fact, are the outcome of a process of natural selection, for ideas (and scientists) compete with one another, so that the body of ideas in a scientific field grows in explanatory content and power (Hull 1988). Considerable scientific research is devoted to imagining ways in which information could be misinterpreted to support a false hypothesis, and devising better safeguards against error. [Among many examples in this book, see Chapters 2 (e.g., pp. 37–38) and 7 (e.g., pp. 163–164).] Science differs in this way from creationism, which does not use evidence to test its claims, does not allow evidence to shake its a priori commitment to certain beliefs, and does not grow in its capacity to explain the natural world. Unshakeable belief despite reason or evidence (i.e., faith) may be considered a virtue in a religious framework, but is precisely antithetical to the practice of science.

How could creationism, in any guise, contribute to scientific understanding? Suppose an advocate of "intelligent design" says that multicellular organisms are so complex, compared with unicellular organisms, that they must have originated by the intervention of an intelligent designer. Unless the ID advocate proposes that extraterrestrial creatures are responsible (which would merely shift the problem a step back), this designer must be a supernatural rather than a material being. So what was this designer, how did it equip organisms with new features, how long did it take, and why did it do this? Natural science can at least imagine ways to address such questions. (For example, we can look for phylogenetic intermediates, analyze sequence differences in genes that encode the relevant character differences, look for fossils, do experiments on the selective advantage of multicellularity; see Chapter 22, p. 617 on such an experiment). But the ID hypothesis generates no research ideas. It stops science dead in its tracks.

Scientific research requires that we have some way of testing hypotheses based on experimental or observational data. *The most important feature of scientific hypotheses is that they are testable,* at least in principle. Sometimes we can test a hypothesis by direct observation, but more often we do not see natural processes or causes directly. (For example, electrons, atoms, hydrogen bonds, molecules, and genes are not directly visible, and we cannot watch the occurrence of mutation during DNA replication. We cannot see DNA replication itself, for that matter.) Rather, we infer such processes by comparing the outcome of observations or experiments with predictions made from competing hypotheses. In order to make such inferences, we must assume that the processes obey **natural laws**: statements that certain patterns of events will always occur if certain conditions hold. That is, science depends on the consistency, or predictability (at least in a statistical sense), of natural phenomena, as exemplified by the laws of physics and chemistry. Because supernatural events or agents are supposed to suspend or violate natural laws, science cannot infer anything about them.

It is important to understand that just as religion does not provide scientific, mechanistic explanations for natural phenomena, science cannot provide answers to any questions that are not about natural phenomena: it cannot tell us what is beautiful or ugly, good or bad, moral or immoral. It cannot tell us what the meaning of life is, and it cannot tell us whether or not supernatural beings exist (see Gould 1999; Pigliucci 2002). On the other hand, religion does not provide any incontrovertible answers to such questions, on which adherents to different religions may honestly disagree (Coyne 2009b, 2012). And ever since Plato, these questions have been addressed by philosophy as well as religion.

Scientists can test and falsify many specific creationist claims, such as the occurrence of a worldwide flood or the claim that Earth and all organisms are less than 10,000 years old, but scientists cannot test the hypothesis that God exists, or that He created anything, because we do not know what consistent patterns these hypotheses might predict. (Try to think of any observation at all that would definitively rule out these supernatural possibilities.) Science must therefore adopt the position that natural causes are responsible for whatever we wish to explain about the natural world. This is not necessarily a commitment to METAPHYSICAL NATURALISM (the assumption that everything *truly* has natural rather than supernatural causes), but it is a commitment to METHODOLOGICAL NATURALISM (the *working principle* that we can entertain only natural causes when we seek scientific explanations). To be sure, some scientists maintain that accepting both scientific and religious assertions (e.g., mechanistic explanations of most phenomena, but also miracles such as people request in intercessory prayer) requires cognitive dissonance and arbitrary decisions about when to require evidence and when not to (Coyne 2009b; Dawkins 2009b).

Evolution as fact and as theory

We have been using the words "hypothesis," "theory," and "fact," and it is imperative that we understand what they mean. A **hypothesis** is a proposition, a supposition. Before 1944, the idea that the genetic material is DNA was a plausible hypothesis that had little evidence to support it. After 1944, this hypothesis became stronger and stronger as evidence grew. Now we consider it a FACT. Many facts in science are, simply, hypotheses that have become so well supported by evidence that we feel safe in acting as if they were true. To use a courtroom analogy, they have been "proven" beyond reasonable doubt. Not beyond any conceivable doubt, but reasonable doubt. (The word "fact" is also used to mean a simple observation of data, such as the color of your hair—which, if you think about it, we accept as if it were true despite the long arguments we could have about the reliability of sense perceptions, the definitions of colors, the accuracy of measurements, and so on.)

A **theory**, as the word is used in science, doesn't mean an unsupported speculation or hypothesis (the popular use of the word). A theory is, instead, a big idea that encompasses other ideas and hypotheses and weaves them into a coherent fabric, as we saw in Chapter 1. Thus atomic theory, quantum theory, and plate tectonic theory are not mere speculations or opinions, but strongly supported ideas that explain a great variety of phenomena. There are few theories in biology, and among them, evolutionary theory is surely the most important.

So is evolution a fact or a theory? In light of these definitions, evolution is a scientific fact. That is, the descent of all species, with modification, from common ancestors is a hypothesis that in the last 150 years or so has been supported by so much evidence, and has so successfully resisted all challenges, that it has become a fact. But this history of evolutionary change is explained by evolutionary theory, the body of statements (about mutation, selection, genetic drift, developmental constraints, and so forth) that together account for the various changes that organisms have undergone.

Because creationist explanations for the diversity and characteristics of living things cannot be evaluated by the methods of science, they should not be given "equal time" in a science classroom. Neither should other hypotheses that are either unscientific or demonstrably wrong. Chemistry teachers do not and should not teach about alchemy (the old idea that one element, such as lead, can be magically transformed into another, such as gold); earth science classes should not even mention the hypothesis that Earth is flat; history and psychology teachers should not consider astrological explanations for historical events or personality traits—even though there are people who believe in all of these pseudoscientific ideas. The ideal of democracy doesn't extend to ideas—some are simply wrong, and as a purely practical matter, it is imperative that we recognize them as such. In everyday life, we assume and depend on natural, not supernatural, explanations. Unlike the Puritans of Salem, Massachusetts, who in 1692 condemned people for witchcraft, we no longer seriously entertain the notion that someone can be victimized by a witch's spell or possessed by devils, and we would be outraged if a criminal successfully avoided conviction because he claimed "the Devil made me do it." Even those who most devoutly believe that God sustains them in the palm of His hand will panic if the airplane's wing flaps malfunction. We depend on scientific explanations, and we know that science has proven its ability—because it works.

The consequences of rejecting science in favor of ideological dogma or folk belief or religious belief can be tragic. Several years ago, news reports described the arrest of an American couple on charges of child abuse. They had allowed their two children to die because they rejected medical help in favor of "faith healing." It has been estimated that the South African government could have prevented the premature death by AIDS of 330,000 adults and 35,000 babies, had it not been for the former president's denial that AIDS is caused by HIV, and his appointment of a like-minded health minister who prescribed garlic, lemon juice, and beetroot as remedies (*New York Times*, November 26, 2008). The perils of ignoring or distorting fundamental science became tragically evident in the Soviet Union between the 1930s and the 1960s under Stalin, when Mendelism and Darwinism came to be viewed as dangerous Western ideologies. Trofim Lysenko, an agronomist with no scientific training, gained Stalin's favor by declaring genetics a capitalist, bourgeois threat to the state, and led a campaign in which geneticists were persecuted and imprisoned and genetics research and training were extinguished. Rejecting the Mendelian-Darwinian view of adaptation, Lysenko preached a Lamarckian doctrine that exposing crop plants to an environmental factor, such as low temperature, would induce inherited adaptive changes in their offspring. Lysenko completely altered the practice of Soviet agriculture and created a disaster for Soviet food production and the Soviet people (Soyfer 1994). Rejection or ideological control of science, the most reliable basis yet developed for knowing about the natural, material world, is profoundly dangerous to the welfare of society.

The Evidence for Evolution

The evidence for evolution has been presented throughout the preceding chapters of this book. The examples provided represent only a very small percentage of the studies that might be cited for each particular line of evidence. In this section, we will simply review the sources of evidence for evolution and refer back to earlier chapters for detailed examples.

The fossil record

The fossil record is extremely incomplete, for reasons that geologists understand well (see Chapter 4). Consequently, the transitional stages that we postulate in the origin of many higher taxa have not (yet) been found. But there is absolutely no truth to the claim, made by many creationists, that the fossil record does not provide any intermediate forms. There are many examples of such forms, both at low and high taxonomic levels; Chapter 4 provides several examples in the evolution of the classes of tetrapod vertebrates. Critically important intermediates are still being found: just in the last few years, several Chinese fossils, including feathered dinosaurs, have greatly expanded the record of the origin of birds, and the discovery of *Tiktaalik* has enlarged our understanding of the origin of tetrapods (see Chapter 4). The fossil record, moreover, documents two important aspects of character evolution: mosaic evolution (e.g., the more or less independent evolution of different features in the evolution of mammals) and gradual change of individual features (e.g., cranial capacity and other features of hominins).

Many discoveries in the fossil record fit predictions based on phylogenetic or other evidence. The earliest fossil ants, for instance, have the wasplike features that had been predicted by entomologists, and the discovery of feathered dinosaurs was to be expected, given the consensus that birds are modified dinosaurs. Likewise, phylogenetic analyses of living organisms imply a sequence of branching events, as well as a sequence of origin of the diagnostic characters of those branches. The fossil record often matches the predicted sequences (as we saw in Chapters 4 and 5): for example, prokaryotes precede eukaryotes in the fossil record, wingless insects precede winged insects, fishes precede tetrapods, amphibians precede amniotes, algae precede vascular plants, ferns and gymnosperms precede flowering plants.

Phylogenetic and comparative studies

Even if we had no fossil record at all, many other kinds of information would provide incontrovertible evidence for evolution. For example, molecular phylogenies support many of the relationships that have long been postulated from morphological data (see Chapter 2). These two data sets are entirely independent (the molecular phylogenies are often based on sequences that have no biological function), so their correspondence justifies confidence that the relationships are real: that the lineages form monophyletic groups and have indeed descended from common ancestors.

The largest monophyletic group encompasses all organisms. Although Darwin allowed that life might have originated from a few original ancestors, we can be confident today that all known living things stem from a single ancestor because of the many features that are universally shared. These features include most of the codons in the genetic code, the machinery of nucleic acid replication, the mechanisms of transcription and translation, proteins composed only of "left-handed" (L-isomer) amino acids, and many aspects of fundamental biochemistry. Many genes are shared among all organisms, including the three major "empires" (Bacteria, Archaea, and Eucarya; see Chapters 4 and 20), and these genes have been successfully used to infer the deepest branches in the tree of life.

Systematists have shown that the differences among related species often form gradual series, ranging from slight differences to great differences with stepwise intermediates (see Chapter 3). Such intermediates often make it difficult to establish clear-cut families or other higher taxa, so classification often becomes a somewhat arbitrary choice between "splitting" species among many taxa and "lumping" them into few. Systematic studies have also demonstrated the common origin, or homology, of characteristics that may differ greatly among taxa—the most familiar examples being the radically different forms of limbs among tetrapod vertebrates. Homology of structures is often more evident in early developmental stages than in adult organisms, and contemporary developmental biology demonstrates that Hox genes and other developmental mechanisms are shared among animal phyla that diverged from common ancestors a billion or more years ago (see Chapter 21).

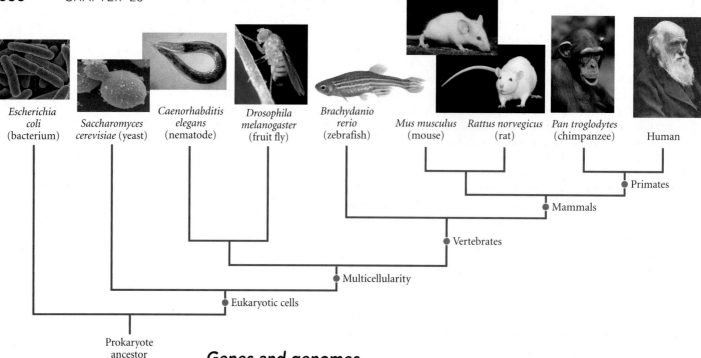

Escherichia coli (bacterium) *Saccharomyces cerevisiae* (yeast) *Caenorhabditis elegans* (nematode) *Drosophila melanogaster* (fruit fly) *Brachydanio rerio* (zebrafish) *Mus musculus* (mouse) *Rattus norvegicus* (rat) *Pan troglodytes* (chimpanzee) Human

Primates
Mammals
Vertebrates
Multicellularity
Eukaryotic cells
Prokaryote ancestor

FIGURE 23.2 Only common ancestry can explain why the U.S. National Institutes of Health should support basic biological research on organisms ranging from bacteria to chimpanzees, on the supposition that it will contribute to understanding human health and disease. (*E. coli* photo courtesy of Rocky Mountain Laboratories, NIAID, NIH; *S. cerevisiae* courtesy of NASA; *C. elegans* courtesy of Amy Pasquinelli/NIH.)

Genes and genomes

The revolution in molecular biology and genomics is yielding data about evolution on a larger scale than ever before. These data increasingly show the extraordinary commonality of all living things. Because of this commonality, the structure and function of genes and genomes can be understood through comparisons among species and evolutionary models. Indeed, it is only because of this common ancestry that there has ever been any reason to think that human biochemistry, physiology, or brain function, much less genome function, could be understood by studying yeast, flies, rats, or monkeys (**Figure 23.2**).

Molecular studies show that the genomes of most organisms have similar elements, such as a great abundance of noncoding pseudogenes and satellite DNA and a plethora of "selfish" transposable elements that generally provide no advantage to the organism. These and other features are readily understandable under evolutionary theory, but would hardly be expected of an intelligent, omnipotent designer (Avise 2010). Molecular evolutionary analyses have shown how new genes arise by processes such as unequal crossing over and divergence of duplicate genes (see Chapter 20). Some DNA polymorphisms are shared between species, so that, for example, some major histocompatibility sequences of humans are more similar and more closely related to chimpanzee sequences than to other human sequences (see Figure 12.32B). What more striking evidence of common ancestry could there be?

Among the many other ways in which molecular studies affirm the reality of evolution, consider just one more: molecular clocks. They are far from perfectly accurate, but sequence differences between species nevertheless are roughly correlated with the time since common ancestry, judged from other evidence such as biogeography or the fossil record (see Chapters 2 and 4). The sequence differences do not simply encode the phenotypic differences that the organisms manifest, and the phenotypic differences themselves are much less correlated with time since common ancestry. No theory but evolution makes sense of these patterns of DNA differences among species.

Biogeography

We noted in Chapter 6 that the geographic distributions of organisms provided Darwin with abundant evidence of evolution, and they have continued to do so. For example, the distributions of many taxa correspond to geological events such as the movement of land masses and the formation and dissolution of connections between them. We saw that the phylogenies of Hawaiian species match the sequence by which the islands came into existence. We saw that diverse ecological niches in a region are typically filled not by the same taxa that occupy similar niches in other parts of the world, but by different mono-phyletic groups that have undergone independent adaptive radiation (such as the anoles of the Greater Antilles). We saw, as did Darwin, that an isolated region such as an island is not populated by all the kinds of organisms that could thrive there, as we might suppose a thoughtful designer could arrange. Instead, whole groups are commonly missing, and human-introduced species often come to dominate.

Failures of the argument from design

Since God cannot be known directly, theologians such as Thomas Aquinas have long attempted to infer His characteristics from His works. Theologians have argued, for instance, that order in the universe, such as the predictable movement of celestial bod-ies, implies that God must be orderly and rational, and that He creates according to a plan. From the observation that organisms have characteristics that serve their survival, it could similarly be inferred that God is a rational, intelligent designer who, furthermore, is benef-icent: He has not only conferred on living things the boon of existence, but has equipped them for all their needs. Such a beneficent God would not create an imperfect world; so, as the philosopher Leibniz said, this must be "the best of all possible worlds." (His phrase was mercilessly ridiculed by Voltaire in his marvelous satire *Candide*.) The adaptive design of organisms, in fact, has long been cited as evidence of an intelligent designer. This was the thrust of William Paley's (1831) famous example: as the design evident in a watch implies a watchmaker, so the design evident in organisms implies a designer of life. This "argument from design" has been renewed in the "intelligent design" version of creation-ism, and it is apparently the most frequently cited reason people give for believing in God (Pigliucci 2002).

Of course, Darwin made this particular theological argument passé by providing a natural mechanism of design: natural selection. Moreover, Darwin and subsequent evolu-tionary biologists have described innumerable examples of biological phenomena that are hard to reconcile with beneficent intelligent design. Just as Voltaire showed (in *Candide*) that cruelties and disasters make a mockery of the idea that this is "the best of all pos-sible worlds," biology has shown that organisms have imperfections and anomalies that can be explained only by the contingencies of history, and characteristics that make sense only if natural selection has produced them. If "good design" were evidence of a kindly, omnipotent designer, would "inferior design" be evidence of an unkind, incompetent, or handicapped designer?

Only evolutionary history can explain vestigial organs—the rudiments of once-func-tional features, such as the tiny, useless pelvis and femur of whales, the reduced wings under the fused wing covers of some flightless beetles, and the nonfunctional stamens or pistils of plants that have evolved separate-sexed flowers from an ancestral hermaph-roditic condition. Likewise, only history can explain why the genome is full of "fossil" genes: pseudogenes that have lost their function. Only the contingencies of history can explain the arbitrary nature of some adaptations. For instance, whereas ectothermic ("cold-blooded") tetrapods have two aortic arches, "warm-blooded" endotherms have only one. This difference is probably adaptive, but can anything except historical chance explain why birds have retained the right arch and mammals the left?

Because characteristics evolve from pre-existing features, often undergoing changes in function, many features are poorly engineered, as anyone who has suffered from lower back pain or wisdom teeth can testify. Once the pentadactyl limb became developmentally canalized, tetrapods could not evolve more than five digits even if they would be useful: the extra "thumb" of the giant panda and of moles is not a true digit at all, and it lacks the flexibility of true fingers because it is not jointed (see Figure 22.10). Similarly, animals would certainly be better off if they could synthesize their own food, and corals do so by harboring endosymbiotic algae—but no animal is capable of photosynthesis. And it is a pity that humans, unlike salamanders, cannot regenerate lost limbs or digits.

If a designer were to equip species with a way to survive environmental change, it might make sense to devise a Lamarckian mechanism, whereby genetic changes would occur in response to need. Instead, adaptation is based on a combination of a random process (mutation) that cannot be trusted to produce the needed variation (and often does not) and a process that is the very epitome of waste and seeming cruelty (natural selection, which requires that great numbers of organisms fail to survive or reproduce). It would be hard to imagine a crueler instance of natural selection than sickle-cell disease, whereby part of the human population is protected against malaria at the expense of countless other people, who are condemned to die because they are homozygous for a gene that happens to be worse for the malarial parasite than for heterozygous carriers (see Chapter 12). Indeed, when Darwin proposed his theory of the cause of evolution, it was widely rejected just because people found it so distasteful, even horrifying, to contemplate. And, of course, this process often does not preserve species in the face of change: more than 99 percent of all species that have ever lived are extinct. Were those species the products of an incompetent designer? Or one that couldn't foresee that species would have to adapt to changing circumstances?

Many species become extinct because of competition, predation, and parasitism. Some of these interactions are so appalling that Darwin was moved to write, "What a book a devil's chaplain might write on the clumsy, wasteful, blundering, low, and horribly cruel works of Nature!" Darwin knew of maggots that work their way up the nasal passages into the brains of sheep, and wasp larvae that, having consumed the internal organs of a living caterpillar, burst out like the monsters in the movie *Alien*. The life histories of parasites, whether parasitic wasp or human immunodeficiency virus (HIV), ill fit our concept of an intelligent, kindly designer, but they are easily explained by natural selection (see Chapter 19).

No one has yet demonstrated a characteristic of any species that serves only to benefit a different species, or only to enhance the so-called balance of nature—for, as Darwin saw, "such could not have been produced through natural selection." Because natural selection consists only of differential reproductive success, it results in "selfish" genes and genotypes, some of which have results that are inexplicable by intelligent design (see Chapter 16). We have seen that genomes are brimming with sequences such as transposable elements that increase their own numbers without benefiting the organism. We have seen maternally transmitted cytoplasmic genes that cause male sterility in many plants, and nuclear genes that have evolved to override them and restore male fertility. Such conflicts among genes in a genome are widespread. Are they predicted by intelligent design theory? Likewise, no theory of design can predict or explain features that we ascribe to sexual selection, such as males that remove the sperm of other males from the female's reproductive tract, or chemicals that enhance a male's reproductive success but shorten his mate's life span. Nor can we rationalize why a beneficent designer would shape the many other selfish behaviors that natural selection explains, such as cannibalism, siblicide, and infanticide.

Evolution, and its mechanisms, observed

Anyone can observe erosion, and geologists can measure the movement of continental plates, which travel at up to 10 centimeters per year. No geologist doubts that these mechanisms, even if they accomplish only slight changes on the scale of human generations, have shaped the Grand Canyon and have separated South America from Africa over the course of millions of years. Likewise, biologists do not expect to see anything like the origin of mammals played out on a human time scale, but they have documented the processes that will yield such grand changes, given enough time.

Evolution requires genetic variation, which originates by mutation. From decades of genetic study of initially homozygous laboratory populations, we know that mutations arise that have effects, ranging from very slight to drastic, on all kinds of phenotypic characters (see Chapter 8). These mutations can provide new variation in quantitative characters, seemingly without limit. This variation has been used by humans for millennia to develop strains of domesticated plants and animals that differ in morphology more than whole families of natural organisms do. In experimental studies of laboratory populations of microorganisms, we have seen new advantageous mutations arise and enable rapid adaptation to temperature changes, toxins, or other environmental stresses. Laboratory studies have documented the occurrence of the same kinds of mutations, at the molecular level, that are found in natural populations and distinguish species. No geneticist or molecular biologist doubts that the differences among species in their genes and genomes originated by natural mutational processes that, by and large, are well understood.

We also know that most natural populations carry a great deal of genetic variation, which can yield rapid responses to artificial or natural selection (see Chapters 9, 11, and 12). We have seen allele frequency differences among recently established populations that can be confidently attributed to genetic drift (see Chapter 10). Evolutionary biologists have documented literally hundreds of examples of natural selection acting on genetic and phenotypic variation (see Chapters 12 and 13). They have described hundreds of cases in which populations have responded to directional selection and have adapted to new environmental factors, ranging from the evolution of resistance to insecticides, herbicides, and antibiotics to the evolution of different diets (see Chapter 13).

Speciation generally takes a very long time, but some processes of speciation can also be observed. Substantial reproductive isolation has evolved in laboratory populations, and species of plants that apparently originated by polyploidy and hybridization have been "re-created" de novo by crossing their suspected parent forms and selecting for the species' diagnostic characters (see Chapter 18).

In summary, the major causes of evolution are known, and they have been extensively documented. The two major processes of long-term evolution, anagenesis (changes in characters within lineages) and cladogenesis (origin of two or more lineages from a common ancestor), are abundantly supported by evidence from every possible source, ranging from molecular biology to paleontology. Over the past century, we have certainly learned of evolutionary processes that were formerly unknown: we now know, for example, that some species may arise from hybridization, and that some DNA sequences are mobile and can cause mutations in other genes. But no scientific observations have ever cast serious doubt on the reality of the basic mechanisms of evolution, such as natural selection, or on the reality of the basic historical patterns of evolution, such as transformation of characters and the origin of all known forms of life from common ancestors. Contrast this mountain of evidence with the evidence for supernatural creation or intelligent design: *there is no such evidence whatever.*

Refuting Creationist Arguments

Creationists attribute the existence of diverse organisms and their characteristics to miracles: direct supernatural intervention. As we have seen, it is impossible to predict miracles or to do experiments on supernatural processes, so creationists do not do original research in support of their theory. Thus, rather than providing positive evidence of divine creation, "creation science" consists entirely of attempts to demonstrate the falsehood or inadequacy of evolutionary science and to show that biological phenomena must, by default, be the products of intelligent design. Here are some of the most commonly encountered creationist arguments, together with capsule counterarguments.

1. *Evolution is outside the realm of science because it cannot be observed.*

   Evolutionary changes have indeed been observed, as we saw earlier in this chapter. In any case, most of science depends not on direct observation, but on testing hypotheses against the predictions they make about what we should observe. Observation of the postulated processes or entities is not required in science, and most of the fundamental processes and objects in physics, chemistry, and cell biology cannot be seen or observed.

2. *Evolution cannot be proved.*

   Nothing in science is ever absolutely proved. "Facts" are hypotheses in which we can have very high confidence because of massive evidence in their favor and the absence of contradictory evidence. Abundant evidence from every area of biology and paleontology supports evolution, and there exists no contradictory evidence.

3. Evolution is not a scientific hypothesis because it is not testable: no possible observations could refute it.

Many conceivable observations could refute or cast serious doubt on evolution, such as finding incontrovertibly mammalian fossils in incontrovertibly Precambrian rocks. In contrast, any puzzling quirk of nature could be attributed to the inscrutable will and infinite power of a supernatural intelligence, so creationism is untestable.

4. The orderliness of the universe, including the order manifested in organisms' adaptations, is evidence of intelligent design.

Order in nature, such as the structure of crystals, arises from natural causes and is not evidence of intelligent design. The order displayed by the correspondence between organisms' structures and their functions is the consequence of natural selection acting on genetic variation, as has been observed in many experimental and natural populations; we have described numerous cases of both throughout this book. Darwin's realization that the combination of a random process (the origin of genetic variation) and a nonrandom process (natural selection) can account for adaptations provided a natural explanation for the apparent design and purpose in the living world and made a supernatural account unnecessary and obsolete.

5. Evolution of greater complexity violates the second law of thermodynamics, which holds that entropy (disorder) increases.

This is a common misrepresentation of one of the most important laws of physics. The second law applies only to closed systems, such as the universe as a whole. Order and complexity can increase in local, open systems as a result of an influx of energy. This is evident in the development of individual organisms, in which biochemical reactions are powered by energy derived ultimately from the Sun.

6. It is almost infinitely improbable that even the simplest life could arise from nonliving matter. The probability of random assembly of a functional nucleotide sequence only 100 bases long is $1/4^{100}$, an exceedingly small number. And scientists have never synthesized life from nonliving matter.

It is true that a fully self-replicating system of nucleic acids and replicase enzymes has not yet arisen from simple organic constituents in the laboratory, but the history of scientific progress shows that it would be foolish and arrogant to assert that what science has not accomplished in a few decades cannot be accomplished. (And even if, given our human limitations, we should never succeed in this endeavor, why should that require us to invoke the supernatural?) Critical steps in the probable origin of life, such as abiotic synthesis of purines, pyrimidines, and amino acids and self-replication of short RNAs, have been demonstrated in the laboratory (see Chapter 5). And there is no reason to think that the first self-replicating or polypeptide-encoding nucleic acids had to have had any particular sequence. If there are many possible sequences with such properties, the probability of their formation rises steeply. Moreover, the origin of life is an entirely different problem from the modification and diversification of life once it has arisen. We do not need to know anything about the origin of life in order to understand and document the evolution of different life forms from their common ancestor.

7. Mutations are harmful and do not give rise to complex new adaptive characteristics.

Most mutations are indeed harmful and are purged from populations by natural selection. Some, however, are beneficial, as shown in many experiments (see Chapters 8 and 13). Complex adaptations are usually based not on single mutations, but on combinations of mutations that jointly or successively increase in frequency as a result of natural selection.

8. *Natural selection merely eliminates unfit mutants, rather than creating new characters.*

"New" characters, in most cases, are modifications of pre-existing characters, which are altered in size, shape, developmental timing, or organization (see Chapters 3, 4, and 21). This is true at the molecular level as well (see Chapter 20). Natural selection "creates" such modifications by increasing the frequencies of alleles at several or many loci so that combinations of alleles, initially improbable because of their rarity, become probable. Observations and experiments on both laboratory and natural populations have demonstrated the efficacy of natural selection.

9. *Chance could not produce complex structures.*

This is true, but natural selection is a deterministic, not a random, process. The random processes of evolution—mutation and genetic drift—do not result in the evolution of complexity, as far as we know. Indeed, when natural selection is relaxed, complex structures, such as the eyes of cave-dwelling animals, slowly degenerate, due in part to the fixation of neutral mutations by genetic drift.

10. *Complex adaptations such as wings, eyes, and biochemical pathways could not have evolved gradually because the first stages would not have been adaptive. The full complexity of such an adaptation is necessary, and it could not arise in a single step by evolution.*

This was one of the first objections that greeted *The Origin of Species*, and it has been christened "irreducible complexity" by advocates of intelligent design. Our answer has two parts. First, many such complex features, such as hemoglobins and eyes, do show various stages of increasing complexity among different organisms (see Chapters 3, 4, 5, and 22). "Half an eye"—an eye capable of discriminating light from dark, but incapable of forming a focused image—is indeed better than no eye at all (see Figure 22.13). Second, many structures have been modified for a new function after being elaborated to serve a different function (see Chapters 3 and 22). The "finished version" of an adaptation that we see today may indeed require precise coordination of many components in order to perform its current function, but the earlier stages, performing different or less demanding functions, and performing them less efficiently, are likely to have been an improvement on the ancestral feature. The evolution of the mammalian skull and jaw (see Chapter 4) provides a good example.

11. *If an altered structure, such as the long neck of the giraffe, is advantageous, why don't all species have that structure?*

This naïve question ignores the fact that different species and populations have different ecological niches and environments, for which different features are adaptive. This principle holds for all features, including "intelligence."

12. *If gradual evolution had occurred, there would be no phenotypic gaps among species, and classification would be impossible.*

Many disparate organisms are connected by intermediate species, and in such cases, classification into higher taxa is indeed rather arbitrary (see Chapter 3). In other cases, gaps exist because of the extinction of intermediate forms (see Chapter 4). Moreover, although much of evolution is gradual, some advantageous mutations with large, discrete effects on the phenotype have probably played a role. Whether or not evolution has been entirely gradual is an empirical question, not a theoretical necessity.

13. ***The fossil record does not contain any transitional forms representing the origin of major new forms of life.***

    This very common claim is flatly false, for there are many such intermediates. Chapter 4 describes a few examples, chosen from the vertebrate groups because those groups are more familiar than many others.

14. ***The fossil record does not objectively represent a time series because rock strata are ordered by their fossil contents and are then assigned different dates on the assumption that evolution has occurred.***

    Even before *The Origin of Species* was published, geologists who did not believe in evolution recognized the temporal order of the fossils that are characteristic of different periods, and had named most of the geological periods. Since then, radiometric dating and other methods have established the absolute ages of geological strata.

15. ***Vestigial structures are not vestigial, but functional.***

    According to creationist thought, an intelligent Creator must have had a purpose, or design, for each element of His creation. Thus all features of organisms must be functional. For this reason, creationists view adaptations as support for their position. However, nonfunctional, imperfect, and even maladaptive structures are expected if evolution is true, especially if a change in an organism's environment or way of life has rendered them superfluous or harmful. As noted earlier, organisms display many features, at both the morphological and molecular levels, that are very unlikely to have any function.

16. ***The classic examples of evolution are false.***

    Some creationists have charged that some of the best-known studies of evolution, cited in almost every textbook, are flawed and that evolutionary biologists have dishonestly perpetuated these supposed falsehoods. For example, H. B. D. Kettlewell, who performed the classic study of industrial melanism in the peppered moth, was accused of having obtained spurious evidence for natural selection by predatory birds because he pinned moths to unnatural resting sites (tree trunks). Kettlewell's conclusions have been strongly supported by later research (see Chapter 12), but suppose for a moment that this and other frequently cited studies were indeed flawed. First, it does not follow that textbook authors and other contemporary biologists have deliberately perpetuated falsehoods; they simply might not have checked and analyzed the original studies, but instead borrowed from other books and "secondary" sources, since no textbook author can check every study in depth. In any case, there is no reason to suspect intellectual dishonesty. Second, whether or not Kettlewell's work was flawed is irrelevant to the validity of the basic claims involved. Both natural selection and rapid evolutionary changes have been demonstrated in so many species that these principles would stand firmly even if the peppered moth story were completely false.

    Of course, the creationists who cite these examples of supposed flaws and frauds realize that the strength of evolutionary biology does not rest on these studies. After all, most creationists accept natural selection and "microevolution," such as changes in moth coloration. Rather, these attacks enable their readers to doubt the truthfulness of evolutionary scientists and to justify their disbelief in evolution. But remember that even if individual scientists are stupid (which a few are) or dishonest (which almost none are), the social process of science uncovers errors and justifies our confidence in its major claims.

17. ***Disagreements among evolutionary biologists show that Darwin was wrong.***

    Disagreements among scientists exist in every field of inquiry and are, in fact, the fuel of scientific progress. They stimulate research and are thus a sign of vitality. Creationists misunderstand or misinterpret evolutionary biologists who have argued (a) that the fossil record displays abrupt shifts rather than gradual change (punctuated equilibrium);

(b) that many characteristics of species may not be adaptations; (c) that evolution may involve mutations with large effects as well as mutations with small effects; and (d) that natural selection does not explain certain major events and trends in the history of life. In fact, none of the evolutionary biologists who hold these positions deny the central proposition that adaptive characteristics evolve by the action of natural selection on random mutations. All these debates arise from differing opinions on the relative frequency and importance of factors known to influence evolution: large-effect versus small-effect mutations, genetic drift versus natural selection, individual selection versus species selection, adaptation versus constraint, and so forth (see Chapters 11, 21, and 22). These arguments about the relative importance of different processes do not at all undermine the strength of the evidence for the historical fact of evolution—that is, descent, with modification, from common ancestors. On this point, there is no disagreement among evolutionary biologists.

18. *There are no fossil intermediates between apes and humans; australopithecines were merely apes. And there exists an unbridgeable gap between humans and all other animals in cognitive abilities.*

This is a claim about one specific detail in evolutionary history, but it is the issue about which creationists care most. This claim is simply false. The array of fossil hominids shows numerous stages in the evolution of posture, hands and feet, teeth, facial structure, brain size, and other features. DNA from Neanderthal remains shows their close relationship to modern humans. Both functional and nonfunctional DNA sequences from humans and African apes are extremely similar and clearly demonstrate that they share a recent common ancestor. The mental abilities of humans are indeed developed to a far greater degree than those of other species, but many of our mental faculties seem to be present in more rudimentary form in other primates and mammals.

19. *As a matter of fairness, alternative theories, such as supernatural creation and intelligent design, should be taught, so that students can make their own decisions.*

This train of thought, if followed to its logical conclusion, would have teachers presenting hundreds of different creation myths, in fairness to the peoples who hold them, and it would compel teachers to entertain supernatural explanations of everything in earth science, astronomy, chemistry, and physics, because anything explained by these sciences, too, could be argued to have a supernatural cause. It would imply teaching students that to do a proper job of investigating an airplane crash, federal agencies should consider the possibility of mechanical failure, a terrorist bomb, a missile impact—and supernatural intervention (Alters and Alters 2001).

Science teachers should be expected to teach the content of contemporary science—which means the hypotheses that have been strongly supported and the ideas that are subjects of ongoing research. That is, they should teach what scientists do. Several scientists have searched the scientific literature for research reports on intelligent design and "creation science" and have found no such reports. Nor is there any evidence that "creation scientists" have carried out scientific research that a biased community of scientists has refused to publish. As noted earlier in this chapter, there is no way of testing hypotheses about the supernatural, so there cannot be any scientific research on the subject. And that means that the subject should not be taught in a science course.

On arguing for evolution

How, and whether or not, to convince people to accept evolution is not a subject for this book, which is devoted to the content of evolutionary science. Brian and Sandra Alters have addressed this topic in *Defending Evolution in the Classroom: A Guide to the Creation/Evolution Controversy* (2001). As these authors point out, for many people, religious beliefs take precedence over scientific evidence, especially among people who believe that their fate for all eternity depends on adhering to their belief system. The points made in this chapter will do little to persuade such people. Presenting the nature of science and the evidence for evolution will, however, have an impact on people who genuinely question what the truth is; but the way in which it is presented is important, as Alters and Alters describe. In this context, anyone who is considering holding a public debate on evolution with a creationist should first become thoroughly familiar with creationist arguments and strategies, and especially with the intelligent design movement, which aggressively markets itself as science when it is not. *Why Evolution Is True*, by Jerry Coyne (2009a), and *The Greatest Show on Earth: The Evidence for Evolution*, by Richard Dawkins (2009), provide the best overall summaries of the argument. Other Internet resources and books are listed at the end of this chapter.

Why Should We Teach Evolution?

If evolution is so controversial, why invite trouble? Why not just drop it from the science curriculum? After all, it's an academic subject that doesn't really affect people's lives, right?

Wrong. Evolution is a foundation of all of biology. Understanding evolution is as relevant to everyday life as understanding physics or chemistry, and research in evolutionary biology affects our lives both directly and indirectly (**Figure 23.3**).

Educating students about evolution should mean teaching them not only the basic principles and facts of evolutionary processes and evolutionary history, but also the concepts

FIGURE 23.3 Cartoonist Gary Trudeau affirms the importance of evolutionary science.

and ways of testing hypotheses that are used in evolutionary science. These approaches have many applications. For example, evolutionary biologists specialize in studying variation both within and among species, and the conceptual approaches and methods they employ are broadly useful. Simply being aware of the importance of genetic and nongenetic variation is a useful lesson; for example, everyone, whether patient or doctor, should be aware that people may vary in their reaction to a drug (or to a disease, for that matter). Understanding the difference between genetic and nongenetic variation is profoundly important for interpreting claims that are made about differences among ethnic or "racial" groups (see Chapter 9). Evolutionary biology shows, moreover, that a trait may have high heritability and nevertheless be readily altered by environmental change (e.g., medical or educational intervention).

Learning about evolution would be important even if it had no practical utility, simply because of the insights and understanding it provides. But evolutionary science does bear directly on applied research and has many practical applications. We can mention only a few of those applications here, but they are treated more fully by David Mindell (2006) and are the subject of two recently founded journals, *Evolutionary Applications* and *Evolution, Medicine, and Public Health*.

Health and medicine

The direct and indirect uses of evolutionary biology are probably more numerous, and more important, in medicine and public health than in any other area (Antolin et al. 2012). Depending on the topic, evolutionary theory may provide new conceptual approaches to medically relevant research (e.g., the evolution of senescence), principles that medical research and practice should take into account (e.g., natural selection for drug resistance in pathogens), or methods for making inferences and discoveries (e.g., phylogenetic methods for tracing the spread of pathogens). Randolph Nesse and Stephen Stearns (2008) explain that *some education in evolutionary biology is essential for every medical researcher*, because it bears directly on almost every field of biological study, and *is useful for every clinical practitioner*, because it deepens one's understanding both of the human body and its ills and of the organisms that cause harm. Michael Antolin and colleagues (2012) cite many applications of evolutionary biology to medicine and public health. The following paragraphs are based largely on these articles.

Two major evolutionary questions can be asked about any trait: what is its evolutionary history, and what has been its evolutionary cause? Often the latter question means, what is its adaptive function? Both questions can lead to deeper understanding of the traits, and the underlying genes, of both humans and pathogens.

For instance, phylogenetic thinking and evolutionary genetics together enable us to understand why some populations have a higher prevalence of lactase persistence, malaria resistance, and other traits (see Chapter 12). They help to explain sickle-cell hemoglobin and the ApoE4 subtype of apolipoprotein, which increases the risk of developing atherosclerosis and Alzheimer's disease, and which has increased in populations that inhabit cold climates, perhaps as an adaptation to a diet high in meat. Population genetic theory enables researchers to identify alleles that have recently increased due to strong selection (see Chapter 12), some of which can increase the risk of infection (Dean et al. 2002). Genomic comparisons of human and other genomes indicate which genes are under strong purifying selection and are therefore likely to cause genetic disease when mutated (Miller and Kumar 2001; see Figure 12.34).

Evolutionary theory helps to explain not only variable human traits, but also universal or nearly universal characteristics, including some that concern health and disease. For example, the theory of genetic conflict can explain the remarkable tug of war between a pregnant woman and her fetus, which is based on the conflict of interest between maternal and imprinted paternal genes (see Chapter 16). Disruption of this balance may underlie some psychiatric disorders, such as autism. Senescence, when viewed through the theory of the evolution of life histories (see Chapter 14), is no longer seen as an inevitable result of wear and tear, but rather as a genetically controlled process based on pleiotropic effects of alleles that are advantageous at younger ages. This theory suggests that genes that reduce oxidative metabolism, and therefore senescence, may lower reproduction—for which there is some evidence. It also suggests that many genes have pleiotropic effects on senescence, which implies that no single genetic alteration would significantly extend the human life span. If we posit that universal human traits are effects of natural selection, we may wonder if the annoying symptoms of infectious disease, such as fever and coughing, might be defenses against pathogens—and whether infection might be prolonged by treating those symptoms. The adaptationist approach leads us naturally to ask about the effects of our modern environment, so unlike the environments we evolved in. Differences between modern Western diets and prehistoric diets have been cited as the cause of high rates of cardiovascular disease, obesity, diabetes, high blood pressure, bowel and autoimmune disorders, and some cancers. It is very likely that the increasing incidence of allergies and other autoimmune disorders is caused by our abnormally sterilized environment, which reduces early exposure to bacteria and parasites that prime the development of immune function. [Remarkably, experimental infection with helminths (worms) has successfully treated Crohn's disease and other immune disorders, and multiple sclerosis develops more slowly in helminth-infected patients.] Humans naturally harbor an enormous diversity of commensal bacteria, the "human microbiome" with which we have coevolved, so it is not surprising that excessive treatment with antibiotics can sometimes have harmful effects.

The evolution of antibiotic resistance by pathogens is one of the greatest threats to public health: infections caused by resistant *Staphylococcus aureus* cause more than 18,000 deaths annually* in the United States, where the total annual cost of antibiotic resistance is estimated to be about $80 billion. It is frustrating that the popular media, and even medical journals, tend not to call this phenomenon by its rightful name—*the evolution of resistance*—but to use euphemisms such as "development" instead (Antonovics et al. 2007). The evolution of antibiotic resistance is greatly encouraged by widespread (and largely unnecessary) uses of antibiotics, including the large quantities administered to farm animals in order to promote their growth. (This use accounts for about 80 percent of the antibiotics sold in the United States.) Population genetic modeling shows that some strategies are better than others in delaying the evolution of drug resistance in human patients, especially applying a mixture of drugs simultaneously, such as the "cocktail" prescribed for people infected with HIV. Pathogens can also evolve to be more or less virulent, and the recognition that natural selection will increase virulence under some conditions will probably have significant implications for devising public health measures in the future. At present, one of the most important applications of evolutionary biology in public health is the use of phylogenetic methods to trace the origin and spread of disease agents, such as bacterial contaminants of food, and the pathways of transmission among human hosts. (Phylogenetic relationships among samples of HIV were used to convict a doctor of attempting to murder his former girlfriend by injecting her with blood from an HIV-infected patient; see Metzker et al. 2002.)

Evolution is the basis of vaccines that use attenuated live virus. A virulent virus is passed through another host species (e.g., horse) for many generations, during which it adapts to the new host by mutation and natural selection, eventually losing its virulence to humans because of the fitness trade-offs required. Evolutionary studies may also prove useful in developing vaccines against new strains of pathogens that are resistant to existing vaccines. Robin Bush and colleagues (1999) tested the hypothesis that new strains of

---

*"At least two million people are sickened and an estimated 99,000 die every year from hospital-acquired infections, the majority of which result from such resistant strains." (*New York Times*, April 12, 2012.)

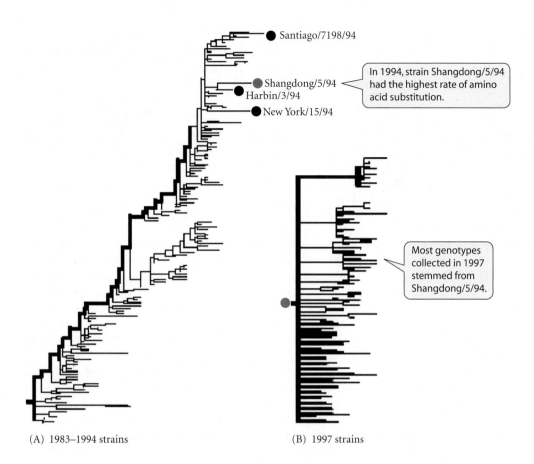

FIGURE 23.4 Prediction of epidemic strains of influenza A virus by phylogenetic analysis. (A) A phylogeny of influenza virus strains from 1983 through the 1993–1994 influenza season. The bold line, or "trunk," traces the single lineage that was successful in leaving descendants through this period. The labeled dots mark lineages that underwent exceptionally high rates of amino acid substitution at positively selected sites in the hemagglutinin gene. Strain Shangdong/5/94 had the highest rate of substitution, and the researchers predicted that future epidemics would stem from this or a closely related lineage. (B) A partial phylogeny of virus strains present in 1997. As researchers had predicted, most genotypes collected in that year were descended from the common ancestor of Shangdong/5/94, as shown here, and its close relatives. (After Bush et al. 1999.)

In 1994, strain Shangdong/5/94 had the highest rate of amino acid substitution.

Most genotypes collected in 1997 stemmed from Shangdong/5/94.

(A) 1983–1994 strains   (B) 1997 strains

influenza A virus spread because their fitness is increased by mutations at codons in the hemagglutinin gene (the main target of human antibodies) that have a history of positive selection. Bush et al. determined the phylogenetic relationships among strains that had been preserved between 1983 and 1994 and identified those strains (e.g., Shangdong/5/94) that had undergone the most amino acid substitutions at positively selected sites in the hemagglutinin gene (**Figure 23.4A**). They predicted that Shangdong/5/94, or its close relatives, would be ancestral to the strains that later spread widely. Their prediction was correct (**Figure 23.4B**). This method may provide a way of estimating which currently rare strains are likely to give rise to epidemic strains, so that vaccines can be developed before those strains spread.

We have seen that phylogenetic methods are a major way of determining the origin of new and emerging infectious diseases (see Chapter 19). The several forms of HIV evolved from viruses carried by chimpanzees and mangabey monkeys (see Figure 3.2), and biologists have recently determined that gorillas may be a reservoir for one form of HIV-1 (van Heuverswyn et al. 2006). The pathogens that cause infectious diseases in humans range from those that are mostly endemic to other animals and only occasionally infect humans (e.g., rabies), through those that are animal-borne but can be transmitted among humans to a greater or lesser extent, to those that are specific to humans, such as the agents that cause smallpox, syphilis, and measles (**Figure 23.5**). Jared Diamond and colleagues (Diamond 1997; Wolfe et al. 2007) note that most human diseases originated in tropical animals, probably because pathogens shift most readily among closely related hosts (see Chapter 19), and most primates have tropical distributions. To an overwhelming degree, the major human pathogens originated in Asia or Africa, rather than in the Western Hemisphere, for several probable reasons: there was much more time for pathogens to transfer to and become adapted to humans in the Old World because hominids evolved there; humans are much more closely related to other Old World (catarrhine) primates than to New World (platyrrhine) primates (see Chapter 6), so pathogen shift was more likely;

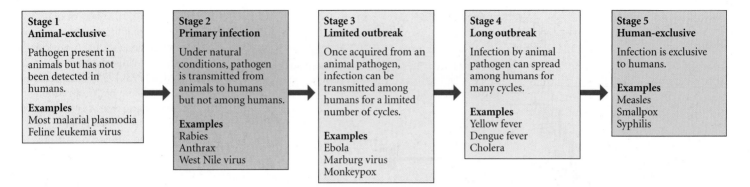

Stage 1 Animal-exclusive	Stage 2 Primary infection	Stage 3 Limited outbreak	Stage 4 Long outbreak	Stage 5 Human-exclusive
Pathogen present in animals but has not been detected in humans.	Under natural conditions, pathogen is transmitted from animals to humans but not among humans.	Once acquired from an animal pathogen, infection can be transmitted among humans for a limited number of cycles.	Infection by animal pathogen can spread among humans for many cycles.	Infection is exclusive to humans.
Examples Most malarial plasmodia Feline leukemia virus	**Examples** Rabies Anthrax West Nile virus	**Examples** Ebola Marburg virus Monkeypox	**Examples** Yellow fever Dengue fever Cholera	**Examples** Measles Smallpox Syphilis

FIGURE 23.5 The ecological relationships between pathogens, humans, and nonhuman hosts can be expressed as five stages of increasing cycling within human populations. Whether or not all pathogens in the higher stages have evolved through the lower stages is not yet known. Some exemplar pathogens are given for each stage. (After Wolfe et al. 2007.)

and most domesticated animals, which have been reservoirs for some pathogens such as influenza, are Old World species.

One form of selection at the level of the gene is segregation distortion, wherein more than half the gametes of a heterozygote carry one of its two alleles, often referred to as a "driver allele" (see Chapter 11). A driver allele can attain high frequency, even fixation, if its pleiotropic costs are not too great. One of the most exciting potential applications of evolutionary biology is GENETIC PEST MANAGEMENT, in which pest insects with genotypes that carry a driver allele, coupled with a gene that has a desirable effect, are released (Gould et al. 2006; James et al. 2011). A desirable gene could, for example, be joined to the *Medea* element, discovered in the flour beetle *Tribolium*, which produces a toxin that kills haploid eggs that lack the element, and which should increase to fixation if it does not have fitness costs (**Figure 23.6**). Such a genetic construct has been experimentally developed in *Drosophila* (Chen et al. 2008). This strategy could be used to control the spread of mosquito-borne diseases by developing mosquitoes with *Medea* linked to a gene that prevents human pathogens, such as those that cause malaria and dengue fever, from surviving and multiplying in the mosquito (Hay et al. 2010). Mating between released and wild mosquitoes would result in the spread of the genetic construct, so that the wild population would become incapable of carrying the pathogen. *Anopheles* mosquitoes with genetic constructs for resistance to malarial infection have recently been developed, although those constructs have not yet been linked to a driver gene (Dong et al. 2011). Similar applications might be used to control insect pests of crops.

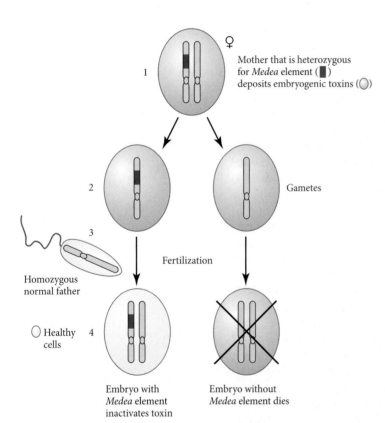

FIGURE 23.6 A driver allele with the potential for use in genetic pest management. The *Medea* element can drive to fixation because it produces a toxin that persists in the mother's eggs. After fertilization, an embryo with the *Medea* element can inactivate the toxin (bottom left), but one that lacks *Medea* dies. Thus all the surviving offspring of *Medea*-carrying mothers carry the element. The *Medea* element can increase to fixation in a population, dragging with it closely linked genes. (From Gould 2010.)

Agriculture and natural resources

The development of improved varieties of domesticated plants and animals is evolution by artificial selection. Evolutionary genetics and plant and animal breeding have had an intimate, mutually beneficial relationship for more than a century. Both theoretical evolutionary methods and experimental studies of *Drosophila* and other model organisms have contributed both to traditional breeding and to modern QTL (quantitative trait locus) analysis, which is used to locate and characterize genes that contribute to traits of interest (see Chapter 13).

Agronomists have learned by bitter experience what evolutionary biologists have long known: that genetic diversity is essential for a population's long-term success. For example, a costly epidemic of the southern corn leaf blight fungus in 1970 was a consequence of widely planting genetically uniform corn with a specific genetic factor that increased susceptibility (see Chapter 19). This example shows why it is important to build "germ plasm" banks of different crop strains, containing genes for various characteristics such as disease resistance. Useful genes can also be found in wild species that are related to crop plants. Genetic and evolutionary studies of the tomato, for example, led to study of the many related species in South America. At least 40 genes for resistance to various diseases have been found in these species, of which about 20 have been crossed into commercial tomato stocks. Studies of wild species' genes for other traits, such as tolerance of salinity and drought, are under way. This approach depends on phylogenetics, evolutionary genetics, and evolutionary ecology.

Genes from wild species can also be transferred to crops by the molecular methods of genetic engineering. Evolutionary biology is aiding this revolution in agronomics by contributing to gene mapping methods, by identifying likely sources of genes for useful characteristics, and by evaluating possible risks posed by transgenic organisms. Genetic engineering is controversial for several reasons; for example, there is concern that transgenes may spread from crop plants to wild species, which then could become more vigorous weeds. Phylogenetic studies can identify wild species with the potential to hybridize with crop plants, and population genetic methods can estimate the fitness effects of transgenes and the chances of gene flow into natural plant populations (Ellstrand 2003).

Insects, weeds, and other organisms cause billions of dollars' worth of crop losses. Much of this loss is caused by the evolution of resistance to chemical insecticides and herbicides by crop pests (see Chapter 12). This resistance not only increases the costs of agriculture but also results in a steady increase in the amount of toxic chemicals sprayed on the landscape (some of which find their way up the food chain, affecting humans and other consumers). For this reason, many crop pests are now managed by INTEGRATED PEST MANAGEMENT (IPM), which uses a combination of chemical pesticides and nonchemical methods, such as release of natural enemies of the pest (BIOLOGICAL CONTROL). In some places, regulations on the use of pesticides follow recommendations made by evolutionary biologists about how to manage pest populations in order to keep them susceptible to pesticides (see Gould 1998; Gould et al. 2002). The development of biological control also benefits from evolutionary analysis. When a new insect species suddenly appears on a crop, phylogenetic systematics is the first approach to identifying the pest and determining where in the world it has come from. That is where entomologists will search for natural enemies, scrutinizing in particular those that are related to known enemies of species that are related to the new pest. Likewise, herbivorous insects used to control weeds or invasive plants must be screened to be sure they will not also attack crops or native plants. A good approach is to see whether they have the potential to feed on or adapt to plants that are related to the target plant species (Futuyma 2000).

Evolutionary biology bears on the management and development of other biological resources as well. The journal *Evolutionary Applications* devoted an entire issue (May 2008) to evolutionary aspects of managing and conserving salmon and trout species. For example, fisheries biologists use genetic markers and population genetic methods to distinguish stocks of fish that migrate from different spawning locations (an issue with great political and economic implications). Resource managers study the evolutionary changes that fishing and hunting have caused in natural populations, especially earlier reproduction at a smaller size (see Figure 11.26). Genetic analyses are similarly important in forestry, for example, in developing commercial stocks of conifers.

Environment and conservation

In a few instances, evolutionary biology has contributed to environmental remediation and restoration of degraded land. For example, populations of grasses that have evolved tolerance to nickel and other heavy metals (see Chapter 13) can be used for revegetating land made barren by mining activities, and some such plants may even be used to remove contaminants from soil and water.

There is little doubt that a major extinction event has been initiated by the huge and accelerating impact of human activities on every aspect of the environment. The most important means of conservation are obvious and require mostly political, legal, and economic expertise: save natural habitats in preserves, establish and enforce limits on the exploitation of fish populations and other biological resources, reduce pollution and global warming. But evolutionary biologists also make indispensable contributions to conservation efforts. They use phylogenetic information to determine where potential nature reserves should be located to protect the greatest variety of biologically different species; they use evolutionary biogeography to identify regions with many endemic species (e.g., Madagascar); they use genetic methods and theory to prevent inbreeding depression in rare species and to distinguish genetically unique populations (Frankham et al. 2002; Allendorf and Luikart 2007); and they use genetic markers to identify illegal traffic in endangered species (see Baker et al. 2000).

Human behavior

No topic in evolutionary biology is more intriguing or more controversial than the evolutionary foundations of human behavior, especially social behavior. Resistance to hypotheses about human behavioral evolution, or even research on the subject, is widespread for several reasons. Many people are emotionally reluctant to see human abilities as extensions of those of other species, and they justify making a sharp distinction by pointing to the immense difference between human mental capacities and those of any other mammal. Almost every aspect of our behavior varies greatly among individuals and among populations because of learning and cultural differences, so the very hypothesis that there are species-specific, genetically based behaviors is viewed with skepticism, especially by many social scientists. Finally, any intimation of biological determinism summons memories of a long, sordid history of political and ideological abuse of evolutionary ideas. Just as the Bible has been used to justify crusades, inquisitions, and witch hunts, scientific ideas have been used to justify social inequity and even persecution. From its earliest days, evolutionary science, under the banner of "social Darwinism," was misappropriated to justify racism and imperialist domination, to exclude women from political and economic power, and to ascribe poverty, illiteracy, and crime not to the social conditions that exclude much of society from education and economic self-sufficiency, but to genetic inferiority (Hofstadter 1955). Evolution was used to justify the American eugenics movement, which advocated encouraging "superior" people to reproduce and discouraging or preventing "inferior" people from doing so; to justify discriminatory quotas in United States immigration policy; and worst of all, to justify the racism that found its most horrifying expression in Nazi Germany. All these abuses were based on misunderstanding or twisting of the data and theory of evolution and genetics (and, to their credit, some evolutionary biologists and geneticists said so). A proper understanding of evolutionary biology, as of any science, is necessary to prevent it from being misused. Research on the evolution of human behavior may as well be used for good as for ill.

Darwin devoted a book, *The Expression of the Emotions in Man and Animals*, to the thesis that rudimentary homologues of human mental abilities and emotions can be seen in other species. It is now clear that many animals are capable of learning, and that some, especially primates, can learn by watching other individuals. Thus cultural differences exist among troops and populations of chimpanzees and some other primates (**Figure 23.7**). Tools are made and used by chimpanzees and, surprisingly, New Caledonian crows (Hunt and Gray 2004). Baboons have been trained to distinguish unfamiliar English words from nonsense combinations of letters (Grainger et al. 2012), and chimpanzees and other apes can learn to use sign language, or other nonverbal symbolic representation, at levels equivalent to those of a young human child. Social interactions, including reciprocal aid and alliances,

FIGURE 23.7 Chimpanzees use tools, and cultural traditions of tool use develop, based on imitation. As a female chimpanzee (*Pan troglodytes*) and her infant watch, a male cracks nuts using a rock as a hammer.

are quite elaborate in some primates, leading researchers to describe "baboon metaphysics" (Cheney and Seyfarth 2007) and "chimpanzee politics" (de Waal 2007).

It would be contrary to everything we know about evolution to suppose that the hominid lineage has diverged so far from other apes that there should remain no trace of homology in the inherited components of our brain organization and behavior, or that the human brain should be a complete tabula rasa, or blank slate, without genetically determined predispositions. Evolutionary psychologists have compiled a long list of supposedly universal, and thus perhaps genetically determined, human behavioral tendencies and capacities, such as fear of snakes and the capacity for language (Pinker 2002), while recognizing that the expression of most such behaviors is culturally highly variable. (All humans speak a grammatical language, for example, but grammar and vocabulary vary greatly.)

Several schools of evolutionary research on human behavior have developed in the last few decades (Laland and Brown 2011). HUMAN SOCIOBIOLOGY, announced by E. O. Wilson in 1975, proposed to interpret a wide range of behaviors (such as conflicts between offspring and parents) as adaptations that had evolved by natural selection, especially kin selection. This approach, which emphasized genetic over cultural explanations of particular behaviors, was hugely controversial. It has been succeeded by several other movements, including HUMAN BEHAVIORAL ECOLOGY, which uses adaptive models to predict and explain a variety of behaviors, including cultural norms. For instance, we may ask why, in polygynous societies, a woman would choose to marry a man who already has one or more wives. Adopting a model that was first developed to explain polygyny in birds, behavioral ecologists have proposed that women decide on the basis of the resources they may expect a man to provide.

Complementing this approach, Leda Cosmides, John Tooby, and others have developed the active (and controversial) field of EVOLUTIONARY PSYCHOLOGY (Barkow et al. 1992; Buss 1999; Pinker 2002; Confer et al. 2010). Evolutionary psychologists propose that over the course of human evolution, especially during the Pleistocene, specific adaptations, conceived as mental "organs" or modules, evolved to solve different classes of problems, especially social challenges such as choosing mates and detecting cheaters. This approach has been criticized on the grounds that it is difficult to separate innate from learned behaviors, to identify discrete behavioral traits, and to demonstrate that traits are adaptations; however, it is credited with developing insights into language, some aspects of sexual behavior, and many other topics. For example, David Buss (1994) reasoned that because reproduction entails a far greater commitment and investment of resources by women than by men, women should have evolved to seek mates who are likely to provide resources, while men, as a consequence of sexual selection, might be expected to place more value on young, physically attractive mates who are likely to be fertile. This sounds like the epitome of the sexism that pervades American culture, yet Buss claimed, based on more than 10,000 interviews in 37 broadly different cultures, that the predicted differences between the sexes are cross-culturally consistent. Many other studies of human sexual behavior and psychology are framed in an evolutionary framework.

Some of the elements of evolutionary theory, such as natural selection and genetic drift, have been applied, mutatis mutandis, to cultural evolution, and some methods developed for evolutionary analysis are being applied to cultural evolution as well (Mace and Holden 2005). Phylogenetic methods are used to trace the evolution of languages and have revealed interesting patterns; for example, frequently used words evolve at slower rates than infrequently used words (Pagel et al. 2007). Cultural evolution and genetic evolution of behavior have been joined in models of GENE-CULTURE COEVOLUTION (Cavalli-Sforza and Feldman 1981) or "dual inheritance" (Boyd and Richerson 1985; Richerson and Boyd 2005). At one level, this field studies how culture affects the selective milieu within which genetic evolution occurs; for example, lactase persistence has evolved in several cultures with milk-based diets, and genome studies are revealing many genes that have undergone recent selection that is thought to stem from changes in diet, the invention of cooking, sexual selection, and migration into novel climates (see Chapter 12; Laland et al. 2010).

Surely the most interesting problem addressed by these models is what accounts for the uniquely hypertrophied organization and cooperation in human societies—far greater than that seen in any other species. Complex social relationships must have required that an individual be able to recognize many other individuals and remember their past interactions; they may have required the ability to interpret the state of mind and intentions of others; they were probably facilitated by punishing cheaters; and they may have been fostered by selection for the ability to form coalitions against stronger "bullies" (Fehr and Fischbacher 2003; Gavrilets 2012). In this context, natural selection may have led to humans' extraordinary capacity for altruism, which extends beyond the limits that kin selection or reciprocity readily explain (Boehm 2012). Many authors have proposed that multilevel selection—selection among groups or tribes, stemming from competition and warfare—played a major role in the evolution of the extraordinary human capacity for cooperation and altruism. Richerson and Boyd (2005) propose that such group selection was facilitated by processes of cultural evolution, such as pressure to conform to the group's norms, that reduce behavioral variation within groups and increase the variance among groups. These processes would make multilevel selection more effective than it usually is in purely genetic models (see Chapters 11 and 16).

Understanding nature and humanity

"All art," said Oscar Wilde, "is perfectly useless." He meant that as high praise: art is a human creation that needs no utilitarian justification, a creation that is justified simply by being an expression—indeed, one of the defining characteristics—of humanity.

Much of what is most meaningful to us is "perfectly useless": music, sunsets, walking on a clean beach, baseball, soccer, movies, gardening, spiritual inspiration—and understanding. Whether the subject be mathematics, the natural world, philosophy, or human nature, attempting to understand is rewarding in itself, quite aside from whatever practical consequences it may yield.

To know about the extraordinary diversity of organisms, about the complexities of the cell, of development, or of our brains, and about how these marvels came to be is deeply rewarding to anyone with a sense of curiosity and wonder. To have achieved such knowledge is, like other advances in science and technology, among humanity's great accomplishments. Likewise, to have some understanding, however imperfect, of what we humans are and how we came into existence is richly rewarding. It is fascinating and ennobling to learn of our 3.5-billion-year-old pedigree, of when and how and possibly why our ancestors evolved the characteristics that led to our present condition, of how and when modern humans emerged from Africa and colonized the rest of Earth, of how genetically unified all humans are with one another, and yet how genetically diverse we are. It is both challenging and important to try to understand "human nature"—to understand how our behavior is shaped by our genes and therefore by our evolutionary past, and how it is shaped by culture, social forces, and our unique individual history of learning and experience.

Evolution is the unifying theme of the biological sciences and an important foundation for the "human sciences" of medicine, psychology, and sociology. Although psychologists and anthropologists may differ among themselves on the role of evolution in determining "human nature," most will agree that some knowledge of evolutionary principles is essential for understanding their subject. And although evolutionary biologists and social scientists do not set social policy, they can speak out against abuses of their science. They can point out misunderstandings of evolutionary theory, such as racist interpretations of differences among human populations, or the "naturalistic fallacy" that what is natural is good and can therefore guide human conduct. Social Darwinism was based on this fallacy; so is the belief that homosexuality is morally wrong because it does not lead to reproduction, and the belief that women should be subservient to men because of their "natural," evolved sex roles. Science can play an important role in the ever-necessary defense of human rights and justice.

Evolution has neither moral nor immoral content, and evolutionary biology provides no philosophical basis for aesthetics or ethics. But evolutionary biology, like other knowledge, can serve the cause of human freedom and dignity by helping to relieve disease and hunger and by helping us to understand and appreciate both the unity and the diversity of humankind. And it can enhance our appreciation of life in all its diversity: in Darwin's phrase, "endless forms most beautiful and most wonderful."

Go to the

EVOLUTION
Companion Website at
sites.sinauer.com/evolution3e

for quizzes, data analysis and simulation exercises, and other study aids.

Summary

1. Evolution is a fact—a hypothesis that is so thoroughly supported that it is extremely unlikely to be false. The theory of evolution is not a speculation, but rather a complex set of well-supported hypotheses that explain how evolution happens.

2. There is a great range of views on whether or not religion and evolution—or religion and science generally—are compatible. Especially in the United States, many reject evolution and instead accept divine creation because they think evolution conflicts with their religious beliefs. The positions taken by creationists on issues such as the age of Earth and of life vary.

3. Science is tentative; it accepts hypotheses provisionally and changes them in the face of convincing new evidence. It is concerned only with testable hypotheses; it depends on empirical studies that are subject to peer scrutiny and that can be verified and repeated by others. Supernatural hypotheses, in contrast, cannot be tested. Creationism has none of the features of science, so it has no claim to be taught in science classes.

4. The evidence for evolution comes from all realms of biology and geology, including comparative studies of morphology, development, life histories, and other features as well as molecular biology, genomics, paleontology, and biogeography. Evolutionary principles can explain features of organisms that would not be expected of a beneficent intelligent designer, such as imperfect adaptation, useless or vestigial features, extinction, selfish DNA, sexually selected characteristics, conflicts among genes within the genome, and infanticide. Furthermore, all the proposed mechanisms of evolution have been thoroughly documented, and evolution has been observed.

5. The arguments used by creationists are all logically refutable in scientific terms and are contradicted by data.

6. It is important to understand evolution not only because it has broad implications for how we think about nature and humanity, but also because it has many practical ramifications. Evolutionary science contributes to many aspects of medicine and public health, agriculture and natural resource management, pest management, and conservation.

7. One of the most difficult and controversial challenges is to join biological and social science in order to understand how the distinctively human cognitive and behavioral characteristics evolved, the extent to which human behaviors have an evolved genetic foundation, how that foundation interacts with cultural and other environmental factors to shape individual behavior, and how genes and culture have coevolved.

Terms and Concepts

creationists	intelligent design (ID)	special creation	theory
hypothesis	natural laws	theistic evolution	

Suggestions for Further Reading

Books on creationism, ID, and defending evolution and science:

Why Evolution Is True, by Jerry A. Coyne (Oxford University Press, New York, 2009), and *The Greatest Show on Earth*, by Richard Dawkins (Free Press, New York, 2009), are outstanding, well-written descriptions of the evidence for evolution and its mechanisms by leading evolutionary biologists.

Evolution vs. Creationism: An Introduction, second edition (Greenwood Press, Westport, CT, 2009), was written by Eugenie C. Scott, director of the National Center for Science Education, who probably knows more about the evolution-creationism controversy than anyone else.

Defending Evolution in the Classroom: A Guide to the Creation/Evolution Controversy, by B. J. Alters and S. M. Alters (Jones and Bartlett, Sudbury, MA, 2001), is a perceptive analysis of how to present and teach evolution.

Denying Evolution: Creationism, Scientism, and the Nature of Science, by M. Pigliucci (Sinauer Associates, Sunderland, MA, 2002), shows why creationism fails as science, but addresses the limits of science as well.

Science, Evolution, and Creationism (National Academy Press, Washington, D.C., 2008) is a 70-page booklet issued by the most prestigious scientific organization in the United States, the National Academy of Sciences, and the Institute of Medicine of the National Academies. It is available at www.nap.edu.

Books and articles that address the relationship between religion and evolution:

Finding Darwin's God: A Scientist's Search for Common Ground between God and Evolution (HarperCollins, New York, 1999), by Kenneth Miller, a cell biologist at a leading university, argues that one can fully accept evolution and reconcile it with religion. See also his later book, *Only a Theory: Evolution and the Battle for America's Soul* (Viking, New York, 2008).

Living with Darwin: Evolution, Design, and the Future of Faith, by Philip Kitcher (Oxford University Press, New York, 2007), is a short reflection by an eminent philosopher on reconciling evolution with our need for meaning in life, whether this is provided by religion or other sources of fulfillment.

"Seeing and believing," *New Republic*, February 4, 2009, 32–41, is an essay and book review by biologist Jerry Coyne, who argues that science and religion cannot be reconciled.

Several excellent websites provide information about evolution and can serve as valuable teaching aids:

Understanding Evolution (http://evolution.berkeley.edu) is an outstanding site, developed by the University of California Museum of Paleontology to provide content and resources for teachers at all grade levels.

www.ncseweb.org is the website of the National Center for Science Education (P.O. Box 9477, Berkeley, CA 94709-0477), which actively supports the teaching of evolution and combats creationism. This is the most comprehensive site on the conflict, and it provides links to a great range of resources.

The Talk Origins Archive (www.talkorigins.org) has a wealth of material on many aspects of evolution and the social controversy. It includes a comprehensive list of creationist claims and rebuttals to them, by Mark Isaak, that is also available in book form [M. Isaak, *The Counter-Creationism Handbook* (University of California Press, Berkeley, 2007)].

Ken Miller's Evolution Resources (www.millerandlevine.com/km/evol/) includes text, video clips of interviews, and other material, especially on debunking creationist claims and on Miller's position that religion and evolution are compatible.

Darwin Day Celebration (www.darwinday.org) describes annual educational events held around the world on or near the anniversary of Darwin's birth (February 12) and includes a link to a free online course on evolution and science.

Applications of evolutionary biology are described by D. P. Mindell in *The Evolving World: Evolution in Everyday Life* (Harvard University Press, Cambridge, MA, 2006). Other sources of information on this topic are "Applied evolution," by J. J. Bull and H. A. Wichman, *Annual Review of Ecology and Systematics* 32: 183–217 (2001), and "Evolution, science, and society," at www.amnat.org or in *American Naturalist* 158 (supplement): S1–S47 (2001). *Principles of Evolutionary Medicine*, by P. Gluckman, A. Beadle, and M. Hanson (Oxford University Press, 2009), is an ideal introduction to this subject, especially for medical students and others in the health sciences.

A huge literature, of variable quality, concerns the evolution of human behavior. A very good introduction to the major contemporary approaches is *Sense and Nonsense: Evolutionary Perspectives on Human Behaviour* (Oxford University Press, Oxford, 2011), by K. N. Laland and G. R. Brown, who are most inclined toward the gene-culture coevolution approach, but provide a fair and lively description of the other schools of evolutionary thought. A very readable, convincing treatment of culture and human evolution is *Not by Genes Alone: How Culture Transformed Human Evolution* (University of Chicago Press, Chicago, 2005), by P. J. Richerson and R. Boyd, leading figures in this field.

Glossary

Most of the terms in this glossary appear at several or many places in the text of this book. Many terms that are used broadly in biology or are used in this book only near their definition in the text are not included here.

A

absolute fitness   *See* **relative fitness**.

active trend   *See* **driven trend**.

adaptation   A process of genetic change in a population whereby, as a result of natural selection, the average state of a character becomes improved with reference to a specific function, or whereby a population is thought to have become better suited to some feature of its environment. Also, *an* adaptation: a feature that has become prevalent in a population because of a selective advantage conveyed by that feature in the improvement in some function. A complex concept; see Chapter 11.

adaptive landscape   A metaphor for the relationship between mean fitness of a population and the allele frequencies at one or more loci that affect fitness. Possible populations with allele frequencies that maximize mean fitness are represented as peaks on the metaphorical landscape.

adaptive peak   That allele frequency, or combination of allele frequencies at two or more loci, at which the mean fitness of a population has a (local) maximum. Also, the mean phenotype (for one or more characters) that maximizes mean fitness. An **adaptive valley** is a set of allele frequencies at which mean fitness has a minimum.

adaptive radiation   Evolutionary divergence of members of a single phylogenetic lineage into a variety of different adaptive forms; usually the taxa differ in the use of resources or habitats, and have diverged over a relatively short interval of geological time. The term **evolutionary radiation** describes a pattern of rapid diversification without assuming that the differences are adaptive.

adaptive valley   *See* **adaptive peak**.

adaptive zone   A set of similar **ecological niches** occupied by a group of (usually) related species, often constituting a higher taxon.

additive effect   The magnitude of the effect of an allele on a character, measured as half the phenotypic difference between homozygotes for that allele compared with homozygotes for a different allele.

additive genetic variance   That component of the **genetic variance** in a character that is attributable to additive effects of alleles.

allele   One of several forms of the same gene, presumably differing by mutation of the DNA sequence. Alleles are usually recognized by their phenotypic effects; DNA sequence variants, which may differ at several or many sites, are usually called **haplotypes**.

allele frequency   The proportion of gene copies in a population that are a given allele; i.e., the probability of finding this allele when a gene is taken randomly from the population; also called **gene frequency**.

allometric growth   Growth of a feature during ontogeny at a rate different from that of another feature with which it is compared.

allopatric   Of a population or species, occupying a geographic region different and separated from that of another population or species. *Cf.* **parapatric**, **sympatric**.

allopatric speciation   Speciation by genetic divergence of allopatric populations of an ancestral species; contrasted with **parapatric** and **sympatric speciation**, in which divergence occurs in **parapatry** or **sympatry** (q.v.).

allopolyploid   A **polyploid** in which the several chromosome sets are derived from more than one species.

allozyme   One of several forms of an enzyme encoded by different alleles at a locus.

alternative splicing   Splicing of different sets of exons from RNA transcripts to form mature transcripts that are translated into different proteins (thus allowing the same gene to encode different proteins).

altruism   Conferral of a benefit on other individuals at an apparent cost to the donor.

anagenesis   Evolutionary change of a feature within a lineage over an arbitrary period of time.

ancestral character state   An evolutionarily older character state, relative to another (derived) state that has evolved from it in one or more lineages.

aneuploid   Of a cell or organism, possessing too many or too few homologous chromosomes, relative to the normal (**euploid**) set.

antagonistic pleiotropy   Contrasting effects of a gene on two different characters, such that the effect of an allele substitution on one character increases fitness, but the effect on the other character decreases fitness.

antagonistic selection   A source of natural selection that opposes another source of selection on a trait.

apomixis   Parthenogenetic reproduction in which an individual develops from one or more mitotically produced cells that have not experienced recombination or syngamy.

apomorphic   Having a **derived** character or state, with reference to another character or state. *See* **synapomorphy**.

aposematic   Coloration or other features that advertise noxious properties; warning coloration.

artificial selection   Selection by humans of a deliberately chosen trait or combination of traits in a (usually captive) population; differing from natural selection in that the criterion for survival and reproduction is the trait chosen, rather than fitness as determined by the entire genotype.

asexual   Pertaining to reproduction that does not entail meiosis and syngamy.

assortative mating   Nonrandom mating on the basis of phenotype; usually refers to positive assortative mating, the propensity to mate with others of like phenotype.

autopolyploid   A **polyploid** in which the several chromosome sets are derived from the same species.

autosome   A chromosome other than a sex chromosome.

B

back mutation   Mutation of an allele back to the allele from which it arose by an earlier mutation.

background extinction   A long-prevailing rate at which taxa become extinct, in contrast to the highly elevated rates that characterize mass extinction.

background selection   Elimination of deleterious mutations in a region of the genome; may explain low levels of neutral sequence variation.

balancing selection   A form of natural selection that maintains polymorphism at a locus within a population.

base pair substitution   As usually used in this book, a base pair that, because of genetic drift or natural selection, has replaced another base pair at a specific DNA site in a population or species.

benthic   Inhabiting the bottom, or substrate, of a body of water. *Cf.* **planktonic**.

behavioral isolation   *See* **sexual isolation**.

biogeographic realm   Major geographic regions of Earth that have characteristic animal and plant taxa.

biogeography   The study of the geographic distribution of organisms.

biological homology   Commonality of different traits, among or within species, based on a shared genetic basis and developmental pathway; the traits are often, but not always, homologous in the usual phylogenetic sense. *See* **homology**.

biological species   A population or group of populations within which genes are actually or potentially exchanged by interbreeding, and which are reproductively isolated from other such groups.

bottleneck   A severe, temporary reduction in population size.

C

Cambrian explosion   The first appearance in the fossil record of many animal phyla, within a relatively short (<20 million years) interval.

canalization   The evolution of internal factors during development that reduce the effect of perturbing environmental and genetic influences, thereby constraining variation and consistently producing a particular (usually wild-type) phenotype.

candidate gene   A gene postulated to be involved in the evolution of a particular trait based on its mutant phenotype or the function of the protein it encodes.

carrying capacity   The population density that can be sustained by limiting resources.

category   In taxonomy, one of the ranks of classification (e.g., genus, family). *Cf.* **taxon**.

cDNA (complementary DNA)   A DNA copy of an mRNA made using reverse transcriptase isolated from a retrovirus.

cDNA library   A collection of cDNAs, representing the transcriptome (all of the mRNAs expressed) of a tissue or whole organism at a particular life history stage, created by isolating

cDNA, cloning it into circular DNA plasmids and propagating it in bacterial cells.

character   A feature, or trait. *Cf.* **character state**.

character displacement   Usually refers to a pattern of geographic variation in which a character differs more greatly between sympatric than between allopatric populations of two species; sometimes used for the evolutionary process of accentuation of differences between sympatric populations of two species as a result of the reproductive or ecological interactions between them.

character state   One of the variant conditions of a character (e.g., yellow versus brown as state of the character "color of snail shell").

chimeric gene   A gene that consists of parts of two or more different ancestral genes.

chronospecies   A segment of an evolving lineage preserved in the fossil record that differs enough from earlier or later members of the lineage to be given a different binomial (name). Not equivalent to biological species.

cis-regulatory element   A noncoding DNA sequence in or near a gene required for proper spatiotemporal expression of that gene, often containing binding sites for transcription factors. *Cf.* **control region**, *trans*-**regulatory element**.

clade   The set of species descended from a particular ancestral species.

cladistic   Pertaining to branching patterns; a cladistic classification classifies organisms on the basis of the historical sequences by which they have diverged from common ancestors.

cladogenesis   Branching of lineages during phylogeny.

cladogram   A branching diagram depicting relationships among taxa; i.e., an estimated history of the relative sequence in which they have evolved from common ancestors. Used by some authors to mean a branching diagram that displays the hierarchical distribution of derived character states among taxa.

cline   A gradual change in an allele frequency or in the mean of a character over a geographic transect.

clone   A lineage of individuals reproduced asexually, by mitotic division.

coadapted gene pool   A population or set of populations in which prevalent genotypes are composed of alleles at two or more loci that confer high fitness in combination with each other, but not with alleles that are prevalent in other such populations.

coalescence   Derivation of the gene copies in one or more populations from a single ancestral copy, viewed retrospectively (from the present back into the past).

codon   A nucleotide triplet that encodes an amino acid or acts as a "stop" signal in translation.

codon bias   Nonrandom usage of synonymous codons to encode a given amino acid.

coefficient of relationship   The proportion of genes that are identical by common descent between two individuals of a species.

coefficient of selection   The proportion by which the average fitness of individuals of one genotype differs from that of a reference genotype.

coevolution   Strictly, the joint evolution of two (or more) ecologically interacting species, each of which evolves in response to selection imposed by the other. Sometimes used loosely to refer to evolution of one species caused by its interaction with another, or simply to a history of joint divergence of ecologically associated species.

commensalism   An ecological relationship between species in which one is benefited but the other is little affected.

common ancestor   A lineage (often designated as a taxon) from which two or more descendant lineages evolved.

common garden   A place in which (usually conspecific) organisms, perhaps from different geographic populations, are reared together, enabling the investigator to ascribe variation among them to genetic rather than environmental differences. Originally applied to plants, but now more generally used to describe any experiment of this design.

comparative genomics   Analysis of similarities and differences between the genomes of different species.

comparative method   A procedure for inferring the adaptive function of a character by correlating its states in various taxa with one or more variables, such as ecological factors hypothesized to affect its evolution.

compartment   A contiguous group of cells, descended from the same progenitor cell, that form a spatially discrete part of a developing organ or structure and often act as a discrete developmental unit. Cells from one compartment typically do not intermix with cells from other compartments.

competition   An interaction between individuals of the same species or different species whereby resources used by one are made unavailable to others.

competitive exclusion   Extinction of a population due to competition with another species.

competitive exclusion principle   The theoretical assertion that one of two ecologically identical species will eventually replace the other by competition.

concerted evolution   Maintenance of a homogeneous nucleotide sequence among the members of a gene family, which evolves over time.

condition-dependent indicator   A characteristic, usually used in behavioral display, that is correlated with, and therefore indicates, the health or physiological vigor ("condition") of an individual.

conservative characters   Features that evolve slowly and are retained with little or no change for long periods of evolutionary time.

conspecific   Belonging to the same species.

constraints   Properties of organisms or their environment that tend to retard evolution of a feature or to direct its evolution along some paths rather than others.

control regions   Untranscribed regions of the genome to which products of other genes bind, and which enhance (**enhancers**) or repress (**repressors**) transcription of specific genes.

convergent evolution (**convergence**)   Evolution of similar features independently in different evolutionary lineages, usually from different antecedent features or by different developmental pathways.

cooperation   Often used to mean activity that benefits both the actor and other individuals.

co-option   The evolution of a function for a gene, tissue, or structure other than the one it was originally adapted for. At the gene level, used interchangeably with **recruitment** and, occasionally, **exaptation**.

Cope's rule   A proposed generalization that individual body size in animals tends to increase during evolution.

copy number variants   Refers to variation among conspecific individuals in the number of duplicates (copies) of a DNA sequence.

correlated selection   Natural selection for specific combinations of traits, such that selection on one trait is correlated with selection on the other.

correlation   A statistical relationship that quantifies the degree to which two variables are associated. For *phenotypic correlation, genetic correlation, environmental correlation* as applied to the relationship between two traits, see Chapter 13.

cost   A reduction in fitness caused by a correlated effect of a feature that provides an increment in fitness (i.e., a benefit).

cost of reproduction   Reduction of an individual's future fitness (survival and/or future reproduction) caused by reproductive activity.

cost of sex   Usually refers to a reduced rate of population growth of a sexual compared to an asexual population, owing to production of males.

creationism   The doctrine that each species (or perhaps higher taxon) was created separately, essentially in its present form, by a supernatural Creator.

crown group   A taxon, distinguished by derived character states, that has descended from an ancestral group (**stem group**) that may bear a different name.

cultural evolution   Changes in the frequency of nongenetic cultural traits within and among populations, based on processes such as nonrandom imitation.

C-value paradox   The lack of correlation between the DNA content of eukaryotic genomes and a given organism's phenotypic complexity (i.e., the genome of a less complex eukaryotic organism, such as a plant, may contain far more DNA than that of a more complex organism, such as a human being). The paradox is explained by the amount of noncoding repetitive DNA sequences in a genome.

D

deme   A local population; usually, a small, panmictic population.

demographic   Pertaining to processes that change the size of a population (i.e., birth, death, dispersal).

density-dependent   Affected by population density.

derived character (**state**)   A character (or character state) that has evolved from an antecedent (ancestral) character or state.

deterministic   Causing a fixed outcome, given initial conditions. *Cf.* **stochastic**.

developmental arrest   A halting of the development of a morphological structure, resulting in a final adult phenotype that lacks the structure or bears an immature form of the structure. This can also refer to developmental arrest at the level of the entire organism, resulting in an adult that resembles the juvenile form of an ancestral or related species (i.e., **paedomorphosis**).

developmental circuit   *See* **developmental pathway**.

developmental constraint   A restriction that prevents the appearance of certain structures or traits due to the inability of an organism's developmental system to produce them.

developmental pathway   A sequence of gene expression through developmental time, involving both gene regulation and the expression of gene products that provide materials for and regulate morphogenesis, resulting in the normal development of a tissue, organ, or other structure. Also called **developmental circuit**.

differential gene expression   Differences in the time, location, and/or quantitative level at which a gene expresses the protein it encodes. Differential gene expression involves differences between species, developmental stages, or physiological states in the specific cells, tissues, structures, or body segments that express a given gene.

dioecious   Of a species, consisting of distinct female and male individuals.

diploid   Of a cell or organism, possessing two chromosome complements. *See also* **haploid**, **polyploid**.

direct development   A life history in which the intermediate larval stage is omitted and development proceeds directly from an embryonic form to an adult-like form. *Cf.* **indirect development**.

direct fitness   *See* **inclusive fitness**.

directional selection   Selection for a value of a character that is higher or lower than its current mean value.

disparity   The magnitude of variation in morphological or other phenotypic characters among species in a clade or taxon.

dispersal   In population biology, movement of individual organisms to different localities; in biogeography, extension of the geographic range of a species by movement of individuals.

disruptive selection   Selection in favor of two or more modal phenotypes and against those intermediate between them; also called **diversifying selection**.

divergence   The evolution of increasing difference between lineages in one or more characters.

diversification   An evolutionary increase in the number of species in a clade, usually accompanied by divergence in phenotypic characters.

diversifying selection   *See* **disruptive selection**.

diversity-dependent factors   Processes that have a stronger effect on per capita rates of speciation or extinction when the diversity of species is greater.

Dobzhansky-Muller (DM) incompatibility   Reduction in the fitness of a hybrid because of interaction between certain alleles in one parent population with specific alleles at other loci in the other parent population.

Dollo's law   A biological generalization positing that complex characters, once lost in evolution, are extremely unlikely to reappear and thus the loss of complex characters is virtually always irreversible.

domain   A relatively small protein segment or module (100 amino acids or less) that can fold into a specific three-dimensional structure independently of other domains.

dominance   Of an allele, the extent to which it produces when heterozygous the same phenotype as when homozygous; may be contrasted with a **recessive** allele, one that is phenotypically detectable only when homozygous. Dominance of a species describes the extent to which it is numerically or otherwise predominant in a community.

driven trend   A prolonged shift in the mean of a character among the species in a clade, owing to more frequent changes within species in one direction than the other. In a **passive trend**, changes in both directions would be equally likely, but are constrained by a boundary in one direction. Also called **active trend**.

duplication   The production of another copy of a locus (or other sequence) that is inherited as an addition to the genome.

E

ecological biogeography   *See* **historical biogeography**.

ecological character displacement   *See* **character displacement**.

ecological niche   The range of combinations of all relevant environmental variables under which a species or population can persist; often more loosely used to describe the "role" of a species, or the resources it utilizes.

ecological release   The expansion of a population's niche (e.g., range of habitats or resources used) where competition with other species is alleviated.

ecological speciation   Speciation caused by divergent selection, by ecological factors, on characteristics that contribute to reproductive isolation.

ecotype   A genetically determined phenotype of a species that is found as a local variant associated with certain ecological conditions.

effective population size   The effective size of a real population is equal to the number of individuals in an ideal population (i.e., a population in which all individuals reproduce equally) that produces the rate of genetic drift seen in the real population.

electrophoresis   A method of separating genetically different forms of a protein, once an important way to detect variation in the encoding genes.

enhancer   A DNA sequence that, when acted on by **transcription factors** controls transcription of an associated gene. *Cf.* ***cis*-regulatory element**, **control region**, **promoter**.

endemic   Of a taxon, restricted to a specified region or locality.

endosymbiont   An organism that resides within the cells of a host species.

environment   Usually, the complex of external physical, chemical, and biotic factors that may affect a population, an organism, or the expression of an organism's genes; more generally, anything external to the object of interest (e.g., a gene, an organism, a population) that may influence its function or activity. Thus, other genes within an organism may be part of a gene's environment, or other individuals in a population may be part of an organism's environment.

environmental correlation (r_E)   *See* **genetic correlation**.

environmental variance   Variation among individuals in a phenotypic trait that is caused by variation in the environment rather than by genetic differences.

epigenetic inheritance   Inherited changes in gene expression or phenotype that are not based on changes in DNA sequence.

epistasis   An effect of the interaction between two or more gene loci on the phenotype or fitness whereby their joint effect differs from the sum of the loci taken separately.

equilibrium   An unchanging condition, as of population size or genetic composition. Also, the value (e.g., of population size, allele frequency) at which this condition occurs. An equilibrium need not be stable. *See* **stability**, **unstable equilibrium**.

ESS   *See* **evolutionarily stable strategy**.

essentialism   The philosophical view that all members of a class of objects (such as a species) share certain invariant, unchanging properties that distinguish them from other classes.

euploid   Of a cell or organism, possessing the normal, balanced, number of chromosomes.

evolution   In a broad sense, the origin of entities possessing different states of one or more characteristics and changes in the proportions of those entities over time. *Organic evolution*, or *biological evolution*, is a change over time in the proportions of individual organisms differing genetically in one or more traits. Such changes transpire by the origin and subsequent alteration of the frequencies of genotypes from generation to generation within populations, by alteration of the proportions of genetically differentiated populations within a species, or by changes in the numbers of species with different characteristics, thereby altering the frequency of one or more traits within a higher taxon.

evolutionarily stable strategy (ESS)   A phenotype such that, if almost all individuals in a population have that phenotype, no alternative phenotype can invade the population or replace it.

evolutionary developmental biology (EDB)   The study of evolutionary changes in the developmental bases of phenotypic characteristics.

evolutionary radiation   *See* **adaptive radiation**.

evolutionary reversal   The evolution of a character from a derived state back toward a condition that resembles an earlier state.

evolutionary synthesis   The reconciliation of Darwin's theory with the findings of modern genetics, which gave rise to a theory that emphasized the coaction of random mutation, selection, genetic drift, and gene flow; also called the **modern synthesis**.

evolutionary trend   A bias in the direction of repeated changes in a character, within one lineage or among multiple lineages, over an extended period of time.

evolvability   Can refer either to a measure of additive genetic variation that enables response to selection, or to the ability of genetic and developmental processes to generate potentially adaptive variation.

exaptation   The evolution of a function of a gene, tissue, or structure other than the one it was originally adapted for; can also refer to the adaptive use of a previously nonadaptive trait.

exon   That part of a gene that is translated into a polypeptide (protein). *Cf.* **intron**.

exon shuffling   The formation of new genes by assembly of exons from two or more preexisting genes. The classical model of exon shuffling generates new combinations of exons mediated via recombination of intervening introns; however, exon shuffling can also come about by retrotransposition of exons into pre-existing genes.

exponential growth   Nonlinear increase (or decrease) of a property (e.g., body size, population size) over time, described by an exponential equation.

F

fecundity   The quantity of gametes (usually eggs) produced by an individual.

fitness   The success of an entity in reproducing; hence, the average contribution of an allele or genotype to the next generation or to succeeding generations. *See also* **relative fitness**.

fixation   Attainment of a frequency of 1 (i.e., 100 percent) by an allele in a population, which thereby becomes **monomorphic** for the allele.

founder effect   The principle that the founders of a new population carry only a fraction of the total genetic variation in the source population.

founder-flush speciation   A hypothesis for speciation, in which genetic change is enhanced in populations that grow rapidly ("flush") after being founded by a few individuals.

frameshift mutation   An insertion or deletion of base pairs in a translated DNA sequence that alters the reading frame, resulting in multiple downstream changes in the gene product.

frequency   In this book, usually used to mean *proportion* (e.g., the frequency of an allele is the proportion of gene copies having that allelic state).

frequency-dependent selection   A mode of natural selection in which the fitness of each genotype varies as a function of its frequency in the population.

function   The way in which a character contributes to the fitness of an organism.

functional constraint   Limitation on the variation expressed in a phenotype (perhaps a protein) because many variants have impaired function and reduce fitness.

G

gametic selection   Natural selection among alleles based on their effects in gametes.

gene   The functional unit of heredity.

gene complex   A group of two or more genes that are members of the same family and in most cases are located in close proximity to one another in the genome, often in tandem separated by various amounts of intergenic, noncoding DNA.

gene conversion   A process involving the unidirectional transfer of DNA information from one gene to another. In a typical conversion event, a gene or part of a gene acquires the same sequence as the other allele at that locus (intralocus or intra-allelic conversion), or the same sequences as a different, usually paralogous, locus (interlocus conversion). One consequence of gene conversion may be the homogenization of sequences among members of a gene family.

gene copy   Refers to one of the representatives of a particular gene in an individual or cell (e.g., one copy in a haploid cell, two copies in a diploid).

gene dispensability   An inverse measure of the degree to which the fitness of an organism is compromised as a result of that gene's being deleted from the organism's genome.

gene duplication The process whereby new genes arise as copies of preexisting gene sequences. The result can be a **gene family**.

gene family Two or more loci with similar nucleotide sequences that have been derived from a common ancestral sequence.

gene flow The incorporation of genes into the gene pool of one population from one or more other populations.

gene frequency *See* **allele frequency**.

gene pool The totality of the genes of a given sexual population.

gene tree A diagram representing the history by which gene copies have been derived from ancestral gene copies in previous generations.

genetic assimilation A process whereby a phenotype whose development is triggered by an environmental stimulus evolves to be constitutively expressed (i.e., to no longer require the stimulus).

genetic conflict Antagonistic fitness relationships between alleles, either at the same locus (intralocus conflict) or at different loci (interlocus conflict).

genetic constraint A restriction that prevents a lineage from evolving along a particular evolutionary trajectory because the genetic variation enabling that trajectory is not available.

genetic correlation Correlated differences among genotypes in two or more phenotypic characters, due to **pleiotropy** or **linkage disequilibrium**. Genetic correlation, together with character correlation caused by different environmental conditions (**environmental correlation**), accounts for the correlation that may be observed between phenotypic characters within a population (**phenotypic correlation**).

genetic distance Any of several measures of the degree of genetic difference between populations, based on differences in allele frequencies.

genetic drift Random changes in the frequencies of two or more alleles or genotypes within a population.

genetic load Any reduction of the mean fitness of a population resulting from the existence of genotypes with a fitness lower than that of the most fit genotype.

genetic marker A readily detected genetic variant (such as a visible mutation or a polymorphic nucleotide) that is used to trace variation and inheritance of a closely linked region that may include a gene of interest.

genetic toolkit The set of genes and proteins, often conserved across distantly related organisms, and the developmental pathways that they comprise, by which multicellular organisms are constructed during development.

genetic variance Variation in a trait within a population, as measured by the variance that is due to genetic differences among individuals.

genic selection A form of selection in which the single gene is the unit of selection, such that the outcome is determined by fitness values assigned to different alleles. *See* **individual selection, kin selection, natural selection**.

genome The entire complement of DNA sequences in a cell or organism. A distinction may be made between the nuclear genome and organelle genomes, such as those of mitochondria and plastids.

genotype The set of genes possessed by an individual organism; often, its genetic composition at a specific locus or set of loci singled out for discussion.

genotype × environment interaction Phenotypic variation arising from the difference in the effect of the environment on the expression of different genotypes.

genotype frequency The proportion of individuals in a population that carry a specific genotype at one or more loci.

geographic variation Variation among spatially distributed populations of a species.

Gondwana The southern of the two large continents that existed in the early Mesozoic.

grade A group of species that have evolved the same state in one or more characters and typically constitute a **paraphyletic** group relative to other species that have evolved further in the same direction.

gradualism The proposition that large differences in phenotypic characters have evolved through many slightly different intermediate states. *See* **phyletic gradualism**.

group selection The differential rate of origination or extinction of whole populations (or species, if the term is used broadly) on the basis of differences among them in one or more characteristics. May also refer to differences among populations in their contribution of genes to the combined gene pool. *See also* **interdemic selection, species selection**.

H

habitat selection The capacity of an organism (usually an animal) to choose a habitat in which to perform its activities. Habitat selection is not a form of natural selection.

habitat tracking The tendency for the geographic range of a species to shift in accordance with changes in the location of its ecological requirements, rather than adapting to environmental changes in its former range.

Haldane's rule The generalization that when only one sex manifests sterility or inviability in hybrids between species, it is the heterogametic sex (with two different sex chromosomes) that does so.

Hamilton's rule The theoretical principle that an altruistic trait can increase if the benefit to recipients, multiplied by their relationship to the altruist, exceeds the fitness cost of the trait to the altruist.

haploid Of a cell or organism, possessing a single chromosome complement, hence a single gene copy at each locus.

haplotype A DNA sequence that differs from homologous sequences at one or more base pair sites.

Hardy-Weinberg Pertaining to the genotype frequencies expected at a locus under ideal equilibrium conditions in a randomly mating population.

heritability The proportion of the **variance** in a trait among individuals that is attributable to differences in genotype. Heritability in the narrow sense is the ratio of **additive genetic variance** to **phenotypic variance**.

hermaphroditic Performance of both female and male sexual functions by a single individual.

heterochrony An evolutionary change in phenotype caused by an alteration of timing of developmental events.

heterokaryotype   A genome or individual that is heterozygous for a chromosomal rearrangement such as an inversion. *Cf.* **homokaryotype**.

heterotopy   Expression of a gene or character in a different location on the body of a descendant than in its ancestor.

heterozygosity   In a population, the proportion of loci at which a randomly chosen individual is heterozygous, on average. Applied to a single locus, it refers to the proportion of heterozygotes in a population. In both senses, **Hardy-Weinberg** equilibrium is often assumed.

heterozygote   An individual organism that possesses different alleles at a locus.

heterozygous advantage   The manifestation of higher fitness by heterozygotes than by homozygotes at a specific locus.

higher taxon   A taxon above the species level, such as a named genus or phylum.

historical biogeography   The study of historical changes in the geographic distribution of organisms, including those that affect their present distribution; **ecological biogeography** addresses current factors that affect present distributions.

hitchhiking   Change in the frequency of an allele due to linkage with a selected allele at another locus.

homeobox genes   A large family of eukaryotic genes that contain a DNA sequence known as the homeobox. The homeobox sequence encodes a protein homeodomain about 60 amino acids in length that binds DNA. Most homeobox genes are transcriptional regulators. *Cf.* **domain**; **Hox genes**.

homeostasis   Maintenance of an equilibrium state by some self-regulating capacity of an individual.

homeotic mutation   A mutation that causes a transformation of one structure into another of the organism's structures.

homeotic selector genes   Genes whose expression is required for the development of an entire organ, segment, or compartment of an organism.

homokaryotype   A genome or individual that is homozygous for a chromosomal rearrangement such as an inversion. *Cf.* **heterokaryotype**.

homologous chromosomes   *See* **homology**.

homology   Possession by two or more species of a character state derived, with or without modification, from their common ancestor. **Homologous chromosomes** are those members of a chromosome complement that bear the same genes.

homonymous   Pertaining to biological structures that occur repeatedly within one segment of the organism, such as teeth or bristles.

homoplasy   Possession by two or more species of a similar or identical character state that has not been derived by both species from their common ancestor; embraces **convergence**, **parallel evolution**, and **evolutionary reversal**.

homozygote   An individual organism that has the same allele at each of its copies of a genetic locus.

horizontal transmission   Movement of genes or symbionts (such as parasites) between individual organisms other than by transmission from parents to their offspring (which is **vertical transmission**). Horizontal transmission of genes is also called **lateral gene transfer**.

Hox genes   A subfamily of **homeobox genes**, conserved in all metazoan animals, that controls anterior-posterior segment identity by regulating the transcription of many genes during development.

hybrid   An individual formed by mating between unlike forms, usually genetically differentiated populations or species.

hybrid zone   A region in which genetically distinct populations come into contact and produce at least some offspring of mixed ancestry.

hybridization   Production of offspring by interbreeding between members of genetically distinct populations.

hypermorphosis   An evolutionary increase in the duration of ontogenetic development, resulting in features that are exaggerated compared to those of the ancestor.

hypothesis   An informed conjecture or proposition of what might be true.

hypothetico-deductive method   A scientific method in which a hypothesis is tested by deducing expected data or observations from it, if it were true, and comparing the deduced predictions with real data.

I

identical by descent   Of two or more gene copies, being derived from a single gene copy in a specified common ancestor of the organisms that carry the copies.

inbreeding   Mating between relatives that occurs more frequently than if mates were chosen at random from a population.

inbreeding coefficient   The probability that a random pair of gene copies, inherited by offspring from two parents, is identical by descent.

inbreeding depression   Reduction, in inbred individuals, of the mean value of a character (usually one correlated with fitness).

inclusive fitness   The fitness of a gene or genotype as measured by its effect on the survival or reproduction of both the organism bearing it (**direct fitness**) and the genes, identical by descent, borne by the organism's relatives (**indirect fitness**).

incomplete lineage sorting   Persistence of a genetic polymorphism through a speciation event, so that fixation occurs only in the descendant species, or in their descendants after subsequent speciation.

indirect development   A life history that includes a larval stage between embryo and adult stages. *Cf.* **direct development**.

indirect fitness   *See* **inclusive fitness**.

individual selection   A form of natural selection consisting of nonrandom differences among different genotypes (or phenotypes) within a population in their contribution to subsequent generations. *See also* **genic selection**, **natural selection**.

individualization   The evolution of distinct form and identity of each of several structures that were not differentiated from one another in an ancestor.

ingroup   *See* **outgroup**.

inheritance of acquired characteristics   The formerly widespread belief that modifications of an individual during its lifetime, due to its behavior or its environment, could be transmitted to its descendants.

intelligent design (ID)   A strain in creationism that claims that the complexity of organisms is too great to have evolved by natural processes and therefore must have been designed by an intelligent being.

inter-, intra- Prefixes meaning, respectively, "between" and "within." For example, "interspecific" differences are differences between species and "intraspecific" differences are differences among individuals within a species.

interaction Strictly, the dependence of an outcome on a combination of causal factors, such that the outcome is not predictable from the average effects of the factors taken separately. More loosely, an interplay between entities that affects one or more of them (as in interactions between species). *See also* **genotype × environment interaction**.

interdemic selection **Group selection** of populations within a species.

intragenic recombination Recombination within a gene.

intrinsic rate of natural increase The potential per capita rate of increase of a population with a stable age distribution whose growth is not depressed by the negative effects of density.

introgression Movement of genes from one species or population into another by hybridization and backcrossing; carries the implication that some genes in a genome undergo such movement, but others do not.

intron A part of a gene that is not translated into a polypeptide. *Cf.* **exon**.

inversion A 180° reversal of the orientation of a part of a chromosome, relative to some standard chromosome.

isolating barrier (isolating mechanism) A genetically determined difference between populations that restricts or prevents gene flow between them. The term does not include spatial segregation by extrinsic geographic or topographic barriers.

isolation by distance A model of population structure in which the likelihood of mating decreases with the geographic distance between individuals, so that local mating causes geographic variation in allele frequencies.

iteroparous Pertaining to a life history in which individuals reproduce more than once. *Cf.* **semelparous**.

K

karyotype The chromosome complement of an individual.

key adaptation An adaptation that provides the basis for using a new, substantially different habitat or resource.

kin selection A form of selection whereby alleles differ in their rate of propagation by influencing the impact of their bearers on the reproductive success of individuals (kin) who carry the same alleles by common descent.

L

Lamarckism The theory that evolution is caused by inheritance of character changes acquired during the life of an individual due to its behavior or to environmental influences.

lateral gene transfer *See* **horizontal transmission**.

Laurasia The northern of the two large continents that existed in the early Mesozoic.

lethal allele An allele (usually recessive) that causes virtually complete mortality, usually early in development.

levels of selection The several kinds of reproducing biological entities (e.g., genes, organisms, species) that can vary in fitness, resulting in potential selection among them.

life history Usually refers to the set of traits that affect changes in numbers of individuals over generations, of the population as a whole or of specific genotypes within the population; these traits include age-specific values of survival, female reproduction, and male reproduction, and may include dispersal as well.

lineage A series of ancestral and descendant populations through time; usually refers to a single evolving species, but may include several species descended from a common ancestor.

lineage sorting The process by which each of several descendant species, carrying several gene lineages inherited from a common ancestral species, acquires a single gene lineage; hence, the derivation of a monophyletic gene tree, in each species, from the paraphyletic gene tree inherited from their common ancestor.

lineage-through-time plot A graph of the apparent change in number of lineages in a clade, often based on a time-calibrated phylogeny.

linkage Occurrence of two loci on the same chromosome: the loci are functionally linked only if they are so close together that they do not segregate independently in meiosis.

linkage disequilibrium The association of two alleles at two or more loci more frequently (or less frequently) than predicted by their individual frequencies.

linkage equilibrium The association of two alleles at two or more loci at the frequency predicted by their individual frequencies.

locus (plural: **loci**) A site on a chromosome occupied by a specific gene; more loosely, the gene itself, in all its allelic states.

logistic equation An equation describing the idealized growth of a population subject to a density-dependent limiting factor. As density increases, the rate of growth gradually declines until population growth stops.

M

macroevolution A vague term, usually meaning the evolution of substantial phenotypic changes, usually great enough to place the changed lineage and its descendants in a distinct genus or higher taxon. *Cf.* **microevolution**.

mass extinction A highly elevated rate of extinction of species, extending over an interval that is relatively short on a geological time scale (although still very long on a human time scale).

maternal effect A nongenetic effect of a mother on the phenotype of her offspring, stemming from factors such as cytoplasmic inheritance, transmission of symbionts from mother to offspring, or nutritional conditions.

maximum likelihood (ML) A statistical method of estimating the parameters of a model (such as the mean and variance, in simple cases) from data.

maximum parsimony *See* **parsimony**.

McDonald-Kreitman (MK) test A test for selection at a locus by comparing DNA sequence variation within species with the variation among species.

mean Usually the arithmetic mean or average; the sum of n values, divided by n. The mean value of x, symbolized as $\bar{x}$, equals $(x_1 + x_2 + \ldots + x_n)/n$.

mean fitness The arithmetic average fitness of all individuals in a population, usually relative to some standard.

meiotic drive Used broadly to denote a preponderance (> 50 percent) of one allele among the gametes produced by a heterozygote, which is more properly called **segregation distortion**; results in **genic selection**.

metapopulation A set of local populations, among which there may be gene flow and patterns of extinction and recolonization.

microarrays Tiny grids of cDNA or genomic DNA fragments blotted onto silicon chips or glass slides that can be exposed to fluorescent DNA or RNA probes in order to assay the presence and quantities of specific genes or mRNAs.

microevolution A vague term, usually referring to slight, short-term evolutionary changes within species. *Cf.* **macroevolution**.

microsatellite A short, highly repeated, untranslated DNA sequence.

migration Used in theoretical population genetics as a synonym for gene flow among populations; in other contexts, refers to directed large-scale movements of organisms that do not necessarily result in gene flow.

mimicry Similarity of certain characters in two or more species due to convergent evolution when there is an advantage conferred by the resemblance. Common types include *Batesian* mimicry, in which a palatable *mimic* experiences lower predation because of its resemblance to an unpalatable *model*; and *Müllerian* mimicry, in which two or more unpalatable species enjoy reduced predation due to their similarity.

modern synthesis *See* **evolutionary synthesis**.

modularity The ability of individual parts of an organism, such as segments or organs, to develop or evolve independently from one another; the ability of developmental regulatory genes and pathways to be regulated independently in different tissues and developmental stages.

molecular clock The concept of a steady rate of change in DNA sequences over time, providing a basis for dating the time of divergence of lineages if the rate of change can be estimated.

monomorphic Having one form; refers to a population in which virtually all individuals have the same genotype at a locus. *Cf.* **polymorphism**.

monophyletic Refers to a taxon, or a branch of a phylogenetic tree or gene tree, that includes all the species (or genes) that descended from a common ancestor. *Cf.* **paraphyletic**, **polyphyletic**.

mosaic evolution Evolution of different characters within a lineage or clade at different rates, hence more or less independently of one another.

multigene family Also called "gene family," a set of distinct loci in a genome that originated from a single locus in an ancestor by duplication and sequence divergence.

multiple stable equilibria *See* **stability**.

multiple-niche polymorphism Stable variation at a locus owing to superior fitness of different genotypes under different conditions of a varying environment.

mutation An error in the replication of a nucleotide sequence, or any other alteration of the genome that is not manifested as reciprocal recombination.

mutational variance The increment in the genetic variance of a phenotypic character caused by new mutations in each generation.

mutualism A symbiotic relation in which each of two species benefits by their interaction.

N

natural laws Consistent natural phenomena, described by statements that certain effects will always occur if specific conditions hold.

natural selection The differential survival and/or reproduction of classes of entities that differ in one or more characteristics. To constitute natural selection, the difference in survival and/or reproduction cannot be due to chance, and it must have the potential consequence of altering the proportions of the different entities. Thus natural selection is also definable as a deterministic difference in the contribution of different classes of entities to subsequent generations. Usually the differences are inherited. The entities may be alleles, genotypes or subsets of genotypes, populations, or, in the broadest sense, species. A complex concept; see Chapter 11. *See also* **genic selection**, **individual selection**, **kin selection**, **group selection**.

naturalistic fallacy A frequently used name for the belief that what is "natural" is morally right or good.

neo-Darwinism Originally, the theory of natural selection of inherited variations, that denied that acquired characteristics might be inherited; often used more broadly to mean the modern theory that natural selection, acting on randomly generated particulate genetic variation, is the major, but not the sole, cause of evolution.

neofunctionalization Divergence of duplicate genes whereby one acquires a new function. *Cf.* **subfunctionalization**.

neoteny Heterochronic evolution whereby development of some or all somatic features is retarded relative to sexual maturation, resulting in sexually mature individuals with juvenile features. *See also* **paedomorphosis**, **progenesis**.

neutral alleles Alleles that do not differ measurably in their effect on fitness.

neutral theory of molecular evolution The hypothesis that most mutations that become fixed do not significantly alter fitness and have become fixed by genetic drift.

nonadaptive evolution Evolution by substitution of neutral alleles.

nonsynonymous substitution A base pair substitution in DNA that results in an amino acid substitution in the protein product; also called **replacement substitution**. *Cf.* **synonymous substitution**.

norm of reaction The set of phenotypic expressions of a genotype under different environmental conditions. *See also* **phenotypic plasticity**.

normal distribution A bell-shaped frequency distribution of a variable; the expected distribution if many factors with independent, small effects determine the value of a variable; the basis for many statistical formulations.

nucleotide substitution The complete replacement of one nucleotide base pair by another within a lineage over evolutionary time.

O

ontogeny The development of an individual organism, from fertilized zygote until death.

optimality theory   Models of adaptive evolution that assume that characters have evolved to nearly their optimum, within limits set by specified constraints.

organism   Usually used in this book to refer to an individual member of a species.

orthologous   Refers to corresponding (homologous) members of a gene family in two or more species. *Cf.* **paralogous**.

outcrossing   Mating with another genetic individual. *Cf.* **selfing**.

outgroup   A taxon that diverged from a group of other taxa (the **ingroup**) before they diverged from one another.

overdominance   The expression by two alleles in heterozygous condition of a phenotypic value for some character that lies outside the range of the two corresponding homozygotes.

P

paedomorphosis   Possession in the adult stage of features typical of the juvenile stage of the organism's ancestor.

Pangaea   The single large "world continent" formed by coalescence of land masses in the late Paleozoic.

panmixia   Random mating among members of a population.

parallel evolution (parallelism)   The evolution of similar or identical features independently in related lineages, thought usually to be based on similar modifications of the same developmental pathways.

paralogous   Refers to the evolutionary relationship between two different members of a gene family, within a species or in a comparison of different species. *Cf.* **orthologous**.

parapatric   Of two species or populations, having contiguous but non-overlapping geographic distributions. *Cf.* **allopatric**, **sympatric**.

parapatric speciation   *See* **allopatric speciation**.

paraphyletic   Refers to a taxon, phylogenetic tree, or gene tree whose members are all derived from a single ancestor, but which does not include all the descendants of that ancestor. *Cf.* **monophyletic**.

parental investment   Parental activities or processes that enhance the survival of existing offspring but whose **costs** reduce the parent's subsequent reproductive success.

parent-offspring conflict   A condition in which a character state that enhances fitness of offspring reduces the fitness of a parent (or vice versa).

parsimony   Economy in the use of means to an end (*Webster's New Collegiate Dictionary*); the principle of accounting for observations by that hypothesis requiring the fewest or simplest assumptions that lack evidence; in systematics, the principle of invoking the minimal number of evolutionary changes to infer phylogenetic relationships.

parthenogenesis   Virgin birth; development from an egg to which there has been no paternal contribution of genes.

passive trend   *See* **driven trend**.

PCR (polymerase chain reaction)   A laboratory technique by which the number of copies of a DNA sequence is increased by replication in vitro.

peak shift   Change in allele frequencies within a population from one to another local maximum of mean fitness by passage through states of lower mean fitness.

peramorphosis   Evolution of a more extreme character state by prolongation of development in the descendant, compared to the ancestor.

peripatric   Of a population, peripheral to most of the other populations of a species.

peripatric speciation   Speciation by evolution of reproductive isolation in peripatric populations as a consequence of a combination of genetic drift and natural selection.

phenetic   Pertaining to phenotypic similarity, as in a phenetic classification.

phenotype   The morphological, physiological, biochemical, behavioral, and other properties of an organism manifested throughout its life; or any subset of such properties, especially those affected by a particular allele or other portion of the **genotype**.

phenotypic correlation (r_P)   *See* **genetic correlation**.

phenotypic integration   Correlation between the state of two or more functionally related characteristics, so that they are advantageously matched in most individuals.

phenotypic plasticity   The capacity of an organism to develop any of several phenotypic states, depending on the environment; usually this capacity is assumed to be adaptive.

phenotypic variance (V_P)   The **variance** (q.v.) in a trait within a population; it may include both **genetic variance** and **environmental variance**.

phyletic gradualism   A term for gradual evolutionary change in features over a long period of time.

phylogenetic niche conservatism   Slow evolution of the ecological requirements of a group of organisms, resulting in long-continued dependence of related species on similar resources and environmental conditions.

phylogenetic species concept (PSC)   Species conceived as groups of populations that are distinguishable from other such groups.

phylogenetic tree   A diagram representing the evolutionary relationships among named groups of organisms, i.e., their history of descent from common ancestors.

phylogeny   The history of descent of a group of taxa such as species from their common ancestors, including the order of branching and sometimes the absolute times of divergence; also applied to the genealogy of genes derived from a common ancestral gene.

phylogeography   Description and analysis of the history and processes that govern the geographic distribution of genes within species and among closely related species, analysis that may shed light on the history of the populations.

physical constraint   A restriction that prevents a lineage from evolving a trait due to the properties of biological materials.

planktonic   Living in open water. *Cf.* **benthic**.

pleiotropy   A phenotypic effect of a gene on more than one character.

ploidy   The number of chromosome complements in an organism.

point mutation   A mutation that maps to a specific gene locus; in a molecular context, usually a change of a single base pair.

polygenic character   A character whose variation is based wholly or in part on allelic variation at more than a few loci.

polymerase chain reaction   *See* **PCR**.

polymorphism   The existence within a population of two or more genotypes, the rarest of which exceeds some arbitrarily low frequency (say, 1 percent); more rarely, the existence of phenotypic variation within a population, whether or not genetically based. *Cf.* **monomorphic**.

polyphenism   The capacity of a species or genotype to develop two or more forms, with the specific form depending on specific environmental conditions or cues, such as temperature or day length. A polyphenism is distinct from a **polymorphism** in that the former is the property of a single genotype, whereas the latter refers to multiple forms encoded by two or more different genotypes.

polyphyletic   Refers to a taxon, phylogenetic tree, or gene tree composed of members derived by evolution from ancestors in more than one ancestral taxon; hence, composed of members that do not share a unique common ancestor. *Cf.* **monophyletic**.

polyploid   Of a cell or organism, possessing more than two chromosome complements.

population   A group of conspecific organisms that occupy a more or less well defined geographic region and exhibit reproductive continuity from generation to generation; ecological and reproductive interactions are more frequent among these individuals than with members of other populations of the same species.

positive selection   Selection for an allele that increases fitness. *Cf.* **purifying selection**.

postzygotic   Occurring after union of the nuclei of uniting gametes; usually refers to inviability or sterility that confers reproductive isolation.

preadaptation   Possession of the necessary properties to permit a shift to a new niche, habitat, or function. A structure is preadapted for a new function if it can assume that function without evolutionary modification.

premating barriers   *See* **prezygotic**.

prezygotic   Occurring before union of nuclei of uniting gametes; usually refers to events in the reproductive process that cause reproductive isolation, including those that occur before mating (**premating barriers**).

primordium   A group of embryonic or larval cells destined to give rise to a particular adult structure.

processed pseudogene   A **pseudogene** that has arisen via the retrotransposition of mRNA into cDNA.

progenesis   A decrease during evolution of the duration of ontogenetic development, resulting in retention of juvenile features in the sexually mature adult. *See also* **neoteny**, **paedomorphosis**.

promoter   Usually refers to the DNA sequences immediately 5′ to (upstream of) a gene that are bound by the RNA polymerase and its cofactors and/or are required in order to transcribe the gene. Sometimes used interchangeably with **enhancer**.

provinciality   The degree to which the taxonomic composition of a biota is differentiated among major geographic regions.

pseudogene   A nonfunctional member of a gene family that has been derived from a functional gene. *Cf.* **processed pseudogene**.

pull of the Recent   An artifact in estimating diversity in the fossil record, whereby taxa that are still alive have apparently longer durations than they would if they had been counted only from fossil data, and so will inflate the count of taxa, compared to the more remote past.

punctuated anagenesis   *See* **punctuated gradualism**.

punctuated equilibria   A pattern of rapid evolutionary change in the phenotype of a lineage separated by long periods of little change; also, a hypothesis intended to explain such a pattern, whereby phenotypic change transpires rapidly in small populations, in concert with the evolution of reproductive isolation.

punctuated gradualism   Alternating periods of slow and more rapid gradual change in a single lineage. Also called **punctuated anagenesis**.

purifying selection   Elimination of deleterious alleles from a population. *Cf.* **positive selection**.

Q

quantitative genetics   Genetic analysis of continuously varying characters, often employing statistical descriptions and estimators of variation.

quantitative trait   A phenotypic character that varies continuously rather than as discretely different character states.

quantitative trait locus/loci (QTL)   A chromosome region containing at least one gene that contributes to variation in a quantitative trait. QTL mapping is a procedure for determining the map positions of QTL on chromosomes.

quantitative variation   *See* **quantitative trait**.

R

race   A poorly defined term for a set of populations occupying a particular region that differ in one or more characteristics from populations elsewhere; equivalent to subspecies. In some writings, a distinctive phenotype, whether or not allopatric from others.

radiation   *See* **adaptive radiation**.

radiometric dating   Estimating ages of geological materials and events by the decay of radioactive elements.

random genetic drift   *See* **genetic drift**.

random walk   A mathematical model of a series of random steps, used to describe random genetic drift and some other biological processes.

realized heritability   The heritability of a trait as calculated retrospectively from the change in a population's mean phenotype, relative to the selection differential that was applied to the character in an artificial selection experiment.

recessive   *See* **dominance**.

reciprocal translocation   A recombinational exchange of parts of two nonhomologous chromosomes.

reciprocity   Cooperation based on reciprocal aid in a succession of encounters between individuals.

recombinational speciation   The origin of a new species by selection among genotypes formed by hybridization between two ancestral species.

recruitment   (1) In evolutionary genetics, the evolution of a new function for a gene other than the function for which that gene was originally adapted. (2) In population biology, refers to the addition of new adult (breeding) individuals to a population via reproduction (i.e., individuals born into the population that reach reproductive age).

recurrent mutation Repeated origin of mutations of a particular kind within a species.

Red Queen hypothesis The proposition that taxa become extinct at an approximately constant rate because they fail to evolve as fast as other species with which they have antagonistic interactions. "Red Queen" more generally refers to averting extinction by evolving as fast as possible.

refugia Locations in which species have persisted while becoming extinct elsewhere.

regression In geology, withdrawal of sea from land, accompanying lowering of sea level; in statistics, a function that best predicts a dependent from an independent variable.

regulatory modularity The property of gene regulation that allows gene expression or protein function to vary in different cells, tissues, or developmental stages of the same organism, without affecting the entire morphology or life history of the organism.

reinforcement Evolution of enhanced reproductive isolation between populations due to natural selection for greater isolation.

relative fitness The fitness of a genotype relative to (as a proportion of) the fitness of a reference genotype, which is often set at 1.0; the fitness values before such standardization are **absolute fitness** values.

relict A species that has been "left behind"; for example, the last survivor of an otherwise extinct group. Sometimes, a species or population left in a locality after extinction throughout most of the region.

replacement substitution *See* **nonsynonymous substitution**.

reporter construct A DNA segment in which a putative *cis*-regulatory sequence is spliced upstream of a gene whose expression can be easily assayed, such as β-galactosidase or green fluorescent protein.

reproductive character displacement *See* **character displacement**.

reproductive effort The proportion of energy or materials that an organism allocates to reproduction rather than to growth and maintenance.

reproductive isolation Reduction of gene exchange between populations by any of several possible factors, usually those arising from biological differences between the populations.

reproductive success The fitness of a genotype or other biological entity, often measured by the average per capita number of offspring that a newly formed zygote will have, or by similar measures.

response to selection The change in the mean value of a character over one or more generations due to selection.

restriction enzyme An enzyme that cuts double-stranded DNA at specific short nucleotide sequences. Genetic variation within a population results in variation in DNA sequence lengths after treatment with a restriction enzyme, or **restriction fragment length polymorphism (RFLP)**.

reticulate evolution Union of different lineages of a clade by hybridization.

reverse transcriptase An enzyme in retroviruses that synthesizes DNA copies of RNA molecules.

RFLP *See* **restriction enzyme**.

rtPCR (reverse transcriptase PCR, real-time PCR) A PCR reaction using mRNA as a template in which an initial step converts the mRNA to cDNA using reverse transcriptase and a subsequent step uses PCR to amplify the cDNA.

runaway sexual selection A model of sexual selection in which a male display character and female preference for the character reinforce one another so that both evolve to be more extreme.

S

saltation A jump; a discontinuous mutational change in one or more phenotypic traits, usually of considerable magnitude.

sampling error The amount of inaccuracy (i.e., random variation) in the estimate of some value of a population, caused by measuring only a portion of the population; by extension, the chance variation in the value of repeated samples from the population.

scala naturae The "scale of nature," or Great Chain of Being: the pre-evolutionary concept that all living things were created in an orderly series of forms, from lower to higher.

scientific theory A coherent body of statements, based on reasoning and evidence, that explains some aspect of nature by recourse to natural laws or processes.

secondary contact Contact and potential interbreeding between formerly allopatric populations, owing to range expansion.

segregation distortion *See* **meiotic drive**.

selection Nonrandom differential survival or reproduction of classes of phenotypically different entities. *See* **natural selection**, **artificial selection**.

selection coefficient The difference between the mean relative fitness of individuals of a given genotype and that of a reference genotype.

selection differential The difference between the mean character value in a population before selection, and in the subset of individuals that survive and reproduce.

selection gradient The slope of the relationship between phenotype and fitness, for a quantitative character, usually taking correlations with other characters into account.

selection plateau The mean character value at which a population ceases to respond to continuing directional selection.

selective advantage The increment in fitness (survival and/or reproduction) provided by an allele or a character state.

selective (or functional) constraint A restriction that prevents a lineage from evolving a particular trait because that trait is always disadvantageous or interferes with the function of another trait.

selective sweep Reduction or elimination of DNA sequence variation in the vicinity of a mutation that has been fixed by natural selection relatively recently.

selfing Self-fertilization; union of female and male gametes produced by the same genetic individual. *Cf.* **outcrossing**.

"selfish DNA" A DNA sequence that has the capacity for its own replication, or replication via other self-replicating elements, but has no immediate function (or is deleterious) for the organism in which it resides.

semelparous   Pertaining to a life history in which individuals (especially females) reproduce only once. *Cf.* **iteroparous**.

semispecies   One of several groups of populations that are partially but not entirely isolated from one another by biological factors (**isolating mechanisms**).

sensory bias   A difference in the ability of an organism to perceive different stimuli (e.g., loud vs. soft sounds).

serial homology   A relationship among repeated, often differentiated, structures of a single organism, defined by their similarity of developmental origin; for example, the several legs and other appendages of an arthropod.

sex-linked   Of a gene, being carried by one of the sex chromosomes; it may be expressed phenotypically in both sexes.

sex ratio   Often described as the proportion of males among offspring, either of an individual ("individual sex ratio") or a population ("population sex ratio").

sexual isolation   Reduction of gene exchange between populations because of aversion to mating with members of the other population; also called **behavioral isolation**.

sexual reproduction   Production of offspring whose genetic constitution is a mixture of those of two potentially genetically different gametes.

sexual selection   Differential reproduction as a result of variation in the ability to obtain mates.

sibling species   Species that are difficult or impossible to distinguish by morphological characters, but may be discerned by differences in ecology, behavior, chromosomes, genetic markers, or other such features.

silent substitution   *See* **synonymous substitution**.

single nucleotide polymorphism (**SNP**)   Variation in the identity of a nucleotide base pair at a single position in a DNA sequence, within or among populations of a species.

sister taxa   Two species or higher taxa that are derived from an immediate common ancestor, and are therefore one another's closest relatives.

special creation   The idea that each species was individually created by God in much its present form.

speciation   Evolution of reproductive isolation within an ancestral species, resulting in two or more descendant species.

species   In the sense of biological species, the members of a group of populations that interbreed or potentially interbreed with one another under natural conditions; a complex concept (see Chapter 17). Also, a fundamental taxonomic category to which individual specimens are assigned, which often but not always corresponds to the biological species. *See also* **biological species**, **phylogenetic species concept**.

species hitchhiking   Increase in the proportion of species with a specific trait because it is correlated with another trait that enhances speciation or reduces extinction.

species selection   A form of **group selection** in which species with different characteristics increase (by speciation) or decrease (by extinction) in number at different rates because of a difference in their characteristics.

species sorting   A correlation between a particular character and the rate of diversification among clades.

stability   Often used to mean constancy; more often in this book, the propensity to return to a condition (a stable equilibrium) or to one of several such conditions (**multiple stable equilibria**) after displacement from that condition.

stabilizing selection   Selection that maintains the mean of a character at or near a constant intermediate value in a population.

standard deviation   The square root of the **variance**.

stasis   Absence of substantial evolutionary change in one or more characters for some period of evolutionary time.

stem group   *See* **crown group**.

stochastic   Random. *Cf.* **deterministic**.

strata   Layers of sedimentary rock that were deposited at different times.

subfunctionalization   Divergence of duplicate genes whereby each retains only a subset of the several functions of the ancestral gene. *Cf.* **neofunctionalization**.

subspecies   A named geographic race; a set of populations of a species that share one or more distinctive features and occupy a different geographic area from other subspecies.

substitution   Usually, the complete replacement of one allele for another within a population or species over evolutionary time (*cf.* **fixation**). Sometimes refers to base pair differences in comparisons of homologous DNA sequences.

superspecies   A group of **semispecies**.

symbiosis   An intimate, usually physical, association between two or more species.

sympatric   Of two species or populations, occupying the same geographic locality so that the opportunity to interbreed is presented. *Cf.* **allopatric**, **parapatric**.

sympatric speciation   *See* **allopatric specation**.

synapomorphy   A derived character state that is shared by two or more taxa and is postulated to have evolved in their common ancestor.

synonymous substitution   Fixation of a base pair change that does not alter the amino acid in the protein product of a gene; also called **silent substitution**. *Cf.* **nonsynonymous substitution**.

T

tandem repeat   One of a group of adjacent duplicate copies of a DNA sequence.

target gene   In developmental genetics, a gene regulated by a transcription factor of interest. This regulation may be direct or indirect.

taxon (plural: **taxa**)   The named taxonomic unit (e.g., *Homo sapiens*, Hominidae, or Mammalia) to which individuals, or sets of species, are assigned. **Higher taxa** are those above the species level. *Cf.* **category**.

teleology   The belief that natural events and objects have purposes and can be explained by their purposes.

territory   An area or volume of habitat defended by an organism or a group of organisms against other individuals, usually of the same species; territorial behavior is the behavior by which the territory is defended.

theistic evolution   The belief that evolution occurs based on natural laws that were established by a deity.

theory   *See* **scientific theory**.

threshold trait A characteristic that is expressed as discrete states, although the genetic variation underlying it is polygenic.

time for speciation (TFS) The amount of time required for reproductive isolation to evolve, once the process starts.

trade-off The existence of both a fitness benefit and a fitness cost of a mutation or character state, relative to another.

transcription factor A protein that interacts with a regulatory DNA sequence and affects the transcription of the associated gene.

transcriptome A term for a specified set of mRNA transcripts, such as those found in a specific cell type or in the organism as a whole. Transcriptomes from the same organ in different species, or from different organs or physiological states of the same species, can reveal how gene expression varies in different contexts.

transition A mutation that changes a nucleotide to another nucleotide in the same class (purine or pyrimidine). *Cf.* **transversion**.

translational robustness The ability of protein to maintain its proper three-dimensional structure and folding in the face of amino acid changes that result from mistranslation of its underlying DNA sequence.

translocation The transfer of a segment of a chromosome to another, nonhomologous, chromosome; the chromosome formed by the addition of such a segment.

transposable element A DNA sequence, copies of which become inserted into various sites in the genome.

transposition Movement of a copy of a transposable element to a different site in the genome.

trans-regulatory element A nucleotide sequence, usually encoding a regulatory protein, that is not closely linked to the structural gene whose expression it regulates. *Cf. cis*-regulatory element.

transversion A mutation that changes a nucleotide to another nucleotide in the opposite class (purine or pyrimidine). *Cf.* **transition**.

trend *See* **evolutionary trend**.

U

ultraconserved elements Regions of the genome that are highly conserved, sometimes 100% identical, between distantly related species. Many ultraconserved elements occur in exons that encode proteins, but a surprising number occur outside of genes and presumably have a regulatory function.

underdominance Lower fitness of a heterozygote than of both of the homozygotes for the same alleles.

unequal crossing over Recombination between nonhomologous sites on two homologous chromosomes.

uniformitarianism The proposition that natural processes that operated in the past are the same as in the present. (The term has usually implied gradual rather than catastrophic change.)

unstable equilibrium An equilibrium to which a system does not return if disturbed.

V

variability Properly, the ability of a system to vary. Often used to mean "variation."

variance (s^2, s^2, V) The average squared deviation of an observation from the arithmetic mean; hence, a measure of variation. $s^2 = [(x_i - \bar{x})^2]/(n - 1)$, where $\bar{x}$ is the mean and n the number of observations.

vegetative propagation Production of offspring from somatic tissues, e.g., by buds.

vertical transmission *See* **horizontal transmission**.

vestigial Occurring in a rudimentary condition as a result of evolutionary reduction from a more elaborated, functional character state in an ancestor.

viability Capacity for survival; often refers to the fraction of individuals surviving to a given age, and is contrasted with inviability due to deleterious genes.

vicariance Separation of a continuously distributed ancestral population or species into separate populations because of the development of a geographic or ecological barrier.

virulence Usually, the damage inflicted on a host by a pathogen or parasite; sometimes, the capacity of a pathogen or parasite to infect and develop in a host.

W

wild-type The allele, genotype, or phenotype that is most prevalent (if there is one) in wild populations; with reference to the wild-type allele, other alleles are often termed mutations.

Z

zygote A single-celled individual formed by the union of gametes. Occasionally used more loosely to refer to an offspring produced by sexual reproduction.

Literature Cited

A

Abbot, P., and 136 others. 2011. Inclusive fitness theory and eusociality. *Nature* 471: E1–E4. [16]

Abbott, R. J. 1992. Plant invasions, interspecific hybridization and the evolution of new plant taxa. *Trends Ecol. Evol.* 7: 401–405. [17]

Abedin, M., and N. King. 2008. The premetazoan ancestry of cadherins. *Science* 319: 946–948. [5, 20]

Abi-Rached, L., and 22 others. 2011. The shaping of modern human immune systems by multiregional admixture with archaic humans. *Science* 333: 89–94. [6]

Abouheif, E., and G. A. Wray. 2002. Evolution of the gene network underlying wing polyphenism in ants. *Science* 297: 249–252. [21]

Abrams, P. A. 1987. Alternative models of character displacement and niche shift. I. Adaptive shifts in resource use when there is competition for nutritionally nonsubstitutable resources. *Evolution* 41: 651–661. [19]

Abrams, P. A. 2000. The evolution of predator-prey interactions: Theory and evidence. *Annu. Rev. Ecol. Syst.* 31: 79–108. [19]

Abrams, P. A., and H. Matsuda. 1997. Fitness minimization and dynamic instability as a consequence of predator-prey coevolution. *Evol. Ecol.* 11: 1–20. [19]

Abzhanov, A., M. Protas, B. R. Grant, P. R. Grant, and C. J. Tabin. 2004. Bmp4 and morphological variation of beaks in Darwin's finches. *Science* 305: 1462–1465. [22]

Ackerly, D. D. 2004. Adaptation, niche conservatism, and convergence: comparative studies of leaf evolution in the California chaparral. *Am. Nat.* 163: 654–671. [19]

Adamowicz, S. J., and A. Purvis. 2006. From more to fewer? Testing an allegedly pervasive trend in the evolution of morphological structure. *Evolution* 60: 1402–1416. [3]

Adamowicz, S. J., A. Purvis, and M. A. Wills. 2008. Increasing morphological complexity in multiple parallel lineages of the Crustacea. *Proc. Natl. Acad. Sci. USA* 105: 4786–4791. [22]

Adams, D. C., and A. Nistri. 2010. Ontogenetic convergence and evolution of foot morphology in European cave salamanders (Family: Plethodontidae). *BMC Evol. Biol.* 10: 216. [21]

Agrawal, A. A. 2005. Natural selection on common milkweed (*Asclepias syriaca*) by a community of specialized insect herbivores. *Evol. Ecol. Res.* 7: 651–667. [19]

Agrawal, A., and C. M. Lively. 2002. Infection genetics: gene-for-gene versus matching-alleles models and all points in between. *Evol. Ecol. Res.* 4: 79–90. [19]

Agrawal, A. A., C. Laforsch, and R. Tollrian. 1999. Transgenerational induction of defences in animals and plants. *Nature* 401: 60–63. [9]

Ahlberg, P. E., and J. A. Clack. 2006. A firm step from water to land. *Nature* 440: 747–749. [4]

Ahlberg, P. E., J. A. Clack, E. Lukševičs, H. Blom, and I. Zupins. 2008. *Ventastega curonica* and the origin of tetrapod morphology. *Nature* 453: 1199–1204. [4]

Alberch, P., and E. A. Gale. 1985. A developmental analysis of an evolutionary trend: Digital reduction in amphibians. *Evolution* 39: 8–23. [21]

Alcock, J. 2013. *Animal Behavior: An Evolutionary Approach*, 10th ed. Sinauer, Sunderland, MA. [16]

Alfaro, M. E., F. Santini, C. Brock, H. Alamillo, A. Dornburg, D. L. Rabosky, G. Carnevale, and L. J. Harmon. 2009. Nine exceptional radiations plus high turnover explain species diversity in jawed vertebrates. *Proc. Natl. Acad. Sci. USA* 106: 13410–13414. [7]

Allen, C. E., P. Beldade, B. J. Zwaan, and P. M. Brakefield. 2008. Differences in the selection response of serially repeated color pattern characters: standing variation, development, and evolution. *BMC Evol. Biol.* 8: 94. [21]

Allen, H. L., and 285 others. 2010. Hundreds of variants clustered in genomic loci and biological pathways affect human height. *Nature* 467: 832–838. [13]

Allendorf, F. W., and G. Luikart. 2007. *Conservation and the Genetics of Populations*. Blackwell, Oxford. [23]

Allison, A. C. 1955. Aspects of polymorphism in man. *Cold Spring Harbor Symp. Quant. Biol.* 20: 239–255. [9]

Alroy, J. 1998. Cope's rule and the dynamics of body mass evolution in North American fossil mammals. *Science* 280: 731–734. [22]

Alroy, J. 2001. A multispecies overkill simulation of the end-Pleistocene megafaunal mass extinction. *Science* 292: 1893–1896. [5]

Alroy, J. 2008. Dynamics of origination and extinction in the marine fossil record. *Proc. Natl. Acad. Sci. USA* 105 (Suppl. 1): 11536–11542. [7]

Alroy, J., P. L. Koch, and J. C. Zachos. 2000. Global climate change and North American mammal evolution. In D. H. Erwin and S. L. Wing (eds.), *Deep Time: Paleobiology's Perspective*, pp. 259–288. *Paleobiology* 25 (4), supplement. [7]

Alroy, J., and 34 others. 2008. Phanerozoic trends in the global diversity of marine invertebrates. *Science* 321: 97–100. [7]

Alters, B. J., and S. M. Alters. 2001. *Defending Evolution: A Guide to the Creation/Evolution Controversy*. Jones and Bartlett, Sudbury, MA. [1, 22, 23]

Alvarez, L. W., W. Alvarez, F. Asaro, and H. V. Michel. 1980. Extraterrestrial cause for the Cretaceous-Tertiary extinction. *Science* 208: 1095–1108. [7]

Ambrose, B. A., D. R. Lerner, P. Ciceri, C. M. Padilla, M. F. Yanofsky, and R. J. Schmidt. 2000. Molecular and genetic analyses of the *silky1* gene reveal conservation in floral organ specification between eudicots and monocots. *Molec. Cell* 5: 569–579. [21]

Anderson, E. 1949. *Introgressive Hybridization*. Wiley, New York. [17]

Anderson, I. A., and J. W. Busch. 2006. Relaxed pollinator-mediated selection weakens floral integration in self-compatible taxa of *Leavenworthia* (Brassicaceae). *Am. J. Bot.* 93: 860–867. [13]

Andersson, M. B. 1982. Female choice selects for extreme tail length in a widowbird. *Nature* 299: 818–820. [11]

Andersson, M. B. 1994. *Sexual Selection*. Princeton University Press, Princeton, NJ. [15]

Andersson, M. B., and Y. Iwasa. 1996. Sexual selection. *Trends Ecol. Evol.* 11: 53–58. [15]

Andersson, M. B., and L. W. Simmons. 2006. Sexual selection and mate choice. *Trends Ecol. Evol.* 21: 296–302. [15]

Andolfatto, P. 2005. Adaptive evolution of non-coding DNA in *Drosophila*. *Nature* 437: 1149–1152. [8]

Andrade, M. C. B. 1996. Sexual selection for male sacrifice in the Australian redback spider. *Science* 271: 70–72. [11]

Andrés, A. M., and 11 others. 2009. Targets of balancing selection in the human genome. *Mol. Biol. Evol.* 26: 2755–2764. [12]

Antolin, M. F. 1993. Genetics of biased sex ratios in subdivided populations: Models, assumptions, and evidence. *Oxford Series in Evolutionary Biology* 9: 239–281. Oxford University Press, Oxford. [15]

Antolin, M. F., and 11 others. 2011. Evolution and medicine in undergraduate education: a prescription for all biology students. *Evolution* 66: 1991–2006. [23]

Antonovics, J. 1968. Evolution in closely adjacent plant populations. V. Evolution of self-fertility. *Heredity* 23: 219–238. [15]

Antonovics, J., A. D. Bradshaw, and R. G. Turner. 1971. Heavy metal tolerance in plants. *Adv. Ecol. Res.* 7: 1–85. [11]

Antonovics, J., and 13 others. 2007. Evolution by any other name: Antibiotic resistance and avoidance of the e-word. *PLoS Biology* 5: e30 doi:10.1371/journal.pbio.0050030. [23]

Apicella, C. L., F. W. Marlowe, J. H. Fowler, and N. A. Christakis. 2012. Social networks and cooperation in hunter-gatherers. *Nature* 481: 497–501. [16]

Arakaki, M., and 8 others. 2011. Contemporaneous and recent radiations of the world's major succulent plant lineages. *Proc. Natl. Acad. Sci. USA* 108: 8379–8384. [5]

Arbeitman, M. N., and 9 others. 2002. Gene expression during the life cycle of *Drosophila melanogaster*. *Science* 297: 2270–2275. [21]

Arendt, J., and D. Reznick. 2008. Convergence and parallelism reconsidered: What have we learned about the genetics of adaptation? *Trend Ecol. Evol.* 23: 26–32. [3]

Armbruster, P., W. E. Bradshaw, and C. M. Holzapfel. 1998. Effects of postglacial range expansion on allozyme and quantitative genetic variation of the pitcher-plant mosquito, *Wyeomyia smithii*. *Evolution* 52: 1697–1704. [9]

Arnegard, M. E., D. J. Zwickl, Y. Lu, and H. H. Zakon. 2010. Old gene duplication facilitates origin and diversification of an innovative communication system—twice. *Proc. Natl. Acad. Sci. USA* 107: 22172–22177. [20]

Arnold, M. J. 1997. *Natural Hybridization and Evolution*. Oxford University Press, Oxford. [17]

Arnold, S. J. 1981. Behavioral variation in natural populations. I. Phenotypic, genetic, and environmental correlations between chemoreceptive responses to prey in the garter snake, *Thamnophis elegans*. *Evolution* 35: 489–509. [13]

Arnqvist, G., and L. Rowe. 2002. Correlated evolution of male and female morphologies in water striders. *Evolution* 56: 936–947. [15]

Arnqvist, G., and L. Rowe. 2005. *Sexual Conflict*. Princeton University Press, Princeton, NJ. [15]

Arnqvist, G., M. Edvardsson, U. Friberg, and T. Nilsson. 2000. Sexual conflict promotes speciation in insects. *Proc. Natl. Acad. Sci. USA* 97: 10460–10464. [18]

Atlan, A., H. Merçot, C. Landre, and C. Montchamp-Moreau. 1997. The sex-ratio trait in *Drosophila simulans*: Geographical distribution of distortion and resistance. *Evolution* 51: 1886–1895. [16]

Atwood, R. C., L. K. Schneider, and F. J. Ryan. 1951. Selective mechanisms in bacteria. *Cold Spring Harbor Symp. Quant. Biol.* 16: 345–355. [11]

Augspurger, C. K. 1984. Demography and life history variation of *Puya dasylirioides*, a long-lived rosette in tropical subalpine bogs. *Oikos* 45: 341–352. [14]

Auld, J. R., A. A. Agrawal, and R. A. Relyea. 2010. Re-evaluating the costs and limits of adaptive phenotypic plasticity. *Proc. R. Soc. Lond. B* 277: 503–511. [13]

Ausich, W. I., and N. G. Lane. 1999. *Life of the past*, 4th ed. Prentice-Hall, Upper Saddle River, NJ. [5]

Averoff, M., and N. H. Patel. 1997. Crustacean appendage evolution associated with changes in Hox gene expression. *Nature* 388: 682–686. [3]

Avise, J. C. 1994. *Molecular Markers, Natural History, and Evolution*. Chapman & Hall, New York. [17, 18]

Avise, J. C. 1998. *The Genetic Gods: Evolution and Belief in Human Affairs*. Harvard University Press, Cambridge, MA. [8]

Avise, J. C. 2000. *Phylogeography*. Harvard University Press, Cambridge, MA. [6]

Avise, J. C. 2004. *Molecular Markers, Natural History, and Evolution*, 2nd ed. Sinauer, Sunderland, MA. [9, 12, 17]

Avise, J. C. 2010. Footprints of nonsentient design inside the human genome. *Proc. Natl. Acad. Sci. USA* 107 (suppl. 2): 8969–8976. [23]

Avise, J. C., and R. M. Ball, Jr. 1990. Principles of genealogical concordance in species concepts and biological taxonomy. *Oxford Surv. Evol. Biol.* 7: 45–67. [17]

Axelrod, R., and W. D. Hamilton. 1981. The evolution of cooperation. *Science* 211: 1390–1396. [16]

Ayala, F. J. 1968. Genotype, environment, and population numbers. *Science* 162: 1453–1459. [14]

Azuma, Y., Y. Kumazawa, M. Miya, K. Mabuchi, and N. Nishida. 2008. Mitogenomic evaluation of the historical biogeography of cichlids toward reliable dating of teleostean divergences. *BMC Evol. Biol.* 8, Art. 215 (doi:10.1186/1471-2148-8-215). [6]

B

Babbitt, C. C., L. R. Warner, O. Fedrigo, C. E. Wall, and G. A. Wray. 2011. Genomic signatures of diet-related shifts during human origins. *Proc. R. Soc. Lond. B* 278: 961–969. [12]

Backwell, P. R. Y., J. H. Christy, S. R. Telford, M. D. Jennions, and N. I. Passmore. 2000. Dishonest signaling in a fiddler crab. *Proc. R. Soc. Lond. B* 267: 719–724. [16]

Baer, C. F., M. M. Niyamoto, and D. R. Denver. 2007. Mutation rate variation in multicellular eukaryotes: causes and consequences. *Nat. Rev. Genet.* 8: 619–631. [8]

Bailey, N. W., and M. Zuk. 2009. Same-sex sexual behavior and evolution. *Trends Ecol. Evol.* 24: 439–446. [16]

Bailey, W. J., D. H. A. Fitch, D. A. Tagle, J. Czelusniak, J. L. Slightom, and M. Goodman. 1991. Molecular evolution of the ψη-globin gene locus: Gibbon phylogeny and the hominoid slowdown. *Mol. Biol. Evol.* 8: 155–184. [2]

Baker, C. S., et al. 2000. Predicted decline of protected whales based on molecular genetic monitoring of Japanese and Korean markets. *Proc. R. Soc. Lond. B* 267: 1191–1199. [23]

Bakewell, M. A., P. Shi, and J. Zhang. 2007. More genes underwent positive selection in chimpanzee evolution than in human evolution. *Proc. Natl. Acad. Sci. USA* 104: 7489–7494. [12]

Bakker, E. G., C. Toomajian, M. Kreitman, and J. Bergelson. 2006. A genome-wide survey of *R* gene polymorphisms in *Arabidopsis*. *Plant Cell* 18: 1803–1818. [19]

Balakrishnan, C. N., and S. V. Edwards. 2009. Nucleotide variation, linkage disequilibrium and founder-facilitated speciation in wild populations of the zebra finch (*Taeniopygia guttata*). *Genetics* 181: 645–660. [18]

Baldauf, S. L., D. Bhattacharya, J. Cockrill, P. Hugenholtz, J. Pawlowski, and A. G. B. Simpson. 2004. The tree of life: An overview. In J. Cracraft and M. J. Donoghue (eds.), *Assembling the Tree of Life*, pp. 44–75. Oxford University Press, New York. [2]

Baldwin, B. G., and M. J. Sanderson. 1998. Age and rate of diversification of the Hawaiian silversword alliance (Compositae). *Proc. Natl. Acad. Sci. USA* 95: 9402–9406. [7]

Bambach, R. K. 1985. Classes and adaptive variety: The ecology of diversification in marine faunas through the Phanerozoic. In J. W. Valentine (ed.), *Phanerozoic Diversity Patterns: Profiles in Macroevolution*, pp. 191–253. Princeton University Press, Princeton, NJ. [7]

Bambach, R. K. 2006. Phanerozoic biodiversity mass extinctions. *Annu. Rev. Earth Planet. Sci.* 34: 127–155. [7]

Bambach, R. K., A. H. Knoll, and J. J. Sepkoski, Jr. 2002. Anatomical and ecological constraints on Phanerozoic animal diversity in the marine realm. *Proc. Natl. Acad. Sci. USA* 99: 6854–6859. [7]

Barber, I., S. A. Arnott, V. A. Braithwaite, J. Andrew, and F. Huntingford. 2001. Indirect fitness consequences of mate choice in sticklebacks: offspring of brighter males grow slowly but resist parasitic infections. *Proc. R. Soc. Lond. B* 268: 71–76. [15]

Barbosa-Morais, N. L., and 16 others. 2012. The evolutionary landscape of alternative splicing in vertebrate species. *Science* 338: 1587–1593. [20]

Barbujani, G. 2005. Human races: classifying people vs. understanding diversity. *Curr. Genom.* 6: 215–226. [9]

Barkow, J. H., L. Cosmides, and J. Tooby (eds.). 1992. *The Adapted Mind: Evolutionary Psychology and the Generation of Culture*. Oxford University Press, New York. [23]

Barmina, O. and A. Kopp. 2007. Sex-specific expression of a Hox gene associated with rapid morphological evolution. *Dev. Biol.* 311: 277–286. [21]

Barnosky, A. D., and 11 others. 2011. Has the Earth's sixth mass extinction already arrived? *Nature* 471: 51–57. [5]

Barraclough, T. G., and A. P. Vogler. 2000. Detecting the geographical pattern of speciation from species–level phylogenies. *Am. Nat.* 155: 419–434. [18]

Barrett, R. D. H., and D. Schluter. 2008. Adaptation from standing genetic variation. *Trends Ecol. Evol.* 23: 38–44. [13]

Barrett, S. C. H. 2002. The evolution of plant sexual diversity. *Nat. Rev. Genet.* 3: 274–284. [15]

Barrett, S. C. H. 2010. Understanding plant reproductive diversity. *Phil. Trans. R. Soc. B* 365: 99–109. [15]

Barrett, S. C. H., and D. Charlesworth. 1991. Effects of a change in the level of inbreeding on the genetic load. *Nature* 352: 522–524. [15]

Bartel, D. P. 2004. MicroRNAs: Genomics, biogenesis, mechanism, and function. *Cell* 116: 281–297. [20]

Bartolomé, C., and B. Charlesworth. 2006. Rates and patterns of chromosomal evolution in *Drosophila pseudoobscura* and *D. miranda*. *Genetics* 173: 779–791. [8]

Barton, N. H., and B. Charlesworth. 1984. Genetic revolutions, founder effects, and speciation. *Annu. Rev. Ecol. Syst.* 15: 133–164. [12]

Barton, N. H., and K. S. Gale. 1993. Genetic analysis of hybrid zones. In R. G. Harrison (ed.), *Hybrid Zones and the Evolutionary Process*, pp. 13–45. Oxford University Press, New York. [17]

Barton, N. H., and G. M. Hewitt. 1985. Analysis of hybrid zones. *Annu. Rev. Ecol. Syst.* 16: 113–148. [18]

Barton, N., and L. Partridge. 2000. Limits to natural selection. *BioEssays* 22: 1075–1084. [9, 13]

Basolo, A. L. 1994. The dynamics of Fisherian sex-ratio evolution: Theoretical and experimental investigations. *Am. Nat.* 144: 473–490. [12]

Basolo, A. L. 1995. Phylogenetic evidence for the role of a preexisting bias in sexual selection. *Proc. R. Soc. Lond.* B 259: 307–311. [15]

Basolo, A. L. 1998. Evolutionary change in a receiver bias: A comparison of female preference functions. *Proc. R. Soc. Lond. B* 265: 2223–2228. [15]

Bateman, A. J. 1947. Contamination of seed crops. II. Wind pollination. *Heredity* 1: 235–246. [9]

Baum, D. A., and K. L. Shaw. 1995. Genealogical perspectives on the species problem. In P. C. Hoch and A. G. Stephenson (eds.), *Experimental and Molecular Approaches to Plant Biosystematics*, pp. 289–303. Monographs in Systematic Botany, Missouri Botanical Garden, St. Louis, MO. [17]

Baxter, S. W., S. E. Johnston, and C. D. Jiggins. 2009. Butterfly speciation and the distribution of gene effect sizes fixed during adaptation. *Heredity* 102: 57–65. [22]

Beaumont, M. A., and R. A. Nichols. 1996. Evaluating loci for use in the genetic analysis of population structure. *Proc. R. Soc. Lond.* B 263: 1619–1626. [12]

Becerra, A., L. Delayre, S. Islas, and A. Lazcano. 2007. The very early stages of biological evolution and the nature of the last common ancestor of the three major cell domains. *Annu. Rev. Ecol. Evol. Syst.* 38: 361–379. [5]

Becerra, J. X. 1997. Insects on plants: Macroevolutionary chemical trends in host use. *Science* 276: 253–256. [19]

Becerra, J. X., and D. L. Venable. 1999. Macroevolution of insect-plant associations: The relevance of host biogeography to host affiliation. *Proc. Natl. Acad. Sci. USA* 96: 12626–12631. [19]

Beck, N. R., M. C. Double, and A. Cockburn. 2003. Microsatellite evolution at two hypervariable loci revealed by extensive avian pedigrees. *Mol. Biol. Evol.* 20: 54–61. [8]

Bégin, M., and D. A. Roff. 2004. From micro- to macroevolution through quantitative genetic variation: Positive evidence from field crickets. *Evolution* 58: 2287–2304. [13]

Behrensmeyer, A. K., J. D. Damuth, W. A. DiMichele, R. Potts, H.-D. Sues, and S. L. Wing (eds.). 1992. *Terrestrial Ecosystems Through Time: Evolutionary Paleoecology of Terrestrial Plants and Animals*. University of Chicago Press, Chicago. [5]

Bejerano, G., M. Pheasant, I. Makurin, S. Stephen, W. J. Kent, J. S. Mattick, and D. Haussler. 2004. Ultraconserved elements in the human genome. *Science* 304: 1321–1325. [12]

Bejerano, G., and 8 others. 2006. A distal enhancer and an ultraconserved exon are derived from a novel retroposon. *Nature* 441: 87–90. [20]

Beldade, P., K. Koops, and P. M. Brakefield. 2002. Developmental constraints versus flexibility in morphological evolution. *Nature* 416: 844–847. [21]

Bell, G., and A. Gonzalez. 2009. Evolutionary rescue can prevent extinction following environmental change. *Ecol. Lett.* 12: 942–948. [13]

Bell, G., and V. Koufopanou. 1986. The cost of reproduction. *Oxford Surv. Evol. Biol.* 3: 83–131. [14]

Bell, M. A., J. V. Baumgartner, and E. C. Olson. 1985. Patterns of temporal change in single morphological characters of a Miocene stickleback fish. *Paleobiology* 11: 258–271. [4]

Benagiano, G., M. Farris, and G. Grudzinskas. 2010. Fate of fertilized human oocytes. *Reprod. Med. Online* 21: 732–741. [12]

Bennett, A. F., R. E. Lenski, and J. E. Mittler. 1992. Evolutionary adaptation to temperature. I. Fitness responses of *Escherichia coli* to changes in its thermal environment. *Evolution* 46: 16–30. [8]

Bennetzen, J. L. 2000. Transposable element contributions to plant gene and genome evolution. *Plant Mol. Biol.* 42: 251–269. [8]

Benton, M. J. (ed.). 1988. *The Phylogeny and Classification of the Tetrapods*. Clarendon Press, Oxford. [5]

Benton, M. J. 1996. On the nonprevalence of competitive replacement in the evolution of tetrapods. In D. Jablonski, D. H. Erwin, and J. Lipps (eds.), *Evolutionary Paleobiology*, pp. 185–210. University of Chicago Press, Chicago. [7]

Benton, M. J. 2009. The Red Queen and the court jester: species diversity and the role of biotic and abiotic factors through time. *Science* 323: 728–732. [7]

Benton, M. J., and R. Hitchin. 1997. Congruence between phylogenetic and stratigraphic data on the history of life. *Proc. R. Soc. Lond. B* 264: 885–890. [4]

Benton, M. J., M. A. Wills, and R. Hitchin. 2000. Quality of the fossil record through time. *Nature* 403: 534–537. [4]

Berenbaum, M. R., and A. R. Zangerl. 1988. Stalemates in the coevolutionary arms race: Synthesis, synergisms, and sundry other sins. In K. C. Spencer (ed.), *Chemical Mediation of Coevolution*, pp. 113–132. Academic Press, San Diego, CA. [19]

Berenbaum, M. R., A. R. Zangerl, and J. K. Nitao. 1986. Constraints on chemical coevolution: Wild parsnips and the parsnip webworm. *Evolution* 40: 1215–1228. [19]

Bergelson, J., M. Kreitman, E. A. Stahl, and D. C. Tian. 2001. Evolutionary dynamics of plant *R*-genes. *Science* 292: 2281–2285. [19]

Bergthorsson, U., D. I. Andersson, and J. R. Roth. 2007. Ohno's dilemma: Evolution of new genes under continuous selection. *Proc. Natl. Acad. Sci. USA* 104: 17004–17009. [20]

Berkman, M. B., and E. Plutzer. 2011. Defeating creationism in the courtroom, but not in the classroom. *Science* 331: 404–405. [23]

Bermingham, E., and J. C. Avise. 1986. Molecular zoogeography of freshwater fishes in the southeastern United States. *Genetics* 113: 939–965. [18]

Bernagiano, G., M. Farris, and G. Grudzinskas. 2010. Fate of fertilized human oocytes. *Reprod. Biomed. Online* 21: 732–741 (doi:10.1026/rbmo.2010.08.011). [12]

Bernays, E. A., and R. F. Chapman. 1994. *Host-Plant Selection by Phytophagous Insects*. Chapman & Hall, London. [19]

Bernays, E. A., and M. Graham. 1988. On the evolution of host specificity by phytophagous arthropods. *Ecology* 69: 886–892. [19]

Bersaglieri, T., and 8 others. 2004. Genetic signatures of strong recent positive selection at the lactase gene. *Am. J. Hum. Genet.* 74: 1111–1120. [12]

Berthold, P., A. J. Heibig, G. Mohr, and U. Querner. 1992. Rapid micro-evolution of migratory behavior in a wild bird species. *Nature* 360: 668–670. [11]

Bharathan, G., and R. Sinha. 2001. The regulation of compound leaf development. *Plant Physiol.* 127: 1–5. [21]

Bikard, D., and 6 others. 2009. Divergent evolution of duplicate genes leads to genetic incompatibilities within *A. thaliana. Science* 323: 623–626. [17]

Birch, L. C., T. Dobzhansky, P. D. Elliott, and R. C. Lewontin. 1963. Relative fitness of geographic races of *Drosophila serrata. Evolution* 17: 72–83. [14]

Birkhead, T. R., and A. P. Møller. 1992. *Sperm Competition in Birds: Evolutionary Causes and Consequences.* Academic Press, London. [15]

Birkhead, T. R., D. J. Hosken, and S. Pitnick (eds.). 2009. *Sperm Biology: An Evolutionary Perspective.* Academic Press, Oxford. [15]

Bishop, J. A. 1981. A neo-Darwinian approach to resistance: Examples from mammals. In J. A. Bishop and L. M. Cook (eds.), *Genetic Consequences of Man Made Change,* pp. 37–51. Academic Press, London. [12]

Blount, Z. D., C. Z. Borland, and R. E. Lenski. 2008. Historical contingency and the evolution of a key innovation in an experimental population of *Escherichia coli. Proc. Natl. Acad. Sci. USA* 105: 7899–7906. [8, 13, 22]

Blows, M. W., and A. A. Hoffmann. 2005. A reassessment of genetic limits to evolutionary change. *Ecology* 86: 1371–1384. [13]

Blows, M. W., S. Chenoweth, and E. Hine. 2004. Orientation of the genetic variance–covariance matrix and the fitness surface for multiple male sexually selected traits. *Am. Nat.* 163: 329–340. [13]

Boag, P. T. 1983. The heritability of external morphology in Darwin's ground finches (*Geospiza*) on Isla Daphne Major, Galápagos. *Evolution* 37: 877–894. [9]

Bodmer, W., and M. Ashburner. 1984. Conservation and change in the DNA sequences coding for alcohol dehydrogenase in sibling species of *Drosophila. Nature* 309: 425–540. [8]

Boehm, C. 2012. *Moral Origins: The Evolution of Virtue, Altruism, and Shame.* Basic Books, New York. [23]

Bolnick, D. I., and B. M. Fitzpatrick. 2007. Sympatric speciation: Models and empirical evidence. *Annu. Rev. Ecol. Evol. Syst.* 38: 459–487. [18]

Bolnick, D. I., R. Svanbäck, J. A. Fordyce, L. H. Yang, J. M. Davis, C. D. Hulsey, and M. L. Forister. 2003. The ecology of individuals: incidence and implications of individual specialization. *Am. Nat.* 161: 1–28. [12, 19]

Bonaparte, J. F. 1978. El Mesozoico de America del Sur y sus tetrapodos. *Opera Lilloana* 26: 1–596. [5]

Bondurianský, R., and S. F. Chenoweth. 2009. Intralocus sexual conflict. *Trends Ecol. Evol.* 24: 280–288. [12, 13]

Bondurianský, R., and T. Day. 2009. Nongenetic inheritance and its evolutionary implications. *Annu. Rev. Ecol. Evol. Syst.* 40: 103–125. [9]

Bonnell, M. L., and R. K. Selander. 1974. Elephant seals: Genetic variation and near extinction. *Science* 184: 908–909. [10]

Bonner, J. T. 1988. *The Evolution of Complexity.* Princeton University Press, Princeton, NJ. [21]

Borgerhoff Mulder, M. 1990. Kipsigis women's preferences for wealthy men: evidence for female choice in mammals? *Behav. Ecol. Sociobiol.* 27: 255–264. [16]

Borries, C., K. Launhardt, C. Epplen, J. T. Epplen, and P. Winkler. 1999. DNA analyses support the hypothesis that infanticide is adaptive in langur monkeys. *Proc. R. Soc. Lond. B* 266: 901–904. [16]

Bouchard, T. J. Jr., D. T. Lykken, M. McGue, N. L. Segal, and A. Tellegen. 1990. Sources of human psychological differences: The Minnesota study of twins reared apart. *Science* 250: 223–228. [9]

Boucot, A. J. 1975. *Evolution and Extinction Rate Controls.* Elsevier, Amsterdam. [7]

Boureau, E. 1964. *Traité de paléobotanique,* Vol. III. Masson, Paris. [5]

Bourke, A. F. G. 2011. *Principles of Social Evolution.* Oxford University Press, Oxford. [5, 16]

Bourke, A. F. G., and N. R. Franks. 1995. *Social Evolution in Ants.* Princeton University Press, Princeton, NJ. [16]

Bowler, P. J. 1989. *Evolution: The History of an Idea.* University of California Press, Berkeley. [1]

Bowler, P. J. 1996. *Life's Splendid Drama: Evolutionary Biology and the Reconstruction of Life's Ancestry 1860–1940.* University of Chicago Press. Chicago. [1]

Bowman, R. I. 1961. *Morphological Differentiation and Adaptation in the Galápagos Finches.* University of California Press, Berkeley. [22]

Boyce, M. S., and C. J. Perrins. 1987. Optimizing great tit clutch size in a fluctuating environment. *Ecology* 68: 142–153. [17]

Boyd, R., and P. J. Richerson. 1985. *Culture and the Evolutionary Process.* University of Chicago Press, Chicago. [23]

Boyd, R., and P. J. Richerson. 2009. Culture and the evolution of human cooperation. *Phil. Trans. R. Soc. B* 364: 3281–3288. [16]

Boyko, A. R., and 13 others. 2008. Assessing the evolutionary impact of amino acid mutations in the human genome. *PLoS Genet.* 4, issue 5, article e1000083, pp. 1–13. [10, 12]

Bradbury, J. W., and S. L. Vehrencamp. 1998. *Principles of Animal Communication.* Sinauer, Sunderland, MA. [16]

Bradshaw, A. D. 1991. Genostasis and the limits to evolution. The Croonian Lecture, 1991. *Phil. Trans. R. Soc. Lond.* B 333: 289–305. [8, 13]

Bradshaw, H. D., and D. W. Schemske. 2003. Allele substitution at a flower colour locus produces a pollinator shift in monkeyflowers. *Nature* 426: 176–178. [17]

Bradshaw, H. D., S. M. Wilbert, K. G. Otto, and D. W. Schemske. 1998. Quantitative trait loci affecting differences in floral morphology between two species of monkeyflower. *Genetics* 149: 367–382 [17]

Bradshaw, W. E., and C. M. Holzapfel. 2001. Genetic shift in photoperiodic response correlated with global warming. *Proc. Natl. Acad. Sci. USA* 98: 14509–14511. [11]

Brakefield, P. M., J. Gates, D. Keys, F. Kesbeke, P. J. Wijngaarden, A. Montelro, V. French, and S. B. Carroll. 1996. Development, plasticity and evolution of butterfly eyespot patterns. *Nature* 384: 236–242. [21]

Branch, G., E. C. Scott, and J. Rosenau. 2010. Dispatches from the evolution wars: shifting tactics and expanding battlefields. *Annu. Rev. Genomics Hum. Genet.* 11: 317–338. [23]

Brayard, A., and 7 others. 2009. Good genes and good luck: ammonoid diversity and the end-Permian mass extinction. *Science* 325: 1118–1121. [7]

Brideau, N. J., H. A. Flores, J. Wang, S. Maheshwari, X. Wang, and D. A. Barbash. 2006. Two Dobzhansky-Muller genes interact to cause hybrid lethality in *Drosophila. Science* 314: 1292–1295. [17]

Bridle, J. R., and T. H. Vines. 2007. Limits to evolution at range margins: When and why does adaptation fail? *Trends Ecol. Evol.* 22: 140–147. [6]

Bridle, J. R., S. Gavaz, and W. J. Kennington. 2009. Testing limits to adaptation along altitudinal gradients in rainforest *Drosophila. Proc. R. Soc. Lond. B* 276: 1507–1515. [12]

Briggs, D. E. G., and P. R. Crowther (eds.). 1990. *Palaeobiology: A Synthesis.* Blackwell Scientific, Oxford. [4]

Briskie, J. V., and M. Mackintosh. 2004. Hatching failure increases with severity of population bottlenecks in birds. *Proc. Natl. Acad. Sci. USA* 101: 558–561. [10]

Brodie, E. D. III. 1992. Correlational selection for color pattern and antipredator behavior in the garter snake *Thamnophis ordinoides. Evolution* 46: 1284–1298. [13]

Brodie, E. D. III. 1993. Homogeneity of the genetic variance-covariance matrix for antipredator traits in two natural populations of the garter snake *Thamnophis ordinoides. Evolution* 47: 844–854. [13, 16]

Brodie, E. D., Jr., B. J. Ridenhour, and E. D. Brodie III. 2002. The evolutionary response of predators to dangerous prey: Hotspots and coldspots in the geographic mosaic of coevolution between garter snakes and newts. *Evolution* 56: 2067–2082. [19]

Bro-Jorgensen, J. 2007. The intensity of sexual selection predicts weapon size in male bovids. *Evolution* 61: 1316–1326. [15]

Bromham, L. 2009. Why do species vary in their rate of molecular evolution? *Biol. Lett.* 5: 401–404. [10]

Brooks, D. R., and D. A. McLennan. 2002. *The Nature of Diversity: An Evolutionary Voyage of Discovery.* University of Chicago Press, Chicago. [3]

Brooks, L. D. 1988. The evolution of recombination rates. In B. R. Levin and R. E. Michod (eds.), *The Evolution of Sex,* pp. 87–105. Sinauer, Sunderland, MA. [15]

Brown, G. P., C. Shilton, B. L. Phillips, and R. Shine. 2007. Invasion, stress, and spinal arthritis in cane toads. *Proc. Natl. Acad. Sci. USA* 104: 17698–17700. [11]

Brown, G. R., T. E. Dickins, R. Sear and K. N. Laland. 2011. Evolutionary accounts of human behavioural diversity. *Phil. Trans. R. Soc. B* 366: 313–324. [16]

Brown, J. H. 1995. *Macroecology.* University of Chicago Press, Chicago. [7]

Brown, J. H., and A. C. Gibson. 1983. *Biogeography.* Mosby, St. Louis. [6]

Brown, J. H., and M. V. Lomolino. 1998. *Biogeography,* 2nd ed. Sinauer, Sunderland, MA [5, 6]

Brown, J. K. M., and W. A. Tellier. 2011. Plant-parasite coevolution: bridging the gap between genetics and ecology. *Annu. Rev. Phytopath.* 49: 345–367. [19]

Brown, W. L. Jr., and E. O. Wilson. 1956. Character displacement. *Syst. Zool.* 5: 49–64. [18, 19]

Brown, W. M., J. M. George, and A. C. Wilson. 1979. Rapid evolution of animal mitochondrial DNA. *Proc. Natl. Acad. Sci. USA* 76: 1967–1971. [10]

Brusca, R. C., and G. J. Brusca. 1990. *Invertebrates.* Sinauer, Sunderland, MA. [3]

Buchanan, K. L., and C. K. Catchpole. 2000. Song as an indicator of male parental effort in the sedge warbler. *Proc. R. Soc. Lond.* 267: 321–326. [16]

Bull, J. J. 1994. Perspective: Virulence. *Evolution* 48: 1423–1437. [19]

Bull, J. J., and E. L. Charnov. 1985. On irreversible evolution. *Evolution* 39: 1149–1155. [22]

Bull, J. J., and W. R. Rice. 1991. Distinguishing mechanisms for the evolution of cooperation. *J. Theor. Biol.* 149: 63–74. [19]

Bull, J. J., and H. A. Wichman. 2001. Applied evolution. *Annu. Rev. Ecol. Syst.* 32:183–217. [23]

Bull, J. J., I. J. Molineaux, and W. R. Rice. 1991. Selection of benevolence in a host–parasite system. *Evolution* 45:875–882. [19]

Burbrink, F. T., R. Lawson, and J. B. Slowinski. 2000. Mitochondrial DNA phylogeography of the polytypic North American rat snake (*Elaphe obsoleta*): a critique of the subspecies concept. *Evolution* 54:2107–2118. [9]

Buri, P. 1956. Gene frequency drift in small population of mutant *Drosophila*. *Evolution* 10: 367–402. [10]

Burt, A., and R. Trivers. 2006. *Genes in Conflict: The Biology of Selfish Genetic Elements.* Harvard University Press, Cambridge, MA. [11, 16]

Bush, A. M., and R. K. Bambach. 2011. Paleoecologic megatrends in marine metazoan. *Annu. Rev. Earth Planet. Sci.* 39: 241–269. [7]

Bush, G. L. 1969. Sympatric host race formation and speciation in frugivorous flies of the genus *Rhagoletis* (Diptera, Tephritidae). *Evolution* 23: 237–251. [18]

Bush, R. M., C. A. Bender, K. Subbarao, N. J. Cox, and W. M. Fitch. 1999. Predicting the evolution of human influenza A. *Science* 286: 1921–1925. [23]

Buss, D. M. 1989. Sex differences in human mate preferences: evolutionary hypotheses tested in 37 cultures. *Behav. Brain Sci.* 12: 1–49. [16]

Buss, D. M. 1994. *The Evolution of Desire: Strategies of Human Mating.* HarperCollins, New York. [23]

Buss, D. M. 1998. Sexual strategies theory: historical origins and current status. *J. Sex Research* 35: 19–31. [16]

Buss, D. M. 1999. Sex differences in human mate preferences: Evolutionary hypotheses tested in 37 cultures. *Behavioral and Brain Sciences* 12: 1–49. [16]

Buss, D. M. 2007. *Evolutionary Psychology: The New Science of the Mind.* Addison Wesley, Boston, Mass. [16]

Buss, L. W. 1987. *The Evolution of Individuality.* Princeton University Press, Princeton, NJ. [16]

Bustamante, C. D., and 13 others. 2005. Natural selection on protein-coding genes in the human genome. *Nature* 437: 1153–1157. [10, 12]

Butlin, R., and 25 others (The Marie Curie Speciation Network). 2012. What do we need to know about speciation? *Trends Ecol. Evol.* 27: 27–39. [18]

Byars, S. G., D. Ewbank, D. R. Govindaraju, and S. C. Stearns. 2010. Natural selection in a contemporary human population. *Proc. Natl. Acad. Sci. USA* 107 (Suppl. 1): 1787–1792. [13]

Byrne, M., M. Timmermans, C. Kidner, and R. Martienssen. 2001. Development of leaf shape *Curr. Opin. Plant Biol.* 4: 38–43. [21]

C

Cai, J., R. Zhao, H. Jiang, and W. Wang. 2008. De novo origination of a new protein-coding gene in *Saccharomyces cerevisiae. Genetics* 179: 487–496. [20]

Calboli, F. C. F., G. W. Gilchrist, and L. Partridge. 2003. Different cell size and cell number contribution in two newly established and one ancient body size cline of *Drosophila subobscura. Evolution* 57: 566–573. [21]

Camperio-Ciani, A., F. Corna, and C. Capiluppi. 2004. Evidence for maternally inherited factors favouring male homosexuality and promoting female fecundity. *Proc. R. Soc. Lond. B* 271: 2217–2221. [16]

Camperio-Ciani, A., P. Cermelli, and G. Zanzotto. 2008. Sexually antagonistic selection in human male homosexuality. *PLoS One* 3: 1–8 (e2282). [16]

Candolin, U., and J. Heuschele. 2008. Is sexual selection beneficial during adaptation to climate change? *Trends Ecol. Evol.* 23: 446–452. [15]

Cann, R. L., M. Stoneking, and A. C. Wilson. 1987. Mitochondrial DNA and human evolution. *Nature* 325: 31–36. [6]

Carbone, M. A., A. Llopart, M. deAngelis, J. A. Coyne, and T. F. Mackay. 2005. Quantitative trait loci affecting the difference in pigmentation between *Drosophila yakuba* and *D. santomea. Genetics* 171: 211–225. [21]

Carlquist, S., B. G. Baldwin, and G. E. Carr (eds.). 2003. *Tarweeds and Silverswords: Evolution of the Madiinae (Asteraceae).* Missouri Botanical Garden Press, St. Louis, MO. [3]

Carmel, L., Y. I. Wolf, I. B. Rogozin, and E. V. Koonin. 2007. Three distinct modes of intron dynamics in the evolution of eukaryotes. *Genome Res.* 17: 1034–1044. [20]

Carreira, V. P., I. M. Soto, J. Mensch, and J. J. Fanara. 2011. Genetic basis of wing morphogenesis in *Drosophila*: sexual dimorphism and non-allometric effects of shape variation. *BMC Dev. Biol.* 11: 32–48. [21]

Carroll, R. L. 1988. *Vertebrate Paleontology and Evolution.* W. H. Freeman, New York. [4, 5]

Carroll, S. B. 2003. Genetics and the making of *Homo sapiens. Nature* 422: 849–857. [21]

Carroll, S. B. 2006. *The Making of the Fittest: DNA as the Ultimate Forensic Record of Evolution.* W. W. Norton, New York. [3]

Carroll, S. B., and C. Boyd. 1992. Host race radiation in the soapberry bug: Natural history with the history. *Evolution* 46: 1052–1069. [11]

Carroll, S. B., S. D. Weatherbee, and J. A. Langeland. 1995. Homeotic genes and the regulation and evolution of insect wing number. *Nature* 375: 58–61. [21]

Carroll, S. B., J. K. Grenier, and S. D. Weatherbee. 2001. *From DNA to Diversity: Molecular Genetics and the Evolution of Animal Design.* Blackwell Science, Malden, MA. [21, 22]

Carroll, S. B., J. K. Grenier, and S. D. Weatherbee. 2005. *From DNA to Diversity: Molecular Genetics and the Evolution of Animal Design,* 2nd ed. Blackwell Science, Malden, MA. [21]

Carroll, S. P., H. Dingle, and S. P. Klassen. 1997. Genetic differentiation of fitness-associated traits among rapidly evolving populations of the soapberry bug. *Evolution* 51: 1182–1188. [11]

Carson, H. L. 1967. Selection for parthenogenesis in *Drosophila mercatorum. Genetics* 55: 157–171. [15]

Carson, H. L. 1975. The genetics of speciation at the diploid level. *Am. Nat.* 109: 73–92. [18]

Carson, H. L., and B. A. Clague. 1995. Geology and biogeography of the Hawaiian Islands. In W. L. Wagner and V. A. Funk (eds.), *Hawaiian Biogeography*, pp. 14–29. Smithsonian Institution Press, Washington, DC. [6]

Carson, H. L., and K. Y. Kaneshiro. 1976. *Drosophila* of Hawaii: Systematics and ecological genetics. *Annu. Rev. Ecol. Syst.* 7: 311–345. [3]

Carstens, B. C., and L. L. Knowles. 2007. Estimating species phylogeny from gene-tree probabilities despite incomplete lineage sorting: an example from *Melanoplus* grasshoppers. *Syst. Biol.* 56: 400–411. [2]

Cartwright, J. 2000. *Evolution and Human Behavior: Darwinian Perspectives on Human Nature.* MIT Press, Cambridge, MA. [16]

Castellanos, M. C., P. Wilson, and J. D. Thomson. 2004. 'Anti-bee' and 'pro-bird' changes during the evolution of hummingbird pollination in *Penstemon* flowers. *J. Evol. Biol.* 17: 876–885. [11]

Cavalier-Smith, T. 2006. Cell evolution and Earth history: Stasis and revolution. *Phil. Trans. R. Soc. Lond. B* 361: 969–1006. [5]

Cavalli-Sforza, L. L., and W. F. Bodmer. 1971. *The Genetics of Human Populations.* W. H. Freeman, San Francisco. [9, 12, 13]

Cavalli-Sforza, L. L., and M. W. Feldman. 1981. *Cultural Transmission and Evolution: A Quantitative Approach.* Princeton University Press, Princeton, NJ. [23]

Cavender-Bares, J., D. D. Ackerly, D. A. Baum, and F. A. Bazzaz. 2004. Phylogenetic overdispersion in Floridian oak communities. *Am. Nat.* 163: 823–843. [19]

Cavender-Bares, J., A. Keen, and B. Miles. 2006. Phylogenetic structure of Floridian plant communities depends on taxonomic and spatial scale. *Ecology* 87: S109–S122. [19]

Cavender-Bares, J., K. H. Kozak, P. V. A. Fine, and S. W. Kembel. 2009. The merging of community ecology and phylogenetic biology. *Ecol. Lett.* 12: 693–715. [19]

Chaline, J., and B. Laurin. 1986. Phyletic gradualism in a European Plio–Pleistocene *Mimomys* lineage (Arvicolidae, Rodentia). *Paleobiology* 12: 203–216. [4]

Chan, Y. F., and 15 others. 2010. Adaptive evolution of pelvic reduction in sticklebacks by recurrent deletion of a *Pitx1* enhancer. *Science* 327: 302–305. [21]

Chanderbali, A. S., and 11 others. 2010. Conservation and canalization of gene expression during angiosperm diversification accompany the origin and evolution of the flower. *Proc. Natl. Acad. Sci. USA* 107: 22570–22575. [21]

Chang, B. S. W., K. Jönsson, M. A. Kazmi, M. J. Donoghue, and T. P. Sakmar. 2002. Recreating a functional ancestral archosaur visual pigment. *Mol. Biol. Evol.* 19: 1483–1489. [3]

Charlesworth, B. 1990. The evolutionary genetics of adaptation. In M. Nitecki (ed.), *Evolutionary Innovations*, pp. 47–70. University of Chicago Press, Chicago. [22]

Charlesworth, B. 1994a. The effect of background selection against deleterious mutations on weakly selected, linked variants. *Genet. Res.* 63: 213–227. [12, 14]

Charlesworth, B. 1994b. *Evolution in Age-Structured Populations.* Cambridge University Press, Cambridge. [14]

Charlesworth, B., and D. Charlesworth. 1978. A model for the evolution of dioecy and gynodioecy. *Am. Nat.* 112: 975–997. [15]

Charlesworth, B., and C. H. Langley. 1989. The population genetics of *Drosophila* transposable elements. *Annu. Rev. Genet.* 23: 251–287. [20]

Charlesworth, B., and S. Rouhani. 1988. The probability of peak shifts in a founder population. II. An additive polygenic trait. *Evolution* 42: 1129–1145. [18]

Charlesworth, B., R. Lande, and M. Slatkin. 1982. A neo-Darwinian commentary on macroevolution. *Evolution* 36: 474–498. [18, 22]

Charlesworth, B., M. T. Morgan, and D. Charlesworth. 1993. The effect of deleterious mutations on neutral molecular variation. *Genetics* 134: 1289–1303. [12]

Charlesworth, B., D. Charlesworth, and N. H. Barton. 2003. The effects of genetic and geographic structure on neutral variation. *Annu. Rev. Ecol. Evol. Syst.* 34: 99–125. [10]

Charnov, E. L. 1982. *The Theory of Sex Allocation.* Princeton University Press, Princeton, NJ. [12, 15]

Cheetham, A. H. 1987. Tempo of evolution in a Neogene bryozoan: Are trends in single morphological characters misleading? *Paleobiology* 13: 286–296. [4]

Chen, C. H., H. Huang, C. M. Ward, J. T. Su, L. V. Schaeffer, M. Guo, and B. A. Hay. 2007. A synthetic maternal-effect selfish genetic element drives population replacement in *Drosophila. Science* 316: 597–600. [23]

Cheney, D. L. 2011. Extent and limits of cooperation in animals. *Proc. Natl. Acad. Sci. USA* 108: 10902–10909. [16]

Cheney, D. L., and R. M. Seyfarth. 2007. *Baboon Metaphysics: The Evolution of a Social Mind.* University of Chicago Press, Chicago. [23]

Cheng, C.-H., and L. Chen. 1999. Evolution of an antifreeze glycoprotein. *Nature* 401: 443–444. [20]

Chevin, L. M., R. Lande, and G. M. Mace. 2010. Adaptation, plasticity, and extinction in a changing environment: towards a predictive theory. *PLoS Biol.* 8(40), article e10000357. doi:10.1371/journal.pbio.10000357. [13]

Chiappe, L. M. 2007. *Glorified Dinosaurs: The Origin and Early Evolution of Birds.* John Wiley and Sons, New York. [4]

Chiappe, L. M., and G. J. Dyke. 2002. The Mesozoic radiation of birds. *Annu. Rev. Ecol. Syst.* 33: 91–124. [4]

Chippindale, P. T., R. M. Bonett, A. S. Baldwin, and J. J. Wiens. 2004. Phylogenetic evidence for a major reversal of life-history evolution in plethodontid salamanders. *Evolution* 58: 2809–2822. [3]

Chouteau, M., and B. Angers. 2012. Wright's shifting balance theory and the diversification of aposematic signals. *PLoS One* 7(3): e34028. doi:10.1371/journal.pone.0034028. [19]

Christiansen, F. B. 1984. The definition and measurement of fitness. In B. Shorrocks (ed.), *Evolutionary Ecology*, pp. 65–79. Blackwell Scientific, Oxford. [12]

Christin, P.-A., T. L. Sage, E. J. Edwards, R. M. Ogburn, R. Khoshravesh, and R. W. Sage. 2011. Complex evolutionary transitions and the significance of C_3-C_4 intermediate forms of photosynthesis in Molluginaceae. *Evolution* 65: 643–660. [3]

Civetta, A., and A. G. Clark. 2000. Correlated effects of sperm competition and postmating female mortality. *Proc. Natl. Acad. Sci. USA* 97: 13162–13165. [15]

Clack, J. A. 2002. *Gaining Ground: The Origin and Evolution of Tetrapods.* Indiana University Press, Bloomington. [4]

Clark, A. G., and D. J. Begun. 1998. Female genotypes affect sperm displacement in *Drosophila. Genetics* 149: 1487–1493. [15]

Clark, A. G. and the *Drosophila* 12 Genomes Consortium. 2007. Evolution of genes and genomes on the *Drosophila* phylogeny. *Nature* 450: 203–218. [20]

Clark, A. G., and 16 others. 2003. Inferring non-neutral evolution from human-chimp-mouse orthologous gene trios. *Science* 302: 1960–1963. [12, 20]

Clausen, J., D. D. Keck, and W. M. Hiesey. 1940. Experimental studies on the nature of species. I. Effect of varied environments on western North American plants. Carnegie Institution of Washington Publication no. 520: 1–452. [9]

Clausen, J., D. D. Keck, and W. M. Hiesey. 1947. Heredity of geographically and ecologically isolated races. *Am. Nat.* 81: 114–133. [9]

Clayton, D. H., S. E. Bush, B. M. Goates, and K. P. Johnson. 2003. Host defense reinforces host–parasite coevolution. *Proc. Natl. Acad. Sci. USA* 100: 15694–15699. [19]

Clegg, S. M., S. M. Degnan, C. Moritz, A. Estoup, J. Kikkawa, and I. P. F. Owens. 2002. Microevolution in island forms: The role of drift and directional selection in morphological divergence of a passerine bird. *Evolution* 56: 2090–2099. [13]

Clements, J., Z. Lu, W. J. Gehring, I. A. Meinertzhagen, and P. Callaerts. 2008. Central projections of photoreceptor axons originating from ectopic eyes in *Drosophila*. *Proc. Natl. Acad. Sci. USA* 105: 8968–8973. [21]

Clutton-Brock, T. H. 1991. *The Evolution of Parental Care*. Princeton University Press, Princeton, NJ. [16]

Clutton-Brock, T. H. 2002. Breeding together: Kin selection and mutualism in cooperative vertebrates. *Science* 296: 69–72. [16]

Clutton-Brock, T. 2007. Sexual selection in males and females. *Science* 318: 1882–1885. [15]

Clutton-Brock, T. 2009. Cooperation between non-kin in animal societies. *Nature* 462: 51–57. [16]

Clutton-Brock, T., and C. Godfray. 1991. Parental investment. In J. R. Krebs and N. B. Davies (eds.), *Behavioural Ecology: An Evolutionary Approach*, third edition, pp. 234–262. Blackwell Scientific, Oxford. [16]

Cockburn, A. 2006. Prevalence of different modes of parental care in birds. *Proc. R. Soc. Lond. B* 273: 1375–1383. [16]

Coen, E. S., and E. M. Meyerowitz. 1991. The war of the whorls: genetic interactions controlling flower development. *Nature* 353: 31–37. [21]

Cohan, F. M. 1984. Can uniform selection retard random genetic divergence between isolated conspecific populations? *Evolution* 38: 495–504. [8]

Colbert, E. H. 1980. *Evolution of the Vertebrates*, 3rd ed. Wiley, New York. [4]

Colbourne, J. K., and 65 others. 2011. The ecoresponsive genome of *Daphnia pulex*. *Science* 331: 555–561. [3]

Cole, C. T. 2003. Genetic variation in rare and common plants. *Annu. Rev. Ecol. Evol. Syst.* 34: 213–237. [10]

Colegrave, N. 2002. Sex releases the speed limit on evolution. *Nature* 420: 664–666. [15]

Coley, P. D., J. P. Bryant, and F. S. Chapin III. 1985. Resource availability and plant antiherbivore defense. *Science* 230: 895–899. [19]

Collin, R., and M. P. Miglietta. 2008. Reversing opinions on Dollo's Law. *Trends Ecol. Evol.* 23: 602–609. [3, 21]

Colosimo, P. F., and 9 others. 2005. Widespread parallel evolution in sticklebacks by repeated fixation of *Ectodysplasin* alleles. *Science* 307: 1928–1933. [13, 21]

Coltman, D. W., P. O'Donoghue, J. T. Jorgenson, J. T. Hogg, C. Strobeck, and M. Festa-Blanchet. 2003. Undesirable evolutionary consequences of trophy hunting. *Nature* 426: 655–658. [11]

Conant, R. 1958. *A Field Guide to Reptiles and Amphibians*. Houghton Mifflin, Boston, MA. [9]

Confer, J. C., and 6 others. 2010. Evolutionary psychology: Controversies, questions, prospects, and limitations. *Amer. Psychol.* 65: 110–126. [23]

Connell, J. H. 1971. On the role of natural enemies in preventing competitive exclusion in some marine animals and in rain forest trees. In P. J. den Boer and G. Gradwell (eds.), *Dynamics in Populations*, pp. 298–312. Centre for Agricultural Publishing and Documentation, Wageningen, the Netherlands. [19]

Conner, J. K., K. Karoly, C. Stewart, V. A. Koelling, H. F. Sahli, and F. H. Shaw. 2011. Rapid independent trait evolution despite a strong pleiotropic genetic correlation. *Am. Nat.* 178: 429–441. [22]

Conway Morris, S. 2003. *Life's Solution: Inevitable Humans in a Lonely Universe*. Cambridge University Press, Cambridge, UK. [22]

Conway Morris, S. (ed.) 2008. *The Deep Structure of Biology: Is Convergence Sufficiently Ubiquitous to Give a Directional Signal?* Templeton Foundation Press, West Conshohocken, Penn. [22]

Cook, C. D. K. 1968. Phenotypic plasticity with particular reference to three amphibious plant species. In V. Heywood (ed.), *Modern Methods in Plant Taxonomy*, pp. 97–111. Academic Press, London. [13]

Cook, L. G., and M. D. Crisp. 2005. Not so ancient: The extant crown group of *Nothofagus* represents a post-Gondwanan radiation. *Proc. R. Soc. Lond. B* 272: 2535–2544. [6]

Cook, L. M. 2003. The rise and fall of the *carbonaria* form of the peppered moth. *Q. Rev. Biol.* 78: 399–417. [12]

Cook, L. M., B. S. Grant, I. J. Saccheri, and J. Mallet. 2012. Selective bird predation on the peppered moth: the last experiment of Michael Majerus. *Biol. Lett.* doi:10.1098/rsbl.2011.1136. [12]

Coop, G., and 9 others. 2009. The role of geography in human adaptation. *PLoS Genet.* 5(6): e1000500. doi:10.1371/journal.pgen.1000500. [12]

Coope, G. R. 1979. Late Cenozoic fossil Coleoptera: Evolution, biogeography, and ecology. *Annu. Rev. Ecol. Syst.* 10: 249–267. [5]

Cooper, S. B. J., K. M. Ibrahim, and G. M. Hewitt. 1995. Postglacial expansion and genome subdivision in the European grasshopper *Chorthippus parallelus*. *Mol. Ecol.* 4: 49–60. [6]

Cooper, T. F. 2007. Recombination speeds adaptation by reducing competition between beneficial mutations in populations of *Escherichia coli*. *PLoS Biol.* 5 (9): e225. doi:10.1371/journal.pbio.0050225. [15]

Cooper, T. F., E. A. Ostrowski, and M. Travisano. 2007. A negative relationship between mutation pleiotropy and fitness effect in yeast. *Evolution* 61: 1495–1499. [8]

Cooper, V. S., and R. E. Lenski. 2000. The population genetics of ecological specialization in evolving *Escherichia coli* populations. *Nature* 407: 736–739. [8]

Cosmides, L., and J. Tooby. 1992. Cognitive adaptations for social exchange. In J. H. Barkow, L. Cosmides, and J. Tooby (eds.), *The Adapted Mind*, pp. 163–228. Oxford University Press, New York. [16]

Cotton, S., K. Fowler, and A. Pomiankowski. 2004. Condition dependence of sexual ornament size and variation in the stalk-eyed fly *Cyrtodiopsis dalmanni* (Diptera: Diopsidae). *Evolution* 58: 1038–1046. [15]

Coyne, J. A. 1974. The evolutionary origin of hybrid inviability. *Evolution* 28: 505–506. [18]

Coyne, J. A. 1984. Genetic basis of male sterility in hybrids between two closely related species of *Drosophila*. *Proc. Natl. Acad. Sci. USA* 81: 4444–4447. [17]

Coyne, J. A. 2009a. *Why Evolution is True*. Viking, New York. [1, 23]

Coyne, J. A. 2009b. Seeing and believing. *The New Republic*, February 4, 2009, pp. 32–41. [23]

Coyne, J. A. 2012. Science, religion, and society: the problem of evolution in America. *Evolution* 66: 2654–2663. [23]

Coyne, J. A., and H. A. Orr. 1989. Patterns of speciation in *Drosophila*. *Evolution* 43: 362–381. [17, 18]

Coyne, J. A., and H. A. Orr. 1997. "Patterns of speciation in *Drosophila*" revisited. *Evolution* 51: 295–303. [17, 18]

Coyne, J. A., and H. A. Orr. 2004. *Speciation*. Sinauer, Sunderland, MA. [17, 18]

Coyne, J. A., and T. D. Price. 2000. Little evidence for sympatric speciation in island birds. *Evolution* 54: 2166–2171. [18]

Cracraft, J. 1989. Speciation and its ontology: The empirical consequences of alternative species concepts for understanding patterns and processes of differentiation. In D. Otte and J. A. Endler (eds.), *Speciation and Its Consequences*, pp. 29–59. Sinauer, Sunderland, MA. [17]

Cracraft, J. 1991. Patterns of diversification within continental biotas: Hierarchical congruence among the areas of endemism of Australian vertebrates. *Austral. Syst. Bot.* 4: 211–227. [6]

Cracraft, J. 2001. Avian evolution, Gondwana biogeography, and the Cretaceous-Tertiary mass extinction event. *Proc. R. Soc. Lond. B* 268: 459–469. [6]

Crawford, D. C., D. T. Akey, and D. A. Nickerson. 2005. The patterns of natural variation in human genes. *Annu. Rev. Genomics Human Genet.* 6: 287–312. [9]

Crawford, N. G., B. C. Faircloth, J. E. McCormack, R. T. Brumfield, K. Winker, and T. G. Glenn. 2012. More than 1000 ultraconserved elements provide evidence that turtles are the sister group of archosaurs. *Biol. Lett.* 8: 783–786. [2]

Creel, S., And P. M. Waser. 1991. Failures of reproductive suppression In dwarf mongooses: Accident or adaptation? *Behav. Ecol.* 2: 7–15. [16]

Crews, D., and 7 others. 2007. Transgenerational epigenetic imprints on mate preference. *Proc. Natl. Acad. Sci. USA* 104: 5942–5946. [9]

Crick, F. H. C. 1968. The origin of the genetic code. *J. Mol. Biol.* 38: 367–379. [5]

Crisp, M. D., S. A. Trewick, and L. G. Cook. 2011. Hypothesis testing in biogeography. *Trends Ecol. Evol.* 26: 66–72. [6]

Cronk, Q. C. 2011. Evolution in reverse gear: the molecular basis of loss and reversal. *Cold Spring Harb. Symp. Quant. Biol.* 74: 259–266. [21]

Crow, J. F. 1993. Mutation, mean fitness, and genetic load. *Oxford Surv. Evol. Biol.* 9: 3–42. [12]

Crow, J. F., and M. Kimura. 1965. Evolution in sexual and asexual populations. *Am. Nat.* 99: 439–450. [15]

Crow, J. F., and M. Kimura. 1970. *An Introduction to Population Genetics Theory*. Harper & Row, New York. [10]

Crow, K. D., P. F. Stadler, V. J. Lynch, C. Amemiya, and G. P. Wagner. 2006. The "fish-specific" Hox cluster duplication is coincident with the origin of teleosts. *Mol. Biol. Evol.* 23: 121–136. [21]

Crozier, R. H., and P. Pamilo. 1996. *Evolution of Social Insect Colonies*. Oxford University Press, Oxford. [16]

Cruzan, M. B., and M. L. Arnold. 1993. Ecological and genetic associations in an *Iris* hybrid zone. *Evolution* 47: 1432–1445. [17]

Cubas, P., C. Vincent, and E. Coen. 1999. An epigenetic mutation responsible for natural variation in floral symmetry. *Nature* 401: 157–161. [9]

Culver, D. C. 1982. *Cave Life: Evolution and Ecology*. Harvard University Press, Cambridge, MA. [14]

Cunningham, C., W.-H. Zhu, and D. M. Hillis. 1998. Best-fit maximum-likelihood models for phylogenetic inference: empirical tests with known phylogenies. *Evolution* 52: 978–987. [2]

Currie, C. R., and 8 others. 2003. Ancient tripartite coevolution in the attine ant–microbe symbiosis. *Science* 299: 386–388. [19]

Currie, C. R., M. Poulsen, J. Mendenhall, J. J. Boomsma, and J. Billen. 2006. Coevolved crypts and exocrine glands support mutualistic bacteria in fungus-growing ants. *Science* 311: 81–83. [19]

Curtsinger, J. W., P. M. Service, and T. Prout. 1994. Antagonistic pleiotropy, reversal of dominance, and genetic polymorphism. *Am. Nat.* 144: 210–228. [12]

Cutter, A. D. 2012. The polymorphic prelude to Bateson-Dobzhansky-Muller incompatibilities. *Trends Ecol. Evol.* 27: 209–218. [18]

D

Daan, S., C. Jijkstra, and J. M. Tinbergen. 1990. Family planning in the kestrel (*Falco tinnunculus*): The ultimate control of covariation of laying date and clutch size. *Behaviour* 114: 83–116. [14]

Dacosta, C. P., and C. M. Jones. 1971. Cucumber beetle resistance and mite susceptibility controlled by the bitter gene in *Cucumis sativus*. *Science* 172: 1145–1146. [19]

Daeschler, E. B., N. H. Shubin, and F. A. Jenkins, Jr. 2006. A Devonian tetrapod-like fish and the evolution of the tetrapod body plan. *Nature* 440: 757–763. [4]

Daly, M., and M. Wilson. 1983. *Sex, Evolution, and Behavior*. Willard Grant Press, Boston, MA. [15, 16]

Damuth, J., and I. L. Heisler. 1988. Alternative formulations of multilevel selection. *Biol. Phil.* 3: 407–430. [11, 16]

Danchin, E., A. Charmantier, F. A. Champagne, A. Masoudi, B. Pujol, and S. Blanchet. 2011. Beyond DNA: integrating inclusive inheritance into an extended theory of evolution. *Nat. Rev. Genet.* 12: 475–486. [9]

Danforth, B. N., L. Conway, and S. Ji. 2003. Phylogeny of eusocial *Lasioglossum* reveals multiple losses of eusociality within a primitively eusocial clade of bees (Hymenoptera: Halictidae). *Syst. Biol.* 52: 23–36. [22]

Dao, D. N., R. H. Kessin, and H. L. Ennis. 2000. Developmental cheating and the evolutionary biology of *Dictyostelium* and *Myxococcus*. *Microbiology* 146: 1505–1512. [16]

Darlington, C. D. 1939. *The Evolution of Genetic Systems*. Cambridge University Press, Cambridge. [18]

Darwin, C. 1854. *A Monograph of the Sub-class Cirripedia, with Figures of All the Species*. The Ray Society, London. [4]

Darwin, C. 1859. *The Origin of Species by Means of Natural Selection, or the Preservation of Favored Races in the Struggle for Life*. Modern Library, New York. [1, 2, 22]

Davies, N. B. 1991. Mating systems. In J. R. Krebs and N. B. Davies (eds.), *Behavioural Ecology: An Ecological Approach*, third edition, pp. 263–294. Blackwell Scientific, Oxford. [16]

Davies, N. B. 1992. *Dunnock Behaviour and Social Evolution*. Oxford University Press, Oxford. [16]

Davies, N. B., and M. de L. Brooke. 1998. Cuckoos versus hosts: Experimental evidence for coevolution. In S. I. Rothstein and S. K. Robinson (eds.), *Parasitic Birds and Their Hosts: Studies in Coevolution*, pp. 59–79. Oxford University Press, New York. [19]

Davis, C. C., and K. J. Wurdack. 2004. Host-to-parasite gene transfer in flowering plants: phylogenetic evidence from Malpighiales. *Science* 305: 676–678. [2]

Davis, D. D. 1964. The giant panda: a morphological study of evolutionary mechanisms. *Fieldiana Memoirs* 3: 1–399. [22]

Davis, J. C., O. Brandman, and D. A. Petrov. 2005. Protein evolution in the context of *Drosophila* development. *J. Mol. Evol.* 60: 774–785. doi:10.1007/s00239-004-0241-2 [20]

Dawkins, R. 1976. *The Selfish Gene*. Oxford University Press, Oxford. [16]

Dawkins, R. 1986. *The Blind Watchmaker*. W. W. Norton, New York. [11]

Dawkins, R. 1989. *The Selfish Gene*. Revised edition. Oxford University Press, Oxford. [11, 16]

Dawkins, R. 2009. *The Greatest Show on Earth: The Evidence for Evolution*. Free Press, New York. [1, 23]

Dawkins, R., and J. R. Krebs. 1979. Arms races between and within species. *Proc. R. Soc. London B* 205: 489–511. [19]

Day, T., and R. Bonduriansky. 2011. A unified approach to the evolutionary consequences of genetic and nongenetic inheritance. *Am. Nat.* 178: E18–E36. [9]

Dayan, T., D. Simberloff, E. Tchernov, and Y. Yom-Tov. 1990. Feline canines: Community-wide character displacement among the small cats of Israel. *Am. Nat.* 136: 39–60. [19]

Dean, A. M., D. E. Dykhuizen, and D. L. Hartl. 1986. Fitness as a function of β-galactosidase activity in *Escherichia coli*. *Genet. Res.* 48: 1–8. [11]

Dean, M., M. Carrington, and S. J. O'Brien. 2002. Balanced polymorphism selected by genetic versus infectious human disease. *Annu. Rev. Genomics Hum. Genet.* 3: 263–292. [23]

Deban, S. M., D. B. Wake, and G. Roth. 1997. Salamander with a ballistic tongue. *Nature* 389: 27–28. [22]

Decaestecker, E., S. Gaba, J. A. M. Raeymaekers, R. Stoks, L. Van Kerckhoven, D. Ebert, and L. De Meester. 2007. Host–parasite 'Red Queen' dynamics archived in pond sediment. *Nature* 450: 870–873. [19]

Degnan, J. H., and N. A. Rosenberg. 2009. Gene tree discordance, phylogenetic inference and the multispecies coalescent. *Trends Ecol. Evol.* 24: 332–340. [2]

Deininger, P. L., and M. A. Batzer. 2002. Mammalian retroelements. *Genome Res* 12: 1455–65. [20]

Delph, L. F., J. C. Steven, I. A. Anderson, C. R. Herlihy, and E. D. Brodie III. 2011. Elimination of a genetic correlation between the sexes via artificial correlational selection. *Evolution* 65: 2872–2880. [13]

de Mazancourt, C., E. Johnson, and T. G. Barraclough. 2008. Biodiversity inhibits species' evolutionary responses to changing environments. *Ecol. Lett.* 11: 380–388. [13]

de Muizon, C. 2001. Walking with whales. *Nature* 413: 259–161. [4]

Dennett, D. C. 1995. *Darwin's Dangerous Idea: Evolution and the Meanings of Life.* Simon & Schuster, New York. [11, 22]

Dennett, D. C. 2006. *Breaking the Spell: Religion as a Natural Phenomenon.* Viking, New York. [16]

Depew, D. J., and B. H. Weber. 1995. *Darwinism Evolving: Systems Dynamics and the Genealogy of Natural Selection.* MIT Press, Cambridge, MA. [21]

de Pontbriand, A., X.-P. Wang, Y. Cavaloc, M.-G. Mattei, and F. Galibert. 2002. Synteny comparison between apes and human using fine-mapping of the genome. *Genomics* 80: 395–401. [8]

de Queiroz, A. 2005. The resurrection of oceanic dispersal in historical biogeography. *Trends Ecol. Evol.* 9: 68–73. [6]

de Queiroz, K. 2007. Species concepts and species delimitation. *Syst. Biol.* 56: 879–886. [17]

de Queiroz, K., and M. J. Donoghue. 1990. Phylogenetic systematics or Nelson's version of cladistics? *Cladistics* 6: 61–75. [17]

de Roode, J. C., and 9 others. 2005. Virulence and competitive ability in genetically diverse malaria infections. *Proc. Natl. Acad. Sci. USA* 102: 7624–7628. [19]

Desmond, A., and J. Moore. 1991. *Darwin.* Warner Books, New York. [1]

Dessain, S., C. T. Gross, M. A. Kuziora, and W. McGinnis. 1992. *Antp*-type homeodomains have distinct DNA-binding specificities that correlate with their different regulatory functions in embryos. *EMBO J.* 11: 991–1002. [21]

Desurmont, G. A., M. J. Donoghue, W. L. Clement, and A. A. Agrawal. 2011. Evolutionary history predicts plant defense against an invasive pest. *Proc. Natl. Acad. Sci. USA* 108: 7070–7074. [19]

de Waal, F. 2007. *Chimpanzee Politics: Power and Sex among Apes.* Johns Hopkins University Press, Baltimore. [23]

DeWitt, T. J., A. Sih, and D. S. Wilson. 1998. Costs and limits of phenotypic plasticity. *Trends Ecol. Evol.* 13: 77–80. [13]

Diamond, J. M. 1975. Assembly of species communities. In M. L. Cody and J. M. Diamond (eds.), *Ecology and Evolution of Communities*, pp. 342–444. Harvard University Press, Cambridge, MA. [6]

Diamond, J. M. 1997. *Guns, Germs, and Steel: The Fates of Human Societies.* Norton, New York. [23]

Dieckmann, U., and M. Doebeli. 1999. On the origin of species by sympatric speciation. *Nature* 400: 354–357. [18]

Diego, F., E. A. Herniou, C. Boschetti, M. Caprioli, G. Melone, C. Ricci, and T. G. Barraclough. 2007. Independently evolving species in asexual bdelloid rotifers. *PLoS Biology* 5(4): e87 [15]

Dilda, C. L., and T. F. C. Mackay. 2002. The genetic architecture of *Drosophila* sensory bristle number. *Genetics* 162: 1655–1674. [13]

Dobzhansky, Th. 1934. Studies on hybrid sterility. I. Spermatogenesis in pure and hybrid *Drosophila pseudoobscura. Z. Zellforsch. Mikrosk. Anat.* 21: 169–221. [17]

Dobzhansky, Th. 1936. Studies on hybrid sterility. II. Localization of sterility factors in *Drosophila pseudoobscura* hybrids. *Genetics* 21: 113–135. [17, 18]

Dobzhansky, Th. 1937. *Genetics and the Origin of Species.* Columbia University Press, New York. [1, 17]

Dobzhansky, Th. 1948. Genetics of natural populations. XVIII. Experiments on chromosomes of *Drosophila pseudoobscura* from different geographic regions. *Genetics* 33: 588–602. [11]

Dobzhansky, Th. 1951. *Genetics and the Origin of Species*, third edition. Columbia University Press, New York. [17, 18]

Dobzhansky, Th. 1955. A review of some fundamental concepts and problems of population genetics. *Cold Spring Harbor Symp. Quant. Biol.* 20: 1–15. [17]

Dobzhansky, Th. 1970. *Genetics of the Evolutionary Process.* Columbia University Press, New York. [8, 9]

Dobzhansky, Th., and O. Pavlovsky. 1953. Indeterminate outcome of certain experiments on *Drosophila* populations. *Evolution* 7: 198–210. [11]

Dobzhansky, Th., and B. Spassky. 1969. Artificial and natural selection for two behavioral traits in *Drosophila pseudoobscura. Proc. Natl. Acad. Sci. USA* 62: 75–80. [9, 13]

Dong, Y. M., S. Das, C. Cirimotich, J. A. Souza-Neto, K. J. McLean, and G. Dimopoulos. 2011. Engineered *Anopheles* immunity to *Plasmodium* infection. *PLoS Pathogens* 7 (12), article e1002458. doi:10.1371/journal.ppat.1002458. [23]

Donoghue, M. J. 1992. Homology. In E. F. Keller and E. A. Lloyd (eds.), *Keywords in Evolutionary Biology*, pp. 170–179. Harvard University Press, Cambridge, MA. [21]

Donoghue, M. J., and S. A. Smith. 2004. Patterns in the assembly of temperate forests around the Northern Hemisphere. *Phil. Trans. R. Soc. Lond. B* 359: 1633–1644. [6]

Donoghue, P. C. J., and M. J. Benton. 2007. Rocks and clocks: calibrating the tree of life using fossils and molecules. *Trends Ecol. Evol.* 22: 424–431. [2]

Doolittle, W. F. 1999. Phylogenetic classification and the universal tree. *Science* 284: 2124–9. [20]

Downs, J. P., E. B. Daeschler, F. A. Jenkins, and N. H. Shubin. 2008. The cranial endoskeleton of *Tiktaalik roseae. Nature* 455: 925–929. [4]

Draghi, J., and G. P. Wagner. 2008. Evolution of evolvability in a developmental model. *Evolution* 62: 301–315. [22]

Drake, J. W., B. Charlesworth, D. Charlesworth, and J. F. Crow. 1998. Rates of spontaneous mutation. *Genetics* 148: 1667–1686. [8]

Drummond, D. A., and C. O. Wilke. 2008. Mistranslation-induced protein misfolding as a dominant constraint on coding-sequence evolution. *Cell* 134: 341–352. [20]

Duberman, M. B., M. Vicinus, and G. Chauncey (eds.). 1989. *Hidden from History: Reclaiming the Gay and Lesbian Past.* Penguin, New York. [16]

Dubrova, Y. E., V. N. Nesterov, N. G. Krouchinsky, V. A. Ostapenko, R. Neumann, D. L. Neil, and A. J. Jeffreys. 1996. Human minisatellite mutation rate after the Chernobyl accident. *Nature* 380: 683–686. [8]

Ducrest, A. L., L. Keller, and A. Roulin. 2008. Pleiotropy in the melanocortin system, coloration and behavioural syndromes. *Trends Ecol. Evol.* 23: 502–510. [21]

Dudley, R., G. Byrnes, S. P. Yanoviak, B. Borrell, R. M. Brown, and J. A. McGuire. 2007. Gliding and the functional origins of flight: Biomechanical novelty or necessity? *Annu. Rev. Ecol. Evol. Syst.* 38: 179–201. [22]

Dudley, S. A., and A. L. File. 2007. Kin recognition in an annual plant. *Biol. Lett.* 3: 435–438. [16]

Dujon, B., and 66 others. 2004. Genome evolution in yeasts. *Nature* 430: 35–44. [20]

Dunbar, R. I. M., and L. Barrett (eds.). 2007. *Oxford Handbook of Evolutionary Psychology.* Oxford University Press, Oxford. [16]

Dunn, C. W., and 17 others. 2008. Broad phylogenetic sampling improves resolution of the animal tree of life. *Nature*: 452: 745–749. [5]

Dyall, S. D., M. T. Brown, and P. J. Johnson. 2004. Ancient invasions: from endosymbionts to organelles. *Science* 304: 253–257. [5]

Dybdahl, M. F., and C. M. Lively. 1998. Host-parasite coevolution: Evidence for rare advantage and time-lagged selection in a natural population. *Evolution* 52: 1057–1066. [19]

Dyer, L. A., and 12 others. 2007. Host specificity of Lepidoptera in tropical and temperate forests. *Nature* 448: 696–699. [7]

E

Eberhard, W. G. 1996. *Female Control: Sexual Selection by Cryptic Female Choice*. Princeton University Press, Princeton, NJ. [17]

Ebert, D. 1994. Virulence and local adaptation of a horizontally transmitted parasite. *Science* 265: 1084–1086. [19]

Edgecombe, G. D., and 8 others. 2011. Higher-level metazoan relationships: recent progress and remaining questions. *Org. Divers. Evol.* 11: 151–172. [5]

Egan, S. P., and D. J. Funk. 2009. Ecologically dependent postmating isolation between sympatric host forms of *Neochlamisus bebbianae* leaf beetles. *Proc. Natl. Acad. Sci. USA* 106: 19426–19431. [18]

Ehrlich, P. R., and P. H. Raven. 1964. Butterflies and plants: A study in coevolution. *Evolution* 18: 586–608. [7, 19]

Eicher, D. L. 1976. *Geologic Time*. Prentice-Hall, Englewood Cliffs, NJ. [4]

Eldakar, O. T., and D. S. Wilson. 2011. Eight criticisms not to make of group selection. *Evolution* 65: 1523–1526. [15, 16]

Eldredge, N. 1989. *Macroevolutionary Patterns and Evolutionary Dynamics: Species, Niches and Adaptive Peaks*. McGraw-Hill, New York. [22]

Eldredge, N., and S. J. Gould. 1972. Punctuated equilibria: An alternative to phyletic gradualism. In T. J. M. Schopf (ed.), *Models in Paleobiology*, pp. 82–115. Freeman, Cooper and Co., San Francisco. [4, 18, 22]

Eldredge, N., and 9 others. 2005. The dynamics of evolutionary stasis. *Paleobiology* 31: 133–145. [18, 22]

Elena, S. F., and R. E. Lenski. 2003. *Evolution* experiments with microorganisms: The dynamics and genetic bases of adaptation. *Nature Rev. Genetics* 4: 457–469. [8]

Elena, S. F., and R. Sanjuán. 2007. Virus evolution: Insights from an experimental approach. *Annu. Rev. Ecol. Evol. Syst.* 38: 27–52. [8, 15]

Ellegren, H. 2008. Comparative genomics and the study of evolution by natural selection. *Mol. Ecol.* 17: 4586–4596. [10]

Ellegren, H., G. Lindgren, C. R. Primmer, and A. P. Møller. 1997. Fitness loss and germline mutations in barn swallows breeding in Chernobyl. *Nature* 389: 593–596. [8]

Ellstrand, N. C. 2003. Current knowledge of gene flow in plants: Implications for transgenic flow. *Phil. Trans. R. Soc. B* 358: 1163–1170 [23]

Ellstrand, N., and J. Antonovics. 1984. Experimental studies of the evolutionary significance of sexual reproduction. I. A test of the frequency-dependent selection hypothesis. *Evolution* 38: 103–115. [12]

Ellstrand, N. C., R. Whitkus, and L. H. Rieseberg. 1996. Distribution of spontaneous plant hybrids. *Proc. Natl. Acad. Sci.* 93: 5090–5093. [17]

Emlen, D. J. 2000. Integrating development with evolution: a case study with beetle horns. *BioScience* 50: 403–418. [9]

Emlen, D. J. 2008. The evolution of animal weapons. *Annu. Rev. Ecol. Evol. Syst.* 39: 387–413. [15]

Emlen, D. J., and H. F. Nijhout. 1999. Hormonal control of male horn length dimorphism in the dung beetle *Onthophagus taurus* (Coleoptera: Scarabaeidae). *J. Insect Physiol.* 45: 45–53. [21]

Emlen, S. T. 1995. An evolutionary theory of the family. *Proc. Natl. Acad. Sci. USA* 92: 8092–8099. [16]

Emlen, S. T., and P. H. Wrege. 1988. The role of kinship in helping decisions among white-fronted bee-eaters. *Behav. Ecol. Sociobiol.* 23: 305–315. [16]

Emlen, S. T., and P. H. Wrege. 1991. Breeding biology of white-fronted bee-eaters at Nakuru—the influence of helpers on breeder fitness. *J. Anim. Ecol.* 60: 309–326. [16]

Emlen, S. T., and P. H. Wrege. 1992. Parent-offspring conflict and the recruitment of helpers among bee-eaters. *Nature* 356: 331–333. [16]

Enard, W., M. Przeworski, S. E. Fisher, C. S. Lai, V. Wiebe, T. Kitano, A. P. Monaco, and S. Pääbo. 2002. Molecular evolution of *FOXP2*, a gene involved in speech and language. *Nature* 418: 869–872. [8]

Endler, J. A. 1973. Gene flow and population differentiation. *Science* 179: 243–250. [12]

Endler, J. A. 1977. *Geographic Variation, Speciation, and Clines*. Princeton University Press, Princeton, NJ. [17, 18]

Endler, J. A. 1980. Natural selection on color patterns in *Poecilia reticulata*. *Evolution* 34: 76–91. [11]

Endler, J. A. 1983. Testing causal hypotheses in the study of geographic variation. In J. Felsenstein (ed.), *Numerical Taxonomy*, pp. 424–443. Springer-Verlag, Berlin. [6]

Endler, J. A. 1986. *Natural Selection in the Wild*. Princeton University Press, Princeton, NJ. [11, 12, 13]

Endler, J. A., and A. L. Basolo. 1998. Sensory ecology, receiver biases and sexual selection. *Trends Ecol. Evol.* 13: 415–420. [18]

Eöry, L., D. L. Halligan, and P. D. Keightley. 2010. Distributions of selectively constrained sites and deleterious mutation rates in the hominid and murid genomes. *Mol. Biol. Evol.* 27: 177–192. [8]

Erbar, C. 2007. Current opinions in flower development and the evo-devo approach in plant phylogeny. *Pl. Syst. Evol.* 269: 107–132. [21]

Erwin, D. H. 1993. *The Great Paleozoic Crisis: Life and Death in the Early Permian*. Columbia University Press, New York. [7]

Erwin, D. H. 2006. *Extinction: How Life on Earth Nearly Ended 250 Million Years Ago*. Princeton University Press, Princeton, NJ. [5, 7]

Erwin, D. H. 2009. Early evolution of the bilaterian developmental toolkit. *Phil. Trans. R. Soc. B* 364: 2253–2261. [5]

Estes, S., B. C. Ajie, M. Lynch, and P. C. Phillips. 2005. Spontaneous mutational correlations for life-history, morphological and behavioral characters in *Caenorhabditis elegans*. *Genetics* 170:645–653. [8]

Etterson, J. R. 2004. Evolutionary potential of *Chamaecrista fasciculata* in relation to climate change. II. Genetic architecture of three populations reciprocally planted along an environmental gradient in the Great Plains. *Evolution* 58: 1549–1471. [13]

Etterson, J. R., and R. G. Shaw. 2001. Constraint to adaptive evolution in response to global warming. *Science* 294: 151–154. [13]

Ewald, P. W. 1994. *Evolution of Infectious Disease*. Oxford University Press, Oxford. [19]

F

Falconer, D. S., and T. F. C. Mackay. 1996. *Introduction to Quantitative Genetics*, 4th ed. Longman, Harlow, U.K. [13]

Farrell, B. D. 1998. "Inordinate fondness" explained: Why are there so many beetles? *Science* 281: 555–559. [7]

Farrell, B. D., and A. S. Sequeira. 2004. Evolutionary rates in the adaptive radiation of beetles on plants. *Evolution* 58: 1984–2001. [7]

Farrell, B., D. Dussourd, and C. Mitter. 1991. Escalation of plant defenses: Do latex and resin canals spur plant diversification? *Am. Nat.* 138: 881–900. [7]

Fay, J. C., and C.-I. Wu. 2003. Sequence divergence, functional constraint, and selection in protein evolution. *Annu. Rev. Genomics Hum. Genet.* 4: 213–235. [12]

Feder, J. L., and A. A. Forbes. 2008. Host fruit-odor discrimination and sympatric host-race formation. pp. 101–116. In K. J. Tilmon (ed.), *Specialization, Speciation, and Radiation: The Evolutionary Biology of Herbivorous Insects*. University of California Press, Berkeley. [18]

Feder, J. L., C. A. Chilcote, and G. L. Bush. 1990. Geographic pattern of genetic differentiation between host-associated populations of *Rhagoletis pomonella* (Diptera: Tephritidae) in the eastern United States and Canada. *Evolution* 44: 570–594. [18]

Feder, J. L., and 8 others. 2005. Mayr, Dobzhansky and Bush and the complexities of sympatric speciation in *Rhagoletis*. *Proc. Natl. Acad. Sci. USA* 102 (Suppl. 1): 6573–6580. [18]

Fedigan, L. M. 1986. The changing role of women in models of human evolution. *Annu. Rev. Anthropol.* 15: 25–66. [4]

Feduccia, A. 1999. *The Origin and Evolution of Birds*, 2nd ed. Yale University Press, New Haven, CT. [22]

Fehér, O., H. Wang, S. Saar, P. P. Mitra, and O. Tchernichovski. 2009. *De novo* establishment of wild-type song culture in the zebra finch. *Nature* 459: 564–568. [9]

Fehr, E., and U. Fischbacher. 2003. The nature of human altruism. *Nature* 425: 785–791. [23]

Felsenstein, J. 1981. Skepticism towards Santa Rosalia, or why are there so few kinds of animals? *Evolution* 35: 124–138. [18]

Felsenstein, J. 1985. Phylogenies and the comparative method. *Am. Nat.* 125: 1–15. [11]

Felsenstein, J. 2004. *Inferring Phylogenies.* Sinauer, Sunderland, MA. [2]

Fenner, F., and F. N. Ratcliffe. 1965. *Myxomatosis.* Cambridge University Press, Cambridge. [19]

ffrench-Constant, R. H., R. T. Roush, D. Mortlock, and G. P. Dively. 1990. Isolation of dieldrin resistance from field populations of *Drosophila melanogaster* (Diptera: Drosophilidae). *J. Econ. Entomol.* 83: 1733–1737. [8]

Field, J., B. Rosenthal, and J. Samuelson. 2000. Early lateral transfer of genes encoding malic enzyme, acetyl-CoA synthetase and alcohol dehydrogenases from anaerobic prokaryotes to *Entamoeba histolytica*. *Molec. Microbiol.* 38: 446–455. [20]

Figueirido, B., C. M. Janis, J. A. Pérez-Claros, M. De Renzi, and P. Palmqvist. 2012. Cenozoic climate change influences mammalian evolutionary dynamics. *Proc. Natl. Acad. Sci. USA* 109: 722–727. [7]

Filchak, K. E., J. B. Roethele, and J. L. Feder. 2000. Natural selection and sympatric divergence in the apple maggot *Rhagoletis pomonella*. *Nature* 407: 739–742. [18]

Fine, P. V. A., and R. H. Ree. 2006. Evidence for a time-integrated species-area effect on the latitudinal gradient in species diversity. *Am. Nat.* 168: 796–804. [6]

Fine, P. V. A., and 8 others. 2006. The growth-defense trade-off and habitat specialization by plants in Amazonian forests. *Ecology* 87: S150–S162. [19]

Finlayson, C. 2005. Biogeography and evolution of the genus *Homo*. *Trends Ecol. Evol.* 20: 457–463. [6]

Fischer, C. S., M. Hout, M. S. Jankowski, S. R. Lucas, A. Swidler, and K. Voss. 1996. *Inequality by Design: Cracking the Bell Curve Myth.* Princeton University Press, Princeton, NJ. [9]

Fisher, D. C. 1986. Progress in organismal design. In D. M. Raup and D. Jablonski (eds.), *Patterns and Processes in the History of Life*, pp. 99–117. Springer-Verlag, Berlin. [22]

Fisher, H. S., and H. E. Hoekstra. 2010. Competition drives cooperation among closely related sperm of deer mice. *Nature* 463: 801–803. [16]

Fisher, R. A. 1930. *The genetical theory of natural selection.* Clarendon Press, Oxford. [12, 15, 18]

Fitzpatrick, B. M., J. A. Fordyce, and S. Gavrilets. 2008. What, if anything, is sympatric speciation? *J. Evol. Biol.* 21: 1452–1459. [18]

Flint, J., and T. F. C. Mackay. 2009. Genetic architecture of quantitative traits in mice, flies, and humans. *Genome Res.* 19: 723–733. [13]

Foote, M. 1988. Survivorship analysis of Cambrian and Ordovician trilobites. *Paleobiology* 14: 258–271. [7]

Foote, M. 1997. The evolution of morphological diversity. *Annu. Rev. Ecol. Syst.* 28: 129–152. [7]

Foote, M. 2000a. Origination and extinction components of diversity: general problems. In D. H. Erwin and S. L. Wing (eds.), *Deep Time: Paleobiology's Perspective*, pp. 74–102. *Paleobiology* 26 (4), supplement. [7]

Foote, M. 2000b. Origination and extinction components of taxonomic diversity: Paleozoic and post-Paleozoic dynamics. *Paleobiology* 26: 578–605. [7]

Foote, M. 2010. The geological history of biodiversity. In M. A. Bell, D. J. Futuyma, W. F. Eanes, and J. S. Levinton (eds.), *Evolution Since Darwin: The First 150 Years*, pp. 479–510. Sinauer, Sunderland, MA. [7]

Foote, M., and A. I. Miller. 2007. *Principles of Paleontology*, 3rd ed. W. H. Freeman, New York. [7]

Force, A., M. Lynch, F. B. Pickett, A. Amores, Y. Yan, and J. Postlethwait. 1999. Preservation of duplicate genes by complementary, degenerate mutations. *Genetics* 151: 1531–1545. [20]

Ford, E. B. 1971. *Ecological Genetics.* Chapman & Hall, London. [9]

Fornoni, J., P. L. Valverde, and J. Núñez-Farfán. 2003. Quantitative genetics of plant tolerance and resistance against natural enemies of two natural populations of *Datura stramonium*. *Evol. Ecol. Res.* 5: 1049–1065. [13]

Forrest, B., and P. R. Gross. 2004. *Creationism's Trojan Horse: The Wedge of Intelligent Design.* Oxford University Press, New York. [23]

Forster, L. M. 1992. The stereotyped behaviour of sexual cannibalism in *Latrodectus hasselti* Thorell (Araneae: Theridiidae), the Australian redback spider. *Aust. J. Zool.* 40: 1–11. [11]

Foster, K. R., T. Wenseleers, and F. L. W. Ratnieks. 2006. Kin selection is the key to altruism. *Trends Ecol. Evol.* 21: 57–60. [16]

Foster, P. L. 2000. Adaptive mutation: implications for evolution. *BioEssays* 22: 1067–1074. [8]

Fournier, G. P., J. Huang, and J. P. Gogarten. 2009. Horizontal gene transfer from extinct and extant lineages: biological innovation and the coral of life. *Phil. Trans. R. Soc. B* 364: 2229–2239. [5]

Fowler, K., and L. Partridge. 1989. A cost of mating in female fruitflies. *Nature* 338: 760–761. [14]

Fowler, N. L., and D. A. Levin. 1984. Ecological constraints on the establishment of a novel polyploid in competition with its diploid progenitor. *Am. Nat.* 124: 703–711. [18]

Fox, C. W., and J. B. Wolf (eds.). *Evolutionary Genetics.* Oxford University Press, New York. [8]

Fox, J. W., and D. A. Vasseur. 2008. Character convergence under competition for nutritionally essential resources. *Am. Nat.* 172: 667–680. [19]

Fraenkel, G. 1959. The *raison d'être* of secondary plant substances. *Science* 129: 1466–1470. [19]

Frank, S. A. 1990. Sex allocation theory for birds and mammals. *Annu. Rev. Ecol. Syst.* 21: 13–55. [17]

Frank, S. A. 1992. Models of plant-pathogen coevolution. *Trends Genet.* 8: 213–219. [19]

Frank, S. A. 1996. Models of parasite virulence. *Q. Rev. Biol.* 71: 37–78. [19]

Frank, S. A. 1997. Models of symbiosis. *Am. Nat.* 150: S80–S99. [16]

Frank, S. A. 1998. *Foundations of Social Evolution.* Princeton University Press, Princeton, NJ. [16]

Frank, S. A. 2003. Perspective: Repression of competition and the evolution of cooperation. *Evolution* 57: 693–705. [16]

Frank, S. A., and B. J. Crespi. 2011. Pathology from evolutionary conflict, with a theory of X chromosome versus autosome conflict over sexually antagonistic traits. *Proc. Natl. Acad. Sci. USA* 108: 10886–10893. [16]

Frankham, R. 1995. Effective population-size: adult population-size ratios in wildlife—A review. *Genet. Res.* 66: 95–107. [10]

Frankham, R., J. D. Ballou, and D. A. Briscoe. 2002. *Introduction to Conservation Genetics.* Cambridge University Press, Cambridge. [9, 10, 23]

Fraser, G. J., C. D. Hulsey, R. F. Bloomquist, K. Uyesugi, N. R. Manley, and J. T. Streelman. 2009. An ancient gene network is co-opted for teeth on old and new jaws. *PLoS Biol.* 7: e31. [21]

Fraser, S. (ed.). 1995. *The Bell Curve Wars: Race, Intelligence, and the Future of America.* Basic Books, New York. [9]

Fricke, C., and G. Arnqvist. 2007. Rapid adaptation to a novel host in a seed beetle (*Callosobruchus maculatus*): the role of sexual selection. *Evolution* 61: 440–454. [15]

Friis, E. M., K. R. Pedersen, and P. R. Crane. 2010. Diversity in obscurity: fossil flowers and the early history of angiosperms. *Phil. Trans. R. Soc. B* 365: 369–382. [5]

Fry, J. D. 2003. Multilocus models of sympatric speciation: Bush vs. Rice vs. Felsenstein. *Evolution* 57: 1735–1746. [18]

Fryer, G., and T. D. Iles. 1972. *The Cichlid Fishes of the Great Lakes of Africa.* T.F.H. Publications, Neptune City, NJ. [3]

Fujita, M. K., A. D. Leaché, F. T. Burbrink, J. A. McGuire, and C. Moritz. 2012. Coalescent-based species delimitation in an integrative taxonomy. *Trends Ecol. Evol.* 27: 480–488. [17]

Fumagalli, M., M. Sironi, U. Pozzoli, A. Ferrer-Admettla, L. Pattini, and R. Nielsen. 2011. Signatures of environmental genetic adaptation pinpoint pathogens as the main selective pressure through human evolution. *PLoS Genet.* 7(11), article e1002355, doi:10.1371/journal.pgen.1002355. [12]

Funk, D. J., and K. E. Omland. 2003. Species-level paraphyly and polyphyly: Frequency, causes, and consequences, with insights from mitochondrial DNA. *Annu. Rev. Ecol. Syst.* 34: 397–423. [17]

Funk, D. J., P. Nosil, and W. J. Etges. 2006. Ecological divergence exhibits consistently positive associations with reproductive isolation across disparate taxa. *Proc. Natl. Acad. Sci. USA* 103: 3209–3213. [18]

Fürsich, F. T., and D. Jablonski. 1984. Lake Triassic naticid drillholes: Carnivorous gastropods gain a major adaptation but fail to radiate. *Science* 224: 78–80. [7]

Futuyma, D. J. 1987. On the role of species in anagenesis. *Am. Nat.* 130: 465–473. [18, 22]

Futuyma, D. J. 1991. A new species of *Ophraella* Wilcox (Coleoptera: Chrysomelidae) from the southeastern United States. *J. NY Entomol. Soc.* 99: 643–653. [17]

Futuyma, D. J. 1995. *Science on Trial: The Case for Evolution*. Sinauer, Sunderland, MA. [3, 4]

Futuyma, D. J. 2000. Some current approaches to the evolution of plant-herbivore interactions. *Plant Species Biol.* 15: 1–9. [23]

Futuyma, D. J. 2004. The fruit of the tree of life: Insights into evolution and ecology. In J. Cracraft and M. J. Donoghue (eds.), *Assembling the Tree of Life*, pp. 25–39. Oxford University Press, New York. [3]

Futuyma, D. J. 2008. Sympatric speciation: Norm or exception? In K. J. Tilmon (ed.), *Specialization, Speciation, and Radiation: The Evolutionary Biology of Herbivorous Insects*, pp. 1136–1148. University of California Press, Berkeley. [18]

Futuyma, D. J. 2010. Evolutionary constraint and ecological consequences. *Evolution* 64: 1865–1884. [13]

Futuyma, D. J., and A. A. Agrawal. 2009. Macroevolution and the biological diversity of plants and herbivores. *Proc. Natl. Acad. Sci. USA* 106: 18054–18061. [19]

Futuyma, D. J., and C. Mitter. 1996. Insect-plant interactions: the evolution of component communities. *Phil. Trans. R. Soc. Lond. B* 351: 1361–1366. [19]

Futuyma, D. J., and M. Slatkin (eds.). 1983. *Coevolution*. Sinauer, Sunderland, MA. [19]

Futuyma, D. J., M. C. Keese, and D. J. Funk. 1995. Genetic constraints on macroevolution: The evolution of host affiliation in the leaf beetle genus *Ophraella*. *Evolution* 49: 797–809. [13, 19]

G

Galiana, A., A. Moya, and F. J. Ayala. 1993. Founder-flush speciation in *Drosophila pseudoobscura*: a large-scale experiment. *Evolution* 47: 432–444. [18]

Galindo, B. E., V. D. Vacquier, and W. J. Swanson. 2003. Positive selection in the egg receptor for abalone sperm lysin. *Proc. Natl. Acad. Sci. USA* 100: 4639–4643. [17]

Galis, F. 1996. The application of functional morphology to evolutionary studies. *Trends Ecol. Evol.* 11: 124–129. [22]

Galis, F. T., and 8 others. 2006. Extreme selection in humans against homeotic transformations of cervical vertebrae. *Evolution* 60: 2643–2654. [12]

Gallup, G. G., Jr., and D. A. Frederick. 2010. The science of sex appeal: an evolutionary perspective. *Rev. Gene. Psychol.* 14: 240–250. [16]

Gao, F., and 11 others. 1999. Origin of HIV–1 in the chimpanzee *Pan troglodytes troglodytes*. *Nature* 397: 436–441. [1]

Gao, L. Z., and H. Innan. 2004. Very low gene duplication rate in the yeast genome. *Science* 306: 1367–1370. [20]

Garland, T., Jr., and M. R. Rose. (ed.). 2009. *Experimental Evolution*. University of California Press, Berkeley. [11]

Garrigan, D., and 11 others. 2007. Inferring human population sizes, divergence times and rates of gene flow from mitochondrial, X and Y chromosome resequencing data. *Genetics* 177: 2195–2207. [10]

Gaston, K. J., and T. M. Blackburn. 2000. *Pattern and Process in Macroecology*. Blackwell Science, Oxford. [7]

Gatesy, J., M. Milinkovitch, V. Waddell, and M. Stanhope. 1999. Stability of cladistic relationships between Cetacea and higher-level artiodactyls taxa. *Syst. Biol.* 48: 6–20. [4]

Gavrilets, S. 2000. Rapid evolution of reproductive barriers driven by sexual conflict. *Nature* 403: 886–889. [15, 18]

Gavrilets, S. 2004. *Fitness Landscapes and the Origin of Species*. Princeton University Press, Princeton, NJ. [18]

Gavrilets, S. 2012. On the evolutionary origins of the egalitarian syndrome. *Proc. Natl. Acad. Sci. USA* 109: 14069–14074. [16]

Gavrilets, S., and A. Hastings. 1996. Founder effect speciation: A theoretical reassessment. *Am. Nat.* 147: 466–491. [18]

Gavrilets, S., and J. B. Losos. 2009. Adaptive radiation: contrasting theory with data. *Science* 323: 732–737. [7]

Gavrilets, S., and W. R. Rice. 2006. Genetic models of homosexuality: Generating testable predictions. *Proc. R. Soc. Lond. B* 273: 3031–3038. [16]

Gavrilets, S., G. Arnqvist, and U. Friberg. 2001. The evolution of female mate choice by sexual conflict. *Proc. R. Soc. Lond. B* 268: 531–539. [15]

Gehring, W. J., and K. Ikeo. 1999. *Pax6* mastering eye morphogenesis and evolution. *Trends Genet.* 15: 371–377. [8]

Gemmell, N. J., V. J. Metcalf, and F. W. Allendorf. 2004. Mother's curse: the effect of mtDNA on individual fitness and population viability. *Trends Ecol. Evol.* 19: 238–244. [16]

Genner, M. J., and 6 others. 2007. Age of cichlids: new dates for ancient lake fish radiations. *Mol. Biol. Evol.* 24: 1269–1282. [6]

Gentles, A. J., and S. Karlin. 1999. Why are human G-protein-coupled receptors predominantly intronless? *Trends Genet.* 15: 47–49. [20]

Gerber, E., and 6 others. 2011. Prospects for biological control of *Ambrosia artemisiifolia* in Europe: learning from the past. *Weed Res.* 51: 559–573. [2]

Gerhart, J., and M. Kirschner. 2007. The theory of facilitated variation. *Proc. Natl. Acad. Sci. USA* 104 (Suppl. 1): 8582–8589. [22]

Gerhold, P., and 8 others. 2011. Phylogenetically poor plant communities receive more alien species, which more easily coexist with natives. *Am. Nat.* 177: 668–680. [19]

Gerrish, P. J., and R. E. Lenski. 1998. The fate of competing beneficial mutations in an asexual population. *Genetica* 102/103: 127–144. [15]

Ghildiyal, M., and P. D. Zamore. 2009. Small silencing RNAs: An expanding universe. *Nature Rev. Genetics* 10: 94–108. [20]

Ghiselin, M. T. 1969. *The Triumph of the Darwinian Method*. University of California Press, Berkeley. [11]

Ghiselin, M. T. 1995. Perspective: Darwin, progress, and economic principles. *Evolution* 49: 1029–1037. [22]

Gibbs, M. J., J. S. Armstrong, and A. J. Gibbs. 2001. The haemagglutinin gene, but not the neuraminidase gene, of 'Spanish flu' was a recombinant. *Phil. Trans. R. Soc. Lond. B* 356: 1845–1855. [19]

Gidaszewski, N. A., M. Baylac, and C. P. Klingenberg. 2009. Evolution of sexual dimorphism of wing shape in the *Drosophila melanogaster* subgroup. *BMC Evol. Biol.* 9: 110–121. [21]

Gignoux, C. R., B. M. Henn, and J. L. Mountain. 2011. Rapid, global demographic expansions after the origins of agriculture. *Proc. Natl. Acad. Sci. USA* 108: 6044–6049. [10]

Gigord, L. D. B., M. R. Macnair, and A. Smithson. 2001. Negative frequency-dependent selection maintains a dramatic color polymorphism in the rewardless orchid *Dactylorhiza sambucina* (L.) Soò. *Proc. Natl. Acad. Sci. USA* 98: 6253–6255. [12]

Gilad, Y., O. Man, S. Pääbo, and D. Lancet. 2003. Human-specific loss of olfactory receptor genes. *Proc. Natl. Acad. Sci. USA* 100: 3324–3327. [20]

Gilbert, S. F, and D. Epel. 2009. *Ecological Developmental Biology*. Sinauer Associates, Sunderland, MA. [21]

Gilbert, S. F., J. Sapp, and A. I. Tauber. 2012. A symbiotic view of life: we have never been individuals. *Q. Rev. Biol.* 87: 325–341. [19]

Gilbert, W. 1987. The exon theory of genes. *Cold Spring Harbor Symp. Quant. Biol.* 52: 901–905. [20]

Gilchrist, G. W., R. B. Huey, and L. Serra. 2001. Rapid evolution of wing size clines in *Drosophila subobscura*. *Genetica* 112: 273–286. [21]

Gilinsky, N. L. 1994. Volatility and the Phanerozoic decline of background extinction. *Paleobiology* 20: 424–444. [7]

Gill, D. E. 1974. Intrinsic rates of increase, saturation density, and competitive ability. II. The evolution of competitive ability. *Am. Nat.* 108: 103–116. [19]

Gill, F. B. 1995. *Ornithology*, 2nd ed. W. H. Freeman, New York. [11]

Gingerich, P. D. 1983. Rates of evolution: Effects of time and temporal scaling. *Science* 222:159–161. [4]

Gingerich, P. D. 2001. Rates of evolution on the time scale of the evolutionary process. *Genetica* 112/113: 127–144. [4]

Gingerich, P. D. 2003. Land-to-sea transition of early whales: Evolution of Eocene Archaeoceti (Cetacea) in relation to skeletal proportions and locomotion of living semiaquatic mammals. *Paleobiology* 29: 429–454. [4]

Gingerich, P. D., M. ul Haq, I. S. Zalmout, I. H. Khan, and M. S., Malkani. 2001. Origin of whales from early artiodactyls: Hands and feet of Eocene Protocetidae from Pakistan. *Science* 293: 2239–2242. [4]

Glor, R. E. 2010. Phylogenetic insights on adaptive radiation. *Annu. Rev. Ecol. Evol. Syst.* 41: 251–270. [7]

Gluckman, P., A. Beedle, and M. Hanson. 2009. *Principles of Evolutionary Medicine*. Oxford University Press, Oxford. [23]

Godfray, H. C. J. 1999. Parent–offspring conflict. In L. Keller (ed.), *Levels of Selection in Evolution*, pp. 100–120. Princeton University Press, Princeton, NJ. [16]

Goldberg, E. E., J. R. Kohn, R. Lande, K. A. Robertson, S. A. Smith, and B. Igić. 2010. Species selection maintains self-incompatibility. *Science* 330: 493–495. [22]

Goldblatt, P. (ed.). 1993. *Biological Relationships Between Africa and South America*. Yale University Press, New Haven, CT. [6]

Goldschmidt, R. B. 1940. *The Material Basis of Evolution*. Yale University Press, New Haven, CT. [17, 22]

Goliber, T., S. Kessler, J.-J. Chen, G. Bharathan, and N. Sinha. 1999. Genetic, molecular, and morphological analysis of compound leaf development. *Curr. Topics Dev. Biol.* 43: 260–290. [21]

Gómez-Zurita, J., T. Hunt, F. Kopliku, and A. P. Vogler. 2007. Recalibrated tree of leaf beetles (Chrysomelidae) indicates independent diversification of angiosperms and their insect herbivores. *PLoS ONE* 2(4): e360. doi:10.1371/journal.pone.0000360.

Gomez, C., E. M. Özbudak, J. Wunderlich, D. Baumann, J. Lewis, and O. Pourquié. 2008. Control of segment number in vertebrate embryos. *Nature* 454: 335–339. [21]

Gompel, N., and S. B. Carroll. 2003. Genetic mechanisms and constraints governing the evolution of correlated traits in drosophilid flies. *Nature* 424: 931–935. [21]

Gomulkiewicz, R., and R. D. Holt. 1995. When does evolution by natural selection prevent extinction? *Evolution* 49: 201–207. [13]

González, J., K. Lenkov, M. Lipatov, J. M. Macpherson, and D. A. Petrov. 2008. High rate of recent transposable element-induced adaptation in *Drosophila melanogaster*. *PLoS Biol.* 6: e251. [8]

Goodman, M., B. F. Koop, J. Czelusniak, D. H. A. Fitch, D. A. Tagle, and J. L. Slightom. 1989. Molecular phylogeny of the family of apes and humans. *Genome* 31: 316–335. [2]

Gould, F. 1998. Sustainability of transgenic insecticidal cultivars: Integrating pest genetics and ecology. *Annu. Rev. Entomol.* 443: 701–726. [23]

Gould, F. 2008. Broadening the application of evolutionarily based genetic pest management. *Evolution* 62: 500–510. [23]

Gould, F. 2010. Applying evolutionary biology: from retrospective analysis to direct manipulation. In M. A. Bell, D. J. Futuyma, W. F. Eanes, and J. S. Levinton (eds.), *Evolution Since Darwin: The First 150 Years*, pp. 591–621. Sinauer, Sunderland, MA. [23]

Gould, F., N. Blair, M. Reid, T. L. Rennie, J. Lopez, and S. Micinski. 2002. *Bacillus thuringiensis*-toxin resistance management: Stable isotope assessment of alternate host use by *Helicoverpa zea*. *Proc. Natl. Acad. Sci. USA* 99: 16581–16586. [23]

Gould, F., K. Magori, and Y. Huang. 2006. Genetic strategies for controlling mosquito-borne diseases. *Am. Sci.* 94: 238–246. [23]

Gould, J. and C. G. Gould. 1989. *Sexual Selection*. Scientific American Library, New York. [15]

Gould, S. J. 1974. The origin and function of "bizarre" structures: Antler size and skull size in the "Irish elk," *Megaloceros giganteus*. *Evolution* 28: 191–220. [3]

Gould, S. J. 1977. *Ontogeny and Phylogeny*. Harvard University Press, Cambridge, MA. [3, 21]

Gould, S. J. 1981. *The Mismeasure of Man*. Norton, New York. [9]

Gould, S. J. 1982. The meaning of punctuated equilibrium and its role in validating a hierarchical approach to macroevolution. In R. Milkman (ed.), *Perspectives on Evolution*, pp. 83–104. Sinauer, Sunderland, MA. [11]

Gould, S. J. 1985. The paradox of the first tier: An agenda for paleobiology. *Paleobiology* 11: 2–12. [7]

Gould, S. J. 1989. *Wonderful Life: The Burgess Shale and the Nature of History*. W. W. Norton, New York. [5, 22]

Gould, S. J. 1999. *Rocks of Ages*. Ballantine, New York. [23]

Gould, S. J. 2002. *The Structure of Evolutionary Theory*. Belknap Press of Harvard University Press, Cambridge, MA. [4, 11, 18, 22, 23]

Gould, S. J. 2007. *Punctuated Equilibrium*. Belknap Press of Harvard University Press, Cambridge, MA. [4, 22]

Gould, S. J., and N. Eldredge. 1993. Punctuated equilibrium comes of age. *Nature* 366: 223–227. [18]

Gould, S. J., and R. C. Lewontin. 1979. The spandrels of San Marco and the Panglossian paradigm. *Proc. R. Soc. Lond. B* 205: 581–598. [14]

Gould, S. J., and E. S. Vrba. 1982. Exaptation: A missing term in the science of form. *Paleobiology* 8: 4–15. [11, 21]

Grace, J. L., and K. L. Shaw. 2011. Coevolution of male mating signal and female preference during early lineage divergence of the Hawaiian cricket, *Laupala cerasina*. *Evolution* 65: 2184–2196. [18]

Gradstein, F., J. Ogg, and A. Smith. 2004. *A Geologic Time Scale 2004*. Cambridge University Press, Cambridge. [4]

Grafen, A. 1991. Modelling in behavioural ecology. In J. R. Krebs and N. B. Davies (eds.), *Behavioural Ecology*, 3rd ed., pp. 5–31. Blackwell Scientific, Oxford. [16]

Grainger, J., S. Dufau, M. Montant, J. C. Zigler, and J. Fagot. 2012. Orthographic processing in baboons (*Papio papio*). *Science* 336: 245–248. [23]

Gramzow, L., M. S. Ritz, and G. Theissen. 2010. On the origin of MADS-domain transcription factors. *Trends Genet.* 26: 149–153. [21]

Grant, B. R., and P. R. Grant. 1989. *Evolutionary Dynamics of a Natural Population: The Large Cactus Finch of the Galápagos*. University of Chicago Press, Chicago. [13]

Grant, B. S. 2002. Sour grapes of wrath. *Science* 297: 940–941. [12]

Grant, B. S. 2012. Industrial melanism. In *Encyclopedia of Life Sciences*, John Wiley & Sons, Ltd, Chichester, UK. doi:10.1002/9780470015902. a0001788.pub3. [12]

Grant, P. R. 1986. *Ecology and Evolution of Darwin's Finches*. Princeton University Press, Princeton, NJ. [9, 13, 19]

Grant, P. R., and B. R. Grant. 2006. Evolution of character displacement in Darwin's finches. *Science* 313: 224–226. [19]

Grant, P. R., and B. R. Grant. 2008. *How and Why Species Multiply: The Radiation of Darwin's Finches*. Princeton University Press, Princeton. [3]

Grant, V. 1966. The selective origin of incompatibility barriers in the plant genus *Gilia*. *Am. Nat.* 100: 99–118. [18]

Grant, V. 1981. *Plant Speciation*. Columbia University Press, New York. [17, 18]

Graur, D., and W.-H. Li. 2000. *Fundamentals of Molecular Evolution*, 2nd ed. Sinauer, Sunderland, MA. [10, 20]

Gravel, S., and 9 others. 2011. Demographic history and rare allele sharing among human populations. *Proc. Natl. Acad. Sci. USA* 108: 11983–11988. [10]

Gray, D. A., and W. H. Cade. 2000. Sexual selection and speciation in field crickets. *Proc. Natl. Acad. Sci. USA* 97: 14449–14454. [18]

Gray, R. D., and Q. D. Atkinson. 2003. Language-tree divergence times support the Anatolian theory of Indo-European origin. *Nature* 426: 435–439. [2]

Gray, R. D., A. J. Drummond, and S. J. Greenhill. 2009. Language phylogenies reveal expansion pulses and pauses in Pacific settlement. *Science* 323: 479–483. [6]

Green, P. M., J. A. Naylor, and F. Giannelli. 1995. The hemophilias. *Adv. Genet.* 32: 99–139. [8]

Green, R. E., and 10 others. 2006. Analysis of one million base pairs of Neanderthal DNA. *Nature* 444: 330–336. [20]

Green, R. E., and 55 others. 2010. A draft sequence of the Neandertal genome. *Science* 328: 710–722. [4, 6]

Gregory, T. 2001. Coincidence, coevolution, or causation? DNA content, cell size, and the C-value enigma. *Biol. Rev.* 76: 65–101. [3]

Grimaldi, D. A. 1987. Phylogenetics and taxonomy of *Zygothrica* (Diptera: Drosophilidae). *Bull. Am. Mus. Nat. Hist.* 186: 103–268. [3]

Grimaldi, D. A., and M. S. Engel. 2005. *Evolution of the Insects*. Cambridge University Press, New York. [5]

Grosberg, R. K., and R. R. Strathmann. 2007. The evolution of multicellularity: A minor major transition? *Annu. Rev. Ecol. Evol. Syst.* 38: 621–654. [5, 22]

Gross, B. L., and L. H. Rieseberg. 2005. The ecological genetics of homoploid hybrid speciation. *J. Hered.* 96: 241–252. [18]

Gross, M. 1984. Sunfish, salmon, and the evolution of alternative reproductive strategies and tactics in fishes. In G. W. Potts and R. J. Wootton (eds.), *Fish Reproduction: Strategies and Tactics*, pp. 55–75. Academic Press, London. [14]

Gruber, J. D., K. Vogel, G. Kalay, and P. J. Wittkopp. 2012. Contrasting properties of gene-specific regulatory, coding, and copy number mutations in *Saccharomyces cerevisiae*: frequency, effects, and dominance. *PLoS Genet.* 8: e1002497. [21]

Gu, X., Y. Wang, and J. Gu. 2002. Age distribution of human gene families shows significant roles of both large- and small-scale duplications in vertebrate evolution. *Nature Genet.* 31: 205–209. [20]

Gupta, A. P., and R. C. Lewontin. 1982. A study of reaction norms in natural populations of *Drosophila pseudoobscura*. *Evolution* 36: 934–948. [9]

Guss, K. A., C. E. Nelson, A. Hudson, M. E. Kraus, and S. B. Carroll. 2001. Control of a genetic regulatory network by a selector gene. *Science* 292: 1164–1167. [21]

Gwynne, D. T. 2008. Sexual conflict over nuptial gifts in insects. *Annu. Rev. Entomol.* 53: 83–101. [15]

H

Haag-Liautard, C., M. Dorris, X. Maside, S. Macaskill, D. L. Halligan, B. Charlesworth, and P. D. Keightley. 2007. Direct estimation of *per* nucleotide and genomic deleterious mutation rates in *Drosophila*. *Nature* 445: 82–85. [8]

Hackett, S. J., and 17 others. 2008. A phylogenomic study of birds reveals their evolutionary history. *Science* 320: 1763–1768. [2]

Haddrath, O., and A. J. Baker. 2001. Complete mitochondrial DNA genome sequences of extinct birds: ratite phylogenetics and the vicariance biogeography hypothesis. *Proc. R. Soc. Lond. B* 268: 939–945. [6]

Hahn, B. H., G. M. Shaw, K. M. De Cock, and P. M. Sharp. 2000. AIDS as a zoonosis: Scientific and public health implications. *Science* 287: 607–614. [3]

Haig, D. 1993. Genetic conflicts in human pregnancy. *Q. Rev. Biol.* 68: 495–532. [16]

Haig, D. 2002. *Genomic Imprinting and Kinship*. Rutgers University Press, New Brunswick, NJ. [16]

Haig, D. 2007. Weismann rules! OK? Epigenetics and the Lamarckian temptation. *Biol. Philos.* 22: 415–428. [9]

Haldane, J. B. S. 1932. *The Causes of Evolution*. Longmans, Green, New York. [12]

Hall, B. G. 1982. Evolution on a petri dish: The evolved β-galactosidase system as a model for studying acquisitive evolution in the laboratory. *Evol. Biol.* 15: 85–150. [8]

Hall, B. G. 2007. *Phylogenetic Trees Made Easy. A How-To Manual for Molecular Biologists*, 3rd ed. Sinauer, Sunderland, MA. [2]

Hall, D. W., and S. B. Joseph. 2010. A high frequency of beneficial mutations across multiple fitness components in *Saccharomyces cerevisiae*. *Genetics* 185: 1397–1409. [8]

Hall, M. C., C. J. Basten, and J. H. Willis. 2006. Pleiotropic quantitative trait loci contribute to population divergence in traits associated with life history variation in *Mimulus guttatus*. *Genetics* 172: 1829–1844. [8]

Halligan, D. L., and P. D. Keightley. 2009. Spontaneous mutation accumulation studies in evolutionary genetics. *Annu. Rev. Ecol. Evol. Syst.* 40: 151–172. [8]

Halligan, D. L., F. Oliver, J. Guthri, K. C. Stemshorn, B. Harr, and P. D. Keightley. 2011. Positive and negative selection in murine ultraconserved noncoding elements. *Mol. Biol. Evol.* 28: 2651–2660. [8]

Hallström, B. M., M. Kullberg, M. A. Nilsson, and A. Janke. 2007. Phylogenomic data analyses provide evidence that Xenarthra and Afrotheria are sister groups. *Mol. Biol. Evol.* 24: 2059–2068. [2]

Hamblin, M. T., E. E. Thompson, and A. Di Rienzo. 2002. Complex signatures of natural selection at the Duffy blood group locus. *Am. J. Hum. Genet.* 70: 369–383. [12]

Hamilton, W. D. 1964. The genetical evolution of social behavior, I and II. *J. Theor. Biol.* 7: 1–52. [11, 16]

Hamilton, W. D. 1967. Extraordinary sex ratios. *Science* 156: 477–488. [15]

Hamilton, W. D. 1971. Geometry for the selfish herd. *J. Theor. Biol.* 31: 295–311. [16]

Hamilton, W. D., and M. Zuk. 1982. Heritable true fitness and bright birds: A role for parasites? *Science* 218: 384–387. [15]

Hammer, M. F. 1995. A recent common ancestry for human Y chromosomes. *Nature* 378: 376–378. [10]

Hammerstein, P. 2003. Why is reciprocity so rare in social animals? A protestant appeal. In P. Hammerstein (ed.), *Genetic and Cultural Evolution of Cooperation*, pp. 481–496. MIT Press, Cambridge, MA. [16]

Hancock, A. M., G. Alkorta-Aranburu, D. B. Witonsky, and A. Di Rienzo. 2010. Adaptations to new environments in humans: the role of subtle allele frequency shifts. *Phil. Trans. R. Soc. B* 365: 2459–2468. [12]

Hanifin, C. T., E. D. Brodie, Jr., and E. D. Brodie III. 2008. Phenotypic mismatches reveal escape from arms-race coevolution. *PLoS Biol.* 6(3): e60, pp. 0471–0482. [19]

Hanken, J. 1984. Miniaturization and its effects on cranial morphology in plethodontid salamanders, genus *Thorius* (Amphibia: Plethodontidae). I. Osteological variation. *Biol. J. Linn. Soc.* 23: 55–75. [3]

Hansen, T. A. 1980. Influence of larval dispersal and geographic distribution on species longevity in neogastropods. *Paleobiology* 6: 193–207. [22]

Hansen, T. F. 2006. The evolution of genetic architecture. *Annu. Rev. Ecol. Evol. Syst.* 37: 123–157. [13]

Hansen, T. F., and D. Houle. 2004. Evolvability, stabilizing selection, and the problem of stasis. In M. Pigliucci and K. Preston (eds.), *Phenotypic*

Integration: Studying the Ecology and Evolution of Complex Phenotypes, pp. 131–150. Oxford University Press, Oxford. [13]

Hansen, T. F., and D. Houle. 2008. Measuring and comparing evolvability and constraint in multivariate characters. *J. Evol. Biol.* 21: 1201–1219. [22]

Hansen, T. F., J. M. Álvarez-Castro, A. J. R. Carter, J. Hermisson, and G. P. Wagner. 2006. Evolution of genetic architecture under directional selection. *Evolution* 60: 1523–1536. [22]

Hansen, T. F., C. Pélabon, and W. S. Armbruster. 2007. Comparing variational properties of homologous floral and vegetative characters in *Dalechampia scandens*: testing the Berg hypothesis. *Evol. Biol.* 34: 86–98. [13]

Harmon, L. J., and 18 others. 2010. Early bursts of body size and shape evolution are rare in comparative data. *Evolution* 64: 2385–2396. [7]

Harris, H., and D. A. Hopkinson. 1972. Average heterozygosity in man. *J. Hum. Genet.* 36: 9–20. [9]

Harris, M. P., S. M. Hasso, M. W. J. Ferguson, and J. F. Fallon. 2006. The development of archosaurian first-generation teeth in a chicken mutant. *Curr. Biol.* 16: 371–377. [3, 21]

Harrison, R. G. 1979. Speciation in North American field crickets: Evidence from electrophoretic comparisons. *Evolution* 33: 1009–1023. [17]

Harrison, R. G. 1990. Hybrid zones: Windows on evolutionary process. *Oxford Surv. Evol. Biol.* 7: 69–128. Oxford University Press, New York. [17]

Harrison, R. G. (ed.). 1993. *Hybrid Zones and the Evolutionary Process.* Oxford University Press, New York. [17]

Harrison, R. G. 1998. Linking evolutionary pattern and process: the relevance of species concepts for the study of speciation. In D. J. Howard and S. H. Berlocher (eds.), *Endless Forms: Species and Speciation*, pp. 19–31. Oxford University Press, New York. [17]

Harrison, T. 2010. Apes among the tangled branches of human origins. *Science* 327: 532–533. [4]

Hartfield, M., and S. P. Otto. 2011. Recombination and hitchhiking of deleterious alleles. *Evolution* 65: 2421–2434. [15]

Hartl, D. L., and A. G. Clark. 1989. *Principles of Population Genetics*, 2nd ed. Sinauer, Sunderland, MA. [9, 10, 12]

Hartl, D. L., and A. G. Clark. 1997. *Principles of Population Genetics*, 3rd ed. Sinauer, Sunderland, MA. [9, 12]

Hartl, D. L., and E. W. Jones. 2001. *Genetics: Analysis of Genes and Genomes.* Jones and Bartlett, Sudbury, MA. [8]

Hartwell, L. H., L. Hood, M. L. Goldberg, A. E. Reynolds, L. M. Silver, and R. C. Veres. 2000. *Genetics: From Genes to Genomes.* McGraw-Hill Higher Education, Boston. [3, 8]

Harvey, P. H., and M. D. Pagel. 1991. *The Comparative Method in Evolutionary Biology.* Oxford University Press, Oxford. [11]

Harvey, P. H., and R. M. May. 1989. Out for the sperm count. *Nature* 337: 508–509. [16]

Hatchwell, B. J. 2009. The evolution of cooperative breeding in birds: kinship, dispersal and life history. *Phil. Trans. R. Soc. B* 364: 3217–3227. [16]

Hattori, D., E. Demir, H. W. Kim, E. Viragh, S. L. Zipursky, and B. J. Dickson. 2007. *Dscam* diversity is essential for neuronal wiring and self-recognition. *Nature* 449: 223–227. [20]

Hauser, M., K. McAuliffe, and P. R. Blake, 2009. Evolving the ingredients for reciprocity and spite. *Phil. Trans. R. Soc. B* 364: 3255–3266. [16]

Hausfater, G., and S. Hrdy (eds.). 1984. *Infanticide: Comparative and Evolutionary Perspectives.* Aldine Publishing Co., New York. [16]

Haussmann, M. F., D. W. Winkler, K. M. O'Reilly, C. E. Huntington, I. C. T. Nisbet, and C. M. Vleck. 2003. Telomeres shorten more slowly in long-lived birds and mammals than in short-lived ones. *Proc. Natl. Acad. Sci. USA* 270: 1387–1392. [14]

Hay, A. J., V. Gregory, A. R. Douglas, and Y. P. Lin. 2001. The evolution of human influenza viruses. *Phil. Trans. R. Soc. Lond. B* 356: 1861–1870. [19]

Hay, B. A., C.-H. Chen, C. M. Ward, H. Huang, J. T. Su, and M. Guo. 2010. Engineering the genomes of wild insect populations: challenges and opportunities provided by synthetic *Medea* selfish genetic elements. *J. Insect Physiol.* 56: 1402–1413. [23]

Haygood, R., C. C. Babbitt, O. Fedrigo, and G. A. Wray. 2010. Contrasts between adaptive coding and noncoding changes during human evolution. *Proc. Natl. Acad. Sci. USA* 107: 7853–7857. [12]

Hayman, P., J. Marchant, and T. Prater. 1986. *Shorebirds: An Identification Guide to the Waders of the World.* Houghton Mifflin, Boston, MA. [3]

Hazen, R. M. 2005. *Genesis: The Scientific Quest for Life's Origin.* Joseph Henry Press, Washington, D.C. [5]

Head, M. L., J. Hunt, M. D. Jennions, and R. Brooks. 2005. The indirect benefits of mating with attractive males outweigh the direct costs. *PLoS Biology* 3: 0289–0294 (doi: 10.1371/journal.pbio.0030033). [15]

Hedrick, P. W. 1986. Genetic polymorphisms in heterogeneous environments: A decade later. *Annu. Rev. Ecol. Syst.* 17: 535–566. [12]

Hellberg, M. E., A. B. Dennis, P. Arbour-Reily, J. E. Aagaard, and W. J. Swanson. 2012. The *Tegula* tango: a coevolutionary dance of interacting, positively selected sperm and egg proteins. *Evolution* 66: 1681–1694. [18]

Hendry, A. P., and M. T. Kinnison. 1999. Perspective: The pace of modern life: Measuring rates of contemporary microevolution. *Evolution* 53: 1637–1653. [11, 13]

Hendry, A. P., P. Nosil, and L. H. Rieseberg. 2007. The speed of ecological speciation. *Func. Ecol.* 21: 455-464. [18]

Henn, B. M., S. Gravel, A. Moreno-Estrada, S. Acevedo-Acevedo, and C. D. Bustamante. 2010. Fine-scale population structure and the era of next-generation sequencing. *Human Mol. Genet.* 19 (Review Issue 2): R221–R226. [9]

Henn, B. M., and 18 others. 2011. Hunter-gatherer genomic diversity suggests a southern African origin for modern humans. *Proc. Natl. Acad. Sci. USA* 108: 5154–5162. [10]

Hennig, W. 1966. *Phylogenetic Systematics.* University of Illinois Press, Urbana. [2]

Herre, E. A., N. Knowlton, U. G. Mueller, and S. A. Rehner. 1999. The evolution of mutualism: Exploring the paths between conflict and cooperation. *Trends Ecol. Evol.* 14: 49–53. [19]

Herrera, C. M., and P. Bazaga. 2011. Untangling individual variation in natural populations: ecological, genetic and epigenetic correlates of long-term inequality in herbivory. *Mol. Ecol.* 20: 1675–1688. [9]

Herrera, C. M., and O. Pellmyr (eds.). 2002. *Plant–Animal Interactions: An Evolutionary Approach.* Blackwell Science, Malden, MA. [19]

Herrnstein, R. J., and C. Murray. 1994. *The Bell Curve: Intelligence and Class Structure in American Life.* The Free Press, New York. [9]

Hewitt, G. M. 2000. The genetic legacy of the Quaternary ice ages. *Nature* 405: 907–913. [6]

Hewitt, G. M. 2004. Genetic consequences of climatic oscillations in the Quaternary. *Phil. Trans. R. Soc. Lond. B.* 359: 183–195. [6]

Hey, J. 2011. Regarding the confusion between the population concept and Mayr's "population thinking." *Q. Rev. Biol.* 86: 253–264. [1]

Hey, J., and C. Pinho. 2012. Population genetics and objectivity in species diagnosis. *Evolution* 66: 1413–1429. [17]

Hickerson, M. J., and 8 others. 2010. Phylogeography's past, present, and future: 10 years after Avise, 2000. *Mol. Phyl. Evol.* 54: 291–301. [6]

Higgie, M., and M. W. Blows. 2007. Are traits that experience reinforcement also under sexual selection? *Am. Nat.* 170: 409–420. [18]

Higgie, M., and M. W. Blows. 2008. The evolution of reproductive character displacement conflicts with how sexual selection operates within a species. *Evolution* 62: 1192–1203. [18]

Hileman, L. C., and P. Cubas. 2009. An expanded evolutionary role for flower symmetry genes. *J. Biol.* 8: 90. [21]

Hill, G. E. 1991. Plumage color is a sexually selected indicator of male quality. *Nature* 350: 337–339. [15]

Hill, W. G., and A. Caballero. 1992. Artificial selection experiments. *Annu. Rev. Ecol. Syst.* 23: 287–310. [13]

Hill, W. G., and A. Robertson. 1966. The effect of linkage on the limits to artificial selection. *Genet. Res.* 8: 269–294. [15]

Hillier, L. W., and the International Chicken Genome Sequencing Consortium. 2004. Sequence and comparative analysis of the chicken genome provide unique perspectives on vertebrate evolution. *Nature* 432: 695–716. [20]

Hillis, D. M. 2010. Phylogenetic progress and applications of the tree of life. In M. A. Bell, D. J. Futuyma, W. F. Eanes, and J. S. Levinton (eds.), *Evolution Since Darwin: The First 150 Years*, pp. 421–449. Sinauer, Sunderland, MA. [5]

Hillis, D. M., J. J. Bull, M. E. White, M. R. Badgett, and I. J. Molineaux. 1992. Experimental phylogenetics: Generation of a known phylogeny. *Science* 255: 589–592. [2]

Hobbs, H. H., M. S. Brown, J. L. Goldstein, and D. W. Russell. 1986. Deletion of exon encoding cysteine-rich repeat of low density lipoprotein receptor alters its binding specificity in a subject with familial hypercholesterolemia. *J. Biol. Chem.* 261: 13114–13120. [8]

Hoekstra, H. E., and J. A. Coyne. 2007. The locus of evolution: Evo devo and the genetics of adaptation. *Evolution* 61: 995–1016. [21]

Hoekstra, H. E., K. F. Drumm, and M. W. Nachman. 2004. Ecological genetics of adaptive color polymorphism in pocket mice: geographic variation in selected and neutral genes. *Evolution* 58: 1329–1341. [12]

Hoekstra, H. E., R. J. Hirschmann, R. A. Bundey, P. A. Insel, and J. P. Crossland. 2006. A single amino acid mutation contributes to adaptive beach mouse color pattern. *Science* 313:101–104. [3]

Hofstadter, R. 1955. *Social Darwinism in American Thought.* Beacon Press, Boston, MA. [1, 23]

Holden, C., and R. A. Mace. 1997. A phylogenetic analysis of the evolution of lactose digestion. *Hum. Biol.* 69: 605–628. [11]

Holland, B. and W. R. Rice. 1998. Chase-away selection: Antagonistic seduction versus resistance. *Evolution* 52: 1–7. [15]

Holland, B. and W. R. Rice. 1999. Experimental removal of sexual selection reverses intersexual antagonistic coevolution and removes a reproductive load. *Proc. Natl. Acad. Sci. USA* 96: 5083–5088. [15]

Hölldobler, B., and E. O. Wilson. 1983. The evolution of communal nest-weaving in ants. *Am. Sci.* 71: 490–499. [11]

Holmes, E. C. 2010. The gorilla connection. *Nature* 467: 404–405. [2]

Holsinger, K. E. 1991. Inbreeding depression and the evolution of plant mating systems. *Trends Ecol. Evol.* 6: 307–308. [15]

Holt, R. D. 1996. Demographic constraints in evolution: Towards unifying the evolutionary theories of senescence and niche conservatism. *Evol. Ecol.* 10: 1–11. [22]

Holt, R. D. 2003. On the evolutionary ecology of species' ranges. *Evol. Ecol. Res.* 5: 159–178. [6]

Holt, R. D., and M. S. Gaines. 1992. Analysis of adaptation in heterogeneous landscapes: Implications for the evolution of fundamental niches. *Evol. Ecol.* 6: 433–447. [6]

Holt, S. B. 1955. Genetics of dermal ridges: Frequency distribution of total finger ridge count. *Ann. Hum. Genet.* 20: 270–281. [9]

Hooper, J. 2002. *Of Moths and Men: Intrigue, Tragedy and the Peppered Moth.* Fourth Estate, London. [12]

Hotopp, J. C. D., and 19 others. 2007. Widespread lateral gene transfer from intracellular bacteria to multicellular eukaryotes. *Science* 317: 1753–1756. [20]

Hougen-Eitzman, D., and M. D. Rausher. 1994. Interactions between herbivorous insects and plant-insect coevolution. *Am. Nat.* 143: 677–697. [19]

Houle, D. 1992. Comparing evolvability and variability of quantitative traits. *Genetics* 130: 195–204. [13, 22]

Houle, D., and A. S. Kondrashov. 2006. Mutation. In C. W. Fox and J. B. Wolf (eds.), *Evolutionary Genetics*, pp. 32–48. Oxford University Press, New York. [8]

Houle, D., B. Morikawa, and M. Lynch. 1996. Comparing mutational variabilities. *Genetics* 143: 1467–1483. [13]

Houston, A. I., T. Székely, and J. M. McNamara. 2005. Conflict between parents over care. *Trends Ecol. Evol.* 20: 33–38. [16]

Howard, D. J. 1999. Conspecific sperm and pollen precedence and speciation. *Annu. Rev. Ecol. Syst.* 30: 109–132. [17]

Howell, F. C. 1978. Hominidae. In V. J. Maglio and H. B. S. Cooke (eds.), *Evolution of African Mammals*, pp. 154–248. Harvard University Press, Cambridge, MA. [4]

Hu, D., L. Hou, L. Zhang, and X. Xu. 2009. A pre-*Archaeopteryx* troodontid theropod from China with long feathers on the metatarsus. *Nature* 461: 640–643. [4]

Huang, Y.-X., K. Magori, A. L. Lloyd, and F. Gould. 2007. Introducing desirable transgenes into insect populations using Y-linked meiotic drive – A theoretical assessment. *Evolution* 61: 717–726. [23]

Hubbell, S. P. 2006. Neutral theory and the evolution of ecological equivalence. *Ecology* 87: 1387–1398. [7]

Hudson, R. R. 1990. Gene genealogies and the coalescent process. In D. J. Futuyma and J. Antonovics (eds.), *Oxford Surv. Evol. Biol.*, pp. 1–44. Oxford University Press, New York. [10]

Hudson, R. R., and J. A. Coyne. 2002. Mathematical consequences of the genealogical species concept. *Evolution* 56: 1557–1565. [17]

Hueffer, K., J. S. L. Parker, W. S. Weichert, R. E. Geisel, J.-Y. Sgro, and C. R. Parrish. 2003. The natural host range shift and subsequent evolution of canine parvovirus resulted from virus-specific binding to the canine transferrin receptor. *J. Virol.* 77: 1718–1726. [19]

Huelsenbeck, J. P., F. Ronquist, R. Nielsen, and J. P. Bollback. 2001. Bayesian inference of phylogeny and its impact on evolutionary biology. *Science* 294: 2310–2314. [2]

Hughes, W. O. H., B. P. Oldroyd, M. Beekman, and F. L. W. Ratnieks. 2008. Ancestral monogamy shows kin selection is key to the evolution of eusociality. *Science* 320: 1213–1216. [16]

Hull, D. L. 1973. *Darwin and his Critics: The Reception of Darwin's Theory of Evolution by the Scientific Community.* Harvard University Press, Cambridge, MA. [1]

Hull, D. L. 1988. *Science as a Process: An Evolutionary Account of the Social and Conceptual Development of Science.* University of Chicago Press, Chicago. [23]

Humphries, C. J., and L. R. Parenti. 1986. *Cladistic biogeography.* Clarendon Press, Oxford. [6]

Hunt, G. 2007a. The relative importance of directional change, random walks, and stasis in the evolution of fossil lineages. *Proc. Natl. Acad. Sci. USA* 104: 18404–18408. [4]

Hunt, G. 2007b. Evolutionary divergence in directions of high phenotypic variance in the ostracode genus *Poseidonamicus*. *Evolution* 61: 1560–1576. [22]

Hunt, G. 2010. Evolution in fossil lineages: Paleontology and *The Origin of Species*. *Am. Nat.* 176 (Suppl.): S61–S76. [4]

Hunt, G. R., and R. D. Gray. 2004. The crafting of hook tools by wild New Caledonian crows. *Proc. R. Soc. Lond. B* 271 (Suppl. 3): S88–S90. [23]

Huntley, J. W., and M. Kowalewski. 2007. Strong coupling of predation intensity and diversity in the Phanerozoic fossil record. *Proc. Natl. Acad. Sci. USA* 104: 15006–15010. [5]

Hurst, G. D. D., and J. H. Werren. 2001. The role of selfish genetic elements in eukaryotic evolution. *Nature Rev. Genet.* 2: 597–606. [11]

Hurst, L. D., A. Atlan, and B. D. Bengtsson. 1996. Genetic conflicts. *Q. Rev. Biol.* 71: 317–364. [16]

Husband, B. C. 2000. Constraints on polyploid evolution: A test of the minority cytotype exclusion principle. *Proc. R. Soc. Lond. B* 267: 217–223. [18]

Huston, M. 1994. Biological diversity: The coexistence of species on changing landscapes. Cambridge University Press, New York. [6]

Hutchinson, G. E. 1957. Concluding remarks. *Cold Spring Harbor Symp. Quant. Biol.* 22: 415–427. [6]

Hutchinson, J. 1969. *Evolution and Phylogeny of Flowering Plants*. Academic Press, New York. [3]

I

Iemmola, F., and A. Camperio-Ciani. 2009. New evidence of genetic factors influencing sexual orientation in men: female fecundity increase in the maternal line. *Arch. Sexual Behav.* 38: 393–399. [16]

Inglis, R. F., A. Gardner, P. Cornelis, and A. Buckling. 2009. Spite and virulence in the bacterium *Pseudomonas aeruginosa*. *Proc. Natl. Acad. Sci. USA* 106: 5703–5707. [16]

Ingman, M., H. Kaessmann, S. Pääbo, and U. Gyllensten. 2000. Mitochondrial genome variation and the origin of modern humans. *Nature* 408: 708–713. [6]

Innocenti, P., E. H. Morrow, and D. K. Dowling. 2011. Experimental evidence supports a sex-specific selective sieve in mitochondrial genome evolution. *Science* 332: 845–848. [1, 16]

International Aphid Genomics Consortium. 2010. Genome sequence of the pea aphid *Acyrthosiphum pisum*. *PLoS Biol.* 8(2): e1000313. [3]

International Human Genome Sequencing Consortium. 2001. Initial sequencing and analysis of the human genome. *Nature* 409: 860–921. [3, 8, 20]

Ioerger, T. R., A. G. Clark, and T.-H. Kao. 1990. Polymorphism at the self-incompatibility locus in Solanaceae predates speciation. *Proc. Natl. Acad. Sci. USA* 87: 9732–9735. [12]

Iyengar, V. K., and T. Eisner. 1999. Female choice increases offspring fitness in an arctiid moth (*Utetheisa ornatrix*). *Proc. Natl. Acad. Sci. USA* 96: 15013–15016. [15]

J

Jablonka, E., and G. Raz. 2009. Transgenerational epigenetic inheritance: prevalence, mechanisms, and implications for the study of heredity and evolution. *Q. Rev. Biol.* 84: 131–175. [9]

Jablonski, D. 1987. Heritability at the species level: Analysis of geographic ranges of Cretaceous molluscs. *Science* 238: 360–363. [7, 11]

Jablonski, D. 1995. Extinctions in the fossil record. In J. H. Lawton and R. M. May (eds.), *Extinction Rates*, pp. 25–44. Oxford University Press, Oxford. [5, 7]

Jablonski, D. 2002. Survival without recovery after mass extinctions. *Proc. Natl. Acad. Sci. USA* 99: 8139–8144. [7]

Jablonski, D. 2005. Mass extinctions and macroevolution. *Paleobiology* 31 (Suppl.): 192–210. [7]

Jablonski, D. 2008a. Biotic interactions and macroevolution: Extensions and mismatches across scales and levels. *Evolution* 62: 715–739. [7]

Jablonski, D. 2008b. Species selection: theory and data. *Annu. Rev. Ecol. Evol. Syst.* 39: 501–524. [7, 11]

Jablonski, D., and D. J. Bottjer. 1990. Onshore-offshore trends in marine invertebrate evolution. In R. M. Ross and W. D. Allmon (eds.), *Causes of Evolution: A Paleontological Perspective*, pp. 21–75. University of Chicago Press, Chicago. [7]

Jablonski, D., and G. Hunt. 2006. Larval ecology, geographic range, and species survivorship in Cretaceous molluscs. *Am. Nat.* 168: 556–564. [11]

Jablonski, D., and R. A. Lutz. 1983. Larval ecology of marine benthic invertebrates: Paleobiological implications. *Biol. Rev.* 58: 21–89. [22]

Jablonski, D., and K. Roy. 2003. Geographical range and speciation in fossil and living molluscs. *Proc. R. Soc. Lond. B* 270: 401–406. [7]

Jablonski, D., S. J. Gould, and D. M. Raup. 1986. The nature of the fossil record: A biological perspective. In D. M. Raup and D. Jablonski (eds.), *Patterns and Processes in the History of Life*, pp. 7–22. Springer-Verlag, Berlin. [4]

Jablonski, D., D. H. Erwin, and J. H. Lipps (eds.). 1996. *Evolutionary Paleobiology: Essays in Honor of James W. Valentine*. University of Chicago Press, Chicago. [4, 7]

Jablonski, D., K. Roy, and J. W. Valentine. 2006. Out of the tropics: Evolutionary dynamics of the latitudinal diversity gradient. *Science* 314: 102–106. [6]

Jablonski, N. G., and G. Chaplin. 2000. The evolution of human skin coloration. *J. Human Evol.* 39: 57–106. [12]

Jackson, A. P., and M. A. Charleston. 2004. A cophylogenetic perspective of RNA-virus evolution. *Mol. Biol. Evol.* 21: 45–57. [19]

Jackson, J. B. C. 1974. Biogeographic consequences of eurytopy and stenotopy among marine bivalves and their biogeographic significance. *Am. Nat.* 104: 541–560. [7]

Jackson, J. B. C. 1995. Constancy and change of life in the sea. In J. H. Lawton and R. M. May (eds.), *Extinction Rates*, pp. 45-54. Oxford University Press, Oxford. [5]

Jackson, J. B. C., and K. G. Johnson. 2000. Life in the last few million years. *Paleobiology* 26 (Suppl.): 221–235. [7]

Jackson, S. T., and J. T. Overpeck. 2000. Responses of plant populations and communities to environmental changes of the late Quaternary. *Paleobiology* 26 (Suppl.): 194–220. [6]

Jackson, S. T., and J. W. Williams. 2004. Modern analogs in Quaternary paleoecology: Here today, gone yesterday, gone tomorrow? *Annu. Rev. Earth Planet. Sci.* 32: 495–537. [5]

Jacob, F. 1977. Evolution and tinkering. *Science* 196: 1161–1166. [21]

Jain, S. K., and D. R. Marshall. 1967. Population studies on predominantly self-pollinating species. X. Variation in natural populations of *Avena fatua* and *A. barbata*. *Am. Nat.* 101: 19–33. [9]

Jakobsson, M., and 23 others. 2008. Genotype, haplotype and copy-number variation in worldwide human populations. *Nature* 451: 998–1005. [10]

James, S., C. P. Simmons, and A. A. James. 2011. Mosquito trials. *Science* 334: 771–772. [23]

Janis, C. M. 1993. Tertiary mammal evolution in the context of changing climates, vegetation, and tectonic events. *Annu. Rev. Ecol. Syst.* 24: 467–500. [7]

Jansson, R., and M. Dynesius. 2002. The fate of clades in a world of recurrent climatic change: Milankovitch oscillations and evolution. *Annu. Rev. Ecol. Syst.* 33: 741–777. [22]

Janz, N., S. Nylin, and N. Wahlberg. 2006. Diversity begets diversity: Host expansions and the diversification of plant-feeding insects. *BMC Evolutionary Biology* 6: 4, doi:10.1186/1471-2148/6/4. (http:www.biomedcentral.com/1471-2148/6/4) [7]

Janzen, D. H. 1970. Herbivores and the number of tree species in tropical forests. *Am. Nat.* 104: 501–528. [19]

Jarne, P., and D. Charlesworth. 1993. The evolution of the selfing rate in functionally hermaphroditic plants and animals. *Annu. Rev. Ecol. Syst.* 24: 441–466. [15]

Jarzebowska, R., and J. Radwan. 2010. Sexual selection counteracts extinction of small populations of the bulb mite. *Evolution* 64: 1283–1289. [15]

Jensen, A. R. 1973. *Educability and Group Differences*. Harper & Row, New York. [9]

Jeong, S., A. Rokas, and S. B. Carroll. 2006. Regulation of body pigmentation by the Abdominal-B Hox protein and its gain and loss in *Drosophila* evolution. *Cell* 125: 1387–1399. [21]

Jeong, S., M. Rebeiz, P. Andolfatto, T. Werner, J. True, and S. B. Carroll. 2008. The evolution of gene regulation underlies a morphological difference between two *Drosophila* sister species. *Cell* 132: 783–793. [21]

Jernvall, J., J. P. Hunter, and M. Fortelius. 1996. Molar tooth diversity, disparity, and ecology in Cenozoic ungulate radiations. *Science* 274: 1469–1492. [7]

Jetz, W., and P. V. A. Fine. 2012. Global gradients in vertebrate diversity predicted by historical area-productivity dynamics and contemporary environment. *PLoS Biol.* 10(3), article e1001292, doi:10.1371/journal.pbio.1001292. [6]

Johnson, J. A., M. R. Bellinger, J. E. Toepfer, and P. Dunn. 2004. Temporal changes in allele frequencies and low effective population size in greater prairie-chickens. *Mol. Ecol.* 13: 2617–2630. [10]

Johnson, L. J. 2007. The genome strikes back: the evolutionary importance of defence against mobile elements. *Evol. Biol.* 34: 121–129. [16]

Johnson, N. K., and C. Cicero. 2004. New mitochondrial DNA data affirm the importance of Pleistocene speciation in North American birds. *Evolution* 58: 1122–1130. [18]

Johnstone, R. A. 1995. Sexual selection, honest advertisement and the handicap principle: Reviewing the evidence. *Biol. Rev.* 70: 1–65. [15]

Johnstone, R. A., and K. Norris. 1993. Badges of status and the cost of aggression. *Behav. Ecol. Sociobiol.* 32: 127–134. [16]

Jokela, J., and C. M. Lively. 1995. Parasites, sex, and early reproduction in a mixed population of freshwater snails. *Evolution* 49: 1268–1271. [15]

Jokela, J., M. F. Dybdahl, and C. M. Lively. 2009. The maintenance of sex, clonal dynamics, and host-parasite coevolution in a mixed population of sexual and asexual snails. *Am. Nat.* 174 (Suppl.): S43–S53. [15]

Jones, A. G., S. J. Arnold, and R. Bürger. 2007. The mutation matrix and the evolution of evolvability. *Evolution* 61: 727–745. [22]

Jones, F. C., and 28 others. 2012. The genomic basis of adaptive evolution in threespine sticklebacks. *Nature* 484: 55–61. [13]

Jones, I. L., and F. M. Hunter. 1999. Experimental evidence for mutual inter- and intrasexual selection favouring a crested auklet ornament. *Animal Behav.* 57: 521–528. [15]

Jones, J. H. 2011. Primates and the evolution of long, slow life histories. *Curr. Biol.* 21: R708–R717. [14]

Jones, S., R. Martin, and D. Pilbeam (eds.). 1992. *The Cambridge Encyclopedia of Human Evolution.* Cambridge University Press, Cambridge. [4]

Jones, T. M., R. J. Quinnell, and A. Balmford. 1998. Fisherian flies: Benefits of female choice in a lekking sandfly. *Proc. R. Soc. Lond. B* 265: 1651–1657. [15]

Judd, W. S., C. S. Campbell, E. A. Kellogg, P. F. Stevens, and M. J. Donoghue. 2002. *Plant Systematics: A Phylogenetic Approach*, 2nd ed. Sinauer, Sunderland, MA. [5]

Judd, W. S., C. S. Campbell, E. A. Kellogg, P. F. Stevens, and M. J. Donoghue. 2007. *Plant Systematics: A Phylogenetic Approach*, 3rd ed. Sinauer, Sunderland, MA. [5]

K

Kaiser, D. 2007. The other evolution wars. *Am. Sci.* 95: 518–525. [23]

Kareiva, P. M., J. G. Kingsolver, and R. B. Huey (eds.). 1993. *Biotic Interactions and Global Change.* Sinauer, Sunderland, MA. [5]

Karn, M. N., and L. S. Penrose. 1951. Birth weight and gestation time in relation to maternal age, parity, and infant survival. *Ann. Eugenics* 16: 147–164. [13]

Kasahara, M. 1997. New insights into the genome organization and origin of the major histocompatibility complex: role of chromosomal (genome) duplication and emergence of the adaptive immune system. *Hereditas* 127: 59–66. [20]

Kassen, R., and T. Bataillon. 2006. Distribution of fitness effects among beneficial mutations before selection in experimental populations of bacteria. *Nature Genetics* 38: 484–488. [8]

Katakura, H., and T. Hosogai. 1994. Performance of hybrid ladybird beetles (*Epilachna*) on the host plants of parental species. *Entomol. Exp. Appl.* 71: 81–85. [17]

Kaufman, T. C., M. A. Seeger, and G. Olsen. 1990. Molecular and genetic organization of the *Antennapedia* gene complex of *Drosophila melanogaster. Adv. Genet.* 27: 309–362. [21]

Kawecki, T. J. 2000. The evolution of genetic canalization under fluctuating selection. *Evolution* 54: 1–12. [13]

Kawecki, T. J., R. E. Lenski, D. Ebert, B. Hollis, I. Olivieri, and M. C. Whitlock. 2012. Experimental evolution. *Trends Ecol. Evol.* 27: 547–560.

Kazazian, H. H., Jr. 2004. Mobile elements: Drivers of genome evolution. *Science* 303: 1626–1632. [8]

Kearsey, M. J., and B. W. Barnes. 1970. Variation for metrical characters in *Drosophila* populations. II. Natural selection. *Heredity* 25: 11–21. [13]

Keeling, P. J. 2004. Diversity and evolutionary history of plastids and their hosts. *Am. J. Bot.* 91: 1481–1493. [5]

Keeling, P. J. 2007. Deep questions in the tree of life. *Science* 317: 1875–1876. [5]

Keightley, P. D. 2012. Rates and fitness consequences of new mutations in humans. *Genetics* 190: 295–304. [8]

Keinan, A., and A. G. Clark. 2012. Recent explosive human population growth has resulted in an excess of rare genetic variants. *Science* 336: 740–743. [10]

Keller, L. (ed.) 1993. *Queen Number and Sociality in Insects.* Oxford University Press, Oxford. [16]

Keller, L. (ed.) 1999. *Levels of Selection in Evolution.* Princeton University Press, Princeton, NJ. [11, 16]

Keller, L., and H. K. Reeve. 1994. Partitioning of reproduction in animal societies. *Trends Ecol. Evol.* 9: 98–102. [16]

Keller, L., and H. K. Reeve. 1999. Dynamics of conflicts within insect societies. In L. Keller (ed.), *Levels of Selection in Evolution*, pp. 153–175. Princeton University Press, Princeton, NJ. [16]

Kellermann, V. M., B. van Heerwaarden, A. A. Hoffmann, and C. M. Sgrò. 2006. Very low additive genetic variance and evolutionary potential in multiple populations of two rainforest *Drosophila* species. *Evolution* 60: 1104–1108. [6]

Kellermann, V., B. van Heerwarden, C. M. Sgrò, and A. A. Hoffmann. 2009. Fundamental evolutionary limits in ecological traits drive *Drosophila* species distribution. *Science* 325: 1244–1246. [13]

Kelley, S. T., and B. D. Farrell. 1998. Is specialization a dead end? The phylogeny of host use in *Dendroctonus* bark beetles (Scolytidae). *Evolution* 52: 1731–1743. [6]

Kemp, T. S. 2005. *The Origin and Evolution of Mammals.* Oxford University Press, Oxford. [4]

Kenrick, P., and P. R. Crane. 1997a. The origin and early evolution of plants on land. *Nature* 389: 33–39. [5]

Kenrick, P., and P. R. Crane. 1997b. *The Origin and Early Diversification of Land Plants.* Smithsonian Institution Press, Washington. [5]

Kerney, R. R., D. C. Blackburn, H. Müller, and J. Hanken. 2012. Do larval traits re-evolve? Evidence from the embryogenesis of a direct-developing salamander, *Plethodon cinereus. Evolution* 66: 252–262. [6]

Khersonsky, O., and D. S. Tawfik. 2010. Enzyme promiscuity: a mechanistic and evolutionary perspective. *Annu. Rev. Biochem.* 79: 471–505. [8]

Kidston, R., and W. H. Lang. 1921. On Old Red Sandstone plants showing structure from the Rhynie chert bed, Aberdeenshire, Part IV. Restorations of the vascular cryptogams, and discussion of their bearing on the general morphology of Pteridophyta and the origin of the organization of land plants. *Trans. R. Soc. Edinburgh* 32: 477–487. [5]

Kiers, E. T., R. A. Rousseau, S. A. West, and R. F. Denison. 2003. Host sanctions and the legume-rhizobium mutualism. *Nature* 425: 78–81. [16]

Kijimoto, T., J. Andrews, and A. P. Moczek. 2010. Programed cell death shapes the expression of horns within and between species of horned beetles. *Evol. Dev.* 12: 449–458. [21]

Kimura, M. 1955. Solution of a process of random genetic drift with a continuous model. *Proc. Natl. Acad. Sci. USA* 41: 144–150. [10]

Kimura, M. 1968. Evolutionary rate at the molecular level. *Nature* 217: 624–626. [10]

Kimura, M. 1983. *The Neutral Theory of Molecular Evolution.* Cambridge University Press, Cambridge. [1, 10]

Kimura, S., D. Koenig, J. Kang, F. Y. Yoong, and N. Sinha. 2008. Natural variation in leaf morphology results from mutation of a novel *KNOX* gene. *Curr. Biol.* 18: 672–677. [21]

King, J. L., and T. H. Jukes. 1969. Non-Darwinian evolution. *Science* 164: 788–798. [10]

King, M. 1993. *Species Evolution: The Role of Chromosome Change.* Cambridge University Press, Cambridge. [17]

King, M.-C., and A. C. Wilson. 1975. Evolution at two levels in humans and chimpanzees. *Science* 188: 107–116. [21]

King, N., and 35 others. 2008. The genome of the choanoflagellate *Monosiga brevicollis* and the origin of metazoans. *Nature* 451: 783–788. [5, 20]

Kingsolver, J. G., and S. E. Diamond. 2011. Phenotypic selection in natural populations: what limits directional selection? *Am. Nat.* 177: 346–357. [13]

Kingsolver, J. G., and D. W. Pfennig. 2004. Individual-level selection as a cause of Cope's rule of phyletic size increase. *Evolution* 58: 1608–1612. [22]

Kinnison, M. T., and N. G. Hairston, Jr. 2007. Eco-evolutionary conservation biology: contemporary evolution and the dynamics of persistence. *Func. Ecol.* 21: 444–454. [13]

Kirk, C. M., S. P. Blomberg, D. L. Duffy, A. C. Heath, I. P. Owens, N. G. Martin. 2001. Natural selection and quantitative genetics of life-history traits in Western women: a twin study. *Evolution* 55: 423–435. [13]

Kirkpatrick, M. 1982. Sexual selection and the evolution of female choice. *Evolution* 3: 1–12. [15]

Kirkpatrick, M. 2009. Patterns of quantitative genetic variation in multiple dimensions. *Genetica* 136: 271–284. [13]

Kirkpatrick, M., and N. H. Barton. 1997a. Evolution of a species' range. *Am. Nat.* 150: 1–23. [6]

Kirkpatrick, M., and N. H. Barton. 1997b. The strength of indirect selection on female mating preferences. *Proc. Natl. Acad. Sci. USA* 94: 1282–1286. [15]

Kirkpatrick, R. C. 2000. The evolution of human homosexual behavior. *Curr. Anthropol.* 41: 385–414. [16]

Kirschner, M., and J. Gerhart. 2005. *The Plausibility of Life.* Yale University Press, New Haven, CT. [22]

Kisel, Y., and T. G. Barraclough. 2010. Speciation has a spatial scale that depends on gene flow. *Am. Nat.* 175: 316–334. [18]

Kitchen, A., C. Ehret, S. Assefa, and C. J. Mulligan. 2009. Bayesian phylogenetic analysis of Semitic languages identifies an Early Bronze Age origin of Semitic in the Near East. *Proc. R. Soc. Lond. B* 276: 2703–2710. [2]

Kitcher, P. 1985. *Vaulting Ambition.* MIT Press, Cambridge, MA. [15, 16]

Kittles, R. A., and K. M. Weiss. 2003. Race, ancestry, and genes: Implications for defining disease risk. *Annu. Rev. Genomics Human Genet.* 4: 33-67. [9]

Klicka, J., and R. M. Zink. 1997. The importance of recent ice ages in speciation: A failed paradigm. *Science* 277: 1666–1669. [18]

Knecht, A. K., K. E. Hosemann, and D. M. Kingsley. 2007. Constraints on utilization of the EDA signaling pathway in threespine stickleback evolution. *Evol. Dev.* 9: 141–154. [21]

Knoll, A. H. 2003. *Life on a Young Planet.* Princeton University Press, Princeton, N.J. [5]

Knoll, A. H. 2011. The multiple origins of multicellularity. *Annu. Rev. Earth Planet. Sci.* 39: 217–239. [5]

Knoll, A. H., and R. K. Bambach. 2000. Directionality in the history of life: Diffusion from the left wall or repeated scaling of the right? In D. H. Erwin and S. L. Wing (eds.), *Deep Time: Paleobiology's Perspective,* pp. 1–14. The Paleontological Society, Allen Press, Lawrence, KS. [22]

Knoll, A. H., R. K. Bambach, J. L. Payne, S. Pruss, and W. Fischer. 2007. Paleophysiology and end-Permian mass extinction. *Earth and Planetary Sci. Lett.* 256: 295–313. [5]

Knowles, L. L. 2009. Statistical phylogeography. *Annu. Rev. Ecol. Evol. Syst.* 40: 593–612. [6]

Knowles, L. L., and B. C. Carstens. 2007. Delimiting species without monophyletic gene trees. *Syst. Biol.* 56: 887–895. [17]

Knowles, L. L., and L. S. Kubatko (eds.). 2010. *Estimating Species Trees: Practical and Theoretical Aspects.* Wiley, New York. [2]

Knowlton, N., L. A. Weigt, L. A. Solórzano, D. K. Mills, and E. Bermingham. 1993. Divergence in proteins, mitochondrial DNA, and reproductive compatibility across the Isthmus of Panama. *Science* 260: 1629–1632. [18]

Koblmüller, S., U. K. Schliewen, N. Duftner, K. M. Sefc, C. Katongo, and C. Sturmbauer. 2008. Age and spread of the haplochromine cichlid fishes in Africa. *Mol. Phyl. Evol.* 49: 153–169. [3]

Koenig, B. A., S. S-J. Lee, and S. S. Richardson. (eds.) 2008. *Revisiting Race in a Genomic Age.* Rutgers University Press, New Brunswick, NJ. [9]

Kojak, K. H., and J. J. Wiens. 2006. Does niche conservatism promote speciation? A case study in North American salamanders. *Evolution* 60: 2604–2621. [18]

Kokko, H. 1998. Should advertising parental care be honest? *Proc. R. Soc. Lond. B* 265: 1871–1878. [16]

Kokko, H. 2007. *Modelling for Field Biologists.* Cambridge University Press, Cambridge. [14]

Kokko, H., R. Brooks, J. M. McNamara, and A. Houston. 2002. The sexual selection continuum. *Proc. R. Soc. Lond. B* 269: 1331–1340. [15]

Kokko, H., M. D. Jennions, and R. Brooks. 2006. Unifying and testing models of sexual selection. *Annu. Rev. Ecol. Evol. Syst.* 37: 43–66. [15]

Kondrashov, A. S., and F. A. Kondrashov. 1999. Interactions among quantitative traits in the course of sympatric speciation. *Nature* 400: 351–354. [18]

Kopp, A., I. Duncan, and S. B. Carroll. 2000. Genetic control and evolution of sexually dimorphic characters in *Drosophila. Nature* 408: 553–559. [21]

Kopp, A., R. M. Graze, S. Xu, S. B. Carroll, and S.V. Nuzhdin. 2003. Quantitative trait loci responsible for variation in sexually dimorphic traits in *Drosophila melanogaster. Genetics* 163: 771–787. [21]

Korber, B., M. Muldoon, J. Theiler, F. Gao, R. Gupta, A. Lapedes, B. H. Hahn, S. Wolinksy, and T. Bhattacharya. 2000. Timing the ancestor of the HIV-1 pandemic strains. *Science* 288: 1789–1796. [1, 10]

Kornegay, J. R., J. W. Schilling, and A. C. Wilson. 1994. Molecular adaptation of a leaf-eating bird: Stomach lysozyme of the hoatzin. *Mol. Biol. Evol.* 11: 921–928. [20]

Koufopanou, V., and G. Bell. 1991. Developmental mutants of *Volvox:* Does mutation recreate patterns of phylogenetic diversity? *Evolution* 45: 1806–1822. [8]

Kozak, K. H., and J. J. Wiens. 2006. Does niche conservatism promote speciation? A case study in North American salamanders. *Evolution* 60: 2604–2621. [18]

Kraaijeveld, K., F. J. L. Kraaijeveld-Smit, and J. Komdeur. 2007. The evolution of mutual ornamentation. *Anim. Behav.* 74: 657–677. [15]

Kramer, E. M., and M. A. Jaramillo. 2005. Genetic basis for innovations in floral organ identity. *J. Exp. Zool. (Mol. Dev. Evol.)* 304B: 526–535. [22]

Krebs, J. R., and R. H. McCleery. 1984. Optimization in behavioural ecology. In J. R. Krebs and N. B. Davies (eds.), *Behavioural Ecology: An Evolutionary Approach,* 2nd ed., pp. 91–121. Sinauer, Sunderland, MA. [14]

Kreitman, M. 1983. Nucleotide polymorphism at the alcohol dehydrogenase locus of *Drosophila melanogaster. Nature* 304: 412–417. [9]

Krug, A. Z., D. Jablonski, and J. W. Valentine. 2009. Signature of the end-Cretaceous mass extinction in the modern biota. *Science* 323: 767–771. [7]

Kucera, M., and B. A. Malmgren. 1998. Differences between evolution of mean form and evolution of new morphotypes: An example from Late Cretaceous planktonic foraminifera. *Paleobiology* 24: 49–63. [4]

Kumar, S., and S. Subramanian. 2002. Mutation rates in mammalian genomes. *Proc. Natl. Acad. Sci. USA* 99: 803–808. [8]

Kuparinen, A., and J. Merilä. 2007. Detecting and managing fisheries-induced evolution. *Trends Ecol. Evol.* 22: 652–659. [11]

Kuraku, S., A. Meyer, and S. Kuratani. 2009. Timing of genome duplications relative to the origin of the vertebrates: did cyclostomes diverge before or after? *Mol. Biol. Evol.* 26: 47–59. [3]

L

Labandeira, C. C., and J. J. Sepkoski Jr. 1993. Insect diversity in the fossil record. *Science* 261: 310–315. [7]

Lack, D. 1954. *The Natural Regulation of Animal Numbers*. Oxford University Press, Oxford. [14]

Lai, C., R. F. Lyman, A. D. Long, C. H. Langley, and T. F. C. Mackay. 1994. Naturally occurring variation in bristle number and DNA polymorphisms at the *scabrous* locus of *Drosophila melanogaster*. *Science* 266: 1697–1702. [13, 21]

Laland, K. N., and G. R. Brown. 2002. *Sense and Nonsense: Evolutionary Perspectives on Human Behaviour*. Oxford University Press, Oxford. [23]

Laland, K. N., and G. R. Brown. 2011. *Sense and Nonsense: Evolutionary Perspectives on Human Behaviour*, 2nd ed. Oxford University Press, Oxford. [23]

Laland, K. N., J. Odling-Smee, and S. Myles. 2010. How culture shaped the human genome: bringing genetics and the human sciences together. *Nat. Rev. Genet.* 11: 137–148. [23]

Lande, R. 1979. Effective deme sizes during long-term evolution estimated from rates of chromosome rearrangement. *Evolution* 33: 234–251. [8]

Lande, R. 1980. Sexual dimorphism, sexual selection and adaptation in polygenic characters. *Evolution* 34: 292–305. [13]

Lande, R. 1981. Models of speciation by sexual selection on polygenic traits. *Proc. Natl. Acad. Sci. USA* 78: 3721–3725. [15, 18]

Lande, R. 1982. Rapid origin of sexual isolation and character divergence in a cline. *Evolution* 36: 213–223. [18]

Lande, R., and S. J. Arnold. 1983. The measurement of selection on correlated characters. *Evolution* 37: 1210–1226. [11, 13]

Lande, R., and D. W. Schemske. 1985. The evolution of self-fertilization and inbreeding depression. I. Genetic models. *Evolution* 39: 24–40. [15]

Landry, C. R., B. Lemos, S. A. Rifkin, W. J. Dickinson, and D. L. Hartl. 2007. Genetic properties influencing the evolvability of gene expression. *Science* 317: 118–121. [8]

Lane, C. E. 2010. The genomic imprint of endosymbiosis. In M. A. Bell, D. J. Futuyma, W. F. Eanes, and J. S. Levinton (eds.), *Evolution Since Darwin: The First 150 Years*, pp. 377–380. Sinauer, Sunderland, MA. [5]

Lanfear, R., J. J. Welch, and L. Bromham. 2010. Watching the clock: studying variation in rates of molecular evolution between species. *Trends Ecol. Evol.* 25: 495–503. [10]

Langley, C. H., and W. M. Fitch. 1974. An examination of the constancy of the rate of molecular evolution. *J. Mol. Evol.* 3: 161–177. [2]

Langstrom, N., O. Rahman, E. Carlstrom, and P. Lichtenstein. 2010. Genetic and environmental effects on same-sex sexual behavior: a population study of twins in Sweden. *Arch. Sexual Behav.* 39: 75–80. [16]

Law, R., A. D. Bradshaw, and P. D. Putwain. 1979. The cost of reproduction in annual meadow grass. *Am. Nat.* 113: 3–16. [14]

Lazcano, A. 2010. The origin and early evolution of life: Did it all start in Darwin's warm little pond? In M. A. Bell, D. J. Futuyma, W. F. Eanes, and J. S. Levinton (eds.), *Evolution Since Darwin: The First 150 Years*, pp. 353–375. Sinauer, Sunderland, MA. [5]

Lederberg, J., and E. M. Lederberg. 1952. Replica plating and indirect selection of bacterial mutants. *J. Bacteriol.* 63: 399–406. [8]

Ledon-Rettig, C. C., D. W. Pfennig, and E. J. Crespi. 2010. Diet and hormonal manipulation reveal cryptic genetic variation: implications for the evolution of novel feeding strategies. *Proc. Biol. Sci.* 277: 3569–3578. [21]

Lee, M. S.Y., T. W. Reeder, J. B. Slowinski, and R. Lawson. 2004. Resolving reptile relationships: Molecular and morphological markers. In J. Cracraft and M. J. Donoghue (eds.), *Assembling the Tree of Life*, pp. 451–467. Oxford University Press, New York. [5]

Lehmann, L., and L. Keller. 2006. The evolution of cooperation and altruism – a general framework and a classification of models. *J. Evol. Biol.* 19: 1365-1376. [16]

Lehtonen, J., M. D. Jennions, and H. Kokko. 2012. The many costs of sex. *Trends Ecol. Evol.* 27: 172–178. [15]

Leigh, E. G. Jr. 1973. The evolution of mutation rates. *Genetics* Suppl. 73: 1–18. [15]

Lemmon, E. M., and A. R. Lemmon. 2010. Reinforcement in chorus frogs: lifetime fitness estimates including intrinsic natural selection and sexual selection against hybrids. *Evolution* 64: 1748–1761. [18]

Lenormand, T. 2002. Gene flow and the limits to natural selection. *Trends Ecol. Evol.* 17: 183–189. [6, 12]

Lessios, H. A. 1998. The first stage of speciation as seen in organisms separated by the Isthmus of Panama. In D. J. Howard and S. H. Berlocher (eds.), *Endless Forms: Species and Speciation*, pp. 186–201. Oxford University Press, Oxford. [6]

Levin, B. R., and R. M. Anderson. 1999. The population biology of anti-infective chemotherapy and the evolution of drug resistance: more questions than answers. In S. C. Stearns (ed.), *Evolution in Health and Disease*, pp. 125–137. Oxford University Press, Oxford. [1]

Levin, D. A. 1983. Polyploidy and novelty in flowering plants. *Am. Nat.* 122: 1–25. [18]

Levin, D. A. 1984. Immigration in plants: An exercise in the subjunctive. In R. Dirzo and J. Sarukhán (eds.), *Perspectives on Plant Population Ecology*, pp. 242–260. Sinauer, Sunderland, MA. [9]

Levine, M. T., C. D. Jones, A. D. Kern, H. A. Lindfors, and D. J. Begun. 2006. Novel genes derived from noncoding DNA in *Drosophila melanogaster* are frequently X-linked and exhibit testis-biased expression. *Proc. Natl. Acad. Sci. USA* 103: 9935–9939. [20]

Levinton, J. S. 2001. *Genetics, Paleontology, and Macroevolution*, 2nd ed. Cambridge University Press, Cambridge. [4, 22]

Lewin, B. 1985. *Genes II*. Wiley, New York. [8]

Lewin, R. 2005. *Human Evolution: An Illustrated Introduction*. Blackwell Publishing, Oxford. [4]

Lewis, H. 1962. Catastrophic speciation as a factor in evolution. *Evolution* 16: 257–271. [18]

Lewontin, R. C. 1970. The units of selection. *Annu. Rev. Ecol. Syst.* 1: 1–18. [11]

Lewontin, R. C. 1974. *The Genetic Basis of Evolutionary Change*. Columbia University Press, New York. [9]

Lewontin, R. C. 2000. *The Triple Helix: Gene, Organism, and Environment*. Harvard University Press, Cambridge, MA. [11, 22]

Lewontin, R. C., and J. L. Hubby. 1966. A molecular approach to the study of genic heterozygosity in natural populations. II. Amount of variation and degree of heterozygosity in natural populations of *Drosophila pseudoobscura*. *Genetics* 54: 595–609. [9, 10]

Lewontin, R. C., S. Rose, and L. Kamin. 1984. *Not in Our Genes: Biology, Ideology, and Human Nature*. Pantheon, New York. [9]

Li, G., J. H. Wang, S. J. Rossiter, G. Jones, and S.Y. Zhang. 2007. Accelerated *FoxP2* evolution in echolocating bats. *PLoS One* 2(9), article e900. doi:10.1371/journal.pone.0000900. [8]

Li, J. Z., and 10 others. 2008. Worldwide human relationships inferred from genome-wide human variation. *Science* 319: 1100–1104. [6, 9]

Li, Q., and 8 others. 2010. Plumage color patterns of an extinct dinosaur. *Science* 327: 1369–1372. [4]

Li, Q., and 9 others. 2012. Reconstruction of *Microraptor* and the evolution of iridescent plumage. *Science* 335: 1215–1219. [4]

Li, W.-H. 1997. *Molecular Evolution*. Sinauer, Sunderland, MA. [2, 3, 8, 9, 10, 20]

Li, W.-H., and D. Graur. 1991. *Fundamentals of Molecular Evolution*. Sinauer, Sunderland, MA. [2]

Li,Y., Z. Liu, P. Shi, and J. Zhang. 2010. The hearing gene *Prestin* unites echolocating bats and whales. *Curr. Biol.* 20: R55–R56. [3]

Lieberman, B. S. 2003. Paleobiogeography: The relevance of fossils to biogeography. *Annu. Rev. Ecol. Evol. Syst.* 34: 51–69. [6]

Liebeskind, B. J., D. M. Hillis, and H. H. Zakon. 2011. Evolution of sodium channels predates the origin of nervous systems in animals. *Proc. Natl. Acad. Sci. USA* 108: 9154–9159. [3]

Lilley, D. M. J., and J. Sutherland. 2011. The chemical origins of life and its early evolution: an introduction. *Phil. Trans. R. Soc. B* 366: 2853–2856. [5]

Lincoln, T. A., and G. F. Joyce. 2009. Self-sustained replication of an RNA enzyme. *Science* 323: 1229–1232. [5]

Linville, S. U., R. Breitwitsch, and A. J. Schilling. 1998. Plumage brightness as an indicator of parental care in northern cardinals. *Anim. Behav.* 55: 119–127. [16]

Lion, S., V. A. A. Jansen, and T. Day. 2011. Evolution in structured populations: beyond the kin versus group debate. *Trends Ecol. Evol.* 26: 193–201. [16]

Liow, L. H., T. B. Quental, and C. R. Marshall. 2010. When can decreasing diversification rates be detected with molecular phylogenies and the fossil record? *Syst. Biol.* 59: 646–659. [7]

Liu, W., and 21 others. 2010. Origin of the human malaria parasite *Plasmodium falciparum* in gorillas. *Nature* 467: 420–425. [2]

Lively, C. M., and M. F. Dybdahl. 2000. Parasite adaptation to locally common host genotypes. *Nature* 405: 679–681. [15]

Lloyd, D. G. 1992. Evolutionarily stable strategies of reproduction in plants: Who benefits and how? In R. Wyatt (ed.), *Ecology and Evolution of Plant Reproduction*, pp. 137–168. Chapman & Hall, New York. [15]

Loehlin, D. W., and J. H. Werren. 2012. Evolution of shape by multiple regulatory changes to a growth gene. *Science* 335: 943–947. [21]

Lohmueller, K. E., and 11 others. 2008. Proportionally more deleterious genetic variation in European than in African populations. *Nature* 451: 994–997. [10]

Lomolino, M. V., B. R. Riddle, R. J. Whittaker, and J. H. Brown. 2010. *Biogeography*, 4th ed. Sinauer, Sunderland, MA. [6]

Long, A. D., R. F. Lyman, C. H. Langley, and T. F. C. Mackay. 1998. Two sites in the *Delta* gene region contribute to naturally occurring variation in bristle number in *Drosophila melanogaster*. *Genetics* 149: 999–1017. [21]

Long, A. D., R. F. Lyman, A. H. Morgan, C. H. Langley, and T. F. C. Mackay. 2000. Both naturally occurring inserts of transposable elements and intermediate frequency polymorphisms at the *achaete-scute* complex are associated with variation in bristle number in *Drosophila melanogaster*. *Genetics* 154: 1255–1269. [21]

Long, M., S. J. de Souza, C. Rosenberg, and W. Gilbert. 1996. Exon shuffling and the origin of the mitochondrial targeting function in plant cytochrome *c1* precursor. *Proc. Natl. Acad. Sci. USA* 93: 7727–7731. [20]

Long, M., E. Betran, K. Thornton, and W. Wang. 2003. The origin of new genes: Glimpses of the young and old. *Nature Rev. Genet.* 4: 865–875. [20]

Longini, I. M., Jr., A. Nizam, S. Xu, K. Ungchusak, W. Hanshaoworakul, D. A. T. Cummings, and M. E. Halloran. 2005. Containing pandemic influenza at the source. *Science* 309: 1083–1087. [19]

López, M. A., and C. López-Fanjul. 1993. Spontaneous mutation for a quantitative trait in *Drosophila melanogaster*. I. Response to artificial selection. *Genet. Res.* 61: 107–116. [13]

Losos, J. B. 1992. The evolution of convergent structure in Caribbean *Anolis* communities. *Syst. Biol.* 41: 403–420. [19]

Losos, J. B. 2009. *Lizards in an Evolutionary Tree: Ecology and Adaptive Radiation of Anoles*. University of California Press, Berkeley. [6]

Losos, J. B. 2010. Adaptive radiation, ecological opportunity, and evolutionary determinism. *Am. Nat.* 175: 623–639. [6, 7, 22]

Losos, J. B., T. R. Jackman, A. Larson, K. de Queiroz, and L. Rodríguez-Schettino. 1998. Contingency and determinism in replicated adaptive radiations of island lizards. *Science* 279: 2115–2118. [6]

Lovejoy, A. O. 1936. *The Great Chain of Being: A Study of the History of an Idea*. Harvard University Press, Cambridge, MA. [1]

Lovejoy, C. O. 1981. The origins of man. *Science* 211: 341–350. [4]

Lovette, I. J. 2005. Glacial cycles and the tempo of avian speciation. *Trends Ecol. Evol.* 20: 57–59. [18]

Lovette, I. J., and E. Bermingham. 1999. Explosive speciation in the New World *Dendroica* warblers. *Proc. R. Soc. Lond. B* 266: 1629–1636. [7]

Ludwig, M. Z., and M. Kreitman. 1995. Evolutionary dynamics of the enhancer region of *even-skipped* in *Drosophila*. *Mol. Biol. Evol.* 12: 1002–1011. [21]

Ludwig, M. Z., N. H. Patel, and M. Kreitman. 1998. Functional analysis of *eve stripe 2* enhancer evolution in *Drosophila*: Rules governing conservation and change. *Development* 125: 949–958. [21]

Ludwig, M. Z., C. Bergman, N. H. Patel, and M. Kreitman. 2000. Evidence for stabilizing selection in a eukaryotic enhancer element. *Nature* 403: 564–567. [21]

Luo, Z.-X. 2011. Developmental patterns in Mesozoic evolution of mammalian ears. *Annu. Rev. Ecol. Evol. Syst.* 42: 355–380. [4]

Luo, Z.-X., A. W. Crompton, and A.-L. Sun. 2001. A new mammaliaform from the early Jurassic and evolution of mammalian characteristics. *Science* 292: 1535–1540. [4]

Lupia, R., S. Lidgard, and P. R. Crane. 1999. Comparing palynological abundance and diversity: implications for biotic replacement during the Cretaceous angiosperm radiation. *Paleobiology* 25: 305–340. [7]

Lyman, R. F., F. Lawrence, S. V. Nuzhdin, and T. F. C. Mackay. 1996. Effects of single *P*-element insertions on bristle number and viability in *Drosophila melanogaster*. *Genetics* 143: 277–292. [8]

Lynch, M. 1988. The rate of polygenic mutation. *Genet. Res.* 51: 137–148. [8, 13]

Lynch, M. 1990. The rate of morphological evolution in mammals from the standpoint of the neutral expectation. *Am. Nat.* 136: 727–741. [13]

Lynch, M. 2007. *The Origins of Genome Architecture*. Sinauer, Sunderland, MA. [3, 20]

Lynch, M., and J. S. Conery. 2000. The evolutionary fate and consequences of duplicate genes. *Science* 290: 1151–1155. [20]

Lynch, M., and J. S. Conery. 2003. The origins of genome complexity. *Science* 302: 1401–1404. [20]

Lynch, M., and R. Lande. 1993. Evolution and extinction in response to environmental change. In P. M. Kareiva, J. G. Kingsolver, and R. B. Huey (eds.), *Biotic Interactions and Global Change*, pp. 234–250. Sinauer, Sunderland, MA. [7, 13, 15]

Lynch, M., and J. B. Walsh. 1998. *Genetics and Analysis of Quantitative Traits*. Sinauer, Sunderland, MA. [9, 13]

Lynch, M., R. Bürger, D. Butcher, and W. Gabriel. 1993. The mutational meltdown in an asexual population. *J. Hered.* 84: 339–344. [15]

Lynch, M., J. Blanchard, D. Houle, T. Kibota, S. Schultz, L. Vassilieva, and J. Willis. 1999. Perspective: Spontaneous deleterious mutation. *Evolution* 53: 645–663. [8]

Lynch, V. J., and G. P. Wagner. 2008. Resurrecting the role of transcription factor change in developmental evolution. *Evolution* 62: 2131–2154. [8, 21]

Lynch, V. J., G. May, and G. P. Wagner. 2011. Regulatory evolution through divergence of a phosphoswitch in the transcription factor CEBPB. *Nature* 480: 383–386. [21]

Lynch, V. J., R. D. Leclerc, G. May, and G. P. Wagner. 2011. Transposon-mediated rewiring of gene regulatory networks contributed to the evolution of pregnancy in mammals. *Nature Genet.* 43: 1154–1159. [8]

Lyon, B. E. 1993. Conspecific brood parasitism as a flexible reproductive tactic in American coots. *Anim. Behav.* 46: 911–928. [16]

Lyon, B. E. 2003. Egg recognition and counting reduce costs of avian conspecific brood parasitism. *Nature* 422: 495–499. [16]

Lyon, B. E., and J. McA. Eadie. 2008. Conspecific brood parasitism in birds: A life-history perspective. *Annu. Rev. Ecol. Evol. Syst.* 39: 343–363. [16]

Lyon, B. E., J. M. Eadie, and L. D. Hamilton. 1994. Parental choice selects for ornamental plumage in American coot chicks. *Nature* 371: 240–243. [16]

M

Ma, H., and C. dePamphilis. 2000. The ABCs of floral evolution. *Cell* 101: 5–8. [21]

MacArthur, R. H., and R. Levins. 1967. The limiting similarity, convergence, and divergence of coexisting species. *Am. Nat.* 101: 377–385. [19]

MacDougall-Shackleton, E. A., and S. A. MacDougall-Shackleton. 2001. Cultural and genetic evolution in mountain white-crowned sparrows: Song dialects are associated with population structure. *Evolution* 55: 2568–2575. [9]

Mace, R., and C. J. Holden. 2005. A phylogenetic approach to cultural evolution. *Trends Ecol. Evol.* 20: 116–123. [2, 23]

Mace, R., and F. M. Jordan. 2011. Macro-evolutionary studies of cultural diversity: a review of empirical studies of cultural transmission and cultural adaptation. *Phil. Trans. R. Soc. B* 366: 402–411. [2]

MacFadden, B. J. 1986. Fossil horses from "Eohippus" (*Hyracotherium*) to *Equus*: Scaling, Cope's law, and the evolution of body size. *Paleobiology* 12: 355–369. [4, 22]

Mack, R. H., D. Simberloff, W. M. Lonsdale, H. Evans, M. Clout, and F. A. Bazzaz. 2000. Biotic invasions: causes, epidemiology, global consequences, and control. *Ecol. Appl.* 10: 689–710. [19]

Mackay, T. D. F., and 51 others. 2012. The *Drosophila melanogaster* genetic reference panel. *Nature* 482: 173–178. [12]

Mackay, T. F. C. 2001a. Quantitative trait loci in *Drosophila*. *Nat. Rev. Genet.* 2: 11–20. [13]

Mackay, T. F. C. 2001b. The genetic architecture of quantitative traits. *Annu. Rev. Genet.* 35: 303–339. [13]

Mackay, T. F. C., and R. F. Lyman. 2005. *Drosophila* bristles and the nature of quantitative genetic variation. *Phil. Trans. R. Soc. Lond. B* 360: 1513–1527. [13]

Mackay, T. F. C., R. F. Lyman, and M. S. Jackson. 1992. Effects of P-element insertion on quantitative traits in *Drosophila melanogaster*. *Genetics* 130: 315–332. [8, 13]

Mackay, T. F. C., J. D. Fry, R. F. Lyman, and S. V. Nuzhdin. 1994. Polygenic mutation in *Drosophila melanogaster*: Estimates from response to selection in inbred strains. *Genetics* 136: 937–951. [13]

Macke, E., S. Magalhães, F. Bach, and I. Olivieri. 2011. Experimental evolution of reduced sex ratio adjustment under local mate competition. *Science* 334: 1127–1129. [15]

Macnair, M. R. 1981. Tolerance of higher plants to toxic materials. In J. A. Bishop and L. M. Cook (eds.), *Genetic Consequences of Man Made Change*, pp. 177–207. Academic Press, New York. [12]

Macnair, M. R. 1997. The evolution of plants in metal-contaminated environments. In R. Bijlsma and V. Loeschcke (eds.), *Environmental Stress, Adaptation and Evolution*, pp. 2–24. Birkhäuser, Basel, Switzerland. [13]

Maddison, W. 1995. Phylogenetic histories within and among species. In P. C. Hoch and A. G. Stevenson (eds.), *Experimental and Molecular Approaches to Plant Biosystematics*, pp. 273–287. Monographs in Systematic Botany 53. Missouri Botanical Garden, St. Louis. [2]

Madsen, T., B. Stille, and R. Shine. 1995. Inbreeding depression in an isolated population of adders *Vipera berus*. *Biol. Conserv.* 75: 113–118. [9]

Madsen, T., R. Shine, M. Olsson, and H. Wittzell. 1999. Restoration of an inbred adder population. *Nature* 402: 34–35. [9]

Magallón, S., and M. J. Sanderson. 2001. Absolute diversification rates in angiosperm clades. *Evolution* 55: 1762–1780. [7]

Mahler, D. L., L. J. Revell, R. E. Glor, and J. B. Losos. 2010. Ecological opportunity and the rate of morphological evolution in the diversification of Greater Antillean anoles. *Evolution* 64: 2731–2745. [7]

Majerus, M. E. 1998. *Melanism: Evolution in Action*. Oxford University Press, Oxford. [12]

Mallarino, R., P. R. Grant, B. R. Grant, A. Herrel, W. P. Kuo, and A. Abzhanov. 2011. Two developmental modules establish 3-D beak shape variation in Darwin's finches. *Proc. Natl. Acad. Sci. USA* 108: 4057–4062. [22]

Mallet, J. 2007. Hybrid speciation. *Nature* 446: 279–283. [17, 18]

Mallet, J. 2010. Shift happens! Shifting balance and the evolution of diversity in warning color and mimicry. *Ecol. Entomol.* 35: 90–104. [19]

Mallet, J., and N. Barton. 1989. Strong natural selection in a warning-color hybrid zone. *Evolution* 43: 421–431. [12]

Mallet, J., and M. Joron. 1999. Evolution of diversity in warning color and mimicry: polymorphisms, shifting balance, and speciation. *Annu. Rev. Ecol. Syst.* 30: 201–233. [3, 18, 19]

Malmgren, B. A., W. A. Berggren, and G. P. Lohmann. 1983. Evidence for punctuated gradualism in the Late Neogene *Globorotalia tumida* lineage of planktonic Foraminifera. *Paleobiology* 9: 377–389. [4]

Manica, A., W. Amos, F. Balloux, and T. Hanihara. 2007. The effect of ancient population bottlenecks on human phenotypic variation. *Nature* 448: 346–348. [10]

Manno, T. G., F. S. Dobson, J. L. Hoogland, and D. W. Foltz. 2007. Social group fission and gene dynamics among black-tailed prairie dogs. *J. Mammal.* 88: 448–456. [16]

Marcot, J. D., and D. W. McShea. 2007. Increasing hierarchical complexity throughout the history of life: Phylogenetic tests of trend mechanisms. *Paleobiology* 33: 182–200. [22]

Margulis, L. 1993. *Symbiosis in Cell Evolution*, 2nd ed. W. H. Freeman, San Francisco. [5]

Margulis, L., and R. Fester (eds.) 1991. *Symbiosis as a Source of Evolutionary Innovation*. MIT Press, Cambridge, MA. [5]

Marshall, C. R. 2006. Explaining the Cambrian "explosion" of animals. *Annu. Rev. Earth Planet. Sci.* 34: 355–384. [5]

Martin, P. S., and R. G. Klein (eds.). 1984. *Quaternary Extinctions: A Prehistoric Revolution*. University of Arizona Press, Tucson. [5]

Martin, W., and 9 others. 2002. Evolutionary analysis of *Arabidopsis*, cyanobacterial, and chloroplast genomes reveals plastid phylogeny and thousands of cyanobacterial genes in the nucleus. *Proc. Natl. Acad. Sci.* 99: 12246–12251. [19]

Martínez Wells, M., and C. S. Henry. 1992a. Behavioural responses of green lacewings (Neuroptera: Chrysopidae: *Chrysoperla*) to synthetic mating songs. *Anim. Behav.* 44: 641–652. [17]

Mather, K. 1949. *Biometrical Genetics: The Study of Continuous Variation*. Methuen, London. [9]

Mathew, S., and R. Boyd. 2011. Punishment sustains large-scale cooperation in prestate warfare. *Proc. Natl. Acad. Sci. USA* 108: 11375–11380. [16]

Matsumura, M. 1996. Genetic analysis of a threshold trait: density-dependent wing dimorphism in *Sogatella furcifera* (Horváth) (Hemptera: Delphacidae), the white-backed planthopper. *Heredity* 76: 229–237. [9]

Mattila, T. M., and F. Bokma. 2008. Extant mammal body masses suggest punctuated equilibrium. *Proc. R. Soc. Lond. B* 275: 2195–2199. [18, 22]

Matton, D. P., N. Nass, A. G. Clark, and E. Newbigin. 1994. Self-incompatibility: How plants avoid illegitimate offspring. *Proc. Natl. Acad. Sci. USA* 91: 1992–1997. [15]

May, R. M., and R. M. Anderson. 1983. Parasite-host coevolution. In D. J. Futuyma and M. Slatkin (eds.), *Coevolution*, pp. 186–206. Sinauer, Sunderland, MA. [19]

Maynard Smith, J. 1978. *The Evolution of Sex*. Cambridge University Press, Cambridge. [15]

Maynard Smith, J. 1980. Selection for recombination in a polygenic model. *Genet. Res.* 35: 269–277. [15]

Maynard Smith, J. 1982. *Evolution and the Theory of Games*. Cambridge University Press, Cambridge. [14, 16]

Maynard Smith, J. 1988. The evolution of recombination. In B. R. Levin and R. E. Michod (eds.), *The Evolution of Sex*, pp. 106–125. Sinauer, Sunderland, MA. [15]

Maynard Smith, J., and J. Haigh. 1974. The hitch-hiking effect of a favourable gene. *Genet. Res.* 23: 23–35. [12]

Maynard Smith, J., and E. Szathmáry. 1995. *The Major Transitions in Evolution*. W. H. Freeman, San Francisco. [5, 16, 22]

Maynard Smith, J., and 8 others. 1985. Developmental constraints and evolution. *Q. Rev. Biol.* 60: 265–287. [21]

Mayr, E. 1942. *Systematics and the Origin of Species*. Columbia University Press, New York. [1, 17, 18]

Mayr, E. 1954. Change of genetic environment and evolution. In J. Huxley, A. C. Hardy, and E. B. Ford (eds.), *Evolution as a Process*, pp. 157–180. Allen and Unwin, London. [4, 18]

Mayr, E. 1960. The emergence of evolutionary novelties. In S. Tax (ed.), *The Evolution of Life*, pp. 157–180. University of Chicago Press, Chicago. [22]

Mayr, E. 1963. *Animal Species and Evolution*. Harvard University Press, Cambridge, MA. [9, 17, 18]

Mayr, E. 1982a. *The Growth of Biological Thought: Diversity, Evolution, and Inheritance*. Harvard University Press, Cambridge, MA. [1]

Mayr, E. 1982b. Processes of speciation in animals. In C. Barigozzi (ed.), *Mechanisms of Speciation*, pp. 1–19. Alan R. Liss, New York. [18]

Mayr, E. 1988a. Cause and effect in biology. In E. Mayr (ed.), *Toward a New Philosophy of Biology*, pp. 24–37. Harvard University Press, Cambridge, MA. [11]

Mayr, E. 1988b. The probability of extraterrestrial intelligent life. In E. Mayr (ed.), *Toward a New Philosophy of Biology*, pp. 67–74. Harvard University Press, Cambridge, MA. [22]

Mayr, E. 2004. *What Makes Biology Unique? Considerations on the Autonomy of a Scientific Discipline*. Cambridge University Press, Cambridge. [3]

Mayr, E., and W. B. Provine (eds.). 1980. *The Evolutionary Synthesis: Perspectives on the Unification of Biology*. Harvard University Press, Cambridge, MA. [1]

McCarrey, J. R., M. Kumari, M. J. Aivaliotis, Z. Q. Wang, P. Zhang, F. Marshall, and J. L. Vandenberg. 1996. Analysis of the cDNA and encoded protein of the human testis-specific PGK-2 gene. *Devel. Genet.* 19: 321–332. [8]

McCauley, D. E. 1993. Evolution in metapopulations with frequent local extinction and recolonization. In D. J. Futuyma and J. Antonovics (eds.), *Oxford Surv. Evol. Biol.*, pp. 109–134. Oxford University Press, Oxford. [9]

McClearn, G. E., B. Johansson, S. Berg, N. L. Pedersen, F. Ahern, S. A. Petrill, and R. Plomin. 1997. Substantial genetic influence on cognitive abilities in twins 80 or more years old. *Science* 276: 1560–1563. [9]

McCommas, S. A., and E. H. Bryant. 1990. Loss of electrophoretic variation in serially bottlenecked populations. *Heredity* 64: 315–321. [10]

McCutcheon, J. P., and N. A. Moran. 2012. Extreme genome reduction in symbiotic bacteria. *Nat. Rev. Microbiol.* 10: 13–26. [22]

McDonald, D. B., and W. K. Potts. 1994. Cooperative display and relatedness among males in a lek-mating bird. *Science* 266: 1030–1032. [16]

McDonald, J. H., and M. Kreitman. 1991. Adaptive protein evolution at the *Adh* locus in *Drosophila*. *Nature* 351: 652–654. [12]

McElwain, J. C., and S. W. Punyasena. 2007. Mass extinction events and the plant fossil record. *Trends Ecol. Evol.* 22: 548–557. [5]

McGaugh, S. E., and M. A. Noor. 2012. Genomic impacts of chromosome inversions in parapatric *Drosophila* species. *Phil. Trans. R. Soc. B* 367: 422–429. [18]

McGinnis, W., and R. Krumlauf. 1992. Homeobox genes and axial patterning. *Cell* 68: 283–302. [21]

McKenzie, J. A., and G. M. Clarke. 1988. Diazinon resistance, fluctuating asymmetry and fitness in the Australian sheep blowfly, *Lucilia cuprina*. *Genetics* 120: 213–220. [13]

McKinney, F. K. 1992. Competitive interactions between related clades: Evolutionary implications of overgrowth interactions between encrusting cyclostome and cheilostome bryozoans. *Mar. Biol.* 114: 645–652. [7]

McKinney, M. L., and K. J. McNamara. 1991. *Heterochrony: The Evolution of Ontogeny*. Plenum, New York. [3]

McLean, C.Y., and 12 others. 2011. Human-specific loss of regulatory DNA and the evolution of human-specific traits. *Nature* 471: 216–219. [20, 21]

McNamara, K. J. 1997. *Shapes of Time*. Johns Hopkins University Press, Baltimore. [3]

McNeilly, T., and J. Antonovics. 1968. Evolution in closely adjacent plant populations. IV. Barriers to gene flow. *Heredity* 23: 205–218. [15]

McPeek, M. A. 2008. The ecological dynamics of clade diversification and community assembly. *Am. Nat.* 172: E270–E284. [7]

McPeek, M. A., and J. M. Brown. 2007. Clade age and not diversification rate explains species richness among animal taxa. *Am. Nat.* 169: E97–E106. [7]

McShea, D. W. 1991. Complexity and evolution: What everybody knows. *Biol. Phil.* 6: 303–324. [22]

McShea, D. W. 1994. Mechanisms of large-scale evolutionary trends. *Evolution* 48: 1747–1763. [22]

McShea, D. W. 1998. Possible largest–scale trends in organismal evolution: eight "live hypotheses." *Annu. Rev. Ecol. Syst.* 29: 293–318. [22]

McShea, D. W. 2001. The hierarchical structure of organisms: a scale and documentation of a trend in the maximum. *Paleobiology* 27: 405–423. [22]

Medawar, P. B. 1952. *An Unsolved Problem of Biology*. H. K. Lewis, London. [14]

Mendelson, T. C., and K. L. Shaw. 2005. Rapid speciation in an arthropod. *Nature* 433: 375. [6]

Meredith, R. W., and 22 others. 2011. Impacts of the Cretaceous terrestrial revolution and KPg extinction on mammal diversification. *Science* 334: 521–524. [5]

Merilä, J., B. C. Sheldon, and H. Ellegren. 1997. Antagonistic natural selection revealed by molecular sex identification of nestling collared flycatchers. *Mol. Ecol.* 6: 1167–1175. [13]

Merilä, J., B. C. Sheldon, and H. Ellegren. 1998. Quantitative genetics of sexual size dimorphism in the collared flycatcher, *Ficedula albicollis*. *Evolution* 52: 870–876. [13]

Messier, W., and C.-B. Stewart. 1997. Episodic adaptive evolution of primate lysozymes. *Nature* 385: 151–154. [20]

Metcalf, J. C., K. E. Rose, and M. Rees. 2003. Evolutionary demography of monocarpic perennials. *Trends Ecol. Evol.* 18: 471–480. [14]

Metcalf, R. L., and W. H. Luckmann (eds.). 1994. *Introduction to Insect Pest Management*, 3rd ed. Wiley, New York. [12]

Metzker, M. L., D. P. Mindell, X.-M. Liu, R. G. Ptak, R. A. Gibbs, and D. M. Hillis. 2002. Molecular evidence of HIV-1 transmission in a criminal case. *Proc. Natl. Acad. Sci. USA* 99: 14292–14297. [23]

Meyer, A., and R. Zardoya. 2003. Recent advances in the (molecular) phylogeny of vertebrates. *Annu. Rev. Ecol. Evol. Syst.* 34: 311–338. [2]

Mezey, J. G., D. Houle, and S. V. Nuzhdin. 2005. Naturally segregating quantitative trait loci affecting wing shape of *Drosophila melanogaster*. *Genetics* 169: 2101–2113. [21]

Michel, A. P., J. Rull, M. Aluja, and J. L. Feder. 2007. The genetic structure of hawthorn-infesting *Rhagoletis pomonella* populations in Mexico: Implications for sympatric speciation. *Mol. Ecol.* 16: 2867–2878. [18]

Michel, A. P., S. Sim, T. H. Q. Powell, M. S. Taylor, P. Nosil, and J. L. Feder. 2010. Widespread genomic divergence during sympatric speciation. *Proc. Natl. Acad. Sci. USA* 107: 9724–9729. [18]

Michod, R. E. 1999. *Darwinian Dynamics: Evolutionary Transitions in Fitness and Individuality*. Princeton University Press, Princeton, NJ. [16, 22]

Michod, R. E. 2007. Evolution of individuality during the transition from unicellular to multicellular life. *Proc. Natl. Acad. Sci. USA* 104: 8613–8618. [5]

Mielke, J. H., L. W. Konigsberg, and J. H. Relethford. 2006. *Human Biological Variation*. Oxford University Press, New York. [9]

Milá, B., D. J. Girman, M. Kimura, and T. B. Smith. 2000. Genetic evidence for the effect of a postglacial population expansion on the phylogeography of a North American songbird. *Proc. R. Soc. Lond. B* 267: 1033–1040. [2, 9]

Millais, J. G. 1897. *British Deer and Their Horns*. Henry Sotheran and Co., London. [3]

Miller, A. I. 2000. Conversations about Phanerozoic global diversity. *Paleobiology* 26 (Suppl.): 53–73. [7]

Miller, G. F. 2000. *The Mating Mind: How Sexual Choice Shaped the Evolution of Human Nature*. Doubleday, New York. [15]

Miller, K. R. 2008. *Only a Theory: Evolution and the Battle for America's Soul*. Viking, New York. [23]

Miller, M. P. and S. Kumar. 2001. Understanding human disease mutations through the use of interspecific genetic variation. *Hum. Molec. Genet.* 21: 2319–2328. [23]

Miller, S. L. 1953. Production of amino acids under possible primitive earth conditions. *Science* 117: 528–529. [5]

Milot, E., F. M. Mayer, D. H. Nussey, M. Boisvert, F. Pelletier, and D. Reale. 2011. Evidence for evolution in response to natural selection in a contemporary human population. *Proc. Natl. Acad. Sci. USA* 108: 17040–17045. [13]

Mindell, D. F., and C. E. Thacker. 1996. Rates of molecular evolution: phylogenetic issues and applications. *Annu. Rev. Ecol. Syst.* 27: 279–303. [2]

Mindell, D. O. 2006. *The Evolving World: Evolution in Everyday Life*. Harvard University Press, Cambridge, MA. [23]

Minelli, A. 2000. Limbs and tails as evolutionarily derived duplicates of the main body axis. *Evol. Devel.* 2: 157–165. [22]

Minelli, A. 2003. *The Development of Animal Form*. Cambridge University Press, Cambridge. [22]

Miralles, R., P. J. Gerrish, A. Moya, and S. F. Elena. 1999. Clonal interference and the evolution of RNA viruses. *Science* 285: 1745–1747. [15]

Mitgutsch, C., M. K. Richardson, R. Jiménez, J. E. Martin, P. Kondrashov, M. A. G. de Bakker, and M. R. Sánchez-Villagra. 2012. Circumventing the polydactyly 'constraint': the mole's 'thumb'. *Biol. Lett.* 8: 74–77. [22]

Mitra, S., H. Landel, and S. Pruett-Jones. 1996. Species richness covaries with mating systems in birds. *Auk* 113: 544–551. [18]

Mittelbach, G. G., and 21 others. 2007. Evolution and the latitudinal diversity gradient: Speciation, extinction and biogeography. *Ecol. Lett.* 10: 315–331. [6]

Mitter, C., and B. D. Farrell. 1991. Macroevolutionary aspects of insect-plant relationships. In E. A. Bernays (ed.), *Insect/Plant Interactions*, vol. 3, pp. 35–78. CRC Press, Boca Raton, FL. [6]

Mitter, C., B. D. Farrell, and B. Wiegmann. 1988. The phylogenetic study of adaptive zones: Has phytophagy promoted insect diversification? *Am. Nat.* 132: 107–128. [7, 22]

Mitton, J. B., and M. C. Grant. 1984. Associations among protein heterozygosity, growth rate, and developmental homeostasis. *Annu. Rev. Ecol. Syst.* 15: 479–499. [12]

Mock, D. W. 2004. *More than Kin and Less than Kind: The Evolution of Family Conflict*. Harvard University Press, Cambridge, MA. [16]

Moczek, A. P. 2006. Pupal remodeling and the development and evolution of sexual dimorphism in horned beetles. *Am. Nat.* 168: 711–729. [21]

Moczek, A. P., J. Hunt, D. J. Emlen, and L. W. Simmons. 2002. Threshold evolution in exotic populations of a polyphenic beetle. *Evol. Ecol. Res.* 4: 587–601. [21]

Monroe, M. J., and F. Bokma. 2010. Punctuated equilibrium in a neontological context. *Theory Biosci.* 129: 103–111. [22]

Monteiro, A., B. Chen, L. C. Scott, L. Vedder, H. J. Prijs, A. Belicha-Villanueva, and P. M. Brakefield. 2007. The combined effect of two mutations that alter serially homologous color pattern elements on the fore and hindwings of a butterfly. *BMC Genet.* 8(22) doi:10.1186/1471-2156-8-22. [22]

Montgomery, S. L. 1982. Biogeography of the moth genus *Eupithecia* in Oceania and the evolution of ambush predation in Hawaiian caterpillars (Lepidoptera: Geometridae). *Entomologia Generalis* 8: 27–34. [7]

Moodley, Y., and 14 others. 2009. The peopling of the Pacific from a bacterial perspective. *Science* 323: 527–530. [6]

Moore, F. B. G., D. E. Rozen, and R. E. Lenski. 2000. Pervasive compensatory adaptation in *Escherichia coli*. *Proc. R. Soc. Lond. B* 267: 515–522. [8]

Moore, W. S., and J. T. Price. 1993. Nature of selection in the northern flicker hybrid zone and its implications for speciation theory. In R. G. Harrison (ed.), *Hybrid Zones and the Evolutionary Process*, pp. 196–225. Oxford University Press, New York. [9]

Moran, N. 1996. Accelerated evolution and Muller's ratchet in endosymbiotic bacteria. *Proc. Natl. Acad. Sci. USA* 93: 2873–2878. [15]

Moran, N. 2003. Tracing the evolution of gene loss in obligate bacterial symbionts. *Curr. Opin. Microbiol.* 6: 512–518. [3, 20]

Moran, N. A. 2007. Symbiosis as an adaptive process and source of phenotypic complexity. *Proc. Natl. Acad. Sci. USA* 104: 8627–8633. [5, 19]

Moran, N. A., and P. Baumann. 1994. Phylogenetics of cytoplasmically inherited microorganisms of arthropods. *Trends Ecol. Evol.* 9: 15–20. [19]

Moran, N. A., H. J. McLaughlin, and R. Sorek. 2009. The dynamics and time scale of ongoing genomic erosion in symbiotic bacteria. *Science* 323: 379–382. [20]

Morjan, C. L., and L. H. Rieseberg. 2004. How species evolve collectively: Implications of gene flow and selection for the spread of advantageous alleles. *Mol. Ecol.* 13: 1341–1356. [18]

Morran, L. T., O. G. Schmidt, I. A. Gelarden, R. C. Parrish II, and C. M. Lively. 2011. Running with the Red Queen: host-parasite coevolution selects for biparental sex. *Science* 333: 216–218. [15]

Morton, W. F., J. F. Crow, and H. J. Muller. 1956. An estimate of the mutational damage in man from data on consanguineous marriages. *Proc. Natl. Acad. Sci. USA* 42: 855–863. [9]

Mousseau, T. A., and D. A. Roff. 1987. Natural selection and the heritability of fitness components. *Heredity* 59: 181–197. [9, 13]

Moyle, P. B., and J. J. Cech Jr. 1983. *Fishes: An Introduction to Ichthyology*. Prentice-Hall, Englewood Cliffs, NJ. [6]

Moyroud, E., E. Kusters, M. Monniaux, R. Koes, and F. Percy. 2010. LEAFY blossoms. *Trends Plant Sci.* 15: 346–352. [21]

Mueller, L. D., P. Guo, and F. J. Ayala. 1991. Density-dependent natural selection and trade-offs in life history traits. *Science* 253: 433–435. [14]

Mukai, T., S. I. Chigusa, L. E. Mettler, and J. F. Crow. 1972. Mutation rate and dominance of genes affecting viability in *Drosophila melanogaster*. *Genetics* 72: 335–355. [8]

Müller, G. B. 1990. Developmental mechanisms at the origin of morphological novelty: A side-effect hypothesis. In M. H. Nitecki (ed.), *Evolutionary Innovations*, pp. 99–130. University of Chicago Press, Chicago. [3, 18]

Müller, G. B., and G. P. Wagner. 1991. Novelty in evolution: Restructuring the concept. *Annu. Rev. Ecol. Syst.* 23: 229–256. [22]

Müller, G. B., and G. P. Wagner. 1996. Homology, *Hox* genes, and developmental integration. *Am. Zool.* 36: 4–13. [3]

Muller, H. J. 1940. Bearing of the *Drosophila* work on systematics. In J. S. Huxley (ed.), *The New Systematics*, pp. 185–268. Clarendon Press, Oxford. [17, 18]

Muller, H. J. 1964. The relation of recombination to mutational advance. *Mutat. Res.* 1: 2–9. [15]

Mundy, N. I. 2005. A window on the genetics of evolution: MC1R and plumage colouration in birds. *Proc. Biol. Sci.* 272: 1633–1640. [21]

Muñoz-Fuentes, V., A. J. Green, M. D. Sorenson, J. J. Negro, and C. Vila. 2006. The ruddy duck *Oxyura jamaicensis* in Europe: Natural colonization or human introduction? *Mol. Ecol.* 15: 1441–1453. [10]

Müntzing, A. 1930. Über Chromosomenvermehrung in Galeopsis-Kreuzungen und ihre phylogenetische Bedeutung. *Hereditas* 14: 153–172. [18]

Myers, A. A., and P. S. Giller (eds.). 1988. *Analytical Biogeography*. Chapman & Hall, London. [6]

N

Nachman, M. W., and S. L. Crowell. 2000. Estimate of the mutation rate per nucleotide in humans. *Genetics* 156: 297–304. [8]

Nachman, M. W., H. E. Hoekstra, and S. L. D'Agostino. 2003. The genetic basis of adaptive melanism in pocket mice. *Proc. Natl. Acad. Sci. USA* 100: 5268–5273. [3]

Nadeau, N. J., and 12 others. 2012. Genomic islands of divergence in hybridizing *Heliconius* butterflies identified by large-scale targeted sequencing. *Phil. Trans. R. Soc. B* 367: 343–353. [18]

Nason, J. D., N. C. Ellstrand, and M. L. Arnold. 1992. Patterns of hybridization and introgression in populations of oaks, manzanitas, and irises. *Am. J. Bot.* 79: 101–111. [17]

Nee, S. 2006. Birth-death models in macroevolution. *Annu. Rev. Ecol. Evol. Syst.* 37: 1–17. [7]

Nei, M., and A. L. Hughes. 1991. Polymorphism and evolution of the major histocompatibility complex loci in mammals. In R. K. Selander, A. G. Clark, and T. S. Whittam (eds.), *Evolution at the Molecular Level*, pp. 222–247. Sinauer, Sunderland, MA. [12]

Nei, M., Y. Suzuki, and M. Nozawa. 2010. The neutral theory of molecular evolution in the genomics era. *Annu. Rev. Genomics Hum. Genet.* 11: 265–289. [10]

Neigel, J. E., and J. C. Avise. 1986. Phylogenetic relationship of mitochondrial DNA under various demographic models of speciation. In E. Nevo and S. Karlin (eds.), *Evolutionary Processes and Theory*, pp. 515–534. Academic Press, London. [17]

Nesse, R., and G. C. Williams. 1994. *Why We Get Sick: The New Science of Darwinian Medicine*. Times Books, New York. [23]

Nesse, R. M., and S. C. Stearns. 2008. The great opportunity: Evolutionary applications to medicine and public health. *Evol. Applications* 1: 28–48. [23]

Nestmann, E. R., and R. F. Hill. 1973. Population genetics in continuously growing mutator cultures of *Escherichia coli*. *Genetics* 73: 41–44. [11]

Nevo, E. 1991. Evolutionary theory and processes of active speciation and adaptive radiation in subterranean mole rats, *Spalax ehrenbergi* superspecies, in Israel. *Evol. Biol.* 25: 1–125. [17]

Newcomb, R. D., P. M. Campbell, D. L. Olles, E. Cheah, R. J. Russell, and J. G. Oakeshott. 1997. A single amino acid substitution converts a carboxylesterase to an organophosphate hydrolase and confers insecticide resistance on a blowfly. *Proc. Natl. Acad. Sci. USA* 94: 7464–7468. [8]

Newman, S. A., G. Forgacs, and G. B. Müller. 2006. Before programs: the physical organization of multicellular forms. *Int. J. Devel. Biol.* 50 (S1): 289–299. [22]

Ng, C. S., and A. Kopp. 2008. Sex combs are important for male mating success in *Drosophila melanogaster*. *Behav. Genet.* 38: 195–201. [21]

Nielsen, R. 2005. Molecular signatures of natural selection. *Annu. Rev. Genet.* 39: 197–218. [12]

Nielsen, R., I. Hellmann, M. Hubisz, C. Bustamante, and A. G. Clark. 2007. Recent and ongoing selection in the human genome. *Nature Rev. Genet.* 8: 857–868. [12]

Niklas, K. J., B. H. Tiffney, and A. H. Knoll. 1983. Patterns in vascular land plant diversification. *Nature* 303: 614–616. [7]

Nilsson, D. E., and S. Pelger. 1994. A pessimistic estimate of the time required for an eye to evolve. *Proc. R. Soc. Lond. B* 256: 59–65. [22]

Nilsson, L. A., L. Jonsson, L. Ralison, and E. Randrianjohany. 1985. Monophily and pollination mechanisms in *Angraecum arachnites* Schltr. (Orchidaceae) in a guild of long-tongued hawkmoths (Sphingidae). *Biol. J. Linn. Soc.* 26: 1–19. [19]

Nisbett, R. 1995. Race, IQ, and scientism. In S. Fraser (ed.), *The Bell Curve Wars*, pp. 36–57. BasicBooks, New York. [9]

Nitecki, M. H. (ed.). 1990. *Evolutionary Innovations*. University of Chicago Press, Chicago. [22]

Noonan, J. P., and 10 others. 2006. Sequencing and analysis of Neanderthal genomic DNA. *Science* 314: 1113–1118. [20]

Noor, M. A. F. 1995. Speciation driven by natural selection in *Drosophila*. *Nature* 375: 674–675. [18]

Noor, M. A. F. 1999. Reinforcement and other consequences of sympatry. *Heredity* 83: 503–508. [18]

Norell, M. A., and M. J. Novacek. 1992. The fossil record and evolution: Comparing cladistic and paleontologic evidence for vertebrate history. *Science* 255: 1690–1693. [4]

Norell, M. A., and X. Xu. 2005. Feathered dinosaurs. *Annu. Rev. Earth & Planetary Sci.* 33: 277–299. [4]

Norenzayan, A., and S. J. Heine. 2005. Psychological universals: what are they and how can we know? *Psychol. Bull.* 131: 763–784. [16]

Normark, B. B., O. P. Judson, and N. A. Moran. 2003. Genomic signatures of ancient asexual lineages. *Biol. J. Linn. Soc.* 79: 69–84. [11, 15]

Norton, H. L., and 9 others. 2007. Genetic evidence for the convergent evolution of light skin in Europeans and East Asians. *Mol. Biol. Evol.* 24: 710–722. [12]

Nosil, P. 2007. Divergent host plant adaptation and reproductive isolation between ecotypes of *Timema cristinae* walking sticks. *Am. Nat.* 169: 151–162. [18]

Nosil, P. 2012. *Ecological Speciation*. Oxford University Press, Oxford. [18]

Nosil, P., and J. L. Feder. 2012. Genomic divergence during speciation: causes and consequences. *Phil. Trans. R. Soc. B* 367: 332–343. [18]

Nosil, P., T. H. Vines, and D. J. Funk. 2005. Perspective: Reproductive isolation caused by natural selection against immigrants from divergent habitats. *Evolution* 59: 705–719. [17, 18]

Novembre, J., and A. Di Rienzo. 2009. Spatial patterns of variation due to natural selection in humans. *Nat. Rev. Genet.* 10: 745–755. [12]

Novembre, J., and S. Ramachandran. 2011. Perspectives on human population structure at the cusp of the sequencing era. *Annu. Rev. Genom. Human Genet.* 12: 245–274. [9]

Novembre, J., and 11 others. 2008. Genes mirror geography within Europe. *Nature* 456: 98–101. [9]

Novotny, V., P. Drozd, S. E. Miller, M. Kulfan, M. Janda, Y. Basset, and G. D. Weiblen. 2006. Why are there so many species of herbivorous insects in tropical forests? *Science* 313: 1115–1118. [7]

Novotny, V., and 15 others. 2007. Low beta diversity of herbivorous insects in tropical forests. *Nature* 448: 692–695. [7]

Nowak, M. A. 2006. Five rules for the evolution of cooperation. *Science* 314: 1560–1563. [16]

Nowak, M. A., C. E. Tarnita, and E. O. Wilson. 2010. The evolution of eusociality. *Nature* 466: 1057–1062. [16]

Nuismer, S. L., B. J. Ridenhour, and B. P. Oswald. 2007. Antagonistic coevolution mediated by phenotypic differences between quantitative traits. *Evolution* 61: 1823–1834. [19]

Numbers, R. L. 2006. *The Creationists: From Scientific Creationism to Intelligent Design*. Harvard University Press, Cambridge, MA. [22, 23]

Nummela, S., J. G. M. Thewissen, S. Bajpal, S. T. Hussain, and K. Kumar. 2004. Eocene evolution of whale hearing. *Nature* 430: 776–778. [4]

Nur, N. 1984. The consequences of brood size for breeding blue tits II. Nestling weight, offspring survival and optimal brood size. *J. Anim. Ecol.* 53: 497–517. [13]

Nuzhdin, S. V., and T. F. C. Mackay. 1994. Direct determination of retrotransposon transposition rates in *Drosophila melanogaster*. *Genet. Res.* 63: 139–144. [8]

Nuzhdin, S. V., C. L. Dilda, and T. F. C. Mackay. 1999. The genetic architecture of selection response: inferences from fine-scale mapping of bristle number quantitative trait loci in *Drosophila melanogaster*. *Genetics* 153: 1317–1331. [13]

O

Oakeshott, J. G., J. B. Gibson, P. R. Anderson, W. R. Knib, D. G. Anderson, and G. K. Chambers. 1982. Alcohol dehydrogenase and glycerol-3-phosphate dehydrogenase clines in *Drosophila melanogaster* on different continents. *Evolution* 36: 86–96. [9]

Oakley, T. H., and M. S. Pankey. 2008. Opening the "black box": The genetic and biochemical basis of eye evolution. *Evolution: Education and Outreach* 1: 390–402. [22]

Ochman, H., J. G. Lawrence, and E. A. Groisman. 2000. Lateral gene transfer and the nature of bacterial innovation. *Nature* 405: 299–304. [2, 20]

Odling-Smee, F. J., K. N. Laland, and M. W. Feldman. 2003. *Niche Construction: The Neglected Process in Evolution*. Princeton University Press, Princeton, NJ. [11, 22]

Ogura, A., K. Ikeo, and T. Gojobori. 2004. Comparative analysis of gene expression for convergent evolution of camera eye between octopus and human. *Genome Res.* 14: 1555–1561. [21]

Ohno, S. 1970. *Evolution by Gene Duplication*. Springer-Verlag, Berlin. [20]

Ohta, T. 1992. The nearly neutral theory of molecular evolution. *Annu. Rev. Ecol. Syst.* 23: 263–286. [10, 12]

Okasha, S. 2006. *Evolution and the Levels of Selection*. Oxford University Press, Oxford. [11]

Oliver, G., A. Mailhos, R. Wehr, N. G. Copeland, N. A. Jenkins, and P. Gruss. 1995. *Six3*, a murine homolog of the *sine oculis* gene, demarcates the most anterior border of the developing neural plate and is expressed during eye development. *Development* 121: 4045–4055. [21]

Oliver, M. J., D. Petrov, D. Ackerley, P. Falkowski, and O. M. Schofield. 2007. The mode and tempo of genome size evolution in eukaryotes. *Genome Research* 17: 594–601. [22]

Olivieri, I., Y. Michalakis, and P.-H. Gouyon. 1995. Metapopulation genetics and the evolution of dispersal. *Am. Nat.* 146: 202–228. [14]

Olsen, E. M., M. Heino, G. R. Lilly, M. J. Morgan, J. Brattley, B. Ernande, and U. Dieckmann. 2004. Maturation trends indicative of rapid evolution preceded the collapse of northern cod. *Nature* 428: 932–935. [11]

Olson, E. C., and R. L. Miller. 1958. *Morphological Integration*. University of Chicago Press, Chicago. [13]

Orr, H. A. 1998. The population genetics of adaptation: The distribution of factors fixed during adaptive evolution. *Evolution* 52: 935–949. [13]

Orr, H. A. 2000. Adaptation and the cost of complexity. *Evolution* 54: 13–20. [13, 22]

Orr, H. A., and J. Coyne. 1992. The genetics of adaptation: A reassessment. *Am. Nat.* 140: 725–742. [22]

Orr, H. A., and S. Irving. 2001. Complex epistasis and the genetic basis of hybrid sterility in the *Drosophila pseudoobscura* Bogotá-USA hybridization. *Genetics* 158: 1089–1100. [17]

Osada, N., and H. Innan. 2008. Duplication and gene conversion in the *Drosophila* melanogaster genome. *PLoS Genetics* 4(12): e1000305. [20]

Osorio, D. 1994. Eye evolution: Darwin's shudder stilled. *Trends Ecol. Evol.* 9: 241–242. [22]

Ostrom, J. H. 1976. On a new specimen of the Lower Cretaceous theropod dinosaur *Deinonychus antirrhopus*. *Breviora* 439: 1–21. [4]

Otto, S. P. 2009. The evolutionary enigma of sex. *Am. Nat.* 174 (Suppl.): S1–S14. [15]

Otto, S. P., and S. L. Nuismer. 2004. Species interactions and the evolution of sex. *Science* 304: 1018–1020. [15]

Otto, S. P., and J. Whitton. 2000. Polyploid incidence and evolution. *Annu. Rev. Genet.* 34: 401–437. [8, 18]

Ownbey, M. 1950. Natural hybridization and amphiploidy in the genus *Tragopogon*. *Am. J. Bot.* 37: 489–499. [18]

Oyama, R. K., and 8 others. 2008. The shrunken genome of *Arabidopsis thaliana*. *Plant Syst. Evol.* 273: 257–271. [22]

P

Paegel, B. M., and G. F. Joyce. 2008. Darwinian evolution on a chip. *PLoS Biol.* 6(4): e85. [5]

Page, R. B., S. R. Voss, A. K. Samuels, J. J. Smith, S. Putta, and C. K. Beachy. 2008. Effect of thyroid hormone concentration on the transcriptional response underlying induced metamorphosis in the Mexican axolotl (*Ambystoma*). *BMC Genomics* 9: 78. [21]

Page, R. B., M. A. Boley, J. J. Smith, S. Putta, and S. R. Voss. 2010. Microarray analysis of a salamander hopeful monster reveals transcriptional signatures of paedomorphic brain development. *BMC Evol. Biol.* 10: 199. [21]

Page, R. D. M. 2002. Introduction. In R. D. M. Page (ed.), *Tangled Trees: Phylogeny, Cospeciation, and Coevolution*, pp. 1–21. University of Chicago Press, Chicago. [19]

Page, S. L., and M. Goodman. 2001. Catarrhine phylogeny: noncoding DNA evidence for a diphyletic origin of the mangabeys and for a human-chimpanzee clade. *Mol. Phyl. Evol.* 18: 14–25. [2]

Pagel, M. 1999. Inferring the historical pattern of biological evolution. *Nature* 401: 877–884. [3]

Pagel, M., C. Venditti, and A. Meade. 2006. Large punctuational contribution of speciation to evolutionary divergence at the molecular level. *Science* 314: 119–121. [18]

Pagel, M., Q. D. Atkinson, and A. Meade. 2007. Frequency of word-use predicts rates of lexical evolution throughout Indo-European history. *Nature* 449: 717–720. [23]

Palmer, A. R. 1982. Predation and parallel evolution: Recurrent parietal plate reduction in balanomorph barnacles. *Paleobiology* 8: 31–44. [4]

Palmer, J. D., and J. M. Logsdon, Jr. 1991. The recent origin of introns. *Curr. Opin. Genet. Dev.* 1: 470–477. [20]

Palmer, J. D., D. E. Soltis, and M. W. Chase. 2004. The plant tree of life: An overview and some points of view. *Am. J. Bot.* 91: 1437–1445. [5]

Palstra, F. P., and D. E. Ruzzante. 2008. Genetic estimates of contemporary effective population size: What can they tell us about the importance of genetic stochasticity for wild population persistence? *Mol. Ecol.* 17: 3428–3447. [10]

Palumbi, S. R. 1998. Species formation and the evolution of gamete recognition loci. In D. J. Howard and S. H. Berlocher (eds.), *Endless Forms: Species and Speciation*, pp. 271–278. Oxford University Press, New York. [17]

Palumbi, S. R. 2001. *The Evolution Explosion: How Humans Cause Rapid Evolutionary Change*. W. W. Norton, New York [1, 11]

Palumbi, S. R. 2009. Speciation and the evolution of gamete recognition genes: pattern and process. *Heredity* 102: 66–76. [18]

Panganiban, G., and 13 others. 1997. The origin and evolution of animal appendages. *Proc. Natl. Acad. Sci. USA* 94: 5162–5166. [21]

Panhuis, T. M., R. Butlin, M. Zuk, and T. Tregenza. 2001. Sexual selection and speciation. *Trends Ecol. Evol.* 16: 364–371. [18]

Panopoulou, G., and A. J. Poustka. 2005. Timing and mechanism of ancient vertebrate genome duplications: The adventure of a hypothesis. *Trends Genet.* 21: 559–567. [3]

Papadopulos, A. S. T., W. J. Baker, D. Crayn, R. K. Butlin, R. G. Kynast, I. Hutton, and V. Savolainen. 2011. Speciation with gene flow on Lord Howe Island. *Proc. Natl. Acad. Sci. USA* 108: 13188–13193. [18]

Paradis, J., and G. C. Williams. 1989. *Evolution and Ethics: T. H. Huxley's Evolution & Ethics with New Essays on its Victorian and Sociobiological Context*. Princeton University Press, Princeton, NJ. [1]

Parfrey, L. W., J. Grant, Y. I. Tekle, E. Lasek-Nesselquist, H. G. Morrison, M. L. Sogin, D. J. Patterson, and L. A. Katz. 2010. Broadly sampled multigene analyses yield a well-resolved eukaryotic tree of life. *Syst. Biol.* 59: 518–533. [5]

Parfrey, L. W., D. J. G. Lahr, A. H. Knoll, and L. A. Katz. 2011. Estimating the timing of early eukaryotic diversification with multigene molecular clocks. *Proc. Natl. Acad. Sci. USA* 108: 13624–12629. [5]

Parker, G. A. 1970. Sperm competition and its evolutionary consequences in the insects. *Biol. Rev.* 45: 525–567. [15]

Parker, G. A. 1979. Sexual selection and sexual conflict. In M. S. Blum and N. A. Blum (eds.), *Sexual Selection and Reproduction*, pp. 123–166. Academic Press, New York. [15]

Parker, G. A. 2006. Sexual conflict over mating and fertilization: an overview. *Phil. Trans. R. Soc. B* 361: 235–259. [15]

Parker, G. A., and J. Maynard Smith. 1990. Optimality theory in evolutionary biology. *Nature* 348: 27–33. [14]

Parmesan, C., and 12 others. 1999. Poleward shift of butterfly species' ranges associated with regional warming. *Nature* 399: 579–583. [6]

Parmesan, C., S. Gaines, L. Gonzalez, D. M. Kaufman, J. Kingsolver, A. T. Peterson, and R. Sagarin. 2005. Empirical perspectives on species borders: from traditional biogeography to global change. *Oikos* 108: 58–75. [6]

Partridge, L. 2001. Evolutionary theories of ageing applied to long-lived organisms. *Exper. Gerontol.* 36: 641–650. [14]

Partridge, L., and L. D. Hurst. 1998. Sex and conflict. *Science* 281: 2003–2008. [15]

Partridge, L., N. Prowse, and P. Pignatelli. 1999. Another set of responses and correlated responses to selection on age at reproduction in *Drosophila melanogaster*. *Proc. R. Soc. Lond. B* 266: 255–261. [14]

Paterson, H. E. H. 1985. The recognition concept of species. In E. S. Vrba (ed.), *Species and Speciation*, pp. 21–29. Transvaal Museum Monograph No. 4, Pretoria, South Africa. [17]

Patterson, J. T. 1943. Studies in the genetics of *Drosophila*. III. The Drosophilidae of the southwest. *Univ. Texas Publ.* 4313: 7–203. [15]

Patton, J. L., and S. Y. Yang. 1977. Genetic variation in *Thomomys bottate* pocket gophers: Macrogeographic patterns. *Evolution* 31: 697–720. [10]

Pavlicev, M., and G. P. Wagner. 2012. A model of developmental evolution: selection, pleiotropy and compensation. *Trends Ecol. Evol.* 27: 316–322. [13, 22]

Peiman, S., and B. W. Robinson. 2007. Heterospecific aggression and adaptive divergence in brook sticklebacks (*Culaea inconstans*). *Evolution* 61: 1327–1338. [19]

Pélabon, C., W. S. Armbruster, and T. F. Hansen. 2011. Experimental evidence for the Berg hypothesis: vegetative traits are more sensitive than pollination traits to environmental variation. *Func. Ecol.* 25: 247–257. [13]

Pellmyr, O., and C. J. Huth. 1994. Evolutionary stability of mutualism between yuccas and yucca moths. *Nature* 372: 257–260. [19]

Pellmyr, O., and J. Leebens-Mack. 1999. Forty million years of mutualism: Evidence for Eocene origin of the yucca-yucca moth association. *Proc. Natl. Acad. Sci. USA* 96: 9178–9183. [19]

Pemberton, R. W. 2000. Predictable risk to native plants in weed biological control. *Oecologia* 125: 489–494. [19]

Pennock, R. T. 1999. *Tower of Babel: The Evidence against the New Creationism*. MIT Press, Cambridge, MA. [1]

Pennock, R. T. 2003. Creationism and intelligent design. *Annu. Rev. Genomics Hum. Genet.* 4: 143–163. [23]

Pepper, J. W., C. S. Findlay, R. Kassen, S. L. Spencer, and C. C. Maley. 2009. Cancer research meets evolutionary biology. *Evol. Appl.* 2: 62–70. [11]

Perry, J. C., and L. Rowe. 2012. Sex role stereotyping and sexual conflict theory. *Anim. Behav.* 83: E10–E13. [15]

Peterson, A. T., J. Soberón, and V. Sánchez-Cordero. 1999. Conservatism of ecological niches in evolutionary time. *Science* 285: 1265–1267. [6]

Peterson, K. J., J. A. Cotton, J. G. Kehling, and D. Pisani. 2008. The Ediacaran emergence of bilaterians: congruence between the genetic and the geological fossil records. *Phil. Trans. R. Soc. Lond. B* 363: 1435–1443. [5]

Petrov, D., and J. F. Wendel. 2006. Evolution of eukaryotic genome structure. In C. W. Fox and J. B. Wolf (eds.), *Evolutionary Genetics*, pp.144–156. Oxford University Press, New York. [8]

Pfennig, D. W., and W. A. Frankino. 1997. Kin-mediated morphogenesis in facultatively cannibalistic tadpoles. *Evolution* 51: 1993–1999. [16]

Pfennig, D. W., and P. J. Murphy. 2000. Character displacement in polyphenic tadpoles. *Evolution* 54: 1738–1749. [21]

Pfennig, D. W., J. P. Collins, and R. E. Ziemba. 1999. A test of alternative hypotheses for kin recognition in cannibalistic tiger salamanders. *Behav. Ecol.* 10: 436–443. [11]

Phillips, B. L., G. P. Brown, J. K. Webb, and R. L. Shine. 2006. Invasion and the evolution of speed in toads. *Nature* 439: 803. [11]

Phillips, P. C. 2008. Epistasis—the essential role of gene interactions in the structure and evolution of genetic systems. *Nat. Rev. Genet.* 9: 855–867. [18]

Piatigorsky, J. 2007. *Gene Sharing and Evolution: The Diversity of Protein Functions*. Harvard University Press, Cambridge, MA. [11]

Pielou, E. C. 1991. *After the Ice Age: The Return of Life to Glaciated North America*. University of Chicago Press, Chicago. [5]

Pigliucci, M. 2001. *Phenotypic Plasticity: Beyond Nature and Nurture*. Johns Hopkins University Press, Baltimore. [9, 13]

Pigliucci, M. 2002. *Denying Evolution: Creationism, Scientism, and the Nature of Science*. Sinauer, Sunderland, MA. [1, 23]

Pigliucci, M., and K. Preston (eds.). 2004. *Phenotypic Integration: Studying the Ecology and Evolution of Complex Phenotypes*. Oxford University Press, Oxford. [13]

Pinho, C., and J. Hey. 2010. Divergence with gene flow: models and data. *Annu. Rev. Ecol. Syst.* 41: 215–230. [18]

Pinker, S. 2002. *The Blank Slate: The Modern Denial of Human Nature*. Penguin Books, New York. [16]

Pires, J. C., and 9 others. 2004. Molecular cytogenetic analysis of recently evolved *Tragopogon* (Asteraceae) allopolyploids reveal a karyotype that is additive of the diploid progenitors. *Am. J. Bot.* 91: 1022–1035. [18]

Pitnick, S., T. A. Markow, and G. S. Spicer. 1999. Evolution of multiple kinds of sperm-storage organs in *Drosophila*. *Evolution* 53: 1804–1822. [15]

Plomin, R., J. C. DeFries, G. E. McClearn, and M. Rutter. 1997. *Behavioral Genetics*, 3rd ed. W. H. Freeman, New York. [9]

Poiani, A. 2010. *Animal Homosexuality: A Biosocial Perspective*. Cambridge University Press, Cambridge. [16]

Pomiankowski, A. 1988. The evolution of female mate preferences for male genetic quality. *Oxford Surv. Evol. Biol.* 5: 136–184. [15]

Pomiankowski, A., and Y. Iwasa. 1998. Runaway ornament diversity caused by Fisherian sexual selection. *Proc. Natl. Acad. Sci. USA* 95: 5106–5111. [15, 18]

Pomiankowski, A., Y. Iwasa, and S. Nee. 1991. The evolution of costly mate preferences. I. Fisher and biased mutation. *Evolution* 45: 1422–1430. [15]

Ponting, C. P., C. Nellaker, and S. Meader. 2011. Rapid turnover of functional sequence in human and other genomes. *Annu. Rev. Genom. Human Genet.* 12: 275–299. [20]

Poole, A. M., and D. Penny. 2006. Evaluating hypotheses for the origin of eukaryotes. *BioEssays* 29: 74–84. [5]

Poole, A. M., D. C. Jeffares, and D. Penny. 1998. The path from the RNA world. *J. Molec. Evol.* 46: 1–17. [20]

Poon, A., and L. Chao. 2004. Drift increases the advantage of sex in RNA bacteriophage Φ6. *Genetics* 166: 19–24. [15]

Popesco, M. C., and 8 others. 2006. Human lineage-specific amplification, selection, and neuronal expression of DUF1220 domains. *Science* 313: 1304–1307. [20]

Porter, K. R. 1972. *Herpetology*. W. B. Saunders, Philadelphia, PA. [11]

Porter, M. L., and K. A. Crandall. 2003. Lost along the way: the significance of evolution in reverse. *Trends Ecol. Evol.* 541–547. [3]

Prabhakar, S., and 12 others. 2008. Human-specific gain of function in a developmental enhancer. *Science* 321: 1346–1350. [21]

Prescott, G. W., D. R. Williams, A. Balmford, R. E. Green, and A. Manica. 2012. Quantitative global analysis of the role of climate and people in explaining late Quaternary megafaunal extinctions. *Proc. Natl. Acad. Sci. USA* 109: 4527–4531. [5]

Presgraves, D. C. 2003. A fine-scale genetic analysis if hybrid incompatibilities in *Drosophila*. *Genetics* 163: 955–972. [17]

Presgraves, D. C. 2010. The molecular evolutionary basis of species formation. *Nat. Rev. Genet.* 11: 175–180. [17]

Prince, V. E. 2002. The Hox paradox: More complex(es) than imagined. *Dev. Biol.* 249: 1–15. [20]

Prince, V. E., and F. B. Pickett. 2002. Splitting pairs: The diverging fates of duplicated genes. *Nature Rev. Genet.* 3: 827–837. [20]

Pritchard, J. K., J. K. Pickrell, and G. Coop. 2010. The genetics of human adaptation: hard sweeps, soft sweeps, and polygenic adaptation. *Curr. Biol.* 20: R208–R215. doi:10.1016/j.cub.2009.11.055. [12]

Prokop, Z. M., Ł. Michalczyk, S. M. Drobniak, M. Herdegen, and J. Radwan. 2012. Meta-analysis suggests choosy females get sexy sons more than "good genes". *Evolution* 66: 2665–2673. [15]

Promislow, D. E. L., and P. H. Harvey. 1991. Mortality rates and the evolution of mammalian life histories. *Acta Oecologica* 220: 417–437. [14]

Prothero, D. R. 2006. *After the Dinosaurs: The Age of Mammals*. Indiana University Press, Bloomington. [5]

Prothero, D. R. 2007. *Evolution: What the Fossils Say and Why it Matters*. Columbia University Press, New York. [4]

Prothero, D. R., and C. Buell. 2007. *Evolution: What the Fossils Say and Why it Matters*. Columbia University Press, New York. [5]

Prud'homme, B., N. Gompel, A. Rokas, V. A. Kassner, T. M. Williams, S.-D. Yeh, J. R. True, and S. B. Carroll. 2006. Repeated morphological evolution through *cis*-regulatory changes in a pleiotropic gene. *Nature* 440: 1050–1053. [21]

Prud'homme, B., N. Gompel, and S. M. Carroll. 2007. Emerging principles of regulatory evolution. *Proc. Natl. Acad. Sci. USA* 104 (Suppl. 1): 8605–8612. [22]

Prum, R. O. 2010. The Lande-Kirkpatrick mechanism is the null model of evolution by intersexual selection: implications for meaning, honesty, and design in intersexual signals. *Evolution* 64: 3085–3100. [15]

Przeworski, M., G. Coop, and J. D. Wall. 2005. The signature of positive selection on standing genetic variation. *Evolution* 59: 2312–2323. [13]

Purugganan, M. D. 1997. The MADS-box floral homeotic gene lineages predate the origin of seed plants: Phylogenetic and molecular clock estimates. *J. Mol. Evol.* 45: 392–396. [21]

Q

Quammen, D. 2006. *The Reluctant Mr. Darwin: An Intimate Portrait of Charles Darwin and Making of His Theory of Evolution*. W. W. Norton, New York. [1]

Queitsch, C., T. A. Sangster, and S. Lindquist. 2002. *Hsp90* as a capacitor of phenotypic variation. *Nature* 417: 618–624. [13]

Queller, D. C. 1994. Genetic relatedness in viscous populations. *Evol. Ecol.* 8: 70–73. [16]

Queller, D. C., F. Zacchi, R. Cervo, S. Turillazzi, M. T. Henshaw, L. A. Santorelli, and J. E. Strassmann. 2000. Unrelated helpers in a social insect. *Nature* 405: 784–787. [16]

Queller, D. C., E. Ponte, S. Bozzaro, and J. E. Strassmann. 2003. Single-gene greenbeard effects in the social amoeba *Dictyostelium discoideum*. *Science* 299: 105–106. [16]

R

Rabosky, D. L. 2009. Ecological limits and diversification rate: alternative paradigms to explain the variation in species richness among clades and regions. *Ecol. Lett.* 12: 735–743. [7]

Rabosky, D. L. 2010. Extinction rates should not be estimated from molecular phylogenies. *Evolution* 64: 1816–1824. [7]

Rabosky, D. L., and I. J. Lovette. 2008. Explosive evolutionary radiations: decreasing speciation or increasing extinction through time? *Evolution* 62: 1866–1875. [7]

Rabosky, D. L., and A. R. McCune. 2010. Reinventing species selection with molecular phylogenies. *Trends Ecol. Evol.* 25: 68–74. [7, 11]

Radinsky, L. B. 1984. Ontogeny and phylogeny in horse skull evolution. *Evolution* 38: 1–15. [21]

Raff, E. C., E. M. Popodi, B. J. Sly, F. R. Turner, J. T. Villinski, and R. A. Raff. 1999. A novel ontogenetic pathway in hybrid embryos between species with different modes of development. *Development* 126: 1937–1945. [21]

Raff, R. A. 1996. *The Shape of Life: Genes, Development, and the Evolution of Animal Form*. University of Chicago Press, Chicago. [3]

Ralls, K., P. H. Harvey, and A. M. Lyles. 1986. Inbreeding in natural populations of birds and mammals. In M. E. Soulé (ed.), *Conservation Biology: The Science of Scarcity and Diversity*, pp. 35–56. Sinauer, Sunderland, MA. [15]

Ramsey, J. 2011. Polyploidy and ecological adaptation in wild yarrow. *Proc. Natl. Acad. Sci. USA* 108: 7096–7101. [8, 18]

Ramsey, J., and D. W. Schemske. 1998. Pathways, mechanisms, and rates of polyploid formation in flowering plants. *Annu. Rev. Ecol. Syst.* 29: 467–501. [8, 18]

Ramsey, J., H. D. Bradshaw, and D. W. Schemske. 2003. Components of reproductive isolation between the monkeyflowers *Mimulus lewisii* and *M. cardinalis* (Scrophulariaceae). *Evolution* 57: 1520–1534. [17]

Randsholt, N. B., and P. Santamaria. 2008. How *Drosophila* change their combs: the Hox gene *Sex combs reduced* and sex comb variation among *Sophophora* species. *Evol. Dev.* 10: 121–133. [21]

Rannala, B., and Z.-H. Yang. 2008. Phylogenetic inference using whole genomes. *Annu. Rev. Genomics Hum. Genet.* 9: 217–231. [2]

Ranz, J. M., F. Casals, and A. Ruiz. 2001. How malleable is the eukaryotic genome? Extreme rate of chromosomal rearrangement in the genus *Drosophila*. *Genome Res.* 11: 230–239. [8]

Rasmussen, M., and 57 others. 2011. An aboriginal Australian genome reveals separate human dispersals into Asia. *Science* 334: 94–98. [6]

Rast, J. P., L. C. Smith, M. Loza-Coll, T. Hibino, and G. W. Litman. 2006. Review: Genomic insights into the immune system of the sea urchin. *Science* 314: 952–956. [20]

Ratcliff, W. C., R. F. Denison, M. Borrello, and M. Travisano. 2012. Experimental evolution of multicellularity. *Proc. Natl. Acad. Sci. USA* 109: 1595–1600. [5, 22]

Ratnieks, F. L. W., and T. Wenseleers. 2007. Altruism in insect societies and beyond: Voluntary or enforced? *Trends Ecol. Evol.* 23: 45–52. [16]

Ratnieks, F. L. W., K. R. Foster, and T. Wenseleers. 2006. Conflict resolution in insect societies. *Annu. Rev. Entomol.* 51: 581–608. [16]

Raup, D. M. 1972. Taxonomic diversity during the Phanerozoic. *Science* 177: 1065–1071. [7]

Raup, D. M., and J. J. Sepkoski Jr. 1982. Mass extinctions in the marine fossil record. *Science* 215: 1501–1503. [7]

Raxworthy, C. J., M. R. J. Forstner, and R. A. Nussbaum. 2002. Chamaeleon radiation by oceanic dispersal. *Nature* 415: 784–787. [6]

Raymond, M., C. Merticat, M. Weill, N. Pasteur, and C. Chevillon. 2001. Insecticide resistance in the mosquito *Culex pipiens*: What have we learned about adaptation? *Genetica* 112-113: 287–296. [13]

Razin, S. 1997. The minimal cellular genome of mycoplasma. *Indian J. Biochem. Biophys.* 34: 124–130. [20]

Rebeiz, M., N. Jikomes, V. A. Kassner, and S. B. Carroll. 2011. Evolutionary origin of a novel gene expression pattern through co-option of the latent activities of existing regulatory sequences. *Proc. Natl. Acad. Sci. USA* 108: 10036–10043. [21]

Rebeiz, M., B. Castro, F. Liu, F. Yue, and J. W. Posakony. 2012. Ancestral and conserved *cis*-regulatory architectures in developmental control genes. *Dev. Biol.* 362: 282–294. [21]

Ree, R. H. 2005. Detecting the historical signature of key innovations using stochastic models of character evolution and cladogenesis. *Evolution* 59: 257–265. [7]

Ree, R. H., and S. A. Smith. 2008. Maximum likelihood inference of geographic range evolution by dispersal, local extinction, and cladogenesis. *Syst. Biol.* 57: 4–14. [6]

Ree, R. H., B. R. Moore, C. O. Webb, and M. J. Donoghue. 2005. A likelihood framework for inferring the evolution of geographic range on phylogenetic trees. *Evolution* 59: 2299–2311. [6]

Reeve, H. K., and P. W. Sherman. 1993. Adaptation and the goals of evolutionary research. *Q. Rev. Biol.* 68: 1–32. [11]

Reich, D., and 27 others. 2010. Genetic history of an archaic hominin group from Denisova Cave in Siberia. *Nature* 468: 1053–1060. [4, 6]

Reimers-Kipping, S., W. Hevers, S. Pääbo, and W. Enard. 2011. Humanized *Foxp2* specifically affects cortico-basal ganglia circuits. *Neuroscience* 175: 75–84. [8]

Relethford, J. H. 1997. Hemispheric difference in human skin color. *Am. J. Phys. Anthropol.* 104: 449–457. [9]

Relethford, J. H. 2008. Genetic evidence and the modern human origins debate. *Heredity* 100: 555–563. [6]

Relyea, R. A. 2002. Costs of phenotypic plasticity. *Am. Nat.* 159: 272–282. [13]

Rendel, J. M., B. L. Sheldon, and D. E. Finlay. 1966. Selection for canalization of the scute phenotype. II. *Am. Nat.* 100: 13–31. [13]

Rensch, B. 1959. *Evolution Above the Species Level*. Columbia University Press, New York. [1, 3, 22]

Reznick, D. 1985. Cost of reproduction: An evaluation of the empirical evidence. *Oikos* 44: 257–267. [14]

Reznick, D., and J. Travis. 2002. Adaptation. In C.W. Fox, D. A. Roff, and D. J. Fairbairn (eds.), *Evolutionary Ecology: Concepts and Case Studies*, pp. 44–57. Oxford University Press, New York. [14]

Reznick, D., H. Bryga, and J. A. Endler. 1990. Experimentally induced life-history evolution in a natural population. *Nature* 346: 357–359. [14]

Rice, W. R. 1992. Sexually antagonistic genes: Experimental evidence. *Science* 256: 1436–1439. [15]

Rice, W. R. 1996. Sexually antagonistic male adaptation triggered by experimental arrest of female evolution. *Nature* 381: 232–234. [15]

Rice, W. R., and E. E. Hostert. 1993. Laboratory experiments on speciation: What have we learned in forty years? *Evolution* 47: 1637–1653. [18]

Rice, W. R., U. Friberg, and S. Gavrilets. 2012. Homosexuality as a consequence of epigenetically canalized development. *Q. Rev. Biol.* 87: 343–368. [16]

Richardson, M. K., J. Hanken, L. Selwood, G. M. Wright, R. J. Richards, C. Pieau, and A. Raynaud. 1998. Haeckel, embryos, and evolution. *Science* 280: 983–984. [3]

Richerson, P. J., and R. Boyd, 2005. *Not by Genes Alone: How Culture Transformed Human Evolution*. University of Chicago Press, Chicago. [11, 16, 23]

Ricklefs, R. E. 2004. A comprehensive framework for global patterns in biodiversity. *Ecol. Lett.* 7: 1–15. [6]

Ricklefs, R. E. 2007. History and diversity: explorations at the intersection of ecology and evolution. *Am. Nat.* 170 (Suppl. 2): S56–S70. [7]

Ricklefs, R. E., and R. E. Latham. 1992. Intercontinental correlation of geographical ranges suggests stasis in ecological traits of relict genera of temperate perennial herbs. *Am. Nat.* 139: 1305–1321. [22]

Rico, P., P. Bouteillon, M. J. H. van Oppen, M. E. Knight, G. M. Hewitt, and G. F. Turner. 2003. No evidence for parallel sympatric speciation in cichlid species of the genus *Pseudotropheus* from northwestern Lake Malawi. *J. Evol. Biol.* 16: 37–46. [18]

Riedl, R. 1978. *Order in Living Organisms: A Systems Analysis of Evolution*. Wiley, New York. [21]

Riehl, C. 2011. Living with strangers: direct benefits favour non-kin cooperation in a communally nesting bird. *Proc. R. Soc. Lond. B* 278: 1728–1735. [16]

Rieseberg, L. H. 1997. Hybrid origins of plant species. *Annu. Rev. Ecol. Syst.* 28: 359–389. [18]

Rieseberg, L. H. 2001. Chromosomal arrangements and speciation. *Trends Ecol. Evol.* 16: 351–358. [17]

Rieseberg, L. H., and J. H. Willis. 2007. Plant speciation. *Science* 317: 910–914. [17]

Rieseberg, L. H., J. Whitten, and K. Gardner. 1999. Hybrid zones and the genetic architecture of a barrier to gene flow between two sunflower species. *Genetics* 152: 713–727. [17]

Rieseberg, L. H., D. M. Raymond, Z. Lai, K. Livingstone, J. L. Durphy, A. E. Schwarzbach, L. A. Donovan, and C. Lexer. 2003. Major ecological transitions in wild sunflowers facilitated by hybridization. *Science* 301: 1211–1216. [18]

Rifkin, S. A., D. Houle, J. Kim, and K. P. White. 2005. A mutation accumulation assay reveals a broad capacity for rapid evolution of gene expression. *Nature* 438: 220–223. [8]

Riley, M. A., and J. E. Wertz. 2002. Bacteriocins: evolution, ecology, and application. *Annu. Rev. Microbiol.* 56: 117–137. [16]

Risch, N., H. Tang, H. Katzenstein, and J. Ekstein. 2003. Geographic distribution of disease mutations in the Ashkenazi Jewish population supports genetic drift over selection. *Am. J. Hum. Genet.* 72: 812–822. [10]

Ritchie, M. G., and S. D. F. Phillips. 1998. The genetics of sexual isolation. In D. J. Howard and S. H. Berlocher (eds.), *Endless Forms: Species and Speciation*, pp. 291–308. Oxford University Press, New York. [17]

Roberts, G. C., and C. W. Smith. 2002. Alternative splicing: Combinatorial output from the genome. *Curr. Opin. Chem. Biol.* 6: 375–383. [20]

Rodríguez, D. J. 1996. A model for the establishment of polyploidy in plants. *Am. Nat.* 147: 33–46. [18]

Roelofs, W. L., and 8 others. 1987. Sex pheromone production and perception in European corn borer moths is determined by both autosomal and sex-linked genes. *Proc. Natl. Acad. Sci. USA* 84: 7585–7589. [17]

Roesti, M., A. P. Hendry, W. Salzburger, and D. Berner. 2012. Genome divergence during evolutionary diversification as revealed in replicate lake-stream stickleback population pairs. *Mol. Ecol.* 21: 2852–2862. [18]

Roff, D. A. 2002. *Life History Evolution*. Sinauer, Sunderland, MA. [14]

Rogers, A. R. 1995. Genetic evidence for a Pleistocene population explosion. *Evolution* 49: 608–615. [10]

Romer, A. S. 1966. *Vertebrate Paleontology*. University of Chicago Press, Chicago. [3, 5]

Romer, A. S., and T. S. Parsons. 1986. *The Vertebrate Body*. Saunders College Publishing, Philadelphia. [5]

Ronquist, F. 1997. Dispersal-vicariance analysis: A new approach to the quantification of historical biogeography. *Syst. Biol.* 46: 195–203. [6]

Ronquist, F., and I. Sanmartín. 2011. Phylogenetic models in biogeography. *Annu. Rev. Ecol. Evol. Syst.* 42: 441–464. [6]

Rose, M. R. 1991. *The Evolutionary Biology of Aging*. Oxford University Press, Oxford. [14]

Rosenzweig, M. L., and R. D. McCord. 1991. Incumbent replacement: Evidence for long-term evolutionary progress. *Paleobiology* 17: 202–213. [7]

Rosindell, J., S. J. Cornell, S. P. Hubbell, and R. S. Etienne. 2010. Protracted speciation revitalizes the neutral theory of biodiversity. *Ecol. Lett.* 13: 716–727. [7]

Rossiter, MC. 1996. Incidence and consequences of inherited environmental effects. *Annu. Rev. Ecol. Syst.* 27: 451–476. [9]

Roth, V. L. 1988. The biological basis of homology. In C. J. Humphries (ed.), *Ontogeny and Systematics*, pp. 1–26. British Museum (Natural History). [21]

Roth, V. L. 1991. Homology and hierarchies: Problems solved and unresolved. *J. Evol. Biol.* 4: 167–194. [21]

Rothschild, L. J., and A. M. Lister (eds.). 2003. *Evolution on Planet Earth: The Impact of the Physical Environment*. Academic Press, San Diego, CA. [7]

Rothstein, S. I., and S. K. Robinson (eds.). 1998. *Parasitic Birds and Their Hosts: Studies in Coevolution*. Oxford University Press, New York. [19]

Roughgarden, J. 1971. Density-dependent natural selection. *Ecology* 52: 453–468. [14]

Roughgarden, J. 2004. *Evolution's Rainbow: Diversity, Gender, and Sexuality in Nature and People*. University of California Press, Berkeley. [15]

Roush, R. T., and B. E. Tabashnik (eds.). 1990. *Pesticide Resistance in Arthropods*. Chapman and Hall, New York. [12]

Rovito, S. M. 2010. Lineage divergence and speciation in the web-toed salamanders (Plethodontidae: *Hydromantes*) of the Sierra Nevada, California. *Mol. Ecol.* 19: 4554–4571. [18]

Rowe, L. E., and D. Houle. 1996. The lek paradox and the capture of genetic variance by condition dependent traits. *Proc. R. Soc. Lond. B* 263: 1415–1421. [15]

Rowe, T. B., T. E. Macrini, and Z.-H. Luo. 2011. Fossil evidence on origin of the mammalian brain. *Science* 332: 955–957. [4]

Royo-Torres, R., A. Cobos, and L. Alcalá. 2006. A giant European dinosaur and a new sauropod clade. *Science* 314: 1925–1927. [4]

Ruhl, M., N. R. Bonis, G.-J. Reichart, J. A. S. Damsté, and W. M. Kürschner. 2011. Atmospheric carbon injection linked to End-Triassic mass extinction. *Science* 333: 430–434. [5]

Rundle, H. D. 2002. A test of ecologically dependent postmating isolation between sympatric sticklebacks. *Evolution* 56: 322–329. [18]

Rundle, H. D. 2003. Divergent environments and population bottlenecks fail to generate premating isolation in *Drosophila pseudoobscura*. *Evolution* 57: 2557–2565. [18]

Rundle, H. D., and P. Nosil. 2005. Ecological speciation. *Ecol. Lett.* 8: 336–352. [18]

Rundle, H. D., L. Nagel, J. W. Boughman, and D. Schluter. 2000. Natural selection and parallel speciation in sympatric sticklebacks. *Science* 287: 306–308. [18]

Ruse, M. 1979. *The Darwinian Revolution*. University of Chicago Press, Chicago. [11]

Ruse, M. 1996. *Monad to Man: The Concept of Progress in Evolutionary Biology*. Harvard University Press, Cambridge, MA. [11, 22]

Russell, C. A., and 18 others. 2012. The potential for respiratory droplet-transmissible A/H5N1 influenza virus to evolve in a mammalian host. *Science* 336: 1541–1547. [19]

Rutherford, S. L., and S. Lindquist. 1998. Hsp90 as a capacitor for morphological evolution. *Nature* 396: 336–342. [13, 22]

Rutledge, R., and 9 others. Characterization of an *AGAMOUS* homologue from the conifer black spruce (*Picea mariana*) that produces floral homeotic conversions when expressed in *Arabidopsis*. *Plant J.* 15: 625–634. [20]

Ruvolo, M. 1997. Molecular phylogeny of the hominoids: Inferences from multiple independent DNA sequence data sets. *Mol. Biol. Evol.* 14: 248–265. [2]

Ryan, M. J. 1998. Sexual selection, receiver biases, and the evolution of sex differences. *Science* 281: 1999–2003. [15]

S

Sabeti, P. C., and 16 others. 2002. Detecting recent positive selection in the human genome from haplotype structure. *Nature* 419: 832–837. [12]

Sabeti, P. C., and 9 others. 2006. Positive natural selection in the human lineage. *Science* 312: 1614–1620. [12]

Sabeti, P. C., 11 others, and The International HapMap Consortium. 2007. Genome-wide detection and characterization of positive selection in human populations. *Nature* 449: 913–918. [12]

Sachs, J. L., U. G. Mueller, T. P. Wilcox, and J. J. Bull. 2004. The evolution of cooperation. *Q. Rev. Biol.* 79: 135–160. [16]

Sakai, A., S. G. Weller, M.-L. Chen, S.-Y. Chou, and C. Tasanont. 1997. Evolution of gynodioecy and maintenance of females: the role of inbreeding depression, outcrossing rates, and resource allocation in *Schiedea adamantis* (Caryophyllaceae). *Evolution* 51: 724–736. [15]

Sala, O. E., and 18 others. 2000. Global biodiversity scenarios for the year 2100. *Science* 287: 1770–1774. [7]

Salvini-Plawen, L. V., and E. Mayr. 1977. On the evolution of photoreceptors and eyes. *Evol. Biol.* 10: 207–263. [22]

Salzberg, S. L., O. White, J. Peterson, and J. A. Eisen. 2001. Microbial genes in the human genome: Lateral transfer or gene loss? *Science* 292: 1903–1906. [20]

Sambatti, J. B. M., J. L. Strasburg, D. Ortiz-Barrientos, E. J. Baack, and L. H. Rieseberg. 2012. Reconciling extremely strong barriers with high levels of gene exchange in annual sunflowers. *Evolution* 66: 1459–1473. [18]

Sanderson, M. J. 2002. Estimating absolute rates of molecular evolution and divergence times: a penalized likelihood approach. *Mol. Biol. Evol.* 19: 101–109. [10]

Sang, T., and Y. Zhong. 2000. Testing hybridization hypotheses based on incongruent gene trees. *Syst. Biol.* 49: 422–434. [2]

Sanmartín, I., and F. Ronquist. 2004. Southern hemisphere biogeography inferred by event-based models: plant versus animal patterns. *Syst. Biol.* 53: 216–243. [6]

Sanmartín, I., H. Enghoff, and F. Ronquist. 2001. Patterns of animal dispersal, vicariance and diversification in the Holarctic. *Biol. J. Linn. Soc.* 73: 345–390. [6]

Sasa, M., P. T. Chippendale, and N. A. Johnson. 1998. Patterns of postzygotic isolation in frogs. *Evolution* 52: 1811–1820. [18]

Satta, Y., and N. Takahata. 2002. Out of Africa with regional interbreeding? Modern human origins. *BioEssays* 24: 871–875. [6]

Savolainen, V., and 9 others. 2006. Sympatric speciation in palms on an oceanic island. *Nature* 441: 210–213. [18]

Sayres, M. A. W., C. Venditti, M. Pagel, and K. D. Makova. 2011. Do variations in substitution rates and male mutation bias correlate with life history traits? A study of 32 mammalian genomes. *Evolution* 65: 2800–2815. [8]

Scally, A., and 70 others. 2012. Insights into hominid evolution from the gorilla genome sequence. *Nature* 483: 169–175. [2]

Schaffer, W. M. 1974. Selection for optimal life histories: effects of age structure. *Ecology* 55: 291–303. [14]

Scharloo, W. 1991. Canalization: Genetic and developmental aspects. *Annu. Rev. Ecol. Syst.* 22: 65–93. [13]

Schemske, D. W., and H. D. Bradshaw, Jr. 1999. Pollinator preference and the evolution of floral traits in monkeyflkowers (*Mimulus*). *Proc. Natl. Acad. Sci. USA* 96: 11910–11915. [17]

Schindewolf, O. H. 1950. *Grundfrage der Paläontologie*. Schweitzerbart, Jena, Germany. [22]

Schlichting, C. D., and M. Pigliucci. 1998. *Phenotypic Evolution: A Reaction Norm Perspective*. Sinauer, Sunderland, MA. [13]

Schliewen, U. K., D. Tautz, and S. Pääbo. 1994. Sympatric speciation suggested by monophyly of crater lake cichlids. *Nature* 368: 629–632. [18]

Schluter, D. 1996. Adaptive radiation along genetic lines of least resistance. *Evolution* 50: 1766–1774. [13]

Schluter, D. 2000. *The Ecology of Adaptive Radiation*. Oxford University Press, Oxford. [3, 13, 18, 19]

Schluter, D. 2009. Evidence for ecological speciation and its alternative. *Science* 323: 737–741. [18]

Schluter, D., and J. D. McPhail. 1992. Ecological character displacement and speciation in sticklebacks. *Am. Nat.* 140: 85–108. [19]

Schluter, D., and L. M. Nagel. 1995. Parallel speciation by natural selection. *Am. Nat.* 146: 292–301. [18]

Schluter, D., and R. E. Ricklefs. 1993. Convergence and the regional component of species diversity. In R. E. Ricklefs and D. Schluter (eds.), *Species Diversity in Ecological Communities: Historical and Geographical Perspectives*, pp. 230-240. University of Chicago Press, Chicago. [6]

Schmalhausen, I. I. 1986. *Factors of Evolution*. University of Chicago Press, Chicago. (Russian publication 1946; English translation 1949.) [13]

Schrag, S. J., G. T. Ndifon, and A. F. Read. 1994. Temperature-determined outcrossing ability in wild populations of a simultaneous hermaphrodite snail. *Ecology* 75: 2066–2077. [9]

Schwenk, K., N. Brede, and B. Streit. 2008. Introduction. Extent, processes and evolutionary impact of interspecific hybridization in animals. *Phil. Trans. R. Soc.* 363: 2805–2811. [17]

Scott, E. C. 1997. Antievolution and creationism in the United States. *Annu. Rev. Anthropol.* 26: 263–289. [23]

Scott, E. C. 2005. *Evolution versus Creationism: An Introduction.* University of California Press, Berkeley. [1]

Scoville, A. G., and M. E. Pfrender. 2010. Phenotypic plasticity facilitates recurrent rapid adaptation to introduced predators. *Proc. Natl. Acad. Sci. USA* 107: 4260–4263. [21]

Seehausen, O. 2006a. African cichlid fish: a model system in adaptive radiation research. *Proc. R. Soc. Lond. B* 273: 1987–1998. [3]

Seehausen, O. 2006b. Losing biodiversity by reverse speciation. *Curr. Biol.* 16: R334–R337. [17]

Seehausen, O., J. J. M. van Alphen, and F. Witte. 1997. Cichlid fish diversity threatened by eutrophication that curbs sexual selection. *Science* 277: 1808–1811. [17]

Seehausen, O., P. J. Mayhew, and J. J. van Alphen. 1999. Evolution of colour patterns in East African cichlid fish. *J. Evol. Biol.* 12: 514–534. [18]

Selander, R. K. 1966. Sexual dimorphism and differential niche utilization in birds. *Condor* 68: 113–151. [19]

Sella, G., D. A. Petrov, M. Przeworski, M., and P. Andolfatto. 2009. Pervasive natural selection in the *Drosophila* genome. *PLoS Genet.* 5: e1000495. [8, 10, 12]

Sepkoski, J. J. Jr. 1984. A kinetic model of Phanerozoic taxonomic diversity. III. Post-Paleozoic families and mass extinctions. *Paleobiology* 10: 246–267. [7]

Sepkoski, J. J. Jr. 1993. Ten years in the library: New data confirm paleontological patterns. *Paleobiology* 19: 43–51. [7]

Sepkoski, J. J. Jr. 1996a. Competition in macroevolution: The double wedge revisited. In D. Jablonski, D. H. Erwin, and J. Lipps (eds.), *Evolutionary Paleobiology*, pp. 211–255. University of Chicago Press, Chicago. [7]

Sepkoski, J. J. Jr. 1996b. Large-scale history of biodiversity. In V. H. Heywood (ed.), *Global Biodiversity Assessment*, pp. 202–212. United Nations Environmental Programme. Cambridge University Press, Cambridge. [7]

Sepkoski, J. J. Jr., F. K. McKinney, and S. Lidgard. 2000. Competitive displacement among post-Paleozoic cyclostome and cheilostome bryozoans. *Paleobiology* 26: 7–18. [7]

Sereno, P. C. 1999. The evolution of dinosaurs. *Science* 284: 2137–2147. [4, 5]

Servedio, M. R., and M. A. F. Noor. 2003. The role of reinforcement in speciation: Theory and data. *Annu. Rev. Ecol. Evol. Syst.* 34: 339–364. [18]

Servedio, M. R., G. S. van Doorn, M. Kopp, A. M. Frame and P. Nosil. 2011. Magic traits in speciation: 'magic' but not rare? *Trends Ecol. Evol.* 26: 389–397. [18]

Sgrò, C. M., and L. Partridge. 1999. A delayed wave of death from reproduction in *Drosophila*. *Science* 286: 2521–2524. [14]

Shabalina S. A., A. Y. Ogurtsov, V. A. Kondrashov, and A. S. Kondrashov. 2001. Selective constraint in intergenic regions of human and mouse genomes. *Trends Genet.* 17: 373–376. [20]

Shaffer, H. B., and S. R. Voss. 1996. Phylogenetic and mechanistic analysis of a developmentally integrated character complex: Alternate life history modes in ambystomatid salamanders. *Am. Zool.* 36: 24–35. [21]

Shapiro, M. D., M. E. Marks, C. L. Peichel, B. K. Blackman, K. S. Nereng, B. Jonsson, D. Schluter, and D. M. Kingsley. 2004. Genetic and developmental basis of evolutionary pelvic reduction in threespine sticklebacks. *Nature* 428: 717–723. [21]

Shapiro, M. D., M. A. Bell, and D. M. Kingsley. 2006. Parallel genetic origins of pelvic reduction in vertebrates. *Proc. Natl. Acad. Sci. USA* 103: 13753–13758. [21]

Sharp, A. J., Z. Cheng, and E. E. Eichler. 2006. Structural variation of the human genome. *Annu. Rev. Genomics Hum. Genet.* 7: 407–442. [8]

Sharp, N. P., and A. F. Agrawal. 2012. Evidence for elevated mutation rates in low-quality genotypes. *Proc. Natl. Acad. Sci. USA* 109: 6142–6146. [8]

Shaver, A. C., P. G. Dombrowski, J. Y. Sweeney, T. Treis, R. M. Zappala, and P. D. Sniegowski. 2002. Fitness evolution and the rise of mutator alleles in experimental *Escherichia coli* populations. *Genetics* 162: 557–566. [15]

Shelby, J. A., R. Madewell, and A. P. Moczek. 2007. Juvenile hormone mediates sexual dimorphism in horned beetles. *J. Exp. Zoolog. B. Mol. Dev. Evol.* 308: 417–427. [21]

Shen, S-Z, and 21 others. 2011. Calibrating the End-Permian mass extinction. *Science* 334: 1367–1372. [5]

Sherman, P. W., H. K. Reeve, and D. W. Pfennig. 1997. Recognition systems. In J. R. Krebs and N. B. Davies (eds.), *Behavioural Ecology: An Evolutionary Approach*, 4th ed., pp. 69–96. Blackwell Scientific, Oxford. [16]

Shine, R., and E. L. Charnov. 1992. Patterns of survival, growth, and maturation in snakes and lizards. *Am. Nat.* 139: 1257–1269. [14]

Shine, R., G. P. Brown, and B. L. Phillips. 2011. An evolutionary process that assembles phenotypes through space rather than through time. *Proc. Natl. Acad. Sci. USA* 108: 5708–5711. [11]

Shirai, L. T., S. V. Saenko, R. A. Keller, M. A. Jeronimo, P. M. Brakefield, H. Descimon, N. Wahlberg, and P. Beldade. 2012. Evolutionary history of the recruitment of conserved developmental genes in association to the formation and diversification of a novel trait. *BMC Evol. Biol.* 12: 21–32. [21]

Short, L. L. 1965. Hybridization in the flickers (*Colaptes*) of North America. *Bull. Am. Mus. Nat. Hist.* 129: 307–428. [9]

Shoshani, J., C. P. Groves, E. L. Simons, and G. F. Gunnell. 1996. Primate phylogeny: Morphological and molecular results. *Mol. Phyl. Evol.* 5: 102–154. [2]

Shu, D.-G., and 10 others. 2003. Head and backbone of the Early Cambrian vertebrate *Haikouichthys*. *Nature* 421: 526–529. [5]

Shubin, N. 2008. *Your Inner Fish: A Journey into the 3.5-Billion-Year History of the Human Body.* Allen Lane/Pantheon, New York. [4]

Shubin, N. H., E. B. Daeschler, and F. A. Jenkins, Jr. 2006. The pectoral fin of *Tiktaalik roseae* and the origin of the tetrapod limb. *Nature* 440: 764–771. [4]

Sidor, C. A. 2001. Simplification as a trend in synapsid cranial evolution. *Evolution* 55: 1419–1442. [4]

Sidor, C. A., and J. A. Hopson. 1998. Ghost lineages and "mammalness:" Assessing the temporal pattern of character acquisition in the Synapsida. *Paleobiology* 24: 254–273. [4]

Siepelski, A. M., and C. W. Benkman. 2007. Convergent patterns in the selection mosaic for two North American bird-dispersed pines. *Ecol. Monogr.* 77: 203–220. [19]

Signor, P. W. III. 1985. Real and apparent trends in species richness through time. In J. W. Valentine (ed.), *Phanerozoic Diversity Patterns: Profiles in Macroevolution*, pp. 129–150. Princeton University Press, Princeton, NJ. [7]

Signor, P. W. III. 1990. The geological history of diversity. *Annu. Rev. Ecol. Syst.* 21: 509–539. [7]

Silk, J. B., and B. R. House. 2011. Evolutionary foundations of human prosocial sentiments. *Proc. Natl. Acad. Sci. USA* 108: 10910–10917. [16]

Simmons, A. D., and C. D. Thomas. 2004. Changes in dispersal during species' range expansions. *Am. Nat.* 164: 378–395. [11]

Simmons, L. W. 2001. *Sperm Competition and its Evolutionary Consequences in the Insects.* Princeton University Press, Princeton, NJ. [15]

Simons, E. L. 1979. The early relatives of man. In G. Isaac and R. E. F Leakey (eds.), *Human Ancestors*, pp. 22–42. W. H Freeman, San Francisco. [5]

Simpson, G. G. 1944. *Tempo and Mode in Evolution.* Columbia University Press, New York. [1, 22]

Simpson, G. G. 1953. *The Major Features of Evolution.* Columbia University Press, New York. [1, 21, 22]

Simpson, G. G. 1964. *This View of Life: The World of an Evolutionist.* Harcourt, Brace and World, New York. [22]

Simpson, P. 2002. Evolution of development in closely related species of flies and worms. *Nat. Rev. Genet.* 3: 907–917. [21]

Sinervo, B. 1990. The evolution of maternal investment in lizards: An experimental and comparative analysis of egg size and its effect on offspring performance. *Evolution* 44: 279–294. [14]

Singh, N. D., A. M. Larracuente, and A. G. Clark. 2008. Contrasting the efficacy of selection on the X and autosomes in *Drosophila. Mol. Biol. Evol.* 25: 454–467. [20]

Skaer, N., D. Pistillo, and P. Simpson. 2002. Transcriptional heterochrony of *scute* and changes in bristle pattern between two closely related species of blow fly. *Dev. Biol.* 252: 31–45. [21]

Skoglund, P., and M. Jakobsson. 2011. Archaic human ancestry in East Asia. *Proc. Natl. Acad. Sci. USA* 108: 18301–18306. [6]

Slatkin, M. 1980. Ecological character displacement. *Ecology* 61: 163–177. [19]

Slatkin, M. 1985. Gene flow in natural populations. *Annu. Rev. Ecol. Syst.* 16: 393–430. [10]

Slatkin, M. 1996. In defense of founder-flush theories of speciation. *Am. Nat.* 147: 493–505. [18]

Slatkin, M. 2004. A population-genetic test of founder effects and implications for Ashkenazi Jewish disease. *Am. J. Hum. Genet.* 75: 282–293. [10]

Slatkin, M., and W. P. Maddison. 1989. A cladistic measure of gene flow inferred from the phylogenies of alleles. *Genetics* 123: 603–613. [10]

Smith, A. B., and K. J. Peterson. 2002. Dating the time of origin of major clades: Molecular clocks and the fossil record. *Annu. Rev. Earth Planet. Sci.* 30: 65–88. [5]

Smith, M. R., and J.-B. Caron. 2010. Primitive soft-bodied cephalopods from the Cambrian. *Nature* 465: 469–472. [5]

Smith, N. G. C., and A. Eyre-Walker. 2002. Adaptive protein evolution in *Drosophila. Nature* 415: 1022–1024. [10]

Smith, T. B. 1993. Disruptive selection and the genetic basis of bill size polymorphism in the African finch *Pyrenestes. Nature* 363: 618–620. [12]

Smith, T. B., and L. Bernatchez. 2008. Evolutionary change in human-altered environments. *Mol. Ecol.* 17: 1–8. [11]

Smith, T. B., and S. Skulason. 1996. Evolutionary significance of resource polymorphisms in fishes, amphibians, and birds. *Annu. Rev. Ecol. Syst.* 27: 111–133. [19]

Smocovitis, V. B. 1996. *Unifying Biology: The Evolutionary Synthesis and Evolutionary Biology.* Princeton University Press, Princeton, NJ. [1]

Sniegowski, P. D., and R. E. Lenski. 1995. Mutation and adaptation: The directed mutation controversy in evolutionary perspective. *Annu. Rev. Ecol. Syst.* 26: 553–578. [8]

Sniegowski, P. D., P. J. Gerrish, T. Johnson, and A. Shaver. 2000. The evolution of mutation rates: Separating causes from consequences. *Bioessays* 22: 1057–1066. [15]

Sobel, J. M., G. F. Chen, L. R. Watt, and D. W. Schemske. 2010. The biology of speciation. *Evolution* 64: 295–315. [18]

Sober, E. 1984. *The Nature of Selection: Evolutionary Theory in Philosophical Focus.* MIT Press, Cambridge, MA. [11]

Sodergren, E., and 232 others. 2006. The genome of the sea urchin *Strongylocentrotus purpuratus. Science* 314: 941–952. [2, 20]

Soltis, D. E., P. S. Soltis, J. C. Pires, A. Kovarik, J. A. Tate, and E. Mavrodiev. 2004. Recent and recurrent polyploidy in *Tragopogon* (Asteraceae): Cytogenetic, genomic and genetic comparisons. *Biol. J. Linn. Soc.* 82: 485–501. [18]

Soltis, D. E., A. B. Morris, J. S. McLachlan, P. M. Manos, and P. S. Soltis. 2006. Comparative phylogeography of unglaciated eastern North America. *Mol. Ecol.* 15: 4261–4293. [6]

Somers, C. M., and D. N. Cooper. 2009. Air pollution and mutations in the germline: are humans at risk? *Hum. Genet.* 125: 119–130. [8]

Somers, C. M., B. E. McCarry, F. Malek, and J. S. Quinn. 2004. Reduction of particulate air pollution lowers the risk of heritable mutations in mice. *Science* 304: 1008–1010. [8]

Soulé, M. 1966. Trends in the insular radiation of a lizard. *Am. Nat.* 100: 47–64. [18]

Soulé, M. E. 1976. Allozyme variation, its determinants in space and time. In F. J. Ayala (ed.), *Molecular Evolution*, pp. 60–77. Sinauer, Sunderland, MA. [10]

Soyfer, V. 1994. *Lysenko and the Tragedy of Soviet Science.* Rutgers University Press, New Brunswick, NJ. [23]

Sparks, J. S., and W. L. Smith. 2004. Phylogeny and biogeography of cichlid fishes (Teleostei: Perciformes: Cichlidae). *Cladistics* 20: 501–507. [6]

Spassky, B., N. Spassky, H. Levene, and Th. Dobzhansky. 1958. Release of genetic variability through recombination. I. *Drosophila pseudoobscura. Genetics* 43: 844–867.

Spiegelman, S. 1970. Extracellular evolution of replicating molecules. In F. O. Schmitt (ed.), *The Neuro Sciences: A Second Study Program*, pp. 927–945. Rockefeller University Press, New York. [5]

Stanley, S. M. 1979. *Macroevolution: Pattern and Process.* W. H. Freeman, San Francisco. [7, 18, 22]

Stanley, S. M. 1986. Anatomy of a regional mass extinction: Plio-Pleistocene decimation of the western Atlantic bivalve fauna. *Palaios* 1: 17–36. [7]

Stanley, S. M. 1990. The general correlation between rate of speciation and rate of extinction: Fortuitous causal linkages. In R. M. Ross and W. D. Allmon (eds.), *Causes of Evolution: A Paleontological Perspective*, pp. 103–127. University of Chicago Press, Chicago. [7]

Stanley, S. M. 1993. *Earth and Life Through Time*, 2nd ed. W.H. Freeman, New York. [5]

Stanley, S. M. 2005. *Earth System History.* W.H. Freeman, New York. [4]

Stanley, S. M., and L. A. Campbell. 1981. Neogene mass extinction of western Atlantic mollusks. *Nature* 293: 457–459. [5]

Stanley, S. M., and X. Yang. 1987. Approximate evolutionary stasis for bivalve morphology over millions of years: A multivariate, multilineage study. *Paleobiology* 13: 113–139. [22]

Stearns, S. C. 1992. *The Evolution of Life Histories.* Oxford University Press, Oxford. [14]

Stearns, S. C., S. G. Byars, D. R. Govindaraju, and D. Ewbank. 2010. Measuring selection in contemporary human populations. *Nat. Rev. Genet.* 11: 611–622. [13]

Stebbins, G. L. 1950. *Variation and Evolution in Plants.* Columbia University Press, New York. [1, 17, 18]

Stebbins, G. L. 1974. *Flowering Plants: Evolution Above the Species Level.* Belknap Press of Harvard University Press, Cambridge, MA. [13]

Stebbins, R. C. 1954. *Amphibians and Reptiles of Western North America.* McGraw-Hill, New York. [17]

Stedman, H. H., and 9 others. 2004. Myosin gene mutation correlates with anatomical changes in the human lineage. *Nature* 428: 415–418. [20]

Steiper, M. E., and N. M. Young. 2006. Primate molecular divergence dates. *Mol. Phyl. Evol.* 41: 384–394. [2]

Steiper, M. E., N. M. Young, and T. Y. Sukarna. 2004. Genomic data support the hominoid slowdown and an Early Oligocene estimate for the hominoid-cercopithecoid divergence. *Proc. Natl. Acad. Sci. USA* 101: 17021–17026. [10]

Stephens J. C., and 27 others. 2001. Haplotype variation and linkage disequilibrium in 313 human genes. *Science* 293: 489–493. [20]

Steppan, S. J., P. C. Phillips, and D. Houle. 2002. Comparative quantitative genetics: evolution of the G matrix. *Trends Ecol. Evol.* 17: 320–327. [13]

Stern, C. 1973. *Principles of Human Genetics*. W. H. Freeman, San Francisco. [9]

Stern, D. L. 1998. A role of Ultrabithorax in morphological differences between *Drosophila* species. *Nature* 396: 463–466. [21]

Stern, D. L., and V. Orgogozo. 2008. The loci of evolution: how predictable is genetic evolution? *Evolution* 62: 2155–2177. [22]

Stern, D. L., and V. Orgogozo. 2009. Is genetic evolution predictable? *Science* 323: 746–751. [8, 21, 22]

Stewart, C.-B., J. W. Schilling, and A. C. Wilson. 1987. Adaptive evolution in the stomach lysozymes of foregut fermenters. *Nature* 330: 401–404. [20]

Stewart, J. R., and C. B. Stringer. 2012. Human evolution out of Africa: the role of refugia and climate change. *Science* 335: 1317–1321. [6]

Stewart, W. N. 1983. *Paleobotany and the Evolution of Plants*. Cambridge University Press, Cambridge. [5]

Stoneking, M., and J. Krause. 2011. Learning about human population history from ancient and modern genomes. *Nat. Rev. Genet.* 12: 603–614. [6]

Storz, J. F. 2005. Using genome scans of DNA polymorphism to infer adaptive population divergence. *Mol. Ecol.* 14: 671–688. [12]

Storz, J. F., G. R. Scott, and Z. A. Cheviron. 2010. Phenotypic plasticity and genetic adaptation to high-altitude hypoxia in vertebrates. *J. Exp. Biol.* 213: 4125–4136. [21]

Strasburg, J. L., N. A. Sherman, K. M. Wright, L. C. Moyle, J. H. Willis, and L. H. Rieseberg. 2012. What can patterns of differentiation across plant genomes tell us about adaptation and speciation? *Phil. Trans. R. Soc. B* 367: 364–373. [18]

Strassmann, J. E., and D. C. Queller. 2007. Insect societies as divided organisms: The complexities of purpose and cross-purpose. *Proc. Natl. Acad. Sci. USA* 104 (Suppl. 1): 8619–8626. [16]

Strassmann, J. E., and D. C. Queller. 2011. Evolution of cooperation and control of cheating in a social microbe. *Proc. Natl. Acad. Sci. USA* 108: 10855–10962. [16]

Strauss, S.Y., C. O. Webb, and N. Salamin. 2006. Exotic taxa less related to native species are more invasive. *Proc. Natl. Acad. Sci. USA* 103: 5841–5845. [19]

Strickberger, M. W. 1968. *Genetics*. Macmillan, New York. [8]

Strobeck, C. 1983. Expected linkage disequilibrium for a neutral locus linked to a chromosomal arrangement. *Genetics* 103: 545–555. [12]

Strömberg, C. A. E. 2006. Evolution of hypsodonty in equids: testing a hypothesis of adaptation. *Paleobiology* 32: 236–258. [4]

Strotz, L. C., and A. P. Allen. 2013. Assessing the role of cladogenesis in macroevolution by integrating fossil and molecular evidence. *Proc. Natl. Acad. Sci. USA*, published online before print February 1, 2013, doi:10.1073/pnas.1208302110 [18, 22]

Stümpke, H. 1957. *Bau und Leben der Rhinogradentia* (English translation: *The Snouters*). Doubleday, New York. (Reprinted 1981, University of Chicago Press.) [22]

Stutt, A. D., and M. T. Siva-Jothy. 2001. Traumatic insemination and sexual conflict in the bed bug *Cimex lectularius*. *Proc. Natl. Acad. Sci. USA* 98: 5683–5687. [15]

Suga, H., P. Tschopp, D. F. Graziussi, M. Stierwald, V. Schmid, and W. J. Gehring. 2010. Flexibly deployed *Pax* genes in eye development at the early evolution of animals demonstrated by studies on a hydrozoan jellyfish. *Proc. Natl. Acad. Sci. USA* 107: 14263–14268. [21]

Sullivan, J., and P. Joyce. 2005. Model selection in phylogenetics. *Annu. Rev. Ecol. Evol. Syst.* 36: 445–466. [2]

Sundström, L., M. Chapuisat, and L. Keller. 1996. Conditional manipulation of sex ratio by ant workers: A test of kin selection theory. *Science* 274: 993–995. [16]

Swanson, W. J., and V. D. Vacquier. 2002. The rapid evolution of reproductive proteins. *Nat. Rev. Genet.* 3: 137–144. [12, 15, 20]

Swofford, D. L., G. J. Olsen, P. J. Waddell, and D. M. Hillis. 1996. Phylogenetic inference. In D. M. Hillis, C. Moritz, and B. K. Mable (eds.), *Molecular Systematics*, 2nd ed., pp. 407–514. Sinauer, Sunderland, MA. [2]

Szathmáry, E. 1993. Coding coenzyme handles: A hypothesis for the origin of the genetic code. *Proc. Natl. Acad. Sci. USA* 90: 9916–9920. [5]

Szymura, J. M. 1993. Analysis of hybrid zones with *Bombina*. In R. G. Harrison (ed.), *Hybrid Zones and the Evolutionary Process*, pp. 261–289. Oxford University Press, New York. [17]

T

Taberlet, P., L. Fumagalli, A.-G. Wust-Saucy, and J.-F. Cosson. 1998. Comparative phylogeography and postglacial colonization routes in Europe. *Mol. Ecol.* 7: 453–464. [6]

Tagle, D. A., J. L. Slightom, R. T. Jones, and M. Goodman. 1991. Concerted evolution led to high expression of a prosimian primate δ-globin gene locus. *J. Biol. Chem.* 266: 7469–7480. [20]

Tajima, F. 1989. Statistical method for testing the neutral mutation hypothesis by DNA polymorphism. *Genetics* 123: 585–595. [12]

Taper, M. L., and T. J. Case. 1992. Coevolution among competitors. *Oxford Surv. Evol. Biol.* 8: 63–109. [19]

Templeton, A. R. 1980. The theory of speciation via the founder principle. *Genetics* 94: 1011–1038. [18]

Templeton, A. R. 1989. The meaning of species and speciation: A genetic perspective. In D. Otte and J. A. Endler (eds.), *Speciation and Its Consequences*, pp. 3–27. Sinauer, Sunderland, MA. [17]

Templeton, A. R. 2008. The reality and importance of founder speciation in evolution. *BioEssays* 30: 470–479. [18]

Theobald, D. L. 2010. A formal test of the theory of universal common ancestry. *Nature* 465: 219–222. [5]

Thewissen, J. G. M., and E. M. Williams. 2002. The early radiations of Cetacea (Mammalia): Evolutionary pattern and developmental correlations. *Annu. Rev. Ecol. Syst.* 33: 73–90. [4]

Thewissen, J. G. M., L. N. Cooper, M. T. Clementz, S. Bajpal, and B. N. Tiwari. 2007. Whales originated from aquatic artiodactyls in the Eocene epoch of India. *Nature* 450: 1190–1194. [4]

Thomas, C. D., and 18 others. 2004. Extinction risk from climate change. *Nature* 427: 145–148. [5, 7]

Thompson, J. N. 1994. *The Coevolutionary Process*. University of Chicago Press, Chicago. [19]

Thompson, J. N. 2004. *The Geographic Mosaic of Coevolution*. University of Chicago Press, Chicago. [22]

Thompson, J. N. 2009. The coevolving web of life. *Am. Nat.* 173: 125–140. [19]

Thomson, J. D., and J. Brunet. 1990. Hypotheses for the evolution of dioecy in seed plants. *Trends Ecol. Evol.* 5: 11–16. [15]

Thornhill, N. W. (ed.). 1993. *The Natural History of Inbreeding and Outbreeding*. University of Chicago Press, Chicago. [15]

Thornhill, R., and J. Alcock. 1983. *The Evolution of Insect Mating Systems*. Harvard University Press, Cambridge, MA. [15]

Thornton, J. W. 2004. Resurrecting ancient genes: Experimental analysis of extinct molecules. *Nat. Rev. Genet.* 5: 366–375. [3]

Thrall, P. H., and J. J. Burdon. 2003. Evolution of virulence in a plant host-pathogen metapopulation. *Science* 299: 1735–1737. [19]

Tilley, S. G., P. A. Verrell, and S. J. Arnold. 1990. Correspondence between sexual isolation and allozyme differentiation: A test in the salamander *Desmognathus ochrophaeus*. *Proc. Natl. Acad. Sci. USA* 87: 2715–2719. [18]

Tilmon, K. J. (ed.) 2008. *Specialization, Speciation, and Radiation: The Evolutionary Biology of Herbivorous Insects.* University of California Press, Berkeley. [19]

Tinbergen, J. M., and S. Daan. 1990. Family planning in the great tit (*Parus major*): optimal clutch size as integration of parent and offspring fitness. *Behaviour* 114: 161–190. [14]

Tishkoff, S. A., and 18 others. 2007. Convergent adaptation of human lactase persistence in Africa and Europe. *Nature Genet.* 39: 31–40. [11, 12]

Tishkoff, S. A., and 24 others. 2009. The genetic structure and history of Africans and African Americans. *Science* 324: 1035–1044. [10]

Tizard, B. 1973. IQ and race. *Nature* 247: 316. [9]

Tobias, J. A., J. Aben, R. T. Brumfield, E. P. Derryberry, W. Halfwerk, H. Slabbekoorn, and N. Seddon. 2010a. Song divergence by sensory drive in Amazonian birds. *Evolution* 64: 2820–2839. [18]

Tobias, J. A., N. Seddon, C. N. Spottiswoode, J. D. Pilgrim, L. D. C. Fishpool, and N. J. Collar. 2010b. Quantitative criteria for species delimitation. *Ibis* 152: 724–746. [17]

Toju, H., and T. Sota. 2006. Imbalance of predator and prey armament: Geographic clines in phenotypic interface and natural selection. *Am. Nat.* 167: 105–117. [19]

Tomoyasu, Y., S. R. Wheeler, and R. E. Denell. 2005. *Ultrabithorax* is required for membranous wing identity in the beetle *Tribolium castaneum*. *Nature* 433: 643–647. [22]

Tomoyasu, Y., Y. Arakane, K. J. Kramer, and R. E. Denell. 2009. Repeated co-options of exoskeleton formation during wing-to-elytron evolution in beetles. *Curr. Biol.* 19: 2057–2065. [21]

Tooby, J., and L. Cosmides. 1992. The psychological foundations of culture. In J. H. Barkow, L. Cosmides, and J. Tooby (eds.), *The Adapted Mind: Evolutionary Psychology and the Generation of Culture*, pp. 19-136. Oxford University Press, New York. [16]

Travis, J. 1989. The role of optimizing selection in natural populations. *Annu. Rev. Ecol. Syst.* 20: 279–296. [13]

Trivers, R. L. 1971. The evolution of reciprocal altruism. *Q. Rev. Biol.* 46: 35–57. [16]

Trivers, R. L. 1972. Parental investment and sexual selection. In B. Campbell (ed.), *Sexual Selection and the Descent of Man*, pp. 136–179. Heinemann, London. [15]

Trivers, R. L. 1974. Parent-offspring conflict. *Am. Zool.* 11: 249–264. [16]

True, J. R., and S. B. Carroll. 2002. Gene co-option in physiological and morphological evolution. *Annu. Rev. Cell. Dev. Biol.* 18: 53–80. [11, 21]

Tsubota, T., K. Saigo, and T. Kojima. 2008. Hox genes regulate the same character by different strategies in each segment. *Mech. Dev.* 125: 894–905. [21]

Turelli, M. 1988. Phenotypic evolution, constant covariances, and the maintenance of additive variance. *Evolution* 42: 1342–1347. [13]

Turelli, M., J. H. Gillespie, and R. Lande. 1988. Rate tests for selection on quantitative characters during macroevolution and microevolution. *Evolution* 42: 1085–1089. [13]

Turelli, M., N. H. Barton, and J. A. Coyne. 2001. Theory and speciation. *Trends Ecol. Evol.* 16: 330–343. [18]

Turner, T. F., J. P. Wares, and J. R. Gold. 2002. Genetic effective size is three orders of magnitude smaller than adult census size in an abundant, estuarine-dependent marine fish (*Sciaenops ocellatus*). *Genetics* 162: 1329–1339. [10]

U

Ullstrup, A. J. 1972. The impacts of the southern corn leaf blight epidemic of 1970–1971. *Annu. Rev. Phytopathol.* 19: 37–50. [19]

Uyeda, J. C., T. F. Hansen, S. J. Arnold, and J. Pienaar. 2011. The million-year wait for evolutionary bursts. *Proc. Natl. Acad. Sci. USA* 108: 15908–15913. [18]

V

Vacquier, V. D., and W. J. Swanson. 2011. Selection in the rapid evolution of gamete recognition proteins in marine invertebrates. *Cold Spring Harbor Persp. Biol.* 3(11): a002931. [15]

Valentine, J. W. 2004. *On the Origin of Phyla.* University of Chicago Press, Chicago. [5]

Valentine, J. W., T. C. Foin, and D. Peart. 1978. A provincial model of Phanerozoic marine diversity. *Paleobiology* 4: 55–66. [7]

van Asch, M., P. H. van Tienderen, L. J. M. Holleman, and M. E. Visser. 2007. Predicting adaptation of phenology in response to climate change, an insect herbivore example. *Global Change Biol.* 13: 1596–1604. [13]

van Ham, R. C., and 15 others. 2003. Reductive genome evolution in *Buchnera aphidicola*. *Proc. Natl. Acad. Sci. USA* 100: 581–586. [3, 20]

Van Heuverswyn, F., and 15 others. 2006. Human immunodeficiency viruses: SIV infection in wild gorillas. *Nature* 444: 164. [23]

Vanhooydonck, B., A. Y. Herrel, R. van Damme, and D. J. Irschick. 2005. Does dewlap size predict male bite performance in Jamaican *Anolis* lizards? *Func. Ecol.* 19: 38–42. [16]

van Noordwijk, A. J., and G. deJong. 1986. Acquisition and allocation of resources: Their influence on variation in life history tactics. *Am. Nat.* 128: 137–142. [14]

van Rheede, T., T. Bastiaans, D. S. Boone, S. B. Hedges, W. W. de Jong, and O. Madsen. 2006. The platypus is in its place: nuclear genes and indels confirm the sister group relation of monotremes and therians. *Mol. Biol. Evol.* 23: 587–597. [2]

Van Valen, L. 1973. A new evolutionary law. *Evol. Theory* 1: 1–30. [7]

Vaughan, T. A. 1986. *Mammalogy*, 3rd ed. Saunders College Publishing, Philadelphia, PA. [3]

Vehrencamp, S. L. 1983. Optimal degree of skew in cooperative societies. *Am. Zool.* 23: 327–335. [16]

Venter, J. C. et al. 2001. The sequence of the human genome. *Science* 291: 1304–1351. [3, 8]

Vermeij, G. J. 1987. *Evolution and Escalation: An Ecological History of Life.* Princeton University Press, Princeton, NJ. [5]

Vigilant, L., M. Stoneking, H. Harpending, K. Hawkes, and A. C. Wilson. 1991. African populations and the evolution of human mitochondrial DNA. *Science* 253: 1503–1507. [6]

Voelker, R. A., H. E. Schaffer, and T. Mukai. 1980. Spontaneous allozyme mutations in *Drosophila melanogaster*: Rate of occurrence and nature of the mutants. *Genetics* 94: 961–968. [10]

Voight, B. F., S. Kudaravalli, X. Wen, and J. K. Pritchard. 2006. A map of recent positive selection in the human genome. *PLoS Biol.* 4: e72 (doi: 10.1371/journal.pbio.0040072). [12]

von Baer, K. E. 1828. *Entwicklungsgeschichte der Thiere: Beobachtung und Reflexion.* Bornträger, Konigsberg. [3]

Vonk, F. J., and M. K. Richardson. 2008. Developmental biology: Serpent clocks tick faster. *Nature* 454: 282–283. [21]

Vonlanthen, P., and 9 others. 2012. Eutrophication causes speciation reversal in whitefish adaptive radiations. *Nature* 482: 357–362. [17]

Voss, S. R., and H. B. Shaffer. 2000. Evolutionary genetics of metamorphic failure using wild-caught vs. laboratory axolotls (*Ambystoma mexicanum*). *Molec. Ecol.* 9: 1401–1407. [21]

Voss, S. R., and J. J. Smith. 2005. Evolution of salamander life cycles: A major-effect quantitative trait locus contributes to discrete and continuous variation for metamorphic timing. *Genetics* 170: 275–281. [21]

Voss, S. R., K. L. Prudic, J. C. Oliver, and H. B. Shaffer. 2003. Candidate gene analysis of metamorphic timing in ambystomatid salamanders. *Molec. Ecol.* 12: 1217–1223. [21]

W

Waddington, C. H. 1942. Canalization of development and the inheritance of acquired characters. *Nature* 150: 563–565. [21]

Waddington, C. H. 1953. Genetic assimilation of an acquired character. *Evolution* 7: 118–126. [13]

Wade, M. J. 1977. An experimental study of group selection. *Evolution* 31: 134–153. [11]

Wade, M. J. 1979. The primary characteristics of *Tribolium* populations group selected for increased and decreased population size. *Evolution* 33: 749–764. [11]

Wade, M. J., and S. J. Arnold. 1980. The intensity of sexual selection in relation to male sexual behavior, female choice, and sperm precedence. *Anim. Behav.* 28: 446–461. [15]

Wagner, A., G. P. Wagner, and P. Similion. 1994. Epistasis can facilitate the evolution of reproductive isolation by peak shifts: A two-locus two-allele model. *Genetics* 138: 533–545. [18]

Wagner, G. P. 1988. The influence of variation and of developmental constraints on the rate of multivariate phenotypic evolution. *J. Evol. Biol.* 1: 45–66. [13]

Wagner, G. P. 1989a. The biological homology concept. *Annu. Rev. Ecol. Syst.* 20: 51–69. [21]

Wagner, G. P. 1989b. The origin of morphological characters and the biological basis of homology. *Evolution* 43: 1157–1171. [21]

Wagner, G. P. 1996. Homologues, natural kinds and the evolution of modularity. *Am. Zool.* 36: 36–43. [3, 22]

Wagner, G. P. 2010. Evolvability: the missing piece in the neo-Darwinian synthesis. In M. A. Bell, D. J. Futuyma, W. F. Eanes, and J. S. Levinton (eds.), *Evolution Since Darwin: The First 150 Years*, pp. 197–213. Sinauer, Sunderland, MA. [22]

Wagner, G. P., and L. Altenberg. 1996. Perspective: Complex adaptations and the evolution of evolvability. *Evolution* 50: 967–976. [13, 22]

Wagner, G. P., and V. J. Lynch. 2008. The gene regulatory logic of transcription factor evolution. *Trends Ecol. Evol.* 23: 377–385. [21]

Wagner, G. P., and J. Zhang. 2011. The pleiotropic structure of the genotype-phenotype map: the evolvability of complex organisms. *Nat. Rev. Genet.* 12: 204–213. [22]

Wagner, G. P., G. Booth, and H. Bagheri-Chaichian. 1997. A population genetic theory of canalization. *Evolution* 51: 329–347. [13]

Wagner, G. P., J. P. Kenney-Hunt, M. Pavlicev, J. R. Peck, D. Waxman, and J. M. Cheverud. 2008. Pleiotropic scaling of gene effects and the 'cost of complexity'. *Nature* 452: 470–472. [13]

Wagner, P. J. 1996. Contrasting the underlying patterns of active trends in morphologic evolution. *Evolution* 50: 990–1007. [21, 22]

Wake, D. B. 1982. Functional and developmental constraints and opportunities in the evolution of feeding systems in urodeles. In D. Mossakowski and G. Roth (eds.), *Environmental Adaptation and Evolution*, pp. 51–66. G. Fischer, Stuttgart. [22]

Wakeley, J., and J. Hey. 1997. Estimating ancestral population parameters. *Genetics* 145: 847–855. [18]

Walsh, B., and M. W. Blows. 2009. Abundant genetic variation + strong selection = multivariate genetic constraints: a geometric view of adaptation. *Annu. Rev. Ecol. Evol. Syst.* 40: 41–59. [13]

Walsh, H. E., I. L. Jones, and V. L. Friesen. 2005. A test of founder effect speciation using multiple loci in the auklets (*Aethia* spp.) *Genetics* 171: 1885–1894. [17]

Wang, J., and M. C. Whitlock. 2003. Estimating effective population size and migration rates from genetic samples over space and time. *Genetics* 163: 429–446. [10]

Wang, Z., B.-Y. Liao, and J. Zhang. 2010. Genomic patterns of pleiotropy and the evolution of complexity. *Proc. Natl. Acad. Sci. USA* 107: 18034–18039. [22]

Wang, Z., R. L. Young, H. L. Xue, and G. P. Wagner. 2011. Transcriptomic analysis of avian digits reveals conserved and derived digit identities in birds. *Nature* 477: 583–586. [22]

Waples, R. S. 1989. A generalized approach for estimating effective population size from temporal changes in allele frequency. *Genetics* 121: 379–391. [10]

Warneken, F., and M. Tomasello. 2009. Varieties of altruism in children and chimpanzees. *Trends Cogn. Sci.* 13: 397–402. [16]

Warner, R. R. 1984. Mating behavior and hermaphroditism in coral reef fishes. *Am. Sci.* 72: 128–136. [14]

Warren, R. W., L. Nagy, J. Selegue, J. Gates, and S. B. Carroll. 1994. Evolution of homeotic gene function in flies and butterflies. *Nature* 372: 458–461. [21]

Weatherbee, S. D., G. Halder, J. Kim, A. Hudson, and S. B. Carroll. 1998 Ultrabithorax regulates genes at several levels of the wing-patterning hierarchy to shape the development of the *Drosophila* haltere. *Genes Devel.* 12: 1474–1482. [21]

Webb, C. O. 2000. Exploring the phylogenetic structure of ecological communities: An example for rain forest trees. *Am. Nat.* 156: 145–155. [19]

Webb, C. O., D. D. Ackerley, M. A. McPeek, and M. J. Donoghue. 2002. Phylogenies and community ecology. *Annu. Rev. Ecol. Evol. Syst.* 33: 475–505. [19]

Webb, C. O., G. S. Gilbert, and M. J. Donoghue. 2006. Phylodiversity-dependent seedling mortality, size, structure, and disease in a Bornean rain forest. *Ecology* 87: S123–S131. [19]

Weber, D. S., B. S. Stewart, J. C. Garza, and N. Lehman. 2000. An empirical genetic assessment of the severity of the northern elephant seal population bottleneck. *Curr. Biol.* 10: 1287–1290. [10]

Weber, K. E. 1996. Large genetic change at small fitness cost in large populations of *Drosophila melanogaster* selected for wind tunnel flight: rethinking fitness surfaces. *Genetics* 144: 205–213. [13]

Weber, K. E., and L. T. Diggins. 1990. Increased selection response in larger populations. II. Selection for ethanol vapor resistance in *Drosophila melanogaster* at two population sizes. *Genetics* 125: 585–597. [13]

Weishampel, D. B., P. Dodson, and H. Osmólska. 2004. *The Dinosauria*, 2nd ed. University of California Press, Berkeley. [5]

Welch, J. J., and L. Bromham. 2005. Molecular dating when rates vary. *Trends Ecol. Evol.* 20: 320–327. [2]

Wellman, C. H., P. L. Osterloff, and U. Mohiuddin. 2003. Fragments of the earliest land plants. *Nature* 425: 282–290. [5]

Wells, J. 2000. *Icons of Evolution: Science or Myth?* Regnery, Washington, DC. [12]

Wen, J. 1999. *Evolution* of eastern Asian and eastern North American disjunct distributions of flowering plants. *Annu. Rev. Ecol. Syst.* 30: 421–455. [6]

Wenseleers, T., and F. L. W. Ratnieks. 2006. Comparative analysis of worker reproduction and policing in eusocial Hymenoptera supports relatedness theory. *Am. Nat.* 168: E163–E179. [16]

Werner, P. A., and W. J. Platt. 1976. Ecological relationships of co-occurring goldenrods (*Solidago*: Compositae). *Am. Nat.* 110: 959–971. [14]

Werner, T. K., and T. W. Sherry. 1987. Behavioral feeding specialization in *Pinaroloxias inornata*, the "Darwin's Finch" of Cocos Island, Costa Rica. *Proc. Natl. Acad. Sci. USA* 84: 5506–5510. [18, 19]

Werren, J. H. 1980. Sex ratio adaptations to local mate competition in a parasitic wasp. *Science* 208: 1157–1159. [15]

Werren, J. H. 1998. *Wolbachia* and speciation. In D. J. Howard and S. H. Berlocher (eds.), *Endless Forms: Species and Speciation*, pp. 245–260. Oxford University Press, New York. [17]

Werren, J. H. 2011. Selfish genetic elements, genetic conflict, and evolutionary innovation. *Proc. Natl. Acad. Sci. USA* 108: 10863–10870. [16]

Werth, C. R., and M. D. Windham. 1991. A model for divergent, allopatric speciation of polyploidy pteridophytes resulting from silencing of duplicate-gene expression. *Am. Nat.* 137: 515–526. [17]

West-Eberhard, M. J. 1979. Sexual selection, social competition, and evolution. *Proc. Am. Phil. Soc.* 123: 222–234. [15]

West-Eberhard, M. J. 1983. Sexual selection, social competition, and speciation. *Q. Rev. Biol.* 58: 155–183. [15, 18]

West-Eberhard, M. J. 2003. *Developmental Plasticity and Evolution*. Oxford University Press, New York. [13]

West, S. 2009. *Sex Allocation*. Princeton University Press, Princeton, N.J. [15]

West, S. A., and A. Gardner. 2010. Altruism, spite, and greenbeards. *Science* 327: 1341–1344. [16]

West, S. A., C. M. Lively, and A. F. Read. 1999. A pluralistic approach to sex and recombination. *J. Evol. Biol.* 12: 1003–1012. [15]

West, S. A., E. T. Kiers, E. L. Simms, and R. F. Denison. 2002. Sanctions and mutualism stability: Why do rhizobia fix nitrogen? *Proc. R. Soc. Lond. B* 269: 685–694. [16]

West, S. A., A. S. Griffin, and A. Gardner. 2007. Social semantics: altruism, cooperation, mutualism, strong reciprocity and group selection. *J. Evol. Biol.* 20: 415–432. [16]

Westneat, D., and C. W. Fox (eds.). 2010. *Evolutionary Behavioral Ecology*. Oxford University Press, Oxford. [16]

Westneat, D. F., and I. R. K. Stewart. 2003. Extra-pair paternity in birds: causes, correlates, and conflict. *Annu. Rev. Ecol. Evol. Syst.* 34: 365–396. [11, 16]

Wheeler, B. M., A. M. Heimberg, V. N. Moy, E. A. Sperling, T. W. Holstein, S. Heber, and K. J. Peterson. 2009. The deep evolution of metazoan microRNAs. *Evol. Dev.* 11: 50–68. [20]

White, B. J., F. H. Collins, and N. J. Besansky. 2011. Evolution of *Anopheles gambiae* in relation to humans and malaria. *Annu. Rev. Ecol. Evol. Syst.* 42: 111–132. [17]

White, M. J. D. 1978. *Modes of Speciation*. W. H. Freeman, San Francisco. [12]

White, T. D., B. Asfaw, Y. Beyene, Y. Haile-Selassie, C. O. Lovejoy, G. Suwa, G. WoldeGabriel. 2009. *Ardipithecus ramidus* and the paleobiology of early hominids. *Science* 326: 75–85. [4]

Whiting, M. F., S. Bradler, and T. Maxwell. 2003. Loss and recovery of wings in stick insects. *Nature* 421: 264–267. [21]

Whitlock, M. C., and A. F. Agrawal. 2009. Purging the genome with sexual selection: reducing mutation load through selection on males. *Evolution* 63: 569–582. [15]

Whittall, J. B., and S. A. Hodges. 2007. Pollinator shifts drive increasingly long nectar spurs in columbine flowers. *Nature* 447: 706–709. [3]

Whittingham, L. A., P. D. Taylor, and R. J. Robertson. 1992. Confidence of paternity and male parental care. *Am. Nat.* 139: 1115–1125. [16]

Wible, J. R., G. W. Rougier, M. J. Novacek, and R. J. Asher. 2007. Cretaceous eutherians and Laurasian origin for placental mammals near the K/T boundary. *Nature* 447: 1003–918. [5]

Wichman, H. A., L. A. Scott, C. D. Yarber, and J. J. Bull. 2000. Experimental evolution recapitulates natural evolution. *Phil. Trans. R. Soc. Lond. B* 355: 1677–1684. [8]

Wiens, J. J. 2001. Widespread loss of sexually selected traits: How the peacock lost its spots. *Trends Ecol. Evol.* 16: 517–523. [15]

Wiens, J. J. 2004. Speciation and ecology revisited: Phylogenetic niche conservatism and the origin of species. *Evolution* 58: 193–197. [18]

Wiens, J. J. 2011. The causes of species richness patterns across space, time, and clades and the role of "ecological limits." *Q. Rev. Biol.* 86: 75–96. [7]

Wiens, J. J., and M. J. Donoghue. 2004. Historical biogeography, ecology and species richness. *Trends Ecol. Evol.* 19: 639–644. [6]

Wiens, J. J., and C. H. Graham. 2005. Niche conservatism: Integrating evolution, ecology, and conservation biology. *Annu. Rev. Ecol. Evol. Syst.* 36: 519–539. [6, 22]

Wiens, J. J., C. H. Graham, D. S. Moen, S. A. Smith, and T. W. Reeder. 2006. Evolutionary and ecological causes of the latitudinal diversity gradient in hylid frogs: Treefrog trees unearth the roots of high tropical diversity. *Am. Nat.* 168: 579–596. [6]

Wildman, D. E., and 9 others. 2007. Genomics, biogeography, and the diversification of placental mammals. *Proc. Natl. Acad. Sci. USA.* 104: 14395–14400. [2]

Wiley, C., C. K. Ellison, and K. L. Shaw. 2012. Widespread genetic linkage of mating signals and female preference in the Hawaiian cricket *Laupala*. *Proc. R. Soc. Lond. B* 279: 1203–1209. [17]

Wiley, E. O. 1978. The evolutionary species concept reconsidered. *Syst. Zool.* 27: 17–26. [17]

Wilkins, A. S. 2002. *The Evolution of Developmental Pathways*. Sinauer, Sunderland, MA. [21]

Wilkinson, G. S. 1988. Reciprocal altruism in bats and other mammals. *Ethol. Sociobiol.* 9: 85–100. [16]

Williams, E. E. 1972. The origin of faunas: Evolution of lizard congeners in a complex island fauna—a trial analysis. *Evol. Biol.* 6: 47–89. [6]

Williams, G. C. 1957. Pleiotropy, natural selection, and the evolution of senescence. *Evolution* 11: 398–411. [14]

Williams, G. C. 1966. *Adaptation and Natural Selection*. Princeton University Press, Princeton, NJ. [11, 14]

Williams, G. C. 1975. *Sex and Evolution*. Princeton University Press, Princeton, NJ. [15]

Williams, G. C. 1992a. *Gaia*, nature worship and biocentric fallacies. *Q. Rev. Biol.* 67: 479–486. [11]

Williams, G. C. 1992b. *Natural Selection: Domains, Levels, and Challenges*. Oxford University Press, New York. [21]

Williams, T. M., J. E. Selegue, T. Werner, N. Gompel, A. Kopp, and S. B. Carroll. 2008. The regulation and evolution of a genetic switch controlling sexually dimorphic traits in *Drosophila*. *Cell* 134: 610–623. [21]

Willig, M. R., D. M. Kaufman, and R. D. Stevens. 2003. Latitudinal gradients of biodiversity: pattern, process, scale, and synthesis. *Annu. Rev. Ecol. Evol. Syst.* 34: 273–309. [6]

Williston, S. W. 1925. *The Osteology of the Reptiles*. Harvard University Press, Cambridge, MA. [5]

Wilson, A. C., S. S. Carlson, and T. J. White. 1977. Biochemical evolution. *Annu. Rev. Biochem.* 46: 573–639. [2]

Wilson, A. M., and K. Thompson. 1989. A comparative study of reproductive allocation in 40 British grasses. *Functional Ecology* 3: 297–302. [14]

Wilson, D. S. 1975. A theory of group selection. *Proc. Natl. Acad. Sci. USA* 72: 143–146. [15, 16]

Wilson, D. S. 1983. The group selection controversy: History and current status. *Annu. Rev. Ecol. Syst.* 14: 159–187. [11]

Wilson, D. S. 1997. Altruism and organism: Disentangling the themes of multilevel selection theory. *Am. Nat.* 150 (Suppl.): S122–S134. [16]

Wilson, D. S. 2002. *Darwin's Cathedral: Evolution, Religion, and the Nature of Society*. University of Chicago Press, Chicago. [16]

Wilson, D. S., and R. K. Colwell. 1981. The evolution of sex ratio in structured demes. *Evolution* 35: 882–897. [15]

Wilson, E. O. 1971. *The Insect Societies*. Harvard University Press, Cambridge, MA. [16]

Wilson, E. O. 1975. *Sociobiology: The New Synthesis*. Harvard University Press, Cambridge, MA. [16, 22, 23]

Wilson, E. O. 1992. *The Diversity of Life*. Harvard University Press, Cambridge, MA. [5, 7]

Wilson, E. O., F. M. Carpenter, and W. L. Brown Jr. 1967. The first Mesozoic ants. *Science* 157: 1038–1040. [4]

Wilson, G., and Q. Rahman. 2005. *Born Gay: The Psychobiology of Sexual Orientation*. Peter Owen Publishers, London. [16]

Wilson, G. P., A. R. Evans, I. J. Corte, P. D. Smits, M. Fortelius, and J. Jernvall. 2012. Adaptive radiation of multituberculate mammals before the extinction of dinosaurs. *Nature* 483: 457–460. [5]

Winkler, I. S., and C. Mitter. 2008. The phylogenetic dimension of insect–plant interactions: A review of recent evidence. In K. J. Tilmon (ed.), *Specialization, Speciation, and Radiation: The Evolutionary Biology of Herbivorous Insects*, pp. 240–263. University of California Press, Berkeley. [6. 7, 19]

Winsor, M. P. 2006. Linnaeus's biology was not essentialist. *Ann. Missouri Bot. Gard.* 93: 2–7. [1]

Winter, K. U., A. Becker, T. Munster, J. T. Kim, H. Saedler, and G. Theissen. 1999. MADS-box genes reveal that gnetophytes are more closely related to conifers than to flowering plants. *Proc. Natl. Acad. Sci. USA* 96: 7342–7347. [21]

Wittkopp, P. J., B. K. Haerum, and A. G. Clark. 2004. Evolutionary changes in *cis* and *trans* gene regulation. *Nature* 430: 85–88. [21]

Woese, C. R. 2000. Interpreting the universal phylogenetic tree. *Proc. Natl. Acad. Sci. USA* 97: 8392–8396. [5]

Wolfe, N. D., C. P. Dunavan, and J. Diamond. 2007. Origins of major human infectious diseases. *Nature* 447: 279–283. [23]

Wood, B. A. 2002. Hominid revelations from Chad. *Nature* 418: 133–135. [4]

Wood, B., and J. Baker. 2011. Evolution in the genus *Homo. Annu. Rev. Ecol. Evol. Syst.* 42: 47–69. [4]

Wood, T., N. Takebayashi, M. S. Barker, I. Mayrose, P. B. Greenspoon, and L. H. Rieseberg. 2009. The frequency of polyploid speciation in vascular plants. *Proc. Natl. Acad. Sci. USA* 106: 13875–13879. [18]

Woolhouse, M. E. J., J. P. Webster, E. Domingo, B. Charlesworth, and B. R. Levin. 2002. Biological and biomedical implications of the co-evolution of pathogens and their hosts. *Nat. Genet.* 32: 569–577. [19]

Woolhouse, M. E. J., D. T. Haydon, and R. Antia. 2005. Emerging pathogens: The epidemiology and evolution of species jumps. *Trends Ecol. Evol.* 20: 238–244. [19]

Wray, G. A. 2007. The evolutionary significance of *cis*-regulatory mutations. *Nat. Rev. Genet.* 8: 206–216. [8, 21]

Wright, S. 1931. Evolution in Mendelian populations. *Genetics* 16: 97–159. [10]

Wright, S. 1935. The analysis of variance and the correlations between relatives with respect to deviations from an optimum. *J. Genet.* 30: 243–256. [10]

Wu, C.-I., and H. Hollocher. 1998. Subtle is nature: Differentiation and speciation. In D. J. Howard and S. H. Berlocher (eds.), *Endless Forms: Species and Speciation*, pp. 339–351. Oxford University Press, New York. [17]

Wu, D., and 12 others. 2006. Metabolic complementarity and genomics of the dual bacterial symbiosis of sharpshooters. *PLoS Biology* 4(6): 1079–1092. [19]

Wu, P., T. X. Jiang, S. Suksaweang, R. B. Widelitz, and C. M. Chuong. 2004. Molecular shaping of the beak. *Science* 305: 1465–1466. [21, 22]

Wyatt, R. 1988. Phylogenetic aspects of the evolution of self-pollination. In L. D. Gottlieb and S. K. Jain (eds.), *Plant Evolutionary Biology*, pp. 109–131. Chapman & Hall, London. [15]

Wyckoff, G. J., W. Wang, and C.-I. Wu. 2000. Rapid evolution of male reproductive genes in the descent of man. *Nature* 403: 304–309. [12]

Wynne-Edwards, V. C. 1962. *Animal Dispersion in Relation to Social Behaviour*. Olver & Boyd, Edinburgh. [11]

X

Xiao, S., and M. Laflamme. 2009. On the eve of animal radiation: phylogeny, ecology and evolution of the Ediacara biota. *Trends Ecol. Evol.* 24: 31–40. [5]

Xu, X., H. You, K. Du, and F. Han. 2011. An *Archaeopteryx*-like theropod from China and the origin of Avialae. *Nature* 475: 465–470. [4]

Xu, X., Z. Zhou, X. Wang, X. Kuang, F. Zhang, and X. Du. 2003. Four-winged dinosaurs from China. *Nature* 421: 335–340. [4]

Xu, Z. P., I. Woo, H. Her, D. R. Beier, and R. L. Maas. 1997. Mouse *Eya* homologues of the *Drosophila eyes absent* gene require *Pax6* for expression in lens and nasal placode. *Development* 124: 219–231. [21]

Y

Yamamoto, S., and T. Sota. 2009. Incipient allochronic speciation by climatic disruption of the reproductive period. *Proc. R. Soc. Lond. B* 276: 2711–2719. [18]

Yang, Z., and B. Rannala. 2010. Bayesian species delimitation using multilocus sequence data. *Proc. Natl. Acad. Sci. USA* 107: 9264–9269. [17]

Yoder, A. D., and M. D. Nowak. 2006. Has vicariance or dispersal been the predominant biogeographic force in Madagascar? Only time will tell. *Annu. Rev. Ecol. Evol. Syst.* 37: 405–431. [6]

Yoder, A. D., and Z. Yang. 2004. Divergence dates for Malagasy lemurs estimated from multiple gene loci: Geological and evolutionary context. *Mol. Ecol.* 13: 757–773. [6]

Yoo, B. H. 1980. Long-term selection for a quantitative character in large replicate populations of *Drosophila melanogaster*. I. Response to selection. II. Lethals and visible mutants with large effects. *Genet. Res.* 35: I. 1–17, II. 19–31. [13]

Young, N. M., G. P. Wagner, and B. Hallgrimsson. 2010. Development and the evolvability of human limbs. *Proc. Natl. Acad. Sci. USA* 107: 3400–3405. [13]

Yu, N., M. I. Jensen-Seaman, L. Chemnick, O. Ryder, and W.-H. Li. 2004. Nucleotide diversity in gorillas. *Genetics* 166: 1375–1383. [10]

Yukilevich, R. 2012. Asymmetrical patterns of speciation uniquely support reinforcement in *Drosophila. Evolution* 66: 1430–1446. [18]

Z

Zahavi, A. 1975. Mate selection: A selection for a handicap. *J. Theor. Biol.* 53: 205–214. [15]

Zangerl, A. R., and M. R. Berenbaum. 2003. Phenotype matching in wild parsnip and parsnip webworm: Causes and consequences. *Evolution* 57: 806–815. [19]

Zangerl, A. R., and M. R. Berenbaum. 2005. Increase in toxicity of an invasive weed after reassociation with its coevolved herbivore. *Proc. Natl. Acad. Sci. USA* 102: 15529–15532. [19]

Zhang, J., and D. M. Webb. 2003. Evolutionary deterioration of the vomeronasal pheromone transduction pathway in catarrhine primates. *Proc. Natl. Acad. Sci. USA* 100: 8337–8341. [20]

Zhang, J., H. F. Rosenberg, and M. Nei. 1998. Positive Darwinian selection after gene duplication in primate ribonuclease genes. *Proc. Natl. Acad. Sci. USA* 95: 3708–3713. [20]

Zhang, J., D. M. Webb, and O. Podlaha. 2002. Accelerated protein evolution and origins of human-specific features: FOXP2 as an example. *Genetics* 162: 1825–1835. [8]

Zhang, J., A. M. Dean, F. Brunet, and M. Long. 2004. Evolving protein functional diversity in new genes of *Drosophila. Proc. Natl. Acad. Sci. USA* 101: 16246–16250. [20]

Zielenski, J., and L.-C. Tsui. 1995. Cystic fibrosis: Genotypic and phenotypic variations. *Annu. Rev. Genet.* 29: 777–807. [8]

Zimmer, C. 2009. On the origin of life on Earth. *Science* 323: 198–199. [5]

Zipursky, S. L., W. M. Wojtowicz, and D. Hattori. 2006. Got diversity? Wiring the fly brain with Dscam. *Trends Biochem. Sci.* 31: 581–588. [20]

Zouros, E. 1987. On the relation between heterozygosity and heterosis: An evaluation of the evidence from marine mollusks. In M. C. Rattazi, J. G. Scandalios, and G. S. Whitt (eds.), *Isozymes: Current Topics in Biological and Medical Research*, pp. 255–270. Alan R. Liss, New York. [12]

Zufall, R. A., and M. D. Rausher. 2004. Genetic changes associated with floral adaptation restrict future evolutionary potential. *Nature* 428: 847–850. [22]

Zuk, M., J. T. Rotenberry, and R. M. Tinghitella. 2006. Silent night: Adaptive disappearance of a sexual signal in a parasitized population of field crickets. *Biol. Lett.* 2: 521–524. [15]

Zwarts, L., M. M. Magwire, M. A. Carbone, M. Verstevena, L. Herteleera, R. R. H. Anholt, P. Callaerts, and T. F. C. Mackay. 2011. Complex genetic architecture of *Drosophila* aggressive behavior. *Proc. Natl. Acad. Sci. USA* 108: 17070–17075. [18]

Illustration Credits

Cover © Bence Máté.

Chapter 1 *Opener*: © Barry Mansell/Superstock/Alamy. Scientist photographs: *Lamarck*, courtesy of the National Library of Medicine; *Darwin*, courtesy of The American Philosphical Library; *Wallace*, from *My Life* by Alfred Russel Wallace; *Fisher*, courtesy of Joan Fisher Box; *Haldane*, courtesy of Dr. K. Patau; *Wright*, courtesy of Doris Marie Provine; *Mayr*, courtesy of the Harvard News Service and E. Mayr; *Stebbins, Simpson, and Dobzhansky*, courtesy of G. L. Stebbins; *Kimura*, courtesy of W. Provine; *Smith, Mayr, and Williams*, photo by Lars Falck, © Royal Swedish Academy of Sciences; *Hamilton*, courtesy of the University of Oxford.

Chapter 2 *Opener*: Courtesy of Douglas Futuyma. 2.7A *left*: © A & J Visage/Alamy. 2.7A *right*: © Holmes Garden Photos/Alamy. 2.7B *right*: © Jarno Gonzalez Zarraonandia/Shutterstock. 2.12A: © Steve Bloom/Alamy Images. 2.12B: © Rolf Nussbaumer Photography/Alamy. 2.12C: © Lars Christensen/istock. 2.12D: © Suzi Eszterhas/Minden Pictures/Corbis. 2.23: © Tim Zurowski/All Canada Photos/Corbis.

Chapter 3 *Opener*: © B. G. Thomson/Photo Researchers, Inc. 3.8A: © fotolincs/Alamy Images. 3.8B: © Photo Resource Hawaii/Alamy Images. 3.8C: © Ivan Kuzmin/Alamy. 3.8D: © Jason Gallier/Alamy. 3.11: David McIntyre. 3.15A: © Roberto Nistri/Alamy. 3.15B: © Chris Mattison/Alamy. 3.15C: © Mark Boulton/Alamy. 3.19: David McIntyre. 3.23A: © Ron Dahlquist/Design Pics/Corbis. 3.23B: © Dani Carlo/Prisma/AGE Fotostock. 3.23C: © Noble Proctor/The National Audubon Society Collection/Science Source. 3.24A: David McIntyre. 3.24B: © Roberto Nistri/Alamy. 3.24C: © Andreas Gradin/Shutterstock. 3.24D: © Roberto Nistri/Alamy.

Chapter 4 *Opener*: © Pascal Goetgheluck/Science Source. 4.6B: © Ted Daeschler/Academy of Natural Sciences/VIREO. 4.8D: © The Natural History Museum, London. 4.9A: © Imaginechina/Corbis. 4.14A: © Julius T. Csotonyi/Science Source. 4.14B: © John Reader/Science Source.

Chapter 5 *Opener*: © John Sibbick/The Natural History Museum, London. 5.3B: David McIntyre. 5.8A: © Sinclair Stammers/Science Source. 5.9A: © Ivan Vdovin/Alamy. 5.9B: David McIntyre. 5.10B: © Custom Life Science Images. 5.13A: © Francois Gohier/Science Source. 5.13B: © Sabena Jane Blackbird/Alamy. 5.13C: David McIntyre. 5.13D: © The Natural History Museum, London. 5.18A: © aasheimmedia/istock. 5.18B, C: David McIntyre. 5.19A: Courtesy of Jane Sinauer. 5.19B: © The Natural History Museum, London. 5.19C: David McIntyre. 5.22A: © The Field Museum, Chicago, negative #GEO84968c. 5.22B: Courtesy of Mitternacht90/Wikipedia. 5.22C: © Andre Maslennikov/AGE Fotostock. 5.26: © Nigel Reed QEDimages/Alamy.

Chapter 6 *Howler monkey*: © Morales/AGE Fotostock. *Douc langur*: © J & C Sohns/AGE Fotostock. 6.1B: © Geoff Bryant/Science Source. 6.3A: © Ron Kacmarcik/istock. 6.3B: © Andy Rouse/Corbis. 6.3C: © Glenn Bartley/All Canada Photos/Corbis. 6.3D: © DK Limited/Corbis. 6.5C: © Papilio/Alamy. 6.5D: © Michael D. Kern/Naturepl.com. 6.7: © FLPA/Alamy. 6.13: © Roberto Nistri/Alamy. 6.15 *Lemur*: © Eric Isselée/istock. 6.15 *Galago*: © DK Limited/Corbis. 6.15 *Chameleon*: © Eric Isselée/istock. 6.29A *Temperate*: © Ryan M. Bolton/Shutterstock. 6.29A *Tropical*: © Mark Kostich/Shutterstock.

Chapter 7 *Opener*: © Jochen Tack/Alamy. 7.2 *inset*: © Ron Dahlquist/Design Pics/Corbis. 7.3A *inset*: © Glenn Bartley/All Canada Photos/Corbis. 7.19B: © Juniors Bildarchiv/Alamy. 7.19C: © Constance McGuire/istock. 7.21 *Asterale*: © Pavel Lebedinsky/istock. 7.21 *Euphorb*: © Piksel/istock. 7.21 *Water lily*: © Yuttasak Jannarong/Shutterstock. 7.22A: © Alex Staroseltsev/Shutterstock. 7.22B: © Eric Isselée/Shutterstock. 7.22C: © Melinda Fawver/istock. 7.22D: © Kim Taylor/Naturepl.com. 7.24: © Vilor/Shutterstock.

Chapter 8 *Opener*: Diego Velázquez, *Las Meninas*, 1656. 8.5: © Greg C Grace/Alamy. 8.27 *top*: © Mike Lane/Alamy Images. 8.27 *bottom*: © FLPA/Alamy.

Chapter 9 *Black-faced finch*: Courtesy of Jürgen Schiller García. *Red-faced finch*: © ANT Photo Library/Science Source. 9.1A *Blue*: © Steve Byland/Shutterstock. 9.1A *White*: © Buschkind/Alamy. 9.1C *clockwise from upper left*: © urosr/Shutterstock; © Tomas Loutocky/Shutterstock; © Gina Smith/Shutterstock; © riekephotos/Shutterstock; © Lucian Coman/Shutterstock; © Gina Smith/Shutterstock. 9.2A: David McIntyre. 9.3: © Nigel Cattlin/Alamy. 9.5 *inset*: © Chad A. Ivany/Shutterstock.

Chapter 10 *Opener*: © ANP Photo/AGE Fotostock. 10.5: © Richard R. Hansen/Science Source. 10.9A: © Mark Carwardine/Naturepl.com.

Chapter 11 *Opener*: © Alex Wild Photography/alexanderwild.com. 11.1C: © Tom McHugh/Science Source. 11.2A: © Martin Fowler/Alamy. 11.2B: © Perennou Nuridsany/Science Source. 11.7A: © Alex Wild Photography/alexanderwild.com. 11.7B: © FLPA/Alamy. 11.10A: © Jason Gallier/Alamy. 11.18A: © Alex Mustard/2020VISION/Naturepl.com. 11.18B: © Christian Musat/Shutterstock. 11.21A: © Pepbaix/Alamy. 11.21B: © Mireille Vautier/Alamy Images. 11.22A *top*: © Robert Shantz/Alamy. 11.22A *bottom*: David McIntyre. 11.26A: © blickwinkel/Alamy. 11.26B: © B. Wilson/Shutterstock.

Chapter 12 *Opener*: © Curt Wiler/Alamy. 12.4: © John Cancalosi/Naturepl.com. 12.5: © Piotr Naskrecki/Minden Pictures/Corbis. 12.17: © blickwinkel/Alamy. 12.22: © Howard Birnstihl/Northside Productions. 12.24A: © Bill Coster/Alamy.

Chapter 13 *Opener*: © Pete Oxford/Minden Pictures/Corbis. 13.23: From Colosimo et al. 2005.

Chapter 14 *Opener*: © David Welling/Naturepl.com. 14.2A: © Carol Walker/Naturepl.com. 14.2B: © INTERFOTO Pressebildagentur/Alamy. 14.3A: © Jim Zipp/Science Source. 14.3B: © Kim Taylor/Naturepl.com. 14.4A *left*: © mauritius images GmbH/Alamy. 14.4A *right*: © Robert Shantz/Alamy. 14.4B *left*: David McIntyre. 14.4B *right*: © Lagui/Shutterstock. 14.10: Courtesy of Douglas Futuyma. 14.15A: © Maslov Dmitry/Shutterstock. 14.15B: © Imago/Zuma Press. 14.17B: © Frederick R. McConnaughey/Science Source.

Index

Numbers in *italic* refer to information in an illustration or illustration caption.

A

Abalones *(Haliotis)*, 422–423, 468, 493–494, 557
ABCDE model, of flower development, 572–574
abdominal A (abdA) gene, 56–57, 545
Abortion
  in plants, 440
  spontaneous, 320
Abouheif, Ehab, 596
Abrams, Peter, 531
Absolute fitness, 312
Acacia karroo, 298
Acanthocephala, 180
Acanthostega, 84
Acheulian technology, 93
Achillea borealis, 506, 507
Achondroplasia, *189*
Acquired immune deficiency syndrome (AIDS), 52, 636
Acrocentric chromosomes, 212, *213*
Acrocephalus schoenobaenus (sedge warbler), *438*
Acrocephalus scirpaceus (reed warbler), *520*
Acromyrmex, 514
Actinopterygii, 115
  See also Ray-finned fishes
Active trends, 621
  See also Driven trends
Adaptation and Natural Selection (Williams), 294
Adaptations
  "adaptive rescue" from extinction, 373–376
  based on new versus standing variation, 343
  concepts of design and mechanism, 284–285
  in Darwin's concept of natural selection, 8
  definitions of, 281, 296–297
  under directional selection in sexual populations, 403–404
  efficiency and, 623–624
  examples of, 282–284
  genetic constraints on, 372–373
  key adaptations, 183, *184*
  methods of recognizing, 297–301
  misconceptions regarding, 304–305
  natural selection and the evolution of, 281 (*see also* Natural selection)
  observations documenting, 301–304
  refuting the creationist arguments regarding, 643
  sexual selection and, 423–424
  to varying environments in sexual populations, 404–405
Adaptive divergence, 485
Adaptive geographic variation, 246
Adaptive landscapes, 329–330, 331, 497
Adaptive peaks, 330, 331
Adaptive radiations
  based on modifications of ancestral traits, 184
  Cambrian explosion, 111–114
  defined, 183
  description and examples of, *68, 69–70*
  ecological opportunity as a factor in, 177, 183
  incumbent replacement as a factor in, 179
  key adaptations, 183, *184*
  mammals in the Cenozoic, 126–128, *129*
  in the Mesozoic, 120
  problems for phylogenetic analysis, 46
  snakes in the Oligocene, 126
Adaptive topographies, 329–330
Adaptive valley, 330
Additive allele effects, 240, 351, 352
Additive genetic variance, 351–352, 364–365, 372, 373
Additive inheritance, 204
Adenine, 190
Adh gene, 235, 246, 335, 555, 556
Admixture, 251
"Advanced" character state. *See* Derived character state
Aethia cristatella (crested auklet), 420–421, 477, *478*
Aethia pusilla (least auklet), 477, *478*
Africa
  Duffy blood group, 335
  modern human origins and, 150–152, *153,* 250, 276
  sickle-cell hemoglobin and, *322,* 323
African-Americans, sickle-cell hemoglobin and, 323
African swallowtail butterfly *(Papilio dardanus),* 218–219, 522
Afrotheria, *45,* 128
Agalychnis callidryas (red-eyed tree frog), *158, 631*
Agave (century plants), 380
"Age of fishes," 115
"Age of mammals," 126
"Age of reptiles," 119
Age rank, correlation with clade rank, 95
Age schedules, of reproduction, 390–392
Age structure, reproductive success and, 386–387
Agelaius phoeniceus (red-winged blackbird), 413
Aggregate properties, 292–293, 295
Aggressive interactions, 429
Aglaiocercus coelestis (violet-tailed sylph hummingbird), *467*
Aglaophyton, 117
Agnathans, 111, *112,* 115, 587
Agouti, 582
Agrawal, Anurag, 523–524
Agriculture
  ecological impact of, 131
  human development of, 94
  significance of evolutionary biology, 650–651
Ailuropoda melanoleuca (giant panda), 66, 613
Air pollution, 201
Alberch, Pere, 596
Alcids, *296,* 297
Alcohol dehydrogenase *(Adh)* gene, 235, 246, 335, 555, 556
Algorithmic methods, in phylogenetic analyses, 36
Alkaline phosphatase, 559
Allele copy, 218
Allele frequency
  adaptive landscapes, 329–330
  additive genetic variance and, 351, 352
  defined, 223
  detecting selection from geographic variation, 333–335
  differences among populations, 248–249
  fixation index, 248
  genetic distance between populations and, 470–472
  genetic drift and, 258–259 (*see also* Genetic drift)
  Hardy-Weinberg principle, 224–228
  heritability and, 352
  heterozygosity, 228, *229*
  heterozygote advantage, 322–323
  interaction of selection and genetic drift, 330–332
  random fluctuations in, 261–262
  selection models, 314, 315–322
  step clines, 320
Alleles
  additive effects, 240, 351, 352
  defined, 11, 193, 218
  deleterious, 319–320
  dominance, 203–204, 228
  effectively neutral, 269
  factors affecting allele distributions in hybrid zones, 479–480
  fixation, 259, 261, *262*
  hitchhiking, 236, 289, 297
  homozygous and heterozygous, 223–224
  inclusive fitness, 434–437
  introgression, 480
  invasion of a population, 315
  lethal, 232, 241, 357
  mutator alleles, 400
  nearly neutral, 330
  neutral, 261 (*see also* Neutral theory of molecular evolution)
  origin of incompatibility, 488–489
  substitution by genetic drift, 263
  turnover, 271
  wild type, 198, 223

About the Book

Editor: Andrew D. Sinauer

Project Editor: Laura Green

Copy Editor: Norma Sims Roche

Photo Editor: David McIntyre

Production Manager: Christopher Small

Book Design and Layout: Jefferson Johnson

Illustration Program: Precision Graphics

Manufacturer: Courier Corporation, Inc.